AF252140

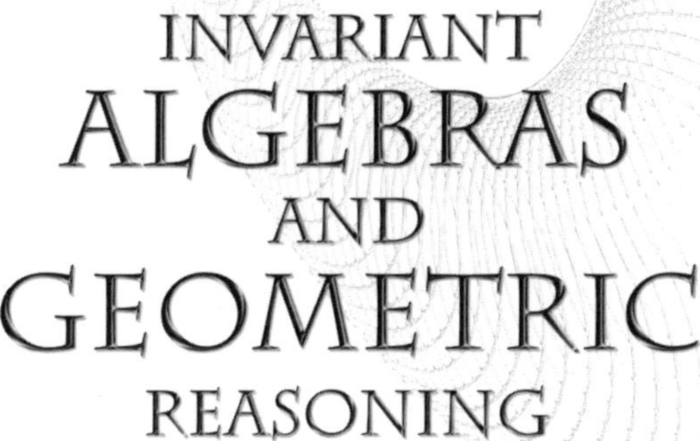

INVARIANT ALGEBRAS AND GEOMETRIC REASONING

INVARIANT ALGEBRAS AND GEOMETRIC REASONING

HONGBO LI

Chinese Academy of Sciences, China

World Scientific

NEW JERSEY · LONDON · SINGAPORE · BEIJING · SHANGHAI · HONG KONG · TAIPEI · CHENNAI

Published by

World Scientific Publishing Co. Pte. Ltd.

5 Toh Tuck Link, Singapore 596224

USA office: 27 Warren Street, Suite 401-402, Hackensack, NJ 07601

UK office: 57 Shelton Street, Covent Garden, London WC2H 9HE

British Library Cataloguing-in-Publication Data
A catalogue record for this book is available from the British Library.

ISBN-13 978-981-270-808-3
ISBN-10 981-270-808-1

Printed in Singapore by World Scientific Printers

Dedicated to my darling Kaiying,
my parents Changlin and Fengqin,
and my angels, Jessie, Bridie and Terry

Foreword

Beginning with now classical ideas of H. Grassmann and W.K. Clifford, *Geometric Algebra* has been developed in recent decades into a unified algebraic framework for geometry and its applications. It is fair to say that no other mathematical system has a broader range of applications from pure mathematics and physics to engineering and computer science.

Geometric computing is the heart of advanced applications. The more complex the application the more evident the need for computing that goes beyond number crunching to generate insight into the structure of systems and processes. This book develops representational and computational tools that enhance the power of *Geometric Algebra* to generate such insight. It demonstrates that power with many examples of automated geometric inference.

Computational geometry began with the invention of coordinate-based *analytic geometry* by Descartes and Fermat, and it was systematized by the invention of matrix algebra in the nineteenth century. Coordinates are the primitives for computer-based computations today, but they are not the natural primitives for most geometric structures. Consequently, the geometry in computations with coordinates is often difficult to divine. Though matrix methods are most common today, alternative approaches to computational geometry have developed almost as separate branches of mathematics. Especially noteworthy is *Invariant Theory*, which evolved from the theory of determinants into a combinatorial calculus called *Grassmann-Cayley algebra*. The present book continues that evolution by integrating the insights, notations and results of Grassmann-Cayley algebra into the more comprehensive Geometric Algebra. The result is a system that combines the geometric insight of classical synthetic geometry with the computational power of analytic geometry based on vectors instead of coordinates. The reader is referred to the text for many surprising examples. I predict that the tools and techniques developed here will ultimately be recognized as standard components of computational geometry.

David Hestenes

Physics & Astronomy Department, ASU

September, 2007

Preface

The demand for more reliable geometric computing in mathematical, physical and computer sciences has revitalized many venerable algebraic subjects in mathematics, and among them, there are Grassmann-Cayley algebra and Geometric Algebra. As distinguished invariant languages for projective, Euclidean, and other classical geometries, the two algebras nowadays have important applications not only in mathematics and physics, but in a variety of areas in computer science such as computer vision, computer graphics and robotics.

This book contains the author and his collaborators' most recent, original development of Grassmann-Cayley algebra and Geometric Algebra and their applications in automated reasoning of classical geometries. It includes the first two of the three advanced invariant algebras: Cayley bracket algebra, conformal geometric algebra, and null bracket algebra, together with their symbolic manipulations, and applications in geometric theorem proving.

The new aspects and mechanisms in integrating the representational simplicity of the advanced invariant algebras with their powerful computational capabilities, form the new theory of classical advanced invariants. It captures the intrinsic beauty of geometric languages and geometric computing, and leads to amazing simplification in algebraic manipulations, in sharp contrast to approaches based on coordinates and basic invariants.

As a treatise offering a detailed and rigorous mathematical exposition of these notions, at the same time offering numerous examples and algorithms that can be implemented in computer algebra systems, this book is meant for both mathematicians and practitioners in invariant algebras and geometric reasoning, for both seasoned professionals and inexperienced readers. It can also be used as a reference book by graduate and undergraduate students in their study of discrete and computational geometry, computer algebra, and other related courses. For the first-time reading, those sections marked with asterisks are suggested to be skipped by beginners.

The author wishes to express his heartfelt gratitude towards his family, for their full support of the author's mathematical career. He warmly thanks his former postdoc supervisors W.-T. Wu and D. Hestenes, for their persistent support and encouragement all these years. He thanks his colleagues H. Shi and X.-S. Gao for their great encouragement to the author in writing this book.

The manuscript of this book has been read by the author's Ph.D. students Y.-L. Shen, L. Huang, Z. Xie, D.-S. Wang, J. Liu, Y.-H. Cao, R.-Y. Sun, L.-X. Zhang, Y.-J. Liu, *et al.*, in a seminar lasting for six months. The author sincerely appreciates their valuable suggestions and proof-reading. He expresses his deep gratitude to Prof. Z.-Q. Xu who recommended this book for publication. His apology for repeatedly postponing the submission of the camera-ready version of the manuscript, and his appreciation for being granted every time to postpone the submission, are both to the editor, Ms. J. Zhang of World Scientific.

Hongbo Li
Beijing
September, 2007

Contents

4. Projective Conic Geometry with Bracket Algebra and Quadratic Grassmann-Cayley Algebra — 151

Chapter 1

Introduction

"In his famous survey of mathematical ideas, F. Klein championed 'the fusion of arithmetic with geometry' as a major unifying principle of mathematics. Klein's seminal analysis of the structure and history of mathematics brings to light two major processes by which mathematics grows and becomes organized. They may be aptly referred to as the algebraic and the geometric. The one emphasizes algebraic structure while the other emphasizes geometric interpretation. Klein's analysis shows one process alternatively dominating the other in the historical development of mathematics. But there is no necessary reason that the two processes should operate in mutual exclusion. Indeed, each process is undoubtedly grounded in one of two great capacities of the human mind: the capacity for language and the capacity for spatial perception. From the psychological point of view, then, the fusion of algebra with geometry is so fundamental that one could well say, 'Geometry without algebra is dumb! Algebra without geometry is blind!' "

— D. Hestenes, 1984.

1.1 Leibniz's dream

The algebraization of geometry started with R. Descartes' introduction of coordinates into geometry. This is one of the greatest achievements in human history, in that it is a key step from qualitative description to quantitative analysis. However, coordinates are sequences of numbers, they have no geometric meaning by themselves.

Co-inventor of calculus, the great mathematician G. Leibniz, once dreamed of having a geometric calculus dealing directly with geometric objects rather than with sequences of numbers. His dream is to have an algebra that is so close to geometry that every expression in it has a clear geometric meaning of being either a geometric object or a geometric relation between geometric objects, that the algebraic

1

manipulations among the expressions, such as addition, subtraction, multiplication and division, correspond to geometric transformations. Such an algebra, if exists, is rightly called *geometric algebra*, and its elements called *geometric numbers*.

Then what is a geometry? In his classic book *Three-dimensional Geometry and Topology* [180], Fields Medalist W. Thurston wrote: " Do we think of a geometry as a space equipped with such notions as lines and planes, or as a space equipped with a notion of congruence, or as a space equipped with either a metric or a Riemannian metric? There are deficiencies in all of these approaches. The best way to think of a geometry, really, is to keep in mind these different points of view all at the same time."

Thurston defined a geometry as a space X equipped with a group G of congruences. Technically, X is a manifold that is connected and simply connected, and G is a Lie group of diffeomorphisms of X, whose action on X is transitive and whose point stabilizers are compact. For a classical geometry, the geometric space X is embedded in a real vector space $\mathcal{V}^n$, and the transformation group G of X is a subgroup of the general linear group $GL(\mathcal{V}^n)$.

To search for a geometric algebra dreamed of by Leibniz, we start with the most fundamental geometry, Euclidean plane geometry. The geometric space is $\mathbb{R}^2 = \mathbb{R} \times \mathbb{R}$. Points are represented by vectors starting from the origin of $\mathbb{R}^2$, so they can be added, and be multiplied with a scalar of $\mathbb{R}$. The transformation group is the *Euclidean group*, where each element can be decomposed into two parts $(\mathbf{R}, \mathbf{t})$, such that

$$\mathbf{x} \mapsto \mathbf{R}\mathbf{x} + \mathbf{t}, \ \forall \mathbf{x} \in \mathbb{R}^2 \tag{1.1.1}$$

is the group action on the geometric space, with $\mathbf{R}$ being a 2D rotation matrix, and $\mathbf{t} \in \mathbb{R}^2$ being a vector of translation.

Two vectors can be multiplied using the complex numbers product, if $(x_1, x_2) \in \mathbb{R}^2$ is identified with $x_1 + ix_2 \in \mathbb{C}$:

$$\mathbf{x}\mathbf{y} = (x_1, x_2)(y_1, y_2) = (x_1 y_1 - x_2 y_2, x_1 y_2 + x_2 y_1). \tag{1.1.2}$$

Although the result is still a vector (complex number), it lacks invariance under the Euclidean group. In other words, the geometric information encoded in the product is unable to be separated from the interference of the reference coordinate frame. Regarding this aspect, we can say that the complex numbers product is *geometrically meaningless*, because its geometric interpretation is always related to the real axis of the specific complex numbers coordinate system of the 2D plane.

If we change the product of $\mathbf{x}, \mathbf{y}$ to

$$\overline{\mathbf{x}}\mathbf{y} := (x_1, -x_2)(y_1, y_2) = (x_1 y_1 + x_2 y_2, x_1 y_2 - x_2 y_1), \tag{1.1.3}$$

then under any 2D rotation $f : \mathbf{x} \mapsto \mathbf{x}e^{i\theta}$ centered at the origin, $(\overline{\mathbf{x}e^{i\theta}})(\mathbf{y}e^{i\theta}) = \overline{\mathbf{x}}\mathbf{y}$ is invariant, in the sense that

$$\overline{f(\mathbf{x})}f(\mathbf{y}) = \overline{\mathbf{x}}\mathbf{y}. \tag{1.1.4}$$

Since any reflection in the plane with respect to a line passing through the origin is the composition of a rotation and the complex conjugate $\mathbf{z} \mapsto \bar{\mathbf{z}}$ for $\mathbf{z} \in \mathbb{C}$, by $\overline{\bar{\mathbf{x}} e^{i\theta}} \left(\bar{\mathbf{y}} e^{i\theta} \right) = \mathbf{x}\bar{\mathbf{y}} = \overline{\bar{\mathbf{x}}\mathbf{y}}$, we get that the product (1.1.3) is conjugate-invariant under any reflection g:

$$\overline{g(\mathbf{x})}g(\mathbf{y}) = \overline{\bar{\mathbf{x}}\mathbf{y}}. \tag{1.1.5}$$

So the product (1.1.3) can be called a *geometric product* of 2D *orthogonal geometry*, where the geometric space is still $\mathbb{R}^2$ but the transformation group is the 2D orthogonal group $O(2)$. For the geometric interpretation, let $\mathbf{x} = |\mathbf{x}|e^{i\theta_\mathbf{x}}$ and $\mathbf{y} = |\mathbf{y}|e^{i\theta_\mathbf{y}}$, then

$$\bar{\mathbf{x}}\mathbf{y} = |\mathbf{x}||\mathbf{y}|e^{i(\theta_\mathbf{y} - \theta_\mathbf{x})} = |\mathbf{x}||\mathbf{y}|e^{i\angle(\mathbf{x},\mathbf{y})}, \tag{1.1.6}$$

where $\angle(\mathbf{x},\mathbf{y})$ is the angle of rotation from vector $\mathbf{x}$ to vector $\mathbf{y}$.

The two components of the geometric product (1.1.3) are also invariant by $O(2)$. They are both geometrically meaningful, and can also be termed as being "geometric". To distinguish among the three products, the real part of the geometric product $\bar{\mathbf{x}}\mathbf{y}$ is called the *inner product* between $\mathbf{x}$ and $\mathbf{y}$, denoted by $\mathbf{x} \cdot \mathbf{y}$; the pure imaginary part of the geometric product is called the *outer product* between $\mathbf{x}$ and $\mathbf{y}$, denoted by $\mathbf{x} \wedge \mathbf{y}$:

$$\begin{aligned} \mathbf{x} \cdot \mathbf{y} &= \frac{1}{2}(\mathbf{x}\bar{\mathbf{y}} + \bar{\mathbf{x}}\mathbf{y}), \\ \mathbf{x} \wedge \mathbf{y} &= \frac{1}{2}(\bar{\mathbf{x}}\mathbf{y} - \mathbf{x}\bar{\mathbf{y}}). \end{aligned} \tag{1.1.7}$$

One can immediately recognize that $\mathbf{x} \cdot \mathbf{y}$ is exactly the inner product of vectors $\mathbf{x}, \mathbf{y}$ in vector algebra, and $\mathbf{x} \times \mathbf{y} = (x_1 y_2 - x_2 y_1)\mathbf{n} = -i(\mathbf{x} \wedge \mathbf{y})\mathbf{n}$ is the cross product of the two vectors in space, where $\mathbf{n}$ is the unit normal direction of the plane described by complex numbers. In trigonometric form,

$$\begin{aligned} \mathbf{x} \cdot \mathbf{y} &= |\mathbf{x}||\mathbf{y}| \cos \angle(\mathbf{x}, \mathbf{y}), \\ \mathbf{x} \wedge \mathbf{y} &= i|\mathbf{x}||\mathbf{y}| \sin \angle(\mathbf{x}, \mathbf{y}). \end{aligned} \tag{1.1.8}$$

All complex numbers form a field, so any nonzero vector in $\mathbb{R}^2$ is invertible. The *geometric division* of $\mathbf{x}$ by $\mathbf{y}$ is just the geometric product of $\mathbf{x}$ and $\mathbf{y}^{-1}$, where the inverse is with respect to the geometric product (1.1.3) instead of the usual complex numbers product:

$$\mathbf{y}^{-1} = \frac{\mathbf{y}}{\mathbf{y}\bar{\mathbf{y}}}, \qquad \bar{\mathbf{x}}\mathbf{y}^{-1} = \frac{\bar{\mathbf{x}}\mathbf{y}}{\mathbf{y}\bar{\mathbf{y}}} = |\mathbf{x}||\mathbf{y}|^{-1}e^{i\angle(\mathbf{x},\mathbf{y})}. \tag{1.1.9}$$

The above analysis leads to the following theorem: The complex numbers equipped with the scalar multiplication, addition, subtraction, geometric product (1.1.3) and geometric division (1.1.9), are a geometric algebra of 2D orthogonal geometry.

For Euclidean plane geometry, Leibniz's dream is partially realized by complex numbers when the transformation group is restricted to the orthogonal group. His dream cannot be fully realized by complex numbers, because neither (1.1.3) nor

(1.1.9) is invariant under translation. With the increase of the dimension of the geometric space and the generalization of the transformation group, realizing Leibniz's dream becomes more and more difficult.

Can Leibniz's dream of geometric algebras be realized for nD classical geometries? The answer is affirmative:

- For nD projective geometry, the geometric algebra is *Grassmann-Cayley algebra*.

- For nD affine geometry, the geometric algebra is *affine Grassmann-Cayley algebra*.

- For nD orthogonal geometry, the geometric algebra is *Clifford algebra*, also called *Geometric Algebra* by Clifford himself.

- For nD Euclidean geometry, nD similarity geometry, nD conformal geometry, nD spherical geometry, nD hyperbolic geometry, and nD elliptic geometry, the geometric algebras are the same. It is *conformal geometric algebra*, which is composed of conformal Grassmann-Cayley algebra and conformal Clifford algebra.

1.2   Development of geometric algebras

In 1844, H. Grassmann published his book *Linear Extension Theory* [68], where he proposed the very original concepts in linear algebra such as linear independence, nD linear space, and linear extension of linear subspaces. Grassmann and A. Cayley are the two co-founders of linear algebra. While Grassmann established the concept of linear space, Cayley set up the matrix representation of linear maps. However, the algebra now bearing both their names, *Grassmann-Cayley algebra*, is not an algebra of matrices. It is an integration of *Grassmann algebra*, also called *exterior algebra*, and the dual of Grassmann algebra called *Cayley algebra* [57].

In Grassmann's vision, a point is zero dimensional, and can be represented by a vector, or a 1D direction. A line is generated by two points on it, and should be represented by a *product* of the two vectors representing the two points. The value of the product of two vectors should be a 2D direction. Two vectors in the same direction cannot span a 2D direction, so their product should be zero. Likewise, a plane is generated by three points on it, and should be represented by the product of the three vectors representing the points. The value of the product of three vectors should be a 3D direction. Three linearly dependent vectors cannot span a 3D direction, so their product should be zero. The product should be associative, as the plane spanned by line **12** and point **3** is identical to the plane spanned by point **1** and line **23**.

For any vector space $\mathcal{V}^n$, there exists a unique associative product, denoted by "$\wedge$", satisfying the requirements that the product of any vector $\mathbf{a} \in \mathcal{V}^n$ with itself is zero, and the product is linear with respect to every participating vector. This is Grassmann's product, nowadays called the *outer product*, or the *exterior product*. Vector space $\mathcal{V}^n$ supplemented with the linear combinations of all kinds of outer product results, becomes a $\mathbb{Z}$-*graded vector space* of dimension 2^n, called the *Grassmann space* over base space $\mathcal{V}^n$.

The $\mathbb{Z}$-grading in a Grassmann space is induced by the outer product. The outer product of r vectors represents an rD direction, and its *grade* is r. The operator extracting the r-graded part is called an r-*grading operator*.

In modern terms, Grassmann's point is a projective point, or a 1D linear subspace. His lines and planes are projective lines and planes respectively, or in other words, 2D and 3D linear subspaces respectively. His product is the direct sum extension of linear subspaces: if the intersection of two linear subspaces is just the zero vector, or in other words, they have *zero intersection*, then their outer product is their *direct sum*; if the two linear subspaces have nonzero intersection, their outer product is zero.

For example, in $\mathbb{R}^2$, the outer product of two vectors $\mathbf{x} = (x_1, x_2)$, $\mathbf{y} = (y_1, y_2)$ is, by (1.1.3) and (1.1.7),

$$\mathbf{x} \wedge \mathbf{y} = i(x_1 y_2 - x_2 y_1). \tag{1.2.1}$$

Grassmann took i as a unit 2D direction representing the complex numbers plane, which is also a real projective line, so that $x_1 y_2 - x_2 y_1$ is the scale, or coefficient, of the outer product $\mathbf{x} \wedge \mathbf{y}$ with respect to the unit i, the latter serving as a basis vector of the one-dimensional linear space of 2D directions.

Since $i(\mathbf{x} \wedge \mathbf{y})$ equals the signed area of the parallelogram spanned by vectors $\mathbf{y}, \mathbf{x}$, it is invariant under any affine transformation of the plane. Under any general linear transformation T of the plane, the outer product changes by a scale independent of $\mathbf{x}, \mathbf{y}$:

$$T(\mathbf{x}) \wedge T(\mathbf{y}) = \det(T)\, \mathbf{x} \wedge \mathbf{y}. \tag{1.2.2}$$

It is called a *relative invariant* in classical invariant theory, and is meaningful in projective geometry.

However, there is no way for one vector to divide another in Grassmann algebra, because no vector in $\mathbb{R}^2$ is invertible with respect to the outer product: for any two vectors $\mathbf{a}, \mathbf{b}$, their outer product is never a scalar.

In $(n-1)$D *projective geometry*, the geometric space is composed of all 1D linear subspaces of an nD vector space $\mathcal{V}^n$, the transformation group is the general linear group $GL(\mathcal{V}^n)$. Addition and subtraction of two generic vectors in $\mathcal{V}^n$ are purely algebraic operations without any meaning in projective geometry, because by arbitrarily rescaling one vector while fixing the other, the sum of the two vectors can be in any 1D subspace of the 2D space spanned by the two vectors.

Grassmann algebra is a geometric algebra of projective geometry where the product of vectors represents the extension of linear subspaces. However, this geometric algebra is incomplete both algebraically and geometrically: algebraically, it lacks the division operation; geometrically, it does not have any operation that represents the intersection of linear subspaces.

To provide the Grassmann algebra with an algebraic operation representing the geometric intersection, Cayley proposed a second product called the *meet product,* which is algebraically dual to the outer product. Geometrically, the meet product of two linear subspaces of $\mathcal{V}^n$, whose dimensions when summed up are not less than n, represents the intersection of the two linear subspaces. A Grassmann space equipped with both the outer product and the meet product is an algebra called *Grassmann-Cayley algebra.*

Grassmann-Cayley algebra is a geometric algebra for nD projective geometry whose two products represent the extension and intersection of linear subspaces.

For example, in $\mathbb{R}^2$, the meet product of two vectors $\mathbf{x} = (x_1, x_2)$, $\mathbf{y} = (y_1, y_2)$ is, using the complex numbers representation,

$$\mathbf{x} \vee \mathbf{y} := x_1 y_2 - x_2 y_1 = \begin{vmatrix} x_1 & y_1 \\ x_2 & y_2 \end{vmatrix} := [\mathbf{xy}]. \tag{1.2.3}$$

The *bracket* $[\mathbf{xy}]$ is the short-hand notation of the determinant formed by the two vectors $\mathbf{x}, \mathbf{y}$. It is just the coefficient of the 2D direction $\mathbf{x} \wedge \mathbf{y}$ with respect to the unit i.

To define a division operation in the Grassmann-Cayley algebra over $\mathbb{R}^2$, we need to find the inverse of vector $\mathbf{y} \in \mathbb{R}^2$. This is possible only for the meet product. For vector

$$\mathbf{z} = (-\frac{y_2}{y_1^2 + y_2^2}, \ \frac{y_1}{y_1^2 + y_2^2}), \tag{1.2.4}$$

since $\mathbf{y} \vee \mathbf{z} = 1$, $\mathbf{z}$ is a *right inverse* of $\mathbf{y}$ with respect to the meet product. It is not a left inverse because $\mathbf{z} \vee \mathbf{y} = -1$. Such a right inverse is not unique, because any $\mathbf{z} + \lambda \mathbf{y}$ for $\lambda \in \mathbb{R}$ is also a right inverse of $\mathbf{y}$. Nevertheless, the right inverse is *unique modulo* $\mathbf{y}$, *i.e.*, any right inverse of $\mathbf{y}$ is of the form $\mathbf{z} + \lambda \mathbf{y}$.

If we require that the right inverse of $\mathbf{y}$ is orthogonal to $\mathbf{y}$, *i.e.*, its inner product with $\mathbf{y}$ equals zero, then (1.2.4) is the unique right inverse of $\mathbf{y}$ with respect to the meet product, called the *orthogonal right inverse* of $\mathbf{y}$, and denoted by $*\mathbf{y}^{-1}$. The *division* of vector $\mathbf{x}$ by vector $\mathbf{y}$ from the right, can be defined by

$$\mathbf{x}/\mathbf{y} := \mathbf{x} \wedge (*\mathbf{y}^{-1}) = \mathbf{x} \wedge \mathbf{z} = i\frac{x_1 y_1 + x_2 y_2}{y_1^2 + y_2^2}, \tag{1.2.5}$$

which congrues with i times the inner product between $\mathbf{x}$ and $\mathbf{y}^{-1} = \mathbf{y}/(\mathbf{y}\overline{\mathbf{y}})$ in complex numbers notation, according to (1.1.3) and (1.1.9).

Similarly, if we define the division of $\mathbf{x}$ by $\mathbf{y}$ from the left as the outer product of the left inverse of $\mathbf{y}$ with $\mathbf{x}$, the result is still (1.2.5). So (1.2.5) can be uniformly called the division of $\mathbf{x}$ by $\mathbf{y}$. This division operation is elegant, but lacks invariance under $GL(\mathbb{R}^2)$, so it is meaningless in projective geometry. However, it is meaningful in orthogonal geometry, because it is invariant under $O(2)$.

Projective geometry does not have a geometric algebra in which the division operation is geometrically meaningful.

The above construction of the $O(n)$-invariant division by the outer product and the meet product, indicates a new way of constructing the geometric algebra $\mathbb{C}$ of 2D orthogonal geometry: by denoting

$$*\mathbf{y} := *(\mathbf{y}^{-1})^{-1} = *(\frac{\mathbf{y}}{|\mathbf{y}|^2})^{-1} = (-y_2, y_1), \tag{1.2.6}$$

we have

$$\begin{aligned}
\overline{\mathbf{x}}\mathbf{y} &= (x_1 y_1 + x_2 y_2, x_1 y_2 - x_2 y_1) \\
&= (x_1 y_1 + x_2 y_2) + i(x_1 y_2 - x_2 y_1) \\
&= -i\mathbf{x}/\mathbf{y}^{-1} + \mathbf{x} \wedge \mathbf{y} \\
&= \mathbf{x} \vee *\mathbf{y} + \mathbf{x} \wedge \mathbf{y} \\
&= \mathbf{x} \cdot \mathbf{y} + \mathbf{x} \wedge \mathbf{y}.
\end{aligned} \tag{1.2.7}$$

The last expression of (1.2.7) as a geometric product of two vectors $\mathbf{x}, \mathbf{y} \in \mathbb{R}^2$ can be extended to more than two dimensions. Historically, in 1843, W.R. Hamilton established the theory of *quaternions*, a geometric algebra of 3D orthogonal geometry as the extension of the complex numbers to space. In 1873, W.K. Clifford extended quaternions to *dual quaternions*, a geometric algebra of 3D Euclidean geometry. Both can be taken as extensions of the left side of (1.2.7) to some geometric products of 3D geometry.

In 1878, Clifford established the general theory of *Clifford algebra*, which he originally called geometric algebra, as "an application of Grassmann's extensive algebra". It is an extension of the last expression of (1.2.7) to a geometric product of nD orthogonal geometry. Quaternions and dual quaternions are both Clifford algebras.

Clifford algebra is a geometric algebra of nD orthogonal geometry in which the division operation is geometrically meaningful. Its product after decomposition also gives the extension and intersection of linear subspaces.

From now on, we only call the product in Clifford algebra the *geometric product*. A Grassmann space equipped with the geometric product is called a *Clifford algebra*. Formally, for any vector space V^n equipped with an inner product structure, the *Clifford algebra* over base space V^n is the universal associative algebra generated by the relation $\mathbf{a}\mathbf{a} = \mathbf{a} \cdot \mathbf{a}$ for any vector $\mathbf{a} \in V^n$, if the product, denoted by juxtaposition, is linear with respect to every participating vector.

Example 1.1. Clifford algebra $\mathcal{CL}(\mathbb{R}^2)$ over the Euclidean plane $\mathbb{R}^2$.

Is it just the algebra $\mathbb{C}$ of complex numbers? Certainly not. The dimension of $\mathcal{CL}(\mathbb{R}^2)$ as a real vector space is $2^2 = 4$, while $\mathbb{C}$ as a real vector space is of dimension 2. Let $\mathbf{e}_1, \mathbf{e}_2$ be an orthonormal basis of $\mathbb{R}^2$. Their geometric product is

$$\mathbf{e}_1 \mathbf{e}_2 = \mathbf{e}_1 \cdot \mathbf{e}_2 + \mathbf{e}_1 \wedge \mathbf{e}_2 = \mathbf{e}_1 \wedge \mathbf{e}_2. \tag{1.2.8}$$

It represents a 2D direction, and so is called a *2-vector*. In complex numbers notation, $\mathbf{e}_1 \mathbf{e}_2 = (1, 0)(0, 1) = i$ is the pure imaginary unit.

As a vector space, $\mathcal{CL}(\mathbb{R}^2)$ is the direct sum of three vector subspaces:

- the subspace $\mathbb{R}$ of scalars, or 0D directions (0-vectors);
- the subspace $\mathbb{R}^2$ of vectors, or 1D directions (1-vectors);
- the subspace of 2-vectors, or 2D directions, spanned by $i = \mathbf{e}_1\mathbf{e}_2$.

The direct sum of the first and the third subspaces is composed of elements of the form $x + y\mathbf{e}_1\mathbf{e}_2 = x + iy$, where $x, y \in \mathbb{R}$. It is a subalgebra of $\mathcal{CL}(\mathbb{R}^2)$ isomorphic to $\mathbb{C}$.

So $\mathcal{CL}(\mathbb{R}^2)$ is composed of both the real representation $\mathbb{R}^2$ and the complex representation $\{x + y\mathbf{e}_1\mathbf{e}_2 \,|\, x, y \in \mathbb{R}\}$ of the Euclidean plane. As an algebra it is isomorphic to the algebra $M_{2\times 2}(\mathbb{R})$ of 2 by 2 real matrices:

$$\begin{pmatrix} a & b \\ c & d \end{pmatrix} = \frac{a+d}{2} + \frac{a-d}{2}\mathbf{e}_1 + \frac{b+c}{2}\mathbf{e}_2 + \frac{b-c}{2}\mathbf{e}_1\mathbf{e}_2. \tag{1.2.9}$$

The algebraic isomorphism between $\mathcal{CL}(\mathbb{R}^2)$ and $M_{2\times 2}(\mathbb{R})$ is not a coincidence. By a theorem of É. Cartan, any real Clifford algebra is isomorphic to a matrix algebra over either the real numbers, or the complex numbers, or the quaternions.

Although created as a geometric algebra for orthogonal geometry, Clifford algebra did not attract more attention than quaternions in the 19th century. At that time, it was generally taken as a mathematical curiosity of being an algebra of "hypercomplex numbers", meaning that its elements, called *Clifford numbers*, are just complex numbers in high dimensional space.

The relation between a Clifford algebra and the corresponding Grassmann-Cayley algebra over the same base space is as follows: first, they are isomorphic as linear spaces, or more accurately, as $\mathbb{Z}$-graded linear spaces; second, both the meet product and the outer product can be expressed as polynomials of the geometric product in Clifford algebra, called the *ungrading* of the two products; third, the geometric product can always be decomposed into a polynomial of the two products in Grassmann-Cayley algebra, called the *grading* of the geometric product. Because of the latter decomposition, Grassmann-Cayley algebra provides a natural representation for Clifford algebra, and can be taken as a version of Clifford algebra if its base space has an inner product structure.

In the 20th century, with the development of spinors and their representations in Clifford algebras in the first half of the century, by great mathematicians É. Cartan, H. Weyl, C. Chevalley, M. Riesz, *et al.*, and by great physicists W. Pauli, P. Dirac, *et al.*, with the applications in index theorems and gauge theory in the second half of the century, by M. Atiyah, I. Singer, N. Seiberg, E. Witten, *et al.*, with the extension of complex analysis to Clifford numbers to form an alternative theory of several complex variables called *Clifford analysis*, by A.C. Dixon, F. Klein, R. Delanghe, L. Ahlfors, *et al.*, with the development into a universal geometric algebra by D. Hestenes and his school, and with many other accomplishments and applications by mathematicians, physicists and computer scientists, Clifford algebra has become a conflux of algebra, analysis and geometry, with wide range of applications in mathematics, theoretical physics, computer science and engineering.

It is hard to imagine that the reason behind so many successful applications of Clifford algebra is other than the intrinsic property that it is a geometric algebra describing the incidence and metric properties of linear subspaces. Just quote one comment from differential geometers:

> "It is a striking (and not commonplace) fact that Clifford algebras and their representations play an important role in many fundamental aspects of differential geometry. These algebras emerge repeatedly at the very core of an astonishing variety of problems in geometry and topology.
>
> Even in discussing Riemannian geometry, the formalism of Clifford multiplication will be used in place of the more conventional exterior tensor calculus. The Clifford multiplication is strictly richer than exterior multiplication; it reflects the inner symmetries and basic identities of the Riemannian structure. The effort invested in becoming comfortable with this algebraic formalism is well worthwhile."
>
> — H. Lawson Jr. and M.-L. Michelson, 1989.

Following the line of developing a universal geometric language out of Clifford algebra, D. Hestenes launched a new approach to Clifford algebra, and formulated a version of this algebra that is more "geometric" than any other version. He called this version *Geometric Algebra*, where the two initial capitals distinguish this specific geometric algebra from other geometric algebras [77], [78], [82], [83].

Hestenes regarded Clifford algebra as a geometric extension of the real number system to provide a complete algebraic representation of the geometric notions of high dimensional directions and magnitudes. The building blocks of *Geometric Algebra* are the outer products of vectors, called *blades*, or *decomposable extensors* in Grassmann algebra. The geometric product of blades describes relations among the spaces they represent. In classical geometries, the primitive elements are points, and geometric objects are point sets with properties. The properties are of two main types: structural and transformational. Geometric objects are characterized by structural relations and compared by transformations. Geometric Algebra provides a unified algebraic framework for both kinds of properties.

There are two main differences between Geometric Algebra and other versions of Clifford algebra. The first difference is "structural", or representational. On one hand, in the versions of hypercomplex numbers and matrix algebra, any expression representing a geometric object or property is a linear combination of a fixed set of basis expressions. In the Grassmann-Cayley algebra version of Clifford algebra, an expression representing an orthogonal transformation of a geometric object is in the form of a polynomial of outer products and meet products.

On the other hand, in Geometric Algebra, any expression representing a geometric object or property is a *graded Clifford monomial*, *i.e.*, a monomial generated by vectors using the geometric product and $\mathbb{Z}$-grading operators. The inner product, outer product and meet product are secondary operations. They can all be expressed as compositions of $\mathbb{Z}$-grading operators and the geometric product.

Structurally, Geometric Algebra is more "geometric" than other versions of Clifford algebra, because geometric objects or properties are expressed more multiplicatively than additively in this algebra. In symbolic form, more addition leads to more algebra, and more multiplication preserves more geometry. While the geometric product and many of the $\mathbb{Z}$-grading operators are geometrically meaningful, the decomposition into a sum of expressions is a way of getting more "algebraic", and often breaks up the high dimensional relations among geometric objects.

To be more specific and exemplary, coordinatization is a typical way of algebraization by decomposing (or "discretizing") a high dimensional geometric object into a sequence of one-dimensional representations. Let $\mathbf{e}_1, \mathbf{e}_2, \ldots, \mathbf{e}_n$ be a basis of $\mathcal{V}^n$, then the coordinate representation of any vector $\mathbf{x} \in \mathcal{V}^n$, $(x_1, x_2, \ldots, x_n) = x_1\mathbf{e}_1 + x_2\mathbf{e}_2 + \cdots + x_n\mathbf{e}_n$, expresses $\mathbf{x}$ by scalars x_i each measuring the affinity of $\mathbf{x}$ with a basis vector $\mathbf{e}_i$. For two vectors $\mathbf{x}, \mathbf{y} \in \mathcal{V}^n$, their relation in $\mathcal{V}^n$ is decomposed into the sum of 1D relations between the x_i and y_i, with the basis vectors $\mathbf{e}_i$ as external agencies.

The second difference between Geometric Algebra and other versions of Clifford algebra is "transformational", or computational. Other versions emphasize the multilinear nature of Clifford algebra, so expansions based on multilinearity are common and frequent in manipulating expressions. The idea behind such manipulations is to normalize the expressions into canonical forms, just like the normalization of the multiplication of two polynomials by expanding it into a sum of monomials. Multilinear expansion is a way of decomposing a high dimensional multiplication into the sum of lower dimensional multiplications. It inevitably decreases the "geometric" feature of Clifford algebra.

Multiplication and division operations are the gist of Geometric Algebra. They are the only two geometrically meaningful operations upon algebraic expressions having geometric meaning, and are always preferred to addition and subtraction. Consequently, as manipulations inverse to expansions and normalizations, factorizations for the multiplicatively decomposed form and contractions to reduce the number of terms are common and frequent in Geometric Algebra. Symbolic computations of geometric problems based on factorizations and contractions prove to be more efficient and effective than those based on expansions and normalizations.

1.3 Conformal geometric algebra

Having surveyed the history of Clifford algebra from the viewpoint of developing a geometric algebra for nD orthogonal geometry, we return to the original problem of Leibniz's dream, *i.e.*, developing a geometric algebra for nD Euclidean geometry.

The Euclidean group $E(n)$ is the semi-direct product of the orthogonal group

$O(n)$ and the translational group $\mathbb{R}^n$. To find a homogeneous space embedded in a real vector space $\mathcal{V}^m$ for $E(n)$ to act on as a transitive transformation group, the orthogonal group $O(\mathcal{V}^m)$ of the surrounding space $\mathcal{V}^m$ should contain $E(n)$ as a subgroup. The only $(n+1)$D real vector space having this property is the space $\mathbb{R}^{n,0,1}$ spanned by $\mathbb{R}^n$ and a *null vector* $\mathbf{e}$ orthogonal to $\mathbb{R}^n$. A nonzero vector is said to be *null* if its inner product with itself is zero.

The orthogonal group of $\mathbb{R}^{n,0,1}$ is isomorphic to the similarity group of $\mathbb{R}^n$, which is generated by the Euclidean group $E(n)$ and dilations

$$D_\lambda : \ \mathbf{x} \longmapsto \lambda\mathbf{x}, \ \forall \mathbf{x} \in \mathbb{R}^n, \tag{1.3.1}$$

for all $\lambda \in \mathbb{R} - \{0\}$. When $n = 3$, the algebra of even-graded elements in $\mathcal{CL}(\mathbb{R}^{n,0,1})$ is isomorphic to the algebra of *dual quaternions*.

The product in dual quaternions is invariant under $E(3)$, so it is a Euclidean geometric product. However, the dual quaternion representations of primitive geometric objects such as points, lines and planes in space are not *covariant*. More accurately, the representations are not *tensors*, they depend upon the position of the origin of the coordinate system irregularly. As a consequence, the compositions of $\mathbb{Z}$-grading operators and the geometric product in $\mathcal{CL}(\mathbb{R}^{3,0,1})$ have rather poor geometric meaning, because they are all related to the origin in use. The algebraic properties of $\mathcal{CL}(\mathbb{R}^{3,0,1})$ are also poor. Because the inner product in $\mathbb{R}^{3,0,1}$ is degenerate, many important invertibilities in non-degenerate Clifford algebras are lost.

To find covariant representations of geometric objects, we need to go up one more dimension. In the $(n+2)$D Minkowski space $\mathbb{R}^{n+1,1}$, there is a well-known result [22], [35], [130], [147], saying that the orthogonal group of $\mathbb{R}^{n+1,1}$ is a double covering of the *conformal group* $M(n)$ of $\mathbb{R}^n$, the latter being composed of angle-preserving diffeomorphisms called *Möbius transformations* in $\mathbb{R}^n$. Since $E(n)$ is a subgroup of $M(n)$, it can be represented by orthogonal transformations in $\mathbb{R}^{n+1,1}$.

The geometric space on which $E(n)$ acts is neither a linear subspace of $\mathbb{R}^{n+1,1}$, nor an affine subspace. Instead, it is a quadric surface of dimension n. This surface, denoted by $\mathcal{N}_{\mathbf{e}}$, is isometric to $\mathbb{R}^n$ as distance space. This model of nD Euclidean geometry has its origin in the work of F. Wachter (1792-1817), a student of Gauss [70]. A revised version of this model, called the *Lie model*, was developed by S. Lie in his Ph. D. dissertation (1872) to study *contact geometry*, also known as *Lie sphere geometry*. In applications, this model is often called the *conformal model* [18], [56], [151], [168].

The Geometric Algebra $\mathcal{CL}(\mathbb{R}^{n+1,1})$ established upon the conformal model is called *conformal geometric algebra*. It is an integration of the previously developed conformal Clifford algebra [5], [44], [72], [80], [130], [136], [184], with the recently developed conformal Grassmann-Cayley algebra [110].

Conformal Clifford algebra studies the conformal transformations in $\mathbb{R}^n$ using various versions of Clifford algebra, ranging from hypercomplex numbers and Clifford matrices to Geometric Algebra. It provides a "transformational" geometric algebra for nD conformal geometry.

Conformal Grassmann-Cayley algebra, on the other hand, provides primitive Euclidean geometric objects such as points, lines, circles, planes and spheres, with covariant representations in $\mathcal{CL}(\mathbb{R}^{n+1,1})$. It represents and computes geometric constructions such as the intersection, extension, contact, and knotting of primitive geometric objects, by combining the meet product, outer product, inner product, and geometric product. It provides a "structural" geometric algebra for nD conformal geometry.

As an integration of conformal Grassmann-Cayley algebra and conformal Clifford algebra, conformal geometric algebra provides the first geometric algebra for nD conformal geometry, including Euclidean geometry as a special case.

Another classical result on conformal groups is that the conformal groups of nD Euclidean space, nD spherical space, and nD hyperbolic space respectively, are isomorphic to each other [35], [147]. Besides the conformal model $\mathcal{N}_{\mathbf{e}}$ of Euclidean geometry, the conformal models $\mathcal{N}_{\mathbf{p}}$ and $\mathcal{N}_{\mathbf{a}}$ of nD spherical geometry and hyperbolic geometry respectively also can be established as nD quadric surfaces in Minkowski space $\mathbb{R}^{n+1,1}$ [111], [113].

To unify the three conformal models, the whole set $\mathcal{N}$ of null vectors in $\mathbb{R}^{n+1,1}$ must be put into service. This space when equipped with a conformal distance function, provides a *universal homogeneous model* for Euclidean, spherical, and hyperbolic geometries [112]. By "homogeneous" we mean that two vectors in $\mathcal{N}$ represent the same point if and only if they differ by scale. Because of the homogeneity, the model is conformal instead of isometric, hence can represent various classical geometries of different metrics.

The Geometric Algebra established upon the universal homogeneous model, still called conformal geometric algebra, is a unified geometric algebra for classical geometries.

1.4 Geometric computing with invariant algebras

The purpose of developing geometric algebras is to use them to compute geometric problems and prove geometric theorems. Now that Leibniz's dream has been completely realized, using geometric algebras to solve difficult unsolved problems in geometry seems highly prospective. However, to change the prospect into reality, a lot of developments are needed.

What is geometric computing? Computing is an algebraic thing. In a suitable algebraic framework, a geometric problem can be translated into an algebraic one, and an algebraic result can be obtained by either symbolic or numerical computing. If the result can be translated into geometric terms, then it is geometrically meaningful. *Geometric computing* refers to a procedure of algebraic manipulations generating a geometrically meaningful result from an input of geometric data.

In the ancient times before algebra was invented, geometric terms such as lengths, areas, ratios and angles were used directly in geometric computing. This synthetic approach to geometry has rather limited capabilities both in representing and in manipulating geometric objects and properties.

By decomposing high dimensional geometry into a sequence of one dimensional geometries, Descartes' introduction of coordinates into geometry greatly facilitates the representation and manipulation of geometric objects. A side-effect is that this factitious decomposition induces two hard-to-solve problems:

(1) The results from algebraic computations based on coordinates are either difficult to explain geometrically, or geometrically meaningless because they are neither invariant nor covariant under coordinate transformations. In other words, their dependencies upon the specific coordinate systems are either difficult or impossible to be separated from the geometric properties they represent.

(2) Even in the most favorable coordinate system, there is the following "middle expression swell" phenomenon: both the input expression and the output expression are small in size, but the expressions in the middle computing steps are huge. Because of this, some computations are possible only theoretically.

We shall see that geometric algebras can alleviate the above difficulties faced by coordinate methods. A geometric algebra can be represented either by coordinates or in a coordinate-free manner. In coordinate form, no result from algebraic computing by a geometric algebra is geometrically meaningless, owning to the invariant nature of the operations in such an algebra. Still the obtaining of a geometric interpretation may be possible only theoretically, because of the difficulty in translating the result from coordinate form to geometric terms.

On one hand, the coordinate-free versions of geometric algebras have obvious representational advantage over coordinates. For example, for two vectors $\mathbf{x}, \mathbf{y} \in \mathbb{R}^2$, by (1.2.3) and (1.2.2), bracket $[\mathbf{xy}]$ is a *projective geometric invariant*. By Laplace expansion, the determinant in (1.2.3) becomes a polynomial of two terms. By the same expansion, a single bracket monomial $[\mathbf{a}_1\mathbf{a}_2][\mathbf{a}_3\mathbf{a}_4] \cdots [\mathbf{a}_{2k-1}\mathbf{a}_{2k}]$, where $\mathbf{a}_1, \mathbf{a}_2, \ldots, \mathbf{a}_{2k} \in \mathbb{R}^2$, becomes a coordinate polynomial of as many as 2^k terms.

Recall that the geometric algebra of projective geometry is Grassmann-Cayley algebra. In this algebra, all brackets form an algebraic system by multiplication, addition and subtraction, called a *bracket algebra*. Bracket algebra is the coordinate-free version of the algebra of determinants. A theorem in classical invariant theory says that all projective geometric invariants are generated algebraically by such brackets. Bracket algebra is in fact the algebra of all projective geometric invariants.

On the other hand, the representational advantage of brackets does not necessarily lead to any manipulational advantage. In bracket algebra, all brackets are indeterminates. Contrary to the coordinates of generic vectors, which are independent of each other, brackets composed of generic vectors are algebraically dependent, and their algebraic relations are called *syzygies*.

In classical invariant theory, a *syzygy* refers to a polynomial of invariants that equals zero when expanded into coordinate form. For example, in $\mathbb{R}^2$, brackets $[\mathbf{a}_i\mathbf{a}_j]$ for $1 \leq i < j \leq 4$ satisfy

$$[\mathbf{a}_1\mathbf{a}_2][\mathbf{a}_3\mathbf{a}_4] - [\mathbf{a}_1\mathbf{a}_3][\mathbf{a}_2\mathbf{a}_4] + [\mathbf{a}_1\mathbf{a}_4][\mathbf{a}_2\mathbf{a}_3] = 0. \qquad (1.4.1)$$

The left side is a syzygy among the six brackets. It becomes zero automatically when expanded into coordinate form using $\mathbf{a}_i = (x_i, y_i)^T$:

$$(x_1y_2 - x_2y_1)(x_3y_4 - x_4y_3) - (x_1y_3 - x_3y_1)(x_2y_4 - x_4y_2)$$
$$+ (x_1y_4 - x_4y_1)(x_2y_3 - x_3y_2) = 0.$$

In the special case where $\mathbf{a}_1 = \mathbf{e}_1$, $\mathbf{a}_2 = \mathbf{e}_2$ are an orthonormal basis of $\mathbb{R}^2$, and $\mathbf{a}_3 = \mathbf{x}$, $\mathbf{a}_4 = \mathbf{y}$ are two generic vectors, this sygyzy is just the Laplace expansion of bracket $[\mathbf{x}\mathbf{y}]$, *i.e.*,

$$[\mathbf{e}_1\mathbf{e}_2][\mathbf{x}\mathbf{y}] - [\mathbf{e}_1\mathbf{x}][\mathbf{e}_2\mathbf{y}] + [\mathbf{e}_1\mathbf{y}][\mathbf{e}_2\mathbf{x}] = \begin{vmatrix} x_1 & y_1 \\ x_2 & y_2 \end{vmatrix} - x_2(-y_1) + y_2(-x_1).$$

The algebraic dependencies among bracket indeterminates pose two long-lasting problems to the invariant theory community: factorization of a bracket polynomial without resorting to coordinates, and reduction of a bracket polynomial to the least number of terms. Having been open for about a century, there is no sign that the two problems can be solved in the near future.

In classical invariant theory, a bracket polynomial is changed into normal form by Young's *straightening algorithm* proposed in 1928. In the straightening procedure, a bracket monomial is transformed many times into bracket polynomials of many terms. This "expansion" procedure does not have any control of the middle expression swell.

Geometric interpretation is also a problem for bracket algebra. Although each bracket, as a determinant of homogeneous coordinates of the constituent points, can be interpreted in affine geometry as the *signed volume* of the simplex spanned by the points as vertices, a bracket polynomial is by no means easily interpretable with geometric terms. If the bracket polynomial can be written as a rational monomial in Grassmann-Cayley algebra, then it can be given a projective geometric interpretation based on the latter form. This translation from bracket algebra to Grassmann-Cayley algebra is called *Cayley factorization*. Except for some special cases [192], [117], this problem is still open.

So in classical invariant theory, the two major problems faced by the coordinate approach are still alive, although in some cases the algebraic manipulations can be simplified because of the simplicity in algebraic representations. Due to the algebraic dependencies among brackets, new difficulties arise, which are by no means easy to handle. In the invariant-theoretic approach, people do not get rid of algebraic dependencies, otherwise it becomes a traditional coordinate approach.

Bracket algebra is the algebra of invariants in projective geometry, and after some revision, also the algebra of invariants in affine geometry. For Euclidean geometry, bracket algebra needs to be supplemented with inner products of vectors.

The resulting algebra, called *inner-product bracket algebra*, is the algebra of invariants in Euclidean geometry. The algebraic relations among inner products of vectors are much more complicated than those among brackets. The task to control middle expression swell is much heavier.

The phenomenon that in classical geometries, the results computed using algebras generated by basic invariants such as determinants and inner products of vectors, are often big rational polynomials of the basic invariants, without any clear geometric meaning, indicates that the generating elements of such algebras are too low-level, *both algebraically and geometrically*. To control the expression size effectively in middle computing steps, at the same time to alleviate the difficulty of translating algebraic results into geometric terms, more advanced invariant algebras are needed.

1.5 From basic invariants to advanced invariants

Advanced invariants are polynomials of basic invariants. By putting them into the algebra of basic invariants as new indeterminates, and treating their polynomial expressions by basic invariants as their defining syzygies, an *algebra of advanced invariants* is obtained.

The purpose of proposing advanced invariants is to simplify algebraic computation and keep geometric meaning. The criteria used in singling out advanced invariants from the polynomials of basic invariants, are the following:

- an advanced invariant should have clear geometric meaning;
- the system of advanced invariants should be hierarchical;
- an advanced invariant should have relatively nice symmetry with respect to its vector constituents.

(1) To keep geometric meaning, we can resort to the geometric algebra of the corresponding geometry. Since the product in such an algebra is always geometrically meaningful, a scalar-valued monomial in such an algebra is naturally an advanced invariant with immediate geometric interpretation. Such monomials occur naturally in representing geometric constructions. If not expanded into polynomials of basic brackets, they can keep the geometric nature within their algebraic structures.

(2) The structure of an algebra of advanced invariants has to be hierarchical: the bottom level is the basic invariants, and each higher level invariant is a polynomial of the lower level ones. The *level* of an advanced invariant is determined by the level of composition of geometric product operations that are used to construct the advanced invariant.

The connections between high-level invariants and low-level ones are through expansions and factorizations. An *expansion* is the transformation of a high-level invariant to a polynomial of low-level ones, by eliminating at least one geometric product operation from the high-level invariant. The reverse procedure to produce a high-level invariant in monomial form is called *factorization*.

To distinguish the expansion and factorization here from those based on multilinearity properties, the terminology usually bears the name of the associated geometric algebra. For example, in projective geometry – *Cayley expansion* and *Cayley factorization*, in orthogonal geometry – *Clifford expansion* and *Clifford factorization*, and in Euclidean geometry – *null expansion* and *null factorization*.

(3) Advanced invariants when put in the form of monomials of the associated geometric algebra, are easy to be given geometric explanations, but are more difficult to manipulate than low-level ones if they do not have nice symmetries within their monomial structures.

Symmetries within advanced invariants, if expressed by basic invariants, are generally complicated syzygies. They provide the most economical way of reining syzygies. Employing the structural symmetries within advanced invariants can drastically, and in some cases, even magically, simplify the manipulation of syzygies.

Example 1.2. Advanced invariants in projective geometry.

All scalar-valued monomials in Grassmann-Cayley algebra generate an advanced invariant algebra, called *Cayley bracket algebra*. The monomials are graded by the number of meet products involved, and each monomial has clear geometric meaning by explaining the outer product as the geometric extension, and the meet product as the geometric intersection.

The role played by symmetries within advanced invariants in simplifying algebraic manipulations can be seen from the following typical example. In $\mathcal{V}^3$, bracket

$$[\mathbf{abc}] := \begin{vmatrix} a_1 & b_1 & c_1 \\ a_2 & b_2 & c_2 \\ a_3 & b_3 & c_3 \end{vmatrix} \tag{1.5.1}$$

is a projective geometric invariant. Vectors $\mathbf{a}, \mathbf{b}, \mathbf{c}$ represent projective points in the plane. If $\mathbf{a} = \mathbf{x}_1\mathbf{x}_2 \cap \mathbf{x}_3\mathbf{x}_4$, $\mathbf{b} = \mathbf{y}_1\mathbf{y}_2 \cap \mathbf{y}_3\mathbf{y}_4$, $\mathbf{c} = \mathbf{z}_1\mathbf{z}_2 \cap \mathbf{z}_3\mathbf{z}_4$ are intersections of lines, then in the Grassmann-Cayley algebra over $\mathcal{V}^3$, the three points have expressions

$$\begin{aligned} \mathbf{a} &= (\mathbf{x}_1 \wedge \mathbf{x}_2) \vee (\mathbf{x}_3 \wedge \mathbf{x}_4), \\ \mathbf{b} &= (\mathbf{y}_1 \wedge \mathbf{y}_2) \vee (\mathbf{y}_3 \wedge \mathbf{y}_4), \\ \mathbf{c} &= (\mathbf{z}_1 \wedge \mathbf{z}_2) \vee (\mathbf{z}_3 \wedge \mathbf{z}_4). \end{aligned} \tag{1.5.2}$$

Substituting them into bracket $[\mathbf{abc}]$, we get, in the notation of Cayley bracket algebra, the following advanced projective geometric invariant:

$$((\mathbf{x}_1\mathbf{x}_2)(\mathbf{x}_3\mathbf{x}_4))\,((\mathbf{y}_1\mathbf{y}_2)(\mathbf{y}_3\mathbf{y}_4))\,((\mathbf{z}_1\mathbf{z}_2)(\mathbf{z}_3\mathbf{z}_4)). \tag{1.5.3}$$

It can be expanded into 16847 different bracket polynomials [117]!

The equality between any two of the 16847 bracket polynomials is a nontrivial syzygy in bracket algebra. In the monomial form (1.5.3) of this advanced invariant, all these equalities are embodied in the following simple symmetries:

(i) the product of each pair of vectors, *e.g.* $\mathbf{x}_1\mathbf{x}_2$ of $\mathbf{x}_1$ and $\mathbf{x}_2$, is antisymmetric;

(ii) the product of each pair of 2-vectors, *e.g.* $(\mathbf{x}_1\mathbf{x}_2)(\mathbf{x}_3\mathbf{x}_4)$ of $\mathbf{x}_1\mathbf{x}_2$ and $\mathbf{x}_3\mathbf{x}_4$, is antisymmetric;

(iii) the product of $(\mathbf{x}_1\mathbf{x}_2)(\mathbf{x}_3\mathbf{x}_4)$, $(\mathbf{y}_1\mathbf{y}_2)(\mathbf{y}_3\mathbf{y}_4)$, $(\mathbf{z}_1\mathbf{z}_2)(\mathbf{z}_3\mathbf{z}_4)$, is both antisymmetric and associative;

(iv) the product is invariant under any duality between vectors and 2-vectors: if

$$\mathbf{x}_1\mathbf{x}_2 \mapsto \mathbf{u}_1, \quad \mathbf{x}_3\mathbf{x}_4 \mapsto \mathbf{u}_2, \quad \mathbf{y}_1\mathbf{y}_2 \mapsto \mathbf{v}_1, \quad \mathbf{y}_3\mathbf{y}_4 \mapsto \mathbf{v}_2, \quad \mathbf{z}_1\mathbf{z}_2 \mapsto \mathbf{w}_1, \quad \mathbf{z}_3\mathbf{z}_4 \mapsto \mathbf{w}_2$$

is a special linear map from the space of 2-vectors to $\mathcal{V}^3$, then (1.5.3) equals $(\mathbf{u}_1\mathbf{u}_2)(\mathbf{v}_1\mathbf{v}_2)(\mathbf{w}_1\mathbf{w}_2)$.

In general, a representation of a geometric property by advanced invariants can be transformed fairly easily into a representation by basic invariants or coordinates. The converse transformation is usually much more difficult to be practically possible. As to computation, a computing procedure based on basic invariants or coordinates generally cannot be translated into one based on advanced invariants. The converse translation is generally also impossible, because basic invariants and coordinates have their own rules of computation, by either straightening or expanding into their own normal forms.

The idea of *advanced invariant computing* is to manipulate advanced invariants by their own mechanism, without resorting to low-level invariants or coordinates. The following *non-commutative diagram* describes the relation among computations based on advanced invariants, basic invariants, and coordinates respectively.

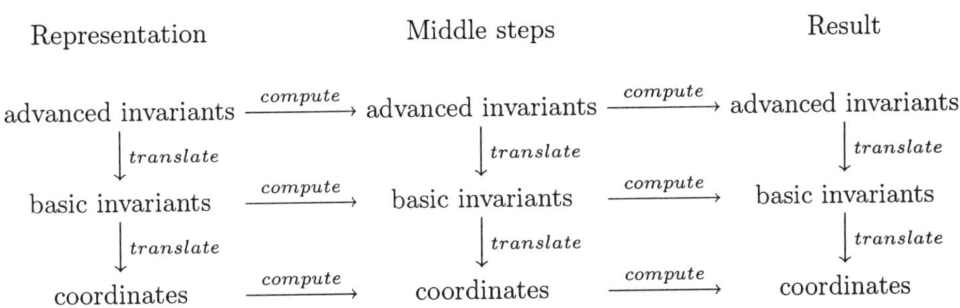

Example 1.3. Advanced invariants in orthogonal geometry.

The algebra of basic invariants in orthogonal geometry is inner-product bracket algebra. Geometrically, the inner product of two vectors represents the cosine of the angle between them. For the cosine of the angle formed by two planes or higher dimensional objects, its representation in inner-product bracket algebra is usually a complicated polynomial of inner products of vectors.

Bracket algebra when supplemented with inner products of high dimensional directions, becomes an algebra of advanced invariants in orthogonal geometry, called *graded inner-product bracket algebra* [50], [51], [136]. In this algebra, the inner products of rD directions are naturally graded by the dimension r. Still this algebra cannot provide simple representations for the cosines of sums of angles. To find monomial representations of such angle objects, we need the geometry of angles, which is the geometric algebra of orthogonal geometry – the Clifford algebra (or Geometric Algebra) over inner-product space $\mathcal{V}^n$.

Clifford bracket algebra [115] is generated by two scalar-valued $\mathbb{Z}$-grading operators acting on the geometric product results of vectors in Clifford algebra, called *angular bracket operator* and *square bracket operator* respectively, together with the graded inner products of vectors. The two brackets are naturally graded by the number of vectors in them.

Algebraically, the angular bracket is the prolongation of the inner product of two vectors to a "hyper-inner-product" of any $2l$ vectors of $\mathcal{V}^n$, for any $l \geq 1$; the square bracket is the prolongation of the determinant of n vectors to a "hyper-determinant" of any $n+2l$ vectors of $\mathcal{V}^n$. They are often referred to as *long brackets*. Geometrically, the two brackets are trigonometric functions of the compositions of directed angles.

Clifford bracket algebra is an advanced invariant algebra of orthogonal geometry. It is not an algebra of invariants in Euclidean geometry, where the transformation group is generated by orthogonal transformations and translations. While a translation is numerically almost trivial, it leads symbolically to the exponential growth of the expression size in the traditional normalization approach. For a single monomial $x_1 x_2 \cdots x_m$ in indeterminates $x_1, x_2, \ldots, x_m$, the translation $x_i \mapsto x_i + t$ changes the monomial into a polynomial of as many as 2^m terms after expansion:

$$(x_1 + t)(x_2 + t) \cdots (x_m + t) = x_1 x_2 \cdots x_m + t x_2 \cdots x_m + \cdots + x_1 t^{m-1} + t^m.$$

To construct advanced invariants in Euclidean geometry, we need the universal geometric algebra of classical geometries – conformal geometric algebra. In this algebra, all vectors representing points in Euclidean space are null vectors. The Clifford bracket algebra generated by null vectors is called *null bracket algebra* (NBA). Being nilpotent, null vectors provide great benefits to algebraic manipulations, and as a result, long brackets composed of null vectors have much richer symmetries than those composed of other vectors.

Null bracket algebra is a universal algebra of advanced invariants for classical metric geometries. It is also an algebra of advanced invariants for a class of non-classical geometry – contact geometry, where all points, hyperplanes and spheres are represented by null vectors.

1.6   Geometric reasoning with advanced invariant algebras

Geometric reasoning is a common task in mathematics education, computer-aided design, computer vision and robot navigation. Traditional geometric reasoning follows either a logical programming approach in artificial intelligence, or a coordinate approach in computer algebra [96], [201], or a hybrid approach based on both basic geometric invariants such as areas, volumes and distances, and dynamic geometric database management [40], [41].

In algebraic approaches to geometric reasoning and theorem proving, the input of a geometric problem is formulated by a set of symbols and their algebraic relations, and the algebraic computing of the conclusion expression, if geometrically meaningful, is called "symbolic geometric computing".

It has been widely known that algebraic methods such as the Shaw Prize winner W.-T. Wu's method of characteristic sets, the Gröbner base method, and the resultant method, are much more efficient than the logical programming approach to mechanical geometric theorem proving. In these methods, coordinates are introduced to formulate the geometric theorems in question. Because of the difficulty in providing geometric interpretations for algebraic expressions in coordinate form, much of the recent research on automated geometric deduction has focused on developing high-level coordinate-free techniques.

Using geometric invariants in symbolic geometric computing has been an active research subject since 1980's. Apart from the benefit of better geometric interpretability when compared with coordinates, geometric invariants have a salient feature of reducing the size of symbolic manipulations.

In the 1990's, several invariant algebraic methods were proposed, including the bracket polynomial straightening method [175], the distance method [70], the area method [40], the biquadratic final polynomial method [148], and the vectorial equation solving method [103]. From the viewpoint of advanced invariants, these methods can all be classified as methods of basic invariants.

The first successful method of advanced invariants for automated theorem proving in projective geometry was proposed in 2003 [117]. Advanced invariants from Cayley bracket algebra are used together with basic invariants in an alternating manner, by eliminations of geometric constructions to get advanced invariants from basic invariants, and by Cayley expansions to get polynomials of basic invariants from advanced invariants. In projective incidence geometry, all the theorems tested by the method of advanced invariants are given *binomial proofs*, which means that at every step of the proving procedure, the conclusion expression as a polynomial of basic and advanced invariants, is always two-termed.

In projective conic geometry, however, many theorems cannot be given binomial proofs. The reason is that the Grassmann-Cayley algebra over $\mathcal{V}^n$ is on intersections and extensions of *linear geometric objects* such as points, lines and planes. For *quadratic geometric objects* such as conics and quadrics, the base space $\mathcal{V}^n$ is too small. Recently (see Chapter 4), the Grassmann-Cayley algebra of nD projective quadric geometry, called *quadratic Grassmann-Cayley algebra*, is established. In the case $n = 3$, this is a geometric algebra of projective conic geometry, by which the algebra of basic conic invariants and the algebra of advanced conic invariants can be set up.

For automated theorem proving in Euclidean geometry, a breakthrough was made by the author in 2007 [124], during his preparation of this book. The author proposed a recipe for symbolic computing with conformal geometric algebra and null bracket algebra. The recipe is composed of three parts:

- use long geometric product to represent and compute multiplicatively;
- use "**breefs**" to control the expression size locally;
- use Clifford factorization for term reduction and translation back to geometry.

(1) Long geometric product:

This strategy has two aspects. In representation, it refers to representing geometric constructions and relations *multiplicatively*, or more accurately, with as few terms as possible. In computation, it refers to eliminating most of the $\mathbb{Z}$-grading operators from a graded Clifford monomial by generating a polynomial of long brackets with as few terms as possible, called *least ungrading*.

Eliminating $\mathbb{Z}$-grading operators is a way of prolonging the geometric product by breaking up the barriers to the associativity of the geometric product. The purpose is to replace complicated syzygy manipulations by simple symmetry manipulations of the long geometric product, with the cost of generating a minimal number of terms out of a monomial in Geometric Algebra.

(2) **Breefs**:

In most cases, the goal of algebraic computing is to make *simplification* to algebraic expressions, *i.e.*, to shorten the expressions by reducing the number of terms, and to make the expressions more decomposed by producing more factors. If at every middle step of the computing, a factored and shortest result is reached, then not only the goal of computing can be realized more easily, but the middle expression swell can be effectively controlled.

The new guideline in invariant computing, called "**breefs**" – **b**racket-oriented **re**presentation, **e**limination, **e**xpansion for **f**actored and **s**hortest result, is contrary to the traditional guideline of computing by normalization, in that it does not transform an expression into normal form, which usually incurs middle expression swell, instead it seeks to squeeze the expression locally to make it more compact. The new guideline leads to a lot of unique techniques. For example, if a vector indeterminate representing a geometric construction occurs several times in an expression, it can be given different algebraic representations at different parts of the same expression.

(3) Clifford factorization:

It includes two new algebras grown out of the interplay of null bracket algebra, Grassmann-Cayley algebra and conformal geometric algebra. They are *null Geometric Algebra* (NGA) and *null Grassmann-Cayley algebra* (NGC). The former is used to make factorizations and term reductions in null bracket algebra, while the latter is used to represent geometric constructions multiplicatively and to make Cayley factorizations.

About one hundred theorems in Euclidean geometry are tested, among which about four-fifth are given monomial or binomial proofs. In particular, more than one third of the theorems are given *monomial proofs*, *i.e.*, the conclusion of each theorem is represented by a single monomial, and keeps to be in monomial form till the end of the proof. This feature makes it impossible to find any analytic proof that is more elegant.

We do not have enough space left in this book to further explore this recipe and its magic effect in simplifying invariant geometric computing, so we advertize the forthcoming sequel [125] of this book for reference.

1.7 Highlights of the chapters

Chapter 2 is an introduction of bracket algebra, Grassmann-Cayley algebra and coalgebra, and the advanced invariant algebra for projective geometry – Cayley bracket algebra. The highlights of this chapter are

- the two new concepts *total meet product* and *reduced meet product*;
- Cayley expansion theory, of which most of the contents are moved to Appendix A to ease reading difficulty;
- Cayley bracket algebra.

Chapter 3 is on simplification techniques in bracket algebra, and applications to automated theorem proving in projective incidence geometry. Highlights:

- rational invariants, which are as important as classical invariants but seem to have been overlooked all the time;
- bracket polynomial divisions, with the discovery that divisions of invariants generate covariants;
- factorization and contraction algorithms.

Chapter 4 is on invariant representations of projective conics, the **breefs** principle, simplification techniques in conic computation, and applications to automated theorem proving. This chapter may be skipped by those not particularly interested in projective conics and quadrics. Highlights:

- conic representations and computations based on Cayley bracket algebra;
- quadratic Grassmann-Cayley algebra and quadratic bracket algebra, proposed by the author when he was an AvH Fellow at Christian-Albrechts Universität zu Kiel in 1998, but never formally published;
- **breefs**.

Chapter 5 is on inner-product bracket algebra, Clifford algebra, and Clifford expansion theory. The highlight is the *Clifford expansion theory* of expanding a monomial of Clifford algebra into a polynomial of inner-product Grassmann algebra. This theory is the foundation of Clifford bracket algebra. The study of Clifford expansions was initiated in the second half of the 20th century, first by physicists

E. Caianiello [32], A. Lotze [129], and later by mathematicians G.-C. Rota and J. Stein [158], A. Brini [29], *et al.*

Driven by the need to find factored and shortest expansions of the geometric product of a sequence of elements into a polynomial of inner products and outer products, and totally ignorant of any existing work on this subject, the author re-derived the whole theory of Clifford expansions in 2001 [115]. In comparison, the author's results contain not only all the formulas discovered earlier by other people, but also many new discoveries that can be taken as simplifications and extensions of the existing results. The author thanks B. Fauser and A. Brini for getting him to know of the earlier work on Clifford expansions.

Chapter 6 is on Geometric Algebra, its main computing techniques, Clifford coalgebra, and Clifford bracket algebra. Among the computing techniques unique to Geometric Algebra, there are three prominent ones that prove to be very powerful: (a) symmetry and commutation, (b) ungrading, (c) monomial simplification. Below we present a short introduction to the latter technique.

Monomial simplification is to reduce the redundancy of participating vectors in the geometric product. It includes *monomial factorization* to generate scalar-valued factors, and *monomial compression* to reduce the number of effective vectors. While monomial factorization techniques are to be developed in [125], monomial compression is the highlight of this chapter. This problem seems rather interesting, and inspires a sequence of conjectures on the algebraic structures of Clifford algebras, *e.g.*, the *maximal grade conjectures*. The author thanks D. Fontijne for putting up the invertible monomial (versor) compression problem to him.

Chapter 7 is on algebraic models of affine, Euclidean, and contact geometries, and on the corresponding geometric algebras of describing their incidence constructions – affine Grassmann-Cayley algebra, conformal Grassmann-Cayley algebra, and the Grassmann-Cayley algebra upon the Lie model. There are two attractive discoveries:

(1) The geometric exploration of the meet products of Minkowski blades in the conformal model, leads to an elegant characterization of the *knotting relation* between two objects.

(2) Positive vectors and negative vectors in the Lie model have the geometric interpretations of representing pencils of *intersecting Lie spheres* and *separating Lie spheres* respectively. So besides the contact problem among points, hyperplanes and spheres, the intersection and separation problems can also be dealt with in the Grassmann-Cayley algebra upon the Lie model.

Chapter 8 is on conformal Clifford algebra, dual Clifford algebra, and the universal homogeneous model of classical geometries. Four kinds of representations of 3D conformal transformations in conformal Clifford algebra are developed in full detail, for the purpose of applying them immediately to engineering problems:

- the geometric product of Minkowski blades [110] is a representation of the conformal transformation by direct geometric constructions;

- the Cayley transform [130] provides a large-scale representation of the conformal transformation by a polynomial of degree four;
- the outer exponential [130] provides a local representation of the conformal transformation by a polynomial of degree two;
- the fractional linear form of nD conformal transformations [184] is a natural extension of the classical complex fractional linear function representation of 2D conformal transformations.

In each chapter, those sections and subsections marked with asterisks at the end of their titles, are suggested to be skipped by inexperienced beginners for their first-time reading.

Chapter 2

Projective Space, Bracket Algebra and Grassmann-Cayley Algebra

Projective geometry is the simplest classical geometry in that it is on the properties of linear subspaces that are invariant under the general linear group. The generators of invariants in this geometry are called brackets, which are determinants of homogeneous coordinates. The geometric algebra of this geometry is called Grassmann-Cayley (GC) algebra. It provides a synthetic description to projective geometry through its two geometrically meaningful products: the outer product representing the extension of linear subspaces, and the meet product representing their intersection.

The revitalization of this century-old mathematical language is due to the Rota school in the 1970-1980's, *cf.* [14], [57], [69], [157], [158], [190]. The translation from GC algebra to bracket algebra is called Cayley expansion. It can be said to be a procedure of translating geometry to algebra. The translation from bracket algebra back to GC algebra is called Cayley factorization. It conglomerates basic invariants to advanced invariants.

This chapter introduces the two algebras and their translations to each other. To make the concepts more easily accessible, we start with a homogeneous coordinate description of invariants and covariants.

2.1 Projective space and classical invariants

Since we only care for classical geometry, throughout this book we assume that the base field $\mathbb{K}$ has characteristic 0.

An $(n-1)$D *projective space* $\mathbb{P}^{n-1}$ is the set of 1D vector subspaces of an nD vector space $\mathcal{V}^n$ over $\mathbb{K}$. A 1D subspace of $\mathcal{V}^n$ is called a projective *point* in $\mathbb{P}^{n-1}$. Let $\{\mathbf{e}_1, \mathbf{e}_2, \ldots, \mathbf{e}_n\}$ be a basis of $\mathcal{V}^n$. Then a projective point can be represented by the following *homogeneous coordinates*:

$$(x_1 : x_2 : \ldots : x_n), \tag{2.1.1}$$

such that vector $x_1\mathbf{e}_1 + x_2\mathbf{e}_2 + \cdots + x_n\mathbf{e}_n$ spans the 1D subspace.

A projective point can be represented by any vector spanning it. Two vectors represent the same projective point if and only if they are equal to each other up to a nonzero scaling, or briefly, *up to scale*. When $\mathbf{x}$ is a vector and we say point $\mathbf{x}$, we mean the 1D subspace spanned by $\mathbf{x}$.

If a geometric object can be represented by an algebraic expression, and the representation is unique up to scale, we say the representation is *homogeneous*. The algebraic representation of any object in projective space is homogeneous. In fact, any homogeneous representation can be written as a *homogeneous function* in its vector variables.

Let $f(\mathbf{x})$ be a function in vector variable $\mathbf{x}$. If

$$f(\lambda \mathbf{x}) = \lambda^r f(\mathbf{x}), \quad \text{for any } \lambda \in \mathbb{K}, \tag{2.1.2}$$

then f is called a *homogeneous function* of degree r in $\mathbf{x}$. An equality in vector variables is valid in projective geometry if and only if it is homogeneous in each vector variable.

In this book, we always use boldfaced digits and boldfaced lower-case letters to denote vectors. We use boldfaced capitals to denote non-scalar and non-vector variables and functions, and use Greek letters to denote scalar parameters.

Let $\mathbf{1}, \mathbf{2}, \cdots, \mathbf{n+1}$ be vectors in $\mathcal{V}^n$ such that no n of them are linearly dependent. Since the $n + 1$ vectors must be linearly dependent, there are scalars $\lambda_1, \lambda_2, \ldots, \lambda_{n+1}$, at least one of which is nonzero, such that

$$\lambda_1 \mathbf{1} + \lambda_2 \mathbf{2} + \cdots + \lambda_{n+1}(\mathbf{n+1}) = 0. \tag{2.1.3}$$

If vector $\mathbf{i}$ is replaced by $\mu_i \mathbf{i}$, the above equality remains valid only when every λ_j for $j \neq i$ is multiplied by μ_i. So every λ_j is a homogeneous function of degree 1 in each vector variable $\mathbf{i} \neq \mathbf{j}$.

Indeed, by linear algebra the equation (2.1.3) can be solved for indeterminates λ's as follows: let $\mathbf{i} = (x_{1i}, x_{2i}, \ldots, x_{ni})^T$, then

$$\lambda_i = (-1)^i \mu \begin{vmatrix} x_{11} & x_{12} & \cdots & x_{1(i-1)} & x_{1(i+1)} & \cdots & x_{1(n+1)} \\ x_{21} & x_{22} & \cdots & x_{2(i-1)} & x_{2(i+1)} & \cdots & x_{2(n+1)} \\ \vdots & \vdots & \vdots & \vdots & \vdots & \ddots & \vdots \\ x_{n1} & x_{n2} & \cdots & x_{n(i-1)} & x_{n(i+1)} & \cdots & x_{n(n+1)} \end{vmatrix}, \tag{2.1.4}$$

where μ is a free parameter independent of i.

Definition 2.1. [Definition of bracket algebra by coordinates] Let $\{\mathbf{e}_1, \mathbf{e}_2, \ldots, \mathbf{e}_n\}$ be a basis of $\mathcal{V}^n$. The *bracket* of a sequence of n vectors $\mathbf{1}, \mathbf{2}, \ldots, \mathbf{n}$ in $\mathcal{V}^n$, denoted by $[\mathbf{12} \cdots \mathbf{n}]$, is the determinant of their coordinates with respect to the basis:

$$[\mathbf{12} \cdots \mathbf{n}] := \begin{vmatrix} x_{11} & x_{12} & \cdots & x_{1n} \\ x_{21} & x_{22} & \cdots & x_{2n} \\ \vdots & \vdots & \ddots & \vdots \\ x_{n1} & x_{n2} & \cdots & x_{nn} \end{vmatrix}. \tag{2.1.5}$$

The nD *bracket algebra* generated by $m > n$ vectors $\mathbf{1}, \mathbf{2}, \ldots, \mathbf{m}$ of $\mathcal{V}^n$, called *atomic vectors*, or *generating vectors*, is the commutative ring generated by the brackets of any n-tuples of the m vectors.

In bracket notation, (2.1.3) can be written as

$$\sum_{i=1}^{n+1}(-1)^i[\mathbf{12}\cdots\check{\mathbf{i}}\cdots(\mathbf{n+1})]\mathbf{i} = 0, \qquad (2.1.6)$$

where $\check{\mathbf{i}}$ denotes that $\mathbf{i}$ does not occur in the sequence. (2.1.6) is the *Cramer's rule* of vectors $\mathbf{1}, \mathbf{2}, \ldots, \mathbf{n+1}$. If $[\mathbf{12}\cdots\mathbf{n}] \neq 0$, then (2.1.6) can be written as

$$(-1)^n[\mathbf{12}\cdots\mathbf{n}](\mathbf{n+1}) = \sum_{i=1}^{n}(-1)^i[\mathbf{12}\cdots\check{\mathbf{i}}\cdots(\mathbf{n+1})]\mathbf{i}. \qquad (2.1.7)$$

It is the *Cramer's rule* of vector $\mathbf{n+1}$ with respect to vectors $\mathbf{1}, \mathbf{2}, \ldots, \mathbf{n}$. By (2.1.7), a bracket can be interpreted as a component of the homogeneous coordinates of a point with respect to a basis of the vector space.

Definition 2.2. Let $f = f(\mathbf{x}_1, \mathbf{x}_2, \ldots, \mathbf{x}_m)$ be a polynomial in the homogeneous coordinates of vector variables $\mathbf{x}_1, \mathbf{x}_2, \ldots, \mathbf{x}_m \in \mathcal{V}^n$. If under any general linear transformation T of $\mathcal{V}^n$,

$$f(T(\mathbf{x}_1), T(\mathbf{x}_2), \ldots, T(\mathbf{x}_m)) = \det(T)^k f(\mathbf{x}_1, \mathbf{x}_2, \ldots, \mathbf{x}_m), \qquad (2.1.8)$$

where k is a nonnegative integer independent of T, then f is called a *classical invariant*. If $k = 0$, then f is called an *absolute invariant*. If for some $1 \le l \le m$,

$$f(T(\mathbf{x}_1), T(\mathbf{x}_2), \ldots, T(\mathbf{x}_l), \mathbf{x}_{l+1}, \mathbf{x}_{l+2}, \ldots, \mathbf{x}_m) = \det(T)^k f(\mathbf{x}_1, \mathbf{x}_2, \ldots, \mathbf{x}_m), \qquad (2.1.9)$$

then f is called a *classical covariant*.

Projective geometry is the subject on the properties of geometric configurations embedded in $\mathcal{V}^n$ that are invariant under the general linear group $GL(\mathcal{V}^n)$. *Classical invariant theory* is the subject on invariants and covariants of homogeneous polynomials [189]. The two subjects are closely related to each other by the homogeneous coordinates of projective geometric objects.

By (2.1.5), under a transformation $T \in GL(\mathcal{V}^n)$, any bracket changes by scale $\det(T)$. So the bracket is a classical invariant. The following theorem characterizes the essential role played by brackets in classical invariant theory.

Theorem 2.3. [First Fundamental Theorem in Classical Invariant Theory] Any classical invariant is a polynomial of brackets, called *bracket polynomial*.

In a bracket algebra, the brackets of the generating vectors are algebraically dependent, *i.e.*, they satisfy some polynomial relations. The polynomial relations satisfied by classical invariants are called *syzygies*. For example, by substituting (2.1.7) into any bracket containing vector $\mathbf{n+1}$, say $[\mathbf{1'2'}\cdots(\mathbf{n-1})'(\mathbf{n+1})]$, we get

$$\sum_{i=1}^{n+1}(-1)^i[\mathbf{12}\cdots\check{\mathbf{i}}\cdots(\mathbf{n+1})][\mathbf{1'2'}\cdots(\mathbf{n-1})'\mathbf{i}] = 0. \qquad (2.1.10)$$

(2.1.10) is called a *Grassmann-Plücker relation* (or *identity*). The left side of (2.1.10) is called a *Grassmann-Plücker* (GP) *syzygy* (or *polynomial*). The following theorem characterizes the essential role played by GP syzygies.

Theorem 2.4. [Second Fundamental Theorem in Classical Invariant Theory] Any syzygy of classical invariants in bracket algebra is in the ideal generated by GP syzygies.

For example, the 3D bracket algebra generated by 5 coplanar points $\mathbf{1}, \mathbf{2}, \mathbf{3}, \mathbf{4}, \mathbf{5}$ is the bracket polynomials of $C_5^3 = 10$ indeterminates

$$[\mathbf{123}], [\mathbf{124}], [\mathbf{125}], \ldots, [\mathbf{345}].$$

The ten brackets are not algebraically independent. They satisfy five GP relations:

$$\begin{aligned}
[\mathbf{123}][\mathbf{145}] - [\mathbf{124}][\mathbf{135}] + [\mathbf{125}][\mathbf{134}] &= 0, \\
[\mathbf{123}][\mathbf{245}] - [\mathbf{124}][\mathbf{235}] + [\mathbf{125}][\mathbf{234}] &= 0, \\
[\mathbf{123}][\mathbf{345}] - [\mathbf{134}][\mathbf{235}] + [\mathbf{135}][\mathbf{234}] &= 0, \\
[\mathbf{124}][\mathbf{345}] - [\mathbf{134}][\mathbf{245}] + [\mathbf{145}][\mathbf{234}] &= 0, \\
[\mathbf{125}][\mathbf{345}] - [\mathbf{135}][\mathbf{245}] + [\mathbf{145}][\mathbf{235}] &= 0.
\end{aligned} \tag{2.1.11}$$

The GP relations are not algebraically independent of each other. In (2.1.11), only three are algebraically independent, *e.g.*, the first three.

For a finite set of generic generating vectors in $\mathcal{V}^n$, the bracket algebra they generate is an integral domain, and even a unique factorization domain (UFD). If the generating vectors are not generic, the bracket algebra is still an integral domain but no longer a UFD. For example, if $\mathbf{1}, \mathbf{2}, \mathbf{3}, \mathbf{4}, \mathbf{5}$ are points in the projective plane, and $\mathbf{1}, \mathbf{2}, \mathbf{3}$ are collinear, then the bracket algebra generated by them is not a UFD, because by the following GP relation,

$$[\mathbf{124}][\mathbf{135}] - [\mathbf{125}][\mathbf{134}] = [\mathbf{123}][\mathbf{145}] = 0. \tag{2.1.12}$$

In the polynomial ring of coordinates, since coordinates are algebraically independent, monomials of coordinates comprise a basis of the ring, when the latter is taken as a $\mathbb{Z}$-*module*. It is common knowledge that any polynomial can be written uniquely as a sum of finitely many monomials with integer coefficients. This property is what we use daily in judging if two polynomials are equal or not. In bracket algebra, however, the algebraic dependencies among brackets make it a nontrivial task to judge whether or not two bracket polynomials are equal.

Definition 2.5. A bracket monomial is said to have *degree d* if it has d and only d bracket factors. A bracket monomial is said to have *degree r* in vector variable $\mathbf{x}$ if vector $\mathbf{x}$ occurs r and only r times in the bracket monomial.

A bracket monomial of degree k and coefficient ± 1 is of the following form:

$$[\mathbf{a}_{11}\mathbf{a}_{12}\cdots\mathbf{a}_{1n}][\mathbf{a}_{21}\mathbf{a}_{12}\cdots\mathbf{a}_{1n}]\cdots[\mathbf{a}_{k1}\mathbf{a}_{k2}\cdots\mathbf{a}_{kn}]. \tag{2.1.13}$$

Here the $\mathbf{a}$'s are generating vectors of the bracket algebra. We may arbitrarily define an order "$\prec$" among the generating vectors, and without loss of generality, we can assume that for any $1 \le i \le k$ in (2.1.13),

$$\mathbf{a}_{i1} \prec \mathbf{a}_{i2} \prec \cdots \prec \mathbf{a}_{in}. \tag{2.1.14}$$

No two elements in (2.1.14) are the same, because this would nullify the corresponding bracket. We write (2.1.13) in the form of a matrix:

$$\begin{bmatrix} \mathbf{a}_{11} & \mathbf{a}_{12} & \cdots & \mathbf{a}_{1n} \\ \mathbf{a}_{21} & \mathbf{a}_{12} & \cdots & \mathbf{a}_{1n} \\ \vdots & \vdots & \ddots & \vdots \\ \mathbf{a}_{k1} & \mathbf{a}_{k2} & \cdots & \mathbf{a}_{kn} \end{bmatrix}. \tag{2.1.15}$$

A $k \times n$ matrix $(\mathbf{a}_{ij})_{k \times n}$ whose entries satisfy (2.1.14) is called a *Young matrix.* A Young matrix is *standard* (or *straight*), if its entries in every column from top to bottom are in ascending order, *i.e.*, for any $1 \le j \le n$,

$$\mathbf{a}_{1j} \preceq \mathbf{a}_{2j} \preceq \cdots \preceq \mathbf{a}_{kj}. \tag{2.1.16}$$

For example, when $n = 3$ and $\mathbf{1} \prec \mathbf{2} \prec \mathbf{3} \prec \mathbf{4} \prec \mathbf{5}$,

$$\begin{bmatrix} \mathbf{125} \\ \mathbf{134} \end{bmatrix} \tag{2.1.17}$$

is not standard, because $\mathbf{5} \succ \mathbf{4}$ in the last column. However, by the first GP relation in (2.1.11),

$$[\mathbf{125}][\mathbf{134}] = -[\mathbf{123}][\mathbf{145}] + [\mathbf{124}][\mathbf{135}], \tag{2.1.18}$$

and the latter two monomials are standard Young matrices. So a nonstandard Young matrix can be "straightened" to a sum of standard ones. This phenomenon discloses a general property of standard Young matrices:

Theorem 2.6. [First Main Theorem in Classical Invariant Theory] Standard Young matrices comprise a basis of bracket algebra as a $\mathbb{Z}$-module.

The procedure leading to linear combination of standard Young matrices is called *straightening* [209]. The following is a brief description of the straightening procedure. Any nonstandard bracket monomial of degree two is of the form

$$\begin{bmatrix} \mathbf{a}_1 \mathbf{a}_2 \cdots \mathbf{a}_s & \mathbf{b}_1 \mathbf{b}_2 \cdots \mathbf{b}_{n-s} \\ \mathbf{c}_1 \mathbf{c}_2 \cdots \mathbf{c}_s & \mathbf{d}_1 \mathbf{d}_2 \cdots \mathbf{d}_{n-s} \end{bmatrix}, \tag{2.1.19}$$

where vectors $\mathbf{a}_i \preceq \mathbf{c}_i$ but $\mathbf{b}_1 \succ \mathbf{d}_1$. So

$$\mathbf{c}_1 \prec \mathbf{c}_2 \prec \cdots \prec \mathbf{c}_s \prec \mathbf{d}_1 \prec \mathbf{b}_1 \prec \mathbf{b}_2 \prec \cdots \prec \mathbf{b}_{n-s}. \tag{2.1.20}$$

To straighten monomial (2.1.19), one of $\mathbf{c}_1, \mathbf{c}_2, \ldots, \mathbf{c}_s, \mathbf{d}_1$ should switch with one of the $\mathbf{b}$'s. Such a switch strictly reduces the order of the sequence formed by the two rows in (2.1.19). Doing this kind of switch recursively, with the strict reduction of the order each time, in finite steps we will get a sum of straight monomials. This is the *straightening algorithm* in classical invariant theory.

The switches in the straightening algorithm are based on the so-called *van der Waerden syzygies* [177]. To describe the syzygies, we introduce a very useful notation in invariant theory and Hopf algebra, called *Sweedler's notation* [179].

Definition 2.7. Let $i_1 + i_2 + \cdots i_r = s$, where each i_j is a non-negative integer. Let S be a sequence of s elements. An *r-partition* of S of *shape* $\lambda = (i_1, i_2, \ldots, i_r)$, denoted by

$$\lambda \vdash S, \tag{2.1.21}$$

is a partition of S into r non-overlapping subsequences of length $i_1, i_2, \ldots, i_r$ respectively. A 2-partition is also called a *bipartition*.

The j-th subsequence is denoted by $S_{(j)}$. If this subsequence undergoes another r'-partition, the j'-th subsequence in the second partition is denoted by $S_{(jj')}$. In a summation $\sum_{\lambda \vdash S} f(S_{(j_1)}, S_{(j_2)}, \ldots, S_{(j_t)})$, where $\{j_1, j_2, \ldots, j_t\} \subseteq \{1, 2, \ldots, r\}$, the first nonzero subsequence in the summand, say $S_{(j_1)}$, is assumed to *carry the sign of permutation* of the new sequence $S_{(1)}, S_{(2)}, \ldots, S_{(r)}$ relative to the original S.

This notation of partition and summation is called *Sweedler's notation*.

Remark: Let $j_1 < j_2 < \ldots < j_r$ be a subsequence of $1, 2, \ldots, s$, and let its remainder in $1, 2, \ldots, s$ be $k_1 < k_2 < \ldots < k_{s-r}$. Then the sign of the permutation of $j_1, \ldots, j_r, k_1, \ldots, k_{s-r}$ is

$$(-1)^{\frac{r(r+1)}{2} + j_1 + j_2 + \cdots + j_r}. \tag{2.1.22}$$

Definition 2.8. Let there be three sequences of vectors $\mathbf{A}_s = \mathbf{a}_1, \mathbf{a}_2, \ldots, \mathbf{a}_s$, and $\mathbf{B}_{n+1} = \mathbf{b}_1, \mathbf{b}_2, \ldots, \mathbf{b}_{n+1}$, and $\mathbf{C}_{n-s-1} = \mathbf{c}_1, \mathbf{c}_2, \ldots, \mathbf{c}_{n-s-1}$, then

$$\sum_{(n-s, s+1) \vdash \mathbf{B}_{n+1}} [\mathbf{A}_s \mathbf{B}_{n+1\,(1)}][\mathbf{B}_{n+1\,(2)} \mathbf{C}_{n-s-1}] = 0 \tag{2.1.23}$$

is called the *van der Waerden* (VW) *relation* of the three sequences. The left side of the equality is called a *van der Waerden* (VW) *syzygy*.

For example, in 2D projective geometry there are only three kinds of VW syzygies, of three, four, and six terms respectively. The first two are GP syzygies, the last one is of the form

$$[\mathbf{123}][\mathbf{456}] - [\mathbf{124}][\mathbf{356}] + [\mathbf{125}][\mathbf{346}] + [\mathbf{134}][\mathbf{256}] - [\mathbf{135}][\mathbf{246}] + [\mathbf{145}][\mathbf{236}]. \tag{2.1.24}$$

Proposition 2.9. VW syzygy (2.1.23) is in the ideal generated by GP syzygies.

Proof. When $s = 0$, (2.1.23) is just the GP relation (2.1.10), now in the form

$$\sum_{(n,1) \vdash \mathbf{B}_{n+1}} [\mathbf{B}_{n+1\,(1)}][\mathbf{B}_{n+1\,(2)} \mathbf{C}_{n-1}] = 0. \tag{2.1.25}$$

Assume that (2.1.23) holds for $s = k-1$, and let $\mathbf{A}_{k-1} = \mathbf{a}_1, \mathbf{a}_2, \ldots, \mathbf{a}_{k-1}$. When $s = k$,

$$\sum_{(n-k,k+1)\vdash\mathbf{B}_{n+1}} [\mathbf{A}_k\mathbf{B}_{n+1(1)}][\mathbf{B}_{n+1(2)}\mathbf{C}_{n-k-1}]$$

$$= \frac{1}{k+1} \sum_{(n-k,k,1)\vdash\mathbf{B}_{n+1}} [\mathbf{A}_{k-1}\mathbf{a}_k\mathbf{B}_{n+1(1)}][\mathbf{B}_{n+1(2)}\mathbf{B}_{n+1(3)}\mathbf{C}_{n-k-1}]$$

$$= \frac{(-1)^{1+(n+1)}}{k+1} \sum_{(n-k+1,k-1,1)\vdash\mathbf{B}_{n+1}} [\mathbf{A}_{k-1}\mathbf{B}_{n+1(1)}][\mathbf{B}_{n+1(2)}\mathbf{B}_{n+1(3)}\mathbf{a}_k\mathbf{C}_{n-k-1}]$$

$$= \frac{(-1)^n k}{k+1} \sum_{(n-k+1,k)\vdash\mathbf{B}_{n+1}} [\mathbf{A}_{k-1}\mathbf{B}_{n+1(1)}][\mathbf{B}_{n+1(2)}\mathbf{a}_k\mathbf{C}_{n-k-1}]$$

$$= 0,$$

where the transfer of $\mathbf{a}_k$ from the first bracket to the second bracket in the second step is caused by the induction hypothesis upon three sequences $\mathbf{A}_{k-1}$, $\mathbf{a}_k\mathbf{B}_{n+1(1)}\mathbf{B}_{n+1(2)}$, and $\mathbf{B}_{n+1(3)}\mathbf{C}_{n-k-1}$, by which the subtotal of the partitions where $\mathbf{a}_k$ is in the first bracket equals negative the subtotal of the partitions where $\mathbf{a}_k$ is in the second bracket. $\square$

We return to the straightening of (2.1.19). By the VW relation among sequences

$$\begin{aligned}
\mathbf{A}_s &= \mathbf{a}_1, \mathbf{a}_2, \ldots, \mathbf{a}_s, \\
\mathbf{B}_{n+1} &= \mathbf{b}_1, \mathbf{b}_2, \ldots, \mathbf{b}_{n-s}, \mathbf{c}_1, \mathbf{c}_2, \ldots, \mathbf{c}_s, \mathbf{d}_1, \\
\mathbf{C}_{n-s-1} &= \mathbf{d}_2, \mathbf{d}_3, \ldots, \mathbf{d}_{n-s},
\end{aligned}$$

we have, by denoting $\mathbf{D} = \mathbf{b}_1, \mathbf{b}_2, \ldots, \mathbf{b}_{n-s}$,

$$\begin{bmatrix} \mathbf{a}_1 \cdots \mathbf{a}_s \ \mathbf{b}_1 \cdots \mathbf{b}_{n-s} \\ \mathbf{c}_1 \cdots \mathbf{c}_s \ \mathbf{d}_1 \cdots \mathbf{d}_{n-s} \end{bmatrix} = - \sum_{\substack{(n-s,s+1)\vdash\mathbf{B}_{n+1}, \\ \mathbf{B}_{n+1(1)}\neq\mathbf{D}}} \begin{bmatrix} \mathbf{A}_s & \mathbf{B}_{n+1(1)} \\ \mathbf{B}_{n+1(2)} & \mathbf{C}_{n-s-1} \end{bmatrix}. \tag{2.1.26}$$

On the right side of the above equality, at least one vector $\mathbf{b}_i$ is moved from the first row to the second row, and because of (2.1.20), each term on the right side has strictly lower order than the term on the left side.

Theorem 2.10. [Second Main Theorem in Classical Invariant Theory] The VW syzygies are a Gröbner base of the ideal generated by GP syzygies. The straightening procedure is the normal form reduction with respect to this base.

The straightening algorithm, unfortunately, generally does not lead to any simplification of a bracket polynomial expression, except for the special case where the bracket polynomial equals zero but is not identical to zero. In the exceptional case, the straightening algorithm definitely simplifies the polynomial to zero because of the basis property of the standard Young matrices.

Remarks on the definitions of "invariant" and "covariant" by Definition 2.2:

In classical invariant theory [189], given a homogeneous polynomial $p = p(\mathbf{x})$ of degree d in vector variable $\mathbf{x} = (x_1, x_2, \ldots, x_n)^T \in \mathcal{V}^n$, let its coefficients in monomial basis $\{x_1^{i_1} x_2^{i_2} \cdots x_n^{i_n} \mid i_1 + i_2 + \cdots + i_n = d\}$ be parameters $\lambda_1, \lambda_2, \ldots, \lambda_s$. A *relative invariant* of *weight* k of polynomial p, refers to a polynomial $g = g(\lambda_1, \lambda_2, \ldots, \lambda_s)$ such that under any general linear transformation T of $\mathcal{V}^n$, by using $\lambda_1', \lambda_2', \ldots, \lambda_s'$ to denote the coefficients of the polynomial $p'(\mathbf{x}) := p(T(\mathbf{x}))$,

$$g(\lambda_1', \lambda_2', \ldots, \lambda_s') = \det(T)^k g(\lambda_1, \lambda_2, \ldots, \lambda_s). \tag{2.1.27}$$

A *relative covariant* of *weight* k of the polynomial p, refers to a polynomial $h = h(\lambda_1, \lambda_2, \ldots, \lambda_s, \mathbf{x})$ such that under any general linear transformation T of $\mathcal{V}^n$,

$$h(\lambda_1', \lambda_2', \ldots, \lambda_s', T(\mathbf{x})) = \det(T)^k h(\lambda_1, \lambda_2, \ldots, \lambda_s, \mathbf{x}). \tag{2.1.28}$$

If there are r homogeneous polynomials $p_1, p_2, \ldots, p_r$ in the same set of vector variables $\mathbf{x}_1, \mathbf{x}_2, \ldots, \mathbf{x}_m$, a *simultaneous* (or *joint*) *relative invariant* of *weight* k of the r polynomials, refers to a polynomial $g = g(\lambda_1, \lambda_2, \ldots, \lambda_t)$, where the λ's are all the coefficients of the r polynomials, such that under any general linear transformation T of $\mathcal{V}^n$, equality (2.1.27) holds if index s is replaced by t.

Similarly, a *simultaneous* (or *joint*) *relative covariant* of *weight* k of the r polynomials, refers to a polynomial $h = h(\lambda_1, \lambda_2, \ldots, \lambda_t, \mathbf{x}_1, \mathbf{x}_2, \ldots, \mathbf{x}_m)$ such that under any general linear transformation T of $\mathcal{V}^n$,

$$h(\lambda_1', \ldots, \lambda_s', T(\mathbf{x}_1), \ldots, T(\mathbf{x}_m)) = \det(T)^k h(\lambda_1, \ldots, \lambda_s, \mathbf{x}_1, \ldots, \mathbf{x}_m). \tag{2.1.29}$$

Traditionally a homogeneous polynomial of degree d is called a *d-form*. A 1-form is also called a *linear form*, and a 2-form called a *quadratic form*.

So in classical invariant theory, an invariant (or covariant) given by Definition 2.2 is a **simultaneous relative invariant** (or **covariant**) of weight k of the following m (or l) **linear forms** dual to vectors $\mathbf{x}_i = (x_{i1}, x_{i2}, \ldots, x_{in})^T$ for $1 \leq i \leq m$:

$$p_i(\mathbf{x}_i) = \lambda_{i1} x_{i1} + \lambda_{i2} x_{i2} + \cdots + \lambda_{in} x_{in}. \tag{2.1.30}$$

However, this does not imply that our definition only covers invariants and covariants of linear forms. As will be shown in Chapter 3, the base vector space $\mathcal{V}^n$ can be enlarged to the vector space of d-forms, and any d-form of the original base space is a linear form of the new base space.

2.2　Brackets from the symbolic point of view

On one hand, because of the algebraic dependencies among brackets, two bracket monomials are equal if and only if they are identical after straightening. On the other hand, after each bracket is changed into a polynomial of coordinates using Laplace expansions of determinants, as every monomial in the coordinates of generic points is algebraically independent, the procedure of normalizing a coordinate-formed bracket monomial is just the procedure of expanding the multiplication of determinants into a polynomial of coordinates. The result of such an expansion is often a polynomial of hundreds of thousands of terms.

For example, let $\mathbf{1}, \mathbf{2}, \mathbf{3}, \mathbf{4}, \mathbf{5}$ be points in the projective plane, with homogeneous coordinates $\mathbf{i} = (x_i, y_i, z_i)$. Then in expanded coordinate polynomial form, the size of a simple bracket binomial can be much larger:

$$[\mathbf{123}][\mathbf{145}] + [\mathbf{124}][\mathbf{135}]$$

$$\begin{aligned}
= \ & -x_1 y_2 z_4 x_3 y_1 z_5 - x_1 y_2 z_4 x_5 z_1 y_3 + x_1^2 z_2 y_4 z_3 y_5 - x_1 z_2 y_4 x_3 z_1 y_5 \\
& - x_1 z_2 y_4 x_5 y_1 z_3 - x_2 y_1 z_4 x_1 y_3 z_5 + x_2 y_1^2 z_4 x_3 z_5 - x_2 y_1 z_4 x_3 z_1 y_5 \\
& + 2x_2 z_1 y_3 x_1 y_4 z_5 - x_2 z_1 y_3 x_1 z_4 y_5 - x_2 z_1 y_3 x_4 y_1 z_5 + x_2 z_1^2 y_3 x_4 y_5 \\
& - x_1 y_2 z_3 x_4 y_1 z_5 + 2x_1 y_2 z_3 x_4 z_1 y_5 + 2x_1 y_2 z_3 x_5 y_1 z_4 - x_1 y_2 z_3 x_5 z_1 y_4 \\
& - x_3 z_1 y_2 x_1 y_4 z_5 + 2x_3 z_1 y_2 x_1 z_4 y_5 + 2x_3 z_1 y_2 x_4 y_1 z_5 - 2x_3 z_1^2 y_2 x_4 y_5 \\
& + x_3 z_1^2 y_2 x_5 y_4 + x_1^2 y_2 z_3 y_4 z_5 - 2x_1^2 y_2 z_3 z_4 y_5 - x_1 z_2 y_3 x_5 y_1 z_4 - x_2 z_1 y_4 x_3 y_1 z_5 \\
& + 2x_1 z_2 y_3 x_5 z_1 y_4 - x_2 y_1 z_3 x_1 y_4 z_5 + 2x_2 y_1 z_3 x_1 z_4 y_5 + x_2 y_1^2 z_3 x_4 z_5 \\
& - x_2 y_1 z_3 x_4 z_1 y_5 - 2x_2 y_1^2 z_3 x_5 z_4 + 2x_2 y_1 z_3 x_5 z_1 y_4 + 2x_2 z_1 y_3 x_5 y_1 z_4 \\
& - 2x_2 z_1^2 y_3 x_5 y_4 + 2x_3 y_1 z_2 x_1 y_4 z_5 - x_3 y_1 z_2 x_1 z_4 y_5 - 2x_3 y_1^2 z_2 x_4 z_5 \\
& + 2x_3 y_1 z_2 x_4 z_1 y_5 + x_3 y_1^2 z_2 x_5 z_4 - x_3 y_1 z_2 x_5 z_1 y_4 - x_4 z_1 y_2 x_1 y_3 z_5 \\
& - x_4 z_1 y_2 x_5 y_1 z_3 + x_4 z_1^2 y_2 x_5 y_3 - 2x_1^2 z_2 y_3 y_4 z_5 + x_1^2 z_2 y_3 z_4 y_5 + x_2 z_1^2 y_4 x_3 y_5 \\
& + 2x_1 z_2 y_3 x_4 y_1 z_5 - x_1 z_2 y_3 x_4 z_1 y_5 - x_3 z_1 y_2 x_5 y_1 z_4 + x_1^2 y_2 z_4 y_3 z_5 \\
& - x_4 y_1 z_2 x_1 z_3 y_5 + x_4 y_1^2 z_2 x_5 z_3 - x_4 y_1 z_2 x_5 z_1 y_3 - x_2 z_1 y_4 x_1 z_3 y_5.
\end{aligned}$$

From the symbolic computation point of view, in order to control the size of middle expression swell, it is very important to use brackets directly in representation and computation. A definition of bracket algebra in terms of bracket symbols and vector symbols, instead of the determinants of homogeneous coordinates, is needed in symbolic manipulation of brackets [192].

Definition 2.11. [Definition of bracket algebra by symbols] Let $\mathbf{a}_1, \ldots, \mathbf{a}_m$ be symbols, and let $m \geq n$. Let the $[\mathbf{a}_{i_1} \cdots \mathbf{a}_{i_n}]$ be indeterminates over $\mathbb{K}$ for each ordered sequence $1 \leq i_1, \ldots, i_n \leq m$, called *brackets*. The nD *bracket algebra* generated by the $\mathbf{a}$'s, is the quotient of the polynomial ring generated by brackets modulo the ideal generated by the following syzygies:

B1. $[\mathbf{a}_{i_1} \mathbf{a}_{i_2} \cdots \mathbf{a}_{i_n}]$ if $i_j = i_k$ for some $j \neq k$.

B2. $[\mathbf{a}_{i_1} \mathbf{a}_{i_2} \cdots \mathbf{a}_{i_n}] - \mathrm{sign}(\sigma)[\mathbf{a}_{i_{\sigma(1)}} \cdots \mathbf{a}_{i_{\sigma(n)}}]$ for any permutation σ of $1, 2, \ldots, n$.

GP. GP syzygy (2.1.10), *i.e.*,

$$\sum_{k=1}^{n+1} (-1)^k [\mathbf{a}_{i_1} \mathbf{a}_{i_2} \cdots \breve{\mathbf{a}}_{i_k} \cdots \mathbf{a}_{i_{n+1}}][\mathbf{a}_{i_k} \mathbf{a}_{j_1} \mathbf{a}_{j_2} \cdots \mathbf{a}_{j_{n-1}}]. \tag{2.2.1}$$

The above definition is coordinate-free, and when supplemented with the following *bracket operator*, which allows for linear operations within a bracket, the definition agrees with Definition 2.1 by homogeneous coordinates [177]:

Definition 2.12. Let $\mathcal{B}$ be the nD bracket algebra generated by symbols $\mathbf{a}_1, \ldots, \mathbf{a}_m$ called *atomic vectors*, or *generating vectors*. Denote by $\mathbb{K}^m$ the mD vector space generated by the m symbols. The *bracket operator*, still denoted by "[ ]", from $\mathcal{A} = \underbrace{\mathbb{K}^m \times \mathbb{K}^m \times \cdots \times \mathbb{K}^m}_{n}$ to $\mathcal{B}$, is defined by

$$
\begin{array}{ccc}
\mathcal{A} & \longrightarrow & \mathcal{B} \\
(\mathbf{a}_{i_1}, \ldots, \mathbf{a}_{i_n}) & \longmapsto & [\mathbf{a}_{i_1} \cdots \mathbf{a}_{i_n}], \\
(\mathbf{x}_1, \ldots, \lambda\mathbf{x}_i, \ldots, \mathbf{x}_n) & \longmapsto & \lambda[\mathbf{x}_1 \cdots \mathbf{x}_n], \\
(\mathbf{x}_1, \ldots, \mathbf{x}_i + \mathbf{y}_i, \ldots, \mathbf{x}_n) & \longmapsto & [\mathbf{x}_1 \cdots \mathbf{x}_n] + [\mathbf{x}_1 \cdots \mathbf{x}_{i-1}\mathbf{y}_i\mathbf{x}_{i+1} \cdots \mathbf{x}_n].
\end{array}
\tag{2.2.2}
$$

Corollary 2.13. In $\mathcal{V}^n$, any n vectors are linearly dependent if and only if their bracket equals 0.

In the symbolic definition of bracket algebra, the base vector space $\mathcal{V}^n$ does not occur, so this definition is in pure scalar form. The most important feature of a vector space is that any $n + 1$ vectors are linearly dependent. This feature is embodied in the scalar form of Cramer's rule: GP syzygy (2.2.1).

Example 2.14. In the 3D bracket algebra generated by symbols $\mathbf{1}, \mathbf{2}, \mathbf{3}, \mathbf{4}, \mathbf{5}$, simplify the bracket polynomial

$$
\begin{aligned}
p = &-[\mathbf{125}][\mathbf{135}][\mathbf{145}][\mathbf{234}]^2 - [\mathbf{124}]^2[\mathbf{135}][\mathbf{235}][\mathbf{345}] \\
&+[\mathbf{125}][\mathbf{134}]^2[\mathbf{235}][\mathbf{245}] + [\mathbf{123}]^2[\mathbf{145}][\mathbf{245}][\mathbf{345}].
\end{aligned}
$$

The answer is $p = 0$. There are various ways to make the simplification. The straightening algorithm definitely can do so, although the procedure is long and boring. In coordinate form, each bracket monomial of p when expanded becomes a coordinate polynomial of 1986 terms. The sum of any two bracket monomials becomes a coordinate polynomial of 2526 terms. Adding any third bracket monomial reduces the size to 1986 terms, and the sum of four bracket monomials is zero. The computing by computer is fast, reliable but unreadable, not to mention pedagogical.

By 3-termed GP syzygies among the five generating elements, the simplification procedure is short, enjoyable and enlightening:

GP relations		p
$\begin{aligned}&[\mathbf{125}][\mathbf{234}]\\&=[\mathbf{124}][\mathbf{235}]-[\mathbf{123}][\mathbf{245}]\end{aligned}$	$\overset{explode}{=}$	$\begin{aligned}&-[\mathbf{124}][\mathbf{135}][\mathbf{145}][\mathbf{234}][\mathbf{235}] - [\mathbf{124}]^2[\mathbf{135}][\mathbf{235}][\mathbf{345}]\\&+[\mathbf{123}][\mathbf{135}][\mathbf{145}][\mathbf{234}][\mathbf{245}] + [\mathbf{123}]^2[\mathbf{145}][\mathbf{245}][\mathbf{345}]\\&+[\mathbf{125}][\mathbf{134}]^2[\mathbf{235}][\mathbf{245}]\end{aligned}$
$\begin{aligned}&[\mathbf{145}][\mathbf{234}]+[\mathbf{124}][\mathbf{345}]\\&=[\mathbf{134}][\mathbf{245}],\\&[\mathbf{135}][\mathbf{234}]+[\mathbf{123}][\mathbf{345}]\\&=[\mathbf{134}][\mathbf{235}]\end{aligned}$	$\overset{contract}{=}$	$\begin{aligned}&-[\mathbf{124}][\mathbf{134}][\mathbf{135}][\mathbf{235}][\mathbf{245}]\\&+[\mathbf{123}][\mathbf{134}][\mathbf{145}][\mathbf{235}][\mathbf{245}]\\&+[\mathbf{125}][\mathbf{134}]^2[\mathbf{235}][\mathbf{245}]\end{aligned}$
$\begin{aligned}&[\mathbf{123}][\mathbf{145}]-[\mathbf{124}][\mathbf{135}]\\&=-[\mathbf{125}][\mathbf{134}]\end{aligned}$	$\overset{contract}{=}$	$0.$

In the first step, monomial $[\mathbf{125}][\mathbf{234}]$ is "exploded" to binomial $[\mathbf{124}][\mathbf{235}] - [\mathbf{123}][\mathbf{245}]$ by a GP relation. $[\mathbf{124}][\mathbf{235}]$ and $[\mathbf{123}][\mathbf{245}]$ are also factors of the second and the last terms of p respectively, and the four terms form two pairs that are respectively "contractible" to single terms by GP relations. When p is changed into a bracket trinomial at the end of the second step, another contraction based on a GP relation changes it to zero. These manipulations are tricky but not accidental. There are systematic ways to generate them by computer programs in Chapter 3.

From the bracket point of view, all homogeneous coordinates are brackets. Let $\mathbf{a}_1, \mathbf{a}_2, \ldots, \mathbf{a}_n$ be linearly independent vectors. Then any vector $\mathbf{a}_{n+1}$ satisfies (2.1.7) with $\mathbf{i}$ denoting $\mathbf{a}_i$. The homogeneous coordinates of $\mathbf{a}_{n+1}$ are brackets

$$\{[\mathbf{a}_1 \mathbf{a}_2 \cdots \breve{\mathbf{a}}_i \cdots \mathbf{a}_{n+1}], \, | \, 1 \le i \le n\}. \tag{2.2.3}$$

When $\mathbf{a}_{n+1}$ varies among different generic points, the brackets in (2.2.3) are algebraically independent. Thus, using coordinates is equivalent to eliminating algebraic dependencies among brackets by reducing all other brackets to polynomials of the brackets of the form (2.2.3). Without resorting to Cramer's rules which are executed to vectors instead of brackets, algebraic dependencies among brackets are eliminated by the following *coordinatization syzygies.*

Proposition 2.15. Let $\mathbf{a}_1, \ldots, \mathbf{a}_m$ be generating vectors of a bracket algebra, and let the first n vectors be linearly independent. Then any GP syzygy, after being multiplied with some power of $[\mathbf{a}_1 \mathbf{a}_2 \cdots \mathbf{a}_n]$, is generated by the following algebraically independent subset of GP syzygies, called the *coordinatization syzygies* with respect to $\mathbf{a}_1, \mathbf{a}_2, \ldots, \mathbf{a}_n$, and where $1 \le i \le n$ is arbitrary:

$$
\begin{aligned}
& [\mathbf{a}_1 \mathbf{a}_2 \cdots \mathbf{a}_n][\mathbf{a}_{j_1} \mathbf{a}_{j_2} \cdots \mathbf{a}_{j_n}] \\
& - \sum_{k=1}^{n} [\mathbf{a}_1 \mathbf{a}_2 \cdots \mathbf{a}_{i-1} \mathbf{a}_{j_k} \mathbf{a}_{i+1} \cdots \mathbf{a}_n][\mathbf{a}_{j_1} \mathbf{a}_{j_2} \cdots \mathbf{a}_{j_{k-1}} \mathbf{a}_i \mathbf{a}_{j_{k+1}} \cdots \mathbf{a}_{j_n}].
\end{aligned}
\tag{2.2.4}
$$

Proof. If at most one of $\mathbf{a}_{j_1}, \mathbf{a}_{j_2}, \ldots, \mathbf{a}_{j_n}$ is not in $\{\mathbf{a}_1, \ldots, \mathbf{a}_n\}$, then (2.2.4) is trivially zero. In the following, this case is not considered.

Assume that the $\mathbf{a}$'s are ordered by their subscripts, and that any bracket of the $\mathbf{a}$'s has its elements rearranged in strictly ascending order. Define an order among the brackets as follows:

(1) $[\mathbf{a}_{j_1} \mathbf{a}_{j_2} \cdots \mathbf{a}_{j_n}] \prec [\mathbf{a}_{l_1} \mathbf{a}_{l_2} \cdots \mathbf{a}_{l_n}]$ if the former bracket contains more elements in $\{\mathbf{a}_1, \ldots, \mathbf{a}_n\}$.
(2) $[\mathbf{a}_{j_1} \mathbf{a}_{j_2} \cdots \mathbf{a}_{j_n}] \prec [\mathbf{a}_{l_1} \mathbf{a}_{l_2} \cdots \mathbf{a}_{l_n}]$ if they contain the same number of elements of $\{\mathbf{a}_1, \ldots, \mathbf{a}_n\}$, but lexicographically $(\mathbf{a}_{j_1}, \mathbf{a}_{j_2}, \ldots, \mathbf{a}_{j_n}) \prec (\mathbf{a}_{l_1}, \mathbf{a}_{l_2}, \cdots, \mathbf{a}_{l_n})$.

By the above ordering, the leading terms of (2.2.4) for different subsets $\{\mathbf{a}_{j_1}, \mathbf{a}_{j_2}, \ldots, \mathbf{a}_{j_n}\} \subset \{\mathbf{a}_1, \mathbf{a}_2, \ldots, \mathbf{a}_m\}$ are different, the leading coefficients are identical and nonzero, so these syzygies are algebraically independent. By repeatedly applying (2.2.4) to replace the leading term with the remaining terms, any bracket $[\mathbf{a}_{j_1} \mathbf{a}_{j_2} \cdots \mathbf{a}_{j_n}]$ after being multiplied with some power of $[\mathbf{a}_1 \mathbf{a}_2 \cdots \mathbf{a}_n]$, is changed into a polynomial of indeterminates

$$\{[\mathbf{a}_1 \mathbf{a}_2 \cdots \mathbf{a}_n], \, [\mathbf{a}_1 \mathbf{a}_2 \cdots \mathbf{a}_{i-1} \mathbf{a}_{j_k} \mathbf{a}_{i+1} \cdots \mathbf{a}_n] \, | \, 1 \le i, k \le n\}.$$

These indeterminates are algebraically independent. In this way, any syzygy of the bracket algebra is reduced to zero by (2.2.4) after being multiplied with some power of $[\mathbf{a}_1\mathbf{a}_2\cdots\mathbf{a}_n]$. $\qquad\square$

A GP syzygy is an r-termed quadratic bracket polynomial, where r ranges from 3 to $n+1$. The following proposition discloses further relations among GP syzygies of different terms.

Proposition 2.16. Any GP syzygy when multiplied by a suitable bracket monomial, is in the ideal generated by 3-termed GP polynomials (or GP *trinomials*).

Proof.   Let r be the number of terms of the GP syzygy. When $r = 2$, the conclusion is trivial. Assume that the conclusion holds for r.

Any $(r+1)$-termed GP polynomial g_{r+1} is of the form

$$g_{r+1} = \sum_{(1,r)\vdash\mathbf{B}_{r+1}} [\mathbf{C}_{n-r}\mathbf{A}_{r-1}\mathbf{B}_{r+1\,(1)}][\mathbf{C}_{n-r}\mathbf{B}_{r+1\,(2)}], \qquad (2.2.5)$$

where $\mathbf{A}_{r-1}, \mathbf{B}_{r+1}, \mathbf{C}_{n-r}$ are sequences of the generating vectors, with length $r-1, r+1, n-r$ respectively. Let $\mathbf{A}_{r-1} = (\mathbf{A}_{r-2}, \mathbf{a})$ be a fixed partition of $\mathbf{A}_{r-1}$ of shape $(r-2, 1)$, and let $\mathbf{B}_{r+1} = (\mathbf{b}_1, \mathbf{b}_2, \mathbf{B}_{r-1})$ be a fixed partition of $\mathbf{B}_{r+1}$ of shape $(1, 1, r-1)$. Denote $\mathbf{C} = \mathbf{C}_{n-r}$.

Let

$$\begin{aligned}
g_r(\mathbf{b}_1) &= \sum_{(1,r)\vdash\mathbf{B}_{r+1}} [\mathbf{C}\mathbf{A}_{r-2}\mathbf{b}_1\mathbf{B}_{r+1\,(1)}][\mathbf{C}\mathbf{B}_{r+1\,(2)}], \\
g_r(\mathbf{b}_2) &= - \sum_{(1,r)\vdash\mathbf{B}_{r+1}} [\mathbf{C}\mathbf{A}_{r-2}\mathbf{b}_2\mathbf{B}_{r+1\,(1)}][\mathbf{C}\mathbf{B}_{r+1\,(2)}], \\
g_3(\mathbf{x}) &= [\mathbf{C}\mathbf{A}_{r-1}\mathbf{b}_1][\mathbf{C}\mathbf{A}_{r-2}\mathbf{b}_2\mathbf{x}] - [\mathbf{C}\mathbf{A}_{r-1}\mathbf{b}_2][\mathbf{C}\mathbf{A}_{r-2}\mathbf{b}_1\mathbf{x}] \\
&\quad + [\mathbf{C}\mathbf{A}_{r-1}\mathbf{x}][\mathbf{C}\mathbf{A}_{r-2}\mathbf{b}_1\mathbf{b}_2].
\end{aligned} \qquad (2.2.6)$$

Then $g_r(\mathbf{b}_1)$ and $g_r(\mathbf{b}_2)$ are r-termed GP polynomials, and $g_3(\mathbf{x})$ is a 3-termed GP polynomial. By induction hypothesis, and the following trivial identity:

$$\begin{aligned}
[\mathbf{C}\mathbf{A}_{r-2}\mathbf{b}_1\mathbf{b}_2]g_{r+1} &= [\mathbf{C}\mathbf{A}_{r-1}\mathbf{b}_2]g_r(\mathbf{b}_1) + [\mathbf{C}\mathbf{A}_{r-1}\mathbf{b}_1]g_r(\mathbf{b}_2) \\
&\quad + \sum_{(1,r)\vdash\mathbf{B}_{r+1}} g_3(\mathbf{B}_{r+1\,(1)})[\mathbf{C}\mathbf{B}_{r+1\,(2)}],
\end{aligned} \qquad (2.2.7)$$

g_{r+1} after being multiplied by some bracket monomials, is generated by GP trinomials. $\qquad\square$

Corollary 2.17. Any $(r+1)$-termed GP polynomial (2.2.5), when multiplied by bracket monomial

$$\prod_{i=1}^{r-2} [\mathbf{C}_{n-r}\mathbf{A}_i\mathbf{B}_{r-i}], \qquad (2.2.8)$$

where $\mathbf{A}_i$ is the first i element of $\mathbf{A}_{r-1}$, and $\mathbf{B}_{r-i}$ is the first $r-i$ elements of $\mathbf{B}_{r+1}$, is in the ideal generated by GP trinomials.

Proof. When (2.2.5) is multiplied by $[\mathbf{C}_{n-r}\mathbf{A}_{r-2}\mathbf{B}_2]$, it is decomposed into two r-termed GP polynomials $g_r(\mathbf{b}_1), g_r(\mathbf{b}_2)$ together with a collection of GP trinomials. In $g_r(\mathbf{b}_1)$, the role of $(\mathbf{C}_{n-r}, \mathbf{A}_{r-1}, \mathbf{B}_{r+1})$ in (2.2.5) is played by $(\mathbf{C}_{n-r}\mathbf{b}_1, \mathbf{A}_{r-2}, \mathbf{B}_{r+1} - \{\mathbf{b}_1\})$, so when multiplied by $[\mathbf{C}_{n-r}\mathbf{A}_{r-3}\mathbf{B}_3]$, $g_r(\mathbf{b}_1)$ is further decomposed into two $(r-1)$-termed GP polynomials together with some GP trinomials. The same multiplier also applies to $g_r(\mathbf{b}_2)$. By induction, the conclusion is proved. $\qquad\square$

Below we introduce an important concept called *deficit bracket*. The number of vectors in a bracket equals the dimension of the base vector space. However, in the base vector space there are various subspaces, and accordingly there are various brackets of different lengths. The concept deficit bracket unifies all such brackets.

First let us recall how this problem is tackled in linear algebra by homogeneous coordinates. Given an rD subspace spanned by vectors $\mathbf{A}_r = \mathbf{a}_1, \mathbf{a}_2, \ldots, \mathbf{a}_r$, we supplement them with another $n - r$ vectors $\mathbf{U}_{n-r} = \mathbf{u}_1, \mathbf{u}_2, \ldots, \mathbf{u}_{n-r}$ to form a basis of $\mathcal{V}^n$. Then any vector $\mathbf{b}$ in the rD subspace has homogeneous coordinates

$$
\begin{aligned}
[\mathbf{U}_{n-r}\mathbf{A}_r]\mathbf{b} &= \sum_{(n-1,1)\vdash\mathbf{U}_{n-r}\mathbf{A}_r} [(\mathbf{U}_{n-r}\mathbf{A}_r)_{(1)}\mathbf{b}](\mathbf{U}_{n-r}\mathbf{A}_r)_{(2)} \\
&= \sum_{(r-1,1)\vdash\mathbf{A}_r} [\mathbf{U}_{n-r}\mathbf{A}_{r(1)}\mathbf{b}]\mathbf{A}_{r(2)},
\end{aligned}
\tag{2.2.9}
$$

where by Corollary 2.13, the coefficients of the vectors $\mathbf{u}$'s on the right side of (2.2.9) are zero. So $[\mathbf{U}_{n-r}\mathbf{A}_r]$ serves as a bracket of the rD bracket algebra. The deficit bracket is a universal extension of this construction to the bracket algebras based on all kinds of subspaces of $\mathcal{V}^n$.

Let $\mathbf{u}_1, \ldots, \mathbf{u}_{n-1}$ be $n - 1$ generic vectors in $\mathcal{V}^n$, called *dummy vectors*. By this we assume that $[\mathbf{u}_1\mathbf{u}_2\cdots\mathbf{u}_{n-r}\mathbf{a}_{i_1}\cdots\mathbf{a}_{i_r}] \neq 0$ for all the generating vectors $\mathbf{a}_1, \mathbf{a}_2, \ldots, \mathbf{a}_m$ of the bracket algebra. Then for any $r \leq n$ vectors $\mathbf{B}_r = \mathbf{b}_1, \mathbf{b}_2, \ldots, \mathbf{b}_r$ in $\mathcal{V}^n$, their r-*deficit bracket* is

$$
[\mathbf{B}_r] := [\mathbf{u}_1\cdots\mathbf{u}_{n-r}\mathbf{B}_r].
\tag{2.2.10}
$$

In fact, we have already met with deficit brackets in (2.2.5). Any $(r+1)$-termed GP polynomial can be viewed as a GP syzygy in an rD subspace by taking $\mathbf{C}_{n-r}$ as dummy vectors.

Because deficit brackets adopt the same notation as regular brackets, later on in this book, a bracket or bracket symbol in which the number of vectors is less than the dimension of the base vector space, is taken as a deficit bracket only when *explicitly announced*.

2.3 Covariants, duality and Grassmann-Cayley algebra

The concept of deficit bracket in the previous section provides the simplest example of a classical covariant. Recall that a covariant is obtained from an invariant of vector variables by deleting several vector variables from it, resulting in a scalar-valued functional in the deleted variables. A covariant is said to be *multilinear* if the original invariant is linear with respect to every vector variable that is deleted later. In this chapter, only multilinear covariants are considered.

In the simplest case of a 3D bracket algebra, a bracket, say [**123**], is the 3×3 determinant of the homogeneous coordinates of the three vector variables. If two vector variables, say **2**, **3**, are deleted from the bracket, the result is a covariant, and is just vector **1**. So any vector is a covariant. If only one vector variable, say **3**, is deleted, the result is a covariant that is *linear* and *anticommutative* with respect to vector variables **1** and **2**. It is denoted by $\mathbf{1} \wedge \mathbf{2}$, or simply by **12** if no other product adopts the juxtaposition notation.

This simple observation leads to the two fundamental concepts *outer product* and *Grassmann algebra*. To make a formal definition, we need a general characterization of an *algebra* over a base field.

Definition 2.18. A $\mathbb{K}$-*algebra* is a vector space $\mathcal{V}$ over $\mathbb{K}$ together with an associative $\mathbb{K}$-linear mapping, called the *product* and denoted by juxtaposition of elements, and a linear mapping $i : \mathbb{K} \longrightarrow \mathcal{V}$ called the *unit map*, such that for any $\lambda \in \mathbb{K}$ and $\mathbf{x} \in \mathcal{V}$,

$$i(\lambda)\mathbf{x} = \mathbf{x}i(\lambda) = \lambda\mathbf{x}. \tag{2.3.1}$$

Usually the i-notation is omitted, and (2.3.1) allows the product of $i(\lambda)$ and $\mathbf{x}$ to be identified with the scaling of vector $\mathbf{x}$ by λ.

Notation.

The symbol "$\otimes$" is the standard notation of the *tensor product* in the tensor algebra generated by a linear space. It is the associative product with and only with the following *multilinear property*:

$$(\lambda_1 \mathbf{A}_1 + \mu_1 \mathbf{B}_1) \otimes (\lambda_2 \mathbf{A}_2 + \mu_2 \mathbf{B}_2) = \begin{aligned} &\lambda_1\lambda_2\mathbf{A}_1 \otimes \mathbf{A}_2 + \mu_1\lambda_2\mathbf{B}_1 \otimes \mathbf{A}_2 \\ &+\lambda_1\mu_2\mathbf{A}_1 \otimes \mathbf{B}_2 + \mu_1\mu_2\mathbf{B}_1 \otimes \mathbf{B}_2, \quad \forall\lambda_i, \mu_j \in \mathbb{K}. \end{aligned}$$

The tensor algebra generated by $\mathcal{V}^n$ is the linear space spanned by all the tensor product results of vectors of $\mathcal{V}^n$, denoted by $\otimes(\mathcal{V}^n)$.

Definition 2.19. The *Grassmann algebra* $\Lambda(\mathcal{V}^n)$ is the $\mathbb{K}$-algebra obtained as the quotient of the tensor algebra $\otimes(\mathcal{V}^n)$ modulo the two-sided ideal generated by elements of the form $\mathbf{x} \otimes \mathbf{x}$ for $\mathbf{x} \in \mathcal{V}^n$. The quotient of the tensor product is called the *outer product*, also known as the *exterior product*, or *Grassmann product*. $\mathbb{K}$ is a 1D subspace of the Grassmann algebra, and the unit map is the identity transformation in $\mathbb{K}$.

When $\Lambda(\mathcal{V}^n)$ is viewed as a vector space, it is called a *Grassmann space*. Its elements are called *multivectors*.

Notation.

In this book we always use the juxtaposition to denote the fundamental product of an algebra, if there is more than one product in this algebra. The fundamental product precedes any other product by default. In the setting of Grassmann algebra and Grassmann-Cayley algebra, it is the outer product that is denoted by juxtaposition, and precedes all other products by default. In other settings, the outer product is denoted by "$\wedge$".

Formally, the Grassmann algebra can be defined from the tensor algebra by defining the outer product as the *complete antisymmetrization* of the tensor product: for any vectors $\mathbf{a}_1, \mathbf{a}_2, \ldots, \mathbf{a}_r$,

$$\mathbf{a}_1 \wedge \mathbf{a}_2 \wedge \cdots \wedge \mathbf{a}_r := \frac{1}{r!} \sum_{\sigma} \operatorname{sign}(\sigma) \mathbf{a}_{\sigma(1)} \otimes \mathbf{a}_{\sigma(2)} \otimes \cdots \otimes \mathbf{a}_{\sigma(r)}, \qquad (2.3.2)$$

where the summation runs over all permutations σ of $1, 2, \ldots, r$. In other words, the Grassmann space is the space of all antisymmetric tensors. Informally, a Grassmann algebra is generated from a base vector space by the generating relation that the product of any vector with itself is zero.

Grassmann space is finite-dimensional. Since any $n + 1$ vectors are linearly dependent, by anticommutativity their outer product is always zero. So the Grassmann space is spanned by the outer products of at most n vectors in $\mathcal{V}^n$.

Definition 2.20. The subspace of $\Lambda(\mathcal{V}^n)$ spanned by the outer products of r-tuples of vectors in $\mathcal{V}^n$ is denoted by $\Lambda^r(\mathcal{V}^n)$. Any element of the space is called an *r-vector*, or *r-extensor*, and r is called the *grade* (or *step*) of the element. A multivector is said to be *homogeneous*, if it is an r-vector for some r. If an r-vector can be decomposed into the outer product of r vectors, then it is called an *r-blade* (or *decomposable r-extensor*). All r-blades form a projective variety, called the *r-Grassmann variety*.

The Grassmann space $\Lambda(\mathcal{V}^n)$ is the direct sum of its r-vector subspaces. The component of a multivector $\mathbf{A}$ in the r-vector subspace is called its *r-graded part*, denoted by $\langle \mathbf{A} \rangle_r$. In particular, the base field $\mathbb{K}$ is the 0-vector subspace of $\Lambda(\mathcal{V}^n)$, and $\mathcal{V}^n$ is the 1-vector subspace. 0-vectors are called *scalars*, 1-vectors are still called *vectors*, and 2-vectors are called *bivectors*, *etc.* In the reverse direction, an n-vector is called a *pseudoscalar*, an $(n-1)$-vector is called a *pseudovector*, an $(n-2)$-vector is called a *pseudobivector*, *etc.* The *r-grading operator* $\langle \ \rangle_r$ returns the r-graded part of a multivector.

Blades have clear geometric meaning. In projective geometry, r points are in the same $(r-2)$D projective subspace if and only if their outer product is zero. So the $(r-1)$D projective subspace spanned by r points can be represented by their outer product, denoted by $\mathbf{A}_r$, in the sense that any point $\mathbf{x}$ is in the subspace if and only if $\mathbf{x}\mathbf{A}_r = 0$. We use the same symbol to denote both a blade and the subspace it represents.

A *Grassmann subspace* of $\Lambda(\mathcal{V}^n)$ is a vector subspace of $\Lambda(\mathcal{V}^n)$ that is closed under the outer product. Any blade generates a Grassmann subspace from the vector subspace it represents. The Grassmann subspace generated by blade $\mathbf{A}_r$ is denoted by $\Lambda(\mathbf{A}_r)$.

Let $\mathbf{e}_1, \mathbf{e}_2, \ldots, \mathbf{e}_n$ be a basis of $\mathcal{V}^n$. The r-vector subspace $\Lambda^r(\mathcal{V}^n)$ is spanned by the *induced basis*

$$\{\mathbf{e}_{i_1}\mathbf{e}_{i_2}\cdots\mathbf{e}_{i_r} \mid 1 \le i_1 < i_2 < \ldots < i_r \le n\}, \tag{2.3.3}$$

so its dimension is C_n^r. The dimension of $\Lambda(\mathcal{V}^n)$ is 2^n.

For any other n-tuple of vectors $\mathbf{a}_1, \mathbf{a}_2, \ldots, \mathbf{a}_n$,

$$\mathbf{a}_1\mathbf{a}_2\cdots\mathbf{a}_n = \det(a_{ij})_{i,j=1..n}\,\mathbf{e}_1\mathbf{e}_2\cdots\mathbf{e}_n, \tag{2.3.4}$$

where $\mathbf{a}_i = (a_{i1}, a_{i2}, \ldots, a_{in})^T$. So bracket $[\mathbf{a}_1\mathbf{a}_2\cdots\mathbf{a}_n]$ is just the coordinate of pseudoscalar $\mathbf{a}_1\mathbf{a}_2\cdots\mathbf{a}_n$. Denote

$$\mathbf{I}_n = \mathbf{e}_1\mathbf{e}_2\cdots\mathbf{e}_n. \tag{2.3.5}$$

Then $[\mathbf{I}_n] = 1$.

The juxtaposition of vectors in a bracket has been taken as the sequence formed by the vectors since Definition 2.1. Now as a corollary of (2.3.4), the juxtaposition can also be taken as the outer product of the sequence of vectors, and the two interpretations agree with each other. The *bracket of a multivector* refers to the bracket of its pseudoscalar part.

The outer product of an r-vector $\mathbf{A}_r$ and an s-vector $\mathbf{B}_s$ is of grade $r + s$, and equals zero if $r + s > n$. It has the following *graded antisymmetric* (or *graded anticommutative*) property:

$$\mathbf{A}_r\mathbf{B}_s = (-1)^{rs}\mathbf{B}_s\mathbf{A}_r. \tag{2.3.6}$$

Let Λ, Λ' denote two Grassmann spaces. Their *tensor product* $\Lambda \otimes \Lambda'$ is the linear space spanned by elements of the form $\mathbf{A} \otimes \mathbf{A}'$, where $\mathbf{A} \in \Lambda$ and $\mathbf{A}' \in \Lambda'$. Vector space $\Lambda \otimes \Lambda'$ is a $\mathbb{Z}$-graded space whose grading is the sum of the gradings of Λ and Λ'. For example, if $\mathbf{A}, \mathbf{A}'$ are elements of grade r, r' in Λ, Λ' respectively, then $\mathbf{A} \otimes \mathbf{A}'$ has grade $r + r'$.

The graded space $\Lambda \otimes \Lambda'$ becomes a Grassmann algebra if endowed with the following outer product: let $\mathbf{A}_r, \mathbf{B}_s \in \Lambda$ and $\mathbf{A}'_{r'}, \mathbf{B}'_{s'} \in \Lambda'$, where the grades of the elements are r, s, r', s' respectively, then

$$(\mathbf{A}_r \otimes \mathbf{A}'_{r'})(\mathbf{B}_s \otimes \mathbf{B}'_{s'}) := \mathbf{A}_r\mathbf{B}_s \otimes \mathbf{A}'_{r'}\mathbf{B}'_{s'}. \tag{2.3.7}$$

$\Lambda \otimes \Lambda'$ equipped with the above outer product is called the *tensor product* of the two Grassmann algebras.

The tensor product of two Grassmann spaces can be endowed with another kind of outer product, called the *twisted outer product*:

$$(\mathbf{A}_r \otimes \mathbf{A}'_{r'})(\mathbf{B}_s \otimes \mathbf{B}'_{s'}) := (-1)^{r's}\mathbf{A}_r\mathbf{B}_s \otimes \mathbf{A}'_{r'}\mathbf{B}'_{s'}. \tag{2.3.8}$$

If we check the sign of permutation of the vector sequence of $\mathbf{A}_r, \mathbf{A}'_{r'}, \mathbf{B}_s, \mathbf{B}'_{s'}$ on the left side of (2.3.8), to the vector sequence of $\mathbf{A}_r, \mathbf{B}_s, \mathbf{A}'_{r'}, \mathbf{B}'_{s'}$ on the right side, we see that it is exactly $(-1)^{r's}$.

The space $\Lambda \otimes \Lambda'$ equipped with the above product is also a Grassmann algebra, called the *twisted tensor product*, or *graded tensor product*, of the two Grassmann algebras, and denoted by $\Lambda \hat{\otimes} \Lambda'$. The term "twisted" refers to the switch of position between $\mathbf{A}'_{r'}$ and $\mathbf{B}_s$ in (2.3.8).

Similarly, the tensor product of k Grassmann spaces can be given a twisted outer product by attaching the sign of permutation to the product. The result is the *twisted tensor product* of the k Grassmann algebras. For example, for vectors $\mathbf{a}$'s, $\mathbf{b}$'s and $\mathbf{c}$'s, in the twisted tensor product of three Grassmann algebras,

$$(\mathbf{a}_1 \otimes \mathbf{b}_1 \otimes \mathbf{c}_1)(\mathbf{a}_2 \otimes \mathbf{b}_2 \otimes \mathbf{c}_2)(\mathbf{a}_3 \otimes \mathbf{b}_3 \otimes \mathbf{c}_3) = -\mathbf{a}_1\mathbf{a}_2\mathbf{a}_3 \otimes \mathbf{b}_1\mathbf{b}_2\mathbf{b}_3 \otimes \mathbf{c}_1\mathbf{c}_2\mathbf{c}_3. \quad (2.3.9)$$

Notations.

(1) The symbol "$\oplus$" is the standard notation of the *direct sum* of two or more linear spaces $\mathcal{V}_1, \ldots, \mathcal{V}_r$, which is defined as the space

$$\{\lambda_1 \mathbf{x}_1 + \ldots + \lambda_r \mathbf{x}_r \mid \mathbf{x}_i \in \mathcal{V}_i, \lambda_i \in \mathbb{K}\},$$

under the assumption that the intersection of any two of the linear subspaces is the set composed of the zero vector.

(2) The symbol "$\cong$" denotes that the two sides are *algebraically isomorphic* to each other. So $\mathcal{A} \cong \mathcal{B}$ for two algebras $\mathcal{A}, \mathcal{B}$ means that there exists an invertible linear mapping $f : \mathcal{A} \longrightarrow \mathcal{B}$ such that $f(\mathbf{ab}) = f(\mathbf{a})f(\mathbf{b})$, for all $\mathbf{a} \in \mathcal{A}$ and $\mathbf{b} \in \mathcal{B}$. If $\mathcal{A}, \mathcal{B}$ are graded, then f is further required to preserve the grade: for any $\mathbf{a} \in \mathcal{A}$, $\mathbf{a}$ and $f(\mathbf{a})$ have the same grade.

Proposition 2.21. The following is an isomorphism of Grassmann algebras:

$$\Lambda(\mathcal{V}^m \oplus \mathcal{V}^n) \cong \Lambda(\mathcal{V}^m) \hat{\otimes} \Lambda(\mathcal{V}^n). \quad (2.3.10)$$

Proof. First, the two sides are isomorphic as *Grassmann spaces*, *i.e.*, there is a linear isomorphism between the two graded spaces that preserves the grading. One such isomorphism can be constructed as follows: for any $\mathbf{A} \in \Lambda(\mathcal{V}^m)$ and $\mathbf{A}' \in \Lambda(\mathcal{V}^n)$,

$$f(\mathbf{A} + \mathbf{A}') = \mathbf{A} \otimes 1 + 1 \otimes \mathbf{A}', \quad f(\mathbf{AA}') = \mathbf{A} \otimes \mathbf{A}'. \quad (2.3.11)$$

The grade of $\mathbf{A} \otimes \mathbf{A}'$ is the sum of the grades of $\mathbf{A}$ and $\mathbf{A}'$. It is easy to verify that f is such an isomorphism.

Second, for any r-blade $\mathbf{A}_r$ and s-blade $\mathbf{B}_s$ in $\Lambda(\mathcal{V}^m)$, for any r'-blade $\mathbf{A}'_{r'}$ and s'-blade $\mathbf{B}'_{s'}$ in $\Lambda(\mathcal{V}^n)$, we need to verify that

$$f((\mathbf{A}_r + \mathbf{A}'_{r'})(\mathbf{B}_s + \mathbf{B}'_{s'})) = f(\mathbf{A}_r + \mathbf{A}'_{r'})f(\mathbf{B}_s + \mathbf{B}'_{s'}). \quad (2.3.12)$$

On one hand,

$$\begin{aligned}
f((\mathbf{A}_r &+ \mathbf{A}'_{r'})(\mathbf{B}_s + \mathbf{B}'_{s'})) \\
&= f(\mathbf{A}_r\mathbf{B}_s + \mathbf{A}_r\mathbf{B}'_{s'} + \mathbf{A}'_{r'}\mathbf{B}_s + \mathbf{A}'_{r'}\mathbf{B}'_{s'}) \\
&= \mathbf{A}_r\mathbf{B}_s \otimes 1 + \mathbf{A}_r \otimes \mathbf{B}'_{s'} + (-1)^{r's}\mathbf{B}_s \otimes \mathbf{A}'_{r'} + 1 \otimes \mathbf{A}'_{r'}\mathbf{B}'_{s'}.
\end{aligned}$$

On the other hand, only under the twisted outer product can (2.3.12) hold:

$$
\begin{aligned}
f(\mathbf{A}_r &+ \mathbf{A}'_{r'})f(\mathbf{B}_s + \mathbf{B}'_{s'}) \\
&= (\mathbf{A}_r \otimes 1 + 1 \otimes \mathbf{A}'_{r'})(\mathbf{B}_s \otimes 1 + 1 \otimes \mathbf{B}'_{s'}) \\
&= (\mathbf{A}_r \otimes 1)(\mathbf{B}_s \otimes 1) + (\mathbf{A}_r \otimes 1)(1 \otimes \mathbf{B}'_{s'}) + (1 \otimes \mathbf{A}'_{r'})(\mathbf{B}_s \otimes 1) \\
&\qquad\qquad\qquad\qquad\qquad\qquad\qquad + (1 \otimes \mathbf{A}'_{r'})(1 \otimes \mathbf{B}'_{s'}) \\
&= \mathbf{A}_r\mathbf{B}_s \otimes 1 + \mathbf{A}_r \otimes \mathbf{B}'_{s'} + (-1)^{r's}\mathbf{B}_s \otimes \mathbf{A}'_{r'} + 1 \otimes \mathbf{A}'_{r'}\mathbf{B}'_{s'}.
\end{aligned}
$$

$\square$

The outer product has clear geometric meaning in projective geometry, and is a fundamental *projective geometric operation*. The first geometric interpretation is that the outer product of two blades is the *extension*, *i.e.*, the direct sum of the two vector subspaces represented by the two blades.

In a projective space, let $\mathbf{1}, \mathbf{2}$ be two points. Their outer product represents the projective *line* passing through them. In general, a projective point is a 1D direction, and a projective r-space is an $(r + 1)$D direction. The outer product of an rD direction and an sD direction is nonzero as long as they have no common 1D direction. When it is nonzero, the outer product represents the direct sum of the two high-dimensional directions, called their *extension*. Grassmann originally called his algebra "the theory of linear extensions" [68].

The second geometric explanation is that the outer product of a blade and any other element is the *perspective projection* of the element with respect to the blade representing the *perspective center*.

Definition 2.22. Let $\mathbf{e}_1, \mathbf{e}_2, \dots, \mathbf{e}_n$ be a basis of $\mathcal{V}^n$. An rD *perspective projection* in $\mathcal{V}^n$ refers to a linear mapping from $\mathcal{V}^n$ to itself in the matrix form

$$
\mathbf{T}^{-1}\mathrm{diag}(1, 1, \dots, 1, \underbrace{0, 0, \dots, 0}_{r})\mathbf{T}, \tag{2.3.13}
$$

where $\mathbf{T}$ is an invertible matrix representing a coordinate transformation in $\mathcal{V}^n$. Let $\mathbf{B}_r$ be the rD vector subspace spanned by basis vectors $\mathbf{e}_{n-r+1}, \mathbf{e}_{n-r+2}, \dots, \mathbf{e}_n$. The rD *perspective center* refers to the preimage of $\mathbf{B}_r$ under the coordinate transformation $\mathbf{T}$.

Let $\mathbf{A}_r$ be an r-blade. For any vector $\mathbf{x} \in \mathcal{V}^n$, the linear mapping

$$
\mathbf{x} \mapsto \mathbf{A}_r\mathbf{x} \tag{2.3.14}
$$

changes $\mathbf{x}$ to a vector in the $(n - r)$D vector space

$$
\mathbf{A}_r\mathcal{V}^n := \{\mathbf{A}_r\mathbf{y} \mid \mathbf{y} \in \mathcal{V}^n\}. \tag{2.3.15}
$$

That (2.3.14) is the rD perspective projection with perspective center $\mathbf{A}_r$ can be easily checked by its action on a fixed basis of $\mathcal{V}^n$. In $\Lambda(\mathcal{V}^n)$, mapping (2.3.14) is called the *outer product operator* induced by $\mathbf{A}_r$.

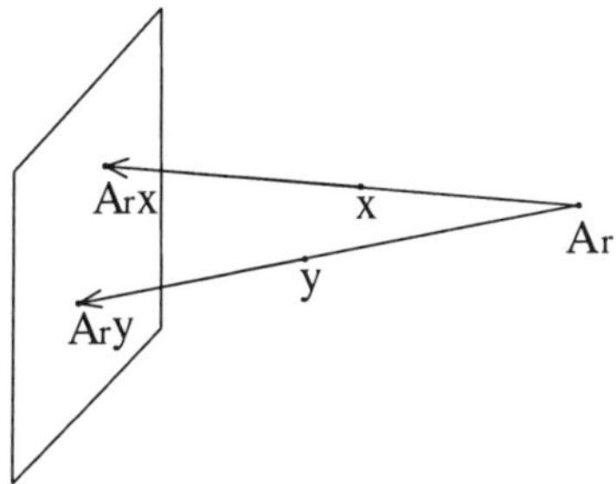

Fig. 2.1   Outer product explained as perspective projection.

The Grassmann algebra generated by vector space $\mathbf{A}_r \mathcal{V}^n$ has its outer product "$\wedge_{\mathbf{A}_r}$" induced from that of $\Lambda(\mathcal{V}^n)$ as follows:

$$(\mathbf{A}_r\mathbf{B}) \wedge_{\mathbf{A}_r} (\mathbf{A}_r\mathbf{C}) := \mathbf{A}_r\mathbf{BC}, \quad \forall\, \mathbf{B}, \mathbf{C} \in \Lambda(\mathcal{V}^n). \tag{2.3.16}$$

The unit map of this algebra is

$$i_{\mathbf{A}_r} : \lambda \in \mathbb{K} \longmapsto \lambda\mathbf{A}_r \in \Lambda^0(\mathbf{A}_r\mathcal{V}^n). \tag{2.3.17}$$

We have already met with perspective projection before. In the notation of deficit brackets with *dummy blade* (also called *dummy extensor*) $\mathbf{U}_{n-2}$, *i.e.*, the outer product of $n - 2$ dummy vectors, any GP trinomial is of the form

$$[\mathbf{U}_{n-2}12][\mathbf{U}_{n-2}34] - [\mathbf{U}_{n-2}13][\mathbf{U}_{n-2}24] + [\mathbf{U}_{n-2}14][\mathbf{U}_{n-2}23]. \tag{2.3.18}$$

It can be viewed as a GP syzygy in the 2D bracket algebra, or the 1D projective geometry, obtained from the $(n - 2)$D perspective projection with center $\mathbf{U}_{n-2}$. Corollary 2.17 says that after $r - 2$ rounds of 1D perspective projections whose centers are selected from $\mathbf{b}_1, \mathbf{b}_2, \ldots, \mathbf{b}_{r-1}$, any GP syzygy boils down to a collection of GP syzygies in 1D projective geometry up to a bracket monomial factor.

Duality is an important concept in projective geometry. For example, in 3D projective space, the dual of a point is a plane, and the dual of a line is a line. In the general case, such a duality, called *Hodge dual*, can be constructed and represented in Grassmann algebra as follows.

First, let the vector sequence $\mathbf{E}_n = \mathbf{e}_1, \mathbf{e}_2, \ldots, \mathbf{e}_n$ be a fixed basis of $\mathcal{V}^n$. Let $\mathcal{V}^{n*}$ be the dual vector space of $\mathcal{V}^n$. The *dual basis* of $\mathbf{E}_n$ in $\mathcal{V}^{n*}$ is denoted by $\mathbf{E}_n{}^* = \mathbf{e}_1^*, \mathbf{e}_2^*, \ldots, \mathbf{e}_n^*$. Under the natural correspondence between $\mathbf{e}_i$ and $\mathbf{e}_i^*$ for $1 \le i \le n$, vector spaces $\Lambda^r(\mathcal{V}^n)$ and $\Lambda^r(\mathcal{V}^{n*}) = (\Lambda^r(\mathcal{V}^n))^*$ are isomorphic for any $0 \le r \le n$. Denote this linear isomorphism from $\Lambda(\mathcal{V}^n)$ to $\Lambda(\mathcal{V}^{n*})$ by J.

Second, there is a natural isomorphism between $\Lambda^r(\mathcal{V}^n)$ and $\Lambda^{n-r}(\mathcal{V}^{n*})$: any $\mathbf{A} \in \Lambda^r(\mathcal{V}^n)$ corresponds to a unique element $\jmath_{\mathbf{A}} \in \Lambda^{n-r}(\mathcal{V}^{n*}) = (\Lambda^{n-r}(\mathcal{V}^n))^*$ as follows:

$$\begin{aligned} \jmath_{\mathbf{A}} : \Lambda^{n-r}(\mathcal{V}^n) &\longrightarrow \mathbb{K} \\ \mathbf{B} &\longmapsto [\mathbf{AB}]. \end{aligned} \tag{2.3.19}$$

Definition 2.23. The composition

$$* := J^{-1} \circ \jmath : \quad \Lambda^r(\mathcal{V}^n) \longrightarrow \Lambda^{n-r}(\mathcal{V}^n), \ 0 \le r \le n \qquad (2.3.20)$$

is a linear isomorphism in $\Lambda(\mathcal{V}^n)$, called the *Hodge dual operator* determined by basis $\mathbf{E}_n$.

Example 2.24. Let there be a partition of $\mathbf{E}_n$ of shape $(r, n - r)$. The outer product of the vectors in $\mathbf{E}_{n(1)}$, denoted by $\wedge \mathbf{E}_{n(1)}$, is a basis vector of $\Lambda^r(\mathcal{V}^n)$. Its Hodge dual is

$$*(\wedge \mathbf{E}_{n(1)}) = \wedge \mathbf{E}_{n(2)}. \qquad (2.3.21)$$

In particular, the dual of a pseudoscalar $\mathbf{A}$ is its bracket $[\mathbf{A}]$, and the dual of a scalar λ is $\lambda \mathbf{I}_n$, where $\mathbf{I}_n = \wedge \mathbf{E}_n$.

The Hodge dual operator is a *grade-dependent involution*:

$$* * \mathbf{A}_r = (-1)^{r(n-r)} \mathbf{A}_r, \quad \forall \, \mathbf{A}_r \in \Lambda^r(\mathcal{V}^n). \qquad (2.3.22)$$

If $\mathbf{A}$ is a blade then so is its dual. This property can be derived as follows. Let $\mathbf{A}$ be the outer product of a sequence of r vectors. The vectors can be extended to a basis of $\mathcal{V}^n$. By (2.3.21), the dual of $\mathbf{A}$ with respect to this basis is an $(n-r)$-blade. When the basis is transformed to $\mathbf{E}_n$, each vector of the $(n-r)$-blade undergoes a linear transformation, and their outer product remains a blade.

Definition 2.25. [Definition of meet product by duality] The *meet product* in a Grassmann space is the unique associative product dual to the outer product:

$$*(\mathbf{A} \vee \mathbf{B}) := (*\mathbf{A})(*\mathbf{B}), \quad \forall \, \mathbf{A}, \mathbf{B} \in \Lambda(\mathcal{V}^n). \qquad (2.3.23)$$

The meet product of an r-vector $\mathbf{A}_r$ and an s-vector $\mathbf{B}_s$ is of grade $r + s - n$, and equals zero if $r + s < n$. It has the following *cograded antisymmetric* (also called *cograded anticommutative*) property:

$$\mathbf{A}_r \vee \mathbf{B}_s = (-1)^{(n-r)(n-s)} \mathbf{B}_s \vee \mathbf{A}_r. \qquad (2.3.24)$$

Below we look for explicit expressions of the meet product.

Let $\mathbf{A}_r$ be an r-blade. Then in vector space $\Lambda^r(\mathcal{V}^n)$, $\mathbf{A}_r$ and the C_n^r induced basis vectors $\{\mathbf{e}_{i_1}\mathbf{e}_{i_2} \cdots \mathbf{e}_{i_r} \,|\, 1 \le i_1 < i_2 < \ldots < i_r \le n\}$ satisfy the following Cramer's rule:

$$\mathbf{A}_r = \sum_{(n-r,r) \vdash \mathbf{E}_n} [\mathbf{E}_{n(1)} \mathbf{A}_r](\wedge \mathbf{E}_{n(2)}). \qquad (2.3.25)$$

Assume that $(\mathbf{E}'_{n(1)}, \mathbf{E}'_{n(2)})$ is a fixed partition of $\mathbf{E}_n$ of shape $(s, n - s)$. If $\mathbf{B}_s = \wedge \mathbf{E}'_{n(1)}$, then by (2.3.21),

$$
\begin{aligned}
(*\mathbf{A}_r)(*\mathbf{B}_s) &= \sum_{(n-r,r) \vdash \mathbf{E}_n} [\mathbf{E}_{n(1)} \mathbf{A}_r](* \wedge \mathbf{E}_{n(2)})(* \wedge \mathbf{E}'_{n(1)}) \\
&= \sum_{(n-r,r) \vdash \mathbf{E}_n} (-1)^{r(n-r)} [\mathbf{E}_{n(1)} \mathbf{A}_r](\wedge \mathbf{E}_{n(1)} \mathbf{E}'_{n(2)}) \\
&= \sum_{(n-r,r+s-n) \vdash \mathbf{E}'_{n(1)}} [\mathbf{A}_r \mathbf{E}'_{n(11)}](\wedge \mathbf{E}'_{n(11)} \mathbf{E}'_{n(2)}).
\end{aligned}
$$

So

$$\mathbf{A}_r \vee \mathbf{B}_s = (-1)^{(r+s)(r+s-n)} * \{(*\mathbf{A}_r)(*\mathbf{B}_s)\}$$

$$= (-1)^{(r+s)(r+s-n)+(n-s)(r+s-n)} \sum_{(n-r,r+s-n)\vdash \mathbf{E}'_{n(1)}} [\mathbf{A}_r \mathbf{E}'_{n(11)}](\wedge \mathbf{E}'_{n(12)}) \tag{2.3.26}$$

$$= \sum_{(r+s-n,n-r)\vdash \mathbf{B}_s} [\mathbf{A}_r \mathbf{B}_{s(2)}]\mathbf{B}_{s(1)}.$$

An s-blade has infinitely many ways to be represented as the outer product of s vectors. Nevertheless, we claim that the result of (2.3.26) is independent of the choice of the sequence of s vectors representing blade $\mathbf{B}_s$. We need the following lemma to prove this.

Lemma 2.26. For any integer $t \geq 0$, define a linear mapping $\Delta_t^\wedge : \Lambda(\mathcal{V}^n) \longrightarrow \Lambda(\mathcal{V}^n) \otimes \Lambda(\mathcal{V}^n)$ as follows: for any sequence of vectors $\mathbf{A}_r = \mathbf{a}_1, \ldots, \mathbf{a}_r$,

$$\Delta_t^\wedge(\wedge \mathbf{A}_r) := \begin{cases} 0, & \text{if } r < t, \\ \displaystyle\sum_{(t,r-t)\vdash \mathbf{A}_r} (\wedge \mathbf{A}_{r(1)}) \otimes (\wedge \mathbf{A}_{r(2)}), & \text{if } r \geq t. \end{cases} \tag{2.3.27}$$

Then $\Delta_t^\wedge$ is well defined.

Proof. We need to prove that if $\wedge \mathbf{B}_r = \wedge \mathbf{A}_r$ for another sequence of vectors $\mathbf{B}_r$, then $\Delta_t^\wedge(\wedge \mathbf{B}_r) = \Delta_t^\wedge(\wedge \mathbf{A}_r)$. Only the case $r > t > 0$ needs to be considered.

First, we have a well-defined linear mapping $\Delta_t^\otimes : \otimes(\mathcal{V}^n) \longrightarrow \otimes(\mathcal{V}^n)$, which is defined for any tensor $\otimes\mathbf{A}_r = \mathbf{a}_1 \otimes \mathbf{a}_2 \otimes \cdots \otimes \mathbf{a}_r$, by

$$\Delta_t^\otimes(\otimes\mathbf{A}_r) := \begin{cases} 0, & \text{if } r < t; \\ \displaystyle\sum_{(t,r-t)\vdash \mathbf{A}_r} (\otimes\mathbf{A}_{r(1)}) \otimes (\otimes \mathbf{A}_{r(2)}), & \text{if } r \geq t. \end{cases} \tag{2.3.28}$$

Second, let $S(r)$ be the permutation group of r integers $1, 2, \ldots, r$. According to (2.3.2), by denoting $\mathbf{A}_{\sigma(r)} = \mathbf{a}_{\sigma(1)}, \mathbf{a}_{\sigma(2)}, \ldots, \mathbf{a}_{\sigma(r)}$ for any permutation $\sigma \in S(r)$,

$$\Delta_t^\otimes(\wedge \mathbf{A}_r)$$

$$= \frac{1}{r!} \sum_{\sigma \in S(r)} \text{sign}(\sigma)\Delta_t^\otimes(\otimes\mathbf{A}_{\sigma(r)})$$

$$= \frac{1}{r!} \sum_{\sigma \in S(r)} \text{sign}(\sigma) \sum_{(t,r-t)\vdash \mathbf{A}_{\sigma(r)}} (\otimes \mathbf{A}_{\sigma(r)(1)}) \otimes (\otimes \mathbf{A}_{\sigma(r)(2)})$$

$$= \frac{t!(r-t)!}{r!} \sum_{(t,r-t)\vdash \mathbf{A}_r} \{\frac{1}{t!} \sum_{\tau \in S(t)} \text{sign}(\tau)(\otimes\mathbf{A}_{\tau(r_{(1)})})\} \otimes \{\frac{1}{(r-t)!} \sum_{\pi \in S(r-t)} \text{sign}(\pi)(\otimes\mathbf{A}_{\pi(r_{(2)})})\}$$

$$= \frac{1}{C_r^t} \Delta_t^\wedge(\wedge \mathbf{A}_r).$$

Thus, $\Delta_t^\wedge$ is a well-defined linear mapping induced from $\Delta_t^\otimes$. $\qquad\square$

Corollary 2.27. The two linear operators $\Delta_t^\wedge$ and $\Delta_t^\otimes$ defined above, when restricted to subspace $\Lambda^r(\mathcal{V}^n)$, satisfy

$$\Delta_t^\otimes = \frac{1}{C_r^t}\,\Delta_t^\wedge. \tag{2.3.29}$$

A tensor is said to be *decomposable*, if it can be written as the tensor product of vectors. By Lemma 2.26 and Corollary 2.27, for decomposable tensors and extensors, their *partitions* are well defined.

Definition 2.28. Let $\mathbf{A}_r = \mathbf{a}_1 \otimes \mathbf{a}_2 \otimes \cdots \otimes \mathbf{a}_r$ be a decomposable tensor of length r, and let $\mathbf{A}_r' = \mathbf{a}_1 \mathbf{a}_2 \cdots \mathbf{a}_r$ be a decomposable extensor of length s. A *partition* of $\mathbf{A}_r$ (or $\mathbf{A}_r'$) of shape λ, refers to a sequence of decomposable tensors (or extensors), which are obtained by first partitioning the sequence of vectors $\mathbf{a}$'s into subsequences of shape λ, then changing each subsequence of vectors by the tensor product (or outer product) into a decomposable tensor (or extensor).

For a partition of shape $(0, r)$ of decomposable tensor or extensor $\mathbf{A}_r$, it is always assumed that $\mathbf{A}_{r(1)} = 1$ and $\mathbf{A}_{r(2)} = \mathbf{A}_r$.

Corollary 2.29. Let $f : \Lambda(\mathcal{V}^n) \times \Lambda(\mathcal{V}^n) \longrightarrow \mathcal{L}$ be a bilinear mapping from $\Lambda(\mathcal{V}^n)$ to a vector space $\mathcal{L}$. Then for any r-blade $\mathbf{A}_r$, any $0 \le t \le r$,

$$f \circ \Delta_t^\wedge(\mathbf{A}_r) = \sum_{(t,r-t) \vdash \mathbf{A}_r} f(\mathbf{A}_{r(1)}, \mathbf{A}_{r(2)}) \tag{2.3.30}$$

is independent of the decomposition of $\mathbf{A}_r$ into any outer product of r vectors.

By Corollary 2.29, the bipartition of blade $\mathbf{B}_s$ in the result of (2.3.26) is meaningful. Furthermore, since brackets are invariants, (2.3.26) holds not only for $\mathbf{B}_s = \wedge\,\mathbf{E}_{n(1)}'$, but for arbitrary s-blade in $\Lambda(\mathcal{V}^n)$.

That the result of (2.3.26) is *independent of the basis* $\mathbf{E}_n$ is an extremely important property, but nowhere explicit in Definition 2.25. A coordinate-free definition of the meet product based on (2.3.26) is necessary.

Definition 2.30. [Definition of meet product by brackets] The *meet product* in a Grassmann algebra is the linear extension of the following product: for any r-blade $\mathbf{A}_r$ and s-blade $\mathbf{B}_s$,

$$\mathbf{A}_r \vee \mathbf{B}_s := \sum_{(r+s-n,\,n-r) \vdash \mathbf{B}_s} [\mathbf{A}_r \mathbf{B}_{s(2)}]\mathbf{B}_{s(1)}. \tag{2.3.31}$$

(2.3.31) is called a *shuffle formula*. The right side is called the *Cayley expansion* of the left side by *distributing* (or *partitioning*, or *separating*) $\mathbf{B}_s$.

The meet product has another *shuffle formula* by partitioning $\mathbf{A}_r$:

$$\mathbf{A}_r \vee \mathbf{B}_s = \sum_{(n-s,\,r+s-n) \vdash \mathbf{A}_r} [\mathbf{A}_{r(1)} \mathbf{B}_s]\mathbf{A}_{r(2)}. \tag{2.3.32}$$

That (2.3.31) and (2.3.32) are equal is equivalent to the associativity of the meet product. It is also equivalent to the cograded antisymmetry (2.3.24) of the meet product.

By induction, (2.3.32) can be extended from two blades to any $t+1$ blades: let the grades of blades $\mathbf{B}_1, \mathbf{B}_2, \ldots, \mathbf{B}_t$ be $b_1, b_2, \ldots, b_t$ respectively, then the meet product of $\mathbf{A}_r$ and these blades is

$$\mathbf{A}_r \vee \mathbf{B}_1 \vee \mathbf{B}_2 \vee \cdots \vee \mathbf{B}_t = \sum_{\substack{(n-b_1,\ldots,n-b_t, \\ r-nt+\sum_{i=1}^{t} b_i) \vdash \mathbf{A}_r}} [\mathbf{A}_{r(1)}\mathbf{B}_1][\mathbf{A}_{r(2)}\mathbf{B}_2] \cdots [\mathbf{A}_{r(t)}\mathbf{B}_t]\mathbf{A}_{r(t+1)}.$$

$$(2.3.33)$$

In terms of meet products, the GP relation (2.1.25) is the Cayley expansion of

$$\mathbf{B}_{n+1} \vee \mathbf{C}_{n-1} = 0, \tag{2.3.34}$$

and the VW relation (2.1.23) is the Cayley expansion of

$$\mathbf{A}_s \vee \mathbf{B}_{n+1} \vee \mathbf{C}_{n-s-1} = 0. \tag{2.3.35}$$

Both equalities are trivial, as any $(n+1)$-vector is zero.

Definition 2.31. A Grassmann space equipped with the outer product and the meet product is called a *Grassmann-Cayley algebra* (GC algebra), also called a *Cayley algebra*. Any monomial in this algebra containing at least one meet product (or no meet product) is called a *Cayley expression* (or *Grassmann expression*), also called a *Cayley monomial* (or *Grassmann monomial*). A *Cayley polynomial* (or *Grassmann polynomial*) is the sum of finitely many Cayley monomials (or Grassmann monomials).

A Grassmann monomial is an outer product of vectors. A blade becomes a Grassmann monomial only after it is decomposed into an outer product of vectors.

While the outer product represents the extension of two subspaces in projective geometry, the meet product represents their intersection. This can be seen clearly from both the duality relation (2.3.23), and the explicit expression (2.3.31) of the meet product. Let $\mathbf{E}_n$ be a basis of $\mathcal{V}^n$. Let $\mathbf{A}_r, \mathbf{B}_s$ be Grassmann monomials composed of basis vector sequences $\mathbf{A}'_r, \mathbf{B}'_s$ respectively, and let the complement of sequence $\mathbf{A}'_r$ in $\mathbf{E}_n$ be $\mathbf{C}'_{n-r}$. By (2.3.31), if the vectors in $\mathbf{A}'_r, \mathbf{B}'_s$ span the whole space $\mathcal{V}^n$, then by denoting with $\mathbf{B}'_s \backslash \mathbf{C}'_{n-r}$ the subsequence of $\mathbf{B}'_s$ after removal of all vectors that also belong to $\mathbf{C}'_{n-r}$, we get

$$\mathbf{A}_r \vee \mathbf{B}_s = [\mathbf{A}'_r \mathbf{C}'_{n-r}] \, (\wedge (\mathbf{B}'_s \backslash \mathbf{C}'_{n-r})) = \wedge (\mathbf{B}'_s \cap \mathbf{A}'_r). \tag{2.3.36}$$

GC algebra provides invariant descriptions of projective geometric statements through its two products. In 2D projective geometry, three lines $\mathbf{12}, \mathbf{1'2'}, \mathbf{1''2''}$ *concur*, i.e., they meet at the same point, if and only if their meet product equals zero:

$$\mathbf{12} \vee \mathbf{1'2'} \vee \mathbf{1''2''} = 0. \tag{2.3.37}$$

In 3D projective geometry, four planes $\mathbf{123}, \mathbf{1'2'3'}, \mathbf{1''2''3''}, \mathbf{1'''2'''3'''}$ are *copunctual*, *i.e.*, they meet at the same point, if and only if their meet product equals zero:

$$\mathbf{123} \vee \mathbf{1'2'3'} \vee \mathbf{1''2''3''} \vee \mathbf{1'''2'''3'''} = 0. \tag{2.3.38}$$

In (2.1.7), we have shown that brackets provide a natural expression of the homogeneous coordinates of a point by means of the Cramer's rule of the point and n basis points. Likewise, scalar-valued meet products provide natural representations for the homogeneous coordinates of higher dimensional geometric objects. For example, in 2D projective geometry, let $\mathbf{12}, \mathbf{1'2'}, \mathbf{1''2''}, \mathbf{1'''2'''}$ be four lines, then their Cramer's rule is

$$\begin{aligned}
(\mathbf{12} \vee \mathbf{1'2'} \vee \mathbf{1''2''})\mathbf{1'''2'''} &- (\mathbf{12} \vee \mathbf{1'2'} \vee \mathbf{1'''2'''})\mathbf{1''2''} \\
+ (\mathbf{12} \vee \mathbf{1''2''} \vee \mathbf{1'''2'''})\mathbf{1'2'} &- (\mathbf{1'2'} \vee \mathbf{1''2''} \vee \mathbf{1'''2'''})\mathbf{12} = 0,
\end{aligned} \tag{2.3.39}$$

where the coefficient of each bivector is the meet product of three lines.

Applications of GC algebra in projective geometry are the main contents of Chapters 3 and 4. As a final remark of this section, in classical invariant theory [69], the meet product is usually denoted by "$\wedge$" to indicate its similarity with the intersection operator "$\cap$" in set theory. We feel that we should reserve the wedge symbol to the outer product in the setting of Clifford algebra, where the outer product is no longer the fundamental product and no longer represented by juxtaposition. Throughout this book, the meet product is always denoted by "$\vee$".

2.4 Grassmann coalgebra

We have seen that the outer product and the meet product represent respectively the extension and intersection of vector subspaces in $\mathcal{V}^n$. Unfortunately, the representations are correct only in some cases. As exceptions,

- if two subspaces have nonzero intersection, then their outer product is always zero, and cannot represent their extension;
- if the sum of their dimensions is less than n, two subspaces may still have nonzero intersection, but their meet product is always zero.

In this section, we present the third definition of the meet product, by which the above exceptions are both covered. The definition is based on *Grassmann coalgebra*. Below we first introduce the concept of *coalgebra* over a base field.

Definition 2.32. A $\mathbb{K}$-*coalgebra* is a vector space $\mathcal{V}$ over $\mathbb{K}$, together with a $\mathbb{K}$-linear mapping $\Delta : \mathcal{V} \longrightarrow \mathcal{V} \otimes_{\mathbb{K}} \mathcal{V}$ called the *coproduct*, and a linear mapping $\epsilon : \mathcal{V} \longrightarrow \mathbb{K}$ called the *counit* map, such that for any $\mathbf{x} \in \mathcal{V}$,

$$\begin{aligned}
(I_{\mathcal{V}} \otimes \Delta)(\Delta \mathbf{x}) &= (\Delta \otimes I_{\mathcal{V}})(\Delta \mathbf{x}), \\
(I_{\mathcal{V}} \otimes \epsilon)(\Delta \mathbf{x}) &= (\epsilon \otimes I_{\mathcal{V}})(\Delta \mathbf{x}) = \mathbf{x},
\end{aligned} \tag{2.4.1}$$

where $I_{\mathcal{V}}$ denotes the identity transformation in $\mathcal{V}$. The first equality in (2.4.1) is called the *coassociative law*.

Because of the coassociativity, we can write $\Delta^2 = (I_{\mathcal{V}} \otimes \Delta) \circ \Delta$, and in the general case, $\Delta^r = (I_{\mathcal{V}} \otimes \Delta) \circ \Delta^{r-1}$. We write $\Delta^0 = I_{\mathcal{V}}$. *The coproduct precedes the tensor product by default.*

Example 2.33. Let the vector space be $\otimes(\mathcal{V}^n)$. For vectors $\mathbf{a}_1, \mathbf{a}_2, \ldots, \mathbf{a}_r \in \mathcal{V}^n$, define the *tensor coproduct* $\Delta^{\otimes}$ by

$$\Delta^{\otimes}(\mathbf{a}_1 \otimes \mathbf{a}_2 \otimes \cdots \otimes \mathbf{a}_r) := \sum_{t=0}^{r} \Delta_t^{\otimes}(\mathbf{a}_1 \otimes \mathbf{a}_2 \otimes \cdots \otimes \mathbf{a}_r), \qquad (2.4.2)$$

where the *t-th part* $\Delta_t^{\otimes}$ of the tensor coproduct is already defined in (2.3.28).

- When $r = 0$, we always set

$$\mathbf{a}_1 \otimes \mathbf{a}_2 \otimes \cdots \otimes \mathbf{a}_r \,|_{r=0} = 1. \qquad (2.4.3)$$

  Then $\Delta^{\otimes}(1) = 1 \otimes 1$.
- When $r = 1$,

$$\Delta^{\otimes}(\mathbf{a}_1) = 1 \times \mathbf{a}_1 + \mathbf{a}_1 \otimes \mathbf{1}. \qquad (2.4.4)$$

- When $r = 2$,

$$\Delta^{\otimes}(\mathbf{a}_1 \otimes \mathbf{a}_2) = 1 \otimes (\mathbf{a}_1 \otimes \mathbf{a}_2) + \mathbf{a}_1 \otimes \mathbf{a}_2 - \mathbf{a}_2 \otimes \mathbf{a}_1 + (\mathbf{a}_1 \otimes \mathbf{a}_2) \otimes 1. \qquad (2.4.5)$$

If the counit map is defined by

$$\epsilon^{\otimes}(\mathbf{a}_1 \otimes \mathbf{a}_2 \otimes \cdots \otimes \mathbf{a}_r) = \delta_r^0 = \begin{cases} 0, & \text{if } r \neq 0, \\ 1, & \text{if } r = 0, \end{cases} \qquad (2.4.6)$$

then the tensor space $\otimes(\mathcal{V}^n)$ equipped with the tensor coproduct and the above counit map, becomes a coalgebra called the *tensor coalgebra* over $\mathcal{V}^n$.

Notation. Let $\mathbf{A}$ be a sequence of elements. In

$$\sum_{\vdash \mathbf{A}} f(\mathbf{A}_{(1)}, \mathbf{A}_{(2)}, \ldots, \mathbf{A}_{(k)}), \qquad (2.4.7)$$

the summation is over *all shapes of partitions* of $\mathbf{A}$ into k non-overlapping subsequences.

Definition 2.34. The *outer coproduct* in Grassmann space $\Lambda(\mathcal{V}^n)$ is the linear extension of the mapping $\Delta^{\wedge} : \Lambda(\mathcal{V}^n) \longrightarrow \Lambda(\mathcal{V}^n) \otimes \Lambda(\mathcal{V}^n)$, which is defined for any r-blade $\mathbf{A}_r \in \Lambda(\mathcal{V}^n)$, by

$$\Delta^{\wedge} \mathbf{A}_r := \sum_{\vdash \mathbf{A}_r} \mathbf{A}_{r(1)} \otimes \mathbf{A}_{r(2)} = \sum_{i=0}^{r} \sum_{(i,r-i)\vdash \mathbf{A}_r} \mathbf{A}_{r(1)} \otimes \mathbf{A}_{r(2)}. \qquad (2.4.8)$$

The *Grassmann coalgebra* over $\mathcal{V}^n$ is the Grassmann space $\Lambda(\mathcal{V}^n)$ equipped with the outer coproduct and the counit mapping $\langle\ \rangle_0\colon \mathbf{A} \mapsto \langle \mathbf{A} \rangle_0$ for $\mathbf{A} \in \Lambda(\mathcal{V}^n)$.

The outer coproduct can be decomposed into $n+1$ parts: $\Delta^\wedge = \sum_{i=0}^n \Delta_i^\wedge$, where the *i-th part*

$$\Delta_i^\wedge \mathbf{A}_r = \sum_{(i,r-i)\vdash \mathbf{A}_r} \mathbf{A}_{r(1)} \otimes \mathbf{A}_{r(2)} \tag{2.4.9}$$

is already defined in (2.3.27).

To understand the algebraic meaning of the outer coproduct, consider the simplest case $r = 2$:

$$\Delta^\wedge(\mathbf{a}_1\mathbf{a}_2) = 1 \otimes \mathbf{a}_1\mathbf{a}_2 + \mathbf{a}_1 \otimes \mathbf{a}_2 - \mathbf{a}_2 \otimes \mathbf{a}_1 + \mathbf{a}_1\mathbf{a}_2 \otimes 1. \tag{2.4.10}$$

Among the four terms on the right side, the outer product of any pair of blades in the same term equals the preimage $\mathbf{a}_1\mathbf{a}_2$ under $\Delta^\wedge$. The four terms include *all the ordered pairs of blades* composed of vectors in $\{\mathbf{a}_1, \mathbf{a}_2\}$ such that the outer product of each pair of blades equals the input $\mathbf{a}_1\mathbf{a}_2$.

For a general grade r, the above observation is still correct. The outer coproduct of blade $\mathbf{A}_r = \mathbf{a}_1\mathbf{a}_2 \cdots \mathbf{a}_r$ is *inverse to the outer product* of the r vectors $\mathbf{a}$'s, in the sense that it outputs all the ordered pairs of blades composed of the $\mathbf{a}$'s such that the outer product of each pair of blades equals $\mathbf{A}_r$. Linear operator $\Delta_i^\wedge$ returns the pair of blades with grade $(i, r - i)$. What is highly remarkable is that when every such pair of blades is multiplied by the tensor product, their sum is *independent of the decomposition* of the input r-blade into any outer product of r vectors.

Outer coproduct should not be a strange thing to any one who knows of Laplace expansions of determinants. For an $n \times n$ determinant $\det(\mathbf{a}_1\mathbf{a}_2 \dots \mathbf{a}_n)$, where $\mathbf{a}_i = (a_{1i}, a_{2i}, \dots, a_{ni})^T$ is the i-th column vector, its $r \times r$ subdeterminant formed by columns $i_1, i_2, \dots, i_r$ and rows $j_1, j_2, \dots, j_r$, can be denoted by

$$(\mathbf{a}_{i_1}\mathbf{a}_{i_2} \dots \mathbf{a}_{i_r}|j_1 j_2 \dots j_r) := \det(a_{j_l i_k})_{k,l=1..r}. \tag{2.4.11}$$

This is the *letter-place notation* of subdeterminants (or minors), also called the *Laplace pairing* [69] between the *places* j_l and the *letters* $\mathbf{a}_i$.

In the letter-place notation, for $\mathbf{A}_n = \mathbf{a}_1\mathbf{a}_2 \cdots \mathbf{a}_n$,

$$[\mathbf{A}_n] = \det(\mathbf{a}_1\mathbf{a}_2 \dots \mathbf{a}_n) = (\mathbf{a}_1\mathbf{a}_2 \dots \mathbf{a}_n|12 \dots n) = (\mathbf{A}_n|12 \dots n). \tag{2.4.12}$$

The *Laplace expansion* of $[\mathbf{A}_n]$ by its first row is

$$\sum_{j=1}^n (-1)^{j+1} a_{1j}(\mathbf{a}_1\mathbf{a}_2 \dots \breve{\mathbf{a}}_j \dots \mathbf{a}_n|2 \dots n) = \sum_{(1,n-1)\vdash \mathbf{A}_n} (\mathbf{A}_{n(1)}|1)(\mathbf{A}_{n(2)}|2 \dots n).$$

In general, for a fixed bipartition (W, W') of $1, 2, \dots, n$ of shape $(r, n - r)$, the Laplace expansion of $[\mathbf{A}_n]$ by its rows W is the following pairing involving the outer coproduct of $\mathbf{A}_n$:

$$\sum_{(r,n-r)\vdash \mathbf{A}_n} (\mathbf{A}_{n(1)}|W)(\mathbf{A}_{n(2)}|W') = (\Delta_r^\wedge \mathbf{A}_n|W \otimes W'). \tag{2.4.13}$$

Proposition 2.35. The tensor coproduct and the outer coproduct are respectively homomorphisms of the following tensor algebras and Grassmann algebras:

$$\begin{aligned} \Delta^\otimes &: \otimes(\mathcal{V}^n) \longrightarrow \otimes(\mathcal{V}^n) \,\hat{\otimes}\, \otimes(\mathcal{V}^n), \\ \Delta^\wedge &: \Lambda(\mathcal{V}^n) \longrightarrow \Lambda(\mathcal{V}^n) \,\hat{\otimes}\, \Lambda(\mathcal{V}^n). \end{aligned} \tag{2.4.14}$$

Proof. We only prove the statement for the outer coproduct. Let $\mathbf{A}_r, \mathbf{B}_s$ be blades in $\Lambda(\mathcal{V}^n)$. Then

$$\Delta^\wedge(\mathbf{A}_r\mathbf{B}_s) = \sum_{\vdash \mathbf{A}_r\mathbf{B}_s} (\mathbf{A}_r\mathbf{B}_s)_{(1)} \otimes (\mathbf{A}_r\mathbf{B}_s)_{(2)}$$

$$= \sum_{\vdash \mathbf{A}_r, \vdash \mathbf{B}_s} (\mathbf{A}_{r(1)} \otimes \mathbf{A}_{r(2)})(\mathbf{B}_{s(1)} \otimes \mathbf{B}_{s(2)})$$

$$= \Delta^\wedge(\mathbf{A}_r)\Delta^\wedge(\mathbf{B}_s).$$

$\square$

Definition 2.36. [Definition of meet product by outer coproduct] The *meet product* in Grassmann space $\Lambda(\mathcal{V}^n)$ is the linear extension of the following product, which is defined for any r-blade $\mathbf{A}_r$ and s-blade $\mathbf{B}_s$ by

$$\mathbf{A}_r \vee \mathbf{B}_s = (I_{\Lambda(\mathcal{V}^n)} \otimes \jmath_{\mathbf{A}_r})(\Delta^\wedge_{r+s-n}\mathbf{B}_s). \tag{2.4.15}$$

Here $I_{\Lambda(\mathcal{V}^n)}$ is the identity transformation in $\Lambda(\mathcal{V}^n)$, and $\jmath_{\mathbf{A}_r}$ is defined by (2.3.19).

This definition provides a clear geometric interpretation of the outer coproduct: for any $\mathbf{B} \in \Lambda(\mathcal{V}^n)$, $\Delta^\wedge\mathbf{B}$ is the linear transformation in $\Lambda(\mathcal{V}^n)$ mapping any $\mathbf{A} \in \Lambda(\mathcal{V}^n)$ to $\mathbf{A} \vee \mathbf{B}$. Similarly, $\Delta^{\wedge k}\mathbf{B}$ is the linear functional

$$(\mathbf{A}_1, \mathbf{A}_2, \dots, \mathbf{A}_k) \mapsto \mathbf{A}_1 \vee \mathbf{A}_2 \vee \cdots \vee \mathbf{A}_k \vee \mathbf{B} \tag{2.4.16}$$

for any $\mathbf{A}$'s and $\mathbf{B}$ in $\Lambda(\mathcal{V}^n)$. From this aspect, the outer coproduct and the meet product are equivalent.

By (2.4.15), the meet product of $\mathbf{A}_r$ and $\mathbf{B}_s$ being zero for $r+s < n$ is caused by the operator $\jmath_{\mathbf{A}_r}$. If we replace this operator with the outer product operator $\mathbf{A}_r\wedge$, we can allow the whole $\Delta^\wedge$ instead of only the part $\Delta^\wedge_{r+s-n}$ to occur in (2.4.15).

Definition 2.37. The *total meet product* of two multivectors in $\Lambda(\mathcal{V}^n)$ is a linear isomorphism in Grassmann algebra $\Lambda(\mathcal{V}^n) \otimes \Lambda(\mathcal{V}^n)$, defined for any r-blade $\mathbf{A}_r$ and s-blade $\mathbf{B}_s$ by

$$\mathbf{A}_r \bar{\vee} \mathbf{B}_s := (1 \otimes \mathbf{A}_r)(\Delta^\wedge\mathbf{B}_s) = \sum_{i=r+s-n}^{s} \sum_{(i,s-i)\vdash\mathbf{B}_s} \mathbf{B}_{s(1)} \otimes \mathbf{A}_r\mathbf{B}_{s(2)}. \tag{2.4.17}$$

Definition 2.38. In the tensor algebra $\otimes(\Lambda(\mathcal{V}^n))$, or $\hat{\otimes}(\Lambda(\mathcal{V}^n))$ if the tensor product is twisted, the *lexicographic order* among the grades of the tensors is defined as follows: for r_i-blades $\mathbf{A}_{r_i}$ and s_j-blades $\mathbf{B}_{s_j}$, $\mathbf{A}_{r_1} \otimes \mathbf{A}_{r_2} \otimes \cdots \otimes \mathbf{A}_{r_u} \prec \mathbf{B}_{s_1} \otimes \mathbf{B}_{s_2} \otimes \cdots \otimes \mathbf{B}_{s_v}$ if one of the following conditions is satisfied:

- $u < v$, and $r_i = s_i$ for $1 \le i \le u$.
- For some $1 \le t < u$, $r_i = s_i$ for all $i < t$, but $r_t < s_t$.

If $(r_1, r_2, \ldots, r_u)$ is the lowest grade of a tensor $\mathbf{A}$ in $\otimes(\Lambda(\mathcal{V}^n))$ or $\hat{\otimes}(\Lambda(\mathcal{V}^n))$, the *lexicographically lowest part* of the tensor, denoted by $\langle \mathbf{A} \rangle_{\min}$, refers to the sum of all the $(r_1, r_2, \ldots, r_u)$-graded terms in the tensor.

Clearly, if the meet product (2.4.15) of $\mathbf{A}_r, \mathbf{B}_s$ is nonzero, then it corresponds to the lexicographically lowest part of their total meet product (2.4.17).

Example 2.39. Let there be a fixed basis in $\mathcal{V}^n$. If $\mathbf{A}_r, \mathbf{B}_s$ are both outer products of some of the basis vectors, let $\mathbf{A}_r = \mathbf{J}_t \mathbf{A}'$, $\mathbf{B}_s = \mathbf{J}_t \mathbf{B}'$, where $\mathbf{A}', \mathbf{B}', \mathbf{J}_t$ are outer products of some non overlapping subsequences of basis vectors. Then the extension of vector subspaces $\mathbf{A}_r, \mathbf{B}_s$ is $\mathbf{J}_t \mathbf{A}' \mathbf{B}'$, and their intersection is $\mathbf{J}_t$. By (2.4.17),

$$\mathbf{A}_r \, \bar{\vee} \, \mathbf{B}_s = \sum_{i=0}^{s-t} \sum_{(i, s-t-i) \vdash \mathbf{B}'} \mathbf{J}_t \mathbf{B}'_{(1)} \otimes \mathbf{J}_t \mathbf{A}' \mathbf{B}'_{(2)}. \tag{2.4.18}$$

The intersection and extension are both in the lexicographically lowest part of (2.4.18): $\mathbf{J}_t \otimes \mathbf{J}_t \mathbf{A}' \mathbf{B}'$.

Definition 2.40. For any r-blade $\mathbf{A}_r$ and s-blade $\mathbf{B}_s$, let $\langle \mathbf{A}_r \, \bar{\vee} \, \mathbf{B}_s \rangle_{\min}$ be the lexicographically lowest part of $\mathbf{A}_r \, \bar{\vee} \, \mathbf{B}_s$. The *intersection product* of $\mathbf{A}_r, \mathbf{B}_s$ is defined by

$$\mathbf{A}_r \sqcup \mathbf{B}_s := (I_{\Lambda(\mathcal{V}^n)} \otimes [\ ])(\langle \mathbf{A}_r \, \bar{\vee} \, \mathbf{B}_s \rangle_{\min}); \tag{2.4.19}$$

their *extension product* is defined by

$$\mathbf{A}_r \sqcap \mathbf{B}_s := ([\ ] \otimes I_{\Lambda(\mathcal{V}^n)})(\langle \mathbf{A}_r \, \bar{\vee} \, \mathbf{B}_s \rangle_{\min}). \tag{2.4.20}$$

Lemma 2.41. If $\mathbf{A}$ is a blade in $\Lambda(\mathcal{V}^n)$, so is $(I_{\Lambda(\mathcal{V}^n)} \otimes [\ ])(\Delta_i^{\wedge}(\mathbf{A}))$ for any $0 \leq i \leq n$, where the symbol "$[\ ]$" denotes the deficit bracket operator in $\Lambda(\mathcal{V}^n)$.

Proof. Let blade $\mathbf{A}$ be of grade r. Let $\mathbf{U}$ be the dummy blade of grade $i + n - r$. Then

$$(I_{\Lambda(\mathcal{V}^n)} \otimes {}_{\mathbf{J}}\mathbf{U})(\Delta_i^{\wedge}(\mathbf{A})) = \sum_{(i, r-i) \vdash \mathbf{A}} \mathbf{A}_{(1)} [\mathbf{U} \mathbf{A}_{(2)}] = \mathbf{U} \vee \mathbf{A}$$

is a blade, as $*(\mathbf{U} \vee \mathbf{A}) = (*\mathbf{U})(*\mathbf{A})$ is. $\qquad\square$

Corollary 2.42. For any r-blade $\mathbf{A}_r$ and s-blade $\mathbf{B}_s$, the intersection and extension of vector subspaces $\mathbf{A}_r, \mathbf{B}_s$ are represented by the intersection product and the extension product of the two blades respectively.

Example 2.43. Let $\mathbf{12}, \mathbf{1'2'}$ be two coplanar lines in nD projective geometry. Then $\mathbf{121'2'} = 0$, and

$$\mathbf{12} \, \bar{\vee} \, \mathbf{1'2'} = \mathbf{1'} \otimes \mathbf{122'} - \mathbf{2'} \otimes \mathbf{121'} + \mathbf{1'2'} \otimes \mathbf{12}. \tag{2.4.21}$$

The result has grades $(1, 3)$ and $(2, 2)$. If the $(1, 3)$-graded part is nonzero, then

$$\mathbf{12} \sqcup \mathbf{1'2'} = \mathbf{1'}[\mathbf{122'}] - \mathbf{2'}[\mathbf{121'}] \tag{2.4.22}$$

is the meet product in 2D projective geometry. If the $(1,3)$-graded part is zero, then the lexicographically lowest part is

$$12 \sqcup 1'2' = 1'2'[12],$$

and the two lines are identical.

Consider the terms of grade $(i, r+s-i)$ in the total meet product (2.4.17). By Lemma 2.41,

$$\mathbf{X}_i = \sum_{(i,s-i)\vdash\mathbf{B}_s} \mathbf{B}_{s(1)}[\mathbf{A}_r\mathbf{B}_{s(2)}] = \mathbf{U}_{i+n-r-s}\mathbf{A}_r \vee \mathbf{B}_s,$$

$$\mathbf{Y}_{r+s-i} = \sum_{(i,s-i)\vdash\mathbf{B}_s} [\mathbf{B}_{s(1)}]\mathbf{A}_r\mathbf{B}_{s(2)} = (-1)^{i(n-i)}\mathbf{A}_r(\mathbf{B}_s \vee \mathbf{U}_{n-i})$$

are both blades. $\mathbf{X}_i$ represents the iD intersection of space $\mathbf{B}_s$ with a generic $(i+n-s)$D space containing $\mathbf{A}_r$ as a subspace. $\mathbf{Y}_{r+s-i}$ represents the $(r+s-i)$D space spanned by subspace $\mathbf{A}_r$ and a generic subspace of codimension i in space $\mathbf{B}_s$.

The meet product of more than two blades has been considered in (2.3.33). Likewise, the total meet product, the intersection product and the extension product can also be extended to more than two blades.

Notation.

The tensor product of i identical algebraic elements $\mathbf{A}$, called the *tensor power* of $\mathbf{A}$ to the i-th, is denoted by

$$\mathbf{A}^{\otimes i} := \underbrace{\mathbf{A} \otimes \mathbf{A} \otimes \cdots \otimes \mathbf{A}}_{i}. \qquad (2.4.23)$$

For example, $[\ ]^{\otimes i} = \underbrace{[\ ] \otimes \cdots \otimes [\ ]}_{i}$, and $1^{\otimes i} = \underbrace{1 \otimes \cdots \otimes 1}_{i}$.

Definition 2.44. For any $\mathbf{A}_1, \mathbf{A}_2, \ldots, \mathbf{A}_s \in \Lambda(\mathcal{V}^n)$, their *total meet product* is

$$\mathbf{A}_1 \bar{\vee} \mathbf{A}_2 \bar{\vee} \cdots \bar{\vee} \mathbf{A}_s := \prod_{i=1}^{s}(1^{\otimes(s-i)} \otimes \Delta^{\wedge(i-1)}\mathbf{A}_i) \in \otimes^s(\Lambda(\mathcal{V}^n)). \qquad (2.4.24)$$

Here "$\prod$" denotes the outer product in $\otimes^s(\Lambda(\mathcal{V}^n)) = \underbrace{\Lambda(\mathcal{V}^n) \otimes \cdots \otimes \Lambda(\mathcal{V}^n)}_{s}.$

Let $\langle \mathbf{A}_1 \bar{\vee} \mathbf{A}_2 \bar{\vee} \cdots \bar{\vee} \mathbf{A}_s\rangle_{\min}$ be the lexicographically lowest part of the total meet product. The *intersection product* of $\mathbf{A}_1, \mathbf{A}_2, \ldots, \mathbf{A}_s$ is defined by

$$\mathbf{A}_1 \sqcup \mathbf{A}_2 \sqcup \cdots \sqcup \mathbf{A}_s := (I_{\Lambda(\mathcal{V}^n)} \otimes [\ ]^{\otimes(s-1)})(\langle \mathbf{A}_1 \bar{\vee} \mathbf{A}_2 \bar{\vee} \cdots \bar{\vee} \mathbf{A}_s\rangle_{\min}); \qquad (2.4.25)$$

their *extension product* is defined by

$$\mathbf{A}_1 \sqcap \mathbf{A}_2 \sqcap \cdots \sqcap \mathbf{A}_s := ([\ ]^{\otimes(s-1)} \otimes I_{\Lambda(\mathcal{V}^n)})(\langle \mathbf{A}_1 \bar{\vee} \mathbf{A}_2 \bar{\vee} \cdots \bar{\vee} \mathbf{A}_s\rangle_{\min}). \qquad (2.4.26)$$

Example 2.45. Let there be a fixed basis in $\mathcal{V}^n$. If $\mathbf{A}, \mathbf{B}, \mathbf{C}$ are blades of grade r, s, t respectively, let

$$
\begin{aligned}
\mathbf{A} &= \mathbf{J}_{ABC}\mathbf{J}_{AB}\mathbf{J}_{AC}\mathbf{A}', \\
\mathbf{B} &= \mathbf{J}_{ABC}\mathbf{J}_{AB}\mathbf{J}_{BC}\mathbf{B}', \\
\mathbf{C} &= \mathbf{J}_{ABC}\mathbf{J}_{AC}\mathbf{J}_{BC}\mathbf{C}',
\end{aligned}
\tag{2.4.27}
$$

where $\mathbf{A}', \mathbf{B}', \mathbf{C}', \mathbf{J}_{AB}, \mathbf{J}_{AC}, \mathbf{J}_{BC}, \mathbf{J}_{ABC}$ are outer products of some non-overlapping sequences of basis vectors. Then

$$
\begin{aligned}
& \mathbf{A} \,\bar{\vee}\, \mathbf{B} \,\bar{\vee}\, \mathbf{C} \\
={}& (1^{\otimes 2} \otimes \mathbf{A})(1 \otimes \Delta^{\wedge}\mathbf{B})(\Delta^{\wedge 2}\mathbf{C}) \\
={}& \sum_{\vdash \mathbf{B}, \vdash \mathbf{C}} \mathbf{C}_{(1)} \otimes \mathbf{B}_{(1)}\mathbf{C}_{(2)} \otimes \mathbf{A}\mathbf{B}_{(2)}\mathbf{C}_{(3)} \\
={}& \sum_{\vdash \mathbf{J}_{AC}, \vdash \mathbf{J}_{BC}\mathbf{B}', \vdash \mathbf{J}_{BC}\mathbf{C}'} \mathbf{J}_{ABC}\mathbf{J}_{AC(1)}(\mathbf{J}_{BC}\mathbf{C}')_{(1)} \otimes \mathbf{J}_{ABC}\mathbf{J}_{AB}(\mathbf{J}_{BC}\mathbf{B}')_{(1)}\mathbf{J}_{AC(2)}(\mathbf{J}_{BC}\mathbf{C}')_{(2)} \\
& \hspace{5.5cm} \otimes \mathbf{J}_{ABC}\mathbf{J}_{AB}\mathbf{J}_{AC}\mathbf{A}'(\mathbf{J}_{BC}\mathbf{B}')_{(2)}(\mathbf{J}_{BC}\mathbf{C}')_{(3)}.
\end{aligned}
$$

Its lexicographically lowest part is

$$
\begin{aligned}
& \operatorname{sign}(\mathbf{J}_{BC(2)}, \mathbf{J}_{BC(1)})\mathbf{J}_{ABC} \otimes \mathbf{J}_{ABC}\mathbf{J}_{AB}\mathbf{J}_{BC(2)}\mathbf{J}_{AC}\mathbf{J}_{BC(1)} \\
& \hspace{3cm} \otimes \mathbf{J}_{ABC}\mathbf{J}_{AB}\mathbf{J}_{AC}\mathbf{A}'\mathbf{J}_{BC(1)}\mathbf{B}'\mathbf{J}_{BC(2)}\mathbf{C}' \tag{2.4.28} \\
={}& \pm\mathbf{J}_{ABC} \otimes \mathbf{J}_{ABC}\mathbf{J}_{AB}\mathbf{J}_{AC}\mathbf{J}_{BC} \otimes \mathbf{J}_{ABC}\mathbf{J}_{AB}\mathbf{J}_{AC}\mathbf{J}_{BC}\mathbf{A}'\mathbf{B}'\mathbf{C}'.
\end{aligned}
$$

The three components of the decomposable tensor in the result of (2.4.28) represent the subspace shared by at least three spaces (the intersection of spaces $\mathbf{A}, \mathbf{B}, \mathbf{C}$), the subspace shared by at least two of the three spaces, and the subspace shared by at least one of the three spaces (the extension), respectively. They can be obtained from $\langle \mathbf{A} \,\bar{\vee}\, \mathbf{B} \,\bar{\vee}\, \mathbf{C} \rangle_{\min}$ by applying the following operators respectively:

$$
I_{\Lambda(\mathcal{V}^n)} \otimes [\ ] \otimes [\ ], \qquad [\ ] \otimes I_{\Lambda(\mathcal{V}^n)} \otimes [\ ], \qquad [\ ] \otimes [\ ] \otimes I_{\Lambda(\mathcal{V}^n)}.
$$

In Section 2.3, we have seen that the outer product operator "$\mathbf{C}_t \wedge$" induced by t-blade $\mathbf{C}_t$ is a perspective projection, and induces a homomorphism from the Grassmann algebra $\Lambda(\mathcal{V}^n)$ to the Grassmann algebra $\Lambda(\mathbf{C}_t\mathcal{V}^n)$. Below we investigate the influence of this map upon the meet products in the two algebras.

Let $\mathbf{A}_r, \mathbf{B}_s, \mathbf{C}_t$ be blades of grade r, s, t respectively, such that $\max(r, s) < n-t < r + s$. Then

$$
\mathbf{C}_t\mathbf{A}_r \vee \mathbf{C}_t\mathbf{B}_s = \sum_{(r+s+t-n,\,n-r-t)\vdash\mathbf{B}_s} [\mathbf{C}_t\mathbf{A}_r\mathbf{B}_{s(2)}]\mathbf{C}_t\mathbf{B}_{s(1)} = \sum_{(n-s-t,\,r+s+t-n)\vdash\mathbf{A}_r} [\mathbf{C}_t\mathbf{A}_{r(1)}\mathbf{B}_s]\mathbf{C}_t\mathbf{A}_{r(2)}.
\tag{2.4.29}
$$

Definition 2.46. The *reduced meet product* of r-blade $\mathbf{A}_r$ and s-blade $\mathbf{B}_s$, with t-blade $\mathbf{C}_t$ as the *base*, is

$$
\mathbf{A}_r \vee_{\mathbf{C}_t} \mathbf{B}_s := \sum_{(r+s+t-n,\,n-r-t)\vdash\mathbf{B}_s} [\mathbf{C}_t\mathbf{A}_r\mathbf{B}_{s(2)}]\mathbf{B}_{s(1)}
\tag{2.4.30}
$$

If $\mathbf{C}_t = \mathbf{U}_t$ is the dummy t-blade, then (2.4.30) is called the *t-deficit meet product* of $\mathbf{A}_r$ and $\mathbf{B}_s$, denoted by $\mathbf{A}_r \vee_t \mathbf{B}_s$.

The *reduced meet product* of blades $\mathbf{A}_{r_1}, \mathbf{A}_{r_2}, \ldots, \mathbf{A}_{r_k}$ with base $\mathbf{C}_t$, and similarly their *t-deficit meet product* if $\mathbf{C}_t$ is the dummy t-blade, is defined by

$$\mathbf{A}_{r_1} \vee_{\mathbf{C}_t} \mathbf{A}_{r_2} \vee_{\mathbf{C}_t} \cdots \vee_{\mathbf{C}_t} \mathbf{A}_{r_k} = \mathbf{A}_{r_1} \vee_{\mathbf{C}_t} (\mathbf{A}_{r_2} \vee_{\mathbf{C}_t} (\cdots (\mathbf{A}_{r_{k-1}} \vee_{\mathbf{C}_t} \mathbf{A}_{r_k}) \cdots)). \quad (2.4.31)$$

Proposition 2.47. Mod-$\mathbf{C}_t$ cograded anticommutativity:

$$\mathbf{C}_t(\mathbf{A}_r \vee_{\mathbf{C}_t} \mathbf{B}_s) = (-1)^{(n-t-r)(n-t-s)} \mathbf{C}_t(\mathbf{B}_s \vee_{\mathbf{C}_t} \mathbf{A}_r). \quad (2.4.32)$$

Mod-$\mathbf{C}_t$ associativity:

$$\mathbf{C}_t(\mathbf{A}_{r_1} \vee_{\mathbf{C}_t} (\mathbf{A}_{r_2} \vee_{\mathbf{C}_t} \mathbf{A}_{r_3})) = \mathbf{C}_t((\mathbf{A}_{r_1} \vee_{\mathbf{C}_t} \mathbf{A}_{r_2}) \vee_{\mathbf{C}_t} \mathbf{A}_{r_3}). \quad (2.4.33)$$

Proposition 2.48. [Collection symmetry] Let $\mathbf{U}, \mathbf{A}, \mathbf{B}, \mathbf{C}$ be blades of grade u, a, b, c respectively, such that $a + b + c = 2(n - u)$. Then for any bipartition $(\mathbf{U}_{(1)}, \mathbf{U}_{(2)})$ of shape (u_1, u_2) of $\mathbf{U}$,

$$\mathbf{UA} \vee \mathbf{U}_{(1)}\mathbf{B} \vee \mathbf{U}_{(2)}\mathbf{C} = (-1)^{u_2(n+a+b+u_1+u_2)}\mathbf{UA} \vee \mathbf{UB} \vee \mathbf{C}. \quad (2.4.34)$$

The result can be further strengthened as follows: (2.4.34) holds for any u-vector $\mathbf{U}$ such that one of $\mathbf{U}_{(1)}, \mathbf{U}_{(2)}$ is a blade factor of $\mathbf{U}$.

Proof. The expansions of both sides by partitioning $\mathbf{UA}$ equal

$$[\mathbf{UA}_{(1)}\mathbf{B}]\mathbf{UA}_{(2)}\mathbf{C} \quad (2.4.35)$$

up to sign. In expanding the left side, blade $\mathbf{UA}$ is split into $\mathbf{U}_{(1)}\mathbf{U}_{(2)}\mathbf{A}_{(1)}\mathbf{A}_{(2)}$, and then $\mathbf{U}_{(2)}\mathbf{A}_{(1)}$ is distributed to $\mathbf{U}_{(1)}\mathbf{B}$. The sign of the result differs from that of (2.4.35) by $(-1)^{a_2 u_2}$, where $a_2 = a - (n - u_1 - u_2 - b)$ is the grade of $\mathbf{A}_{(2)}$.

Similarly, in expanding the right side of (2.4.34), blade $\mathbf{UA}$ is split into $\mathbf{UA}_{(1)}\mathbf{A}_{(2)}$, and then $\mathbf{A}_{(1)}$ is distributed to $\mathbf{UB}$. The sign of the result is identical to that of (2.4.35). $\qquad\square$

Corollary 2.49. [Commutation symmetry] With notations as above,

$$\mathbf{UA} \vee \mathbf{UB} \vee \mathbf{C} = (-1)^{u(n+a+b+u)}\mathbf{UA} \vee \mathbf{B} \vee \mathbf{UC}. \quad (2.4.36)$$

Proposition 2.50. [Separation symmetry] Let $\mathbf{U}, \mathbf{A}, \mathbf{B}, \mathbf{A}', \mathbf{B}'$ be blades of grade u, a, b, a', b' respectively, such that $a + b + u = n$ and $a + b = a' + b'$. Then

$$\mathbf{UA} \vee \mathbf{UB} \vee \mathbf{A}'\mathbf{B}' = (-1)^{ua'}\mathbf{UA}' \vee \mathbf{AB} \vee \mathbf{UB}'. \quad (2.4.37)$$

Proof. Expanding $\mathbf{UA} \vee \mathbf{UB} \vee \mathbf{A}'\mathbf{B}'$ by distributing $\mathbf{UA}$, we get $[\mathbf{UAB}]\mathbf{UA}'\mathbf{B}'$. Expanding $\mathbf{UA}' \vee \mathbf{AB} \vee \mathbf{UB}'$ by distributing $\mathbf{UA}'$, we get $(-1)^{ua'}[\mathbf{UAB}]\mathbf{UA}'\mathbf{B}'$. $\qquad\square$

2.5 Cayley expansion

Given a Cayley expression generated by vectors, its *Cayley expansion* refers to the procedure of eliminating all the meet products from it, so that the result contains only brackets and outer products of vectors. Cayley expansion is the procedure of changing multiplication to addition. Since multiplication preserves geometric meaning while addition breaks it up, Cayley expansion is a procedure of "algebraization". It is an indispensable step in making geometric deduction from geometric constructions using GC algebra and bracket algebra.

Despite its importance, Cayley expansion is one of the least studied topics in GC algebra. The first systematic investigation of this topic was made in [117], where the *Cayley expansion theory* on the *complete classification* of all the factored and binomial expansions of layer-one Cayley expressions in 2D and 3D projective geometries was established.

In this section, we first investigate basic Cayley expansions, then introduce the classification of basic Cayley expansions of layer-one Cayley expressions, and finally discuss general Cayley expansions.

2.5.1 *Basic Cayley expansions*

Definition 2.51. The *Cayley space* over $\mathcal{V}^n$ is the linear space spanned by the meet product results of pseudovectors in $\Lambda(\mathcal{V}^n)$. The Cayley space is still denoted by $\Lambda(\mathcal{V}^n)$. An *r-coblade* in the Cayley space refers to the meet product of r pseudovectors. Any r-coblade is an $(n-r)$-blade, and still denoted by $\mathbf{A}_{n-r}$.

Notation.

The symbol "$\vdash_\vee$" denotes a partition of a coblade into a sequence of coblades, by first partitioning the sequence of pseudovectors of the coblade, then changing each subsequence of pseudovectors into a coblade by the meet product.

Definition 2.52. The *meet coproduct* in Cayley space $\Lambda(\mathcal{V}^n)$ is the linear extension of the mapping $\Delta^\vee : \Lambda(\mathcal{V}^n) \longrightarrow \Lambda(\mathcal{V}^n) \otimes \Lambda(\mathcal{V}^n)$, which is defined for any r-coblade $\mathbf{A}_{n-r} \in \Lambda(\mathcal{V}^n)$ by partitioning $\mathbf{A}_{n-r}$ as the meet product of r pseudovectors:

$$\Delta^\vee \mathbf{A}_{n-r} := \sum_{\vdash_\vee \mathbf{A}_{n-r}} \mathbf{A}_{n-r\,(1)} \otimes \mathbf{A}_{n-r\,(2)} = \sum_{i=0}^{r} \sum_{(i,r-i)\vdash_\vee \mathbf{A}_{n-r}} \mathbf{A}_{n-r\,(1)} \otimes \mathbf{A}_{n-r\,(2)}.$$

$$(2.5.1)$$

Definition 2.53. The *Grassmann-Cayley coalgebra* over $\mathcal{V}^n$ is the Grassmann space $\Lambda(\mathcal{V}^n)$ equipped with the outer coproduct, the meet coproduct, the counit mapping $\langle\ \rangle_0$ of the outer coproduct, and the counit mapping $\langle\ \rangle_n$ of the meet coproduct:

$$\langle\ \rangle_n : \mathbf{A} \mapsto \langle \mathbf{A} \rangle_n, \quad \forall \mathbf{A} \in \Lambda(\mathcal{V}^n).$$

$$(2.5.2)$$

Corresponding to the shuffle formulas (2.3.31) and (2.3.32) for the outer coproduct, there are the following *shuffle formulas* for the meet coproduct: for any r-coblade $\mathbf{A}_{n-r}$ and s-coblade $\mathbf{B}_{n-s}$ in the Cayley space,

$$\mathbf{A}_{n-r}\mathbf{B}_{n-s} = \sum_{(r+s-n,n-r)\vdash_\vee \mathbf{B}_{n-s}} \left(\mathbf{A}_{n-r} \vee \mathbf{B}_{n-s(2)}\right) \mathbf{B}_{n-s(1)}, \tag{2.5.3}$$

and

$$\mathbf{A}_{n-r}\mathbf{B}_{n-s} = \sum_{(n-s,r+s-n)\vdash_\vee \mathbf{A}_{n-r}} \left(\mathbf{A}_{n-r(1)} \vee \mathbf{B}_{n-s}\right) \mathbf{A}_{n-r(2)}. \tag{2.5.4}$$

The right sides of the above two formulas are still called the *Cayley expansions* of the left sides. To distinguish between the two kinds of Cayley expansions, the expansions based on partitioning vector sequences are called *outer coproduct expansions*, or *meet product expansion*; the expansions based on partitioning pseudovector sequences are called *meet coproduct expansions*, or *outer product expansion*. Cayley expansions are realized by changing meet products and their outer products into outer coproducts and meet coproducts respectively.

Definition 2.54. Given a Cayley expression, its *basic Cayley expansion* is the procedure of changing the expression into an equal expression involving only brackets and outer products of vectors, by using the four shuffle formulas (2.3.31), (2.3.32), (2.5.3) and (2.5.4)

Besides basic Cayley expansions, there are also other Cayley expansions, *e.g.*, (2.5.33) and (2.5.38) to be introduced in Subsection 2.5.3. However, general Cayley expansions produce more terms than the basic ones. They do not appear to be useful in practice. With the basic Cayley expansions at hand, we are ready to prove some simple projective geometric theorems.

Example 2.55. [Fano's axiom] In the projective plane, a *complete quadrilateral* is composed of four points called *vertices*, no three of which are collinear, together with the six line segments connecting each pair of vertex, called *sides*. There are three pairs of sides that do not meet any vertex. Their intersections are the three *diagonal points*.

Fano's axiom says that there is no complete quadrilateral whose diagonal points are collinear.

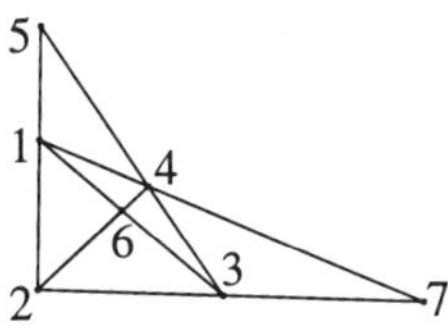

Fig. 2.2   Fano's axiom.

The configuration (Figure 2.2) of Fano's axiom can be constructed as follows:

Free points (vertices): $\mathbf{1,2,3,4}$, satisfying $[\mathbf{123}],[\mathbf{124}],[\mathbf{134}],[\mathbf{234}] \neq 0$.
Intersections (diagonal points):

$$\mathbf{5 = 12 \cap 34}, \quad \mathbf{6 = 13 \cap 24}, \quad \mathbf{7 = 14 \cap 23}.$$

Conclusion: $[\mathbf{567}] \neq 0$.

Proof. The three intersections have Cayley expressions

$$\mathbf{5 = 12 \vee 34}, \quad \mathbf{6 = 13 \vee 24}, \quad \mathbf{7 = 14 \vee 23}.$$

Substituting them into bracket $[\mathbf{567}]$, we get

$$[\mathbf{567}] = [(\mathbf{12 \vee 34})(\mathbf{13 \vee 24})(\mathbf{14 \vee 23})]. \tag{2.5.5}$$

We need to prove that after expanding the meet products, the result is always not equal to zero.

In (2.5.5), the first meet product has two different expansions:

$$\mathbf{12 \vee 34} = [\mathbf{134}]\mathbf{2} - [\mathbf{234}]\mathbf{1} = [\mathbf{124}]\mathbf{3} - [\mathbf{123}]\mathbf{4}.$$

Substituting any of them, say the first one, into (2.5.5), we get

$$[\mathbf{567}] = [\mathbf{134}][\mathbf{2}(\mathbf{13 \vee 24})(\mathbf{14 \vee 23})] - [\mathbf{234}][\mathbf{1}(\mathbf{13 \vee 24})(\mathbf{14 \vee 23})]. \tag{2.5.6}$$

The first term in (2.5.6) has two meet products to be expanded. If we expand $\mathbf{13 \vee 24}$ by separating $\mathbf{2,4}$, *i.e.*,

$$\mathbf{13 \vee 24} = [\mathbf{134}]\mathbf{2} + [\mathbf{123}]\mathbf{4}, \tag{2.5.7}$$

and substituting the result into $[\mathbf{2}(\mathbf{13 \vee 24})(\mathbf{14 \vee 23})]$, then vector $\mathbf{2}$ on the right side of (2.5.7) is canceled by the same vector $\mathbf{2}$ in $[\mathbf{2}(\mathbf{13 \vee 24})(\mathbf{14 \vee 23})]$. The result after the expansion remains a Cayley monomial. On the contrary, if we use the other expansion

$$\mathbf{13 \vee 24} = [\mathbf{124}]\mathbf{3} + [\mathbf{234}]\mathbf{1},$$

and substitute it into $[\mathbf{2}(\mathbf{13 \vee 24})(\mathbf{14 \vee 23})]$, the result is a Cayley binomial. The monomial result is obviously better for later manipulations.

For the same reason, in the second term of (2.5.6), the first meet product $\mathbf{13 \vee 24}$ should be expanded by separating $\mathbf{1,3}$. This is the simplest case of *bracket-oriented expansion for factored and shortest result*.

After making the expansions of $\mathbf{13 \vee 24}$ in both terms of (2.5.6), we get

$$[\mathbf{567}] = [\mathbf{134}][\mathbf{123}][\mathbf{24}(\mathbf{14 \vee 23})] - [\mathbf{234}][\mathbf{124}][\mathbf{13}(\mathbf{14 \vee 23})]. \tag{2.5.8}$$

The meet product $\mathbf{14 \vee 23}$ can be expanded by splitting either $\mathbf{1,4}$ or $\mathbf{2,3}$, and the results are identical.

$$[\mathbf{24}(\mathbf{14 \vee 23})] = -[\mathbf{234}][\mathbf{124}],$$
$$[\mathbf{13}(\mathbf{14 \vee 23})] = [\mathbf{123}][\mathbf{134}].$$

After all the expansions, we get
$$[(\mathbf{12} \vee \mathbf{34})(\mathbf{13} \vee \mathbf{24})(\mathbf{14} \vee \mathbf{23})] = -2\,[\mathbf{123}][\mathbf{124}][\mathbf{134}][\mathbf{234}]. \qquad (2.5.9)$$
The right side is nonzero by the given hypotheses. $\qquad\Box$

This example discloses that Cayley expansions have strong combinatorial nature. If a Cayley expression contains two meet products, then if one meet product generates k terms by expansion, the other meet product can have different expansions in each of the k terms produced by the first expansion.

Since Cayley expansion is the procedure of changing multiplication to addition, it inevitably leads to the growth of the expression size. For the purpose of simplifying algebraic manipulations, a Cayley expansion leading to a monomial result is the most desired. If such a result is unavailable, then a result containing bracket monomial factors, called *factored result*, is preferred. If no factored result is available, then a Cayley expansion generating the least number of terms among all the expansions of the expression is optimal.

The classification of all the Cayley expansions of a Cayley expression is purposed to provide all the available optimal Cayley expansions. By making optimal Cayley expansions, the size growth induced by Cayley expansions can be effectively controlled. In classifying Cayley expansions, we do not consider general Cayley expansions (2.5.33) and (2.5.38), because they usually generate a lot more terms than basic Cayley expansions. The *Cayley expansion theory*, which is on the classification of all optimal Cayley expansions, is key to simplifying symbolic computations involving meet products.

2.5.2 *Cayley expansion theory*

Cayley expansions can be divided into two categories according to the algebraic constraints among the constituent vectors of the Cayley expression to be expanded. If all the distinct vectors in the expression are independent generic vectors, then every Cayley expansion of the expression is said to be *generic*. If there are *linear constraints* among distinct vectors in the expression, *e.g.*, collinearity and coplanarity constraints, then every Cayley expansion of the expression is said to be *semifree*.

The difference in classifying generic expansions and semifree expansions is solely caused by the *rewriting rules* of the bracket operator. They are the rules that are *automatically applied* to all brackets in a computer program implementation of bracket algebra. In Cayley expansion theory, it is assumed that only the following two rewriting rules are applied:

Rule of antisymmetry. The antisymmetry within a bracket, which is composed of the two syzygies **B1**, **B2** in Definition 2.11, is automatically applied whenever a bracket occurs. This means that if the input is $[\mathbf{112}]$, then it is automatically identified as 0, and if the input is $[\mathbf{213}]$, then it is automatically replaced by $-[\mathbf{123}]$.

Rule of evaluation. The evaluations of specific brackets demanded by linear constraints, are automatically applied whenever the specific brackets occur. In 2D and 3D projective geometries, the specific brackets are

- if points $\mathbf{1}, \mathbf{2}, \ldots, \mathbf{r}$ are collinear, then for any $1 \leq i, j, k \leq r$, $[\mathbf{ijk}] = 0$, and for any point $\mathbf{x}$ in the projective space, $[\mathbf{ijkx}] = 0$;
- if the points are coplanar in space, then for any $1 \leq i, j, k, l \leq r$, $[\mathbf{ijkl}] = 0$.

There are two rules that are not applied to brackets automatically. The multilinearity property of the bracket operator can be used in both directions, either for expanding a bracket, or for factorizing a bracket polynomial. A GP syzygy has at least three terms, and can be used to replace any term with negative the rest terms. The ambiguities in the aim and target of manipulations determine that the two syzygies cannot be applied unconditionally to all brackets.

Definition 2.56. By taking a bracket as a Cayley expression of *layer 0*, a Cayley expression is said to be of *layer 1*, if it can be obtained from a bracket by replacing one or several vectors in the bracket by either meet products of Grassmann monomials or linear combinations of vectors.

Recursively, a Cayley expression is said to be of *layer i*, if it can be obtained from a Cayley expression of layer $i - 1$ by similar replacement of one or several vectors of it, but cannot be obtained from any Cayley expression of layer lower than $i - 1$ by replacements of vectors.

For example, in 2D projective geometry, starting from a bracket $[\mathbf{abc}]$, by replacing vector $\mathbf{a}$ with meet product $\mathbf{12} \vee \mathbf{34}$, we get a layer-1 expression $[(\mathbf{12} \vee \mathbf{34})\mathbf{bc}] = \mathbf{12} \vee \mathbf{34} \vee \mathbf{bc}$. Continuing to replace $\mathbf{b}$ by $\mathbf{1'2'} \vee \mathbf{3'4'}$, we get an expression $[(\mathbf{12} \vee \mathbf{34})(\mathbf{1'2'} \vee \mathbf{3'4'})\mathbf{c}]$, which is still of layer one, because it can be obtained directly from bracket $[\mathbf{abc}]$ by making the two replacements simultaneously. If we replace $\mathbf{1}$ by $\mathbf{1'2'} \vee \mathbf{3'4'}$, the result $(\mathbf{1'2'} \vee \mathbf{3'4'})\mathbf{2} \vee \mathbf{34} \vee \mathbf{bc} = [(\mathbf{1'2'} \vee \mathbf{3'4'})\mathbf{2}(\mathbf{34} \vee \mathbf{bc})]$ is still of layer one. A layer-2 expression is shown in Figure 2.3.

Since the expansion result of a Cayley expression is a bracket polynomial, we classify all Cayley expressions into different layers by their affinities to brackets. Then Cayley expansions reduce the layer of a Cayley expression by one each time, and recursively they change the expression into a bracket polynomial.

In 2D geometry, there are three layer-1 Cayley expressions caused by replacements of vectors with meet products, or equivalently, with intersections of lines:

$$
\begin{aligned}
p_I &= \mathbf{12} \vee \mathbf{1'2'} \vee \mathbf{1''2''}, \\
p_{III} &= [\mathbf{1}(\mathbf{1'2'} \vee \mathbf{3'4'})(\mathbf{1''2''} \vee \mathbf{3''4''})], \\
p_{IV} &= [(\mathbf{12} \vee \mathbf{34})(\mathbf{1'2'} \vee \mathbf{3'4'})(\mathbf{1''2''} \vee \mathbf{3''4''})].
\end{aligned}
\tag{2.5.10}
$$

There are three other layer-1 Cayley expressions caused by replacements of vectors with at least one meet product and one linear combination:

$$
\begin{aligned}
q_I &= [\mathbf{1}(\mathbf{1'2'} \vee \mathbf{3'4'})\mathbf{a}_{\mathbf{5''6''}}], \\
q_{II} &= [(\mathbf{12} \vee \mathbf{34})\mathbf{a}_{\mathbf{5'6'}}\mathbf{a}_{\mathbf{5''6''}}], \\
q_{III} &= [(\mathbf{12} \vee \mathbf{34})(\mathbf{1'2'} \vee \mathbf{3'4'})\mathbf{a}_{\mathbf{5''6''}}],
\end{aligned}
\tag{2.5.11}
$$

Layer 0:   [abc]

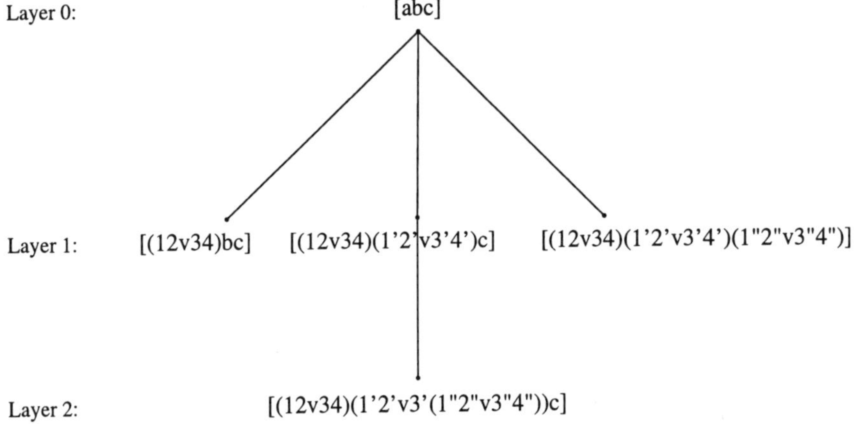

Fig. 2.3   layers of a Cayley expression.

where

$$\begin{aligned}
\mathbf{a_{56}} &= \lambda_6 \mathbf{5} + \lambda_5 \mathbf{6}, \\
\mathbf{a_{5'6'}} &= \lambda_{6'} \mathbf{5'} + \lambda_{5'} \mathbf{6'}, \\
\mathbf{a_{5''6''}} &= \lambda_{6''} \mathbf{5''} + \lambda_{5''} \mathbf{6''}
\end{aligned} \tag{2.5.12}$$

represent three generic points on lines $\mathbf{56}$, $\mathbf{5'6'}$, $\mathbf{5''6''}$ respectively, with the λ's as generic scalar parameters.

In 3D projective geometry, there are much more layer-1 Cayley expressions. While a single vector can be replaced by the meet product of either three 3-vectors or a 2-vector and a 3-vector, two vectors can be put together to be replaced by the meet product of two 3-vectors. If there is only one replacement, and it is by a point of intersection (meet product), then there are the following two layer-1 expressions:

$$\begin{aligned}
r_I &= [(\mathbf{12} \vee \mathbf{1'2'3'})\mathbf{1''2''3''}] &&= \mathbf{12} \vee \mathbf{1'2'3'} \vee \mathbf{1''2''3''}, \\
r_{II} &= [(\mathbf{123} \vee \mathbf{1'2'3'} \vee \mathbf{1''2''3''})\mathbf{1'''2'''3'''}] &&= \mathbf{123} \vee \mathbf{1'2'3'} \vee \mathbf{1''2''3''} \vee \mathbf{1'''2'''3'''}.
\end{aligned} \tag{2.5.13}$$

In this subsection and in Appendix A, we classify the basic Cayley expansions of the above layer-1 Cayley expressions. Below we use the term *Cayley expansion* to denote a basic one. We say that an expansion is *monomial* (or *binomial*, or *i-termed*), if it leads to a bracket polynomial of one term (or two terms, or i terms). We use the term *factored expansion* to refer to an expansion leading to a bracket polynomial in factored form, and use the term *shortest expansion* of an expression to refer to an expansion leading to a bracket polynomial whose number of terms is the smallest among all the expansions of the expression. A Cayley expression is said to have *unique expansion*, if all its Cayley expansions lead to the same result.

We illustrate the arguments used in the classifications by classifying all the generic expansions, semifree expansions and unique expansions of the simplest expression: p_I of (2.5.10).

The Cayley expansion of $p_I = \mathbf{12} \vee \mathbf{1'2'} \vee \mathbf{1''2''}$ by distributing $\mathbf{1,2}$ is

$$p_I = [\mathbf{11'2'}][\mathbf{21''2''}] - [\mathbf{21'2'}][\mathbf{11''2''}]. \qquad (2.5.14)$$

There are two other expansions of p_I, which distribute $\mathbf{1',2'}$ and $\mathbf{1'',2''}$ respectively.

According to the number of terms generated, p_I has three different expansion results: zero, bracket monomial and bracket binomial. Geometrically, if two points are identical then their outer product is zero, if two lines are identical then their meet product is zero, and if three lines have an explicit common point then their meet product is zero. These are the only three *trivially zero* cases.

Notice that when using GC algebra to represent geometric objects and their constraints, we usually do not distinguish between the geometric descriptions and their algebraic representations, and use them in a mixed manner.

If (2.5.14) is not trivially zero, then it is a bracket monomial if and only if one of the four brackets equals zero, *i.e.*, $\mathbf{1}$ or $\mathbf{2}$ is on one of the lines $\mathbf{1'2'}, \mathbf{1''2''}$. In all other cases, (2.5.14) is a bracket binomial.

Below we find all the unique expansions. If the results from the three expansions of p_I are identical, then each result has to be a bracket monomial.

Case 1. Generic expansions:

If different points in p_I are generic ones, then (2.5.14) is a monomial if and only if point $\mathbf{1}$ or $\mathbf{2}$ belongs to one of the pairs $\mathbf{1'2'}$ and $\mathbf{1''2''}$. By symmetry, we can assume that $\mathbf{1} = \mathbf{1'}$. Since p_I is not trivially zero, no other point of p_I can be identical to $\mathbf{1}$. The expansion by splitting $\mathbf{1''}$ and $\mathbf{2''}$ is

$$[\mathbf{121''}][\mathbf{12'2''}] - [\mathbf{122''}][\mathbf{12'1''}].$$

It is a monomial if and only if one of $\mathbf{2}, \mathbf{2'}$ is identical with one of $\mathbf{1''}, \mathbf{2''}$. By symmetry, we can assume that $\mathbf{2'} = \mathbf{2''}$. Then the expansion by splitting $\mathbf{1''}$ and $\mathbf{2''} = \mathbf{2'}$ must be a monomial: $-[\mathbf{122'}][\mathbf{12'1''}]$.

So in generic expansions, p_I has a unique expansion if and only if $\mathbf{1} = \mathbf{1'}$ and $\mathbf{2'} = \mathbf{2''}$ up to symmetry.

Case 2. Semifree expansions:

If there are collinearity constraints among different points in p_I, they only evaluate some brackets to zero, and have no influence upon the identification of two brackets. The conclusion in Case 1 is still valid for semifree expansions.

Proposition 2.57. Semifree expansions of $p_I = \mathbf{12} \vee \mathbf{1'2'} \vee \mathbf{1''2''}$:

(1) (Trivially zero) p_I is trivially zero if one of the following conditions is satisfied: (a) one of the three pairs $\mathbf{12}, \mathbf{1'2'}, \mathbf{1''2''}$, is a pair of identical points; (b) two of the three pairs are collinear; (C) one of the six points is on all the three lines.

In the following, assume that p_I is not trivially zero.

(2) (Monomial expansion) p_I has a monomial expansion if and only if one of the six points is on two of the three lines $\mathbf{12}, \mathbf{1'2'}, \mathbf{1''2''}$.

(3) (Unique expansion) A *double point* is a point that occurs twice in p_I. p_I has a unique expansion if and only if it has two double points. The unique expansion is

$$\mathbf{12} \vee \mathbf{12'} \vee \mathbf{22''} = [\mathbf{122'}][\mathbf{122''}]. \tag{2.5.15}$$

Cayley expansion theory is established by using the technique of case-by-case symmetric analysis of all possible results. The procedure is long and tedious. To make the text more readable, we move most of the classification results together with their proofs to Appendix A.

The multiplication of two p_I-typed Cayley expressions is of the form

$$p_{II} = (\mathbf{12} \vee \mathbf{34} \vee \mathbf{56})(\mathbf{1'2'} \vee \mathbf{3'4'} \vee \mathbf{5'6'}). \tag{2.5.16}$$

At first glance, it seems that the classification of p_{II} should be direct from that of p_I, if we expand both p_I-typed Cayley expressions in p_{II} simultaneously and then multiply the results together. This naive idea is incorrect.

Indeed, the only way to expand p_{II} is to expand the two p_I-typed Cayley expressions. However, the two expressions need not be expanded simultaneously. For example, the Cayley expansion of p_{II} by separating $\mathbf{1}, \mathbf{2}$ as the first step, is

$$p_{II} = [\mathbf{134}][\mathbf{256}]\mathbf{1'2'} \vee \mathbf{3'4'} \vee \mathbf{5'6'} - [\mathbf{234}][\mathbf{156}]\mathbf{1'2'} \vee \mathbf{3'4'} \vee \mathbf{5'6'}. \tag{2.5.17}$$

In each term on the right side, expression $\mathbf{1'2'} \vee \mathbf{3'4'} \vee \mathbf{5'6'}$ can have a different expansion. Besides, there are six different first-step expansions of p_{II}, called the *initial expansions*.

Proposition 2.58. If the 12 points in p_{II} are different and are generic points in the plane, then p_{II} has 45 different expansion results of bracket polynomials.

Proof. If starting with the expansion (2.5.17), there are 9 expansions into bracket polynomials. So p_{II} has $2 \times 3 \times 9 = 54$ expansions, among which $3 \times 3 = 9$ are counted twice. All together there are $54 - 9 = 45$ different expansions of p_{II} into bracket polynomials, no two of which produce the same result. $\qquad\square$

Any Cayley expansion of p_{II} into a bracket polynomial can be decomposed into three steps:

(1) initial expansion on one p_I-typed Cayley expression,
(2) successive expansions of the other p_I-typed Cayley expression in the two newly generated terms respectively,
(3) combining the *like terms* generated by the two successive expansions.

In Cayley expansion theory, if two terms are identical up to an integer scale, they are called *like terms*. They can either cancel each other or merge into a single nonzero term. This situation makes the classification of Cayley expansions much more complicated than expected. Since like terms can only be generated from different terms produced by the initial Cayley expansion, they are given a special name, called *intergroup like terms*.

The proof of the following proposition is moved to Section A.1 of Appendix A.

Proposition 2.59. Semifree expansions of $p_{II} = (\mathbf{12} \vee \mathbf{34} \vee \mathbf{56})(\mathbf{1'2'} \vee \mathbf{3'4'} \vee \mathbf{5'6'})$:

(1) (Zero) $p_{II} = 0$ if and only if one of its meet products is zero.

   In the following, assume that $p_{II} \neq 0$.

(2) (Inner point) If $\mathbf{1}, \mathbf{2}, \mathbf{3}$ are collinear, point $\mathbf{3}$ is called an *inner point* of p_{II}. Then

$$p_{II} = [\mathbf{124}][\mathbf{356}]\mathbf{1'2'} \vee \mathbf{3'4'} \vee \mathbf{5'6'} \qquad (2.5.18)$$

is a shortest expansion. If $\mathbf{3} = \mathbf{1}$ or $\mathbf{2}$, and all distinct points in p_{II} are generic ones, then (2.5.18) is the unique shortest expansion.

(3) (Like terms) If p_{II} has no inner point, and its two meet products are not identical, then it has only one pattern having an expansion that generates intergroup like terms:

$$(\mathbf{12} \vee \mathbf{34} \vee \mathbf{56})(\mathbf{13} \vee \mathbf{24} \vee \mathbf{56}) = [\mathbf{124}][\mathbf{134}][\mathbf{256}][\mathbf{356}] - [\mathbf{123}][\mathbf{234}][\mathbf{156}][\mathbf{456}].$$
$$(2.5.19)$$

Furthermore, the left side of (2.5.19) has only one expansion that generates intergroup like terms; the result of the expansion is the right side of (2.5.19).

(4) p_{II} has a monomial expansion if and only if both of its meet products have monomial expansions. The shortest expansions of p_{II} are 2-termed if either it is of the pattern (2.5.19), or a meet product of p_{II} has an inner point. The shortest expansions of p_{II} are 3-termed if the two meet products are identical. In other cases, the shortest expansions are 4-termed.

As a surprising byproduct, the third item (2.5.19) in the classification of the expansions of p_{II} leads to the following elegant Cayley factorization formula:

$$[\mathbf{124}][\mathbf{134}][\mathbf{256}][\mathbf{356}] - [\mathbf{123}][\mathbf{234}][\mathbf{156}][\mathbf{456}] = (\mathbf{12} \vee \mathbf{34} \vee \mathbf{56})(\mathbf{13} \vee \mathbf{24} \vee \mathbf{56}). \quad (2.5.20)$$

This is a typical benefit from Cayley expansion theory.

Cayley factorization is much more difficult than its reverse procedure Cayley expansion. By classifying all factored expansions and binomial expansions of low-layer Cayley expressions, we can get a complete list of Cayley factorizations for low-degree bracket binomials, by equalizing the factored expansions and other expansions of the same Cayley expressions. In this way, at least the Cayley factorization problem for low-degree bracket binomials can be solved completely. For general bracket polynomials, since the classification of Cayley expansions is very complicated, this approach of Cayley factorization no longer works.

Besides the benefit produced for Cayley factorization, Cayley expansion theory also provides a list of optimal expansion results for low-layer Cayley expansions. They can be used to evaluate the simplicity of a Cayley expression, in order to make selections among different Cayley expressions representing the same construction or constraint. In Chapter 4, there are plenty of such examples arising from geometric computing involving projective conics.

The following theorem provides a complete classification of all generic expansions of p_{III}. Its proof can be found in Section A.2 of Appendix A.

Theorem 2.60. Generic expansions of $p_{III} = [\mathbf{1}(\mathbf{1'2'} \vee \mathbf{3'4'})(\mathbf{1''2''} \vee \mathbf{3''4''})]$:

(1) (Trivially zero) If one of the following conditions is satisfied, p_{III} is trivially zero: (a) one of the four pairs, $\mathbf{1'2'}, \mathbf{3'4'}, \mathbf{1''2''}, \mathbf{3''4''}$, is a pair of identical points; (b) one of the two pairs, $\{\mathbf{1'2'}, \mathbf{3'4'}\}$ and $\{\mathbf{1''2''}, \mathbf{3''4''}\}$, is a pair of identical lines; (c) the two sets $\{\mathbf{1'2'}, \mathbf{3'4'}\}$ and $\{\mathbf{1''2''}, \mathbf{3''4''}\}$ are identical as sets of lines.

In the following, assume that p_{III} is not trivially zero.

(2) (Inner intersection) If $\mathbf{1'} = \mathbf{3'}$, point $\mathbf{1'}$ is called an *inner intersection* in p_{III}. The following is a shortest expansion in this case:

$$[\mathbf{1}(\mathbf{1'2'} \vee \mathbf{1'4'})(\mathbf{1''2''} \vee \mathbf{3''4''})] = [\mathbf{1'2'4'}]\mathbf{11'} \vee \mathbf{1''2''} \vee \mathbf{3''4''}. \qquad (2.5.21)$$

The right side is the unique factored expansion of the left side. Furthermore, the right side is the unique shortest expansion if and only if on the left side, points $\mathbf{2'}, \mathbf{4'}$ are not in $\{\mathbf{1''}, \mathbf{2''}\}, \{\mathbf{3''}, \mathbf{4''}\}$ respectively. In the exceptional case, the other shortest expansion is

$$[\mathbf{1}(\mathbf{1'2'} \vee \mathbf{1'4'})(\mathbf{2'2''} \vee \mathbf{4'4''})] = [\mathbf{11'4'}][\mathbf{1'2'2''}][\mathbf{2'4'4''}] - [\mathbf{11'2'}][\mathbf{1'4'4''}][\mathbf{2'4'2''}].$$
$$(2.5.22)$$

(3) (Double line) If $\mathbf{1'} = \mathbf{1''}$, $\mathbf{2'} = \mathbf{2''}$, line $\mathbf{1'2'}$ is called a *double line* in p_{III}. The following is the unique shortest expansion in this case:

$$[\mathbf{1}(\mathbf{1'2'} \vee \mathbf{3'4'})(\mathbf{1'2'} \vee \mathbf{3''4''})] = [\mathbf{11'2'}]\mathbf{1'2'} \vee \mathbf{3'4'} \vee \mathbf{3''4''}. \qquad (2.5.23)$$

The right side is also the unique factored expansion of the left side.

(4) (Recursion of **1**) If $\mathbf{1'} = \mathbf{1}$, point $\mathbf{1}$ is said to *recur* in p_{III}. The following is a shortest expansion in this case:

$$[\mathbf{1}(\mathbf{12'} \vee \mathbf{3'4'})(\mathbf{1''2''} \vee \mathbf{3''4''})] = [\mathbf{13'4'}]\mathbf{12'} \vee \mathbf{1''2''} \vee \mathbf{3''4''}. \qquad (2.5.24)$$

The right side is the unique factored expansion of the left side. Furthermore, the right side is the unique shortest expansion if and only if on the left side, points $\mathbf{3'}, \mathbf{4'}$ are not in $\{\mathbf{1''}, \mathbf{2''}\}, \{\mathbf{3''}, \mathbf{4''}\}$ respectively. In the exceptional case, the other shortest expansion is

$$[\mathbf{1}(\mathbf{12'} \vee \mathbf{3'4'})(\mathbf{3'2''} \vee \mathbf{4'4''})] = [\mathbf{12'4'}][\mathbf{13'2''}][\mathbf{3'4'4''}] - [\mathbf{12'3'}][\mathbf{14'4''}][\mathbf{3'4'2''}].$$
$$(2.5.25)$$

(5) (Other cases) If p_{III} has neither inner intersection nor double line, and $\mathbf{1}$ does not recur, then p_{III} has no factored expansion. The shortest expansions of p_{III} are two, three, or four termed if and only if the set $\mathcal{M} = \{\mathbf{1'}, \mathbf{2'}, \mathbf{3'}, \mathbf{4'}\} \cap \{\mathbf{1''}, \mathbf{2''}, \mathbf{3''}, \mathbf{4''}\}$ has at least two elements, has only one element, or is empty.

Definition 2.61. A vector variable in a homogeneous bracket polynomial or Cayley polynomial p is said to be a *single point*, *double point*, or *triple point*, if it occurs once, twice, or three times in every term of p.

The following theorem provides a complete classification of all factored expansions and binomial expansions of p_{IV}. Its proof can be found in Section A.3.

Theorem 2.62. Generic expansions of $p_{IV} = [(\mathbf{12} \vee \mathbf{34})(\mathbf{1'2'} \vee \mathbf{3'4'})(\mathbf{1''2''} \vee \mathbf{3''4''})]$:

(1) (Trivially zero) If one of the following conditions is satisfied, p_{IV} is trivially zero: (a) one of the six pairs, $\mathbf{12}, \mathbf{34}, \mathbf{1'2'}, \mathbf{3'4'}, \mathbf{1''2''}, \mathbf{3''4''}$, is a pair of identical points; (b) one of the three pairs, $\{\mathbf{12}, \mathbf{34}\}, \{\mathbf{1'2'}, \mathbf{3'4'}\}, \{\mathbf{1''2''}, \mathbf{3''4''}\}$, is a pair of identical lines; (c) two of the three sets $\{\mathbf{12}, \mathbf{34}\}, \{\mathbf{1'2'}, \mathbf{3'4'}\}, \{\mathbf{1''2''}, \mathbf{3''4''}\}$, are identical as sets of lines.

In the following, assume that p_{IV} is not trivially zero.

(2) (Inner intersection) If $\mathbf{1} = \mathbf{3}$, point $\mathbf{1}$ is called an *inner intersection* in p_{IV}. The following is the unique factored expansion in this case:

$$[(\mathbf{12} \vee \mathbf{14})(\mathbf{1'2'} \vee \mathbf{3'4'})(\mathbf{1''2''} \vee \mathbf{3''4''})] = [\mathbf{124}][\mathbf{1}(\mathbf{1'2'} \vee \mathbf{3'4'})(\mathbf{1''2''} \vee \mathbf{3''4''})].$$
$$(2.5.26)$$

The right side also leads to a shortest expansion of the left side.

(3) (Double line) If $\mathbf{12} = \mathbf{1'2'}$, line $\mathbf{12}$ is called a *double line* in p_{IV}. The following is the unique factored expansion in this case:

$$[(\mathbf{12} \vee \mathbf{34})(\mathbf{12} \vee \mathbf{3'4'})(\mathbf{1''2''} \vee \mathbf{3''4''})] = (\mathbf{12} \vee \mathbf{34} \vee \mathbf{3'4'})(\mathbf{12} \vee \mathbf{1''2''} \vee \mathbf{3''4''}). \quad (2.5.27)$$

The right side leads to a shortest expansion of the left side if either it has an expansion of two terms, or p_{IV} has another double line, or the set $\mathcal{N} = \{\mathbf{3}, \mathbf{4}, \mathbf{3'}, \mathbf{4'}\} \cap \{\mathbf{1''}, \mathbf{2''}, \mathbf{3''}, \mathbf{4''}\}$ has at most two elements. When neither is satisfied, the shortest expansions of the left side have three terms if and only if $\mathcal{N}$ has three elements.

(4) (Triangle) If $\mathbf{1'} = \mathbf{1}$ and $\mathbf{1''2''} = \mathbf{22'}$, then $\mathbf{122'}$ is called a *triangle* in p_{IV}. The following is a factored expansion in this case:

$$\begin{aligned} &[(\mathbf{12} \vee \mathbf{34})\,(\mathbf{12'} \vee \mathbf{3'4'})\,(\mathbf{22'} \vee \mathbf{3''4''})] \\ &= [\mathbf{122'}]([\mathbf{134}][\mathbf{23''4''}][\mathbf{2'3'4'}] - [\mathbf{13'4'}][\mathbf{234}][\mathbf{2'3''4''}]). \end{aligned} \quad (2.5.28)$$

When there is neither inner intersection nor double line, the triangle pattern has three subpatterns where more bracket factors can be generated:

(a) (Complete quadrilateral) If $\{\mathbf{1}, \mathbf{2}, \mathbf{3}, \mathbf{4}\} = \{\mathbf{1'}, \mathbf{2'}, \mathbf{3'}, \mathbf{4'}\} = \{\mathbf{1''}, \mathbf{2''}, \mathbf{3''}, \mathbf{4''}\}$, then

$$[(\mathbf{12} \vee \mathbf{34})(\mathbf{13} \vee \mathbf{24})(\mathbf{14} \vee \mathbf{23})] = -2[\mathbf{123}][\mathbf{124}][\mathbf{134}][\mathbf{234}]. \quad (2.5.29)$$

The monomial expression is symmetric with respect to $\mathbf{1}, \mathbf{2}, \mathbf{3}, \mathbf{4}$.

(b) (Quadrilateral) If p_{IV} has two triangles sharing the same side, *i.e.*, they share a pair of common points, for example it is triangles **123** and **134**, then the only pattern for the two triangles to coexist in p_{IV} is $[(\mathbf{12} \vee \mathbf{34})\,(\mathbf{13} \vee \mathbf{24})\,(\mathbf{14} \vee \mathbf{3''4''})]$. We have

$$[(\mathbf{12}\vee\mathbf{34})\,(\mathbf{13}\vee\mathbf{24})\,(\mathbf{14}\vee\mathbf{3''4''})] = -[\mathbf{124}][\mathbf{134}]([\mathbf{123}][\mathbf{43''4''}]+[\mathbf{13''4''}][\mathbf{234}]). \tag{2.5.30}$$

The pair of 4-tuple and 2-tuple $(\mathbf{1234}, \mathbf{14})$ on the left side of (2.5.30) is called a *quadrilateral* in p_{IV}.

(c) (Triangle pair) If p_{IV} has two disentangled triangles, *i.e.*, the two triangles have no point in common, the only pattern for the two triangles to coexist in p_{IV} is

$$[(\mathbf{12} \vee \mathbf{34})\,(\mathbf{12'} \vee \mathbf{34'})\,(\mathbf{22'} \vee \mathbf{44'})] = -[\mathbf{122'}][\mathbf{344'}]\mathbf{13} \vee \mathbf{24} \vee \mathbf{2'4'}. \tag{2.5.31}$$

The pair of 3-tuples $(\mathbf{122'}, \mathbf{344'})$ on the left side of (2.5.31) is called a *triangle pair* in p_{IV}. This pattern corresponds to Desargues Theorem, see (3.1.7) in Chapter 3.

(5) If p_{IV} is not trivially zero, and has neither inner intersection, nor double line, nor triangle, then it has no factored expansion.

In the following items, the above hypothesis is always assumed.

(6) (Two triple points) If p_{IV} has two triple points, then it has a binomial expansion.

(7) (One triple point and two double points) If p_{IV} has only one triple point **1**, then it has a binomial expansion if and only if there are two double points **2, 3**, such that p_{IV} is of the pattern $[(\mathbf{14} \vee \mathbf{23})(\mathbf{12'} \vee \mathbf{3'4'})(\mathbf{12''} \vee \mathbf{3''4''})]$, where $\mathbf{2} \in \{\mathbf{2'}, \mathbf{3'}, \mathbf{4'}\}$ and $\mathbf{3} \in \{\mathbf{2''}, \mathbf{3''}, \mathbf{4''}\}$.

(8) (Four double points) If p_{IV} has no triple point, then it has a binomial expansion if and only if there are four double points **1, 2, 3, 4**, such that p_{IV} is of one of the following two patterns:

(a) (4-2-2 pattern) $[(\mathbf{12} \vee \mathbf{34})(\mathbf{1'2'} \vee \mathbf{3'4'})(\mathbf{1''2''} \vee \mathbf{3''4''})]$, where $\{\mathbf{1}, \mathbf{3}\} \subset \{\mathbf{1'}, \ldots, \mathbf{4'}\}$ and $\{\mathbf{2}, \mathbf{4}\} \subset \{\mathbf{1''}, \ldots, \mathbf{4''}\}$. This pattern corresponds to Pascal's Conic Theorem, see (4.1.1) of Chapter 4.

(b) (3-3-2 pattern) $[(\mathbf{12} \vee \mathbf{56})(\mathbf{13} \vee \mathbf{46'})(\mathbf{24} \vee \mathbf{36''})]$, where $\mathbf{5}, \mathbf{6}, \mathbf{6'}, \mathbf{6''}$ are either single or double points.

Remark: (2.5.30) comes from the following Cayley expansions:

$$\begin{aligned}(\mathbf{12} \vee \mathbf{34})\,(\mathbf{13} \vee \mathbf{24}) &= -[\mathbf{134}][\mathbf{234}]\mathbf{12} + [\mathbf{123}][\mathbf{124}]\mathbf{34} \\ &= -[\mathbf{124}][\mathbf{234}]\mathbf{13} + [\mathbf{123}][\mathbf{134}]\mathbf{24}.\end{aligned} \tag{2.5.32}$$

(2.5.32) is antisymmetric with respect to **1, 2, 3, 4**. Its geometric meaning in conic geometry will be explained at the end of Chapter 4.

2.5.3   *General Cayley expansions*

In the two formulas (2.3.31) and (2.3.32) for basic Cayley expansions of the meet product $\mathbf{A}_r \vee \mathbf{B}_s$, either $\mathbf{A}_r$ or $\mathbf{B}_s$ disappears from the non-scalar part of the expansion result. In fact, both $\mathbf{A}_r$ and $\mathbf{B}_s$ can have some of their vectors remain in the non-scalar part of the expansion result.

Proposition 2.63. For any r-blade $\mathbf{A}_r$ and s-blade $\mathbf{B}_s$ in $\Lambda(\mathcal{V}^n)$, for any integers r', s' between 0 and $r + s - n$,

$$\mathbf{A}_r \vee \mathbf{B}_s = \frac{(-1)^{r'(n-s)}}{C_{r+s-n}^{r'}} \sum_{(r',r-r') \vdash \mathbf{A}_r} \mathbf{A}_{r(1)} (\mathbf{A}_{r(2)} \vee \mathbf{B}_s) = \frac{1}{C_{r+s-n}^{s'}} \sum_{(s',s-s') \vdash \mathbf{B}_s} \mathbf{B}_{s(1)} (\mathbf{A}_r \vee \mathbf{B}_{s(2)}).$$

$$(2.5.33)$$

Proof. First, we prove the first equality. When $r' = 1$,

$$\mathbf{A}_r \vee \mathbf{B}_s = \frac{1}{r} \sum_{(1,r-1) \vdash \mathbf{A}_r} \mathbf{A}_{r(1)} \mathbf{A}_{r(2)} \vee \mathbf{B}_s$$

$$= \frac{1}{r} \sum_{(1,r-1) \vdash \mathbf{A}_r} \left(\sum_{(n-s,r+s-n-1) \vdash \mathbf{A}_{r(2)}} (-1)^{n-s} [\mathbf{A}_{r(21)} \mathbf{B}_s] \mathbf{A}_{r(1)} \mathbf{A}_{r(22)} \right.$$

$$\left. + \sum_{(n-s-1,r+s-n) \vdash \mathbf{A}_{r(2)}} [\mathbf{A}_{r(1)} \mathbf{A}_{r(21)} \mathbf{B}_s] \mathbf{A}_{r(22)} \right)$$

$$= \frac{1}{r} \left\{ \sum_{(1,r-1) \vdash \mathbf{A}_r} (-1)^{n-s} \mathbf{A}_{r(1)} (\mathbf{A}_{r(2)} \vee \mathbf{B}_s) \right.$$

$$\left. + \sum_{(n-s,r+s-n) \vdash \mathbf{A}_r} \sum_{(1,n-s-1) \vdash \mathbf{A}_{r(1)}} [\mathbf{A}_{r(11)} \mathbf{A}_{r(12)} \mathbf{B}_s] \mathbf{A}_{r(2)} \right\}$$

$$= \frac{1}{r} \left\{ \sum_{(1,r-1) \vdash \mathbf{A}_r} (-1)^{n-s} \mathbf{A}_{r(1)} (\mathbf{A}_{r(2)} \vee \mathbf{B}_s) + (n - s) \mathbf{A}_r \vee \mathbf{B}_s \right\}.$$

Assume that the first equality holds for $r' - 1$. Then for r',

$$\mathbf{A}_r \vee \mathbf{B}_s = \frac{(-1)^{n-s}}{r + s - n} \sum_{(1,r-1) \vdash \mathbf{A}_r} \mathbf{A}_{r(1)} (\mathbf{A}_{r(2)} \vee \mathbf{B}_s)$$

$$= \frac{(-1)^{r'(n-s)}}{(r + s - n) C_{r+s-n-1}^{r'-1}} \sum_{(1,r-1) \vdash \mathbf{A}_r} \sum_{(r'-1,r-r') \vdash \mathbf{A}_{r(2)}} \mathbf{A}_{r(1)} \mathbf{A}_{r(21)} (\mathbf{A}_{r(22)} \vee \mathbf{B}_s)$$

$$= \frac{(-1)^{r'(n-s)}}{(r + s - n) C_{r+s-n-1}^{r'-1}} \sum_{(r',r-r') \vdash \mathbf{A}_r} \sum_{(1,r'-1) \vdash \mathbf{A}_{r(1)}} \mathbf{A}_{r(11)} \mathbf{A}_{r(12)} (\mathbf{A}_{r(2)} \vee \mathbf{B}_s)$$

$$= \frac{(-1)^{r'(n-s)}}{C_{r+s-n}^{r'}} \sum_{(r',r-r') \vdash \mathbf{A}_r} \mathbf{A}_{r(1)} (\mathbf{A}_{r(2)} \vee \mathbf{B}_s).$$

Second, the second equality is proved as follows:

$$\mathbf{A}_r \vee \mathbf{B}_s = (-1)^{(n-r)(n-s)} \mathbf{B}_s \vee \mathbf{A}_r$$
$$= \frac{(-1)^{(n-r)(n-s)+s'(n-r)}}{C_{r+s-n}^{s'}} \sum_{(s',s-s')\vdash \mathbf{B}_s} \mathbf{B}_{s(1)} \left(\mathbf{B}_{s(2)} \vee \mathbf{A}_r \right)$$
$$= \frac{1}{C_{r+s-n}^{s'}} \sum_{(s',s-s')\vdash \mathbf{B}_s} \mathbf{B}_{s(1)} \left(\mathbf{A}_r \vee \mathbf{B}_{s(2)} \right).$$

$\square$

Example 2.64. When $n = 3$, $r = s = 2$, then $r + s - n = 1$. The only nontrivial case is $r' = s' = 1$, where (2.5.33) are two basic expansion formulas. So in 2D projective geometry, there are no Cayley expansions other than basic ones.

Example 2.65. When $n = 4$, if $r = 2$ and $s = 3$, still (2.5.33) are two basic expansion formulas. If $r = s = 3$, $r' = s' = 1$, then (2.5.33) becomes

$$\begin{aligned} \mathbf{123} \vee \mathbf{1'2'3'} &= -\frac{1}{2}\{\mathbf{1}(\mathbf{23} \vee \mathbf{1'2'3'}) - \mathbf{2}(\mathbf{13} \vee \mathbf{1'2'3'}) + \mathbf{3}(\mathbf{12} \vee \mathbf{1'2'3'})\} \\ &= \frac{1}{2}\{\mathbf{1'}(\mathbf{123} \vee \mathbf{2'3'}) - \mathbf{2'}(\mathbf{123} \vee \mathbf{1'3'}) + \mathbf{3'}(\mathbf{123} \vee \mathbf{1'2'})\}. \end{aligned} \tag{2.5.34}$$

It is the only non-basic Cayley expansion in 3D geometry provided by (2.5.33).

Notation.

The *number of partitions* of a set of k elements into r non-overlapping subsets of $k_1, k_2, \ldots, k_r$ elements respectively, where $k_1 + k_2 + \cdots + k_r = k$, is

$$C_k^{k_1, k_2, \ldots, k_{r-1}} := \frac{k!}{k_1! k_2! \cdots k_r!}. \tag{2.5.35}$$

When $r = 2$, this is just the usual combinatorial number $C_k^{k_1}$.

Corollary 2.66. For any $0 \le r' + s' \le r + s - n$,

$$\mathbf{A}_r \vee \mathbf{B}_s = \frac{(-1)^{r'(n-s)}}{C_{r+s-n}^{r',s'}} \sum_{\substack{(r',r-r')\vdash \mathbf{A}_r, \\ (s',s-s')\vdash \mathbf{B}_s}} \mathbf{A}_{r(1)} \mathbf{B}_{s(1)} \left(\mathbf{A}_{r(2)} \vee \mathbf{B}_{s(2)} \right). \tag{2.5.36}$$

Example 2.67. When $n = 4$, $r = s = 3$, $r' = s' = 1$, (2.5.36) becomes

$$\begin{aligned} \mathbf{123} \vee \mathbf{1'2'3'} = -\frac{1}{2}\{ &\mathbf{11'}[\mathbf{232'3'}] - \mathbf{12'}[\mathbf{231'3'}] + \mathbf{13'}[\mathbf{231'2'}] \\ &-\mathbf{21'}[\mathbf{132'3'}] + \mathbf{22'}[\mathbf{131'3'}] - \mathbf{23'}[\mathbf{131'2'}] \\ &+\mathbf{31'}[\mathbf{122'3'}] - \mathbf{32'}[\mathbf{121'3'}] + \mathbf{33'}[\mathbf{121'2'}]\}. \end{aligned} \tag{2.5.37}$$

It is the only non-basic Cayley expansion in 3D projective geometry provided by (2.5.36). It can also be obtained from (2.5.34) by basic Cayley expansions.

Proposition 2.68. Let $\mathbf{A}_{r_1}, \mathbf{A}_{r_2}, \ldots, \mathbf{A}_{r_k}$ be blades in $\Lambda(\mathcal{V}^n)$ of grade $r_1, r_2, \ldots, r_k$ respectively. Then for any $0 \le r_1' + r_2' + \cdots r_k' \le s_k = r_1 + r_2 + \cdots r_k - (k-1)n$,

$$
\mathbf{A}_{r_1} \vee \mathbf{A}_{r_2} \vee \cdots \vee \mathbf{A}_{r_k} = \frac{(-1)^{\sum_{1 \le i < j \le k} r_i'(n-r_j)}}{C_{s_k}^{r_1', r_2', \ldots, r_k'}} \sum_{i=1}^{k} \sum_{(r_i', r_i - r_i') \vdash \mathbf{A}_{r_i}}
$$
$$
\mathbf{A}_{r_1(1)} \mathbf{A}_{r_2(1)} \cdots \mathbf{A}_{r_k(1)} \left(\mathbf{A}_{r_1(2)} \vee \mathbf{A}_{r_2(2)} \vee \cdots \vee \mathbf{A}_{r_k(2)} \right). \tag{2.5.38}
$$

Proof. For $1 \le i \le k$, let

$$
s_{i1} = (r_1 - r_1') + (r_2 - r_2') + \cdots + (r_{i-1} - r_{i-1}') - (i-2)n,
$$
$$
s_{i2} = r_{i+1} + r_{i+2} + \cdots + r_k - (k-i-1)n,
$$

then

$$
\mathbf{A}_{r_1(2)} \vee \mathbf{A}_{r_2(2)} \vee \cdots \vee \mathbf{A}_{r_{i-1}(2)} \vee \mathbf{A}_{r_i} \vee \mathbf{A}_{r_{i+1}} \vee \cdots \vee \mathbf{A}_{r_k}
$$
$$
= (-1)^{(n-r_i)(n-s_{i1})} \mathbf{A}_{r_i} \vee \mathbf{A}_{r_1(2)} \vee \mathbf{A}_{r_2(2)} \vee \cdots \vee \mathbf{A}_{r_{i-1}(2)} \vee \mathbf{A}_{r_{i+1}} \vee \cdots \vee \mathbf{A}_{r_k}
$$
$$
= (-1)^{(n-r_i)(n-s_{i1}) + r_i'(n - s_{i1} - s_{i2} + n)} (C_{s_{i1} + s_{(i-1)2} - n}^{r_i'})^{-1} \sum_{(r_i', r_i - r_i') \vdash \mathbf{A}_{r_i}}
$$
$$
\mathbf{A}_{r_i(1)} (\mathbf{A}_{r_i(2)} \vee \mathbf{A}_{r_1(2)} \vee \mathbf{A}_{r_2(2)} \vee \cdots \vee \mathbf{A}_{r_{i-1}(2)} \vee \mathbf{A}_{r_{i+1}} \vee \cdots \vee \mathbf{A}_{r_k})
$$
$$
= (-1)^{(n-r_i)(n-s_{i1}) + r_i'(s_{i1} + s_{i2}) + (n - r_i + r_i')(n - s_{i1})} (C_{s_{i1} + s_{(i-1)2} - n}^{r_i'})^{-1}
$$
$$
\sum_{(r_i', r_i - r_i') \vdash \mathbf{A}_{r_i}} \mathbf{A}_{r_i(1)} (\mathbf{A}_{r_1(2)} \vee \mathbf{A}_{r_2(2)} \vee \cdots \vee \mathbf{A}_{r_i(2)} \vee \mathbf{A}_{r_{i+1}} \vee \cdots \vee \mathbf{A}_{r_k})
$$
$$
= (-1)^{r_i'(n + s_{i2})} (C_{s_{i1} + s_{(i-1)2} - n}^{r_i'})^{-1}
$$
$$
\sum_{(r_i', r_i - r_i') \vdash \mathbf{A}_{r_i}} \mathbf{A}_{r_i(1)} (\mathbf{A}_{r_1(2)} \vee \mathbf{A}_{r_2(2)} \vee \cdots \vee \mathbf{A}_{r_i(2)} \vee \mathbf{A}_{r_{i+1}} \vee \cdots \vee \mathbf{A}_{r_k}).
$$

When i goes from 1 to up k, we get (2.5.38). $\qquad \square$

2.6 Grassmann factorization*

In this section, the outer product of a multivector $\mathbf{A}$ with itself is denoted by $\mathbf{A}^2$.

Definition 2.69. Given a multivector $\mathbf{A} \in \Lambda(\mathcal{V}^n)$, if there exist two non-scalar multivectors $\mathbf{B}, \mathbf{C} \in \Lambda(\mathcal{V}^n)$ such that $\mathbf{A} = \mathbf{BC}$, then $\mathbf{A}$ is said to be *divisible* by $\mathbf{B}, \mathbf{C}$, and $\mathbf{B}, \mathbf{C}$ are called nontrivial *factors* or *divisors* of $\mathbf{A}$. The decomposition is called a *Grassmann factorization* of $\mathbf{A}$.

Grassmann factorization is one of the oldest topics in Grassmann algebra. The problem of finding all vector factors is completely solved. However, the problem of finding a *complete factorization* of a multivector $\mathbf{A}$, *i.e.*, finding $\mathbf{B}_1, \mathbf{B}_2, \ldots, \mathbf{B}_r$ such that $\mathbf{A} = \mathbf{B}_1 \mathbf{B}_2 \cdots \mathbf{B}_r$, and each $\mathbf{B}_i$ does not have any nontrivial factor, remains an open problem. The problem of finding a common divisor of maximal grade for two multivectors, called their *Grassmann GCD*, is solved only for the special case where the GCD is a blade.

At the first appearance, the factorization problem is irrelevant to the meet product and Cayley expansion. However, this problem can be effectively dealt with only in the framework of GC algebra, by using meet products and various Cayley expansions extensively. In this section, we present some basic results on finding vector factors of a homogeneous multivector. Most of the results in this section can be found in [69].

Definition 2.70. The *span* of a k-vector $\mathbf{A} \in \Lambda^k(\mathcal{V}^n)$ is the following subspace in $\mathcal{V}^n$:

$$\mathrm{span}(\mathbf{A}) := \{\mathbf{A} \vee \mathbf{U}_{n-k+1} \,|\, \mathbf{U}_{n-k+1} \in \Lambda^{n-k+1}(\mathcal{V}^n)\}. \tag{2.6.1}$$

The *rank* of $\mathbf{A}$ is the dimension of $\mathrm{span}(\mathbf{A})$.

For example, let $\mathbf{1}, \mathbf{2}, \mathbf{3}, \mathbf{4}$ be four linearly independent vectors. Let $\mathbf{A} = \mathbf{12} + \mathbf{34}$. Then $\mathrm{rank}(\mathbf{A}) = 4$, because for any four different generic $(n-1)$-blades $\mathbf{U}_1, \mathbf{U}_2, \mathbf{U}_3, \mathbf{U}_4$ in $\Lambda(\mathcal{V}^n)$,

$$(\mathbf{A} \vee \mathbf{U}_1)(\mathbf{A} \vee \mathbf{U}_2)(\mathbf{A} \vee \mathbf{U}_3)(\mathbf{A} \vee \mathbf{U}_4) = (\det([\mathbf{U}_i\mathbf{j}])_{i,j=1..4})\,\mathbf{1234} \neq 0.$$

Lemma 2.71. Let $\mathbf{A}_k$ be a nonzero k-vector. Then $\mathrm{rank}(\mathbf{A}_k) \geq k$, and the equality holds if and only if $\mathbf{A}_k$ is a blade. Furthermore, if $\mathrm{rank}(\mathbf{A}_k) > k$, then $\mathrm{rank}(\mathbf{A}_k) \geq k + 2$.

Proof. Since $*(\mathbf{A}_k \vee \mathbf{U}_{n-k+1}) = (*\mathbf{A}_k)(*\mathbf{U}_{n-k+1})$, we consider the following outer product operator:

$$
\begin{aligned}
f : \Lambda^{k-1}(\mathcal{V}^n) &\longrightarrow \Lambda^{n-1}(\mathcal{V}^n) \\
\mathbf{W}_{k-1} &\longmapsto (*\mathbf{A}_k)\mathbf{W}_{k-1}.
\end{aligned}
\tag{2.6.2}
$$

On one hand, $\mathrm{span}(\mathbf{A}_k)$ is isomorphic to the image space of f, so

$$\mathrm{rank}(\mathbf{A}_k) = \dim(\Lambda^{k-1}(\mathcal{V}^n)) - \dim\ker(f).$$

On the other hand, $\mathbf{W}_{k-1} \in \ker(f)$ if and only if for any $\mathbf{x} \in \mathcal{V}^n$, $[(*\mathbf{A}_k)\mathbf{x}\mathbf{W}_{k-1}] = 0$. Since the bracket operator is nondegenerate, by duality,

$$\dim\ker(f) = \dim(\Lambda^{k-1}(\mathcal{V}^n)) - \dim(\{(*\mathbf{A}_k)\mathbf{x} \,|\, \mathbf{x} \in \mathcal{V}^n\}).$$

So

$$\mathrm{rank}(\mathbf{A}_k) = \dim((*\mathbf{A}_k)\mathcal{V}^n) = n - \ker((*\mathbf{A}_k) \wedge |_{\mathcal{V}^n}). \tag{2.6.3}$$

The kernel of $(*\mathbf{A}_k) \wedge |_{\mathcal{V}^n}$ is all vectors dividing $*\mathbf{A}_k$, so its dimension is $\leq n - k$, and the equality holds if and only if $\mathbf{A}_k$ is a blade. So $\mathrm{rank}(\mathbf{A}_k) \geq n - (n-k) = k$.

If $\mathrm{rank}(\mathbf{A}_k) > k$, then the kernel of $(*\mathbf{A}_k) \wedge |_{\mathcal{V}^n}$ has dimension $< n - k$. If the dimension is $n - k - 1$, then $n - k - 1$ linearly independent vectors divide $*\mathbf{A}_k$, so $*\mathbf{A}_k$ is a blade and $\mathrm{rank}(\mathbf{A}_k) = k$. Contradiction. $\qquad\square$

Lemma 2.72. Let $\mathbf{A}_k$ be a nonzero k-vector where $k \geq 2$. For any $(n-1)$-blade $\mathbf{U}_{n-1}$ such that $\mathbf{A}_k \vee \mathbf{U}_{n-1} \neq 0$, there exists a vector $\mathbf{x} \in \mathrm{span}(\mathbf{A}_k)$ such that

$$(-1)^k \mathbf{A}_k = \mathbf{A}_k \mathbf{x} \vee \mathbf{U}_{n-1} - (\mathbf{A}_k \vee \mathbf{U}_{n-1})\mathbf{x}, \qquad (2.6.4)$$

$$\mathrm{span}(\mathbf{A}_k) = \mathrm{span}(\mathbf{x}) \oplus (\mathrm{span}(\mathbf{A}_k \mathbf{x} \vee \mathbf{U}_{n-1}) + \mathrm{span}(\mathbf{A}_k \vee \mathbf{U}_{n-1})). \quad (2.6.5)$$

Proof. Let $\mathbf{U}'_{n-k+1}$ be an $(n-k+1)$-blade such that $\mathbf{A}_k \vee \mathbf{U}'_{n-k+1} \vee \mathbf{U}_{n-1} = 1$. Let

$$\mathbf{x} = \mathbf{A}_k \vee \mathbf{U}'_{n-k+1}. \qquad (2.6.6)$$

Then (2.6.4) is just the Cayley expansion of $\mathbf{A}_k \mathbf{x} \vee \mathbf{U}_{n-1}$ by distributing $\mathbf{A}_k \mathbf{x}$.

For any $(n-k+1)$-blade $\mathbf{U}''$, by (2.6.4), vector

$$\begin{aligned}
\mathbf{A}_k \mathbf{x} \vee \mathbf{U}_{n-1} \vee \mathbf{U}'' &= (-1)^k \mathbf{A}_k \vee \mathbf{U}'' + (\mathbf{A}_k \vee \mathbf{U}_{n-1})\mathbf{x} \vee \mathbf{U}'' \\
&= (-1)^k \mathbf{A}_k \vee \mathbf{U}'' + (\mathbf{A}_k \vee \mathbf{U}_{n-1} \vee \mathbf{U}'')\mathbf{x} - \mathbf{A}_k \vee \mathbf{U}_{n-1} \vee \mathbf{x} \mathbf{U}''
\end{aligned}$$

is in $\mathrm{span}(\mathbf{A}_k)$, so $\mathrm{span}(\mathbf{A}_k \mathbf{x} \vee \mathbf{U}_{n-1}) \subseteq \mathrm{span}(\mathbf{A}_k)$. By (2.6.4),

$$\mathrm{span}(\mathbf{A}_k) = \mathrm{span}(\mathbf{x}) + \mathrm{span}(\mathbf{A}_k \mathbf{x} \vee \mathbf{U}_{n-1}) + \mathrm{span}(\mathbf{A}_k \vee \mathbf{U}_{n-1}).$$

On one hand, $\mathbf{x} \vee \mathbf{U}_{n-1} = 1$. On the other hand, obviously $(\mathbf{A}_k \mathbf{x} \vee \mathbf{U}_{n-1}) \vee \mathbf{U}_{n-1} = 0$ and $(\mathbf{A}_k \vee \mathbf{U}_{n-1}) \vee \mathbf{U}_{n-1} = 0$. So $\mathbf{x} \notin \mathrm{span}(\mathbf{A}_k \mathbf{x} \vee \mathbf{U}_{n-1}) + \mathrm{span}(\mathbf{A}_k \vee \mathbf{U}_{n-1})$. $\qquad \square$

Proposition 2.73. [Fundamental lemma of bivectors] For any bivector $\mathbf{A}_2$, its rank must be an even number $2r$, and there exists a basis $\mathbf{e}_1, \mathbf{e}_2, \ldots, \mathbf{e}_{2r}$ of $\mathrm{span}(\mathbf{A}_2)$ such that $\mathbf{A}_2$ is in the following *standard form* (or *canonical form*):

$$\mathbf{A}_2 = \mathbf{e}_1 \mathbf{e}_2 + \mathbf{e}_3 \mathbf{e}_4 + \cdots + \mathbf{e}_{2r-1} \mathbf{e}_{2r}. \qquad (2.6.7)$$

Proof. Induction on r. If $r = 1$, then $\mathbf{A}_2$ is a blade, and the conclusion is trivial.

Let $\mathrm{rank}(\mathbf{A}_2) = k$. The induction assumption is that any bivector whose rank is less than k must have an even rank, and the bivector must be in the standard form (2.6.7). We need to prove the same conclusion for $\mathbf{A}_2$, whose rank is k.

For any $(n-1)$-blade $\mathbf{U}_{n-1}$ such that $\mathbf{A}_2 \vee \mathbf{U}_{n-1} \neq 0$, by (2.6.4) and (2.6.6), vector $\mathbf{x} = \mathbf{A}_2 \vee \mathbf{U}'_{n-1}$ satisfies

$$\mathbf{A}_2 = \mathbf{A}_2 \mathbf{x} \vee \mathbf{U}_{n-1} + \mathbf{x}(\mathbf{A}_2 \vee \mathbf{U}_{n-1}).$$

Since $\mathbf{A}_2 \mathbf{x} \vee \mathbf{U}_{n-1}$ is a bivector whose rank is strictly lower than k, by induction hypothesis, its rank is even, denoted by $2l$. Then $k = 2l + 2$ if and only if vector $\mathbf{A}_2 \vee \mathbf{U}_{n-1}$ is not in $\mathrm{span}(\mathbf{A}_2 \mathbf{x} \vee \mathbf{U}_{n-1})$.

As in the proof of Lemma 2.72, vector $\mathbf{A}_2 \vee \mathbf{U}_{n-1}$ satisfies $\mathbf{A}_2 \vee \mathbf{U}'_{n-1} \vee \mathbf{U}_{n-1} = 1$. For any $(n-1)$-blade $\mathbf{U}''$, by (2.6.6),

$$\begin{aligned}
\mathbf{A}_2 \mathbf{x} \vee \mathbf{U}_{n-1} \vee \mathbf{U}'' \vee \mathbf{U}'_{n-1} &= \mathbf{A}_2(\mathbf{A}_2 \vee \mathbf{U}'_{n-1}) \vee \mathbf{U}_{n-1} \vee \mathbf{U}'' \vee \mathbf{U}'_{n-1} \\
&= \mathbf{A}_2(\mathbf{A}_2 \vee \mathbf{U}'_{n-1}) \vee \mathbf{U}'_{n-1} \vee \mathbf{U}_{n-1} \vee \mathbf{U}'' \\
&= 0,
\end{aligned}$$

so vector $\mathbf{A}_2 \vee \mathbf{U}_{n-1}$ is not in $\mathrm{span}(\mathbf{A}_2\mathbf{x} \vee \mathbf{U}_{n-1})$, and

$$\mathrm{span}(\mathbf{A}_k) = \mathrm{span}(\mathbf{x}) \oplus \mathrm{span}(\mathbf{A}_k\mathbf{x} \vee \mathbf{U}_{n-1}) \oplus \mathrm{span}(\mathbf{A}_k \vee \mathbf{U}_{n-1}).$$

Since $\mathbf{A}_2\mathbf{x} \vee \mathbf{U}_{n-1}$ has standard form $\mathbf{e}_1\mathbf{e}_2 + \mathbf{e}_3\mathbf{e}_4 + \cdots + \mathbf{e}_{2l-1}\mathbf{e}_{2l}$, $\mathbf{A}_2$ is also in standard form by setting $\mathbf{e}_{2l+1} = \mathbf{x}$ and $\mathbf{e}_{2l+2} = \mathbf{A}_2 \vee \mathbf{U}_{n-1}$. $\square$

Theorem 2.74. [Grassmann reduction] For any k-vector $\mathbf{A}_k$, $\mathbf{A}_k \in \Lambda(\mathrm{span}(\mathbf{A}_k))$.

Proof.   Induction on k and $r = \mathrm{rank}(\mathbf{A}_k)$. When $k = 1$, the conclusion is trivial. When $k = 2$, the conclusion is direct from (2.73). Assume that the conclusion holds for $k = m - 1 \geq 2$. When $k = m$, if $r = m$, then $\mathbf{A}_k$ is a blade and the conclusion is trivial. Assume that the conclusion holds for $k = m$ and $r < s$, where $s > m$.

When $r = s$, by (2.6.4), $(-1)^m \mathbf{A}_m = \mathbf{A}_m\mathbf{x} \vee \mathbf{U}_{n-1} - (\mathbf{A}_m \vee \mathbf{U}_{n-1})\mathbf{x}$. Since $\mathrm{rank}(\mathbf{A}_m\mathbf{x} \vee \mathbf{U}_{n-1}) < s$, the conclusion holds for $\mathbf{A}_m\mathbf{x} \vee \mathbf{U}_{n-1}$. Since $\mathbf{A}_m \vee \mathbf{U}_{n-1}$ has grade $m - 1$, the conclusion holds for $\mathbf{A}_m \vee \mathbf{U}_{n-1}$. By (2.6.5), the conclusion is proved for $k = m$ and $r = s$. $\square$

Lemma 2.75. Let $\mathbf{C}_k$ be a k-blade, and let $\mathbf{A}_r$, $\mathbf{C}'_{r-k}$ be r-vector, $(r-k)$-vector respectively. If $\mathbf{A}_r = \mathbf{C}_k\mathbf{C}'_{r-k}$, then for any s-blade $\mathbf{B}_s$ where $s > n - k$, $\mathbf{C}_k \vee \mathbf{B}_s$ is either zero or a blade factor of $\mathbf{A}_r \vee \mathbf{B}_s$.

Proof.   Since $\mathbf{B}_s$ can always be written as $\mathbf{U}_1 \vee \mathbf{U}_2 \vee \cdots \vee \mathbf{U}_{n-s}$ where the $\mathbf{U}$'s are $(n-1)$-blades, we only need to prove the lemma for the case $s = n - 1$. Let vector $\mathbf{c} \in \mathbf{C}_k$ satisfy $\mathbf{C}_k = (\mathbf{C}_k \vee \mathbf{B}_{n-1})\mathbf{c}$. Then

$$
\begin{aligned}
\mathbf{C}_k\mathbf{C}'_{r-k} \vee \mathbf{B}_{n-1} &= (\mathbf{C}_k \vee \mathbf{B}_{n-1})\mathbf{C}'_{r-k} + (-1)^k\mathbf{C}_k(\mathbf{C}'_{r-k} \vee \mathbf{B}_{n-1}) \\
&= (\mathbf{C}_k \vee \mathbf{B}_{n-1})\{\mathbf{C}'_{r-k} + (-1)^k\mathbf{c}(\mathbf{C}'_{r-k} \vee \mathbf{B}_{n-1})\}.
\end{aligned}
$$

   $\square$

Lemma 2.76. Let $\mathbf{A}_k$ be a nonzero k-vector.

(1) If $k \leq n - 3$, then $\mathbf{A}_k$ has a maximal blade factor of grade r if and only if for every vector $\mathbf{x} \in \mathcal{V}^n$, $\mathbf{A}_k\mathbf{x}$ is either zero or has a blade factor of grade at least $r + 1$, and the equality holds for some $\mathbf{x}$.

(2) If $k \geq 3$, then $\mathbf{A}_k$ has a maximal blade factor of grade r if and only if for every pseudovector $\mathbf{U} \in \Lambda^{n-1}(\mathcal{V}^n)$, $\mathbf{A}_k \vee \mathbf{U}$ is either zero or has a blade factor of grade at most $r - 1$, and the equality holds for some $\mathbf{U}$.

Proof.   We only need to prove the sufficiency statement in Case (1).

Induction on r. If $r = 0$ but $\mathbf{A}_k$ has a vector factor, then for any vector $\mathbf{x}$, $\mathbf{A}_k\mathbf{x}$ either is zero or has at least two linearly independent vector factors, contradicting with the assumption.

If $r = 1$ but $\mathbf{A}_k$ has no vector factor, by the sufficiency assumption, there is a vector $\mathbf{x}$ such that $\mathbf{A}_k\mathbf{x}$ has a 2-blade factor $\mathbf{b}\mathbf{x}$. Then $\mathbf{b} + \lambda\mathbf{x}$ for some scalar λ is a factor of $\mathbf{A}_k$. Contradiction.

Assume that the sufficiency statement holds for $1 < r < s$. When $r = s$, by similar argument, $\mathbf{A}_k$ cannot have any blade factor of grade $> s$. Assume that a maximal blade factor of $\mathbf{A}_k$ has grade $g < s$. Let $\mathbf{A}_k = \mathbf{B}_g \mathbf{C}_{k-g}$ where $\mathbf{B}_g$ is a blade and $\mathbf{C}_{k-g}$ does not have any vector factor. By extending a basis $\mathbf{e}_1, \mathbf{e}_2, \ldots, \mathbf{e}_g$ of $\mathbf{B}_g$ to a basis $\mathbf{e}_1, \mathbf{e}_2, \ldots, \mathbf{e}_n$ of $\mathcal{V}^n$, and denoting $\mathbf{D}_{n-g} = \mathbf{e}_{g+1}\mathbf{e}_{g+2}\cdots\mathbf{e}_n$, we can assume that $\mathbf{C}_{k-g} \in \Lambda^{k-g}(\mathbf{D}_{n-g})$. Then for any $\mathbf{x} = \mathbf{x}_b + \mathbf{x}_d \in \mathcal{V}^n$ such that $\mathbf{x}_b \in \mathbf{B}_g$ and $\mathbf{x}_d \in \mathbf{D}_{n-g}$,

$$\mathbf{A}_k\mathbf{x} = \mathbf{B}_g(\mathbf{C}_{k-g}\mathbf{x}_d).$$

So any maximal blade factor $\mathbf{E}_h$ of $\mathbf{A}_k\mathbf{x}$ can be decomposed into two parts: $\mathbf{E}_h = \mathbf{B}_g\mathbf{F}_{h-g}$, where $\mathbf{F}_{h-g} \in \Lambda^{h-g}(\mathbf{D}_{n-g})$. By the sufficiency assumption, $\mathbf{C}_{k-g}\mathbf{x}_d$ has a blade factor of grade $\geq s+1-g \geq 2$, for any $\mathbf{x}_d \in \mathbf{D}_{n-g}$ such that $\mathbf{C}_{k-g}\mathbf{x}_d \neq 0$. By what we have proved for $r = 1$, $\mathbf{C}_{k-g}$ has a vector factor. Contradiction. $\quad\square$

Lemma 2.77. Let $\mathbf{A}_k, \mathbf{B}_k$ be nonzero k-vectors. Then $\mathbf{A}_k \vee \mathbf{B}_k\mathbf{U}_{n-k-1} = 0$ for any generic $(n-k-1)$-blade $\mathbf{U}_{n-k-1}$ if and only if $\mathbf{A}_k, \mathbf{B}_k$ are blades and are equal to each other up to scale.

Proof. If $\mathbf{A}_k \vee \mathbf{B}_k\mathbf{U}_{n-k-1} = 0$, then vector $*(\mathbf{B}_k\mathbf{U}_{n-k-1}) = (*\mathbf{B}_k) \vee (*\mathbf{U}_{n-k-1})$ divides $*\mathbf{A}_k$. On one hand, the dimension of subspace

$$\{*(\mathbf{B}_k\mathbf{U}_{n-k-1}) \,|\, \mathbf{U}_{n-k-1} \in \Lambda^{n-k-1}(\mathcal{V}^n)\}$$

is $\mathrm{rank}(*\mathbf{B}_k) \geq n-k$. On the other hand, $*\mathbf{A}_k$ has at most $n-k$ linearly independent vector factors, and the equality holds if and only if $*\mathbf{A}_k$ is a blade. This proves that both $*\mathbf{A}_k, *\mathbf{B}_k$ are blades spanned by the same set of vectors. $\quad\square$

Theorem 2.78. Let $\mathbf{A}_k$ be a nonzero k-vector, where $2 \leq k \leq n - 2$. Let

$$\mathbf{C}_{k-2} = \mathbf{A}_k \vee \mathbf{A}_k\mathbf{U}_{n-k-2}, \tag{2.6.8}$$

where $\mathbf{U}_{n-k-2}$ denotes a generic $(n-k-2)$-blade. Then $\mathbf{A}_k$ is a blade if and only if $\mathbf{C}_{k-2} = 0$.

Proof. We only need to prove the sufficiency statement.

Induction on k. When $k = n - 2$, then $*\mathbf{C}_{n-4} = (*\mathbf{A}_{n-2})((*\mathbf{A}_{n-2}) \vee (*\mathbf{U}_0)) = \mathbf{U}_0(*\mathbf{A}_{n-2})^2 = 0$. It is a classical result called *Plücker Theorem* that a bivector $\mathbf{B}_2$ is a blade if and only if $\mathbf{B}_2^2 = 0$. Plücker Theorem can also be obtained from Proposition 2.73.

Assume that the conclusion holds for $k > m$, where $2 \leq m \leq n - 3$. When $k = m$, we only need to prove that $\mathbf{A}_k\mathbf{x}$ is a blade for any vector $\mathbf{x}$. By Corollary 2.49, with $\mathbf{x}$ playing the role of $\mathbf{U}$ there, for any $(n - m + 1)$-blade $\mathbf{U}'$,

$$\mathbf{A}_m\mathbf{x} \vee \mathbf{A}_m\mathbf{x}\mathbf{U}_{n-m-3} \vee \mathbf{U}' = \mathbf{A}_m \vee \mathbf{A}_m\mathbf{x}\mathbf{U}_{n-m-3} \vee \mathbf{x}\mathbf{U}' = 0.$$

By induction hypothesis, $\mathbf{A}_m\mathbf{x}$ is a blade. $\quad\square$

Let $\mathbf{e}_1, \mathbf{e}_2, \ldots, \mathbf{e}_n$ be a fixed basis of V^n. Let N be the sequence $1, 2, \ldots, n$, and let $\mathbf{e}_{i_1 i_2 \ldots i_r} = \mathbf{e}_{i_1} \mathbf{e}_{i_2} \cdots \mathbf{e}_{i_r}$. Then any k-vector has *Plücker coordinates*

$$\mathbf{A}_k = \sum_{(k, n-k) \vdash N} \lambda_{N_{(1)}} \mathbf{e}_{N_{(1)}}. \tag{2.6.9}$$

If $\mathbf{A}_k$ is a blade, then for any partition (N', N'') of N of shape $(n - k + 1, k - 1)$, the following vector is in $\mathbf{A}_k$, and all such vectors span $\mathbf{A}_k$:

$$\mathbf{e}_{N'} \vee \mathbf{A}_k = \sum_{(n-k, 1) \vdash N'} \lambda_{\overline{N'_{(2)} N''}} \, \mathbf{e}_{N'_{(2)}} = \sum_{(1, n-k) \vdash N'} \lambda_{\overline{N'_{(1)} N''}} \, \mathbf{e}_{N'_{(1)}}. \tag{2.6.10}$$

For example, when $n = 4$, $k = 2$, $\mathbf{A}_2$ has Plücker coordinates

$$\mathbf{A}_2 = \lambda_{12} \mathbf{e}_1 \mathbf{e}_2 + \lambda_{13} \mathbf{e}_1 \mathbf{e}_3 + \lambda_{14} \mathbf{e}_1 \mathbf{e}_4 + \lambda_{23} \mathbf{e}_2 \mathbf{e}_3 + \lambda_{24} \mathbf{e}_2 \mathbf{e}_4 + \lambda_{34} \mathbf{e}_3 \mathbf{e}_4. \tag{2.6.11}$$

$\mathbf{A}_2$ is a blade if and only if $\mathbf{A}_2^2 = 0$, *i.e.*,

$$\lambda_{12} \lambda_{34} - \lambda_{13} \lambda_{24} + \lambda_{14} \lambda_{23} = 0. \tag{2.6.12}$$

If $\mathbf{A}_2$ is a blade, then it contains vectors

$$\begin{aligned}
-\lambda_{12} \mathbf{e}_2 - \lambda_{13} \mathbf{e}_3 - \lambda_{14} \mathbf{e}_4, &\quad \text{for } N', N'' = 234, 1; \\
\lambda_{12} \mathbf{e}_1 - \lambda_{23} \mathbf{e}_3 - \lambda_{24} \mathbf{e}_4, &\quad \text{for } N', N'' = 134, 2; \\
\lambda_{13} \mathbf{e}_1 + \lambda_{23} \mathbf{e}_2 - \lambda_{34} \mathbf{e}_4, &\quad \text{for } N', N'' = 124, 3; \\
\lambda_{14} \mathbf{e}_1 + \lambda_{24} \mathbf{e}_2 + \lambda_{34} \mathbf{e}_3, &\quad \text{for } N', N'' = 123, 4.
\end{aligned} \tag{2.6.13}$$

If $\lambda_{ij} \neq 0$ for some i, j, then the two vectors in (2.6.13) containing λ_{ij} span $\mathbf{A}_2$.

Corollary 2.79. Let $\mathbf{A} = \mathbf{A}_1 + \mathbf{A}_2 + \cdots + \mathbf{A}_r$, where each $\mathbf{A}_i$ is a k-blade, and for every $i \neq j$, $\mathbf{A}_i$ and $\mathbf{A}_j$ differ only by a vector factor, then $\mathbf{A}$ is a blade. If $r = 2$, then the converse is also true.

Proof. Only the second part needs proof. If $\mathbf{A}_1 = \mathbf{B}_s \mathbf{A}_1'$ and $\mathbf{A}_2 = \mathbf{B}_s \mathbf{A}_2'$, where s-blade $\mathbf{B}_s$ is a maximal common blade of $\mathbf{A}_1, \mathbf{A}_2$, and $\mathbf{A}_1', \mathbf{A}_2'$ are $(k - s)$-blades, where $k - s \geq 2$, then $\mathbf{B}_s \mathbf{A}_1' \mathbf{A}_2' \neq 0$. For any generic $(n - k - 2)$-blade $\mathbf{U}$,

$$\begin{aligned}
&\mathbf{B}_s(\mathbf{A}_1' + \mathbf{A}_2') \vee \mathbf{B}_s(\mathbf{A}_1' + \mathbf{A}_2') \mathbf{U}_{n-r-2} \\
&= \mathbf{B}_s \mathbf{A}_1' \vee \mathbf{B}_s \mathbf{A}_2' \mathbf{U} + \mathbf{B}_s \mathbf{A}_2' \vee \mathbf{B}_s \mathbf{A}_1' \mathbf{U} \\
&= \mathbf{B}_s \Big(\sum_{(2, k-s-2) \vdash \mathbf{A}_1'} [\mathbf{B}_s \mathbf{A}_{1(1)}' \mathbf{A}_2' \mathbf{U}] \mathbf{A}_{1(2)}' + \sum_{(2, k-s-2) \vdash \mathbf{A}_2'} [\mathbf{B}_s \mathbf{A}_{2(1)}' \mathbf{A}_1' \mathbf{U}] \mathbf{A}_{2(2)}' \Big) \\
&\neq 0,
\end{aligned}$$

So $\mathbf{A}_1 + \mathbf{A}_2$ is not a blade. $\qquad\qquad\square$

In Corollary 2.79, if $r = 3$, then the converse is not true. To see this we set $n = 4$, $k = 2$. A bivector $\mathbf{A}_2$ is a blade if and only if its Plücker coordinates satisfy (2.6.12). If we change λ_{12} to $\mu \lambda_{12}$, at the same time change λ_{34} to $\mu^{-1} \lambda_{34}$, where $\mu \notin \{0, 1\}$, then (2.6.12) is unchanged, so the result remains a blade. The new blade is $\mathbf{A}_2 + (\mu - 1) \lambda_{12} \mathbf{e}_1 \mathbf{e}_2 + (\mu^{-1} - 1) \lambda_{34} \mathbf{e}_3 \mathbf{e}_4$, where the last two terms differ by two vector factors.

Theorem 2.80. Let $\mathbf{A}_k, \mathbf{B}_k$ be nonzero k-vectors, where $2 \leq k \leq n - 2$. Further assume that $\mathbf{A}_k$ is a blade. Let

$$\mathbf{C}'_{k-2} = \mathbf{A}_k \vee \mathbf{B}_k \mathbf{U}_{n-k-2},$$
$$\mathbf{D}'_{k-2} = \mathbf{A}_k \vee \mathbf{B}_k \mathbf{U}_{1(n-k-1)} \vee \mathbf{B}_k \mathbf{U}_{2(n-k-1)},$$
$$(2.6.14)$$

where $\mathbf{U}_{1(n-k-1)}, \mathbf{U}_{2(n-k-1)}$ are two independent generic $(n - k - 1)$-blades.

(1) If $\mathbf{B}_k$ is a blade, then $\mathbf{A}_k, \mathbf{B}_k$ differ from each other by exactly one vector factor if and only if $\mathbf{A}_k \vee \mathbf{B}_k \mathbf{U}_{n-k-1} \neq 0$ but $\mathbf{C}'_{k-2} = 0$.

(2) $\mathbf{A}_k, \mathbf{B}_k$ are blades and differ from each other by exactly one vector factor if and only if $\mathbf{A}_k \vee \mathbf{B}_k \mathbf{U}_{n-k-1} \neq 0$ but $\mathbf{D}'_{k-2} = 0$.

In both cases, $\mathbf{B}_k \sqcup \mathbf{A}_k = \mathbf{A}_k \vee \mathbf{B}_k \mathbf{U}_{n-k-1}$ is a maximal common blade factor of $\mathbf{A}_k, \mathbf{B}_k$.

Proof.   We only need to prove the sufficiency statement in both cases.

Induction on k. Denote $\mathbf{A}'_{n-k} = *\mathbf{A}_k$ and $\mathbf{B}'_{n-k} = *\mathbf{B}_k$. When $k = n - 2$, in Case (1), $*\mathbf{C}'_{n-4} = \mathbf{A}'_2(\mathbf{B}'_2 \vee (*\mathbf{U}_0)) = \mathbf{U}_0 \mathbf{A}'_2 \mathbf{B}'_2 = 0$. Since $\mathbf{A}'_2, \mathbf{B}'_2$ are both blades, they share a common vector.

When $k = n - 2$, in Case (2), $*\mathbf{D}'_{n-4} = \mathbf{A}'_2(\mathbf{B}'_2 \vee \mathbf{U}_{1(n-1)})(\mathbf{B}'_2 \vee \mathbf{U}_{2(n-1)}) = 0$. If $\mathbf{B}'_2$ is a blade, then $\mathbf{A}'_2, \mathbf{B}'_2$ share a common vector. If $\mathbf{B}'_2$ is not a blade, let $\mathbf{B}'_2$ be in the canonical form (2.6.7). Then $r > 1$ and $\mathbf{A}'_2 \mathbf{e}_i \mathbf{e}_j = 0$ for any $1 \leq i < j \leq 2r$. So $\mathbf{A}'_2 = 0$, a contradiction.

Assume that the conclusion holds for $k > m$, where $2 \leq m \leq n - 3$. When $k = m$, we only need to prove that for any vector $\mathbf{x} \in \mathcal{V}^n$, $\mathbf{A}_m \mathbf{x}$ and $\mathbf{B}_m \mathbf{x}$ are blades having a maximal common blade factor of grade m. The proof is much the same with that of Theorem 2.78.

Finally, we prove that blade $\mathbf{A}_k \vee \mathbf{B}_k \mathbf{U}_{n-k-1}$ is a factor of $\mathbf{B}_k$. This is obvious in Case (2). In Case (1), for any $(n - k + 2)$-blade $\mathbf{U}'$,

$$\mathbf{C}'_{k-2} \vee \mathbf{U}' = \mathbf{A}_k \vee \mathbf{U}' \vee \mathbf{B}_k \mathbf{U}_{n-k-2} = [(\mathbf{A}_k \vee \mathbf{U}')\mathbf{B}_k \mathbf{U}_{n-k-2}] = 0,$$

so $(\mathbf{A}_k \vee \mathbf{U}')\mathbf{B}_k = 0$. By the following argument, any vector factor of $\mathbf{A}_k \vee \mathbf{B}_k \mathbf{U}_{n-k-1}$ is also a factor of $\mathbf{B}_k$: for any $(n - k - 1)$-blade $\mathbf{U}''$,

$$[(\mathbf{A}_k \vee \mathbf{B}_k \mathbf{U}_{n-k-1} \vee \mathbf{U}')\mathbf{B}_k \mathbf{U}''] = (-1)^k (\mathbf{A}_k \vee \mathbf{U}') \vee \mathbf{B}_k \mathbf{U}_{n-k-1} \vee \mathbf{B}_k \mathbf{U}''$$
$$= (-1)^{nk} \mathbf{B}_k (\mathbf{A}_k \vee \mathbf{U}') \vee \mathbf{U}_{n-k-1} \vee \mathbf{B}_k \mathbf{U}''$$
$$= 0.$$

$\square$

One can also construct

$$\mathbf{E}'_{k-2} = \mathbf{A}_k \vee \mathbf{A}_k \mathbf{U}_{1(n-k-1)} \vee \mathbf{B}_k \mathbf{U}_{2(n-k-1)}, \qquad (2.6.15)$$

and check if $\mathbf{E}'_{k-2} = 0$ discloses any Grassmann factorization property. If $\mathbf{A}_k$ is a blade, then $\mathbf{E}'_{k-2} = 0$ trivially. If $\mathbf{A}_k$ is not a blade, then for the simplest case $k = n - 2$, $\mathbf{B}'_2 \mathbf{A}'_2(\mathbf{A}'_2 \vee \mathbf{U}_{n-1}) = 0$ for any $(n - 1)$-blade $\mathbf{U}_{n-1}$. Let $\mathbf{A}'_2$ be in the canonical form (2.6.7). By choosing different $\mathbf{U}_{n-1}$'s, it turns out that if $r > 2$ then $\mathbf{B}_2' = 0$, if $r = 2$ then $\mathbf{B}_2' \in \Lambda^2(\mathbf{e}_1\mathbf{e}_2\mathbf{e}_3\mathbf{e}_4)$ is arbitrary. So $\mathbf{E}_{k-2}'$ does not contain interesting factorization information.

Corollary 2.81. The following statements on k-vector $\mathbf{A}_k$ are equivalent:

(1) $\mathbf{A}_k$ is a blade.
(2) $\sum_{(1,k-1)\vdash\mathbf{A}_k}[\mathbf{A}_k\mathbf{A}_{k(1)}]\mathbf{A}_{k(2)} = 0$.
(3) $\sum_{(2,k-2)\vdash\mathbf{A}_k}[\mathbf{A}_k\mathbf{A}_{k(1)}]\mathbf{A}_{k(2)} = 0$.
(4) $\mathbf{A}_k \vee \mathbf{A}_k\mathbf{U}_{1(n-k+1)} \vee \mathbf{A}_k\mathbf{U}_{2(n-k+1)} = 0$ for two independent generic $(n-k+1)$-blades $\mathbf{U}_{1(n-k+1)}, \mathbf{U}_{2(n-k+1)}$.

Theorem 2.82. [Maximal blade factor criterion] Let $\mathbf{A}_k \in \Lambda^k(\mathcal{V}^n)$, where $2 \leq k \leq n-2$. For any $2 \leq r \leq k-1$, define

$$\begin{aligned}
\mathbf{C}_{k-r-1} &= \mathbf{C}_{k-2} \vee \mathbf{A}_k\mathbf{U}_{1(n-k-1)} \vee \cdots \vee \mathbf{A}_k\mathbf{U}_{(r-1)(n-k-1)}, \\
\mathbf{D}_{k-r-1} &= \mathbf{A}_k \vee \mathbf{A}_k\mathbf{U}_{1(n-k-1)} \vee \cdots \vee \mathbf{A}_k\mathbf{U}_{(r+1)(n-k-1)},
\end{aligned} \tag{2.6.16}$$

where the $\mathbf{U}_{i(n-k-1)}$ are independent generic $(n-k-1)$-blades, and where $\mathbf{C}_{k-2}$ is defined by (2.6.8). Then the following statements are equivalent:

- $\mathbf{A}_k$ has a maximal blade factor of grade $k-r$.
- $\mathbf{C}_{k-i-1} \neq 0$ for $1 \leq i \leq r-1$, but $\mathbf{C}_{k-r-1} = 0$.
- $\mathbf{D}_{k-i-1} \neq 0$ for $1 \leq i \leq r-1$, but $\mathbf{D}_{k-r-1} = 0$.

Proof. We only prove that the second statement implies the first. For any $(n-k+r+1)$-blade $\mathbf{U}'$,

$$\begin{aligned}
0 &= \mathbf{C}_{k-r-1} \vee \mathbf{U}' \\
&= \mathbf{C}_{k-r} \vee \mathbf{A}_k\mathbf{U}_{(r-1)(n-k-1)} \vee \mathbf{U}' \\
&= (-1)^{k-r-1}\mathbf{C}_{k-r} \vee \mathbf{U}' \vee \mathbf{A}_k\mathbf{U}_{(r-2)(n-k-1)} \\
&= (-1)^{k-r-1}[(\mathbf{C}_{k-r} \vee \mathbf{U}')\mathbf{A}_k\mathbf{U}_{(r-2)(n-k-1)}].
\end{aligned}$$

So $(\mathbf{C}_{k-r} \vee \mathbf{U}')\mathbf{A}_k = 0$. Since $\mathbf{C}_{k-r} \neq 0$, vector $\mathbf{C}_{k-r} \vee \mathbf{U}'$ is in $\text{span}(\mathbf{C}_{k-r})$. It divides $\mathbf{A}_k$, and all such vectors span a subspace in $\mathcal{V}^n$ of dimension $\text{rank}(\mathbf{C}_{k-r}) \geq k-r$.

Now do induction on r. When $r = 2$, $\mathbf{A}_k$ is not a blade, so its blade factor has grade at most $k-2$. Then $\text{rank}(\mathbf{C}_{k-2}) = k-2$, and $\mathbf{C}_{k-2}$ is a maximal blade factor.

Assume that the conclusion holds for $r < m$, where $3 \leq m \leq k-1$. When $r = m$, then $\mathbf{A}_k$ has a blade factor of grade at most $k-m$. So $\text{rank}(\mathbf{C}_{k-m}) = k-m$, and $\mathbf{C}_{k-m}$ is a maximal blade factor. $\quad\square$

Corollary 2.83. With the same notation as in Theorem 2.82, if $\mathbf{A}_k$ has any maximal blade factor of grade $k-r$, then $\mathbf{C}_{k-r}$ and $\mathbf{D}_{k-r}$ defined by (2.6.16) are both maximal blade factors of $\mathbf{A}_k$. The following is also a maximal blade factor if it is nonzero:

$$\mathbf{A}_k \vee \mathbf{A}_k\mathbf{U}_{n-r} = \sum_{(r,k-r)\vdash\mathbf{A}_k}[\mathbf{A}_k\mathbf{A}_{k(1)}]\mathbf{A}_{k(2)}. \tag{2.6.17}$$

One may expect that the series $\mathbf{A}_k \vee \mathbf{A}_k \mathbf{U}_{n-k-i}$ for $2 \leq i \leq k-1$, should also be used as a criterion for blade factors. For the simplest nontrivial case $i = 3$ and $k = n - 3$, by denoting $\mathbf{A}_3' = *\mathbf{A}_k$ we get that the equality $\mathbf{A}_k \vee \mathbf{A}_k \mathbf{U}_{n-k-i} = 0$ is equivalent to the equality $\mathbf{A}_3'^2 = 0$. Unfortunately, the latter is an identity for all 3-vectors. So the series $\mathbf{A}_k \vee \mathbf{A}_k \mathbf{U}_{n-k-i}$ cannot be used as a criterion. Nevertheless, it indeed can be used to construct a maximal blade factor if the result is nonzero. The operator "$\vee \mathbf{A}_k \mathbf{U}_{n-k-1}$" removes a vector factor from $\mathbf{A}_k$, while the operator "$\vee \mathbf{A}_k \mathbf{U}_{n-k-i}$" removes a blade factor of grade i.

Theorem 2.84. [Maximal common blade factor criterion] Let $\mathbf{A}_k, \mathbf{B}_k \in \Lambda^k(\mathcal{V}^n)$, where $2 \leq k \leq n - 2$. Further assume that $\mathbf{A}_k$ is a blade. For any $2 \leq r \leq k - 1$, define

$$\begin{aligned}
\mathbf{C}_{k-r-1}' &= \mathbf{C}_{k-2}' \vee \mathbf{B}_k \mathbf{U}_{1(n-k-1)} \vee \cdots \vee \mathbf{B}_k \mathbf{U}_{(r-1)(n-k-1)}, \\
\mathbf{D}_{k-r-1}' &= \mathbf{D}_{k-2}' \vee \mathbf{B}_k \mathbf{U}_{3(n-k-1)} \vee \cdots \vee \mathbf{B}_k \mathbf{U}_{(r+1)(n-k-1)},
\end{aligned} \qquad (2.6.18)$$

where the $\mathbf{U}_{i(n-k-1)}$ are independent generic $(n-k-1)$-blades.

(1) If $\mathbf{B}_k$ is a blade, then $\mathbf{A}_k, \mathbf{B}_k$ have a maximal blade factor of grade $k - r$ if and only if $\mathbf{C}_{k-i-1}' \neq 0$ for $1 \leq i \leq r-1$ but $\mathbf{C}_{k-r-1}' = 0$.
(2) $\mathbf{A}_k, \mathbf{B}_k$ have a maximal blade factor of grade $k - r$ if and only if $\mathbf{D}_{k-i-1}' \neq 0$ for $1 \leq i \leq r-1$ but $\mathbf{D}_{k-r-1}' = 0$.

When the conditions in the two cases are satisfied, $\mathbf{C}_{k-r}'$ and $\mathbf{D}_{k-r}'$ are both maximal common blade factors of $\mathbf{A}_k, \mathbf{B}_k$.

Proof. Much the same with that of Theorem 2.82. $\qquad\square$

Proposition 2.85. If n is odd, then any $(n-2)$-vector $\mathbf{A}_{n-2}$ in $\Lambda(\mathcal{V}^n)$ has a vector factor.

Proof. Denote $\mathbf{B}_2 = *\mathbf{A}_{n-2}$, and let $\mathrm{rank}(\mathbf{B}_2) = 2s$. Let $n = 2l+1$. Then $s \leq l$.

By (2.6.8), $\mathbf{C}_{n-4} = \lambda(\mathbf{A}_{n-2} \vee \mathbf{A}_{n-2})$, where λ is a generic scalar. By (2.6.16), $\mathbf{C}_{n-3-r} = \mathbf{C}_{n-4} \vee \mathbf{A}_{n-2}\mathbf{u}_1 \vee \cdots \vee \mathbf{A}_{n-2}\mathbf{u}_{r-1}$, where the $\mathbf{u}_i$ are generic vectors in $\mathcal{V}^n$, and $1 \leq r \leq n - 3$. Then

$$\begin{aligned}
\mathbf{C}_{n-3-r} &= \lambda(\mathbf{A}_{n-2})(*\mathbf{A}_{n-2})(*(\mathbf{A}_{n-2}\mathbf{u}_1)) \cdots (*(\mathbf{A}_{n-2}\mathbf{u}_{r-1})) \\
&= \lambda \mathbf{B}_2^2 (\mathbf{B}_2 \vee (*\mathbf{u}_1)) \cdots (\mathbf{B}_2 \vee (*\mathbf{u}_{r-1})) \\
&= \lambda \mathbf{B}_2^2 \mathbf{x}_1 \cdots \mathbf{x}_{r-1},
\end{aligned}$$

where the $\mathbf{x}_i$ are generic vectors in $\mathrm{span}(\mathbf{B}_2)$. When $r = 2s - 2 \leq n - 3$, then in $\Lambda(\mathrm{span}(\mathbf{B}_2))$, $*\mathbf{C}_{n-3-r} = 0$. By Theorem 2.82, the conclusion is proved. $\qquad\square$

Proposition 2.86. If $\mathbf{B}_s$ is a blade factor of $\mathbf{A}_k$, then $\mathbf{A}_k = \mathbf{B}_s((*\mathbf{B}_s) \vee \mathbf{A}_k)$ up to scale.

Proof. Let blade $\mathbf{B}_s$ be composed of basis vectors. Since $\mathbf{B}_s$ is a blade factor of $\mathbf{A}_k$, $\mathbf{B}_s((*\mathbf{B}_s) \vee \mathbf{A}_k) = (\mathbf{B}_s * \mathbf{B}_s) \vee \mathbf{A}_k = \mathbf{A}_k$. $\qquad\square$

Given a basis $\mathbf{e}_1, \mathbf{e}_2, \ldots, \mathbf{e}_n$ of $\mathcal{V}^n$, any r-vector $\mathbf{W}_r$ can be written as a linear combination of the induced basis $\{\mathbf{e}_{i_1}\mathbf{e}_{i_2}\cdots\mathbf{e}_{i_r} \,|\, 1 \leq i_1 < i_2 < \ldots < i_r \leq n\}$ of $\Lambda^r(\mathcal{V}^n)$. The coefficients are the Plücker coordinates (2.6.9). For the r-vector to be a blade, the condition in Theorem 2.78 needs to be translated into constraints on the Plücker coordinates.

Sometimes only part of the basis of $\mathcal{V}^n$ is given, for example, only $\mathbf{e}_1, \mathbf{e}_2, \ldots, \mathbf{e}_m$ where $0 < m < n$ is given. It is called a *partial basis* of $\mathcal{V}^n$. Denote

$$\mathbf{E}_{n-m} = \mathbf{e}_{m+1}\mathbf{e}_{m+2}\cdots\mathbf{e}_n. \tag{2.6.19}$$

Then any r-vector $\mathbf{W}_r$ can be decomposed into the following sum uniquely:

$$\mathbf{W}_r = \mathbf{A} + \sum_{i=1}^{m} \mathbf{e}_i \mathbf{A}_i + \sum_{1 \leq i < j \leq m} \mathbf{e}_i \mathbf{e}_j \mathbf{A}_{ij} + \cdots + \mathbf{e}_1 \mathbf{e}_2 \cdots \mathbf{e}_m \mathbf{A}_{12\ldots m}, \tag{2.6.20}$$

where $\mathbf{A} \in \Lambda^r(\mathbf{E}_{n-m})$, $\mathbf{A}_{i_1 i_2 \ldots i_k} \in \Lambda^{r-k}(\mathbf{E}_{n-m})$ if $k \leq r$, and $\mathbf{A}_{i_1 i_2 \ldots i_k} = 0$ if $k > r$.

As an exercise, let us translate the condition in Theorem 2.78 into constraints on the "Grassmann coefficients" $\mathbf{A}$'s of (2.6.20).

The simplest case is $m = 1$. We have

$$\mathbf{W}_r = \mathbf{A}_r + \mathbf{e}_1 \mathbf{B}_{r-1}. \tag{2.6.21}$$

Obviously, $\mathbf{A}_r = \mathbf{e}_1 \mathbf{W}_r \vee \mathbf{E}_{n-1}$ and $\mathbf{B}_{r-1} = \mathbf{W}_r \vee \mathbf{E}_{n-1}$. If $\mathbf{W}_r$ is a blade, then both $\mathbf{A}_r$ and $\mathbf{B}_{r-1}$ are blades. Furthermore, if $\mathbf{A}_r \neq 0$, then $\mathbf{B}_{r-1} \in \Lambda(\mathbf{A}_r)$. The two conditions are also sufficient for $\mathbf{W}_r$ to be a blade, according to Theorem 2.78 and

$$\mathbf{W}_r \vee \mathbf{W}_r \mathbf{U}_{n-r-2} = \mathbf{A}_r \vee \mathbf{e}_1 \mathbf{B}_{r-1} \mathbf{U}_{n-r-2} + \mathbf{e}_1 \mathbf{B}_{r-1} \vee \mathbf{A}_r \mathbf{U}_{n-r-2} = 0.$$

The first nontrivial case is $m = 2$.

Proposition 2.87. Let r-vector $\mathbf{W}_r$ be of the form

$$\mathbf{A}_r + \mathbf{e}_1 \mathbf{B}_{r-1} + \mathbf{e}_2 \mathbf{C}_{r-1} + \mathbf{e}_1 \mathbf{e}_2 \mathbf{D}_{r-2}, \tag{2.6.22}$$

where $\mathbf{A}_r, \mathbf{B}_{r-1}, \mathbf{C}_{r-1}, \mathbf{D}_{r-2} \in \Lambda(\mathbf{e}_3 \mathbf{e}_4 \cdots \mathbf{e}_n)$ are of grade $r, r-1, r-1, r-2$ respectively. $\mathbf{W}_r$ is a blade if and only if all the following conditions are satisfied:

(1) $\mathbf{A}_r, \mathbf{B}_{r-1}, \mathbf{C}_{r-1}, \mathbf{D}_{r-2}$ are blades;
(2) if $\mathbf{A}_r \neq 0$, then $\mathbf{B}_{r-1}, \mathbf{C}_{r-1}, \mathbf{D}_{r-2} \in \Lambda(\mathbf{A}_r)$, and in $\Lambda(\mathbf{A}_r)$,

$$\mathbf{B}_{r-1} \vee \mathbf{C}_{r-1} = [\mathbf{A}_r]\mathbf{D}_{r-2}; \tag{2.6.23}$$

(3) if $\mathbf{A}_r = 0$, then $\mathbf{B}_{r-1}$ and $\mathbf{C}_{r-1}$ are linearly dependent;
(4) if $\mathbf{B}_{r-1} \neq 0$, then $\mathbf{D}_{r-2} \in \Lambda(\mathbf{B}_{r-1})$;
(5) if $\mathbf{C}_{r-1} \neq 0$, then $\mathbf{D}_{r-2} \in \Lambda(\mathbf{C}_{r-1})$.

Proof. Let $\mathbf{E}_{n-2} = \mathbf{e}_3\mathbf{e}_4\cdots\mathbf{e}_n$. If $\mathbf{W}_r$ is a blade, then

$$
\begin{aligned}
\mathbf{A}_r &= \mathbf{e}_1\mathbf{e}_2\mathbf{W}_r \vee \mathbf{E}_{n-2}, \\
\mathbf{B}_{r-1} &= -\mathbf{e}_2\mathbf{W}_r \vee \mathbf{E}_{n-2} = \mathbf{W}_r \vee \mathbf{e}_2\mathbf{E}_{n-2}, \\
\mathbf{C}_{r-1} &= \mathbf{e}_1\mathbf{W}_r \vee \mathbf{E}_{n-2} = -\mathbf{W}_r \vee \mathbf{e}_1\mathbf{E}_{n-2}, \\
\mathbf{D}_{r-2} &= \mathbf{W}_r \vee \mathbf{E}_{n-2}.
\end{aligned}
$$

So $\mathbf{A}_r = 0$ if and only if $\mathbf{e}_1\mathbf{e}_2\mathbf{W}_r = 0$, *i.e.*, $\mathbf{W}_r$ has a common vector factor with $\mathbf{e}_1\mathbf{e}_2$. $\mathbf{B}_{r-1} = 0$ if and only if $\mathbf{W}_r \in \Lambda(\mathbf{e}_2\mathbf{E}_{n-2})$, and $\mathbf{C}_{r-1} = 0$ if and only if $\mathbf{W}_r \in \Lambda(\mathbf{e}_1\mathbf{E}_{n-2})$. $\mathbf{D}_{r-2} = 0$ if and only if $\mathbf{W}_r$ has a common $(r-1)$-blade factor with $\mathbf{E}_{n-2}$.

When $\mathbf{W}_r$ is a blade, only (2.6.23) needs proof. When all other conditions are satisfied, we prove that this equality is equivalent to $\mathbf{W}_r \vee \mathbf{W}_r\mathbf{U} = 0$ for generic $(n-r-2)$-blade $\mathbf{U}$.

Case 1. $\mathbf{D}_{r-2} \neq 0$. Let $\mathbf{A}_r = \mathbf{A}_2'\mathbf{D}_{r-2}$, $\mathbf{B}_{r-1} = \mathbf{b}\mathbf{D}_{r-2}$ and $\mathbf{C}_{r-1} = \mathbf{c}\mathbf{D}_{r-2}$, where blade $\mathbf{A}_2' \in \Lambda^2(\mathbf{E}_{n-2})$ and vectors $\mathbf{b}, \mathbf{c} \in \mathbf{E}_{n-2}$. Then

$$\mathbf{W}_r \vee \mathbf{W}_r\mathbf{U}$$

$$
\begin{aligned}
= {}& \sum_{(2,r-2)\vdash\mathbf{A}_r} [\mathbf{A}_{r(1)}\mathbf{e}_1\mathbf{e}_2\mathbf{D}_{r-2}\mathbf{U}]\mathbf{A}_{r(2)} + \sum_{(1,r-2)\vdash\mathbf{B}_{r-1}} [\mathbf{e}_1\mathbf{B}_{r-1(1)}\mathbf{e}_2\mathbf{C}_{r-1}\mathbf{U}]\mathbf{B}_{r-1(2)} \\
& + \sum_{(2,r-3)\vdash\mathbf{B}_{r-1}} [\mathbf{B}_{r-1(1)}\mathbf{e}_2\mathbf{C}_{r-1}\mathbf{U}]\mathbf{e}_1\mathbf{B}_{r-1(2)} + \sum_{(1,r-2)\vdash\mathbf{C}_{r-1}} [\mathbf{e}_2\mathbf{C}_{r-1(1)}\mathbf{e}_1\mathbf{B}_{r-1}\mathbf{U}]\mathbf{C}_{r-1(2)} \\
& + \sum_{(2,r-3)\vdash\mathbf{C}_{r-1}} [\mathbf{C}_{r-1(1)}\mathbf{e}_1\mathbf{B}_{r-1}\mathbf{U}]\mathbf{e}_2\mathbf{C}_{r-1(2)} + [\mathbf{e}_1\mathbf{e}_2\mathbf{A}_r]\mathbf{D}_{r-2}
\end{aligned}
$$

$$= 2\left([\mathbf{e}_1\mathbf{e}_2\mathbf{A}_2'\mathbf{D}_{r-2}\mathbf{U}] - [\mathbf{e}_1\mathbf{e}_2\mathbf{b}\mathbf{c}\mathbf{D}_{r-2}\mathbf{U}]\right)\mathbf{D}_{r-2}.$$

It equals zero if and only if

$$\mathbf{A}_r = \mathbf{b}\mathbf{c}\mathbf{D}_{r-2}. \tag{2.6.24}$$

When $\mathbf{A}_r \neq 0$, then (2.6.24) is just (2.6.23). When $\mathbf{A}_r = 0$, since $\mathbf{B}_{r-1}, \mathbf{C}_{r-1}$ are linearly dependent, it must be that $\mathbf{b}\mathbf{c}\mathbf{D}_{r-2} = 0$.

Case 2. $\mathbf{D}_{r-2} = 0$ but one of $\mathbf{B}_{r-1}, \mathbf{C}_{r-1}$ is nonzero, say $\mathbf{B}_{r-1} \neq 0$. Let $\mathbf{A}_r = \mathbf{a}\mathbf{B}_{r-1}$, and $\mathbf{C}_{r-1} = \lambda\mathbf{B}_{r-1}$. Then $\mathbf{W}_r = (\mathbf{a} + \mathbf{e}_1 + \lambda\mathbf{e}_2)\mathbf{B}_{r-1}$ is a blade, and both sides of (2.6.23) are trivially zero.

Case 3. $\mathbf{B}_{r-1}, \mathbf{C}_{r-1}, \mathbf{D}_{r-2}$ are all zero, but $\mathbf{A}_r \neq 0$. Then $\mathbf{W}_r = \mathbf{A}_r$ is a blade, and both sides of (2.6.23) are trivially zero. $\qquad\square$

Corollary 2.88. The standard form of an r-blade with respect to a partial basis $\mathbf{e}_1, \mathbf{e}_2$ of $\mathcal{V}^n$ is either (1) $(\mathbf{a}_1 + \lambda_1\mathbf{e}_1)(\mathbf{a}_2 + \lambda_2\mathbf{e}_2)\mathbf{A}_{r-2}$, or (2) $(\mathbf{a}_1 + \lambda_1\mathbf{e}_1 + \lambda_2\mathbf{e}_2)\mathbf{A}_{r-1}$, where $\mathbf{A}_k$ is a k-blade in $\Lambda(\mathbf{E}_{n-2})$ in which $\mathbf{E}_{n-2} = \mathbf{e}_3\mathbf{e}_4\cdots\mathbf{e}_n$, and the $\mathbf{a}$'s are vectors in $\mathbf{E}_{n-2}$, and λ_1, λ_2 are scalars.

The above criterion for the $m = 2$ case can be easily extended to the general case of any $m > 0$.

Proposition 2.89. The standard form of an r-blade with respect to a partial basis $\mathbf{e}_1, \mathbf{e}_2, \ldots, \mathbf{e}_m$ of $\mathcal{V}^n$ is one of the following:

(1) $(\mathbf{a}_1 + \lambda_1 \mathbf{e}_1)(\mathbf{a}_2 + \lambda_2 \mathbf{e}_2) \cdots (\mathbf{a}_m + \lambda_m \mathbf{e}_m)\mathbf{A}_{r-m};$

(2) $(\mathbf{a}_1 + \lambda_1 \mathbf{e}_1 + \lambda_2 \mathbf{e}_2 + \cdots + \lambda_m \mathbf{e}_m)\mathbf{A}_{r-1};$

(3) for some integer $1 < j < m$, and some indices $\{i_1, i_2, \ldots, i_j\} \subset \{1, 2, \ldots, m\}$, for $I = \{1, 2, \ldots, m\} - \{i_1, i_2, \ldots, i_j\}$,

$$(\mathbf{a}_1 + \lambda_1 \mathbf{e}_{i_1} + \sum_{l \in I} \lambda_{1l} \mathbf{e}_l)(\mathbf{a}_2 + \lambda_2 \mathbf{e}_{i_2} + \sum_{l \in I} \lambda_{2l} \mathbf{e}_l) \cdots (\mathbf{a}_j + \lambda_j \mathbf{e}_{i_j} + \sum_{l \in I} \lambda_{jl} \mathbf{e}_l)\mathbf{A}_{r-j};$$

$$(2.6.25)$$

where $\mathbf{A}_k$ is a k-blade in $\Lambda(\mathbf{E}_{n-m})$ in which $\mathbf{E}_{n-m} = \mathbf{e}_{m+1}\mathbf{e}_{m+2}\cdots\mathbf{e}_n$, and the $\mathbf{a}$'s are vectors in $\mathbf{E}_{n-m}$, and the λ's are scalars.

2.7 Advanced invariants and Cayley bracket algebra

We have talked a lot about Cayley expansions. The inverse procedure of Cayley expansion is called *Cayley factorization*. It aims at eliminating all additions in an expression to obtain a monomial result in Grassmann-Cayley algebra. It is a transformation from addition to multiplication. An expression in the Grassmann algebra is said to be *Cayley factorizable*, if it equals a Cayley expression when both expressions are expanded into polynomials of homogeneous coordinates.

In the special case where the input expression is a Cayley factorizable bracket polynomial, the output is a scalar-valued Cayley expression. A bracket polynomial is said to be an *implicit advanced invariant* if it is Cayley factorizable. A scalar-valued Cayley expression is called an *explicit advanced invariant*.

From the viewpoint of geometric interpretation, Cayley expansion serves as the procedure of translating geometry into algebra, while Cayley factorization serves as the reverse procedure of translating algebra back to geometry. From the viewpoint of invariant theory, Cayley expansion is the procedure of representing advanced invariants by basic invariants, while Cayley factorization is the procedure of conglomerating basic invariants to advanced invariants.

There is a remarkable phenomenon illustrating the advantages of advanced invariants in symbolic manipulations. The phenomenon is that all transformations based on GP and VW syzygies in bracket algebra can be realized by combining Cayley factorizations and Cayley expansions, and the transformations are almost trivial in terms of Cayley expressions. Below we explore this phenomenon.

We start with an analysis of the transformations based on GP syzygies. Let $\mathbf{A}_{n+1} = \mathbf{a}_1 \mathbf{a}_2 \cdots \mathbf{a}_{n+1}$ and $\mathbf{B}_{n-1} = \mathbf{b}_1 \mathbf{b}_2 \cdots \mathbf{b}_{n-1}$ be two sequences of vectors in $\mathcal{V}^n$. Fix an integer r between 1 and $n+1$, and fix a partition $(\mathbf{A}_{n+1(1)}, \mathbf{A}_{n+1(2)})$ of $\mathbf{A}_{n+1}$ of shape $(r, n+1-r)$. The same symbols $\mathbf{A}_{n+1(1)}, \mathbf{A}_{n+1(2)}, \mathbf{B}_{n-1}$ also denote three outer products of vectors in $\Lambda(\mathcal{V}^n)$. When expanding

$$\mathbf{A}_{n+1(1)} \vee \mathbf{B}_{n-1} \vee \mathbf{A}_{n+1(2)}$$

by distributing $\mathbf{A}_{n+1(1)}$, we get an r-termed bracket polynomial, which is part of a GP polynomial. If $r = n+1$, we get a whole GP polynomial.

Given any r-termed GP polynomial g_r, we can divide its terms arbitrarily into two parts $g_{r,r'} + g_{r,r-r'}$ of r' and $r - r'$ terms respectively. The transformation

$$g_{r,r'} = -g_{r,r-r'} \tag{2.7.1}$$

is called a *Grassmann-Plücker* (GP) *transformation*.

Proposition 2.90. Any GP transformation can be realized by a Cayley factorization followed by a Cayley expansion.

Proof. By (2.2.5), any r'-termed part of an r-termed GP polynomial is of the form

$$g_{r,r'} = \sum_{(1,r'-1)\vdash \mathbf{B}_{r'}} [\mathbf{C}_{n-r+1}\mathbf{A}_{r-2}\mathbf{B}_{r'(1)}][\mathbf{C}_{n-r+1}\mathbf{D}_{r-r'}\mathbf{B}_{r'(2)}], \tag{2.7.2}$$

where $\mathbf{A}_{r-2}, \mathbf{B}_{r'}, \mathbf{C}_{n-r+1}, \mathbf{D}_{r-r'}$ are blades of grade $r - 2, r', n - r + 1, r - r'$ respectively. By (2.4.30) and (2.4.31), $g_{r,r'}$ has the following Cayley factorization:

$$\begin{aligned}
g_{r,r'} &= (-1)^{r'(n-r')}\mathbf{B}_{r'} \vee \mathbf{C}_{n-r+1}\mathbf{A}_{r-2} \vee \mathbf{C}_{n-r+1}\mathbf{D}_{r-r'} \\
&= \mathbf{A}_{r-2} \vee_{\mathbf{C}_{n-r+1}} \mathbf{D}_{r-r'} \vee_{\mathbf{C}_{n-r+1}} \mathbf{B}_{r'}.
\end{aligned} \tag{2.7.3}$$

The right side of (2.7.1) is obtained from (2.7.3) by splitting $\mathbf{D}_{r-r'}$. $\square$

In the above proof, if $r' = 1$, then (2.7.3) becomes $\mathbf{C}_{n-r+1}\mathbf{A}_{r-2} \vee \mathbf{B}_1 \vee \mathbf{C}_{n-r+1}\mathbf{D}_{r-1}$. So in GC algebra, the GP transformation for $r' = 1$ is the rewriting

$$\begin{aligned}
[\mathbf{C}_{n-r+1}\mathbf{D}_{r-1}][\mathbf{C}_{n-r+1}\mathbf{A}_{r-2}\mathbf{B}_1] &= 1 \vee \mathbf{C}_{n-r+1}\mathbf{D}_{r-1} \vee \mathbf{C}_{n-r+1}\mathbf{A}_{r-2}\mathbf{B}_1 \\
&= (-1)^{n-1}\mathbf{B}_1 \vee \mathbf{C}_{n-r+1}\mathbf{D}_{r-1} \vee \mathbf{C}_{n-r+1}\mathbf{A}_{r-2},
\end{aligned} \tag{2.7.4}$$

followed by expanding the last expression of (2.7.4) by distributing $\mathbf{C}_{n-r+1}\mathbf{D}_{r-1}$.

(2.7.4) is a special case of the following *straightening transformation*. Let $\mathbf{A}_{r-1} = \mathbf{a}_1\mathbf{a}_2\cdots\mathbf{a}_{r-1}$, $\mathbf{B}_n = \mathbf{b}_1\mathbf{b}_2\cdots\mathbf{b}_n$, and $\mathbf{C}_{n-r} = \mathbf{a}_{r+1}\mathbf{a}_{r+2}\cdots\mathbf{a}_n$ be outer products of vectors. Then

$$\mathbf{A}_{r-1} \vee \mathbf{B}_n\mathbf{a}_r \vee \mathbf{C}_{n-r} = 0. \tag{2.7.5}$$

Any VW syzygy is obtained by expanding the left side of (2.7.5) by splitting $\mathbf{B}_n\mathbf{a}_r$. The terms generated by distributing $\mathbf{a}_r$ to $\mathbf{A}_{r-1}$ are $(-1)^{n-r}\mathbf{A}_{r-1}\mathbf{a}_r \vee \mathbf{B}_n \vee \mathbf{C}_{n-r}$, the terms generated by distributing $\mathbf{a}_r$ to $\mathbf{C}_{n-r}$ are $(-1)^{n+1-r}\mathbf{A}_{r-1} \vee \mathbf{B}_n \vee \mathbf{a}_r\mathbf{C}_{n-r}$. The expansion changes (2.7.5) to

$$\mathbf{A}_{r-1} \vee \mathbf{B}_n \vee \mathbf{a}_r\mathbf{C}_{n-r} = \mathbf{A}_{r-1}\mathbf{a}_r \vee \mathbf{B}_n \vee \mathbf{C}_{n-r}. \tag{2.7.6}$$

(2.7.6) is called a *straightening transformation* of grade r. (2.7.4) is a straightening transformation of grade 1.

Recall that in straightening (2.1.19), *i.e.*,

$$\begin{bmatrix} \mathbf{a}_1\mathbf{a}_2\cdots\mathbf{a}_s \ \mathbf{b}_1\mathbf{b}_2\cdots\mathbf{b}_{n-s} \\ \mathbf{c}_1\mathbf{c}_2\cdots\mathbf{c}_s \ \mathbf{d}_1\mathbf{d}_2\cdots\mathbf{d}_{n-s} \end{bmatrix}, \tag{2.7.7}$$

where $\mathbf{a}_i \preceq \mathbf{c}_i$ but $\mathbf{b}_1 \succ \mathbf{d}_1$, one of $\mathbf{c}_1, \mathbf{c}_2, \ldots, \mathbf{c}_s, \mathbf{d}_1$ should switch with one of the $\mathbf{b}$'s. This switch is relatively difficult to understand and manipulate in bracket algebra. In the form (2.7.6) by Cayley expressions, this switch is very easy to understand and manipulate. (2.7.6) clearly indicates how vector $\mathbf{a}_r$ is moved to the front of $\mathbf{B}_n$ by straightening. The correspondence between (2.7.6) and (2.7.7) is based on expanding both sides of (2.7.6) by splitting $\mathbf{B}_n$, and is composed of

$$
\begin{aligned}
r &: \quad s+1, \\
\mathbf{A}_{r-1} &: \quad \mathbf{a}_1 \mathbf{a}_2 \cdots \mathbf{a}_s, \\
\mathbf{B}_n &: \quad \mathbf{b}_1 \mathbf{b}_2 \cdots \mathbf{b}_{n-s} \mathbf{c}_1 \mathbf{c}_2 \cdots \mathbf{c}_s, \\
\mathbf{a}_r &: \quad \mathbf{d}_1, \\
\mathbf{C}_{n-r} &: \quad \mathbf{d}_2 \cdots \mathbf{d}_{n-s}.
\end{aligned}
$$

Cayley factorization is a difficult task. Only the simplest case where a bracket polynomial is linear with respect to every vector variable of it is solved [192]. Even the following seemingly simple question remains open: Let $\mathbf{1}, \mathbf{2}, \cdots \mathbf{k}$ and $\mathbf{1}', \mathbf{2}', \cdots \mathbf{k}'$ be points in the projective plane. Is the following *Crapo's binomial*

$$[\mathbf{12}'\mathbf{3}'][\mathbf{23}'\mathbf{4}'] \cdots [\mathbf{k1}'\mathbf{2}'] + (-1)^{k-1}[\mathbf{11}'\mathbf{2}'][\mathbf{22}'\mathbf{3}'] \cdots [\mathbf{kk}'\mathbf{1}'] \qquad (2.7.8)$$

Cayley factorizable?

In [175], the problem of "rational Cayley factorizability" was investigated. The problem is as follows: Given a bracket polynomial that is not Cayley factorizable, is it Cayley factorizable after being multiplied with a suitable bracket monomial? The technique *rational Cayley factorization* is very important in projective geometric computing, as will be shown in Chapter 4.

Example 2.91. For six points $\mathbf{1}, \mathbf{2}, \mathbf{3}, \mathbf{4}, \mathbf{5}, \mathbf{6}$ in the projective plane,

$$[\mathbf{134}][\mathbf{256}] - [\mathbf{234}][\mathbf{156}] = \mathbf{12} \vee \mathbf{34} \vee \mathbf{56} \qquad (2.7.9)$$

is a Cayley factorization. If the minus sign on the left side is changed to plus sign, the bracket binomial is no longer Cayley factorizable. Instead, it is rationally Cayley factorizable:

$$
\begin{aligned}
[\mathbf{134}][\mathbf{256}] + [\mathbf{234}][\mathbf{156}] &= -\frac{[(\mathbf{12} \vee \mathbf{56})(\mathbf{13} \vee \mathbf{24})(\mathbf{14} \vee \mathbf{23})]}{[\mathbf{123}][\mathbf{124}]} \\
&= -\frac{[(\mathbf{12} \vee \mathbf{34})(\mathbf{15} \vee \mathbf{26})(\mathbf{16} \vee \mathbf{25})]}{[\mathbf{125}][\mathbf{126}]}.
\end{aligned}
\qquad (2.7.10)
$$

It can be proved that in any rational Cayley factorization of the bracket binomial in (2.7.10), the degree of the denominator cannot be lower than two.

All the advanced invariants in projective incidence geometry form an algebra under addition and multiplication, called *Cayley bracket algebra*. This algebra is the key to Cayley factorization. We make a formal investigation of this algebra.

A Cayley expression, like any other expression in mathematics, has a natural tree structure. The tree structure is determined by the meet product operations

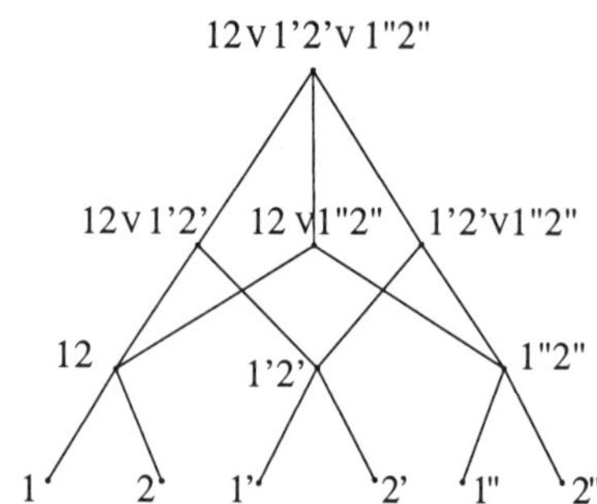

Fig. 2.4   Tree structure of $\mathbf{12} \vee \mathbf{1'2'} \vee \mathbf{1''2''}$.

and outer product operations in the expression. Any nontrivial subtree is called a *proper subexpression*. For example, Cayley expression $\mathbf{12} \vee \mathbf{1'2'} \vee \mathbf{1''2''}$ has the tree structure shown in Figure 2.4, where each node denotes a subtree.

Definition 2.92. [Definition of Cayley bracket by Cayley expressions] An nD *Cayley bracket* is a scalar-valued Cayley expression in $\Lambda(\mathcal{V}^n)$ whose every proper subexpression has grade greater than 0 and less than n. In a Cayley bracket, the outer product and the meet product can be denoted by the same symbol, called the *Cayley product*, because the type of product between any two subexpressions is unambiguously determined by their grades.

Notation.

In the setting of Cayley brackets in this chapter, the Cayley product is denoted by *juxtaposition*. The outer product is denoted by "$\wedge$".

All brackets are Cayley brackets. For example, in 3D bracket algebra, $[\mathbf{123}] = \mathbf{1} \vee (\mathbf{2} \wedge \mathbf{3}) = \mathbf{1}(\mathbf{23})$ is a Cayley bracket, and $\mathbf{12} \vee \mathbf{1'2'} \vee \mathbf{1''2''} = (\mathbf{12})(\mathbf{1'2'})(\mathbf{1''2''})$ is another Cayley bracket.

Any Cayley bracket can be written in the following *nested form*:

$$\mathbf{C}_r(\mathbf{C}_{r-1}(\cdots(\mathbf{C}_1))\cdots) := \mathbf{C}_r \vee (\mathbf{C}_{r-1} \wedge (\mathbf{C}_{r-2} \vee (\mathbf{C}_{r-3} \wedge (\cdots(\mathbf{C}_1))\cdots))), \quad (2.7.11)$$

where the $\mathbf{C}$'s are subexpressions, and where the meet products alternate with the outer products. The parentheses are used as delimiters of the nonassociativity of the Cayley product. Since the result of (2.7.11) is of grade 0, the outermost product can only be the meet product.

Lemma 2.93. As Cayley brackets, $\mathbf{C}_3(\mathbf{C}_2\mathbf{C}_1) = (\mathbf{C}_3\mathbf{C}_2)\mathbf{C}_1$.

Proof. $\mathbf{C}_3 \vee (\mathbf{C}_2 \wedge \mathbf{C}_1) = [\mathbf{C}_3\mathbf{C}_2\mathbf{C}_1] = (\mathbf{C}_3 \wedge \mathbf{C}_2) \vee \mathbf{C}_1$.   □

Proposition 2.94. [131] [Reversion symmetry] As Cayley brackets,

$$\mathbf{C}_r(\mathbf{C}_{r-1}(\cdots(\mathbf{C}_2\mathbf{C}_1))\cdots) = ((\cdots(\mathbf{C}_r\mathbf{C}_{r-1})\cdots\cdots)\mathbf{C}_2)\mathbf{C}_1. \quad (2.7.12)$$

Proof. Induction on r. Lemma 2.93 is on the case $r = 3$. Assume that the conclusion holds for $r-1$. For r, denote $\mathbf{A} = \mathbf{C}_2\mathbf{C}_1$, and $\mathbf{B} = (\cdots(\mathbf{C}_r\mathbf{C}_{r-1})\cdots)\mathbf{C}_3$. Then

$$\mathbf{C}_r(\mathbf{C}_{r-1}(\cdots(\mathbf{C}_3\mathbf{A})\cdots)) = ((\cdots(\mathbf{C}_r\mathbf{C}_{r-1})\cdots)\mathbf{C}_3)\mathbf{A} = \mathbf{B}(\mathbf{C}_2\mathbf{C}_1) = (\mathbf{B}\mathbf{C}_2)\mathbf{C}_1.$$

$\square$

Proposition 2.95. In Cayley bracket (2.7.11), if $\mathbf{C}_{r-2}$ is a Grassmann monomial, then

$$\mathbf{C}_r(\mathbf{C}_{r-1}(\cdots(\mathbf{C}_1))\cdots) = \sum_{\vdash\mathbf{C}_{r-2}} [\mathbf{C}_r\mathbf{C}_{r-1}\mathbf{C}_{r-2(2)}]\,\mathbf{C}_{r-2(1)}(\mathbf{C}_{r-3}(\cdots(\mathbf{C}_1))\cdots).$$

$$(2.7.13)$$

Proof. By Lemma 2.93,

$$\mathbf{C}_r(\mathbf{C}_{r-1}(\mathbf{C}_{r-2}(\mathbf{C}_{r-3}(\cdots(\mathbf{C}_1))\cdots)))) = (\mathbf{C}_r \wedge \mathbf{C}_{r-1}) \vee \mathbf{C}_{r-2} \vee (\mathbf{C}_{r-3}(\cdots(\mathbf{C}_1))\cdots).$$

Expanding the right side by splitting $\mathbf{C}_{r-2}$, we get (2.7.13). $\square$

Corollary 2.96. In Cayley bracket (2.7.11), if all the $\mathbf{C}$'s are Grassmann monomials, then

$$\mathbf{C}_r(\mathbf{C}_{r-1}(\cdots(\mathbf{C}_1))\cdots) = \sum_{i=1}^{[\frac{r}{2}]-1} \sum_{\vdash\mathbf{C}_{r-2i}} \prod_{j=0}^{[\frac{r}{2}]-1} [\mathbf{C}_{r-2j(1)}\mathbf{C}_{r-2j-1}\mathbf{C}_{r-2j-2(2)}], \quad (2.7.14)$$

with the understanding that $\mathbf{C}_{r(1)} = \mathbf{C}_r$, $\mathbf{C}_{1(2)} = \mathbf{C}_1$, and $\mathbf{C}_{0(2)} = 1$ on the right side of (2.7.14).

The following is a constructive definition of Cayley bracket and Cayley product.

Definition 2.97. [Recursive definition of Cayley product and Cayley bracket] Let $\mathbf{a}_1, \ldots, \mathbf{a}_m$ be atomic vectors generating an nD bracket algebra. In the GC algebra associated with the bracket algebra, let $M(\mathbf{a})$ be the *multiplicative set* generated from the atomic vectors by the *Cayley product*, which is defined as follows:

- The Cayley product of any $r < n$ atomic vectors is their outer product, called a *Cayley r-vector*.
- The Cayley product of any $r = n$ atomic vectors is their bracket, called a *Cayley bracket*.
- The Cayley product of any $r > n$ atomic vectors is zero.
- The Cayley product of any Cayley bracket with any element in $M(\mathbf{a})$ is zero.
- For $r + s > n$, the Cayley product of any Cayley r-vector and Cayley s-vector is their meet product, and the result is called a *Cayley $(r + s - n)$-vector*.
- For $r + s = n$, the Cayley product of any Cayley r-vector and Cayley s-vector is their meet product, and the result is called a *Cayley bracket*.

- For $r + s < n$, The Cayley product of any Cayley r-vectors and Cayley s-vector is their outer product, and the result is called a *Cayley $(r + s)$-vector*.

For any element in $M(\mathbf{a})$, the number of atomic vectors it contains is called its *length*. An element in $M(\mathbf{a})$ whose length equals 0 mod n is called a *Cayley bracket*. An element of $M(\mathbf{a})$ which is a subexpression of a Cayley bracket and whose length equals i mod n, where $i \neq 0$ mod n, is called a *Cayley i-vector*. A Cayley 1-vector is called a *Cayley vector*, a Cayley 2-vector is called a *Cayley bivector, etc.*

Proposition 2.98. Any Cayley i-vector is an i-blade in GC algebra.

Proof. Induction on the length l of the subexpression of a Cayley bracket. The case $l < n$ is obvious, as only outer product is allowed among the atomic vectors. Assume that for $l < kn$ the conclusion is true. Let $l = kn + i$, where $0 < i < n$. If $i = 1$, then the subexpression must be of the form $\mathbf{A}_r(\mathbf{B}_{n+1-r})$ for some $r > 1$, where $\mathbf{A}_r$ and $\mathbf{B}_{n+1-r}$ are Cayley r-vector and Cayley $(n+1-r)$-vector respectively. Since both of them have length $< kn$, by induction hypothesis, the grades of $\mathbf{A}_r$ and $\mathbf{B}_{n+1-r}$ are r and $n+1-r$ respectively, and the Cayley product between them must be the meet product. The result is a vector.

Assume that for $l = kn + i$ where $i < j < n$, the conclusion is true. When $i = j$, any Cayley j-vector must take one of the following two forms:

- $\mathbf{A}_r(\mathbf{B}_{j-r})$ for some $r < j$, where $\mathbf{A}_r$ and $\mathbf{B}_{j-r}$ are Cayley r-vector and Cayley $(j - r)$-vector respectively;
- $\mathbf{A}_r(\mathbf{B}_{n+j-r})$ for some $r > j$, where $\mathbf{A}_r$ and $\mathbf{B}_{n+j-r}$ are Cayley r-vector and Cayley $(n + j - r)$-vector respectively.

Both $\mathbf{A}_r$ and $\mathbf{B}_{j-r}$ or $\mathbf{B}_{n+j-r}$ satisfy the induction hypothesis, so in the case of $\mathbf{B}_{j-r}$, the Cayley product is the outer product, the result is a j-blade; in the case of $\mathbf{B}_{n+j-r}$, the Cayley product is the meet product, the result is also a j-blade. $\square$

Definition 2.99. *Cayley bracket algebra* is the ring generated by all the Cayley brackets of a set of atomic vectors, modulo the ideal generated by syzygies **B1**, **B2**, **GP** in Definition 2.11 for basic brackets, together with the four basic Cayley expansion formulas (2.3.31), (2.3.32), (2.5.3), (2.5.4) for atomic vectors after changing each formula into a polynomial equality of Cayley brackets by making Cayley products on both sides with some Cayley i-vectors.

By the above definition, the transition from a Cayley bracket to a polynomial of basic brackets is realized by Cayley expansions starting from the innermost meet product of atomic vectors. This recursive definition is based on Cayley expansions from the bottom up. By the reversion symmetry (2.7.12), Cayley bracket algebra can also be defined in a top-down manner, so that the Cayley expansions in a Cayley bracket always start with the outermost Cayley product. Below we present the top-down definition for $n = 3$ and 4.

Lemma 2.100. When $n = 3$, any $\mathbf{C}_{r-2i}$ in the nested form (2.7.11) where $i \geq 1$, is the outer product of two vectors, and all other $\mathbf{C}$'s in (2.7.11) are vectors.

Proof. Since the outermost product in (2.7.11) is the meet product, the next outermost product must be the outer product. Since $r \geq 3$, $\mathbf{C}_r$ must be a vector, so is $\mathbf{C}_{r-1}$. For $i < r - 1$, if the right side of $\mathbf{C}_i$ in (2.7.11) is the meet product, then $\mathbf{C}_i$ must be a bivector; if its right side is the outer product, then $\mathbf{C}_i$ must be a vector. $\qquad\square$

A 3D Cayley vector is either one of the atomic vectors $\mathbf{a}$'s, or an element in $M(\mathbf{a})$ of the form $(\mathbf{A}_1\mathbf{A}_2)(\mathbf{A}_3\mathbf{A}_4)$, where the $\mathbf{A}_i$ are 3D Cayley vectors. Any 3D Cayley bracket is of the form $\mathbf{A}_1(\mathbf{A}_2\mathbf{A}_3)$, denoted by $[\mathbf{A}_1\mathbf{A}_2\mathbf{A}_3]$.

Definition 2.101. [Top-down definition] The 3D *Cayley bracket algebra* generated by the atomic vectors $\mathbf{a}$'s, is the bracket ring generated by the 3D Cayley brackets in the form $[\mathbf{A}_i\mathbf{A}_j\mathbf{A}_k]$, where the $\mathbf{A}$'s are Cayley vectors, modulo the ideal generated by the following elements:

$$[\mathbf{A}_1\mathbf{A}_2\{(\mathbf{A}_3\mathbf{A}_4)(\mathbf{A}_5\mathbf{A}_6)\}] - [\mathbf{A}_1\mathbf{A}_3\mathbf{A}_4][\mathbf{A}_2\mathbf{A}_5\mathbf{A}_6] + [\mathbf{A}_2\mathbf{A}_3\mathbf{A}_4][\mathbf{A}_1\mathbf{A}_5\mathbf{A}_6]. \quad (2.7.15)$$

(2.7.15) is the Cayley expansion of $\mathbf{A}_1\mathbf{A}_2 \vee \mathbf{A}_3\mathbf{A}_4 \vee \mathbf{A}_5\mathbf{A}_6$ by separating $\mathbf{A}_1, \mathbf{A}_2$. By applying this syzygy recursively, any Cayley bracket can be expanded in a top-down manner to a polynomial of basic brackets.

A major difference between a 4D Cayley bivector and a 3D Cayley bivector is that the latter is always in the form of the outer product of two vectors, while the former can take the form of the meet product of two 3-vectors. In the following, let the $\mathbf{A}_i$ and $\mathbf{B}_j$ be 4D Cayley vectors and Cayley bivectors respectively.

- A 4D Cayley vector is either one of the atomic vectors, or an element in $M(\mathbf{a})$ of the form $\mathbf{B}_1(\mathbf{A}\mathbf{B}_2)$, which includes also $\mathbf{B}_1(\mathbf{B}_2\mathbf{A})$, $(\mathbf{A}\mathbf{B}_2)\mathbf{B}_1$ and $(\mathbf{B}_2\mathbf{A})\mathbf{B}_1$.
- A 4D Cayley bivector is an element in the form of either $\mathbf{A}_1\mathbf{A}_2$ or $(\mathbf{A}_1\mathbf{B}_1)(\mathbf{A}_2\mathbf{B}_2)$, where the latter also includes $(\mathbf{A}_1\mathbf{B}_1)(\mathbf{B}_2\mathbf{A}_2)$, $(\mathbf{B}_1\mathbf{A}_1)(\mathbf{A}_2\mathbf{B}_2)$ and $(\mathbf{B}_1\mathbf{A}_1)(\mathbf{B}_2\mathbf{A}_2)$.
- A 4D Cayley 3-vector is an element of the form $\mathbf{A}_1\mathbf{B}_2$, including $\mathbf{B}_2\mathbf{A}_1$.
- A 4D Cayley bracket is of the form $\mathbf{A}_1(\mathbf{A}_2\mathbf{B})$, including also $\mathbf{A}_1(\mathbf{B}\mathbf{A}_2)$, $(\mathbf{A}_1\mathbf{B})\mathbf{A}_2$ and $(\mathbf{B}\mathbf{A}_1)\mathbf{A}_2$, and is denoted by $[\mathbf{A}_1\mathbf{A}_2\mathbf{B}]$.

Definition 2.102. [Top-down definition] The 4D *Cayley bracket algebra* generated by the atomic vectors $\mathbf{a}$'s, is the bracket ring generated by the 4D Cayley brackets in the form $[\mathbf{A}_i\mathbf{A}_j\mathbf{B}_k]$ where the $\mathbf{A}$'s are Cayley vectors, and the $\mathbf{B}$'s are Cayley bivectors, modulo the ideal generated by the left sides of the following three identities:

$$[\mathbf{A}_1\mathbf{A}_2\{\mathbf{A}_3(\mathbf{B}_1(\mathbf{A}_4\mathbf{B}_2))\}] - [\mathbf{A}_1\mathbf{A}_2\mathbf{B}_1][\mathbf{A}_3\mathbf{A}_4\mathbf{B}_2]$$
$$+[\mathbf{A}_1\mathbf{A}_3\mathbf{B}_1][\mathbf{A}_2\mathbf{A}_4\mathbf{B}_2] - [\mathbf{A}_2\mathbf{A}_3\mathbf{B}_1][\mathbf{A}_1\mathbf{A}_4\mathbf{B}_2] = 0, \qquad (2.7.16)$$

$$[\mathbf{A}_1\mathbf{A}_2\{(\mathbf{A}_3\mathbf{B}_1)(\mathbf{A}_4\mathbf{B}_2)\}] - [\mathbf{A}_1\mathbf{A}_3\mathbf{B}_1][\mathbf{A}_2\mathbf{A}_4\mathbf{B}_2]$$
$$+[\mathbf{A}_2\mathbf{A}_3\mathbf{B}_1][\mathbf{A}_1\mathbf{A}_4\mathbf{B}_2] = 0, \qquad (2.7.17)$$

$$[\mathbf{A}_1\{\mathbf{B}_1(\mathbf{A}_2\mathbf{B}_2)\}(\mathbf{A}_3\mathbf{A}_4)] - [\mathbf{A}_3\mathbf{A}_4\{\mathbf{A}_1(\mathbf{B}_1(\mathbf{A}_2\mathbf{B}_2))\}] = 0. \qquad (2.7.18)$$

(2.7.16) is the Cayley expansion of $\mathbf{A}_1\mathbf{A}_2\mathbf{A}_3 \vee \mathbf{B}_1 \vee \mathbf{A}_4\mathbf{B}_2$ by distributing $\mathbf{A}_1, \mathbf{A}_2, \mathbf{A}_3$, and (2.7.17) is the Cayley expansion of $\mathbf{A}_1\mathbf{A}_2 \vee \mathbf{A}_3\mathbf{B}_1 \vee \mathbf{A}_4\mathbf{B}_2$ by distributing $\mathbf{A}_1, \mathbf{A}_2$.

In Chapters 3 and 4, there are many applications of Cayley bracket algebra in projective geometric computing.

Chapter 3

Projective Incidence Geometry with Cayley Bracket Algebra

An *incidence constraint* in projective geometry refers to either a collinearity or coplanarity relation of various dimensions, or a concurrency relation (*i.e.*, linear spaces meeting at the same linear subspace), or a multilinear equality on the cross-ratios of collinear or coplanar objects.

While Grassmann-Cayley algebra and bracket algebra provide an invariant language for describing incidence relations, Cayley bracket algebra provides the toolkit for advanced invariant computing in projective incidence geometry. This chapter develops the kernel computational tools in Cayley bracket algebra, including various factorization, contraction, and division techniques.

The power of the techniques is demonstrated in two geometric applications: first, in machine proving of incidence geometric theorems with Cayley bracket algebra, without a single exception each theorem can be given a binomial proof, *i.e.*, throughout the proving procedure, any bracket polynomial has at most two terms; second, Cayley bracket algebra has superior symbolic manipulation performance in attacking challenging open problems in enumerative geometry.

3.1 Symbolic methods for projective incidence geometry

The time-honored method for symbolic computing in projective geometry is the method of homogeneous coordinates. In this method, the homogeneous coordinates of generic geometric objects are algebraically independent. From the bracket algebra point of view, let $\mathbf{a}_1, \mathbf{a}_2, \ldots, \mathbf{a}_n$ be fixed points in projective space $\mathbb{P}^{n-1}$ such that they form a basis of the underlying vector space. Then for any point $\mathbf{a}_{n+1}$, its homogeneous coordinates are those brackets composed of $\mathbf{a}_{n+1}$ and $n-1$ basis points. Because of the algebraic independence among such brackets, the method of homogeneous coordinates is the most general one in both algebraic representation and manipulation. The cost is that there is no control of middle expression swell.

We take a look at a concrete example.

Example 3.1. [2D Desargues Theorem] In the projective plane there are two triangles $\mathbf{123}$ and $\mathbf{1'2'3'}$. If the three lines $\mathbf{11'}, \mathbf{22'}, \mathbf{33'}$ concur, then the three pairs of lines $(\mathbf{12}, \mathbf{1'2'})$, $(\mathbf{13}, \mathbf{1'3'})$, $(\mathbf{23}, \mathbf{2'3'})$ intersect at three points that are collinear.

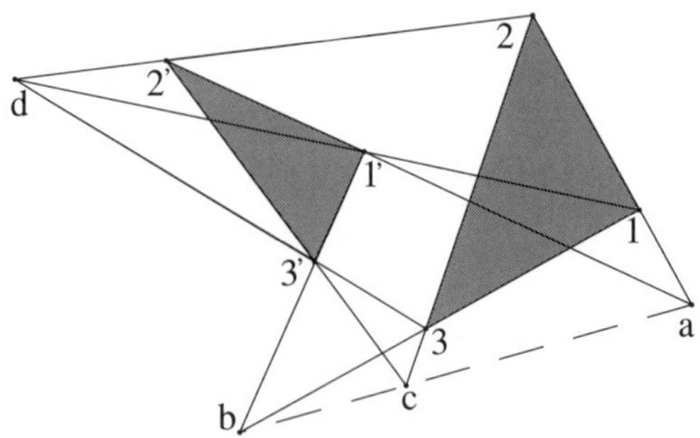

Fig. 3.1   Desargues Theorem.

In Figure 3.1, let $\mathbf{a} = \mathbf{12} \cap \mathbf{1'2'}$, $\mathbf{b} = \mathbf{13} \cap \mathbf{1'3'}$ and $\mathbf{c} = \mathbf{23} \cap \mathbf{2'3'}$; let $\mathbf{d}$ be the point where lines $\mathbf{11'}, \mathbf{22'}, \mathbf{33'}$ concur. The configuration can be constructed as follows:

Free points: $\mathbf{1}, \mathbf{2}, \mathbf{3}, \mathbf{d}$.
Free collinear points: $\mathbf{1'}$ on line $\mathbf{d1}$, $\mathbf{2'}$ on line $\mathbf{d2}$, $\mathbf{3'}$ on line $\mathbf{d3}$.
Intersections: $\mathbf{a} = \mathbf{12} \cap \mathbf{1'2'}$, $\mathbf{b} = \mathbf{13} \cap \mathbf{1'3'}$, $\mathbf{c} = \mathbf{23} \cap \mathbf{2'3'}$.
Conclusion: $\mathbf{a}, \mathbf{b}, \mathbf{c}$ are collinear.

Using coordinate method to prove this theorem is very easy. Every intersection has its coordinates explicitly expressed by the coordinates of those points with which to construct it. Every free collinear point is parametrized by two fixed points collinear with it. When all the constructed points in the reverse order of the sequence of constructions are eliminated, the conclusion expression ends up with zero, and the theorem is proved.

In coordinate form, the input conclusion expression $[\mathbf{abc}]$ is a polynomial of 6 terms. The size grows quickly in the process of elimination, but in the end suddenly reduces to zero. This is a typical example of *middle expression swell*. For difficult problems, the computing is usually stuck in some middle stage because the polynomial size becomes too big to handle.

To reduce the size of middle expressions to manipulable size, geometric invariants such as brackets are needed. Classical invariant-theoretic method follows much the same procedure as the coordinate method: the conclusion expression is in the form of a bracket polynomial, and each intersection is expressed by a Cayley expression; after the substitution of the Cayley expression into the conclusion, the latter needs to be expanded into a bracket polynomial. Each free collinear point is expressed as a linear combination of two fixed points collinear with it by using their Cramer's rule. After all the eliminations and expansions, the conclusion expression is changed into a bracket polynomial containing brackets of different sizes, as any Cramer's rule of three collinear points involves brackets of length two, while the brackets produced by Cayley expansions are of length three. The result is then straightened to zero.

In straightening a bracket polynomial, all kinds of VW syzygies are used, in sharp contrast to simplifying a coordinate polynomial, where there is no syzygy at all. Because straightening a bracket polynomial is less efficient than simplifying the corresponding coordinate polynomial, and because size of a bracket polynomial still grows fast during the straightening, usually classical invariant-theoretic method is less efficient than coordinate method.

In the 1990's, two methods were proposed to curtail the size of middle bracket polynomials. The first is the method of *biquadratic final polynomials* [148]. It is based on the algebraic theory of final polynomials [27], [173]. A proof of Desargues Theorem by this method is as follows:

$$
\begin{array}{llll}
\mathbf{3'c, 1'a, 2d} & \text{concur} \implies & [\mathbf{23'd}][\mathbf{1'ac}] & = -[\mathbf{2cd}][\mathbf{1'3'a}] \\
\mathbf{1'd, 2a, 3b} & \text{concur} \implies & [\mathbf{2ab}][\mathbf{31'd}] & = [\mathbf{23a}][\mathbf{1'bd}] \\
\mathbf{3, 3', d} & \text{collinear} \implies & [\mathbf{23d}][\mathbf{1'3'd}] & = -[\mathbf{23'd}][\mathbf{31'd}] \\
\mathbf{1', 3', b} & \text{collinear} \implies & [\mathbf{1'bd}][\mathbf{1'3'a}] & = -[\mathbf{1'ab}][\mathbf{1'3'd}] \\
\mathbf{2, 3, c} & \text{collinear} \implies & [\mathbf{23a}][\mathbf{2cd}] & = -[\mathbf{23d}][\mathbf{2ac}]
\end{array}
\qquad (3.1.1)
$$

$$
\begin{array}{cc}
\times & \times \\
\Downarrow & \Downarrow
\end{array}
$$

$$
\mathbf{a, b, c} \quad \text{collinear} \impliedby \quad [\mathbf{2ab}][\mathbf{1'ac}] \;=\; [\mathbf{2ac}][\mathbf{1'ab}].
$$

The algorithm searches for all kinds of geometric constraints that can be expressed by *biquadratic equalities*, *i.e.*, equalities each side of which is a bracket monomial of degree two. If a subset of such equalities is found with the property that after multiplying each side together and canceling common bracket factors, the result is a biquadratic equality representation of the conclusion, then the theorem is proved. If the proof is successful then clearly it is very elegant, because only two terms are produced at each step.

In (3.1.1), two out of the six concurrency constraints in the hypotheses are selected, and for each meet-product expression representing a concurrency constraint, *e.g.*, $\mathbf{3'c} \vee \mathbf{1'a} \vee \mathbf{2d} = 0$, one out of its three Cayley expansions is selected, *e.g.*, $[\mathbf{23'd}][\mathbf{1'ac}] = -[\mathbf{2cd}][\mathbf{1'3'a}]$. Furthermore, three out of the nine collinearity constraints in the hypotheses are selected, and for each bracket equality representing a collinearity constraint, *e.g.*, $[\mathbf{33'd}] = 0$, a specific bracket $[\mathbf{21'd}]$ is selected to multiply with it in order to generate a biquadratic equality as follows:

$$
0 = [\mathbf{21'd}][\mathbf{33'd}] = [\mathbf{23d}][\mathbf{1'3'd}] + [\mathbf{23'd}][\mathbf{31'd}].
$$

This equality is a GP relation. The multiplication of the five biquadratic equalities in (3.1.1) produces a biquadratic binomial representation of the conclusion $[\mathbf{abc}] = 0$.

Although some strategies can be applied to reduce the size of the set of biquadratic binomial candidates, the set is still too large to search effectively, restricting the general applicability of this method.

The second method is the *area method* [40]. It uses much the same strategy as the coordinate method and invariant-theoretic method in elimination, only the appearance is different. In the determinant form of a bracket $[\mathbf{123}]$, if the first homogeneous coordinate of each of the three points is set to be 1, then the three points are on the affine plane, and the determinant equals twice the signed area of triangle $\mathbf{123}$. The sign is positive if the orientation of the triangle follows that of the affine plane, and negative otherwise. Although the area method is not directly applicable to projective geometry, after transforming the area equalities into bracket ones by homogenization, the method can be translated into the language of bracket algebra.

In the area method, to represent the conclusion of Example 3.1, two more intersections $\mathbf{z}_1 = \mathbf{12} \cap \mathbf{bc}$, $\mathbf{z}_2 = \mathbf{1'2'} \cap \mathbf{bc}$ need to be introduced. The conclusion is then expressed by the following *ratio-formed* equality:

$$\frac{\mathbf{cz}_1}{\mathbf{bz}_1} \frac{\mathbf{bz}_2}{\mathbf{cz}_2} = 1, \tag{3.1.2}$$

where each ratio is understood to be the ratio of two collinear vectors, or signed line segments. The proof of (3.1.2) by the area method, when translated into bracket algebra, is as follows:

$$
\begin{array}{ccc}
\text{Rules} & & \dfrac{\mathbf{cz}_1}{\mathbf{bz}_1}\dfrac{\mathbf{bz}_2}{\mathbf{cz}_2} \\[2ex]
\boxed{\begin{array}{l} \mathbf{cz}_1/\mathbf{bz}_1 = \quad [\mathbf{12c}]/[\mathbf{12b}] \\ \mathbf{bz}_2/\mathbf{cz}_2 = [\mathbf{1'2'b}]/[\mathbf{1'2'c}] \end{array}} & \overset{\mathbf{z}_1,\mathbf{z}_2}{=\!=} & \dfrac{[\mathbf{12c}][\mathbf{1'2'b}]}{[\mathbf{12b}][\mathbf{1'2'c}]} \\[3ex]
\boxed{\begin{array}{l} [\mathbf{12b}] = \quad [\mathbf{123}][\mathbf{11'3'}] \\ [\mathbf{12c}] = \quad [\mathbf{123}][\mathbf{22'3'}] \\ [\mathbf{1'2'b}] = -[\mathbf{131'}][\mathbf{1'2'3'}] \\ [\mathbf{1'2'c}] = -[\mathbf{232'}][\mathbf{1'2'3'}] \end{array}} & \overset{\mathbf{b},\mathbf{c}}{=\!=} & \dfrac{[\mathbf{131'}][\mathbf{22'3'}]}{[\mathbf{11'3'}][\mathbf{232'}]} \\[3ex]
\boxed{\begin{array}{l} [\mathbf{11'3'}] = -[\mathbf{131'}]\mathbf{d3'}/\mathbf{d3} \\ [\mathbf{22'3'}] = -[\mathbf{232'}]\mathbf{d3'}/\mathbf{d3} \end{array}} & \overset{\mathbf{3'}}{=\!=} & 1.
\end{array}
\tag{3.1.3}
$$

The proof is very short and elegant. The method grows out of the geometric consideration of the ratios of signed areas and line segments in the setting of incidence constraints. For example, the rule of eliminating $\mathbf{z}_1$ from $\mathbf{cz}_1/\mathbf{bz}_1$ is $[\mathbf{12c}]/[\mathbf{12b}]$, whose derivation is shown in Figure 3.2: since $\mathbf{z}_1$ is where line $\mathbf{12}$ meets line $\mathbf{bc}$, the ratio $\mathbf{cz}_1/\mathbf{bz}_1$ of the two line segments $\mathbf{cz}_1, \mathbf{bz}_1$ must be equal to the ratio of the signed distances from $\mathbf{c}, \mathbf{b}$ to line $\mathbf{12}$, and hence be equal to the ratio of the signed areas of triangles $\mathbf{12c}$ and $\mathbf{12b}$.

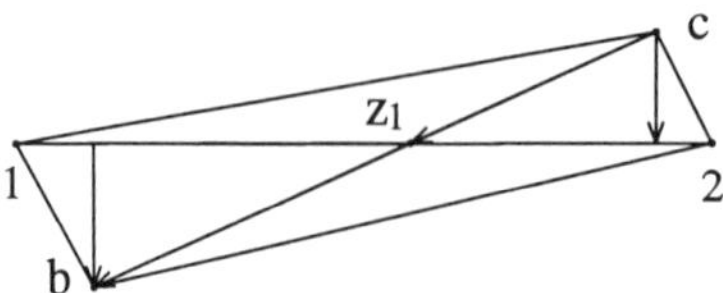

Fig. 3.2   Derivation of elimination rule $\mathbf{cz}_1/\mathbf{bz}_1 = [\mathbf{12c}]/[\mathbf{12b}]$ geometrically.

A spectacular phenomenon in the proof (3.1.3) is that no matter what point is eliminated, every middle result in the right column of (3.1.3) remains a rational monomial. The area method has very good control of the result of elimination. As long as there are sufficiently many elimination rules, the method is applicable to general geometric problems.

From the viewpoint of Cayley bracket algebra,

$$\mathbf{cz_1} \overset{\mathbf{z_1}}{=} \mathbf{c}(\mathbf{12} \vee \mathbf{bc}) \overset{expand}{=} -[\mathbf{12c}]\mathbf{bc},$$
$$\mathbf{bz_1} \overset{\mathbf{z_1}}{=} \mathbf{b}(\mathbf{12} \vee \mathbf{bc}) \overset{expand}{=} -[\mathbf{12b}]\mathbf{bc}. \tag{3.1.4}$$

We see that the elimination rule $\mathbf{cz_1}/\mathbf{bz_1} = [\mathbf{12c}]/[\mathbf{12b}]$ can be obtained by direct algebraic computing, instead of by geometric argument based on incidence constraints. In fact, by monomial expansions of meet-product expressions such as

$$[\mathbf{12b}] \overset{\mathbf{b}}{=} \mathbf{12} \vee \mathbf{13} \vee \mathbf{1'3'} \overset{expand}{=} [\mathbf{123}][\mathbf{11'3'}], \tag{3.1.5}$$

all other rules in the left column of (3.1.3) for the elimination of intersections of lines can be generated *purely algebraically*.

The transition from geometric consideration to algebraic computation has the significant benefit of being capable of generating as many new elimination rules as possible, in a totally mechanical manner, without worrying about the shortage of elimination rules. Indeed, in the area method there are many different elimination rules based on a big list of special geometric constructions.

Ratio-formed representation and computation are another feature of the area method. Many elimination rules can only be expressed in the ratio form. For example, because neither covariants nor deficit brackets exist in the method, the expansion results in (3.1.4) can only be put in the ratio form. A noticeable drawback of the ratio form is its inefficiency in representing geometric constraints. For example, (3.1.3) shows that only after two supplementary points of intersection are constructed can a conclusion as simple as $[\mathbf{abc}] = 0$ be represented in the ratio form.

In Cayley bracket algebra, on the contrary, without resorting to either the construction of supplementary points or special elimination rules, the result after eliminating $\mathbf{z_1}, \mathbf{z_2}$ in the second line of (3.1.3) can be obtained directly from the input conclusion expression by a binomial expansion:

$$[\mathbf{abc}] \overset{\mathbf{a}}{=} \mathbf{12} \vee \mathbf{1'2'} \vee \mathbf{bc} \overset{expand}{=} [\mathbf{12b}][\mathbf{1'2'c}] - [\mathbf{12c}][\mathbf{1'2'b}].$$

The ratio form prohibits the combination of the two groups of terms in the numerator and denominator from making both usual polynomial and bracket polynomial manipulations. In particular, the latter manipulation involves syzygies of at least three terms, which is impossible to implement without breaking up rational bracket monomials. Indeed, the area method does not have any GP syzygy-based simplification.

The above investigations indicate that there is still a lot of room for further improvement of the already very elegant proofs (3.1.1) and (3.1.3). This is exactly what we are going to do next. We show that using the same old strategy of the coordinate method and the invariant-theoretic method, by purely algebraic simplifications we can easily derive nD Desargues Theorem and its converse for the symbolic value n. Using Cayley bracket algebra, the geometric theorem and its converse are represented by a single algebraic identity (3.1.7) below, which discloses not only the qualitative relationship between the hypotheses and the conclusion, but also the quantitative relationship between them. Desargues Theorem and its converse represented in this form can be directly used in algebraic computing as a Cayley expansion formula. This is a much higher level algebraization of the geometric theorem.

First we consider the theorem in 2D case. The hypothesis is $\mathbf{11'} \vee \mathbf{22'} \vee \mathbf{33'} = 0$, and the conclusion is

$$[(\mathbf{12} \vee \mathbf{1'2'})(\mathbf{13} \vee \mathbf{1'3'})(\mathbf{23} \vee \mathbf{2'3'})] = 0. \tag{3.1.6}$$

By (2.5.31), the left side of (3.1.6) has factored Cayley expansion

$$[(\mathbf{12} \vee \mathbf{1'2'})(\mathbf{13} \vee \mathbf{1'3'})(\mathbf{23} \vee \mathbf{2'3'})] = -[\mathbf{123}][\mathbf{1'2'3'}]\mathbf{11'} \vee \mathbf{22'} \vee \mathbf{33'}, \tag{3.1.7}$$

so Desargues Theorem and its converse in 2D case are direct corollaries of (3.1.7).

Without resorting to formula (2.5.31), the conclusion expression in (3.1.6) can be expanded step by step, using only basic Cayley expansions, in two different ways that lead to the same result:

(1) Outer product expansion:

$$\begin{aligned}
& [(\mathbf{12} \vee \mathbf{1'2'})(\mathbf{13} \vee \mathbf{1'3'})(\mathbf{23} \vee \mathbf{2'3'})] \\
= \;& (\mathbf{12} \vee \mathbf{13} \vee \mathbf{1'3'})(\mathbf{1'2'} \vee \mathbf{23} \vee \mathbf{2'3'}) - (\mathbf{12} \vee \mathbf{23} \vee \mathbf{2'3'})(\mathbf{1'2'} \vee \mathbf{13} \vee \mathbf{1'3'}) \\
= \;& [\mathbf{123}][\mathbf{1'2'3'}](-[\mathbf{11'3'}][\mathbf{232'}] + [\mathbf{131'}][\mathbf{22'3'}]) \\
\overset{factor}{=} \;& -[\mathbf{123}][\mathbf{1'2'3'}]\mathbf{11'} \vee \mathbf{22'} \vee \mathbf{33'}.
\end{aligned}$$

The last step is the Cayley factorization reverse to expansion (2.5.14).

(2) Meet product expansion:

$$\begin{aligned}
& [(\mathbf{12} \vee \mathbf{1'2'})(\mathbf{13} \vee \mathbf{1'3'})(\mathbf{23} \vee \mathbf{2'3'})] \\
= \;& [\mathbf{11'2'}][\mathbf{2}(\mathbf{13} \vee \mathbf{1'3'})(\mathbf{23} \vee \mathbf{2'3'})] - [\mathbf{21'2'}][\mathbf{1}(\mathbf{13} \vee \mathbf{1'3'})(\mathbf{23} \vee \mathbf{2'3'})] \\
= \;& [\mathbf{11'2'}][\mathbf{22'3'}][\mathbf{2}(\mathbf{13} \vee \mathbf{1'3'})\mathbf{3}] - [\mathbf{21'2'}][\mathbf{11'3'}][\mathbf{13}(\mathbf{23} \vee \mathbf{2'3'})] \\
= \;& [\mathbf{123}]([\mathbf{11'2'}][\mathbf{22'3'}][\mathbf{31'3'}] - [\mathbf{21'2'}][\mathbf{11'3'}][\mathbf{32'3'}]) \\
\overset{factor}{=} \;& -[\mathbf{123}][\mathbf{1'2'3'}]\mathbf{11'} \vee \mathbf{22'} \vee \mathbf{33'}.
\end{aligned}$$

The last step is the Cayley factorization obtained from expansions (2.5.21) and (2.5.22).

In the above computing, after the elimination of the three intersections, everything is left to the algebraic simplification of the Cayley bracket obtained. The proving strategy is identical to that of the coordinate method and the invariant-theoretic method. The difference lies in the computing techniques. In Cayley expansion it is the *factored and shortest result* that is acquired, following the classification theorems in Chapter 2. In simplification it is the Cayley factorization that is used, which not only simplifies the expression, but makes the computing robust against different Cayley expansions. The syzygies among brackets are integrated into the symmetries within a Cayley bracket. This is a typical feature of advanced invariants.

In sharp contrast, even for the simplest case $n = 2$, the expression on the right side of (3.1.7) when changed into a polynomial of coordinates, contains as many as 1290 terms. After polynomial factorization, the coordinate form of factor $\mathbf{11'} \vee \mathbf{22'} \vee \mathbf{33'}$ has 48 terms. The effect of controlling middle expression swell by Cayley expansion and factorization is obvious.

(3.1.7) remains valid in nD projective space. In the GC algebra over $\mathcal{V}^{n+1}$, $[\mathbf{123}]$ represents deficit bracket $[\mathbf{U}_{n-2}\mathbf{123}]$, where $\mathbf{U}_{n-2}$ is the dummy $(n-2)$-blade, and $\mathbf{11'} \vee \mathbf{22'} \vee \mathbf{33'}$ is the abbreviation of

$$\mathbf{U}_{n-2}\mathbf{11'} \vee \mathbf{U}_{n-2}\mathbf{22'} \vee \mathbf{33'}. \tag{3.1.8}$$

(3.1.7) is just

$$\begin{aligned}
&[\mathbf{U}_{n-2}(\mathbf{U}_{n-2}\mathbf{12} \vee \mathbf{1'2'})(\mathbf{U}_{n-2}\mathbf{13} \vee \mathbf{1'3'})(\mathbf{U}_{n-2}\mathbf{23} \vee \mathbf{2'3'})] \\
&= -[\mathbf{U}_{n-2}\mathbf{123}][\mathbf{U}_{n-2}\mathbf{1'2'3'}]\mathbf{U}_{n-2}\mathbf{11'} \vee \mathbf{U}_{n-2}\mathbf{22'} \vee \mathbf{33'},
\end{aligned} \tag{3.1.9}$$

whose validity does not need any further proof.

Theorem 3.2. [nD Desargues Theorem and its converse] In the nD projective space there are two triangles $\mathbf{123}$ and $\mathbf{1'2'3'}$. Assume that line pairs $(\mathbf{12}, \mathbf{1'2'})$, $(\mathbf{13}, \mathbf{1'3'})$, $(\mathbf{23}, \mathbf{2'3'})$ intersect in the nD space at points $\mathbf{a}, \mathbf{b}, \mathbf{c}$ respectively.

(1) [Desargues] If lines $\mathbf{11'}, \mathbf{22'}, \mathbf{33'}$ concur, then points $\mathbf{a}, \mathbf{b}, \mathbf{c}$ are collinear.
(2) [Converse Desargues] If points $\mathbf{a}, \mathbf{b}, \mathbf{c}$ are collinear, then under the assumption that triangles $\mathbf{123}, \mathbf{1'2'3'}$ are not degenerate, lines $\mathbf{11'}, \mathbf{22'}, \mathbf{33'}$ concur.

Besides Cayley factorization, there is another technique for algebraic simplification: reducing the number of terms of a bracket polynomial by GP transformations, called *contraction*. For example, immediately before $\mathbf{3'}$ is eliminated in (3.1.3), the ratio-formed expression can be written as a bracket binomial

$$[\mathbf{131'}][\mathbf{22'3'}] - [\mathbf{11'3'}][\mathbf{232'}], \tag{3.1.10}$$

which can be contracted to zero by $\mathbf{11'} \vee \mathbf{22'} \vee \mathbf{33'} = 0$, using Cayley factorization and the concurrency of the three lines. The elimination of $\mathbf{3'}$ in (3.1.3) can be replaced by contraction.

Compared with previous methods based on homogeneous coordinates and brackets, the Cayley bracket algebra approach has much richer algebraic structures: Grassmann-Cayley expressions as covariants, brackets as basic invariants, deficit brackets as rational invariants, and Cayley brackets as advanced invariants. In this new framework, typical algebraic manipulations include representation, elimination, expansion and simplification. The purpose of algebraic manipulations is to produce *factored and shortest result*, by which the middle expression swell can be tightly controlled.

3.2   Factorization techniques in bracket algebra

A bracket factor can occur either explicitly or implicitly in a bracket polynomial. It is said to be *explicit*, if it occurs in every term of the bracket polynomial, otherwise it is said to be *implicit*. Implicit bracket factors come from two sources:

(1) GP syzygies of bracket algebra, where each bracket polynomial having an implicit bracket factor has at least two terms, *e.g.*,

$$[\mathbf{123}][\mathbf{145}] - [\mathbf{124}][\mathbf{135}] = -[\mathbf{125}][\mathbf{134}].$$

(2) Incidence constraints among the atomic vectors of bracket algebra, which destroy the unique factorization property of this algebra. With the presence of incidence constraints, even a bracket monomial can have an implicit bracket factor, *e.g.*, if $[\mathbf{123}] = 0$, then

$$[\mathbf{124}][\mathbf{135}] = [\mathbf{125}][\mathbf{134}].$$

In this section, we introduce two corresponding groups of factorization techniques in bracket algebra. The first group consists of two Cayley factorization formulas obtained from Cayley expansion theory by equalizing a factored result and a binomial result of the same Cayley expression. The equality of the two results is caused by GP relations. The second group consists of two transformations based on the collinearity, coplanarity, and concurrency constraints among atomic vectors.

3.2.1   *Factorization based on GP relations*

Vectors in the same bracket are called *bracket mates* of each other. In a bracket monomial, the *bracket mates* of a vector refer to all the bracket mates of the vector in the monomial.

By (2.5.22) and (2.5.25), for any points $\mathbf{1}, \mathbf{2}, \mathbf{3}, \mathbf{1'}, \mathbf{2'}, \mathbf{3'}$ in the projective plane,

$$[\mathbf{121'}][\mathbf{232'}][\mathbf{133'}] - [\mathbf{122'}][\mathbf{233'}][\mathbf{131'}] = [\mathbf{123}]\mathbf{11'} \vee \mathbf{22'} \vee \mathbf{33'}. \tag{3.2.1}$$

Algorithm 3.3. Degree-3 bracket binomial factorization based on (3.2.1).

Input: A polynomial p composed of brackets and meet products of type p_I, and already factorized in the polynomial ring of these elements.

Output: A polynomial q of brackets and meet products of type p_I.

Procedure. For every factor f of p, do the following.

    Step 1. If f does not satisfy any of the following conditions, put it into q:

        (1) f is a bracket binomial of degree three, the coefficients are ± 1.

        (2) f has six points, three of which are double points, denoted by $\mathbf{1}, \mathbf{2}, \mathbf{3}$; the other three are single ones, denoted by $\mathbf{1'}, \mathbf{2'}, \mathbf{3'}$.

Step 2. Let the two terms of f be p_1, p_2. If p_1 is of the form $\epsilon[\mathbf{121'}][\mathbf{232'}][\mathbf{133'}]$, where $\epsilon \in \{\pm 1\}$, then it defines the matching between points $\mathbf{i}$ and $\mathbf{i'}$ for $\mathbf{i} = \mathbf{1, 2, 3}$;

(1) if $p_2 = -\epsilon[\mathbf{122'}][\mathbf{233'}][\mathbf{131'}]$, put into q the factor $\epsilon[\mathbf{123}]\mathbf{11'} \vee \mathbf{22'} \vee \mathbf{33'}$;

(2) if $p_2 = -\epsilon[\mathbf{123'}][\mathbf{231'}][\mathbf{132'}]$, put into q the factor $-\epsilon[\mathbf{123}]\mathbf{13'} \vee \mathbf{21'} \vee \mathbf{32'}$.

In all other cases, put f into q.

By (2.5.19), for any six points $\mathbf{1}$ to $\mathbf{6}$ in the plane,

$$[\mathbf{125}][\mathbf{126}][\mathbf{345}][\mathbf{346}] - [\mathbf{123}][\mathbf{124}][\mathbf{563}][\mathbf{564}] = (\mathbf{12} \vee \mathbf{35} \vee \mathbf{46})(\mathbf{12} \vee \mathbf{36} \vee \mathbf{45}). \quad (3.2.2)$$

Algorithm 3.4. Degree-4 bracket binomial factorization based on (3.2.2).

Input: A polynomial p composed of brackets and meet products of type p_I, and already factorized in the polynomial ring of these elements.

Output: A polynomial q of brackets and meet products of type p_I.

Procedure. For every factor f of p, do the following.

Step 1. If f does not satisfy any of the following conditions, put it into q:

(1) f is a bracket binomial of degree four, the coefficients are ± 1.

(2) There are six points in f, and they are all double points.

Step 2. Count the number of bracket mates for each point in f. In the first term p_1 of f, there are two points, denoted by $\mathbf{5, 6}$, each having four bracket mates, while the other four points each have three bracket mates. In the second term p_2 of f, there are two other points, denoted by $\mathbf{3, 4}$, each having four bracket mates, while the other four points each have three bracket mates. The two points left in f are denoted by $\mathbf{1, 2}$.

Step 3. If $p_1 = \epsilon[\mathbf{125}][\mathbf{126}][\mathbf{345}][\mathbf{346}]$ and $p_2 = -\epsilon[\mathbf{123}][\mathbf{124}][\mathbf{563}][\mathbf{564}]$, put into q the factor $\epsilon(\mathbf{12} \vee \mathbf{35} \vee \mathbf{46})(\mathbf{12} \vee \mathbf{36} \vee \mathbf{45})$, else put f into q.

3.2.2 *Factorization based on collinearity constraints*

In 2D projective geometry, if three lines $\mathbf{12}, \mathbf{1'2'}, \mathbf{1''2''}$ concur, then their meet product equals zero. Expanding the meet product by splitting $\mathbf{12}$, we get

$$[\mathbf{11'2'}][\mathbf{21''2''}] = [\mathbf{11''2''}][\mathbf{21'2'}]. \quad (3.2.3)$$

If no two of the three lines are collinear, then (3.2.3) is called a *concurrency transformation* of monomial $[\mathbf{11'2'}][\mathbf{21''2''}]$. If $\mathbf{1', 2', 1'', 2''}$ are collinear, then (3.2.3) is called a *collinearity transformation*. In a collinearity transformation, at least three of the four collinear points are distinct from each other.

A projective line is said to be *long*, if it contains at least three atomic points, *i.e.*, there is at least one collinearity constraint about the line. For a bracket $[\mathbf{123}]$, the pair **12** is called a *line* of the bracket, and **3** is called the *bracket mate* of the line. For long line **1′2′1″2″** in two different forms **1′2′** and **1″2″**, the collinearity transformation (3.2.3) is nothing but the interchange of the respective bracket mates of the two forms.

Example 3.5. In the projective plane, let $(1, 2', 3', 4)$, $(2, 3', 4', 5)$, $(3, 4', 5', 1)$ be collinear quadruplets of points. Simplify $p = ([12'5'][243'])/([122'][43'5'])$.

The two brackets in the numerator of p each contain two long lines. The first bracket contains long lines **142′3′**, **134′5′**, and the second bracket contains long lines **253′4′**, **142′3′**. Their common long line is **142′3′**, so the two bracket mates **5′, 2** of the common long line can be interchanged: $[12'5'][243'] = -[122'][43'5']$, yielding $p = -1$.

Example 3.6. [Leisening's Theorem] Let **126**, **347** be two lines in the plane. Let $5 = 27 \cap 36$, $9 = 24 \cap 13$, $0 = 17 \cap 46$, and $8 = 12 \cap 34$. Then the three intersections $85 \cap 14$, $89 \cap 67$, $80 \cap 23$ are collinear.

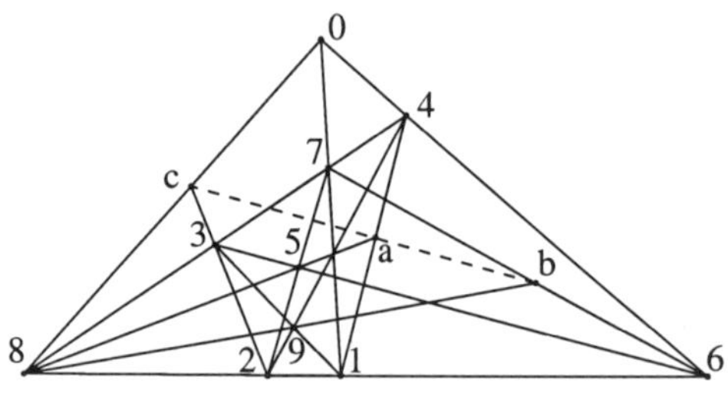

Fig. 3.3   Leisening's Theorem.

Hypotheses:
Free points: $1, 2, 3, 4$.
Free collinear points: **6** on line **12**; **7** on line **34**.
Intersections: $5 = 27 \cap 36$, $9 = 24 \cap 13$, $0 = 17 \cap 46$, $8 = 12 \cap 34$.

Conclusion: $58 \cap 14$, $67 \cap 89$, $23 \cap 80$ are collinear.

Proof.

$$[(58 \vee 14)(67 \vee 89)(23 \vee 80)]$$
$$\overset{expand}{=} (58 \vee 67 \vee 89)(14 \vee 23 \vee 80) - (14 \vee 67 \vee 89)(58 \vee 23 \vee 80)$$
$$\overset{expand}{=} -[678][589][80(14 \vee 23)] + [238][580][89(14 \vee 67)]$$
$$\overset{5,8,9,0}{=} 0,$$

where the last step is simultaneous elimination of four points:

$$[678] \qquad \overset{8}{=} \quad 67 \vee 12 \vee 34$$
$$\overset{expand}{=} -[127][346],$$

$$[238] \quad \overset{8}{=} \quad 23 \vee 12 \vee 34$$
$$\overset{expand}{=} \quad -[123][234],$$

$$[589] \quad \overset{5,8,9}{=} \quad [(12 \vee 34)(24 \vee 13)(27 \vee 36)]$$
$$\overset{expand}{=} \quad (12 \vee 24 \vee 13)(34 \vee 27 \vee 36) - (34 \vee 24 \vee 13)(12 \vee 27 \vee 36)$$
$$\overset{expand}{=} \quad [123][234]([124][367] - [134][267])$$
$$\overset{factor}{=} \quad [123][234]14 \vee 23 \vee 67,$$

$$[580] \quad \overset{5,8,0}{=} \quad [(12 \vee 34)(17 \vee 46)(27 \vee 36)]$$
$$\overset{expand}{=} \quad (12 \vee 17 \vee 46)(34 \vee 27 \vee 36) - (34 \vee 17 \vee 46)(12 \vee 27 \vee 36)$$
$$\overset{expand}{=} \quad [127][346]([146][237] - [147][236])$$
$$\overset{factor}{=} \quad [127][346]14 \vee 23 \vee 67,$$

$$[80(14 \vee 23)] \quad \overset{8,0}{=} \quad [(12 \vee 34)(17 \vee 46)(14 \vee 23)]$$
$$\overset{expand}{=} \quad (12 \vee 17 \vee 46)(34 \vee 14 \vee 23) - (34 \vee 17 \vee 46)(12 \vee 14 \vee 23)$$
$$\overset{expand}{=} \quad [124][134]([167][234] + [123][467]),$$

$$[89(14 \vee 67)] \quad \overset{8,9}{=} \quad [(12 \vee 34)(24 \vee 13)(14 \vee 67)]$$
$$\overset{expand}{=} \quad (12 \vee 24 \vee 13)(34 \vee 14 \vee 67) - (34 \vee 24 \vee 13)(12 \vee 14 \vee 67)$$
$$\overset{expand}{=} \quad [124][134]([123][467] + [167][234]).$$

$$\square$$

A careful check of the above proof, in particular the Cayley expansions immediately after $5, 8, 9, 0$ are eliminated, shows that the selection of Cayley expansions is extremely delicate. As an example, if expanding $[589] = [(12 \vee 34)(24 \vee 13)(27 \vee 36)]$ by separating 27 and 36, then

$$(27 \vee 12 \vee 34)(36 \vee 24 \vee 13) - (36 \vee 12 \vee 34)(27 \vee 24 \vee 13)$$
$$= [127][136][234]^2 - [123]^2[247][346]. \tag{3.2.4}$$

The result is not Cayley factorizable if the points are generic ones in the plane.

The computing procedure is abundant with such traps of "bad" Cayley expansions. As one more example, in expanding

$$[80(14 \vee 23)] = (12 \vee 17 \vee 46)(34 \vee 14 \vee 23) - (34 \vee 17 \vee 46)(12 \vee 14 \vee 23),$$

If $12 \vee 17 \vee 46$ is expanded by separating $1, 2$, and $34 \vee 17 \vee 46$ is expanded by separating $3, 4$, then

$$(12 \vee 17 \vee 46)(34 \vee 14 \vee 23) - (34 \vee 17 \vee 46)(12 \vee 14 \vee 23)$$
$$= -[127][134][146][234] - [123][124][147][346], \tag{3.2.5}$$

which is not Cayley factorizable if the points are generic ones.

So the previous proof is, although elegant, extremely sensitive to Cayley expansions and thus too difficult to obtain. The reason for this delicacy is that there are several free collinear points in the construction, which hide the common bracket factors by breaking up the unique factorization property of bracket polynomials in generic vector variables.

The proof can be made robust by collinearity transformations. Some technical terms are needed to explain the factorization algorithm. A *line form* of a long line l in a bracket polynomial p, refers to a pair of points $\mathbf{x}, \mathbf{y}$ on the line such that $[\mathbf{xyz}]$ is a bracket of p for some point $\mathbf{z}$. If a point occurs in every term of p as a bracket mate of l, it is called a *common bracket mate* of l in p. If a line form of l occurs in every term of p, it is called a *common line form* of l in p.

By (3.2.3), a common bracket factor hidden by collinearity constraints must be composed of a common line form $\mathbf{L}$ of a long line l and a common bracket mate of l. If both exist then they comprise a common bracket factor. If they are not in the same bracket of a bracket monomial, they can be put together by a collinearity transformation.

Algorithm 3.7. Finding implicit bracket factors by collinearity transformations.

Input: A bracket polynomial p without explicit common bracket factors.

Output: p in factored form.

Step 1. Find all the long lines occurring at least twice in each term of p.

For each long line l found in this way, if there exists a common line form $\mathbf{L}$ of l and a common bracket mate $\mathbf{a}$ of l, then for every $(\mathbf{L}, \mathbf{a})$, do the following two steps; if there is no more long line, then output p.

Step 2. In every term of p, for any pair of brackets containing $\mathbf{L}$ and $\mathbf{a}$ separately, interchange $\mathbf{a}$ and the bracket mate of $\mathbf{L}$ in the pair of brackets.

Step 3. Output the explicit common bracket factor $[\mathbf{aL}]^r$ of p to maximal r. Delete $[\mathbf{aL}]^r$ from each term of p. Return to Step 1.

Example 3.8. Simplify the result of (3.2.4).

By the following collinearity transformations based on long lines $\mathbf{126}$ and $\mathbf{347}$ respectively,

$$[\mathbf{127}][\mathbf{136}] = -[\mathbf{123}][\mathbf{167}], \quad [\mathbf{247}][\mathbf{346}] = -[\mathbf{234}][\mathbf{467}],$$

we get

$$[\mathbf{127}][\mathbf{136}][\mathbf{234}]^2 - [\mathbf{123}]^2[\mathbf{247}][\mathbf{346}] = [\mathbf{123}][\mathbf{234}](-[\mathbf{167}][\mathbf{234}] + [\mathbf{123}][\mathbf{467}])$$
$$\overset{factor}{=} [\mathbf{123}][\mathbf{234}]\, \mathbf{14} \vee \mathbf{23} \vee \mathbf{67}.$$

$$(3.2.6)$$

On the left side of (3.2.6), the bracket mates of long line $\mathbf{347}$ in the two terms are respectively $\mathbf{2,2}$ and $\mathbf{2,6}$. The common line form is $\mathbf{34}$, and the common bracket mate is **2**. In the second term, **2** and **6** in **[247][346]** are switched, and an explicit common bracket **[234]** is obtained.

Similarly, the bracket mates of long line **126** in the two terms are respectively **7, 3** and **3, 3**, and a common factor **[123]** is produced by a collinearity transformation based on long line **126**.

The result of (3.2.5) can be simplified similarly:

$$-[127][134][146][234] - [123][124][147][346]$$
$$= [134][124]([167][234] + [123][467]);$$

the collinearity transformations are

$$[127][146] = -[124][167], \quad [147][346] = -[134][467].$$

The collinearity transformations can certainly be extended to 3D and higher dimensions as coplanarity transformations of various dimensions. In 3D projective geometry, let points $1, 2, 3, 1', 2', 3'$ be coplanar, then for any points $4, 4'$ in space,

$$[1234][1'2'3'4'] = [1234'][1'2'3'4] \tag{3.2.7}$$

is called a *coplanarity transformation*.

Example 3.9. Let there be six points **1** to **6** in space, among which $1, 2, 4, 5$ and $2, 3, 4, 6$ are separately coplanar. Simplify

$$p = [1256][1345][1346][2356] + [1235][1236][1456][3456].$$

In the first term of p, the first two brackets each contain plane **1245**, the last two brackets each contain plane **2346**. The two coplanarity transformations

$$[1256][1345] = -[1235][1456], \quad [1346][2356] = [1236][3456]$$

change p to zero.

3.2.3 *Factorization based on concurrency constraints*

Assume that three lines concur at a point **3**. Let p be the left side of (3.2.3), Then point **3** either does not occur in p, or is a single or double point of p.

If **3** is a double point of p, let $2' = 2'' = 3$, then (3.2.3) becomes

$$[11'3][21''3] = [21'3][11''3].$$

It is just a collinearity transformation based on long line **123**.

If **3** is a single point of p, then p cannot contain any double point. Among the three lines passing through **3**, two lines must be long. Let **3** be on lines **12** and $1'2'$. Then **3** occurs explicitly in the pair $1''2''$. Let $2'' = 3$. Then (3.2.3) becomes

$$[11'2'][21''3] = [21'2'][11''3]. \tag{3.2.8}$$

Let $p = [11'2'][21''3]$ be the left side of (3.2.8). Bracket $[11'2']$ of p contains a long line $1'2'3$, whose bracket mate is **1**. Bracket $[21''3]$ of p contains a line

$1''3$ whose bracket mate 2 is on long line 123. The concurrency transformation interchanges the bracket mates $1, 2$ of lines $1'2'3$ and $1''3$ respectively.

If 3 does not occur in p, it is called an *implicit point* of p. Then all the three lines $12, 1'2', 1''2''$ are long. In $p = [11'2'][21''2'']$, the long lines in the two brackets are $1'2'3$, $1''2''3$ respectively, whose bracket mates are $1, 2$ respectively. The two bracket mates are both on the third long line 123, and are interchanged by the concurrency transformation.

Algorithm 3.10. Finding concurrency transformations of a bracket monomial.

Input: A bracket monomial p.

Output: All concurrency transformations of p that are not collinearity ones.

Step 1. Find all long lines in the brackets of p.

For each pair of brackets sharing no common point, do the following:

Step 2. If the two brackets each contain a long line, and the two long lines share a common point, denoted by 3, then

(1) [Common single point] if the bracket mate of one long line is on the other long line, then the two brackets match pattern (3.2.8):

The point of concurrency is 3, the bracket mate 1 of long line $1'2'3$ is on long line 123. The output transformation interchanges $1, 2$.

(2) [Common implicit point] Else if point 3 is collinear with the bracket mates of the two long lines, then the two brackets match pattern (3.2.3):

The point of concurrency is 3, the two long lines are $1'2'3, 1''2''3$, their bracket mates are $1, 2$. The output transformation interchanges $1, 2$.

Example 3.11. Let $1', 2', 3'$ be points on sides $23, 31, 12$ of triangle 123 respectively. The following two meet products each have three monomial expansions:

$$\begin{aligned}
1'2 \vee 2'3 \vee 3'1 &= -[121'][32'3'] = \ \ [133'][21'2'] = [11'3'][232'], \\
23' \vee 31' \vee 12' &= \ \ [122'][31'3'] = -[11'2'][233'] = [131'][22'3'].
\end{aligned} \qquad (3.2.9)$$

These expansions are interchangeable by concurrency transformations.

For example, let $p = [121'][32'3']$. The long lines of p are $123'$, $21'3$, $32'1$, with bracket mates $1', 1, 3'$ respectively. The first two long lines are in the same bracket.

(1) Since $3'$ is on long line $123'$, a concurrency transformation interchanges $2, 3'$:

$$[121'][32'3'] = [13'1'][32'2] = -[11'3'][232'].$$

(2) Similarly, since 1 is on long line $32'1$, a concurrency transformation interchanges $1, 2'$, and changes p to $-[133'][21'2']$.

3.3 Contraction techniques in bracket computing

In bracket algebra, GP relations can be used to reduce the number of terms of a bracket polynomial. This idea leads to a series of powerful techniques for simplifying bracket computation.

In a GP transformation defined by (2.7.1), if the right side has fewer terms than the left side after the bracket rewriting rules are applied, we say the left side is *contractible*, and the GP transformation is a *direct contraction*, or *regular contraction*. Usually we use the term *contraction* to indicate a direct contraction.

(2.7.1) comes from the decomposition of an r-termed GP polynomial arbitrarily into two parts $g_{r,r'} + g_{r,r-r'}$ of r' and $r - r'$ terms respectively:

$$g_{r,r'} = -g_{r,r-r'}. \tag{3.3.1}$$

For (3.3.1) to be a contraction, it is not necessary that $r' > r/2$. A natural question is, given the left side of (3.3.1), is the right side unique? By (2.7.3),

$$g_{r,r'} = (-1)^{r'(n-r')} \mathbf{B}_{r'} \vee \mathbf{C}_{n-r+1} \mathbf{A}_{r-2} \vee \mathbf{C}_{n-r+1} \mathbf{D}_{r-r'}. \tag{3.3.2}$$

(1) If $r' = r$, then $g_{r,r} = 0$.

(2) If $r' = r - 1$, then

$$g_{r,r-1} = (-1)^{(r-1)(n-r+1)} [\mathbf{B}_{r-1}\mathbf{C}_{n-r+1}][\mathbf{C}_{n-r+1}\mathbf{A}_{r-2}\mathbf{D}_1] \tag{3.3.3}$$

is the unique monomial expansion of (3.3.2).

(3) If $r' = 1$, then

$$\begin{aligned}
g_{r,1} &= (-1)^{n-1} [\mathbf{B}_1\mathbf{C}_{n-r+1}\mathbf{A}_{r-2}][\mathbf{C}_{n-r+1}\mathbf{D}_{r-1}] \\
&= (-1)^r \sum_{(r-2,1)\vdash\mathbf{D}_{r-1}} [\mathbf{C}_{n-r+1}\mathbf{B}_1\mathbf{D}_{r-1\,(1)}][\mathbf{C}_{n-r+1}\mathbf{A}_{r-2}\mathbf{D}_{r-1\,(2)}]
\end{aligned} \tag{3.3.4}$$

are the only two basic Cayley expansions of (3.3.2) where $r' = 1$.

If $r' = 1$ and $r > 3$, then the right side of (3.3.2) has $2r - 2$ possibilities, because by the first expansion in (3.3.4), there are two choices in selecting an $(r - 1)$-blade $\mathbf{B}_1\mathbf{A}_{r-2}$ from two $(r - 1)$-blades composed of single points of the bracket monomial $[\mathbf{B}_1\mathbf{C}_{n-r+1}\mathbf{A}_{r-2}][\mathbf{C}_{n-r+1}\mathbf{D}_{r-1}]$, and by the second expansion in (3.3.4), there are $r - 1$ choices in selecting a vector $\mathbf{B}_1$ from the $(r - 1)$-blade $\mathbf{B}_1\mathbf{A}_{r-2}$. If $r' = 1$ and $r = 3$, then the right side of (3.3.2) is unique, because the second expansion in (3.3.4) is independent of the choices of both 2-blade $\mathbf{B}_1\mathbf{A}_1$ and vector $\mathbf{B}_1$.

So in the three special cases $(r = 3, r' = 1)$, $r' = r - 1$, and $r' = r$, the right side of (3.3.1) is unique. In the general case $2 \leq r' \leq r - 2$, since (3.3.2) has three different expansions, the right side of (3.3.1) has two possibilities if the Cayley factorization (3.3.2) is fixed.

In practice, it is sufficient to choose $r' = 2$. The corresponding Cayley factorization is

$$\begin{aligned}
&[\mathbf{C}_{n-r+1}\mathbf{A}_{r-2}\mathbf{b}_1][\mathbf{C}_{n-r+1}\mathbf{D}_{r-2}\mathbf{b}_2] - [\mathbf{C}_{n-r+1}\mathbf{A}_{r-2}\mathbf{b}_2][\mathbf{C}_{n-r+1}\mathbf{D}_{r-2}\mathbf{b}_1] \\
&= \mathbf{C}_{n-r+1}\mathbf{A}_{r-2} \vee \mathbf{C}_{n-r+1}\mathbf{D}_{r-2} \vee \mathbf{b}_1\mathbf{b}_2.
\end{aligned} \tag{3.3.5}$$

3.3.1 *Contraction*

A group of GP transformations are said to be *of the same level,* if they are independent of each other and can be carried out in parallel. The following algorithm realizes contractions in nD bracket algebra.

Algorithm 3.12. nD contraction.

Input: A homogeneous bracket polynomial p of degree at least two. Assume that p is already factorized in the polynomial ring of brackets.

Output: A homogeneous bracket polynomial q.

Procedure: Move all numerical factor and single bracket factors of p to q. While p is not empty, for every factor f of p, do the following.

 Step 1. [Contractions of the same level]

  Set $g = 0$. For every pair of terms $p_1 + p_2$ in f, do the following:

  1.1. Let $p_1 = cd_1$, $p_2 = cd_2$, where c is their common factors. If one of d_1, d_2 is not a monomial of two brackets with coefficient ± 1, skip to the next pair of terms.

  1.2. Count the degrees of the points in d_1. Let $\mathbf{C}_{n-r+1}$ be the set of double points in d_1, whose cardinality is denoted by $n-r+1$. By this we obtain the number r.

  1.3. If $r = 3$, then there are only four single points in d_1. Do the following:

  (1) Fix a single point $\mathbf{a}$ in d_1. Let the single-point bracket mates of $\mathbf{a}$ in d_1, d_2 be $\mathbf{b}_1$, $\mathbf{b}_2$ respectively. Let the fourth single point in d_1 be $\mathbf{d}$.

  (2) Now d_1 is of the form $\epsilon[\mathbf{C}_{n-2}\mathbf{ab}_1][\mathbf{C}_{n-2}\mathbf{db}_2]$, where $\epsilon \in \{\pm 1\}$. If $d_2 \neq -\epsilon[\mathbf{C}_{n-2}\mathbf{ab}_2][\mathbf{C}_{n-2}\mathbf{db}_1]$, skip to the next pair of terms.

  (3) According to (3.3.3), set $g = g + \epsilon c[\mathbf{C}_{n-2}\mathbf{b}_1\mathbf{b}_2][\mathbf{C}_{n-2}\mathbf{ad}]$. Remove $p_1 + p_2$ from f.

  1.4. If $r > 3$, do the following:

  (1) If the following case does not occur, skip to the next pair of terms: when counting the single-point bracket mates of the $2r - 2$ single points in d_1, there are two points, denoted by $\mathbf{b}_1, \mathbf{b}_2$, each having $2r - 4$ such mates, while the other $2r - 4$ single points each have $r - 1$ such mates.

  (2) d_1 must be of the form $\epsilon[\mathbf{C}_{n-r+1}\mathbf{A}_{r-2}\mathbf{b}_1][\mathbf{C}_{n-r+1}\mathbf{D}_{r-2}\mathbf{b}_2]$. The $2r - 4$ single points are thus separated into two groups: $\mathbf{A}_{r-2}$ and $\mathbf{D}_{r-2}$. If $d_2 \neq -\epsilon[\mathbf{C}_{n-r+1}\mathbf{D}_{r-2}\mathbf{b}_1][\mathbf{C}_{n-r+1}\mathbf{A}_{r-2}\mathbf{b}_2]$, then skip to the next pair of terms.

  (3) If by replacing $p_1 + p_2$ with ϵcx, where x is one of the two expansions of the result of (3.3.5) by splitting either $\mathbf{C}_{n-r+1}\mathbf{A}_{r-2}$ or $\mathbf{C}_{n-r+1}\mathbf{D}_{r-2}$,

the number of terms of f is reduced, then set $g = g + \epsilon cx$, and remove $p_1 + p_2$ from f.

Step 2. [Contractions of deeper level] Set $f = f + g$. All contractions recorded in g are of the same level.

If $f = 0$, then return $q = 0$; else if either f is a monomial or no contraction occurs in Step 1, then move f to q; else, go back to Step 1.

Example 3.13. Let **1** to **5** be five points in space, where point **5** is on plane **123**. Contract $p = [\mathbf{135}][\mathbf{2345}] - [\mathbf{235}][\mathbf{1345}]$, where $[\mathbf{135}] = [\mathbf{U}_1\mathbf{135}]$ is a deficit bracket.

In $p = [\mathbf{U}_1\mathbf{135}][\mathbf{2345}] - [\mathbf{U}_1\mathbf{235}][\mathbf{1345}]$, the six points with their degrees are $\mathbf{1}^1, \mathbf{2}^1, \mathbf{3}^2, \mathbf{4}^1,\ \mathbf{5}^2, \mathbf{U}_1^1$. So in Step 1.2 of Algorithm 3.12, $\mathbf{C}_2 = \mathbf{35}$ and $r = 3$.

Fix a single-point **1**. Its single-point bracket mates in p are $\mathbf{U}_1$ and **4**. The fourth single point is **2**. By writing the first term of p as $-[\mathbf{35U}_1\mathbf{1}][\mathbf{3524}]$, and writing the second term as $[\mathbf{35U}_1\mathbf{2}][\mathbf{3514}]$, we get, by Step 1.3 of the algorithm, $p = [\mathbf{35U}_1\mathbf{4}][\mathbf{3512}] = 0$.

3.3.2 *Level contraction*

In the GP transformation (3.3.1) where $r' = 2$, if both sides have the same number of terms after the bracket rewriting rules are applied, then (3.3.1) is not a contraction. However, the result of the transformation may be contractible when combined with other terms of the bracket polynomial.

A *level contraction* of a non-contractible bracket polynomial is composed of one or several GP transformations and one or several successive contractions, the latter being of the same level, by which the number of terms of the polynomial is decreased.

Algorithm 3.14. Level contraction.

Input: A bracket polynomial p of degree at least two. Assume that p is neither factorizable in the polynomial ring of brackets, nor contractible.

Output: p after level contractions.

Step 1. [Single level-contraction] For every GP transformable pair of terms $p_1 + p_2$ in p, which can be detected by Step 1.4 of Algorithm 3.12, there are two different GP transformation results p' and p''.

If when p' or p'' is substituted into p, the latter can be directly contracted to a bracket polynomial of fewer terms, then do the transformation and the contractions, output p and exit.

Step 2. [Combined level-contraction] It requires that p has at least two GP transformable pairs.

For the GP transformable pairs in p, combine their identity transformations and GP transformations in different ways. Each combination should contain at least two GP transformations. If there is any combination that makes the transformation result contractible, do the transformations and contractions.

Example 3.15. Let **1** to **7** be seven points in the plane. Do level contraction to

$$p = [124][125][127][136][236][357][457] + [125][126][127][134][234][357][567]$$
$$-[124][126][135][137][234][257][567] - [124][126][135][137][236][257][457]$$
$$+[124][126][134][157][235][237][567] + [124][126][136][157][235][237][457].$$

There are two pairs of terms in p each having five common bracket factors. Each pair is GP transformable.

(1) Terms 3 and 4:

$$-[124][126][135][137][257]([234][567] + [236][457])$$

$$\overset{GP}{=} \begin{cases} -[124][126][135][137][235][257][467] - [124][126][135][137][237][257][456], \\ -[124][126][135][137][257][246][357] - [124][126][135][137][257]^2[346]. \end{cases}$$

$$(3.3.6)$$

None of the four new terms has five common bracket factors with any of the remaining terms of p. So this GP transformable pair is neither contractible nor leading to any contraction.

(2) Terms 5 and 6:

$$[124][126][157][235][237]([134][567] + [136][457])$$

$$\overset{GP}{=} \begin{cases} [124][126][135][157][235][237][467] + [124][126][137][157][235][237][456], \\ [124][126][146][157][235][237][357] - [124][126][157]^2[235][237][346]. \end{cases}$$

$$(3.3.7)$$

Similarly, this GP transformable pair is neither contractible nor leading to any contraction.

Now we do combination to the GP transformations. For each pair of the two pairs of GP transformations, we only need to consider the four combinations of the four newly produced terms. The conclusion is that the first transformations in (3.3.6) and (3.3.7) produce two contractible pairs:

$$-[137][257]+[157][237] = -[127][357], \quad -[135][257]+[157][235] = -[125][357].$$

So

$$p = [357]\{[124][125][127][136][236][457] + [125][126][127][134][234][567]$$
$$-[124][125][126][137][237][456] - [124][126][127][135][235][467]\},$$

$$(3.3.8)$$

which has two terms fewer than the original p. The other three pairs of transformations cannot produce any pair of terms with five common bracket factors.

The level contractions and the strong contractions to be introduced in the next subsection are more complicated than direct contractions. They are used primarily to bracket polynomials in generic vector variables.

3.3.3 *Strong contraction*

(3.3.1) is called an *explosion* if $r' = 1$. A *strong contraction* of a bracket polynomial is an explosion followed by several contractions of the same level, by which the number of terms of the bracket polynomial is decreased.

Proposition 3.16. Let p be a bracket polynomial of generic points.

(1) If p has no GP transformation but has a strong contraction induced by an r-termed GP polynomial, then the strong contraction induces $r - 1$ successive contractions of the same level. Each contraction reduces the sum of a term generated by the explosion and an original term of p to a bracket monomial.

(2) If every pair of terms in p has at least $i \geq 3$ brackets left in each term after removal of their common bracket factors, then p has no GP transformation.

    (a) If $i \geq 5$, then p has no strong contraction.

    (b) If $i = 4$, then for each contractible pair of terms in a strong contraction, the two brackets generated by explosion are common factors of the pair.

    (c) If $i = 3$, then for each contractible pair of terms in a strong contraction, one of the two brackets generated by explosion is a common factor of the pair.

Proof. (1) The explosion produces $r - 1$ new terms. If there is any contraction among an original term of p and more than one new term, the original term of p must form a GP transformable pair with the exploded term of p, contradicting with the hypothesis that p has no GP transformation. (2) Direct corollary of (1). $\qquad\square$

The following algorithm realizes strong contractions for $n = 3$. The explosions are

$$
\begin{aligned}
[123][456] = \ & [124][356] - [125][346] + [126][345] \\
= \ & -[134][256] + [135][246] - [136][245] \\
= \ & [234][156] - [235][146] + [236][145] \\
= \ & [145][236] - [245][136] + [345][126] \\
= \ & [146][235] - [246][135] + [346][125] \\
= \ & [156][234] - [256][134] + [356][124],
\end{aligned}
\tag{3.3.9}
$$

and

$$
[123][145] = [124][135] - [125][134].
\tag{3.3.10}
$$

Algorithm 3.17. Strong contraction.

Input: A bracket polynomial p of at least three terms and five different points. Assume that p is not factorizable in the polynomial ring of brackets.

Output: p after strong contraction.

Step 1. For all pairs of terms in p, compute the degree of the remaining bracket polynomial after removal of the common bracket factors from the pair. Let i be the lowest of such degrees. If $i \geq 5$, then there is no strong contraction.

Step 2. For every bracket in p, compute its *total degree* in p, which is the sum of the degrees in p over the points contained in the bracket. The total degree defines an order among the brackets in p.

Step 3. [Explosion based on (3.3.9)] If p has at least four terms and involves at least six different points, then start from a bracket with the lowest total degree in p, say $[\mathbf{123}]$, do the following:

Let S be the set of points in p other than $\mathbf{1, 2, 3}$. Let P be the set of terms of p containing $[\mathbf{123}]$. For any element $p' \in P$, start from a bracket $[\mathbf{456}]$ in p' with the lowest total degree and where $\{\mathbf{4, 5, 6}\} \subseteq S$, let $p' = [\mathbf{123}][\mathbf{456}]p''$ where p'' is a bracket monomial, do the following.

    Case 1. If $i < 3$, let q_1 to q_6 be the six results obtained by multiplying p'' with the terms of the six bracket trinomials in (3.3.9) respectively. For j from 1 to 6, for each term of bracket trinomial q_j, check if there is any term in the bracket polynomial $p - p'$ that can form a contractible pair with it, using Step 1.3 of Algorithm 3.12. If this is true for each term of q_j, replace p' and the three involved terms of $p - p'$ in the contractions by the contraction results, exit.

    Case 2. If $i = 3$, find in (3.3.9) such bracket trinomials where each term has a common bracket factor with a term of $p - p'$, and the three involved terms of $p - p'$ are pairwise different. This establishes a set of correspondences, each being from the three terms of a bracket trinomial in (3.3.9) multiplied by p'', to the three involved terms of $p - p'$. For every such a correspondence, check if the three pairs of terms in correspondence are all contractible, using Step 1.3 of Algorithm 3.12, and if so, replace p' and the three involved terms of $p - p'$ by the contraction results, exit.

    Case 3. If $i = 4$, find in (3.3.9) such bracket trinomials where each term is a factor of a term in $p - p'$, and the three involved terms in $p - p'$ are pairwise different. This establishes a set of correspondences, each being from the three terms of a bracket trinomial in (3.3.9) multiplied by p'', to the three involved terms of $p - p'$. For every such a correspondence, check if the three pairs of terms in correspondence are all contractible, using Step 1.3 of Algorithm 3.12, and if so, replace p' and the three involved terms of $p - p'$ by the contraction results, exit.

Step 4. [Explosion based on (3.3.10)] Start from a bracket q with the lowest total degree in p, do the following:

Let S be the set of points in p but not in q. Let P be the set of terms of p containing bracket q. For any element $p' \in P$, start from a bracket $[\mathbf{145}]$ of p' with the lowest total degree, where $\{\mathbf{4, 5}\} \subseteq S$ and $\mathbf{1} \in q$, let $q = [\mathbf{123}]$ and $p' = [\mathbf{123}][\mathbf{145}]p''$, where p'' is a bracket monomial, do the following.

For each term of bracket binomial $r = p''[124][135] - p''[125][134]$, check if there is any term of $p - p'$ that can form a contractible pair with it, using Step 1.3 of Algorithm 3.12. If this is true for both terms in r, replace p' and the two involved terms of $p - p'$ by the contraction results, exit.

Example 3.18. The polynomial factor on the right side of (3.3.8) is as follows:

$$p = \ [124][125][127][136][236][457] + [125][126][127][134][234][567]$$
$$-[124][125][126][137][237][456] - [124][126][127][135][235][467].$$

Do strong contraction to it.

By Step 1 of Algorithm 3.17, we get $i = 4$. The points in p with their degrees are $1^4, 2^4, 3^2, 4^2, 5^2, 6^2, 7^2$. The brackets in p with their total degrees are $[12j]^{10}, [13j]^8, [23j]^8$ for $4 \le j \le 7$, and $[jkl]^6$ for $4 \le j < k < l \le 7$.

We start from any bracket with total degree 6, for example $[567]$. In Step 3, $\mathcal{S} = \{1, 2, 3, 4\}$. The second term p' of p is the only one that contains $[567]$. There are two brackets in p' with total degree 8 and composed of points in $\mathcal{S}$: $[134], [234]$. Start from any of them, say $[134]$. Then $p'' = [125][126][127][234]$.

The six explosions of $[567][134]$ are

$$
\begin{aligned}
[134][567] =\ & [135][467] - [136][457] + [137][456] \\
=\ & -[145][367] + [146][357] - [147][356] \\
=\ & [345][167] - [346][157] + [347][156] \\
=\ & [167][345] - [367][145] + [467][135] \\
=\ & -[157][346] + [357][146] - [457][136] \\
=\ & [156][347] - [356][147] + [456][137].
\end{aligned}
\qquad (3.3.11)
$$

Only the first explosion has its three terms in the three terms of $p - p'$ respectively. The correspondence discovered in Case (3) of Step 3 is

$$
\begin{aligned}
p''[135][467] &\Longleftrightarrow -[124][126][127][135][235][467], \\
-p''[136][457] &\Longleftrightarrow\ [124][125][127][136][236][457], \\
p''[137][456] &\Longleftrightarrow -[124][125][126][137][237][456].
\end{aligned}
$$

The three corresponding pairs are all contractible. The contractions are

$$
\begin{aligned}
[125][234] - [124][235] &= -[123][245], \\
[124][236] - [126][234] &=\ [123][246], \\
[127][234] - [124][237] &= -[123][247].
\end{aligned}
$$

So

$$p = [123]\{-[126][127][135][245][467] + [125][127][136][246][457]$$
$$- [125][126][137][247][456]\}.$$

It is also possible to use VW syzygies and successive contractions to reduce the number of terms. A VW syzygy which is not a GP syzygy contains at least six terms. In 2D projective geometry there is only one such syzygy:

$$
\begin{aligned}
[121'][342'] &- [131'][242'] + [141'][232'] \\
&+ [231'][142'] - [241'][132'] + [341'][122'] = 0.
\end{aligned}
\qquad (3.3.12)
$$

Among all pairs of terms in (3.3.12), there is only one kind of combination that is not contractible:

$$[121'][342']+[122'][341'] = [131'][242']-[141'][232']-[231'][142']+[241'][132'].$$
$$(3.3.13)$$

(3.3.13) is called a *van der Waerden* (VW) *transformation*. The right side of (3.3.13) is uniquely determined by the left side.

Assume that there is a bracket polynomial that is not contractible, but contains the left side of (3.3.13). When (3.3.13) is applied to the polynomial, the number of terms is increased by two, so three contractions to monomial results are required to reduce the number of terms. This kind of simplification cannot be realized by either strong contraction or level contraction. It is neither encountered in our bracket computing practice.

3.4   Exact division and pseudodivision

In the polynomial ring of coordinates, division and pseudodivision are major algebraic manipulations. In bracket algebra, algebraic dependencies among brackets make the two manipulations much more complicated. In this section, the two manipulations are explored at a preliminary level, and some surprising phenomena are discovered.

3.4.1   *Exact division by brackets without common vectors*

Given a bracket polynomial p and a set of bracket polynomials $q_1, q_2, \ldots, q_k$, if no two q_i have any vector variable in common, then an *exact division* of p by the q_i refers to a decomposition of the form

$$p = q_1 r_1 + q_2 r_2 + \cdots q_k r_k, \qquad (3.4.1)$$

where the r_i are bracket polynomials. It is not required that the r_i be unique.

Exact division can be easily done by means of homogeneous coordinates, *i.e.*, by eliminating all algebraic dependencies among brackets. Without resorting to coordinatization, an exact division can usually be realized by explosions and successive contractions. Cayley factorizations (3.2.1) and (3.2.2) are typical examples of division by either a basic bracket or a Cayley bracket.

Example 3.19. Divide by bracket [**123**] the left side of (3.2.1):

$$p = [121'][232'][133'] - [122'][233'][131'].$$

From the following explosion in the first term of p:

$$[121'][133'] = [123][11'3'] + [123'][131'],$$

the divisor $[\mathbf{123}]$ is generated. So

$$p \overset{explode}{=} [\mathbf{123}][\mathbf{11'3'}][\mathbf{232'}] + [\mathbf{131'}]([\mathbf{123'}][\mathbf{232'}] - [\mathbf{122'}][\mathbf{233'}])$$
$$\overset{contract}{=} [\mathbf{123}]([\mathbf{11'3'}][\mathbf{232'}] - [\mathbf{131'}][\mathbf{22'3'}]).$$

Example 3.20. Divide by Cayley bracket $\mathbf{12} \vee \mathbf{35} \vee \mathbf{46}$ the left side of (3.2.2):

$$p = [\mathbf{125}][\mathbf{126}][\mathbf{345}][\mathbf{346}] - [\mathbf{123}][\mathbf{124}][\mathbf{356}][\mathbf{456}].$$

From the following explosion in the first term of p:

$$[\mathbf{125}][\mathbf{346}] = \mathbf{12} \vee \mathbf{35} \vee \mathbf{46} - [\mathbf{123}][\mathbf{456}],$$

the divisor $\mathbf{12} \vee \mathbf{35} \vee \mathbf{46}$ is generated. So

$$p \overset{explode}{=} [\mathbf{126}][\mathbf{345}]\,\mathbf{12} \vee \mathbf{35} \vee \mathbf{46} - [\mathbf{123}][\mathbf{456}]([\mathbf{126}][\mathbf{345}] + [\mathbf{124}][\mathbf{356}])$$
$$\overset{factor}{=} ([\mathbf{126}][\mathbf{345}] - [\mathbf{123}][\mathbf{456}])\,\mathbf{12} \vee \mathbf{35} \vee \mathbf{46}.$$

In the previous two examples, using coordinates leads to the same result although in a different form. The following example shows that using coordinates can produce drastically different results from using brackets, and the results from the coordinate approach may be geometrically inexplicable.

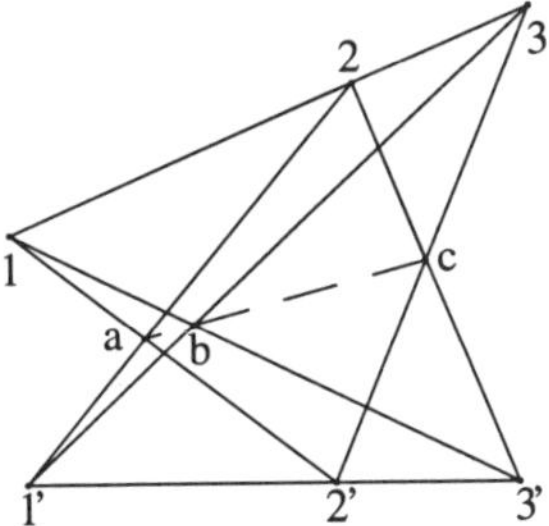

Fig. 3.4   Pappus Theorem.

Example 3.21. [Pappus Theorem] In the projective plane there are two lines $\mathbf{123}$ and $\mathbf{1'2'3'}$ each passing through three points. Then the intersections $\mathbf{a} = \mathbf{12'} \cap \mathbf{1'2}$, $\mathbf{b} = \mathbf{13'} \cap \mathbf{1'3}$, $\mathbf{c} = \mathbf{23'} \cap \mathbf{2'3}$ are collinear.

Hypotheses: $[\mathbf{123}] = [\mathbf{1'2'3'}] = 0$.
Conclusion: $[(\mathbf{12'} \vee \mathbf{1'2})(\mathbf{13'} \vee \mathbf{1'3})(\mathbf{23'} \vee \mathbf{2'3})] = 0$.

We remove the two hypotheses and check how they influence the conclusion expression.

$$[(\mathbf{12'} \vee \mathbf{1'2})(\mathbf{13'} \vee \mathbf{1'3})(\mathbf{23'} \vee \mathbf{2'3})]$$
$$\overset{expand}{=} -[\mathbf{122'}][\mathbf{1'}(\mathbf{13'} \vee \mathbf{1'3})(\mathbf{23'} \vee \mathbf{2'3})] + [\mathbf{11'2'}][\mathbf{2}(\mathbf{13'} \vee \mathbf{1'3})(\mathbf{23'} \vee \mathbf{2'3})]$$
$$\overset{expand}{=} -[\mathbf{122'}][\mathbf{11'3'}][\mathbf{1'3}(\mathbf{23'} \vee \mathbf{2'3})] - [\mathbf{11'2'}][\mathbf{232'}][\mathbf{2}(\mathbf{13'} \vee \mathbf{1'3})\mathbf{3'}]$$
$$\overset{expand}{=} -[\mathbf{122'}][\mathbf{11'3'}][\mathbf{233'}][\mathbf{31'2'}] + [\mathbf{123'}][\mathbf{11'2'}][\mathbf{232'}][\mathbf{31'3'}].$$

(3.4.2)

The expansion is finished but the result has neither $[\mathbf{123}]$ nor $[\mathbf{1'2'3'}]$. Division of the result of (3.4.2) by $[\mathbf{123}]$ is needed to find out both brackets.

Any GP relation containing $[\mathbf{123}]$ is one of the following forms:

$$[\mathbf{123}][\mathbf{145}] - [\mathbf{124}][\mathbf{135}] + [\mathbf{125}][\mathbf{134}] = 0,$$
$$[\mathbf{123}][\mathbf{456}] - [\mathbf{124}][\mathbf{356}] + [\mathbf{125}][\mathbf{346}] - [\mathbf{126}][\mathbf{345}] = 0,$$
$$[\mathbf{123}][\mathbf{456}] - [\mathbf{124}][\mathbf{356}] + [\mathbf{134}][\mathbf{256}] - [\mathbf{234}][\mathbf{156}] = 0.$$

A bracket monomial can generate $[\mathbf{123}]$ through explosion if and only if it contains two such brackets: one contains two of the three vectors $\mathbf{1, 2, 3}$, say it is bracket $[\mathbf{124}]$; the other contains the third vector $\mathbf{3}$ but not the bracket mate $\mathbf{4}$ of the first two vectors in the previous bracket.

In the result of (3.4.2), the first term has two brackets $[\mathbf{122'}][\mathbf{233'}]$ that are the only pair of brackets that can generate $[\mathbf{123}]$ by explosion. The explosion is

$$[\mathbf{122'}][\mathbf{233'}] = [\mathbf{123}][\mathbf{22'3'}] + [\mathbf{123'}][\mathbf{232'}].$$

Immediately after the explosion we need to do simplification to the "remainder of the division". The technique is contraction:

$$-[\mathbf{11'3'}][\mathbf{31'2'}] + [\mathbf{11'2'}][\mathbf{31'3'}] = [\mathbf{131'}][\mathbf{1'2'3'}].$$

Now the result of (3.4.2) becomes

$$-[\mathbf{122'}][\mathbf{11'3'}][\mathbf{233'}][\mathbf{31'2'}] + [\mathbf{123'}][\mathbf{11'2'}][\mathbf{232'}][\mathbf{31'3'}]$$
$$\overset{explode}{=} -[\mathbf{123}][\mathbf{11'3'}][\mathbf{22'3'}][\mathbf{31'2'}] - [\mathbf{123'}][\mathbf{232'}]([\mathbf{11'3'}][\mathbf{31'2'}] - [\mathbf{11'2'}][\mathbf{31'3'}])$$
$$\overset{contract}{=} -[\mathbf{123}][\mathbf{11'3'}][\mathbf{22'3'}][\mathbf{31'2'}] + [\mathbf{123'}][\mathbf{131'}][\mathbf{232'}][\mathbf{1'2'3'}],$$

$$(3.4.3)$$

where both brackets in the hypotheses are produced.

In comparison, using homogeneous coordinates $\mathbf{i} = (x_i, y_i, z_i)^T$ and $\mathbf{i'} = (x_i', y_i', z_i')^T$ for $\mathbf{i} = \mathbf{1, 2, 3}$ to represent the result of (3.4.2) leads to a polynomial f of 720 terms. Do pseudodivision to f by

$$g = [\mathbf{123}] = (y_2 z_3 - y_3 z_2)x_1 + z_1 x_2 y_3 - z_1 x_3 y_2 - y_1 x_2 z_3 + y_1 x_3 z_2,$$

with x_1 as the leading variable, then we get

$$f = -\frac{h[\mathbf{123}]}{(y_2 z_3 - y_3 z_2)^2} + \frac{(y_1 z_3 - z_1 y_3)(y_1 z_2 - y_2 z_1)[\mathbf{1'2'3'}][\mathbf{231'}][\mathbf{232'}][\mathbf{233'}]}{(y_2 z_3 - y_3 z_2)^2},$$

where h is an irreducible polynomial of 894 terms, and where the brackets represent their corresponding coordinate polynomial expressions. Polynomial h has no clear geometric meaning.

The difference of the above result with (3.4.3) is caused by the complete elimination of the coordinate x_1 of point $\mathbf{1}$ from the result of (3.4.2). In coordinate form, vector $\mathbf{1}$ cannot stay in brackets $[\mathbf{11'3'}], [\mathbf{123'}], [\mathbf{131'}]$ of the result of (3.4.3) after pseudodivision.

One may ask whether the equality between the first and the last expressions in (3.4.3) has any geometric meaning. Indeed the equality is meaningful only in projective conic geometry:

Lines **123** and **1'2'3'** form a conic in the plane. By Pascal's Conic Theorem, the three intersections **a, b, c** are collinear if and only if points **1, 2, 3, 1', 2', 3'** are on the same conic. The function

$$\text{conic}(\mathbf{123456}) = [\mathbf{135}][\mathbf{245}][\mathbf{126}][\mathbf{346}] - [\mathbf{125}][\mathbf{345}][\mathbf{136}][\mathbf{246}] \qquad (3.4.4)$$

is antisymmetric with respect to the six points **1** to **6** in the plane, and equals zero if and only if the six points are on the same conic. The equality between the first and the last expressions in (3.4.3) is just $-\text{conic}(\mathbf{121'32'3'}) = \text{conic}(\mathbf{121'2'33'})$, *i.e.*, the antisymmetry of the function with respect to **3, 2'**. It shows that $\text{conic}(\mathbf{123456})$ is an advanced invariant, where the nontrivial identity (3.4.3) is translated into an easy antisymmetry within the advanced invariant.

3.4.2 *Pseudodivision by brackets with common vectors*

Division by a set of bracket polynomials having common vector variables is much more complicated. The result is rather surprising: the pseudo-coefficient, the quotients and the remainder are no longer bracket polynomials, instead they are symmetric tensor products of vectors.

In 2D projective geometry, if two bracket polynomial equations have a common vector variable, then the common vector can be completely determined by the two equations. The remainder of a bracket polynomial by the two bracket polynomials, with the common vector as the leading vector variable, is the result of eliminating the common vector from the dividend.

For example, let **3** be the intersection of lines **12** and **1'2'** in the plane. By substituting $\mathbf{3} = \mathbf{12} \vee \mathbf{1'2'}$ into bracket [**345**], we get that the remainder of [**345**] by [**123**] and [**1'2'3**], with **3** as the leading vector variable, contains a factor **12** $\vee$ **1'2'** $\vee$ **45**. This factor is all we can get by elimination alone.

If **3** is not on the two lines, how does bracket [**345**] vary with [**123**] and [**1'2'3**]? This question can only be answered by pseudodivision in bracket algebra.

For the current example, the *pseudodivision* of [**345**] by [**123**] and [**1'2'3**], with **3** as the leading vector variable, refers to the finding of bracket polynomials c, q, q', r, where c is called the *pseudo-coefficient*, q and q' are called the first and second *quotients* respectively, and r is called the *remainder*, such that

(i) $c[\mathbf{345}] = q[\mathbf{123}] + q'[\mathbf{1'2'3}] + r$,

(ii) r contains a factor **12** $\vee$ **1'2'** $\vee$ **45**,

(iii) c does not contain **3**,

(iv) the degree of **c** is the lowest among all the c's satisfying requirement (i).

In coordinate form, let $\mathbf{i} = (x_i, y_i, z_i)^T$ for $i = 1, 2, 3$, and let $\mathbf{j'} = (x_{jp}, y_{jp}, z_{jp})^T$ for $j = 1, 2$. The following Maple session computes c, q, q', r:

```
>  f1:=det([[x1,y1,z1],[x2,y2,z2],[x3,y3,z3]])-a:
>  f2:=det([[x1p,y1p,z1p],[x2p,y2p,z2p],[x3,y3,z3]])-b:
>  g:=det([[x4,y4,z4],[x5,y5,z5],[x3,y3,z3]]):
```

```
> h:=solve({f1,f2},{x3,y3}):
> fin:=simplify(subs(h,g)):
> collect(fin,[a,b]);
```

$$-\frac{(-z_4 y_5 x_{1p} z_{2p} + y_4 z_5 x_{1p} z_{2p} + x_5 z_4 y_{1p} z_{2p} - x_5 z_4 z_{1p} y_{2p} - x_4 z_5 y_{1p} z_{2p} + x_4 z_5 z_{1p} y_{2p} - y_4 z_5 x_{2p} z_{1p} + z_4 y_5 x_{2p} z_{1p})\,a}{-x_{1p} z_{2p} y_1 z_2 + x_{1p} z_{2p} z_1 y_2 + x_{2p} z_{1p} y_1 z_2 - x_{2p} z_{1p} z_1 y_2 + y_{1p} z_{2p} x_1 z_2 - y_{1p} z_{2p} x_2 z_1 - z_{1p} y_{2p} x_1 z_2 + z_{1p} y_{2p} x_2 z_1}$$

$$-\frac{(-x_4 z_5 z_1 y_2 - y_4 z_5 x_1 z_2 + x_4 z_5 y_1 z_2 - x_5 z_4 y_1 z_2 + x_5 z_4 z_1 y_2 + y_4 z_5 x_2 z_1 + z_4 y_5 x_1 z_2 - z_4 y_5 x_2 z_1)\,b}{-x_{1p} z_{2p} y_1 z_2 + x_{1p} z_{2p} z_1 y_2 + x_{2p} z_{1p} y_1 z_2 - x_{2p} z_{1p} z_1 y_2 + y_{1p} z_{2p} x_1 z_2 - y_{1p} z_{2p} x_2 z_1 - z_{1p} y_{2p} x_1 z_2 + z_{1p} y_{2p} x_2 z_1}$$

$$-\frac{1}{-x_{1p} z_{2p} y_1 z_2 + x_{1p} z_{2p} z_1 y_2 + x_{2p} z_{1p} y_1 z_2 - x_{2p} z_{1p} z_1 y_2 + y_{1p} z_{2p} x_1 z_2 - y_{1p} z_{2p} x_2 z_1 - z_{1p} y_{2p} x_1 z_2 + z_{1p} y_{2p} x_2 z_1}$$

$$\begin{aligned}
(&x_4 y_5 z_3 x_{1p} z_{2p} y_1 z_2 - x_4 y_5 z_3 x_{1p} z_{2p} z_1 y_2 - x_4 y_5 z_3 x_{2p} z_{1p} y_1 z_2 + x_4 y_5 z_3 x_{2p} z_{1p} z_1 y_2 - x_4 y_5 z_3 y_{1p} z_{2p} x_1 z_2 \\
&+ x_4 y_5 z_3 y_{1p} z_{2p} x_2 z_1 + x_4 y_5 z_3 z_{1p} y_{2p} x_1 z_2 - x_4 y_5 z_3 z_{1p} y_{2p} x_2 z_1 - x_4 z_5 x_{1p} y_{2p} z_3 y_1 z_2 + x_4 z_5 x_{1p} y_{2p} z_3 z_1 y_2 \\
&+ x_4 z_5 x_{2p} y_{1p} z_3 y_1 z_2 - x_4 z_5 x_{2p} y_{1p} z_3 z_1 y_2 + x_4 z_5 y_{1p} z_{2p} x_1 y_2 z_3 - x_4 z_5 y_{1p} z_{2p} x_2 y_1 z_3 - x_4 z_5 z_{1p} y_{2p} x_1 y_2 z_3 \\
&+ x_4 z_5 z_{1p} y_{2p} x_2 y_1 z_3 - x_5 y_4 z_3 x_{1p} z_{2p} y_1 z_2 + x_5 y_4 z_3 x_{1p} z_{2p} z_1 y_2 + x_5 y_4 z_3 x_{2p} z_{1p} y_1 z_2 - x_5 y_4 z_3 x_{2p} z_{1p} z_1 y_2 \\
&+ x_5 y_4 z_3 y_{1p} z_{2p} x_1 z_2 - x_5 y_4 z_3 y_{1p} z_{2p} x_2 z_1 - x_5 y_4 z_3 z_{1p} y_{2p} x_1 z_2 + x_5 y_4 z_3 z_{1p} y_{2p} x_2 z_1 + x_5 z_4 x_{1p} y_{2p} z_3 y_1 z_2 \\
&- x_5 z_4 x_{1p} y_{2p} z_3 z_1 y_2 - x_5 z_4 x_{2p} y_{1p} z_3 y_1 z_2 + x_5 z_4 x_{2p} y_{1p} z_3 z_1 y_2 - x_5 z_4 y_{1p} z_{2p} x_1 y_2 z_3 + x_5 z_4 y_{1p} z_{2p} x_2 y_1 z_3 \\
&+ x_5 z_4 z_{1p} y_{2p} x_1 y_2 z_3 - x_5 z_4 z_{1p} y_{2p} x_2 y_1 z_3 + y_4 z_5 x_{1p} y_{2p} z_3 x_1 z_2 - y_4 z_5 x_{1p} y_{2p} z_3 x_2 z_1 - y_4 z_5 x_{1p} z_{2p} x_1 y_2 z_3 \\
&+ y_4 z_5 x_{1p} z_{2p} x_2 y_1 z_3 - y_4 z_5 x_{2p} y_{1p} z_3 x_1 z_2 + y_4 z_5 x_{2p} y_{1p} z_3 x_2 z_1 + y_4 z_5 x_{2p} z_{1p} x_1 y_2 z_3 - y_4 z_5 x_{2p} z_{1p} x_2 y_1 z_3 \\
&- z_4 y_5 x_{1p} y_{2p} z_3 x_1 z_2 + z_4 y_5 x_{1p} y_{2p} z_3 x_2 z_1 + z_4 y_5 x_{1p} z_{2p} x_1 y_2 z_3 - z_4 y_5 x_{1p} z_{2p} x_2 y_1 z_3 + z_4 y_5 x_{2p} y_{1p} z_3 x_1 z_2 \\
&- z_4 y_5 x_{2p} y_{1p} z_3 x_2 z_1 - z_4 y_5 x_{2p} z_{1p} x_1 y_2 z_3 + z_4 y_5 x_{2p} z_{1p} x_2 y_1 z_3)
\end{aligned}$$

It takes us great efforts to decipher the geometric meaning of the above result. Let $\mathbf{x}, \mathbf{y}$ be the basis vectors along the x-axis and y-axis respectively. Then the above computing returns the pseudo-coefficient c:

$$\begin{aligned}
\mathbf{12} \vee \mathbf{1'2'} \vee \mathbf{xy} = {}&-x_{1p} z_{2p} y_1 z_2 + x_{1p} z_{2p} z_1 y_2 + x_{2p} z_{1p} y_1 z_2 - x_{2p} z_{1p} z_1 y_2 \\
&+ y_{1p} z_{2p} x_1 z_2 - y_{1p} z_{2p} x_2 z_1 - z_{1p} y_{2p} x_1 z_2 + z_{1p} y_{2p} x_2 z_1;
\end{aligned} \tag{3.4.5}$$

the quotient q with respect to $a = [\mathbf{123}]$:

$$\begin{aligned}
-\mathbf{1'2'} \vee \mathbf{45} \vee \mathbf{xy} = {}&-(-z_4 y_5 x_{1p} z_{2p} + y_4 z_5 x_{1p} z_{2p} + x_5 z_4 y_{1p} z_{2p} - x_5 z_4 z_{1p} y_{2p} \\
&- x_4 z_5 y_{1p} z_{2p} + x_4 z_5 z_{1p} y_{2p} - y_4 z_5 x_{2p} z_{1p} + z_4 y_5 x_{2p} z_{1p});
\end{aligned} \tag{3.4.6}$$

the quotient q' with respect to $b = [\mathbf{1'2'3}]$:

$$\begin{aligned}
\mathbf{12} \vee \mathbf{45} \vee \mathbf{xy} = {}&-(-x_4 z_5 z_1 y_2 - y_4 z_5 x_1 z_2 + x_4 z_5 y_1 z_2 - x_5 z_4 y_1 z_2 \\
&+ x_5 z_4 z_1 y_2 + y_4 z_5 x_2 z_1 + z_4 y_5 x_1 z_2 - z_4 y_5 x_2 z_1);
\end{aligned} \tag{3.4.7}$$

and the remainder r:

$$z_3\, \mathbf{12} \vee \mathbf{1'2'} \vee \mathbf{45} = [\mathbf{3xy}]\, \mathbf{12} \vee \mathbf{1'2'} \vee \mathbf{45}, \tag{3.4.8}$$

as a coordinate polynomial of 48 terms.

We consider how to obtain the above results using brackets instead of coordinates.

(1) To divide $[\mathbf{345}]$ by $[\mathbf{123}]$ is to find bracket polynomials x, y, z of lowest degree such that

$$x[\mathbf{345}] = y[\mathbf{123}] + z. \tag{3.4.9}$$

The GP relation in the form of (3.4.9) is

$$[\mathbf{12v_1}][\mathbf{345}] = [\mathbf{45v_1}][\mathbf{123}] - [\mathbf{124}][\mathbf{35v_1}] + [\mathbf{125}][\mathbf{34v_1}], \qquad (3.4.10)$$

where vector $\mathbf{v_1}$ is arbitrary. This explosion is just the pseudodivision of $[\mathbf{345}]$ by $[\mathbf{123}]$, because the pseudo-coefficient $[\mathbf{12v_1}]$ is a bracket monomial of lowest degree. The remainder is

$$z = -[\mathbf{124}][\mathbf{35v_1}] + [\mathbf{125}][\mathbf{34v_1}]. \qquad (3.4.11)$$

(2) To divide (3.4.11) by $[\mathbf{1'2'3}]$, we only need to divide those brackets containing vector $\mathbf{3}$.

$$x'[\mathbf{34v_1}] = y'[\mathbf{1'2'3}] + z' \implies$$
$$[\mathbf{1'2'v_2}][\mathbf{34v_1}] = -[\mathbf{4v_2v_1}][\mathbf{1'2'3}] + [\mathbf{1'2'4}][\mathbf{3v_2v_1}] + [\mathbf{1'2'v_1}][\mathbf{34v_2}]. \qquad (3.4.12)$$

Similarly,

$$x''[\mathbf{35v_1}] = y''[\mathbf{1'2'3}] + z'' \implies$$
$$-[\mathbf{1'2'v_2}][\mathbf{35v_1}] = [\mathbf{5v_2v_1}][\mathbf{1'2'3}] - [\mathbf{1'2'5}][\mathbf{3v_2v_1}] - [\mathbf{1'2'v_1}][\mathbf{35v_2}]. \qquad (3.4.13)$$

In (3.4.12) and (3.4.13), the same arbitrary vector $\mathbf{v_2}$ is used to control the degree of the pseudo-coefficient to one.

(3) Putting together the results of the previous steps, we get

$$
\begin{aligned}
&[\mathbf{12v_1}][\mathbf{1'2'v_2}][\mathbf{345}] \\
&= [\mathbf{45v_1}][\mathbf{1'2'v_2}][\mathbf{123}] + ([\mathbf{124}][\mathbf{5v_2v_1}] - [\mathbf{125}][\mathbf{4v_2v_1}])[\mathbf{1'2'3}] \\
&\qquad\qquad\qquad - ([\mathbf{124}][\mathbf{1'2'5}] - [\mathbf{125}][\mathbf{1'2'4}])[\mathbf{3v_2v_1}] \\
&\qquad\qquad\qquad - ([\mathbf{124}][\mathbf{35v_2}] - [\mathbf{125}][\mathbf{34v_2}])[\mathbf{1'2'v_1}] \\
&= [\mathbf{45v_1}][\mathbf{1'2'v_2}][\mathbf{123}] - \mathbf{12} \vee \mathbf{45} \vee \mathbf{v_2v_1}\,[\mathbf{1'2'3}] \\
&\qquad\qquad - \mathbf{12} \vee \mathbf{1'2'} \vee \mathbf{45}\,[\mathbf{3v_2v_1}] - \mathbf{12} \vee \mathbf{45} \vee \mathbf{3v_2}\,[\mathbf{1'2'v_1}].
\end{aligned}
\qquad (3.4.14)
$$

(4) The last term in the result of (3.4.14) is not part of the remainder. It is a combination of the dividend $[\mathbf{345}]$ and the divisor $[\mathbf{123}]$, according to the expansion

$$\mathbf{12} \vee \mathbf{45} \vee \mathbf{3v_2} = [\mathbf{123}][\mathbf{45v_2}] - [\mathbf{12v_2}][\mathbf{345}].$$

By Cayley factorizations

$$
\begin{aligned}
[\mathbf{12v_1}][\mathbf{1'2'v_2}] - [\mathbf{12v_2}][\mathbf{1'2'v_1}] &= -\mathbf{12} \vee \mathbf{1'2'} \vee \mathbf{v_2v_1}, \\
[\mathbf{45v_1}][\mathbf{1'2'v_2}] - [\mathbf{45v_2}][\mathbf{1'2'v_1}] &= \mathbf{1'2'} \vee \mathbf{45} \vee \mathbf{v_2v_1},
\end{aligned}
$$

we get the final result of the pseudodivision:

$$
\begin{aligned}
\mathbf{12} \vee \mathbf{1'2'} \vee \mathbf{v_2v_1}\,[\mathbf{345}] = &-\mathbf{1'2'} \vee \mathbf{45} \vee \mathbf{v_2v_1}\,[\mathbf{123}] + \mathbf{12} \vee \mathbf{45} \vee \mathbf{v_2v_1}\,[\mathbf{1'2'3}] \\
&+ \mathbf{12} \vee \mathbf{1'2'} \vee \mathbf{45}\,[\mathbf{3v_2v_1}].
\end{aligned}
\qquad (3.4.15)
$$

On one hand, the final result (3.4.15) agrees with the results (3.4.5) to (3.4.8) from coordinate computing, by setting $\mathbf{v}_2, \mathbf{v}_1$ to be the basis vectors along the x-axis and y-axis respectively. On the other hand, the following peculiarity of pseudodivision in bracket algebra is disclosed: since (3.4.15) depends on arbitrary vectors $\mathbf{v}_2, \mathbf{v}_1 \in \mathcal{V}^3$, *the result of pseudodivision is covariant instead of invariant.*

In purely covariant form, the pseudodivision result (3.4.15) is

$$[\mathbf{345}]\,\mathbf{12} \vee \mathbf{1'2'} = [\mathbf{123}]\,\mathbf{45} \vee \mathbf{1'2'} + [\mathbf{1'2'3}]\,\mathbf{12} \vee \mathbf{45} + (\mathbf{12} \vee \mathbf{1'2'} \vee \mathbf{45})\mathbf{3}. \quad (3.4.16)$$

It is nothing but the Cramer's rule of vector $\mathbf{12} \vee \mathbf{1'2'}$ with respect to vectors $\mathbf{45} \vee \mathbf{1'2'}$, $\mathbf{12} \vee \mathbf{45}$ and $\mathbf{3}$ in $\mathcal{V}^3$, if multiplied by $-\mathbf{12} \vee \mathbf{1'2'} \vee \mathbf{45}$:

$$\begin{aligned}
[\mathbf{3}(\mathbf{12} \vee \mathbf{45})(\mathbf{45} \vee \mathbf{1'2'})]\,\mathbf{12} \vee \mathbf{1'2'} = &-[\mathbf{3}(\mathbf{12} \vee \mathbf{1'2'})(\mathbf{12} \vee \mathbf{45})]\,\mathbf{45} \vee \mathbf{1'2'} \\
&+[\mathbf{3}(\mathbf{12} \vee \mathbf{1'2'})(\mathbf{45} \vee \mathbf{1'2'})]\,\mathbf{12} \vee \mathbf{45} \\
&+[(\mathbf{12} \vee \mathbf{1'2'})(\mathbf{12} \vee \mathbf{45})(\mathbf{45} \vee \mathbf{1'2'})]\,\mathbf{3}.
\end{aligned}$$
$$(3.4.17)$$

Similarly, if we do pseudodivision to degree-2 bracket monomial $[\mathbf{345}][\mathbf{34'5'}]$ by brackets $[\mathbf{123}]$ and $[\mathbf{1'2'3}]$, with $\mathbf{3}$ as the leading vector variable, we get

$$\begin{aligned}
[\mathbf{345}][\mathbf{34'5'}]\,(\mathbf{12} \vee \mathbf{1'2'}) \otimes (\mathbf{12} \vee \mathbf{1'2'}) = &(\mathbf{12} \vee \mathbf{1'2'} \vee \mathbf{45})(\mathbf{12} \vee \mathbf{1'2'} \vee \mathbf{4'5'})\,\mathbf{3} \otimes \mathbf{3} \\
&+u[\mathbf{123}] + v[\mathbf{1'2'3}],
\end{aligned}$$
$$(3.4.18)$$

where u, v are each a symmetric tensor product of two vectors.

3.5 Rational invariants

A projective space has various subspaces, and each subspace has its own invariants. An invariant of a projective subspace is no longer an invariant in the whole space, instead it becomes a covariant. For example, on a projective line, the bracket of two points is an invariant, but no longer an invariant in the projective plane. It becomes the covariant representing the line: the outer product of the two points.

Nevertheless, the ratio of two invariants of a projective subspace is always an invariant, called an *invariant ratio*. For example, the ratio of two brackets $[\mathbf{12}]$ and $[\mathbf{34}]$ for collinear points $\mathbf{1}, \mathbf{2}, \mathbf{3}, \mathbf{4}$ is an invariant: for any point $\mathbf{5}$ not collinear with the four points,

$$\frac{[\mathbf{12}]}{[\mathbf{34}]} = \frac{[\mathbf{125}]}{[\mathbf{345}]} \quad (3.5.1)$$

is independent of the choice of dummy vector $\mathbf{5}$. This is a ubiquitous property of all invariant ratios in projective geometry.

Invariant ratios are as fundamental as brackets in projective geometry. They are the direct heritage of low dimensional invariants. The fundamental absolute invariant in projective geometry, the *cross-ratio* of four collinear points $\mathbf{1}, \mathbf{2}, \mathbf{3}, \mathbf{4}$, is the following absolute invariant ratio:

$$(\mathbf{12}; \mathbf{34}) = \frac{[\mathbf{13}]\,[\mathbf{24}]}{[\mathbf{14}]\,[\mathbf{23}]}. \quad (3.5.2)$$

Definition 3.22. A *rational invariant* is a rational monomial composed of invariant ratios and rational bracket monomials. The *algebra of rational invariants* is generated by invariant ratios over the quotient field of the bracket ring.

Despite its indispensable role played in projective geometry, the algebra of rational invariants is far from being well studied in the literature. In this section, we investigate rational invariants by two approaches: partial antisymmetrization and partial symmetrization.

3.5.1　*Antisymmetrization of rational invariants*

The transformation from a rational invariant to an equal rational bracket monomial is called the *completion* of the rational invariant. Since this is always possible by adding dummy extensors to the components of invariant ratios, the aim should be to find the *simplest completion* of a rational invariant.

There are two aspects for a completion to be the simplest:

(1) The degree of the resulting rational bracket monomial should be the lowest.
(2) When the degree is the lowest, sometimes it is required that the sum or difference of the numerator and the denominator of the rational bracket monomial be Cayley factorizable.

Like other symbolic geometric computing problems, the completion problem of a rational invariant whose vector variables satisfy incidence constraints, can be handled by the elimination approach. Following the reverse order of a construction sequence of the geometric configuration, the constrained points in the rational invariant are eliminated, and after each elimination, the Cayley expressions produced in the rational invariant are expanded and simplified, leading ultimately to a rational bracket polynomial that contains no constrained points. If each Cayley expansion results in a bracket monomial, then the final result is a rational bracket monomial.

Example 3.23. In the projective plane, let the rows and columns of the following table of points be collinear, while any other triplets of points in the table are non-collinear ones:

$$
\begin{array}{ccc}
 & 2\ 3 & \\
4 & 5 & 6 \\
7 & 8 & 9
\end{array}
\tag{3.5.3}
$$

Write equation

$$
\frac{79}{78}\frac{28}{25}\frac{36}{39}\frac{45}{46} = 1
\tag{3.5.4}
$$

as an incidence constraint in projective geometry.

By the table of incidence relations (3.5.3), the geometric configuration can be constructed as follows:

Free points: $\mathbf{2, 3, 4, 5, 7, 9}$.

Intersections: $\mathbf{6} = \mathbf{45} \cap \mathbf{39}$, $\mathbf{8} = \mathbf{25} \cap \mathbf{79}$.

The eliminations of the two intersections from the left side of (3.5.4) are by straightforward substitutions of their Cayley expressions followed by Cayley expansions:

$$\frac{79}{78} \stackrel{8}{=} \frac{79}{7(25 \vee 79)} \stackrel{expand}{=} -\frac{1}{[257]}, \qquad \frac{28}{25} \stackrel{8}{=} \frac{2(25 \vee 79)}{25} \stackrel{expand}{=} [279],$$

$$\frac{36}{39} \stackrel{6}{=} \frac{3(45 \vee 39)}{39} \stackrel{expand}{=} -[345], \qquad \frac{45}{46} \stackrel{6}{=} \frac{45}{4(45 \vee 39)} \stackrel{expand}{=} -\frac{1}{[349]}. \tag{3.5.5}$$

After the eliminations, equality (3.5.4) is changed into

$$[279][345] + [257][349] \stackrel{factor}{=} 27 \vee 59 \vee 34 = 0, \tag{3.5.6}$$

whose geometric meaning is obvious: lines $\mathbf{27, 59, 34}$ concur.

The elimination approach requires a construction sequence of the points in a rational invariant. It guarantees neither the rational monomial form of the final result, nor the simplicity of the completion. Techniques for the specific purpose of simplest completion need to be developed.

Since the numerator and denominator of an invariant ratio are both covariants, they can multiply with other covariants by either the outer product or the meet product. A monomial of invariant ratios is then changed into a monomial of ratios of outer products and meet products. Since the two products are antisymmetric, this transformation is called a partial *antisymmetrization* of the rational invariant.

After the antisymmetrization, by monomial expansions, a monomial of ratios of meet products and outer products can be changed into a rational bracket monomial. This is the *antisymmetrization approach* to the completion of rational invariants.

Example 3.23 can be handled in the antisymmetrization approach as follows: let p be the left side of (3.5.4), where each invariant ratio is taken as a ratio of two blades that differ by scale only. The 2-blades in the numerator and denominator of p can be separately multiplied using the meet product as follows:

$$p = \frac{79}{78}\frac{28}{25}\frac{36}{39}\frac{45}{46} = \frac{79 \vee 28}{78 \vee 25}\frac{36 \vee 45}{39 \vee 46} \stackrel{expand}{=} \frac{-[279]8}{[257]8}\frac{[345]6}{[349]6} = -\frac{[279][345]}{[257][349]}. \tag{3.5.7}$$

Although the result is still the same with that of (3.5.5), where points $\mathbf{6, 8}$ are also eliminated from the rational invariant, the antisymmetrization approach does not demand any construction sequence for elimination.

In (3.5.7), the key to reducing the degree of the numerator (and also the denominator) of p from four to two, is the generation of two common vectors $\mathbf{6, 8}$ in both the numerator and the denominator by Cayley expansions of meet products. Since $\mathbf{6}$ is the intersection of lines $\mathbf{456}$ and $\mathbf{369}$, vector $\mathbf{6}$ has to occur in both $\mathbf{36} \vee \mathbf{45}$

and $\mathbf{39 \vee 46}$ as a common factor after their Cayley expansions. Similarly, vector $\mathbf{8 = 789 \cap 258}$ occurs as a common factor of $\mathbf{79 \vee 28}$ and $\mathbf{78 \vee 25}$ after their Cayley expansions.

Definition 3.24. A *line* in a rational invariant q refers to a 2-blade occurring in an invariant ratio of q. Symbolically, a line is represented by all atomic points on it, so it may have more than two atomic points. An *explicit intersection* in q, refers to the common atomic point of two different lines in q.

The left side of (3.5.4) contains four lines $\mathbf{789, 258, 369, 456}$. Besides $\mathbf{6, 8}$, there are two other explicit intersections: $\mathbf{5 = 456 \cap 258}$, $\mathbf{9 = 789 \cap 369}$. If using the latter two intersections instead, then in the antisymmetrization approach,

$$p = \frac{\mathbf{79}}{\mathbf{78}} \frac{\mathbf{28}}{\mathbf{25}} \frac{\mathbf{36}}{\mathbf{39}} \frac{\mathbf{45}}{\mathbf{46}} = \frac{\mathbf{79 \vee 36}}{\mathbf{78 \vee 39}} \frac{\mathbf{28 \vee 45}}{\mathbf{25 \vee 46}} \overset{expand}{=} \frac{[\mathbf{367}]\mathbf{9}}{-[\mathbf{378}]\mathbf{9}} \frac{[\mathbf{248}]\mathbf{5}}{[\mathbf{246}]\mathbf{5}},$$

and (3.5.4) is changed into

$$[\mathbf{248}][\mathbf{367}] + [\mathbf{246}][\mathbf{378}] \overset{factor}{=} \mathbf{37 \vee 68 \vee 24} = 0. \tag{3.5.8}$$

(3.5.6) and (3.5.8) are related as follows: Substituting $\mathbf{5 = 46 \vee 28}$, $\mathbf{9 = 36 \vee 78}$ into meet product $\mathbf{27 \vee 59 \vee 34}$, we get

$$\begin{aligned}
& \mathbf{27 \vee 59 \vee 34} \\
&= -[(\mathbf{27 \vee 34})(\mathbf{46 \vee 28})(\mathbf{36 \vee 78})] \\
&= -(\mathbf{27 \vee 46 \vee 28})(\mathbf{34 \vee 36 \vee 78}) + (\mathbf{34 \vee 46 \vee 28})(\mathbf{27 \vee 36 \vee 78}) \\
&= [\mathbf{246}][\mathbf{278}][\mathbf{346}][\mathbf{378}] + [\mathbf{346}][\mathbf{248}][\mathbf{367}][\mathbf{278}] \\
&= [\mathbf{278}][\mathbf{346}] \, \mathbf{37 \vee 68 \vee 24}.
\end{aligned}$$

Conversely, substituting $\mathbf{6 = 45 \vee 39}$, $\mathbf{8 = 25 \vee 79}$ into meet product $\mathbf{37 \vee 68 \vee 24}$, we get $\mathbf{27 \vee 59 \vee 34}$ up to two bracket factors.

Completion strategy 1. To reduce the degree of the completion of a rational invariant, multiply by the meet product pairs of lines having explicit intersections.

Example 3.25. Let $\mathbf{1', 2', 3'}$ be points collinear with sides $\mathbf{23, 13, 12}$ of triangle $\mathbf{123}$ respectively.

(1) [Ceva's Theorem and its converse] Lines $\mathbf{11', 22', 33'}$ concur if and only if

$$\frac{\mathbf{1'2}}{\mathbf{31'}} \frac{\mathbf{2'3}}{\mathbf{12'}} \frac{\mathbf{3'1}}{\mathbf{23'}} = 1. \tag{3.5.9}$$

(2) [Menelaus' Theorem and its converse] Points $\mathbf{1', 2', 3'}$ are collinear if and only if

$$\frac{\mathbf{1'2}}{\mathbf{31'}} \frac{\mathbf{2'3}}{\mathbf{12'}} \frac{\mathbf{3'1}}{\mathbf{23'}} = -1. \tag{3.5.10}$$

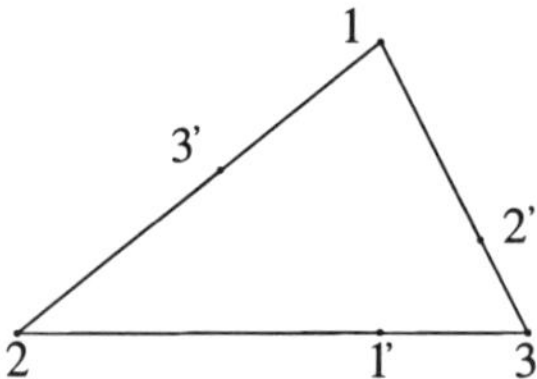

Fig. 3.5   Ceva's Theorem and Menelaus' Theorem.

Proof.   (1) Denote the left side of (3.5.9) by p. Since the degree of the numerator of p is three, a natural antisymmetrization is

$$p = \frac{\mathbf{1'2} \vee \mathbf{2'3} \vee \mathbf{3'1}}{\mathbf{31'} \vee \mathbf{12'} \vee \mathbf{23'}}. \tag{3.5.11}$$

The following expansion of (3.5.11) leads to Ceva's Theorem:

$$\frac{(\mathbf{1'2} \vee \mathbf{2'3}) \vee \mathbf{3'1}}{(\mathbf{23'} \vee \mathbf{31'}) \vee \mathbf{12'}} = \frac{[\mathbf{21'2'}][\mathbf{133'}]}{[\mathbf{31'3'}][\mathbf{122'}]}.$$

It changes (3.5.9) into

$$[\mathbf{21'2'}][\mathbf{133'}] - [\mathbf{122'}][\mathbf{31'3'}] = \mathbf{11'} \vee \mathbf{22'} \vee \mathbf{33'} = 0.$$

(2) The following expansion of (3.5.11) leads to Menelaus' Theorem:

$$\frac{(\mathbf{1'2} \vee \mathbf{2'3}) \vee \mathbf{3'1}}{(\mathbf{31'} \vee \mathbf{12'}) \vee \mathbf{23'}} = -\frac{[\mathbf{21'2'}][\mathbf{133'}]}{[\mathbf{11'2'}][\mathbf{233'}]}.$$

It changes (3.5.10) into

$$[\mathbf{21'2'}][\mathbf{133'}] - [\mathbf{11'2'}][\mathbf{233'}] = -\mathbf{12} \vee \mathbf{1'2'} \vee \mathbf{33'} = -[\mathbf{123}][\mathbf{1'2'3'}] = 0.$$

□

Remark: (3.5.11) has nine different Cayley expansions, leading to nine different results of the form s/t, where s, t are degree-2 bracket monomials. So there are eighteen expressions of the form $s \pm t$. Half of them are not Cayley factorizable. Among the other half, six have factorization $[\mathbf{123}][\mathbf{1'2'3'}]$, leading to Menelaus' Theorem; the other three have factorization $\mathbf{11'} \vee \mathbf{22'} \vee \mathbf{33'}$, leading to Ceva's Theorem.

By concurrency transformations, all the eighteen expressions of the form $s \pm t$ can be changed into Cayley factorizable expressions, see Example 3.11. Concurrency transformations make the derivation of the simplest completion robust against different Cayley expansions.

Completion strategy 2. To make the derivation of the simplest completion robust against different Cayley expansions and antisymmetrizations, use collinearity and concurrency transformations.

Example 3.26. [Menelaus' Theorem for quadrilateral] A line cuts the four sides $\mathbf{12}, \mathbf{23}, \mathbf{34}, \mathbf{41}$ of quadrilateral $\mathbf{1234}$ at points $\mathbf{1'}, \mathbf{2'}, \mathbf{3'}, \mathbf{4'}$ respectively. Then

$$\frac{\mathbf{11'}}{\mathbf{21'}} \frac{\mathbf{22'}}{\mathbf{32'}} \frac{\mathbf{33'}}{\mathbf{43'}} \frac{\mathbf{44'}}{\mathbf{14'}} = 1. \tag{3.5.12}$$

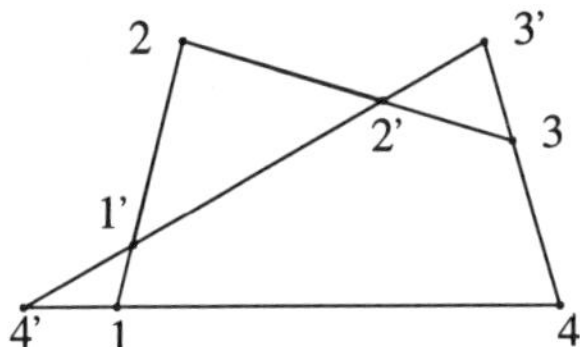

Fig. 3.6 Menelaus' Theorem for quadrilateral.

Proof.   We remove the collinearity of points $\mathbf{1}', \mathbf{2}', \mathbf{3}', \mathbf{4}'$, and see how the conclusion depends on the removed hypothesis. It turns out that this hypothesis can be replaced by a more general one: lines $\mathbf{13}, \mathbf{1}'\mathbf{2}', \mathbf{3}'\mathbf{4}'$ are concurrent. It is a sufficient and necessary condition for the conclusion to be true.

The left side of (3.5.12), denoted by p, has four explicit intersections:

$$\mathbf{1} = \mathbf{121}' \cap \mathbf{144}', \quad \mathbf{2} = \mathbf{121}' \cap \mathbf{232}', \quad \mathbf{3} = \mathbf{232}' \cap \mathbf{343}', \quad \mathbf{4} = \mathbf{144}' \cap \mathbf{343}'.$$

Using explicit intersections $\mathbf{2}, \mathbf{4}$,

$$p = \frac{\mathbf{11}' \vee \mathbf{22}'}{\mathbf{21}' \vee \mathbf{32}'} \frac{\mathbf{33}' \vee \mathbf{44}'}{\mathbf{43}' \vee \mathbf{14}'} = \frac{[\mathbf{11}'\mathbf{2}']}{[\mathbf{31}'\mathbf{2}']} \frac{[\mathbf{33}'\mathbf{4}']}{[\mathbf{13}'\mathbf{4}']},$$

so (3.5.12) can be written as

$$[\mathbf{11}'\mathbf{2}'][\mathbf{33}'\mathbf{4}'] - [\mathbf{31}'\mathbf{2}'][\mathbf{13}'\mathbf{4}'] = \mathbf{13} \vee \mathbf{1}'\mathbf{2}' \vee \mathbf{3}'\mathbf{4}' = 0. \tag{3.5.13}$$

Alternatively, using explicit intersections $\mathbf{1}, \mathbf{3}$,

$$p = \frac{\mathbf{11}' \vee \mathbf{44}'}{\mathbf{21}' \vee \mathbf{14}'} \frac{\mathbf{22}' \vee \mathbf{33}'}{\mathbf{32}' \vee \mathbf{43}'} = \frac{[\mathbf{41}'\mathbf{4}']}{[\mathbf{21}'\mathbf{4}']} \frac{[\mathbf{22}'\mathbf{3}']}{[\mathbf{42}'\mathbf{3}']},$$

so (3.5.12) can be written as

$$[\mathbf{22}'\mathbf{3}'][\mathbf{41}'\mathbf{4}'] - [\mathbf{21}'\mathbf{4}'][\mathbf{42}'\mathbf{3}'] = -\mathbf{24} \vee \mathbf{1}'\mathbf{4}' \vee \mathbf{2}'\mathbf{3}' = 0. \tag{3.5.14}$$

Its equivalence with (3.5.13) can be proved as in Example 3.23. $\square$

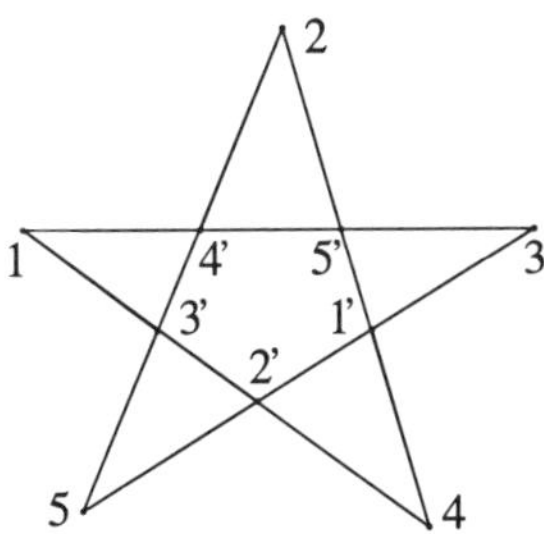

Fig. 3.7 Five-star in Example 3.27.

Example 3.27. Let there be a five-star with vertices $\mathbf{1}, \mathbf{2}, \mathbf{3}, \mathbf{4}, \mathbf{5}$ and concave points $\mathbf{1}', \mathbf{2}', \mathbf{3}', \mathbf{4}', \mathbf{5}'$, as shown in Figure 3.7. Then

$$\frac{\mathbf{43}'}{\mathbf{12}'} \frac{\mathbf{54}'}{\mathbf{23}'} \frac{\mathbf{15}'}{\mathbf{34}'} \frac{\mathbf{21}'}{\mathbf{45}'} \frac{\mathbf{32}'}{\mathbf{51}'} = -1. \tag{3.5.15}$$

Analysis: Denote the left side of (3.5.15) by p. The explicit intersections in p are

$$1 = 134'5' \cap 142'3', \quad 2 = 241'5' \cap 253'4', \quad 3 = 134'5' \cap 351'2',$$
$$4 = 142'3' \cap 241'5', \quad 5 = 253'4' \cap 351'2'.$$

So any antisymmetrization of p leads to a ratio of two bracket monomials of degree three. The difference lies in the number of common factors of the two bracket monomials. The more the number of common bracket factors, the lower the degree of the completion result.

We check how a common bracket factor is produced. In the numerator of p, the meet product $43' \vee 54'$ of the first two bivectors produces a bracket by monomial expansion: $[454']$. This bracket cannot be reproduced by any Cayley expression of the denominator of p, as points $4, 5, 4'$ are scattered in three different bivectors. On the contrary, the meet product $43' \vee 15'$ of the first and the third bivectors in the numerator produces a bracket $[43'5']$ by monomial expansion. This bracket can be reproduced in the denominator by the meet product $23' \vee 45' = [43'5']2$.

First proof. Produce one common bracket factor, add a ratio of two identical vectors to produce two other common bracket factors.

$$p = \frac{43' \vee 15'}{12' \vee 34'} \frac{54' \vee 21'}{23' \vee 45'} \frac{32'}{51'} = -\frac{[43'5']}{[32'4']} \frac{[51'4']}{[43'5']} \frac{32'}{51'} \frac{4'}{4'} = -\frac{[51'4']}{[32'4']} \frac{[32'4']}{[51'4']} = -1.$$

□

Second proof. Produce two (and hence three) common bracket factors. In the denominator of p, the meet product $12' \vee 34'$ of the first two bivectors produces a bracket $[32'4']$. This bracket can be reproduced in the numerator by the meet product $54' \vee 32' = -[32'4']5$.

$$p = -\frac{43' \vee 15'}{12' \vee 34'} \frac{(54' \vee 32') \vee 21'}{(23' \vee 45') \vee 51'} = -\frac{[43'5']}{[32'4']} \frac{[32'4'][251']}{[43'5'][251']} = -1.$$

□

Completion strategy 3. In the completion of a rational invariant, commute stepwise between the numerator and the denominator of the rational invariant, in order to reproduce at each step the bracket factor that is produced in the previous step.

Despite the above strategies, concurrency transformations can make the derivation of the simplest completion robust against different antisymmetrizations. For example, let p be the left side of (3.5.15), then by concurrency transformations, any antisymmetrization of p can be simplified to -1, no matter what the degree of the antisymmetrization result is.

A typical antisymmetrization of p is

$$\frac{43' \vee 54' \vee 15'}{12' \vee 23' \vee 34'} \frac{21' \vee 32'}{45' \vee 51'}, \tag{3.5.16}$$

where the two long meet products each have three bracket monomial expansions:

$$43' \vee 54' \vee 15' = -[13'5'][454'] = -[155'][43'4'] = -[154'][43'5'],$$
$$12' \vee 23' \vee 34' = -[123'][32'4'] = -[122'][33'4'] = [233'][12'4']. \tag{3.5.17}$$

If choosing arbitrarily among the above monomial expansions, say choosing in (3.5.17) the first expansion of each meet-product expression, then (3.5.16) becomes

$$\frac{[13'5'][454']}{[123'][32'4']}\frac{[232']}{[455']}. \tag{3.5.18}$$

The numerator of (3.5.18) has three concurrency transformations:

$$[13'5'][454'] = -[155'][43'4'] = -[154'][43'5'],$$
$$[454'][232'] = [245][32'4'].$$

The last transformation produces a bracket $[32'4']$ common to the denominator of (3.5.18), changing (3.5.18) into

$$\frac{[13'5'][245]}{[123'][455']}. \tag{3.5.19}$$

The two brackets in the denominator of (3.5.19) have two concurrency transformations:

$$[123'][455'] = [125][43'5'] = -[13'5'][245]. \tag{3.5.20}$$

The last transformation changes (3.5.19) to -1.

3.5.2 *Symmetrization of rational invariants*

Deficit brackets and deficit meet products occur naturally in algebraic representations of low-dimensional geometric constraints. For example, In $(n-1)$D projective space $\mathcal{V}^n$, points $\mathbf{a}, \mathbf{b}, \mathbf{c}$ are collinear if and only if $\mathbf{a}$ is on hyperplane $\mathbf{U}_{n-3}\mathbf{bc}$, where $\mathbf{U}_{n-3}$ is a generic $(n-3)$-blade representing a generic $(n-4)$D projective plane:

$$[\mathbf{U}_{n-3}\mathbf{abc}] = [\mathbf{abc}] = 0. \tag{3.5.21}$$

Here $[\mathbf{abc}]$ is a deficit bracket.

As another example, if lines $\mathbf{12}$ and $\mathbf{1'2'}$ intersect at point $\mathbf{a}$ in the $(n-1)$D projective space, then $\mathbf{a}$ is where hyperplane $\mathbf{U}_{n-3}\mathbf{12}$ and line $\mathbf{1'2'}$ meet, so

$$\mathbf{a} = \mathbf{U}_{n-3}\mathbf{12} \vee \mathbf{1'2'} = \mathbf{12} \vee \mathbf{1'2'}, \tag{3.5.22}$$

where the latter is an $(n-3)$-deficit meet product.

In a projective subspace, rational invariants are invariants, and so can be used to represent geometric constraints among low dimensional objects. In the whole projective space, such a representation takes the form of deficit brackets and deficit meet products. Since the two deficit algebraic operators are both scalar-valued and thus commutative, transforming a rational invariant into a ratio of deficit brackets and deficit meet products is called a partial *symmetrization* of the rational invariant.

In the case of 2D projective geometry, let $\mathbf{u}_1, \mathbf{u}_2$ be generic points outside lines $\mathbf{12}, \mathbf{34}$ respectively. By (3.5.1), for collinear points $\mathbf{1}, \mathbf{2}, \mathbf{1'}, \mathbf{2'}$, ratio $\mathbf{12}/\mathbf{1'2'} = [\mathbf{12u_1}]/[\mathbf{1'2'u_1}]$ is independent of the vector variable $\mathbf{u}_1$. Similarly, for collinear points $\mathbf{3}, \mathbf{4}, \mathbf{3'}, \mathbf{4'}$, ratio $\mathbf{34}/\mathbf{3'4'} = [\mathbf{34u_2}]/[\mathbf{3'4'u_2}]$ is independent of the vector variable $\mathbf{u}_2$. So

$$\frac{\mathbf{12}}{\mathbf{1'2'}}\frac{\mathbf{34}}{\mathbf{3'4'}} = \frac{[\mathbf{12u_1}]}{[\mathbf{1'2'u_1}]}\frac{[\mathbf{34u_2}]}{[\mathbf{3'4'u_2}]} = \frac{[\mathbf{12u_2}]}{[\mathbf{1'2'u_2}]}\frac{[\mathbf{34u_1}]}{[\mathbf{3'4'u_1}]}. \tag{3.5.23}$$

To change $\mathbf{u}_1, \mathbf{u}_2$ into the same dummy vector $\mathbf{U}_1$, we need to make *symmetrization* to the two results in (3.5.23):

$$\begin{aligned}
\frac{\mathbf{12}}{\mathbf{1'2'}}\frac{\mathbf{34}}{\mathbf{3'4'}} &= \frac{[\mathbf{12u_1}][\mathbf{34u_2}] + [\mathbf{12u_2}][\mathbf{34u_1}]}{[\mathbf{1'2'u_1}][\mathbf{3'4'u_2}] + [\mathbf{1'2'u_2}][\mathbf{3'4'u_1}]} \\
&= \frac{[\mathbf{12(u_1+u_2)}][\mathbf{34(u_1+u_2)}] - [\mathbf{12u_1}][\mathbf{34u_1}] - [\mathbf{12u_2}][\mathbf{34u_2}]}{[\mathbf{1'2'(u_1+u_2)}][\mathbf{3'4'(u_1+u_2)}] - [\mathbf{1'2'u_1}][\mathbf{3'4'u_1}] - [\mathbf{1'2'u_2}][\mathbf{3'4'u_2}]} \\
&= \frac{[\mathbf{U_1 12}][\mathbf{U_1 34}]}{[\mathbf{U_1 1'2'}][\mathbf{U_1 3'4'}]}.
\end{aligned}$$

$$\tag{3.5.24}$$

where the last step is based on the property that if $a_1 : b_1 = a_2 : b_2 = a_3 : b_3$ for scalars a_i, b_j, then $(a_1 + a_2 + a_3) : (b_1 + b_2 + b_3) = a_1 : b_1$.

In comparison, the antisymmetrization of the two results in (3.5.23) yields

$$\begin{aligned}
\frac{\mathbf{12}}{\mathbf{1'2'}}\frac{\mathbf{34}}{\mathbf{3'4'}} &= \frac{[\mathbf{12u_2}][\mathbf{34u_1}] - [\mathbf{12u_1}][\mathbf{34u_2}]}{[\mathbf{1'2'u_2}][\mathbf{3'4'u_1}] - [\mathbf{1'2'u_1}][\mathbf{3'4'u_2}]} \\
&= \frac{\mathbf{12} \vee \mathbf{34} \vee \mathbf{u_1 u_2}}{\mathbf{1'2'} \vee \mathbf{3'4'} \vee \mathbf{u_1 u_2}} = \frac{\mathbf{12} \vee \mathbf{34}}{\mathbf{1'2'} \vee \mathbf{3'4'}}.
\end{aligned}$$

$$\tag{3.5.25}$$

By (3.5.24), the symmetrization of a rational invariant can be done by simply appending the same dummy vector to each bivector in the invariant ratios of the rational invariant. The difficulty lies in the geometric part of the symmetrization: explaining the symmetrization result in high-dimensional projective geometry.

Proposition 3.28. By means of deficit brackets and $(n-3)$-deficit meet products, the incidence relations among points and lines in $(n-1)$D projective space have the same algebraic representations as in the projective plane.

Proof. We only prove that any three lines $\mathbf{11'}, \mathbf{22'}, \mathbf{33'}$, no two of which are collinear, concur in the $(n-1)$D projective space $\mathcal{V}^n$ if and only if for a generic $(n-3)$-blade $\mathbf{U}_{n-3} \in \Lambda(\mathcal{V}^n)$, hyperplanes $\mathbf{U}_{n-3}\mathbf{11'}$, $\mathbf{U}_{n-3}\mathbf{22'}$ and line $\mathbf{33'}$ concur, *i.e.*,

$$\mathbf{U}_{n-3}\mathbf{11'} \vee \mathbf{U}_{n-3}\mathbf{22'} \vee \mathbf{33'} = \mathbf{11'} \vee \mathbf{22'} \vee \mathbf{33'} = 0. \tag{3.5.26}$$

If any two of the three lines, say $\mathbf{22'}, \mathbf{33'}$, are coplanar, then $\mathbf{a} = \mathbf{U}_{n-3}\mathbf{22'} \vee \mathbf{33'}$ represents the point of intersection of the two lines. The representation is independent of $\mathbf{U}_{n-3}$ in the sense that the vector changes only by scale for different $\mathbf{U}_{n-3}$'s. So $\mathbf{U}_{n-3}\mathbf{11'} \vee \mathbf{a} = [\mathbf{11'a}] = 0$ if and only if points $\mathbf{1}, \mathbf{1'}, \mathbf{a}$ are collinear.

If no two of the three lines are coplanar, the dimension d of the vector space spanned by vectors $\mathbf{1}, \mathbf{1}', \mathbf{2}, \mathbf{2}', \mathbf{3}, \mathbf{3}'$ ranges from 4 to 6. Without loss of generality, assume $n = d$.

(1) If $d = 6$, then the six vectors are all basis vectors. For any generic 3-blade $\mathbf{U} \in \Lambda(\mathcal{V}^6)$,

$$\mathbf{U11}' \vee \mathbf{U22}' \vee \mathbf{33}' = [\mathbf{U22}'\mathbf{3}'][\mathbf{U11}'\mathbf{3}] - [\mathbf{U22}'\mathbf{3}][\mathbf{U11}'\mathbf{3}']. \tag{3.5.27}$$

The right side of (3.5.27) is a polynomial of 4 different coordinate components of $\mathbf{U}$, so the left side cannot be zero for generic $\mathbf{U}$.

(2) If $d = 5$, suppose $\mathbf{1}', \mathbf{2}, \mathbf{2}', \mathbf{3}, \mathbf{3}'$ form a basis of $\mathcal{V}^5$. For any generic 2-blade $\mathbf{U} \in \Lambda(\mathcal{V}^5)$, if (3.5.27) equals zero, then the coordinate component $[\mathbf{U22}'\mathbf{3}']$ of $\mathbf{U}$ is a factor of $[\mathbf{U11}'\mathbf{3}']$, but this is impossible because by Cramer's rule

$$[\mathbf{1}'\mathbf{22}'\mathbf{33}']\mathbf{1} = [\mathbf{122}'\mathbf{33}']\mathbf{1}' - [\mathbf{11}'\mathbf{2}'\mathbf{33}']\mathbf{2} + [\mathbf{11}'\mathbf{233}']\mathbf{2}' - [\mathbf{11}'\mathbf{22}'\mathbf{3}']\mathbf{3} + [\mathbf{11}'\mathbf{22}'\mathbf{3}]\mathbf{3}',$$

the right side of (3.5.27) is changed into a rational polynomial that does not have the coordinate variable $[\mathbf{U22}'\mathbf{3}']$:

$$[\mathbf{U22}'\mathbf{3}'][\mathbf{U11}'\mathbf{3}] - \frac{[\mathbf{U22}'\mathbf{3}]}{[\mathbf{1}'\mathbf{22}'\mathbf{33}']}([\mathbf{11}'\mathbf{2}'\mathbf{33}'][\mathbf{U1}'\mathbf{23}'] - [\mathbf{11}'\mathbf{233}'][\mathbf{U1}'\mathbf{2}'\mathbf{3}']$$
$$+ [\mathbf{11}'\mathbf{22}'\mathbf{3}'][\mathbf{U1}'\mathbf{33}']).$$

(3) If $d = 4$, suppose that $\mathbf{2}, \mathbf{2}', \mathbf{3}, \mathbf{3}'$ form a basis of $\mathcal{V}^4$. For any generic vector $\mathbf{U} \in \mathcal{V}^4$, if (3.5.27) equals zero, then $[\mathbf{U22}'\mathbf{3}']$ is a factor of $[\mathbf{U11}'\mathbf{3}']$ when $\mathbf{11}'$ is written as a linear combination of bivector basis $\mathbf{22}', \mathbf{23}, \mathbf{23}', \mathbf{2}'\mathbf{3}, \mathbf{2}'\mathbf{3}', \mathbf{33}'$. So the coordinate components of $\mathbf{11}'$ with respect to $\mathbf{23}, \mathbf{2}'\mathbf{3}$ are both zero. Similarly, if (3.5.27) equals zero, then $[\mathbf{U22}'\mathbf{3}]$ is a factor of $[\mathbf{U11}'\mathbf{3}]$, and the coordinate components of $\mathbf{11}'$ with respect to $\mathbf{23}', \mathbf{2}'\mathbf{3}'$ are both zero. So $\mathbf{11}'$ is a linear combination of $\mathbf{22}', \mathbf{33}'$. Since $\mathbf{11}'$ is a blade, from

$$0 = (\mathbf{11}')(\mathbf{11}') = (\lambda\mathbf{22}' + \mu\mathbf{33}')(\lambda\mathbf{22}' + \mu\mathbf{33}') = 2\lambda\mu\,\mathbf{22}'\mathbf{33}',$$

we get $\lambda\mu = 0$, so $\mathbf{11}'$ must be in the 1D space spanned by $\mathbf{22}'$ or $\mathbf{33}'$, *i.e.*, either $\mathbf{1}, \mathbf{1}', \mathbf{2}, \mathbf{2}'$ or $\mathbf{1}, \mathbf{1}', \mathbf{3}, \mathbf{3}'$ are collinear. Both are against the noncoplanarity assumption of any two of the three lines $\mathbf{11}', \mathbf{22}', \mathbf{33}'$. $\qquad\square$

Example 3.29. Example 3.23 is valid in 3D, where the four points $\mathbf{5}, \mathbf{6}, \mathbf{8}, \mathbf{9}$ need not be coplanar. Similarly, Example 3.26 is valid in 3D, where the four points $\mathbf{1}', \mathbf{2}', \mathbf{3}', \mathbf{4}'$ are not only noncollinear, they are even noncoplanar. However, any lift of the geometric configuration of Example 3.6 or 3.25 to nD space lies in a 2D plane, because the configuration is completely determined by a triangle, which has only 2D structure.

To extend a 2D projective incidence theorem to nD, the key is to explain in nD projective geometry the components of the Cayley algebraic identity representing the theorem. The translation from the Cayley bracket algebra of deficit brackets and deficit meet products to nD projective incidence geometry, is related to the high dimensional projective reconstruction of an nD incidence geometric configuration from the 2D data obtained from it by a sequence of 1D perspective projections [122].

Example 3.30. Find the condition for the configuration of Figure 3.8, drawn as a 2D wireframe, to be realizable in space as a solid figure by the reverse of a general perspective projection, such that the five planes **123**, **1'2'3'**, **11'22'**, **11'33'**, **22'33'** are pairwise noncoplanar.

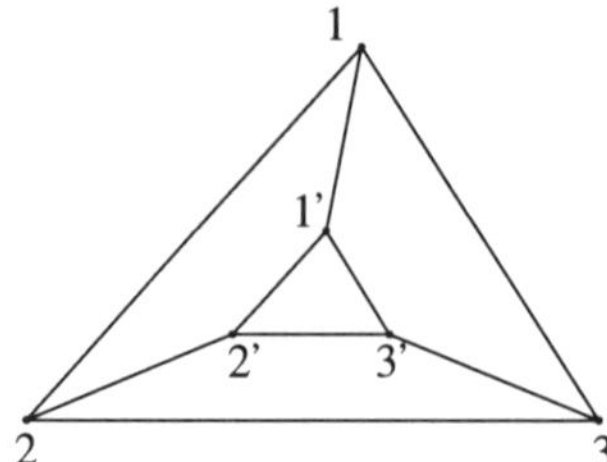

Fig. 3.8   3D realizability of a truncated pyramid.

The 3D configuration can be constructed as follows:

Free points: **1, 2, 3, 1'**.
Free coplanar point: **2'** on plane **11'2**, *i.e.*, $[\mathbf{11'22'}] = 0$.
Free collinear point: **3'** on line **311'** ∩ **322'**, *i.e.*, $\mathbf{3'(311'} \vee \mathbf{322')} = 0$.

By a perspective projection onto the plane, the first constraint $[\mathbf{11'22'}] = 0$ is changed into a trivial equality, because the outer product of any four points in the projective plane is always zero.

The second constraint can be written as

$$\mathbf{33'(11'} \vee_3 \mathbf{22')} = 0. \tag{3.5.28}$$

Since points **1, 1', 2, 2'** are coplanar in space, the reduced meet product $\mathbf{11'} \vee_3 \mathbf{22'}$ is up to scale independent of the choice of **3**, and can be replaced by $\mathbf{11'} \vee_{\mathbf{U}_1} \mathbf{22'}$. Then (3.5.28) becomes

$$\mathbf{U_1 33'(11'} \vee_{\mathbf{U}_1} \mathbf{22')} = \mathbf{U_1 33'} \vee \mathbf{U_1 11'} \vee \mathbf{22'} = \mathbf{33'} \vee \mathbf{11'} \vee \mathbf{22'} = 0. \tag{3.5.29}$$

This equality is changed into a nontrivial 2D condition by perspective projection: the images of lines **33', 11', 22'** in the image plane concur. It is the sufficient and necessary condition for the configuration of Figure 3.8 to be realizable in space as a truncated pyramid [178].

3.6   Automated theorem proving

The *characteristic set* method [201] of automated geometric theorem proving is based on triangulating a set of polynomials. Given a set S of polynomials in free parameters $u_1, u_2, \ldots, u_p$ and indeterminates $x_1 \prec x_2 \prec \ldots \prec x_q$, where $u_k \prec x_1$ for all $1 \leq k \leq p$, a *triangulation* of S is a procedure of changing S into the following *triangular form*:

$$\begin{aligned}
&\{f_{0i}(u_1, u_2, \ldots, u_p) \mid i \in I\}, \\
&f_1(x_1, u_1, u_2, \ldots, u_p), \\
&f_2(x_1, x_2, u_1, u_2, \ldots, u_p), \\
&\qquad\qquad\vdots \\
&f_q(x_1, x_2, \ldots, x_q, u_1, u_2, \ldots, u_p),
\end{aligned} \qquad (3.6.1)$$

where I is an index set, the f's are polynomials, and for any $1 \leq j \leq q$, the leading variable of f_j is x_j.

Pseudodivision is the most important technique in triangulation. Under some inequality conditions called *nondegeneracy conditions*, the original set of polynomials is equivalent to the new set of polynomials (3.6.1) when only their zero sets are taken into account, and the verification of a conclusion in polynomial form based on the set of hypotheses S, can be made by reducing the conclusion polynomial to zero with (3.6.1).

In the setting of Grassmann-Cayley algebra, the triangular form needs to be revised as follows:

Definition 3.31. Given a set S of polynomials in GC algebra $\Lambda(\mathcal{V}^n)$, with free atomic vectors $\mathbf{u}_1, \mathbf{u}_2, \ldots, \mathbf{u}_p$ and vector indeterminates $\mathbf{x}_1 \prec \mathbf{x}_2 \prec \ldots \prec \mathbf{x}_q$, where $\mathbf{u}_k \prec \mathbf{x}_1$ for all $1 \leq k \leq p$, a *triangulation* of S is a procedure of changing S into the following *triangular form*:

$$\begin{aligned}
&\{f_{0i}(\mathbf{u}_1, \mathbf{u}_2, \ldots, \mathbf{u}_p) \mid i \in I_0\}, \\
&\{f_{1i}(\mathbf{x}_1, \mathbf{u}_1, \mathbf{u}_2, \ldots, \mathbf{u}_p) \mid i \in I_1\}, \\
&\{f_{2i}(\mathbf{x}_1, \mathbf{x}_2, \mathbf{u}_1, \mathbf{u}_2, \ldots, \mathbf{u}_p) \mid i \in I_2\}, \\
&\qquad\qquad\vdots \\
&\{f_{qi}(\mathbf{x}_1, \mathbf{x}_2, \ldots, \mathbf{x}_q, \mathbf{u}_1, \mathbf{u}_2, \ldots, \mathbf{u}_p) \mid i \in I_q\},
\end{aligned} \qquad (3.6.2)$$

where the I_j are index sets, the f_{ji} are GC polynomials, such that for any $1 \leq j \leq q$,

- the leading vector variable of f_{ji} is $\mathbf{x}_j$, for all $i \in I_j$;
- the GC polynomials in $\{f_{ji}(\mathbf{x}_1, \mathbf{x}_2, \ldots, \mathbf{x}_j, \mathbf{u}_1, \mathbf{u}_2, \ldots, \mathbf{u}_p) \mid i \in I_j\}$, after being decomposed with respect to a basis of $\Lambda(\mathcal{V}^n)$, are a set of scalar-valued GC polynomials among which at most $n-1$ are algebraically independent modulo the f_{ki} for all $k < j$ and $i \in I_k$.

If under some inequality conditions called *nondegeneracy conditions*, the original set of GC polynomials S is equivalent to the new set of GC polynomials (3.6.2) when only their zero sets are taken into account, the verification of a conclusion in GC polynomial form based on the set of hypotheses S, can be made by reducing the conclusion expression to zero with (3.6.2).

In projective incidence geometry with Cayley bracket algebra, the most important method of triangulation is *multivector equation solving* [103], [107]. As a simple example, let there be four points $\mathbf{1}, \mathbf{2}, \mathbf{3}, \mathbf{4}$ in the plane, and let point $\mathbf{x}$ be collinear with lines $\mathbf{12}$ and $\mathbf{34}$. Then under the nondegeneracy conditions $\mathbf{12} \neq 0$ and $\mathbf{34} \neq 0$, the GC polynomial equations

$$[\mathbf{x12}] = 0, \qquad [\mathbf{x34}] = 0, \tag{3.6.3}$$

can be triangulated into the following covariant form, or equivalently, a form of deficit bracket and deficit meet product:

$$\mathbf{x} = \mathbf{12} \vee \mathbf{34}, \quad i.e., \quad [\mathbf{U}_2\mathbf{x}] = \mathbf{U}_2 \vee \mathbf{12} \vee \mathbf{34}. \tag{3.6.4}$$

Obviously, the most efficient way of solving for $\mathbf{x}$ from (3.6.3) is simply by representing the input geometric constraint $\mathbf{x} = \mathbf{12} \cap \mathbf{34}$ algebraically as (3.6.4). There are others that cannot be so easily solved. They require more advanced multivector equation solving techniques [107].

The result of multivector equation solving in projective incidence geometry is usually a Cayley expression. The verification of a conclusion is usually reduced to eliminating the vector indeterminates from the conclusion expression by substituting them sequentially with their Cayley expressions, and then simplifying the result using techniques like Cayley expansions, factorizations, contractions, *etc.*

There are two main issues in elimination: the order by which to eliminate the vector indeterminates, and the nondegeneracy conditions under which the original geometric constraints are equivalent to the Cayley expression representations of the vector indeterminates. In this section, we first investigate the two main problems, then present an algorithm of automated theorem proving based on Cayley expansions and bracket polynomial simplifications, and use it to prove theorems in both 2D and 3D projective incidence geometries.

3.6.1 *Construction sequence and elimination sequence*

In classical geometry, a geometric configuration is usually determined by a *sequence of geometric constructions*, and each construction determines a point of the configuration. Example 3.1 at the beginning of this chapter is a typical example. The construction sequence is usually composed of three parts: the first part is the set of free points in the geometric space; the second part is the set of *semifree points*, *i.e.*, non-free points having nonzero degree of freedom, *e.g.*, free collinear points in 2D projective geometry, free coplanar points in 3D geometry, *etc.*; the third part is the set of *constrained points*, *i.e.*, points with no degree of freedom.

The *parents* and *children* of a constructed point are respectively those points used directly in its construction and those points constructed directly with it. A constructed point usually has more than one parent. A construction without any child is called an *end*. The parents-children relations among the constructions form a diagram, called the *parents-children diagram*. The *ancestors* and *descendents* of a constructed point can be defined recursively by the diagram.

The construction sequence defines a *total order* among the constructed points. However, constructions are generally *in batch* instead of one by one. If several points can be constructed independently and in parallel, they are said to be in the same *construction batch*. In Example 3.1, free points **1, 2, 3, d** are constructed in the same batch. The order among the four points are arbitrary, and in our opinion, totally unnecessary for bracket computing without resorting to either coordinatization or straightening. The construction sequence in fact defines a total order among the batches of constructed points.

In theorem proving by Cayley bracket algebra, the elimination of the constructions in batch, called *batch elimination*, is needed to control middle expression swell. The most often used order in batch elimination is the reverse of the order of construction batches. However, this order is not necessarily the most appropriate for the best performance of batch elimination, as to be shown in Example 3.36 of this section.

We use an example to demonstrate a *dynamic order of batch elimination*, which is generally not the reverse of the order of construction batches.

Example 3.32. [Nehring's Theorem] Let **18, 27, 36** be three lines in triangle **123** concurrent at point **4**, and let point **5** be on line **12**. Let **9 = 13 ∩ 58, 0 = 23 ∩ 69, a = 12 ∩ 70, b = 13 ∩ 8a, c = 23 ∩ 6b**. Then points **5, 7, c** are collinear.

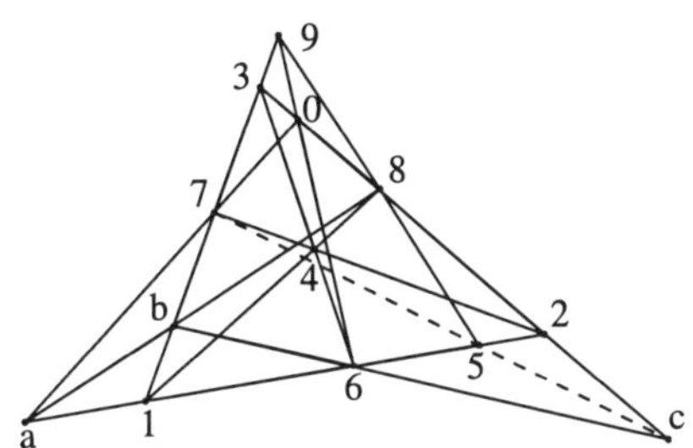

Fig. 3.9　Nehring's Theorem.

Construction sequence:

Free points:　**1, 2, 3, 4.**
Free collinear point: **5** on line **12**.
Intersections:

$$6 = 12 \cap 34, \quad 7 = 13 \cap 24, \quad 8 = 14 \cap 23, \quad 9 = 13 \cap 58,$$
$$0 = 23 \cap 69, \quad a = 12 \cap 70, \quad b = 13 \cap 8a, \quad c = 23 \cap 6b.$$

Conclusion: **5, 7, c** are collinear.

The construction sequence defines the following order of construction batches:

$$1, 2, 3, 4 \prec 5 \prec 6, 7, 8 \prec 9 \prec 0 \prec a \prec b \prec c. \tag{3.6.5}$$

The order is unable to reflect the following parents-children structure of the constructions:

$$1, 2, 3, 4 \longrightarrow 5 \longrightarrow \begin{cases} 6, 7 \\ 8 \longrightarrow 9 \xrightarrow{+6} 0 \xrightarrow{+7} a \longrightarrow b \xrightarrow{+6} c, \end{cases} \tag{3.6.6}$$

where $9 \xrightarrow{+6} 0$ means that the construction of **0** requires, besides point **9**, another point **6** that is constructed in parallel with point **8** in another branch.

The *dynamic order of batch elimination* should be

$$\mathbf{c} \prec \mathbf{b} \prec \mathbf{a} \prec \mathbf{7, 0} \prec \mathbf{6, 9} \prec \mathbf{8} \prec \mathbf{5} \prec \mathbf{1, 2, 3, 4}. \tag{3.6.7}$$

The reason is as follows: First, **c** is an end of the whole constructions, so the elimination starts from **c**. After the elimination, **c** disappears from the parents-children diagram (3.6.6), and **b** becomes the new end. After eliminating **b** and then **a**, we find that both **7** and **0** become the new ends. They can be eliminated at the same time. After the eliminations of **7** and **0**, the new ends are **6** and **9**. Hence we get (3.6.7).

In elimination, if a vector indeterminate representing a construction to be eliminated from an expression, no longer has any descendent in the expression, then it is called an *end* of the expression. A batch elimination is always carried out to the ends of the expression, and after each batch elimination, those points having been eliminated from the expression are removed from the parents-children diagram. Such a diagram is called a *dynamic parents-children diagram*, and the order of batch elimination guided by the ends of the expression is called a *dynamic order*.

To maintain the dynamic parents-children diagram is very easy: before a batch elimination, we only need to scan the set of constructed points in search of those without children; after the batch elimination, we only need to update the children information for the parents of the eliminated points. There is no tracing beyond two generations for grandparents or grandchildren.

The dynamic order of batch elimination is generally more efficient than the reverse of the order of construction batches, because Cayley expansion is more efficient in the former order of elimination. There are exceptions. For Nehring's Theorem, whose proof immediately follows this paragraph, the efficiency is much the same for both orders.

Proof of Example 3.32.

<table>
<tr><td>Rules</td><td>$[57\mathbf{c}]$</td></tr>
<tr><td></td><td>$\overset{\mathbf{c}}{=} \ \mathbf{57} \vee \mathbf{23} \vee \mathbf{6b}$</td></tr>
</table>

$$\boxed{\begin{aligned} &\mathbf{57 \vee 23 \vee 6b} \\ &= [(\mathbf{57} \vee \mathbf{23})\mathbf{6}(\mathbf{13} \vee \mathbf{8a})] \\ &= -[\mathbf{235}]\mathbf{67} \vee \mathbf{13} \vee \mathbf{8a} - [\mathbf{237}]\mathbf{56} \vee \mathbf{13} \vee \mathbf{8a} \\ &= -[\mathbf{235}][\mathbf{136}][\mathbf{78a}] - [\mathbf{237}][\mathbf{13a}][\mathbf{568}] \end{aligned}}$$

$$\overset{\mathbf{b}}{=} \ -[\mathbf{136}][\mathbf{235}][\mathbf{78a}] - [\mathbf{13a}][\mathbf{237}][\mathbf{568}]$$

$$\begin{array}{ll} [78a] = -[127][780] \\ [13a] = -[123][170] \end{array} \qquad \overset{a}{=} \ [127][136][235][780]+[123][170][237][568]$$

$$\begin{array}{ll} [780] = -[237][689] \\ [170] = [127][369] \end{array} \qquad \overset{0}{=} \ \underbrace{[127][237]}(-[136][235][689] \\ \phantom{\overset{0}{=}} \qquad\qquad\qquad + [123][369][568])$$

$$\begin{array}{ll} [689] = [138][568] \\ [369] = -[136][358] \end{array} \qquad \overset{9}{=} \ \underbrace{[136][568]}(-[138][235] - [123][358])$$

$$\overset{contract}{=} 0.$$

The brackets with underbraces in the proof are explicit common bracket factors that occur at each step, and are removed once they are detected. They are moved to a set containing all common bracket and p_I-typed meet-product factors of the conclusion expression. Generally no elements of the set can be zero, and by removing them from the main procedure the proof can be significantly simplified. If by chance, one element of the set is exactly the only factor nullifying the conclusion expression, then after the main procedure finishes, the removed factors will be further computed.

If by the dynamic order of batch eliminations (3.6.7), $\mathbf{7}, \mathbf{0}$ are eliminated simultaneously, and then $\mathbf{6}, \mathbf{9}$ also simultaneously, the proof is almost the same. In the last step of the above proof, by the GP relation $[\mathbf{138}][\mathbf{235}] + [\mathbf{123}][\mathbf{358}] = [\mathbf{135}][\mathbf{238}]$, since $[\mathbf{238}] = 0$ by collinearity, the program simply evaluates the result to zero. Thus, in the reverse order of the construction batches, the proof finishes even before points $\mathbf{5}, \mathbf{6}, \mathbf{7}, \mathbf{8}$ are eliminated. The disappearance of points $\mathbf{6}, \mathbf{7}$ before their eliminations is caused by the removal of four common bracket factors.

3.6.2 *Geometric constructions and nondegeneracy conditions*

Each geometric construction is associated with one or several inequality constraints as the prerequisite for the existence of the construction, called the *associated nondegeneracy conditions*. For example, if $\mathbf{3}$ is the intersection of lines $\mathbf{12}$ and $\mathbf{1'2'}$, then both lines must exist, and the four points are not collinear, so that the two lines intersect; the associated nondegeneracy condition is $\mathbf{12} \vee \mathbf{1'2'} \neq 0$.

The following is a list of projective incidence constructions in 2D and 3D geometries, together with their associated nondegeneracy conditions:

(1) $\mathbf{x}$ is a free point: no inequality constraint.

(2) $\mathbf{x}$ is a free collinear point on line $\mathbf{12}$: $\mathbf{12} \neq 0$.

(3) $\mathbf{x}$ is a free coplanar point in plane $\mathbf{123}$: $\mathbf{123} \neq 0$.

(4) $\mathbf{x}$ is a free collinear point on line $\mathbf{123} \cap \mathbf{1'2'3'}$: $\mathbf{123} \vee \mathbf{1'2'3'} \neq 0$.

Let $\mathbf{c}$ be a point on line $\mathbf{ab}$. Then $[\mathbf{ab}]\mathbf{c} = [\mathbf{ac}]\mathbf{b} - [\mathbf{bc}]\mathbf{a}$. The *harmonic conjugate*, or simply called *conjugate*, of $\mathbf{c}$ with respect to $\mathbf{a}, \mathbf{b}$, is a point $\mathbf{d}$ on line $\mathbf{ab}$ such that the cross-ratio

$$(\mathbf{ab};\mathbf{cd}) = \frac{[\mathbf{ac}][\mathbf{bd}]}{[\mathbf{ad}][\mathbf{bc}]} = -1. \tag{3.6.8}$$

The conjugate of $\mathbf{c}$ with respect to $\mathbf{a},\mathbf{b}$ has the following expression:

$$\mathrm{conj}_{\mathbf{ab}}(\mathbf{c}) := [\mathbf{ac}]\mathbf{b} + [\mathbf{bc}]\mathbf{a}. \tag{3.6.9}$$

(5) $\mathbf{x}$ is the conjugate of a point $\mathbf{3}$ on line $\mathbf{12}$: $\mathbf{12} \neq 0$.

(6) $\mathbf{x}$ is the point of intersection of two lines $\mathbf{12}, \mathbf{1'2'}$ in the plane: $\mathbf{12} \vee \mathbf{1'2'} \neq 0$.

(7) $\mathbf{x}$ is the point of intersection of line $\mathbf{12}$ and plane $\mathbf{1'2'3'}$ in space: $\mathbf{12}\vee\mathbf{1'2'3'} \neq 0$.

(8) $\mathbf{x}$ is the point of intersection of three planes $\mathbf{123}, \mathbf{1'2'3'}, \mathbf{1''2''3''}$ in space: $\mathbf{123} \vee \mathbf{1'2'3'} \vee \mathbf{1''2''3''} \neq 0$.

In the procedure of proving a geometric theorem, some inequality requirements that are not associated nondegeneracy conditions may occur. They are called the *additional nondegeneracy conditions*. Such inequalities are not needed by the geometric constructions, but are required by the algebraic proof of the theorem.

For example, to eliminate a free collinear point $\mathbf{x}$ on line $\mathbf{12}$, we can use the Cramer's rule $[\mathbf{12}]\mathbf{x} = [\mathbf{1x}]\mathbf{2} - [\mathbf{2x}]\mathbf{1}$. If $\mathbf{3}$ is another free collinear point on line $\mathbf{12}$, and the conclusion expression is relevant to $\mathbf{1},\mathbf{3}$ but irrelevant to $\mathbf{2}$, then using $\mathbf{1},\mathbf{3}$ to represent $\mathbf{x}$ may reduce the number of terms, with the cost of the additional nondegeneracy condition $\mathbf{13} \neq 0$.

If the expression of $\mathbf{x}$ by $\mathbf{1},\mathbf{3}$ is used in the proof of a geometric theorem, then the proof is incomplete without considering the degenerate case $\mathbf{13} = 0$. This does not indicate that the theorem is incorrect if $\mathbf{13} = 0$. The additional nondegeneracy condition demands an additional proof for the degenerate case, only after this supplement is the algebraic proof of the theorem finished.

Proposition 3.33. If $\mathbf{x}$ is a free collinear point on the line of intersection of planes $\mathbf{123}, \mathbf{1'2'3'}$ in space, then in elimination, $\mathbf{x}$ can be replaced by any of the following two expressions:

$$\begin{aligned}
\mathbf{x} &= ([\mathbf{x2}][\mathbf{31'2'3'}] - [\mathbf{x3}][\mathbf{21'2'3'}])\mathbf{1} - ([\mathbf{x1}][\mathbf{31'2'3'}] - [\mathbf{x3}][\mathbf{11'2'3'}])\mathbf{2} \\
&\qquad\qquad\qquad\qquad + ([\mathbf{x1}][\mathbf{21'2'3'}] - [\mathbf{x2}][\mathbf{11'2'3'}])\mathbf{3} \\
&= ([\mathbf{x2'}][\mathbf{1233'}] - [\mathbf{x3'}][\mathbf{1232'}])\mathbf{1'} - ([\mathbf{x1'}][\mathbf{1233'}] - [\mathbf{x3'}][\mathbf{1231'}])\mathbf{2'} \\
&\qquad\qquad\qquad\qquad + ([\mathbf{x1'}][\mathbf{1232'}] - [\mathbf{x2'}][\mathbf{1231'}])\mathbf{3'}.
\end{aligned} \tag{3.6.10}$$

In particular, if $\mathbf{1} = \mathbf{1'}$, then

$$\mathbf{x} = [\mathbf{x1}]\mathbf{23} \vee_1 \mathbf{2'3'} - [\mathbf{x}(\mathbf{23} \vee_1 \mathbf{2'3'})]\mathbf{1} = -[\mathbf{x1}]\mathbf{2'3'} \vee_1 \mathbf{23} + [\mathbf{x}(\mathbf{2'3'} \vee_1 \mathbf{23})]\mathbf{1}. \tag{3.6.11}$$

Proof. For dummy bivector $\mathbf{U}_2$, by $[\mathbf{123x}] = [\mathbf{1'2'3'x}] = 0$,

$$\mathbf{U}_2\mathbf{x} \vee \mathbf{123} \vee \mathbf{1'2'3'}$$

$$\overset{expand}{=} \left\{ \begin{aligned}
&(\mathbf{U}_2 \vee \mathbf{123} \vee \mathbf{1'2'3'})\mathbf{x}, \\
&(\mathbf{U}_2\mathbf{x} \vee \mathbf{123} \vee \mathbf{2'3'})\mathbf{1'} - (\mathbf{U}_2\mathbf{x} \vee \mathbf{123} \vee \mathbf{1'3'})\mathbf{2'} + (\mathbf{U}_2\mathbf{x} \vee \mathbf{123} \vee \mathbf{1'2'})\mathbf{3'}, \\
&-(\mathbf{U}_2\mathbf{x} \vee \mathbf{1'2'3'} \vee \mathbf{23})\mathbf{1} + (\mathbf{U}_2\mathbf{x} \vee \mathbf{1'2'3'} \vee \mathbf{13})\mathbf{2} - (\mathbf{U}_2\mathbf{x} \vee \mathbf{1'2'3'} \vee \mathbf{12})\mathbf{3}.
\end{aligned} \right.$$

Continuing to expand the meet products, we get (3.6.10), where the coefficient of $\mathbf{x}$ should have been $\mathbf{U}_2 \vee \mathbf{123} \vee \mathbf{1'2'3'}$, which is nonzero by the associated nondegeneracy condition of $\mathbf{x}$, and is removed because the representation is homogeneous.

If $\mathbf{1} = \mathbf{1'}$, then points $\mathbf{x}, \mathbf{1}, \mathbf{23} \vee_1 \mathbf{2'3'}$ are collinear, and (3.6.11) is their Cramer's rule, where the coefficient of $\mathbf{x}$ is removed. $\qquad\square$

3.6.3 *Theorem proving algorithm and practice*

Algorithm 3.34. Automated theorem proving in projective incidence geometry.

Input: (1) A sequence of batches of constructed points $\mathbf{x}_i$, (2) a conclusion expression *conc* in the Cayley bracket algebra generated by the $\mathbf{x}_i$.

Output: (1) The proving procedure, including eliminations, expansions, various contractions and factorizations; (2) additional nondegeneracy conditions.

Step 1. [Collection] Collect planes and lines. A line is composed of all points (at least three) collinear with each other; a plane is composed of all points (at least four) coplanar with each other.

Step 2. [Dynamic batch elimination] Start from the ends $\mathbf{x}_i$ of *conc* in the dynamic parents-children diagram, while *conc* $\neq 0$ and the $\mathbf{x}_i$ are not free points, eliminate the $\mathbf{x}_i$ from *conc* by substituting their GC algebraic expressions into *conc*, expand and simplify the result.

The algorithm is tested by more than 50 theorems in 2D and 3D projective incidence geometries. All the theorems tested are given binomial proofs. Several examples are presented below for illustration. In the proofs, explicit common bracket and meet-product factors are marked with underbraces and then removed. Their detection does not require any multivariate polynomial factorization.

Example 3.35. [Saam's Theorem]
    Free points: **1, 2, 3, 4, 5, 6**.
    Free collinear point: **7** on line **12**.

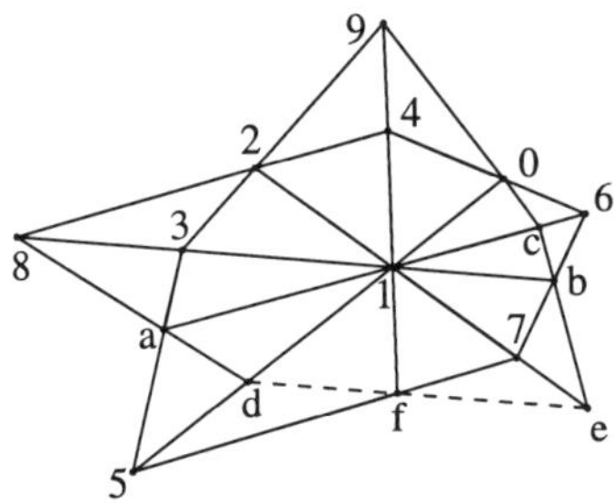

Fig. 3.10   Saam's Theorem.

Intersections:

$$8 = 13 \cap 24, \quad 9 = 23 \cap 14, \quad 0 = 15 \cap 46, \quad a = 35 \cap 16, \quad b = 13 \cap 67,$$
$$f = 57 \cap 14, \quad c = 16 \cap 90, \quad d = 15 \cap 8a, \quad e = 12 \cap bc.$$

Conclusion: **d, e, f** are collinear.

Proof.

<table>
<tr><td colspan="2" align="center">Rules</td><td align="center">[def]</td></tr>
<tr>
<td colspan="2">

[def]= [(15∨8a)(12∨bc)(57∨14)]

 = [18a][5(12∨bc)(57∨14)]

  −[58a][1(12∨bc)(57∨14)]

 = [125]([147][1bc][58a]−[145][18a][7bc])

</td>
<td>

d, e, f $\underset{=}{}$ [125]([147][1bc][58a] −[145][18a][7bc])

</td>
</tr>
<tr>
<td colspan="2">

[18a]= [135][168]

[58a]= [156][358]

[7bc]=−[137]67∨16∨90=[137][167][690]

[1bc]= [167]13∨16∨90=[136][167][190]

</td>
<td>

a, b, c $\underset{=}{}$ [167]([136][147][156][190][358] −[135][137][145][168][690])

</td>
</tr>
<tr>
<td colspan="2">

[168]=−[124][136]

[358]= [135][234]

[690]= [156]23∨14∨46=[146][156][234]

[190]=−[123]14∨15∨46=−[123][145][146]

</td>
<td>

8, 9, 0 $\underset{=}{}$ [135][136][145][146][156][234] (−[123][147] + [124][137])

</td>
</tr>
</table>

$$\overset{contract}{=} 0.$$

Additional nondegeneracy condition: none.   □

Remark: $[(\mathbf{15} \vee \mathbf{8a})(\mathbf{12} \vee \mathbf{bc})(\mathbf{57} \vee \mathbf{14})]$ is the *perspective pattern* (A.3.49) in Appendix A. If expanding the first meet product by separating **1, 5**, or expanding the last meet product by separating **5, 7**, then a unique factored result can be obtained. For other expansions, no common bracket factor can be produced. With the aid of collinearity transformations, the common bracket factor **[125]** can be produced from any binomial expansion of the Cayley expression. For example,

$$[(\mathbf{15} \vee \mathbf{8a})(\mathbf{12} \vee \mathbf{bc})(\mathbf{57} \vee \mathbf{14})]$$
$$\overset{expand}{=} [\mathbf{18a}][\mathbf{5}(\mathbf{12} \vee \mathbf{bc})(\mathbf{57} \vee \mathbf{14})] - [\mathbf{58a}][\mathbf{1}(\mathbf{12} \vee \mathbf{bc})(\mathbf{57} \vee \mathbf{14})]$$
$$\overset{expand}{=} -[\mathbf{125}][\mathbf{145}][\mathbf{18a}][\mathbf{7bc}] + [\mathbf{124}][\mathbf{157}][\mathbf{1bc}][\mathbf{58a}]$$
$$\overset{collinear}{=} [\mathbf{125}](-[\mathbf{145}][\mathbf{18a}][\mathbf{7bc}] + [\mathbf{147}][\mathbf{1bc}][\mathbf{58a}]),$$

where the collinearity transformation in the last step is $[\mathbf{124}][\mathbf{157}] = [\mathbf{125}][\mathbf{147}]$.

Example 3.36. Free points: **1, 2, 3, 4, 5, 6, 7, 8, 9.**

Semifree point: **0** on line **19**.

Intersections:

$$a = 13 \cap 24, \quad b = 24 \cap 35, \quad c = 35 \cap 46, \quad d = 46 \cap 57, \quad e = 57 \cap 68,$$
$$f = 68 \cap 17, \quad g = 17 \cap 28, \quad h = 28 \cap 13, \quad a_1 = 29 \cap 0h, \quad b_1 = 39 \cap aa_1,$$
$$c_1 = 49 \cap bb_1, \quad d_1 = 59 \cap cc_1, \quad e_1 = 69 \cap dd_1, \quad f_1 = 79 \cap ee_1, \quad g_1 = 89 \cap ff_1.$$

Conclusion: **0, g, g_1** are collinear.

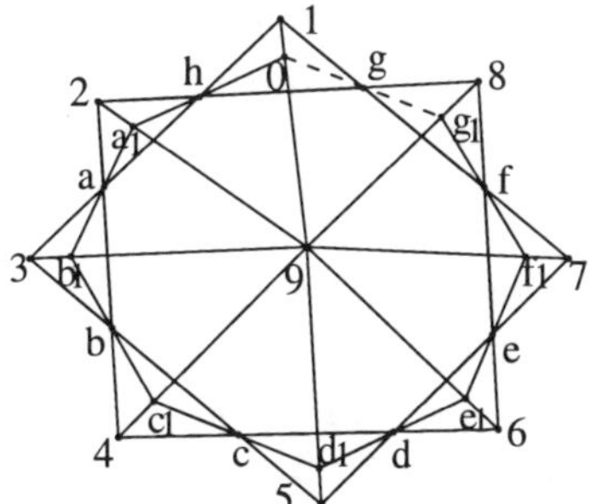

Fig. 3.11   Example 3.36.

Proof.

<table>
<tr>
<td>

Rules

$[0gg_1] = [0(17\vee 28)(89\vee ff_1)]$
$\quad = [028]17\vee 89\vee ff_1 - [017]28\vee 89\vee ff_1$
$\quad = -[028][17f_1][89f] - [017][289][8ff_1]$

$[17f_1] = [179][7ee_1]$
$[89f] = -[178][689]$
$[8ff_1] = [178]68\vee 79\vee ee_1 = -[178][68e_1][79e]$

$[68e_1] = [689][6dd_1]$
$[79e] = [579][678]$
$[7ee_1] = -[678]57\vee 69\vee dd_1 = [57d_1][678][69d]$

$[57d_1] = [579][5cc_1]$
$[69d] = [469][567]$
$[6dd_1] = -[567]46\vee 59\vee cc_1 = [46c_1][567][59c]$

$[46c_1] = [469][4bb_1]$
$[59c] = [359][456]$
$[5cc_1] = -[456]35\vee 49\vee bb_1 = [35b_1][456][49b]$

$[35b_1] = [359][3aa_1]$
$[49b] = [249][345]$
$[4bb_1] = -[345]24\vee 39\vee aa_1 = [24a_1][345][39a]$

$[24a_1] = -[02h][249]$
$[39a] = [139][234]$
$[3aa_1] = -[234]13\vee 29\vee 0h = -[013][234][29h]$

$[013][179] = [017][139]$

</td>
<td>

$[0gg_1]$

$\underset{g,g_1}{=}\ -[028][17f_1][89f]$
$\qquad -[017][289][8ff_1]$

$\underset{f,f_1}{=}\ [178]\underbrace{([028][179][689][7ee_1]}$
$\qquad +[017][289][68e_1][79e])$

$\underset{e,e_1}{=}\ [678][689]\underbrace{([028][179][57d_1][69d]}$
$\qquad +[017][289][579][6dd_1])$

$\underset{d,d_1}{=}\ [567][579]\underbrace{([028][179][469][5cc_1]}$
$\qquad +[017][289][46c_1][59c])$

$\underset{c,c_1}{=}\ [456][469]\underbrace{([028][179][35b_1][49b]}$
$\qquad +[017][289][359][4bb_1])$

$\underset{b,b_1}{=}\ [345][359]\underbrace{([028][179][249][3aa_1]}$
$\qquad +[017][289][24a_1][39a])$

$\underset{a,a_1}{=}\ -[234][249]\underbrace{([013][028][179][29h]}$
$\qquad +[017][02h][139][289])$

$\overset{collinear}{=} [017][139]\underbrace{([028][29h] + [02h][289])}$

$\overset{contract}{=} 0.$

</td>
</tr>
</table>

Additional nondegeneracy condition: none.
$\qquad\qquad\qquad\qquad\qquad\qquad\qquad\qquad\qquad\qquad\qquad$ $\square$

If the reverse of the order of construction batches is used in elimination, the proof is still binomial, but the maximal degree among the bracket binomials is nine instead of four here. In this example, considerable simplifications are achieved by using the dynamic order of batch eliminations.

Example 3.37. [148] [A non-realizable torus]

  Free points in space: **1, 2, 3, 4, 5**.

  Free coplanar points: **6** on plane **134**, **7** on plane **125**.

  Intersections: $\mathbf{8 = 124 \cap 236 \cap 457}$, $\mathbf{9 = 237 \cap 456 \cap 678}$.

  Conclusion: **1, 3, 5, 9** are coplanar.

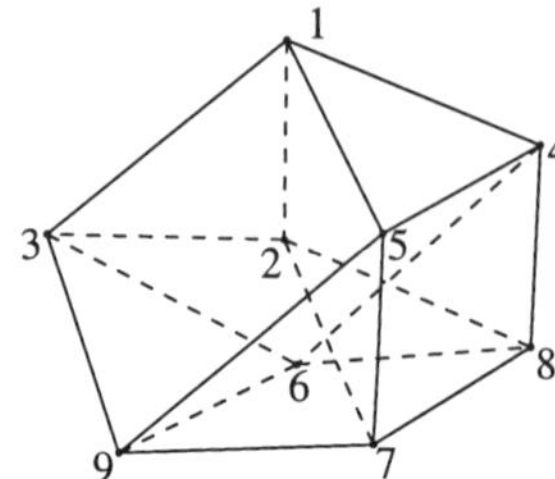

Fig. 3.12 A non-realizable torus in space.

Proof.

Rules   **[1359]**

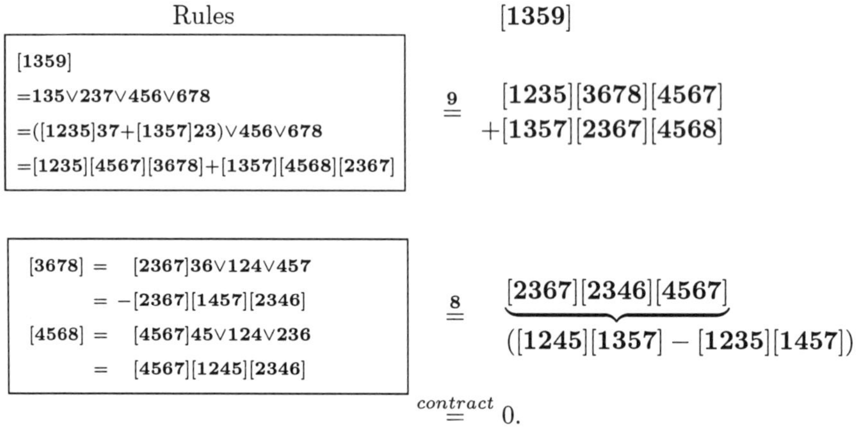

$$[1359]$$
$$=135 \vee 237 \vee 456 \vee 678$$
$$=([1235]37+[1357]23) \vee 456 \vee 678$$
$$=[1235][4567][3678]+[1357][4568][2367]$$

$$\overset{9}{=} \quad [1235][3678][4567] + [1357][2367][4568]$$

$$[3678] = [2367]36 \vee 124 \vee 457$$
$$= -[2367][1457][2346]$$
$$[4568] = [4567]45 \vee 124 \vee 236$$
$$= [4567][1245][2346]$$

$$\overset{8}{=} \quad \underbrace{[2367][2346][4567]}_{([1245][1357] - [1235][1457])}$$

$$\overset{contract}{=} 0.$$

Additional nondegeneracy condition: none.   □

Example 3.38. [148] [Sixteen-Point Theorem] Let there be two groups of lines in space, each group being composed of four lines. When selecting one line from each group and put them together, there are sixteen pairs of lines. If fifteen pairs are coplanar ones, so is the sixteenth pair.

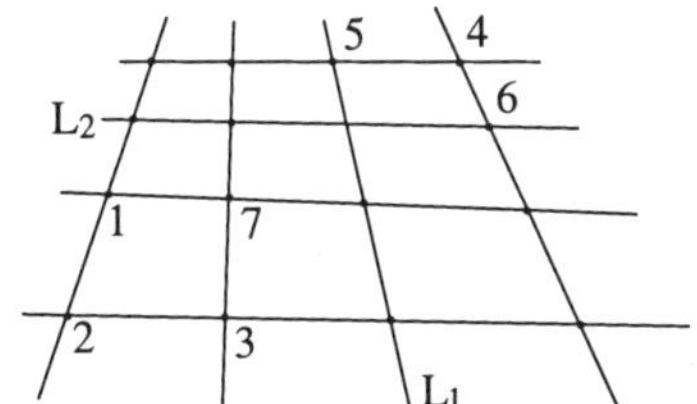

Fig. 3.13　3D Sixteen-Point Theorem.

In Figure 3.13, the sixteen pairs of lines are

$$\{12,17\}\ \{12,23\}\ \{23,37\}\ \{37,17\}\ \{45,46\}\ \{45,\mathbf{L_1}\}\ \{46,\mathbf{L_2}\}\ \{12,45\}$$
$$\{23,46\}\ \{17,46\}\ \{37,45\}\ \{17,\mathbf{L_1}\}\ \{23,\mathbf{L_1}\}\ \{12,\mathbf{L_2}\}\ \{37,\mathbf{L_2}\}\ \{\mathbf{L_1},\mathbf{L_2}\}.$$

Free points: $\mathbf{1,2,3,4}$.

Free coplanar points: $\mathbf{5}$ on plane $\mathbf{124}$, $\mathbf{6}$ on plane $\mathbf{234}$.

Free collinear point: $\mathbf{7}$ on line $\mathbf{416} \cap \mathbf{435}$.

Conclusion: $\mathbf{L_1 = 157 \cap 235}$ and $\mathbf{L_2 = 126 \cap 367}$ are coplanar.

Proof.

Rules　　　　　　　　　　　　　$\mathbf{157 \vee 235 \vee 126 \vee 367}$

$$\boxed{\begin{aligned}
&\mathbf{157\vee235=[1257]35-[1357]25}\\
&\mathbf{35\vee126\vee367=-[1236][3567]}\\
&\mathbf{25\vee126\vee367=-[1256][2367]}
\end{aligned}}$$

$$= \quad [1256][1357][2367] - [1236][1257][3567]$$

$$\boxed{\begin{aligned}
&\mathbf{7=-[47]16\vee_435-[7(16\vee_435)]4}\\
&\mathbf{[1357]=\ [7(16\vee_435)][1345]}\\
&\mathbf{[2367]=\ [47][1236][3456]}\\
&\mathbf{[1257]=\ [47][1235][1456]}\\
&\mathbf{[3567]=-[7(16\vee_435)][3456]}
\end{aligned}}$$

$$\overset{7}{=}\quad \underbrace{[47][7(16 \vee_4 35)][1236][3456]}_{([1256][1345] + [1235][1456])}$$

$$\overset{contract}{=}\quad 0.$$

Additional nondegeneracy condition: none.　　　　　　　　　　　　□

Remark: In eliminating $\mathbf{7}$, two brackets containing $\mathbf{7}$ each have a unique monomial expansion:

$$\begin{aligned}
[\mathbf{1357}] &=\ [7(16 \vee_4 35)][1345],\\
[\mathbf{3567}] &= -[7(16 \vee_4 35)][3456];
\end{aligned}$$

two other brackets containing $\mathbf{7}$ each have two monomial expansions:

$$\begin{aligned}
[\mathbf{2367}] &= -[47][236(16 \vee_4 35)] =\ [47][1236][3456] = [47][1346][2356],\\
[\mathbf{1257}] &= -[47][125(16 \vee_4 35)] = -[47][1256][1345] = [47][1235][1456].
\end{aligned}$$

$$(3.6.12)$$

So there are four different combinations of the monomial expansions in (3.6.12), leading to four proofs different in their final steps. The first monomial expansions in each line of (3.6.12) are the best combination, and the proof finishes immediately after the elimination of **7**. The last monomial expansions in each line of (3.6.12) are the worst combination, the proof cannot be finished without using either Cramer's rule to eliminate **6**, or two coplanarity transformations as shown in Example 3.9 to make factorization. The other two combinations each need a contraction to finish, one of which is already in the above proof.

3.7 Erdös' consistent 5-tuples*

Around 1994, Erdös *et al.* proposed the following challenging problem in enumerative projective incidence geometry:

"For ten points $\mathbf{a}_{ij}$, $1 \leq i < j \leq 5$, in the projective plane, if there are five points $\mathbf{a}_k$, $1 \leq k \leq 5$, in which at least two points are different, such that $\mathbf{a}_i, \mathbf{a}_j, \mathbf{a}_{ij}$ are collinear for all $1 \leq i < j \leq 5$, we say the five points form a *consistent 5-tuple*. Now assume that no three of the $\mathbf{a}_{ij}$ are collinear. Is it true that there are only finitely many consistent 5-tuples?"

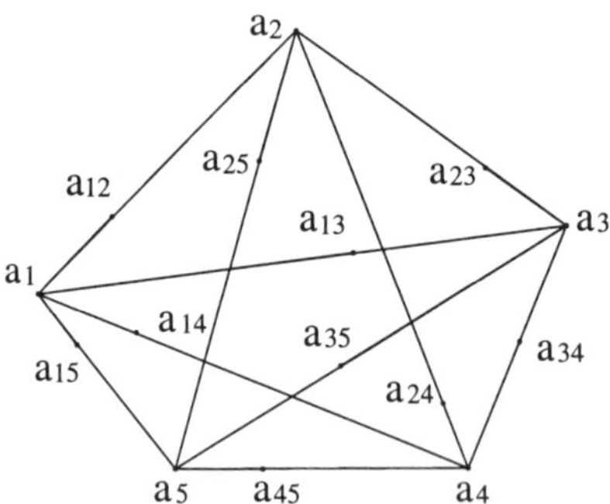

Fig. 3.14 Erdös' consistent 5-tuple configuration.

This is Erdös' consistent 5-tuple problem. It still remains open today. The number of consistent 5-tuples is called the number of *solutions of Erdös' problem*. In [105], the following two theorems were established:

Theorem 3.39. For ten generic points $\{\mathbf{a}_{ij} \,|\, 1 \leq i < j \leq 5\}$ in the plane, any consistent 5-tuple $\{\mathbf{a}_k \,|\, 1 \leq k \leq 5\}$ satisfies:

(1) $\mathbf{a}_i \neq \mathbf{a}_j$ for $i \neq j$.
(2) $\mathbf{a}_i \neq \mathbf{a}_{ij}$ for $i \neq j$.
(3) $\mathbf{a}_i \neq \mathbf{a}_{jk}$ for $i \neq j \neq k$.
(4) $\mathbf{a}_i, \mathbf{a}_{ij}, \mathbf{a}_{ik}$ are noncollinear for $i \neq j \neq k$.
(5) $\mathbf{a}_i, \mathbf{a}_{ij}, \mathbf{a}_{jk}$ are noncollinear for $i \neq j \neq k$.

If any of the above conditions is violated by the ten input points $\mathbf{a}_{ij}$, then there are only finitely many solutions.

Theorem 3.40. For ten generic points $\{\mathbf{a}_{ij} \mid 1 \leq i < j \leq 5\}$ in the plane, there are at most six solutions.

In this section, we show that Cayley expansion and Cayley factorization techniques contribute to not only significant improvements over the proof of [105], but also new results on this open problem.

3.7.1 *Derivation of the fundamental equations*

We shall prove Theorem 3.39 in Subsection 3.7.3. In this subsection, we use the theorem directly.

The ten constraints on the five points $\mathbf{a}_k$ are

$$[\mathbf{a}_i \mathbf{a}_j \mathbf{a}_{ij}] = 0, \ \forall i \neq j, \tag{3.7.1}$$

where when $i > j$, we set $\mathbf{a}_{ij} = \mathbf{a}_{ji}$. To solve (3.7.1) for the $\mathbf{a}_k$, we do triangulation to the ten equations by the order of vector variables $\{\mathbf{a}_{ij}\} \prec \mathbf{a}_5 \prec \mathbf{a}_4 \prec \mathbf{a}_3 \prec \mathbf{a}_2 \prec \mathbf{a}_1$.

Step 1. The equations with leading vector $\mathbf{a}_1$ are

$$\begin{cases} [\mathbf{a}_1 \mathbf{a}_2 \mathbf{a}_{12}] = 0, \\ [\mathbf{a}_1 \mathbf{a}_3 \mathbf{a}_{13}] = 0, \\ [\mathbf{a}_1 \mathbf{a}_4 \mathbf{a}_{14}] = 0, \\ [\mathbf{a}_1 \mathbf{a}_5 \mathbf{a}_{15}] = 0. \end{cases} \tag{3.7.2}$$

From the last two equations we get $\mathbf{a}_1$, whose existence for generic data $\{\mathbf{a}_{ij}\}$ is guaranteed by Theorem 3.39:

$$\mathbf{a}_1 = \mathbf{a}_4 \mathbf{a}_{14} \vee \mathbf{a}_5 \mathbf{a}_{15}. \tag{3.7.3}$$

Step 2. The equations with leading vector $\mathbf{a}_2$ are

$$\begin{cases} [\mathbf{a}_2 \mathbf{a}_3 \mathbf{a}_{23}] = 0, \\ [\mathbf{a}_2 \mathbf{a}_4 \mathbf{a}_{24}] = 0, \\ [\mathbf{a}_2 \mathbf{a}_5 \mathbf{a}_{25}] = 0. \end{cases} \tag{3.7.4}$$

From the last two equations we get

$$\mathbf{a}_2 = \mathbf{a}_4 \mathbf{a}_{24} \vee \mathbf{a}_5 \mathbf{a}_{25}. \tag{3.7.5}$$

Step 3. The equations with leading vector $\mathbf{a}_3$ are

$$\begin{cases} [\mathbf{a}_3 \mathbf{a}_4 \mathbf{a}_{34}] = 0, \\ [\mathbf{a}_3 \mathbf{a}_5 \mathbf{a}_{35}] = 0, \end{cases} \tag{3.7.6}$$

from which we get

$$\mathbf{a}_3 = \mathbf{a}_4 \mathbf{a}_{34} \vee \mathbf{a}_5 \mathbf{a}_{35}. \tag{3.7.7}$$

Step 4. The equation with leading element $\mathbf{a}_4$ is

$$[\mathbf{a}_4 \mathbf{a}_5 \mathbf{a}_{45}] = 0. \tag{3.7.8}$$

Substituting the obtained Cayley expressions of a_1, a_2, a_3 into the three unused equations in (3.7.2) and (3.7.4), we get

$$[(a_4a_{14} \vee a_5a_{15})(a_4a_{24} \vee a_5a_{25})a_{12}] = 0, \tag{3.7.9}$$

$$[(a_4a_{14} \vee a_5a_{15})(a_4a_{34} \vee a_5a_{35})a_{13}] = 0, \tag{3.7.10}$$

$$[(a_4a_{24} \vee a_5a_{25})(a_4a_{34} \vee a_5a_{35})a_{23}] = 0. \tag{3.7.11}$$

We need to solve equations (3.7.8) to (3.7.11) for a_4. To this end we first change (3.7.9) to (3.7.11) into bracket polynomial equations by Cayley expansions. By interchanging subscripts 1 and 2, (3.7.10) and (3.7.11) are switched, and by interchanging subscripts 1 and 3, (3.7.9) and (3.7.11) are switched. So we only need to concentrate on one equation, say (3.7.11).

The left side of (3.7.11) has all together two different binomial expansions, which are obtained by distributing a_4a_{24}, a_5a_{25} and a_4a_{34}, a_5a_{35} respectively. After the expansions, we get two equations both equivalent to (3.7.11):

$$\begin{aligned}
[a_4a_{23}a_{34}][a_4a_5a_{24}][a_5a_{25}a_{35}] &= [a_4a_{24}a_{34}][a_4a_5a_{25}][a_5a_{23}a_{35}], \\
[a_4a_{23}a_{24}][a_4a_5a_{34}][a_5a_{25}a_{35}] &= [a_4a_{24}a_{34}][a_4a_5a_{35}][a_5a_{23}a_{25}].
\end{aligned} \tag{3.7.12}$$

Generically $a_4 \neq a_5$ and $a_5 \neq a_{45}$ by Theorem 3.39, so blade a_4a_5 in the brackets of (3.7.12) can be replaced by blade a_5a_{45}, because they differ only by scale according to (3.7.8). This *collinearity-like transformation* will be used frequently in this section.

Replacing a_4a_5 by a_5a_{45} eliminates a vector a_4 from each side of the two equations in (3.7.12), changing (3.7.12) into two equations linear in a_4:

$$a_4a_{34}([a_5a_{24}a_{45}][a_5a_{25}a_{35}]a_{23} - [a_5a_{23}a_{35}][a_5a_{25}a_{45}]a_{24}) = 0, \tag{3.7.13}$$

$$a_4a_{24}([a_5a_{25}a_{35}][a_5a_{34}a_{45}]a_{23} + [a_5a_{23}a_{25}][a_5a_{35}a_{45}]a_{34}) = 0. \tag{3.7.14}$$

By changing subscript 2 to 1 in (3.7.13) and (3.7.14), we get two equations both equivalent to (3.7.10):

$$a_4a_{34}([a_5a_{14}a_{45}][a_5a_{15}a_{35}]a_{13} - [a_5a_{13}a_{35}][a_5a_{15}a_{45}]a_{14}) = 0, \tag{3.7.15}$$

$$a_4a_{14}([a_5a_{15}a_{35}][a_5a_{34}a_{45}]a_{13} + [a_5a_{13}a_{15}][a_5a_{35}a_{45}]a_{34}) = 0. \tag{3.7.16}$$

By changing subscript 3 to 1 in (3.7.13) and (3.7.14), we get two equations both equivalent to (3.7.9):

$$a_4a_{14}([a_5a_{15}a_{25}][a_5a_{24}a_{45}]a_{12} + [a_5a_{12}a_{15}][a_5a_{25}a_{45}]a_{24}) = 0, \tag{3.7.17}$$

$$a_4a_{24}([a_5a_{14}a_{45}][a_5a_{15}a_{25}]a_{12} - [a_5a_{12}a_{25}][a_5a_{15}a_{45}]a_{14}) = 0. \tag{3.7.18}$$

In (3.7.13) and (3.7.15), since generically $a_4 \neq a_{34}$ by Theorem 3.39, by the collinearity of the four points: a_4, a_{34}, and

$$\begin{aligned}
&[a_5a_{24}a_{45}][a_5a_{25}a_{35}]a_{23} - [a_5a_{23}a_{35}][a_5a_{25}a_{45}]a_{24}, \\
&[a_5a_{14}a_{45}][a_5a_{15}a_{35}]a_{13} - [a_5a_{13}a_{35}][a_5a_{15}a_{45}]a_{14},
\end{aligned}$$

we get the following equation in a_5:

$$\begin{aligned}
[a_{34}(&[a_5a_{24}a_{45}][a_5a_{25}a_{35}]a_{23} - [a_5a_{23}a_{35}][a_5a_{25}a_{45}]a_{24}) \\
&([a_5a_{14}a_{45}][a_5a_{15}a_{35}]a_{13} - [a_5a_{13}a_{35}][a_5a_{15}a_{45}]a_{14})] = 0.
\end{aligned} \tag{3.7.19}$$

Conversely, if (3.7.19) holds, then by the noncollinearities of a_{34}, a_{23}, a_{24} and a_{34}, a_{13}, a_{14}, (3.7.13) and (3.7.15) are equivalent.

Notice that (3.7.19) is antisymmetric with respect to subscripts 1 and 2. By interchanging subscripts 2 and 3 in (3.7.19), we get the second equation in a_5:

$$[a_{24}([a_5a_{25}a_{35}][a_5a_{34}a_{45}]a_{23} + [a_5a_{23}a_{25}][a_5a_{35}a_{45}]a_{34}) \atop ([a_5a_{14}a_{45}][a_5a_{15}a_{25}]a_{12} - [a_5a_{12}a_{25}][a_5a_{15}a_{45}]a_{14})] = 0; \tag{3.7.20}$$

by interchanging subscripts 1 and 3 in (3.7.19), we get the third equation in a_5:

$$[a_{14}([a_5a_{15}a_{25}][a_5a_{24}a_{45}]a_{12} + [a_5a_{12}a_{15}][a_5a_{25}a_{45}]a_{24}) \atop ([a_5a_{15}a_{35}][a_5a_{34}a_{45}]a_{13} + [a_5a_{13}a_{15}][a_5a_{35}a_{45}]a_{34})] = 0. \tag{3.7.21}$$

The three equations (3.7.19), (3.7.20) and (3.7.21) are called the *fundamental equations of Erdös' problem* in a_5. Any two of them deduce the third.

Triangulation result:

Under some inequality conditions stated in Theorem 3.39, the original ten equations (3.7.1) are equivalent to the three Cayley expression representations of a_1, a_2, a_3 (they are equivalent to six scalar equations), plus equation (3.7.8), plus equations (3.7.13) and (3.7.14) (they are equivalent to each other), and plus the three fundamental equations (two of them are algebraically independent).

Proposition 3.41. Under the five groups of inequality conditions in Theorem 3.39, the number of solutions of Erdös' problem is equal to the number of solutions of the fundamental equations in vector indeterminate a_5.

Clearly $a_5 = a_{45}$ satisfies all three equations, $a_5 = a_{15}$ satisfies (3.7.19) and (3.7.21), $a_5 = a_{25}$ satisfies (3.7.20) and (3.7.21), and $a_5 = a_{35}$ satisfies (3.7.19) and (3.7.20). They are all solutions of the fundamental equations.

Proposition 3.42. For any given data $\{a_{ij}\}$ satisfying the noncollinearity assumption, the left side of any fundamental equation is not identically zero, and no two fundamental equations represent the same quartic curve in a_5.

Proof. We only need to consider the first two fundamental equations (3.7.19) and (3.7.20). If they represent the same quartic in a_5, then a_{25} should be a double point of (3.7.19), *i.e.*, the derivative of the left side of (3.7.19) with respect to a_5 at $a_5 = a_{25}$ should be zero. If (3.7.19) is an identity for all points a_5 in the plane, this derivative should also be zero. Denote

$$x = [a_{14}a_{25}a_{45}][a_{15}a_{25}a_{35}]a_{13} - [a_{13}a_{25}a_{35}][a_{15}a_{25}a_{45}]a_{14}. \tag{3.7.22}$$

Then the above-mentioned derivative equals the dual of the following bivector:

$$a_{25}(-[a_{24}a_{25}a_{45}][xa_{23}a_{34}]a_{35} + [a_{23}a_{25}a_{35}][xa_{24}a_{34}]a_{45}), \tag{3.7.23}$$

which is nonzero because of the noncollinearities of a_{25}, a_{35}, a_{45} and a_{13}, a_{14}, a_{34}.
$\square$

Proposition 3.43. Line $\mathbf{a}_{35}\mathbf{a}_{45}$ is a branch of quartic (3.7.19) in $\mathbf{a}_5$ if and only if

$$\mathbf{a}_{13}\mathbf{a}_{14} \vee \mathbf{a}_{23}\mathbf{a}_{24} \vee \mathbf{a}_{35}\mathbf{a}_{45} = 0. \tag{3.7.24}$$

It cannot be a common branch of (3.7.19) and any of (3.7.20), (3.7.21).

Proof. Substituting $\mathbf{a}_5 = \lambda\mathbf{a}_{35} + \mu\mathbf{a}_{45}$ into (3.7.19), we get, by denoting $\tau = \lambda^2\tau^2[\mathbf{a}_{15}\mathbf{a}_{35}\mathbf{a}_{45}][\mathbf{a}_{25}\mathbf{a}_{35}\mathbf{a}_{45}]$,

$$\begin{aligned}
0 &= \tau([\mathbf{a}_{13}\mathbf{a}_{23}\mathbf{a}_{34}][\mathbf{a}_{14}\mathbf{a}_{35}\mathbf{a}_{45}][\mathbf{a}_{24}\mathbf{a}_{35}\mathbf{a}_{45}] - [\mathbf{a}_{13}\mathbf{a}_{24}\mathbf{a}_{34}][\mathbf{a}_{14}\mathbf{a}_{35}\mathbf{a}_{45}][\mathbf{a}_{23}\mathbf{a}_{35}\mathbf{a}_{45}] \\
&\quad -[\mathbf{a}_{13}\mathbf{a}_{35}\mathbf{a}_{45}][\mathbf{a}_{14}\mathbf{a}_{23}\mathbf{a}_{34}][\mathbf{a}_{24}\mathbf{a}_{35}\mathbf{a}_{45}] + [\mathbf{a}_{13}\mathbf{a}_{35}\mathbf{a}_{45}][\mathbf{a}_{14}\mathbf{a}_{24}\mathbf{a}_{34}][\mathbf{a}_{23}\mathbf{a}_{35}\mathbf{a}_{45}]) \\
&= \tau[(\mathbf{a}_{13}\mathbf{a}_{14} \vee \mathbf{a}_{35}\mathbf{a}_{45})(\mathbf{a}_{23}\mathbf{a}_{24} \vee \mathbf{a}_{35}\mathbf{a}_{45})\mathbf{a}_{34}] \\
&= \tau[\mathbf{a}_{34}\mathbf{a}_{35}\mathbf{a}_{45}]\,\mathbf{a}_{13}\mathbf{a}_{14} \vee \mathbf{a}_{23}\mathbf{a}_{24} \vee \mathbf{a}_{35}\mathbf{a}_{45}.
\end{aligned}$$

When line $\mathbf{a}_{35}\mathbf{a}_{45}$ is a branch of quartic (3.7.19), $\mathbf{a}_{13}\mathbf{a}_{14} \vee \mathbf{a}_{23}\mathbf{a}_{24} \vee \mathbf{a}_{35}\mathbf{a}_{45} = 0$. Assume that the three lines $\mathbf{a}_{13}\mathbf{a}_{14}, \mathbf{a}_{23}\mathbf{a}_{24}, \mathbf{a}_{35}\mathbf{a}_{45}$ concur at point $\mathbf{o}$. Then $\mathbf{o} \neq \mathbf{a}_{ij}$ for any $i \neq j$. By symmetry we only need to prove that line $\mathbf{a}_{35}\mathbf{a}_{45}$ is not a branch of (3.7.21).

Assume that the line is a branch of the quartic. Substituting $\mathbf{a}_5 = \lambda\mathbf{a}_{35} + \mu\mathbf{a}_{45}$ into (3.7.21), we get

$$\begin{aligned}
&[\mathbf{a}_{12}\mathbf{a}_{13}\mathbf{a}_{14}][\mathbf{a}_{24}\mathbf{a}_{35}\mathbf{a}_{45}](\lambda[\mathbf{a}_{15}\mathbf{a}_{25}\mathbf{a}_{35}] + \mu[\mathbf{a}_{15}\mathbf{a}_{25}\mathbf{a}_{45}]) \\
&+[\mathbf{a}_{13}\mathbf{a}_{14}\mathbf{a}_{24}][\mathbf{a}_{25}\mathbf{a}_{35}\mathbf{a}_{45}](\lambda[\mathbf{a}_{12}\mathbf{a}_{15}\mathbf{a}_{35}] + \mu[\mathbf{a}_{12}\mathbf{a}_{15}\mathbf{a}_{45}]) = 0.
\end{aligned}$$

The coefficients of λ, μ respectively should be both zero, so

$$\begin{aligned}
[\mathbf{a}_{12}\mathbf{a}_{13}\mathbf{a}_{14}][\mathbf{a}_{15}\mathbf{a}_{25}\mathbf{a}_{35}][\mathbf{a}_{24}\mathbf{a}_{35}\mathbf{a}_{45}] &= -[\mathbf{a}_{12}\mathbf{a}_{15}\mathbf{a}_{35}][\mathbf{a}_{13}\mathbf{a}_{14}\mathbf{a}_{24}][\mathbf{a}_{25}\mathbf{a}_{35}\mathbf{a}_{45}], \\
[\mathbf{a}_{12}\mathbf{a}_{13}\mathbf{a}_{14}][\mathbf{a}_{15}\mathbf{a}_{25}\mathbf{a}_{45}][\mathbf{a}_{24}\mathbf{a}_{35}\mathbf{a}_{45}] &= -[\mathbf{a}_{12}\mathbf{a}_{15}\mathbf{a}_{45}][\mathbf{a}_{13}\mathbf{a}_{14}\mathbf{a}_{24}][\mathbf{a}_{25}\mathbf{a}_{35}\mathbf{a}_{45}].
\end{aligned} \tag{3.7.25}$$

By collinearity, replacing $\mathbf{a}_{13}\mathbf{a}_{14}$ by $\mathbf{o}\mathbf{a}_{14}$, and replacing $\mathbf{a}_{35}\mathbf{a}_{45}$ by $\mathbf{o}\mathbf{a}_{45}$ in the brackets of (3.7.25), we get

$$\begin{aligned}
[\mathbf{o}\mathbf{a}_{12}\mathbf{a}_{14}][\mathbf{a}_{15}\mathbf{a}_{25}\mathbf{a}_{35}][\mathbf{o}\mathbf{a}_{24}\mathbf{a}_{45}] &= -[\mathbf{a}_{12}\mathbf{a}_{15}\mathbf{a}_{35}][\mathbf{o}\mathbf{a}_{14}\mathbf{a}_{24}][\mathbf{o}\mathbf{a}_{25}\mathbf{a}_{45}], \\
[\mathbf{o}\mathbf{a}_{12}\mathbf{a}_{14}][\mathbf{a}_{15}\mathbf{a}_{25}\mathbf{a}_{45}][\mathbf{o}\mathbf{a}_{24}\mathbf{a}_{45}] &= -[\mathbf{a}_{12}\mathbf{a}_{15}\mathbf{a}_{45}][\mathbf{o}\mathbf{a}_{14}\mathbf{a}_{24}][\mathbf{o}\mathbf{a}_{25}\mathbf{a}_{45}],
\end{aligned}$$

which can be written in the following ratio form:

$$-\frac{[\mathbf{o}\mathbf{a}_{12}\mathbf{a}_{14}][\mathbf{o}\mathbf{a}_{24}\mathbf{a}_{45}]}{[\mathbf{o}\mathbf{a}_{14}\mathbf{a}_{24}][\mathbf{o}\mathbf{a}_{25}\mathbf{a}_{45}]} = \frac{[\mathbf{a}_{12}\mathbf{a}_{15}\mathbf{a}_{35}]}{[\mathbf{a}_{15}\mathbf{a}_{25}\mathbf{a}_{35}]} = \frac{[\mathbf{a}_{12}\mathbf{a}_{15}\mathbf{a}_{45}]}{[\mathbf{a}_{15}\mathbf{a}_{25}\mathbf{a}_{45}]}. \tag{3.7.26}$$

From the last equality in (3.7.26), we get

$$0 = [\mathbf{a}_{12}\mathbf{a}_{15}\mathbf{a}_{35}][\mathbf{a}_{15}\mathbf{a}_{25}\mathbf{a}_{45}] - [\mathbf{a}_{12}\mathbf{a}_{15}\mathbf{a}_{45}][\mathbf{a}_{15}\mathbf{a}_{25}\mathbf{a}_{35}] \overset{contract}{=} [\mathbf{a}_{12}\mathbf{a}_{15}\mathbf{a}_{25}][\mathbf{a}_{15}\mathbf{a}_{35}\mathbf{a}_{45}],$$

violating the noncollinearity assumption. $\qquad\qquad\square$

Now let us find one more special solution of the fundamental equations. In [105], a special solution in coordinate form is found on line $\mathbf{a}_{14}\mathbf{a}_{45}$. Below we deduce its expression in GC algebra.

In (3.7.19), if $[\mathbf{a}_5\mathbf{a}_{14}\mathbf{a}_{45}] = 0$ (generically this is not true according to Theorem 3.39), then after removing bracket factors $[\mathbf{a}_5\mathbf{a}_{13}\mathbf{a}_{35}][\mathbf{a}_5\mathbf{a}_{15}\mathbf{a}_{45}]$, we get from (3.7.19) the following:

$$[\mathbf{a}_5\mathbf{a}_{24}\mathbf{a}_{45}][\mathbf{a}_5\mathbf{a}_{25}\mathbf{a}_{35}][\mathbf{a}_{14}\mathbf{a}_{23}\mathbf{a}_{34}] = [\mathbf{a}_5\mathbf{a}_{23}\mathbf{a}_{35}][\mathbf{a}_5\mathbf{a}_{25}\mathbf{a}_{45}][\mathbf{a}_{14}\mathbf{a}_{24}\mathbf{a}_{34}]. \tag{3.7.27}$$

Replacing $a_5 a_{45}$ by $a_{14} a_{45}$ in the brackets of (3.7.27), we get

$$[a_5 a_{35}([a_{14}a_{23}a_{34}][a_{14}a_{24}a_{45}]a_{25} - [a_{14}a_{24}a_{34}][a_{14}a_{25}a_{45}]a_{23})] = 0. \qquad (3.7.28)$$

Combining (3.7.28) and $[a_5 a_{14} a_{45}] = 0$, we get

$$\begin{aligned}
a_5 &= a_{14}a_{45} \vee ([a_{14}a_{23}a_{34}][a_{14}a_{24}a_{45}]a_{25} - [a_{14}a_{24}a_{34}][a_{14}a_{25}a_{45}]a_{23})a_{35} \\
&= \quad [a_{14}a_{24}a_{34}][a_{14}a_{25}a_{45}][a_{14}a_{35}a_{45}]a_{23} \\
&\quad -[a_{14}a_{23}a_{34}][a_{14}a_{24}a_{45}][a_{14}a_{35}a_{45}]a_{25} \\
&\quad +[a_{14}a_{23}a_{24}][a_{14}a_{25}a_{45}][a_{14}a_{34}a_{45}]a_{35},
\end{aligned}$$
$$(3.7.29)$$

where we have used the contraction

$$[a_{14}a_{23}a_{34}][a_{14}a_{24}a_{45}] - [a_{14}a_{24}a_{34}][a_{14}a_{23}a_{45}] = [a_{14}a_{23}a_{24}][a_{14}a_{34}a_{45}].$$

Denote by $t_{14,45}$ the point represented by (3.7.29). It is the fourth point of intersection of line $a_{14}a_{45}$ with quartic curve (3.7.19) in a_5, the other three being point a_{45} counted twice and point $a_{14}a_{45} \vee a_{13}a_{35}$.

Similarly, in (3.7.20), if $[a_5 a_{14} a_{45}] = 0$, then after removing bracket factors $[a_5 a_{12} a_{25}][a_5 a_{15} a_{45}]$, we get from (3.7.20) the following:

$$[a_5 a_{34} a_{45}][a_5 a_{25} a_{35}][a_{14} a_{23} a_{24}] = [a_5 a_{23} a_{25}][a_5 a_{35} a_{45}][a_{14} a_{24} a_{34}]. \qquad (3.7.30)$$

Replacing $a_5 a_{45}$ by $a_{14} a_{45}$ in the brackets of (3.7.30), we get

$$[a_5 a_{25}([a_{14}a_{23}a_{24}][a_{14}a_{34}a_{45}]a_{35} - [a_{14}a_{24}a_{34}][a_{14}a_{35}a_{45}]a_{23})] = 0. \qquad (3.7.31)$$

Combining (3.7.31) and $[a_5 a_{14} a_{45}] = 0$, we get

$$\begin{aligned}
a_5 &= \quad a_{14}a_{45} \vee a_{25}([a_{14}a_{23}a_{24}][a_{14}a_{34}a_{45}]a_{35} + [a_{14}a_{24}a_{34}][a_{14}a_{35}a_{45}]a_{23}) \\
&= \quad [a_{14}a_{25}a_{45}]([a_{14}a_{23}a_{24}][a_{14}a_{34}a_{45}]a_{35} + [a_{14}a_{24}a_{34}][a_{14}a_{35}a_{45}]a_{23}) \\
&\quad -[a_{14}a_{35}a_{45}]([a_{14}a_{23}a_{45}][a_{14}a_{24}a_{34}] + [a_{14}a_{23}a_{24}][a_{14}a_{34}a_{45}])a_{25} \\
&\overset{contract}{=} \quad [a_{14}a_{24}a_{34}][a_{14}a_{25}a_{45}][a_{14}a_{35}a_{45}]a_{23} \\
&\quad -[a_{14}a_{23}a_{34}][a_{14}a_{24}a_{45}][a_{14}a_{35}a_{45}]a_{25} \\
&\quad +[a_{14}a_{23}a_{24}][a_{14}a_{25}a_{45}][a_{14}a_{34}a_{45}]a_{35} \\
&= \quad t_{14,45}.
\end{aligned}$$
$$(3.7.32)$$

It is the fourth point of intersection of line $a_{14}a_{45}$ with quartic curve (3.7.20) in a_5, the other three being point a_{45} counted twice and point $a_{14}a_{45} \vee a_{12}a_{25}$.

From (3.7.29) and (3.7.32), we get

Proposition 3.44. $a_5 = t_{14,45}$ is a solution of the two fundamental equations (3.7.19) and (3.7.20).

3.7.2   *Proof of Theorem 3.40*

We write (3.7.19) and (3.7.20) as

$$\begin{cases} g_1(\mathbf{a}_5) = 0, \\ g_2(\mathbf{a}_5) = 0. \end{cases} \tag{3.7.33}$$

The following is a classical result in the theory of algebraic curves [10]:

Lemma 3.45. Let $f(x,y) = 0, g(x,y) = 0$ be two algebraic curves in $\mathbb{C}^2$. Let the lowest total degrees of $f(x,y), g(x,y)$ be m, n respectively. Then $(0,0)$ is an intersection of the two curves with multiplicity at least mn.

Corollary 3.46. Let $f(x,y), g(x,y)$ be two complex polynomials of lowest total degree m, n respectively. Then x^{mn} is a factor of the resultant of $f(x,y), g(x,y)$ with respect to y.

In (3.7.19), the degrees of the vector variables are

$$\mathbf{a}_5^4 \mathbf{a}_{12}^0 \mathbf{a}_{13}^1 \mathbf{a}_{14}^1 \mathbf{a}_{15}^1 \mathbf{a}_{23}^1 \mathbf{a}_{24}^1 \mathbf{a}_{25}^1 \mathbf{a}_{34}^1 \mathbf{a}_{35}^2 \mathbf{a}_{45}^2.$$

In (3.7.20), the degrees of the vector variables are

$$\mathbf{a}_5^4 \mathbf{a}_{12}^1 \mathbf{a}_{13}^0 \mathbf{a}_{14}^1 \mathbf{a}_{15}^1 \mathbf{a}_{23}^2 \mathbf{a}_{24}^1 \mathbf{a}_{25}^2 \mathbf{a}_{34}^1 \mathbf{a}_{35}^1 \mathbf{a}_{45}^2.$$

When $\mathbf{a}_5 = \mathbf{a}_{15}, \mathbf{a}_{25}, \mathbf{a}_{35}, \mathbf{a}_{45}, \mathbf{t}_{14,45}$ respectively, both g_1 and g_2 are zero, with multiplicity $1, 1, 2, 2, 1$ and $1, 2, 1, 2, 1$ respectively. So the five points are solutions of (3.7.33) in $\mathbf{a}_5$, with multiplicity $\geq 1 \times 1 = 1$, $1 \times 2 = 2$, $2 \times 1 = 2$, $2 \times 2 = 4$, $1 \times 1 = 1$ respectively.

By the classical Bezout's Theorem [2], if two complex quartic curves intersect at isolated points, then there are sixteen points of intersection. For the two quartic curves g_1 and g_2, if the condition of Bezout's Theorem is satisfied, then after removal of the above ten points of intersection, there are six points of intersection left. So to prove Theorem 3.40, We only need to prove that for generic $\{\mathbf{a}_{ij}\}$, the two quartic curves intersect at isolated points. We need the method of proving a **generically negative** conclusion by a single **integer** instance, called the *Gnein method*, to overcome the symbolic computation difficulty in proving such a conclusion.

This method is based on the following obvious facts on complex polynomials of one or two variables. Let u be a free complex parameter, x, y be complex variables, and m be an integer. All the polynomials are assumed to have complex coefficients.

(i) Let $f(u)$ be a polynomial. If $f(m) \neq 0$, then for generic u, $f(u) \neq 0$.

(ii) Let $f(x, u)$ be a polynomial whose leading coefficient in x is $l(u)$. If $l(m) \neq 0$ and $f(x, m)$ is irreducible, then for generic u, $f(x, u)$ is irreducible.

(iii) Let $f(x, u), g(x, u)$ be polynomials whose leading coefficients in x are $l_1(u), l_2(u)$ respectively. If $l_1(m) \neq 0, l_2(m) \neq 0$, then

$$(resultant(f(x, u), g(x, u), x))|_{u=m} = resultant(f(x, m), g(x, m), x).$$

By the Gnein method, replacing a free parameter u by an integer instance m can be used to prove that a complex polynomial with free parameter u is generically nonzero or irreducible, and can evaluate the resultant of two complex polynomials with common free parameter u at the instance $u = m$ by first evaluating the two polynomials at instance m. This is a method of indirect inference. It is used when a computer algebra software is unable to complete the symbolic computation needed to verify a generically negative conclusion directly.

To apply the Gnein method to Erdös' problem, we need to introduce homogeneous coordinates. Since $\mathbf{a}_{45}, \mathbf{a}_{25}, \mathbf{a}_{35}$ are not collinear, they form an affine basis of the affine plane in space, where the last coordinate of every point is 1. Their homogeneous coordinates can be set to

$$\mathbf{a}_{45} = (0,0,1), \quad \mathbf{a}_{25} = (1,0,1), \quad \mathbf{a}_{35} = (0,1,1). \tag{3.7.34}$$

Let

- $\mathbf{a}_\alpha = (x_\alpha, y_\alpha, 1)$ for any other double index $\alpha \in \{ij \,|\, i \neq j\} - \{25, 35, 45\}$;
- $\mathbf{a}_5 = (x, y, z)$;
- $\mathbf{t}_{14,45} = (x_t, y_t, z_t)$. By (3.7.29), generically $z_t \neq 0$, so we may assume $z_t = 1$.

Proposition 3.47. For generic $\{\mathbf{a}_{ij}\}$, any point in a consistent 5-tuple is an affine point. In particular, $\mathbf{a}_5$ is affine, *i.e.*, $z \neq 0$.

Proof. If $z = 0$, by Theorem 3.39, generically x, y cannot be both zero. Assume $y \neq 0$. Since g_1, g_2 are homogeneous multivariate polynomials, we can set $y = 1$. The resultant of $g_1(x, 1, 0), g_2(x, 1, 0)$ with respect to x is a polynomial h in variables x_α, y_α. Using the Gnein method, by replacing the x_α, y_α with special integers generated randomly between 1 and 100, such that the noncollinearity conditions of the $\mathbf{a}_\alpha$ are satisfied, we get that h is nonzero in such an instance of the x_α, y_α, so for generic $\{x_\alpha, y_\alpha\}$, the resultant $h \neq 0$, and polynomial equations $g_1(x, 1, 0) = g_2(x, 1, 0) = 0$ have no solution. $\qquad\square$

Choose $z = 1$. The resultant of $g_1(x, y, 1), g_2(x, y, 1)$ with respect to y is a polynomial $h'(x)$ of degree sixteen in x. This polynomial cannot be factorized by Maple 10. By Corollary 3.46 and the five special solutions of (3.7.33) in $\mathbf{a}_5$ together with their multiplicities, we get that $h'(x)$ has factors

$$x^4(x - x_{15})(x - x_{25})^2(x - x_{35})^2(x - x_t). \tag{3.7.35}$$

Removing these factors from $h'(x)$, we get a polynomial of degree six in x, denoted by $f(x)$.

By the Gnein method, we find that $f(x)$ is irreducible for generic $\{x_\alpha, y_\alpha\}$. Its leading coefficient is generically nonzero. The triangulation result of (3.7.33) is of the form

$$\begin{cases} l(x)y - r(x), \\ f(x), \end{cases}$$

where $l(x), r(x)$ are nonzero polynomials of degree less than six. Again by the Gnein method, we get that when $f(x) = 0$, generically $l(x) \neq 0$. This finishes the proof of Theorem 3.40.

3.7.3 *Proof of Theorem 3.39*

There are five cases (1) to (5) in the theorem. We prove them one by one.

(1). By symmetry we only need to consider the case $\mathbf{a}_4 = \mathbf{a}_5$. Since at least two of the five points are different, without loss of generality, let $\mathbf{a}_1 \neq \mathbf{a}_4$.

If $\mathbf{a}_2 = \mathbf{a}_4$, then $\mathbf{a}_1, \mathbf{a}_4, \mathbf{a}_{14}, \mathbf{a}_{15}, \mathbf{a}_{12}$ are collinear, violating the assumption that the latter three points are not collinear. So $\mathbf{a}_2 \neq \mathbf{a}_4$. Similarly, $\mathbf{a}_3 \neq \mathbf{a}_4$.

So lines $\mathbf{a}_1\mathbf{a}_4 = \mathbf{a}_{14}\mathbf{a}_{15}$, $\mathbf{a}_2\mathbf{a}_4 = \mathbf{a}_{24}\mathbf{a}_{25}$, and $\mathbf{a}_3\mathbf{a}_4 = \mathbf{a}_{34}\mathbf{a}_{35}$. The three lines $\mathbf{a}_{14}\mathbf{a}_{15}, \mathbf{a}_{24}\mathbf{a}_{25}, \mathbf{a}_{34}\mathbf{a}_{35}$ concur (at point $\mathbf{a}_4$), which is generically not true. This proves that generically $\mathbf{a}_4 \neq \mathbf{a}_5$.

We further consider the non-generic case $\mathbf{a}_4 = \mathbf{a}_5$, which is caused by $\mathbf{a}_{14}\mathbf{a}_{15} \vee \mathbf{a}_{24}\mathbf{a}_{25} \vee \mathbf{a}_{34}\mathbf{a}_{35} = 0$.

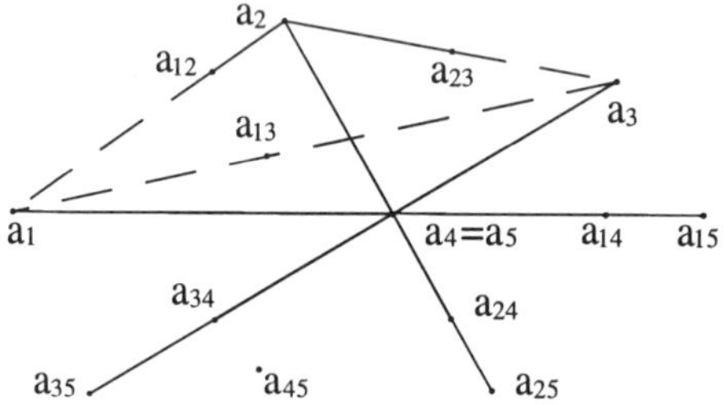

Fig. 3.15   Consistent 5-tuple in the non-generic case $\mathbf{a}_4 = \mathbf{a}_5$.

If $\mathbf{a}_1, \mathbf{a}_2, \mathbf{a}_4$ are collinear, so are $\mathbf{a}_{14}, \mathbf{a}_{15}, \mathbf{a}_{24}$, violating the noncollinearity assumption. So $\mathbf{a}_1, \mathbf{a}_2, \mathbf{a}_4$ are not collinear. Similarly, $\mathbf{a}_1, \mathbf{a}_3, \mathbf{a}_4$ and $\mathbf{a}_2, \mathbf{a}_3, \mathbf{a}_4$ are separately not collinear. If $\mathbf{a}_3 = \mathbf{a}_{13}$, then $\mathbf{a}_{13}, \mathbf{a}_4, \mathbf{a}_{34}, \mathbf{a}_{35}$ are collinear, violating the noncollinearity assumption. So $\mathbf{a}_3 \neq \mathbf{a}_{13}$. Similarly, $\mathbf{a}_i \neq \mathbf{a}_{ij}$ for all $1 \leq i < j \leq 3$. The geometric configuration is shown in Figure 3.15.

The following intersections exist:

$$\mathbf{a}_1 = \mathbf{a}_3\mathbf{a}_{13} \vee \mathbf{a}_4\mathbf{a}_{14}, \quad \mathbf{a}_2 = \mathbf{a}_3\mathbf{a}_{23} \vee \mathbf{a}_4\mathbf{a}_{24}. \tag{3.7.36}$$

Substituting them into $[\mathbf{a}_1\mathbf{a}_2\mathbf{a}_{12}] = 0$ and expanding the result, we get

$$[\mathbf{a}_3\mathbf{a}_4\mathbf{a}_{23}][\mathbf{a}_3\mathbf{a}_{12}\mathbf{a}_{13}][\mathbf{a}_4\mathbf{a}_{14}\mathbf{a}_{24}] = [\mathbf{a}_3\mathbf{a}_4\mathbf{a}_{24}][\mathbf{a}_3\mathbf{a}_{13}\mathbf{a}_{23}][\mathbf{a}_4\mathbf{a}_{12}\mathbf{a}_{14}]. \tag{3.7.37}$$

Replacing $\mathbf{a}_3\mathbf{a}_4$ by $\mathbf{a}_{34}\mathbf{a}_{35}$ in the brackets of (3.7.37), and combining with equation $[\mathbf{a}_3\mathbf{a}_{34}\mathbf{a}_{35}] = 0$, we get

$$\mathbf{a}_3 = \mathbf{a}_{34}\mathbf{a}_{35} \vee \mathbf{a}_{13}([\mathbf{a}_{23}\mathbf{a}_{34}\mathbf{a}_{35}][\mathbf{a}_4\mathbf{a}_{14}\mathbf{a}_{24}]\mathbf{a}_{12} + [\mathbf{a}_{24}\mathbf{a}_{34}\mathbf{a}_{35}][\mathbf{a}_4\mathbf{a}_{12}\mathbf{a}_{14}]\mathbf{a}_{23}), \tag{3.7.38}$$

where the meet product is nonzero because $\mathbf{a}_{13}, \mathbf{a}_{12}, \mathbf{a}_{23}$ and $\mathbf{a}_{34}, \mathbf{a}_{35}, \mathbf{a}_{13}$ are separately noncollinear.

So points $\mathbf{a}_1, \mathbf{a}_2$ are determined by (3.7.36), point $\mathbf{a}_3$ is determined by (3.7.37), and points $\mathbf{a}_4 = \mathbf{a}_5 = \mathbf{a}_{14}\mathbf{a}_{15} \vee \mathbf{a}_{24}\mathbf{a}_{25}$. *The solution of Erdös' problem is unique if* $\mathbf{a}_4 = \mathbf{a}_5$.

(2). By case (1), we can assume that $\mathbf{a}_i \neq \mathbf{a}_j$ for all $i \neq j$. By symmetry, we only need to consider the case $\mathbf{a}_5 = \mathbf{a}_{45}$. Then $\mathbf{a}_4 \neq \mathbf{a}_{45}$.

If $\mathbf{a}_4 = \mathbf{a}_{14}$, then

$$\begin{aligned}
\mathbf{a}_2 &= \mathbf{a}_{14}\mathbf{a}_{24} \vee \mathbf{a}_{25}\mathbf{a}_{45}, \\
\mathbf{a}_3 &= \mathbf{a}_{14}\mathbf{a}_{34} \vee \mathbf{a}_{35}\mathbf{a}_{45}, \\
\mathbf{a}_1 &= \mathbf{a}_{15}\mathbf{a}_{45} \vee \mathbf{a}_{12}(\mathbf{a}_{14}\mathbf{a}_{24} \vee \mathbf{a}_{25}\mathbf{a}_{45}).
\end{aligned} \tag{3.7.39}$$

Substituting the expressions of $\mathbf{a}_2, \mathbf{a}_3$ into $[\mathbf{a}_2\mathbf{a}_3\mathbf{a}_{23}] = 0$, we get

$$[\mathbf{a}_{23}(\mathbf{a}_{14}\mathbf{a}_{24} \vee \mathbf{a}_{25}\mathbf{a}_{45})(\mathbf{a}_{14}\mathbf{a}_{34} \vee \mathbf{a}_{35}\mathbf{a}_{45})] = 0, \tag{3.7.40}$$

which is generically false.

In the non-generic case where $\mathbf{a}_4 = \mathbf{a}_{14}$, by (3.7.39), the solution is unique. The cases $\mathbf{a}_4 = \mathbf{a}_{24}$ and $\mathbf{a}_4 = \mathbf{a}_{34}$ are similar.

Now assume $\mathbf{a}_4 \notin \{\mathbf{a}_{14}, \mathbf{a}_{24}, \mathbf{a}_{34}\}$. Then

$$\begin{aligned}
\mathbf{a}_1 &= \mathbf{a}_{15}\mathbf{a}_{45} \vee \mathbf{a}_4\mathbf{a}_{14}, \\
\mathbf{a}_2 &= \mathbf{a}_{25}\mathbf{a}_{45} \vee \mathbf{a}_4\mathbf{a}_{24}, \\
\mathbf{a}_3 &= \mathbf{a}_{35}\mathbf{a}_{45} \vee \mathbf{a}_4\mathbf{a}_{34}.
\end{aligned} \tag{3.7.41}$$

Substituting them into $[\mathbf{a}_i\mathbf{a}_j\mathbf{a}_{ij}] = 0$ for $1 \leq i < j \leq 3$ and expanding the results, we get two sets of equations by different Cayley expansions. The first set is

$$[\mathbf{a}_4\mathbf{a}_{14}\mathbf{a}_{24}][\mathbf{a}_4\mathbf{a}_{25}\mathbf{a}_{45}][\mathbf{a}_{12}\mathbf{a}_{15}\mathbf{a}_{45}] = [\mathbf{a}_4\mathbf{a}_{12}\mathbf{a}_{14}][\mathbf{a}_4\mathbf{a}_{24}\mathbf{a}_{45}][\mathbf{a}_{15}\mathbf{a}_{25}\mathbf{a}_{45}], \tag{3.7.42}$$

$$[\mathbf{a}_4\mathbf{a}_{14}\mathbf{a}_{34}][\mathbf{a}_4\mathbf{a}_{35}\mathbf{a}_{45}][\mathbf{a}_{13}\mathbf{a}_{15}\mathbf{a}_{45}] = [\mathbf{a}_4\mathbf{a}_{13}\mathbf{a}_{14}][\mathbf{a}_4\mathbf{a}_{34}\mathbf{a}_{45}][\mathbf{a}_{15}\mathbf{a}_{35}\mathbf{a}_{45}], \tag{3.7.43}$$

$$[\mathbf{a}_4\mathbf{a}_{24}\mathbf{a}_{34}][\mathbf{a}_4\mathbf{a}_{35}\mathbf{a}_{45}][\mathbf{a}_{23}\mathbf{a}_{25}\mathbf{a}_{45}] = [\mathbf{a}_4\mathbf{a}_{23}\mathbf{a}_{24}][\mathbf{a}_4\mathbf{a}_{34}\mathbf{a}_{45}][\mathbf{a}_{25}\mathbf{a}_{35}\mathbf{a}_{45}]. \tag{3.7.44}$$

The second set is

$$[\mathbf{a}_4\mathbf{a}_{14}\mathbf{a}_{24}][\mathbf{a}_4\mathbf{a}_{15}\mathbf{a}_{45}][\mathbf{a}_{12}\mathbf{a}_{25}\mathbf{a}_{45}] = [\mathbf{a}_4\mathbf{a}_{12}\mathbf{a}_{24}][\mathbf{a}_4\mathbf{a}_{14}\mathbf{a}_{45}][\mathbf{a}_{15}\mathbf{a}_{25}\mathbf{a}_{45}], \tag{3.7.45}$$

$$[\mathbf{a}_4\mathbf{a}_{14}\mathbf{a}_{34}][\mathbf{a}_4\mathbf{a}_{15}\mathbf{a}_{45}][\mathbf{a}_{13}\mathbf{a}_{35}\mathbf{a}_{45}] = [\mathbf{a}_4\mathbf{a}_{13}\mathbf{a}_{34}][\mathbf{a}_4\mathbf{a}_{14}\mathbf{a}_{45}][\mathbf{a}_{15}\mathbf{a}_{35}\mathbf{a}_{45}], \tag{3.7.46}$$

$$[\mathbf{a}_4\mathbf{a}_{24}\mathbf{a}_{34}][\mathbf{a}_4\mathbf{a}_{25}\mathbf{a}_{45}][\mathbf{a}_{23}\mathbf{a}_{35}\mathbf{a}_{45}] = [\mathbf{a}_4\mathbf{a}_{23}\mathbf{a}_{34}][\mathbf{a}_4\mathbf{a}_{24}\mathbf{a}_{45}][\mathbf{a}_{25}\mathbf{a}_{35}\mathbf{a}_{45}]. \tag{3.7.47}$$

They are equivalent to the three equations in the first set one by one sequentially.

Obviously, $\mathbf{a}_4 = \mathbf{a}_{14}, \mathbf{a}_{24}, \mathbf{a}_{45}$ each satisfy (3.7.42), $\mathbf{a}_4 = \mathbf{a}_{14}, \mathbf{a}_{34}, \mathbf{a}_{45}$ each satisfy (3.7.43), and $\mathbf{a}_4 = \mathbf{a}_{24}, \mathbf{a}_{34}, \mathbf{a}_{45}$ each satisfy (3.7.44).

Introduce homogeneous coordinates:

- $\mathbf{a}_{45} = (0,0,1)$, $\mathbf{a}_{14} = (1,0,1)$, $\mathbf{a}_{24} = (0,1,1)$;
- $\mathbf{a}_\alpha = (x_\alpha, y_\alpha, 1)$ for all other $\alpha \in \{ij \mid i \neq j\} - \{14, 24, 45\}$;
- $\mathbf{a}_4 = (x, y, z)$. If $z = 0$, then we may assume $y \neq 0$ and choose $y = 1$. Then (3.7.42), (3.7.43), (3.7.44) are univariate polynomials with free parameters $\{x_\alpha, y_\alpha\}$. By the Gnein method, the resultant of (3.7.42) and (3.7.43) is generically nonzero. So generically $z \neq 0$, and we can choose $z = 1$.

Denote (3.7.42), (3.7.43), (3.7.44) by

$$f_{12}(x,y) = 0, \quad f_{13}(x,y) = 0, \quad f_{23}(x,y) = 0,$$

and let

$$resultant(f_{13}, f_{23}, y) = h_1(x),$$
$$resultant(f_{12}, f_{23}, y) = h_2(x).$$

Then $x(x - x_{34})$ is a factor of $h_1(x)$, and $x(x - x_{24})$ is a factor of $h_2(x)$. Removing these factors from h_1, h_2, we get two irreducible polynomials h_1', h_2' of degree two in x. By the Gnein method, for generic $\{x_\alpha, y_\alpha\}$, the resultant of h_1', h_2' with respect to x is nonzero. This proves that generically $\mathbf{a}_5 \neq \mathbf{a}_{45}$.

Next we prove that in the non-generic case $\mathbf{a}_5 = \mathbf{a}_{45}$, the number of solutions is finite. (3.7.42) taken as an equation in $\mathbf{a}_4$, represents a conic passing through three explicit points $\mathbf{a}_{14}, \mathbf{a}_{24}, \mathbf{a}_{45}$. Let us find more points on the conic.

If $[\mathbf{a}_4\mathbf{a}_{12}\mathbf{a}_{14}] = 0$, then (3.7.42) is changed into $[\mathbf{a}_4\mathbf{a}_{25}\mathbf{a}_{45}] = 0$. So

$$\mathbf{a}_4 = \mathbf{a}_{12}\mathbf{a}_{14} \vee \mathbf{a}_{25}\mathbf{a}_{45} \tag{3.7.48}$$

is a fourth point on the conic. If $[\mathbf{a}_4\mathbf{a}_{15}\mathbf{a}_{45}] = 0$, then the same conic (3.7.42) but in a different algebraic form (3.7.45) is changed into $[\mathbf{a}_4\mathbf{a}_{12}\mathbf{a}_{24}] = 0$. So

$$\mathbf{a}_4 = \mathbf{a}_{12}\mathbf{a}_{24} \vee \mathbf{a}_{15}\mathbf{a}_{45}. \tag{3.7.49}$$

is a fifth point on the conic.

Among the five points

$$\mathbf{a}_{14}, \ \mathbf{a}_{24}, \ \mathbf{a}_{45}, \ \mathbf{a}_{12}\mathbf{a}_{14} \vee \mathbf{a}_{25}\mathbf{a}_{45}, \ \mathbf{a}_{12}\mathbf{a}_{24} \vee \mathbf{a}_{15}\mathbf{a}_{45}, \tag{3.7.50}$$

no three are collinear, so they determine a unique *nondegenerate conic*, *i.e.*, no three points on the conic are collinear. Similarly, by changing subscript 2 to 3 in (3.7.50), we get that (3.7.43) represents the unique nondegenerate conic determined by points

$$\mathbf{a}_{14}, \ \mathbf{a}_{34}, \ \mathbf{a}_{45}, \ \mathbf{a}_{13}\mathbf{a}_{14} \vee \mathbf{a}_{35}\mathbf{a}_{45}, \ \mathbf{a}_{13}\mathbf{a}_{34} \vee \mathbf{a}_{15}\mathbf{a}_{45}; \tag{3.7.51}$$

by changing subscript 1 to 2 in (3.7.51), we get that (3.7.44) represents the unique nondegenerate conic determined by points

$$\mathbf{a}_{24}, \ \mathbf{a}_{34}, \ \mathbf{a}_{45}, \ \mathbf{a}_{23}\mathbf{a}_{24} \vee \mathbf{a}_{35}\mathbf{a}_{45}, \ \mathbf{a}_{23}\mathbf{a}_{34} \vee \mathbf{a}_{25}\mathbf{a}_{45}. \tag{3.7.52}$$

Erdös' problem has infinitely many solutions if and only if the above three nondegenerate conics are identical. Since any line in the plane has at most two points of intersection with a nondegenerate conic, if the two conics determined by (3.7.50) and (3.7.52) respectively are identical, then among the three collinear points $\mathbf{a}_{45}, \mathbf{a}_{12}\mathbf{a}_{14} \vee \mathbf{a}_{25}\mathbf{a}_{45}, \mathbf{a}_{23}\mathbf{a}_{34} \vee \mathbf{a}_{25}\mathbf{a}_{45}$, two must be identical. It can only be the latter two points that are identical, by the noncollinearity assumption. So

$$\mathbf{a}_{12}\mathbf{a}_{14} \vee \mathbf{a}_{23}\mathbf{a}_{34} \vee \mathbf{a}_{25}\mathbf{a}_{45} = 0.$$

By similar arguments, we get that the three conics determined by (3.7.50), (3.7.51) and (3.7.52) respectively are identical if and only if the following conditions are satisfied:

(i)

$$\mathbf{a}_{12}\mathbf{a}_{14} \vee \mathbf{a}_{23}\mathbf{a}_{34} \vee \mathbf{a}_{25}\mathbf{a}_{45} = 0,$$
$$\mathbf{a}_{13}\mathbf{a}_{14} \vee \mathbf{a}_{23}\mathbf{a}_{24} \vee \mathbf{a}_{35}\mathbf{a}_{45} = 0, \qquad (3.7.53)$$
$$\mathbf{a}_{12}\mathbf{a}_{24} \vee \mathbf{a}_{13}\mathbf{a}_{34} \vee \mathbf{a}_{15}\mathbf{a}_{45} = 0;$$

(ii) points $\mathbf{a}_{14}, \mathbf{a}_{24}, \mathbf{a}_{34}, \mathbf{a}_{45}$ are on the same conic with points

$$\mathbf{p} = \mathbf{a}_{12}\mathbf{a}_{14} \vee \mathbf{a}_{23}\mathbf{a}_{34}, \quad \mathbf{q} = \mathbf{a}_{13}\mathbf{a}_{14} \vee \mathbf{a}_{23}\mathbf{a}_{24}, \quad \mathbf{r} = \mathbf{a}_{12}\mathbf{a}_{24} \vee \mathbf{a}_{13}\mathbf{a}_{34}. \quad (3.7.54)$$

By (3.7.54) and the nondegeneracy of the conic, we get

$$\mathbf{a}_{12} = \mathbf{p}\mathbf{a}_{14} \cap \mathbf{r}\mathbf{a}_{24}, \quad \mathbf{a}_{13} = \mathbf{q}\mathbf{a}_{14} \cap \mathbf{r}\mathbf{a}_{34}, \quad \mathbf{a}_{23} = \mathbf{p}\mathbf{a}_{34} \cap \mathbf{q}\mathbf{a}_{24}. \qquad (3.7.55)$$

By Pascal's Conic Theorem, points $\mathbf{p}, \mathbf{q}, \mathbf{r}, \mathbf{a}_{14}, \mathbf{a}_{24}, \mathbf{a}_{34}$ are on the same conic if and only if $[\mathbf{a}_{12}\mathbf{a}_{13}\mathbf{a}_{23}] = 0$. Since the latter is forbidden, the conic cannot pass through the six points simultaneously. This proves that when $\mathbf{a}_5 = \mathbf{a}_{45}$, there are finitely many solutions to Erdös' problem, or more accurately, there at most $3 + 2 = 5$ solutions, three from $\mathbf{a}_4 = \mathbf{a}_{i4}$ for $i = 1, 2, 3$, and two from the common intersection of the three nondegenerate conics, as any two conics already have two explicit points of intersection $\mathbf{a}_{45}$ and $\mathbf{a}_{j4}$, for some $1 \leq j \leq 3$.

(3). Assume that $\mathbf{a}_i \neq \mathbf{a}_{ij}$. By symmetry we only need to prove $\mathbf{a}_5 \neq \mathbf{a}_{34}$. If $\mathbf{a}_5 = \mathbf{a}_{34}$, by $[\mathbf{a}_{34}\mathbf{a}_3\mathbf{a}_4] = [\mathbf{a}_{34}\mathbf{a}_3\mathbf{a}_{35}] = [\mathbf{a}_{34}\mathbf{a}_4\mathbf{a}_{45}] = 0$, points $\mathbf{a}_3, \mathbf{a}_4, \mathbf{a}_{34}, \mathbf{a}_{35}, \mathbf{a}_{45}$ are collinear, violating the noncollinearity assumption.

(4). We only need to consider the case where $\mathbf{a}_5, \mathbf{a}_{35}, \mathbf{a}_{45}$ are collinear. Assume that $\mathbf{a}_i \neq \mathbf{a}_j$, $\mathbf{a}_i \neq \mathbf{a}_{ij}$, and $\mathbf{a}_i \neq \mathbf{a}_{jk}$ for $i \neq j \neq k$. Then $\mathbf{a}_5, \mathbf{a}_{35}, \mathbf{a}_3, \mathbf{a}_{45}, \mathbf{a}_4, \mathbf{a}_{34}$ are collinear, violating the noncollinearity assumption.

(5). We only need to consider the case where $\mathbf{a}_5, \mathbf{a}_{45}, \mathbf{a}_{34}$ are collinear. Again assume that $\mathbf{a}_i \neq \mathbf{a}_j$, $\mathbf{a}_i \neq \mathbf{a}_{ij}$, and $\mathbf{a}_i \neq \mathbf{a}_{jk}$ for $i \neq j \neq k$. Then $\mathbf{a}_5, \mathbf{a}_{45}, \mathbf{a}_4, \mathbf{a}_{34}, \mathbf{a}_3, \mathbf{a}_{35}$ are collinear, violating the noncollinearity assumption.

Chapter 4

Projective Conic Geometry with Bracket Algebra and Quadratic Grassmann-Cayley Algebra

Because of its nonlinear nature, projective conic geometry is more complicated than incidence geometry. For a basic geometric relation like six points being on the same conic, it is well known that this can be represented by an equality of degree-four bracket binomial, which is the bracket algebraic representation of Pascal's Conic Theorem. It is less well known that there are fifteen such equalities representing the same relation, although in homogeneous coordinates these equalities are identical. It is much less well known how to employ these different equalities to simplify bracket computing involving conic points.

The first part of this chapter is devoted to the development of methods and algorithms for the representation and simplification of conic geometric objects, constraints and their computing based on Cayley bracket algebra. The second part is on a very efficient algorithm based on the "**breefs**" principle, for generating hand-checkable proofs for theorems in projective conic geometry. While the majority of the theorems can be given two-termed proofs, those theorems on intersections of conics are always outside the scope of two-termed proofs.

This phenomenon indicates that the Grassmann-Cayley algebra $\Lambda(\mathcal{V}^3)$ and the associated bracket algebra may not be the intrinsic language for projective conic geometry. They are invariant algebras only for linear forms on $\mathcal{V}^3$. To represent nonlinear forms the base vector space $\mathcal{V}^3$ needs to be enlarged to the space of symmetric tensors generated by vectors in $\mathcal{V}^3$. The Grassmann-Cayley algebra and bracket algebra established upon the space of symmetric tensors of step two, called *quadratic Grassmann-Cayley algebra* and *quadratic bracket algebra* respectively, are the intrinsic language for describing and computing conic geometric problems. The third part of this chapter is an exploration of the two new algebras and their connections with the two old algebras based on $\mathcal{V}^3$.

4.1 Conics with bracket algebra

Conic geometry has a prominent feature of many different representations for the same geometric constraint in the GC algebra over $\mathcal{V}^3$. The first necessary work is to

151

represent in this algebra basic geometric constructions such as conics, intersections, conjugates, poles, polars and tangents.

4.1.1 *Conics determined by points*

There are three kinds of projective conics: (1) a line and itself, called a *double-line conic*, (2) two different lines, called a *line-pair conic*, (3) a conic without any line branch, called a *nondegenerate conic*. The intersection of the two lines in a line-pair conic is the *double point* of the conic. In this chapter, the term "conic" refers to the latter two kinds of conics by default.

Any point on a given conic is called a *conic point*. By Pascal's Conic Theorem, six points $1, 2, 3, 4, 5, 6 \in \mathcal{V}^3$ are on the same conic, called *coconic*, if and only if the intersections $12 \cap 56, 13 \cap 45, 24 \cap 36$ are collinear. Expanding the left side of

$$[(12 \vee 56)(13 \vee 45)(24 \vee 36)] = 0 \tag{4.1.1}$$

into a bracket binomial by splitting $5, 6$ in $12 \vee 56$, we get

$$\mathrm{conic}(123456) := [135][245][126][346] - [125][345][136][246]. \tag{4.1.2}$$

Let $1, 2, 3, 4, 5$ be five points in the projective plane. Assume that no four of them are collinear. This inequality condition is denoted by $\exists 12345$. A classical conclusion is that such five points determine a unique conic, denoted by $\mathbf{12345}$, such that any point $\mathbf{x}$ is on the conic if and only if

$$\mathrm{conic}(12345\mathbf{x}) = 0. \tag{4.1.3}$$

The juxtaposition here does not denote the outer product of vectors, although the antisymmetry property still exists. Such a conic is called a *point-conic*, meaning that it is determined by points.

Proposition 4.1. For any six points $1, \ldots, 6$ in the plane, the expression $\mathrm{conic}(123456)$ is antisymmetric with respect to the six points. Furthermore, for any point $6'$ in the plane,

$$\begin{aligned} &[126'][346']\mathrm{conic}(123456) + [125][345]\mathrm{conic}(123466') \\ &= [126][346]\,\mathrm{conic}(123456'). \end{aligned} \tag{4.1.4}$$

Proof. Obviously, $\mathrm{conic}(123456)$ is antisymmetric within each of the pairs $(1, 4)$, $(2, 3)$ and $(5, 6)$. We only need to prove the antisymmetries within $(1, 2)$ and within $(1, 5)$. This can be verified by contractions:

$$\begin{aligned} &\mathrm{conic}(123456) + \mathrm{conic}(213456) \\ =\ &[126][346]([135][245] - [235][145]) - [125][345]([136][246] - [236][146]) \\ \overset{contract}{=}\ &0, \end{aligned}$$

$$\begin{aligned} &\mathrm{conic}(123456) + \mathrm{conic}(523416) \\ =\ &[135][346]([245][126] + [124][256]) - [125][246]([345][136] + [134][356]) \\ \overset{contract}{=}\ &0. \end{aligned}$$

(4.1.4) is a trivial identity obtained from (4.1.2). $\qquad\square$

Corollary 4.2. If points $1, \dots, 5, 1', \dots, 5'$ are coconic, then for any point $\mathbf{a}$ in the plane,

$$
\begin{aligned}
\frac{\mathrm{conic}(\mathbf{a}12345)}{\mathrm{conic}(\mathbf{a}12345')} &= \frac{[125][345]}{[125'][345']}, \\
\frac{\mathrm{conic}(\mathbf{a}12345)}{\mathrm{conic}(\mathbf{a}1234'5')} &= \frac{[124][125][345]}{[124'][125'][34'5']}, \\
\frac{\mathrm{conic}(\mathbf{a}12345)}{\mathrm{conic}(\mathbf{a}123'4'5')} &= \frac{[123][124][125][345]}{[123'][124'][125'][3'4'5']}.
\end{aligned}
\tag{4.1.5}
$$

Proof. The first formula is a corollary of (4.1.4). The second formula comes from

$$
\frac{\mathrm{conic}(\mathbf{a}12345)}{\mathrm{conic}(\mathbf{a}1234'5')} = \frac{\mathrm{conic}(\mathbf{a}12345)}{\mathrm{conic}(\mathbf{a}12345')} \frac{\mathrm{conic}(\mathbf{a}12345')}{\mathrm{conic}(\mathbf{a}1234'5')}.
$$

The third formula can be obtained similarly. $\qquad\square$

Recall that if there are more than two points on the same line, then the line can be represented by the outer product of any two points. The ratio of the two representations is a rational invariant. Similarly, if there are more than five points on the same conic, then the conic can be represented by any five points. Different representations differ by a ratio of bracket monomials as in (4.1.5), which is a rational invariant of the conic.

Definition 4.3. In the GC algebra of 2D projective geometry, let $p(\mathcal{A})$ be a polynomial containing vector variables $\mathcal{A} = \{\mathbf{a}_1, \mathbf{a}_2, \mathbf{a}_3, \mathbf{a}_4, \mathbf{a}_5\}$ which represent five points on the same conic. If for some $1 \le i \le 5$, for any two points $1, 1'$ on the conic,

$$
\frac{p(\mathcal{A})\,|_{\mathbf{a}_i=1}}{p(\mathcal{A})\,|_{\mathbf{a}_i=1'}} = \frac{[\mathbf{1}\mathbf{a}_{k_1}\mathbf{a}_{k_2}][\mathbf{1}\mathbf{a}_{k_3}\mathbf{a}_{k_4}]}{[\mathbf{1'}\mathbf{a}_{k_1}\mathbf{a}_{k_2}][\mathbf{1'}\mathbf{a}_{k_3}\mathbf{a}_{k_4}]}
\tag{4.1.6}
$$

for any permutation k_1, k_2, k_3, k_4 of $\{1, 2, 3, 4, 5\} - \{i\}$ such that the right side of (4.1.6) is nonzero, then $p(\mathcal{A})$ is said to satisfy the *point-conic transformation rule* in $\mathbf{a}_i$.

The right side of (4.1.6) is called the *transformation coefficient* of $p(\mathcal{A})$ in $\mathbf{a}_i$ from 1 to $1'$. It is independent of the permutation of $\mathbf{a}_{k_1}, \mathbf{a}_{k_2}, \mathbf{a}_{k_3}, \mathbf{a}_{k_4}$, according to (4.1.2).

Besides (4.1.2), a point conic has a degree-five bracket trinomial representation as follows:

Proposition 4.4. Any point $\mathbf{x}$ on conic 12345 satisfies the following equation:

$$
\begin{aligned}
\mathrm{conic}_{123,45}(\mathbf{x}) := {}& [145][234][235][\mathbf{x}12][\mathbf{x}13] - [134][135][245][\mathbf{x}12][\mathbf{x}23] \\
& + [124][125][345][\mathbf{x}13][\mathbf{x}23] \\
={}& 0.
\end{aligned}
\tag{4.1.7}
$$

The expression $\text{conic}_{123,45}(\mathbf{x})$ is symmetric with respect to $\mathbf{1}, \mathbf{2}, \mathbf{3}$ but antisymmetric with respect to $\mathbf{4}, \mathbf{5}$. It satisfies the point-conic transformation rules in $\mathbf{4}$ and $\mathbf{5}$. Furthermore, if $\mathbf{1}, \mathbf{2}, \mathbf{3}$ are not collinear, then (4.1.7) is equivalent to (4.1.2).

Proof. Applying the following GP relations to (4.1.7),

$$[\mathbf{134}][\mathbf{x23}] = [\mathbf{123}][\mathbf{x34}] + [\mathbf{234}][\mathbf{x13}],$$
$$[\mathbf{124}][\mathbf{x23}] = [\mathbf{123}][\mathbf{x24}] + [\mathbf{234}][\mathbf{x12}],$$

we get

$$\text{conic}_{123,45}(\mathbf{x}) = [\mathbf{123}]([\mathbf{125}][\mathbf{345}][\mathbf{13x}][\mathbf{24x}] - [\mathbf{135}][\mathbf{245}][\mathbf{12x}][\mathbf{34x}])$$
$$+ [\mathbf{x12}][\mathbf{x13}][\mathbf{234}]([\mathbf{145}][\mathbf{235}] + [\mathbf{125}][\mathbf{345}] - [\mathbf{135}][\mathbf{245}])$$
$$\overset{contract}{=} -[\mathbf{123}]\,\text{conic}(\mathbf{12345x}).$$

$\square$

Polars and tangents:

For four distinct points $\mathbf{a}, \mathbf{b}, \mathbf{c}, \mathbf{d}$ on the same line, pairs $(\mathbf{a}, \mathbf{b})$ and $(\mathbf{c}, \mathbf{d})$ are said to be *conjugate* with respect to each other, if the cross-ratio $(\mathbf{ab}; \mathbf{cd}) = -1$. If $\mathbf{a} = \mathbf{b}$, then the pair $(\mathbf{a}, \mathbf{a})$ is *conjugate* with respect to any pair $(\mathbf{a}, \mathbf{d})$ where $\mathbf{d}$ is any point on the line.

Two distinct points $\mathbf{a}, \mathbf{b}$ are said to be *conjugate* with respect to a conic, if either they are conjugate with respect to the points $\mathbf{c}, \mathbf{d}$ where line $\mathbf{ab}$ meets the conic, or line $\mathbf{ab}$ is part of the conic. A point is *conjugate* to itself with respect to a conic if it is on the conic.

If $\mathbf{a}$ is not a double point of a conic, *i.e.*, not the intersection of the two lines of a line-pair conic, then the conjugates of $\mathbf{a}$ with respect to the conic form a line, called the *polar* of $\mathbf{a}$. In particular, if $\mathbf{a}$ is on the conic, its polar is the *tangent* at $\mathbf{a}$. Dually, the points on a line L which is not part of a conic, have a unique common conjugate with respect to the conic, called the *pole* of L. When L is tangent to the conic, its pole is the *point of tangency*. The term *tangency* agrees with the geometric intuition that a line is tangent to a conic if and only if the point of tangency is where they meet twice, *i.e.*, with multiplicity two.

Definition 4.5. The *polarization* (or *derivative*) of the function $\text{conic}(\mathbf{12345a})$ of $\mathbf{a}$ by $\mathbf{b}$ is the polynomial

$$\text{conic}(\mathbf{12345(ab)}) := \frac{1}{2}\{[\mathbf{135}][\mathbf{245}]([\mathbf{12a}][\mathbf{34b}] + [\mathbf{12b}][\mathbf{34a}]) \tag{4.1.8}$$
$$- [\mathbf{125}][\mathbf{345}]([\mathbf{13a}][\mathbf{24b}] + [\mathbf{13b}][\mathbf{24a}])\}.$$

Proposition 4.6. Two points $\mathbf{a}, \mathbf{b}$ are conjugate with respect to conic $\mathbf{12345}$ if and only if

$$\text{conic}(\mathbf{12345(ab)}) = 0. \tag{4.1.9}$$

When $\mathbf{a}$ is not a double point of the conic, its polar has the following bivector expression:

$$\text{polar}_{\mathbf{a}}(\mathbf{12345}) := [135][245]([12\mathbf{a}]\,34 + [34\mathbf{a}]\,12) - [125][345]([13\mathbf{a}]\,24 + [24\mathbf{a}]\,13).$$
$$(4.1.10)$$

The expressions $\text{conic}(\mathbf{12345}(\mathbf{ab}))$ and $\text{polar}_{\mathbf{a}}(\mathbf{12345})$ are both antisymmetric in $\mathbf{1, 2, 3, 4, 5}$, and follow the point-conic transformation rules in the five points.

Proof. If $\mathbf{a} = \mathbf{b}$, we have

$$\text{conic}(\mathbf{12345}(\mathbf{aa})) = \text{conic}(\mathbf{12345a}). \qquad (4.1.11)$$

If $\mathbf{a} \neq \mathbf{b}$ and $\mathbf{a}$ is not a conic point, let $\mathbf{x}$ be a point of intersection of line $\mathbf{ab}$ with the conic. Let $\lambda = -[\mathbf{bx}]/[\mathbf{ax}]$. Substituting $\mathbf{x} = \mathbf{b} + \lambda\mathbf{a}$ into $\text{conic}(\mathbf{12345x})$ and expanding the result, we get

$$\begin{aligned}
\text{conic}(\mathbf{12345x}) = \; & [135][245]([12\mathbf{b}] + \lambda[12\mathbf{a}])([34\mathbf{b}] + \lambda[34\mathbf{a}]) \\
& -[125][345]([13\mathbf{b}] + \lambda[13\mathbf{a}])([24\mathbf{b}] + \lambda[24\mathbf{a}]) \\
= \; & \text{conic}(\mathbf{12345b}) + 2\lambda\,\text{conic}(\mathbf{12345}(\mathbf{ab})) + \lambda^2\,\text{conic}(\mathbf{12345a}).
\end{aligned}$$
$$(4.1.12)$$

Let $\mathbf{c}, \mathbf{d}$ be the points of intersection of line $\mathbf{ab}$ with the conic. They are conjugate with respect to $\mathbf{a}, \mathbf{b}$, *i.e.*, $[\mathbf{bc}]/[\mathbf{ac}] = -[\mathbf{bd}]/[\mathbf{ad}]$, if and only if the sum of the two roots of (4.1.12) in λ is zero, *i.e.*, the linear part of the polynomial is zero, which is just (4.1.9).

If $\mathbf{a} \neq \mathbf{b}$ and $\mathbf{b}$ is not a conic point, we still have (4.1.9). If $\mathbf{a} \neq \mathbf{b}$ and both are conic points, they are conjugate with respect to the conic if and only if line $\mathbf{ab}$ is part of the conic, *i.e.*, if and only if (4.1.12) equals zero for any scalar λ. Since the quadratic and constant parts of the polynomial are already zero, the linear part must also be zero.

By removing $\mathbf{b}$ from both sides of (4.1.8) we get (4.1.10), *i.e.*,

$$[\text{polar}_{\mathbf{a}}(\mathbf{12345})\mathbf{x}] = 2\,\text{conic}(\mathbf{12345}(\mathbf{ax})), \quad \forall \mathbf{x} \in \mathcal{V}^3. \qquad (4.1.13)$$

$\square$

Proposition 4.7. The tangent of conic $\mathbf{12345}$ at point $\mathbf{5}$ which is assumed not being a double point of the conic, has the following bivector expression:

$$\text{tangent}_{\mathbf{5,1234}} := \text{polar}_{\mathbf{5}}(\mathbf{12345}) = -[134][235][245]\,15 + [135][145][234]\,25.$$
$$(4.1.14)$$

The expression is antisymmetric with respect to $\mathbf{1, 2, 3, 4}$ and satisfies the point-conic transformation rules in the four points.

Proof. Setting $\mathbf{a} = \mathbf{5}$ in (4.1.10), we get, for any $\mathbf{x} \in \mathcal{V}^3$,

$$\begin{aligned}
[\text{polar}_{\mathbf{5}}(\mathbf{12345})\mathbf{x}] \; = \; & [125][135]([245][34\mathbf{x}] - [24\mathbf{x}][345]) \\
& +[245][345]([12\mathbf{x}][135] - [125][13\mathbf{x}]) \\
\overset{contract}{=} \; & [125][135][234][45\mathbf{x}] - [123][15\mathbf{x}][245][345] \\
= \; & -[\text{tangent}_{\mathbf{5,1432}}\,\mathbf{x}].
\end{aligned}$$
$$(4.1.15)$$

$\square$

Poles:

The pole of line $\mathbf{ab}$ with respect to a conic is the intersection of the polars of points $\mathbf{a}, \mathbf{b}$ respectively. Direct expansion of the meet product of the polars leads to a complicated expression of the pole. To obtain a succinct expression, we consider constructing the pole by two tangents.

Let $\mathbf{1}, \mathbf{2}$ be the points of intersection of line $\mathbf{ab}$ and conic $\mathbf{12345}$. Then

$$
\begin{aligned}
-\text{tangent}_{\mathbf{1,2345}} \vee \text{tangent}_{\mathbf{2,1345}} = \ & ([\mathbf{134}][\mathbf{135}][\mathbf{245}]\mathbf{12} - [\mathbf{124}][\mathbf{125}][\mathbf{345}]\mathbf{13}) \\
& \vee ([\mathbf{145}][\mathbf{234}][\mathbf{235}]\mathbf{12} + [\mathbf{124}][\mathbf{125}][\mathbf{345}]\mathbf{23}) \\[4pt]
= \ & [\mathbf{123}][\mathbf{124}][\mathbf{125}][\mathbf{345}] \, ([\mathbf{145}][\mathbf{234}][\mathbf{235}]\mathbf{1} \\
& +[\mathbf{134}][\mathbf{135}][\mathbf{245}]\mathbf{2} - [\mathbf{124}][\mathbf{125}][\mathbf{345}]\mathbf{3}).
\end{aligned}
\tag{4.1.16}
$$

From this we get the following representation of a pole:

Proposition 4.8. If line $\mathbf{12}$ is not a branch of conic $\mathbf{12345}$, then the pole of line $\mathbf{12}$ with respect to the conic is

$$
\text{pole}_{\mathbf{12,345}} := [\mathbf{145}][\mathbf{234}][\mathbf{235}]\mathbf{1} + [\mathbf{134}][\mathbf{135}][\mathbf{245}]\mathbf{2} - [\mathbf{124}][\mathbf{125}][\mathbf{345}]\mathbf{3}.
\tag{4.1.17}
$$

It is symmetric with respect to $\mathbf{1}, \mathbf{2}$, antisymmetric with respect to $\mathbf{3}, \mathbf{4}, \mathbf{5}$, and follows the point-conic transformation rules in $\mathbf{3}, \mathbf{4}, \mathbf{5}$.

Corollary 4.9. For points $\mathbf{3}', \mathbf{4}', \mathbf{5}'$ on conic $\mathbf{12345}$,

$$
\text{tangent}_{\mathbf{1,2345}} \vee \text{tangent}_{\mathbf{2,13'4'5'}} = [\mathbf{123'}][\mathbf{124'}][\mathbf{125'}][\mathbf{3'4'5'}]\,\text{pole}_{\mathbf{12,345}},
\tag{4.1.18}
$$

$$
\text{pole}_{\mathbf{12,345}}\,\text{pole}_{\mathbf{13,24'5'}} = -2\,[\mathbf{14'5'}][\mathbf{234'}][\mathbf{235'}]\,\text{tangent}_{\mathbf{1,2345}}.
\tag{4.1.19}
$$

Proof. (4.1.18) is from (4.1.16) by the point-conic transformation rules of $\text{tangent}_{\mathbf{2,1345}}$ in $\mathbf{3}, \mathbf{4}, \mathbf{5}$. When $\mathbf{4} = \mathbf{4}'$ and $\mathbf{5} = \mathbf{5}'$, (4.1.19) is straightforward from (4.1.17). By the point-conic transformation rules of $\text{pole}_{\mathbf{13,245}}$ in $\mathbf{4}, \mathbf{5}$,

$$
\begin{aligned}
\text{pole}_{\mathbf{12,345}}\,\text{pole}_{\mathbf{13,24'5'}} &= \frac{[\mathbf{14'5'}][\mathbf{234'}][\mathbf{235'}]}{[\mathbf{145}][\mathbf{234}][\mathbf{235}]}\,\text{pole}_{\mathbf{12,345}}\,\text{pole}_{\mathbf{13,245}} \\
&= -2\,[\mathbf{14'5'}][\mathbf{234'}][\mathbf{235'}]\,\text{tangent}_{\mathbf{1,2345}}.
\end{aligned}
$$

$\qquad\square$

Corollary 4.9 characterizes the relations among different Grassmann-Cayley representations of poles and tangents when there are more than five coconic points. The proof shows the power of point-conic transformation rules. The following proposition provides more elegant results on the brackets formed by poles and tangents.

Proposition 4.10. Let $\mathbf{1}, \ldots, \mathbf{5}; \mathbf{2}', \ldots, \mathbf{5}'; \mathbf{1}'', \ldots, \mathbf{5}''$ be coconic points, then

(1)

$$[\text{pole}_{45,123}\ \text{tangent}_{1,2'3'4'5}] = 2\,[124][12'5][134][13'5][14'5][2'3'4'][235].\quad (4.1.20)$$

(2)

$$[\text{pole}_{12,34''5''}\ \text{pole}_{13,24'5'}\ \text{pole}_{45,1''23}]$$
$$= -4\,[1''24][125][134][1''35][14'5'][14''5''][234'][234''][235'][235''],$$
$$[\text{pole}_{12,34''5''}\ \text{pole}_{13,245'}\ \text{pole}_{45,12''3''}]$$
$$= -4\,[124][12''5][134][13''5][14''5''][145'][2''3''4][234''][235''][235'].$$
$$\quad (4.1.21)$$

(3)

$$[1\,\text{pole}_{12,3'45}\ \text{pole}_{3'4',12'5'}] = 2\,[123'][12'4'][13'4][13'5][14'5'][2'3'5'][245],$$
$$\quad (4.1.22)$$

and

$$\frac{[1\,\text{pole}_{12,345}\ \text{pole}_{34,125}]}{[2\,\text{pole}_{12,345}\ \text{pole}_{34,125}]} = \frac{[134]}{[234]},\qquad \frac{[1\,\text{pole}_{12,345}\ \text{pole}_{34,125}]}{[3\,\text{pole}_{12,345}\ \text{pole}_{34,125}]} = -\frac{[124]}{[234]}.\quad (4.1.23)$$

Proof. As the proofs are similar, we only prove (4.1.20). First remove all the primes. We have

$$[\text{pole}_{45,123}\ \text{tangent}_{1,2345}]$$
$$= -[124][125][135]\{[134][235]([124][345] - [134][245])$$
$$\qquad\qquad -[134][234]([125][345] - [135][245])\}$$
$$= 2\,[124][125][134][135][145][234][235].$$

By the point-conic transformation rules of $\text{tangent}_{1,2345}$ in $2, 3, 4$, we get (4.1.20). $\qquad\square$

Intersections:

In the projective plane, if a line and a conic intersect at a given point, then they must intersect at a second point, disregard of the base field of the geometry. The reason is that the *second point of intersection* satisfies a linear equation by removing a factor representing the given point of intersection as a root.

Proposition 4.11. If line **ab** is not part of conic **a1234**, then their second point of intersection always exists, denoted by $\mathbf{x} = \mathbf{ab} \cap \mathbf{a1234}$. It has the following representation in GC algebra:

$$\mathbf{x}_{\mathbf{ab},1234} := [134][24a][3ab]\,12 \vee \mathbf{ab} - [124][34a][2ab]\,13 \vee \mathbf{ab}.\quad (4.1.24)$$

(4.1.24) is antisymmetric with respect to $\mathbf{1, 2, 3, 4}$ and satisfies the point-conic transformation rules in the four points.

Proof. Expanding (4.1.24) by separating $\mathbf{a}, \mathbf{b}$ in the meet products, we get

$$
\begin{aligned}
\mathbf{x}_{\mathbf{ab},1234} =\;& ([12\mathbf{b}][134][24\mathbf{a}][3\mathbf{ab}] - [124][13\mathbf{b}][2\mathbf{ab}][34\mathbf{a}])\mathbf{a} \\
&+ ([124][13\mathbf{a}][2\mathbf{ab}][34\mathbf{a}] - [12\mathbf{a}][134][24\mathbf{a}][3\mathbf{ab}])\mathbf{b} \\
=\;& c\mathbf{a} + t\mathbf{b},
\end{aligned}
\tag{4.1.25}
$$

where by (4.1.13) and (4.1.15),

$$
c = \mathrm{conic}(\mathbf{123ab4}), \qquad t = -[\mathbf{b}\,\mathrm{tangent}_{\mathbf{a},2314}] = -2\,\mathrm{conic}(\mathbf{1234(ab)}).
$$

Since line $\mathbf{ab}$ is not part of the conic, c and t cannot be both zero. Substituting (4.1.25) into $\mathrm{conic}(\mathbf{4321ax}) = [12\mathbf{x}][13\mathbf{a}][24\mathbf{a}][34\mathbf{x}] - [12\mathbf{a}][13\mathbf{x}][24\mathbf{x}][34\mathbf{a}]$, we get

$$
\begin{aligned}
\mathrm{conic}(\mathbf{4321ax}_{\mathbf{ab},1234}) =\;& ct\{[12\mathbf{a}][24\mathbf{a}]([13\mathbf{a}][34\mathbf{b}] - [13\mathbf{b}][34\mathbf{a}]) \\
&+ [13\mathbf{a}][34\mathbf{a}]([12\mathbf{b}][24\mathbf{a}] - [12\mathbf{a}][24\mathbf{b}])\} \\
&+ t^2([12\mathbf{b}][13\mathbf{a}][24\mathbf{a}][34\mathbf{b}] - [12\mathbf{a}][13\mathbf{b}][24\mathbf{b}][34\mathbf{a}]) \\
=\;& ct([12\mathbf{a}][134][24\mathbf{a}][3\mathbf{ab}] - [124][13\mathbf{a}][2\mathbf{ab}][34\mathbf{a}]) + ct^2 \\
=\;& 0.
\end{aligned}
$$

So $\mathbf{x}$ is a point on both line $\mathbf{ab}$ and conic $\mathbf{a1234}$. $\qquad\qquad\square$

Corollary 4.12. Let points $\mathbf{1}, \mathbf{2}, \mathbf{a}$ be noncollinear. If intersections $\mathbf{1'} = \mathbf{12345} \cap \mathbf{1a}$ and $\mathbf{2'} = \mathbf{12345} \cap \mathbf{2a}$ have their expressions as (4.1.24), then

$$
\mathrm{polar}_{\mathbf{a}}(\mathbf{12345}) = \mathbf{12}'_{\mathbf{2a},1345} - \mathbf{21}'_{\mathbf{1a},2345}.
\tag{4.1.26}
$$

Likewise, in the projective plane, if two conics intersect at three given points, then after removing three factors representing the given points of intersection as roots, the *fourth point of intersection* satisfies a linear equation, disregard of the base field of the geometry. For two distinct conics $\mathbf{12345}$ and $\mathbf{1234'5'}$, if none of lines $\mathbf{12}$, $\mathbf{13}$, $\mathbf{23}$ is part of the two conics, then $\mathbf{1}, \mathbf{2}, \mathbf{3}$ are not collinear, and the two conics do not have a line in common. So they have four points of intersection. Denote the fourth point of intersection by $\mathbf{12345} \cap \mathbf{1234'5'}$.

Proposition 4.13. $\mathbf{12345} \cap \mathbf{1234'5'}$ has the following coordinate representation:

$$
\lambda_2\lambda_3\,\mathbf{1} + \lambda_1\lambda_3\,\mathbf{2} + \lambda_1\lambda_2\,\mathbf{3},
\tag{4.1.27}
$$

where in $\mathbb{K}^3$ ($\mathbb{K}$ is the base field of the geometry),

$$
\frac{1}{[123]}
\begin{pmatrix} \lambda_1 \\ \lambda_2 \\ \lambda_3 \end{pmatrix}
=
\begin{pmatrix} [145][234][235] \\ [134][135][245] \\ [124][125][345] \end{pmatrix}
\times
\begin{pmatrix} [14'5'][234'][235'] \\ [134'][135'][24'5'] \\ [124'][125'][34'5'] \end{pmatrix}.
\tag{4.1.28}
$$

Here "$\times$" is the cross product in the vector algebra over $\mathbb{K}^3$.

Furthermore, for any permutation $\mathbf{i}, \mathbf{j}, \mathbf{k}$ of $\mathbf{1}, \mathbf{2}, \mathbf{3}$,

$$
\begin{aligned}
\lambda_i =\;& [\mathbf{i}\,\mathrm{pole}_{\mathbf{jk},i45}\,\mathrm{pole}_{\mathbf{jk},i4'5'}] \\
=\;& -[\mathbf{i}\,\mathrm{pole}_{\mathbf{ij},k45}\,\mathrm{pole}_{\mathbf{ij},k4'5'}] \\
=\;& [\mathbf{i}\,\mathrm{pole}_{\mathbf{ij},k45}\,\mathrm{pole}_{\mathbf{ik},j4'5'}].
\end{aligned}
\tag{4.1.29}
$$

Proof. Denote the right side of (4.1.28) by $\mathbf{v}$. Let the fourth point of intersection be $\mathbf{x}$. Since $[\mathbf{123}] \neq 0$, a point $\mathbf{y}$ is on conic $\mathbf{12345}$ if and only if $\mathrm{conic}_{\mathbf{123},\mathbf{45}}(\mathbf{y}) = 0$. By (4.1.7),

$$
\begin{aligned}
\mathrm{conic}_{\mathbf{123},\mathbf{45}}(\mathbf{x}) &= [\mathbf{145}][\mathbf{234}][\mathbf{235}][\mathbf{x12}][\mathbf{x13}] - [\mathbf{134}][\mathbf{135}][\mathbf{245}][\mathbf{x12}][\mathbf{x23}] \\
&\qquad\qquad + [\mathbf{124}][\mathbf{125}][\mathbf{345}][\mathbf{x13}][\mathbf{x23}] \\
&= 0,
\end{aligned}
$$

$$
\begin{aligned}
\mathrm{conic}_{\mathbf{123},\mathbf{4'5'}}(\mathbf{x}) &= [\mathbf{14'5'}][\mathbf{234'}][\mathbf{235'}][\mathbf{x12}][\mathbf{x13}] - [\mathbf{134'}][\mathbf{135'}][\mathbf{24'5'}][\mathbf{x12}][\mathbf{x23}] \\
&\qquad\qquad + [\mathbf{124'}][\mathbf{125'}][\mathbf{34'5'}][\mathbf{x13}][\mathbf{x23}] \\
&= 0.
\end{aligned}
$$

$$(4.1.30)$$

So $\mathbf{v} = 0$ if and only if the above two equations differ by a constant scale, *i.e.*, if and only if the two conics are identical. Since the latter is not true, this proves $\mathbf{v} \neq 0$.

By (4.1.30), $\mathbf{v}$ is parallel to vector $(\lambda'_1, \lambda'_2, \lambda'_3)^T$, where

$$
\lambda'_1 = [\mathbf{x12}][\mathbf{x13}], \quad \lambda'_2 = -[\mathbf{x12}][\mathbf{x23}], \quad \lambda'_3 = [\mathbf{x13}][\mathbf{x23}].
$$

So vectors $(\lambda'_1, \lambda'_2, \lambda'_3)^T$ and $(\lambda_1, \lambda_2, \lambda_3)^T$ are parallel. By this and the Cramer's rule $[\mathbf{123}]\mathbf{x} = [\mathbf{x23}]\mathbf{1} - [\mathbf{x13}]\mathbf{2} + [\mathbf{x12}]\mathbf{3}$, we get that vector $\mathbf{x}$ is parallel to vector $(\lambda'_2\lambda'_3, \lambda'_1\lambda'_3, \lambda'_1\lambda'_2)^T$ in the coordinate system $\{\mathbf{1}, \mathbf{2}, \mathbf{3}\}$. Since vectors $(\lambda_2\lambda_3, \lambda_1\lambda_3, \lambda_1\lambda_2)^T$ and $(\lambda'_2\lambda'_3, \lambda'_1\lambda'_3, \lambda'_1\lambda'_2)^T$ are parallel, (4.1.27) follows.

(4.1.29) can be easily deduced from (4.1.17) and (4.1.28). $\qquad\square$

The representation of the fourth point of intersection is very complicated. Its compact representation in GC algebra by meet products is still not found.

4.1.2 Conics determined by tangents and points

A conic can be determined by any five linear objects incident to it: conic points or tangents. In this section, we consider conics determined by one or two points of tangency and the corresponding tangent lines, together with three or one conic point respectively.

Tangent-point conic:

Proposition 4.14. There exists a unique conic passing through points $\mathbf{1}, \mathbf{2}, \mathbf{3}, \mathbf{4}$ and tangent to line $\mathbf{45}$, denoted by $\mathbf{1234(45)}$, if the following set of nondegeneracy conditions, denoted by $\exists\mathbf{1234(45)}$, are satisfied:

- $\mathbf{1}, \mathbf{2}, \mathbf{3}, \mathbf{4}$ are distinct and noncollinear,
- $\mathbf{4}, \mathbf{5}$ are distinct,
- either (a) points $\mathbf{1}, \mathbf{2}, \mathbf{3}$ are not on line $\mathbf{45}$, and point $\mathbf{4}$ is not on any of the lines $\mathbf{12}, \mathbf{13}, \mathbf{23}$, or (b) only one of the three points $\mathbf{1}, \mathbf{2}, \mathbf{3}$ is on line $\mathbf{45}$.

A point $\mathbf{x}$ is on the conic if and only if for the polarization function (4.1.8),

$$2\,\mathrm{conic}(\mathbf{x1234(45)}) = [\mathbf{134}][\mathbf{245}][\mathbf{14x}][\mathbf{23x}] - [\mathbf{234}][\mathbf{145}][\mathbf{13x}][\mathbf{24x}] = 0. \quad (4.1.31)$$

Proof. By the given nondegeneracy conditions, if $[\mathbf{145}] = 0$, then $[\mathbf{134}], [\mathbf{245}]$ are nonzero, and (4.1.31) is the equation of the line-pair conic $(\mathbf{14}, \mathbf{23})$. If $[\mathbf{245}] = 0$, then (4.1.31) is the equation of the line-pair conic $(\mathbf{13}, \mathbf{24})$. If $[\mathbf{345}] = 0$, by collinearity transformation $[\mathbf{234}][\mathbf{145}] = [\mathbf{134}][\mathbf{245}]$, (4.1.31) becomes

$$
\begin{aligned}
0 &= [\mathbf{134}][\mathbf{245}]([\mathbf{14x}][\mathbf{23x}] - [\mathbf{13x}][\mathbf{24x}]) \\
&\overset{contract}{=} -[\mathbf{134}][\mathbf{245}][\mathbf{12x}][\mathbf{34x}],
\end{aligned}
$$

which is the equation of the line-pair conic $(\mathbf{12}, \mathbf{34})$. The conclusion is true in the three degenerate cases.

Below we assume that points $\mathbf{1}, \mathbf{2}, \mathbf{3}$ are not on line $\mathbf{45}$. By the given hypotheses, $[\mathbf{124}], [\mathbf{134}], [\mathbf{234}]$ are all nonzero. If $[\mathbf{123}] = 0$, by collinearity transformation $[\mathbf{234}][\mathbf{13x}] = [\mathbf{134}][\mathbf{23x}]$, (4.1.31) becomes

$$
\begin{aligned}
0 &= [\mathbf{134}][\mathbf{23x}]([\mathbf{245}][\mathbf{14x}] - [\mathbf{145}][\mathbf{24x}]) \\
&\overset{contract}{=} -[\mathbf{124}][\mathbf{134}][\mathbf{23x}][\mathbf{45x}],
\end{aligned}
$$

which is the equation of the line-pair conic $(\mathbf{23}, \mathbf{45})$. The conclusion also holds in this degenerate case.

So we further assume $[\mathbf{123}] \neq 0$. Then no three of the four points $\mathbf{1}, \mathbf{2}, \mathbf{3}, \mathbf{4}$ are collinear. Let $\mathbf{x}$ be any point in the plane distinct from the four points, then no four of the five points $\mathbf{1}, \mathbf{2}, \mathbf{3}, \mathbf{4}, \mathbf{x}$ are collinear, so they determine a unique conic $\mathbf{1234x}$. Obviously none of the four points $\mathbf{1}, \mathbf{2}, \mathbf{3}, \mathbf{4}$ can be a double point of the conic, and the tangent at $\mathbf{4}$ exists. Comparing (4.1.31) with (4.1.14), we find that (4.1.31) is exactly

$$
0 = 2\,\mathrm{conic}(\mathbf{x1234}(\mathbf{45})) = -2\,\mathrm{conic}(\mathbf{123x4}(\mathbf{45})) = -[\mathrm{tangent}_{\mathbf{4,123x}}\mathbf{5}].
$$

$\square$

The conic in Proposition 4.14 is called a *tangent-point conic*, as besides points, it is determined by a tangent and the point of tangency.

Corollary 4.15. Any point $\mathbf{x}$ on conic $\mathbf{1234}(\mathbf{45})$ satisfies

$$
[\mathbf{234}]^2[\mathbf{145}][\mathbf{12x}][\mathbf{13x}] - [\mathbf{134}]^2[\mathbf{245}][\mathbf{12x}][\mathbf{23x}] + [\mathbf{124}]^2[\mathbf{345}][\mathbf{13x}][\mathbf{23x}] = 0.
$$

$$\tag{4.1.32}$$

If $[\mathbf{123}] \neq 0$, point $\mathbf{x}$ is on the conic if and only if (4.1.32) holds.

Proof. By Cramer's rule $[\mathbf{123}]\mathbf{x} = [\mathbf{23x}]\mathbf{1} - [\mathbf{13x}]\mathbf{2} + [\mathbf{12x}]\mathbf{3}$,

$$
\begin{aligned}
&2\,[\mathbf{123}]\,\mathrm{conic}(\mathbf{x1234}(\mathbf{45})) \\
&\overset{cramer}{=} [\mathbf{234}]^2[\mathbf{145}][\mathbf{12x}][\mathbf{13x}] - [\mathbf{134}]^2[\mathbf{245}][\mathbf{12x}][\mathbf{23x}] \\
&\qquad + [\mathbf{124}][\mathbf{13x}][\mathbf{23x}]([\mathbf{134}][\mathbf{245}] - [\mathbf{145}][\mathbf{234}]) \\
&\overset{contract}{=} [\mathbf{234}]^2[\mathbf{145}][\mathbf{12x}][\mathbf{13x}] - [\mathbf{134}]^2[\mathbf{245}][\mathbf{12x}][\mathbf{23x}] \\
&\qquad\qquad + [\mathbf{124}]^2[\mathbf{345}][\mathbf{13x}][\mathbf{23x}].
\end{aligned}
$$

$$\tag{4.1.33}$$

$\square$

Similar to the point-conic case, it can be proved that for any points $1, \ldots, 5, \mathbf{x}, 1'$ in the plane,

$$[1'34][1'4\mathbf{x}]\,\text{conic}(\mathbf{x}1234(45)) - [134][14\mathbf{x}]\,\text{conic}(\mathbf{x}1'234(45))$$
$$= [234][24\mathbf{x}]\,\text{conic}(\mathbf{x}11'34(45)). \tag{4.1.34}$$

A direct corollary is the following *tangent-point conic transformation rules*:

(1) If $1', 2', 3'$ are points on conic $1234(45)$, then for any point $\mathbf{x}$ in the plane,

$$\frac{\text{conic}(\mathbf{x}1234(45))}{\text{conic}(\mathbf{x}1'234(45))} = \frac{[124][134]}{[1'24][1'34]} = \frac{[123][145]}{[1'23][1'45]},$$

$$\frac{\text{conic}(\mathbf{x}1234(45))}{\text{conic}(\mathbf{x}1'2'34(45))} = \frac{[124][134][234]}{[1'2'4][1'34][2'34]} = \frac{[123][145][245]}{[1'2'3][1'45][2'45]},$$

$$\frac{\text{conic}(\mathbf{x}1234(45))}{\text{conic}(\mathbf{x}1'2'3'4(45))} = \frac{[124][134][234]}{[1'2'4][1'3'4][2'3'4]} = \frac{[123][145][245][345]}{[1'2'3'][1'45][2'45][3'45]}. \tag{4.1.35}$$

(2) If line $4'5'$ is tangent to conic $1234(45)$ at point $4'$, where point 1 is not a double point, then for any point $\mathbf{x}$ in the plane,

$$\frac{\text{conic}(\mathbf{x}1234(45))}{\text{conic}(\mathbf{x}1234'(4'5'))} = \frac{[145][234]^2}{[14'5'][234']^2}. \tag{4.1.36}$$

Proof. (1) The second equality in the first line of (4.1.35) is by the incidence of point 2 and conic $11'34(45)$, *i.e.*, $[124][134][1'23][1'45] = [123][145][1'24][1'34]$. The other equalities in (4.1.35) are easy to derive.

(2) If $1, 2, 3$ are not collinear, then if $4, 5$ are replaced by $4', 5'$, from (4.1.33) and the independence of the coordinate variables $[12\mathbf{x}], [13\mathbf{x}], [23\mathbf{x}]$ of vector $\mathbf{x}$, we get (4.1.36).

If $1, 2, 3$ are collinear, then $4, 5, 4', 5'$ must also be collinear, and (4.1.31) becomes $-[124][134][23\mathbf{x}][45\mathbf{x}] = 0$, which is the equation of the line-pair conic $(23, 45)$. Since 1 is not a double point, it is not on line 45. Then (4.1.36) is the result of

$$\frac{[45\mathbf{x}]}{[4'5'\mathbf{x}]} = \frac{[145]}{[14'5']}, \qquad \frac{[234']^2}{[124'][134']} = \frac{[234]^2}{[124][134]}.$$

$\square$

Tangency:

Proposition 4.16. If point $1 \neq 23 \cap 45$, then the tangent of conic $1234(45)$ at point 1 is

$$\text{tangent}_1((1234(45)) := [134]^2[245]12 - [124]^2[345]13$$
$$= [134][145][234]12 - [123][124][345]14 \tag{4.1.37}$$
$$= [124][145][234]13 - [123][134][245]14.$$

It is antisymmetric with respect to $2, 3$ and satisfies the tangent-point conic transformation rules (4.1.35) in the two points.

Proof. The equalities among the three bivector expressions in (4.1.37) can be proved by contractions. For example, for any point $\mathbf{x}$ in the plane,

$$([134]^2[245][12\mathbf{x}] - [124]^2[345][13\mathbf{x}]) - ([134][145][234][12\mathbf{x}]$$
$$-[123][124][345][14\mathbf{x}])$$
$$= \quad [12\mathbf{x}][134]([134][245] - [145][234]) - [124][345]([124][13\mathbf{x}][-[123][14\mathbf{x}])$$
$$\overset{contract}{=} 0.$$

We prove that the first bivector expression in (4.1.37), denoted by $\mathbf{B}$, represents the tangent at $\mathbf{1}$. For conic $\mathbf{1234(45)}$, if $[245] = 0$, by Proposition 4.14, $[345]$, $[124]$ are both nonzero. By symmetry, if $[345] = 0$, then $[134][245] \neq 0$. Similarly, if $[134] = 0$, then $[124][345] \neq 0$; also by symmetry, if $[124] = 0$, then $[134][245] \neq 0$. So the two coefficients in $\mathbf{B}$ cannot be both zero.

Assume $[124][345] \neq 0$. Let $\mathbf{x}$ be a point on conic $\mathbf{1234(45)}$. By (4.1.14), the tangent of conic $\mathbf{1234x}$ at $\mathbf{1}$ is

$$[134][13\mathbf{x}][24\mathbf{x}]\mathbf{12} - [124][12\mathbf{x}][34\mathbf{x}]\mathbf{13}. \tag{4.1.38}$$

By writing $\text{conic}(\mathbf{x3214(45)}) = 0$ as

$$\frac{[13\mathbf{x}][24\mathbf{x}]}{[12\mathbf{x}][34\mathbf{x}]} = \frac{[134][245]}{[124][345]},$$

and substituting the ratio equality into (4.1.38), we get $\mathbf{B}$ up to scale. $\qquad\square$

Intersection:

Proposition 4.17. Let $\mathbf{x} = \mathbf{a234(45)} \cap \mathbf{ab}$ be the second point of intersection of conic $\mathbf{a234(45)}$ and line $\mathbf{ab}$, then it has the following expressions in GC algebra:

$$\mathbf{x}_{\mathbf{ab,a234(45)}} := [234][45a][3ab]\mathbf{24} \vee \mathbf{ab} - [245][34a][4ab]\mathbf{23} \vee \mathbf{ab}$$
$$= [234][24a][3ab]\mathbf{45} \vee \mathbf{ab} - [245][23a][4ab]\mathbf{34} \vee \mathbf{ab} \tag{4.1.39}$$
$$= [345][24a][3ab]\mathbf{24} \vee \mathbf{ab} - [245][34a][2ab]\mathbf{34} \vee \mathbf{ab}.$$

Proof. First we have

$$2\,\text{conic}(\mathbf{ax234(45)}) = [34\mathbf{x}][245][4a\mathbf{x}][23a] - [234][45\mathbf{x}][3a\mathbf{x}][24a] = 0. \tag{4.1.40}$$

Substituting $\mathbf{x} = \lambda\mathbf{a} + \mu\mathbf{b}$ into it, we get

$$\frac{\lambda}{\mu} = \frac{[234][24b][3ab][45a] - [23b][245][34a][4ab]}{[23a][245][34a][4ab] - [234][24a][3ab][45a]}.$$

The expression

$$\mathbf{x} = \quad ([234][24b][3ab][45a] - [23b][245][34a][4ab])\mathbf{a}$$
$$+([23a][245][34a][4ab] - [234][24a][3ab][45a])\mathbf{b}$$

equals the first expression in (4.1.39) if the latter is expanded by splitting $\mathbf{a}, \mathbf{b}$ in the two meet products. Similarly, from $\text{conic}(\mathbf{a2x34(45)}) = \text{conic}(\mathbf{a23x4(45)}) = 0$, we get the other two expressions in (4.1.39). The equalities among the three expressions can be proved by contractions. $\qquad\square$

Bitangent-point conic:

When expression $2\,\mathrm{conic}(\mathbf{x1234(45)})$ is taken as a function of vector variable $\mathbf{3}$, its *polarization* (or *derivative*) by vector $\mathbf{a}$ is

$$2\,\mathrm{conic}(\mathbf{x12(3a)4(45)}) := \frac{1}{2}\{[\mathbf{14x}][\mathbf{245}]([\mathbf{134}][\mathbf{2ax}] - [\mathbf{14a}][\mathbf{23x}]) \\ -[\mathbf{145}][\mathbf{24x}](-[\mathbf{13x}][\mathbf{24a}] + [\mathbf{1ax}][\mathbf{234}])\}. \tag{4.1.41}$$

If $\mathbf{3} = \mathbf{2}$, then

$$4\,\mathrm{conic}(\mathbf{x12(2a)4(45)}) = [\mathbf{14x}][\mathbf{124}][\mathbf{245}][\mathbf{2ax}] + [\mathbf{12x}][\mathbf{145}][\mathbf{24a}][\mathbf{24x}]. \tag{4.1.42}$$

By the last expression of $\mathrm{tangent}_2(\mathbf{2x14(45)})$ in (4.1.37), using contractions

$$[\mathbf{12a}][\mathbf{24x}] - [\mathbf{124}][\mathbf{2ax}] = [\mathbf{12x}][\mathbf{24a}], \quad [\mathbf{124}][\mathbf{45x}] - [\mathbf{145}][\mathbf{24x}] = -[\mathbf{14x}][\mathbf{245}],$$

we get from (4.1.42) the following relation:

$$4\,\mathrm{conic}(\mathbf{x12(2a)4(45)}) = [\mathrm{tangent}_2(\mathbf{2x14(45)})\mathbf{a}] = [\mathrm{tangent}_2(\mathbf{x124(45)})\mathbf{a}]. \tag{4.1.43}$$

Corollary 4.18.

$$4\,\mathrm{conic}(\mathbf{x12(24)3(35)}) = -[\mathbf{123}]^2[\mathbf{24x}][\mathbf{35x}] + [\mathbf{124}][\mathbf{135}][\mathbf{23x}]^2 \\ = [\mathbf{123}][\mathbf{13x}][\mathbf{235}][\mathbf{24x}] + [\mathbf{12x}][\mathbf{135}][\mathbf{234}][\mathbf{23x}] \tag{4.1.44} \\ = [\mathbf{124}][\mathbf{13x}][\mathbf{235}][\mathbf{23x}] + [\mathbf{123}][\mathbf{12x}][\mathbf{234}][\mathbf{35x}].$$

Proposition 4.19. There exists a unique conic passing through points $\mathbf{1, 2, 3}$ and tangent to lines $\mathbf{24, 35}$, denoted by $\mathbf{12(24)3(35)}$, if the following set of nondegeneracy conditions, denoted by $\exists\mathbf{12(24)3(35)}$, are satisfied:

- $\mathbf{1, 2, 3}$ are not collinear,
- $\mathbf{24, 35}$ are lines and are distinct,
- either (a) $\mathbf{1}$ is on one of the lines $\mathbf{24, 35}$, or (b) $\mathbf{1, 3}$ are not on line $\mathbf{24}$, and $\mathbf{1, 2}$ are not on line $\mathbf{35}$.

A point $\mathbf{x}$ is on the conic if and only if $\mathrm{conic}(\mathbf{x12(24)3(35)}) = 0$.

Proof. Denote the first expression on the right side of (4.1.44) by $p = p(\mathbf{x})$. If $[\mathbf{124}]$ or $[\mathbf{135}]$ is zero, then $p(\mathbf{x}) = 0$ represents the line-pair conic $(\mathbf{24, 35})$. If $[\mathbf{234}] = 0$ or $[\mathbf{235}] = 0$, then $[\mathbf{135}] = 0$, and $p(\mathbf{x}) = 0$ still represents $(\mathbf{24, 35})$. The conclusion is true in these degenerate cases.

Below we assume $[\mathbf{124}][\mathbf{135}][\mathbf{234}][\mathbf{235}] \neq 0$. Then points $\mathbf{1, 3}$ are not on line $\mathbf{24}$. Let $\mathbf{x}$ be any point not on lines $\mathbf{12}$ and $\mathbf{23}$. Then point $\mathbf{2}$ is not on any of the three lines $\mathbf{13}, \mathbf{1x}, \mathbf{3x}$. Since (1) $\mathbf{3}, \mathbf{1}, \mathbf{x}, \mathbf{2}$ are pairwise distinct and noncollinear, (2) $\mathbf{2}, \mathbf{4}$ are distinct, (3) either it is only point $\mathbf{x}$ among $\{\mathbf{1}, \mathbf{3}, \mathbf{x}\}$ that is on line $\mathbf{24}$, or it is all three points $\mathbf{1}, \mathbf{3}, \mathbf{x}$ that are not on line $\mathbf{24}$, by Proposition 4.14, there exists a unique conic $\mathbf{31x2(24)}$. Since $\mathbf{3}$ is not on line $\mathbf{24}$, it cannot be a double point of the conic, and the tangent at $\mathbf{3}$ exists, whose equation is just $p = 0$ where it is $\mathbf{5}$ instead of $\mathbf{x}$ that is the vector indeterminate.   □

By the first expression in (4.1.44), it can be easily proved that for any points $\mathbf{1}, \ldots, \mathbf{5}, \mathbf{x}, \mathbf{1'}$ in the plane,

$$[\mathbf{1'23}]^2 \text{conic}(\mathbf{x12(24)3(35)}) - [\mathbf{123}]^2 \text{conic}(\mathbf{x1'2(24)3(35)})$$
$$+ [\mathbf{x23}]^2 \text{conic}(\mathbf{11'2(24)3(35)}) = 0. \tag{4.1.45}$$

The conic in Proposition 4.19 is called a *bitangent-point conic*. There are the following *bitangent-point conic transformation rules*:

(1) If point $\mathbf{1'}$ is on conic $\mathbf{12(24)3(35)}$, then for any point $\mathbf{x}$ in the plane,

$$\frac{\text{conic}(\mathbf{x12(24)3(35)})}{\text{conic}(\mathbf{x1'2(24)3(35)})} = \frac{[\mathbf{123}]^2}{[\mathbf{1'23}]^2}. \tag{4.1.46}$$

(2) If lines $\mathbf{2'4'}, \mathbf{3'5'}$ are tangent to the conic at points $\mathbf{2'}, \mathbf{3'}$, then

$$\frac{\text{conic}(\mathbf{x12(24)3(35)})}{\text{conic}(\mathbf{x12(24)3'(3'5')})} = \frac{[\mathbf{123}]^2[\mathbf{235}]}{[\mathbf{123'}]^2[\mathbf{23'5'}]} = \frac{[\mathbf{234}]^2[\mathbf{135}]}{[\mathbf{23'4}]^2[\mathbf{13'5'}]},$$
$$\frac{\text{conic}(\mathbf{x12(24)3(35)})}{\text{conic}(\mathbf{x12'(2'4')3'(3'5')})} = \frac{[\mathbf{123}]^2[\mathbf{23'4}][\mathbf{235}]}{[\mathbf{12'3'}]^2[\mathbf{2'3'4'}][\mathbf{23'5'}]}. \tag{4.1.47}$$

Tangency:

Proposition 4.20. The tangent of conic $\mathbf{12(24)3(35)}$ at point $\mathbf{1} \neq \mathbf{24} \cap \mathbf{35}$ is

$$\text{tangent}_1(\mathbf{12(24)3(35)}) := [\mathbf{135}][\mathbf{234}]\mathbf{12} + [\mathbf{124}][\mathbf{235}]\mathbf{13}. \tag{4.1.48}$$

In particular if $\mathbf{4} = \mathbf{5}$, *i.e.*, $\mathbf{4}$ is the pole of line $\mathbf{23}$, the tangent of conic $\mathbf{12(24)3(34)}$ at point $\mathbf{1}$ is

$$\text{tangent}_1'(\mathbf{12(24)3(34)}) := [\mathbf{134}]\mathbf{12} + [\mathbf{124}]\mathbf{13}. \tag{4.1.49}$$

Proof. For conic $\mathbf{12(24)3(35)}$, by Proposition 4.19, if $[\mathbf{124}]$ or $[\mathbf{235}]$ is zero, then $[\mathbf{135}], [\mathbf{234}]$ are both nonzero, and (4.1.48) represents line $\mathbf{12}$. If $[\mathbf{135}]$ or $[\mathbf{234}]$ is zero, then (4.1.48) represents line $\mathbf{13}$. The conclusion is true in these degenerate cases.

When none of $[\mathbf{124}], [\mathbf{135}], [\mathbf{234}], [\mathbf{235}]$ is zero, let $\mathbf{x}$ be a point on conic $\mathbf{12(24)3(35)}$. By the second expression in (4.1.37), the tangent of conic $\mathbf{12x3(35)}$ at point $\mathbf{1}$ is

$$[\mathbf{135}][\mathbf{13x}][\mathbf{23x}]\mathbf{12} - [\mathbf{123}][\mathbf{12x}][\mathbf{35x}]\mathbf{13}. \tag{4.1.50}$$

The last expression of $4\,\mathrm{conic}(\mathbf{x12(24)3(35)}) = 0$ in (4.1.44) can be written as

$$[\mathbf{123}][\mathbf{12x}][\mathbf{35x}] = -[\mathbf{124}][\mathbf{13x}][\mathbf{235}][\mathbf{23x}]/[\mathbf{234}].$$

Substituting it into (4.1.50), we get (4.1.48) multiplied by $[\mathbf{13x}][\mathbf{23x}]/[\mathbf{234}]$. $\quad\square$

The following are some transformation properties of the tangent expressions:

(1) If lines $\mathbf{2'4'}, \mathbf{3'5'}$ are tangent to conic $\mathbf{12(24)3(35)}$ at points $\mathbf{2'}, \mathbf{3'}$ respectively, then

$$\begin{aligned}
\frac{\mathrm{tangent}_1(\mathbf{12(24)3(35)})}{\mathrm{tangent}_1(\mathbf{12(24)3'(3'5')})} &= \frac{[\mathbf{123}][\mathbf{235}]}{[\mathbf{123'}][\mathbf{23'5'}]}, \\
\frac{\mathrm{tangent}_1(\mathbf{123},\mathbf{24},\mathbf{35})}{\mathrm{tangent}_1(\mathbf{12'(2'4')3'(3'5')})} &= \frac{[\mathbf{123}][\mathbf{23'4}][\mathbf{235}]}{[\mathbf{12'3'}][\mathbf{2'3'4'}][\mathbf{23'5'}]}.
\end{aligned}$$
(4.1.51)

(2) If $\mathbf{4} = \mathbf{5}$ and $\mathbf{4'} = \mathbf{5'}$ in (4.1.51), then

$$\begin{aligned}
\frac{\mathrm{tangent}'_1(\mathbf{12(24)3(34)})}{\mathrm{tangent}'_1(\mathbf{12(24')3'(3'4')})} &= \frac{[\mathbf{123}][\mathbf{124}]}{[\mathbf{123'}][\mathbf{124'}]} = \frac{[\mathbf{123}][\mathbf{23'4}]}{[\mathbf{123'}][\mathbf{23'4'}]}, \\
\frac{\mathrm{tangent}'_1(\mathbf{12(24)3(34)})}{\mathrm{tangent}'_1(\mathbf{12'(2'4')3'(3'4')})} &= \frac{[\mathbf{123}][\mathbf{23'4}]}{[\mathbf{12'3'}][\mathbf{23'4'}]}.
\end{aligned}$$
(4.1.52)

Intersection:

The second point of intersection $\mathbf{ab} \cap \mathbf{a2(24)3(35)}$ of line $\mathbf{ab}$ and conic $\mathbf{a2(24)3(35)}$ has the following expression in GC algebra:

$$[\mathbf{234}][\mathbf{2ab}][\mathbf{35a}]\mathbf{ab} \vee \mathbf{23} + [\mathbf{235}][\mathbf{23a}][\mathbf{3ab}]\mathbf{ab} \vee \mathbf{24}. \qquad (4.1.53)$$

4.2 Bracket-oriented representation

We have seen from the previous section that for the same conic geometric object or constraint, usually there are several different representations in GC algebra or bracket algebra. Although the representations are equal up to scale, such equalities are difficult to establish from the syzygy relations and coconic constraints. It is an important task to select *suitable algebraic representations* for both geometric constructions and conclusions.

The idea *bracket-oriented representation* refers to determining optimal algebraic representations of *all points in a bracket* at the same time, by substituting the representations into the bracket and setting the goal of optimization as producing a factored and shortest bracket polynomial result out of the expansions of the Cayley polynomials obtained by different substitutions.

4.2.1 *Representations of geometric constructions*

The following is a list of typical geometric constructions in projective conic geometry, together with their associated nondegeneracy conditions. Points constructed in items (3), (4) below are called *incidence points*.

(1) $\mathbf{x}$ is a free point in the plane: no nondegeneracy condition.

(2) $\mathbf{x}$ is a free collinear point on line $\mathbf{12}$: $\mathbf{12} \neq 0$.

(3) $\mathbf{x}$ is the conjugate of point $\mathbf{3}$ on line $\mathbf{12}$: $\mathbf{12} \neq 0$.

(4) $\mathbf{x}$ is the intersection of two lines $\mathbf{12}, \mathbf{34}$: $\mathbf{12} \vee \mathbf{34} \neq 0$.

(5) $\mathbf{x}$ is a free point on a conic.
Nondegeneracy condition: $\exists$conic, where the conic is one of $\mathbf{12345}$, $\mathbf{1234(45)}$, and $\mathbf{12(24)3(35)}$.

(6) Conic $\mathbf{12\ldots i}$, where the number of points $i \geq 6$.
It means that five of the i points are free and determine a conic, and the others are free points on the conic. All i points are called *free conic points*.
Nondegeneracy condition: there exist five points $\{\mathbf{j_1}, \ldots, \mathbf{j_5}\} \subset \{\mathbf{1}, \mathbf{2}, \ldots, \mathbf{i}\}$ such that $\exists \mathbf{j_1} \ldots \mathbf{j_5}$; denoted by $\exists \mathbf{12 \ldots i}$.

(7) $\mathbf{L}$ is the polar (including tangent) of point $\mathbf{a}$ with respect to a conic.
Nondegeneracy conditions: $\exists$conic, and $\mathbf{a}$ is not a double point; denoted by $\exists \text{polar}_{\mathbf{a}}(\text{conic})$.

(8) $\mathbf{x}$ is a free collinear point on the tangent at point $\mathbf{1}$ of a conic: $\exists \text{polar}_{\mathbf{1}}(\text{conic})$.

(9) $\mathbf{x}$ is the intersection of line $\mathbf{ab}$ and the polar of point $\mathbf{1}$ with respect to a conic.
Nondegeneracy conditions: $\mathbf{ab} \neq 0$, $\exists \text{polar}_{\mathbf{1}}(\text{conic})$, either $\mathbf{a}$ or $\mathbf{b}$ is not conjugate to $\mathbf{1}$ with respect to the conic.

(10) $\mathbf{x}$ is the pole of line $\mathbf{ab}$ with respect to a conic.
Nondegeneracy conditions: $\mathbf{ab} \neq 0$, $\exists$conic, $\mathbf{ab}$ is not part of the conic; denoted by $\exists \text{pole}_{\mathbf{ab}}(\text{conic})$.

(11) $\mathbf{x}$ is the second point of intersection of line $\mathbf{ab}$ and conic $\mathbf{a1234}$: $\exists \text{pole}_{\mathbf{ab}}(\text{conic})$.

(12) $\mathbf{x}$ is the fourth point of intersection of conics $\mathbf{12345}$ and $\mathbf{1234'5'}$.
Nondegeneracy conditions: $\mathbf{123} \neq 0$, points $\mathbf{1}, \mathbf{2}, \mathbf{3}, \mathbf{4}, \mathbf{5}, \mathbf{4'}, \mathbf{5'}$ are not coconic, and $\exists \text{pole}_{\mathbf{ij}}(\mathbf{12345})$, $\exists \text{pole}_{\mathbf{ij}}(\mathbf{1234'5'})$ for $\mathbf{ij} = \mathbf{12}, \mathbf{13}, \mathbf{23}$.

Definition 4.21. Let $\mathbf{x}$ be a point or line constructed from a set of points $\mathcal{A}$. Usually there are more than one way of construction, so $\mathbf{x}$ can be constructed by a subset of $\mathcal{A}$. If a point $\mathbf{y}$ in $\mathcal{A}$ has the property that in every construction of $\mathbf{x}$ by a subset of $\mathcal{A}$, in every GC algebraic expression of $\mathbf{x}$ by points of the subset, $\mathbf{y}$ always occurs in the vector part of the expression instead of the scalar coefficient part, then $\mathbf{y}$ is an *essential point* of $\mathbf{x}$. If no such point exists for $\mathbf{x}$, then $\mathbf{x}$ is its own *essential point*.

In the above twelve constructions, the corresponding essential points are

$$(1)\ \mathbf{x} \quad (2)\ \mathbf{1}, \mathbf{2} \quad (3)\ \mathbf{1}, \mathbf{2} \quad (4)\ \mathbf{1}, \mathbf{2}, \mathbf{3}, \mathbf{4} \quad (5)\ \mathbf{x} \quad (6)\ \mathbf{1}, \mathbf{2}, \ldots, \mathbf{i}$$

$$(7)\ \mathbf{a} \quad (8)\ \mathbf{1} \quad (9)\ \mathbf{a}, \mathbf{b}, \mathbf{1} \quad (10)\ \mathbf{a}, \mathbf{b} \quad (11)\ \mathbf{a}, \mathbf{b} \quad (12)\ \mathbf{1}, \mathbf{2}, \mathbf{3}.$$

In constructions (5) and (6), there are no essential points other than the constructed points themselves, because a free conic point can be represented by three arbitrarily selected conic points.

The difference among various representations of the same geometric construction lies not only in the 5-tuple of points and tangents representing the conic, but also in the order of elements within the 5-tuple. Below we check an example.

Example 4.22. Let there be five points in the plane: $\mathbf{1, 2, 3, 4, 5}$. Let $\mathbf{6} = \mathbf{34} \cap \mathbf{15}$, $\mathbf{7} = \mathbf{14} \cap \mathbf{35}$, $\mathbf{8} = \mathbf{13} \cap \mathbf{45}$. Conic $\mathbf{12345}$ intersects line $\mathbf{26}$ at points $\mathbf{2, 9}$, and lines $\mathbf{25, 19}$ intersect at point $\mathbf{0}$. Then points $\mathbf{7, 8, 0}$ are collinear.

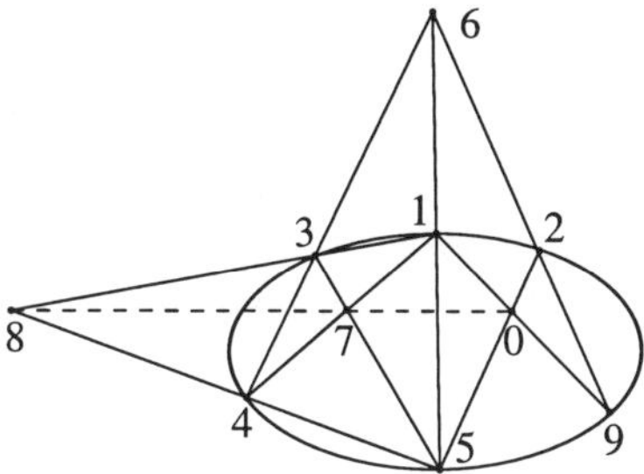

Fig. 4.1   Example 4.22: optimal representation of $9 =$ conic $\mathbf{12345} \cap$ line $\mathbf{26}$.

Free points: $\mathbf{1, 2, 3, 4, 5}$.
Intersections:

$$\mathbf{6} = \mathbf{34} \cap \mathbf{15}, \qquad \mathbf{7} = \mathbf{14} \cap \mathbf{35}, \quad \mathbf{8} = \mathbf{13} \cap \mathbf{45},$$
$$\mathbf{9} = \mathbf{12345} \cap \mathbf{26}, \quad \mathbf{0} = \mathbf{25} \cap \mathbf{19}.$$

Conclusion: $[\mathbf{780}] = 0$.

Proof.

Rules $\qquad\qquad\qquad\qquad\qquad\qquad\qquad$ $[\mathbf{780}]$

$$
\begin{aligned}
[\mathbf{780}] &= [(14 \vee 35)(13 \vee 45)(25 \vee 19)] \\
&= (13 \vee 45 \vee 19)(14 \vee 35 \vee 25) \\
&\quad - (13 \vee 45 \vee 25)(14 \vee 35 \vee 19) \\
&= [139][145]^2[235] - [245][135]^2[149]
\end{aligned}
$$

$$9_{26,1345} = -[145][235][246]13 \vee 26$$
$$+ [135][236][245]14 \vee 26$$

$[\mathbf{780}] \overset{7,8,0}{=} [135]^2[149][245] - [139][145]^2[235]$

$$\overset{9}{=} \frac{[126][134][135][145][235][245]}{([135][246] - [145][236])}$$

$$\overset{contract}{=} 0.$$

Remark: In the above proof, the first Cayley expansion produces a bracket binomial in which $\mathbf{9}$ occurs in two brackets: $[\mathbf{139}]$ and $[\mathbf{149}]$. To eliminate $\mathbf{9}$, since

1 occurs in both brackets, **3** and **4** each occur in a bracket, an ideal representation is to let points **1, 3, 4** occur in the vector part of **9**. By (4.1.24), the unique optimal representation for **9** is $\mathbf{9}_{26,1345}$, as **3, 4** are trivially antisymmetric in the expression of $\mathbf{9}_{26,1345}$.

The order $\mathbf{2} \succ \mathbf{1} \succ \mathbf{3}, \mathbf{4} \succ \mathbf{5}$ of conic points in the representation of **9** can be explained as follows: **2** is the essential point of the construction; **1** occurs twice as a bracket mate of **9** in the conclusion expression, while **3, 4** each occur once; **5** is not a bracket mate of **9**.

In this example, conic points **1, 3, 4** occur explicitly in the brackets containing the element **9** to be represented. In many examples, the bracket mates of the element to be represented are not conic points, instead they are constructed by some conic points as essential points. After such bracket mates are eliminated, their essential conic points become explicit bracket mates of the element to be represented. Taking into consideration both the explicit and the implicit conic bracket mates, we propose the following *conic points selection algorithm* for optimal representation of a geometric construction involving conic points.

Algorithm 4.23. Conic points selection.

Input: (1) A construction $\mathbf{x} = \mathbf{x}(\mathcal{P})$ related to a conic, where $\mathcal{P}$ is the set of points used in the construction,

(2) $\mathcal{C}$, the set of existing conic points before $\mathbf{x}$ is constructed,

(3) $p(\mathbf{x})$, a Cayley polynomial where $\mathbf{x}$ occurs in brackets (including also Cayley brackets and deficit brackets).

Output: A set of pairs $(q(\mathbf{x}), s_q)$, each q being a bracket in $p(\mathbf{x})$ containing $\mathbf{x}$, and each s_q being a sequence of elements in $\mathcal{C}$.

Step 1. Let $\mathcal{C}_\mathbf{x}$ be the elements of $\mathcal{C}$ that are not essential to $\mathbf{x}$. For each bracket $q(\mathbf{x})$ in $p(\mathbf{x})$ containing $\mathbf{x}$, do the following.

(1) For every bracket mate $\mathbf{y}$ of $\mathbf{x}$, find all its essential points $\mathcal{E}_\mathbf{y}$ in $\mathcal{C}_\mathbf{x}$. Set the *essential weight* of each element in $\mathcal{E}_\mathbf{y}$ to be $(\#(\mathcal{E}_\mathbf{y}))^{-1}$, where $\#(\mathcal{E}_\mathbf{y})$ is the number of elements in $\mathcal{E}_\mathbf{y}$. The more essential conic points $\mathbf{y}$ has, the less important each of them is in representing $\mathbf{x}$.

(2) Let the union of all the $\mathcal{E}_\mathbf{y}$'s be $\mathcal{E}_q$. Compute the sum of the essential weights for each element of $\mathcal{E}_q$. Order the elements by their essential weights, denote the descending sequence by the same symbol $\mathcal{E}_q$.

Step 2. If there is only one $q(\mathbf{x})$ in $p(\mathbf{x})$, then return $(q(\mathbf{x}), \mathcal{E}_q, \mathcal{C}_\mathbf{x} - \mathcal{E}_q)$; else, the order of elements in $\mathcal{C}_\mathbf{x} - \mathcal{E}_q$ should be refined as follows by the principle of minimizing the change of conic representations for different q's, in order to minimize the size of transformation coefficient.

Let $\mathcal{E}$ be the union of all the $\mathcal{E}_q$'s. Compute the sum of the essential weights for each element in $\mathcal{E}$. Order the elements by their essential weights, denote the descending sequence by the same symbol $\mathcal{E}$.

Step 3. For every $q(\mathbf{x})$, return $(q(\mathbf{x}), \mathcal{E}_q, \mathcal{E} - \mathcal{E}_q, \mathcal{C}_\mathbf{x} - \mathcal{E})$.

Conic points selection has already been used tacitly in the proofs of Corollary 4.9 and Proposition 4.10:

(i) In the proof of (4.1.18), when computing the intersection of the tangents at points $\mathbf{1}, \mathbf{2}$ of a conic, we selected the representations $\text{tangent}_{\mathbf{1},\mathbf{2345}}$ and $\text{tangent}_{\mathbf{2},\mathbf{1345}}$. The result has four bracket factors. For other representations, usually no bracket factor can be obtained.

(ii) In the proof of (4.1.19), when computing the line connecting the poles of $\mathbf{12}$ and $\mathbf{13}$ with respect to a conic, we selected $\text{pole}_{\mathbf{12},\mathbf{345}}$ and $\text{pole}_{\mathbf{13},\mathbf{245}}$ to represent the poles, and obtained three bracket factors.

(iii) In the proof of (4.1.20), when computing the bracket composed of the tangent at point $\mathbf{1}$ of a conic and the pole of line $\mathbf{45}$, we selected the representations $\text{tangent}_{\mathbf{1},\mathbf{2345}}$ and $\text{pole}_{\mathbf{45},\mathbf{123}}$ by setting $\mathbf{2}' = \mathbf{2}$, $\mathbf{3}' = \mathbf{3}$ and $\mathbf{4}' = \mathbf{4}$, otherwise the result is very difficult to factorize.

(iv) In the proof of (4.1.22), when computing the bracket composed of conic point $\mathbf{1}$ and the poles of $\mathbf{12}, \mathbf{3}'\mathbf{4}'$, we first computed $[\mathbf{1}\,\text{pole}_{\mathbf{12},\mathbf{345}}\,\text{pole}_{\mathbf{34},\mathbf{125}}]$ by setting $\mathbf{3} = \mathbf{3}'$, $\mathbf{4} = \mathbf{4}'$, $\mathbf{5} = \mathbf{5}'$, and then applied the point-conic transformation rules. The computing based on the original representations is very difficult.

Representation of the second point of intersection of a line and a conic.

As in Example 4.22, in a bracket polynomial $p(\mathbf{x})$, the second point of intersection $\mathbf{x}$ of line $\mathbf{ab}$ with a conic passing through point $\mathbf{a}$ should be represented as follows:

(1) Use the conic points selection algorithm to find for each bracket $q(\mathbf{x})$ containing $\mathbf{x}$ a sequence of conic points s_q. Substitute the first four elements of s_q into $\mathbf{1}, \mathbf{2}, \mathbf{3}, \mathbf{4}$ of (4.1.24).

(2) Fix a $q(\mathbf{x})$, substitute the corresponding representation of $\mathbf{x}$ into it. For any other bracket containing $\mathbf{x}$, substitute into it the corresponding representation of $\mathbf{x}$, and multiply the result with the corresponding transformation coefficient.

Representations of poles, polars and tangents.

By (4.1.10), for conic $\mathbf{12345}$, the polar of point $\mathbf{a}$ has three different forms:

$$\text{polar}_\mathbf{a}(\mathbf{12345})$$
$$= [\mathbf{135}][\mathbf{245}]([\mathbf{12a}]\,\mathbf{34} + [\mathbf{34a}]\,\mathbf{12}) - [\mathbf{125}][\mathbf{345}]([\mathbf{13a}]\,\mathbf{24} + [\mathbf{24a}]\,\mathbf{13})$$
$$= [\mathbf{145}][\mathbf{235}]([\mathbf{13a}]\,\mathbf{24} + [\mathbf{24a}]\,\mathbf{13}) - [\mathbf{135}][\mathbf{245}]([\mathbf{14a}]\,\mathbf{23} + [\mathbf{23a}]\,\mathbf{14})$$
$$= [\mathbf{145}][\mathbf{235}]([\mathbf{12a}]\,\mathbf{34} + [\mathbf{34a}]\,\mathbf{12}) - [\mathbf{125}][\mathbf{345}]([\mathbf{14a}]\,\mathbf{23} + [\mathbf{23a}]\,\mathbf{14}). \tag{4.2.1}$$

For conic $\mathbf{1234(45)}$, the tangent has three different forms (4.1.37).

The representations of poles, polars and tangents in a bracket polynomial or Cayley polynomial follow much the same procedure as the above representation of the second point of intersection of a line and a conic. If there is any difference, it is that if there are several equivalent forms of the same representation, *e.g.*, (4.2.1) and (4.1.37), then the forms are substituted one by one into the same place of the polynomial, and the simplest among the computing results is selected.

Example 4.24. [Brianchon's Theorem] Let there be six points $\mathbf{1, 2, 3, 4, 5, 6}$ on a conic. Draw tangent lines of the conic at the six points, such that the neighboring tangents meet at points $\mathbf{7, 8, 9, 0, a, b}$ respectively. Then lines $\mathbf{9b, 8a, 70}$ concur.

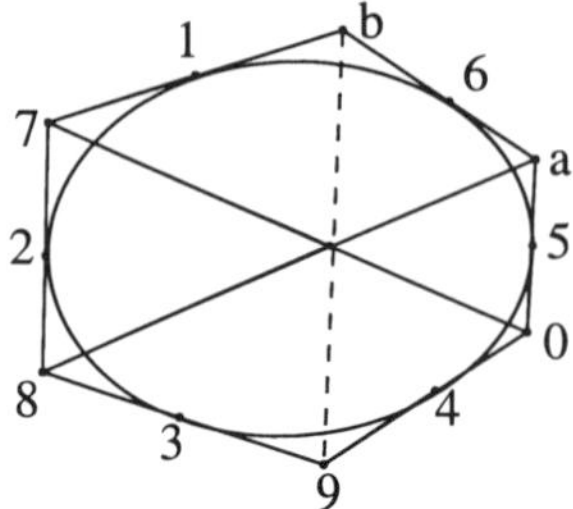

Fig. 4.2 Brianchon's Theorem.

Free conic points: $\mathbf{1, 2, 3, 4, 5, 6}$.
Poles:

$$\mathbf{7} = \mathrm{pole}_{\mathbf{12}}(\mathbf{123456}), \quad \mathbf{8} = \mathrm{pole}_{\mathbf{23}}(\mathbf{123456}), \quad \mathbf{9} = \mathrm{pole}_{\mathbf{34}}(\mathbf{123456}),$$
$$\mathbf{0} = \mathrm{pole}_{\mathbf{45}}(\mathbf{123456}), \quad \mathbf{a} = \mathrm{pole}_{\mathbf{56}}(\mathbf{123456}), \quad \mathbf{b} = \mathrm{pole}_{\mathbf{61}}(\mathbf{123456}).$$

Conclusion: $\mathbf{9b} \vee \mathbf{8a} \vee \mathbf{70} = 0$.

Proof.

$$\mathbf{9b} \vee \mathbf{8a} \vee \mathbf{70} \ \overset{expand}{=}\ [780][9ab] - [70a][89b]$$

$$\overset{represent}{=}\ [7_{21,345}\, 8_{23,145}\, 0_{45,213}][9_{34,651}\, a_{65,134}\, b_{61,534}]$$
$$-[7_{12,546}\, 0_{54,612}\, a_{56,412}][8_{32,461}\, 9_{34,261}\, b_{61,324}]$$

$$\frac{7_{21,345}\, 8_{23,145}\, 9_{34,651}\, 0_{45,213}\, a_{65,134}\, b_{61,534}}{7_{12,546}\, 8_{32,461}\, 9_{34,261}\, 0_{54,612}\, a_{56,412}\, b_{61,324}}$$

$$\overset{7,8,9,0,a,b}{=}\ 16\ \underbrace{[124][125][134]^2[136][145]^2[146][235]}$$

$$\underbrace{[245][346][356][126]^{-4}[234]^{-2}[456]^{-2}}$$

$$\{[126]^4[135]^4[234]^3[245][346][456]^3$$
$$-[123]^3[125][136][156]^3[246]^4[345]^4\}$$

$$\overset{conic}{=} \underbrace{[123]^3[156]^3[246]^3[345]^3}\{[126][135][245][346]$$
$$-[125][136][246][345]\}$$
$$\overset{conic}{=} 0.$$

Additional nondegeneracy conditions:

$$\exists 12345, \quad \exists 12346, \quad \exists 12456, \quad \exists 13456, \quad 126 \neq 0, \quad 234 \neq 0, \quad 456 \neq 0. \quad (4.2.2)$$

$\square$

Remarks:

(1) $9b \vee 8a \vee 70$ has three different expansions, leading to much the same proofs.

(2) The second step is choosing representations for all the points in the brackets for batch elimination. For example, in $[780]$, the three points have essential points $1^1, 2^2, 3^1, 4^1, 5^1$, where the exponents denote the multiplicities of the points as essential points of $7, 8, 0$. So 2^2 denotes that point 2 is an essential point of two of $7, 8, 0$. Definitely it should be used in the representation. $0_{45,213}$ is the unique optimal representation for pole 0. For poles 7 and 8, the representations $7_{21,345}$ and $8_{23,145}$ minimize the size of transformation coefficient.

(3) The four brackets in the first line of the proof can be computed by applying (4.1.21):

$$\begin{aligned}
[7_{21,345}\, 8_{23,145}\, 0_{45,213}] &= -4\,[124][125][134]^2[135]^2[234][235][245]^2, \\
[9_{34,651}\, a_{65,134}\, b_{61,534}] &= -4\,[135]^2[136][145]^2[146][346]^2[356][456], \\
[7_{12,546}\, 0_{54,612}\, a_{56,412}] &= -4\,[125]^2[145][146]^2[156][245][246]^2[256], \\
[8_{32,461}\, 9_{34,261}\, b_{61,324}] &= -4\,[123][124]^2[134][136]^2[236][246]^2[346].
\end{aligned} \quad (4.2.3)$$

(4) The six points $7, 8, 9, 0, a, b$ each need two different representations, so the second term in the second line of the proof is multiplied by the transformation coefficient

$$k = \frac{7_{21,345}\, 8_{23,145}\, 9_{34,651}\, 0_{45,213}\, a_{65,134}\, b_{61,534}}{7_{12,546}\, 8_{32,461}\, 9_{34,261}\, 0_{54,612}\, a_{56,412}\, b_{61,324}}$$

$$= \frac{[3(\breve{3}\breve{6})_{(1)}][3(\breve{3}\breve{6})_{(2)}]}{[6(\breve{3}\breve{6})_{(1)}][6(\breve{3}\breve{6})_{(2)}]} \frac{[5(\breve{5}\breve{6})_{(1)}][5(\breve{5}\breve{6})_{(2)}]}{[6(\breve{5}\breve{6})_{(1)}][6(\breve{5}\breve{6})_{(2)}]} \frac{[5(\breve{5}\breve{2})_{(1)}][5(\breve{5}\breve{2})_{(2)}]}{[2(\breve{5}\breve{2})_{(1)}][2(\breve{5}\breve{2})_{(2)}]} \quad (4.2.4)$$

$$\frac{[3(\breve{3}\breve{6})'_{(1)}][3(\breve{3}\breve{6})'_{(2)}]}{[6(\breve{3}\breve{6})'_{(1)}][6(\breve{3}\breve{6})'_{(2)}]} \frac{[3(\breve{3}\breve{2})_{(1)}][3(\breve{3}\breve{2})_{(2)}]}{[2(\breve{3}\breve{2})_{(1)}][2(\breve{3}\breve{2})_{(2)}]} \frac{[5(\breve{5}\breve{2})'_{(1)}][5(\breve{5}\breve{2})'_{(2)}]}{[2(\breve{5}\breve{2})'_{(1)}][2(\breve{5}\breve{2})'_{(2)}]},$$

where $\breve{i}\breve{j}$ denotes the 4-tuple of points after removal of ij from 123456, and each $\breve{i}\breve{j}$ is partitioned into two parts $(\breve{i}\breve{j})_{(1)}$ and $(\breve{i}\breve{j})_{(2)}$ of equal size. Some 4-tuples have two ways of partition, and to distinguish between them, the prime symbol is used.

(5) (4.2.4) shows that there are many different representations of the transformation coefficient. In the proof, the partitions of 4-tuples in (4.2.4) are automatically selected according to the representations of the poles. For example,

$$\frac{7_{21,345}}{7_{12,546}} = -\frac{[123][453]}{[126][456]}, \quad (4.2.5)$$

because the partition $\mathbf{12}, \mathbf{45}$ of 4-tuple $\mathbf{\check{3}\check{6}} = \mathbf{1245}$ is already provided in the representations of $\mathbf{7}$ on the left side of (4.2.5) by the commas in the subscripts. The negative sign on the right side of (4.2.5) is caused by

$$\frac{\mathbf{7}_{\mathbf{21,345}}}{\mathbf{7}_{\mathbf{12,546}}} = -\frac{\mathbf{7}_{\mathbf{12,345}}}{\mathbf{7}_{\mathbf{12,645}}}.$$

(6) In (4.2.4), there is a lot of freedom in choosing the partitions – we do not need to follow the partitions provided by the representations such as (4.2.5). The following is the unique set of partitions for the result after the elimination of $\mathbf{7}, \mathbf{8}, \mathbf{9}, \mathbf{0}, \mathbf{a}, \mathbf{b}$ to be a bracket polynomial instead of a rational bracket polynomial:

$$k = \frac{[\mathbf{315}][\mathbf{324}]}{[\mathbf{615}][\mathbf{624}]} \frac{[\mathbf{314}][\mathbf{325}]}{[\mathbf{614}][\mathbf{625}]} \frac{[\mathbf{513}][\mathbf{546}]}{[\mathbf{213}][\mathbf{246}]} \frac{[\mathbf{514}][\mathbf{536}]}{[\mathbf{214}][\mathbf{236}]} \frac{[\mathbf{513}][\mathbf{524}]}{[\mathbf{613}][\mathbf{624}]} \frac{[\mathbf{315}][\mathbf{346}]}{[\mathbf{215}][\mathbf{246}]}. \quad (4.2.6)$$

There are two approaches to find the best representation (4.2.6). After eliminating the six poles by (4.2.3) and removing common bracket factors, the conclusion expression becomes

$$\begin{array}{c} [\mathbf{134}][\mathbf{135}]^4[\mathbf{145}][\mathbf{234}][\mathbf{235}][\mathbf{245}][\mathbf{346}][\mathbf{356}][\mathbf{456}] \\ -[\mathbf{123}][\mathbf{124}][\mathbf{125}][\mathbf{136}][\mathbf{146}][\mathbf{156}][\mathbf{236}][\mathbf{246}]^4[\mathbf{256}]\, k. \end{array} \quad (4.2.7)$$

Denote

$$\begin{aligned} c' &= [\mathbf{134}][\mathbf{135}]^4[\mathbf{145}][\mathbf{234}][\mathbf{235}][\mathbf{245}][\mathbf{346}][\mathbf{356}][\mathbf{456}], \\ c &= [\mathbf{123}][\mathbf{124}][\mathbf{125}][\mathbf{136}][\mathbf{146}][\mathbf{156}][\mathbf{236}][\mathbf{246}]^4[\mathbf{256}]. \end{aligned}$$

The first approach is to consider the brackets of c that can occur in the denominator of the result of (4.2.4) in one and only one manner. There are four such brackets:

$$\begin{aligned} [\mathbf{123}] &: \text{ occurs only in } \frac{[\mathbf{153}]}{[\mathbf{123}]} \frac{[\mathbf{546}]}{[\mathbf{246}]} \text{ of type } \frac{[\mathbf{5*}]}{[\mathbf{2*}]} \frac{[\mathbf{5*}]}{[\mathbf{2*}]}, \\[4pt] [\mathbf{125}] &: \text{ occurs only in } \frac{[\mathbf{135}]}{[\mathbf{125}]} \frac{[\mathbf{346}]}{[\mathbf{246}]} \text{ of type } \frac{[\mathbf{3*}]}{[\mathbf{2*}]} \frac{[\mathbf{3*}]}{[\mathbf{2*}]}, \\[4pt] [\mathbf{136}] &: \text{ occurs only in } \frac{[\mathbf{135}]}{[\mathbf{136}]} \frac{[\mathbf{245}]}{[\mathbf{246}]} \text{ of type } \frac{[\mathbf{5*}]}{[\mathbf{6*}]} \frac{[\mathbf{5*}]}{[\mathbf{6*}]}, \\[4pt] [\mathbf{156}] &: \text{ occurs only in } \frac{[\mathbf{153}]}{[\mathbf{156}]} \frac{[\mathbf{243}]}{[\mathbf{246}]} \text{ of type } \frac{[\mathbf{3*}]}{[\mathbf{6*}]} \frac{[\mathbf{3*}]}{[\mathbf{6*}]}, \end{aligned} \quad (4.2.8)$$

where every "$*$" is an abbreviation of a 2-blade.

The denominators in (4.2.8) cancel eight of the twelve brackets in c. The four brackets left in c, $[\mathbf{124}][\mathbf{146}][\mathbf{236}][\mathbf{256}]$, can be generated by the denominators of the two remaining ratios in (4.2.4):

$$\text{type } \frac{[\mathbf{3*}]}{[\mathbf{6*}]} \frac{[\mathbf{3*}]}{[\mathbf{6*}]} : \frac{[\mathbf{143}][\mathbf{253}]}{[\mathbf{146}][\mathbf{256}]}; \qquad \text{type } \frac{[\mathbf{5*}]}{[\mathbf{2*}]} \frac{[\mathbf{5*}]}{[\mathbf{2*}]} : \frac{[\mathbf{514}][\mathbf{536}]}{[\mathbf{214}][\mathbf{236}]}.$$

The second approach is to consider the ratio $k' = c'/c$ and prove that it is a representation of k. Start with the bracket of the highest degree in the denominator (or numerator) of k'. Distribute it among the brackets in the numerator (or denominator) of k' to form ratios of type **3/6, 5/2, 5/6, 3/2** respectively. Then continue to the brackets of lower degree in the denominator (or numerator). In this example, starting with $[246]^4$ (or $[135]^4$), we get the same pairing as (4.2.8).

The benefit of representation (4.2.6) is obvious: substituting it into (4.2.7) we get zero immediately, without resorting to conic transformations as in the above proof. Furthermore, the last three nondegeneracy conditions in (4.2.2) are avoided.

(7) Is it indispensable that all the six poles must change their representations in order to obtain a bracket polynomial proof rather than a rational one? According to (4.1.21), we only need to change the representations of two points instead of six:

$$
\begin{aligned}
[9bc] \;=\; & [7_{21,345}\, 8_{23,145}\, 0_{45,261}][9_{34,612}\, a_{65,134}\, b_{61,534}] \\
& -[7_{12,534}\, 0_{54,612}\, a_{56,412}][8_{32,461}\, 9_{34,261}\, b_{61,345}]\frac{8_{23,145}\, a_{65,134}}{8_{32,461}\, a_{56,412}} \\[2mm]
\;=\; & 16\,\underbrace{[124][125][134]^2[135]^2[136][145]^2[146][235][245][246]^2[346][356]}_{\{[123][156][245][346]-[125][136][234][456]\}}
\end{aligned}
$$

$$
\overset{conic}{=}\; 0.
$$

(8) The last two steps are *conic transformations* to be studied in the next section.

(9) The additional nondegeneracy conditions include all the existence conditions of the 5-point conics used in the proof, together with the bracket factors in the denominator of the transformation coefficient that cannot be canceled by the bracket monomial being multiplied with the transformation coefficient.

Representation of the fourth point of intersection of two conics.

Let $\mathbf{x} = 12345 \cap 1234'5'$. The coordinate representation (4.1.27) – (4.1.28) of $\mathbf{x}$ has twelve terms, where the bracket coefficients are of degree twelve. Some simplifications must be carried out to the representation when substituted into a bracket polynomial $p(\mathbf{x})$ containing $\mathbf{x}$:

(i) First apply Cramer's rule $[123]\mathbf{x} = [\mathbf{x}23]1 - [\mathbf{x}13]2 + [\mathbf{x}12]3$ to $p(\mathbf{x})$. It is the coordinatization of $\mathbf{x}$ with respect to basis $1, 2, 3$. Since the computing is always homogeneous, the coefficient $[123]$ of $\mathbf{x}$ is not needed.

(ii) Compute

$$
\begin{aligned}
\mu_1 &= [145][234][235], & \mu_2 &= [134][135][245], & \mu_3 &= [124][125][345], \\
\mu_1' &= [14'5'][234'][235'], & \mu_2' &= [134'][135'][24'5'], & \mu_3' &= [124'][125'][34'5']
\end{aligned}
$$

$$\tag{4.2.9}$$

by eliminating all the incidence points from the brackets of the μ's.

(iii) Compute $\lambda_1 = \mu_2\mu_3' - \mu_3\mu_2'$, $\lambda_2 = \mu_3\mu_1' - \mu_1\mu_3'$, $\lambda_3 = \mu_1\mu_2' - \mu_2\mu_1'$. Substitute

$$
[\mathbf{x}23] = \lambda_2\lambda_3, \quad [\mathbf{x}13] = -\lambda_1\lambda_3, \quad [\mathbf{x}12] = \lambda_1\lambda_2
$$

into $p(\mathbf{x})$ after removing their common bracket factors.

4.2.2 *Representations of geometric conclusions*

There are four typical conclusions in conic geometry:

(a) points $1, 2, 3$ are collinear;

(b) lines $12, 1'2', 1''2''$ concur;

(c) six points $1, \ldots, 6$ are on the same conic;

(d) points $\mathbf{a}, \mathbf{b}$ are conjugate with respect to a conic.

The first two conclusions have unique algebraic representations. The latter two are to be considered in this subsection.

Representation of the coconic conclusion.

The conclusion that points $1, 2, 3, 4, 5, 6$ are coconic has fifteen different representations: without using GP relations, the expression

$$\mathrm{conic}(\mathbf{123456}) = [135][245][126][346] - [125][345][136][246]$$

is antisymmetric within each of the pairs $\mathbf{14}, \mathbf{23}, \mathbf{56}$, and is symmetric among the three pairs, so the number of different representations is $C_6^2 \times C_4^2/3! = 15$. They are listed as follows:

> **123456 123546 123645 124356 124536 124635 125346 125436**
> **125634 126345 126435 126534 134256 135246 136245**.

The six points form $C_6^3 = 20$ brackets. Without computing the brackets any representation is just as good as any other one. To compute them means to eliminate from them all the incidence points by substituting into them the corresponding Cayley expressions, and then make Cayley expansions.

An obvious criterion for a good representation is that the degree of the explicit common factors obtained from the above bracket computing, called the *common degree* of the representation, is maximal among the fifteen representations.

The following is a list of formulas on generic factored Cayley expansions of some layer-1 Cayley expressions, called *factorizable Cayley brackets*. Their derivation can be found in Section 2.5 and Appendix A. In the list, $\mathbf{a_{ij}} = \lambda_{\mathbf{j}}\mathbf{i} + \lambda_{\mathbf{i}}\mathbf{j}$ and $\mathbf{b_{kl}} = \mu_{\mathbf{l}}\mathbf{k} + \mu_{\mathbf{k}}\mathbf{l}$ are points on lines $\mathbf{ij}$ and $\mathbf{kl}$ respectively, where the coefficients λ's and μ's are generic polynomials linear in the vector variables marked as boldfaced subscripts.

Generic factored expansions of factorizable Cayley brackets

Double line: Line $1'2'$ or 12 or $5'6'$.

$$[1(1'2' \vee 3'4')(1'2' \vee 3''4'')] = [11'2']1'2' \vee 3'4' \vee 3''4'',$$

$$[(12 \vee 34)(12 \vee 3'4')(1''2'' \vee 3''4'')] = (12 \vee 34 \vee 3'4')(12 \vee 1''2'' \vee 3''4''),$$

$$[1(1'2' \vee 3'4')\mathbf{a_{1'2'}}] = [11'2'][3'4'\mathbf{a_{1'2'}}],$$

$$[(12 \vee 34)\mathbf{a_{5'6'}}\mathbf{b_{5'6'}}] = (\lambda_{\mathbf{6'}}\mu_{\mathbf{5'}} - \lambda_{\mathbf{5'}}\mu_{\mathbf{6'}})12 \vee 34 \vee 5'6',$$

$$[(12 \vee 34)\mathbf{a_{12}}\mathbf{b_{5''6''}}] = [12\mathbf{b_{5''6''}}][34\mathbf{a_{12}}],$$

$$[(12 \vee 34)(12 \vee 3'4')\mathbf{a_{5''6''}}] = [12\mathbf{a_{5''6''}}]12 \vee 34 \vee 3'4',$$

$$[(12 \vee 34)(1'2' \vee 3'4')\mathbf{a_{12}}] = [34\mathbf{a_{12}}]12 \vee 1'2' \vee 3'4'.$$

Recursion: Point **1** recurs.

$$
\begin{aligned}
\mathbf{12} \vee \mathbf{12'} \vee \mathbf{1''2''} &= [\mathbf{122'}][\mathbf{11''2''}], \\
[\mathbf{1}(\mathbf{12'} \vee \mathbf{3'4'})\mathbf{a}_{5''6''}] &= [\mathbf{13'4'}][\mathbf{12'a}_{5''6''}], \\
[\mathbf{1}(\mathbf{1'2'} \vee \mathbf{3'4'})\mathbf{a}_{16''}] &= -\lambda_1 \mathbf{16''} \vee \mathbf{1'2'} \vee \mathbf{3'4'}, \\
[\mathbf{1}(\mathbf{12'} \vee \mathbf{3'4'})(\mathbf{1''2''} \vee \mathbf{3''4''})] &= [\mathbf{13'4'}]\mathbf{12'} \vee \mathbf{1''2''} \vee \mathbf{3''4''}.
\end{aligned}
$$

Complete quadrilateral: **1234**.

$$[(\mathbf{12} \vee \mathbf{34})(\mathbf{13} \vee \mathbf{24})(\mathbf{14} \vee \mathbf{23})] = -2\,[\mathbf{123}][\mathbf{124}][\mathbf{134}][\mathbf{234}].$$

Triangle pair: $(\mathbf{122'}, \mathbf{344'})$.

$$[(\mathbf{12} \vee \mathbf{34})\,(\mathbf{12'} \vee \mathbf{34'})\,(\mathbf{22'} \vee \mathbf{44'})] = -[\mathbf{122'}][\mathbf{344'}]\mathbf{13} \vee \mathbf{24} \vee \mathbf{2'4'}.$$

Quadrilateral: $(\mathbf{1234}, \mathbf{14})$.

$$
\begin{aligned}
[(\mathbf{12} \vee \mathbf{34})\,(\mathbf{13} \vee \mathbf{24})\,(\mathbf{14} \vee \mathbf{3''4''})] &= -[\mathbf{124}][\mathbf{134}]([\mathbf{123}][\mathbf{43''4''}] + [\mathbf{13''4''}][\mathbf{234}]), \\
[(\mathbf{12} \vee \mathbf{34})(\mathbf{13} \vee \mathbf{24})\mathbf{a}_{14}] &= \ \ [\mathbf{124}][\mathbf{134}](\lambda_4[\mathbf{123}] - \lambda_1[\mathbf{234}]).
\end{aligned}
$$

Triangle: $\mathbf{122'}$.

$$
\begin{aligned}
[(\mathbf{12} \vee \mathbf{34})\,(\mathbf{12'} \vee \mathbf{3'4'})\,(\mathbf{22'} \vee \mathbf{3''4''})] &= [\mathbf{122'}]([\mathbf{134}][\mathbf{23''4''}][\mathbf{2'3'4'}] \\
&\qquad\qquad -[\mathbf{13'4'}][\mathbf{234}][\mathbf{2'3''4''}]), \\
[(\mathbf{12} \vee \mathbf{34})\mathbf{a}_{12'}\mathbf{b}_{22'}] &= [\mathbf{122'}](\lambda_1\mu_{2'}[\mathbf{234}] - \lambda_{2'}\mu_2[\mathbf{134}]), \\
[(\mathbf{12} \vee \mathbf{34})(\mathbf{12'} \vee \mathbf{3'4'})\mathbf{a}_{22'}] &= [\mathbf{122'}](\lambda_{2'}[\mathbf{13'4'}][\mathbf{234}] + \lambda_2[\mathbf{134}][\mathbf{2'3'4'}]).
\end{aligned}
$$

Except for the first triangle pattern $p_{IV} = [(\mathbf{12} \vee \mathbf{34})\,(\mathbf{12'} \vee \mathbf{3'4'})\,(\mathbf{22'} \vee \mathbf{3''4''})]$, each factorizable Cayley bracket in the above list has a unique factored expansion. In the exceptional case, there are three factored results, which can be obtained by either (a) the three outer product expansions, or (b) the three meet product expansions by splitting $\mathbf{1}, \mathbf{2}$ in the first meet product, or $\mathbf{1}, \mathbf{2'}$ in the second meet product, or $\mathbf{2}, \mathbf{2'}$ in the third meet product:

$$
\begin{aligned}
p_{IV} &= [\mathbf{122'}]([\mathbf{23''4''}]\mathbf{12'} \vee \mathbf{34} \vee \mathbf{3'4'} - [\mathbf{13'4'}]\mathbf{22'} \vee \mathbf{34} \vee \mathbf{3''4''}) \\
&= [\mathbf{122'}]([\mathbf{2'3''4''}]\mathbf{12} \vee \mathbf{34} \vee \mathbf{3'4'} - [\mathbf{134}]\mathbf{22'} \vee \mathbf{3'4'} \vee \mathbf{3''4''}) \qquad (4.2.10) \\
&= [\mathbf{122'}]([\mathbf{2'3'4'}]\mathbf{12} \vee \mathbf{34} \vee \mathbf{3''4''} - [\mathbf{234}]\mathbf{12'} \vee \mathbf{3'4'} \vee \mathbf{3''4''}).
\end{aligned}
$$

Algorithm 4.25. Optimal representation of the six-point coconic conclusion.

Input: The constructions of points $\mathbf{1}, \mathbf{2}, \mathbf{3}, \mathbf{4}, \mathbf{5}, \mathbf{6}$.

Output: A set of sequences of the six points.

Step 1. Let $\mathcal{B}$ be the set of twenty brackets formed by the six points. Find and compute all the factorizable Cayley brackets in $\mathcal{B}$.

Step 2. If there are two triangle p_{IV}-typed Cayley brackets generating two different bracket binomial factors of degree three, say p_1 and p_2, compare them by contracting $p_1 \pm p_2$. If they are equal up to sign, then identify the two factors up to scale. This step is necessary because the expansion results of such Cayley brackets may not be unique.

Step 3. Let $\mathcal{C}$ be the fifteen representations of the coconic conclusion. For every element c in $\mathcal{C}$, first substitute into it the computed results of the factorizable Cayley brackets, then collect explicit common factors of the expression of c to count the common degree.

Step 4. Output the elements of $\mathcal{C}$ of maximal common degree.

Example 4.26. If points $1, 2, 3, 4, 5, 6$ are on a conic, then intersections $12 \cap 34$, $13 \cap 24$, $14 \cap 23$, $34 \cap 56$, $35 \cap 46$, $45 \cap 36$ are coconic.

Free conic points: $1, 2, 3, 4, 5, 6$.
Intersections:

$$7 = 12 \cap 34, \quad 8 = 13 \cap 24, \quad 9 = 14 \cap 23,$$
$$0 = 34 \cap 56, \quad a = 35 \cap 46, \quad b = 36 \cap 45.$$

Conclusion: $7, 8, 9, 0, a, b$ are coconic.

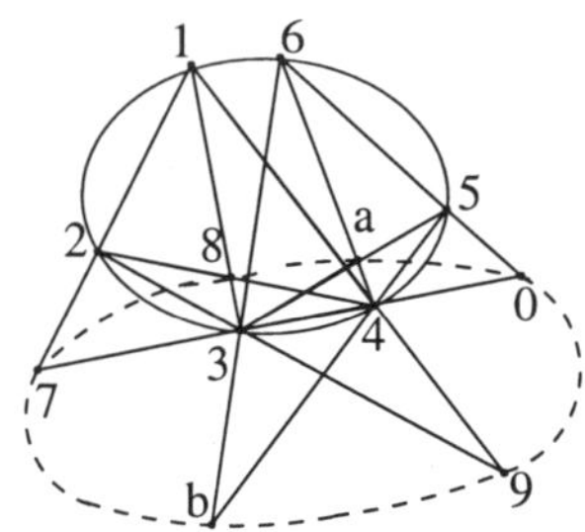

Fig. 4.3   Example 4.26: representation of the coconic conclusion.

Procedure of representing the conclusion:

Steps 1–2. There are eight factorizable Cayley brackets formed by $7, 8, 9, 0, a, b$:

1 line: **70** of line **34**. There are four associated brackets:

$$[780] = -[134][234]\,12 \vee 34 \vee 56, \quad [790] = [134][234]\,12 \vee 34 \vee 56,$$
$$[70a] = -[345][346]\,12 \vee 34 \vee 56, \quad [70b] = [345][346]\,12 \vee 34 \vee 56.$$

2 complete quadrilaterals: $[789]$ of 1234, and $[0ab]$ of 3456.

$$[789] = -2[123][124][134][234], \quad [0ab] = -2[345][346][356][456].$$

2 quadrilaterals: $[890]$ of $(1234, 34)$, and $[7ab]$ of $(3456, 34)$.

$$[7ab] = [345][346]([123][456] + [124][356]),$$
$$[890] = [134][234]([123][456] + [124][356]).$$

Steps 3–4. The fifteen representations with their common degrees are

$$789a0b^{(4)} \quad 789b0a^{(4)} \quad 7809ab^{(4)} \quad \underline{780a9b^{(6)}} \quad \underline{780b9a^{(6)}}$$
$$\underline{78a90b^{(6)}} \quad \underline{78a09b^{(6)}} \quad \underline{78ab90^{(6)}} \quad \underline{78b90a^{(6)}} \quad \underline{78b09a^{(6)}}$$
$$\underline{78ba90^{(6)}} \quad 7908ab^{(4)} \quad \underline{79a80b^{(6)}} \quad \underline{79b80a^{(6)}} \quad 7890ab^{(0)}.$$

Ten representations have maximal common degree 6.

Let us see how the proof goes on when choosing one of the ten representations, say $\mathbf{78a09b}^{(6)}$. In this example, a very nice property of the twelve non-factorizable Cayley brackets is that their binomial expansion results are all unique. For example, all the binomial expansion results of $[\mathbf{80b}]$ are identical:

$$[(\mathbf{13} \vee \mathbf{24})(\mathbf{34} \vee \mathbf{56})(\mathbf{36} \vee \mathbf{45})] = [\mathbf{134}][\mathbf{245}][\mathbf{346}][\mathbf{356}] - [\mathbf{136}][\mathbf{234}][\mathbf{345}][\mathbf{456}].$$

Proof of Example 4.26.

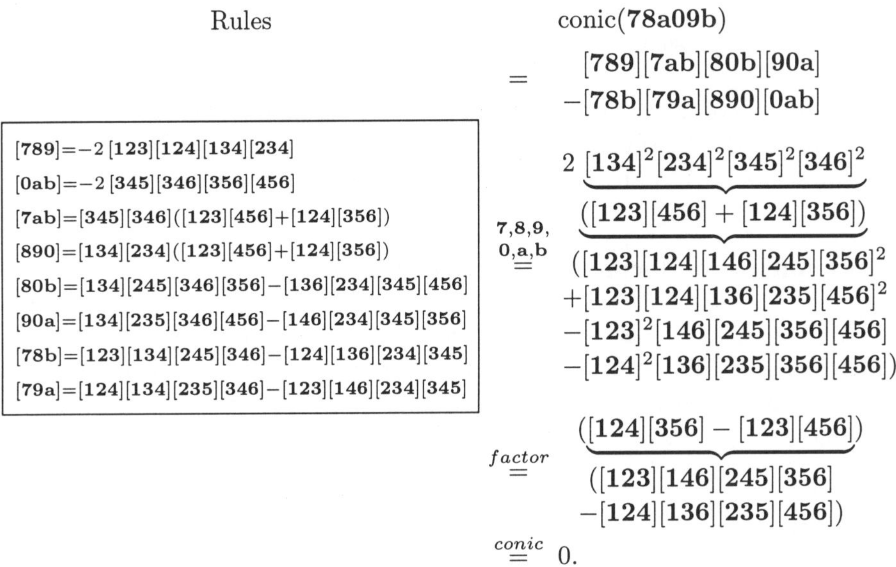

The next to the last step is a factorization in the polynomial ring of brackets. The last step is a conic transformation to be introduced in the next section. □

Representation of the conic conjugacy conclusion.

The guideline is to reduce the number of terms in (4.1.8). This is only possible when some brackets involving the two conjugate points $\mathbf{a}, \mathbf{b}$ with respect to the conic are zero. If any two conic points are collinear with $\mathbf{a}$ (or $\mathbf{b}$), they should be used in the representation of the conic conjugacy conclusion. In the general case, for points $\mathbf{a}$ and $\mathbf{b}$, their essential conic points with bigger essential weights should be in the conclusion representation.

Example 4.27. Let $\mathbf{7}, \mathbf{9}$ be points not on conic $\mathbf{12}(\mathbf{24})\mathbf{3}(\mathbf{34})$. Let $\mathbf{8}$ be the intersection of line $\mathbf{23}$ and the tangent of the conic at $\mathbf{1}$. Let $\mathbf{a} = \mathbf{23} \cap \mathbf{47}$, $\mathbf{b} = \mathbf{12} \cap \mathbf{79}$. Represent the conclusion that $\mathbf{a}, \mathbf{b}$ are conjugate with respect to the conic.

The conic has three points $\mathbf{1}, \mathbf{2}, \mathbf{3}$ and three corresponding tangents $\mathbf{18}, \mathbf{24}, \mathbf{34}$. The essential conic points of $\mathbf{a}, \mathbf{b}$ together with their essential weights are $\mathbf{2}^1, \mathbf{1}^{\frac{1}{2}}, \mathbf{3}^{\frac{1}{2}}$.

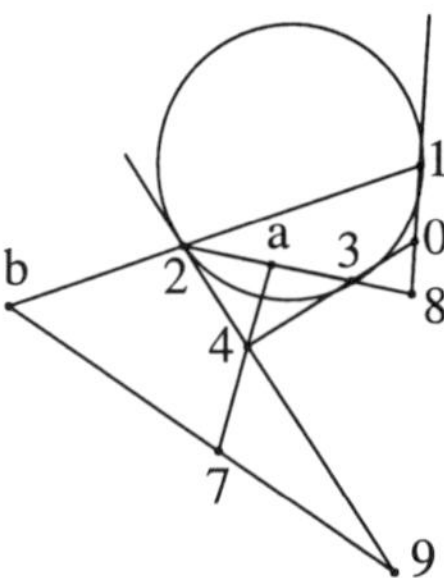

Fig. 4.4　Example 4.27: representation of the conic conjugacy conclusion.

So there are two optimal representations of the conclusion based on the two conic representations **32(24)1(18)** and **12(24)3(34)** respectively. The proofs following these representations can be found in Example 4.51 of Subsection 4.5.2.

4.3　Simplification techniques in conic computing

The six-point-on-conic constraint is the basic structure in conic geometry. Just like the 2D collinearity and concurrency transformations in projective incidence geometry, the 2D coconic constraint can be used to simplify bracket computing involving conic points as a transformation of bracket polynomials. The major difference is that the coconic constraint is of degree four, and besides the fifteen equivalent forms (4.1.2), it has another group of equivalent forms (4.1.4). In this section, we present three transformations based on (4.1.2) and (4.1.4) for contractions and factorizations of bracket polynomials involving conic points.

4.3.1　*Conic transformation*

Definition 4.28. Let p be a bracket polynomial which is neither contractible nor factorizable in the polynomial ring of brackets. For any six conic points $\mathbf{a}, \mathbf{b}, \mathbf{c}, \mathbf{d}, \mathbf{x}, \mathbf{y}$ in p, if the transformation

$$[\mathbf{xab}][\mathbf{xcd}][\mathbf{yac}][\mathbf{ybd}] = [\mathbf{xac}][\mathbf{xbd}][\mathbf{yab}][\mathbf{ycd}] \tag{4.3.1}$$

either reduces the number of terms of p, or makes it factorizable in the polynomial ring of brackets, or makes it contractible, the transformation is called a *conic transformation*.

No matter if (4.3.1) is a conic transformation or not, the term of p containing the left side of (4.3.1) is said to be *conic transformable*.

In some sense, (4.3.1) and the pseudoconic transformation (4.3.4) to be introduced in the next subsection are the counterpart of the collinearity transformation in incidence geometry. Transformation (4.3.1) is performed only to one term of a bracket polynomial. The simplest way to realize the transformation is to divide the term by the left side of the equality, and multiply the quotient with the right side.

Proposition 4.29. Transformation (4.3.1) has the following properties:

(1) Any three brackets on the left side determine a unique transformation.

(2) There are only two transformations on $[\mathbf{xab}][\mathbf{xcd}]$ when multiplied by two brackets containing $\mathbf{y}$. The two brackets are $[\mathbf{yac}][\mathbf{ybd}]$ or $[\mathbf{yad}][\mathbf{ybc}]$.

(3) For a conic point $\mathbf{x}$, if there is a point that occurs in every bracket containing $\mathbf{x}$, then there is no transformation involving $\mathbf{x}$.

(4) Let t be the term of p containing the left side of (4.3.1). Then (4.3.1) is a conic transformation if and only if one of the following conditions is satisfied:

(a) a bracket on the right side of (4.3.1) is in every term of $p - t$;
(b) t after the transformation becomes a like term of another term of p;
(c) t after the transformation forms a contractible pair with another term of p.

In particular, if p has only two terms, then (4.3.1) is a conic transformation if and only if a bracket on the right side of (4.3.1) is in $p - t$.

Algorithm 4.30. Conic transformation.

Input: A bracket polynomial p of degree at least four and involving at least six conic points. Assume that p is already factorized in the polynomial ring of brackets, and that all factors of degree less than four or involving fewer than six conic points have been moved into a set q.

Output: A set q composed of bracket polynomial factors.

Step 1. Do the following steps for each factor f of p, for each term t of f, by setting $\mathcal{C}$ to be all conic points in t.

Step 2. Let p' be the square-free bracket factors of t formed by points in $\mathcal{C}$. Count the degree of each point in p', which is the number of occurrences of the point in p'. Denote by $\mathcal{C}^2$ the points with degree at least two.

If $\#(\mathcal{C}^2) < 6$, then move f to q and go back to Step 1, else if $\mathcal{C}^2 \neq \mathcal{C}$, then set $\mathcal{C} = \mathcal{C}^2$ and go back to the beginning of Step 2.

Step 3. Let $\mathbf{x}$ be a point in $\mathcal{C}$ whose degree in p' is the lowest. Let $b(\mathbf{x})$ be the brackets of p' containing $\mathbf{x}$, and let $\breve{b}(\mathbf{x})$ be the brackets not containing $\mathbf{x}$.

Find from $b(\mathbf{x})$ all the bracket pairs $[\mathbf{xab}][\mathbf{xcd}]$ such that $\{\mathbf{a}, \mathbf{b}\} \cap \{\mathbf{c}, \mathbf{d}\}$ is empty. If there is no such pair, go to Step 5.

Step 4. For each pair $[\mathbf{xab}][\mathbf{xcd}]$ found in Step 3, set

$$\mathcal{R}_{\mathbf{c}} = \{\mathbf{y} \in \mathcal{C} - \{\mathbf{x}, \mathbf{a}, \mathbf{b}, \mathbf{c}, \mathbf{d}\} \mid [\mathbf{yac}][\mathbf{ybd}] \in \breve{b}(\mathbf{x})\},$$
$$\mathcal{R}_{\mathbf{d}} = \{\mathbf{y} \in \mathcal{C} - \{\mathbf{x}, \mathbf{a}, \mathbf{b}, \mathbf{c}, \mathbf{d}\} \mid [\mathbf{yad}][\mathbf{ybc}] \in \breve{b}(\mathbf{x})\}.$$

For every $\mathbf{y} \in \mathcal{R_c}$, let the biggest power of $[\mathbf{xab}][\mathbf{xcd}][\mathbf{yac}][\mathbf{ybd}]$ in t be m. For every $\mathbf{y}' \in \mathcal{R_d}$, let the biggest power of $[\mathbf{xab}][\mathbf{xcd}][\mathbf{y}'\mathbf{ad}][\mathbf{y}'\mathbf{bc}]$ in t be m'.

For i from m down to 1, for i' from m' down to 1, if one of

$$([\mathbf{xab}][\mathbf{xcd}][\mathbf{yac}][\mathbf{ybd}])^{i} = ([\mathbf{xac}][\mathbf{xbd}][\mathbf{yab}][\mathbf{ycd}])^{i},$$
$$([\mathbf{xab}][\mathbf{xcd}][\mathbf{y}'\mathbf{ad}][\mathbf{y}'\mathbf{bc}])^{i'} = ([\mathbf{xad}][\mathbf{xbc}][\mathbf{y}'\mathbf{ab}][\mathbf{y}'\mathbf{cd}])^{i'},$$

is a conic transformation, then perform it, contract and factorize the result, replace f by the factors, and go back to Step 1.

Step 5. If $\#(\mathcal{C}) > 6$, delete $\mathbf{x}$ from $\mathcal{C}$, go back to Step 2; else, skip to the next term of f, and if f has no more terms, move f into q.

Example 4.31. Let $\mathbf{1, 2, 3, 4, 5, 6}$ be conic points. Simplify the following bracket binomial occurred in the proof of Example 4.24:

$$p = [\mathbf{126}]^4[\mathbf{135}]^4[\mathbf{234}]^3[\mathbf{245}][\mathbf{346}][\mathbf{456}]^3 - [\mathbf{123}]^3[\mathbf{125}][\mathbf{136}][\mathbf{156}]^3[\mathbf{246}]^4[\mathbf{345}]^4.$$

Steps 1–2. Let t be the first term. Then $p' = [\mathbf{126}][\mathbf{135}][\mathbf{234}][\mathbf{245}][\mathbf{346}][\mathbf{456}]$. The conic points with their degrees are $\mathcal{C} = \mathcal{C}^2 = \{\mathbf{1}^2, \mathbf{2}^3, \mathbf{3}^3, \mathbf{4}^4, \mathbf{5}^3, \mathbf{6}^3\}$.

Step 3. Set $\mathbf{x} = \mathbf{1}$. Then $b(\mathbf{x}) = [\mathbf{126}][\mathbf{135}]$ and $\breve{b}(\mathbf{x}) = [\mathbf{234}][\mathbf{245}][\mathbf{346}][\mathbf{456}]$.

Steps 4-5. $\mathbf{y} = \mathbf{4}$ is the only conic point not in $b(\mathbf{x})$. $[\mathbf{234}][\mathbf{456}]$ and $[\mathbf{245}][\mathbf{346}]$ are both in $\breve{b}(\mathbf{x})$, with multiplicity $m = 3$ and $m' = 1$ respectively.

If we choose $[\mathbf{234}][\mathbf{456}]$, then the transformation

$$([\mathbf{126}][\mathbf{135}][\mathbf{234}][\mathbf{456}])^3 = ([\mathbf{123}][\mathbf{156}][\mathbf{246}][\mathbf{345}])^3$$

changes t to $[\mathbf{123}]^3[\mathbf{126}][\mathbf{135}][\mathbf{156}]^3[\mathbf{245}][\mathbf{246}]^3[\mathbf{346}][\mathbf{345}]^3$, which has common factors $[\mathbf{123}]^3[\mathbf{156}]^3[\mathbf{246}]^3[\mathbf{345}]^3$ with the second term of p. After removing the common factors, we get

$$p = [\mathbf{126}][\mathbf{135}][\mathbf{245}][\mathbf{346}] - [\mathbf{125}][\mathbf{136}][\mathbf{246}][\mathbf{345}].$$

Another conic transformation changes p to 0.

If we choose the other pair $[\mathbf{245}][\mathbf{346}]$, then the simplification procedure is much the same and only the order between the above two transformations is reversed.

Example 4.32. Let $\mathbf{1, 2, 3, 4, 5, 6}$ be conic points. Simplify

$$\begin{aligned} p = \ & [\mathbf{124}]^3[\mathbf{135}]^2[\mathbf{136}][\mathbf{256}][\mathbf{346}] - [\mathbf{124}]^2[\mathbf{125}][\mathbf{134}]^2[\mathbf{136}][\mathbf{256}][\mathbf{356}] \\ & +[\mathbf{124}][\mathbf{126}]^2[\mathbf{134}][\mathbf{135}]^2[\mathbf{245}][\mathbf{346}] - [\mathbf{125}][\mathbf{126}]^2[\mathbf{134}]^3[\mathbf{245}][\mathbf{356}]. \end{aligned}$$

The first, the third and the last terms of p each have one conic transformation:

$$\begin{aligned} [\mathbf{124}][\mathbf{135}][\mathbf{256}][\mathbf{346}] &= [\mathbf{125}][\mathbf{134}][\mathbf{246}][\mathbf{356}], \\ [\mathbf{126}][\mathbf{135}][\mathbf{245}][\mathbf{346}] &= [\mathbf{125}][\mathbf{136}][\mathbf{246}][\mathbf{345}], \qquad\qquad (4.3.2) \\ [\mathbf{126}][\mathbf{134}][\mathbf{245}][\mathbf{356}] &= [\mathbf{124}][\mathbf{136}][\mathbf{256}][\mathbf{345}]. \end{aligned}$$

The second term has no conic transformation, because according to Proposition 4.29, among the conic points $\mathcal{C} = \{\mathbf{1}^4, \mathbf{2}^3, \mathbf{3}^3, \mathbf{4}^2, \mathbf{5}^3, \mathbf{6}^3\}$, for point $\mathbf{4}$ which has the lowest degree in the set, another point **1** occurs in every bracket containing **4** in the second term.

The first transformation in (4.3.2) produces a common factor $[\mathbf{134}]$ of p. After removing it we get

$$p = \;[\mathbf{124}]^2[\mathbf{125}][\mathbf{135}][\mathbf{136}][\mathbf{246}][\mathbf{356}] - [\mathbf{124}]^2[\mathbf{125}][\mathbf{134}][\mathbf{136}][\mathbf{256}][\mathbf{356}]$$
$$+[\mathbf{124}][\mathbf{126}]^2[\mathbf{135}]^2[\mathbf{245}][\mathbf{346}] - [\mathbf{125}][\mathbf{126}]^2[\mathbf{134}]^2[\mathbf{245}][\mathbf{356}].$$

The second transformation produces another common factor $[\mathbf{125}]$ of p, and after removing it we get

$$p = \;[\mathbf{124}]^2[\mathbf{135}][\mathbf{136}][\mathbf{246}][\mathbf{356}] - [\mathbf{124}]^2[\mathbf{134}][\mathbf{136}][\mathbf{256}][\mathbf{356}]$$
$$+[\mathbf{124}][\mathbf{126}][\mathbf{135}][\mathbf{136}][\mathbf{246}][\mathbf{345}] - [\mathbf{126}]^2[\mathbf{134}]^2[\mathbf{245}][\mathbf{356}].$$

The last conic transformation produces two common factors $[\mathbf{124}][\mathbf{136}]$. Finally, with all the common factors retrieved, we have

$$p = [\mathbf{124}][\mathbf{125}][\mathbf{134}][\mathbf{136}]\{[\mathbf{124}][\mathbf{135}][\mathbf{246}][\mathbf{356}] - [\mathbf{124}][\mathbf{134}][\mathbf{256}][\mathbf{356}]$$
$$+[\mathbf{126}][\mathbf{135}][\mathbf{246}][\mathbf{345}] - [\mathbf{126}][\mathbf{134}][\mathbf{256}][\mathbf{345}]\}. \tag{4.3.3}$$

4.3.2 *Pseudoconic transformation*

Definition 4.33. Let p be a bracket polynomial which is neither contractible, nor factorizable in the polynomial ring of brackets, nor conic transformable. For any six conic points $\mathbf{a}, \mathbf{b}, \mathbf{c}, \mathbf{d}, \mathbf{x}, \mathbf{y}$ in p, if by the transformation

$$[\mathbf{xab}][\mathbf{xcd}][\mathbf{yac}] = \frac{[\mathbf{xac}][\mathbf{xbd}][\mathbf{yab}][\mathbf{ycd}]}{[\mathbf{ybd}]}, \tag{4.3.4}$$

after removal of common rational bracket monomial factors from p, either the degree of p is decreased, or p becomes contractible, then the transformation is called a *pseudoconic transformation*.

Proposition 4.34. Let p be a bracket polynomial which is neither contractible, nor factorizable in the polynomial ring of brackets, nor conic transformable. Let t be the term of p containing the left side of (4.3.4). Let λ be the remainder of t after removal of the left side of (4.3.4). Let r be the numerator of the right side of (4.3.4).

(i) The transformation (4.3.4) cannot reduce the number of terms of p.

(ii) (4.3.4) is a pseudoconic transformation if and only if one of the following conditions is satisfied:

    (a) Two brackets in r are in every term of $p - t$.

    (b) The numerator of t after the transformation forms a contractible pair with another term of p multiplied by $[\mathbf{ybd}]$.

(iii) If p is a degree-3 binomial, then it has no pseudoconic transformation.

(iv) If p is degree-3 and has at least three terms, then (4.3.4) is a pseudoconic transformation if and only if one of the following terms is in $p - t$:

(a) $-\lambda[\mathbf{xac}][\mathbf{xbd}][\mathbf{yac}]$, (b) $-\lambda[\mathbf{xac}][\mathbf{xcd}][\mathbf{yab}]$, (c) $-\lambda[\mathbf{xab}][\mathbf{xac}][\mathbf{ycd}]$.

Proof.   (i). If p has a term which when multiplied by $[\mathbf{ybd}]$ becomes a like term of λr, then $[\mathbf{ybd}]$ has to be in λr, contradicting with the assumption that p is not conic transformable.

(ii). Obvious.

(iii). Let u be the other term of p. Then (4.3.4) is a pseudoconic transformation if and only if $[\mathbf{ybd}]u$ and $[\mathbf{xac}][\mathbf{xbd}][\mathbf{yab}][\mathbf{ycd}]$ have two common bracket factors. Then two of the latter four brackets must be in u, contradicting with the assumption that p is not factorizable.

(iv). Let u be a term of p containing two brackets of $[\mathbf{xac}][\mathbf{xbd}][\mathbf{yab}][\mathbf{ycd}]$. Since p is homogeneous, the third bracket in u is unique. Up to coefficient there are only the three cases given in the proposition. Because every two of the three cases have only one common bracket factor, no two brackets of $[\mathbf{xac}][\mathbf{xbd}][\mathbf{yab}][\mathbf{ycd}]$ can be common to every term different from t. So one term, say $u[\mathbf{ybd}]$, must form a GP transformable pair with $\lambda[\mathbf{xac}][\mathbf{xbd}][\mathbf{yab}][\mathbf{ycd}]$. The negative signs in the three cases come from the GP transformation requirement. $\square$

Algorithm 4.35. Pseudoconic transformation.

Input: A bracket polynomial p of degree at least three and involving at least six conic points. Assume that p is already factorized in the polynomial ring of brackets, and whose factors do not have conic transformations. Further assume that all the factors of p satisfying one of the following conditions are moved into a set q:

(1) the degree is less than three,
(2) the factor contains fewer than six conic points,
(3) the factor is a degree-3 bracket binomial.

Output: A set q composed of rational bracket polynomial factors.

Steps 1 to 3. Identical to those of Algorithm 4.30 for conic transformation, except that the second paragraph of Step 2 is revised to the following:

If either $\#(\mathcal{C}) < 6$ or $\#(\mathcal{C}^2) < 3$, then move f into q and go back to Step 1.

Step 4. For each pair $[\mathbf{xab}][\mathbf{xcd}]$ found in Step 3, set

$$\mathcal{R} = \{\, [\mathbf{yij}] \in \breve{b}(\mathbf{x}) \mid \mathbf{ij} \in \{\mathbf{ac}, \mathbf{ad}, \mathbf{bc}, \mathbf{bd}\},\ \mathbf{y} \in \mathcal{C} - \{\mathbf{x}, \mathbf{a}, \mathbf{b}, \mathbf{c}, \mathbf{d}\} \,\}.$$

For every bracket in $\mathcal{R}$, say $[\mathbf{yac}]$, let the biggest power of $[\mathbf{xab}][\mathbf{xcd}][\mathbf{yac}]$ in t be m. For i from m down to 1, if

$$([\mathbf{xab}][\mathbf{xcd}][\mathbf{yac}])^i = ([\mathbf{xac}][\mathbf{xbd}][\mathbf{yab}][\mathbf{ycd}][\mathbf{ybd}]^{-1})^i$$

is a pseudoconic transformation, then perform it, contract and factorize the result, put the result into q, delete f from p and go back to Step 1.

Step 5. Identical to that of Algorithm 4.30.

Example 4.36. Simplify the expression $\mathrm{conic}_{123,\,45}(\mathbf{x})$ in (4.1.7):

$$p = [145][234][235][\mathbf{x}12][\mathbf{x}13] - [134][135][245][\mathbf{x}12][\mathbf{x}23]$$
$$+ [124][125][345][\mathbf{x}13][\mathbf{x}23].$$

Let t be the first term of p. The conic points with their degrees are $\mathcal{C} = \mathcal{C}^2 = \{1^3, 2^3, 3^3, 4^2, 5^2, \mathbf{x}^2\}$. For point $\mathbf{x}$, since $b(\mathbf{x}) = [\mathbf{x}12][\mathbf{x}13]$, there is no pseudoconic transformation. For point $\mathbf{5}$, since $b(\mathbf{5}) = [145][235]$ and $\mathcal{R} = [\mathbf{x}12][\mathbf{x}13]$, there is a transformation

$$[145][235][\mathbf{x}12] = [125][345][14\mathbf{x}][23\mathbf{x}][\mathbf{x}34]^{-1},$$

which changes p into

$$[125][234][345][\mathbf{x}13][\mathbf{x}14] - [134][135][245][\mathbf{x}12][\mathbf{x}34] \tag{4.3.5}$$
$$+ [124][125][345][\mathbf{x}13][\mathbf{x}34]$$

multiplied by factor $[\mathbf{x}23][\mathbf{x}34]^{-1}$. Then a contraction between the first and the third terms of (4.3.5) changes it to

$$[134]([125][345][\mathbf{x}13][\mathbf{x}24] - [135][245][\mathbf{x}12][\mathbf{x}34]). \tag{4.3.6}$$

A conic transformation changes (4.3.6) to zero.

Example 4.37. Let $\mathbf{1}, \ldots, \mathbf{6}$ be conic points. Simplify

$$p = [123][134][145][356] + [124][135][136][345].$$

Let t be the first term of p. The conic points with their degrees are $\mathcal{C} = \{1^3, 2^1, 3^3, 4^2, 5^2, 6^1\}$, so $\mathcal{C}^2 = \{1^3, 3^3, 4^2, 5^2\}$.

For $\mathbf{x} = \mathbf{4}$, there is no pseudoconic transformation. For $\mathbf{x} = \mathbf{5}, \mathbf{1}$ or $\mathbf{3}$, the unique pseudoconic transformation is

$$[123][145][356] = [124][135][236][456][246]^{-1}.$$

Performing it to t, we get

$$p = [124][135][246]^{-1}([134][236][456] + [136][246][345]). \tag{4.3.7}$$

Similarly, for the second term of p, there is only one pseudoconic transformation:

$$[124][136][345] = [126][134][245][356][256]^{-1}.$$

It yields another result different from (4.3.7):

$$p = [134][356][256]^{-1}([123][145][256] + [126][135][245]). \tag{4.3.8}$$

This example shows that different pseudoconic transformations produce different simplification results. To produce a unique result irrelevant to the pseudoconic transformations, the factorization techniques to be introduced in Section 4.4 are required.

4.3.3 *Conic contraction*

The degree-5 bracket polynomial equality $\text{conic}_{\mathbf{234,56}}(\mathbf{1}) = 0$ can be written as

$$[\mathbf{123}][\mathbf{245}][\mathbf{246}][\mathbf{356}] - [\mathbf{124}][\mathbf{235}][\mathbf{236}][\mathbf{456}] = [\mathbf{123}][\mathbf{124}][\mathbf{256}][\mathbf{345}][\mathbf{346}][\mathbf{134}]^{-1}.$$
$$(4.3.9)$$

As shown in Example 4.36, deriving (4.3.9) from the left to the right by previous simplification techniques can be made by a pseudoconic transformation, a contraction and a conic transformation. To save the effort, it is better that (4.3.9) be used directly in bracket polynomial simplification, called *conic contraction*.

Algorithm 4.38. Conic contraction.

Input: A bracket polynomial p of degree at least four and involving at least six conic points. Assume that p is already factorized in the polynomial ring of brackets, and that all the factors of p of degree less than four are moved into a set q.

Output: A set q composed of rational bracket polynomial factors.

Step 1. Do the following steps for each factor f of p, for each pair of terms $t_1 + t_2$ of f.

Step 2. Let c be the explicit common factors of $t_1 + t_2$, and let $t_i = ct_i'$ for $i = 1, 2$. If any of the following conditions is not satisfied, skip to the next pair of terms, and if f has no more pair of terms, move f to q and go back to Step 1:

(1) Each t_i' has four brackets, all of which are square-free.
(2) The coefficient of t_i' is ± 1.
(3) t_1' has six points, all of which are on the same conic.
(4) The six points, denoted by $\mathbf{1}$ to $\mathbf{6}$ respectively, have degree $1, 3, 2, 2, 2, 2$ in t_1'. This fixes $\mathbf{1}, \mathbf{2}$.
(5) Bivector $\mathbf{12}$ occurs once and only once in each t_i'. Denote the brackets containing $\mathbf{12}$ in t_1', t_2' by $[\mathbf{123}], [\mathbf{124}]$ respectively. This fixes $\mathbf{3}, \mathbf{4}$.
(6) $t_1 = \epsilon[\mathbf{123}][\mathbf{245}][\mathbf{246}][\mathbf{356}]$ and $t_2 = -\epsilon[\mathbf{124}][\mathbf{235}][\mathbf{236}][\mathbf{456}]$, where $\epsilon \in \{\pm 1\}$. This fixes $\mathbf{5}, \mathbf{6}$.

Step 3. Substitute

$$t_1 + t_2 = \epsilon c[\mathbf{123}][\mathbf{124}][\mathbf{256}][\mathbf{345}][\mathbf{346}][\mathbf{134}]^{-1}$$

into f, contract and factorize the result, put the result into q and remove f from p.

Example 4.39. Let $\mathbf{1}, \ldots, \mathbf{6}$ be conic points. Simplify

$$p = [\mathbf{126}][\mathbf{234}][\mathbf{245}][\mathbf{356}] - [\mathbf{124}][\mathbf{236}][\mathbf{256}][\mathbf{345}].$$

The degrees of the points are $1^1, 2^3, 3^2, 4^2, 5^2, 6^2$. The two brackets $[\mathbf{126}]$ and $[\mathbf{124}]$ each occur in a term of p. Point sequence $\mathbf{126453}$ of p matches $\mathbf{123456}$ of (4.3.9), so

$$p = -[\mathbf{124}][\mathbf{126}][\mathbf{235}][\mathbf{346}][\mathbf{456}][\mathbf{146}]^{-1}.$$

Algorithm 4.40. *Conic simplification:*
Bracket polynomial simplification based on coconic constraints.

Input: A bracket polynomial p involving at least six conic points. Assume that p is already factorized in the polynomial ring of brackets.

Output: p.

Procedure. For every factor f of p, do repeatedly (a) contractions, (b) conic transformations, (c) conic contractions, (d) pseudoconic transformations, until the result no longer changes.

Every time there is any change in steps (b), (c), (d), do contractions.

4.4 Factorization techniques in conic computing

There is one more technique in computing bracket polynomials involving conic points, without which the computing is often not only very difficult, but extremely sensitive to different Cayley expansions and (pseudo)conic transformations, as already disclosed by Example 4.37.

The technique is called *conic Cayley factorization*. It is an integration of the Cayley factorization techniques developed in Section 3.2 and the conic simplification techniques developed in the previous section. For maximally factorizing a bracket polynomial involving conic points, bracket monomials must be allowed to occur in the denominator, which is the feature of this factorization. This is the second form of *rational Cayley factorization*, besides the one discussed in Section 2.7.

4.4.1 *Bracket unification*

In conic geometry, the implicit bracket factors caused by coconic constraints can be found by conic transformations and pseudoconic transformations, through the following *bracket unification* algorithm, the counterpart of Algorithm 3.7.

Algorithm 4.41. Bracket unification.

Input: Two bracket polynomials p_1, p_2. Assume that they are already factorized in the polynomial ring of brackets and their explicit common factors are already removed.

Output: p_1, p_2.

Procedure. Let $p_i = d_i c_i$ for $i = 1, 2$, where d_i is all the bracket factors of p_i. Set $d = d_1 + \lambda d_2$ for transcendental element λ. Set $b = 1$.

Do the following to d until the result no longer changes, then output $p_1 = b d_1 c_1$ and $p_2 = b d_2 c_2$.

(1) Do conic transformations, move the bracket factors to b.

(2) Do pseudoconic transformations, move the rational bracket factors to b.

Example 4.42. If points $\mathbf{1}, \mathbf{2}, \mathbf{3}, \mathbf{4}, \mathbf{5}, \mathbf{6}$ are on a conic, then intersections $\mathbf{12} \cap \mathbf{34}$, $\mathbf{14} \cap \mathbf{35}$, $\mathbf{35} \cap \mathbf{26}$, $\mathbf{12} \cap \mathbf{56}$, $\mathbf{13} \cap \mathbf{46}$, $\mathbf{25} \cap \mathbf{46}$ are also on a conic.

Free conic points: $\mathbf{1}, \mathbf{2}, \mathbf{3}, \mathbf{4}, \mathbf{5}, \mathbf{6}$.

Intersections:

$$\mathbf{7} = \mathbf{12} \cap \mathbf{34}, \quad \mathbf{8} = \mathbf{14} \cap \mathbf{35}, \quad \mathbf{9} = \mathbf{26} \cap \mathbf{35},$$
$$\mathbf{0} = \mathbf{12} \cap \mathbf{56}, \quad \mathbf{a} = \mathbf{13} \cap \mathbf{46}, \quad \mathbf{b} = \mathbf{25} \cap \mathbf{46}.$$

Conclusion: $\mathbf{7}, \mathbf{8}, \mathbf{9}, \mathbf{0}, \mathbf{a}, \mathbf{b}$ are coconic.

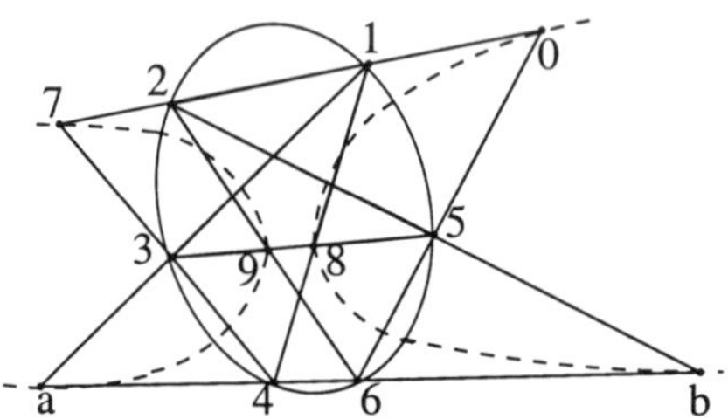

Fig. 4.5   Example 4.42: the role of bracket unification.

Proof.   The following are factorizable Cayley brackets of $\mathbf{7}, \mathbf{8}, \mathbf{9}, \mathbf{0}, \mathbf{a}, \mathbf{b}$:

3 lines: $\mathbf{70}$ on line $\mathbf{12}$, $\mathbf{89}$ on line $\mathbf{35}$, and $\mathbf{ab}$ on line $\mathbf{46}$. There are twelve brackets:

$$
\begin{aligned}
[\mathbf{780}] &= -[124][135]\,\mathbf{12} \vee \mathbf{34} \vee \mathbf{56}, & [\mathbf{790}] &= -[126][235]\,\mathbf{12} \vee \mathbf{34} \vee \mathbf{56}, \\
[\mathbf{70a}] &= [123][146]\,\mathbf{12} \vee \mathbf{34} \vee \mathbf{56}, & [\mathbf{70b}] &= [125][246]\,\mathbf{12} \vee \mathbf{34} \vee \mathbf{56}, \\
[\mathbf{789}] &= [123][345]\,\mathbf{14} \vee \mathbf{26} \vee \mathbf{35}, & [\mathbf{890}] &= -[125][356]\,\mathbf{14} \vee \mathbf{26} \vee \mathbf{35}, \\
[\mathbf{89a}] &= -[135][346]\,\mathbf{14} \vee \mathbf{26} \vee \mathbf{35}, & [\mathbf{89b}] &= [235][456]\,\mathbf{14} \vee \mathbf{26} \vee \mathbf{35}, \\
[\mathbf{7ab}] &= [124][346]\,\mathbf{13} \vee \mathbf{25} \vee \mathbf{46}, & [\mathbf{8ab}] &= [146][345]\,\mathbf{13} \vee \mathbf{25} \vee \mathbf{46}, \\
[\mathbf{9ab}] &= -[246][356]\,\mathbf{13} \vee \mathbf{25} \vee \mathbf{46}, & [\mathbf{0ab}] &= -[126][456]\,\mathbf{13} \vee \mathbf{25} \vee \mathbf{46}.
\end{aligned}
$$

2 triangles: $[\mathbf{78a}]$ of $\mathbf{134}$, and $[\mathbf{90b}]$ of $\mathbf{256}$.

$$[\mathbf{78a}] = -[134]([123][146][345] + [124][135][346]),$$
$$[\mathbf{90b}] = -[256]([125][246][356] + [126][235][456]).$$

The fifteen representations together with their common degrees are

$$\mathbf{7890ab}^{(0)} \quad \mathbf{789a0b}^{(4)} \quad \mathbf{789b0a}^{(4)} \quad \mathbf{7809ab}^{(4)} \quad \underline{\mathbf{780a9b}}^{(6)} \quad \underline{\mathbf{780b9a}}^{(6)}$$
$$\underline{\mathbf{78a90b}}^{(6)} \quad \mathbf{78a09b}^{(4)} \quad \underline{\mathbf{78ab90}}^{(6)} \quad \underline{\mathbf{78b90a}}^{(6)} \quad \mathbf{78b09a}^{(4)} \quad \underline{\mathbf{78ba90}}^{(6)}$$
$$\mathbf{7908ab}^{(4)} \quad \underline{\mathbf{79a80b}}^{(6)} \quad \underline{\mathbf{79b80a}}^{(6)}.$$

There are eight representations with maximal common degree 6. In particular, in each of the two representations $\mathbf{78b90a}^{(6)}$, $\mathbf{780b9a}^{(6)}$, the eight involved Cayley brackets are all factorizable. Choosing any of the two representations, say the first one, we get

$$\text{conic}(\mathbf{78b90a}) = [780][7ab][89a][90b] - [78a][70b][890][9ab]$$
$$= \underbrace{(12 \vee 34 \vee 56)(14 \vee 26 \vee 35)(13 \vee 25 \vee 46)}$$
$$\{[125]^2[134][246]^2[356]^2([123][146][345] + [124][135][346])$$
$$-[124]^2[135]^2[256][346]^2([125][246][356] + [126][235][456])\}.$$
$$(4.4.1)$$

Now let us see how the bracket unification works. For

$$b_1 = [125]^2[134][246]^2[356]^2, \qquad b_2 = [124]^2[135]^2[256][346]^2,$$

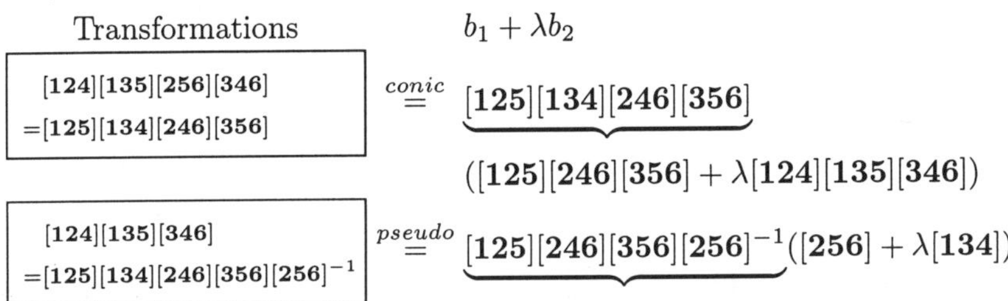

$$\text{Transformations} \qquad\qquad b_1 + \lambda b_2$$

$$\boxed{\begin{array}{c}[124][135][256][346] \\ =[125][134][246][356]\end{array}} \quad \overset{conic}{=} \quad \underbrace{[125][134][246][356]}$$

$$([125][246][356] + \lambda[124][135][346])$$

$$\boxed{\begin{array}{c}[124][135][346] \\ =[125][134][246][356][256]^{-1}\end{array}} \quad \overset{pseudo}{=} \quad \underbrace{[125][246][356][256]^{-1}}([256] + \lambda[134]).$$

Essentially b_1 and b_2 are simplified to $[256]$ and $[134]$ respectively. Substituting them into (4.4.1) and removing common factors, we get

$$\begin{aligned}
\text{conic}(\mathbf{78b90a}) \;=\;\; & [123][146][256][345] + [124][135][256][346] \\
& -[125][134][246][356] - [126][134][235][456] \qquad (4.4.2)
\end{aligned}$$
$$\overset{conic}{=} 0.$$

Additional nondegeneracy condition: $\mathbf{256} \neq 0.$ $\qquad\qquad\qquad\square$

4.4.2 *Conic Cayley factorization*

In computing a coconic conclusion expression, generally it is not difficult to find binomial expansions for non-factorizable Cayley brackets, what is difficult is that there are often several binomial results for the same Cayley bracket, and they can form a large number of combinations. It is a common phenomenon that succeeding bracket manipulations work well for one particular combination of expansions, but not for any other one. Thus, the computing is very fragile.

Example 4.43. In Example 4.42, instead of choosing a representation with maximal common degree, we choose one with common degree 4, for example

$$\text{conic}(\mathbf{78b09a}) = [78a][79b][890][0ab] - [789][7ab][80a][90b]. \qquad (4.4.3)$$

Prove the coconic conclusion again.

Proof. (4.4.3) has only two non-factorizable Cayley brackets: $[\mathbf{79b}], [\mathbf{80a}]$. They each have six binomial expansions:

$$\begin{aligned}
[\mathbf{79b}] &= [\mathbf{124}][\mathbf{235}][\mathbf{236}][\mathbf{456}] - [\mathbf{123}][\mathbf{245}][\mathbf{246}][\mathbf{356}] \\
&= [\mathbf{125}][\mathbf{234}][\mathbf{236}][\mathbf{456}] - [\mathbf{123}][\mathbf{245}][\mathbf{256}][\mathbf{346}] \\
&= [\mathbf{125}][\mathbf{234}][\mathbf{246}][\mathbf{356}] - [\mathbf{124}][\mathbf{235}][\mathbf{256}][\mathbf{346}] \\
&= [\mathbf{126}][\mathbf{234}][\mathbf{245}][\mathbf{356}] - [\mathbf{124}][\mathbf{236}][\mathbf{256}][\mathbf{345}] \\
&= [\mathbf{126}][\mathbf{234}][\mathbf{235}][\mathbf{456}] - [\mathbf{123}][\mathbf{246}][\mathbf{256}][\mathbf{345}] \\
&= [\mathbf{126}][\mathbf{235}][\mathbf{245}][\mathbf{346}] - [\mathbf{125}][\mathbf{236}][\mathbf{246}][\mathbf{345}],
\end{aligned}$$

$$\begin{aligned}
[\mathbf{80a}] &= [\mathbf{125}][\mathbf{136}][\mathbf{146}][\mathbf{345}] - [\mathbf{126}][\mathbf{135}][\mathbf{145}][\mathbf{346}] \\
&= [\mathbf{123}][\mathbf{146}][\mathbf{156}][\mathbf{345}] - [\mathbf{126}][\mathbf{134}][\mathbf{135}][\mathbf{456}] \\
&= [\mathbf{123}][\mathbf{145}][\mathbf{156}][\mathbf{346}] - [\mathbf{125}][\mathbf{134}][\mathbf{136}][\mathbf{456}] \\
&= [\mathbf{124}][\mathbf{136}][\mathbf{156}][\mathbf{345}] - [\mathbf{126}][\mathbf{134}][\mathbf{145}][\mathbf{356}] \\
&= [\mathbf{124}][\mathbf{135}][\mathbf{156}][\mathbf{346}] - [\mathbf{125}][\mathbf{134}][\mathbf{146}][\mathbf{356}] \\
&= [\mathbf{123}][\mathbf{145}][\mathbf{146}][\mathbf{356}] - [\mathbf{124}][\mathbf{135}][\mathbf{136}][\mathbf{456}].
\end{aligned}$$

$$(4.4.4)$$

The conic simplification works well for the first expansions of the two brackets, but not for any other combination. The proof based on the first expansions goes on smoothly as in Example 4.42, while the proofs based on other expansions are very difficult to continue.

How to overcome the difficulty of the extreme sensitivity to different binomial expansions of non-factorizable Cayley brackets? In this example, a very nice property of the non-factorizable Cayley brackets is that all their binomial expansion results can be conic-contracted to rational bracket monomials. By means of conic contractions and bracket unifications, any of the fifteen representations of the coconic conclusion is just as good as any other one.

For the expansions in (4.4.4), the conic contractions give

$$\begin{aligned}
[\mathbf{79b}] &= -[\mathbf{123}][\mathbf{124}][\mathbf{256}][\mathbf{345}][\mathbf{346}][\mathbf{134}]^{-1} \\
&= -[\mathbf{123}][\mathbf{125}][\mathbf{246}][\mathbf{345}][\mathbf{356}][\mathbf{135}]^{-1} \\
&= -[\mathbf{124}][\mathbf{125}][\mathbf{236}][\mathbf{345}][\mathbf{456}][\mathbf{145}]^{-1} \\
&= -[\mathbf{124}][\mathbf{126}][\mathbf{235}][\mathbf{346}][\mathbf{456}][\mathbf{146}]^{-1} \\
&= -[\mathbf{123}][\mathbf{126}][\mathbf{245}][\mathbf{346}][\mathbf{356}][\mathbf{136}]^{-1} \\
&= -[\mathbf{125}][\mathbf{126}][\mathbf{234}][\mathbf{356}][\mathbf{456}][\mathbf{156}]^{-1},
\end{aligned}$$

$$\begin{aligned}
[\mathbf{80a}] &= -[\mathbf{125}][\mathbf{126}][\mathbf{134}][\mathbf{356}][\mathbf{456}][\mathbf{256}]^{-1} \\
&= -[\mathbf{123}][\mathbf{126}][\mathbf{145}][\mathbf{346}][\mathbf{356}][\mathbf{236}]^{-1} \\
&= -[\mathbf{123}][\mathbf{125}][\mathbf{146}][\mathbf{345}][\mathbf{356}][\mathbf{235}]^{-1} \\
&= -[\mathbf{124}][\mathbf{126}][\mathbf{135}][\mathbf{346}][\mathbf{456}][\mathbf{246}]^{-1} \\
&= -[\mathbf{124}][\mathbf{125}][\mathbf{136}][\mathbf{345}][\mathbf{456}][\mathbf{245}]^{-1} \\
&= -[\mathbf{123}][\mathbf{124}][\mathbf{156}][\mathbf{345}][\mathbf{346}][\mathbf{234}]^{-1}.
\end{aligned}$$

$$(4.4.5)$$

If we choose the first expansions in (4.4.4), we get their corresponding conic contractions as the first expressions in (4.4.5), by which we directly get (4.4.2) without making bracket unification. The proof has no additional nondegeneracy condition. This combination is the best.

If we choose the worst combination, which is the last expansions of the two brackets in (4.4.4) and hence in (4.4.5), then

$$\text{conic}(\textbf{78b09a}) = \underbrace{[\textbf{156}]^{-1}[\textbf{234}]^{-1}}$$

$$\{[\textbf{125}]^2[\textbf{126}]^2[\textbf{134}][\textbf{234}]^2[\textbf{356}]^2[\textbf{456}]^2([\textbf{123}][\textbf{146}][\textbf{345}] + [\textbf{124}][\textbf{135}][\textbf{346}])$$
$$-[\textbf{123}]^2[\textbf{124}]^2[\textbf{156}]^2[\textbf{256}][\textbf{345}]^2[\textbf{346}]^2([\textbf{125}][\textbf{246}][\textbf{356}] + [\textbf{126}][\textbf{235}][\textbf{456}])\}.$$

For

$$b_1 = [\textbf{125}]^2[\textbf{126}]^2[\textbf{134}][\textbf{234}]^2[\textbf{356}]^2[\textbf{456}]^2,$$
$$b_2 = [\textbf{123}]^2[\textbf{124}]^2[\textbf{156}]^2[\textbf{256}][\textbf{345}]^2[\textbf{346}]^2,$$

the bracket unification goes as follows:

Transformations		$b_1 + \lambda b_2$
$([\textbf{125}][\textbf{234}][\textbf{356}])^2$ $=([\textbf{124}][\textbf{156}][\textbf{235}][\textbf{346}][\textbf{146}]^{-1})^2$	$\overset{pseudo}{=}$	$\underbrace{[\textbf{124}]^2[\textbf{156}]^2[\textbf{346}]^2[\textbf{146}]^{-2}}$ $([\textbf{126}]^2[\textbf{134}][\textbf{235}]^2[\textbf{456}]^2$ $+\lambda[\textbf{123}]^2[\textbf{146}]^2[\textbf{256}][\textbf{345}]^2)$
$[\textbf{126}][\textbf{134}][\textbf{235}][\textbf{456}]$ $=[\textbf{123}][\textbf{146}][\textbf{256}][\textbf{345}]$	$\overset{conic}{=}$	$\underbrace{[\textbf{123}][\textbf{146}][\textbf{256}][\textbf{345}]}([\textbf{126}][\textbf{235}][\textbf{456}]$ $+\lambda[\textbf{123}][\textbf{146}][\textbf{345}])$
$[\textbf{126}][\textbf{235}][\textbf{456}]$ $=[\textbf{123}][\textbf{146}][\textbf{256}][\textbf{345}][\textbf{134}]^{-1}$	$\overset{pseudo}{=}$	$\underbrace{[\textbf{123}][\textbf{146}][\textbf{345}][\textbf{134}]^{-1}}([\textbf{256}] + \lambda[\textbf{134}]).$

So even in the worst case we still get the same bracket coefficients [**256**] and [**134**] as in the best combination. The difference is the increase of additional nondegeneracy conditions: $\textbf{134} \neq 0$, $\textbf{146} \neq 0$, $\textbf{156} \neq 0$, $\textbf{234} \neq 0$. $\qquad\square$

This example suggests the application of conic simplification immediately after the expansions of non-factorizable Cayley brackets, because the coconic constraints have not been used to make any simplification during Cayley expansions. For more complicated computing, for instance Example 4.44 below, bracket unification alone is not sufficient to make the computing robust, and must be complemented with Cayley factorization techniques to produce common bracket and meet product factors.

Example 4.44. [Steiner's Theorem] Let points $1, 2, 3, 4, 5, 6$ be on a conic, and let $\textbf{7}, \textbf{8}, \textbf{9}, \textbf{0}, \textbf{a}, \textbf{b}$ be the intersections $12 \cap 35$, $13 \cap 45$, $14 \cap 25$, $13 \cap 26$, $12 \cap 46$, $14 \cap 36$ respectively, then lines $\textbf{7b}, \textbf{8a}, \textbf{90}$ are concurrent.

Free conic points: $1, 2, 3, 4, 5, 6$.
Intersections:

$$7 = 12 \cap 35, \quad 8 = 13 \cap 45, \quad 9 = 14 \cap 25,$$
$$0 = 13 \cap 26, \quad a = 12 \cap 46, \quad b = 14 \cap 36.$$

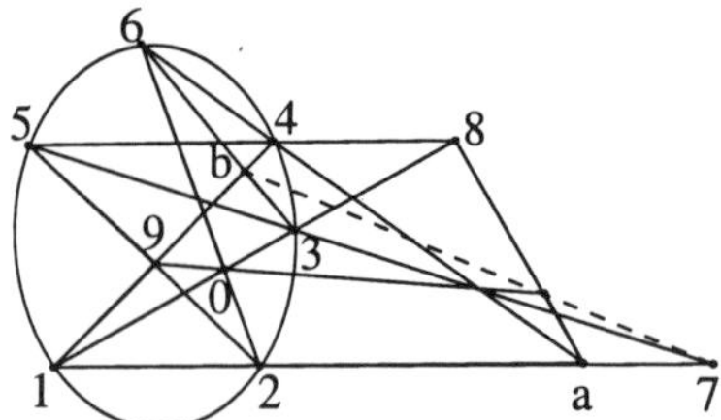

Fig. 4.6   Steiner's Theorem.

Conclusion: **7b, 8a, 90** are concurrent.

Proof.

Rules   **7b $\vee$ 90 $\vee$ 8a**

$$\overset{expand}{=} \mathbf{[7ab][890] - [78b][90a]}$$

[7ab]=[125][136]12$\vee$35$\vee$46
[890]=[125][134]13$\vee$26$\vee$45
[78b]=[125][134][356][256]$^{-1}$13$\vee$26$\vee$45
[90a]=[126][134][245][345]$^{-1}$12$\vee$35$\vee$46

$$\overset{7,8,9,0,a,b}{=} \frac{\mathbf{[125][134][256]^{-1}[345]^{-1}}}{\underbrace{\mathbf{(12 \vee 35 \vee 46)(13 \vee 26 \vee 45)}}}$$

$$\underbrace{}_{\begin{array}{c}\mathbf{([124][125][256][345]} \\ \mathbf{-[126][134][245][356])}\end{array}}$$

$$\overset{conic}{=} 0.$$

Procedure of computing brackets **[78b]** and **[90a]**:

Rules   **[78b]**

[(12$\vee$35)(13$\vee$45)(14$\vee$36)]
=   **[134][5(12$\vee$35)(14$\vee$36)]**
−[135][4(12$\vee$35)(14$\vee$36)]
=   **[124][135]^2[346]−[125][134]^2[356]**
[124][135][346]
=[125][134][246][356][256]$^{-1}$

$$\overset{7,8,b}{=} \mathbf{[124][135]^2[346] - [125][134]^2[356]}$$

$$\overset{pseudo}{=} \mathbf{[125][134][356][256]^{-1}([135][246]}$$
$$\mathbf{-[134][256])}$$

$$\overset{factor}{=} \mathbf{[125][134][356][256]^{-1}13 \vee 26 \vee 45,}$$

[90a]

[(14$\vee$25)(13$\vee$26)(12$\vee$46)]
=   **[126][4(14$\vee$25)(13$\vee$26)]**
−[124][6(14$\vee$25)(13$\vee$26)]
=   **[124]^2[136][256]+[126]^2[134][245]**
[124][136][256]
=[126][134][245][356][345]$^{-1}$

$$\overset{9,0,a}{=} \mathbf{[124]^2[136][256] + [126]^2[134][245]}$$

$$\overset{pseudo}{=} \mathbf{[126][134][245][345]^{-1}([124][356]}$$
$$\mathbf{+[126][345])}$$

$$\overset{factor}{=} \mathbf{[126][134][245][345]^{-1}12 \vee 35 \vee 46.}$$

Additional nondegeneracy conditions: $\mathbf{256} \neq 0$, $\mathbf{345} \neq 0$. $\qquad\square$

In the above proof, brackets $[\mathbf{78b}]$, $[\mathbf{90a}]$ each have three binomial expansions:

$$\begin{aligned}
[\mathbf{78b}] &= [124][135]^2[346] - [125][134]^2[356] \\
&= [123][134][145][356] + [124][135][136][345] \\
&= [123][135][145][346] + [125][134][136][345],
\end{aligned}$$

$$\begin{aligned}
[\mathbf{90a}] &= [124]^2[136][256] + [126]^2[134][245] \\
&= [123][124][146][256] + [125][126][134][246] \\
&= [124][125][136][246] - [123][126][146][245].
\end{aligned}$$

$$(4.4.6)$$

If the conic simplification is not carried out immediately after Cayley expansions, then it works well only for the first expansions in (4.4.6):

$$\begin{aligned}
[\mathbf{78b}][\mathbf{90a}] = \ &[124]^3[135]^2[136][256][346] - [124]^2[125][134]^2[136][256][356] \\
&+[124][126]^2[134][135]^2[245][346] - [125][126]^2[134]^3[245][356].
\end{aligned}$$

The details have been provided in Example 4.32. By (4.3.3), after removing common bracket factors, $[\mathbf{78b}][\mathbf{90a}]$ can be simplified to

$$\begin{aligned}
&[124][135][246][356] - [124][134][256][356] \\
+&[126][135][246][345] - [126][134][256][345].
\end{aligned}$$

A Cayley factorization then changes it to $(\mathbf{12} \vee \mathbf{35} \vee \mathbf{46})(\mathbf{13} \vee \mathbf{26} \vee \mathbf{45})$, which agrees with the meet product factors of $[\mathbf{7ab}][\mathbf{890}]$ after eliminating $\mathbf{7, 8, 9, 0, a, b}$. The proof finishes without invoking any additional nondegeneracy condition.

However, the proving is too fragile in that a proof based on any other combination of the expansions of $[\mathbf{78b}]$, $[\mathbf{90a}]$ in (4.4.6) is very difficult to continue.

Let us see how conic Cayley factorization makes the proving robust. Similar to Example 4.37, we get that each expansion in (4.4.6) has two pseudoconic transformations, so there are all together twelve different results from conic simplification:

$$\begin{aligned}
[\mathbf{78b}] &= [125][134][356]([135][246] - [134][256])[256]^{-1} \\
&= [124][135][346]([135][246] - [134][256])[246]^{-1} \\
&= [124][135]([134][236][456] + [136][246][345])[246]^{-1} \\
&= [134][356]([123][145][256] + [126][135][245])[256]^{-1} \\
&= [125][134]([136][256][345] + [135][236][456])[256]^{-1} \\
&= [135][346]([126][134][245] + [123][145][246])[246]^{-1},
\end{aligned}$$

$$\begin{aligned}
[\mathbf{90a}] &= [126][134][245]([124][356] + [126][345])[345]^{-1} \\
&= [124][136][256]([124][356] + [126][345])[356]^{-1} \\
&= [126][134]([124][235][456] + [125][246][345])[345]^{-1} \\
&= [124][256]([126][135][346] + [123][146][356])[356]^{-1} \\
&= [126][245]([124][135][346] - [123][146][345])[345]^{-1} \\
&= [124][136]([125][246][356] - [126][235][456])[356]^{-1}.
\end{aligned}$$

$$(4.4.7)$$

After degree-2 and degree-3 Cayley factorizations, there are only four different results:

$$[\mathbf{78b}] = [\mathbf{125}][\mathbf{134}][\mathbf{356}][\mathbf{256}]^{-1}\mathbf{13} \vee \mathbf{26} \vee \mathbf{45}$$
$$= [\mathbf{124}][\mathbf{135}][\mathbf{346}][\mathbf{246}]^{-1}\mathbf{13} \vee \mathbf{26} \vee \mathbf{45},$$

$$[\mathbf{90a}] = [\mathbf{126}][\mathbf{134}][\mathbf{245}][\mathbf{345}]^{-1}\mathbf{12} \vee \mathbf{35} \vee \mathbf{46}$$
$$= [\mathbf{124}][\mathbf{136}][\mathbf{256}][\mathbf{356}]^{-1}\mathbf{12} \vee \mathbf{35} \vee \mathbf{46}.$$

Thus, after bracket-wise conic simplification and Cayley factorization, the proofs based on different combinations of Cayley expansions are much the same.

In conic computing, there is a need to factorize a polynomial composed of brackets and meet products of type p_I to maximal extent, with brackets allowed in the denominator. Such a polynomial generally occurs after bracket-wise eliminations and expansions, and is a linear combination of some multiplications of polynomials. Owning to their invariant inheritance from the eliminated brackets, the polynomial components of the multiplications are generally much easier to be factorized, but not so after the multiplications are expanded. Conic Cayley factorization before merging the Cayley expansion results of different Cayley brackets is mandatory.

The above observation is developed into the idea *bracket-oriented simplification*, and is integrated into the following *conic Cayley factorization* algorithm.

Algorithm 4.45. Conic Cayley factorization.

Input: A polynomial p generated by a set $\mathcal{C}$ of polynomials of brackets and meet products of type p_I.

Output: p.

Step 1. For each element c of $\mathcal{C}$ involving at least six conic points, do (1) conic simplification, (2) bracket unification, (3) Cayley factorization.

After this step, change p into a factored form by collecting all its explicit common factors.

Step 2. Do expansion within each factor of p.

Step 3. Do the following to each factor of p: (1) bracket polynomial simplification, (2) conic simplification, (3) bracket unification, (4) Cayley factorization.

Example 4.46. Unify the results (4.3.7) and (4.3.8) of Example 4.37, which are just the expressions in the third and fourth lines of (4.4.7) respectively.

Steps 1–2. Do nothing.

Steps 3. (1)–(2). Do nothing.

(3). Bracket unification: The two bracket factors in (4.3.7) and (4.3.8) are simplified to [**125**] and [**346**] respectively:

$$[124][135][246]^{-1} + \lambda[134][356][256]^{-1}$$
$$= [246]^{-1}[256]^{-1}([124][135][256] + \lambda[134][246][356])$$
$$\overset{pseudo}{=} [134][246][356][346]^{-1}([125] + \lambda[346]).$$

(4). Cayley factorization:

$$[134][236][456] + [136][246][345] = [346]13 \vee 26 \vee 45,$$
$$[123][145][256] + [126][135][245] = [125]13 \vee 26 \vee 45.$$

So $p = [125][134][356][256]^{-1}13 \vee 26 \vee 45$ is independent of the pseudodivisions used in Example 4.37.

4.5 Automated theorem proving

The techniques developed so far for geometric computing in both projective incidence and projective conic geometries, when put together, form a brand new toolkit featuring *factored and shortest computing.*

As analyzed at the beginning of this chapter, the traditional algebraic approach to geometric theorem proving is the normalization of the conclusion expression by either a characteristic set or a Gröbner base of the hypotheses expressions. The normalization is carried out stepwise by either polynomial pseudodivision or bracket polynomial straightening, and at the middle steps the expression size often increases considerably.

In order to control the middle expression swell, factored and shortest computing is necessary. By factorization, a large system of hypotheses expressions is decomposed into smaller ones, among which generally only one leads to the conclusion, and the others function as masks of the truth. By term reduction, the expression size is further controlled.

A novel guideline in symbolic geometric computing – **breefs**: bracket-oriented representation, elimination and expansion for factored and shortest result, appears to be natural in carrying out factored and shortest computing. It is already embodied in the following computing techniques:

Representation: The representations of the coconic conclusion and the conic conjugacy conclusion are optimized by maximizing the common degree and by reducing the number of terms respectively.

The representation of a constructed point is optimized based on its bracket mates. For the same point occurring in different parts of the same expression, completely different algebraic representations can be used, with appropriately selected transformation coefficients as the trade-off. In contrast, all previous elimination methods follow the same unanimous representation style.

Elimination: The batch elimination enables several or all elements in a bracket to be eliminated at the same time, preparing the ground for factored and shortest expansions of layer-1 Cayley expressions.

On the contrary, all previous elimination methods follow the one-by-one style; since they are unable to make full use of the representation information of the bracket mates, it is very difficult for them to produce factored and shortest results.

Expansion: If one or several points are eliminated simultaneously from a bracket, the result is a layer-1 Cayley expression. Generally there are many Cayley expansion results, among which the factored and shortest ones are the most desired.

The classification of factored and shortest expansions is used in optimal representation of the coconic conclusion. Factored expansions can be implemented directly into the Cayley expansion algorithm to speed up computing.

Simplification: It includes factorization and term reduction. In bracket polynomial factorization, to find implicit bracket factors, collinearity and concurrency transformations in incidence geometry, conic and pseudoconic transformations in conic geometry, are proposed. In bracket polynomial term reduction, three general techniques are proposed: contraction, level contraction and strong contraction. In conic geometry, three more techniques are added: conic transformation, pseudoconic transformation, and conic contraction.

Bracket-oriented simplification refers to simplifying the Cayley expansion result of a layer-1 Cayley bracket in an expression before merging the result into the expression. It also includes subsequent Cayley factorizations and eliminations of all the incidence points in the result, as is done in (4.2.9) for the representation of the fourth point of intersection of two conics. This again is in sharp contrast to all previous elimination methods.

Factorization: Cayley factorization is a very important means of making the computing robust against different Cayley expansions and transformations. Bracket-oriented Cayley factorization as the last stage of bracket-oriented simplification, is used to find common bracket and meet product factors.

The following theorem proving algorithm is based on the **breefs** principle.

Algorithm 4.47. Hand-checkable machine proof in projective conic geometry.

Input: A sequence of points, lines and conics together with their constructions; a conclusion statement.

Output: Representation and computation procedure of the conclusion expression, including all kinds of representations, eliminations, expansions, contractions, transformations, factorizations, and additional nondegeneracy conditions.

Step 1. [Registration] Collect points, lines, conics, polars and tangents.

(1) A line is composed of at least three points.

(2) A conic is composed of the construction and all its points and tangents.

(3) A polar is composed of the conic, the pole, and all points on it.

(4) A tangent is composed of the conic, the tangent point, and all points on it.

Step 2. [Conclusion representation] If it is a coconic or conic conjugacy conclusion, find an optimal conclusion expression. Let $conc = 0$ be the conclusion, where $conc$ is an expression in GC algebra.

Step 3. [Dynamic batch elimination] Start from the ends $\mathbf{x}_i$ of $conc$ in the dynamic parents-children diagram, while $conc \neq 0$ and the $\mathbf{x}_i$ are constrained points or free collinear points, eliminate the $\mathbf{x}_i$ from $conc$ by the following procedure:

(1) find the optimal expressions of the $\mathbf{x}_i$ in every different bracket of $conc$,

(2) do Cayley expansions after substituting the expressions of the $\mathbf{x}_i$ into $conc$,

(3) simplify each Cayley expansion result immediately,

(4) simplify $conc$, remove common bracket or meet product factors.

Step 4. [Level and strong contractions] Expand $conc$ into a bracket polynomial.

If $conc \neq 0$, then do repeatedly (1) level contractions, (2) strong contractions, until the result no longer changes.

At the end of each operation, do contractions and remove common bracket factors.

Step 5. [Additional nondegeneracy conditions] There are two sources:

(1) the denominators which are produced by transformation rules, Cramer's rules, conic contractions and pseudoconic transformations, and which are not canceled after substitutions;

(2) the associated nondegeneracy conditions of the geometric constructions not included in the input.

Theoretically, Step 4 does not guarantee that $conc = 0$ can be reached; in practice, however, all but one example need this step. It is Example 4.54 later in this section, whose conclusion is proved by a level contraction and a strong contraction. Practically, no theorem needs to go through the elimination of any free point or free conic point.

The efficiency of **breefs** is also revealed by the statistics that among the 40 difficult theorems tested by the algorithm, 32 are given binomial proofs. In particular, almost all the theorems involving only free conic points, tangent points, and poles constructed by tangents are given binomial proofs. The next three subsections are devoted to some typical examples and their proofs.

4.5.1   *Almost incidence geometry*

If the constructions of a geometric problem involve only free points, free conic points and incidence points, we say the problem belongs to *almost incidence geometry*. Such problems are among the simplest in conic geometry, and usually can be solved in a binomial manner.

Example 4.48. If points $1, 2, 3, 4, 5, 6$ are on a conic, then $12 \cap 34$, $13 \cap 24$, $14 \cap 25$, $14 \cap 36$, $15 \cap 46$, $45 \cap 16$ are also on a conic.

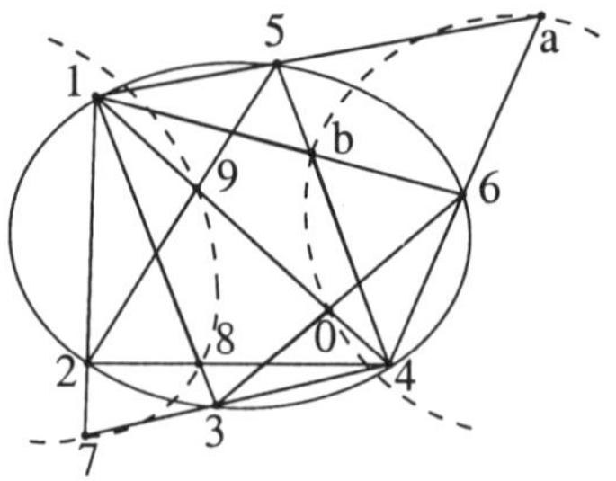

Fig. 4.7   Example 4.48.

Free conic points: $1, 2, 3, 4, 5, 6$.
Intersections:

$$7 = 12 \cap 34, \quad 8 = 13 \cap 24, \quad 9 = 14 \cap 25,$$
$$0 = 14 \cap 36, \quad a = 15 \cap 46, \quad b = 16 \cap 45.$$

Conclusion: $7, 8, 9, 0, a, b$ are coconic.

Proof.   The following are factorizable Cayley brackets:

1 line: **90** on line **14**. There are four associated brackets:

$$[790] = [890] = -[124][134]14 \vee 25 \vee 36,$$
$$[90a] = [90b] = [145][146]14 \vee 25 \vee 36.$$

2 quadrilaterals: $[789]$ and $[780]$ of $(1234, 14)$, $[9ab]$ and $[0ab]$ of $(1456, 14)$.

$$[789] = [124][134]([123][245] - [125][234]),$$
$$[780] = [124][134]([123][346] - [136][234]),$$
$$[9ab] = [145][146]([156][245] - [125][456]),$$
$$[0ab] = [145][146]([156][346] - [136][456]).$$

4 triangles: $[79a]$ of **125**, $[70a]$ of **346**, $[89b]$ of **245**, and $[80b]$ of **136**. Each bracket has two different factored expansions.

$$[70a] = [346]([123][145][146] - [124][134][156])$$
$$= [346]([124][136][145] - [125][134][146]),$$

$$[89b] = -[245]([123][145][146] - [124][134][156])$$
$$= -[245]([124][136][145] - [125][134][146]),$$
$$[79a] = [125]([124][134][456] - [145][146][234])$$
$$= [125]([124][145][346] + [134][146][245]),$$
$$[80b] = -[136]([124][134][456] - [145][146][234])$$
$$= -[136]([124][145][346] + [134][146][245]).$$

The conclusion can be represented by conic($\mathbf{78ab90}$).

$$\text{conic}(\mathbf{78ab90})$$
$$= [780][79a][89b][0ab] - [789][70a][80b][9ab]$$
$$\overset{7,8,9,0,a,b}{=} [124][134][145][146]([123][145][146] - [124][134][156])$$
$$\underbrace{([124][134][456] - [145][146][234])([125][346] - [136][245])}$$
$$([125][136][234][456] - [123][156][245][346])$$
$$\overset{conic}{=} 0.$$

Additional nondegeneracy condition: none. $\qquad\square$

Example 4.49. [Nine-point Conic Theorem] Let $\mathbf{1234}$ be a quadrilateral, and let $\mathbf{7,8,9}$ be the three intersections $\mathbf{12} \cap \mathbf{34}$, $\mathbf{13} \cap \mathbf{24}$ and $\mathbf{23} \cap \mathbf{14}$. A line intersects the six sides of complete quadrilateral $\mathbf{1234789}$ at points $\mathbf{5,6,0,a,b,c}$ respectively. Then the six conjugate points of $\mathbf{5,6,0,a,b,c}$ with respect to the corresponding collinear pairs of vertices of quadrilateral $\mathbf{1234}$, are on the same conic with points $\mathbf{7,8,9}$.

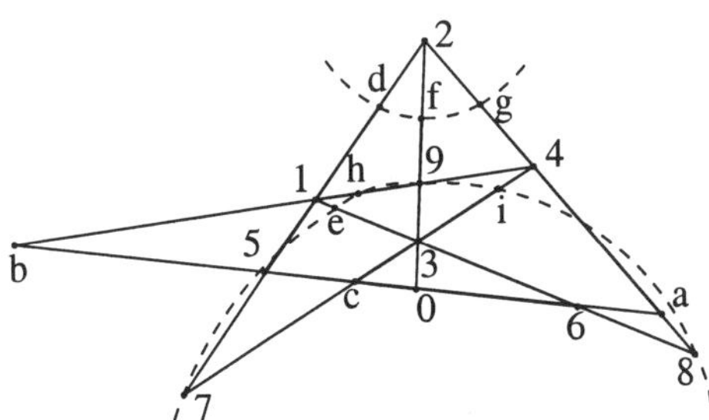

Fig. 4.8   Nine-point Conic Theorem.

Free points: $\mathbf{1,2,3,4}$.
Free collinear points: $\mathbf{5}$ on line $\mathbf{12}$, $\mathbf{6}$ on line $\mathbf{13}$.
Intersections:
$$\mathbf{7} = \mathbf{12} \cap \mathbf{34}, \quad \mathbf{8} = \mathbf{13} \cap \mathbf{24}, \quad \mathbf{9} = \mathbf{23} \cap \mathbf{14}, \quad \mathbf{0} = \mathbf{23} \cap \mathbf{56},$$
$$\mathbf{a} = \mathbf{24} \cap \mathbf{56}, \quad \mathbf{b} = \mathbf{14} \cap \mathbf{56}, \quad \mathbf{c} = \mathbf{34} \cap \mathbf{56}.$$

Conjugates:

$$d = \text{conj}_{12}(5), \quad e = \text{conj}_{13}(6), \quad f = \text{conj}_{23}(0),$$
$$g = \text{conj}_{24}(a), \quad h = \text{conj}_{14}(b), \quad i = \text{conj}_{34}(c).$$

Conclusion: $7, 8, 9, d, e, f, g, h, i$ are coconic.

Proof. If it can be proved that $7, 8, 9, d, f, i$ are coconic, then by symmetry, the following 6-tuples are coconic points: $\{7, 8, 9, d, g, i\}$, $\{7, 8, 9, d, e, i\}$, $\{7, 8, 9, d, h, i\}$. Under the additional nondegeneracy condition $\exists 789di$, the nine points are on the same conic.

The six points $7, 8, 9, d, f, i$ have the following factorizable Cayley brackets, according to the representations

$$d = [25]1 + [15]2,$$
$$f = [30]2 + [20]3,$$
$$i = [4c]3 + [3c]4.$$

3 lines: $7d$ on line 12, $9f$ on line 23, and $7i$ on line 34. There are eleven brackets:

$$[78d] = [123][124]([15][234] + [25][134]),$$
$$[79d] = [123][124]([15][234] + [25][134]),$$
$$[7df] = -[20][123]([15][234] + [25][134]),$$
$$[7di] = -([3c][124] + [4c][123])([15][234] + [25][134]),$$
$$[79f] = -[123][234]([20][134] + [30][124]),$$
$$[89f] = [123][234]([20][134] + [30][124]),$$
$$[9df] = -[25][123]([20][134] + [30][124]),$$
$$[9fi] = [3c][234]([20][134] + [30][124]),$$
$$[78i] = -[134][234]([3c][124] + [4c][123]),$$
$$[79i] = -[134][234]([3c][124] + [4c][123]),$$
$$[7fi] = -[30][234]([3c][124] + [4c][123]).$$

1 complete quadrilateral: $[789]$ of 1234.

$$[789] = -2\,[123][124][134][234].$$

3 quadrilaterals: $[78f]$ of $(1234, 23)$, $[89d]$ of $(1234, 12)$, and $[89i]$ of $(1234, 34)$.

$$[78f] = [123][234]([30][124] - [20][134]),$$
$$[89d] = [123][124]([15][234] - [25][134]),$$
$$[89i] = [134][234]([4c][123] - [3c][124]).$$

2 triangles: $[8df]$ of 123, and $[8fi]$ of 234.

$$[8df] = [123]([25][30][124] + [15][20][234]),$$
$$[8fi] = [234]([20][3c][134] + [30][4c][123]).$$

The conclusion can be represented by conic(**78f9di**).

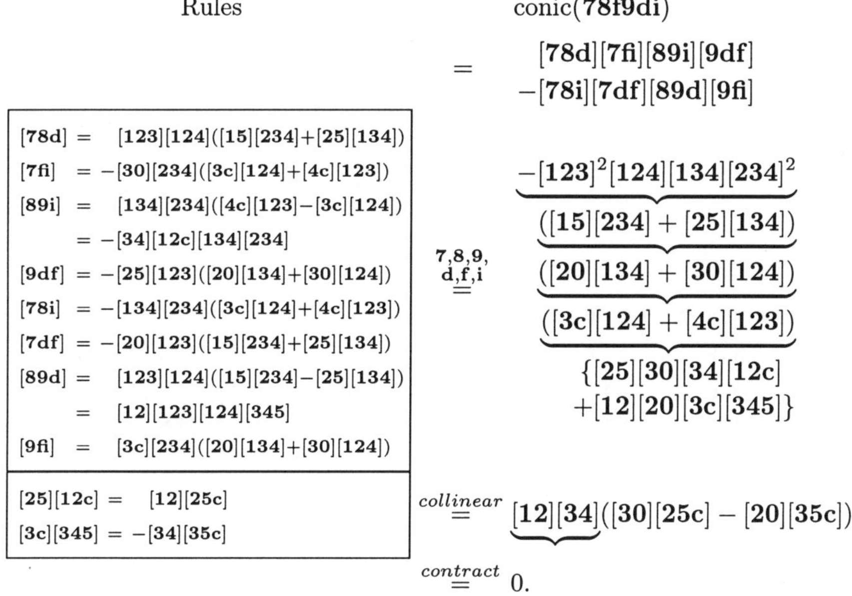

$$\text{Rules} \qquad\qquad \text{conic(\textbf{78f9di})}$$

$$= \;\; [78d][7fi][89i][9df] - [78i][7df][89d][9fi]$$

$$
\begin{aligned}
[78d] &= [123][124]([15][234]+[25][134])\\
[7fi] &= -[30][234]([3c][124]+[4c][123])\\
[89i] &= [134][234]([4c][123]-[3c][124])\\
&= -[34][12c][134][234]\\
[9df] &= -[25][123]([20][134]+[30][124])\\
[78i] &= -[134][234]([3c][124]+[4c][123])\\
[7df] &= -[20][123]([15][234]+[25][134])\\
[89d] &= [123][124]([15][234]-[25][134])\\
&= [12][123][124][345]\\
[9fi] &= [3c][234]([20][134]+[30][124])
\end{aligned}
$$

$$
\begin{aligned}
\overset{7,8,9,}{\underset{d,f,i}{=}}\;\; &-[123]^2[124][134][234]^2\\
&\underbrace{([15][234]+[25][134])}\\
&\underbrace{([20][134]+[30][124])}\\
&\underbrace{([3c][124]+[4c][123])}\\
&\{[25][30][34][12c]\\
&\;+[12][20][3c][345]\}
\end{aligned}
$$

$$
\begin{aligned}
{[25][12c]} &= [12][25c]\\
[3c][345] &= -[34][35c]
\end{aligned}
$$

$$\overset{collinear}{=} \underbrace{[12][34]}([30][25c]-[20][35c])$$

$$\overset{contract}{=} 0.$$

Additional nondegeneracy condition: $\exists$**789di**. $\qquad\square$

Remark: Two contractions are made immediately after the Cayley expansions of the eight brackets in the conclusion expression. By collinearity transformations, the eliminations of $0, \mathbf{c}$ are avoided.

4.5.2 *Tangency and polarity*

For theorems on tangents and poles that can be constructed by tangents, generally binomial proofs can be found. For more general polars and poles, finding a binomial proof is difficult.

Example 4.50. When the two tangents to a conic from each vertex of a triangle intersect at two points with the opposite side of the triangle respectively, the six points of intersection are on a common conic.

Free conic points: $\mathbf{1, 2, 3, 4, 5, 6}$.
Poles: $\mathbf{7} = \text{pole}_{\mathbf{14}}(\mathbf{123456})$, $\mathbf{8} = \text{pole}_{\mathbf{25}}(\mathbf{123456})$, $\mathbf{9} = \text{pole}_{\mathbf{36}}(\mathbf{123456})$.
Intersections:

$$
\begin{aligned}
\mathbf{0} &= \mathbf{89} \cap \mathbf{17}, & \mathbf{a} &= \mathbf{89} \cap \mathbf{47}, & \mathbf{b} &= \mathbf{79} \cap \mathbf{28},\\
\mathbf{c} &= \mathbf{79} \cap \mathbf{58}, & \mathbf{d} &= \mathbf{39} \cap \mathbf{78}, & \mathbf{e} &= \mathbf{69} \cap \mathbf{78}.
\end{aligned}
$$

Conclusion: $\mathbf{0, a, b, c, d, e}$ are coconic.

Proof. All the brackets of $\mathbf{0, a, b, c, d, e}$ are factorizable Cayley brackets:

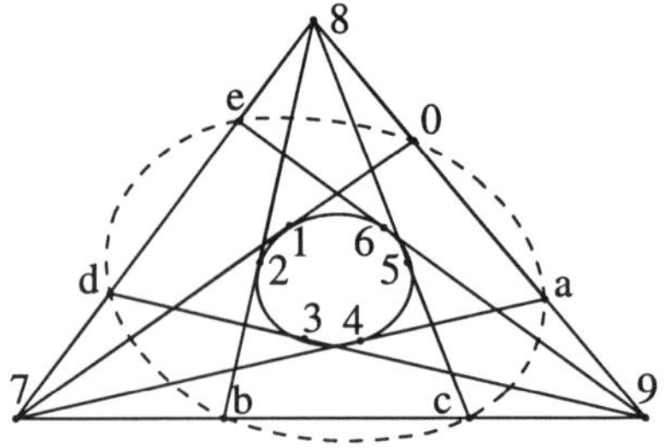

Fig. 4.9   Example 4.50.

3 lines: **0a** on line **89**, **de** on line **78**, and **bc** on line **79**. There are twelve brackets:

$$[0ab] = [147][289][789]^2, \qquad [0ac] = -[147][589][789]^2,$$
$$[0ad] = -[147][389][789]^2, \qquad [0ae] = -[147][689][789]^2,$$
$$[0de] = -[178][369][789]^2, \qquad [ade] = [369][478][789]^2,$$
$$[bde] = -[278][369][789]^2, \qquad [cde] = [369][578][789]^2,$$
$$[0bc] = -[179][258][789]^2, \qquad [abc] = [258][479][789]^2,$$
$$[bcd] = [258][379][789]^2, \qquad [bce] = [258][679][789]^2.$$

1 triangle: The other eight brackets are all of **789**.

$$[0bd] = -[789]([178][289][379] + [179][278][389]),$$
$$[0be] = -[789]([178][289][679] + [179][278][689]),$$
$$[0cd] = -[789]([178][379][589] + [179][389][578]),$$
$$[0ce] = -[789]([178][589][679] + [179][578][689]),$$
$$[abd] = -[789]([278][389][479] + [289][379][478]),$$
$$[abe] = -[789]([278][479][689] + [289][478][679]),$$
$$[acd] = -[789]([379][478][589] + [389][479][578]),$$
$$[ace] = -[789]([478][589][679] + [479][578][689]).$$

The conclusion can be represented by conic(**0aecbd**).

conic(**0aecbd**)

$$= [abc][cde][0ad][0be] - [0ab][0de][acd][bce]$$

$$\overset{0,a,b,\\c,d,e}{=} [147][258][369][789]^9([179][278][389][479][578][689]$$
$$\qquad\qquad\qquad -[178][289][379][478][589][679])$$

$$\overset{7,8,9}{=} 64\,[123]^6[124][125][126]^4[134][135]^4[136][145][146][156]^2[234]^4[235]$$
$$[236][245][246]^2[256][345]^2[346][356]([125][134][146][236][245][356]$$
$$\qquad\qquad\qquad -[124][136][145][235][256][346])$$

$$\overset{pseudo}{=} [134][236][245][234]^{-1}([125][146][234][356] - [124][156][235][346])$$

$$\overset{conic}{=} 0.$$

The eliminations of $\mathbf{7},\mathbf{8},\mathbf{9}$ follow (4.1.22):

$$[\mathbf{1}\,7_{14,253}\,8_{25,143}] = 2\,[123][124][125][135][145][234][345],$$
$$[\mathbf{4}\,7_{14,253}\,8_{25,143}] = 2\,[123][124][135][145][234][245][345],$$
$$[\mathbf{2}\,7_{14,253}\,8_{25,143}] = 2\,[123][124][125][135][234][245][345],$$
$$[\mathbf{5}\,7_{14,253}\,8_{25,143}] = 2\,[123][125][135][145][234][245][345],$$
$$[\mathbf{1}\,7_{14,362}\,9_{36,142}] = 2\,[123][126][134][136][146][234][246],$$
$$[\mathbf{4}\,7_{14,362}\,9_{36,142}] = 2\,[123][126][134][146][234][246][346],$$
$$[\mathbf{3}\,7_{14,362}\,9_{36,142}] = 2\,[123][126][134][136][234][246][346],$$
$$[\mathbf{6}\,7_{14,362}\,9_{36,142}] = 2\,[123][126][136][146][234][246][346],$$
$$[\mathbf{2}\,8_{25,361}\,9_{36,251}] = 2\,[123][126][135][156][235][236][256],$$
$$[\mathbf{5}\,8_{25,361}\,9_{36,251}] = 2\,[123][126][135][156][235][256][356],$$
$$[\mathbf{3}\,8_{25,361}\,9_{36,251}] = 2\,[123][126][135][156][235][236][356],$$
$$[\mathbf{6}\,8_{25,361}\,9_{36,251}] = 2\,[123][126][135][156][236][256][356].$$

The pseudoconic transformation in the next to the last step is

$$[136][145][256] = [134][156][236][245][234]^{-1}.$$

Additional nondegeneracy conditions: $\exists\,\mathbf{12345},\ \exists\,\mathbf{12346},\ \exists\,\mathbf{12356},\ \mathbf{234}\neq 0.$ $\quad\Box$

Example 4.51. A conic touches the three sides $\mathbf{90},\mathbf{49},\mathbf{40}$ of a triangle $\mathbf{490}$ at points $\mathbf{1},\mathbf{2},\mathbf{3}$ respectively. Show that the three points $\mathbf{12}\cap\mathbf{40}$, $\mathbf{13}\cap\mathbf{49}$, $\mathbf{23}\cap\mathbf{90}$ lie on a line. If the lines joining $\mathbf{4},\mathbf{9},\mathbf{0}$ to any point $\mathbf{7}$ of this line meet $\mathbf{23},\mathbf{12},\mathbf{13}$ at points $\mathbf{a},\mathbf{b},\mathbf{c}$ respectively, prove that triangle $\mathbf{abc}$ is *self-polar* relative to the conic, *i.e.*, any vertex of the triangle is the pole of the opposite side.

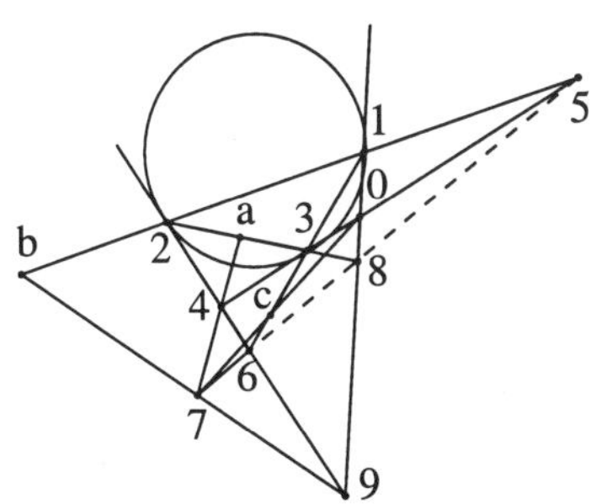

Fig. 4.10   Example 4.51.

Free points: $\mathbf{1},\mathbf{2},\mathbf{3},\mathbf{4}$.

Intersections and free collinear points:

$\mathbf{5} = \mathbf{12}\cap\mathbf{34}$,	$\mathbf{6} = \mathbf{13}\cap\mathbf{24}$,	$\mathbf{7}$ on line $\mathbf{56}$,
$\mathbf{8} = \mathbf{23}\cap\mathrm{tangent}_1(\mathbf{12(24)3(34)})$,	$\mathbf{9} = \mathbf{24}\cap\mathbf{18}$,	$\mathbf{0} = \mathbf{34}\cap\mathbf{18}$,
$\mathbf{a} = \mathbf{23}\cap\mathbf{47}$,	$\mathbf{b} = \mathbf{12}\cap\mathbf{79}$,	$\mathbf{c} = \mathbf{13}\cap\mathbf{70}$.

Conclusion: (1) $\mathbf{5},\mathbf{6},\mathbf{8}$ are collinear;

(2) $\mathbf{a},\mathbf{b}$ are conjugate with respect to the conic;

202   *Invariant Algebras and Geometric Reasoning*

(3) $\mathbf{a}, \mathbf{c}$ are conjugate with respect to the conic;

(4) $\mathbf{b}, \mathbf{c}$ are conjugate with respect to the conic.

Proof.   (1). By (4.1.49), $\mathrm{tangent}_1(\mathbf{12(24)3(34)}) = [\mathbf{134}]\mathbf{12} + [\mathbf{124}]\mathbf{13}$, so

$$\mathbf{8} = [\mathbf{134}]\mathbf{23} \vee \mathbf{12} + [\mathbf{124}]\mathbf{23} \vee \mathbf{13} = -[\mathbf{123}]([\mathbf{134}]\mathbf{2} + [\mathbf{124}]\mathbf{3}), \qquad (4.5.1)$$

and

$$[\mathbf{568}] \stackrel{\mathbf{5,6,8}}{=} [\mathbf{134}][\mathbf{2}(\mathbf{12} \vee \mathbf{34})(\mathbf{13} \vee \mathbf{24})] + [\mathbf{124}][\mathbf{3}(\mathbf{12} \vee \mathbf{34})(\mathbf{13} \vee \mathbf{24})] \stackrel{expand}{=} 0.$$

Additional nondegeneracy condition: none.

(2). By Example 4.27, the conclusion has two optimal representations. The following proof is based on the conic representation $\mathbf{12(24)3(34)}$ and the polarization of the first expression in (4.1.44):

Rules $\mathrm{conj}_{\mathbf{12(24)3(34)}}(\mathbf{a}, \mathbf{b})$

$$= [\mathbf{123}]^2([\mathbf{34a}][\mathbf{24b}] + [\mathbf{24a}][\mathbf{34b}])$$

Rules	
$[\mathbf{24a}]=-[\mathbf{234}][\mathbf{247}],\ [\mathbf{24b}]=-[\mathbf{124}][\mathbf{279}]$ $[\mathbf{34a}]=-[\mathbf{234}][\mathbf{347}],\ [\mathbf{34b}]=-\mathbf{12}\vee\mathbf{34}\vee\mathbf{79}$	$\stackrel{\mathbf{a,b}}{=} [\mathbf{234}]([\mathbf{124}][\mathbf{279}][\mathbf{347}] + [\mathbf{247}]\mathbf{12} \vee \mathbf{34} \vee \mathbf{79})$
$[\mathbf{279}]=[\mathbf{128}][\mathbf{247}],$ $\mathbf{12}\vee\mathbf{34}\vee\mathbf{79}$ $=[\mathbf{127}]\mathbf{24}\vee\mathbf{18}\vee\mathbf{34}-[\mathbf{347}]\mathbf{24}\vee\mathbf{18}\vee\mathbf{12}$ $=[\mathbf{127}][\mathbf{148}][\mathbf{234}]+[\mathbf{347}][\mathbf{124}][\mathbf{128}]$	$\stackrel{\mathbf{9}}{=} [\mathbf{247}](2\,[\mathbf{124}][\mathbf{128}][\mathbf{347}] + [\mathbf{127}][\mathbf{148}][\mathbf{234}])$
$\mathbf{7}=[\mathbf{57}]\mathbf{6}-[\mathbf{67}]\mathbf{5},\quad \mathbf{8}=[\mathbf{134}]\mathbf{2}+[\mathbf{124}]\mathbf{3}$ $[\mathbf{127}]=[\mathbf{57}][\mathbf{126}],\quad [\mathbf{347}]=[\mathbf{57}][\mathbf{346}]$ $[\mathbf{128}]=[\mathbf{123}][\mathbf{124}],\ [\mathbf{148}]=-2\,[\mathbf{124}][\mathbf{134}]$	$\stackrel{\mathbf{7,8}}{=} 2\,[\mathbf{57}][\mathbf{124}]([\mathbf{123}][\mathbf{124}][\mathbf{346}]$ $\qquad\qquad -[\mathbf{126}][\mathbf{134}][\mathbf{234}])$
$[\mathbf{126}]=[\mathbf{123}][\mathbf{124}],\qquad [\mathbf{346}]=[\mathbf{134}][\mathbf{234}]$	$\stackrel{\mathbf{6}}{=} 0.$

Additional nondegeneracy condition: none.

The following proof is based on the conic representation $\mathbf{31(18)2(24)}$ and the polarization of the first expression in (4.1.44):

Rules $\mathrm{conj}_{\mathbf{31(18)2(24)}}(\mathbf{a}, \mathbf{b})$

$$= [\mathbf{123}]^2([\mathbf{24a}][\mathbf{18b}] + [\mathbf{18a}][\mathbf{24b}])$$

Rules	
$[\mathbf{18a}]=[\mathbf{123}][\mathbf{478}],\quad [\mathbf{18b}]=-[\mathbf{129}][\mathbf{178}]$ $[\mathbf{24a}]=-[\mathbf{234}][\mathbf{247}],\ [\mathbf{24b}]=[\mathbf{129}][\mathbf{247}]$	$\stackrel{\mathbf{a,b}}{=} [\mathbf{129}][\mathbf{247}]([\mathbf{123}][\mathbf{478}] + [\mathbf{178}][\mathbf{234}])$
$\mathbf{7}=[\mathbf{57}]\mathbf{6}-[\mathbf{67}]\mathbf{5},\quad \mathbf{8}=[\mathbf{134}]\mathbf{2}+[\mathbf{124}]\mathbf{3}$ $[\mathbf{478}]=[\mathbf{57}][\mathbf{124}][\mathbf{346}]-[\mathbf{67}][\mathbf{134}][\mathbf{245}]$ $[\mathbf{178}]=-[\mathbf{57}][\mathbf{126}][\mathbf{134}]+[\mathbf{67}][\mathbf{124}][\mathbf{135}]$	$\stackrel{\mathbf{7,8}}{=} -[\mathbf{67}][\mathbf{123}][\mathbf{134}][\mathbf{245}] + [\mathbf{57}][\mathbf{123}][\mathbf{124}][\mathbf{346}]$ $\qquad -[\mathbf{57}][\mathbf{126}][\mathbf{134}][\mathbf{234}] + [\mathbf{67}][\mathbf{124}][\mathbf{135}][\mathbf{234}]$
$[\mathbf{245}]=-[\mathbf{124}][\mathbf{234}],\ [\mathbf{346}]=[\mathbf{134}][\mathbf{234}]$ $[\mathbf{135}]=-[\mathbf{123}][\mathbf{134}],\ [\mathbf{126}]=[\mathbf{123}][\mathbf{124}]$	$\stackrel{\mathbf{5,6}}{=} 0.$

Additional nondegeneracy condition: $\exists 31(18)2(24)$.

(3) and (4): similar to (2). $\qquad\qquad\qquad\square$

4.5.3 *Intersection*

For theorems on intersections of lines and conics, and intersections of conics, generally no binomial proof can be found (Example 4.22 is an exception).

Example 4.52. Let K and G be two conics through four points $1, 2, 3, 4$. Let 5 and 7 be two points of K that do not lie on G, and are such that 5 does not lie on the tangent to G at 3, and 7 does not lie on the tangent to G at 1. Then 35 intersects G at a point 8 other than 3, and 71 intersects G at a point 9 other than 1, and 24 intersects 57 at a point 0 collinear with $8, 9$.

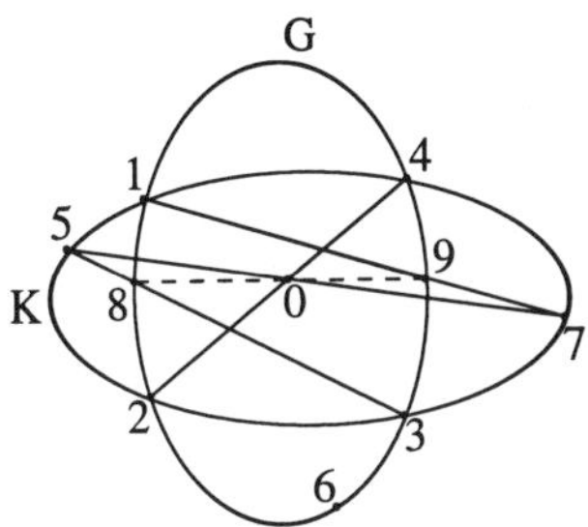

Fig. 4.11   Example 4.52.

Free conic points: $1, 2, 3, 4, 5, 7$.
Free point: 6.
Intersections: $8 = 35 \cap 12346$, $9 = 17 \cap 12346$, $0 = 24 \cap 57$.
Conclusion: $8, 9, 0$ are collinear.

Proof.

$[890] \overset{8,9,0}{=} \quad [126][146][147][236][345][346]\,[(12 \vee 35)\,(23 \vee 17)\,(24 \vee 57)]$
$\qquad\qquad -[126]^2[147][235][346]^2\,[(14 \vee 35)\,(23 \vee 17)\,(24 \vee 57)]$
$\qquad\qquad -[127][146]^2[236]^2[345]\,[(12 \vee 35)\,(34 \vee 17)\,(24 \vee 57)]$
$\qquad\qquad +[126][127][146][235][236][346]\,[(14 \vee 35)\,(34 \vee 17)\,(24 \vee 57)]$

$\qquad \overset{expand}{=} \; -[126][146][147][236][345][346]([127][135][234][257]$
$\qquad\qquad\qquad\qquad\qquad\qquad\qquad\qquad\qquad + [123][157][235][247])$
$\qquad\qquad -[126]^2[147][235][346]^2([123][157][247][345] - [127][135][234][457])$
$\qquad\qquad -[127][146]^2[236]^2[345]([135][147][234][257] - [134][157][235][247])$
$\qquad\qquad +[126][127][146][235][236][346]([135][147][234][457]$
$\qquad\qquad\qquad\qquad\qquad\qquad\qquad\qquad\qquad + [134][157][247][345])$

$$\overset{csimp}{=} \underbrace{[126][146][157][236][245][247][346][357][257]^{-1}[457]^{-1}}$$
$$([127][134][235][457] - [123][147][257][345])$$
$$\overset{conic}{=} 0.$$

The following are representations of $\mathbf{8, 9}$:

$$\mathbf{8} = \mathbf{8}_{35,1246} = [146][236][345]\,\mathbf{12} \vee \mathbf{35} - [126][235][346]\,\mathbf{14} \vee \mathbf{35},$$
$$\mathbf{9} = \mathbf{9}_{17,3246} = [126][147][346]\,\mathbf{23} \vee \mathbf{17} - [127][146][236]\,\mathbf{34} \vee \mathbf{17}.$$

The conic simplification (*csimp*) in the next to the last step contains two conic contractions (*cct*) and two conic transformations:

$$[127][135][234][257] + [123][157][235][247] \overset{cct}{=} [123][157][245][247][357]/[457],$$
$$[123][157][247][345] - [127][135][234][457] \overset{conic}{=} 0,$$
$$[135][147][234][257] - [134][157][235][247] \overset{conic}{=} 0,$$
$$[135][147][234][457] + [134][157][247][345] \overset{cct}{=} [134][157][245][247][357]/[257].$$

Additional nondegeneracy conditions: $\mathbf{257} \neq 0$, $\mathbf{457} \neq 0$. $\qquad\square$

Example 4.53. (Figure 4.12)

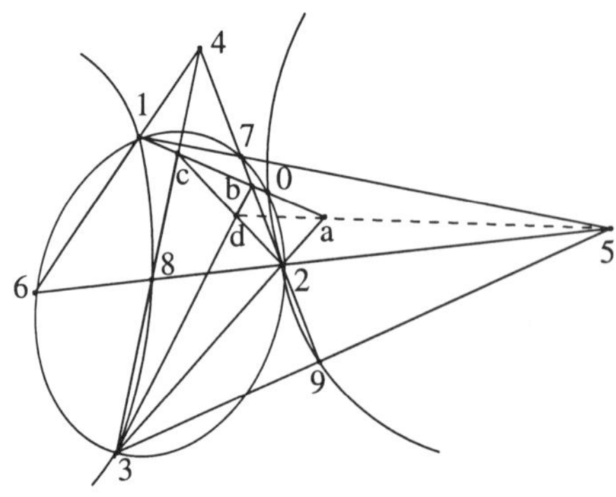

Fig. 4.12   Example 4.53.

Free points: $\mathbf{1, 2, 3, 4, 5}$.
Intersections:

$\mathbf{6} = \mathbf{14} \cap \mathbf{25}$, $\mathbf{7} = \mathbf{15} \cap \mathbf{24}$, $\mathbf{8} = \mathbf{25} \cap \mathbf{34}$, $\mathbf{9} = \mathbf{24} \cap \mathbf{35}$, $\mathbf{0} = \mathbf{12367} \cap \mathbf{12389}$,
$\mathbf{a} = \mathbf{10} \cap \mathbf{23}$, $\mathbf{b} = \mathbf{10} \cap \mathbf{24}$, $\mathbf{c} = \mathbf{10} \cap \mathbf{34}$, $\mathbf{d} = \mathbf{3b} \cap \mathbf{2c}$.

Conclusion: $\mathbf{5, a, d}$ are collinear.

Proof.

Rules

$[5(10 \vee 23)(3b \vee 2c)]$ $= [235]10 \vee 3b \vee 2c + [150]23 \vee 3b \vee 2c$ $= -[235][120][3bc] - [150][23b][23c]$

$[23c] = [130][234],\ [23b] = [120][234]$ $[3bc] = -[130]10 \vee 24 \vee 34$ $\quad\quad = [130][140][234]$

$0 = [230]1 - [130]2 + [120]3$

$[120] = \quad [124][145][235]^2$ $\quad\quad + [125][134][235][245]$ $\quad\quad + [125]^2[234][345]$ $[130] = [135][245]([134][235]$ $\quad\quad\quad\quad + [135][234])$

$[124][135] - [125][134] = [123][145]$ $[135][245] - [125][345] = [145][235]$

$[5ad]$

$$\overset{a,\,d}{=} \ -[120][235][3bc] - [150][23b][23c]$$

$$\overset{b,\,c}{=} \ -\underbrace{[120][130][234]([140][235] + [150][234])}$$

$$\overset{cramer}{=} \ \begin{aligned}&[130]([124][235] + [125][234]) \\ &-[120]([134][235] + [135][234])\end{aligned}$$

$$\overset{0}{=} \ \begin{aligned}&\underbrace{([134][235] + [135][234])} \\ &\{[124][135][235][245] \\ &-[124][145][235]^2 + [125][135][234][245] \\ &-[125]^2[234][345] - [125][134][235][245]\}\end{aligned}$$

$$\overset{contract}{=} \ \underbrace{[145][235]}([123][245] - [124][235] \\ + [125][234])$$

$$\overset{contract}{=} \ 0.$$

Procedure of deriving the elimination rules of $\mathbf{0}$:

$$\mu_1 = [167][236][237] \overset{6,7}{=} \underbrace{[124]^2[125]^2[145][234][235]},$$

$$\mu_2 = [136][137][267] \overset{6,7}{=} -\underbrace{[124]^2[125]^2[134][135][245]},$$

$$\mu_3 = [126][127][367] \overset{6,7}{=} \underbrace{[124]^2[125]^2(-[124][145][235] - [125][134][245])},$$

$$\mu_1' = [189][238][239] \overset{8,9}{=} \underbrace{[234]^2[235]^2(-[134][235][245] - [125][234][345])},$$

$$\mu_2' = [138][139][289] \overset{8,9}{=} -\underbrace{[234]^2[235]^2[134][135][245]},$$

$$\mu_3' = [128][129][389] \overset{8,9}{=} \underbrace{[234]^2[235]^2[124][125][345]}.$$

$$\begin{aligned}
\lambda_1 &= \mu_2\mu_3' - \mu_2'\mu_3 \\
&= -[134][135][245]([124][125][345] + [124][145][235] + [125][134][245]) \\
&\overset{contract}{=} -[134][245]^2[135]([124][135] + [125][134]),
\end{aligned}$$

$$\begin{aligned}
\lambda_2 &= \mu_3\mu_1' - \mu_3'\mu_1 \\
&= [134][245]([124][145][235]^2 + [125][134][235][245] + [125]^2[234][345]),
\end{aligned}$$

$$\begin{aligned}
\lambda_3 &= \mu_1\mu_2' - \mu_1'\mu_2 \\
&= -[134][135][245]([145][234][235] + [134][235][245] + [125][234][345]) \\
&\overset{contract}{=} -[134][245]^2[135]([134][235] + [135][234]).
\end{aligned}$$

$$[\mathbf{120}] = \quad \lambda_1\lambda_2 = -\underbrace{[\mathbf{134}]^2[\mathbf{135}][\mathbf{245}]^3([\mathbf{124}][\mathbf{135}] + [\mathbf{125}][\mathbf{134}])}$$
$$([\mathbf{124}][\mathbf{145}][\mathbf{235}]^2 + [\mathbf{125}][\mathbf{134}][\mathbf{235}][\mathbf{245}] + [\mathbf{125}]^2[\mathbf{234}][\mathbf{345}]),$$

$$[\mathbf{130}] = -\lambda_1\lambda_3 = -\underbrace{[\mathbf{134}]^2[\mathbf{135}][\mathbf{245}]^3([\mathbf{124}][\mathbf{135}] + [\mathbf{125}][\mathbf{134}])}$$
$$[\mathbf{135}][\mathbf{245}]([\mathbf{134}][\mathbf{235}] + [\mathbf{135}][\mathbf{234}]).$$

Additional nondegeneracy condition: none. $\qquad\qquad\square$

Example 4.54. A *chord of a conic* is a line segment connecting two points of the conic. Let there be three conics sharing a common chord C. If the three conics are taken in pairs such that the common chord of each pair opposite to C is drawn, then the resulting three lines are concurrent.

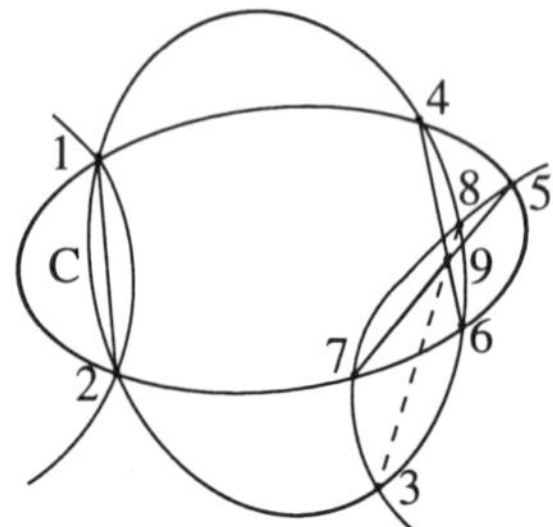

Fig. 4.13   Three conics $\mathbf{123468}, \mathbf{123578}, \mathbf{124567}$ with a common chord $C = \mathbf{12}$.

Free conic points: $\mathbf{1, 2, 4, 5, 6, 7}$.
Free point: $\mathbf{3}$.
Intersections: $\mathbf{8} = \mathbf{12346} \cap \mathbf{12357}, \quad \mathbf{9} = \mathbf{46} \cap \mathbf{57}$.
Conclusion: $\mathbf{3, 8, 9}$ are collinear.

Proof.

Rules

$$[\mathbf{389}]$$
$$\overset{\mathbf{9}}{=} \quad \mathbf{38} \vee \mathbf{46} \vee \mathbf{57}$$

$\mathbf{8} = [\mathbf{238}]\mathbf{1} - [\mathbf{138}]\mathbf{2} + [\mathbf{128}]\mathbf{3}$

$$\overset{cramer}{=} \quad [\mathbf{138}]\mathbf{23} \vee \mathbf{46} \vee \mathbf{57} - [\mathbf{238}]\mathbf{13} \vee \mathbf{46} \vee \mathbf{57}$$

$[\mathbf{138}] = [\mathbf{125}][\mathbf{127}][\mathbf{134}][\mathbf{136}][\mathbf{246}][\mathbf{357}]$
$\quad - [\mathbf{124}][\mathbf{126}][\mathbf{135}][\mathbf{137}][\mathbf{257}][\mathbf{346}]$
$[\mathbf{238}] = [\mathbf{125}][\mathbf{127}][\mathbf{146}][\mathbf{234}][\mathbf{236}][\mathbf{357}]$
$\quad - [\mathbf{124}][\mathbf{126}][\mathbf{157}][\mathbf{235}][\mathbf{237}][\mathbf{346}]$
$\mathbf{23} \vee \mathbf{46} \vee \mathbf{57} = [\mathbf{234}][\mathbf{567}] + [\mathbf{236}][\mathbf{457}]$
$\mathbf{13} \vee \mathbf{46} \vee \mathbf{57} = [\mathbf{134}][\mathbf{567}] + [\mathbf{136}][\mathbf{457}]$

$$\begin{aligned}
&\quad [\mathbf{125}][\mathbf{127}][\mathbf{134}][\mathbf{136}][\mathbf{234}][\mathbf{246}][\mathbf{357}][\mathbf{567}] \\
&- [\mathbf{124}][\mathbf{126}][\mathbf{135}][\mathbf{137}][\mathbf{234}][\mathbf{257}][\mathbf{346}][\mathbf{567}] \\
&+ [\mathbf{125}][\mathbf{127}][\mathbf{134}][\mathbf{136}][\mathbf{236}][\mathbf{246}][\mathbf{357}][\mathbf{457}] \\
\overset{\mathbf{8}}{=}\; &- [\mathbf{124}][\mathbf{126}][\mathbf{135}][\mathbf{137}][\mathbf{236}][\mathbf{257}][\mathbf{346}][\mathbf{457}] \\
&+ [\mathbf{124}][\mathbf{126}][\mathbf{134}][\mathbf{157}][\mathbf{235}][\mathbf{237}][\mathbf{346}][\mathbf{567}] \\
&- [\mathbf{125}][\mathbf{127}][\mathbf{134}][\mathbf{146}][\mathbf{234}][\mathbf{236}][\mathbf{357}][\mathbf{567}] \\
&+ [\mathbf{124}][\mathbf{126}][\mathbf{136}][\mathbf{157}][\mathbf{235}][\mathbf{237}][\mathbf{346}][\mathbf{457}] \\
&- [\mathbf{125}][\mathbf{127}][\mathbf{136}][\mathbf{146}][\mathbf{234}][\mathbf{236}][\mathbf{357}][\mathbf{457}]
\end{aligned}$$

$$\underbrace{[346]}$$

$$\{[124][125][127][136][236][357][457]$$
$$+[125][126][127][134][234][357][567]$$
$$-[124][126][135][137][234][257][567]$$
$$-[124][126][135][137][236][257][457]$$
$$+[124][126][134][157][235][237][567]$$
$$+[124][126][136][157][235][237][457]\}$$

Left box (contract):
$$[136][246]-[146][236]=[126][346]$$
$$[134][246]-[146][234]=[124][346]$$

$$\stackrel{contract}{=}$$

$$\underbrace{[357]}\{[124][125][127][136][236][457]$$
$$+[125][126][127][134][234][567]$$
$$-[124][125][126][137][237][456]$$
$$-[124][126][127][135][235][467]\}$$

Left box (level):
$$[234][567]+[236][457]$$
$$=[235][467]+[237][456]$$
$$[134][567]+[136][457]$$
$$=[135][467]+[137][456]$$
$$-[137][257]+[157][237]=-[127][357]$$
$$-[135][257]+[157][235]=-[125][357]$$

$$\stackrel{level}{=}$$

$$\underbrace{[123]}\{-[126][127][135][245][467]$$
$$+[125][127][136][246][457]$$
$$-[125][126][137][247][456]\}$$

Left box (strong):
$$[134][567]=[135][467]-[136][457]$$
$$+[137][456]$$
$$[125][234]-[124][235]=-[123][245]$$
$$[124][236]-[126][234]=[123][246]$$
$$[127][234]-[124][237]=-[123][247]$$

$$\stackrel{strong}{=}$$

$$\underbrace{[125][157]^{-1}}\{-[127][135][167][246][457]$$
$$+[127][136][157][246][457]$$
$$-[126][137][157][247][456]\}$$

Left box (pseudo):
$$[126][245][467]$$
$$=[125][167][246][457][157]^{-1}$$

$$\stackrel{pseudo}{=}$$

$$\underbrace{[137]}([127][156][246][457]$$
$$-[126][157][247][456])$$

Left box (contract):
$$-[135][167]+[136][157]=[137][156]$$

$$\stackrel{contract}{=}$$

$$\stackrel{conic}{=}\ 0.$$

Procedure of deriving the elimination rules of **8**:

$$\mu_1 = [146][234][236],$$
$$\mu_2 = [134][136][246],$$
$$\mu_3 = [124][126][346],$$
$$\mu_1' = [157][235][237],$$
$$\mu_2' = [135][137][257],$$
$$\mu_3' = [125][127][357];$$
$$\lambda_1 = [125][127][134][136][246][357]-[124][126][135][137][257][346],$$
$$\lambda_2 = [124][126][157][235][237][346]-[125][127][146][234][236][357],$$
$$\lambda_3 = [135][137][146][234][236][257]-[134][136][157][235][237][246];$$
$$[238] = \underline{\lambda_3\lambda_2},$$
$$[138] = -\underline{\lambda_3\lambda_1}.$$

Additional nondegeneracy condition: $\mathbf{157} \neq 0$. $\square$

Remarks:

(1) After the elimination of **8** and the succeeding contraction, there exists no conic simplification before the degree of **3** is reduced to one. The reason is that **3** takes too many brackets, which often makes the number of conic points fewer than six after removal of the brackets containing **3**. The level contraction and strong contraction each produce a common bracket factor containing **3**, thus reducing the degree of **3** by two. Only after these bracket simplifications can the conic simplification take place, which gets rid of the last degree of **3**.

(2) This is the only example we have met in conic geometry in which the level and strong contractions both take place and are indispensable for a step-by-step hand-checkable verification of the conclusion.

The difficulty in representing and computing the intersections of conics suggests that GC algebra and bracket algebra may not be the intrinsic language for describing and manipulating conic geometry. This suspicion is correct, leading to the new language composed of *quadratic Grassmann-Cayley algebra* and *quadratic bracket algebra*, which is the content of the next section.

4.6   Conics with quadratic Grassmann-Cayley algebra*

A conic is determined by five points on it. Six points $\mathbf{1}, \ldots, \mathbf{6}$ are on the same conic if and only if conic($\mathbf{123456}$) = 0, where the left side is a bracket binomial of degree four. This kind of representation, however, is not satisfactory for several reasons:

(1) there are fifteen different bracket binomials,
(2) the antisymmetry among the six points is nontrivial,
(3) the intersections of conics are difficult to represent,
(4) the extension to quadrics in 3D geometry, or more generally, to hyperquadrics in nD geometry, is highly complicated.

We make clear the last point. A 0D conic in 1D projective geometry is just a pair of points. In $\mathbb{P}^1$, three points $\mathbf{1}, \mathbf{2}, \mathbf{3}$ are coconic if and only if the antisymmetrization of bracket monomial

$$[\mathbf{12}][\mathbf{13}][\mathbf{23}] \tag{4.6.1}$$

with respect to the three points equals zero. Since the monomial itself is already antisymmetric, it is unchanged by the antisymmetrization.

In $\mathbb{P}^2$, six points $\mathbf{1}$ to $\mathbf{6}$ are coconic if and only if the antisymmetrization of

$$[\underline{\mathbf{124}}][\underline{\mathbf{135}}][\underline{\mathbf{236}}][\mathbf{456}] \tag{4.6.2}$$

with respect to the six points equals zero. The underlined pieces are exactly the contents of the brackets in (4.6.1). So (4.6.2) is obtained simply by attaching three new points $\mathbf{4}, \mathbf{5}, \mathbf{6}$ to the three brackets in (4.6.1) respectively, and then adding a fourth bracket composed of the three new points. The shortest form of the antisymmetrization has only two terms, *e.g.*, any of the fifteen expressions of conic($\mathbf{123456}$).

In $\mathbb{P}^3$, ten points $\mathbf{1}, \mathbf{2}, \ldots, \mathbf{9}, \mathbf{0}$ are on the same quadric if and only if the antisymmetrization of

$$[\underline{\mathbf{1247}}][\underline{\mathbf{1358}}][\underline{\mathbf{2369}}][\underline{\mathbf{4560}}][\mathbf{7890}] \tag{4.6.3}$$

with respect to the ten points equals zero. Again notice that the underlined pieces are just (4.6.2). One form of the antisymmetrization, known as the *Turnbull-Young invariant*, has 240 terms, and after straightening, is shortened to 138 terms [191]. It is unknown whether or not this bracket polynomial can be further shortened, and whether or not it is Cayley factorizable.

Continuing in this way, if the antisymmetrization of $[\mathbf{A}_1][\mathbf{A}_2]\cdots[\mathbf{A}_{n+1}]$ being zero is a bracket polynomial representation for a sequence of C_{n+1}^2 points to be on the same $(n-2)$D quadric in $\mathbb{P}^{n-1}$, then by adding $n+1$ new points $\mathbf{b}_1$ to $\mathbf{b}_{n+1}$ to the end of the sequence of points, the antisymmetrization of

$$[\mathbf{A}_1\mathbf{b}_1][\mathbf{A}_2\mathbf{b}_2]\cdots[\mathbf{A}_{n+1}\mathbf{b}_{n+1}][\mathbf{b}_1\mathbf{b}_2\cdots\mathbf{b}_{n+1}] \tag{4.6.4}$$

being zero is a bracket polynomial representation for the $C_{n+2}^2 = C_{n+1}^2 + (n+1)$ points to be on the same *hyperquadric* in $\mathbb{P}^n$, which is an $(n-1)$D algebraic surface defined by a degree-2 homogeneous polynomial equation.

Theoretically, the constraint that C_{n+2}^2 points in nD projective geometry are on the same hyperquadric can always be represented by a bracket polynomial equality of degree $n+2$. In practice, however, the size becomes unmanageable even for small values of n. We have to seek for more efficient algebraic representation of the constraint, so that on one hand it should reflect the antisymmetry among the point variables, and on the other hand it is still equal to the above bracket polynomial representation.

4.6.1 *Quadratic Grassmann space and quadratic bracket algebra*

The GC algebra $\Lambda(\mathcal{V}^3)$ is essentially the algebra of linear objects in $\mathbb{P}^2$. As conics are quadratic objects in $\mathbb{P}^2$, representing conics by multivectors in $\Lambda(\mathcal{V}^3)$ is impossible. The base space of the GC algebra needs to be enlarged in order to represent nonlinear objects in a linearly manner such as by multivectors.

Definition 4.55. The *symmetrization* in the tensor algebra $\otimes(\mathcal{V}^n)$ is the linear transformation defined by

$$\mathbf{1}\otimes\mathbf{2}\otimes\cdots\otimes\mathbf{r} \longmapsto \mathbf{1}\odot\mathbf{2}\odot\cdots\odot\mathbf{r} := \frac{1}{r!}\sum_{\sigma}\sigma(\mathbf{1})\otimes\sigma(\mathbf{2})\otimes\cdots\otimes\sigma(\mathbf{r}), \ \ \forall \mathbf{i}\in\mathcal{V}^n, \tag{4.6.5}$$

where the summation runs over all permutations σ of $\mathbf{1}, \mathbf{2}, \ldots, \mathbf{r}$.

The product "$\odot$" is called the *symmetric tensor product*. It is multilinear and associative, and generates an algebra over $\mathcal{V}^n$, called the *symmetric tensor product algebra*, denoted by $\odot(\mathcal{V}^n)$. Its elements are called *symmetric tensors*. Its $\mathbb{Z}$-grading is induced from $\otimes(\mathcal{V}^n)$. Its *r-graded subspace*, denoted by $\odot^r(\mathcal{V}^n)$, is the space of symmetric tensors of grade r.

For example, let $\mathbf{a}, \mathbf{b}$ be two vectors in the vector space $\mathcal{V}^n$ realizing $\mathbb{P}^{n-1}$, then

$$\mathbf{a} \odot \mathbf{b} = \frac{\mathbf{a} \otimes \mathbf{b} + \mathbf{b} \otimes \mathbf{a}}{2}. \tag{4.6.6}$$

It is called a *point pair*.

Notation.

In this section, we use the juxtaposition of elements to denote their symmetric tensor product, instead, we use the symbol "$\wedge$" to denote the outer product. For any $\mathbf{a} \in \mathcal{V}^n$, we write

$$\mathbf{a}^2 = \mathbf{a}\mathbf{a} = \mathbf{a} \odot \mathbf{a} = \mathbf{a} \otimes \mathbf{a}, \tag{4.6.7}$$

and call it a *quadratic point*.

Definition 4.56. The Grassmann space over base space $\odot^2(\mathcal{V}^n)$, denoted by $\Lambda(\odot^2(\mathcal{V}^n))$, is called the *quadratic Grassmann space* over $\mathcal{V}^n$. Its outer product is still denoted by "$\wedge$" as in $\Lambda(\mathcal{V}^n)$. The Grassmann-Cayley algebra and bracket algebra on the quadratic Grassmann space are called *quadratic Grassmann-Cayley algebra* and *quadratic bracket algebra* respectively.

The dimension of vector space $\odot^2(\mathcal{V}^n)$ is $n(n+1)/2$. Let $\mathbf{e}_1, \mathbf{e}_2, \ldots, \mathbf{e}_n$ be a basis of $\mathcal{V}^n$. Then the *induced basis* of $\odot^2(\mathcal{V}^n)$ is

$$\{\mathbf{e}_i^2, \ 2(\mathbf{e}_j\mathbf{e}_k) \,|\, 1 \le i \le n, \ 1 \le j < k \le n\}. \tag{4.6.8}$$

The coefficient 2 in front of $\mathbf{e}_j\mathbf{e}_k$ is used as a convention, so that for any vector $\mathbf{x} = \sum_i x_i\mathbf{e}_i \in \mathcal{V}^n$, $\mathbf{x}^2$ has coordinate $x_i x_j$ in the (i,j)-th basis vector of $\odot^2(\mathcal{V}^n)$:

$$\mathbf{x}^2 = \sum_i x_i^2 \mathbf{e}_i^2 + \sum_{j<k} x_j x_k (2(\mathbf{e}_j\mathbf{e}_k)).$$

Proposition 4.57. Let $\mathbf{1}, \mathbf{2}, \mathbf{3}$ be a basis of $\mathcal{V}^3$. Let

$$\mathbf{A} = \lambda_{11}\mathbf{1}^2 + 2\lambda_{12}\mathbf{12} + 2\lambda_{13}\mathbf{13} + \lambda_{22}\mathbf{2}^2 + 2\lambda_{23}\mathbf{23} + \lambda_{33}\mathbf{3}^2 \tag{4.6.9}$$

be an element of $\odot^2(\mathcal{V}^3)$. Then $\mathbf{A}$ is a quadratic point (up to scale if $\mathbb{K} \ne \mathbb{C}$) if and only if

$$\lambda_{11} : \lambda_{12} : \lambda_{13} = \lambda_{12} : \lambda_{22} : \lambda_{23} = \lambda_{13} : \lambda_{23} : \lambda_{33}. \tag{4.6.10}$$

When (4.6.10) holds, let $\mathbf{x} = (\lambda_{11}, \lambda_{12}, \lambda_{13})^T \in \mathcal{V}^3$, then $\mathbf{A} = \mathbf{x}^2$ up to scale.

We investigate blades in $\Lambda(\odot^2(\mathcal{V}^n))$ and their geometric explanations. First, $\mathbf{1}^2 \wedge \mathbf{2}^2$ equals zero if and only if $\mathbf{1}$ and $\mathbf{2}$ are linearly dependent. When it is nonzero, the blade represents two points $\mathbf{1}, \mathbf{2}$, in the sense that a point pair $\mathbf{34}$ satisfies

$$\mathbf{1}^2 \wedge \mathbf{2}^2 \wedge 2(\mathbf{34}) = 0 \tag{4.6.11}$$

if and only if $\mathbf{3} = \mathbf{4}$ up to scale, and $\mathbf{3} = \mathbf{1}$ or $\mathbf{2}$ up to scale.

Second, $1^2 \wedge 2(12) \wedge 2^2$ equals zero if and only if 1 and 2 are linearly dependent. When it is nonzero, then a point 3 satisfies

$$1^2 \wedge 2(12) \wedge 2^2 \wedge 3^2 = 0 \tag{4.6.12}$$

if and only if 3 is on line 12.

Proposition 4.58. In $\Lambda(\odot^2(\mathcal{V}^n))$, $1^2 \wedge 2^2$ represents the 0D conic composed of points $1, 2$, and $1^2 \wedge 2(12) \wedge 2^2$ represents the double-line conic composed of line 12, in the sense that a projective point $\mathbf{x} \in \mathcal{V}^n$ is on a projective quadric represented by a blade $\mathbf{A} \in \Lambda(\odot^2(\mathcal{V}^n))$ if and only if $\mathbf{x}^2 \wedge \mathbf{A} = 0$.

Proof. We only need to prove the double-line conic statement. If $3 = \lambda_1 1 + \lambda_2 2$, then (4.6.12) is obvious. If $1, 2, 3$ are part of a basis of $\mathcal{V}^n$, then $1^2, 2^2, 3^2, 2(12)$ are part of a basis of $\odot^2(\mathcal{V}^n)$, and (4.6.12) cannot hold. $\square$

The following lemma can be proved similarly.

Lemma 4.59. (1) Three points $1, 2, 3$ are collinear if and only if

$$1^2 \wedge 2(12) \wedge 2(13) \wedge 2^2 \wedge 2(23) \wedge 3^2 = 0. \tag{4.6.13}$$

(2) If $1, 2, 3$ are collinear, then

$$[12]^2 1^2 \wedge 2^2 \wedge 3^2 = [13][23] 1^2 \wedge 2(12) \wedge 2^2, \tag{4.6.14}$$

where the brackets are deficit ones in $\Lambda(\mathcal{V}^n)$.

Corollary 4.60. (1) If no two of the four points $1, 2, 3, 4$ are identical, then the four points are collinear if and only if

$$1^2 \wedge 2^2 \wedge 3^2 \wedge 4^2 = 0. \tag{4.6.15}$$

(2) $1^2 \wedge 2^2 \wedge 3^2$ represents either a straight line or three noncollinear points, depending on whether $1, 2, 3$ are collinear or not.

Lemma 4.61. If no three of $1, 2, 3, 4$ are collinear, then for any point 5 different from the four points,

$$1^2 \wedge 2^2 \wedge 3^2 \wedge 4^2 \wedge 5^2 \neq 0. \tag{4.6.16}$$

As a corollary, $1^2 \wedge 2^2 \wedge 3^2 \wedge 4^2$ represents either a straight line and a point outside the line, or four points in which no three are collinear.

Proof. First assume $n = 3$, and assume $1^2 \wedge 2^2 \wedge 3^2 \wedge 4^2 \wedge 5^2 = 0$. With respect to the basis $1, 2, 3$, let

$$4 = \lambda_1 1 + \lambda_2 2 + \lambda_3 3, \quad 5 = \mu_1 1 + \mu_2 2 + \mu_3 3,$$

where the λ_i are nonzero, and at most one μ_j is zero.

If $\mu_1 = 0$, then the other two μ_j are nonzero; however, by

$$
\begin{aligned}
\mathbf{1}^2 \wedge \mathbf{2}^2 \wedge \mathbf{3}^2 \wedge \mathbf{4}^2 \wedge \mathbf{5}^2 =\ & \lambda_1\mu_1(\lambda_2\mu_3 - \lambda_3\mu_2)\mathbf{1}^2 \wedge \mathbf{2}^2 \wedge \mathbf{3}^2 \wedge 2(\mathbf{12}) \wedge 2(\mathbf{13}) \\
& + \lambda_2\mu_2(\lambda_1\mu_3 - \lambda_3\mu_1)\mathbf{1}^2 \wedge \mathbf{2}^2 \wedge \mathbf{3}^2 \wedge 2(\mathbf{12}) \wedge 2(\mathbf{23}) \\
& + \lambda_3\mu_3(\lambda_1\mu_2 - \lambda_2\mu_1)\mathbf{1}^2 \wedge \mathbf{2}^2 \wedge \mathbf{3}^2 \wedge 2(\mathbf{13}) \wedge 2(\mathbf{23}),
\end{aligned}
$$
$$(4.6.17)$$

it must be that $\lambda_2(\lambda_1\mu_2\mu_3) = \lambda_3(\lambda_1\mu_2\mu_3) = 0$. Contradiction. So none of the three μ_j is zero. Again by (4.6.17), we get $\lambda_1 : \mu_1 = \lambda_2 : \mu_2 = \lambda_3 : \mu_3$, so points $\mathbf{4}$ and $\mathbf{5}$ are identical, contradiction. This proves (4.6.16) for the case $n = 3$.

Second, if $\mathbf{1}, \mathbf{2}, \mathbf{3}, \mathbf{4}$ are linearly independent vectors, then (4.6.16) obviously holds for any point $\mathbf{5}$ different from the four points. $\qquad\square$

Theorem 4.62. [Fundamental theorem of quadratic bracket algebra over $\mathcal{V}^3$] Let $\mathbf{1}, \mathbf{2}, \mathbf{3}, \mathbf{4}, \mathbf{5}, \mathbf{6}$ be points in the projective plane. Then

$$[\mathbf{1}^2\mathbf{2}^2\mathbf{3}^2\mathbf{4}^2\mathbf{5}^2\mathbf{6}^2] = 0 \qquad (4.6.18)$$

if and only if the six points are coconic.

Proof.   If four of the six points, say $\mathbf{1}, \mathbf{2}, \mathbf{3}, \mathbf{4}$, are collinear, then $\mathbf{1}^2 \wedge \mathbf{2}^2 \wedge \mathbf{3}^2 \wedge \mathbf{4}^2 = 0$. When this happens, lines $\mathbf{56}$ and $\mathbf{1234}$ form a line-pair conic, and the conclusion is true.

Assume that no four of the six points are collinear. Let $\mathbf{e}_1, \mathbf{e}_2, \mathbf{e}_3$ be a basis of $\mathcal{V}^3$. Denote the dual basis in $\mathcal{V}^{3*}$ by $\mathbf{e}_1^*, \mathbf{e}_2^*, \mathbf{e}_3^*$. Then the induced basis in $\odot^2(\mathcal{V}^3)$ is

$$\mathbf{e}_1{}^2,\ 2\mathbf{e}_1\mathbf{e}_2,\ 2\mathbf{e}_1\mathbf{e}_3,\ \mathbf{e}_2{}^2,\ 2\mathbf{e}_2\mathbf{e}_3,\ \mathbf{e}_3{}^2; \qquad (4.6.19)$$

the *induced dual basis* in $\odot^2(\mathcal{V}^{3*})$ is

$$\mathbf{e}_1^{*2},\ \mathbf{e}_1^*\mathbf{e}_2^*,\ \mathbf{e}_1^*\mathbf{e}_3^*,\ \mathbf{e}_2^{*2},\ \mathbf{e}_2^*\mathbf{e}_3^*,\ \mathbf{e}_3^{*2}. \qquad (4.6.20)$$

Let the coordinates of $\mathbf{i}$ with respect to basis $\mathbf{e}_1, \mathbf{e}_2, \mathbf{e}_3$ be (x_i, y_i, z_i). Then

$$[\mathbf{123}] = \frac{\mathbf{1} \wedge \mathbf{2} \wedge \mathbf{3}}{\mathbf{e}_1 \wedge \mathbf{e}_2 \wedge \mathbf{e}_3} = \frac{(\mathbf{1} \wedge \mathbf{2} \wedge \mathbf{3}) \cdot (\mathbf{e}_1^* \wedge \mathbf{e}_2^* \wedge \mathbf{e}_3^*)}{(\mathbf{e}_1 \wedge \mathbf{e}_2 \wedge \mathbf{e}_3) \cdot (\mathbf{e}_1^* \wedge \mathbf{e}_2^* \wedge \mathbf{e}_3^*)} = \begin{vmatrix} x_1 & x_2 & x_3 \\ y_1 & y_2 & y_3 \\ z_1 & z_2 & z_3 \end{vmatrix}, \qquad (4.6.21)$$

where the dot symbol denotes the pairing between $\Lambda(\mathcal{V}^3)$ and $\Lambda(\mathcal{V}^{3*})$. Similarly,

$$
\begin{aligned}
& [\mathbf{1}^2\mathbf{2}^2\mathbf{3}^2\mathbf{4}^2\mathbf{5}^2\mathbf{6}^2] \\[4pt]
& = \frac{(\mathbf{1}^2 \wedge \mathbf{2}^2 \wedge \mathbf{3}^2 \wedge \mathbf{4}^2 \wedge \mathbf{5}^2 \wedge \mathbf{6}^2) \cdot (\mathbf{e}_1^{*2} \wedge \mathbf{e}_1^*\mathbf{e}_2^* \wedge \mathbf{e}_1^*\mathbf{e}_3^* \wedge \mathbf{e}_2^{*2} \wedge \mathbf{e}_2^*\mathbf{e}_3^* \wedge \mathbf{e}_3^{*2})}{(\mathbf{e}_1^2 \wedge 2\mathbf{e}_1\mathbf{e}_2 \wedge 2\mathbf{e}_1\mathbf{e}_3 \wedge \mathbf{e}_2^2 \wedge 2\mathbf{e}_2\mathbf{e}_3 \wedge \mathbf{e}_3^2) \cdot (\mathbf{e}_1^{*2} \wedge \mathbf{e}_1^*\mathbf{e}_2^* \wedge \mathbf{e}_1^*\mathbf{e}_3^* \wedge \mathbf{e}_2^{*2} \wedge \mathbf{e}_2^*\mathbf{e}_3^* \wedge \mathbf{e}_3^{*2})} \\[6pt]
& = \begin{vmatrix}
x_1^2 & x_1 y_1 & x_1 z_1 & y_1^2 & y_1 z_1 & z_1^2 \\
x_2^2 & x_2 y_2 & x_2 z_2 & y_2^2 & y_2 z_2 & z_2^2 \\
x_3^2 & x_3 y_3 & x_3 z_3 & y_3^2 & y_3 z_3 & z_3^2 \\
x_4^2 & x_4 y_4 & x_4 z_4 & y_4^2 & y_4 z_4 & z_4^2 \\
x_5^2 & x_5 y_5 & x_5 z_5 & y_5^2 & y_5 z_5 & z_5^2 \\
x_6^2 & x_6 y_6 & x_6 z_6 & y_6^2 & y_6 z_6 & z_6^2
\end{vmatrix},
\end{aligned}
$$

where the dot symbol denotes the pairing between $\Lambda(\odot^2(\mathcal{V}^3))$ and $\Lambda(\odot^2(\mathcal{V}^{3*}))$. By expanding the 6×6 determinant, we get a polynomial of 720 terms, which equals the coordinate polynomial form of

$$\text{conic}(\mathbf{123456}) = [\mathbf{125}][\mathbf{345}][\mathbf{136}][\mathbf{246}] - [\mathbf{126}][\mathbf{346}][\mathbf{135}][\mathbf{245}] \qquad (4.6.22)$$

by using (4.6.21). Thus

$$[\mathbf{1}^2\mathbf{2}^2\mathbf{3}^2\mathbf{4}^2\mathbf{5}^2\mathbf{6}^2] = \text{conic}(\mathbf{123456}). \qquad (4.6.23)$$

$\square$

Corollary 4.63.

$$\begin{aligned}
[\mathbf{1}^2\mathbf{2}^2\mathbf{3}^2\mathbf{4}^2\mathbf{5}^2(\mathbf{66'})] &= \text{conic}(\mathbf{12345(66')}), \\
[\mathbf{1}^2\mathbf{2}^2\mathbf{3}^2(\mathbf{44'})\mathbf{5}^2(\mathbf{66'})] &= \text{conic}(\mathbf{123(44')5(66')}).
\end{aligned} \qquad (4.6.24)$$

Lemma 4.61 and Theorem 4.62 lead to the following classical result:

Corollary 4.64. Let $\mathbf{1}, \mathbf{2}, \mathbf{3}, \mathbf{4}, \mathbf{5}$ be points in the plane. Then $\mathbf{1}^2 \wedge \mathbf{2}^2 \wedge \mathbf{3}^2 \wedge \mathbf{4}^2 \wedge \mathbf{5}^2$ represents a conic if and only if no four of the five points are collinear.

Lemma 4.65. For any four points $\mathbf{1}, \mathbf{2}, \mathbf{3}, \mathbf{4}$ in the plane,

$$2(\mathbf{14}) \wedge 2(\mathbf{24}) \wedge 2(\mathbf{34}) \wedge \mathbf{4}^2 = 0, \qquad (4.6.25)$$
$$\mathbf{1}^2 \wedge 2(\mathbf{12}) \wedge \mathbf{2}^2 \wedge \mathbf{3}^2 \wedge 2(\mathbf{34}) \wedge \mathbf{4}^2 = 0, \qquad (4.6.26)$$
$$2(\mathbf{12}) \wedge 2(\mathbf{23}) \wedge \mathbf{2}^2 \wedge 2(\mathbf{14}) \wedge 2(\mathbf{34}) \wedge \mathbf{4}^2 = 0. \qquad (4.6.27)$$

Proof. (4.6.25): Obvious from the linear dependence of $\mathbf{1}, \mathbf{2}, \mathbf{3}, \mathbf{4}$. (4.6.26): If $\mathbf{1}, \mathbf{2}, \mathbf{3}$ are collinear, then $\mathbf{1}^2 \wedge 2(\mathbf{12}) \wedge \mathbf{2}^2 \wedge \mathbf{3}^2 = 0$. If they are not collinear, then $\mathbf{4} = \lambda_1 \mathbf{1} + \lambda_2 \mathbf{2} + \lambda_3 \mathbf{3}$, and the proof is by direct computing. (4.6.27): similar. $\square$

Corollary 4.66. Let $\mathbf{3}$ be a point outside line $\mathbf{12}$. The double-line conic composed of line $\mathbf{12}$ is represented by

$$\mathbf{1}^2 \wedge 2(\mathbf{12}) \wedge \mathbf{2}^2 \wedge 2(\mathbf{13}) \wedge 2(\mathbf{23}). \qquad (4.6.28)$$

The line-pair conic composed of lines $\mathbf{12}, \mathbf{13}$ is represented by

$$\mathbf{1}^2 \wedge 2(\mathbf{12}) \wedge \mathbf{2}^2 \wedge 2(\mathbf{13}) \wedge \mathbf{3}^2. \qquad (4.6.29)$$

The line-pair conic composed of lines $\mathbf{12}, \mathbf{34}$ is represented by

$$\mathbf{1}^2 \wedge 2(\mathbf{12}) \wedge \mathbf{2}^2 \wedge 2(\mathbf{34}) \wedge \mathbf{3}^2. \qquad (4.6.30)$$

Fix a basis $\mathbf{e}_1, \mathbf{e}_2, \mathbf{e}_3$ of $\mathcal{V}^3$. In Section 2.3, a Hodge dual operator "$*$" is defined in the Grassmann space $\Lambda(\mathcal{V}^3)$. The induced basis of $\odot^2(\mathcal{V}^3)$ also defines a Hodge dual operator, denoted by the same symbol "$*$", in the quadratic Grassmann space $\Lambda(\odot^2(\mathcal{V}^3))$. The two duals are related geometrically as follows:

Proposition 4.67. For any two distinct points $\mathbf{x}, \mathbf{y}$ in the projective plane, $*(\mathbf{x}^2)$ is the double-line conic composed of line $*\mathbf{x}$, and $*(\mathbf{xy})$ is the line-pair conic composed of lines $*\mathbf{x}$ and $*\mathbf{y}$.

Proof.   Without loss of generality, let $\mathbf{x} = \mathbf{e}_1$. Then

$$*(\mathbf{x}^2) = 2\mathbf{e}_1\mathbf{e}_2 \wedge 2\mathbf{e}_1\mathbf{e}_3 \wedge \mathbf{e}_2^2 \wedge 2\mathbf{e}_2\mathbf{e}_3 \wedge \mathbf{e}_3^2$$

represents the double-line conic composed of line $\mathbf{e}_2 \wedge \mathbf{e}_3 = *\mathbf{x}$, according to (4.6.28). Similarly, let $\mathbf{y} = \mathbf{e}_2$, then

$$*(2\mathbf{x}\mathbf{y}) = -\mathbf{e}_1^2 \wedge 2\mathbf{e}_1\mathbf{e}_3 \wedge \mathbf{e}_2^2 \wedge 2\mathbf{e}_2\mathbf{e}_3 \wedge \mathbf{e}_3^2$$

represents the line-pair conic composed of lines $\mathbf{e}_2 \wedge \mathbf{e}_3 = *\mathbf{x}$ and $\mathbf{e}_1 \wedge \mathbf{e}_3 = -(*\mathbf{y})$, according to (4.6.29). $\square$

4.6.2   *Extension and Intersection*

In Grassmann space $\Lambda(\mathcal{V}^n)$, the extension product $\mathbf{A}_r \sqcap \mathbf{B}_s$ of two blades $\mathbf{A}_r, \mathbf{B}_s$ defined by (2.4.20), represents the sum of the two vector subspaces represented by the two blades:

$$\{\mathbf{x}_1 + \mathbf{x}_2 \in \mathcal{V}^n \,|\, \mathbf{x}_1 \in \mathbf{A}_r, \ \mathbf{x}_2 \in \mathbf{B}_s\}. \tag{4.6.31}$$

In quadratic Grassmann space $\Lambda(\odot^2(\mathcal{V}^n))$, the extension product of two blades is defined just the same. For example, for quadratic points $\mathbf{1}^2$ and $\mathbf{2}^2$, their extension product is

$$\mathbf{1}^2 \sqcap \mathbf{2}^2 = \mathbf{1}^2 \wedge \mathbf{2}^2, \tag{4.6.32}$$

which represents the 0D conic composed of the two points. In contrast, the extension product of $\mathbf{1}, \mathbf{2}$ in $\Lambda(\mathcal{V}^n)$ is $\mathbf{1} \wedge \mathbf{2}$, which represents the line through the two points.

In $\Lambda(\mathcal{V}^3)$, point $\mathbf{3}$ is on line $\mathbf{12}$ if and only if $[\mathbf{123}] = 0$. Squaring this equation, we get the following double-line conic equation, with $\mathbf{3}$ as the vector indeterminate:

$$[\mathbf{123}]^2 = 0. \tag{4.6.33}$$

The equation should agree with the two equations (4.6.12) and (4.6.13) representing the same constraint. Since $[\mathbf{123}]^2$ is the square of $[\mathbf{123}]$, it is natural to expect that $\mathbf{1}^2 \wedge 2(\mathbf{12}) \wedge \mathbf{2}^2$ is the "square" of $\mathbf{1} \wedge \mathbf{2}$, and (4.6.13) is the "square" of $\mathbf{1} \wedge \mathbf{2} \wedge \mathbf{3}$, where the "square" is a natural extension of the symmetrization from vectors to blades in $\Lambda(\mathcal{V}^3)$. This extension, however, is neither the extension product in $\Lambda(\odot^2(\mathcal{V}^3))$, nor the symmetrization operation in $\otimes(\mathcal{V}^3)$.

Definition 4.68. The *quadratic extension product* between two blades in $\Lambda(\mathcal{V}^n)$, denoted by the same symbol "$\odot$" as the symmetrization, or simply by the juxtaposition of elements, is a blade in $\Lambda(\odot^2(\mathcal{V}^n))$ that is defined for any r-blade $\mathbf{A}_r = \mathbf{a}_1 \wedge \mathbf{a}_2 \wedge \cdots \mathbf{a}_r$ and s-blade $\mathbf{B}_s = \mathbf{b}_1 \wedge \mathbf{b}_2 \wedge \cdots \wedge \mathbf{b}_s$ of $\Lambda(\mathcal{V}^n)$ as follows:

$$\begin{aligned} \mathbf{a}_i\mathbf{B}_s &= \mathbf{B}_s\mathbf{a}_i := \mathbf{a}_i\mathbf{b}_1 \wedge \mathbf{a}_i\mathbf{b}_2 \wedge \cdots \mathbf{a}_i\mathbf{b}_s, \\ \mathbf{A}_r\mathbf{B}_s &:= \mathbf{a}_1\mathbf{B}_s \sqcap \mathbf{a}_2\mathbf{B}_s \sqcap \cdots \sqcap \mathbf{a}_r\mathbf{B}_s. \end{aligned} \tag{4.6.34}$$

It is easy to verify that the quadratic extension product $\mathbf{A}_r\mathbf{B}_s$ is well-defined, *i.e.*, independent of the decomposition of $\mathbf{A}_r$ into an outer product of vectors. $\mathbf{A}_r\mathbf{B}_s$ represents the vector subspace of $\odot^2(\mathcal{V}^n)$ spanned by

$$\{\mathbf{x}_1\mathbf{x}_2 \in \odot^2(\mathcal{V}^n)\,|\,\mathbf{x}_1 \in \mathbf{A}_r,\ \mathbf{x}_2 \in \mathbf{B}_s\}.$$

It is homogeneous of degree s in $\mathbf{A}_r$ and of degree r in $\mathbf{B}_s$, and satisfies the following commutation rule:

$$\mathbf{A}_r\mathbf{B}_s = (-1)^{\frac{r(r-1)s(s-1)}{4}}\mathbf{B}_s\mathbf{A}_r, \tag{4.6.35}$$

where the sign is caused by the scanning by rows of an $r \times s$ matrix versus the scanning by columns, both starting from the upper-left corner of the matrix.

Examples. Let $\mathbf{1},\ldots,\mathbf{r},\mathbf{1}',\ldots,\mathbf{r}' \in \mathcal{V}^3$, where $\mathbf{1},\mathbf{2},\mathbf{3}$ are not collinear.

(1) The quadratic extension product of vectors $\mathbf{1},\mathbf{2}$ is the symmetrization $\mathbf{12}$.

(2) $\mathbf{1}(\mathbf{1}' \wedge \mathbf{2}' \wedge \cdots \wedge \mathbf{r}') = \mathbf{11}' \wedge \mathbf{12}' \wedge \cdots \wedge \mathbf{1r}'$.

(3) $(\mathbf{1} \wedge \mathbf{2})(\mathbf{1} \wedge \mathbf{2}) = (\mathbf{1}^2 \wedge \mathbf{12}) \sqcap (\mathbf{12} \wedge \mathbf{2}^2) = [\mathbf{12}]\mathbf{1}^2 \wedge \mathbf{12} \wedge \mathbf{2}^2$ as expected.

(4) Let $\mathbf{A} = \mathbf{12} \wedge \mathbf{2}^2 \wedge \mathbf{23}$, $\mathbf{B} = \mathbf{13} \wedge \mathbf{23} \wedge \mathbf{3}^2$, then as expected,

$$(\mathbf{1} \wedge \mathbf{2} \wedge \mathbf{3})(\mathbf{1} \wedge \mathbf{2} \wedge \mathbf{3}) = (\mathbf{1}^2 \wedge \mathbf{12} \wedge \mathbf{13}) \sqcap (\mathbf{12} \wedge \mathbf{2}^2 \wedge \mathbf{23}) \sqcap (\mathbf{13} \wedge \mathbf{23} \wedge \mathbf{3}^2)$$

$$= \sum_{(1,2)\vdash\mathbf{A},(2,1)\vdash\mathbf{B}} [\mathbf{A}_{(1)} \wedge \mathbf{B}_{(1)}]\mathbf{1}^2 \wedge \mathbf{12} \wedge \mathbf{13} \wedge \mathbf{A}_{(2)} \wedge \mathbf{B}_{(2)}$$

$$+ \sum_{(2,1)\vdash\mathbf{A},(1,2)\vdash\mathbf{B}} [\mathbf{A}_{(1)} \wedge \mathbf{B}_{(1)}]\mathbf{1}^2 \wedge \mathbf{12} \wedge \mathbf{13} \wedge \mathbf{A}_{(2)} \wedge \mathbf{B}_{(2)}$$

$$= 2\,[\mathbf{12} \wedge \mathbf{13} \wedge \mathbf{23}]\mathbf{1}^2 \wedge \mathbf{12} \wedge \mathbf{13} \wedge \mathbf{2}^2 \wedge \mathbf{23} \wedge \mathbf{3}^2.$$

(5) $(\mathbf{1} \wedge \mathbf{2} \wedge \mathbf{4})(\mathbf{1} \wedge \mathbf{3} \wedge \mathbf{4}) = (\mathbf{1}^2 \wedge \mathbf{13} \wedge \mathbf{14}) \sqcap (\mathbf{12} \wedge \mathbf{23} \wedge \mathbf{24}) \sqcap (\mathbf{14} \wedge \mathbf{34} \wedge \mathbf{4}^2)$. To obtain the line-pair conic representation (4.6.29), we need to eliminate $\mathbf{4}$ from $\mathbf{14},\mathbf{24},\mathbf{34}$. By the Cramer's rule of $\mathbf{4}$ with respect to $\mathbf{1},\mathbf{2},\mathbf{3}$,

$$\begin{aligned}
[\mathbf{123}]\mathbf{14} &= [\mathbf{234}]\mathbf{1}^2 - [\mathbf{134}]\mathbf{12} + [\mathbf{124}]\mathbf{13}, \\
[\mathbf{123}]\mathbf{24} &= [\mathbf{234}]\mathbf{12} - [\mathbf{134}]\mathbf{2}^2 + [\mathbf{124}]\mathbf{23}, \\
[\mathbf{123}]\mathbf{34} &= [\mathbf{234}]\mathbf{13} - [\mathbf{134}]\mathbf{23} + [\mathbf{124}]\mathbf{3}^2.
\end{aligned} \tag{4.6.36}$$

So

$$(\mathbf{1} \wedge \mathbf{2} \wedge \mathbf{4})(\mathbf{1} \wedge \mathbf{3} \wedge \mathbf{4})$$

$$= \underbrace{[\mathbf{123}]^{-2}[\mathbf{134}]^2}\,(\mathbf{1}^2 \wedge \mathbf{13} \wedge \mathbf{12}) \sqcap (\mathbf{12} \wedge \mathbf{23} \wedge \mathbf{2}^2) \sqcap (\mathbf{14} \wedge \mathbf{34} \wedge \mathbf{4}^2)$$

$$= (\mathbf{14} \wedge \mathbf{34} \wedge \mathbf{4}^2) \sqcap (\mathbf{1}^2 \wedge \mathbf{13} \wedge \mathbf{12}) \sqcap (\mathbf{12} \wedge \mathbf{23} \wedge \mathbf{2}^2)$$

$$\begin{aligned}
= \ &2\,[\mathbf{2}^2 \wedge \mathbf{13} \wedge \mathbf{12}]\mathbf{14} \wedge \mathbf{34} \wedge \mathbf{4}^2 \wedge \mathbf{1}^2 \wedge \mathbf{12} \wedge \mathbf{23} \\
&+2\,[\mathbf{1}^2 \wedge \mathbf{12} \wedge \mathbf{23}]\mathbf{14} \wedge \mathbf{34} \wedge \mathbf{4}^2 \wedge \mathbf{13} \wedge \mathbf{12} \wedge \mathbf{2}^2 \\
&-2\,[\mathbf{13} \wedge \mathbf{12} \wedge \mathbf{23}]\mathbf{14} \wedge \mathbf{34} \wedge \mathbf{4}^2 \wedge \mathbf{1}^2 \wedge \mathbf{12} \wedge \mathbf{2}^2 \\
&-2\,[\mathbf{1}^2 \wedge \mathbf{12} \wedge \mathbf{2}^2]\mathbf{14} \wedge \mathbf{34} \wedge \mathbf{4}^2 \wedge \mathbf{13} \wedge \mathbf{12} \wedge \mathbf{23}.
\end{aligned} \tag{4.6.37}$$

The above result seems terrible, far away from the outer product of 4^2 with (4.6.29). However, the *advantage of deficit brackets* in homogeneous computing is that the dummy extensor can be chosen arbitrarily, as long as the final result is not identical to zero. Different choices of the dummy extensor lead to equal-up-to-scale results.

Notice that $\{1^2, 12, 13, 2^2, 23, 3^3\}$ is a basis of $\odot^2(\mathcal{V}^3)$. In the result of (4.6.37), the four deficit brackets are all composed of basis vectors. For the first deficit bracket $[2^2 \wedge 13 \wedge 12]$, if choosing the dummy extensor $\mathbf{U}_3$ to be the trivector $1^2 \wedge 3^2 \wedge 23$ dual to $2^2 \wedge 12 \wedge 13$, then only the first term in the result of (4.6.37) is nonzero. Similar arguments go through to the other deficit brackets by choosing $\mathbf{U}_3$ to be the corresponding dual trivectors.

Thus the result of (4.6.37) is four *equal-up-to-scale* forms of $(1 \wedge 2 \wedge 4)(1 \wedge 3 \wedge 4)$:

$$
\begin{aligned}
&14 \wedge 34 \wedge 4^2 \wedge 1^2 \wedge 12 \wedge 23, \\
&14 \wedge 34 \wedge 4^2 \wedge 13 \wedge 12 \wedge 2^2, \\
&14 \wedge 34 \wedge 4^2 \wedge 1^2 \wedge 12 \wedge 2^2, \\
&14 \wedge 34 \wedge 4^2 \wedge 13 \wedge 12 \wedge 23.
\end{aligned}
\tag{4.6.38}
$$

Eliminating $\mathbf{14}, \mathbf{34}$ from (4.6.38) by (4.6.36), and removing common factors $-[\mathbf{123}]^{-2}$, we get

$$
\begin{aligned}
&[124]^2 13 \wedge 3^2 \wedge 1^2 \wedge 12 \wedge 23 \wedge 4^2, \\
&[234] 1^2 \wedge (-[134]23 + [124]3^2) \wedge 13 \wedge 12 \wedge 2^2 \wedge 4^2, \\
&[124] 13 \wedge (-[134]23 + [124]3^2) \wedge 1^2 \wedge 12 \wedge 2^2 \wedge 4^2, \\
&[124][234] 1^2 \wedge 3^2 \wedge 13 \wedge 12 \wedge 23 \wedge 4^2,
\end{aligned}
\tag{4.6.39}
$$

from which and by symmetry we get the following three equivalent representations of the line-pair conic $\mathbf{12}, \mathbf{13}$:

$$
\begin{aligned}
&1^2 \wedge 3^2 \wedge 12 \wedge 13 \wedge 23, \\
&1^2 \wedge 2^2 \wedge 12 \wedge 13 \wedge 23, \\
&1^2 \wedge 2^2 \wedge 3^2 \wedge 12 \wedge 13.
\end{aligned}
\tag{4.6.40}
$$

In GC algebra $\Lambda(\mathcal{V}^3)$, the representation of the intersection of two lines is very easy, but very complicated when it comes to two conics. In quadratic GC algebra $\Lambda(\odot^2(\mathcal{V}^n))$, the situation is reversed.

First, consider two lines $\mathbf{12}$ and $\mathbf{34}$. In $\Lambda(\odot^2(\mathcal{V}^3))$, the lines are represented by $1^2 \wedge 2(12) \wedge 2^2$ and $3^2 \wedge 2(34) \wedge 4^2$ respectively. By (4.6.26), their outer product is zero. Their intersection is

$$
\begin{aligned}
(1^2 \wedge 12 \wedge 2^2) \sqcup (3^2 \wedge 34 \wedge 4^2) = \; &[1^2 \wedge (12) \wedge 3^2 \wedge (34) \wedge 4^2] 2^2 \\
&-[1^2 \wedge 2^2 \wedge 3^2 \wedge (34) \wedge 4^2] \, 12 \\
&+[(12) \wedge 2^2 \wedge 3^2 \wedge (34) \wedge 4^2] 1^2.
\end{aligned}
\tag{4.6.41}
$$

According to (4.6.10), with respect to the basis $\mathbf{1}, \mathbf{2}$ of line $\mathbf{12}$, the intersection is

$$
[2(12) \wedge 2^2 \wedge 3^2 \wedge (34) \wedge 4^2] 1 - [1^2 \wedge 2^2 \wedge 3^2 \wedge (34) \wedge 4^2] 2.
\tag{4.6.42}
$$

In contrast, in $\Lambda(\mathcal{V}^3)$ the intersection has a very simple expression:

$$(\mathbf{1} \wedge \mathbf{2}) \vee (\mathbf{3} \wedge \mathbf{4}) = [\mathbf{134}]\mathbf{2} - [\mathbf{234}]\mathbf{1}. \tag{4.6.43}$$

To verify that it equals (4.6.42) up to scale, we need to eliminate one point, say point $\mathbf{1}$, from the two deficit brackets of (4.6.42) by Cramer's rule $[\mathbf{234}]\mathbf{1} = [\mathbf{134}]\mathbf{2} - [\mathbf{124}]\mathbf{3} + [\mathbf{123}]\mathbf{4}$. Complementing the deficit brackets by selecting dummy point-pair $\mathbf{U}_1 = \mathbf{24}$, we get the equality.

Next, consider the intersection of two conics:

$$(\mathbf{1}^2 \wedge \mathbf{2}^2 \wedge \mathbf{3}^2 \wedge \mathbf{4}^2 \wedge \mathbf{5}^2) \vee (\mathbf{1'}^2 \wedge \mathbf{2'}^2 \wedge \mathbf{3'}^2 \wedge \mathbf{4'}^2 \wedge \mathbf{5'}^2). \tag{4.6.44}$$

If $\mathbf{1} = \mathbf{1'}, \mathbf{2} = \mathbf{2'}, \mathbf{3} = \mathbf{3'}$, then (4.6.44) equals

$$\mathbf{1}^2 \wedge \mathbf{2}^2 \wedge \mathbf{3}^2 \wedge ([\mathbf{5}^2\mathbf{1}^2\mathbf{2}^2\mathbf{3}^2\mathbf{4'}^2\mathbf{5'}^2]\mathbf{4}^2 - [\mathbf{4}^2\mathbf{1}^2\mathbf{2}^2\mathbf{3}^2\mathbf{4'}^2\mathbf{5'}^2]\mathbf{5}^2). \tag{4.6.45}$$

Let the fourth point of intersection of the two conics be $\mathbf{x} \in \mathcal{V}^3$, where

$$\mathbf{x} = \lambda_2\lambda_3\mathbf{1} + \lambda_1\lambda_3\mathbf{2} + \lambda_1\lambda_2\mathbf{3}. \tag{4.6.46}$$

Then (4.6.45) should be equal to $\mathbf{1}^2 \wedge \mathbf{2}^2 \wedge \mathbf{3}^2 \wedge \mathbf{x}^2$ up to scale. Eliminating $\mathbf{x}$ from $\mathbf{1}^2 \wedge \mathbf{2}^2 \wedge \mathbf{3}^2 \wedge \mathbf{x}^2$ by (4.6.46), we get that (4.6.45) equals

$$\mathbf{1}^2 \wedge \mathbf{2}^2 \wedge \mathbf{3}^2 \wedge (\lambda_1\mathbf{23} + \lambda_2\mathbf{13} + \lambda_3\mathbf{12}) \tag{4.6.47}$$

up to scale, which is another form of (4.1.28).

Third, consider the intersection of line $\mathbf{12}$ and conic $\mathbf{1'2'3'4'5'}$:

$$\begin{aligned}
(\mathbf{1}^2 \wedge 2(\mathbf{12}) \wedge \mathbf{2}^2) \vee (\mathbf{1'}^2 \wedge \mathbf{2'}^2 \wedge \mathbf{3'}^2 \wedge \mathbf{4'}^2 \wedge \mathbf{5'}^2) \\
= [\mathbf{2}^2\mathbf{1'}^2\mathbf{2'}^2\mathbf{3'}^2\mathbf{4'}^2\mathbf{5'}^2]\mathbf{1}^2 \wedge 2(\mathbf{12}) - [2(\mathbf{12})\mathbf{1'}^2\mathbf{2'}^2\mathbf{3'}^2\mathbf{4'}^2\mathbf{5'}^2]\mathbf{1}^2 \wedge \mathbf{2}^2 \\
+ [\mathbf{1}^2\mathbf{1'}^2\mathbf{2'}^2\mathbf{3'}^2\mathbf{4'}^2\mathbf{5'}^2]2(\mathbf{12}) \wedge \mathbf{2}^2.
\end{aligned} \tag{4.6.48}$$

If point $\mathbf{1}$ is on the conic, then by (4.6.10), the second point of intersection is

$$[\mathbf{2}^2\mathbf{1'}^2\mathbf{2'}^2\mathbf{3'}^2\mathbf{4'}^2\mathbf{5'}^2]\mathbf{1} - [2(\mathbf{12})\mathbf{1'}^2\mathbf{2'}^2\mathbf{3'}^2\mathbf{4'}^2\mathbf{5'}^2]\mathbf{2}. \tag{4.6.49}$$

In the general case, let $\mathbf{x} = \lambda\mathbf{1} + \mu\mathbf{2}$ be a point at the intersection (4.6.48). Then the outer product of $\mathbf{x}^2$ and (4.6.48) equals zero, *i.e.*,

$$\lambda^2[\mathbf{1}^2\mathbf{1'}^2\mathbf{2'}^2\mathbf{3'}^2\mathbf{4'}^2\mathbf{5'}^2] + \lambda\mu[2(\mathbf{12})\mathbf{1'}^2\mathbf{2'}^2\mathbf{3'}^2\mathbf{4'}^2\mathbf{5'}^2] + \mu^2[\mathbf{2}^2\mathbf{1'}^2\mathbf{2'}^2\mathbf{3'}^2\mathbf{4'}^2\mathbf{5'}^2] = 0.$$

The discriminant of this quadratic equation is

$$[(\mathbf{12})\mathbf{1'}^2\mathbf{2'}^2\mathbf{3'}^2\mathbf{4'}^2\mathbf{5'}^2]^2 - [\mathbf{1}^2\mathbf{1'}^2\mathbf{2'}^2\mathbf{3'}^2\mathbf{4'}^2\mathbf{5'}^2][\mathbf{2}^2\mathbf{1'}^2\mathbf{2'}^2\mathbf{3'}^2\mathbf{4'}^2\mathbf{5'}^2]. \tag{4.6.50}$$

It equals zero if and only if the line is tangent to the conic.

There are two remarks to make before ending this chapter. In the remark after Theorem 2.62, there is a Cayley expression $(\mathbf{12} \vee \mathbf{34})(\mathbf{13} \vee \mathbf{24})$ that is antisymmetric in $\mathbf{1}, \mathbf{2}, \mathbf{3}, \mathbf{4}$, and it is not clear why it should be so in the setting of incidence geometry. In conic geometry, the expression represents the line connecting the two double-points of line-pair conics $\mathbf{1}^2 \wedge \mathbf{2}^2 \wedge \mathbf{3}^2 \wedge \mathbf{4}^2 \wedge 2(\mathbf{12})$ and $\mathbf{1}^2 \wedge \mathbf{2}^2 \wedge \mathbf{3}^2 \wedge \mathbf{4}^2 \wedge 2(\mathbf{13})$ respectively, so it must be a linear functional of the pencil of conics represented by

$\mathbf{1}^2 \wedge \mathbf{2}^2 \wedge \mathbf{3}^2 \wedge \mathbf{4}^2$. The antisymmetry comes from this 4-blade representation of the pencil of conics.

Cubics in the plane can certainly be described and computed with the *cubic Grassmann-Cayley algebra* $\Lambda(\odot^3(\mathcal{V}^3))$ and the *cubic bracket algebra* over $\mathcal{V}^3$. Formally, all degree-m algebraic curves and surfaces of various dimensions can be described and manipulated with the *m-ic Grassmann-Cayley algebra* $\Lambda(\odot^m(\mathcal{V}^n))$ and the *m-ic bracket algebra* over $\mathcal{V}^n$. For $m > 2$, it is still not clear how the expansion of brackets from m-ic bracket algebra to l-ic bracket algebra for $l < m$ should proceed.

Chapter 5

Inner-product Bracket Algebra and Clifford Algebra

When a vector space is equipped with an inner product structure, the linear transformations preserving this structure form the orthogonal group of the vector space. Orthogonal geometry is on the properties of the vector space that are invariant under the orthogonal group. Its geometric algebra and algebra of basic invariants are inner-product Grassmann algebra and inner-product bracket algebra respectively.

To establish a complete system of advanced orthogonal invariants, a natural extension of inner-product Grassmann algebra called Clifford algebra is needed. The associativity and almost invertibility of the geometric product make Clifford algebra an ideal tool in solving multivector equations, including both algebraic and differential ones.

The translation from Clifford algebra to inner-product Grassmann algebra is called Clifford expansion. The translation back to Clifford algebra is called Clifford factorization. This chapter introduces the two geometric algebras, with emphasis on the representations of Clifford algebra and the translation from Clifford algebra to inner-product Grassmann algebra.

From this chapter on, the outer product is always denoted by "$\wedge$". The juxtaposition of elements always denotes the geometric product in Clifford algebra.

5.1 Inner-product bracket algebra

5.1.1 *Inner-product space*

An *inner product*, or *scalar product*, in $\mathcal{V}^n$, denoted by the dot symbol, is a bilinear symmetric function over $\mathcal{V}^n \times \mathcal{V}^n$, with values in the base field $\mathbb{K}$. Vector space $\mathcal{V}^n$ equipped with an inner product is called an *inner-product space*.

An inner product naturally induces a quadratic form:

$$Q(\mathbf{x}) = \mathbf{x} \cdot \mathbf{x}, \ \forall \mathbf{x} \in \mathcal{V}^n. \tag{5.1.1}$$

Conversely, a quadratic form determines an inner product by polarization, or more explicitly,

$$\mathbf{x} \cdot \mathbf{y} = \frac{Q(\mathbf{x}+\mathbf{y}) - Q(\mathbf{x}) - Q(\mathbf{y})}{2}, \ \forall \mathbf{x}, \mathbf{y} \in \mathcal{V}^n. \tag{5.1.2}$$

219

An *inner product space* is also called a *quadratic space*, when equipped with the quadratic form induced from the inner product.

A linear isomorphism between two inner-product spaces, if preserving the inner product structure, is called an *isometry*. An isometry from an inner-product space to itself is called an *orthogonal transformation* of the inner-product space.

Two vectors in $\mathcal{V}^n$ are said to be *orthogonal* if their inner product equals zero. An inner product is said to be *degenerate*, or $\mathcal{V}^n$ is said to be *degenerate*, if there exists a nonzero vector in $\mathcal{V}^n$ that is orthogonal to every vector in $\mathcal{V}^n$, including itself.

If $\mathcal{V}^n$ is real, a vector in it is said to be *positive*, *null*, or *negative*, if the inner product of the vector with itself is positive, zero, or negative respectively. The sign of the inner product result is called the *signature* of the vector. If a subspace of $\mathcal{V}^n$ has the property that any two vectors in it are orthogonal to each other, the subspace is called a *null subspace*, or *totally isotropic subspace*.

Definition 5.1. In $\mathcal{V}^n$, the set

$$\mathrm{rad}(\mathcal{V}^n) := \{\mathbf{x} \in \mathcal{V}^n \,|\, \mathbf{x} \cdot \mathbf{y} = 0, \ \forall \mathbf{y} \in \mathcal{V}^n\} \tag{5.1.3}$$

is called the *radical* of $\mathcal{V}^n$. For a subspace represented by a blade $\mathbf{A}$, its radical is denoted by $\mathrm{rad}(\mathbf{A})$.

A classical theorem in linear algebra [48] says that there always exists a subspace $\mathcal{V}'$ in $\mathcal{V}^n$, such that

$$\mathcal{V}^n = \mathrm{rad}(\mathcal{V}^n) \oplus \mathcal{V}'. \tag{5.1.4}$$

Let $\mathbf{e}_1, \mathbf{e}_2, \ldots, \mathbf{e}_n$ be a basis of $\mathcal{V}^n$. The matrix

$$(\mathbf{e}_i \cdot \mathbf{e}_j)_{i,j=1..n}$$

is called the *matrix of the inner product*, or the *Gram matrix*, of the basis. If the matrix is diagonal, the basis is called an *orthogonal basis*. If the inner product is nondegenerate, the *reciprocal basis* of the basis $\mathbf{e}$'s is another basis $\mathbf{e}^*$'s of $\mathcal{V}^n$, such that $\mathbf{e}_i \cdot \mathbf{e}_j^* = \delta_{ij}$ for all $1 \leq i, j \leq n$. Here δ_{ij} is the *Kronecker symbol*, it equals 1 if $i = j$ and equals 0 otherwise.

In the case $\mathbb{K} = \mathbb{R}$, the classical Sylvester's Theorem in linear algebra says that by changing the basis, the Gram matrix can be diagonized such that the diagonal elements are all in $\{1, -1, 0\}$. The respective numbers of 1, -1 and 0 in the diagonal, denoted by (p, q, r), are independent of the basis chosen. The triplet is called the *signature*, or *index*, of the inner product. A basis whose Gram matrix is in this form is called an *orthonormal basis*. If $\{\mathbf{e}_1, \mathbf{e}_2, \ldots, \mathbf{e}_n\}$ is an orthonormal basis of $\mathcal{V}^n$, then $\mathbf{e}_i^* = \epsilon_i \mathbf{e}_i$ for all $1 \leq i \leq n$, where $\epsilon_i = \mathbf{e}_i \cdot \mathbf{e}_i$ is the signature of $\mathbf{e}_i$.

In the case $\mathbb{K} = \mathbb{C}$, again Sylvester's Theorem says that by changing the basis, the Gram matrix can be diagonized such that its diagonal elements are all in $\{0, 1\}$. The respective numbers of 1 and 0, denoted by (p, r), are independent of the basis chosen. The pair is called the *signature*, or *index*, of the inner product. A basis whose Gram matrix is in this form is called an *orthonormal basis*.

A real vector space of signature (p, q, r) is often denoted by $\mathbb{R}^{p,q,r}$. If $r = 0$, the inner product is nondegenerate, and the space can be denoted by $\mathbb{R}^{p,q}$. For example, an nD *Minkowski space* is a space $\mathbb{R}^{n-1,1}$. If furthermore $q = 0$, the space is an nD *Euclidean vector space*, denoted by $\mathbb{R}^n$, else if $p = 0$, the space is an nD *anti-Euclidean vector space*, denoted by $\mathbb{R}^{-n}$. A complex vector space of signature (p, r) is often denoted by $\mathbb{C}^{p,r}$, and when $r = 0$, $\mathbb{C}^{p,0}$ is usually denoted by $\mathbb{C}^p$.

Example 5.2. In Euclidean space $\mathbb{R}^3$, the inner product of two vectors $\mathbf{a} = (a_1, a_2, a_3)^T$ and $\mathbf{b} = (b_1, b_2, b_3)^T$ is

$$\mathbf{a} \cdot \mathbf{b} = a_1 b_1 + a_2 b_2 + a_3 b_3.$$

In anti-Euclidean space $\mathbb{R}^{-3}$, the inner product is

$$\mathbf{a} \cdot \mathbf{b} = -a_1 b_1 - a_2 b_2 - a_3 b_3.$$

The space-time in special relativity is the 4D Minkowski space $\mathbb{R}^{3,1}$. For two vectors $\mathbf{a} = (a_1, a_2, a_3, a_4)^T$ and $\mathbf{b} = (b_1, b_2, b_3, b_4)^T$ in $\mathbb{R}^{3,1}$, their inner product is

$$\mathbf{a} \cdot \mathbf{b} = a_1 b_1 + a_2 b_2 + a_3 b_3 - a_4 b_4.$$

A subspace of $\mathcal{V}^n$ is naturally equipped with the inner product structure of $\mathcal{V}^n$. When the subspace is represented by a blade, the *signature of the blade* refers to the signature of the inner product in the subspace. Accordingly, if a blade is said to be null, or degenerate, or Euclidean, or Minkowski, or others, the meaning is that the inner product in the corresponding subspace has the indicated property.

Definition 5.3. A *Witt pair*, or *hyperbolic pair*, refers to a pair of null vectors $\mathbf{a}, \mathbf{b}$ such that $\mathbf{a} \cdot \mathbf{b} = -1$. A *Witt basis* refers to a basis

$$\{\mathbf{a}_i, \mathbf{b}_i, \mathbf{c}_j \mid 1 \leq i \leq u, \, 1 \leq j \leq v\} \tag{5.1.5}$$

of $\mathcal{V}^n$, where $2u = n - v$, such that for any $1 \leq i \leq u$, $(\mathbf{a}_i, \mathbf{b}_i)$ is a Witt pair, $\mathbf{a}_i$ is orthogonal to all other basis vectors except $\mathbf{b}_i$, and $\mathbf{b}_i$ is orthogonal to all other basis vectors except $\mathbf{a}_i$, and $\mathbf{c}_j$ is orthogonal to all other basis vectors.

An orthogonal basis is a Witt basis without any Witt pair. For $\mathcal{V}^n = \mathbb{R}^{p,q,r}$, the maximal number of Witt pairs allowed in a Witt basis is $u = \min(p, q)$; for $\mathcal{V}^n = \mathbb{C}^{p,r}$, the maximal number is $u = [p/2]$.

Corollary 5.4. The dimension of any maximal null subspace of $\mathbb{R}^{p,q,r}$ is $r + \min(p, q)$; the dimension of any maximal null subspace of $\mathbb{C}^{p,r}$ is $r + [p/2]$.

The classical *Witt Theorem* [48] states that for a nondegenerate inner-product space, a Witt basis of the whole space can be built by extending a Witt basis of any subspace. In particular, if $\mathbf{a}_1, \mathbf{a}_2, \ldots, \mathbf{a}_k$ is a basis of a null subspace of $\mathcal{V}^n$, then there exist null vectors $\mathbf{b}_1, \mathbf{b}_2, \ldots, \mathbf{b}_k \in \mathcal{V}^n$ such that each $(\mathbf{a}_i, \mathbf{b}_i)$ is a Witt pair, and all such Witt pairs are part of a Witt basis of $\mathcal{V}^n$. In contrast, an orthogonal basis of a subspace can be extended to an orthogonal basis of the whole space only when the inner product of the subspace is nondegenerate. Of course, any orthogonal basis of a subspace can be extended to a Witt basis of the whole space.

The inner product in $\mathcal{V}^n$ can be extended to an inner product in the Grassmann space $\Lambda(\mathcal{V}^n)$ by Laplace expansions of determinants, as follows:

Definition 5.5. The following scalar-valued product in $\Lambda(\mathcal{V}^n)$, denoted by "$\rfloor$", is called the *Hodge scalar product*: for vectors $\mathbf{a}$'s and $\mathbf{b}$'s in $\mathcal{V}^n$,

$$(\mathbf{a}_1 \wedge \mathbf{a}_2 \wedge \cdots \wedge \mathbf{a}_r) \rfloor (\mathbf{b}_1 \wedge \mathbf{b}_2 \wedge \cdots \wedge \mathbf{b}_s) := \delta_{rs} (\mathbf{a}_1 \mathbf{a}_2 \ldots \mathbf{a}_r \,|\, \mathbf{b}_1 \mathbf{b}_2 \ldots \mathbf{b}_r)$$
$$:= \delta_{rs} \, \det(\mathbf{a}_i \cdot \mathbf{b}_j)_{i,j=1..r}. \tag{5.1.6}$$

Here $(\mathbf{a}_1 \mathbf{a}_2 \ldots \mathbf{a}_r \,|\, \mathbf{b}_1 \mathbf{b}_2 \ldots \mathbf{b}_r)$ is the *Gram determinant* $\det(\mathbf{a}_i \cdot \mathbf{b}_j)_{i,j=1..r}$, in letter-place notation (2.4.11), of blades $\mathbf{a}_1 \wedge \mathbf{a}_2 \wedge \cdots \wedge \mathbf{a}_r$ and $\mathbf{b}_1 \wedge \mathbf{b}_2 \wedge \cdots \wedge \mathbf{b}_s$.

The *reverse scalar product* in $\Lambda(\mathcal{V}^n)$, denoted by the dot symbol, is defined by

$$(\mathbf{a}_1 \wedge \mathbf{a}_2 \wedge \cdots \wedge \mathbf{a}_r) \cdot (\mathbf{b}_1 \wedge \mathbf{b}_2 \wedge \cdots \wedge \mathbf{b}_s) := (\mathbf{a}_r \wedge \mathbf{a}_{r-1} \wedge \cdots \wedge \mathbf{a}_1) \rfloor (\mathbf{b}_1 \wedge \mathbf{b}_2 \wedge \cdots \wedge \mathbf{b}_s).$$
$$\tag{5.1.7}$$

The (Hodge, reverse) scalar product of $\lambda, \mu \in \mathbb{K}$ is defined as $\lambda\mu$. The (Hodge, reverse) scalar product of any two elements in $\Lambda(\mathcal{V}^n)$ is the linear extension of the (Hodge, reverse) scalar product of blades.

Definition 5.6. The *magnitude* of r-vector $\mathbf{A}_r$ is defined by

$$|\mathbf{A}_r| = \sqrt{|\mathbf{A}_r \cdot \mathbf{A}_r|} = \sqrt{|\mathbf{A}_r \rfloor \mathbf{A}_r|}. \tag{5.1.8}$$

In particular, if an r-blade has magnitude 1, it is called a *unit blade*. The *inverse* of a blade $\mathbf{A}_r$ is

$$\mathbf{A}_r^{-1} = \frac{\mathbf{A}_r}{\mathbf{A}_r \cdot \mathbf{A}_r}, \tag{5.1.9}$$

if the denominator is nonzero.

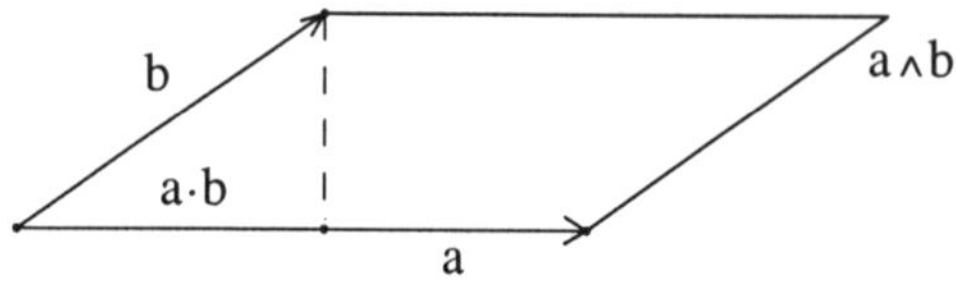

Fig. 5.1 $\mathbf{a} \cdot \mathbf{b}$ and $\mathbf{a} \wedge \mathbf{b}$.

Example 5.7. In $\mathbb{R}^n$, the magnitude of a vector is its length. For two vectors $\mathbf{a}, \mathbf{b}$, let $\mathbf{a} = \mathbf{d} + \mathbf{c}$ where $\mathbf{d}$ is parallel to $\mathbf{b}$ and $\mathbf{c}$ is orthogonal to $\mathbf{b}$. Then

$$\mathbf{a} \cdot \mathbf{b} = \mathbf{d} \cdot \mathbf{b} = \epsilon |\mathbf{d}| \, |\mathbf{b}|,$$
$$(\mathbf{a} \wedge \mathbf{b}) \cdot (\mathbf{a} \wedge \mathbf{b}) = (\mathbf{c} \wedge \mathbf{b}) \cdot (\mathbf{c} \wedge \mathbf{b}) = -(\mathbf{b} \cdot \mathbf{b})(\mathbf{c} \cdot \mathbf{c}),$$

where $\epsilon = 1$ if $\mathbf{b}, \mathbf{d}$ are in the same direction, and -1 otherwise. So

- $|\mathbf{a} \cdot \mathbf{b}|$ is the multiplication of the length of vector $\mathbf{a}$ and the length of the orthogonal projection of vector $\mathbf{b}$ into the 1D subspace $\mathbf{a}$.
- $|\mathbf{a} \wedge \mathbf{b}| = |\mathbf{b}|\,|\mathbf{c}|$ is the area of the parallelogram spanned by vectors $\mathbf{a}$, $\mathbf{b}$.
- In general, the magnitude of r-blade $\mathbf{a}_1 \wedge \cdots \wedge \mathbf{a}_r$ is the volume of the rD parallelotope spanned by the vectors $\mathbf{a}$'s.
- The absolute value of the scalar product $(\mathbf{a}_1 \wedge \cdots \wedge \mathbf{a}_r) \cdot (\mathbf{b}_1 \wedge \cdots \wedge \mathbf{b}_r)$, is the multiplication of the volume of the rD parallelotope spanned by the $\mathbf{b}$'s, and the volume of the rD orthogonal projection of the rD parallelotope spanned by the $\mathbf{a}$'s into rD subspace $\mathbf{b}_1 \wedge \cdots \wedge \mathbf{b}_r$.

The following is a reformulation of (2.4.13) in the setting of Gram determinants.

Definition 5.8. Let $\mathbf{A}_r = \mathbf{a}_{i_1}, \mathbf{a}_{i_2}, \ldots, \mathbf{a}_{i_r}$ be a sequence of vectors. For $0 < s < r$, let there be a fixed partition $(\mathbf{B}_{r(1)}, \mathbf{B}_{r(2)})$ of sequence $\mathbf{B}_r = \mathbf{a}_{j_1}, \mathbf{a}_{j_2}, \ldots, \mathbf{a}_{j_r}$ of shape $(s, r-s)$. The *Laplace expansion* of the Gram determinant $(\mathbf{A}_r | \mathbf{B}_r)$ by this fixed partition of places, is

$$(\mathbf{A}_r | \mathbf{B}_r) = \sum_{(s,r-s) \vdash \mathbf{A}_r} (\mathbf{A}_{r(1)} | \mathbf{B}_{r(1)}) (\mathbf{A}_{r(2)} | \mathbf{B}_{r(2)}). \tag{5.1.10}$$

Lemma 5.9. [Laplace expansions of scalar products] For r-blades $\mathbf{A}_r$ and $\mathbf{B}_r$, and for any $1 \leq l \leq r$, let $(\mathbf{A}'_{r(1)}, \mathbf{A}'_{r(2)})$ be a fixed partition of $\mathbf{A}_r$ of shape $(r-l, l)$, and let $(\mathbf{B}'_{r(1)}, \mathbf{B}'_{r(2)})$ be a fixed partition of $\mathbf{B}_r$ of shape $(l, r-l)$. Then

$$\begin{aligned}
\mathbf{A}_r \cdot \mathbf{B}_r &= \sum_{(r-l,l) \vdash \mathbf{A}_r} (\mathbf{A}_{r(2)} \cdot \mathbf{B}'_{r(1)}) (\mathbf{A}_{r(1)} \cdot \mathbf{B}'_{r(2)}) \\
&= \sum_{(l,r-l) \vdash \mathbf{B}_r} (\mathbf{A}'_{r(2)} \cdot \mathbf{B}_{r(1)}) (\mathbf{A}'_{r(1)} \cdot \mathbf{B}_{r(2)}).
\end{aligned} \tag{5.1.11}$$

Proof. When each inner product $\mathbf{A}_i \cdot \mathbf{B}_i$ is replaced by the corresponding Gram determinant $(\mathbf{A}_i | \mathbf{B}_i)$ in letter-place notation, the vectors in $\mathbf{A}_i$ serve as the column indices (letters), and the vectors in $\mathbf{B}_i$ serve as the row indices (places). The two equalities in (5.1.11) are just the Laplace expansions of the determinant $(\mathbf{A}_r | \mathbf{B}_r)$ by l rows, and by l columns, respectively. $\qquad\square$

Proposition 5.10. [Grassmann-Plücker identity of scalar products] Let $\mathbf{A}_{r+1}$, $\mathbf{B}_{r-1}$ be two sequences of vectors of length $r+1, r-1$ respectively. Then

$$\sum_{(r,1) \vdash \mathbf{A}_{r+1}} (\wedge \mathbf{A}_{r+1(1)}) \cdot (\wedge \mathbf{A}_{r+1(2)} \mathbf{B}_{r-1}) = 0. \tag{5.1.12}$$

It can also be written as an identity of Gram determinants:

$$\sum_{(r,1) \vdash \mathbf{A}_{r+1}} (\mathbf{A}_{r+1(1)} | \mathbf{A}_{r+1(2)} \mathbf{B}_{r-1}) = 0. \tag{5.1.13}$$

Proof. The left side of (5.1.13) can be written as the following more explicit form:

$$\sum_{i=1}^{r+1}(-1)^{i+1+r}(\mathbf{a}_1\ldots\breve{\mathbf{a}}_i\ldots\mathbf{a}_{r+1}|\,\mathbf{a}_i\mathbf{b}_1\ldots\mathbf{b}_{r-1}).$$

Expanding the determinants in the sum by their first rows, we get

$$\Big(\sum_{j<i}(-1)^{i+j+r}+\sum_{j>i}(-1)^{i+j+1+r}\Big)(\mathbf{a}_j\cdot\mathbf{a}_i)\det(\mathbf{a}_k\cdot\mathbf{b}_l)_{\substack{k\neq i,j,\\ l=1..r-1}}=0.$$

$\square$

Corollary 5.11. [van der Waerden identity of scalar products] For $r>s$, let $\mathbf{A}_s,\mathbf{B}_{r+1},\mathbf{C}_{r-s-1}$ be sequences of vectors of length $s,r+1,r-s-1$ respectively. Then

$$\sum_{(r-s,s+1)\vdash\mathbf{B}_{r+1}}(\wedge\,\mathbf{A}_s\mathbf{B}_{r+1(1)})\cdot(\wedge\,\mathbf{B}_{r+1(2)}\mathbf{C}_{r-s-1})=0. \tag{5.1.14}$$

It can also be written as an identity of Gram determinants:

$$\sum_{(r-s,s+1)\vdash\mathbf{B}_{r+1}}(\mathbf{A}_s\mathbf{B}_{r+1(1)}|\,\mathbf{B}_{r+1(2)}\mathbf{C}_{r-s-1})=0. \tag{5.1.15}$$

Definition 5.12. The *Hodge interior product* in $\Lambda(\mathcal{V}^n)$, also called the *Hodge inner product*, or the *two-sided contraction*, still denoted by "$\rfloor$", is a bilinear mapping from $\Lambda(\mathcal{V}^n)\times\Lambda(\mathcal{V}^n)$ to $\Lambda(\mathcal{V}^n)$ as follows: for any vectors **a**'s and **b**'s, for any r-blade $\mathbf{A}_r$ and s-blade $\mathbf{B}_s$, where $r\leq s$, for any $\lambda\in\mathbb{K}$,

$$\mathbf{a}\,\rfloor\,(\mathbf{b}_1\wedge\cdots\wedge\mathbf{b}_s):=\sum_{i=1}^{s}(-1)^{i+1}(\mathbf{a}\cdot\mathbf{b}_i)\,\mathbf{b}_1\wedge\cdots\wedge\breve{\mathbf{b}}_i\wedge\cdots\wedge\mathbf{b}_s, \tag{5.1.16}$$

$$(\mathbf{a}_1\wedge\cdots\wedge\mathbf{a}_r)\,\rfloor\,\mathbf{B}_s:=(\mathbf{a}_2\wedge\cdots\wedge\mathbf{a}_r)\,\rfloor\,(\mathbf{a}_1\rfloor\mathbf{B}_s), \tag{5.1.17}$$

$$\mathbf{B}_s\,\rfloor\,\mathbf{A}_r:=(-1)^{\frac{(s-r)(s-r-1)}{2}}\mathbf{A}_r\,\rfloor\mathbf{B}_s, \tag{5.1.18}$$

$$\lambda\,\rfloor\,\mathbf{A}_r:=\lambda\mathbf{A}_r. \tag{5.1.19}$$

The *reverse inner product* (or *reverse interior product*) in $\Lambda(\mathcal{V}^n)$, henceforth always called the *inner product*, and still denoted by the dot symbol, is a bilinear mapping from $\Lambda(\mathcal{V}^n)\times\Lambda(\mathcal{V}^n)$ to $\Lambda(\mathcal{V}^n)$ defined by

$$(\mathbf{a}_1\wedge\mathbf{a}_2\wedge\cdots\wedge\mathbf{a}_r)\cdot\mathbf{B}_s:=(\mathbf{a}_r\wedge\mathbf{a}_{r-1}\wedge\cdots\wedge\mathbf{a}_1)\,\rfloor\,\mathbf{B}_s. \tag{5.1.20}$$

It has the following symmetry induced from (5.1.18):

$$\mathbf{A}_r\cdot\mathbf{B}_s=(-1)^{rs-\min(r,s)}\mathbf{B}_s\cdot\mathbf{A}_r. \tag{5.1.21}$$

Two multivectors are said to be *orthogonal*, if their inner product equals zero. Two blades are said to be *completely orthogonal*, if any vector of one blade is orthogonal to any vector of the other blade.

Remark: (1) In this book, when we use the scalar product of two multivectors, we always assume that they are homogeneous multivectors of the same grade. In such a setting, the scalar product always agrees with the inner product, so there is no need to introduce any different notation.

(2) The sign in (5.1.18) is introduced in order to make the relation (5.1.20) sign-free. A more natural definition is to set $\mathbf{B}_s \rfloor \mathbf{A}_r = \mathbf{A}_r \rfloor \mathbf{B}_s$, with the trade-off that (5.1.20) is changed into

$$(\mathbf{a}_1 \wedge \mathbf{a}_2 \wedge \cdots \wedge \mathbf{a}_r) \cdot \mathbf{B}_s = \begin{cases} (\mathbf{a}_r \wedge \mathbf{a}_{r-1} \wedge \cdots \wedge \mathbf{a}_1) \rfloor \mathbf{B}_s, & \text{if } r \leq s, \\ (-1)^{\frac{(r-s)(r-s-1)}{2}} (\mathbf{a}_r \wedge \mathbf{a}_{r-1} \wedge \cdots \wedge \mathbf{a}_1) \rfloor \mathbf{B}_s, & \text{if } r > s. \end{cases}$$

In comparison, (5.1.20) is more convenient in the translation between the Hodge inner product and the reverse inner product.

(3) In Clifford algebra, conventionally the inner product of a scalar and a multivector is set to be zero. We feel that it brings much more disadvantages than advantages in symbolic computation, so we have to abandon it. In the above definition, the inner product of a scalar and a multivector is the scaling of the multivector by the scalar. An immediate corollary is that the inner product when restricted to the set of scalars and pseudoscalars, is always associative and commutative.

From Corollary 5.11 and Definition 5.12, we get

Corollary 5.13. [van der Waerden identity of inner products] For any r-blade $\mathbf{A}_r$ where $r > 1$, for any $1 \leq s \leq r - 1$,

$$\sum_{(r-s,s) \vdash \mathbf{A}_r} \mathbf{A}_{r(1)} \cdot \mathbf{A}_{r(2)} = 0. \tag{5.1.22}$$

Proposition 5.14. The inner product of any two blades is a blade.

Proof. By Definition 5.12, we only need to prove that $\mathbf{a} \cdot \mathbf{B}_s$ is a blade for any vector $\mathbf{a}$ and s-blade $\mathbf{B}_s$, where $s > 2$. On the right side of (5.1.16), any two blades differ by a vector factor. By Corollary 2.79, $\mathbf{a} \cdot \mathbf{B}_s$ is a blade. $\square$

Proposition 5.15. [Laplace expansions of inner products]

(1) Let $\mathbf{A}_r, \mathbf{B}_s$ be blades of grade r, s respectively, where $r \leq s$, then

$$\mathbf{A}_r \cdot \mathbf{B}_s = \sum_{(r,s-r) \vdash \mathbf{B}_s} (\mathbf{A}_r \cdot \mathbf{B}_{s(1)}) \mathbf{B}_{s(2)}, \tag{5.1.23}$$

$$\mathbf{A}_r \rfloor \mathbf{B}_s = \sum_{(r,s-r) \vdash \mathbf{B}_s} (\mathbf{A}_r \rfloor \mathbf{B}_{s(1)}) \mathbf{B}_{s(2)}. \tag{5.1.24}$$

(2) Let $\mathbf{A}_r, \mathbf{B}_s, \mathbf{C}_t$ be blades of grade r, s, t respectively, where $r + s \leq t$, then

$$(\mathbf{A}_r \wedge \mathbf{B}_s) \cdot \mathbf{C}_t = \mathbf{A}_r \cdot (\mathbf{B}_s \cdot \mathbf{C}_t), \tag{5.1.25}$$

$$(\mathbf{A}_r \cdot \mathbf{C}_t) \cdot \mathbf{B}_s = \mathbf{A}_r \cdot (\mathbf{C}_t \cdot \mathbf{B}_s), \tag{5.1.26}$$

$$(\mathbf{A}_r \wedge \mathbf{B}_s) \rfloor \mathbf{C}_t = \mathbf{B}_s \rfloor (\mathbf{A}_r \rfloor \mathbf{C}_t). \tag{5.1.27}$$

Proof. (1) Induction on r. When $r = 1$, (5.1.23) is just (5.1.16). Assume that (5.1.23) holds for $r - 1 < s$. Let $\mathbf{A}_r = \mathbf{A}_{r-1} \wedge \mathbf{a}_r$, then by induction hypothesis and (5.1.11),

$$\mathbf{A}_r \cdot \mathbf{B}_s = \mathbf{A}_{r-1} \cdot (\mathbf{a}_r \cdot \mathbf{B}_s)$$

$$= \mathbf{A}_{r-1} \cdot \sum_{(1,s-1)\vdash \mathbf{B}_s} \left((\mathbf{a}_r \cdot \mathbf{B}_{s(1)}) \, \mathbf{B}_{s(2)} \right)$$

$$= \sum_{(1,r-1,s-r)\vdash \mathbf{B}_s} (\mathbf{a}_r \cdot \mathbf{B}_{s(1)}) \, (\mathbf{A}_{r-1} \cdot \mathbf{B}_{s(2)}) \, \mathbf{B}_{s(3)}$$

$$= \sum_{(r,s-r)\vdash \mathbf{B}_s} (\mathbf{A}_r \cdot \mathbf{B}_{s(1)}) \, \mathbf{B}_{s(2)}.$$

(5.1.24) is another form of (5.1.23).

(2) By (5.1.17), if $r \le s$, then

$$(\mathbf{a}_1 \wedge \cdots \wedge \mathbf{a}_r) \cdot \mathbf{B}_s = (\mathbf{a}_1 \wedge \cdots \wedge \mathbf{a}_{r-1}) \cdot (\mathbf{a}_r \cdot \mathbf{B}_s). \tag{5.1.28}$$

Continuing the transfer of $\mathbf{a}_i$ to $\mathbf{B}_s$ in this way, we get, for any $1 \le k \le r$,

$$(\mathbf{a}_1 \wedge \cdots \wedge \mathbf{a}_k \wedge \cdots \wedge \mathbf{a}_r) \cdot \mathbf{B}_s = (\mathbf{a}_1 \wedge \cdots \wedge \mathbf{a}_k) \cdot ((\mathbf{a}_{k+1} \wedge \cdots \wedge \mathbf{a}_r) \cdot \mathbf{B}_s), \tag{5.1.29}$$

which is exactly (5.1.25) if we replace $\mathbf{a}_1 \wedge \cdots \wedge \mathbf{a}_k$ by $\mathbf{A}_r$, replace $\mathbf{a}_{k+1} \wedge \cdots \wedge \mathbf{a}_r$ by $\mathbf{B}_s$, and replace $\mathbf{B}_s$ in (5.1.29) by $\mathbf{C}_t$. (5.1.26) and (5.1.27) are rewritings of (5.1.25). $\qquad\square$

Remark: If $r + s = t$, by substituting (5.1.23) into (5.1.25), we get (5.1.11). This justifies that the above proposition is the extension of Laplace expansions from scalar products to interior products.

Corollary 5.16. Let $\mathbf{A}_r, \mathbf{B}_s$ be r-blade and s-blade respectively. Then for any $0 \le i \le \min(r, s)$,

$$\mathbf{A}_r \cdot \mathbf{B}_s = \frac{1}{C^i_{\min(r,s)}} \sum_{\substack{(r-i,i)\vdash \mathbf{A}_r, \\ (i,s-i)\vdash \mathbf{B}_s}} (\mathbf{A}_{r(2)} \cdot \mathbf{B}_{s(1)}) \, (\mathbf{A}_{r(1)} \cdot \mathbf{B}_{s(2)}). \tag{5.1.30}$$

Example 5.17. Let $\mathcal{V}^n$ be a nondegenerate inner-product space. For any vectors $\mathbf{u}_1, \mathbf{u}_2$ and bivector $\mathbf{A}_2$, let

$$\mathbf{B}_2 = (\mathbf{A}_2 \cdot (\mathbf{u}_1 \wedge \mathbf{u}_2)) \, \mathbf{A}_2 + (\mathbf{A}_2 \cdot \mathbf{u}_1) \wedge (\mathbf{A}_2 \cdot \mathbf{u}_2).$$

Prove that the two equations $(\mathbf{A}_2 \cdot \mathbf{u}_i) \cdot \mathbf{B}_2 = 0$ for $i = 1, 2$ are equivalent to the following n equations in bivector variable $\mathbf{u}_1 \wedge \mathbf{u}_2$:

$$(\mathbf{u}_1 \wedge \mathbf{u}_2) \cdot (\mathbf{A}_2 \cdot (\mathbf{A}_2 \cdot (\mathbf{A}_2 \cdot (\mathbf{u}_1 \wedge \mathbf{u}_2 \wedge \mathbf{v})))) = 0, \ \forall \text{ basis vector } \mathbf{v} \in \mathcal{V}^n. \tag{5.1.31}$$

Proof. First, the vectorial equation $(\mathbf{A}_2 \cdot \mathbf{u}_i) \cdot \mathbf{B}_2 = 0$ is equivalent to a set of n scalar equations of the form

$$((\mathbf{A}_2 \cdot \mathbf{u}_i) \cdot \mathbf{B}_2) \cdot \mathbf{v} = 0, \tag{5.1.32}$$

where $\mathbf{v} \in \mathcal{V}^n$ is any basis vector. We have

$$
\begin{aligned}
((\mathbf{A}_2 \cdot \mathbf{u}_1) \cdot \mathbf{B}_2) \cdot \mathbf{v} &= -\mathbf{B}_2 \cdot ((\mathbf{A}_2 \cdot \mathbf{u}_1) \wedge \mathbf{v}) \\
&= -(\mathbf{A}_2 \cdot (\mathbf{u}_1 \wedge \mathbf{u}_2))\, \mathbf{A}_2 \cdot ((\mathbf{A}_2 \cdot \mathbf{u}_1) \wedge \mathbf{v}) \\
&\quad -((\mathbf{A}_2 \cdot \mathbf{u}_1) \cdot \mathbf{v})\,(\mathbf{A}_2 \cdot \mathbf{u}_1) \cdot (\mathbf{A}_2 \cdot \mathbf{u}_2) \\
&\quad +(\mathbf{A}_2 \cdot \mathbf{u}_1)^2\,(\mathbf{A}_2 \cdot \mathbf{u}_2) \cdot \mathbf{v} \\
&= \ \ (\mathbf{A}_2 \cdot (\mathbf{u}_1 \wedge \mathbf{u}_2))\,(\mathbf{A}_2 \cdot \mathbf{v}) \cdot (\mathbf{A}_2 \cdot \mathbf{u}_1) \\
&\quad -(\mathbf{A}_2 \cdot (\mathbf{u}_1 \wedge \mathbf{v}))\,(\mathbf{A}_2 \cdot \mathbf{u}_1) \cdot (\mathbf{A}_2 \cdot \mathbf{u}_2) \\
&\quad +(\mathbf{A}_2 \cdot \mathbf{u}_1)^2\,\mathbf{A}_2 \cdot (\mathbf{u}_2 \wedge \mathbf{v}) \\
&= \ \ (\mathbf{A}_2 \cdot \mathbf{u}_1) \cdot (\mathbf{A}_2 \cdot (\mathbf{A}_2 \cdot (\mathbf{u}_1 \wedge \mathbf{u}_2 \wedge \mathbf{v}))) \\
&= -\mathbf{u}_1 \cdot (\mathbf{A}_2 \cdot (\mathbf{A}_2 \cdot (\mathbf{A}_2 \cdot (\mathbf{u}_1 \wedge \mathbf{u}_2 \wedge \mathbf{v})))).
\end{aligned}
\tag{5.1.33}
$$

If $\mathbf{u}_1 \wedge \mathbf{u}_2 = 0$, then $\mathbf{B}_2 = 0$, and the input equations are trivial. Assume that $\mathbf{u}_1 \wedge \mathbf{u}_2 \neq 0$. By writing $((\mathbf{A}_2 \cdot \mathbf{u}_i) \cdot \mathbf{B}_2) \cdot \mathbf{v} = 0$ as $\mathbf{u}_{3-i}(((\mathbf{A}_2 \cdot \mathbf{u}_i) \cdot \mathbf{B}_2) \cdot \mathbf{v}) = 0$ for $i = 1, 2$, we get, by (5.1.33),

$$
\begin{aligned}
&\mathbf{u}_2(((\mathbf{A}_2 \cdot \mathbf{u}_1) \cdot \mathbf{B}_2) \cdot \mathbf{v}) - \mathbf{u}_1(((\mathbf{A}_2 \cdot \mathbf{u}_2) \cdot \mathbf{B}_2) \cdot \mathbf{v}) \\
&= (\mathbf{u}_1 \wedge \mathbf{u}_2) \cdot (\mathbf{A}_2 \cdot (\mathbf{A}_2 \cdot (\mathbf{A}_2 \cdot (\mathbf{u}_1 \wedge \mathbf{u}_2 \wedge \mathbf{v})))).
\end{aligned}
$$

$$\square$$

5.1.2 *Inner-product Grassmann algebra*

The interior product enables us to define a basis-free dual operator, in contrast to the basis-dependent dual operator "$*$" in Definition 2.23. Assume that $\mathcal{V}^n$ is a nondegenerate inner-product space, and $\mathbf{I}_n$ is a unit pseudoscalar in $\Lambda(\mathcal{V}^n)$.

Definition 5.18. The *reverse dual* operator in $\Lambda(\mathcal{V}^n)$, henceforth called the *dual operator* with respect to $\mathbf{I}_n$, denoted by "$\sim$", is a linear operator defined by

$$\mathbf{A}^\sim := \mathbf{A} \cdot \mathbf{I}_n^{-1}, \ \forall \mathbf{A} \in \Lambda(\mathcal{V}^n). \tag{5.1.34}$$

The *Hodge dual operator*, still denoted by "$*$", is a linear operator in $\Lambda(\mathcal{V}^n)$ defined for any r-blade $\mathbf{A}_r$ by

$$((*\mathbf{A}_r) \rfloor \mathbf{B}_{n-r})\,\mathbf{I}_n := \mathbf{A}_r \wedge \mathbf{B}_{n-r}, \ \forall \mathbf{B}_{n-r} \in \Lambda^{n-r}(\mathcal{V}^n). \tag{5.1.35}$$

The Hodge inner product and the reverse inner product each generate a dual operator. Their difference is only the sign of the result. The Hodge formalism is convenient in geometric interpretation, but inconvenient in algebraic manipulation. It is used primarily in differential geometry. The reverse formalism has the opposite property. Because of its algebraic advantage, the reverse formalism is adopted in this book as the default one.

Example 5.19. Let $\mathbf{E}_n = \mathbf{e}_1, \mathbf{e}_2, \ldots, \mathbf{e}_n$ be a fixed orthonormal basis of $\mathbb{R}^n$, such that $\mathbf{I}_n = \mathbf{e}_1 \wedge \mathbf{e}_2 \wedge \cdots \wedge \mathbf{e}_n$. Then the reciprocal basis is $\mathbf{e}_i^* = \mathbf{e}_i$ for $1 \leq i \leq n$. For any fixed bipartition $(\mathbf{E}_{n(1)}, \mathbf{E}_{n(2)})$ of $\mathbf{E}_n$, it is easy to verify that

$$*(\wedge \mathbf{E}_{n(1)}) = \wedge \mathbf{E}_{n(2)}. \tag{5.1.36}$$

(5.1.36) congrues with (2.3.21), showing that when $\mathcal{V}^n = \mathbb{R}^n$, the Hodge dual operator defined by (5.1.35) agrees with its earlier version in Definition 2.23. When the basis undergoes a special orthogonal transformation, (5.1.36) is unchanged. From this aspect, we can say that in $\Lambda(\mathbb{R}^n)$, the basis-dependent Hodge dual operator given in Definition 2.23 is independent of the selected orthonormal basis of $\mathbb{R}^n$.

Definition 5.20. The *inverse dual operator* in $\Lambda(\mathcal{V}^n)$, denoted by "$-\sim$", is a linear operator defined by

$$\mathbf{A}^{-\sim} := \mathbf{A} \cdot \mathbf{I}_n, \ \forall \mathbf{A} \in \Lambda(\mathcal{V}^n). \tag{5.1.37}$$

Lemma 5.21. For any integers n, r,

$$\begin{aligned}
(-1)^{\frac{r(r+1)}{2} + \frac{n(n+1)}{2} + \frac{(n-r)(n-r+1)}{2} + r(n-r)} &= 1, \\
(-1)^{\frac{r(r-1)}{2} + \frac{n(n-1)}{2} + \frac{(n-r)(n-r-1)}{2} + r(n-r)} &= 1.
\end{aligned} \tag{5.1.38}$$

Proposition 5.22. For any $\mathbf{A} \in \Lambda(\mathcal{V}^n)$,

$$(\mathbf{A}^\sim)^{-\sim} = (\mathbf{A}^{-\sim})^\sim = \mathbf{A}. \tag{5.1.39}$$

Proof. Let $\mathbf{e}_1, \mathbf{e}_2, \ldots, \mathbf{e}_n$ be an orthonormal basis of $\mathcal{V}^n$. Let $\mathbf{e}_i \cdot \mathbf{e}_i = \epsilon_i$, where $\epsilon_i^2 = 1$. By linearity, we only need to consider the case where $\mathbf{A} = \mathbf{e}_1 \wedge \mathbf{e}_2 \wedge \cdots \wedge \mathbf{e}_r$ and $1 \leq r \leq n - 1$. Since $\mathbf{I}_n = \mathbf{e}_1 \wedge \mathbf{e}_2 \wedge \cdots \wedge \mathbf{e}_n$,

$$\mathbf{I}_n^{-1} = (-1)^{\frac{n(n-1)}{2}} \epsilon_1 \epsilon_2 \cdots \epsilon_n \mathbf{I}_n. \tag{5.1.40}$$

By Lemma 5.21,

$$\begin{aligned}
(\mathbf{e}_1 \wedge \mathbf{e}_2 \wedge \cdots \wedge \mathbf{e}_r)^{-\sim} &= (-1)^{\frac{r(r-1)}{2}} \epsilon_1 \epsilon_2 \cdots \epsilon_r \, \mathbf{e}_{r+1} \wedge \mathbf{e}_{r+2} \wedge \cdots \wedge \mathbf{e}_n, \\
(\mathbf{e}_{r+1} \wedge \mathbf{e}_{r+2} \wedge \cdots \wedge \mathbf{e}_n)^\sim &= (-1)^{\frac{n(n-1)}{2} + \frac{(n-r)(n-r-1)}{2} + r(n-r)} \epsilon_1 \epsilon_2 \cdots \epsilon_r \epsilon_{r+1}^2 \epsilon_{r+2}^2 \cdots \epsilon_n^2 \\
&\qquad\qquad\qquad\qquad\qquad\qquad\qquad\qquad\qquad \mathbf{e}_1 \wedge \mathbf{e}_2 \wedge \cdots \wedge \mathbf{e}_r \\
&= (-1)^{\frac{r(r-1)}{2}} \epsilon_1 \epsilon_2 \cdots \epsilon_r \, \mathbf{e}_1 \wedge \mathbf{e}_2 \wedge \cdots \wedge \mathbf{e}_r.
\end{aligned}$$

So $(\mathbf{A}^{-\sim})^\sim = \mathbf{A}$. The other equality can be proved similarly. $\qquad\square$

In $\Lambda(\mathbb{R}^{p,q})$, any two nonzero pseudoscalars differ by a nonzero scale, and the sign of the scale divides the set of pseudoscalars into two equivalent classes. Pseudoscalars differing by positive scales represent the same *orientation* of $\mathbb{R}^{p,q}$. Let $\mathbf{I}_n$ be a unit pseudoscalar in $\Lambda(\mathbb{R}^{p,q})$. Then $\mathbb{R}^{p,q}$ together with orientation $\mathbf{I}_n$ is called an *oriented real vector space*.

In fact, the oriented space can be compactly represented by a single algebraic element, the pseudoscalar $\mathbf{I}_n$, for both the vector space and its orientation. There are two unit pseudoscalars in $\Lambda(\mathbb{R}^{p,q})$. They differ by sign, and represent the two opposite orientations of $\mathbb{R}^{p,q}$. An orthogonal transformation keeping the orientation invariant is called a *special orthogonal transformation*. The special orthogonal group of $\mathbb{R}^{p,q}$ is denoted by $SO(p,q)$, while the orthogonal group of $\mathbb{R}^{p,q}$ is denoted by $O(p,q)$.

Definition 5.23. In the Grassmann algebra over a nondegenerate inner-product space $\mathcal{V}^n$ with a fixed unit pseudoscalar $\mathbf{I}_n$, the *bracket* of a pseudoscalar $\mathbf{A}_n$ is defined by

$$[\mathbf{A}_n] := \mathbf{A}_n^{\sim} = \mathbf{A}_n \cdot \mathbf{I}_n^{-1}. \tag{5.1.41}$$

Notation. Let $\mathbf{I}_n$ be the unit pseudoscalar defining the bracket by (5.1.41). Set

$$\iota := \mathbf{I}_n \rfloor \mathbf{I}_n. \tag{5.1.42}$$

For example, in (5.1.40),

$$\iota = \epsilon_1 \epsilon_2 \cdots \epsilon_n. \tag{5.1.43}$$

If $\mathcal{V}^n = \mathbb{C}^n$, then $\iota = 1$; if $\mathcal{V}^n = \mathbb{R}^{p,q}$, then $\iota = (-1)^q$.

The above definition of a bracket by the reverse dual operator congrues with the classical definition by the determinant of the homogeneous coordinates, in both the real case and the complex case. The Hodge dual operator, however, defines a bracket different from the classical one in the real case $\mathbb{R}^{p,q}$, by sign $(-1)^q$.

A *linear involution*, or simply called *involution*, refers to an invertible linear transformation having the property that its composition with itself is the identity transformation. The reverse dual has the advantages that it is a *grade-independent* involution, is clean both in representing the duality between the inner product and the outer product, and in defining the meet product with the inner product, as to be seen in the following proposition.

Proposition 5.24. In $\Lambda(\mathcal{V}^n)$, where $\mathcal{V}^n$ is a nondegenerate inner-product space, the inner product and the outer product are dual to each other, in the sense that for any r-blade $\mathbf{A}_r$ and s-blade $\mathbf{B}_s$,

$$(\mathbf{A}_r \cdot \mathbf{B}_s)^{\sim} = \mathbf{A}_r \wedge \mathbf{B}_s^{\sim}, \text{ if } r \leq s; \tag{5.1.44}$$

$$(\mathbf{A}_r \wedge \mathbf{B}_s)^{\sim} = \mathbf{A}_r \cdot \mathbf{B}_s^{\sim}, \text{ if } r + s \leq n. \tag{5.1.45}$$

Furthermore,

$$(\mathbf{A}_r^{\sim})^{\sim} = (-1)^{\frac{n(n-1)}{2}} \iota \, \mathbf{A}_r, \tag{5.1.46}$$

$$\mathbf{A}_r^{\sim} \cdot \mathbf{B}_s^{\sim} = (-1)^{\frac{n(n-1)}{2} + s(n-1)} \iota \, \mathbf{A}_r \cdot \mathbf{B}_s. \tag{5.1.47}$$

Proof. (5.1.46) is direct from (5.1.40) and (5.1.43). (5.1.45) is direct from (5.1.25). Setting $\mathbf{C}_{n-s} = \mathbf{B}_s^{\sim}$ in (5.1.45), we get $(\mathbf{A}_r \cdot \mathbf{C}_{n-s})^{-\sim} = \mathbf{A}_r \wedge \mathbf{C}_{n-s}^{-\sim}$. By (5.1.46), we get (5.1.44).

If $r \geq s$, then $n - r \leq n - s$. By (5.1.45), (5.1.44) and (5.1.46),

$$
\begin{aligned}
\mathbf{A}_r^\sim \cdot \mathbf{B}_s^\sim &= (\mathbf{A}_r^\sim \wedge \mathbf{B}_s)^\sim \\
&= (-1)^{s(n-r)} (\mathbf{B}_s \wedge \mathbf{A}_r^\sim)^\sim \\
&= (-1)^{s(n-r)} ((\mathbf{B}_s \cdot \mathbf{A}_r)^\sim)^\sim \\
&= (-1)^{s(n-r)+\frac{n(n-1)}{2}} \iota\, \mathbf{B}_s \cdot \mathbf{A}_r \\
&= (-1)^{\frac{n(n-1)}{2}+s(n-1)} \iota\, \mathbf{A}_r \cdot \mathbf{B}_s.
\end{aligned}
$$

If $r < s$, by (5.1.21), we still get (5.1.47). $\qquad\square$

Lemma 5.25. For any r-blade $\mathbf{A}_r$,

$$
*\mathbf{A}_r = (-1)^{\frac{n(n-1)}{2}+\frac{r(r-1)}{2}} \mathbf{A}_r^\sim = \iota\, \mathbf{A}_r \rfloor \mathbf{I}_n. \tag{5.1.48}
$$

Proof. For any $(n-r)$-blade $\mathbf{B}_{n-r}$, by (5.1.35), (5.1.38), (5.1.40) and (5.1.43),

$$
\begin{aligned}
(*\mathbf{A}_r) \rfloor \mathbf{B}_{n-r} &= (\mathbf{A}_r \wedge \mathbf{B}_{n-r})^\sim \\
&= (-1)^{r(n-r)} (\mathbf{B}_{n-r} \wedge \mathbf{A}_r)^\sim \\
&= (-1)^{r(n-r)} \mathbf{B}_{n-r} \cdot \mathbf{A}_r^\sim \\
&= (-1)^{r(n-r)+\frac{(n-r)(n-r-1)}{2}} \mathbf{A}_r^\sim \rfloor \mathbf{B}_{n-r} \\
&= (-1)^{\frac{n(n-1)}{2}+\frac{r(r-1)}{2}} (\mathbf{A}_r \cdot \mathbf{I}_n^{-1}) \rfloor \mathbf{B}_{n-r} \\
&= \iota\, (\mathbf{A}_r \rfloor \mathbf{I}_n) \rfloor \mathbf{B}_{n-r}.
\end{aligned}
$$

$\qquad\square$

The following proposition is a rewriting of Proposition 5.24 from the reverse formalism to the Hodge formalism. In comparison, the signs in the formulas are more complicated than those in Proposition 5.24.

Proposition 5.26. For any r-blade $\mathbf{A}_r$ and s-blade $\mathbf{B}_s$,

$$
\begin{aligned}
(\mathbf{A}_r \rfloor \mathbf{B}_s) &= (-1)^{r(s-r)} \mathbf{A}_r \wedge (\mathbf{B}_s), && \text{if } r \leq s; \\
(\mathbf{A}_r \wedge \mathbf{B}_s) &= (-1)^{\frac{n(n-1)}{2}+rs} \mathbf{A}_r \rfloor (\mathbf{B}_s), && \text{if } r+s \leq n; \\
(\mathbf{A}_r) &= (-1)^{\frac{n(n-1)}{2}+r(n-r)} \iota\, \mathbf{A}_r; \\
(*\mathbf{A}_r) \rfloor (*\mathbf{B}_s) &= (-1)^{\frac{r(r+1)}{2}+\frac{s(s+1)}{2}+n(r+s)} \iota\, \mathbf{A}_r \rfloor \mathbf{B}_s.
\end{aligned} \tag{5.1.49}
$$

We have the following clean result:

Proposition 5.27. If $r + s \geq n$, then

$$
(\mathbf{B}_s^\sim \cdot \mathbf{A}_r) = (\mathbf{A}_r) \wedge (*\mathbf{B}_s). \tag{5.1.50}
$$

Proof. Replacing $\mathbf{A}_r$ in the first formula of (5.1.49) by $*\mathbf{A}_r$, we get

$$
((\mathbf{A}_r) \rfloor \mathbf{B}_s) = (-1)^{(n-r)(s-n+r)} (*\mathbf{A}_r) \wedge (*\mathbf{B}_s) = (-1)^{r(n-r)} (*\mathbf{B}_s) \wedge (*\mathbf{A}_r).
$$

Alternatively, by (5.1.48) and Lemma 5.21,

$$(*\mathbf{A}_r)\rfloor\mathbf{B}_s = (-1)^{\frac{n(n-1)+r(r-1)+(n-r)(n-r-1)}{2}}\mathbf{A}_r^{\sim}\cdot\mathbf{B}_s = (-1)^{r(n-r)}\mathbf{A}_r^{\sim}\cdot\mathbf{B}_s.$$

$\square$

In Definition 2.25, the meet product is defined by $*(\mathbf{A}_r\vee\mathbf{B}_s) = (*\mathbf{A}_r)\wedge(*\mathbf{B}_s)$. Comparing this with (5.1.50), we get the following beautiful dualities:

Proposition 5.28. For r-blade $\mathbf{A}_r$ and s-blade $\mathbf{B}_s$, where $r+s\geq n$,

$$\begin{aligned}
\mathbf{A}_r\vee\mathbf{B}_s &= \mathbf{B}_s^{\sim}\cdot\mathbf{A}_r,\\
(\mathbf{A}_r\vee\mathbf{B}_s)^{\sim} &= \mathbf{B}_s^{\sim}\wedge\mathbf{A}_r^{\sim}.
\end{aligned} \tag{5.1.51}$$

In fact, (5.1.51) can be obtained directly from the Laplace expansion (5.1.23) of the inner product and the shuffle formula (2.3.32) of the meet product. For example,

$$\mathbf{B}_s^{\sim}\cdot\mathbf{A}_r = \sum_{(n-s,r+s-n)\vdash\mathbf{A}_r}(\mathbf{B}_s^{\sim}\cdot\mathbf{A}_{r(1)})\mathbf{A}_{r(2)} = \sum_{(n-s,r+s-n)\vdash\mathbf{A}_r}[\mathbf{A}_{r(1)}\mathbf{B}_s]\mathbf{A}_{r(2)}. \tag{5.1.52}$$

In [77], the (reverse) dual is defined by $\mathbf{A}_r\vee\mathbf{B}_s = \mathbf{A}_r^{\sim}\cdot\mathbf{B}_s$. It is incompatible with the shuffle formula of the meet product, so it has to be revised to (5.1.51).

Definition 5.29. The Grassmann space over a nondegenerate inner product space $\mathcal{V}^n$, when equipped with the outer product, the inner product, and the dual operator with respect to a fixed unit pseudoscalar, is called the *inner-product Grassmann algebra* over $\mathcal{V}^n$, still denoted by $\Lambda(\mathcal{V}^n)$.

Since the meet product can be defined by the inner product and the dual operator, the inner-product Grassmann algebra generates a Grassmann-Cayley algebra. It enables us to do all projective geometric computing by inner products, outer products and dual operators.

For a vector space $\mathcal{V}^n$ without any inner product structure, by fixing a nonzero pseudoscalar in $\Lambda(\mathcal{V}^n)$, we can define the bracket operator and the meet product. They are basis-independent operators, and are meaningful in projective geometry. By further fixing a basis of $\mathcal{V}^n$, we can define the dual operator with respect to the basis, and define an inner product structure in $\mathcal{V}^n$ as follows:

$$\mathbf{a}\cdot\mathbf{b} := [\mathbf{a}\wedge(*\mathbf{b})] = \mathbf{a}\vee(*\mathbf{b}). \tag{5.1.53}$$

This inner product has Euclidean signature, and changes $\mathcal{V}^n$ into $\mathbb{R}^n$. Thus, by fixing a basis of $\mathcal{V}^n$, we can generate a Euclidean inner-product Grassmann algebra from the GC algebra over $\mathcal{V}^n$.

For a nondegenerate inner-product space $\mathcal{V}^n$, by fixing a nonzero pseudoscalar in $\Lambda(\mathcal{V}^n)$, we can define a GC algebra over $\mathcal{V}^n$. This algebra does not include the inner product. Introducing the dual operator based on an orthonormal basis of $\mathcal{V}^n$, is equivalent to introducing the inner product into the GC algebra and changing it into an inner-product Grassmann algebra.

So no matter whether or not $\mathcal{V}^n$ is equipped with an inner-product structure,

$$\text{GC algebra} + \text{dual operator} = \text{inner-product Grassmann algebra.} \qquad (5.1.54)$$

Since the dual operator, or equivalently, the inner product, is invariant only by a change of basis in the orthogonal group of $\mathcal{V}^n$, for invariant computing in projective geometry, inner-product Grassmann algebra is not the correct language.

Below we investigate the geometric meaning of the dual operator and the inner product in inner-product Grassmann algebra.

Proposition 5.30. Let $\mathbf{A}_r$ be an r-blade in $\Lambda(\mathcal{V}^n)$. Then $\mathbf{A}_r^{\sim}$ represents the orthogonal complement of the rD subspace $\mathbf{A}_r$ in $\mathcal{V}^n$. If $\mathcal{V}^n = \mathbb{R}^n$, then the orientation of $\mathbf{A}_r$ succeeded by the orientation of $*\mathbf{A}_r$ is the orientation of $\mathcal{V}^n$.

Proof. For any vector $\mathbf{a} \in \mathbf{A}_r$, $\mathbf{a} \cdot \mathbf{A}_r^{\sim} = (\mathbf{a} \wedge \mathbf{A}_r)^{\sim} = 0$. If $\mathcal{V}^n = \mathbb{R}^n$, then $\mathbf{A}_r \rfloor \mathbf{A}_r > 0$. By the first equality in (5.1.49), $\mathbf{A}_r \wedge (*\mathbf{A}_r) = *(\mathbf{A}_r \rfloor \mathbf{A}_r) = (\mathbf{A}_r \rfloor \mathbf{A}_r)\mathbf{I}_n$ represents the same orientation of $\mathcal{V}^n$ as $\mathbf{I}_n$. $\qquad\square$

For example, in the Euclidean plane $\mathbb{R}^2$, the Hodge dual of any vector $\mathbf{a}$ is obtained by rotating $\mathbf{a}$ $90°$ in the orientation of the plane. In the Euclidean space $\mathbb{R}^3$, the Hodge dual of any bivector $\mathbf{a} \wedge \mathbf{b}$ is, by the right hand rule, a normal vector of the oriented plane $\mathbf{a} \wedge \mathbf{b}$.

In $\Lambda(\mathcal{V}^n)$, let $\mathbf{A}_r, \mathbf{B}_s$ be r-blade and s-blade respectively, where $r \leq s$. Then $\mathbf{A}_r \cdot \mathbf{B}_s$ is an $(s - r)$-blade in $\Lambda(\mathbf{B}_s)$. When $\mathcal{V}^n = \mathbb{R}^n$, let $\mathbf{A}_r = \mathbf{a}_1 \wedge \cdots \wedge \mathbf{a}_r$, and let $\mathbf{a}_i = \mathbf{d}_i + \mathbf{c}_i$, where $\mathbf{d}_i \in \mathbf{B}_s$, $\mathbf{c}_i \perp \mathbf{B}_s$. Then

$$(\mathbf{a}_1 \wedge \cdots \wedge \mathbf{a}_r) \cdot \mathbf{B}_s = (\mathbf{d}_1 \wedge \cdots \wedge \mathbf{d}_r) \cdot \mathbf{B}_s. \qquad (5.1.55)$$

Clearly, $\mathbf{d}_1 \wedge \cdots \wedge \mathbf{d}_r$ is the rD orthogonal projection of $\mathbf{A}_r$ into subspace $\mathbf{B}_s$. By this and Proposition 5.30, we get

Proposition 5.31. In $\Lambda(\mathbb{R}^n)$, for r-blade $\mathbf{A}_r$ and s-blade $\mathbf{B}_s$ where $r < s$, blade $\mathbf{A}_r \cdot \mathbf{B}_s$ represents an $(s - r)$D subspace in sD space $\mathbf{B}_s$, which is the orthogonal complement of the rD orthogonal projection of space $\mathbf{A}_r$ into space $\mathbf{B}_s$. Its magnitude equals the multiplication of the volume of the sD parallelotope $\mathbf{B}_s$ with the volume of the rD orthogonal projection of the rD parallelotope $\mathbf{A}_r$ into space $\mathbf{B}_s$.

Let $\mathbf{B}_s$ be an s-blade having nonzero magnitude. Then $\mathbf{B}_s^{-1} = \mathbf{B}_s/(\mathbf{B}_s \cdot \mathbf{B}_s)$ exists. Again by Proposition 5.30 and (5.1.55), when $r \leq s$, the dual of $\mathbf{A}_r \cdot \mathbf{B}_s$ in $\Lambda(\mathbf{B}_s)$ is the orthogonal projection of $\mathbf{A}_r$ into $\mathbf{B}_s$:

$$P_{\mathbf{B}_s}(\mathbf{A}_r) := (\mathbf{A}_r \cdot \mathbf{B}_s) \cdot \mathbf{B}_s^{-1}. \qquad (5.1.56)$$

When $r > s$, define $P_{\mathbf{B}_s}(\mathbf{A}_r) = 0$. The linear operator $P_{\mathbf{B}_s}$ does not depend on the scale of $\mathbf{B}_s$, so it is an operator determined by vector space $\mathbf{B}_s$, called the *orthogonal projection operator* into $\mathbf{B}_s$. Its orthogonal complement

$$P_{\mathbf{B}_s}^{\perp}(\mathbf{A}) := \mathbf{A} - P_{\mathbf{B}_s}(\mathbf{A}), \quad \forall \mathbf{A} \in \Lambda(\mathcal{V}^n), \qquad (5.1.57)$$

is called the *orthogonal rejection operator* from $\mathbf{B}_s$.

Proposition 5.32. Let $\mathbf{a}$ be a vector and $\mathbf{B}_s$ be an s-blade, where $0 < s < n$. Then

$$P^{\perp}_{\mathbf{B}_s}(\mathbf{a}) = (\mathbf{a} \wedge \mathbf{B}_s) \cdot \mathbf{B}_s^{-1}. \tag{5.1.58}$$

Proof. By (5.1.45) and (5.1.47),

$$P^{\perp}_{\mathbf{B}_s}(\mathbf{a}) = P_{\mathbf{B}_s^{\sim}}(\mathbf{a}) = (\mathbf{a} \cdot \mathbf{B}_s^{\sim}) \cdot (\mathbf{B}_s^{\sim})^{-1} = \frac{(\mathbf{a} \wedge \mathbf{B}_s)^{\sim} \cdot \mathbf{B}_s^{\sim}}{\mathbf{B}_s^{\sim} \cdot \mathbf{B}_s^{\sim}} = \frac{(\mathbf{a} \wedge \mathbf{B}_s) \cdot \mathbf{B}_s}{\mathbf{B}_s \cdot \mathbf{B}_s}.$$

$\square$

As a corollary, the orthogonal decomposition of any vector $\mathbf{a}$ with respect to subspace $\mathbf{B}_s$, where $s > 0$, is

$$\mathbf{a} = P_{\mathbf{B}_s}(\mathbf{a}) + P^{\perp}_{\mathbf{B}_s}(\mathbf{a}) = (\mathbf{a} \cdot \mathbf{B}_s + \mathbf{a} \wedge \mathbf{B}_s) \cdot \mathbf{B}_s^{-1}. \tag{5.1.59}$$

Proposition 5.33. $P_{\mathbf{B}_s}$ is a homomorphism of Grassmann algebras from $\Lambda(\mathcal{V}^n)$ to $\Lambda(\mathbf{B}_s)$.

Proof. For two vectors $\mathbf{a}$ and $\mathbf{b}$, by (5.1.39) and the fact that $P_{\mathbf{B}_s}(\mathbf{a})$ and $P_{\mathbf{B}_s}(\mathbf{b})$ are respectively the orthogonal projections of $\mathbf{a}$ and $\mathbf{b}$ into subspace $\mathbf{B}_s$, we get

$$\begin{aligned} P_{\mathbf{B}_s}(\mathbf{a} \wedge \mathbf{b}) &= ((\mathbf{a} \wedge \mathbf{b}) \cdot \mathbf{B}_s) \cdot \mathbf{B}_s^{-1} \\ &= ((P_{\mathbf{B}_s}(\mathbf{a}) \wedge P_{\mathbf{B}_s}(\mathbf{b})) \cdot \mathbf{B}_s) \cdot \mathbf{B}_s^{-1} \\ &= P_{\mathbf{B}_s}(\mathbf{a}) \wedge P_{\mathbf{B}_s}(\mathbf{b}). \end{aligned}$$

The proof of the general case $P_{\mathbf{B}_s}(\mathbf{a}_1 \wedge \cdots \wedge \mathbf{a}_r) = P_{\mathbf{B}_s}(\mathbf{a}_1) \wedge \cdots \wedge P_{\mathbf{B}_s}(\mathbf{a}_r)$ is similar.

$\square$

5.1.3 *Algebras of basic invariants and advanced invariants*

Brackets and inner products of vectors are two basic invariants in orthogonal geometry [189]. In Euclidean orthogonal geometry, inner products of vectors are equivalent to squared distances of points in that one can be represented by the other. It is well known that distances are the basic invariants in Euclidean geometry, so why do we resort to two other invariants and call them "basic invariants"?

First, let $\mathbf{a} = (a_1, \ldots, a_n)^T$ be a vector in $\mathbb{R}^n$. The distance between the origin and the end of the vector is

$$|\mathbf{a}| = \sqrt{a_1^2 + \cdots + a_n^2}, \tag{5.1.60}$$

which is not a polynomial function of the Cartesian coordinates. An *algebraic invariant* refers to a polynomial of coordinates that is invariant under the group of coordinate transformations. Only when the distance is squared can we treat it as an algebraic invariant.

Second, a bracket is another algebraic invariant. It cannot be represented as a polynomial function of the inner products of vectors, so it is another generator of the algebra of orthogonal invariants. It is a classical result in invariant theory [189] that inner products of vectors and brackets generate all algebraic invariants in orthogonal geometry.

The two basic invariants are related as follows. First, starting from the Cramer's rule for $n+1$ vectors $\mathbf{A}_{n+1} = \mathbf{a}_1, \mathbf{a}_2, \ldots, \mathbf{a}_{n+1}$,

$$\sum_{(1,n) \vdash \mathbf{A}_{n+1}} \mathbf{A}_{n+1\,(1)} [\mathbf{A}_{n+1\,(2)}] = 0, \tag{5.1.61}$$

and making inner product at both sides with a vector $\mathbf{b}$, we get

$$\sum_{(1,n) \vdash \mathbf{A}_{n+1}} (\mathbf{A}_{n+1\,(1)} \cdot \mathbf{b}) [\mathbf{A}_{n+1\,(2)}] = 0. \tag{5.1.62}$$

The left side of (5.1.62) is called an *inner-product Grassmann-Plücker syzygy* (IGP).

Second, consider the multiplication of two brackets:

$$\begin{aligned}
[\mathbf{a}_1 \ldots \mathbf{a}_n][\mathbf{b}_1 \ldots \mathbf{b}_n] &= (\mathbf{a}_1 \wedge \cdots \wedge \mathbf{a}_n) \cdot \mathbf{I}_n^{-1} \cdot (\mathbf{b}_1 \wedge \cdots \wedge \mathbf{b}_n) \cdot \mathbf{I}_n^{-1} \\
&= ((\mathbf{a}_1 \wedge \cdots \wedge \mathbf{a}_n) \rfloor (\mathbf{b}_1 \wedge \cdots \wedge \mathbf{b}_n)) \cdot (\mathbf{I}_n^{-1} \rfloor \mathbf{I}_n^{-1}) \\
&= \iota\,(\mathbf{a}_1 \ldots \mathbf{a}_n \,|\, \mathbf{b}_1 \ldots \mathbf{b}_n) \\
&= \iota \det(\mathbf{a}_i \cdot \mathbf{b}_j)_{i,j=1..n}.
\end{aligned} \tag{5.1.63}$$

It is called the *Laplace expansion* of the two brackets. When written as

$$[\mathbf{a}_1 \ldots \mathbf{a}_n][\mathbf{b}_1 \ldots \mathbf{b}_n] - \iota(\mathbf{a}_1 \ldots \mathbf{a}_n \,|\, \mathbf{b}_1 \ldots \mathbf{b}_n) = 0, \tag{5.1.64}$$

the left side is called a *bracket Laplace expansion syzygy* (BL).

By (5.1.64), (5.1.13) and (5.1.15), we get immediately

Proposition 5.34. All GP and VW syzygies among brackets in $\Lambda(\mathcal{V}^n)$, where $\mathcal{V}^n$ is an inner-product space, are generated by BL syzygies.

It is another classical result in invariant theory that the two kinds of syzygies, IGP and BL, generate all syzygies among brackets and inner products of vectors. Based on this result, we can define the following *inner-product bracket algebra*.

Definition 5.35. [Definition of inner-product bracket algebra] Let $\mathbf{a}_1, \ldots, \mathbf{a}_m$ be symbols, called *atomic vectors*, and let $m \geq n$.

- Let the $[\mathbf{a}_{i_1} \cdots \mathbf{a}_{i_n}]$ be indeterminates over $\mathbb{K}$ for each sequence of indices $1 \leq i_1, \ldots, i_n \leq m$, called *brackets*.
- Let the $\mathbf{a}_{j_1} \cdot \mathbf{a}_{j_2}$ be indeterminates over $\mathbb{K}$ for each ordered pair $1 \leq j_1, j_2 \leq m$, called *inner products* of vectors.

The nD *inner-product bracket algebra* generated by the $\mathbf{a}$'s, is the quotient of the polynomial ring generated by the brackets and inner products of vectors, modulo the ideal generated by the following syzygies:

B1. $[\mathbf{a}_{i_1}\cdots\mathbf{a}_{i_n}]$ if $i_j = i_k$ for some $j \neq k$.

B2. $[\mathbf{a}_{i_1}\cdots\mathbf{a}_{i_n}] - \mathrm{sgn}(\sigma)[\mathbf{a}_{i_{\sigma(1)}}\cdots\mathbf{a}_{i_{\sigma(n)}}]$ for any permutation σ of $1, 2, \ldots, n$.

IS. (Inner-product symmetry) $\mathbf{a}_i \cdot \mathbf{a}_j - \mathbf{a}_j \cdot \mathbf{a}_i$ if $i \neq j$.

IGP. Inner-product Grassmann-Plücker syzygy (5.1.62), with $\mathbf{a}_i, \mathbf{b}$ denoting $\mathbf{a}_{j_i}, \mathbf{a}_k$.

BL. Bracket Laplace expansion syzygy (5.1.64), with $\mathbf{a}_i, \mathbf{b}_j$ denoting $\mathbf{a}_{k_i}, \mathbf{a}_{l_j}$, and $(\mathbf{a}_1 \ldots \mathbf{a}_n \,|\, \mathbf{b}_1 \ldots \mathbf{b}_n)$ denoting the completely expanded form of the determinant.

In (2.3.35), it is shown that any VW syzygy in bracket algebra can be obtained from the following meet product expansion:

$$\mathbf{A}_r \vee \mathbf{B}_{n+1} \vee \mathbf{C}_{n-r-1} = \sum_{(r+1,n-r)\vdash\mathbf{B}_{n+1}} [\mathbf{A}_r\mathbf{B}_{n+1(2)}][\mathbf{B}_{n+1(1)}\mathbf{C}_{n-r-1}]. \quad (5.1.65)$$

In inner-product bracket algebra, there is a similar result. Since an inner-product Grassmann algebra is just a GC algebra equipped with a dual operator, we only need to consider the influence of the dual operator upon the right side of (5.1.65). There are two other cases besides (5.1.65).

Case 1. $\mathbf{C}_{n-r-1} = \mathbf{D}_{r+1}^{\sim}$. Then

$$
\begin{aligned}
[\mathbf{B}_{n+1(1)}\mathbf{C}_{n-r-1}] &= (\mathbf{B}_{n+1(1)} \wedge \mathbf{D}_{r+1}^{\sim})^{\sim} \\
&= \mathbf{B}_{n+1(1)} \cdot (\mathbf{D}_{r+1}^{\sim})^{\sim} \\
&= (-1)^{\frac{n(n-1)+r(r+1)}{2}} \iota\, \mathbf{B}_{n+1(1)} \rfloor \mathbf{D}_{r+1} \\
&= (-1)^{\frac{n(n-1)+r(r+1)}{2}} \iota\, (\mathbf{B}_{n+1(1)}|\mathbf{D}_{r+1}).
\end{aligned}
\quad (5.1.66)
$$

Substituting (5.1.66) into (5.1.65), we get the following *inner-product van der Waerden syzygy* (IVW) for $0 \le r < n$:

$$\sum_{(r+1,n-r)\vdash\mathbf{B}_{n+1}} [\mathbf{A}_r\mathbf{B}_{n+1(2)}]\,(\mathbf{B}_{n+1(1)}|\mathbf{D}_{r+1}). \quad (5.1.67)$$

Clearly the IGP syzygy (5.1.62) is an IVW syzygy for $r = 0$.

Case 2. $\mathbf{C}_{n-r-1} = \mathbf{D}_{r+1}^{\sim}$ and $\mathbf{A}_r = \mathbf{E}_{n-r}^{\sim}$. Let $\mathbf{F}_{n+1}$ be the sequence of vectors of $\mathbf{D}_{r+1}$ followed by those of $\mathbf{E}_{n-r}$. By similar argument, and using Laplace expansions of determinants, we get that (5.1.65) equals, up to scale,

$$\sum_{(r+1,n-r)\vdash\mathbf{B}_{n+1}} (\mathbf{B}_{n+1(2)}|\mathbf{E}_{n-r})\,(\mathbf{B}_{n+1(1)}|\mathbf{D}_{r+1}) = (\mathbf{B}_{n+1}|\mathbf{F}_{n+1}). \quad (5.1.68)$$

The right side of (5.1.68) is called an *inner-product Laplace expansion syzygy* (IL).

Proposition 5.36. Any IVW syzygy is generated by IGP syzygies, and any IL syzygy is generated by IGP and BL syzygies.

Proof. The second statement is obvious from the Laplace expansion of $(\mathbf{B}_{n+1}|\mathbf{F}_{n+1})$ by any row (or column). For the first statement, assume that it is true for (5.1.67) where $r = k - 1 < n - 1$. For $r = k$, let $\mathbf{D}_{k+1} = \mathbf{D}_k \wedge \mathbf{d}$, where $\mathbf{d}$ is a vector. By Laplace expansions, IGP relation (5.1.62) and induction hypothesis,

$$\sum_{(k+1,n-k)\vdash\mathbf{B}_{n+1}} [\mathbf{A}_k\mathbf{B}_{n+1\,(2)}]\,(\mathbf{B}_{n+1\,(1)}|\mathbf{D}_{k+1})$$

$$= \sum_{(k,1,n-k)\vdash\mathbf{B}_{n+1}} [\mathbf{A}_k\mathbf{B}_{n+1\,(3)}]\,(\mathbf{B}_{n+1\,(2)}\cdot\mathbf{d})\,(\mathbf{B}_{n+1\,(1)}|\mathbf{D}_k)$$

$$= \sum_{\substack{(k,n-k+1)\vdash\mathbf{B}_{n+1},\\(k-1,1)\vdash\mathbf{A}_k}} (\mathbf{A}_{k\,(2)}\cdot\mathbf{d})\,[\mathbf{A}_{k\,(1)}\mathbf{B}_{n+1\,(2)}]\,(\mathbf{B}_{n+1\,(1)}|\mathbf{D}_k)$$

$$= 0.$$

$\square$

Compared with bracket algebra, inner-product bracket algebra is much more complicated in that the defining syzygy BL contains as many as $n! + 1$ terms, all but one of which come from the complete expansion of the determinant $(\mathbf{a}_1\ldots\mathbf{a}_n|\mathbf{b}_1\ldots\mathbf{b}_n)$. The inner-products of vectors satisfy only two syzygies, the symmetry syzygy IS and the inner-product Laplace expansion syzygy IL. The latter is a polynomial of $(n+1)!$ terms in inner products of vectors.

To effectively employ such syzygies, we need to introduce the Hodge interior products of blades, or equivalently, minors of the inner-product matrix of the generating vectors in letter-place notation, $(\mathbf{a}_{i_1}\ldots\mathbf{a}_{i_r}|\mathbf{a}_{j_1}\ldots\mathbf{a}_{j_r})$ for all $0 < r < n$. On one hand, as polynomials of inner products of vectors, they are *advanced invariants*; on the other hand, they can be used to slow down the size explosion of Laplace expansions by replacing complete expansions with incomplete ones.

Definition 5.37. [Definition of graded inner-product bracket algebra] Let $\mathbf{a}_1,\ldots,$ $\mathbf{a}_m$ be symbols, called *atomic vectors*, and let $m \geq n$.

- Let the $[\mathbf{a}_{i_1}\cdots\mathbf{a}_{i_n}]$ be indeterminates over $\mathbb{K}$ for each sequence of indices $1 \leq i_1,\ldots,i_n \leq m$, called *brackets*.
- For any $1 \leq r \leq n$, let the $(\mathbf{a}_{j_1}\ldots\mathbf{a}_{j_r}|\mathbf{a}_{k_1}\ldots\mathbf{a}_{k_r})$ be indeterminates over $\mathbb{K}$ for every ordered pair of sequences $1 \leq j_1,\ldots,j_r \leq m$ and $1 \leq k_1,\ldots,k_r \leq m$, called *r-graded inner products*, or *Gram r-minors*.

The nD *graded inner-product bracket algebra* generated by the $\mathbf{a}$'s, is the quotient of the polynomial ring generated by the brackets and graded inner products, modulo the ideal generated by syzygies B1, B2, IGP in Definition 5.35, where $\mathbf{a}_i \cdot \mathbf{a}_j$ is replaced by $(\mathbf{a}_i|\mathbf{a}_j)$, together with the following syzygies:

GI1. $(\mathbf{a}_{j_1}\ldots\mathbf{a}_{j_r}|\mathbf{a}_{k_1}\ldots\mathbf{a}_{k_r}) - (\mathbf{a}_{k_1}\ldots\mathbf{a}_{k_r}|\mathbf{a}_{j_1}\ldots\mathbf{a}_{j_r})$ if the two sequences are not identical.

GI2. $(\mathbf{a}_{j_1} \ldots \mathbf{a}_{j_r} | \mathbf{a}_{k_1} \ldots \mathbf{a}_{k_r})$ if $j_p = j_q$ for some $p \neq q$.

GI3. $(\mathbf{a}_{j_1} \ldots \mathbf{a}_{j_r} | \mathbf{a}_{k_1} \ldots \mathbf{a}_{k_r}) - \mathrm{sgn}(\sigma)\,(\mathbf{a}_{j_{\sigma(1)}} \ldots \mathbf{a}_{j_{\sigma(r)}} | \mathbf{a}_{k_1} \ldots \mathbf{a}_{k_r})$ for any permutation σ of $1, \ldots, r$.

GIL. Graded inner-product Laplace expansion syzygy by (5.1.10), with $\mathbf{A}_r, \mathbf{B}_r$ denoting two subsequences of the $\mathbf{a}$'s of length r:

$$(\mathbf{A}_r | \mathbf{B}_r) - \sum_{(s,r-s) \vdash \mathbf{A}_r} (\mathbf{A}_{r(1)} | \mathbf{B}_{r(1)})\,(\mathbf{A}_{r(2)} | \mathbf{B}_{r(2)}).$$

Geometrically, graded inner-product bracket algebra introduces the cosines of the angles formed by high dimensional linear objects. The cosine of a sum of finitely many such angles, in general, can only be represented by a complicated polynomial in this algebra. To complete the system by allowing sums of angles to be advanced invariants, the inner-product Grassmann algebra must be extended to the *Clifford algebra* based on the same inner-product space.

5.2 Clifford algebra

Definition 5.38. The *Clifford algebra* $\mathcal{CL}(\mathcal{V}^n)$ over an inner-product space $\mathcal{V}^n$, is the $\mathbb{K}$-algebra obtained as the quotient of the tensor algebra $\otimes(\mathcal{V}^n)$ modulo the two-sided ideal, called *generating ideal*, generated by elements of the form $\mathbf{x} \otimes \mathbf{x} - \mathbf{x} \cdot \mathbf{x}$, for all $\mathbf{x} \in \mathcal{V}^n$. The numbers field $\mathbb{K}$ is a 1D subspace of the algebra, and the unit map in this algebra is the identity transformation in $\mathbb{K}$.

The quotient of the tensor product modulo the generating ideal is called the *geometric product*, also known as the *Clifford product*, or *Clifford multiplication*. When $\mathcal{CL}(\mathcal{V}^n)$ is viewed as a vector space, it is called a *Clifford space*. Its elements are called *Clifford polynomials*. A *Clifford monomial* is the geometric product of finitely many vectors.

Equivalently, the generators of the generating ideal can be enlarged to $\mathbf{x} \otimes \mathbf{y} + \mathbf{y} \otimes \mathbf{x} - 2\mathbf{x} \cdot \mathbf{y}$, for all $\mathbf{x}, \mathbf{y} \in \mathcal{V}^n$. The procedure of changing a tensor into its equivalent class modulo the generating ideal is called *Cliffordization*. It is the canonical quotient map from tensor algebra to Clifford algebra.

Notations.

From now on, the geometric product is always denoted by juxtaposition of elements, and precedes all other products by default. The geometric product of r identical elements is denoted by the r-th power of the element. The mapping from several elements to their geometric product is denoted by a multilinear map cl from $\mathcal{CL}(\mathcal{V}^n) \times \mathcal{CL}(\mathcal{V}^n) \times \cdots \times \mathcal{CL}(\mathcal{V}^n)$ to $\mathcal{CL}(\mathcal{V}^n)$, which can also be taken as a linear map from $\otimes(\mathcal{CL}(\mathcal{V}^n))$ to $\mathcal{CL}(\mathcal{V}^n)$:

$$\begin{aligned}
\mathsf{cl}: \quad &\otimes(\mathcal{CL}(\mathcal{V}^n)) \quad \longrightarrow \quad \mathcal{CL}(\mathcal{V}^n) \\
&\mathbf{A}_1 \otimes \mathbf{A}_2 \otimes \cdots \otimes \mathbf{A}_k \longmapsto \mathbf{A}_1 \mathbf{A}_2 \cdots \mathbf{A}_k.
\end{aligned} \tag{5.2.1}$$

Informally, a Clifford algebra is generated from an inner-product space by the geometric product under the generating relation that the geometric product of any vector with itself is their inner product. Formally, $\mathcal{CL}(\mathcal{V}^n)$ is the unique associative and multilinear algebra with the following *universal property*: if there is any isometry f from $\mathcal{V}^n$ into an inner-product $\mathbb{K}$-algebra $\mathcal{A}$, then f can be uniquely extended to an isometry from $\mathcal{CL}(\mathcal{V}^n)$ into $\mathcal{A}$.

When the inner product is completely degenerate, *i.e.*, the inner product of any two vectors is zero, the corresponding Clifford algebra is just the Grassmann algebra. In other cases, the two algebras are different mainly by their gradings.

Notation. For integers p, q, $[p/q]$ is the standard notation of the biggest integer that is less than or equal to p/q. For example, $[1/2] = 0$, $[0] = 0$, $[-1/2] = -1$.

The $\mathbb{Z}$-grading in Grassmann algebra is identical to that in tensor algebra, by imbedding the Grassmann space into the tensor space as the subspace of antisymmetric tensors. The grading in Clifford algebra is influenced by the nonzero inner products of vectors: a tensor $\mathbf{a}_1 \otimes \mathbf{a}_2 \otimes \cdots \otimes \mathbf{a}_r$ of grade r is equivalent by Cliffordization to a tensor of grade ranging from $r - 2[r/2]$ to r. In Clifford algebra, tensors of different grades can represent the same element, and there is no natural heritage from the $\mathbb{Z}$-grading of tensor algebra. In contrast, a tensor $\mathbf{a}_1 \otimes \mathbf{a}_2 \otimes \cdots \otimes \mathbf{a}_r$ of grade r modulo the generating relations $\mathbf{a} \otimes \mathbf{a} = 0$ for all $\mathbf{a} \in \mathcal{V}^n$, becomes either zero or an antisymmetric tensor of grade r. In Grassmann algebra, the heritage from the $\mathbb{Z}$-grading of tensor algebra is natural.

Example 5.39. [130] Based on $\mathbb{R}^2$, a whole family of isomorphic Clifford algebras can be generated. Let $\mathbf{e}_1, \mathbf{e}_2$ be an orthonormal basis of $\mathbb{R}^2$, and let $\mathbf{e}_{12} = \mathbf{e}_1 \wedge \mathbf{e}_2$ in the corresponding Grassmann algebra. Define

$$
\begin{aligned}
\mathbf{e}_1\mathbf{e}_1 &= \mathbf{e}_2\mathbf{e}_2 = 1, & \mathbf{e}_1\mathbf{e}_2 &= -\mathbf{e}_2\mathbf{e}_1 = \mathbf{e}_{12} + \lambda, \\
\mathbf{e}_1\mathbf{e}_{12} &= \mathbf{e}_2 - \lambda\mathbf{e}_1, & \mathbf{e}_2\mathbf{e}_{12} &= -\mathbf{e}_1 - \lambda\mathbf{e}_2, \\
\mathbf{e}_{12}\mathbf{e}_1 &= -\mathbf{e}_2 - \lambda\mathbf{e}_1, & \mathbf{e}_{12}\mathbf{e}_2 &= \mathbf{e}_1 - \lambda\mathbf{e}_2, \\
\mathbf{e}_{12}\mathbf{e}_{12} &= -2\lambda\mathbf{e}_{12} - \lambda^2 - 1, &&
\end{aligned}
\tag{5.2.2}
$$

where $\lambda \in \mathbb{R}$ is a parameter. It is easy to verify that for different λ's, the Clifford algebras defined by the multiplication table (5.2.2) are isomorphic to each other. They are different representations of the same Clifford algebra.

The $\mathbb{Z}$-grading in Grassmann algebra,

$$
\mathrm{grade}(1) = 0, \quad \mathrm{grade}(\mathbf{e}_1) = \mathrm{grade}(\mathbf{e}_2) = 1, \quad \mathrm{grade}(\mathbf{e}_{12}) = 2, \tag{5.2.3}
$$

leads to different $\mathbb{Z}$-gradings in different representations of the same Clifford algebra, by choosing different λ's.

Although Clifford algebra does not have a canonical $\mathbb{Z}$-grading, it does have a canonical $\mathbb{Z}_2$-grading. In the generating relations $\mathbf{a} \otimes \mathbf{a} = \mathbf{a} \cdot \mathbf{a}$ for all $\mathbf{a} \in \mathcal{V}^n$, the two sides of each equality are *even-graded tensors*. The *parity* of a tensor is even if the tensor is even-graded, and odd if the tensor is odd-graded. The parity is unchanged by Cliffordization. Thus $\mathcal{CL}(\mathcal{V}^n)$ is naturally decomposed into two subspaces: the even-graded subspace $\mathcal{CL}^+(\mathcal{V}^n)$, and the odd-graded subspace $\mathcal{CL}^-(\mathcal{V}^n)$. The former is also a *Clifford subalgebra*, i.e., it is a linear subspace that is closed under the geometric product.

Definition 5.40. An *even* (or *odd*) Clifford monomial is the geometric product of an even (or odd) number of vectors. The linear operators "$\langle\ \rangle_+$" and "$\langle\ \rangle_-$" extract from a Clifford polynomial its even-graded part and odd-graded part respectively. They are called the *even grading* operator and *odd grading* operator respectively.

The non-existence of a canonical $\mathbb{Z}$-grading determines the non-existence of a canonical linear isomorphism from the Clifford space to the Grassmann space. The canonical linear isomorphism in the reverse direction does exist. Recall that in the generating relations $\mathbf{x} \otimes \mathbf{y} + \mathbf{y} \otimes \mathbf{x} = 2\mathbf{x} \cdot \mathbf{y}$ for all $\mathbf{x}, \mathbf{y} \in \mathcal{V}^n$, the two sides of each equality are symmetric tensors. On one hand, the complete antisymmetrization eliminates any symmetric part of a tensor, so it eliminates both sides of the generating relations. On the other hand, an antisymmetric tensor does not have any symmetric part, so it is unchanged by the Cliffordization.

The above arguments show that the two sides of the following equality are in the same equivalent class of Cliffordization:

$$\mathbf{a}_1 \wedge \mathbf{a}_2 \wedge \cdots \wedge \mathbf{a}_r = \frac{1}{r!} \sum_{\sigma} \text{sign}(\sigma)\mathbf{a}_{\sigma(1)}\mathbf{a}_{\sigma(2)} \cdots \mathbf{a}_{\sigma(r)}, \tag{5.2.4}$$

where the $\mathbf{a}$'s are vectors, and where the summation runs over all permutations σ of $1, 2, \ldots, r$. By setting the two sides to be equal, a representative of the equivalent class is specified. The map from the left side of (5.2.4) to the right side, is the *canonical linear isomorphism* from the Grassmann space to the Clifford space. As a corollary, if the $\mathbf{a}$'s are mutually orthogonal, then

$$\mathbf{a}_1 \wedge \mathbf{a}_2 \wedge \cdots \wedge \mathbf{a}_r = \mathbf{a}_1\mathbf{a}_2 \cdots \mathbf{a}_r. \tag{5.2.5}$$

Definition 5.41. The *grade involution*, or *main involution* in Clifford algebra, denoted by the overhat symbol, is a linear operator defined by

$$\widehat{\mathbf{A}} = \langle\mathbf{A}\rangle_+ - \langle\mathbf{A}\rangle_-, \quad \forall \mathbf{A} \in \mathcal{CL}(\mathcal{V}^n). \tag{5.2.6}$$

The *reversion* in Clifford algebra, denoted by the dagger symbol, is a linear operator defined by

$$(\mathbf{a}_1\mathbf{a}_2 \cdots \mathbf{a}_r)^\dagger = \mathbf{a}_r\mathbf{a}_{r-1} \cdots \mathbf{a}_1, \quad \forall \mathbf{a}_i \in \mathcal{V}^n. \tag{5.2.7}$$

The *conjugate* in Clifford algebra, denoted by the overbar symbol, is the composition of the grade involution and the reversion:

$$\overline{\mathbf{A}} = \left(\widehat{\mathbf{A}}\right)^\dagger = \widehat{\mathbf{A}^\dagger}, \quad \forall \mathbf{A} \in \mathcal{CL}(\mathcal{V}^n). \tag{5.2.8}$$

Example 5.42. Let the **a**'s be vectors in $\mathcal{V}^n$, then

$$\widehat{\mathbf{a}_1\mathbf{a}_2\cdots\mathbf{a}_r} = (-1)^r\mathbf{a}_1\mathbf{a}_2\cdots\mathbf{a}_r,$$
$$\overline{\mathbf{a}_1\mathbf{a}_2\cdots\mathbf{a}_r} = (-1)^r\mathbf{a}_r\mathbf{a}_{r-1}\cdots\mathbf{a}_1,$$
$$(\mathbf{a}_1\wedge\mathbf{a}_2\wedge\cdots\wedge\mathbf{a}_r)^\dagger = \mathbf{a}_r\wedge\mathbf{a}_{r-1}\wedge\cdots\wedge\mathbf{a}_1.$$

That the Clifford space is isomorphic to the Grassmann space by (5.2.4) is easy to understand. For any two vectors $\mathbf{a}, \mathbf{b} \in \mathcal{V}^n$, their tensor product can be decomposed into the symmetric part and the antisymmetric part:

$$\mathbf{a}\otimes\mathbf{b} = \frac{\mathbf{a}\otimes\mathbf{b}+\mathbf{b}\otimes\mathbf{a}}{2} + \frac{\mathbf{a}\otimes\mathbf{b}-\mathbf{b}\otimes\mathbf{a}}{2}. \tag{5.2.9}$$

By Cliffordization, the symmetric part is identified with the scalar $\mathbf{a}\cdot\mathbf{b}$, the antisymmetric part is identified with $\mathbf{a}\wedge\mathbf{b}$, and the tensor product is changed into the geometric product. So (5.2.9) becomes

$$\mathbf{ab} = \mathbf{a}\cdot\mathbf{b}+\mathbf{a}\wedge\mathbf{b}. \tag{5.2.10}$$

The geometric product of two vectors is composed of two parts, the 0-graded part and the 2-graded part. They are respectively the inner product and the outer product of the two vectors, and the $\mathbb{Z}$-grading is induced from the Grassmann algebra via the isomorphism (5.2.4). For $r \leq n$ vectors, similar arguments show that their geometric product is composed of $[r/2] + 1$ parts of different grades:

$$\mathbf{a}_1\mathbf{a}_2\cdots\mathbf{a}_r = \sum_{i=0}^{[\frac{r}{2}]} \langle\mathbf{a}_1\mathbf{a}_2\cdots\mathbf{a}_r\rangle_{r-2i}. \tag{5.2.11}$$

The $\mathbb{Z}$-grading is again from the Grassmann algebra.

When $r > n$, the complete antisymmetrization of any tensor of grade $i > n$ is identified to zero by Cliffordization, so only those partial antisymmetrizations of grade $\leq n$ remain. The result is always contained in the Grassmann space $\Lambda(\mathcal{V}^n)$. Since the Grassmann space is embedded in the Clifford space by (5.2.4), the two linear spaces are isomorphic.

Thus, under the isomorphism (5.2.4), the Clifford space and the Grassmann space are identified, and the Clifford space becomes $\mathbb{Z}$-graded. The *i-grading operators* $\langle\ \rangle_i$ for all $0 \leq i \leq n$, are identical to those in Grassmann space.

By the symmetry of the inner product and the antisymmetry of the outer product, we get from (5.2.10) the following equalities:

$$\mathbf{a}\cdot\mathbf{b} = \frac{\mathbf{ab}+\mathbf{ba}}{2}, \qquad \mathbf{a}\wedge\mathbf{b} = \frac{\mathbf{ab}-\mathbf{ba}}{2}. \tag{5.2.12}$$

So both the inner product and the outer product are included in and can be derived from the geometric product. The inner product of two vectors can be extended to the inner product of any two elements in $\mathcal{CL}(\mathcal{V}^n)$ via (5.2.4). If starting from the geometric product, then other products in Grassmann algebra, inner-product Grassmann algebra, GC algebra, *etc.*, can be defined through the geometric product. From this aspect, Clifford algebra unifies and contains all previous invariant algebras.

Conversely, (5.2.10) suggests that the geometric product of two vectors is nothing but a deformation of the outer product by attaching a scalar term, the inner product of the two vectors. In the general case, the geometric product of any r-vector $\mathbf{A}_r$ and s-vector $\mathbf{B}_s$ in $\Lambda(\mathcal{V}^n)$ has the following decomposition by grade:

$$\mathbf{A}_r\mathbf{B}_s = \sum_{i=\max(0,[\frac{r+s+1-n}{2}])}^{\min(r,s)} \langle \mathbf{A}_r\mathbf{B}_s \rangle_{r+s-2i}. \tag{5.2.13}$$

In Section 5.4, it will be shown that any term on the right side of (5.2.13) is a polynomial in the inner-product Grassmann algebra $\Lambda(\mathcal{V}^n)$. In the rest of this section, we only investigate the two outermost terms in (5.2.13): $\mathbf{A}_r{\cdot}\mathbf{B}_s$ and $\mathbf{A}_r{\wedge}\mathbf{B}_s$.

Lemma 5.43. Let $r > 0$. For an r-blade $\mathbf{A}_r$ and any vector $\mathbf{a}$,

$$\begin{aligned} \mathbf{a}\mathbf{A}_r &= \mathbf{a}\cdot\mathbf{A}_r + \mathbf{a}\wedge\mathbf{A}_r, \\ \mathbf{A}_r\mathbf{a} &= \mathbf{A}_r\cdot\mathbf{a} + \mathbf{A}_r\wedge\mathbf{a}. \end{aligned} \tag{5.2.14}$$

Proof. We only prove the first equality. Let $\mathbf{e}_1,\ldots,\mathbf{e}_r$ be an orthogonal basis of the rD space $\mathbf{A}_r$. In $\otimes(\mathcal{V}^n)$, tensor $\mathbf{a}\otimes\mathbf{A}_r$ can be decomposed into a linear combination of tensor products of the partial symmetrizations and partial antisymmetrizations of vectors $\mathbf{a},\mathbf{e}_1,\ldots,\mathbf{e}_r$. During Cliffordization, all partial symmetrizations among the $\mathbf{e}$'s are reduced to zero, so only the partial symmetrizations between $\mathbf{a}$ and $\mathbf{e}_i$, for all $1 \leq i \leq r$, are left, and they are changed into inner products $\mathbf{a}\cdot\mathbf{e}_i$. Since the $\mathbf{e}$'s are antisymmetric before Cliffordization, they keep this property afterwards. Therefore, the only remaining terms from the partial symmetrizations of $\mathbf{a}\otimes\mathbf{A}_r$ are

$$\sum_{i=1}^{r}(-1)^{i+1}(\mathbf{a}\cdot\mathbf{e}_i)\,\mathbf{e}_1\wedge\cdots\wedge\check{\mathbf{e}}_i\wedge\cdots\wedge\mathbf{e}_r, \tag{5.2.15}$$

where the signs are caused by the antisymmetry among $\mathbf{e}_1,\ldots,\mathbf{e}_r$.

If $\mathbf{a}$ is in any partial antisymmetrization, then the antisymmetry among $\mathbf{e}_1,\ldots,\mathbf{e}_r$ requires that $\mathbf{a}$ and $\mathbf{e}_i$, for all $1 \leq i \leq r$, must be antisymmetric. After Cliffordization, $\mathbf{a}$ and $\mathbf{e}_1,\ldots,\mathbf{e}_r$ must form a complete antisymmetrization. The result is $\mathbf{a}\wedge\mathbf{e}_1\wedge\cdots\wedge\mathbf{e}_r$. $\qquad\square$

Proposition 5.44. Let $\mathbf{A}_r$ and $\mathbf{B}_s$ be r-blade and s-blade in $\Lambda(\mathcal{V}^n)$ respectively, where $0 \leq r,s \leq n$. Then

$$\mathbf{A}_r \cdot \mathbf{B}_s = \langle \mathbf{A}_r\mathbf{B}_s \rangle_{|r-s|}, \tag{5.2.16}$$

$$\mathbf{A}_r \wedge \mathbf{B}_s = \langle \mathbf{A}_r\mathbf{B}_s \rangle_{r+s}. \tag{5.2.17}$$

Proof. The leading grade of $\mathbf{A}_r\mathbf{B}_s$ is $r+s$, because the tensor product $\mathbf{A}_r\otimes\mathbf{B}_s$ itself is of grade $r+s$. The $(r+s)$-graded part of $\mathbf{A}_r\mathbf{B}_s$ must be the result of the complete antisymmetrization of $\mathbf{A}_r\otimes\mathbf{B}_s$ by Cliffordization, as any symmetric part is always down-graded by Cliffordization. This proves (5.2.17).

Within each of the blades $\mathbf{A}_r, \mathbf{B}_s$, the constituent vectors are completely anti-symmetric, so the grade reduction in Cliffordization can only be between the two blades. Since $|r - s|$ is the difference of the grades of the two blades, it is the lowest grade of $\mathbf{A}_r\mathbf{B}_s$, and is obtained by the complete symmetrization between the two blades. When $r = s$, the complete symmetrization is obviously the inner product $\mathbf{A}_r \cdot \mathbf{B}_s$. When $r \neq s$, by (5.1.23), the complete symmetrization is still the inner product.

An alternative proof of (5.2.16) by (5.2.14) without resorting to Cliffordization is available after Theorem 5.4.33 in Subsection 5.4.3 is established. $\qquad\square$

Corollary 5.45. For any vectors a's in $\mathcal{V}^n$,

$$\mathbf{a}_1 \wedge \cdots \wedge \mathbf{a}_r = \langle \mathbf{a}_1 \cdots \mathbf{a}_r \rangle_r. \tag{5.2.18}$$

Corollary 5.46. For any r-blade $\mathbf{A}_r$,

$$\mathbf{A}_r\mathbf{A}_r = \mathbf{A}_r \cdot \mathbf{A}_r. \tag{5.2.19}$$

Proof. Choose an orthogonal basis $\mathbf{e}_1, \ldots, \mathbf{e}_r$ of the rD space $\mathbf{A}_r$. Then $\mathbf{e}_i\mathbf{e}_i = \mathbf{e}_i \cdot \mathbf{e}_i$, and for $i \neq j$, $\mathbf{e}_i\mathbf{e}_j = \mathbf{e}_i \wedge \mathbf{e}_j = -\mathbf{e}_j \wedge \mathbf{e}_i = -\mathbf{e}_j\mathbf{e}_i$. So

$$\mathbf{A}_r\mathbf{A}_r = \mathbf{e}_1 \cdots \mathbf{e}_r\mathbf{e}_1 \cdots \mathbf{e}_r = (-1)^{r-1}\mathbf{e}_1\mathbf{e}_1(\mathbf{e}_2 \cdots \mathbf{e}_r\mathbf{e}_2 \cdots \mathbf{e}_r).$$

Continuing to move together pairs of identical basis vectors in this way, we get

$$\mathbf{A}_r\mathbf{A}_r = (-1)^{\frac{r(r-1)}{2}}(\mathbf{e}_1\mathbf{e}_1) \cdots (\mathbf{e}_r\mathbf{e}_r) = (-1)^{\frac{r(r-1)}{2}}(\mathbf{e}_1 \cdot \mathbf{e}_1) \cdots (\mathbf{e}_r \cdot \mathbf{e}_r) \in \mathbb{K},$$

so $\mathbf{A}_r\mathbf{A}_r = \langle \mathbf{A}_r\mathbf{A}_r \rangle_0 = \mathbf{A}_r \cdot \mathbf{A}_r$. $\qquad\square$

Proposition 5.47. For any r-blade $\mathbf{A}_r$ and n-blade $\mathbf{B}_n$ in $\Lambda(\mathcal{V}^n)$,

$$\mathbf{A}_r\mathbf{B}_n = \mathbf{A}_r \cdot \mathbf{B}_n. \tag{5.2.20}$$

In particular, if the inner product in $\mathcal{V}^n$ is nondegenerate, then

$$\mathbf{A}^\sim = \mathbf{A}\mathbf{I}_n^{-1}, \ \forall \mathbf{A} \in \mathcal{CL}(\mathcal{V}^n). \tag{5.2.21}$$

Proof. If the inner product in the subspace $\mathbf{A}_r$ is nondegenerate, then any orthogonal basis of the subspace can be extended to an orthogonal basis of the whole space $\mathcal{V}^n$. A procedure similar to the proof of Corollary 5.46 leads to (5.2.20). Below we assume that the inner product in subspace $\mathbf{A}_r$ is degenerate.

Let $\mathbf{v}_1, \ldots, \mathbf{v}_r$ be a basis of $\mathbf{A}_r$. By decomposition (5.1.4), $\mathbf{v}_i = \mathbf{x}_i + \mathbf{y}_i$, where $\mathbf{x}_i \in \mathrm{rad}(\mathcal{V}^n)$ and $\mathbf{y}_i \in \mathcal{V}'$. r-blade $\mathbf{A}_r$ is decomposed into a sum of blades in which only one blade $\mathbf{y}_1 \wedge \cdots \wedge \mathbf{y}_r$ does not contain any $\mathbf{x}_i$.

If some $\mathbf{x}_i$ is nonzero, say $\mathbf{x}_1$, then it can be extended to a basis $\mathbf{x}_1, \mathbf{e}_2, \ldots, \mathbf{e}_n$ of $\mathcal{V}^n$, by first extending to a basis of $\mathrm{rad}(\mathcal{V}^n)$, and then appending an orthogonal basis of $\mathcal{V}'$. Since $\mathbf{x}_1$ is orthogonal to any vector $\mathbf{z}$ in $\mathcal{V}^n$, $\mathbf{x}_1\mathbf{z} = -\mathbf{z}\mathbf{x}_1 = \mathbf{x}_1 \wedge \mathbf{z}$. For any vectors $\mathbf{z}_2, \ldots, \mathbf{z}_r \in \mathcal{V}^n$,

$$\begin{aligned}
(\mathbf{x}_1 \wedge \mathbf{z}_2 \wedge \cdots \wedge \mathbf{z}_r)\mathbf{B}_n &= \mathbf{x}_1(\mathbf{z}_2 \wedge \cdots \wedge \mathbf{z}_r)(\mathbf{x}_1 \wedge \mathbf{e}_2 \wedge \cdots \wedge \mathbf{e}_n) \\
&= (-1)^{r-1}(\mathbf{z}_2 \wedge \cdots \wedge \mathbf{z}_r)\mathbf{x}_1\mathbf{x}_1(\mathbf{e}_2 \wedge \cdots \wedge \mathbf{e}_n) \\
&= 0.
\end{aligned}$$

Thus in $\mathbf{A}_r\mathbf{B}_n$, only the component $\mathbf{Y}_r = \mathbf{y}_1 \wedge \cdots \wedge \mathbf{y}_r$ of $\mathbf{A}_r$ is left with $\mathbf{B}_n$. If $\mathbf{Y}_r = 0$, then $\mathbf{A}_r\mathbf{B}_n = 0$, and the conclusion is trivial. So we assume $\mathbf{Y}_r \neq 0$. In the nondegenerate inner-product subspace $\mathcal{V}'$, if the inner product in subspace $\mathbf{Y}_r$ is nondegenerate, then the conclusion can be proved in a way similar to Corollary 5.46. So we further assume that $\mathbf{Y}_r$ is a degenerate inner-product space.

Let $\mathbf{a}_1, \ldots, \mathbf{a}_k, \mathbf{c}_1, \ldots, \mathbf{c}_{r-k}$ be an orthogonal basis of $\mathbf{Y}_r$, in which only the $\mathbf{a}$'s are null. By Witt Theorem, there exists the following Witt basis of $\mathcal{V}'$:

$$(\mathbf{a}_1, \mathbf{b}_1), \ldots, (\mathbf{a}_k, \mathbf{b}_k), \ (\mathbf{a}_{k+1}, \mathbf{b}_{k+1}), \ldots, (\mathbf{a}_u, \mathbf{b}_u), \ \mathbf{c}_1, \ldots, \mathbf{c}_{r-k}, \ \mathbf{c}_{r-k+1}, \ldots, \mathbf{c}_v.$$

Let $\mathbf{D}_{n-2u-v}$ be a pseudoscalar in $\Lambda(\mathrm{rad}(\mathcal{V}^n))$. By the following set of computational formulas:

- for any i, j, $\mathbf{c}_i(\mathbf{a}_j \wedge \mathbf{b}_j) = (\mathbf{a}_j \wedge \mathbf{b}_j)\mathbf{c}_i$, because

$$\mathbf{c}_i(\mathbf{a}_j \wedge \mathbf{b}_j) = \mathbf{c}_i \wedge \mathbf{a}_j \wedge \mathbf{b}_j = \mathbf{a}_j \wedge \mathbf{b}_j \wedge \mathbf{c}_i;$$

- for any $i \neq j$, $\mathbf{a}_i(\mathbf{a}_j \wedge \mathbf{b}_j) = (\mathbf{a}_j \wedge \mathbf{b}_j)\mathbf{a}_i$;
- for any i, $\mathbf{a}_i(\mathbf{a}_i \wedge \mathbf{b}_i) = \mathbf{a}_i$, because $\mathbf{a}_i(\mathbf{a}_i \wedge \mathbf{b}_i) = \mathbf{a}_i \cdot (\mathbf{a}_i \wedge \mathbf{b}_i) = -(\mathbf{a}_i \cdot \mathbf{b}_i)\mathbf{a}_i$;

we get

$$
\begin{aligned}
\mathbf{A}_r\mathbf{B}_n &= \mathbf{Y}_r\mathbf{B}_n \\
&= (\mathbf{a}_1 \wedge \cdots \wedge \mathbf{a}_k \wedge \mathbf{c}_1 \wedge \cdots \wedge \mathbf{c}_{r-k}) \\
&\qquad ((\mathbf{a}_1 \wedge \mathbf{b}_1) \wedge \cdots \wedge (\mathbf{a}_u \wedge \mathbf{b}_u) \wedge \mathbf{c}_1 \wedge \cdots \wedge \mathbf{c}_v \wedge \mathbf{D}_{n-2u-v}) \\[6pt]
&= (\mathbf{a}_1\mathbf{a}_2 \cdots \mathbf{a}_k)(\mathbf{c}_1\mathbf{c}_2 \cdots \mathbf{c}_{r-k}) \\
&\qquad (\mathbf{a}_1 \wedge \mathbf{b}_1)(\mathbf{a}_2 \wedge \mathbf{b}_2) \cdots (\mathbf{a}_u \wedge \mathbf{b}_u)(\mathbf{c}_1\mathbf{c}_2 \cdots \mathbf{c}_v)\mathbf{D}_{n-2u-v} \\[6pt]
&= \{\mathbf{a}_1(\mathbf{a}_1 \wedge \mathbf{b}_1)\} \{\mathbf{a}_2(\mathbf{a}_2 \wedge \mathbf{b}_2)\} \cdots \{\mathbf{a}_k(\mathbf{a}_k \wedge \mathbf{b}_k)\} \\
&\qquad (\mathbf{a}_{k+1} \wedge \mathbf{b}_{k+1}) \cdots (\mathbf{a}_u \wedge \mathbf{b}_u)(\mathbf{c}_1 \cdots \mathbf{c}_{r-k})^2(\mathbf{c}_{r-k+1} \cdots \mathbf{c}_v)\mathbf{D}_{n-2u-v} \\[6pt]
&= (\mathbf{a}_1\mathbf{a}_2 \cdots \mathbf{a}_k)(\mathbf{a}_{k+1} \wedge \mathbf{b}_{k+1}) \cdots (\mathbf{a}_u \wedge \mathbf{b}_u) \\
&\qquad (\mathbf{c}_1 \cdots \mathbf{c}_{r-k})^2(\mathbf{c}_{r-k+1} \cdots \mathbf{c}_v)\mathbf{D}_{n-2u-v} \\[6pt]
&= (\mathbf{c}_1 \cdots \mathbf{c}_{r-k})^2 (\mathbf{a}_1 \wedge \cdots \wedge \mathbf{a}_k) \wedge (\mathbf{a}_{k+1} \wedge \mathbf{b}_{k+1}) \wedge \cdots \wedge (\mathbf{a}_u \wedge \mathbf{b}_u) \\
&\qquad \wedge(\mathbf{c}_{r-k+1} \wedge \cdots \wedge \mathbf{c}_v) \wedge \mathbf{D}_{n-2u-v}.
\end{aligned}
$$

So the grade of $\mathbf{A}_r\mathbf{B}_n$ is $n - r$. Then $\mathbf{A}_r\mathbf{B}_n = \langle \mathbf{A}_r\mathbf{B}_n \rangle_{n-r} = \mathbf{A}_r \cdot \mathbf{B}_n$. $\qquad\square$

Proposition 5.44 indicates that the geometric product is the completion of the outer product and the inner product, not a mere union of the two products. The significance of the completion lies in two aspects: first, the geometric product is associative while the inner product and the outer product are not; second, the geometric product is *almost invertible*, while the inner product and the outer product are almost never invertible.

Definition 5.48. For any $\mathbf{A} \in \mathcal{CL}(\mathcal{V}^n)$, if there exists another element $\mathbf{B} \in \mathcal{CL}(\mathcal{V}^n)$ such that $\mathbf{AB} = \mathbf{BA} = 1$, then $\mathbf{B}$ is called the *inverse* of $\mathbf{A}$, denoted by $\mathbf{A}^{-1}$.

According to Corollary 5.54 to be introduced in Subsection 5.3.2, almost all elements in $\mathcal{CL}(\mathcal{V}^n)$ are invertible, if the inner product in $\mathcal{V}^n$ is nondegenerate. For example, an r-blade $\mathbf{A}_r$ is invertible if and only if $\mathbf{A}_r \cdot \mathbf{A}_r \neq 0$, and $\mathbf{A}_r^{-1}$ is provided by (5.1.9). In particular, all nonzero blades in $\Lambda(\mathbb{R}^n)$ and $\Lambda(\mathbb{R}^{-n})$ are invertible.

That the inner product and the outer product are almost never invertible is easy to understand. They are respectively the grade-lowering and the grade-raising operations. The geometric product executes both operations simultaneously, thus allowing almost all elements to be invertible.

The associativity and almost invertibility of the geometric product determine that Clifford algebra plays a key role in solving *multivector equations*. For example, let $\mathbf{B}_s$ be an invertible s-blade. Then (5.1.59) provides an orthogonal decomposition of vector $\mathbf{a}$ with respect to sD space $\mathbf{B}_s$:

$$\mathbf{a} = (\mathbf{a}\cdot\mathbf{B}_s)\cdot\mathbf{B}_s^{-1} + (\mathbf{a}\wedge\mathbf{B}_s)\cdot\mathbf{B}_s^{-1} = (\mathbf{a}\cdot\mathbf{B}_s)\mathbf{B}_s^{-1} + (\mathbf{a}\wedge\mathbf{B}_s)\mathbf{B}_s^{-1} = (\mathbf{a}\mathbf{B}_s)\mathbf{B}_s^{-1}. \quad (5.2.22)$$

The decomposition is trivial by associativity: $(\mathbf{a}\mathbf{B}_s)\mathbf{B}_s^{-1} = \mathbf{a}(\mathbf{B}_s\mathbf{B}_s^{-1}) = \mathbf{a}$. By invertibility, if the geometric product $\mathbf{a}\mathbf{B}_s$ is known, then $\mathbf{a}$ is completely determined by $\mathbf{B}_s^{-1}$ and $\mathbf{a}\mathbf{B}_s$. We say $\mathbf{a}\mathbf{B}_s$ (or $\mathbf{B}_s\mathbf{a}$) is a left (or right) *faithful representation* of $\mathbf{a}$ by the geometric product with $\mathbf{B}_s$.

In general, any invertible element $\mathbf{B} \in \mathcal{CL}(\mathcal{V}^n)$ provides a left (or right) faithful representation of $\mathcal{CL}(\mathcal{V}^n)$ by mapping any $\mathbf{A} \in \mathcal{CL}(\mathcal{V}^n)$ to $\mathbf{AB}$ (or $\mathbf{BA}$). The geometric product of two invertible elements contains all the algebraic relations between them, so that one element can completely determine the other by their given geometric product. If the two elements represent geometric objects, then the geometric product contains *all the geometric relations* between them.

Example 5.49. In $\mathbb{R}^2$, let $\mathbf{e}_1, \mathbf{e}_2$ be an orthonormal basis, called the x-axis and the y-axis respectively. Any point $\mathbf{a}$ is represented by a vector $a_1\mathbf{e}_1 + a_2\mathbf{e}_2$. The right faithful representation of $\mathbf{a}$ with respect to the x-axis $\mathbf{e}_1$ is

$$\mathbf{e}_1\mathbf{a} = \mathbf{e}_1 \cdot \mathbf{a} + \mathbf{e}_1 \wedge \mathbf{a} = a_1 + a_2\,\mathbf{e}_1 \wedge \mathbf{e}_2. \quad (5.2.23)$$

Since bivector $\mathbf{e}_1 \wedge \mathbf{e}_2$ functions as the pure imaginary unit i, $\mathbf{e}_1\mathbf{a}$ is nothing but the complex numbers representation of point $\mathbf{a}$. The right faithful representation changes the 2D real space $\mathbb{R}^2$ into the complex numbers field $\mathbb{C}$, and the latter has much better algebraic property.

5.3 Representations of Clifford algebras

Clifford algebra has five major representations, according to the general appearance of the single-termed elements in each representation.

(1) Hypercomplex Clifford numbers:

In 1878, Clifford defined the prototype of a Clifford algebra [43]. If translated into modern mathematical language, what he defined is $\mathcal{CL}(\mathbb{R}^{-n})$. Let $\mathbf{e}_1, \ldots, \mathbf{e}_n$

be an orthonormal basis of $\mathbb{R}^{-n}$. The n basis vectors, together with an additional element "-1", generate a finite group F_n, subject to the following multiplication relations:

- -1 commutes with the basis vectors;
- $(-1)^2 = 1$, where 1 is the unit of $\mathbb{R}$;
- $(\mathbf{e}_i)^2 = -1$ for $i = 1, \ldots, n$;
- $\mathbf{e}_i \mathbf{e}_j = (-1)\mathbf{e}_j \mathbf{e}_i$ for $i \neq j$.

Clifford's geometric algebra is the quotient of the group algebra $\mathbb{R}F_n$ modulo $\mathbb{R}\{1 + (-1)\}$. Its elements are later called *Clifford numbers*.

The all-negative signature of the base space comes from Clifford's intention to include and extend the complex numbers and quaternions to a high dimensional numbers system. This explains the term "hypercomplex Clifford numbers".

By revising the generating relations and allowing other numbers fields, Clifford's original definition can be readily extended to cover all kinds of Clifford algebra. The definition is not basis free, nor are the algebraic maniputations based on this definition.

By extending real harmonic analysis, complex analysis and quaternionic analysis to high dimensional Clifford algebras, a branch of modern analysis, called *Clifford analysis*, was created. Clifford analysis has its origin traced back to A. C. Dixon [53] and F. Klein [97].

(2) Multilinear tensor-formed Clifford algebra:

Definition 5.38 gives the standard form of a Clifford algebra. It was originally proposed by Chevalley [36]. Elements in this representation are in the form of Clifford polynomials.

Multilinear Clifford algebra has been used extensively by algebraists, analysts, theoretical physicists, differential geometers, *et al.*, in algebraic representation theory, spin geometry, Atiyah-Singer index theory, Seiberg-Witten theory, *etc.* Spinors are naturally represented as elements in the minimal left (or right) ideals of this algebra.

In practice, however, it rarely happens that elements in the form of Clifford polynomials are used directly in algebraic computation. Matrix representations are usually resorted to.

(3) Matrix representation of Clifford algebra:

The complete classifications of all real and complex Clifford algebras were made by É. Cartan [33]. The result is that any Clifford algebra over the real numbers or complex numbers is isomorphic to a matrix algebra. Matrix representations of Clifford algebras are direct consequences of Cartan's classification. They are very convenient for the understanding, characterizing and numerical computing of Clifford algebras.

However, matrix representations are not basis free. For invariant geometric computing with Clifford algebras, matrix representations do not prove to be efficient.

(4) Geometric Algebra:

The version of Clifford algebra called *Geometric Algebra* was first proposed by D. Hestenes [76], originally called *Space-Time algebra* in the setting of special relativity. In this representation, a single-termed element is generated from atomic vectors by the compositions of geometric products, duals and $\mathbb{Z}$-gradings, called a *graded Clifford monomial.*

The contents of Geometric Algebra are composed of three parts: the part of philosophy that advocates the use of Geometric Algebra as a universal language for describing and manipulating problems in mathematics and physics related to geometry; the part of usage that prefers the geometric product to the addition operation and other products in algebraic computation; the part of language body that is on symbolic manipulations of symmetries and $\mathbb{Z}$-grading operators.

The benefit of graded Clifford monomial representation can be seen from the simple example of deriving (5.1.25) by symmetries and $\mathbb{Z}$-grading operators [77]. For blades $\mathbf{A}_r, \mathbf{B}_s, \mathbf{C}_t$ of grade r, s, t respectively such that $r + s \leq t$, to prove

$$(\mathbf{A}_r \wedge \mathbf{B}_s) \cdot \mathbf{C}_t = \mathbf{A}_r \cdot (\mathbf{B}_s \cdot \mathbf{C}_t), \tag{5.3.1}$$

we first change the three inner products and one outer product into $\mathbb{Z}$-gradings of geometric products, then eliminate most of the $\mathbb{Z}$-grading operators, and finally apply the associativity of the geometric product:

$$\begin{aligned}
(\mathbf{A}_r \wedge \mathbf{B}_s) \cdot \mathbf{C}_t &= \langle\langle \mathbf{A}_r \mathbf{B}_s\rangle_{r+s}\mathbf{C}_t\rangle_{t-(r+s)} = \langle(\mathbf{A}_r\mathbf{B}_s)\mathbf{C}_t\rangle_{t-r-s}; \\
\mathbf{A}_r \cdot (\mathbf{B}_s \cdot \mathbf{C}_t) &= \langle \mathbf{A}_r\langle \mathbf{B}_s\mathbf{C}_t\rangle_{t-s}\rangle_{(t-s)-r} = \langle \mathbf{A}_r(\mathbf{B}_s\mathbf{C}_t)\rangle_{t-s-r}.
\end{aligned} \tag{5.3.2}$$

Besides applications in mechanics and theoretical physics, Geometric Algebra has found a lot of applications in computer science for modeling and manipulating geometric objects in a covariant and coordinate-free manner.

(5) Clifford deformation of Grassmann algebra:

The geometric product can be defined directly on the Grassmann space $\Lambda(\mathcal{V}^n)$, based on the fact that the geometric product of any two vectors is nothing but a deformation of the outer product by attaching a scalar term. This new product in the Grassmann space can always be written as a linear combination of outer products with inner product coefficients. All elements in a Clifford algebra can be written as multivectors in the corresponding Grassmann algebra with inner product coefficients. In this formulation, a Clifford algebra is just a Grassmann algebra equipped with a "deformed outer product".

This viewpoint was proposed by physicists [32], [129] in the 1960-70s, and further developed by mathematicians and physicists [29], [62], [116], [140], [158]. In this viewpoint, Clifford algebra is closely related to Hopf algebra, and is a topic of interest in algebraic combinatorics.

Dissidence: The inner product of a scalar and a multivector.

In the literature, some people insist that $\lambda \cdot \mathbf{A} = \mathbf{A} \cdot \lambda = 0$, so that *Chevalley's formula*

$$\mathbf{a}\mathbf{A} = \mathbf{a} \cdot \mathbf{A} + \mathbf{a} \wedge \mathbf{A} \tag{5.3.3}$$

holds for any multivector $\mathbf{A}$ and vector $\mathbf{a}$. Some insist on $\lambda \wedge \mathbf{A} = \mathbf{A} \wedge \lambda = 0$, so that $\lambda \cdot \mathbf{A} = \mathbf{A} \cdot \lambda = \lambda\mathbf{A}$ for the same reason. Some others insist on

$$\mathbf{a} \cdot \mathbf{A}_r = \langle \mathbf{a}\mathbf{A}_r \rangle_{|r-1|}, \tag{5.3.4}$$

so that the grading formula (5.2.16) holds for any $r, s \geq 0$, although the result

$$\mathbf{a} \cdot \lambda = \mathbf{a} \wedge \lambda = \lambda\mathbf{a} \tag{5.3.5}$$

invalidates Chevalley's formula. We adopt (5.3.4) and hence (5.3.5) for two main reasons: first, this definition of the inner product can simplify a lot of formulas without excluding the scalar case; second, it is fully compatible with the relation (5.1.51) between the meet product and the inner product, so that the resulting Clifford algebra is fully compatible with GC algebra.

5.3.1 *Clifford numbers*

In this subsection, we consider two typical Clifford numbers other than the complex numbers: quaternions and dual quaternions.

The extension of complex numbers to a higher dimensional numbers system was first achieved by Hamilton in 1843. Before this discovery, it had taken him ten years to realize that there is no 3D numbers system at all. Quaternions are 4D numbers, with basis $\{1, \mathbf{i}, \mathbf{j}, \mathbf{k}\}$ such that

$$\begin{aligned}
\mathbf{i}^2 &= \mathbf{j}^2 = \mathbf{k}^2 = -1, \\
\mathbf{i}\mathbf{j} &= -\mathbf{j}\mathbf{i} = \mathbf{k}, \\
\mathbf{j}\mathbf{k} &= -\mathbf{k}\mathbf{j} = \mathbf{i}, \\
\mathbf{k}\mathbf{i} &= -\mathbf{i}\mathbf{k} = \mathbf{j}.
\end{aligned} \tag{5.3.6}$$

The basis elements $\mathbf{i}, \mathbf{j}, \mathbf{k}$ form an orthonormal basis of $\mathbb{R}^{-3}$. In a general quaternion $x_0 + x_1\mathbf{i} + x_2\mathbf{j} + x_3\mathbf{k}$, the *scalar part* refers to x_0, the *vector part* refers to $x_1\mathbf{i} + x_2\mathbf{j} + x_3\mathbf{k}$. The multiplication $\mathbf{A}\mathbf{B}$ of two quaternions is naturally decomposed into two parts: the scalar part $\mathbf{A} \cdot \mathbf{B}$, called the *inner product*, and the vector part $[\mathbf{A}, \mathbf{B}]$, called the *vector product*. So

$$\mathbf{A}\mathbf{B} = \mathbf{A} \cdot \mathbf{B} + [\mathbf{A}, \mathbf{B}]. \tag{5.3.7}$$

The algebraic of quaternions $\mathbb{Q}$ is isomorphic to the Clifford algebra $\mathcal{CL}(\mathbb{R}^{-2})$. Another commonly used algebraic isomorphism is $\mathbb{Q} \cong \mathcal{CL}^+(\mathbb{R}^3)$: let $\{\mathbf{e}_1, \mathbf{e}_2, \mathbf{e}_2\}$ be an orthonormal basis of $\mathbb{R}^3$, then

$$\mathbf{i} = \mathbf{e}_1^{\sim}, \quad \mathbf{j} = \mathbf{e}_2^{\sim}, \quad \mathbf{k} = \mathbf{e}_3^{\sim}. \tag{5.3.8}$$

In history, Gibbs' 3D *vector algebra* was developed out of quaternions by the duality (5.3.8): for any vectors $\mathbf{a}, \mathbf{b} \in \mathbb{R}^3$, their *cross product*, also called *vector product*, denoted by $\mathbf{a} \times \mathbf{b}$, is defined by

$$(\mathbf{a} \times \mathbf{b})^{\sim} := [\mathbf{a}^{\sim}, \mathbf{b}^{\sim}]. \tag{5.3.9}$$

An alternative explanation is that via the vector product isomorphism induced by the dual operator, the quaternions form a 3D vector algebra over $\mathbb{R}^{-3}$.

The product (5.3.7) does not congrue with the geometric product if $\mathbf{A}, \mathbf{B}$ are taken as vectors. It congrues with the geometric product only when $\mathbf{A}, \mathbf{B}$ are taken as bivectors. In either case, the vector product $[\mathbf{A}, \mathbf{B}]$ does not congrue with the outer product. This may be the reason why Grassmann had been rejecting quaternions. It is Clifford who, by integrating Grassmann algebra with Hamilton's quaternions, formed a unified system in which the two algebras are fully compatible.

Proposition 5.50. For any $\mathbf{a}, \mathbf{b} \in \mathbb{R}^3$,

$$\begin{aligned} \mathbf{a} \times \mathbf{b} &= (\mathbf{a} \wedge \mathbf{b})^{\sim}, \\ [\mathbf{a}^{\sim}, \mathbf{b}^{\sim}] &= \langle \mathbf{a}^{\sim} \mathbf{b}^{\sim} \rangle_2. \end{aligned} \tag{5.3.10}$$

Proof. The second equality comes from the isomorphism $\mathbb{Q} \cong \mathcal{CL}^+(\mathbb{R}^3)$ under the correspondence (5.3.8). For the first equality, by (5.3.9),

$$(\mathbf{a} \times \mathbf{b})^{\sim} = [\mathbf{a}^{\sim}, \mathbf{b}^{\sim}] = \langle \mathbf{a}^{\sim} \mathbf{b}^{\sim} \rangle_2 = \langle \mathbf{a} \mathbf{I}_3^{-1} \mathbf{b} \mathbf{I}_3^{-1} \rangle_2 = \langle \mathbf{a} \mathbf{b} (\mathbf{I}_3^{-1})^2 \rangle_2 = -\mathbf{a} \wedge \mathbf{b}.$$

$$\square$$

Clifford algebra $\mathcal{CL}(\mathbb{R}^2)$ is four dimensional. It includes both the real vector space $\mathbb{R}^2 = \mathcal{CL}^-(\mathbb{R}^2)$ as its odd subspace, and the complex numbers $\mathbb{C} \cong \mathcal{CL}^+(\mathbb{R}^2)$ as its even subalgebra. Is this redundancy of having both the real and the complex representations for the same Euclidean plane necessary?

The complex numbers representation has the huge advantage that the product is commutative and invertible, but is inconvenient in representing the inner product and the vector product, and the representation (5.2.23) is basis dependent. The vector algebra representation changes the previous disadvantages into its own advantages, but also changes the previous advantages into its disadvantages. It is the Clifford algebra $\mathcal{CL}(\mathbb{R}^2)$ that includes both representations by summing them up to a 4D representation, where the two representations are unified with the geometric product. The Clifford algebraic representation inherits the advantages of both representations.

Quaternions are suited to represent points and their relations in 3D Euclidean geometry, but not for lines and planes. Again it is Clifford who proposed an algebra called *biquaternions*, or *dual quaternions*, to solve this problem.

Definition 5.51. Let ϵ be a nilpotent algebraic element commuting with every other algebraic element. An element in $\mathbb{R} + \epsilon\mathbb{R}$ is called a *dual number*, and an element in $\mathbb{Q} + \epsilon\mathbb{Q}$ is called a *dual quaternion*. An element in $\mathbb{R}^3 + \epsilon\mathbb{R}^3$ is called a *dual vector.* The 3D *dual vector algebra* over $\mathbb{R}^3$ is the set of dual vectors equipped with the inner product and the vector product induced from the 3D vector algebra over $\mathbb{R}^3$.

In the dual vector algebra over $\mathbb{R}^3$,

- point $\mathbf{a} \in \mathbb{R}^3$ is represented by $1 + \epsilon \mathbf{a}$;
- the line $(\mathbf{a}, \mathbf{l})$ passing through point $\mathbf{a} \in \mathbb{R}^3$ and following direction $\mathbf{l}$, is represented by $\mathbf{l} + \epsilon \mathbf{a} \times \mathbf{l}$;
- the plane normal to vector $\mathbf{n}$ and passing through point $\mathbf{a}$, is represented by $\mathbf{n} + \epsilon \mathbf{n} \cdot \mathbf{a}$.

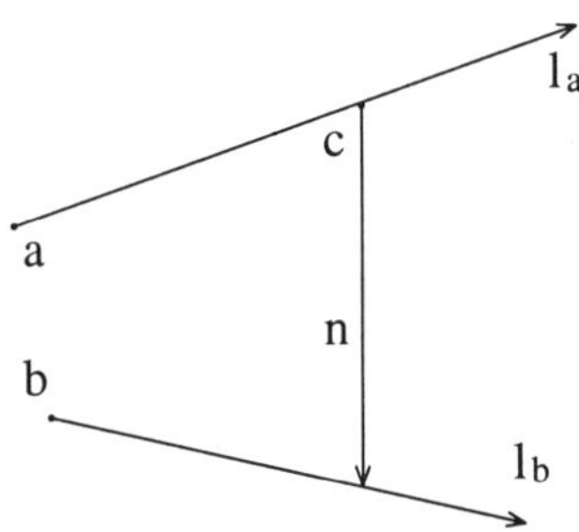

Fig. 5.2   Relationship between two spatial lines.

Dual vector algebra has been used in mechanical design and robot kinematics in computing the spatial relationship of two non-coplanar lines [209]. For two such lines $(\mathbf{a}, \mathbf{l}_a), (\mathbf{b}, \mathbf{l}_b)$, each represented by an incident point and a direction, let $\mathbf{n}$ be the unit vector along direction $\mathbf{l}_a \times \mathbf{l}_b$. Let δ be the signed distance from line $(\mathbf{a}, \mathbf{l}_a)$ to line $(\mathbf{b}, \mathbf{l}_b)$ along direction $\mathbf{n}$, let θ be the angle of rotation from $\mathbf{l}_a$ to $\mathbf{l}_b$, and let $\mathbf{c}$ be any point on the line of common perpendicular of the two lines. Denote

$$\theta_\delta := \theta + \epsilon\delta,$$
$$\mathbf{n_c} := \mathbf{n} + \epsilon \mathbf{c} \times \mathbf{n}, \tag{5.3.11}$$

and define

$$\cos\theta_\delta := \cos\theta - \epsilon\delta\sin\theta,$$
$$\sin\theta_\delta := \sin\theta + \epsilon\delta\cos\theta, \tag{5.3.12}$$
$$e^{\theta_\delta \mathbf{n_c}} := \cos\theta_\delta + \mathbf{n_c}\sin\theta_\delta.$$

Proposition 5.52. [172] With notations as above,

$$(\mathbf{l}_a + \epsilon\mathbf{a} \times \mathbf{l}_a)(\mathbf{l}_b + \epsilon\mathbf{b} \times \mathbf{l}_b) = e^{\theta_\delta \mathbf{n_c}}. \tag{5.3.13}$$

Proof.   Since $\mathbf{n_c}$ is irrelevant to the position of point $\mathbf{c}$ on the common perpendicular of the two lines, we can choose $\mathbf{c}$ to be on line $(\mathbf{a}, \mathbf{l}_a)$. Expanding both sides of (5.3.13), then using (5.3.10) to compare separately the scalar and vector terms, each being decomposed into a part with ϵ and a part without ϵ, we get

$$\mathbf{l}_a \cdot \mathbf{l}_b = \cos\theta,$$

$$\mathbf{l}_a \times \mathbf{l}_b = \mathbf{n}\sin\theta,$$

$$\mathbf{l}_a \cdot (\mathbf{b} \times \mathbf{l}_b) + (\mathbf{a} \times \mathbf{l}_a) \cdot \mathbf{l}_b = (\mathbf{a} - \mathbf{b}) \cdot (\mathbf{l}_a \times \mathbf{l}_b) = -\delta \sin\theta,$$

$$\begin{aligned}
\mathbf{l}_a \times (\mathbf{b} \times \mathbf{l}_b) + (\mathbf{a} \times \mathbf{l}_a) \times \mathbf{l}_b &= -\mathbf{l}_a \cdot (\mathbf{b} \wedge \mathbf{l}_b) + \mathbf{l}_b \cdot (\mathbf{a} \wedge \mathbf{l}_a)\\
&= -\mathbf{l}_a \cdot ((\mathbf{c} + \mathbf{n}\delta) \wedge \mathbf{l}_b) + \mathbf{l}_b \cdot (\mathbf{c} \wedge \mathbf{l}_a)\\
&= \mathbf{n}\delta (\mathbf{l}_a \cdot \mathbf{l}_b) - \mathbf{c} \cdot (\mathbf{l}_a \wedge \mathbf{l}_b)\\
&= \mathbf{n}\delta \cos\theta + \mathbf{c} \times \mathbf{n}\sin\theta.
\end{aligned}$$

$\square$

The set of dual quaternions equipped with its inner product and vector product, is isomorphic to the dual vector algebra over $\mathbb{R}^{-3}$. In Chapter 8, dual quaternions and dual vector algebra are extended to *dual Clifford algebra* for nD Euclidean geometry.

5.3.2　*Matrix-formed Clifford algebras*

É. Cartan made a complete classification of all real and complex Clifford algebras in 1908 [33]. He further discovered a variety of periodicity symmetries in the set of real and complex Clifford algebras.

Let $M(\mathbb{R}), M(\mathbb{C}), M(\mathbb{Q})$ denote matrix algebras of certain dimension, with real, complex and quaternion components respectively. Let $^2M(\mathbb{R}), ^2M(\mathbb{C}), ^2M(\mathbb{Q})$ be respectively the direct sum of two identical matrix algebras, whose components are in $\mathbb{R}, \mathbb{C}, \mathbb{Q}$ respectively.

Theorem 5.53. [Cartan's classification and period-8 theorem] For real Clifford algebras,

$$
\mathcal{CL}(\mathbb{R}^{p,q}) \cong
\begin{cases}
M(\mathbb{R}), & \text{if } p - q \equiv 0 \text{ or } 2 \quad \text{mod } 8,\\
^2M(\mathbb{R}), & \text{if } p - q \equiv 1 \qquad\quad \text{mod } 8,\\
M(\mathbb{C}), & \text{if } p - q \equiv 3 \text{ or } 7 \quad \text{mod } 8,\\
M(\mathbb{Q}), & \text{if } p - q \equiv 4 \text{ or } 6 \quad \text{mod } 8,\\
^2M(\mathbb{Q}), & \text{if } p - q \equiv 5 \qquad\quad \text{mod } 8.
\end{cases}
\tag{5.3.14}
$$

For complex Clifford algebras,

$$\mathcal{CL}(\mathbb{C}^{2r}) \cong M(\mathbb{C}), \quad \mathcal{CL}(\mathbb{C}^{2r+1}) \cong {}^2M(\mathbb{C}). \tag{5.3.15}$$

The dimensions of the matrix algebras in the table can be easily fixed. For example, $\mathcal{CL}(\mathbb{R}^{a+4b+3,a-4b-2})$ has real dimension 2^{2a+1}, so each matrix algebra component in $^2M(\mathbb{Q}) \cong \mathcal{CL}(\mathbb{R}^{a+4b+3,a-4b-2})$ has real dimension 2^{2a} and quaternionic dimension 2^{2a-2}, and the matrix size is $2^{a-1} \times 2^{a-1}$.

Corollary 5.54. If the inner product in $\mathcal{V}^n$ is nondegenerate, then almost all elements in $\mathcal{CL}(\mathcal{V}^n)$ are invertible.

Theorem 5.55. [Cartan's first group of periodicity theorems]

$$\mathcal{CL}(\mathbb{R}^{r,s}) \cong \mathcal{CL}(\mathbb{R}^{s+1,r-1}), \tag{5.3.16}$$

$$\mathcal{CL}(\mathbb{R}^{r,s}) \cong \mathcal{CL}(\mathbb{R}^{r+4,s-4}). \tag{5.3.17}$$

Example 5.56. $\mathcal{CL}(\mathbb{R}^2) \cong \mathcal{CL}(\mathbb{R}^{1,1}) \cong M_{2\times 2}(\mathbb{R})$.

Let $\mathbf{e}_1, \mathbf{e}_2$ be an orthonormal basis of $\mathbb{R}^2$. Since $(\mathbf{e}_1\mathbf{e}_2)^2 = -1$, $\mathbf{e}_1$ and $\mathbf{e}_1\mathbf{e}_2$ span a 2D real vector space that is isomorphic to $\mathbb{R}^{1,1}$. Since $\mathbf{e}_1(\mathbf{e}_1\mathbf{e}_2) = \mathbf{e}_2$, $\mathcal{CL}(\mathbb{R}^2) \cong \mathcal{CL}(\mathbb{R}^{1,1})$ as Clifford algebras.

The following correspondence between two bases of $\mathcal{CL}(\mathbb{R}^2)$ and $M_{2\times 2}(\mathbb{R})$ respectively is unique up to an orthogonal transformation in $\mathbb{R}^2$:

$$1 = \begin{pmatrix} 1 & \\ & 1 \end{pmatrix}, \quad \mathbf{e}_1 = \begin{pmatrix} 1 & \\ & -1 \end{pmatrix}, \quad \mathbf{e}_2 = \begin{pmatrix} & 1 \\ 1 & \end{pmatrix}, \quad \mathbf{e}_1\mathbf{e}_2 = \begin{pmatrix} & 1 \\ -1 & \end{pmatrix}. \tag{5.3.18}$$

In the matrix representation of $\mathcal{CL}(\mathbb{R}^2)$,

- 0-vectors are scalar multiples of the identity matrix;
- 1-vectors are symmetric matrices of trace zero;
- 2-vectors are antisymmetric matrices;
- the reversion operation is the matrix transpose.

Example 5.57. $\mathcal{CL}(\mathbb{R}^3) \cong \mathcal{CL}(\mathbb{R}^{1,2}) \cong M_{2\times 2}(\mathbb{C})$.

Let $\mathbf{e}_1, \mathbf{e}_2, \mathbf{e}_3$ be an orthonormal basis of $\mathbb{R}^3$. Then $\{\mathbf{e}_1, \mathbf{e}_1\mathbf{e}_2, \mathbf{e}_1\mathbf{e}_3\}$ is an orthonormal basis of $\mathbb{R}^{1,2}$. Consequently, $\mathcal{CL}(\mathbb{R}^3) \cong \mathcal{CL}(\mathbb{R}^{1,2})$.

The following correspondence is unique up to an orthogonal transformation in $\mathbb{R}^3$:

$$1 = \begin{pmatrix} 1 & \\ & 1 \end{pmatrix}, \quad \mathbf{e}_1 = \begin{pmatrix} & 1 \\ 1 & \end{pmatrix}, \quad \mathbf{e}_2 = \begin{pmatrix} & -i \\ i & \end{pmatrix}, \quad \mathbf{e}_3 = \begin{pmatrix} 1 & \\ & -1 \end{pmatrix},$$

$$\mathbf{e}_1\mathbf{e}_2\mathbf{e}_3 = \begin{pmatrix} i & \\ & i \end{pmatrix}, \quad \mathbf{e}_2\mathbf{e}_3 = \begin{pmatrix} & i \\ i & \end{pmatrix}, \quad \mathbf{e}_3\mathbf{e}_1 = \begin{pmatrix} & 1 \\ -1 & \end{pmatrix}, \quad \mathbf{e}_1\mathbf{e}_2 = \begin{pmatrix} i & \\ & -i \end{pmatrix}.$$

The matrices corresponding to $\mathbf{e}_1, \mathbf{e}_2, \mathbf{e}_3$ are called *Pauli spin matrices* [141]. In the matrix representation of $\mathcal{CL}(\mathbb{R}^3)$,

- 0-vectors are real multiples of the identity matrix;
- 1-vectors are Hermitian symmetric matrices of trace zero;
- 2-vectors are Hermitian antisymmetric matrices of trace zero;
- 3-vectors are pure imaginary multiples of the identity matrix;
- the reversion operation is the matrix Hermitian-transpose;
- the dual operation is the scaling by $-i$.

The tensor product of two Clifford algebras $\mathcal{A}, \mathcal{B}$ as $\mathbb{Z}$-graded spaces, if endowed with the geometric product defined by (2.3.7) (or by (2.3.8)), with the understanding that the juxtaposition there denotes the geometric product instead of the outer product, becomes a Clifford algebra, called the *tensor product* (or *twisted tensor product*) of the two Clifford algebras, denoted by $\mathcal{A} \otimes \mathcal{B}$ (or $\mathcal{A} \,\hat{\otimes}\, \mathcal{B}$).

The following is the counterpart of Proposition 2.21:

Proposition 5.58. If inner-product spaces $\mathcal{V}^n, \mathcal{W}^m$ are orthogonal, then

$$\mathcal{CL}(\mathcal{V}^n \oplus \mathcal{W}^m) \cong \mathcal{CL}(\mathcal{V}^n) \,\hat{\otimes}\, \mathcal{CL}(\mathcal{W}^m). \tag{5.3.19}$$

Theorem 5.59. [Cartan's second group of periodicity theorems]

$$\mathcal{CL}(\mathbb{R}^{r+1,s+1}) \cong \mathcal{CL}(\mathbb{R}^{r,s}) \otimes \mathcal{CL}(\mathbb{R}^{1,1}), \tag{5.3.20}$$

$$\mathcal{CL}(\mathbb{R}^{r+2,s}) \cong \mathcal{CL}(\mathbb{R}^{s,r}) \otimes \mathcal{CL}(\mathbb{R}^{2,0}), \tag{5.3.21}$$

$$\mathcal{CL}(\mathbb{R}^{r,s+2}) \cong \mathcal{CL}(\mathbb{R}^{s,r}) \otimes \mathcal{CL}(\mathbb{R}^{0,2}), \tag{5.3.22}$$

$$\mathcal{CL}(\mathbb{C}^{r+2}) \cong \mathcal{CL}(\mathbb{C}^r) \otimes \mathcal{CL}(\mathbb{C}^2). \tag{5.3.23}$$

Proof. The derivations of these periodicities are much the same, so we only consider the first one. Let $\mathbb{R}^{r+1,s+1} = \mathbb{R}^{r,s} \oplus \mathbb{R}^{1,1}$ be a fixed orthogonal decomposition. Let $\mathbf{I}_2$ be a unit pseudoscalar in $\Lambda(\mathbb{R}^{1,1})$. Then $\mathbf{I}_2^2 = 1$. Any vector $\mathbf{x} \in \mathbb{R}^{r+1,s+1}$ has a unique orthogonal decomposition with respect to plane $\mathbf{I}_2$: $\mathbf{x} = P_{\mathbf{I}_2}(\mathbf{x}) + P_{\mathbf{I}_2}^{\perp}(\mathbf{x})$.

Define a linear isomorphism f from $\mathbb{R}^{r+1,s+1}$ into $\mathcal{CL}(\mathbb{R}^{r,s}) \otimes \mathcal{CL}(\mathbb{R}^{1,1})$ as follows [48]:

$$f: \quad \mathbf{x} \longmapsto P_{\mathbf{I}_2}^{\perp}(\mathbf{x}) \otimes \mathbf{I}_2 + 1 \otimes P_{\mathbf{I}_2}(\mathbf{x}). \tag{5.3.24}$$

Since $(f(\mathbf{x}))^2 = (P_{\mathbf{I}_2}(\mathbf{x}))^2 + (P_{\mathbf{I}_2}^{\perp}(\mathbf{x}))^2 = \mathbf{x}^2$, by the universal property of Clifford algebra, f induces an isomorphism of Clifford algebras from $\mathcal{CL}(\mathbb{R}^{r+1,s+1})$ into $\mathcal{CL}(\mathbb{R}^{r,s}) \otimes \mathcal{CL}(\mathbb{R}^{1,1})$. That this isomorphism is also surjective is because the two sides of it are vector spaces of the same dimension. $\quad\square$

By $\mathcal{CL}(\mathbb{R}^{1,1}) \cong M_{2\times 2}(\mathbb{R})$, (5.3.20) can be written as

$$\mathcal{CL}(\mathbb{R}^{r+1,s+1}) \cong M_{2\times 2}(\mathcal{CL}(\mathbb{R}^{r,s})). \tag{5.3.25}$$

The right side of (5.3.25) is an instance of *Clifford matrix algebra*, i.e., an algebra of matrices whose components are Clifford numbers. It is an important tool in the study of nD Möbius transformations, and will be investigated in Chapter 8.

From (5.3.24) and (5.3.18), we get

Corollary 5.60. [145] Let $\mathbf{e}_1, \ldots, \mathbf{e}_{r+s}$ be an orthonormal basis of $\mathbb{R}^{r,s}$. Then

$$\begin{pmatrix} \mathbf{e}_1 & 0 \\ 0 & -\mathbf{e}_1 \end{pmatrix}, \begin{pmatrix} \mathbf{e}_2 & 0 \\ 0 & -\mathbf{e}_2 \end{pmatrix}, \ldots, \begin{pmatrix} \mathbf{e}_{r+s} & 0 \\ 0 & -\mathbf{e}_{r+s} \end{pmatrix}, \begin{pmatrix} 0 & 1 \\ 1 & 0 \end{pmatrix}, \begin{pmatrix} 0 & -1 \\ 1 & 0 \end{pmatrix} \tag{5.3.26}$$

are an orthonormal basis of $\mathbb{R}^{r+1,s+1}$ under the isomorphism (5.3.25).

5.3.3 *Groups in Clifford algebra*

Definition 5.61. A *versor* refers to a Clifford monomial composed of invertible vectors. It is called a *rotor*, or *spinor*, if the number of vectors is even. It is called a *unit versor* if its magnitude is 1.

All versors in $\mathcal{CL}(\mathcal{V}^n)$ form a group under the geometric product, called the *versor group*, also known as the *Clifford group*, or *Lipschitz group*. All rotors form a subgroup, called the *rotor group*. All unit versors form a subgroup, called the *pin group*, and all unit rotors form a subgroup, called the *spin group*, denoted by $\mathrm{Spin}(\mathcal{V}^n)$.

The left *graded adjoint action* (or *twisted adjoint representation*) of a versor $\mathbf{V}$ upon $\mathcal{CL}(\mathcal{V}^n)$, denoted by $Ad_{\mathbf{V}}$, is defined by

$$Ad_{\mathbf{V}}(\mathbf{A}) := \mathbf{V}\mathbf{A}\hat{\mathbf{V}}^{-1}, \ \forall \mathbf{A} \in \mathcal{CL}(\mathcal{V}^n), \tag{5.3.27}$$

where the overhat symbol denotes the grade involution (5.2.6).

It is a classical result that when $\mathbb{K} = \mathbb{R}$ or $\mathbb{C}$, an invertible element $\mathbf{A} \in \mathcal{CL}(\mathcal{V}^n)$ is a versor if and only if $Ad_{\mathbf{A}}(\mathbf{x}) \in \mathcal{V}^n$ for any $\mathbf{x} \in \mathcal{V}^n$. The latter is often used as an alternative definition of versors.

In the simplest case where $\mathbf{V} = \mathbf{a}$ is a vector, for any vector $\mathbf{x} \in \mathcal{V}^n$,

$$\begin{aligned}
Ad_{\mathbf{a}}(\mathbf{x}) &= -\mathbf{a}\mathbf{x}\mathbf{a}^{-1}\\
&= -(\mathbf{a}\mathbf{x}\mathbf{a})\mathbf{a}^{-2}\\
&= -\{\mathbf{a}(\mathbf{x}\cdot\mathbf{a}) + \mathbf{a}\cdot(\mathbf{x}\wedge\mathbf{a})\}\mathbf{a}^{-2}\\
&= -\{2(\mathbf{a}\cdot\mathbf{x})\mathbf{a} - \mathbf{a}^2\mathbf{x}\}\mathbf{a}^{-2}\\
&= \mathbf{x} - 2(\mathbf{a}^{-1}\cdot\mathbf{x})\mathbf{a}.
\end{aligned} \tag{5.3.28}$$

So if $\mathbf{x}\wedge\mathbf{a} = 0$, then $Ad_{\mathbf{a}}(\mathbf{x}) = -\mathbf{x}$; if $\mathbf{x}\cdot\mathbf{a} = 0$, then $Ad_{\mathbf{a}}(\mathbf{x}) = \mathbf{x}$. Geometrically, $Ad_{\mathbf{a}}$ is the *mirror reflection* with respect to the hyperplane normal to $\mathbf{a}$.

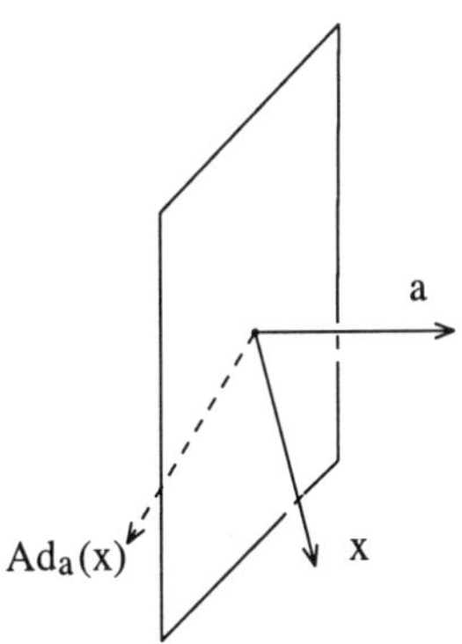

Fig. 5.3   Mirror reflection with respect to the hyperplane normal to **a**.

For any versor $\mathbf{V} = \mathbf{a}_1\mathbf{a}_2\cdots\mathbf{a}_k$, since

$$Ad_{\mathbf{V}}(\mathbf{x}) = (-1)^k\mathbf{a}_1\cdots\mathbf{a}_k\mathbf{x}\mathbf{a}_k^{-1}\cdots\mathbf{a}_1^{-1}, \tag{5.3.29}$$

the result is the composition of r reflections. $Ad_{\mathbf{V}}$ is an orthogonal transformation because $(Ad_{\mathbf{V}}(\mathbf{x}))^2 = \mathbf{V}\mathbf{x}\mathbf{V}^{-1}\mathbf{V}\mathbf{x}\mathbf{V}^{-1} = \mathbf{x}^2$. Furthermore, for any $\mathbf{A}, \mathbf{B} \in \mathcal{CL}(\mathcal{V}^n)$, since

$$Ad_{\mathbf{V}}(\mathbf{AB}) = (-1)^k \mathbf{VAV}^{-1}\mathbf{VBV}^{-1} = (-1)^k Ad_{\mathbf{V}}(\mathbf{A})\, Ad_{\mathbf{V}}(\mathbf{B}), \qquad (5.3.30)$$

$Ad_{\mathbf{V}}$ is a Clifford algebraic automorphism in $\mathcal{CL}(\mathcal{V}^n)$ if $\mathbf{V}$ is even, and an *anti-automorphism* if $\mathbf{V}$ is odd, the meaning of the latter title being clear from (5.3.30).

Since $Ad_{\mathbf{V}}$ is an orthogonal transformation of $\mathcal{V}^n$, it preserves the $\mathbb{Z}$-grading in $\mathcal{CL}(\mathcal{V}^n)$. $Ad_{\mathbf{V}}$ is also a Grassmann algebraic automorphism (or anti-automorphism) in $\Lambda(\mathcal{V}^n)$, if $\mathbf{V}$ is even (or odd).

For unit vectors $\mathbf{a}, \mathbf{b}$ in $\mathbb{R}^n$, let $\mathbf{I}_2$ be the unit 2-blade representing the oriented plane $\mathbf{a} \wedge \mathbf{b}$, and let θ be the angle of rotation from $\mathbf{a}$ to $\mathbf{b}$. Then

$$\begin{aligned}
\mathbf{ab} \quad &= \mathbf{a}\cdot\mathbf{b} + \mathbf{a}\wedge\mathbf{b} = \cos\theta + \mathbf{I}_2\sin\theta = e^{\mathbf{I}_2\theta}, \\
(\mathbf{ab})^{-1} = \quad \mathbf{ba} \quad &= \cos\theta - \mathbf{I}_2\sin\theta = e^{-\mathbf{I}_2\theta}.
\end{aligned} \qquad (5.3.31)$$

For any vector $\mathbf{x} \in \mathbb{R}^n$,

$$Ad_{\mathbf{ba}}(\mathbf{x}) = e^{-\mathbf{I}_2\theta}(P_{\mathbf{I}_2}(\mathbf{x}) + P_{\mathbf{I}_2}^{\perp}(\mathbf{x}))e^{\mathbf{I}_2\theta} = P_{\mathbf{I}_2}(\mathbf{x})e^{\mathbf{I}_2(2\theta)} + P_{\mathbf{I}_2}^{\perp}(\mathbf{x}). \qquad (5.3.32)$$

So $Ad_{\mathbf{ba}}$ is the rotation of angle 2θ in oriented plane $\mathbf{a} \wedge \mathbf{b}$. In this way, a rotor realizes a sequence of rotations by the graded adjoint action. This justifies the term "rotor". (5.3.32) also shows that a rotor is a half-angle representation of a rotation.

Conversely, if a 2D rotation changes vector $\mathbf{a}$ to vector $\mathbf{b}$, then a rotor realizing this rotation is

$$(\mathbf{a} + \mathbf{b})\mathbf{a}, \qquad (5.3.33)$$

as $\mathbf{a} + \mathbf{b}$ is the half-way direction between $\mathbf{a}$ and $\mathbf{b}$.

Theorem 5.62. [Cartan-Dieudonné Theorem] Let $\mathcal{V}^n$ be a nondegenerate inner-product space. Then any orthogonal transformation in $\mathcal{V}^n$ can be realized by the graded adjoint action of a versor of length $\leq n$ in $\mathcal{CL}(\mathcal{V}^n)$. Conversely, any versor realizes an orthogonal transformation. Two versors realize the same orthogonal transformation if and only if they are equal up to scale.

Example 5.63. [171] (Steenrod 1974, pp. 119-121) Let $\mathbb{S}^n$ be the unit sphere of $\mathbb{R}^{n+1}$. The unit point on the i-th axis of $\mathbb{R}^{n+1}$ is denoted by $\mathbf{x}_{i-1}$. The antipodal point of $\mathbf{x} \in \mathbb{S}^n$ is denoted by $-\mathbf{x}$. The *normal form* of the principle bundle $p : SO(n+1) \longrightarrow \mathbb{S}^n$, where $p(\rho) = \rho(\mathbf{x}_n)$ for $\rho \in SO(n+1)$, is defined as follows:

- The coordinate neighborhoods in $\mathbb{S}^n$ are $V_1 = \mathbb{S}^n\backslash(-\mathbf{x}_n)$ and $V_2 = \mathbb{S}^n\backslash(\mathbf{x}_n)$.
- Define $\phi : V_1 \longrightarrow SO(n+1)$ to be the map which assigns to $\mathbf{x} \neq -\mathbf{x}_n$ the rotation which leaves invariant all points orthogonal to both $\mathbf{x}$ and $\mathbf{x}_n$, and rotates the great circle through the two points so as to carry $\mathbf{x}_n$ to $\mathbf{x}$.
- Define $\psi = (\phi(\mathbf{x}_{n-1}))^2$. The juxtaposition and power here denote the group multiplication in $SO(n+1)$.

- It is assumed that $SO(n)$ operates trivially on the $(n+1)$-th axis of $\mathbb{R}^{n+1}$. The coordinate functions of the principle bundle are

$$\phi_1(\mathbf{x}, \rho) = \phi(\mathbf{x})\,\rho, \qquad \text{for } \mathbf{x} \in V_1,\ \rho \in SO(n);$$
$$\phi_2(\mathbf{x}, \rho) = \psi\,\phi(\psi(\mathbf{x}))\,\rho, \quad \text{for } \mathbf{x} \in V_2,\ \rho \in SO(n).$$

Compute the coordinate transformation in the bundle:

$$g_{12}(\mathbf{x}) = (\phi(\mathbf{x}))^{-1}\,\psi\,\phi(\psi(\mathbf{x})). \tag{5.3.34}$$

The computing can be based on orthogonal matrices, but is more complicated and geometrically much less intuitive than the following rotor approach.

Bear in mind that the scale of a rotor is removable in representing an orthogonal transformation. By (5.3.33), $\phi(\mathbf{x}) = (\mathbf{x}_n+\mathbf{x})\mathbf{x}_n$ for $\mathbf{x} \neq -\mathbf{x}_n$. Then $(\phi(\mathbf{x}))^{-1}$ equals, up to scale, $(\phi(\mathbf{x}))^{\dagger} = \mathbf{x}_n(\mathbf{x}_n + \mathbf{x})$. By

$$(\phi(\mathbf{x}_{n-1}))^2 = ((\mathbf{x}_n + \mathbf{x}_{n-1})\mathbf{x}_n)^2 = 2\mathbf{x}_{n-1}\mathbf{x}_n,$$

we can let $\psi = \mathbf{x}_{n-1}\mathbf{x}_n$. Then $\psi(\mathbf{x}) = \mathbf{x}_{n-1}\mathbf{x}_n\mathbf{x}\mathbf{x}_n\mathbf{x}_{n-1}$, and

$$\phi(\psi(\mathbf{x})) = (\mathbf{x}_n + \mathbf{x}_{n-1}\mathbf{x}_n\mathbf{x}\mathbf{x}_n\mathbf{x}_{n-1})\mathbf{x}_n = 1 + \mathbf{x}_n\mathbf{x}_{n-1}\mathbf{x}\mathbf{x}_{n-1}.$$

Substituting the above expressions into (5.3.34), we get

$$\begin{aligned}
g_{12}(\mathbf{x}) &= \mathbf{x}_n(\mathbf{x}_n + \mathbf{x})\mathbf{x}_{n-1}\mathbf{x}_n(1 + \mathbf{x}_n\mathbf{x}_{n-1}\mathbf{x}\mathbf{x}_{n-1}) \\
&= \mathbf{x}\mathbf{x}_{n-1} - \mathbf{x}_n\mathbf{x}\mathbf{x}_n\mathbf{x}_{n-1} \\
&= P^{\perp}_{\mathbf{x}_n}(\mathbf{x})\,\mathbf{x}_{n-1}.
\end{aligned} \tag{5.3.35}$$

5.4   Clifford expansion theory

Changing an expression of geometric products in Clifford algebra into an equal expression of inner products and outer products in inner-product Grassmann algebra is called *Clifford expansion*. *Clifford expansion theory* studies different Clifford expansions of the same expression in Clifford algebra. This theory exhibits a strong flavor of combinatorics.

Notation.
From now on, the 0-grading operator $\langle\ \rangle_0$ is always denoted by the *angular bracket* $\langle\ \rangle$. It has the following obvious property, according to (5.2.13) and (5.2.16):

$$\langle \mathbf{AB} \rangle = \langle \mathbf{BA} \rangle, \quad \forall \mathbf{A}, \mathbf{B} \in \mathcal{CL}(\mathcal{V}^n). \tag{5.4.1}$$

In this section, boldfaced letters $\mathbf{a}_i, \mathbf{b}_j$ always denote vectors in $\mathcal{V}^n$.

5.4.1   *Expansion of the geometric product of vectors*

Theorem 5.64. [32] [Fundamental Clifford expansion of the geometric product of vectors] Let $\mathbf{A}_k = \mathbf{a}_1\mathbf{a}_2\cdots\mathbf{a}_k$. Then for any $1 \leq l \leq [k/2]$,

$$\langle \mathbf{A}_k \rangle_{k-2l} = \sum_{(2l,\,k-2l)\vdash \mathbf{A}_k} \langle \mathbf{A}_{k(1)} \rangle\,\langle \mathbf{A}_{k(2)} \rangle_{k-2l}. \tag{5.4.2}$$

Theorem 5.65. With the same notation as above,

$$\langle \mathbf{a}_1 \mathbf{a}_2 \cdots \mathbf{a}_{2l} \rangle = \sum_{i=2}^{2l} (-1)^i (\mathbf{a}_1 \cdot \mathbf{a}_i) \langle \mathbf{a}_2 \cdots \breve{\mathbf{a}}_i \cdots \mathbf{a}_{2l} \rangle. \qquad (5.4.3)$$

Proof. We prove the two theorems simultaneously.

When $l = 1$, (5.4.3) is trivial; when $k = 2$, (5.4.2) is trivial. Assume that for $l = 1$, $k = m - 1$, (5.4.2) is true. For $l = 1$, $k = m$,

$$\langle \mathbf{A}_m \rangle_{m-2} = \langle \mathbf{A}_{m-1} \rangle_{m-1} \cdot \mathbf{a}_m + \langle \mathbf{A}_{m-1} \rangle_{m-3} \wedge \mathbf{a}_m$$

$$= \sum_{(m-2,1)\vdash \mathbf{A}_{m-1}} (\mathbf{A}_{m-1\,(2)} \cdot \mathbf{a}_m) \langle \mathbf{A}_{m-1\,(1)} \rangle_{m-2}$$

$$+ \sum_{(2,m-3)\vdash \mathbf{A}_{m-1}} \langle \mathbf{A}_{m-1\,(1)} \rangle \langle \mathbf{A}_{m-1\,(2)} \rangle_{m-3} \wedge \mathbf{a}_m$$

$$= \Big(\sum_{\substack{(2,m-2)\vdash \mathbf{A}_m, \\ \mathbf{a}_m \in \mathbf{A}_{m\,(1)}}} + \sum_{\substack{(2,m-2)\vdash \mathbf{A}_m, \\ \mathbf{a}_m \notin \mathbf{A}_{m\,(1)}}} \Big) \langle \mathbf{A}_{m\,(1)} \rangle \langle \mathbf{A}_{m\,(2)} \rangle_{m-2}.$$

So (5.4.2) is true for $l = 1$ and any k. Now assume that for $l = r - 1$ and any k, (5.4.2), (5.4.3) are true. Then

$$\langle \mathbf{a}_1 \cdots \mathbf{a}_{2r} \rangle = \mathbf{a}_1 \cdot \langle \mathbf{a}_2 \cdots \mathbf{a}_{2r} \rangle_1 = \mathbf{a}_1 \cdot \sum_{i=2}^{2r} (-1)^i \langle \mathbf{a}_2 \cdots \breve{\mathbf{a}}_i \cdots \mathbf{a}_{2r} \rangle \mathbf{a}_i.$$

So (5.4.3) is true for $l = r$. This proves (5.4.3).

For $l = r$, (5.4.2) is true for $k = 2l$. Assume that (5.4.2) is true for $k = m - 1 \geq 2l$. For $k = m$,

$$\langle \mathbf{A}_m \rangle_{m-2r} = \langle \mathbf{A}_{m-1} \rangle_{m-2r+1} \cdot \mathbf{a}_m + \langle \mathbf{A}_{m-1} \rangle_{m-2r-1} \wedge \mathbf{a}_m$$

$$= \sum_{(2r-2,m-2r+1)\vdash \mathbf{A}_{m-1}} \langle \mathbf{A}_{m-1\,(1)} \rangle \langle \mathbf{A}_{m-1\,(2)} \rangle_{m+1-2r} \cdot \mathbf{a}_m$$

$$+ \sum_{(2r,m-2r-1)\vdash \mathbf{A}_{m-1}} \langle \mathbf{A}_{m-1\,(1)} \rangle \langle \mathbf{A}_{m-1\,(2)} \rangle_{m-1-2r} \wedge \mathbf{a}_m$$

$$= \sum_{(2r-2,m-2r,1)\vdash \mathbf{A}_{m-1}} (\mathbf{A}_{m-1\,(3)} \cdot \mathbf{a}_m) \langle \mathbf{A}_{m-1\,(1)} \rangle \langle \mathbf{A}_{m-1\,(2)} \rangle_{m-2r}$$

$$+ \sum_{(2r,m-2r-1)\vdash \mathbf{A}_{m-1}} \langle \mathbf{A}_{m-1\,(1)} \rangle \langle \mathbf{A}_{m-1\,(2)} \rangle_{m-2r-1} \wedge \mathbf{a}_m$$

$$= \sum_{(m-2r,2r-2,1)\vdash \mathbf{A}_{m-1}} \langle \mathbf{A}_{m-1\,(1)} \rangle_{m-2r} (\mathbf{A}_{m-1\,(3)} \cdot \mathbf{a}_m) \langle \mathbf{A}_{m-1\,(2)} \rangle$$

$$+ \sum_{(2r,m-2r-1)\vdash \mathbf{A}_{m-1}} \langle \mathbf{A}_{m-1\,(1)} \rangle \langle \mathbf{A}_{m-1\,(2)} \rangle_{m-2r-1} \wedge \mathbf{a}_m$$

$$= \Big(\sum_{\substack{(2r,m-2r)\vdash \mathbf{A}_m, \\ \mathbf{a}_m \in \mathbf{A}_{m\,(1)}}} + \sum_{\substack{(2r,m-2r)\vdash \mathbf{A}_m, \\ \mathbf{a}_m \notin \mathbf{A}_{m\,(1)}}} \Big) \langle \mathbf{A}_{m\,(1)} \rangle \langle \mathbf{A}_{m\,(2)} \rangle_{m-2r}.$$

So (5.4.2) is true for $l = r$ and $k = m$. $\qquad\qquad\qquad\qquad\qquad\qquad\square$

Corollary 5.66. For $\mathbf{A}_{2l} = \mathbf{a}_1 \mathbf{a}_2 \cdots \mathbf{a}_{2l}$,

$$\langle \mathbf{A}_{2l} \rangle = \frac{1}{l!} \sum_{\underbrace{(2,2,\ldots,2)}_{l} \vdash \mathbf{A}_{2l}} \prod_{i=1}^{l} \langle \mathbf{A}_{2l\,(i)} \rangle. \tag{5.4.4}$$

The right side is the *Pfaffian* of the Gram matrix $(\mathbf{a}_i \cdot \mathbf{a}_j)_{i,j=1..2l}$.

Remark: (1) For a partition of shape $(0, k)$ of Clifford monomial $\mathbf{A}_k$, it is always assumed that $\mathbf{A}_{k\,(1)} = 1$ and $\mathbf{A}_{k\,(2)} = \mathbf{A}_k$.

(2) In linear algebra, the *Pfaffian* of a matrix $A = (a_{ij})_{i,j=1..2l}$ is defined as follows: let I be the sequence of indices $1, 2, \ldots, 2l$, then

$$\mathrm{pf}(A) := \frac{1}{l!} \sum_{\underbrace{(2,2,\ldots,2)}_{l} \vdash I} \prod_{i=1}^{l} a_{I_{(i)}}. \tag{5.4.5}$$

In particular, for a fixed basis $\mathbf{e}_1, \mathbf{e}_2, \ldots, \mathbf{e}_{2l}$ of $\mathcal{V}^{2l}$, the Plücker coordinates of a bivector $\mathbf{B}_2$ can be written as

$$\mathbf{B}_2 = \sum_{1 \leq i < j \leq n} (\mathbf{B}_2 \,|\, ij)\, \mathbf{e}_i \wedge \mathbf{e}_j, \tag{5.4.6}$$

where minor $(\mathbf{B}_2 \,|\, ij)$ is the coordinate of $\mathbf{B}_2$ with respect to $\mathbf{e}_i \wedge \mathbf{e}_j$. The Plücker coordinate matrix of $\mathbf{B}_2$ is $((\mathbf{B}_2 \,|\, ij))_{i,j=1..2l}$, where $(\mathbf{B}_2 \,|\, ii) = 0$, and $(\mathbf{B}_2 \,|\, ji) = -(\mathbf{B}_2 \,|\, ij)$. The Pfaffian of the matrix is called the *Pfaffian* of $\mathbf{B}_2$ with respect to the basis of $\mathcal{V}^{2l}$:

$$\mathrm{pf}(\mathbf{B}_2) := \frac{1}{l!} \sum_{\underbrace{(2,2,\ldots,2)}_{l} \vdash I} \prod_{i=1}^{l} (\mathbf{B}_2 \,|\, I_{(i)}) = \frac{1}{l!} (\underbrace{\mathbf{B}_2 \mathbf{B}_2 \ldots \mathbf{B}_2}_{l} \,|\, \mathbf{e}_1 \mathbf{e}_2 \ldots \mathbf{e}_{2l})$$

$$= \frac{1}{l!} [\underbrace{\mathbf{B}_2 \mathbf{B}_2 \ldots \mathbf{B}_2}_{l}]. \tag{5.4.7}$$

A classical theorem in invariant theory says that $\mathrm{pf}(\mathbf{B}_2)$ is a homogeneous invariant polynomial over $\Lambda^2(\mathcal{V}^{2l})$, with $\mathbf{B}_2$ as the bivector indeterminate, and any homogeneous invariant polynomial over $\Lambda^2(\mathcal{V}^{2l})$ is a nonnegative integer power of the Pfaffian up to a constant scale [69].

Example 5.67. For vectors $\mathbf{1}, \mathbf{2}, \mathbf{3}, \mathbf{4}, \mathbf{5}, \mathbf{6} \in \mathcal{V}^n$,

$$\begin{aligned}
\langle \mathbf{123456} \rangle = \;& (1 \cdot 2)(3 \cdot 4)(5 \cdot 6) - (1 \cdot 2)(3 \cdot 5)(4 \cdot 6) + (1 \cdot 2)(3 \cdot 6)(4 \cdot 5) \\
& -(1 \cdot 3)(2 \cdot 4)(5 \cdot 6) + (1 \cdot 3)(2 \cdot 5)(4 \cdot 6) - (1 \cdot 3)(2 \cdot 6)(4 \cdot 5) \\
& +(1 \cdot 4)(2 \cdot 3)(5 \cdot 6) - (1 \cdot 4)(2 \cdot 5)(3 \cdot 6) + (1 \cdot 4)(2 \cdot 6)(3 \cdot 5) \\
& -(1 \cdot 5)(2 \cdot 3)(4 \cdot 6) + (1 \cdot 5)(2 \cdot 4)(3 \cdot 6) - (1 \cdot 5)(2 \cdot 6)(3 \cdot 4) \\
& +(1 \cdot 6)(2 \cdot 3)(4 \cdot 5) - (1 \cdot 6)(2 \cdot 4)(3 \cdot 5) + (1 \cdot 6)(2 \cdot 5)(3 \cdot 4),
\end{aligned}$$

$$\langle \mathbf{123456} \rangle_4 = \quad (1 \cdot 2)3 \wedge 4 \wedge 5 \wedge 6 - (1 \cdot 3)2 \wedge 4 \wedge 5 \wedge 6 + (1 \cdot 4)2 \wedge 3 \wedge 5 \wedge 6$$
$$-(1 \cdot 5)2 \wedge 3 \wedge 4 \wedge 6 + (1 \cdot 6)2 \wedge 3 \wedge 4 \wedge 5 + (2 \cdot 3)1 \wedge 4 \wedge 5 \wedge 6$$
$$-(2 \cdot 4)1 \wedge 3 \wedge 5 \wedge 6 + (2 \cdot 5)1 \wedge 3 \wedge 4 \wedge 6 - (2 \cdot 6)1 \wedge 3 \wedge 4 \wedge 5$$
$$+(3 \cdot 4)1 \wedge 2 \wedge 5 \wedge 6 - (3 \cdot 5)1 \wedge 2 \wedge 4 \wedge 6 + (3 \cdot 6)1 \wedge 2 \wedge 4 \wedge 5$$
$$+(4 \cdot 5)1 \wedge 2 \wedge 3 \wedge 6 - (4 \cdot 6)1 \wedge 2 \wedge 3 \wedge 5 + (5 \cdot 6)1 \wedge 2 \wedge 3 \wedge 4.$$

Corollary 5.68. Let $\mathbf{A}_r = \mathbf{a}_1 \mathbf{a}_2 \cdots \mathbf{a}_r$, $\mathbf{B}_s = \mathbf{b}_1 \mathbf{b}_2 \cdots \mathbf{b}_s$. Then

$$\langle \mathbf{A}_r \mathbf{B}_s \rangle_{r+s-2l} = \sum_{\substack{i+j=2l, \\ 0 \le i \le r, 0 \le j \le s}} \sum_{\substack{(r-i,i) \vdash \mathbf{A}_r, \\ (j,s-j) \vdash \mathbf{B}_s}} \langle \mathbf{A}_{r(2)} \mathbf{B}_{s(1)} \rangle \langle \mathbf{A}_{r(1)} \mathbf{B}_{s(2)} \rangle_{r+s-2l}, \tag{5.4.8}$$

$$\langle \mathbf{B}_s \mathbf{A}_r \rangle_{r+s-2l} = \sum_{\substack{i+j=2l, \\ 0 \le i \le r, 0 \le j \le s}} \sum_{\substack{(r-i,i) \vdash \mathbf{A}_r, \\ (j,s-j) \vdash \mathbf{B}_s}} (-1)^{rs-i} \langle \mathbf{A}_{r(2)} \mathbf{B}_{s(1)} \rangle \langle \mathbf{A}_{r(1)} \mathbf{B}_{s(2)} \rangle_{r+s-2l}. \tag{5.4.9}$$

Proof. (5.4.8) is direct from (5.4.2). (5.4.9) is from (5.4.8) as follows:

$$\langle \mathbf{B}_s \mathbf{A}_r \rangle_{r+s-2l}$$

$$= \sum_{\substack{i+j=2l, \\ 0 \le i \le r, 0 \le j \le s}} \sum_{\substack{(i,r-i) \vdash \mathbf{A}_r, \\ (s-j,j) \vdash \mathbf{B}_s}} \langle \mathbf{B}_{s(2)} \mathbf{A}_{r(1)} \rangle \langle \mathbf{B}_{s(1)} \mathbf{A}_{r(2)} \rangle_{r+s-2l}$$

$$= \sum_{\substack{i+j=2l, \\ 0 \le i \le r, 0 \le j \le s}} \sum_{\substack{(r-i,i) \vdash \mathbf{A}_r, \\ (j,s-j) \vdash \mathbf{B}_s}} (-1)^{i(r-i)+j(s-j)+(r-i)(s-j)} \langle \mathbf{A}_{r(2)} \mathbf{B}_{s(1)} \rangle \langle \mathbf{A}_{r(1)} \mathbf{B}_{s(2)} \rangle_{r+s-2l}.$$

$\square$

In (5.4.2), $\mathbf{A}_k$ is partitioned into shape $(2l, k-2l)$. There can be other partitions. The following series of propositions show that the influence of the shape is only a scale by combinatorial numbers.

Lemma 5.69. Let $\mathbf{A}_{2l} = \mathbf{a}_1 \mathbf{a}_2 \cdots \mathbf{a}_{2l}$. Then for any $0 \le r \le l$,

$$\langle \mathbf{A}_{2l} \rangle = \frac{1}{C_l^r} \sum_{(2r,2l-2r) \vdash \mathbf{A}_{2l}} \langle \mathbf{A}_{2l(1)} \rangle \langle \mathbf{A}_{2l(2)} \rangle. \tag{5.4.10}$$

Proof. Expand both sides by (5.4.4). $\square$

Corollary 5.70. For any $0 \le l \le [k/2]$, any $0 \le r \le l$,

$$\langle \mathbf{A}_k \rangle_{k-2l} = \frac{1}{C_l^r} \sum_{(2r,k-2r) \vdash \mathbf{A}_k} \langle \mathbf{A}_{k(1)} \rangle \langle \mathbf{A}_{k(2)} \rangle_{k-2l}. \tag{5.4.11}$$

Lemma 5.71. For any $0 \le l \le [k/2]$, any $0 \le m \le k - 2l$,

$$\langle \mathbf{A}_k \rangle_{k-2l} = \frac{1}{C_{k-2l}^m} \sum_{(m,k-m) \vdash \mathbf{A}_k} \langle \mathbf{A}_{k(1)} \rangle_m \wedge \langle \mathbf{A}_{k(2)} \rangle_{k-2l-m}. \tag{5.4.12}$$

Proof. Direct from (5.4.2) and the trivial identity that for any $0 \leq j \leq k$,

$$\langle \mathbf{A}_k \rangle_k = \frac{1}{C_k^j} \sum_{(j,k-j) \vdash \mathbf{A}_k} \langle \mathbf{A}_{k(1)} \rangle_j \wedge \langle \mathbf{A}_{k(2)} \rangle_{k-j}. \tag{5.4.13}$$

$\square$

By (5.4.2) and the above two lemmas, we get

Theorem 5.72. [General Clifford expansion of the geometric product of vectors] For any $0 \leq l \leq [k/2]$, any $0 \leq m \leq k - 2l$ and $0 \leq r \leq l$,

$$\langle \mathbf{A}_k \rangle_{k-2l} = \frac{1}{C_l^r C_{k-2l}^m} \sum_{(m+2r,k-m-2r) \vdash \mathbf{A}_k} \langle \mathbf{A}_{k(1)} \rangle_m \wedge \langle \mathbf{A}_{k(2)} \rangle_{k-m-2l}. \tag{5.4.14}$$

5.4.2 *Expansion of square bracket**

In $\mathcal{CL}(\mathcal{V}^n)$, Theorems 5.64 and 5.72 are valid for any i-grading operator, where $1 \leq i \leq n$. However, the expansions they generate are not necessarily the shortest. When $i = n$, by employing the syzygies in inner-product bracket algebra, the number of terms generated by a Clifford expansion can be significantly reduced.

For example, when $n = 4$, the following table displays the number of terms generated by expanding $\langle \mathbf{A}_{4+2l} \rangle_4$ using Theorem 5.64, versus the result obtained by using Theorem 5.79 to be introduced in this subsection:

$$
\begin{array}{lcccc}
\text{Integer } l: & 1 & 2 & 3 & \\
\# \text{ terms by (5.4.2):} & 15 & 210 & 3150 & (5.4.15) \\
\# \text{ terms by (5.4.24) for } m = l+2: & 6 & 42 & 465 &
\end{array}
$$

Notation.
From now on, for the n-graded part of an element $\mathbf{A} \in \mathcal{CL}(\mathcal{V}^n)$, its bracket in $\Lambda^n(\mathcal{V}^n)$ is referred to as the *square bracket* of $\mathbf{A}$:

$$[\mathbf{A}] := [\langle \mathbf{A} \rangle_n]. \tag{5.4.16}$$

According to (5.2.18), in $[\mathbf{a}_1 \mathbf{a}_2 \dots \mathbf{a}_n]$, the juxtaposition of vectors $\mathbf{a}$'s can be understood as any of the geometric product, the outer product, and the sequence of vectors, because they all agree with each other.

Lemma 5.73. Let $\mathbf{A}_{r+2l+s} = \mathbf{a}_1 \mathbf{a}_2 \cdots \mathbf{a}_{r+2l+s}$, and let $\mathbf{B}_s$ be an s-blade. Then for any $0 \leq m \leq l$,

$$\langle \mathbf{A}_{r+2l+s} \rangle_{r+s} \cdot \mathbf{B}_s = (C_l^m)^{-1} \sum_{(r+2l-2m,2m+s) \vdash \mathbf{A}_{r+2l+s}} \langle \mathbf{A}_{r+2l+s(1)} \rangle_r \langle \mathbf{A}_{r+2l+s(2)} \mathbf{B}_s \rangle. \tag{5.4.17}$$

Proof. Direct from Proposition 5.15 and (5.4.11). $\square$

Corollary 5.74. Let $\mathbf{A}_r = \mathbf{a}_1\mathbf{a}_2\cdots\mathbf{a}_r$, where $r \geq 1$. Then for any vector $\mathbf{b}$, any $0 \leq l \leq [\frac{r-1}{2}]$,

$$\mathbf{b}\cdot\langle\mathbf{A}_r\rangle_{r-2l} = \sum_{(2l+1,r-2l-1)\vdash\mathbf{A}_r} \langle\mathbf{b}\mathbf{A}_{r(1)}\rangle\langle\mathbf{A}_{r(2)}\rangle_{r-2l-1}. \tag{5.4.18}$$

Corollary 5.75. Let $\mathbf{B}_{n+2l-1} = \mathbf{a}_2\mathbf{a}_3\cdots\mathbf{a}_{n+2l}$. Then for any $1 \leq i \leq l$,

$$\sum_{(2i-1,n+2l-2i)\vdash\mathbf{B}_{n+2l-1}} \langle\mathbf{a}_1\mathbf{B}_{n+2l-1(1)}\rangle[\mathbf{B}_{n+2l-1(2)}] = 0. \tag{5.4.19}$$

Lemma 5.76. [Generalized Cramer's rule]

$$\sum_{i=1}^{n+2l+1} (-1)^{i+1}[\mathbf{a}_1\cdots\check{\mathbf{a}}_i\cdots\mathbf{a}_{n+2l+1}]\,\mathbf{a}_i = 0. \tag{5.4.20}$$

Proof. Direct from Theorem 5.64 and the Cramer's rule of $n+1$ vectors. $\qquad\square$

Lemma 5.77. For any $0 \leq m \leq n+2l$, let $\mathbf{A}_m = \mathbf{a}_1\mathbf{a}_2\cdots\mathbf{a}_m$ and $\mathbf{B}_{n+2l-m} = \mathbf{a}_{m+1}\mathbf{a}_{m+2}\cdots\mathbf{a}_{n+2l}$. Then

$$\begin{aligned}
[\mathbf{A}_{n+2l}] = \frac{1}{l}\Big(&\sum_{(2,m-2)\vdash\mathbf{A}_m} \langle\mathbf{A}_{m(1)}\rangle[\mathbf{A}_{m(2)}\mathbf{B}_{n+2l-m}] \\
&+ \sum_{(2,n+2l-m-2)\vdash\mathbf{B}_{n+2l-m}} \langle\mathbf{B}_{n+2l-m(1)}\rangle[\mathbf{A}_m\mathbf{B}_{n+2l-m(2)}]\Big).
\end{aligned} \tag{5.4.21}$$

Proof. By (5.4.11), the equality holds for $m = 0$. Assume that (5.4.21) holds for $m = k - 1$. Denote $\mathbf{B} = \mathbf{B}_{n+2l-k+1}$ and $\mathbf{C} = \mathbf{B}_{n+2l-k}$. By (5.4.19),

$$\begin{aligned}
\sum_{(2,n+2l-k-1)\vdash\mathbf{B}} \langle\mathbf{B}_{(1)}\rangle[\mathbf{A}_{k-1}\mathbf{B}_{(2)}] =\ & \sum_{(2,n+2l-k-2)\vdash\mathbf{C}} \langle\mathbf{C}_{(1)}\rangle[\mathbf{A}_{k-1}\mathbf{a}_k\mathbf{C}_{(2)}] \\
&+ \sum_{i=k+1}^{n+2l} (-1)^{i+k+1}(\mathbf{a}_k\cdot\mathbf{a}_i)[\mathbf{a}_1\cdots\check{\mathbf{a}}_k\cdots\check{\mathbf{a}}_i\cdots\mathbf{a}_{n+2l}] \\
=\ & \sum_{(2,n+2l-k-2)\vdash\mathbf{C}} \langle\mathbf{C}_{(1)}\rangle[\mathbf{A}_k\mathbf{C}_{(2)}] \\
&+ \sum_{i=1}^{k-1} (-1)^{i+k+1}(\mathbf{a}_k\cdot\mathbf{a}_i)[\mathbf{a}_1\cdots\check{\mathbf{a}}_i\cdots\check{\mathbf{a}}_k\cdots\mathbf{a}_{n+2l}].
\end{aligned}$$

Then

$$\begin{aligned}
l\,[\mathbf{A}_{n+2l}] &= \sum_{(2,k-3)\vdash\mathbf{A}_{k-1}} \langle\mathbf{A}_{k-1(1)}\rangle[\mathbf{A}_{k-1(2)}\mathbf{B}] + \sum_{(2,n+2l-k-1)\vdash\mathbf{B}} \langle\mathbf{B}_{(1)}\rangle[\mathbf{A}_{k-1}\mathbf{B}_{(2)}] \\
&= \Big(\sum_{\substack{(2,k-2)\vdash\mathbf{A}_k,\\ \mathbf{a}_k\notin\mathbf{A}_{k(1)}}} + \sum_{\substack{(2,k-2)\vdash\mathbf{A}_k,\\ \mathbf{a}_k\in\mathbf{A}_{k(1)}}}\Big)\langle\mathbf{A}_{k(1)}\rangle[\mathbf{A}_{k(2)}\mathbf{C}] + \sum_{(2,n+2l-k-2)\vdash\mathbf{C}} \langle\mathbf{C}_{(1)}\rangle[\mathbf{A}_k\mathbf{C}_{(2)}].
\end{aligned}$$

So (5.4.21) holds for $m = k$. $\qquad\square$

Lemma 5.78. For any $1 \leq k \leq l$,

$$[\mathbf{A}_{n+2l}] = (C_l^k)^{-1} \sum_{(2k,n+2l-2k-1)\vdash \mathbf{A}_{n+2l-1}} \langle \mathbf{A}_{n+2l-1(1)} \rangle [\mathbf{A}_{n+2l-1(2)}\mathbf{a}_{n+2l}]. \tag{5.4.22}$$

Proof. By (5.4.11),

$$
\begin{aligned}
C_l^k \, [\mathbf{A}_{n+2l}] &= \sum_{(2k,n+2l-2k-1)\vdash \mathbf{A}_{n+2l-1}} \langle \mathbf{A}_{n+2l-1(1)} \rangle [\mathbf{A}_{n+2l-1(2)}\mathbf{a}_{n+2l}] \\
&\quad + \sum_{(2k-1,n+2l-2k)\vdash \mathbf{A}_{n+2l-1}} (-1)^n \langle \mathbf{A}_{n+2l-1(1)}\mathbf{a}_{n+2l} \rangle [\mathbf{A}_{n+2l-1(2)}].
\end{aligned}
\tag{5.4.23}
$$

By (5.4.3), (5.4.11) and (5.4.19), the second term in the result of (5.4.23) equals

$$
\begin{aligned}
&\sum_{(1,2k-2,n+2l-2k)\vdash \mathbf{A}_{n+2l-1}} (-1)^n (\mathbf{a}_{n+2l} \cdot \mathbf{A}_{n+2l-1(1)}) \langle \mathbf{A}_{n+2l-1(2)} \rangle [\mathbf{A}_{n+2l-1(3)}] \\
&= \sum_{(1,n+2l-2)\vdash \mathbf{A}_{n+2l-1}} (-1)^n C_{l-1}^{k-1}(\mathbf{a}_{n+2l} \cdot \mathbf{A}_{n+2l-1(1)}) [\mathbf{A}_{n+2l-1(2)}] \\
&= \qquad 0.
\end{aligned}
$$

$\qquad\square$

Theorem 5.79. Let $0 \leq m \leq n + 2l$, let $\mathbf{A}_m = \mathbf{a}_1\mathbf{a}_2\cdots\mathbf{a}_m$ and $\mathbf{B}_{n+2l-m} = \mathbf{a}_{m+1}\mathbf{a}_{m+2}\cdots\mathbf{a}_{n+2l}$. Then for any $1 \leq k \leq l$,

$$
\begin{aligned}
[\mathbf{A}_{n+2l}] &= \frac{1}{C_l^k} \sum_{(2k,m-2k)\vdash \mathbf{A}_m} \langle \mathbf{A}_{m(1)} \rangle [\mathbf{A}_{m(2)}\mathbf{B}_{n+2l-m}] \\
&\quad + \sum_{i=1}^{k} (-1)^{i+1} \frac{C_k^i}{C_l^i} \sum_{(2i,n+2l-2i-m)\vdash \mathbf{B}_{n+2l-m}} \langle \mathbf{B}_{n+2l-m(1)} \rangle [\mathbf{A}_m \mathbf{B}_{n+2l-m(2)}]
\end{aligned}
\tag{5.4.24}
$$

$$
\begin{aligned}
&= \frac{1}{C_l^k} \sum_{(2k,n+2l-2k-m)\vdash \mathbf{B}_{n+2l-m}} \langle \mathbf{B}_{n+2l-m(1)} \rangle [\mathbf{A}_m \mathbf{B}_{n+2l-m(2)}] \\
&\quad + \sum_{i=1}^{k} (-1)^{i+1} \frac{C_k^i}{C_l^i} \sum_{(2i,m-2i)\vdash \mathbf{A}_m} \langle \mathbf{A}_{m(1)} \rangle [\mathbf{A}_{m(2)} \mathbf{B}_{n+2l-m}].
\end{aligned}
\tag{5.4.25}
$$

It can also be written as the following formula of "summation by part":

$$
\begin{aligned}
&\sum_{(2k,m-2k)\vdash \mathbf{A}_m} \langle \mathbf{A}_{m(1)} \rangle [\mathbf{A}_{m(2)}\mathbf{B}_{n+2l-m}] \\
&= \sum_{i=0}^{k} \sum_{(2i,n+2l-2i-m)\vdash \mathbf{B}_{n+2l-m}} (-1)^i C_{l-i}^{k-i} \langle \mathbf{B}_{n+2l-m(1)} \rangle [\mathbf{A}_m \mathbf{B}_{n+2l-m(2)}].
\end{aligned}
\tag{5.4.26}
$$

Proof. By symmetry we only need to prove (5.4.24). By (5.4.21), the equality holds for $k = 1$ and any m. When $m = n + 2l$, by (5.4.11), the equality holds for any k. When $m = n + 2l - 1$, by (5.4.22), the equality holds for any k.

Assume that (5.4.24) is true for $k = r - 1 \leq l - 1$ and any $0 \leq m \leq n + 2l$. We prove that it is true for $k = r$ and any $0 \leq m \leq n + 2l - 2$. There are three cases:

Case 1. $2r \leq m$. By induction hypothesis,

$$
\begin{aligned}
[\mathbf{A}_{n+2l}] = {} & \frac{1}{C_l^{r-1}} \sum_{(2r-2,m-2r+2)\vdash \mathbf{A}_m} \langle \mathbf{A}_{m(1)} \rangle \, [\mathbf{A}_{m(2)}\mathbf{B}_{n+2l-m}] \\
& + \sum_{i=2}^{r-1} (-1)^{i+1} \frac{C_{r-1}^i}{C_l^i} \sum_{(2i,n+2l-2i-m)\vdash \mathbf{B}_{n+2l-m}} \langle \mathbf{B}_{n+2l-m(1)} \rangle \, [\mathbf{A}_m \mathbf{B}_{n+2l-m(2)}] \\
& + \frac{r}{l} \sum_{(2,n+2l-2-m)\vdash \mathbf{B}_{n+2l-m}} \langle \mathbf{B}_{n+2l-m(1)} \rangle \, [\mathbf{A}_m \mathbf{B}_{n+2l-m(2)}] \\
& - \frac{1}{l} \sum_{(2,n+2l-2-m)\vdash \mathbf{B}_{n+2l-m}} \langle \mathbf{B}_{n+2l-m(1)} \rangle \, [\mathbf{A}_m \mathbf{B}_{n+2l-m(2)}].
\end{aligned}
\tag{5.4.27}
$$

First, for partition $(2, n + 2l - 2 - m) \vdash \mathbf{B}_{n+2l-m}$ in the last term of (5.4.27),

$$
\begin{aligned}
& [\mathbf{A}_m \mathbf{B}_{n+2l-m(2)}] \\
= {} & \frac{1}{C_{l-1}^{r-1}} \sum_{(2r-2,m-2r+2)\vdash \mathbf{A}_m} \langle \mathbf{A}_{m(1)} \rangle \, [\mathbf{A}_{m(2)}\mathbf{B}_{n+2l-m(2)}] \\
& + \sum_{i=1}^{r-1} (-1)^{i+1} \frac{C_{r-1}^i}{C_{l-1}^i} \sum_{(2i,n+2l-2i-2-m)\vdash \mathbf{B}_{n+2l-m(2)}} \langle \mathbf{B}_{n+2l-m(21)} \rangle \, [\mathbf{A}_m \mathbf{B}_{n+2l-m(22)}].
\end{aligned}
\tag{5.4.28}
$$

Second, by (5.4.21), for partition $(2r - 2, m - 2r + 2) \vdash \mathbf{A}_m$ in the first term of (5.4.27),

$$
\begin{aligned}
& \sum_{(2,n+2l-2-m)\vdash \mathbf{B}_{n+2l-m}} \langle \mathbf{B}_{n+2l-m(1)} \rangle \, [\mathbf{A}_{m(2)}\mathbf{B}_{n+2l-m(2)}] \\
= {} & (l - r + 1) [\mathbf{A}_{m(2)}\mathbf{B}_{n+2l-m}] - \sum_{(2,m-2r)\vdash \mathbf{A}_{m(2)}} \langle \mathbf{A}_{m(21)} \rangle \, [\mathbf{A}_{m(22)}\mathbf{B}_{n+2l-m}].
\end{aligned}
\tag{5.4.29}
$$

Combining (5.4.28) and (5.4.29), we get that the last term of (5.4.27) equals

$$
\begin{aligned}
& - \frac{1}{l C_{l-1}^{r-1}} \sum_{\substack{(2r-2,m-2r+2)\vdash \mathbf{A}_m, \\ (2,n+2l-2-m)\vdash \mathbf{B}_{n+2l-m}}} \langle \mathbf{A}_{m(1)} \rangle \langle \mathbf{B}_{n+2l-m(1)} \rangle \, [\mathbf{A}_{m(2)}\mathbf{B}_{n+2l-m(2)}] \\
& - \frac{1}{l} \sum_{i=1}^{r-1} (-1)^{i+1} \frac{C_{r-1}^i}{C_{l-1}^i} \sum_{\substack{(2,2i,n+2l-2i-2-m) \\ \vdash \mathbf{B}_{n+2l-m}}} \langle \mathbf{B}_{n+2l-m(1)} \rangle \langle \mathbf{B}_{n+2l-m(2)} \rangle \, [\mathbf{A}_m \mathbf{B}_{n+2l-m(3)}]
\end{aligned}
$$

$$
\begin{aligned}
&= -\frac{l-r+1}{lC_{l-1}^{r-1}} \sum_{(2r-2,m-2r+2)\vdash \mathbf{A}_m} \langle \mathbf{A}_{m(1)}\rangle [\mathbf{A}_{m(2)}\mathbf{B}_{n+2l-m}] \\
&\quad + \frac{1}{lC_{l-1}^{r-1}} \sum_{(2r-2,2,m-2r)\vdash \mathbf{A}_m} \langle \mathbf{A}_{m(1)}\rangle \langle \mathbf{A}_{m(2)}\rangle [\mathbf{A}_{m(3)}\mathbf{B}_{n+2l-m}] \\
&\quad + \sum_{i=1}^{r-1} (-1)^i \frac{(i+1)C_{r-1}^i}{lC_{l-1}^i} \sum_{(2i+2,n+2l-2i-2-m)\vdash \mathbf{B}_{n+2l-m}} \langle \mathbf{B}_{n+2l-m(1)}\rangle [\mathbf{A}_m\mathbf{B}_{n+2l-m(2)}] \\[2ex]
&= -\frac{1}{C_l^{r-1}} \sum_{(2r-2,m-2r+2)\vdash \mathbf{A}_m} \langle \mathbf{A}_{m(1)}\rangle [\mathbf{A}_{m(2)}\mathbf{B}_{n+2l-m}] \\
&\quad + \frac{1}{C_l^r} \sum_{(2r,m-2r)\vdash \mathbf{A}_m} \langle \mathbf{A}_{m(1)}\rangle [\mathbf{A}_{m(2)}\mathbf{B}_{n+2l-m}] \\
&\quad + \sum_{i=2}^{r} (-1)^{i+1} \frac{C_{r-1}^{i-1}}{C_l^i} \sum_{(2i,n+2l-2i-m)\vdash \mathbf{B}_{n+2l-m}} \langle \mathbf{B}_{n+2l-m(1)}\rangle [\mathbf{A}_m\mathbf{B}_{n+2l-m(2)}].
\end{aligned}
$$

$$(5.4.30)$$

Replacing the last term of (5.4.27) with the above result, we get (5.4.24) for $k=r$.

Case 2. $2r-2 \le m$ but $2r > m$.

In this case, the right side of (5.4.29) contains only the first term. The second term in the result of (5.4.30) vanishes. The previous proof is still valid.

Case 3. $2r-2 > m$.

In this case, the first term on the right side of (5.4.27) vanishes, so does the first term on the right side of (5.4.28). The proof can be finished by substituting (5.4.28) into the last term of (5.4.27). $\qquad\square$

Corollary 5.80. With the same notation as Theorem 5.79,

$$
\begin{aligned}
[\mathbf{A}_{n+2l}] &= \sum_{(2l,n-m)\vdash \mathbf{B}_{n+2l-m}} \langle \mathbf{B}_{n+2l-m(1)}\rangle [\mathbf{A}_m\mathbf{B}_{n+2l-m(2)}] \\
&\quad + \sum_{i=1}^{l} \sum_{(2i,m-2i)\vdash \mathbf{A}_m} (-1)^{i+1} \langle \mathbf{A}_{m(1)}\rangle [\mathbf{A}_{m(2)}\mathbf{B}_{n+2l-m}].
\end{aligned}
$$

$$(5.4.31)$$

Proof. Set $k=l$ in (5.4.25). $\qquad\square$

Example 5.81. When $n=4$, the expansion of $[\mathbf{123456}]$ by Theorem 5.64 produces $C_6^2 = 15$ terms, see Example 5.67. Using (5.4.24) for $k=l=1$ and $m=3$, we get

$$
\begin{aligned}
[\mathbf{123456}] = \ & 1\cdot 2[\mathbf{3456}] - 1\cdot 3[\mathbf{2456}] + 2\cdot 3[\mathbf{1456}] \\
& + 4\cdot 5[\mathbf{1236}] - 4\cdot 6[\mathbf{1235}] + 5\cdot 6[\mathbf{1234}].
\end{aligned}
$$

$$(5.4.32)$$

5.4.3 *Expansion of the geometric product of blades*[*]

The formulas provided by Theorems 5.64 and 5.72 on expanding the geometric product of vectors can be extended to blades. When there are only two blades, the expansion formula was first discovered by Caianiello in 1973 [32], and re-discovered by Rota and Stein in 1986 [158]. For the general case of any finitely many blades, the expansion formula was discovered by the author in 2001 [116].

In this subsection, symbols $\mathbf{A}_r, \mathbf{B}_s$, and $\mathbf{C}_t$ always denote r-blade, s-blade, and t-blade each being in the explicit form of the outer product of r, s, and t vectors. If two blades in a sequence named by $\mathbf{A}$ have the same grade r, they must be distinguished by introducing a second index: $\mathbf{A}_{1r}, \mathbf{A}_{2r}$. The first index is used to distinguish different blades in the sequence, the second index denotes the grade.

Theorem 5.82. For r-blade $\mathbf{A}_r$ and s-blade $\mathbf{B}_s$,

$$\langle \mathbf{A}_r \mathbf{B}_s \rangle_{r+s-2l} = \sum_{\substack{(r-l,l) \vdash \mathbf{A}_r, \\ (l,s-l) \vdash \mathbf{B}_s}} \langle \mathbf{A}_{r\,(2)} \mathbf{B}_{s\,(1)} \rangle \, \mathbf{A}_{r\,(1)} \wedge \mathbf{B}_{s\,(2)}. \tag{5.4.33}$$

Proof. When $r = 0$ or $s = 0$ or $l = 0$, (5.4.33) is always true. When $r = 1$, (5.4.33) is true for any s and l. Assume that (5.4.33) is true for r up to $m - 1 > 0$, and for any s and l. Let $\mathbf{C}_{m-1} = \mathbf{a}_2 \wedge \mathbf{a}_3 \wedge \cdots \wedge \mathbf{a}_m$. Then $\langle \mathbf{A}_m \mathbf{B}_s \rangle_{m+s-2l}$ equals

$$\langle \mathbf{a}_1 \mathbf{C}_{m-1} \mathbf{B}_s \rangle_{m+s-2l} - \sum_{(1,m-2) \vdash \mathbf{C}_{m-1}} (\mathbf{a}_1 \cdot \mathbf{C}_{m-1\,(1)}) \langle \mathbf{C}_{m-1\,(2)} \mathbf{B}_s \rangle_{m+s-2l}$$

$$= \mathbf{a}_1 \cdot \langle \mathbf{C}_{m-1} \mathbf{B}_s \rangle_{(m-1)+s-2(l-1)} + \mathbf{a}_1 \wedge \langle \mathbf{C}_{m-1} \mathbf{B}_s \rangle_{(m-1)+s-2l}$$

$$- \sum_{(1,m-2) \vdash \mathbf{C}_{m-1}} (\mathbf{a}_1 \cdot \mathbf{C}_{m-1\,(1)}) \langle \mathbf{C}_{m-1\,(2)} \mathbf{B}_s \rangle_{(m-2)+s-2(l-1)}$$

$$= \sum_{\substack{(m-l,l-1) \vdash \mathbf{C}_{m-1} \\ (l-1,s-l+1) \vdash \mathbf{B}_s}} (\mathbf{C}_{m-1\,(2)} \cdot \mathbf{B}_{s\,(1)}) \, \mathbf{a}_1 \cdot (\mathbf{C}_{m-1\,(1)} \wedge \mathbf{B}_{s\,(2)})$$

$$+ \sum_{\substack{(m-l-1,l) \vdash \mathbf{C}_{m-1} \\ (l,s-l) \vdash \mathbf{B}_s}} (\mathbf{C}_{m-1\,(2)} \cdot \mathbf{B}_{s\,(1)}) \, \mathbf{a}_1 \wedge \mathbf{C}_{m-1\,(1)} \wedge \mathbf{B}_{s\,(2)}$$

$$- \sum_{\substack{(1,m-l-1,l-1) \vdash \mathbf{C}_{m-1} \\ (l-1,s-l+1) \vdash \mathbf{B}_s}} (\mathbf{a}_1 \cdot \mathbf{C}_{m-1\,(1)}) (\mathbf{C}_{m-1\,(3)} \cdot \mathbf{B}_{s\,(1)}) \, \mathbf{C}_{m-1\,(2)} \wedge \mathbf{B}_{s\,(2)}. \tag{5.4.34}$$

The first term in the result of (5.4.34) can be decomposed into

$$\sum_{\substack{(1,m-l-1,l-1) \vdash \mathbf{C}_{m-1}, \\ (l-1,s-l+1) \vdash \mathbf{B}_s}} (\mathbf{C}_{m-1\,(3)} \cdot \mathbf{B}_{s\,(1)}) (\mathbf{a}_1 \cdot \mathbf{C}_{m-1\,(1)}) \, \mathbf{C}_{m-1\,(2)} \wedge \mathbf{B}_{s\,(2)}$$

$$+ \sum_{\substack{(m-l,l-1) \vdash \mathbf{C}_{m-1}, \\ (l-1,1,s-l) \vdash \mathbf{B}_s}} (-1)^{m-l} (\mathbf{C}_{m-1\,(2)} \cdot \mathbf{B}_{s\,(1)}) (\mathbf{a}_1 \cdot \mathbf{B}_{s\,(2)}) \, \mathbf{C}_{m-1\,(1)} \wedge \mathbf{B}_{s\,(3)}. \tag{5.4.35}$$

Substituting it into (5.4.34), and using (5.1.11), we get

$$\langle \mathbf{A}_m \mathbf{B}_s \rangle_{m+s-2l} = \Big(\sum_{\substack{(m-l,l)\vdash \mathbf{A}_m,\, \mathbf{a}_1 \in \mathbf{A}_{m(2)},\\ (l,s-l)\vdash \mathbf{B}_s}} + \sum_{\substack{(m-l,l)\vdash \mathbf{A}_m,\, \mathbf{a}_1 \notin \mathbf{A}_{m(2)},\\ (l,s-l)\vdash \mathbf{B}_s}} \Big) \langle \mathbf{A}_{m(2)} \mathbf{B}_{s(1)} \rangle \, \mathbf{A}_{m(1)} \wedge \mathbf{B}_{s(2)}.$$

$\square$

Example 5.83. For $r = s = 3$ and $l = 2$,

$$\langle (1 \wedge 2 \wedge 3)(4 \wedge 5 \wedge 6) \rangle_2$$
$$= \;\; (1 \wedge 2) \cdot (4 \wedge 5)\; 3 \wedge 6 - (1 \wedge 2) \cdot (4 \wedge 6)\; 3 \wedge 5 + (1 \wedge 2) \cdot (5 \wedge 6)\; 3 \wedge 4$$
$$-(1 \wedge 3) \cdot (4 \wedge 5)\; 2 \wedge 6 + (1 \wedge 3) \cdot (4 \wedge 6)\; 2 \wedge 5 - (1 \wedge 3) \cdot (5 \wedge 6)\; 2 \wedge 4$$
$$+(2 \wedge 3) \cdot (4 \wedge 5)\; 1 \wedge 6 - (2 \wedge 3) \cdot (4 \wedge 6)\; 1 \wedge 5 + (2 \wedge 3) \cdot (5 \wedge 6)\; 1 \wedge 4.$$

Corollary 5.84. For r-blade $\mathbf{A}_r$ and s-blade $\mathbf{B}_s$, for any $0 \le l \le \min(r, s)$,

$$\langle \mathbf{A}_r \mathbf{B}_s \rangle_{r+s-2l} = \sum_{(r-l,l)\vdash \mathbf{A}_r} \mathbf{A}_{r(1)} \wedge (\mathbf{A}_{r(2)} \cdot \mathbf{B}_s) = \sum_{(l,s-l)\vdash \mathbf{B}_s} (\mathbf{A}_r \cdot \mathbf{B}_{s(1)}) \wedge \mathbf{B}_{s(2)}. \tag{5.4.36}$$

Corollary 5.85. For r-blade $\mathbf{A}_r$ and s-blade $\mathbf{B}_s$, for any $0 \le l \le \min(r, s)$ and $0 \le m \le l$,

$$\langle \mathbf{A}_r \mathbf{B}_s \rangle_{r+s-2l} = \frac{1}{C_l^m} \sum_{\substack{(r-m,m)\vdash \mathbf{A}_r,\\ (m,r-m)\vdash \mathbf{B}_s}} \langle \mathbf{A}_{r(2)} \mathbf{B}_{s(1)} \rangle \langle \mathbf{A}_{r(1)} \mathbf{B}_{s(2)} \rangle_{r+s-2l}. \tag{5.4.37}$$

Proof. Direct from (5.4.33) and (5.1.11). $\square$

In the geometric product of a sequence of blades, when vectors in the blades are aligned sequentially, they form a *sequence of sequences*, or *bi-sequence*. The *interior sequences* are vectors in the same blade, the *exterior sequence* is the sequence of blades. A partition of a blade results in a sequence of smaller blades whose outer product equals the blade. In partitioning the geometric product of blades, each blade is partitioned into smaller blades, then the smaller blades are reordered into a new exterior sequence.

Definition 5.86. Let the $\mathbf{A}_i$ be blades for all $1 \le i \le k$. A *partition* of the geometric product $\mathbf{A}_1 \mathbf{A}_2 \cdots \mathbf{A}_k$ of *shape* $(\lambda_1, \lambda_2, \ldots, \lambda_k)$, refers to a sequence of blades obtained by first partitioning each $\mathbf{A}_i$ into shape λ_i, and then reordering the new blades. The *sign of partition* is the sign of permutation of the whole sequence of vectors constituting the exterior sequence.

In *Sweedler's notation*, if the k-th new blade is the j-th part from the partition of $\mathbf{A}_i$, then the new blade is denoted by the double indexed notation $\mathbf{A}_{i(j)}^{(k)}$, where k is the *exterior index* of the whole sequence, and j is the *interior index* within a blade. The sign of partition is attached to the first blade in the new exterior sequence. If the first blade does not occur in a summand, then the sign is attached to the blade with the lowest exterior index in the summand.

Example 5.87. Let $\mathbf{A}_3 = 1 \wedge 2 \wedge 3$ and $\mathbf{B}_3 = 4 \wedge 5 \wedge 6$ be 3-blades. Then $2, 5 \wedge 6, 1 \wedge 3, 4$ form a partition of the geometric product $\mathbf{A}_3\mathbf{B}_3$ of shape $((1,2),(2,1))$. The sign of partition is the sign of permutation of $2, 5, 6, 1, 3, 4$ to the original vector sequence, so it is -1. In Sweedler's notation, $((1,2),(2,1)) \vdash \mathbf{A}_3\mathbf{B}_3$, and

$$-2 = \mathbf{A}_{3\,(1)}^{(1)}, \quad 5 \wedge 6 = \mathbf{B}_{3\,(1)}^{(2)}, \quad 1 \wedge 3 = \mathbf{A}_{3\,(2)}^{(3)}, \quad 4 = \mathbf{B}_{3\,(2)}^{(4)}. \tag{5.4.38}$$

The sign -1 is attached to the first blade $\mathbf{A}_{3\,(1)}^{(1)}$ of the partition result.

Notice that in the above example, the partition is also of shape $(\sigma(1,2), \tau(2,1))$, for any two permutations σ, τ of $1, 2$. The permutations change the lower indices accordingly. This is a common property of all partitions of bi-sequences.

Lemma 5.88. For $1 \le i \le k$, let $s_i = r_1 + r_2 + \cdots + r_i$, and let $\mathbf{A}_{ir_i} = \mathbf{a}_{s_{i-1}+1} \wedge \mathbf{a}_{s_{i-1}+2} \wedge \cdots \wedge \mathbf{a}_{s_i}$. For l from 1 to k, let $\mathbf{A}_{lr_l(1)}, \mathbf{A}_{lr_l(2)}$ be a fixed partition of $\mathbf{A}_{lr_l}$ of shape (i_l, j_l). Then

$$\mathrm{sign}(\mathbf{A}_{1r_1(1)}, \ldots, \mathbf{A}_{kr_k(1)}, \mathbf{A}_{1r_1(2)}, \ldots, \mathbf{A}_{kr_k(2)}) = (-1)^{\sum_{1 \le q < p \le k} i_p j_q}. \tag{5.4.39}$$

Proof. First notice that $\mathrm{sign}(\mathbf{A}_{(1)}, \mathbf{A}_{(2)}) = 1$ for any sequence $\mathbf{A}$. By

$$\mathrm{sign}(\mathbf{A}_{1r_1(1)}, \ldots, \mathbf{A}_{kr_k(1)}, \mathbf{A}_{1r_1(2)}, \ldots, \mathbf{A}_{kr_k(2)}) = (-1)^{j_1(i_2+\cdots+i_k)}\mathrm{sign}(\mathbf{A}_{1r_1(1)},$$
$$\mathbf{A}_{1r_1(2)}, \mathbf{A}_{2r_2(1)}, \mathbf{A}_{3r_3(1)}, \ldots, \mathbf{A}_{kr_k(1)}, \mathbf{A}_{2r_2(2)}, \ldots, \mathbf{A}_{kr_k(2)}),$$

and deduction on k, we get (5.4.39). $\qquad\square$

Theorem 5.89. For r-blade $\mathbf{A}_r$ and s-blade $\mathbf{B}_s$, for any $0 \le l \le \min(r,s)$, any $0 \le m \le l$, any $0 \le i \le r - l$, and any $0 \le j \le s - l$,

$$\langle \mathbf{A}_r\mathbf{B}_s \rangle_{r+s-2l} = \frac{1}{C_l^m C_{r-l}^i C_{s-l}^j} \sum_{\substack{((m+i,r-m-i), \\ (m+j,s-m-j)) \vdash \mathbf{A}_r\mathbf{B}_s}} \langle \mathbf{A}_{r\,(1)}^{(1)}\mathbf{B}_{s\,(1)}^{(2)} \rangle_{i+j} \wedge \langle \mathbf{A}_{r\,(2)}^{(3)}\mathbf{B}_{s\,(2)}^{(4)} \rangle_{r-i+s-j-2l}.$$
$$\tag{5.4.40}$$

Proof. By (5.4.33), (5.4.37) and (5.4.39),

$$\langle \mathbf{A}_r\mathbf{B}_s \rangle_{r+s-2l}$$

$$= \frac{1}{C_l^m} \sum_{\substack{(r-m,m) \vdash \mathbf{A}_r, \\ (m,s-m) \vdash \mathbf{B}_s}} \langle \mathbf{A}_{r(2)}\mathbf{B}_{s(1)} \rangle \langle \mathbf{A}_{r(1)}\mathbf{B}_{s(2)} \rangle_{(r-m)+(s-m)-2(l-m)}$$

$$= \frac{1}{C_l^m} \sum_{\substack{(r-l,l-m,m) \vdash \mathbf{A}_r, \\ (m,l-m,s-l) \vdash \mathbf{B}_s}} \langle \mathbf{A}_{r(3)}\mathbf{B}_{s(1)} \rangle \langle \mathbf{A}_{r(2)}\mathbf{B}_{s(2)} \rangle \mathbf{A}_{r(1)} \wedge \mathbf{B}_{s(3)}$$

$$= \frac{1}{C_l^m C_{r-l}^i C_{s-l}^j} \sum_{\substack{(i,r-l-i,l-m,m) \vdash \mathbf{A}_r, \\ (m,l-m,j,s-l-j) \vdash \mathbf{B}_s}} \langle \mathbf{A}_{r(4)}\mathbf{B}_{s(1)} \rangle \langle \mathbf{A}_{r(3)}\mathbf{B}_{s(2)} \rangle \mathbf{A}_{r(1)} \wedge \mathbf{A}_{r(2)} \wedge \mathbf{B}_{s(3)} \wedge \mathbf{B}_{s(4)}$$

$$= \frac{(-1)^{m(r-m-i)+j(l-m)}}{C_l^m C_{r-l}^i C_{s-l}^j} \sum_{\substack{(i,m,r-l-i,l-m)\vdash \mathbf{A}_r, \\ (m,j,l-m,s-l-j)\vdash \mathbf{B}_s}} \langle \mathbf{A}_{r(2)}\mathbf{B}_{s(1)}\rangle \langle \mathbf{A}_{r(4)}\mathbf{B}_{s(3)}\rangle \mathbf{A}_{r(1)} \wedge \mathbf{A}_{r(3)} \wedge \mathbf{B}_{s(2)} \wedge \mathbf{B}_{s(4)}$$

$$= \frac{(-1)^{(r-m-i)(m+j)}}{C_l^m C_{r-l}^i C_{s-l}^j} \sum_{\substack{(i,m,r-l-i,l-m)\vdash \mathbf{A}_r, \\ (m,j,l-m,s-l-j)\vdash \mathbf{B}_s}} \underline{\langle \mathbf{A}_{r(2)}\mathbf{B}_{s(1)}\rangle \mathbf{A}_{r(1)} \wedge \mathbf{B}_{s(2)}} \wedge \underline{\langle \mathbf{A}_{r(4)}\mathbf{B}_{s(3)}\rangle \mathbf{A}_{r(3)} \wedge \mathbf{B}_{s(4)}}$$

$$= \frac{1}{C_l^m C_{r-l}^i C_{s-l}^j} \sum_{\substack{((m+i,r-m-i), \\ (m+j,s-m-j))\vdash \mathbf{A}_r\mathbf{B}_s}} \langle \mathbf{A}_{r\,(1)}^{(1)}\mathbf{B}_{s\,(1)}^{(2)}\rangle_{i+j} \wedge \langle \mathbf{A}_{r\,(2)}^{(3)}\mathbf{B}_{s\,(2)}^{(4)}\rangle_{r-i+s-j-2l}.$$

$\square$

Example 5.90. In expanding $\langle(1\wedge2\wedge3)(4\wedge5\wedge6)\rangle_2$ by (5.4.40), we have $r = s = 3$ and $l = 2$. If choosing $m = j = 1$ and $i = 0$, we get that the summation in the expansion is over all $((1,2),(2,1))$-typed partitions of the geometric product:

$$\begin{aligned}
2\,\langle(\mathbf{1} \wedge \mathbf{2} \wedge \mathbf{3})(\mathbf{4} \wedge \mathbf{5} \wedge \mathbf{6})\rangle_2 \\
= \quad &\{\mathbf{1}\cdot(\mathbf{4}\wedge\mathbf{5})\} \wedge \{(\mathbf{2}\wedge\mathbf{3})\cdot\mathbf{6}\} - \{\mathbf{1}\cdot(\mathbf{4}\wedge\mathbf{6})\} \wedge \{(\mathbf{2}\wedge\mathbf{3})\cdot\mathbf{5}\} \\
&+\{\mathbf{1}\cdot(\mathbf{5}\wedge\mathbf{6})\} \wedge \{(\mathbf{2}\wedge\mathbf{3})\cdot\mathbf{4}\} - \{\mathbf{2}\cdot(\mathbf{4}\wedge\mathbf{5})\} \wedge \{(\mathbf{1}\wedge\mathbf{3})\cdot\mathbf{6}\} \\
&+\{\mathbf{2}\cdot(\mathbf{4}\wedge\mathbf{6})\} \wedge \{(\mathbf{1}\wedge\mathbf{3})\cdot\mathbf{5}\} - \{\mathbf{2}\cdot(\mathbf{5}\wedge\mathbf{6})\} \wedge \{(\mathbf{1}\wedge\mathbf{3})\cdot\mathbf{4}\} \\
&+\{\mathbf{3}\cdot(\mathbf{4}\wedge\mathbf{5})\} \wedge \{(\mathbf{1}\wedge\mathbf{2})\cdot\mathbf{6}\} - \{\mathbf{3}\cdot(\mathbf{4}\wedge\mathbf{6})\} \wedge \{(\mathbf{1}\wedge\mathbf{2})\cdot\mathbf{5}\} \\
&+\{\mathbf{3}\cdot(\mathbf{5}\wedge\mathbf{6})\} \wedge \{(\mathbf{1}\wedge\mathbf{2})\cdot\mathbf{4}\}.
\end{aligned}$$

Proposition 5.91. Let $1 \leq r, s, t \leq m$ be integers such that $r + s + t = 2m$. Then for any r-blade $\mathbf{A}_r$, s-blade $\mathbf{B}_s$ and t-blade $\mathbf{C}_t$,

$$\langle \mathbf{A}_r\mathbf{B}_s\mathbf{C}_t\rangle = \sum_{\substack{(m-s,m-t)\vdash \mathbf{A}_r, \\ (m-t,m-r)\vdash \mathbf{B}_s, (m-r,m-s)\vdash \mathbf{C}_t}} \langle \mathbf{A}_{r(2)}\mathbf{B}_{s(1)}\rangle \langle \mathbf{B}_{s(2)}\mathbf{C}_{t(1)}\rangle \langle \mathbf{C}_{t(2)}\mathbf{A}_{r(1)}\rangle. \qquad (5.4.41)$$

Proof. Direct from

$$\langle \mathbf{A}_r\mathbf{B}_s\mathbf{C}_t\rangle = \langle \mathbf{A}_r\mathbf{B}_s\rangle_{r+s-2(m-t)} \cdot \mathbf{C}_t = \sum_{\substack{(m-s,m-t)\vdash \mathbf{A}_r \\ (m-t,m-r)\vdash \mathbf{B}_s}} \langle \mathbf{A}_{r(2)}\mathbf{B}_{s(1)}\rangle (\mathbf{A}_{r(1)} \wedge \mathbf{B}_{s(2)}) \cdot \mathbf{C}_t.$$

$\square$

Theorem 5.92. Let $1 \leq r, s, t \leq m$ be integers such that $r + s + t = 2m$. For any r-blade $\mathbf{A}_r$, s-blade $\mathbf{B}_s$ and t-blade $\mathbf{C}_t$, for any $0 \leq m' \leq m$ and $0 \leq r', s', t' \leq m'$, such that $r' + s' + t' = 2m'$ and $0 \leq r - r', s - s', t - t' \leq m - m'$,

$$\langle \mathbf{A}_r\mathbf{B}_s\mathbf{C}_t\rangle = \frac{1}{C_{m-r}^{m'-r'} C_{m-s}^{m'-s'} C_{m-t}^{m'-t'}} \sum_{\substack{((r-r',r'),(s-s',s'), \\ (t-t',t'))\vdash \mathbf{A}_r\mathbf{B}_s\mathbf{C}_t}} \langle \mathbf{A}_{r\,(2)}^{(1)}\mathbf{B}_{s\,(2)}^{(2)}\mathbf{C}_{t\,(2)}^{(3)}\rangle \langle \mathbf{A}_{r\,(1)}^{(4)}\mathbf{B}_{s\,(1)}^{(5)}\mathbf{C}_{t\,(1)}^{(6)}\rangle.$$

$$(5.4.42)$$

Proof. For any $0 \le k \le m - t$, any $0 \le i \le m - s$, and any $0 \le j \le m - r$,

$$\langle \mathbf{A}_r \mathbf{B}_s \mathbf{C}_t \rangle$$

$$= \langle \mathbf{A}_r \mathbf{B}_s \rangle_{r+s-2(m-t)} \cdot \mathbf{C}_t$$

$$= \frac{1}{C_{m-t}^k C_{m-s}^i C_{m-r}^j} \sum_{\substack{((k+i,r-k-i), \\ (k+j,s-k-j)) \vdash \mathbf{A}_r \mathbf{B}_s}} \left(\langle \mathbf{A}_{r\,(1)}^{(1)} \mathbf{B}_{s\,(1)}^{(2)} \rangle_{i+j} \wedge \langle \mathbf{A}_{r\,(2)}^{(3)} \mathbf{B}_{s\,(2)}^{(4)} \rangle_{t-i-j} \right) \cdot \mathbf{C}_t$$

$$= \frac{1}{C_{m-t}^k C_{m-s}^i C_{m-r}^j} \sum_{\substack{(t-i-j,i+j) \vdash \mathbf{C}_t, \\ ((k+i,r-k-i),(k+j,s-k-j)) \vdash \mathbf{A}_r \mathbf{B}_s}} \langle \mathbf{A}_{r\,(1)}^{(1)} \mathbf{B}_{s\,(1)}^{(2)} \mathbf{C}_{t\,(2)} \rangle \, \langle \mathbf{A}_{r\,(2)}^{(3)} \mathbf{B}_{s\,(2)}^{(4)} \mathbf{C}_{t\,(1)} \rangle$$

$$= \frac{1}{C_{m-t}^k C_{m-s}^i C_{m-r}^j} \sum_{\substack{((k+i,r-k-i),(k+j,s-k-j), \\ (i+j,t-i-j)) \vdash \mathbf{A}_r \mathbf{B}_s \mathbf{C}_t}} \langle \mathbf{A}_{r\,(1)}^{(1)} \mathbf{B}_{s\,(1)}^{(2)} \mathbf{C}_{t\,(1)}^{(3)} \rangle \langle \mathbf{A}_{r\,(2)}^{(4)} \mathbf{B}_{s\,(2)}^{(5)} \mathbf{C}_{t\,(2)}^{(6)} \rangle.$$

The last step is because $(-1)^{(r-k-i+s-k-j+t-i-j)(i+j)} = 1$. Setting $r' = k+i$, $s' = k+j$, $t' = i+j$, we get $k = m' - t'$, $i = m' - s'$ and $j = m' - r'$, hence (5.4.42). $\square$

Lemma 5.93. Let $\mathbf{A}_{ir_i}$ be an r_i-blade for $1 \le i \le k$. Then $\langle \mathbf{A}_{1r_1} \mathbf{A}_{2r_2} \cdots \mathbf{A}_{kr_k} \rangle_g = 0$ for any grade g satisfying $g < 2\max(r_1, \ldots, r_k) - \sum_{i=1}^{k} r_i$.

Proof. Let $r_m = \max(r_1, \ldots, r_k)$. For $1 \le i \le k$, let $s_i = \sum_{h=1}^{i} r_h$. If $2r_m - s_k > 0$, then $r_m - s_{m-1} = 2r_m - s_m > s_k - s_m \ge 0$. Since

$$\mathbf{A}_{1r_1} \mathbf{A}_{2r_2} \cdots \mathbf{A}_{kr_k}$$

$$= \left(\sum_{j=r_m-s_{m-1}}^{s_m} \langle \mathbf{A}_{1r_1} \cdots \mathbf{A}_{mr_m} \rangle_j \right) \left(\sum_{l=0}^{s_k-s_m} \langle \mathbf{A}_{(m+1)r_{m+1}} \cdots \mathbf{A}_{kr_k} \rangle_l \right)$$

$$= \sum_{h=r_m-s_{m-1}-s_k+s_m}^{s_k} \left\langle \left(\sum_{j=r_m-s_{m-1}}^{s_m} \langle \mathbf{A}_{1r_1} \cdots \mathbf{A}_{mr_m} \rangle_j \right) \left(\sum_{l=0}^{s_k-s_m} \langle \mathbf{A}_{(m+1)r_{m+1}} \cdots \mathbf{A}_{kr_k} \rangle_l \right) \right\rangle_h,$$

if $g < 2r_m - s_k = r_m - s_{m-1} - s_k + s_m$, then $\langle \mathbf{A}_{1r_1} \mathbf{A}_{2r_2} \cdots \mathbf{A}_{kr_k} \rangle_g = 0$. $\square$

Proposition 5.94. Let $r_1, \ldots, r_k$ be a sequence of positive integers, among which there are three integers $r_{i_1} \ge r_{i_2} \ge r_{i_3}$, such that $r_{i_3} \ge r_j$ for all $j \notin \{i_1, i_2, i_3\}$. Let $r_1 + \cdots + r_k = s_k$. Let $\mathbf{A}_{ir_i}$ be an r_i-blade for $1 \le i \le k$. Then $\mathbf{B}_{k,l} = \langle \mathbf{A}_{1r_1} \mathbf{A}_{2r_2} \cdots \mathbf{A}_{kr_k} \rangle_{s_k-2l}$ equals zero if l is not within the following range:

$$r_{i_1} + r_{i_2} - n \le l \le \min(s_k - r_{i_1}, s_k + n - r_{i_1} - r_{i_2} - r_{i_3}). \tag{5.4.43}$$

Proof. When $k = 2$, if $r + s > n$, then for any l such that $\mathbf{B}_{2,l}$ is nonzero, $r + s - 2l \le n - r + n - s$. So

$$\max(0, r + s - n) \le l \le \min(r, s).$$

This proves (5.4.43) for $k = 2$.

Assume that (5.4.43) is true for $k = m - 1$. On one hand, by

$$\mathbf{B}_{m,l} = \sum_{h=0}^{l} \left\langle \langle \mathbf{B}_{m-1,h} \rangle_{s_{m-1}-2h} \mathbf{A}_{m r_m} \right\rangle_{s_{m-1}-2h+r_m-2(l-h)},$$

we have

$$\max(0, s_m - 2h - n) \leq l - h \leq \min(s_{m-1} - 2h, r_m). \tag{5.4.44}$$

On the other hand, by induction hypothesis,

$$\begin{aligned} \max(0, r_i + r_j - n \,|\, 1 \leq i < j < m) &\leq h \\ \leq \min(s_{m-1} - r_i, s_{m-1} + n - r_i - r_j - r_k \,|\, 1 \leq i < j < k < m). \end{aligned} \tag{5.4.45}$$

Combining (5.4.44) and (5.4.45), we get

$$\begin{aligned} \max(l - r_m, s_m - n - l, 0, r_i + r_j - n \,|\, 1 \leq i < j < m) &\leq h \\ \leq \min(l, s_{m-1} - l, s_{m-1} - r_i, s_{m-1} + n - r_i - r_j - r_k \,|\, 1 \leq i < j < k < m). \end{aligned}$$

So

$$\max(0, \frac{s_m - n}{2}, r_m + r_i - n, r_i + r_j - n, r_m + r_i + r_j + r_k - 2n \,\Big|\, {\scriptstyle 1 \leq i < j < k < m}) \leq l$$
$$\leq \min(\frac{s_m}{2}, s_{m-1}, s_m - r_i, s_m + n - r_i - r_j - r_k, s_{m-1} + n - r_i - r_j \,\Big|\, {\scriptstyle 1 \leq i < j < k < m}),$$

which is a subrange of (5.4.43) for l, when setting $k = m$. $\qquad\square$

Theorem 5.95. [Fundamental Clifford expansion of the geometric product of blades] For $1 \leq i \leq k$, let $\mathbf{A}_{i r_i}$ be an r_i-blade, let $s_i = r_1 + \cdots + r_i$. Then for any $l \geq 0$,

$$\langle \mathbf{A}_{1 r_1} \mathbf{A}_{2 r_2} \cdots \mathbf{A}_{k r_k} \rangle_{s_k - 2l}$$
$$= \sum_{\substack{i_1 + \cdots + i_k = 2l, \\ 0 \leq i_j \leq \min(l, r_j),\, 1 \leq j \leq k}} \sum_{\substack{((i_1, r_1 - i_1), \ldots, (i_k, r_k - i_k)) \\ \vdash \mathbf{A}_{1 r_1} \cdots \mathbf{A}_{k r_k}}} \langle \mathbf{A}_{1 r_1}{}_{(1)}^{(1)} \cdots \mathbf{A}_{k r_k}{}_{(1)}^{(k)} \rangle \, \mathbf{A}_{1 r_1}{}_{(2)}^{(k+1)} \wedge \cdots \wedge \mathbf{A}_{k r_k}{}_{(2)}^{(2k)}.$$

$$\tag{5.4.46}$$

Theorem 5.96. With the same notation as above, if $s_k = 2n$ and $r_1 > 0$, then

$$\langle \mathbf{A}_{1 r_1} \mathbf{A}_{2 r_2} \cdots \mathbf{A}_{k r_k} \rangle = \sum_{\substack{i_2 + \cdots + i_k = r_1, \\ 0 \leq i_j \leq r_j,\, 2 \leq j \leq k}} \sum_{\substack{((i_2, r_2 - i_2), \ldots, (i_k, r_k - i_k)) \\ \vdash \mathbf{A}_{2 r_2} \cdots \mathbf{A}_{k r_k}}}$$
$$\mathbf{A}_{1 r_1} \cdot (\mathbf{A}_{2 r_2}{}_{(1)}^{(1)} \wedge \cdots \wedge \mathbf{A}_{k r_k}{}_{(1)}^{(k-1)}) \, \langle \mathbf{A}_{2 r_2}{}_{(2)}^{(k)} \cdots \mathbf{A}_{k r_k}{}_{(2)}^{(2k-2)} \rangle.$$

$$\tag{5.4.47}$$

Remark: When $r_1 = r_2 = \ldots = r_k = 1$, Theorems 5.95 and 5.96 become Theorems 5.64 and 5.65 respectively.

Proof. We prove Theorem 5.95 and Theorem 5.96 at the same time.

When $k = 2$, (5.4.47) is obviously true, and (5.4.46) is just (5.4.33). Assume that both (5.4.46) and (5.4.47) hold for $k = m - 1$, where $m \geq 3$.

When $k = m$, if $s_m = 2n$, since $r_1 = r_2 + \cdots + r_m - 2(n - r_1)$,

$$\langle \mathbf{A}_{1r_1} \mathbf{A}_{2r_2} \cdots \mathbf{A}_{mr_m} \rangle = \mathbf{A}_{1r_1} \cdot \langle \mathbf{A}_{2r_2} \cdots \mathbf{A}_{mr_m} \rangle_{r_1}$$

$$= \sum_{\substack{i_2 + \cdots + i_m = 2n - 2r_1, \\ 0 \le i_j \le \min(n - r_1, r_j),\, 2 \le j \le m}} \sum_{\substack{((i_2, r_2 - i_2), \ldots, (i_k, r_k - i_k)) \\ \vdash \mathbf{A}_{2r_2} \cdots \mathbf{A}_{kr_k}}} \langle \mathbf{A}_{2r_2\,(1)}^{\,(1)} \cdots \mathbf{A}_{kr_k\,(1)}^{\,(k-1)} \rangle$$

$$\left(\mathbf{A}_{1r_1} \cdot \left(\mathbf{A}_{2r_2\,(2)}^{\,(k)} \wedge \cdots \wedge \mathbf{A}_{kr_k\,(2)}^{\,(2k-2)} \right) \right).$$

Setting $h_j = r_j - i_j$, then using Lemma 5.93, we get (5.4.47).

We prove (5.4.46) for $k = m$.

$$\langle \mathbf{A}_{1r_1} \mathbf{A}_{2r_2} \cdots \mathbf{A}_{mr_m} \rangle_{s_m - 2l}$$

$$= \sum_{h=0}^{l} \langle \langle \mathbf{A}_{1r_1} \cdots \mathbf{A}_{(m-1)r_{m-1}} \rangle_{s_{m-1} - 2h} \, \mathbf{A}_{mr_m} \rangle_{s_{m-1} - 2h + r_m - 2(l - h)}$$

$$= \sum_{\substack{0 \le h \le l,\, i_1 + \cdots + i_{m-1} = 2h, \\ 0 \le i_j \le \min(h, r_j),\, 1 \le j \le m-1}} \sum_{\substack{((i_1, r_1 - i_1), \ldots, (i_{m-1}, r_{m-1} - i_{m-1})) \\ \vdash \mathbf{A}_{1r_1} \cdots \mathbf{A}_{(m-1)r_{m-1}}}} \langle \mathbf{A}_{1r_1\,(1)}^{\,(1)} \cdots \mathbf{A}_{(m-1)r_{m-1}\,(1)}^{\,(m-1)} \rangle$$

$$\langle \left(\mathbf{A}_{1r_1\,(2)}^{\,(m)} \wedge \cdots \wedge \mathbf{A}_{(m-1)r_{m-1}\,(2)}^{\,(2m-2)} \right) \mathbf{A}_{mr_m} \rangle_{s_{m-1} - 2h + r_m - 2(l - h)}$$

$$= \sum_{\substack{0 \le h \le l,\, i_1 + \cdots + i_{m-1} = 2h, \\ 0 \le i_j \le \min(h, r_j),\, 1 \le j \le m-1}} \sum_{\substack{((i_1, r_1 - i_1), \ldots, (i_{m-1}, r_{m-1} - i_{m-1})) \\ \vdash \mathbf{A}_{1r_1} \cdots \mathbf{A}_{(m-1)r_{m-1}},\, (l - h, r_m - l + h) \vdash \mathbf{A}_{mr_m}}} \langle \mathbf{A}_{1r_1\,(1)}^{\,(1)} \cdots \mathbf{A}_{(m-1)r_{m-1}\,(1)}^{\,(m-1)} \rangle$$

$$\left\{ \left(\mathbf{A}_{1r_1\,(2)}^{\,(m)} \wedge \cdots \wedge \mathbf{A}_{(m-1)r_{m-1}\,(2)}^{\,(2m-2)} \right) \cdot \mathbf{A}_{mr_m\,(1)} \right\} \wedge \mathbf{A}_{mr_m\,(2)}.$$
$$(5.4.48)$$

In the result of (5.4.48),

$$\left(\mathbf{A}_{1r_1\,(2)}^{\,(m)} \wedge \mathbf{A}_{2r_2\,(2)}^{\,(m+1)} \wedge \cdots \wedge \mathbf{A}_{(m-1)r_{m-1}\,(2)}^{\,(2m-2)} \right) \cdot \mathbf{A}_{mr_m\,(1)}$$

$$= \sum_{\substack{h_1 + \cdots + h_{m-1} = l - h, \\ 0 \le h_j \le r_j - i_j,\, 1 \le j \le m-1}} \sum_{\substack{((r_1 - i_1 - h_1, h_1), \ldots, \\ (r_{m-1} - i_{m-1} - h_{m-1}, h_{m-1})) \\ \vdash \mathbf{A}_{1r_1\,(2)}^{\,(m)} \cdots \mathbf{A}_{(m-1)r_{m-1}\,(2)}^{\,(2m-2)}}} \qquad (5.4.49)$$

$$\mathbf{A}_{1r_1\,(21)}^{\,(m)} \wedge \mathbf{A}_{2r_2\,(21)}^{\,(m+1)} \wedge \cdots \wedge \mathbf{A}_{(m-1)r_{m-1}\,(21)}^{\,(2m-2)}$$

$$\left\{ \left(\mathbf{A}_{1r_1\,(22)}^{\,(2m-1)} \wedge \mathbf{A}_{2r_2\,(22)}^{\,(2m)} \wedge \cdots \wedge \mathbf{A}_{(m-1)r_{m-1}\,(22)}^{\,(3m-3)} \right) \cdot \mathbf{A}_{mr_m\,(1)} \right\}.$$

For $1 \le j \le m - 1$, let $g_j = i_j + h_j$. Let $g_m = l - h$. Then $\sum_{i=1}^{m} g_i = 2l$. By substituting (5.4.49) into (5.4.48), then using (5.4.47) and Lemma 5.93, we get

$$\langle \mathbf{A}_{1r_1}\mathbf{A}_{2r_2}\cdots \mathbf{A}_{mr_m}\rangle_{s_m-2l}$$

$$= \sum_{\substack{g_1+\cdots+g_m=2l,\\ g_j\geq 0,\, 1\leq j\leq m}} \sum_{\substack{h_1+\cdots+h_{m-1}=g_m,\\ h_j\geq 0,\, 1\leq j\leq m-1}} \sum_{\substack{((g_1-h_1,r_1-g_1,h_1),\ldots,\\ (g_{m-1}-h_{m-1},r_{m-1}-g_{m-1},h_{m-1}),\\ (g_m,r_m-g_m))\vdash \mathbf{A}_{1r_1}\cdots \mathbf{A}_{(m-1)r_{m-1}}\mathbf{A}_{mr_m}}} \langle \mathbf{A}_{1r_1\,(1)}^{(1)}\cdots \mathbf{A}_{(m-1)r_{m-1}\,(1)}^{(m-1)}\rangle$$

$$\mathbf{A}_{1r_1\,(2)}^{(m)} \wedge\cdots\wedge \mathbf{A}_{(m-1)r_{m-1}\,(2)}^{(2m-2)} \wedge \mathbf{A}_{mr_m\,(2)}^{(3m-1)}$$

$$\{(\mathbf{A}_{1r_1\,(3)}^{(2m-1)} \wedge\cdots\wedge \mathbf{A}_{(m-1)r_{m-1}\,(3)}^{(3m-3)}) \cdot \mathbf{A}_{mr_m\,(1)}^{(3m-2)}\}$$

$$= \sum_{\substack{g_1+\cdots+g_m=2l,\\ g_j\geq 0,\, 1\leq j\leq m}} \sum_{\substack{h_1+\cdots+h_{m-1}=g_m,\\ h_j\geq 0,\, 1\leq j\leq m-1}} \sum_{\substack{((g_1-h_1,h_1,r_1-g_1),\ldots,\\ (g_{m-1}-h_{m-1},h_{m-1},r_{m-1}-g_{m-1}),\\ (g_m,r_m-g_m))\vdash \mathbf{A}_{1r_1}\cdots \mathbf{A}_{(m-1)r_{m-1}}\mathbf{A}_{mr_m}}} \langle \mathbf{A}_{1r_1\,(1)}^{(1)}\cdots \mathbf{A}_{(m-1)r_{m-1}\,(1)}^{(m-1)}\rangle$$

$$\{(\mathbf{A}_{1r_1\,(2)}^{(m)} \wedge\cdots\wedge \mathbf{A}_{(m-1)r_{m-1}\,(2)}^{(2m-2)}) \cdot \mathbf{A}_{mr_m\,(1)}^{(2m-1)}\}$$

$$\mathbf{A}_{1r_1\,(3)}^{(2m)} \wedge\cdots\wedge \mathbf{A}_{(m-1)r_{m-1}\,(3)}^{(3m-2)} \wedge \mathbf{A}_{mr_m\,(2)}^{(3m-1)}$$

$$= \sum_{\substack{g_1+\cdots+g_m=2l,\\ 0\leq g_j\leq \min(l,r_j),\, 1\leq j\leq m}} \sum_{\substack{((g_1,r_1-g_1),\ldots,(g_m,r_m-g_m))\\ \vdash \mathbf{A}_{1r_1}\cdots \mathbf{A}_{mr_m}}} \langle \mathbf{A}_{1r_1\,(1)}^{(1)}\cdots \mathbf{A}_{mr_m\,(1)}^{(m)}\rangle \mathbf{A}_{1r_1\,(2)}^{(m+1)} \wedge\cdots\wedge \mathbf{A}_{mr_m\,(2)}^{(2m)}.$$

$\square$

Example 5.97. Let $\mathbf{A}_3, \mathbf{B}_3, \mathbf{C}_3$ be 3-blades. In expanding $\langle \mathbf{A}_3\mathbf{B}_3\mathbf{C}_3\rangle_7$ by (5.4.46), we have $k=3$, $r_1=r_2=r_3=3$, and $l=1$, so one of the three i_j's must be zero. The expansion result is

$$\langle \mathbf{A}_3\mathbf{B}_3\mathbf{C}_3\rangle_7 = \sum_{((1,2),(1,2))\vdash \mathbf{A}_3\mathbf{B}_3} (\mathbf{A}_{3\,(1)}^{(1)} \cdot \mathbf{B}_{3\,(1)}^{(2)})\, \mathbf{A}_{3\,(2)}^{(3)} \wedge \mathbf{B}_{3\,(2)}^{(4)} \wedge \mathbf{C}_3$$

$$- \sum_{((1,2),(1,2))\vdash \mathbf{A}_3\mathbf{C}_3} (\mathbf{A}_{3\,(1)}^{(1)} \cdot \mathbf{C}_{3\,(1)}^{(2)})\, \mathbf{A}_{3\,(2)}^{(3)} \wedge \mathbf{C}_{3\,(2)}^{(4)} \wedge \mathbf{B}_3$$

$$+ \sum_{((1,2),(1,2))\vdash \mathbf{B}_3\mathbf{C}_3} (\mathbf{B}_{3\,(1)}^{(1)} \cdot \mathbf{C}_{3\,(1)}^{(2)})\, \mathbf{B}_{3\,(2)}^{(3)} \wedge \mathbf{C}_{3\,(2)}^{(4)} \wedge \mathbf{A}_3$$

$$= \langle \mathbf{A}_3\mathbf{B}_3\rangle_4 \wedge \mathbf{C}_3 - \langle \mathbf{A}_3\mathbf{C}_3\rangle_4 \wedge \mathbf{B}_3 + \langle \mathbf{B}_3\mathbf{C}_3\rangle_4 \wedge \mathbf{A}_3.$$

Chapter 6

Geometric Algebra

While the other versions of Clifford algebra emphasize the linear nature of the algebra, *i.e.*, care more for addition than for multiplication, Geometric Algebra prefers the usage of multiplication to addition.

Geometrically, the geometric product conglomerates all geometric relations within itself and is geometrically meaningful. Algebraically, exploiting the associativity of the geometric product and the commutations within the grading operators, has incomparable superiority over all other usages of Clifford algebra and inner-product Grassmann algebra in symbolic geometric computing.

This chapter investigates major computing techniques in Geometric Algebra, Clifford coalgebra and Clifford bracket algebra. In particular, three new computing techniques are developed for Geometric Algebra: ungrading, monomial compression, and Clifford factorization. They not only are powerful tools in algebraic manipulations, but also inspire a sequence of explorations and conjectures on the fundamental Grassmann structure of Clifford algebra. We start with a persuasive example from differential geometry as a prelude.

6.1 Major techniques in Geometric Algebra

Example 6.1. [197] In studying nD conformal manifold $(M, [g])$, F. Pedit proposed a triple product T as follows:

$$
\begin{aligned}
T: \quad TM \times T^*M \times TM &\longrightarrow \qquad\qquad TM \\
(\mathbf{x}, \alpha, \mathbf{y}) &\longmapsto \alpha(\mathbf{x})\mathbf{y} + \alpha(\mathbf{y})\mathbf{x} - g(\mathbf{x}, \mathbf{y})\alpha^\#,
\end{aligned}
\tag{6.1.1}
$$

where $\alpha^\# \in TM$ is defined by $g(\alpha^\#, \mathbf{z}) = \alpha(\mathbf{z})$, $\forall\, \mathbf{z} \in TM$. Let T' be the dual of T: for any $\alpha, \beta \in T^*M$ and $\mathbf{z} \in TM$, $T' : T^*M \times TM \times T^*M \longrightarrow T^*M$ is defined by

$$
T'(\alpha, \mathbf{z}, \beta)(\mathbf{u}) = \alpha(T(\mathbf{z}, \beta, \mathbf{u})), \ \forall\, \mathbf{u} \in TM.
\tag{6.1.2}
$$

Prove the following identity satisfied by T and T':

$$
T(\mathbf{x}, \alpha, T(\mathbf{y}, \beta, \mathbf{z})) - T(\mathbf{x}, T'(\alpha, \mathbf{z}, \beta), \mathbf{y}) = T(T(\mathbf{x}, \alpha, \mathbf{y}), \beta, \mathbf{z}) - T(T(\mathbf{x}, \beta, \mathbf{z}), \alpha, \mathbf{y}).
\tag{6.1.3}
$$

273

Willmore wrote the following in p.110 of his book [197]: "The only way I have succeeded in establishing this identity is by the time-honored method of tediously showing the left-hand side and the right-hand side when expanded do in fact lead to identical expressions."

We present an elegant proof with Geometric Algebra. In the Clifford bundle over M, T and T' have the following graded monomial expressions:

$$T(\mathbf{x}, \alpha, \mathbf{y}) = \langle \mathbf{x}\alpha^{\#}\mathbf{y}\rangle_1, \qquad T'(\alpha, \mathbf{z}, \beta) = (\langle \alpha^{\#}\mathbf{z}\beta^{\#}\rangle_1)^{\flat}, \tag{6.1.4}$$

where $\mathbf{x}^{\flat} \in T^*M$ is defined by $\mathbf{x}^{\flat}(\mathbf{y}) = g(\mathbf{x}, \mathbf{y})$, $\forall\, \mathbf{y} \in TM$.

Denote $\mathbf{x} = \mathbf{1}$, $\alpha^{\#} = \mathbf{2}$, $\mathbf{y} = \mathbf{3}$, $\beta^{\#} = \mathbf{4}$, $\mathbf{z} = \mathbf{5}$. Then (6.1.3) becomes

$$\langle \mathbf{12}\langle \mathbf{345}\rangle_1\rangle_1 - \langle \mathbf{1}\langle \mathbf{254}\rangle_1\mathbf{3}\rangle_1 = \langle\langle \mathbf{123}\rangle_1\mathbf{45}\rangle_1 - \langle\langle \mathbf{145}\rangle_1\mathbf{23}\rangle_1. \tag{6.1.5}$$

Proof. The strategy is to eliminate all interior 1-grading operators from both sides of the equality, called *ungrading*, and then simplify the result by symmetries within the 1-grading operator and by the associativity of the geometric product.

The ungrading follows the simple formula

$$\langle \mathbf{123}\rangle_1 = \frac{1}{2}(\mathbf{123} + \mathbf{321}), \tag{6.1.6}$$

which is valid for any vectors $\mathbf{1}, \mathbf{2}, \mathbf{3} \in V^n$. The left side of (6.1.5), after ungrading, becomes

$$\frac{1}{2}\langle \mathbf{12345} + \underline{\mathbf{12543}} - \underline{\mathbf{12543}} - \mathbf{14523}\rangle_1 = \frac{1}{2}\langle \mathbf{12345} - \mathbf{14523}\rangle_1. \tag{6.1.7}$$

The right side of (6.1.5) becomes

$$\frac{1}{2}\langle \mathbf{12345} + \underline{\mathbf{32145}} - \mathbf{14523} - \underline{\mathbf{54123}}\rangle_1 = \frac{1}{2}\langle \mathbf{12345} - \mathbf{14523}\rangle_1. \tag{6.1.8}$$

The associativity is automatically applied to both sides. The equality of (6.1.8) is based on the *reversion symmetry* inside the 1-grading operator:

$$\langle \mathbf{32145}\rangle_1 = \langle (\mathbf{32145})^{\dagger}\rangle_1 = \langle \mathbf{54123}\rangle_1. \tag{6.1.9}$$

$$\square$$

Definition 6.2. In $\mathcal{CL}(V^n)$, a *graded Clifford polynomial* is a $\mathbb{K}$-linear combination of graded Clifford monomials defined in Section 5.3. The space of graded Clifford polynomials is called the *graded Clifford space*, denoted by $\mathcal{G}(V^n)$.

The graded Clifford space equipped with the geometric product, the dual operator and the $\mathbb{Z}$-grading operators, is called a *Geometric Algebra*, denoted by the same symbol $\mathcal{G}(V^n)$. It is a representation of Clifford algebra.

The number of atomic vectors in a graded Clifford monomial is called the *length* of the monomial. A graded Clifford monomial does not allow the addition operation, but allows all the algebraic products defined so far. The following are typical graded Clifford monomials:

- Clifford monomial: $\mathbf{a}_1\mathbf{a}_2\cdots\mathbf{a}_r$;
- Grassmann monomial: $\langle\mathbf{a}_1\mathbf{a}_2\cdots\mathbf{a}_r\rangle_r$;
- integer-graded part of a Clifford monomial, called *single-graded Clifford monomial*: $\langle\mathbf{a}_1\mathbf{a}_2\cdots\mathbf{a}_r\rangle_{r-2i}$, where $0 \le i \le [r/2]$;
- inner product, outer product and meet product of graded Clifford monomials;
- Cayley monomial.

The elegant proof in Example 6.1 reveals the superiority of high-level computing by multiplications (long geometric product) over low-level computing by additions (Clifford expansion). It is based on the first two of the following five major manipulations in Geometric Algebra.

(1) Symmetries within $\mathbb{Z}$-grading operators:

(6.1.9) is a typical example. Following such symmetries, elements can change their positions within a grading operator. The associativity of the geometric product is another kind of symmetry. The use of symmetries in Geometric Algebra is the most economical way of employing syzygies in the corresponding inner-product bracket algebra.

(2) Commutations:

Definition 6.3. The *commutator*, also called *cross product*, of two elements $\mathbf{A}, \mathbf{B} \in \mathcal{CL}(\mathcal{V}^n)$, is

$$[\mathbf{A}, \mathbf{B}] := \frac{1}{2}(\mathbf{AB} - \mathbf{BA}). \tag{6.1.10}$$

Notation.
In [77], the commutator is denoted by the cross symbol "$\times$". In this book we preserve this symbol for the vector product in 3D vector algebra, because $\mathbf{a} \times \mathbf{b} \ne [\mathbf{a}, \mathbf{b}]$ for $\mathbf{a}, \mathbf{b} \in \mathbb{R}^3$.

In general, a grading operator does not have many symmetries. Switching two elements in the same grading operator can be realized by one or two commutators: for any $\mathbf{A}, \mathbf{B}, \mathbf{C} \in \mathcal{CL}(\mathcal{V}^n)$,

$$\mathbf{ABC} - \mathbf{CBA} = \mathbf{ABC} - \mathbf{ACB} + \mathbf{ACB} - \mathbf{CBA} = 2\mathbf{A}[\mathbf{B}, \mathbf{C}] + 2[\mathbf{A}, \mathbf{CB}]. \tag{6.1.11}$$

The commutator is important for two more reasons: the commutator with a bivector commutes with any grading operator, and all bivectors together with the commutator form the Lie algebra of $\mathrm{Spin}(\mathcal{V}^n)$.

(3) Ungrading:

In $\mathcal{G}(\mathcal{V}^n)$, the procedure of eliminating $\mathbb{Z}$-grading operators is called *ungrading*. Just like Cayley expansions in GC algebra, in Geometric algebra there are various ungradings of the same graded Clifford monomial, and finding the shortest ungrading is an important task. For the simplest case, 3-blade $\mathbf{1} \wedge \mathbf{2} \wedge \mathbf{3} \in \mathcal{G}(\mathcal{V}^n)$ has, by the very definition of complete antisymmetrization, an ungrading of $3! = 6$ terms:

$$\mathbf{1} \wedge \mathbf{2} \wedge \mathbf{3} = \frac{1}{6}(\mathbf{123} - \mathbf{132} + \mathbf{231} - \mathbf{213} + \mathbf{312} - \mathbf{321}). \qquad (6.1.12)$$

However, its shortest ungrading is 2-termed:

$$\mathbf{1} \wedge \mathbf{2} \wedge \mathbf{3} = \frac{1}{2}(\mathbf{123} - \mathbf{321}). \qquad (6.1.13)$$

This phenomenon discloses the sharp difference in antisymmetrization between Clifford algebra and its ancestor tensor algebra.

(4) Monomial simplification:

Clifford monomial simplification refers to monomial factorizations and compressions inside a graded Clifford monomial.

Clifford monomial factorization is to transform a graded Clifford monomial into the geometric product of two or more commutative (up to sign) graded Clifford monomials. For example, if a monomial contains an adjoining pair of identical vectors such as $\mathbf{aa}$, then by Cliffordization, $\mathbf{aa} = \langle \mathbf{aa} \rangle$ becomes a scalar. It can move freely within the whole sequence. For a Clifford monomial of generic vectors, there are few monomial factorization techniques. For special vectors such as null vectors, there are a lot more such techniques [125].

Given a Clifford monomial $\mathbf{A}_k$ of length k, if there is another sequence of vectors of length strictly less than k but whose geometric product equals $\mathbf{A}_k$, then $\mathbf{A}_k$ is said to be *compressed*, or its *effective vectors* are said to be reduced. The original monomial form of $\mathbf{A}_k$ is said to be *compressible*.

For example, if vectors $\mathbf{1}, \mathbf{2}, \mathbf{3}$ are linearly dependent, then $\mathbf{123} = \langle \mathbf{123} \rangle_1$ is also a vector by Cliffordization, so the monomial is automatically compressed to a vector, or equivalently, the number of effective vectors in the monomial is reduced from three to one.

(5) Rational Clifford expansion and Clifford factorization:

Rational Clifford expansion changes a graded Clifford monomial into a rational polynomial in inner-product bracket algebra. The purpose is to reduce the size of a Clifford expansion result. For example, when $n = 4$, for null vectors $\mathbf{1}$ to $\mathbf{6}$ in $\mathcal{V}^4$, the expansion of $[\mathbf{123456}]$ by Theorem 5.64 produces a polynomial of fifteen terms, the shortest expansion by Theorem 5.79 produces a polynomial of six terms (5.4.15), while the shortest rational expansion has only two terms [125]:

$$[\mathbf{123456}] = -2\frac{2 \cdot 3[\mathbf{1256}][\mathbf{3456}] + 5 \cdot 6\,[\mathbf{1236}]\,[\mathbf{2345}]}{[\mathbf{2356}]}. \qquad (6.1.14)$$

The reverse procedure of (rational) Clifford expansion is called (*rational*) *Clifford factorization*. It intends to change a multivector of inner-product coefficients into a graded Clifford monomial. A (rational) Cayley factorization is naturally a (rational) Clifford factorization. For example, in $\Lambda(\mathcal{V}^4)$ where $\mathcal{V}^4$ is an inner-product space, for vectors $\mathbf{0}, \mathbf{1}, \mathbf{2}, \mathbf{3}, \mathbf{4}, \mathbf{5}$, the following is a Clifford factorization:

$$0 \cdot 1[2345] - 0 \cdot 2[1345] = -0 \cdot ((1 \wedge 2) \vee (3 \wedge 4 \wedge 5)). \qquad (6.1.15)$$

The commutator can also be used to make Clifford factorization. For example, both sides of (6.1.5) can be changed into the following graded Clifford monomial:

$$\frac{1}{2}\langle 12345 - 14523 \rangle_1 = \langle 1\,[23, 45]\,\rangle_1 = \langle 1\,\langle (2 \wedge 3)(4 \wedge 5) \rangle_2 \,\rangle_1. \qquad (6.1.16)$$

The last step in the above computing will be explained soon in Subsection 6.1.2.

Graded Clifford space is an intermediate stage between Clifford space and inner-product Grassmann space. Because of this, there are three ways to make algebraic manipulations in Geometric Algebra:

1. **Direct way:** change a graded Clifford monomial to another graded Clifford polynomial by symmetries and commutations within each grading operator.

2. **Upward way:** get rid of interior grading operators by ungrading operations, use syzygies among high-level invariants in Clifford bracket algebra to make simplifications. In the extreme case, the result of ungrading is a Clifford polynomial, and a monomial normalization procedure similar to the straightening of bracket monomials is necessary.

3. **Downward way:** increase the number of grading operators by (rational) Clifford expansions, use syzygies among low-level invariants in Clifford bracket algebra to make simplifications. In the extreme case, the result of expansion is a multivector in Grassmann algebra, and the corresponding algebra of invariants is inner-product bracket algebra.

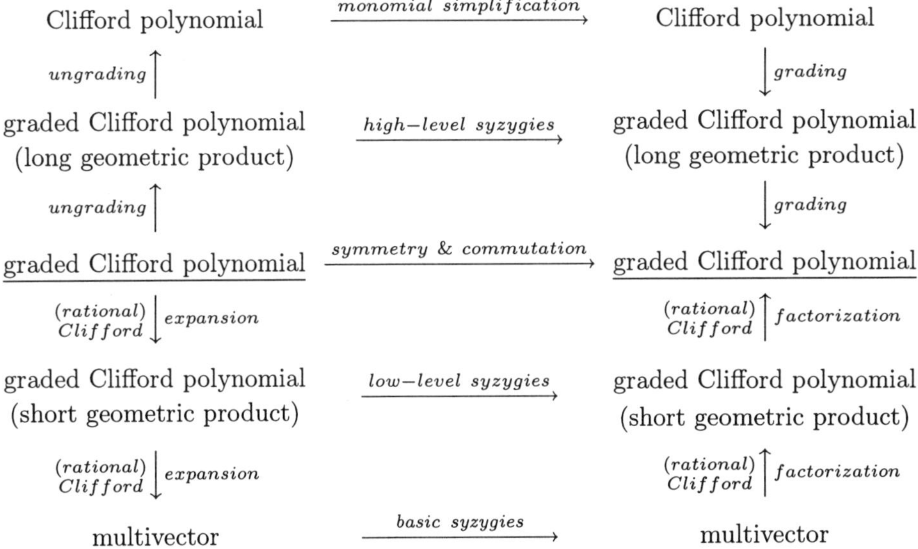

6.1.1   *Symmetry*

Proposition 6.4. The grading operators commute with each of the grade involution, reversion and conjugate operators.

Proof. Only the commutation of $\langle\ \rangle_i$ with reversion needs to be proved. For $\mathbf{A}_k = \mathbf{a}_1 \mathbf{a}_2 \cdots \mathbf{a}_k$, we need to prove

$$\langle \mathbf{A}_k^\dagger \rangle_{k-2l} = (\langle \mathbf{A}_k \rangle_{k-2l})^\dagger = (-1)^{l+\frac{k(k-1)}{2}} \langle \mathbf{A}_k \rangle_{k-2l}. \tag{6.1.17}$$

When $l = 0$, (6.1.17) is trivial for any k. Assume that it is true for $l = m - 1$ and for any k. For $l = m$ and any k,

$$\begin{aligned}
\langle \mathbf{A}_k^\dagger \rangle_{k-2m} &= \frac{1}{m} \sum_{(2,k-2) \vdash \mathbf{A}_k^\dagger} \langle (\mathbf{A}_k^\dagger)_{(1)} \rangle \langle (\mathbf{A}_k^\dagger)_{(2)} \rangle_{k-2-2(m-1)} \\
&= \frac{1}{m} (-1)^{m-1+\frac{(k-2)(k-3)}{2}} \sum_{(2,k-2) \vdash \mathbf{A}_k^\dagger} \langle ((\mathbf{A}_k^\dagger)^\dagger)_{(1)} \rangle \langle ((\mathbf{A}_k^\dagger)^\dagger)_{(2)} \rangle_{k-2m} \\
&= \frac{1}{m} (-1)^{m+\frac{k(k-1)}{2}} \sum_{(2,k-2) \vdash \mathbf{A}_k} \langle \mathbf{A}_{k(1)} \rangle \langle \mathbf{A}_{k(2)} \rangle_{k-2m} \\
&= (-1)^{m+\frac{k(k-1)}{2}} \langle \mathbf{A}_k \rangle_{k-2m}.
\end{aligned}$$

$\square$

Corollary 6.5. For any r-vector $\mathbf{A}_r$ and s-vector $\mathbf{B}_s$,

$$\langle \mathbf{A}_r \mathbf{B}_s \rangle_{r+s-2l} = (-1)^{rs-l} \langle \mathbf{B}_s \mathbf{A}_r \rangle_{r+s-2l}. \tag{6.1.18}$$

Proof.

$$\begin{aligned}
\langle \mathbf{B}_s \mathbf{A}_r \rangle_{r+s-2l} &= (-1)^{\frac{s(s-1)+r(r-1)}{2}} \langle \mathbf{B}_s^\dagger \mathbf{A}_r^\dagger \rangle_{r+s-2l} \\
&= (-1)^{\frac{s(s-1)+r(r-1)+(r+s-2l)(r+s-2l-1)}{2}} \{ \langle (\mathbf{A}_r \mathbf{B}_s)^\dagger \rangle_{r+s-2l} \}^\dagger \\
&= (-1)^{rs-l} \langle \mathbf{A}_r \mathbf{B}_s \rangle_{r+s-2l}.
\end{aligned}$$

$\square$

Proposition 6.6. For vectors $\mathbf{a}$'s in $\mathcal{V}^n$,

$$\langle \mathbf{a}_1 \mathbf{a}_2 \cdots \mathbf{a}_{2l} \rangle = \langle \mathbf{a}_{2l} \mathbf{a}_1 \mathbf{a}_2 \cdots \mathbf{a}_{2l-1} \rangle \tag{6.1.19}$$

$$= \langle \mathbf{a}_{2l} \mathbf{a}_{2l-1} \cdots \mathbf{a}_1 \rangle, \tag{6.1.20}$$

$$[\mathbf{a}_1 \mathbf{a}_2 \cdots \mathbf{a}_{n+2l}] = (-1)^{n-1} [\mathbf{a}_{n+2l} \mathbf{a}_1 \mathbf{a}_2 \cdots \mathbf{a}_{n+2l-1}] \tag{6.1.21}$$

$$= (-1)^{\frac{n(n-1)}{2}} [\mathbf{a}_{n+2l} \mathbf{a}_{n+2l-1} \cdots \mathbf{a}_1]. \tag{6.1.22}$$

Proof.

$$\langle \mathbf{a}_1 \mathbf{a}_2 \cdots \mathbf{a}_{2l} \rangle = \langle \mathbf{a}_1 \mathbf{a}_2 \cdots \mathbf{a}_{2l-1} \rangle_1 \cdot \mathbf{a}_{2l}$$
$$= \mathbf{a}_{2l} \cdot \langle \mathbf{a}_1 \mathbf{a}_2 \cdots \mathbf{a}_{2l-1} \rangle_1$$
$$= \langle \mathbf{a}_{2l} \mathbf{a}_1 \mathbf{a}_2 \cdots \mathbf{a}_{2l-1} \rangle,$$

$$[\mathbf{a}_1 \mathbf{a}_2 \cdots \mathbf{a}_{n+2l}] = [\langle \mathbf{a}_1 \mathbf{a}_2 \cdots \mathbf{a}_{n+2l-1} \rangle_{n-1} \wedge \mathbf{a}_{n+2l}]$$
$$= (-1)^{n-1} [\mathbf{a}_{n+2l} \wedge \langle \mathbf{a}_1 \mathbf{a}_2 \cdots \mathbf{a}_{n+2l-1} \rangle_{n-1}]$$
$$= (-1)^{n-1} [\mathbf{a}_{n+2l} \mathbf{a}_1 \cdots \mathbf{a}_{n+2l-1}].$$

(6.1.20) and (6.1.22) are direct corollaries of Proposition 6.4. $\qquad\square$

Lemma 6.7. For any vector $\mathbf{a}$ and multivector $\mathbf{A}$,

$$\mathbf{a} \wedge \mathbf{A} = \frac{1}{2}(\mathbf{a}\mathbf{A} + \hat{\mathbf{A}}\mathbf{a}) = \hat{\mathbf{A}} \wedge \mathbf{a}, \tag{6.1.23}$$

$$\mathbf{a} \cdot \mathbf{A} = \frac{1}{2}(\mathbf{a}\mathbf{A} - \hat{\mathbf{A}}\mathbf{a}) + \mathbf{a}\langle \mathbf{A} \rangle = -\hat{\mathbf{A}} \cdot \mathbf{a} + 2\mathbf{a}\langle \mathbf{A} \rangle, \tag{6.1.24}$$

$$\mathbf{a}\mathbf{A} = \mathbf{a} \cdot \mathbf{A} + \mathbf{a} \wedge \mathbf{A} - \mathbf{a}\langle \mathbf{A} \rangle. \tag{6.1.25}$$

Proof. (6.1.25) is the sum of (6.1.23) and (6.1.24). The second equality in (6.1.23) is trivial. Expanding $\mathbf{a} \cdot (\mathbf{A} - \langle \mathbf{A} \rangle) = -(\hat{\mathbf{A}} - \langle \mathbf{A} \rangle) \cdot \mathbf{a}$, we get the second equality in (6.1.24).

For the first equalities in (6.1.23) and (6.1.24) respectively, by linearity we can assume that $\mathbf{A}$ is an r-vector. If $r = 0$, then $\mathbf{a} \cdot \mathbf{A} = \mathbf{a} \wedge \mathbf{A} = \mathbf{a}\mathbf{A} = \mathbf{A}\mathbf{a}$, and the equalities are obvious. If $r \geq 1$, then

$$\mathbf{a} \cdot \mathbf{A} = \langle \mathbf{a}\mathbf{A} \rangle_{r-1} = (-1)^{\frac{(r-1)(r-2)}{2}} \langle (\mathbf{a}\mathbf{A})^\dagger \rangle_{r-1} = (-1)^{r-1} \langle \mathbf{A}\mathbf{a} \rangle_{r-1} = -\langle \hat{\mathbf{A}}\mathbf{a} \rangle_{r-1},$$
$$\mathbf{a} \wedge \mathbf{A} = \langle \mathbf{a}\mathbf{A} \rangle_{r+1} = (-1)^{\frac{(r+1)r}{2}} \langle (\mathbf{a}\mathbf{A})^\dagger \rangle_{r+1} = (-1)^r \langle \mathbf{A}\mathbf{a} \rangle_{r+1} = \langle \hat{\mathbf{A}}\mathbf{a} \rangle_{r+1}.$$

By $\mathbf{a}\mathbf{A} = \langle \mathbf{a}\mathbf{A} \rangle_{r-1} + \langle \mathbf{a}\mathbf{A} \rangle_{r+1}$ and $\mathbf{A}\mathbf{a} = \langle \mathbf{A}\mathbf{a} \rangle_{r-1} + \langle \mathbf{A}\mathbf{a} \rangle_{r+1}$, we get the two equalities. $\qquad\square$

The geometric product is no longer associative when composed with grading operators. For example, for vector $\mathbf{a}$ and blades $\mathbf{B}, \mathbf{C}$, in general $\langle \mathbf{a}\langle \mathbf{B}\mathbf{C} \rangle_i \rangle_k \neq \langle \langle \mathbf{a}\mathbf{B} \rangle_j \mathbf{C} \rangle_k$ (nonassociative), and $\langle \mathbf{a}\langle \mathbf{B}\mathbf{C} \rangle_i \rangle_l \neq \langle \langle \mathbf{a}\mathbf{B} \rangle_j \mathbf{C} \rangle_l \pm \langle \mathbf{B}\langle \mathbf{a}\mathbf{C} \rangle_k \rangle_l$ (nonderivative). However, for special values of i, j, k, l, there do exist "pseudo-derivative" and "pseudo-associative" properties.

The following proposition is direct from Lemma 6.7:

Proposition 6.8. For any vector $\mathbf{a}$ and blades $\mathbf{B}, \mathbf{C}$ of nonzero grade,

$$\mathbf{a} \cdot (\mathbf{B}\mathbf{C} - \langle \mathbf{B}\mathbf{C} \rangle) = (\mathbf{a} \cdot \mathbf{B})\mathbf{C} + \hat{\mathbf{B}}(\mathbf{a} \cdot \mathbf{C})$$
$$= (\mathbf{a} \wedge \mathbf{B})\mathbf{C} - \hat{\mathbf{B}}(\mathbf{a} \wedge \mathbf{C}), \tag{6.1.26}$$
$$\mathbf{a} \wedge (\mathbf{B}\mathbf{C}) = (\mathbf{a} \wedge \mathbf{B})\mathbf{C} - \hat{\mathbf{B}}(\mathbf{a} \cdot \mathbf{C})$$
$$= (\mathbf{a} \cdot \mathbf{B})\mathbf{C} + \hat{\mathbf{B}}(\mathbf{a} \wedge \mathbf{C}). \tag{6.1.27}$$

Corollary 6.9. Let $\mathbf{B}_s, \mathbf{C}_t$ be blades of nonzero grade s, t respectively. Then for any vector $\mathbf{a} \in \mathbf{C}_t$,

$$
\begin{aligned}
(\mathbf{a} \cdot \mathbf{B}_s)\mathbf{C}_t &= \mathbf{a} \wedge (\mathbf{B}_s \mathbf{C}_t); \\
(\mathbf{a} \wedge \mathbf{B}_s)\mathbf{C}_t &= \mathbf{a} \cdot (\mathbf{B}_s \mathbf{C}_t - \langle \mathbf{B}_s \mathbf{C}_t \rangle); \\
(\mathbf{a} \cdot \mathbf{B}_s) \cdot \mathbf{C}_t &= \mathbf{a} \wedge (\mathbf{B}_s \cdot \mathbf{C}_t), && \text{if } s \le t; \\
(\mathbf{a} \wedge \mathbf{B}_s) \cdot \mathbf{C}_t &= \mathbf{a} \cdot (\mathbf{B}_s \cdot \mathbf{C}_t), && \text{if } s \le t - 1.
\end{aligned}
\tag{6.1.28}
$$

Proof. The first two equalities are special cases of (6.1.27) and (6.1.26) respectively. The third is the $(t - s + 1)$-graded part of the first, and the last is the $(t - s - 1)$-graded part of the second. $\qquad\square$

6.1.2   *Commutation*

The commutator measures the difference caused by the switch of positions between two neighboring elements. For example, for vectors $\mathbf{a}, \mathbf{b}$ and bivector $\mathbf{B}_2$,

$$
[\mathbf{a}, \mathbf{b}] = \mathbf{a} \wedge \mathbf{b}, \qquad [\mathbf{a}, \mathbf{B}_2] = \mathbf{a} \cdot \mathbf{B}_2.
\tag{6.1.29}
$$

In general, for any vector $\mathbf{a}$ and multivector $\mathbf{A}$, by (6.1.23) and (6.1.24),

$$
\begin{aligned}
[\mathbf{a}, \mathbf{A}] &= [\mathbf{a}, \langle \mathbf{A} \rangle_+] + [\mathbf{a}, \langle \mathbf{A} \rangle_-] \\
&= \frac{1}{2}\{(\mathbf{a}\langle \mathbf{A} \rangle_+ - \langle \hat{\mathbf{A}} \rangle_+ \mathbf{a}) + (\mathbf{a}\langle \mathbf{A} \rangle_- + \langle \hat{\mathbf{A}} \rangle_- \mathbf{a})\} \\
&= \mathbf{a} \cdot (\langle \mathbf{A} \rangle_+ - \langle \mathbf{A} \rangle) + \mathbf{a} \wedge \langle \mathbf{A} \rangle_-.
\end{aligned}
\tag{6.1.30}
$$

Proposition 6.10.

(1) For any multivectors $\mathbf{A}, \mathbf{C}_1, \ldots, \mathbf{C}_r \in \mathcal{CL}(\mathcal{V}^n)$,

$$
[\mathbf{A}, \mathbf{C}_1 \mathbf{C}_2 \cdots \mathbf{C}_r] = \sum_{i=1}^{r} \mathbf{C}_1 \cdots \mathbf{C}_{i-1}[\mathbf{A}, \mathbf{C}_i]\mathbf{C}_{i+1} \cdots \mathbf{C}_r.
\tag{6.1.31}
$$

(2) $\mathcal{CL}(\mathcal{V}^n)$ is a Lie algebra under the commutator operation.

(3) For any bivector $\mathbf{B}_2$ and vectors $\mathbf{a}_i \in \mathcal{V}^n$,

$$
[\mathbf{B}_2, \mathbf{a}_1 \wedge \cdots \wedge \mathbf{a}_r] = \sum_{i=1}^{r} \mathbf{a}_1 \wedge \cdots \wedge \mathbf{a}_{i-1} \wedge [\mathbf{B}_2, \mathbf{a}_i] \wedge \mathbf{a}_{i+1} \wedge \cdots \wedge \mathbf{a}_r.
\tag{6.1.32}
$$

(4) The mapping $\mathbf{C} \mapsto [\mathbf{B}_2, \mathbf{C}]$, $\forall \mathbf{C} \in \mathcal{CL}(\mathcal{V}^n)$, is a grade-preserving operator, *i.e.*, it commutes with $\langle \ \rangle_i$.

(5) $\Lambda^2(\mathcal{V}^n)$ is the Lie algebra of $\mathrm{Spin}(\mathcal{V}^n)$. The exponential map is

$$
e^{\mathbf{A}} := \sum_{i=0}^{\infty} \frac{(\mathbf{A})^i}{i!}, \quad \forall \mathbf{A} \in \mathcal{CL}(\mathcal{V}^n).
\tag{6.1.33}
$$

Proof. (1) Direct extension of the identity $[\mathbf{A}, \mathbf{C}_1 \mathbf{C}_2] = [\mathbf{A}, \mathbf{C}_1]\mathbf{C}_2 + \mathbf{C}_1[\mathbf{A}, \mathbf{C}_2]$.
(2) Corollary of the Jacobi identity

$$[\mathbf{A}, [\mathbf{B}, \mathbf{C}]] = [[\mathbf{A}, \mathbf{B}], \mathbf{C}] + [\mathbf{B}, [\mathbf{A}, \mathbf{C}]]. \tag{6.1.34}$$

(3) Direct from (6.1.31) and (6.1.29). (4) Corollary of (6.1.32). (5) Classical result in Lie group theory. $\quad\square$

Example 6.11. For the 2D rotation induced by rotor $\mathbf{ab}$, since $\mathbf{ab} = e^{\mathbf{I}_2 \frac{\theta}{2}}$, where θ is the angle of rotation, bivector $\mathbf{I}_2 \theta/2$ is the generator of the rotor in Lie algebra $\Lambda^2(\mathcal{V}^n)$.

Proposition 6.12. For any vector $\mathbf{a}$ and bivectors $\mathbf{A}_2, \mathbf{B}_2$,
$$\begin{aligned}
\mathbf{a} \cdot [\mathbf{A}_2, \mathbf{B}_2] &= (\mathbf{a} \cdot \mathbf{A}_2) \cdot \mathbf{B}_2 - (\mathbf{a} \cdot \mathbf{B}_2) \cdot \mathbf{A}_2, \\
\mathbf{a} \cdot [\mathbf{A}_2, \mathbf{B}_2] &= (\mathbf{a} \wedge \mathbf{A}_2) \cdot \mathbf{B}_2 - (\mathbf{a} \wedge \mathbf{B}_2) \cdot \mathbf{A}_2.
\end{aligned} \tag{6.1.35}$$

Proof. By (6.1.29), (6.1.32) and (6.1.26),
$$\begin{aligned}
\mathbf{a} \cdot [\mathbf{A}_2, \mathbf{B}_2] &= [\mathbf{a}, [\mathbf{A}_2, \mathbf{B}_2]] \\
&= [[\mathbf{a}, \mathbf{A}_2], \mathbf{B}_2] + [\mathbf{A}_2, [\mathbf{a}, \mathbf{B}_2]] \\
&= (\mathbf{a} \cdot \mathbf{A}_2) \cdot \mathbf{B}_2 - (\mathbf{a} \cdot \mathbf{B}_2) \cdot \mathbf{A}_2, \\
\mathbf{a} \cdot [\mathbf{A}_2, \mathbf{B}_2] &= \frac{1}{2}(\mathbf{a} \cdot (\mathbf{A}_2 \mathbf{B}_2) - \mathbf{a} \cdot (\mathbf{B}_2 \mathbf{A}_2)) \\
&= \frac{1}{2}\{(\mathbf{a} \wedge \mathbf{A}_2)\mathbf{B}_2 - \hat{\mathbf{A}}_2(\mathbf{a} \wedge \mathbf{B}_2) - (\mathbf{a} \wedge \mathbf{B}_2)\mathbf{A}_2 + \hat{\mathbf{B}}_2(\mathbf{a} \wedge \mathbf{A}_2)\} \\
&= (\mathbf{a} \wedge \mathbf{A}_2) \cdot \mathbf{B}_2 - (\mathbf{a} \wedge \mathbf{B}_2) \cdot \mathbf{A}_2.
\end{aligned}$$

$\quad\square$

Proposition 6.13. Let $\mathbf{B}_2$ be a bivector, and $\mathbf{A}_r$ be an r-vector. Then
$$\langle \mathbf{B}_2 \mathbf{A}_r \rangle_r = [\mathbf{B}_2, \mathbf{A}_r]. \tag{6.1.36}$$

Proof. By linearity we can assume $\mathbf{A}_r = \mathbf{a}_1 \wedge \mathbf{a}_2 \wedge \cdots \wedge \mathbf{a}_r$ and $\mathbf{B}_2 = \mathbf{b}_1 \wedge \mathbf{b}_2$, where $\mathbf{b}_1 \cdot \mathbf{b}_2 = 0$. The left side of (6.1.36) equals
$$\begin{aligned}
\langle \mathbf{b}_1 \mathbf{b}_2 \mathbf{A}_r \rangle_r &= \mathbf{b}_1 \cdot (\mathbf{b}_2 \wedge \mathbf{A}_r) + \mathbf{b}_1 \wedge (\mathbf{b}_2 \cdot \mathbf{A}_r) \\
&= -\mathbf{b}_2 \wedge \sum_{i=1}^{r} (-1)^{i+1}(\mathbf{b}_1 \cdot \mathbf{a}_i)\mathbf{a}_1 \wedge \cdots \wedge \breve{\mathbf{a}}_i \wedge \cdots \wedge \mathbf{a}_r \\
&\quad + \mathbf{b}_1 \wedge \sum_{i=1}^{r} (-1)^{i+1}(\mathbf{b}_2 \cdot \mathbf{a}_i)\mathbf{a}_1 \wedge \cdots \wedge \breve{\mathbf{a}}_i \wedge \cdots \wedge \mathbf{a}_r \\
&= \sum_{i=1}^{r} \mathbf{a}_1 \wedge \cdots \wedge \mathbf{a}_{i-1} \wedge ((\mathbf{b}_2 \cdot \mathbf{a}_i)\mathbf{b}_1 - (\mathbf{b}_1 \cdot \mathbf{a}_i)\mathbf{b}_2) \wedge \mathbf{a}_{i+1} \wedge \cdots \wedge \mathbf{a}_r \\
&= \sum_{i=1}^{r} \mathbf{a}_1 \wedge \cdots \wedge \mathbf{a}_{i-1} \wedge [\mathbf{b}_1 \wedge \mathbf{b}_2, \mathbf{a}_i] \wedge \mathbf{a}_{i+1} \wedge \cdots \wedge \mathbf{a}_r \\
&= [\mathbf{b}_1 \wedge \mathbf{b}_2, \mathbf{a}_1 \wedge \mathbf{a}_2 \wedge \cdots \wedge \mathbf{a}_r].
\end{aligned}$$

$\quad\square$

The commutator of two blades generally does not produce a blade. For example,

$$[\mathbf{a} \wedge \mathbf{b}, \mathbf{c} \wedge \mathbf{d}] = [\mathbf{a}, [\mathbf{b}, \mathbf{c} \wedge \mathbf{d}]] - [\mathbf{b}, [\mathbf{a}, \mathbf{c} \wedge \mathbf{d}]]$$
$$= \mathbf{a} \wedge (\mathbf{b} \cdot (\mathbf{c} \wedge \mathbf{d})) - \mathbf{b} \wedge (\mathbf{a} \cdot (\mathbf{c} \wedge \mathbf{d})). \qquad (6.1.37)$$

Since

$$[\mathbf{a} \wedge \mathbf{b}, \mathbf{c} \wedge \mathbf{d}] \wedge [\mathbf{a} \wedge \mathbf{b}, \mathbf{c} \wedge \mathbf{d}] = -2\,\mathbf{a} \wedge (\mathbf{b} \cdot (\mathbf{c} \wedge \mathbf{d})) \wedge \mathbf{b} \wedge (\mathbf{a} \cdot (\mathbf{c} \wedge \mathbf{d}))$$
$$= 2\left\{(\mathbf{a} \cdot \mathbf{d})(\mathbf{b} \cdot \mathbf{c}) - (\mathbf{a} \cdot \mathbf{c})(\mathbf{b} \cdot \mathbf{d})\right\} \mathbf{a} \wedge \mathbf{b} \wedge \mathbf{c} \wedge \mathbf{d}$$
$$= 2\left\{(\mathbf{a} \wedge \mathbf{b}) \cdot (\mathbf{c} \wedge \mathbf{d})\right\} \mathbf{a} \wedge \mathbf{b} \wedge \mathbf{c} \wedge \mathbf{d},$$

we get, by Theorem 2.78, the following conclusion:

Proposition 6.14. The commutator of two 2-blades is a blade if and only if the 2D subspaces represented by the two blades are either perpendicular to each other, or intersecting in a 1D subspace, or identical.

Besides commutations of the form $\mathbf{AB} - \mathbf{BA}$, there are other commutations of the form $\mathbf{AB} + \mathbf{BA}$, where $\mathbf{A}, \mathbf{B} \in \mathcal{G}(\mathcal{V}^n)$. The simplest case is where $\mathbf{A} = \mathbf{a}$ is a vector and $\mathbf{B}$ is a Clifford monomial.

Proposition 6.15. [Vector commutation formula] For any vectors $\mathbf{a}$ and $\mathbf{b}_i$,

$$\mathbf{a}\mathbf{b}_1\mathbf{b}_2 \cdots \mathbf{b}_r + (-1)^{r-1}\mathbf{b}_1\mathbf{b}_2 \cdots \mathbf{b}_r\mathbf{a} = 2 \sum_{i=1}^{r} (-1)^{i+1}(\mathbf{a} \cdot \mathbf{b}_i)\, \mathbf{b}_1\mathbf{b}_2 \cdots \check{\mathbf{b}}_i \cdots \mathbf{b}_r.$$

$$(6.1.38)$$

Proof. Denote $\mathbf{B}_r = \mathbf{b}_1\mathbf{b}_2 \cdots \mathbf{b}_r$. Then

$$\langle \mathbf{a}\mathbf{B}_r \rangle_{r+1-2l}$$

$$= \sum_{(2l-1,\,r+1-2l) \vdash \mathbf{B}_r} \langle \mathbf{a}\mathbf{B}_{r(1)} \rangle \langle \mathbf{B}_{r(2)} \rangle_{r+1-2l} + \sum_{(2l,\,r-2l) \vdash \mathbf{B}_r} \langle \mathbf{B}_{r(1)} \rangle \langle \mathbf{a}\mathbf{B}_{r(2)} \rangle_{r+1-2l}$$

$$= \sum_{(2l-1,\,r+1-2l) \vdash \mathbf{B}_r} \langle \mathbf{B}_{r(1)}\mathbf{a} \rangle \langle \mathbf{B}_{r(2)} \rangle_{r+1-2l} + \sum_{(2l,\,r-2l) \vdash \mathbf{B}_r} (-1)^r \langle \mathbf{B}_{r(1)} \rangle \langle \mathbf{B}_{r(2)}\mathbf{a} \rangle_{r+1-2l}.$$

Similarly,

$$\langle \mathbf{B}_r\mathbf{a} \rangle_{r+1-2l} = \sum_{(2l-1,\,r+1-2l) \vdash \mathbf{B}_r} (-1)^{r+1} \langle \mathbf{B}_{r(1)}\mathbf{a} \rangle \langle \mathbf{B}_{r(2)} \rangle_{r+1-2l}$$

$$+ \sum_{(2l,\,r-2l) \vdash \mathbf{B}_r} \langle \mathbf{B}_{r(1)} \rangle \langle \mathbf{B}_{r(2)}\mathbf{a} \rangle_{r+1-2l}.$$

So

$$\langle \mathbf{a}\mathbf{B}_r + (-1)^{r-1}\mathbf{B}_r\mathbf{a} \rangle_{r+1-2l} = \sum_{(2l-1,\,r+1-2l) \vdash \mathbf{B}_r} 2\,\langle \mathbf{a}\mathbf{B}_{r(1)} \rangle \langle \mathbf{B}_{r(2)} \rangle_{r+1-2l}$$

$$= \sum_{(1,\,2l-2,\,r+1-2l) \vdash \mathbf{B}_r} 2\,(\mathbf{a} \cdot \mathbf{B}_{r(1)}) \langle \mathbf{B}_{r(2)} \rangle \langle \mathbf{B}_{r(3)} \rangle_{r+1-2l}$$

$$= \sum_{(1,\,r-1) \vdash \mathbf{B}_r} 2\,(\mathbf{a} \cdot \mathbf{B}_{r(1)}) \langle \mathbf{B}_{r(2)} \rangle_{r+1-2l}.$$

$$\square$$

Remark: In contrast, by (6.1.23), we have

$$\mathbf{ab}_1\mathbf{b}_2\cdots\mathbf{b}_r + (-1)^r\mathbf{b}_1\mathbf{b}_2\cdots\mathbf{b}_r\mathbf{a} = 2\,\mathbf{a}\wedge(\mathbf{b}_1\mathbf{b}_2\cdots\mathbf{b}_r). \tag{6.1.39}$$

Corollary 6.16. For any vectors $\mathbf{a}$ and $\mathbf{b}_i$,

$$\begin{aligned}
\mathbf{a}\cdot(\mathbf{b}_1\mathbf{b}_2\cdots\mathbf{b}_r) &= \mathbf{a}\,\langle\mathbf{b}_1\mathbf{b}_2\cdots\mathbf{b}_r\rangle + \sum_{i=1}^{r}(-1)^{i+1}(\mathbf{a}\cdot\mathbf{b}_i)\,\mathbf{b}_1\mathbf{b}_2\cdots\check{\mathbf{b}}_i\cdots\mathbf{b}_r, \\
(\mathbf{b}_1\mathbf{b}_2\cdots\mathbf{b}_r)\cdot\mathbf{a} &= \mathbf{a}\,\langle\mathbf{b}_1\mathbf{b}_2\cdots\mathbf{b}_r\rangle + \sum_{i=1}^{r}(-1)^{r-i}(\mathbf{a}\cdot\mathbf{b}_i)\,\mathbf{b}_1\mathbf{b}_2\cdots\check{\mathbf{b}}_i\cdots\mathbf{b}_r.
\end{aligned} \tag{6.1.40}$$

By repeatedly applying (6.1.38), we get

Corollary 6.17. For any vectors $\mathbf{a}_i$ and $\mathbf{b}_j$,

$$\begin{aligned}
&\mathbf{a}_1\mathbf{a}_2\cdots\mathbf{a}_r\mathbf{b}_1\mathbf{b}_2\cdots\mathbf{b}_s + (-1)^{rs-1}\mathbf{b}_1\mathbf{b}_2\cdots\mathbf{b}_s\mathbf{a}_1\mathbf{a}_2\cdots\mathbf{a}_r \\
&= \; 2\sum_{i=1}^{r}\sum_{j=1}^{s}(-1)^{(r+s)i+s-j}(\mathbf{a}_i\cdot\mathbf{b}_j)\mathbf{a}_{i+1}\cdots\mathbf{a}_r\mathbf{b}_1\cdots\check{\mathbf{b}}_j\cdots\mathbf{b}_s\mathbf{a}_1\cdots\mathbf{a}_{i-1} \\
&\quad +2\sum_{1\le i<j\le r}(\mathbf{a}_i\cdot\mathbf{a}_j)\{(-1)^{(r+s)(i-1)+j}\mathbf{a}_{i+1}\cdots\check{\mathbf{a}}_j\cdots\mathbf{a}_r\mathbf{b}_1\cdots\mathbf{b}_s\mathbf{a}_1\cdots\mathbf{a}_{i-1} \\
&\qquad\qquad\qquad +(-1)^{(r+s)j+i}\mathbf{a}_{j+1}\cdots\mathbf{a}_r\mathbf{b}_1\cdots\mathbf{b}_s\mathbf{a}_1\cdots\check{\mathbf{a}}_i\cdots\mathbf{a}_{j-1}\}.
\end{aligned} \tag{6.1.41}$$

In particular, when $r = 2$,

$$\begin{aligned}
&\mathbf{a}_1\mathbf{a}_2\mathbf{b}_1\mathbf{b}_2\cdots\mathbf{b}_s - \mathbf{b}_1\mathbf{b}_2\cdots\mathbf{b}_s\mathbf{a}_1\mathbf{a}_2 \\
&= 2\sum_{k=1}^{s}\{(-1)^k(\mathbf{a}_1\cdot\mathbf{b}_k)\mathbf{a}_2\mathbf{b}_1\cdots\check{\mathbf{b}}_k\cdots\mathbf{b}_s + (-1)^{s-k}(\mathbf{a}_2\cdot\mathbf{b}_k)\mathbf{b}_1\cdots\check{\mathbf{b}}_k\cdots\mathbf{b}_s\mathbf{a}_1\} \\
&= 2\sum_{k=1}^{s}\{(\mathbf{a}_2\cdot\mathbf{b}_k)\mathbf{b}_1\cdots\mathbf{b}_{k-1}\mathbf{a}_1\mathbf{b}_{k+1}\cdots\mathbf{b}_s - (\mathbf{a}_1\cdot\mathbf{b}_k)\mathbf{b}_1\cdots\mathbf{b}_{k-1}\mathbf{a}_2\mathbf{b}_{k+1}\cdots\mathbf{b}_s\}.
\end{aligned} \tag{6.1.42}$$

Lemma 6.18. For any vectors $\mathbf{a}_i$ and $\mathbf{b}_j$, let $\mathbf{B}_s = \mathbf{b}_1\mathbf{b}_2\cdots\mathbf{b}_s$, then for any $l \ge 0$ such that $2l < s$,

$$\frac{1}{2}\langle\mathbf{a}_1\mathbf{a}_2\mathbf{B}_s - \mathbf{B}_s\mathbf{a}_1\mathbf{a}_2\rangle_{s-2l} = \mathbf{a}_1\wedge(\mathbf{a}_2\cdot\langle\mathbf{B}_s\rangle_{s-2l}) - \mathbf{a}_2\wedge(\mathbf{a}_1\cdot\langle\mathbf{B}_s\rangle_{s-2l}). \tag{6.1.43}$$

Proof. The left side equals

$$\begin{aligned}
\langle[\mathbf{a}_1\mathbf{a}_2,\mathbf{B}_s]\rangle_{s-2l} &= [\mathbf{a}_1\wedge\mathbf{a}_2,\langle\mathbf{B}_s\rangle_{s-2l}] \\
&= \sum_{(2l,s-2l)\vdash\mathbf{B}_s}\langle\mathbf{B}_{s(1)}\rangle[\mathbf{a}_1\wedge\mathbf{a}_2,\langle\mathbf{B}_{s(2)}\rangle_{s-2l}] \\
&= \sum_{(2l,1,s-2l-1)\vdash\mathbf{B}_s}\langle\mathbf{B}_{s(1)}\rangle((\mathbf{a}_1\wedge\mathbf{a}_2)\cdot\mathbf{B}_{s(2)})\wedge\langle\mathbf{B}_{s(3)}\rangle_{s-2l-1}
\end{aligned}$$

$$= \sum_{(1,2l,s-2l-1)\vdash \mathbf{B}_s} ((\mathbf{a}_2 \cdot \mathbf{B}_{s(1)})\mathbf{a}_1 - (\mathbf{a}_1 \cdot \mathbf{B}_{s(1)})\mathbf{a}_2) \wedge (\langle \mathbf{B}_{s(2)} \rangle \langle \mathbf{B}_{s(3)} \rangle_{s-2l-1})$$

$$= \sum_{(1,s-1)\vdash \mathbf{B}_s} ((\mathbf{a}_2 \cdot \mathbf{B}_{s(1)})\mathbf{a}_1 - (\mathbf{a}_1 \cdot \mathbf{B}_{s(1)})\mathbf{a}_2) \wedge \langle \mathbf{B}_{s(2)} \rangle_{s-2l-1}$$

$$= \mathbf{a}_1 \wedge (\mathbf{a}_2 \cdot \langle \mathbf{B}_s \rangle_{s-2l}) - \mathbf{a}_2 \wedge (\mathbf{a}_1 \cdot \langle \mathbf{B}_s \rangle_{s-2l}).$$

$\square$

Proposition 6.19. [Bivector commutation formula] For any vectors $\mathbf{a}_1, \mathbf{a}_2$ and multivector $\mathbf{B}$,

$$\frac{1}{2}(\mathbf{a}_1 \mathbf{a}_2 \mathbf{B} - \mathbf{B}\mathbf{a}_1 \mathbf{a}_2) = \mathbf{a}_1 \wedge (\mathbf{a}_2 \cdot \mathbf{B}) - \mathbf{a}_2 \wedge (\mathbf{a}_1 \cdot \mathbf{B}) - 2\,(\mathbf{a}_1 \wedge \mathbf{a}_2)\langle \mathbf{B} \rangle. \quad (6.1.44)$$

Proof. Direct from either (6.1.42) or (6.1.43) for multivector $\mathbf{B} - \langle \mathbf{B} \rangle$, which has no 0-graded part. $\square$

Proposition 6.20. For any vectors $\mathbf{a}, \mathbf{c}$ and multivector $\mathbf{B}$,

$$\frac{1}{2}(\mathbf{a}\mathbf{B}\mathbf{c} - \mathbf{c}\mathbf{B}\mathbf{a}) = \mathbf{a} \wedge \mathbf{c} \wedge (\hat{\mathbf{B}} + \langle \mathbf{B} \rangle) - (\mathbf{a} \wedge \mathbf{c}) \cdot (\hat{\mathbf{B}} + \langle \mathbf{B} \rangle_1). \quad (6.1.45)$$

Proof. The left side equals

$$\mathbf{a}[\mathbf{B}, \mathbf{c}] + [\mathbf{a}, \mathbf{c}\mathbf{B}]$$

$$= \mathbf{a}((\langle \mathbf{B} \rangle_+ - \langle \mathbf{B} \rangle) \cdot \mathbf{c}) + \mathbf{a}(\langle \mathbf{B} \rangle_- \wedge \mathbf{c}) + \mathbf{a} \cdot (\langle \mathbf{c}\mathbf{B} \rangle_+ - \langle \mathbf{c}\mathbf{B} \rangle) + \mathbf{a} \wedge \langle \mathbf{c}\mathbf{B} \rangle_-$$

$$= \mathbf{a} \cdot ((\langle \mathbf{B} \rangle_+ - \langle \mathbf{B} \rangle) \cdot \mathbf{c}) + \mathbf{a} \wedge ((\langle \mathbf{B} \rangle_+ - \langle \mathbf{B} \rangle) \cdot \mathbf{c}) + \mathbf{a} \cdot (\langle \mathbf{B} \rangle_- \wedge \mathbf{c}) + \mathbf{a} \wedge (\langle \mathbf{B} \rangle_- \wedge \mathbf{c})$$

$$+ \mathbf{a} \cdot (\mathbf{c} \cdot (\langle \mathbf{B} \rangle_- - \langle \mathbf{B} \rangle_1)) + \mathbf{a} \cdot (\mathbf{c} \wedge \langle \mathbf{B} \rangle_-) + \mathbf{a} \wedge (\mathbf{c} \cdot (\langle \mathbf{B} \rangle_+ - \langle \mathbf{B} \rangle) + \mathbf{c} \wedge \langle \mathbf{B} \rangle_+)$$

$$= -(\mathbf{a} \wedge \mathbf{c}) \cdot (\langle \mathbf{B} \rangle_+ - \langle \mathbf{B} \rangle - \langle \mathbf{B} \rangle_- + \langle \mathbf{B} \rangle_1) + \mathbf{a} \wedge \mathbf{c} \wedge (\langle \mathbf{B} \rangle_+ - \langle \mathbf{B} \rangle_-).$$

$\square$

Proposition 6.21. For any vectors $\mathbf{a}, \mathbf{b}$ and multivector $\mathbf{C}$,

$$\frac{1}{2}(\mathbf{a}\mathbf{b}\mathbf{C} - \mathbf{C}\mathbf{b}\mathbf{a}) = (\mathbf{a} \wedge \mathbf{b}) \cdot (\mathbf{C} - \langle \mathbf{C} \rangle_1 - \langle \mathbf{C} \rangle) + \mathbf{a} \wedge \mathbf{b} \wedge \mathbf{C}. \quad (6.1.46)$$

Proof. By linearity we can assume that $\mathbf{C}$ is an r-blade, where $r > 1$. The left side of (6.1.46) equals $((\mathbf{a} \wedge \mathbf{b})\mathbf{C} + \mathbf{C}(\mathbf{a} \wedge \mathbf{b}))/2$. By

$$(\mathbf{a} \wedge \mathbf{b})\mathbf{C} = (\mathbf{a} \wedge \mathbf{b}) \cdot \mathbf{C} + [\mathbf{a} \wedge \mathbf{b}, \mathbf{C}] + \mathbf{a} \wedge \mathbf{b} \wedge \mathbf{C}$$

$$= \mathbf{C} \cdot (\mathbf{a} \wedge \mathbf{b}) - [\mathbf{C}, \mathbf{a} \wedge \mathbf{b}] + \mathbf{C} \wedge \mathbf{a} \wedge \mathbf{b},$$

and $\mathbf{C}(\mathbf{a} \wedge \mathbf{b}) = \mathbf{C} \cdot (\mathbf{a} \wedge \mathbf{b}) + [\mathbf{C}, \mathbf{a} \wedge \mathbf{b}] + \mathbf{C} \wedge \mathbf{a} \wedge \mathbf{b}$, we get (6.1.46). $\square$

6.1.3 *Ungrading*

Transforming a graded Clifford monomial into a Clifford polynomial of minimal number of terms is called *minimal ungrading*, also called *shortest ungrading*, or *least ungrading*. This is a very important technique in Geometric Algebra, but very few results are known.

Proposition 6.22. Let $\mathbf{A} = \mathbf{a}_1\mathbf{a}_2\cdots\mathbf{a}_{2k+1}$, $\mathbf{B} = \mathbf{a}_1\mathbf{a}_2\cdots\mathbf{a}_{2k}$, where $\mathbf{a}_i \in \mathcal{V}^n$. Then

$$
\begin{aligned}
2\langle\mathbf{A}\rangle_1 &= \mathbf{A} + \mathbf{A}^\dagger, \ \text{if}\ n < 5;\\
2\langle\mathbf{A}\rangle_3 &= \mathbf{A} - \mathbf{A}^\dagger, \ \text{if}\ n < 7;\\
2\langle\mathbf{B}\rangle\ &= \mathbf{B} + \mathbf{B}^\dagger, \ \text{if}\ n < 4;\\
2\langle\mathbf{B}\rangle_2 &= \mathbf{B} - \mathbf{B}^\dagger, \ \text{if}\ n < 6.
\end{aligned}
\tag{6.1.47}
$$

Proof. Direct from $\langle\mathbf{A}^\dagger\rangle_i = (\langle\mathbf{A}\rangle_i)^\dagger = (-1)^{i(i-1)/2}\langle\mathbf{A}\rangle_i$. $\qquad\square$

Proposition 6.23. When $n = 4$, let $\mathbf{A} = \mathbf{a}_1\mathbf{a}_2\cdots\mathbf{a}_{2k-1}$, $\mathbf{B} = \mathbf{a}_1\mathbf{a}_2\cdots\mathbf{a}_{2k}$, and $\mathbf{C} = \mathbf{a}_2\mathbf{a}_3\cdots\mathbf{a}_{2k}$, where $\mathbf{a}_i \in \mathcal{V}^4$. Then

$$
\begin{aligned}
4\langle\mathbf{B}\rangle\ &= \mathbf{a}_1\mathbf{C} + \mathbf{C}^\dagger\mathbf{a}_1 + \mathbf{C}\mathbf{a}_1 + \mathbf{a}_1\mathbf{C}^\dagger\\
&= \mathbf{A}\mathbf{a}_{2k} + \mathbf{a}_{2k}\mathbf{A}^\dagger + \mathbf{a}_{2k}\mathbf{A} + \mathbf{A}^\dagger\mathbf{a}_{2k};\\[4pt]
4\langle\mathbf{B}\rangle_4 &= \mathbf{a}_1\mathbf{C} + \mathbf{C}^\dagger\mathbf{a}_1 - \mathbf{C}\mathbf{a}_1 - \mathbf{a}_1\mathbf{C}^\dagger\\
&= \mathbf{A}\mathbf{a}_{2k} + \mathbf{a}_{2k}\mathbf{A}^\dagger - \mathbf{a}_{2k}\mathbf{A} - \mathbf{A}^\dagger\mathbf{a}_{2k}.
\end{aligned}
\tag{6.1.48}
$$

Proof. Direct from the commutativity between the reversion and the grading operators, the reversion invariance of $\langle\mathbf{A}\rangle$ and $\langle\mathbf{A}\rangle_4$, and the shift symmetries (6.1.19), (6.1.21). $\qquad\square$

Proposition 6.24. In $\mathcal{CL}(\mathcal{V}^n)$,

$$
\begin{aligned}
4\langle\mathbf{a}_1\mathbf{a}_2\mathbf{a}_3\mathbf{a}_4\mathbf{a}_5\rangle_5 &= \mathbf{a}_1\mathbf{a}_2\mathbf{a}_3\mathbf{a}_4\mathbf{a}_5 + \mathbf{a}_1\mathbf{a}_5\mathbf{a}_4\mathbf{a}_3\mathbf{a}_2 - \mathbf{a}_5\mathbf{a}_2\mathbf{a}_3\mathbf{a}_4\mathbf{a}_1 - \mathbf{a}_4\mathbf{a}_3\mathbf{a}_2\mathbf{a}_5\mathbf{a}_1,\\
4\langle\mathbf{a}_1\mathbf{a}_2\mathbf{a}_3\mathbf{a}_4\mathbf{a}_5\rangle_1 &= \mathbf{a}_1\mathbf{a}_2\mathbf{a}_3\mathbf{a}_4\mathbf{a}_5 - \mathbf{a}_1\mathbf{a}_5\mathbf{a}_4\mathbf{a}_3\mathbf{a}_2 + \mathbf{a}_5\mathbf{a}_2\mathbf{a}_3\mathbf{a}_4\mathbf{a}_1 + \mathbf{a}_4\mathbf{a}_3\mathbf{a}_2\mathbf{a}_5\mathbf{a}_1\\
&\qquad\qquad\qquad\qquad\qquad\qquad\qquad + 2\,\mathbf{a}_5\mathbf{a}_4\mathbf{a}_3\mathbf{a}_2\mathbf{a}_1.
\end{aligned}
\tag{6.1.49}
$$

Proof. Let $\mathbf{A} = \mathbf{a}_1\mathbf{a}_2\mathbf{a}_3\mathbf{a}_4\mathbf{a}_5$ and $\mathbf{B} = \mathbf{a}_2\mathbf{a}_3\mathbf{a}_4\mathbf{a}_5$. By (6.1.48),

$$
\begin{aligned}
\mathbf{a}_1\mathbf{a}_2\mathbf{a}_3\mathbf{a}_4\mathbf{a}_5 + \mathbf{a}_5\mathbf{a}_4\mathbf{a}_3\mathbf{a}_2\mathbf{a}_1 &= 2(\langle\mathbf{A}\rangle_1 + \langle\mathbf{A}\rangle_5),\\
\langle\mathbf{a}_1\mathbf{a}_2\mathbf{a}_3\mathbf{a}_4\mathbf{a}_5 + \mathbf{a}_2\mathbf{a}_3\mathbf{a}_4\mathbf{a}_5\mathbf{a}_1\rangle_1 &= 2\,\langle\mathbf{B}\rangle\mathbf{a}_1,\\
\mathbf{a}_1\mathbf{a}_2\mathbf{a}_3\mathbf{a}_4\mathbf{a}_5 + \mathbf{a}_5\mathbf{a}_4\mathbf{a}_3\mathbf{a}_2\mathbf{a}_1 + \mathbf{a}_2\mathbf{a}_3\mathbf{a}_4\mathbf{a}_5\mathbf{a}_1 + \mathbf{a}_1\mathbf{a}_5\mathbf{a}_4\mathbf{a}_3\mathbf{a}_2 &= 4(\langle\mathbf{A}\rangle_5 + \langle\mathbf{B}\rangle\mathbf{a}_1),\\
\mathbf{a}_2\mathbf{a}_3\mathbf{a}_4\mathbf{a}_5\mathbf{a}_1 + \mathbf{a}_5\mathbf{a}_4\mathbf{a}_3\mathbf{a}_2\mathbf{a}_1 + \mathbf{a}_5\mathbf{a}_2\mathbf{a}_3\mathbf{a}_4\mathbf{a}_1 + \mathbf{a}_4\mathbf{a}_3\mathbf{a}_2\mathbf{a}_5\mathbf{a}_1 &= 4\,\langle\mathbf{B}\rangle\mathbf{a}_1,
\end{aligned}
$$

from which we get the conclusion. $\qquad\square$

Corollary 6.25. In $\mathcal{CL}(\mathcal{V}^5)$,

$$
\begin{aligned}
4\langle\mathbf{a}_1\mathbf{a}_2\mathbf{a}_3\mathbf{a}_4\mathbf{a}_5\mathbf{a}_6\rangle = \ & \mathbf{a}_1\mathbf{a}_2\mathbf{a}_3\mathbf{a}_4\mathbf{a}_5\mathbf{a}_6 + \mathbf{a}_6\mathbf{a}_5\mathbf{a}_4\mathbf{a}_3\mathbf{a}_2\mathbf{a}_1 + \mathbf{a}_1\mathbf{a}_6\mathbf{a}_5\mathbf{a}_4\mathbf{a}_3\mathbf{a}_2\\
& - \mathbf{a}_2\mathbf{a}_6\mathbf{a}_5\mathbf{a}_4\mathbf{a}_3\mathbf{a}_1 + \mathbf{a}_6\mathbf{a}_3\mathbf{a}_4\mathbf{a}_5\mathbf{a}_2\mathbf{a}_1 + \mathbf{a}_5\mathbf{a}_4\mathbf{a}_3\mathbf{a}_6\mathbf{a}_2\mathbf{a}_1.
\end{aligned}
\tag{6.1.50}
$$

Proof. Let $\mathbf{A} = \mathbf{a}_1\mathbf{a}_2\mathbf{a}_3\mathbf{a}_4\mathbf{a}_5\mathbf{a}_6$ and $\mathbf{B} = \mathbf{a}_2\mathbf{a}_3\mathbf{a}_4\mathbf{a}_5\mathbf{a}_6$. Then

$$\mathbf{a}_1\mathbf{a}_2\mathbf{a}_3\mathbf{a}_4\mathbf{a}_5\mathbf{a}_6 + \mathbf{a}_6\mathbf{a}_5\mathbf{a}_4\mathbf{a}_3\mathbf{a}_2\mathbf{a}_1 = 2(\langle\mathbf{A}\rangle + \langle\mathbf{A}\rangle_4),$$
$$\langle\mathbf{a}_1\mathbf{a}_2\mathbf{a}_3\mathbf{a}_4\mathbf{a}_5\mathbf{a}_6 + \mathbf{a}_2\mathbf{a}_3\mathbf{a}_4\mathbf{a}_5\mathbf{a}_6\mathbf{a}_1\rangle_4 = 2\,\mathbf{a}_1\cdot\langle\mathbf{B}\rangle_5,$$

so

$$\mathbf{a}_1\mathbf{a}_2\mathbf{a}_3\mathbf{a}_4\mathbf{a}_5\mathbf{a}_6 + \mathbf{a}_6\mathbf{a}_5\mathbf{a}_4\mathbf{a}_3\mathbf{a}_2\mathbf{a}_1 + \mathbf{a}_2\mathbf{a}_3\mathbf{a}_4\mathbf{a}_5\mathbf{a}_6\mathbf{a}_1 + \mathbf{a}_1\mathbf{a}_6\mathbf{a}_5\mathbf{a}_4\mathbf{a}_3\mathbf{a}_2$$
$$= 4(\langle\mathbf{A}\rangle + \langle\mathbf{B}\rangle_5\mathbf{a}_1). \tag{6.1.51}$$

By (6.1.49),

$$\mathbf{a}_2\mathbf{a}_3\mathbf{a}_4\mathbf{a}_5\mathbf{a}_6\mathbf{a}_1 + \mathbf{a}_2\mathbf{a}_6\mathbf{a}_5\mathbf{a}_4\mathbf{a}_3\mathbf{a}_1 - \mathbf{a}_6\mathbf{a}_3\mathbf{a}_4\mathbf{a}_5\mathbf{a}_2\mathbf{a}_1 - \mathbf{a}_5\mathbf{a}_4\mathbf{a}_3\mathbf{a}_6\mathbf{a}_2\mathbf{a}_1 = 4\langle\mathbf{B}\rangle_5\mathbf{a}_1. \tag{6.1.52}$$

Combining (6.1.51) and (6.1.52), we get the conclusion. $\qquad\square$

Proposition 6.26. In $\mathcal{CL}(\mathcal{V}^n)$,

$$8\,\langle\mathbf{a}_1\mathbf{a}_2\mathbf{a}_3\mathbf{a}_4\mathbf{a}_5\mathbf{a}_6\rangle_6$$
$$= \ \ \mathbf{a}_1\mathbf{a}_2\mathbf{a}_3\mathbf{a}_4\mathbf{a}_5\mathbf{a}_6 - \mathbf{a}_2\mathbf{a}_3\mathbf{a}_4\mathbf{a}_5\mathbf{a}_6\mathbf{a}_1 + \mathbf{a}_1\mathbf{a}_2\mathbf{a}_6\mathbf{a}_5\mathbf{a}_4\mathbf{a}_3 - \mathbf{a}_2\mathbf{a}_6\mathbf{a}_5\mathbf{a}_4\mathbf{a}_3\mathbf{a}_1$$
$$\quad -\mathbf{a}_1\mathbf{a}_5\mathbf{a}_4\mathbf{a}_3\mathbf{a}_6\mathbf{a}_2 + \mathbf{a}_5\mathbf{a}_4\mathbf{a}_3\mathbf{a}_6\mathbf{a}_2\mathbf{a}_1 - \mathbf{a}_1\mathbf{a}_6\mathbf{a}_3\mathbf{a}_4\mathbf{a}_5\mathbf{a}_2 + \mathbf{a}_6\mathbf{a}_3\mathbf{a}_4\mathbf{a}_5\mathbf{a}_2\mathbf{a}_1,$$

$$8\,\langle\mathbf{a}_1\mathbf{a}_2\mathbf{a}_3\mathbf{a}_4\mathbf{a}_5\mathbf{a}_6\rangle_2 \tag{6.1.53}$$
$$= 3\,\mathbf{a}_1\mathbf{a}_2\mathbf{a}_3\mathbf{a}_4\mathbf{a}_5\mathbf{a}_6 + \mathbf{a}_2\mathbf{a}_3\mathbf{a}_4\mathbf{a}_5\mathbf{a}_6\mathbf{a}_1 - \mathbf{a}_1\mathbf{a}_2\mathbf{a}_6\mathbf{a}_5\mathbf{a}_4\mathbf{a}_3 + \mathbf{a}_2\mathbf{a}_6\mathbf{a}_5\mathbf{a}_4\mathbf{a}_3\mathbf{a}_1$$
$$\quad +\mathbf{a}_1\mathbf{a}_5\mathbf{a}_4\mathbf{a}_3\mathbf{a}_6\mathbf{a}_2 - \mathbf{a}_5\mathbf{a}_4\mathbf{a}_3\mathbf{a}_6\mathbf{a}_2\mathbf{a}_1 + \mathbf{a}_1\mathbf{a}_6\mathbf{a}_3\mathbf{a}_4\mathbf{a}_5\mathbf{a}_2 - \mathbf{a}_6\mathbf{a}_3\mathbf{a}_4\mathbf{a}_5\mathbf{a}_2\mathbf{a}_1$$
$$-4\,\mathbf{a}_6\mathbf{a}_5\mathbf{a}_4\mathbf{a}_3\mathbf{a}_2\mathbf{a}_1.$$

Proof. Let $\mathbf{A} = \mathbf{a}_1\mathbf{a}_2\mathbf{a}_3\mathbf{a}_4\mathbf{a}_5\mathbf{a}_6$ and $\mathbf{B} = \mathbf{a}_2\mathbf{a}_3\mathbf{a}_4\mathbf{a}_5\mathbf{a}_6$. Then

$$\mathbf{a}_1\mathbf{a}_2\mathbf{a}_3\mathbf{a}_4\mathbf{a}_5\mathbf{a}_6 - \mathbf{a}_6\mathbf{a}_5\mathbf{a}_4\mathbf{a}_3\mathbf{a}_2\mathbf{a}_1 = 2(\langle\mathbf{A}\rangle_6 + \langle\mathbf{A}\rangle_2),$$
$$\langle\mathbf{a}_1\mathbf{a}_2\mathbf{a}_3\mathbf{a}_4\mathbf{a}_5\mathbf{a}_6 - \mathbf{a}_2\mathbf{a}_3\mathbf{a}_4\mathbf{a}_5\mathbf{a}_6\mathbf{a}_1\rangle_2 = 2\,\mathbf{a}_1\wedge\langle\mathbf{B}\rangle_1 \tag{6.1.54}$$
$$= \mathbf{a}_1\langle\mathbf{B}\rangle_1 - \langle\mathbf{B}\rangle_1\mathbf{a}_1,$$

so

$$2(\mathbf{a}_1\mathbf{a}_2\mathbf{a}_3\mathbf{a}_4\mathbf{a}_5\mathbf{a}_6 - \mathbf{a}_6\mathbf{a}_5\mathbf{a}_4\mathbf{a}_3\mathbf{a}_2\mathbf{a}_1 - \mathbf{a}_2\mathbf{a}_3\mathbf{a}_4\mathbf{a}_5\mathbf{a}_6\mathbf{a}_1 + \mathbf{a}_1\mathbf{a}_6\mathbf{a}_5\mathbf{a}_4\mathbf{a}_3\mathbf{a}_2)$$
$$= 8\langle\mathbf{A}\rangle_6 + 4(\mathbf{a}_1\langle\mathbf{B}\rangle_1 - \langle\mathbf{B}\rangle_1\mathbf{a}_1). \tag{6.1.55}$$

By (6.1.49),

$$4\langle\mathbf{B}\rangle_1 = \mathbf{a}_2\mathbf{a}_3\mathbf{a}_4\mathbf{a}_5\mathbf{a}_6 - \mathbf{a}_2\mathbf{a}_6\mathbf{a}_5\mathbf{a}_4\mathbf{a}_3 + \mathbf{a}_6\mathbf{a}_3\mathbf{a}_4\mathbf{a}_5\mathbf{a}_2 + \mathbf{a}_5\mathbf{a}_4\mathbf{a}_3\mathbf{a}_6\mathbf{a}_2 + 2\,\mathbf{a}_6\mathbf{a}_5\mathbf{a}_4\mathbf{a}_3\mathbf{a}_2.$$
$$\tag{6.1.56}$$

Combining (6.1.54), (6.1.55) and (6.1.56), we get the conclusion. $\qquad\square$

Proposition 6.27. In $\mathcal{CL}(\mathcal{V}^n)$, the outer product of $r > 4$ generic vectors can be expressed as a Clifford polynomial of at most 2^{r-3} terms.

Proof. By induction and (6.1.23), (6.1.49). $\qquad\square$

The ungrading of r-blades provided by Proposition 6.27 may not be minimal. There are various syzygies among Clifford polynomials, making the minimal ungrading problem very complicated. For example, for any $\mathbf{1},\mathbf{2},\mathbf{3},\mathbf{4} \in \mathcal{V}^n$, since $[\mathbf{12},\mathbf{34}] = [\mathbf{21},\mathbf{43}]$, it must be that

$$\mathbf{1234} + \mathbf{4321} - \mathbf{2143} - \mathbf{3412} = 0. \tag{6.1.57}$$

This simple syzygy identity is by no means obvious from the generating relations $\mathbf{x}^2 = \mathbf{x} \cdot \mathbf{x}$ for all $\mathbf{x} \in \mathcal{V}^n$.

By Proposition 6.27, the number of terms generated by ungrading a monomial of the form $\langle \mathbf{a}_1 \mathbf{a}_2 \cdots \mathbf{a}_r \rangle_r$ may grow exponentially with r. Even if the dimension n of the base space $\mathcal{V}^n$ is fixed, ungrading $\langle \mathbf{a}_1 \mathbf{a}_2 \cdots \mathbf{a}_{r+2k} \rangle_r$ for fixed r but varying k, can lead to a Clifford polynomial whose number of terms tends to infinity with the increase of k.

Similar to rational Clifford factorization, if the ungrading result is allowed to be a rational Clifford polynomial whose denominator is scalar-valued, the number of terms may be significantly reduced. This kind of ungrading is called *rational ungrading*.

For example, when $n = 5$, Theorem 6.30 below provides for $\langle \mathbf{a}_1 \mathbf{a}_2 \cdots \mathbf{a}_{5+2k} \rangle_5$ a rational ungrading of constant number of terms. In contrast, with the growth of k, it does not seem possible that any ungrading of this graded monomial can generate a Clifford polynomial of bounded number of terms.

Lemma 6.28. Let $\mathbf{A} = \mathbf{a}_1 \mathbf{a}_2 \cdots \mathbf{a}_{2k-1}$, $\mathbf{B} = \mathbf{a}_1 \mathbf{a}_2 \cdots \mathbf{a}_{2k}$, where $\mathbf{a}_i \in \mathcal{V}^5$. Then

$$\begin{aligned}
\mathbf{aA} + \mathbf{A}^\dagger \mathbf{a} + \mathbf{Aa} + \mathbf{aA}^\dagger &= 4\left(\langle \mathbf{aA} \rangle + \mathbf{a}\langle \mathbf{A} \rangle_5\right), \\
\mathbf{aB} + \mathbf{B}^\dagger \mathbf{a} + \mathbf{Ba} + \mathbf{aB}^\dagger &= 4\left(\mathbf{a}\langle \mathbf{B} \rangle + \langle \mathbf{aB} \rangle_5\right).
\end{aligned} \tag{6.1.58}$$

Proof. We only prove the first equality. First,

$$\mathbf{aA} + \mathbf{A}^\dagger \mathbf{a} + \mathbf{Aa} + \mathbf{aA}^\dagger = 2(\langle \mathbf{aA} + \mathbf{Aa} \rangle + \langle \mathbf{aA} + \mathbf{Aa} \rangle_4) = 4\langle \mathbf{aA} \rangle + 2\langle \mathbf{aA} + \mathbf{Aa} \rangle_4.$$

Second, $\langle \mathbf{aA} \rangle_4 + \langle \mathbf{Aa} \rangle_4 = \mathbf{a} \cdot \langle \mathbf{A} \rangle_5 + \mathbf{a} \wedge \langle \mathbf{A} \rangle_3 + \langle \mathbf{A} \rangle_5 \cdot \mathbf{a} + \langle \mathbf{A} \rangle_3 \wedge \mathbf{a} = 2\,\mathbf{a} \cdot \langle \mathbf{A} \rangle_5.$ $\quad\square$

Lemma 6.29. Let $\mathbf{A} = \mathbf{a}_1 \mathbf{a}_2 \cdots \mathbf{a}_{2k-1}$, $\mathbf{B} = \mathbf{a}_1 \mathbf{a}_2 \cdots \mathbf{a}_{2k}$, where $\mathbf{a}_i \in \mathcal{V}^5$. Then for any vector $\mathbf{b} \in \mathcal{V}^5$,

$$\begin{aligned}
8\,\langle \mathbf{A} \rangle_5 \mathbf{a} \wedge \mathbf{b} &= \mathbf{aAb} + \mathbf{A}^\dagger \mathbf{ab} - \mathbf{bAa} - \mathbf{baA}^\dagger - \mathbf{bA}^\dagger \mathbf{a} - \mathbf{Aba} + \mathbf{abA} + \mathbf{aA}^\dagger \mathbf{b}, \\
8\,\langle \mathbf{B} \rangle\,\mathbf{a} \wedge \mathbf{b} &= \mathbf{aBb} + \mathbf{B}^\dagger \mathbf{ab} - \mathbf{bBa} - \mathbf{baB}^\dagger - \mathbf{bB}^\dagger \mathbf{a} - \mathbf{Bba} + \mathbf{abB} + \mathbf{aB}^\dagger \mathbf{b}.
\end{aligned} \tag{6.1.59}$$

Proof. We only prove the second equality. Applying the second equality of (6.1.58) to monomial $\mathbf{bB}$, we get

$$\mathbf{bB} + \mathbf{B}^\dagger \mathbf{b} + \mathbf{Bb} + \mathbf{bB}^\dagger = 4\langle \mathbf{B} \rangle \mathbf{b} + 4\langle \mathbf{Bb} \rangle_5. \tag{6.1.60}$$

Substituting the expression of $\langle \mathbf{Bb} \rangle_5$ from (6.1.60) into the first equality of (6.1.58) for $\mathbf{A} = \mathbf{Bb}$, we get

$$4\left(\langle \mathbf{aBb} \rangle - \langle \mathbf{B} \rangle \mathbf{ab}\right) = \mathbf{bB}^\dagger \mathbf{a} - \mathbf{abB} + \mathbf{Bba} - \mathbf{aB}^\dagger \mathbf{b}. \tag{6.1.61}$$

Now that (6.1.61) is obtained from (6.1.58) as an equality of sequence $\mathbf{aBb}$, if reversing the sequence to $\mathbf{bB}^\dagger \mathbf{a}$, we can get from (6.1.61) the following:

$$4\left(\langle \mathbf{aBb} \rangle - \langle \mathbf{B} \rangle \mathbf{ba}\right) = \mathbf{aBb} - \mathbf{baB}^\dagger + \mathbf{B}^\dagger \mathbf{ab} - \mathbf{bBa}. \tag{6.1.62}$$

Subtracting (6.1.61) from (6.1.62) gives (6.1.59). □

Making geometric product to both sides of (6.1.59) with $\mathbf{ab} - \mathbf{ba}$, then using the following ungrading:

$$(\mathbf{a} \wedge \mathbf{b})^2 = \frac{1}{4}(\mathbf{ab} - \mathbf{ba})(\mathbf{ab} - \mathbf{ba}) = \frac{1}{4}(\mathbf{abab} + \mathbf{baba} - 2\mathbf{a}^2\mathbf{b}^2), \qquad (6.1.63)$$

we get

Theorem 6.30. Let $\mathbf{A} \in \mathcal{G}(\mathcal{V}^5)$, and let $\mathbf{a}, \mathbf{b} \in \mathcal{V}^5$ such that $(\mathbf{a} \wedge \mathbf{b})^2 \neq 0$. If $\mathbf{A}$ is odd, then for $r = 5$,

$$\begin{aligned}
4(\mathbf{abab} + \mathbf{baba} - 2\mathbf{a}^2\mathbf{b}^2)\langle\mathbf{A}\rangle_r = \ & \mathbf{ababA} + \mathbf{babaA}^\dagger - \mathbf{a}^2\mathbf{b}^2\mathbf{A} - \mathbf{a}^2\mathbf{b}^2\mathbf{A}^\dagger \\
& + \mathbf{abaAb} + \mathbf{babAa} + \mathbf{abaA}^\dagger\mathbf{b} + \mathbf{babA}^\dagger\mathbf{a} \\
& - \mathbf{a}^2\mathbf{bAb} - \mathbf{a}^2\mathbf{bA}^\dagger\mathbf{b} - \mathbf{b}^2\mathbf{aAa} - \mathbf{b}^2\mathbf{aA}^\dagger\mathbf{a} \\
& + \mathbf{abA}^\dagger\mathbf{ab} - \mathbf{baA}^\dagger\mathbf{ab} - \mathbf{abAba} + \mathbf{baAba}.
\end{aligned}$$
$$(6.1.64)$$

If $\mathbf{A}$ is even, then for $r = 0$, (6.1.64) is still correct.

In practical computing, it often occurs that scalars and pseudoscalars do not need ungrading. Then the minimal ungrading problem becomes much easier.

Proposition 6.31. For any $\mathbf{a}_i \in \mathcal{V}^n$,

$$\begin{aligned}
2\langle\mathbf{a}_1\mathbf{a}_2\cdots\mathbf{a}_{2k+1}\rangle_1 &= \mathbf{a}_1\mathbf{a}_2\cdots\mathbf{a}_{2k+1} + \mathbf{a}_{2k+1}\cdots\mathbf{a}_2\mathbf{a}_1 - 2\langle\mathbf{a}_1\mathbf{a}_2\cdots\mathbf{a}_{2k+1}\rangle_5, && \text{if } n = 5; \\
2\langle\mathbf{a}_1\mathbf{a}_2\cdots\mathbf{a}_{2k}\rangle_2 &= \mathbf{a}_1\mathbf{a}_2\cdots\mathbf{a}_{2k} - \mathbf{a}_{2k}\cdots\mathbf{a}_2\mathbf{a}_1 - 2\langle\mathbf{a}_1\mathbf{a}_2\cdots\mathbf{a}_{2k}\rangle_6, && \text{if } n = 6; \\
2\langle\mathbf{a}_1\mathbf{a}_2\cdots\mathbf{a}_{2k}\rangle_4 &= \mathbf{a}_1\mathbf{a}_2\cdots\mathbf{a}_{2k} + \mathbf{a}_{2k}\cdots\mathbf{a}_2\mathbf{a}_1 - 2\langle\mathbf{a}_1\mathbf{a}_2\cdots\mathbf{a}_{2k}\rangle, && \text{if } n < 8.
\end{aligned}$$
$$(6.1.65)$$

Lemma 6.32. [29] [Ungrading of blades] Let $\mathbf{A}_k = \mathbf{a}_1\mathbf{a}_2\cdots\mathbf{a}_k$. Then

$$\mathbf{a}_1 \wedge \mathbf{a}_2 \wedge \cdots \wedge \mathbf{a}_k = \sum_{i=0}^{[\frac{k}{2}]} \sum_{(2i,k-2i)\vdash\mathbf{A}_k} (-1)^i \langle\mathbf{A}_{k(1)}\rangle \mathbf{A}_{k(2)}. \qquad (6.1.66)$$

For example, the result of applying (6.1.66) to blade $\mathbf{1} \wedge \mathbf{2} \wedge \mathbf{3}$ is

$$\mathbf{123} - \langle\mathbf{123}\rangle_1 = \mathbf{123} - (\mathbf{1}\cdot\mathbf{2})\mathbf{3} + (\mathbf{1}\cdot\mathbf{3})\mathbf{2} - (\mathbf{2}\cdot\mathbf{3})\mathbf{1}.$$

Proof. The conclusion is trivial for $k = 1, 2$. Assume that it is true for $k < m$, where $m \geq 3$. For $k = m$,

$$\begin{aligned}
\mathbf{a}_1 \wedge \cdots \wedge \mathbf{a}_m = \ & (\mathbf{a}_1 \wedge \cdots \wedge \mathbf{a}_{m-1})\mathbf{a}_m - (\mathbf{a}_1 \wedge \cdots \wedge \mathbf{a}_{m-1})\cdot\mathbf{a}_m \\
= \ & \sum_{i=0}^{[\frac{m-1}{2}]} \sum_{(2i,m-2i-1)\vdash\mathbf{A}_{m-1}} (-1)^i\langle\mathbf{A}_{m-1(1)}\rangle \mathbf{A}_{m-1(2)}\mathbf{a}_m \\
& - \sum_{i=0}^{[\frac{m}{2}]-1} \sum_{(2i,m-2i-2,1)\vdash\mathbf{A}_{m-1}} (-1)^i(\mathbf{A}_{m-1(3)} \cdot \mathbf{a}_m) \langle\mathbf{A}_{m-1(1)}\rangle \mathbf{A}_{m-1(2)}
\end{aligned}$$

$$
= \sum_{i=0}^{[\frac{m-1}{2}]} \sum_{\substack{(2i,m-2i)\vdash \mathbf{A}_m, \\ \mathbf{a}_m \notin \mathbf{A}_{m(1)}}} (-1)^i \langle \mathbf{A}_{m(1)} \rangle \, \mathbf{A}_{m(2)} - \sum_{i=0}^{[\frac{m}{2}]-1} \sum_{\substack{(2i+2,m-2i-2)\vdash \mathbf{A}_m, \\ \mathbf{a}_m \in \mathbf{A}_{m(1)}}} (-1)^i \langle \mathbf{A}_{m(1)} \rangle \, \mathbf{A}_{m(2)}
$$

$$
= \sum_{i=0}^{[\frac{m-1}{2}]} \sum_{\substack{(2i,m-2i)\vdash \mathbf{A}_m, \\ \mathbf{a}_m \notin \mathbf{A}_{m(1)}}} (-1)^i \langle \mathbf{A}_{m(1)} \rangle \, \mathbf{A}_{m(2)} + \sum_{i=1}^{[\frac{m}{2}]} \sum_{\substack{(2i,m-2i)\vdash \mathbf{A}_m, \\ \mathbf{a}_m \in \mathbf{A}_{m(1)}}} (-1)^i \langle \mathbf{A}_{m(1)} \rangle \, \mathbf{A}_{m(2)}
$$

$$
= \sum_{i=0}^{[\frac{m}{2}]} \sum_{(2i,m-2i)\vdash \mathbf{A}_m} (-1)^i \langle \mathbf{A}_{m(1)} \rangle \, \mathbf{A}_{m(2)}.
$$

$\square$

Corollary 6.33. With notations as above,

$$
\langle \mathbf{A}_k \rangle_{k-2l} = \sum_{i=0}^{[\frac{k}{2}]-l} \sum_{(2l+2i,k-2l-2i)\vdash \mathbf{A}_k} (-1)^i C_{l+i}^l \langle \mathbf{A}_{k(1)} \rangle \, \mathbf{A}_{k(2)}. \tag{6.1.67}
$$

The following theorem is direct from Lemma 6.32 for $\langle \mathbf{A}_m \rangle_m$:

Theorem 6.34. For any $0 \le m \le 2l$, let $\mathbf{A}_m = \mathbf{a}_1 \mathbf{a}_2 \cdots \mathbf{a}_m$ and $\mathbf{B}_{2l-m} = \mathbf{a}_{m+1} \mathbf{a}_{m+2} \cdots \mathbf{a}_{2l}$. Then

$$
\langle \mathbf{A}_{2l} \rangle = \langle \langle \mathbf{A}_m \rangle_m \mathbf{B}_{2l-m} \rangle + \sum_{i=1}^{[\frac{m}{2}]} \sum_{(2i,m-2i)\vdash \mathbf{A}_m} (-1)^{i+1} \langle \mathbf{A}_{m(1)} \rangle \, \langle \mathbf{A}_{m(2)} \mathbf{B}_{2l-m} \rangle. \tag{6.1.68}
$$

It can also be written in the following form: for any $0 \le m \le 2l$,

$$
\langle \mathbf{A}_m \rangle_m \cdot \langle \mathbf{B}_{2l-m} \rangle_m = \sum_{i=0}^{[\frac{m}{2}]} \sum_{(2i,m-2i)\vdash \mathbf{A}_m} (-1)^i \langle \mathbf{A}_{m(1)} \rangle \, \langle \mathbf{A}_{m(2)} \mathbf{B}_{2l-m} \rangle. \tag{6.1.69}
$$

Remark: Theorem 6.34 is the dual of Corollary 5.80. If $l = m$, the left side of (6.1.69) is an m-graded inner product, and the right side is its Pfaffian expansion. Conversely, by using (6.1.68) recursively, any Pfaffian $\langle \mathbf{A}_{2l} \rangle$ can be expanded into a polynomial of graded inner products whose leading grade is $\min(n, l)$.

6.2 Versor compression

A versor is a Clifford monomial of invertible vectors. *Versor compression* refers to changing a versor into an equal versor but of shorter length. The problem has its origin in Cartan-Dieudonné Theorem 5.62. Since an orthogonal transformation can be represented by a versor of up to length n, given a versor of arbitrary length, a natural requirement is to reduce to minimum the number of effective vectors in the versor, which is equal to the number of effective reflections composing the orthogonal transformation.

Versor compression problem was studied in [56] from the numerical computation point of view. In this section, we study the problem from the symbolic computation point of view. The problem includes three parts: criteria for the compressibility, constructions of the compressions, and characterization of the obstructions to further compression.

We assume that the vectors $\mathbf{a}_i, \mathbf{b}_j$ in this section are invertible by default.

Proposition 6.35. [Commutation within a versor] Let $\mathbf{A}_k = \mathbf{a}_1\mathbf{a}_2\cdots\mathbf{a}_k$. Then for any subsequence $p_1, p_2, \ldots, p_m$ of $1, 2, \ldots, k$, there exist Clifford monomials $\mathbf{B}_{u_m}$, $\mathbf{B}'_v$ and $\mathbf{B}''_{w_m}$ of length $u_m = p_m - p_1 - m + 1$, $v = p_1 - 1$, $w_m = k - p_1 - m + 1$ respectively, such that

$$\begin{aligned}
\mathbf{A}_k &= (\mathbf{a}_1\mathbf{a}_2\cdots\mathbf{a}_{p_1-1})(\mathbf{a}_{p_1}\mathbf{a}_{p_2}\cdots\mathbf{a}_{p_m})\mathbf{B}_{u_m}(\mathbf{a}_{p_m+1}\mathbf{a}_{p_m+2}\cdots\mathbf{a}_k)\\
&= (\mathbf{a}_{p_1}\mathbf{a}_{p_2}\cdots\mathbf{a}_{p_m})\mathbf{B}'_v\mathbf{B}_{u_m}(\mathbf{a}_{p_m+1}\mathbf{a}_{p_m+2}\cdots\mathbf{a}_k) \qquad (6.2.1)\\
&= (\mathbf{a}_1\mathbf{a}_2\cdots\mathbf{a}_{p_1-1})\mathbf{B}''_{w_m}(\mathbf{a}_{p_1}\mathbf{a}_{p_2}\cdots\mathbf{a}_{p_m}).
\end{aligned}$$

Proof. We only prove the first equality. When $m = 0$, the conclusion is trivial. Assume that the conclusion holds for $m \leq r$. For $m = r + 1$, let $p_1, p_2, \ldots, p_{r+1}$ be a subsequence of $1, 2, \ldots, k$. By induction hypothesis, Clifford monomial $\mathbf{a}_{p_1}\mathbf{a}_{p_2}\cdots\mathbf{a}_{p_r}$ can be moved next to $\mathbf{a}_1\mathbf{a}_2\cdots\mathbf{a}_{p_1-1}$, changing the monomial $(\mathbf{a}_{p_1}\mathbf{a}_{p_1+1}\cdots\mathbf{a}_{p_r})(\mathbf{a}_{p_r+1}\mathbf{a}_{p_r+2}\cdots\mathbf{a}_{p_{r+1}})$ of length $p_{r+1} - p_1 + 1$ into

$$(\mathbf{a}_{p_1}\mathbf{a}_{p_2}\cdots\mathbf{a}_{p_r})(\mathbf{b}_1\mathbf{b}_2\cdots\mathbf{b}_{u_r})(\mathbf{a}_{p_r+1}\mathbf{a}_{p_r+2}\cdots\mathbf{a}_{p_{r+1}}), \qquad (6.2.2)$$

where $u_r = p_r - p_1 - r + 1$.

In (6.2.2), vector $\mathbf{x} = \mathbf{a}_{p_{r+1}}$ can move forward across all the $\mathbf{b}_j$ to be next to $\mathbf{a}_{p_r}$, as follows:

$$\begin{aligned}
&(\mathbf{b}_1\mathbf{b}_2\cdots\mathbf{b}_{u_r})(\mathbf{a}_{p_r+1}\mathbf{a}_{p_r+2}\cdots\mathbf{a}_{p_{r+1}})\\
&= \mathbf{x}(\mathbf{x}^{-1}\mathbf{b}_1\mathbf{x})(\mathbf{x}^{-1}\mathbf{b}_2\mathbf{x})\cdots(\mathbf{x}^{-1}\mathbf{b}_{u_r}\mathbf{x})(\mathbf{x}^{-1}\mathbf{a}_{p_r+1}\mathbf{x})(\mathbf{x}^{-1}\mathbf{a}_{p_r+2}\mathbf{x})\cdots(\mathbf{x}^{-1}\mathbf{a}_{p_{r+1}-1}\mathbf{x})\\
&= \mathbf{a}_{p_{r+1}}\mathbf{b}'_1\mathbf{b}'_2\cdots\mathbf{b}'_{u_{r+1}},
\end{aligned}$$

where $u_{r+1} = u_r + p_{r+1} - p_r - 1 = p_{r+1} - p_1 - (r+1) + 1$. So $\mathbf{B}_{u_{r+1}} = \mathbf{b}'_1\mathbf{b}'_2\cdots\mathbf{b}'_{u_{r+1}}$, and the conclusion is proved. $\qquad\square$

Corollary 6.36. Any vector in a versor can move anywhere within the versor, and alters only those vectors that block its way on the left (or right), by its right (or left) adjoint action upon them.

Given a versor $\mathbf{A}_k = \mathbf{a}_1\mathbf{a}_2\cdots\mathbf{a}_k$ of length k, if there are any two linearly dependent vectors in $\mathbf{A}_k$, then they can always be moved together to form a scalar, reducing the length of the versor by two. If there are any three linearly dependent vectors in $\mathbf{A}_k$, then when moved together, they form a single vector by their geometric product. This reduction of the number of effective vectors is called *3-tuple compression*. A vector obtained by 3-tuple compression is called a *3-compressed vector*.

If monomial $\mathbf{A}_{2k} = \mathbf{a}_1\mathbf{a}_2\cdots\mathbf{a}_k$ is compressible, then the $2k$ vectors are always linearly dependent. A monomial composed of linearly independent vectors is said to be *rigid*. Any rigid monomial is incompressible.

When there are m vectors in $\mathbf{A}_k$ that are linearly dependent, by commutation we only need to consider the case $k = m$. The m vectors span a vector space of dimension at most $m - 1$. By Cartan-Dieudonné Theorem, if the inner product in this space is nondegenerate, the length of $\mathbf{A}_m$ can be strictly reduced to within the dimension of the space. In fact, even if the inner product is degenerate, the compression is still possible.

The versor compression of the geometric product of an m-tuple of linearly dependent invertible vectors that span an $(m - 1)$D vector space, is called *m-tuple compression*. We first consider the simplest nontrivial case $m = 4$.

6.2.1 *4-tuple compression*

In $\mathbf{A}_4 = \mathbf{a}_1\mathbf{a}_2\mathbf{a}_3\mathbf{a}_4$, assume that the four vectors are linearly dependent, but any three of them are linearly independent. Denote by $\mathcal{W}^3$ the 3D vector space spanned by them.

If $\mathbf{A}_4$ can be compressed to $\mathbf{b}_1\mathbf{b}_2$, then

$$\mathbf{b}_1 = \mathbf{A}_4\mathbf{b}_2^{-1}. \tag{6.2.3}$$

So if we can find an invertible vector $\mathbf{b}_2$ such that $\mathbf{A}_4\mathbf{b}_2$ is a vector, we can reduce the length of $\mathbf{A}_4$ by two. In fact, we only need to solve the following equation in $\mathbf{b}_2^{-1}$ for an invertible vector solution:

$$\langle\mathbf{A}_4\mathbf{b}_2^{-1}\rangle_3 = \langle\mathbf{A}_4\rangle_2 \wedge \mathbf{b}_2^{-1} = 0. \tag{6.2.4}$$

The solution is all invertible vectors in 2D subspace $\langle\mathbf{A}_4\rangle_2$. The 2-blade $\langle\mathbf{A}_4\rangle_2$ is nonzero because

$$\mathbf{a}_1 \wedge \langle\mathbf{A}_4\rangle_2 = \langle\mathbf{a}_1^2\mathbf{a}_2\mathbf{a}_3\mathbf{a}_4\rangle_3 = \mathbf{a}_1^2\langle\mathbf{a}_2\mathbf{a}_3\mathbf{a}_4\rangle_3 \neq 0.$$

Proposition 6.37. If the vectors in $\mathbf{A}_4 = \mathbf{a}_1\mathbf{a}_2\mathbf{a}_3\mathbf{a}_4$ span a 3D vector space $\mathcal{W}^3$, then $\mathbf{A}_4$ is compressible if and only if the 2D subspace $\langle\mathbf{A}_4\rangle_2$ is not null.

If $\mathcal{W}^3$ has a null 2D subspace, then in $\mathcal{G}(\mathcal{W}^3)$ there are incompressible versors of length 4.

Example 6.38. In $\mathcal{W}^3 = \mathbb{R}^{1,0,2}$, let $\mathbf{e}_1, \mathbf{e}_2, \mathbf{e}_3$ be an orthonormal basis such that $\mathbf{e}_1^2 = 1$, then for vectors $\mathbf{a}_i = \mathbf{e}_1 + \lambda_i\mathbf{e}_2 + \mu_i\mathbf{e}_3$, where $1 \leq i \leq 4$, we have $\mathbf{a}_i \cdot \mathbf{a}_j = 1$ for any $1 \leq i, j \leq 4$, and

$$\begin{aligned}
\langle\mathbf{a}_1\mathbf{a}_2\mathbf{a}_3\mathbf{a}_4\rangle_2 = {}&-(\lambda_1 - \lambda_2 + \lambda_3 - \lambda_4)\mathbf{e}_1\mathbf{e}_2 - (\mu_1 - \mu_2 + \mu_3 - \mu_4)\mathbf{e}_1\mathbf{e}_3 \\
&+\{\lambda_1(\mu_2 - \mu_3 + \mu_4) - \lambda_2(\mu_1 - \mu_3 + \mu_4) \\
&+\lambda_3(\mu_1 - \mu_2 + \mu_4) - \lambda_4(\mu_1 - \mu_2 + \mu_3)\}\mathbf{e}_2\mathbf{e}_3.
\end{aligned}$$

If

$$\lambda_1 - \lambda_2 + \lambda_3 - \lambda_4 = \mu_1 - \mu_2 + \mu_3 - \mu_4 = 0, \quad i.e., \quad \mathbf{a}_4 = \mathbf{a}_1 - \mathbf{a}_2 + \mathbf{a}_3, \tag{6.2.5}$$

and $\mathbf{a}_1, \mathbf{a}_2, \mathbf{a}_3$ are linearly independent, then

$$\langle \mathbf{a}_1\mathbf{a}_2\mathbf{a}_3\mathbf{a}_4 \rangle_2 = \{\lambda_1(\mu_2 - \mu_3) + \lambda_2(\mu_3 - \mu_1) + \lambda_3(\mu_1 - \mu_2)\}\mathbf{e}_2\mathbf{e}_3 = [\mathbf{a}_1\mathbf{a}_2\mathbf{a}_3]\mathbf{e}_2\mathbf{e}_3 \neq 0,$$

blade $\langle \mathbf{a}_1\mathbf{a}_2\mathbf{a}_3\mathbf{a}_4 \rangle_2$ is null, and versor $\mathbf{a}_1\mathbf{a}_2\mathbf{a}_3\mathbf{a}_4$ is incompressible.

If the space $\mathcal{W}^3$ is changed to $\mathbb{R}^{0,1,2}$ or $\mathbb{C}^{1,2}$, the construction of incompressible non-rigid versors is just the same.

Example 6.39. In $\mathcal{W}^3 = \mathbb{R}^{1,1,1}$, let $\mathbf{e}_1, \mathbf{e}_2, \mathbf{e}_3$ be a Witt basis such that $(\mathbf{e}_1, \mathbf{e}_2)$ is a Witt pair, then for vectors $\mathbf{a}_i = \mathbf{e}_1 + \lambda_i\mathbf{e}_2 + \mu_i\mathbf{e}_3$, where $1 \leq i \leq 4$ and $\lambda_i \neq 0$, we have $\mathbf{a}_i \cdot \mathbf{a}_j = -(\lambda_i + \lambda_j)$ for any $1 \leq i, j \leq 4$. Furthermore,

$$-\frac{1}{2}\langle \mathbf{a}_1\mathbf{a}_2\mathbf{a}_3\mathbf{a}_4 \rangle_2 = (\lambda_2\lambda_4 - \lambda_1\lambda_3)\mathbf{e}_1 \wedge \mathbf{e}_2 + \{\lambda_2(\mu_4 - \mu_3) + \lambda_3(\mu_2 - \mu_1)\}\mathbf{e}_1 \wedge \mathbf{e}_3$$
$$+\{\lambda_1(\lambda_3\mu_4 - \lambda_4\mu_3) + \lambda_4(\lambda_1\mu_2 - \lambda_2\mu_1)\}\mathbf{e}_2 \wedge \mathbf{e}_3.$$
$$(6.2.6)$$

In the following two cases, by assuming that $\mathbf{a}_1, \mathbf{a}_2, \mathbf{a}_3$ are linearly independent, blade $\langle \mathbf{a}_1\mathbf{a}_2\mathbf{a}_3\mathbf{a}_4 \rangle_2$ is null, and versor $\mathbf{a}_1\mathbf{a}_2\mathbf{a}_3\mathbf{a}_4$ is incompressible:

Case 1.

$$\lambda_1\lambda_3 = \lambda_2\lambda_4 \quad \text{and} \quad \lambda_2(\mu_3 - \mu_4) = -\lambda_3(\mu_1 - \mu_2). \qquad (6.2.7)$$

Then

$$\begin{aligned}
\lambda_2(\mathbf{a}_3 - \mathbf{a}_4) &= -\lambda_3(\mathbf{a}_1 - \mathbf{a}_2), \\
[\mathbf{a}_1\mathbf{a}_2\mathbf{a}_3] &= \lambda_1(\mu_2 - \mu_3) + \lambda_2(\mu_4 - \mu_1), \\
2\lambda_2\mathbf{a}_4 &= -\langle \mathbf{a}_1\mathbf{a}_2\mathbf{a}_3 \rangle_1 + [\mathbf{a}_1\mathbf{a}_2\mathbf{a}_3]\mathbf{e}_3, \\
\langle \mathbf{a}_1\mathbf{a}_2\mathbf{a}_3\mathbf{a}_4 \rangle_2 &= -2\lambda_4[\mathbf{a}_1\mathbf{a}_2\mathbf{a}_3]\mathbf{e}_2 \wedge \mathbf{e}_3 \neq 0.
\end{aligned} \qquad (6.2.8)$$

Shifting the four vectors in $\mathbf{a}_1\mathbf{a}_2\mathbf{a}_3\mathbf{a}_4$ leads to

$$\begin{aligned}
\langle \mathbf{a}_2\mathbf{a}_3\mathbf{a}_4\mathbf{a}_1 \rangle_2 &= 2\lambda_3\lambda_2^{-1}[\mathbf{a}_1\mathbf{a}_2\mathbf{a}_3]\mathbf{e}_1 \wedge \mathbf{e}_3, \\
\langle \mathbf{a}_3\mathbf{a}_4\mathbf{a}_1\mathbf{a}_2 \rangle_2 &= -2\lambda_3[\mathbf{a}_1\mathbf{a}_2\mathbf{a}_3]\mathbf{e}_2 \wedge \mathbf{e}_3, \\
\langle \mathbf{a}_4\mathbf{a}_1\mathbf{a}_2\mathbf{a}_3 \rangle_2 &= 2[\mathbf{a}_1\mathbf{a}_2\mathbf{a}_3]\mathbf{e}_1 \wedge \mathbf{e}_3.
\end{aligned} \qquad (6.2.9)$$

Case 2.

$$\lambda_1\lambda_3 = \lambda_2\lambda_4 \quad \text{and} \quad \lambda_1(\mu_2 - \mu_3) = -\lambda_2(\mu_4 - \mu_1). \qquad (6.2.10)$$

Then

$$\begin{aligned}
\lambda_1(\mathbf{a}_2 - \mathbf{a}_3) &= -\lambda_2(\mathbf{a}_4 - \mathbf{a}_1), \\
[\mathbf{a}_1\mathbf{a}_2\mathbf{a}_3] &= -\{\lambda_2(\mu_4 - \mu_3) + \lambda_3(\mu_2 - \mu_1)\}, \\
2\lambda_2\mathbf{a}_4 &= -\langle \mathbf{a}_1\mathbf{a}_2\mathbf{a}_3 \rangle_1 - [\mathbf{a}_1\mathbf{a}_2\mathbf{a}_3]\mathbf{e}_3, \\
\langle \mathbf{a}_1\mathbf{a}_2\mathbf{a}_3\mathbf{a}_4 \rangle_2 &= 2\lambda_4[\mathbf{a}_1\mathbf{a}_2\mathbf{a}_3]\mathbf{e}_1 \wedge \mathbf{e}_3 \neq 0.
\end{aligned} \qquad (6.2.11)$$

If the space $\mathcal{W}^3$ is changed to $\mathbb{C}^{2,1}$, the construction of incompressible non-rigid versors is the same.

To find an explicit symbolic solution of (6.2.4), we need to intersect the 2D plane $\langle \mathbf{A}_4 \rangle_2$ with another 2D plane of $\mathcal{W}^3$. The simplest solution is

$$\mathbf{b}_2^{-1} = \mathbf{a}_4^{-1} \mathbf{a}_3^{-1} \{ (\mathbf{a}_1 \wedge \mathbf{a}_2) \vee (\mathbf{a}_3 \wedge \mathbf{a}_4) \}, \tag{6.2.12}$$

as long as the 1D intersection of 2D planes $\mathbf{a}_1 \wedge \mathbf{a}_2$ and $\mathbf{a}_3 \wedge \mathbf{a}_4$ is not a null space. Unfortunately, the latter cannot be guaranteed by the hypothesis of Proposition 6.37.

Lemma 6.40. Any non-null 2D space has at most two null 1D subspaces.

Proof. $\mathbb{C}^2$ has two null 1D subspaces. Let $\mathbf{e}_1, \mathbf{e}_2$ be an orthonormal basis of $\mathbb{C}^2$, then $\mathbf{e}_1 \pm i\mathbf{e}_2$ are two null vectors spanning the two null 1D subspaces respectively. If a complex 2D space is degenerate but not null, it has only one null 1D subspace.

In the real case, all possible signatures of a 2D space are $(2,0,0)$, $(0,2,0)$, $(1,1,0), (1,0,1), (0,1,1)$. The first two signatures do not allow any null vector to exist in the 2D space, the last two signatures each allow only one null 1D subspace, and the third signature allows two null 1D subspaces. $\qquad\square$

Definition 6.41. In the $(n-1)$D projective space represented by an nD inner-product space $\mathcal{V}^n$, a projective point represented by a null vector (or invertible vector), is called a *null point* (or *invertible point*).

Lemma 6.42. If $\mathbf{A}_4 = \mathbf{a}_1 \mathbf{a}_2 \mathbf{a}_3 \mathbf{a}_4$ is compressible, and the four vectors span a 3D vector space $\mathcal{W}^3$, then in the 2D projective geometry of $\mathcal{W}^3$, line $\langle \mathbf{A}_4 \rangle_2$ meets one of the three lines $\mathbf{a}_1 \mathbf{a}_4, \mathbf{a}_2 \mathbf{a}_4, \mathbf{a}_3 \mathbf{a}_4$ at an invertible point.

Proof. By Lemma 6.40, projective line $\langle \mathbf{A}_4 \rangle_2$ has at most two null points. If the three lines $\mathbf{a}_1 \mathbf{a}_4, \mathbf{a}_2 \mathbf{a}_4, \mathbf{a}_3 \mathbf{a}_4$ each intersect with line $\langle \mathbf{A}_4 \rangle_2$ at a null point, then at least two of the three lines meet at a null point. However, the intersection of any two such lines is obviously the invertible point $\mathbf{a}_4$. Contradiction. $\qquad\square$

By Lemma 6.42, assume that the following vector in $\mathcal{W}^3$ is not null:

$$\mathbf{b}_2^{-1} = (\mathbf{a}_1 \wedge \mathbf{a}_4) \vee \langle \mathbf{A}_4 \rangle_2. \tag{6.2.13}$$

By Proposition 6.35, suppose $\mathbf{a}_1 \mathbf{a}_2 \mathbf{a}_3 \mathbf{a}_4 = \mathbf{b}_1' \mathbf{b}_2' \mathbf{a}_1 \mathbf{a}_4$. Substituting (6.2.13) into (6.2.3), we get

$$\mathbf{b}_1 = \mathbf{b}_1' \mathbf{b}_2' \{ \mathbf{a}_1 \mathbf{a}_4 ((\mathbf{a}_1 \wedge \mathbf{a}_4) \vee \langle \mathbf{b}_1' \mathbf{b}_2' \mathbf{a}_1 \mathbf{a}_4 \rangle_2)) \} = \mathbf{b}_1' \mathbf{b}_2' \mathbf{b}_3'. \tag{6.2.14}$$

That $\mathbf{b}_3'$ is a 3-compressed vector is because of the linear dependence among vectors $\mathbf{a}_1, \mathbf{a}_4$, and $(\mathbf{a}_1 \wedge \mathbf{a}_4) \vee \langle \mathbf{b}_1' \mathbf{b}_2' \mathbf{a}_1 \mathbf{a}_4 \rangle_2$; that $\mathbf{b}_1$ is a 3-compressed vector is because

$$\langle \mathbf{b}_1' \mathbf{b}_2' \mathbf{a}_1 \mathbf{a}_4 ((\mathbf{a}_1 \wedge \mathbf{a}_4) \vee \langle \mathbf{b}_1' \mathbf{b}_2' \mathbf{a}_1 \mathbf{a}_4 \rangle_2)) \rangle_3$$
$$= \langle \mathbf{b}_1' \mathbf{b}_2' \mathbf{a}_1 \mathbf{a}_4 \rangle_2 \wedge ((\mathbf{a}_1 \wedge \mathbf{a}_4) \vee \langle \mathbf{b}_1' \mathbf{b}_2' \mathbf{a}_1 \mathbf{a}_4 \rangle_2)$$
$$= 0.$$

Lemma 6.43. In $\mathcal{W}^3$, $(\mathbf{a}_1 \wedge \mathbf{a}_4) \vee (\mathbf{a}_2 \wedge \mathbf{a}_3)$ is a null vector if and only if so is $(\mathbf{a}_1 \wedge \mathbf{a}_4) \vee \langle \mathbf{A}_4 \rangle_2$. The two points are identical if and only if they are both null.

Proof.   Let $\mathbf{a}_1\mathbf{a}_2\mathbf{a}_3\mathbf{a}_4 = \mathbf{b}_1'\mathbf{b}_2'\mathbf{a}_1\mathbf{a}_4$ by virtue of Proposition 6.35. Then

$$
\begin{aligned}
(\mathbf{a}_1 \wedge \mathbf{a}_4) \vee \langle \mathbf{b}_1'\mathbf{b}_2'\mathbf{a}_1\mathbf{a}_4 \rangle_2
&= [\mathbf{a}_1\langle \mathbf{b}_1'\mathbf{b}_2'\mathbf{a}_1\mathbf{a}_4\rangle_2]\mathbf{a}_4 - [\mathbf{a}_4\langle \mathbf{b}_1'\mathbf{b}_2'\mathbf{a}_1\mathbf{a}_4\rangle_2]\mathbf{a}_1 \\
&= [\mathbf{a}_1\mathbf{b}_1'\mathbf{b}_2'\mathbf{a}_1\mathbf{a}_4]\mathbf{a}_4 - [\mathbf{a}_4\mathbf{b}_1'\mathbf{b}_2'\mathbf{a}_1\mathbf{a}_4]\mathbf{a}_1 \\
&= [\mathbf{a}_1\mathbf{a}_4\mathbf{a}_1\mathbf{b}_1'\mathbf{b}_2']\mathbf{a}_4 - [\mathbf{a}_4^2\mathbf{b}_1'\mathbf{b}_2'\mathbf{a}_1]\mathbf{a}_1 \\
&= \{2(\mathbf{a}_1 \cdot \mathbf{a}_4)[\mathbf{a}_1\mathbf{b}_1'\mathbf{b}_2'] - \mathbf{a}_1^2[\mathbf{a}_4\mathbf{b}_1'\mathbf{b}_2']\}\mathbf{a}_4 - \mathbf{a}_4^2[\mathbf{a}_1\mathbf{b}_1'\mathbf{b}_2']\mathbf{a}_1.
\end{aligned}
$$
$$(6.2.15)$$

So

$$
\begin{aligned}
((\mathbf{a}_1 &\wedge \mathbf{a}_4) \vee \langle \mathbf{b}_1'\mathbf{b}_2'\mathbf{a}_1\mathbf{a}_4 \rangle_2)^2 \\
&= \mathbf{a}_1^2\mathbf{a}_4^2(\mathbf{a}_1^2[\mathbf{a}_4\mathbf{b}_1'\mathbf{b}_2']^2 - 2(\mathbf{a}_1 \cdot \mathbf{a}_4)[\mathbf{a}_1\mathbf{b}_1'\mathbf{b}_2'][\mathbf{a}_4\mathbf{b}_1'\mathbf{b}_2'] + \mathbf{a}_4^2[\mathbf{a}_1\mathbf{b}_1'\mathbf{b}_2']^2) \\
&= \mathbf{a}_1^2\mathbf{a}_4^2(\mathbf{a}_1[\mathbf{a}_4\mathbf{b}_1'\mathbf{b}_2'] - \mathbf{a}_4[\mathbf{a}_1\mathbf{b}_1'\mathbf{b}_2'])^2 \\
&= \mathbf{a}_1^2\mathbf{a}_4^2((\mathbf{a}_1 \wedge \mathbf{a}_4) \vee (\mathbf{b}_1' \wedge \mathbf{b}_2'))^2.
\end{aligned}
$$
$$(6.2.16)$$

Since $\mathbf{b}_1' = \mathbf{a}_1\mathbf{a}_2\mathbf{a}_1^{-1}$ and $\mathbf{b}_2' = \mathbf{a}_1\mathbf{a}_3\mathbf{a}_1^{-1}$,

$$\mathbf{a}_1(\mathbf{a}_1 \wedge \mathbf{a}_4)\mathbf{a}_1^{-1} = \mathbf{a}_1 \wedge \mathbf{a}_1\mathbf{a}_4\mathbf{a}_1^{-1} = -\mathbf{a}_1 \wedge \mathbf{a}_4.$$

The mapping $\mathbf{A} \mapsto \mathbf{a}_1\mathbf{A}\mathbf{a}_1^{-1}$ for all $\mathbf{A} \in \mathcal{CL}(\mathcal{W}^3)$, is a special orthogonal transformation when restricted to $\mathcal{W}^3$, so

$$
\begin{aligned}
\mathbf{a}_1\{(\mathbf{a}_1 \wedge \mathbf{a}_4) \vee (\mathbf{a}_2 \wedge \mathbf{a}_3)\}\mathbf{a}_1^{-1} &= \{\mathbf{a}_1(\mathbf{a}_1 \wedge \mathbf{a}_4)\mathbf{a}_1^{-1}\} \vee \{\mathbf{a}_1(\mathbf{a}_2 \wedge \mathbf{a}_3)\mathbf{a}_1^{-1}\} \\
&= -(\mathbf{a}_1 \wedge \mathbf{a}_4) \vee (\mathbf{b}_1' \wedge \mathbf{b}_2'),
\end{aligned}
$$
$$(6.2.17)$$

and $((\mathbf{a}_1 \wedge \mathbf{a}_4) \vee (\mathbf{a}_2 \wedge \mathbf{a}_3))^2 = ((\mathbf{a}_1 \wedge \mathbf{a}_4) \vee (\mathbf{b}_1' \wedge \mathbf{b}_2'))^2$. Combining this and (6.2.16), we get the first statement of the lemma.

For the second statement, (6.2.15) equals $(\mathbf{a}_1 \wedge \mathbf{a}_4) \vee (\mathbf{a}_2 \wedge \mathbf{a}_3) = [\mathbf{a}_1\mathbf{a}_2\mathbf{a}_3]\mathbf{a}_4 - [\mathbf{a}_2\mathbf{a}_3\mathbf{a}_4]\mathbf{a}_1$ if and only if

$$\frac{[\mathbf{a}_1\mathbf{a}_2\mathbf{a}_3]}{[\mathbf{a}_2\mathbf{a}_3\mathbf{a}_4]} = \frac{2(\mathbf{a}_1 \cdot \mathbf{a}_4)[\mathbf{a}_1\mathbf{b}_1'\mathbf{b}_2'] - \mathbf{a}_1^2[\mathbf{a}_4\mathbf{b}_1'\mathbf{b}_2']}{\mathbf{a}_4^2[\mathbf{a}_1\mathbf{b}_1'\mathbf{b}_2']},$$

which after simplification, becomes

$$(\mathbf{a}_1[\mathbf{a}_4\mathbf{b}_1'\mathbf{b}_2'] - \mathbf{a}_4[\mathbf{a}_1\mathbf{b}_1'\mathbf{b}_2'])^2 = ((\mathbf{a}_1 \wedge \mathbf{a}_4) \vee (\mathbf{b}_1' \wedge \mathbf{b}_2'))^2 = 0.$$

$\square$

Proposition 6.44. [4-tuple compression] If $\mathbf{A}_4 = \mathbf{a}_1\mathbf{a}_2\mathbf{a}_3\mathbf{a}_4$ is compressible, and the four vectors span a 3D vector space $\mathcal{W}^3$, then there is a partition $(\mathbf{A}_{3(1)}, \mathbf{A}_{3(2)})$ of blade $\mathbf{A}_3 = \mathbf{a}_1 \wedge \mathbf{a}_2 \wedge \mathbf{a}_3$ of shape $(2, 1)$, such that

(1) vector $\mathbf{A}_{3(1)} \vee (\mathbf{A}_{3(2)} \wedge \mathbf{a}_4) \in \mathcal{W}^3$ is invertible;

(2) by setting $\mathbf{a}_1\mathbf{a}_2\mathbf{a}_3\mathbf{a}_4 = \mathbf{b}_1'\mathbf{b}_2'\mathbf{A}_{3(2)}\mathbf{a}_4$, there are the following 3-compressed vectors:

$$
\begin{aligned}
\mathbf{b}_2^{-1} &= \mathbf{a}_4^{-1}\mathbf{A}_{3(2)}^{-1}\{(\mathbf{b}_1' \wedge \mathbf{b}_2') \vee (\mathbf{A}_{3(2)} \wedge \mathbf{a}_4)\}, \\
\mathbf{b}_3' &= \mathbf{A}_{3(2)}\mathbf{a}_4\mathbf{b}_2^{-1}, \\
\mathbf{b}_1 &= \mathbf{b}_1'\mathbf{b}_2'\mathbf{b}_3';
\end{aligned}
$$

(3) $\mathbf{a}_1\mathbf{a}_2\mathbf{a}_3\mathbf{a}_4 = \mathbf{b}_1\mathbf{b}_2$.

6.2.2 *5-tuple compression*

We proceed to the next case $m = 5$. In $\mathbf{A}_5 = \mathbf{a}_1\mathbf{a}_2\mathbf{a}_3\mathbf{a}_4\mathbf{a}_5$, assume that the five vectors are linearly dependent, and that they span a 4D space $\mathcal{W}^4$. By commutation within the versor, we can assume that the last four vectors $\mathbf{a}_2, \mathbf{a}_3, \mathbf{a}_4, \mathbf{a}_5$ are linearly independent.

If $\mathbf{A}_5 = \mathbf{b}_1\mathbf{b}_2\mathbf{b}_3$, then $\mathbf{b}_1\mathbf{b}_2 = \mathbf{A}_5\mathbf{b}_3^{-1}$. For the existence of $\mathbf{b}_1\mathbf{b}_2$, it is necessary (but not sufficient) that

$$\langle \mathbf{A}_5\mathbf{b}_3^{-1}\rangle_4 = \langle \mathbf{A}_5\rangle_3 \wedge \mathbf{b}_3^{-1} = 0. \tag{6.2.18}$$

The 3-blade $\langle \mathbf{A}_5\rangle_3$ is nonzero because

$$\mathbf{a}_1 \wedge \langle \mathbf{A}_5\rangle_3 = \langle \mathbf{a}_1\mathbf{A}_5\rangle_4 = \mathbf{a}_1^2\langle \mathbf{a}_2\mathbf{a}_3\mathbf{a}_4\mathbf{a}_5\rangle_4 \neq 0.$$

Lemma 6.45. For arbitrary vectors $\mathbf{d}_i \in \mathcal{V}^4$, let $\mathbf{D} = \mathbf{d}_1\mathbf{d}_2\cdots\mathbf{d}_{2k+1}$, then

$$[\langle \mathbf{D}\rangle_1\langle \mathbf{D}\rangle_3] = 0. \tag{6.2.19}$$

Proof.

$$[\langle \mathbf{D}\rangle_1\langle \mathbf{D}\rangle_3] = \frac{1}{4}[(\mathbf{D} + \mathbf{D}^\dagger)(\mathbf{D} - \mathbf{D}^\dagger)] = \frac{1}{4}([\mathbf{D}\mathbf{D}] - [\mathbf{D}^\dagger\mathbf{D}^\dagger]) = 0.$$

$\square$

Lemma 6.46. Any non-null 3D space has at most two null 2D subspaces.

Proof. We only consider the real case, as the complex case is much the same. In the nondegenerate case, no 3D space has any null 2D subspace, because by Witt Theorem, a null 2D subspace requires a 4D nondegenerate space to embed it.

In the degenerate case, by Corollary 5.4, a non-null 3D space has a null 2D subspace if and only if its signature (p, q, r) satisfies

$$\begin{aligned} \min(p, q) + \max(p, q) + r &= 3, \\ r + \min(p, q) &= 2. \end{aligned} \tag{6.2.20}$$

By simple argument, it can be shown that (6.2.20) is equivalent to $p, q \leq 1$.

When one of p, q is zero, then $r = 2$, the 3D space is semi-definite, and has only one null 2D subspace. When $p = q = 1$, let $\mathbf{e}_1, \mathbf{e}_2$ be a Witt basis of a Minkowski plane of the 3D space, and let $\mathbf{e}_3$ be orthogonal to the Minkowski plane, then $\mathbf{e}_1 \wedge \mathbf{e}_3$ and $\mathbf{e}_2 \wedge \mathbf{e}_3$ are the only two null 2D subspaces. $\square$

Proposition 6.47. In $\mathbf{A}_5 = \mathbf{a}_1\mathbf{a}_2\mathbf{a}_3\mathbf{a}_4\mathbf{a}_5$, if the five vectors span a 4D vector space $\mathcal{W}^4$, and no three of them are linearly dependent, then $\mathbf{A}_5$ is compressible if and only if $\langle \mathbf{A}_5\rangle_3$ is not null.

Proof. We only prove the sufficiency statement. Assume that 3D subspace $\langle \mathbf{A}_5\rangle_3$ is not null. Then almost all vectors in it are invertible. If $\langle \mathbf{A}_5\rangle_1 = 0$, then $\mathbf{A}_5$ is a 3-blade, so the length of $\mathbf{A}_5$ can be reduced to three. So we assume $\langle \mathbf{A}_5\rangle_1 \neq 0$.

By (6.2.19), $\langle \mathbf{A}_5 \rangle_1 \in \langle \mathbf{A}_5 \rangle_3$. For any invertible vector $\mathbf{c} \in \langle \mathbf{A}_5 \rangle_3$, since

$$\langle \mathbf{A}_5 \mathbf{c} \rangle_2 = \langle \mathbf{A}_5 \rangle_3 \cdot \mathbf{c} + \langle \mathbf{A}_5 \rangle_1 \wedge \mathbf{c}, \tag{6.2.21}$$

$\langle \mathbf{A}_5 \mathbf{c} \rangle_2$ is a 2-blade in $\Lambda(\langle \mathbf{A}_5 \rangle_3)$.

As vector space $\{ \langle \mathbf{A}_5 \rangle_1 \wedge \mathbf{c} \,|\, \mathbf{c} \in \langle \mathbf{A}_5 \rangle_3 \}$ has dimension two, by (6.2.21), vector space $\mathcal{B} = \{ \langle \mathbf{A}_5 \mathbf{c} \rangle_2 \,|\, \mathbf{c} \in \langle \mathbf{A}_5 \rangle_3 \}$ has dimension at least two. Let $\langle \mathbf{A}_5 \mathbf{c}_1 \rangle_2$, $\langle \mathbf{A}_5 \mathbf{c}_2 \rangle_2$ be two linearly independent bivectors in $\mathcal{B}$, where $\mathbf{c}_1, \mathbf{c}_2$ are invertible vectors in $\langle \mathbf{A}_5 \rangle_3$. Then for any invertible vector $\mathbf{c}$ that is a linear combination of $\mathbf{c}_1, \mathbf{c}_2$, $\langle \mathbf{A}_5 \mathbf{c} \rangle_2 \in \Lambda(\langle \mathbf{A}_5 \rangle_3)$. There are infinitely many such invertible vectors.

By Lemma 6.46, $\langle \mathbf{A}_5 \rangle_3$ has at most two null 2D subspaces. There must be an invertible vector $\mathbf{c}' \in \langle \mathbf{A}_5 \rangle_3$ such that $\langle \mathbf{A}_5 \mathbf{c}' \rangle_2$ is not null. Then $\mathbf{A}_5 \mathbf{c}' = \langle \mathbf{A}_5 \mathbf{c}' \rangle + \langle \mathbf{A}_5 \mathbf{c}' \rangle_2$ equals a rotor in $\mathcal{G}(\langle \mathbf{A}_5 \mathbf{c}' \rangle_2)$, and accordingly, $\mathbf{A}_5 = (\mathbf{A}_5 \mathbf{c}') \mathbf{c}'^{-1}$ is compressible to a versor of length three. $\qquad \square$

The following result is in sharp contrast to the $m = 4$ case.

Proposition 6.48. *Versor $\mathbf{A}_5$ described in Proposition 6.47 is always compressible.*

Proof. Assume that $\mathbf{A}_5$ is incompressible. Then $\langle \mathbf{A}_5 \rangle_3$ is null. Since the 4D space has a null 3D subspace, $\mathcal{W}^4$ must be one of $\mathbb{R}^{1,0,3}$, $\mathbb{R}^{0,1,3}$, $\mathbb{R}^{1,1,2}$, $\mathbb{C}^{1,3}$, $\mathbb{C}^{2,2}$.

Case 1. $\mathcal{W}^4 = \mathbb{R}^{1,0,3}$. Two other cases $\mathbb{R}^{0,1,3}, \mathbb{C}^{1,3}$ are similar.

Let $\mathbf{e}_1, \mathbf{e}_2, \mathbf{e}_3, \mathbf{e}_4$ be an orthonormal basis of $\mathbb{R}^{1,0,3}$ such that $\mathbf{e}_1^2 = 1$. Since $\mathbf{a}_i$ is invertible, the coordinate component of $\mathbf{a}_i$ with respect to $\mathbf{e}_1$ is nonzero. Let $\mathbf{a}_i = \mathbf{e}_1 + \lambda_i \mathbf{e}_2 + \mu_i \mathbf{e}_3 + \tau_i \mathbf{e}_4$ for $1 \leq i \leq 5$. Then $\mathbf{a}_i \cdot \mathbf{a}_j = 1$ for any $1 \leq i, j \leq 5$. As the null 3D subspace of $\mathcal{W}^4$ is unique, $\langle \mathbf{A}_5 \rangle_3$ and $\mathbf{e}_2 \mathbf{e}_3 \mathbf{e}_4$ must be equal up to scale, so

$$[\mathbf{a}_1 \langle \mathbf{A}_5 \rangle_3] = [\mathbf{a}_2 \langle \mathbf{A}_5 \rangle_3] = [\mathbf{a}_3 \langle \mathbf{A}_5 \rangle_3] = [\mathbf{a}_4 \langle \mathbf{A}_5 \rangle_3] = [\mathbf{a}_5 \langle \mathbf{A}_5 \rangle_3] \neq 0. \tag{6.2.22}$$

The $[\mathbf{a}_i \langle \mathbf{A}_5 \rangle_3]$ can be computed as follows:

$$\begin{cases} [\mathbf{a}_1 \langle \mathbf{A}_5 \rangle_3] = \mathbf{a}_1^2 [\mathbf{a}_2 \mathbf{a}_3 \mathbf{a}_4 \mathbf{a}_5], \\[2pt] [\mathbf{a}_2 \langle \mathbf{A}_5 \rangle_3] = 2(\mathbf{a}_1 \cdot \mathbf{a}_2)[\mathbf{a}_2 \mathbf{a}_3 \mathbf{a}_4 \mathbf{a}_5] - \mathbf{a}_2^2 [\mathbf{a}_1 \mathbf{a}_3 \mathbf{a}_4 \mathbf{a}_5], \\[2pt] [\mathbf{a}_3 \langle \mathbf{A}_5 \rangle_3] = 2(\mathbf{a}_1 \cdot \mathbf{a}_3)[\mathbf{a}_2 \mathbf{a}_3 \mathbf{a}_4 \mathbf{a}_5] - 2(\mathbf{a}_2 \cdot \mathbf{a}_3)[\mathbf{a}_1 \mathbf{a}_3 \mathbf{a}_4 \mathbf{a}_5] + \mathbf{a}_3^2 [\mathbf{a}_1 \mathbf{a}_2 \mathbf{a}_4 \mathbf{a}_5] \\[2pt] \qquad\quad = 2(\mathbf{a}_3 \cdot \mathbf{a}_4)[\mathbf{a}_1 \mathbf{a}_2 \mathbf{a}_3 \mathbf{a}_5] - 2(\mathbf{a}_3 \cdot \mathbf{a}_5)[\mathbf{a}_1 \mathbf{a}_2 \mathbf{a}_3 \mathbf{a}_4] - \mathbf{a}_3^2 [\mathbf{a}_1 \mathbf{a}_2 \mathbf{a}_4 \mathbf{a}_5], \\[2pt] [\mathbf{a}_4 \langle \mathbf{A}_5 \rangle_3] = \mathbf{a}_4^2 [\mathbf{a}_1 \mathbf{a}_2 \mathbf{a}_3 \mathbf{a}_5] - 2(\mathbf{a}_4 \cdot \mathbf{a}_5)[\mathbf{a}_1 \mathbf{a}_2 \mathbf{a}_3 \mathbf{a}_4], \\[2pt] [\mathbf{a}_5 \langle \mathbf{A}_5 \rangle_3] = -\mathbf{a}_5^2 [\mathbf{a}_1 \mathbf{a}_2 \mathbf{a}_3 \mathbf{a}_4]. \end{cases}$$
$$\tag{6.2.23}$$

By $[\mathbf{a}_1 \langle \mathbf{A}_5 \rangle_3] = [\mathbf{a}_2 \langle \mathbf{A}_5 \rangle_3]$, we get $[\mathbf{a}_2 \mathbf{a}_3 \mathbf{a}_4 \mathbf{a}_5] = [\mathbf{a}_1 \mathbf{a}_3 \mathbf{a}_4 \mathbf{a}_5]$. By $[\mathbf{a}_1 \langle \mathbf{A}_5 \rangle_3] = [\mathbf{a}_5 \langle \mathbf{A}_5 \rangle_3] = [\mathbf{a}_4 \langle \mathbf{A}_5 \rangle_3]$, we get $[\mathbf{a}_2 \mathbf{a}_3 \mathbf{a}_4 \mathbf{a}_5] = -[\mathbf{a}_1 \mathbf{a}_2 \mathbf{a}_3 \mathbf{a}_5] = -[\mathbf{a}_1 \mathbf{a}_2 \mathbf{a}_3 \mathbf{a}_4]$. Substituting them into the two results of $[\mathbf{a}_3 \langle \mathbf{A}_5 \rangle_3]$ in (6.2.23), we get $[\mathbf{a}_1 \mathbf{a}_2 \mathbf{a}_4 \mathbf{a}_5] = -[\mathbf{a}_1 \mathbf{a}_2 \mathbf{a}_4 \mathbf{a}_5] = [\mathbf{a}_1 \mathbf{a}_2 \mathbf{a}_3 \mathbf{a}_4] = 0$, which is impossible because the five vectors span $\mathcal{W}^4$.

Case 2. $\mathcal{W}^4 = \mathbb{R}^{1,1,2}$. The other case $\mathbb{C}^{2,2}$ is similar.

Let $\mathbf{e}_1, \mathbf{e}_2, \mathbf{e}_3, \mathbf{e}_4$ be a Witt basis of $\mathbb{R}^{1,1,1}$ such that $(\mathbf{e}_1, \mathbf{e}_2)$ is a Witt pair. Since $\mathbf{a}_i$ is invertible, the coordinate components of $\mathbf{a}_i$ with respect to $\mathbf{e}_1, \mathbf{e}_2$ are both nonzero. Let $\mathbf{a}_i = \mathbf{e}_1 + \lambda_i \mathbf{e}_2 + \mu_i \mathbf{e}_3 + \tau_i \mathbf{e}_4$ for $1 \leq i \leq 4$ and $\lambda_i \neq 0$. Then $\mathbf{a}_i \cdot \mathbf{a}_j = -(\lambda_i + \lambda_j)$ for any $1 \leq i, j \leq 5$, and

$$\mathbf{a}_i^{-1} = \frac{\mathbf{a}_i}{\mathbf{a}_i \cdot \mathbf{a}_i} = -\frac{1}{2}\mathbf{e}_2 - \frac{1}{2\lambda_i}(\mathbf{e}_1 + \mu_i \mathbf{e}_3 + \tau_i \mathbf{e}_4). \tag{6.2.24}$$

For any $1 \leq i < j \leq 5$, by the expressions of $\mathbf{a}_i$ and $\mathbf{a}_i^{-1}$, we get

$$\begin{aligned} (\mathbf{a}_i - \mathbf{a}_j) \wedge (\mathbf{e}_2 \wedge \mathbf{e}_3 \wedge \mathbf{e}_4) &= 0, \\ (\mathbf{a}_i^{-1} - \mathbf{a}_j^{-1}) \wedge (\mathbf{e}_1 \wedge \mathbf{e}_3 \wedge \mathbf{e}_4) &= 0. \end{aligned} \tag{6.2.25}$$

Null subspace $\langle \mathbf{A}_5 \rangle_3$ must be identical to one of $\mathbf{e}_2 \wedge \mathbf{e}_3 \wedge \mathbf{e}_4$, $\mathbf{e}_1 \wedge \mathbf{e}_3 \wedge \mathbf{e}_4$, which are the only two null 3D subspaces of $\mathcal{W}^4$, so there are two cases:

Case 2a. $[(\mathbf{a}_i - \mathbf{a}_j)\langle \mathbf{a}_1 \mathbf{a}_2 \mathbf{a}_3 \mathbf{a}_4 \mathbf{a}_5 \rangle_3] = 0$ for any $1 \leq i < j \leq 5$.

Case 2b. $[(\mathbf{a}_i^{-1} - \mathbf{a}_j^{-1})\langle \mathbf{a}_1^{-1} \mathbf{a}_2^{-1} \mathbf{a}_3^{-1} \mathbf{a}_4^{-1} \mathbf{a}_5^{-1} \rangle_3] = 0$ for any $1 \leq i < j \leq 5$.

The two cases are similar, so we only check the ten equations in Case 2a. They are identical to (6.2.22), where the inequality is caused by the linear independence among any three of the five vectors. (6.2.23) provides the same computing results of the $[\mathbf{a}_i \langle \mathbf{A}_5 \rangle_3]$. Substituting them into (6.2.22), we get

$$\begin{aligned} \lambda_1 \lambda_4 &= -\lambda_2 \lambda_5, \\ [\mathbf{a}_2 \mathbf{a}_3 \mathbf{a}_4 \mathbf{a}_5] &= [\mathbf{a}_1 \mathbf{a}_3 \mathbf{a}_4 \mathbf{a}_5], \\ [\mathbf{a}_1 \mathbf{a}_2 \mathbf{a}_4 \mathbf{a}_5] &= \lambda_2 \lambda_3^{-1} [\mathbf{a}_1 \mathbf{a}_3 \mathbf{a}_4 \mathbf{a}_5], \\ [\mathbf{a}_1 \mathbf{a}_2 \mathbf{a}_3 \mathbf{a}_4] &= -\lambda_1 \lambda_5^{-1} [\mathbf{a}_1 \mathbf{a}_3 \mathbf{a}_4 \mathbf{a}_5] = [\mathbf{a}_1 \mathbf{a}_2 \mathbf{a}_3 \mathbf{a}_5]. \end{aligned} \tag{6.2.26}$$

Substituting (6.2.26) into the Cramer's rule of the five vectors, *i.e.*,

$$[\mathbf{a}_2 \mathbf{a}_3 \mathbf{a}_4 \mathbf{a}_5]\mathbf{a}_1 - [\mathbf{a}_1 \mathbf{a}_3 \mathbf{a}_4 \mathbf{a}_5]\mathbf{a}_2 + [\mathbf{a}_1 \mathbf{a}_2 \mathbf{a}_4 \mathbf{a}_5]\mathbf{a}_3 - [\mathbf{a}_1 \mathbf{a}_2 \mathbf{a}_3 \mathbf{a}_5]\mathbf{a}_4 + [\mathbf{a}_1 \mathbf{a}_2 \mathbf{a}_3 \mathbf{a}_4]\mathbf{a}_5 = 0,$$

we get $\mathbf{a}_1 - \mathbf{a}_2 + \lambda_2 \lambda_3^{-1} \mathbf{a}_3 + \lambda_1 \lambda_5^{-1}(\mathbf{a}_4 - \mathbf{a}_5) = 0$. However, the $\mathbf{e}_1$-component of the left side is $\lambda_2 \lambda_3^{-1} \neq 0$. Contradiction. $\qquad \square$

In Proposition 6.47, it is assumed that four of the five vectors are linearly independent. If this condition is not satisfied, we have the following result:

Proposition 6.49. In $\mathbf{A}_5 = \mathbf{a}_1 \mathbf{a}_2 \mathbf{a}_3 \mathbf{a}_4 \mathbf{a}_5$, if any three vectors are linearly independent, but any four vectors are linearly dependent, then there exists a partition $(\mathbf{A}_{5(1)}, \mathbf{A}_{5(2)})$ of $\mathbf{A}_5$ of shape $(4, 1)$, such that versor $\mathbf{A}_{5(1)}$ is compressible.

Proof. We only consider the real case, as the complex case is much the same.

If the conclusion is false, then in the 3D space $\mathcal{W}^3$ spanned by the five vectors, $\langle \mathbf{a}_1 \mathbf{a}_2 \mathbf{a}_3 \mathbf{a}_4 \rangle_2, \langle \mathbf{a}_1 \mathbf{a}_2 \mathbf{a}_3 \mathbf{a}_5 \rangle_2, \langle \mathbf{a}_1 \mathbf{a}_2 \mathbf{a}_4 \mathbf{a}_5 \rangle_2, \langle \mathbf{a}_1 \mathbf{a}_3 \mathbf{a}_4 \mathbf{a}_5 \rangle_2, \langle \mathbf{a}_2 \mathbf{a}_3 \mathbf{a}_4 \mathbf{a}_5 \rangle_2$ are all null 2D subspaces. So $\mathcal{W}^3$ must be one of $\mathbb{R}^{1,0,2}, \mathbb{R}^{0,1,2}, \mathbb{R}^{1,1,1}$. When $\mathcal{W}^3 = \mathbb{R}^{1,0,2}$ or $\mathbb{R}^{0,1,2}$, by (6.2.5), the five vectors are equal to each other up to scale, contradicting with the linear independence assumption of any three of the five vectors.

When $\mathcal{W}^3 = \mathbb{R}^{1,1,1}$, let $\mathbb{R}^{1,1,1} = \mathbb{R}^{1,1,0} \oplus \mathbb{R}^{0,0,1}$ be a fixed orthogonal decomposition. By (6.2.8) and (6.2.11), the components of $\mathbf{a}_4, \mathbf{a}_5$ in $\mathbb{R}^{1,1,0}$ are both equal to $\langle \mathbf{a}_1\mathbf{a}_2\mathbf{a}_3 \rangle_1 \neq 0$ up to scale. For $\langle \mathbf{a}_1\mathbf{a}_2\mathbf{a}_4\mathbf{a}_5 \rangle_2$ to be null, the component of $\mathbf{a}_5$ in $\mathbb{R}^{1,1,0}$ must be equal to $\langle \mathbf{a}_1\mathbf{a}_2\mathbf{a}_4 \rangle_1 = \langle \mathbf{a}_1\mathbf{a}_2\langle \mathbf{a}_1\mathbf{a}_2\mathbf{a}_3 \rangle_1 \rangle_1 \neq 0$ up to scale. So $\langle \mathbf{a}_1\mathbf{a}_2\mathbf{a}_3 \rangle_1$ and $\langle \mathbf{a}_1\mathbf{a}_2\langle \mathbf{a}_1\mathbf{a}_2\mathbf{a}_3 \rangle_1 \rangle_1$ are equal up to scale.

Since

$$
\begin{aligned}
\langle \mathbf{a}_1\mathbf{a}_2\mathbf{a}_3 \rangle_1 \quad &= \quad (\mathbf{a}_2 \cdot \mathbf{a}_3)\mathbf{a}_1 - (\mathbf{a}_1 \cdot \mathbf{a}_3)\mathbf{a}_2 + (\mathbf{a}_1 \cdot \mathbf{a}_2)\mathbf{a}_3, \\
\langle \mathbf{a}_1\mathbf{a}_2\langle \mathbf{a}_1\mathbf{a}_2\mathbf{a}_3 \rangle_1 \rangle_1 &= \quad \{3(\mathbf{a}_1 \cdot \mathbf{a}_2)(\mathbf{a}_2 \cdot \mathbf{a}_3) - \mathbf{a}_2^2(\mathbf{a}_1 \cdot \mathbf{a}_3)\}\mathbf{a}_1 \\
&\quad -\{\mathbf{a}_1^2(\mathbf{a}_2 \cdot \mathbf{a}_3) + (\mathbf{a}_1 \cdot \mathbf{a}_2)(\mathbf{a}_1 \cdot \mathbf{a}_3)\}\mathbf{a}_2 + (\mathbf{a}_1 \cdot \mathbf{a}_2)^2\mathbf{a}_3,
\end{aligned}
\tag{6.2.27}
$$

the two expressions on left side of (6.2.27) are equal up to scale if and only if on the right side, the ratios of the coefficients of $\mathbf{a}_i$ for $i = 1, 2, 3$ are equal to each other. By simplification, the latter equalities are changed into $\mathbf{a}_2 \cdot \mathbf{a}_3 = \mathbf{a}_1 \cdot \mathbf{a}_3 = 0$.

So $\langle \mathbf{a}_1\mathbf{a}_2\mathbf{a}_3 \rangle_1 = (\mathbf{a}_1 \cdot \mathbf{a}_2)\mathbf{a}_3 \neq 0$. Now that the components of $\mathbf{a}_3, \mathbf{a}_4, \mathbf{a}_5$ in $\mathbb{R}^{1,1,0}$ are equal to each other up to scale, vectors $\mathbf{a}_3, \mathbf{a}_4, \mathbf{a}_5$ must be in the same 2D subspace of $\mathcal{W}^3$, contradicting with the linear independence assumption. $\qquad\square$

Combining Propositions 6.48 and 6.49, we get

Theorem 6.50. Any non-rigid versor of length 5 is compressible.

Below we investigate the problem of constructing a 5-tuple compression explicitly. For $\mathbf{A}_5 = \mathbf{a}_1\mathbf{a}_2\mathbf{a}_3\mathbf{a}_4\mathbf{a}_5$ in which no three vectors are linearly dependent, assume that the last four vectors span a 4D vector space $\mathcal{W}^4$. The idea is to use lines $\mathbf{a}_i\mathbf{a}_j$ to intersect plane $\langle \mathbf{A}_5 \rangle_3$ for an invertible intersection in the 3D projective space $\mathcal{W}^4$. The intersection, being on line $\mathbf{a}_i\mathbf{a}_j$, will form a 3-compressed vector with $\mathbf{a}_i, \mathbf{a}_j$.

Lemma 6.51. If $\mathbf{A}_5$ can be compressed to $\mathbf{b}_1\mathbf{b}_2\mathbf{b}_3$, and $\mathbf{a}_4\mathbf{a}_5\mathbf{b}_3$ is a 3-compressed vector, then $\mathbf{a}_4\mathbf{a}_5\mathbf{b}_3$ equals $(\mathbf{a}_1 \wedge \mathbf{a}_2 \wedge \mathbf{a}_3) \vee (\mathbf{a}_4 \wedge \mathbf{a}_5) \in \mathcal{W}^4$ up to scale.

Proof. By the hypotheses, point $\mathbf{b}_3$ is on both line $\mathbf{a}_4\mathbf{a}_5$ and plane $\langle \mathbf{A}_5 \rangle_3$, so

$$
0 = \langle \mathbf{b}_1\mathbf{b}_2\mathbf{b}_3^2 \rangle_4 = \langle \mathbf{A}_5\mathbf{b}_3 \rangle_4 = \langle (\mathbf{a}_1\mathbf{a}_2\mathbf{a}_3)(\mathbf{a}_4\mathbf{a}_5\mathbf{b}_3) \rangle_4 = \mathbf{a}_1 \wedge \mathbf{a}_2 \wedge \mathbf{a}_3 \wedge (\mathbf{a}_4\mathbf{a}_5\mathbf{b}_3).
$$

Then point $\mathbf{a}_4\mathbf{a}_5\mathbf{b}_3$ is on both line $\mathbf{a}_4\mathbf{a}_5$ and plane $\mathbf{a}_1 \wedge \mathbf{a}_2 \wedge \mathbf{a}_3$. $\qquad\square$

Lemma 6.52. In $\mathcal{W}^4$, for any permutation σ of indices $2, 3, 4, 5$, vector $(\mathbf{a}_{\sigma(4)} \wedge \mathbf{a}_{\sigma(5)}) \vee (\mathbf{a}_1 \wedge \mathbf{a}_{\sigma(2)} \wedge \mathbf{a}_{\sigma(3)})$ is null if and only if so is vector $(\mathbf{a}_{\sigma(4)} \wedge \mathbf{a}_{\sigma(5)}) \vee \langle \mathbf{A}_5 \rangle_3$. The two projective points are identical if and only if they are both null.

Proof. Similar to the proof of Lemma 6.43. $\qquad\square$

Lemma 6.53. In the 3D projective space $\mathcal{W}^4$, plane $\langle \mathbf{A}_5 \rangle_3$ meets one of the lines $\mathbf{a}_i\mathbf{a}_j$, where $2 \leq i < j \leq 5$, at an invertible point.

Proof. By Lemma 6.46, plane $\langle \mathbf{A}_5 \rangle_3$ has at most two null lines, say lines L_1, L_2. If the three lines $\mathbf{a}_2\mathbf{a}_5, \mathbf{a}_3\mathbf{a}_5, \mathbf{a}_4\mathbf{a}_5$ each intersect with plane $\langle \mathbf{A}_5 \rangle_3$ at a null point, then at least two of the three lines meet the same null line of the plane. Say $\mathbf{a}_2\mathbf{a}_5$ and $\mathbf{a}_3\mathbf{a}_5$ meet line L_1. If $\mathbf{a}_4\mathbf{a}_5$ also meets line L_1, then $\mathbf{a}_2, \mathbf{a}_3, \mathbf{a}_4, \mathbf{a}_5$ are in the same plane determined by point $\mathbf{a}_5$ and line L_1, contradicting with the assumption that they are linearly independent vectors. So $\mathbf{a}_4\mathbf{a}_5$ meets line L_2 but not line L_1.

Now points $\mathbf{a}_2, \mathbf{a}_3, \mathbf{a}_5$ are coplanar with line L_1, and point $\mathbf{a}_4$ is outside the plane on which lie the points $\mathbf{a}_2, \mathbf{a}_3, \mathbf{a}_5$ and line L_1. If lines $\mathbf{a}_2\mathbf{a}_4, \mathbf{a}_3\mathbf{a}_4$ each intersect with plane $\langle \mathbf{A}_5 \rangle_3$ at a null point, then they cannot meet line L_1 because of the noncoplanarity. So they all meet line L_2. Then $\mathbf{a}_2, \mathbf{a}_3, \mathbf{a}_4, \mathbf{a}_5$ are in the same plane determined by point $\mathbf{a}_4$ and line L_2, contradicting with the linear independence assumption of the four vectors. $\qquad\square$

Lemma 6.53 can be further strengthened as follows:

Proposition 6.54. Among the ten lines $\mathbf{a}_i\mathbf{a}_j$ for $1 \leq i < j \leq 5$, there are at least two lines intersecting plane $\langle \mathbf{A}_5 \rangle_3$ at non-null points.

Proof. We only consider the case where plane $\langle \mathbf{A}_5 \rangle_3$ has two null lines L_1, L_2. By Lemma 6.53, assume that the intersection of line $\mathbf{a}_4\mathbf{a}_5$ and plane $\langle \mathbf{A}_5 \rangle_3$ is not null. Then one of $\mathbf{a}_4, \mathbf{a}_5$ must be outside plane $\langle \mathbf{A}_5 \rangle_3$. Assume that it is $\mathbf{a}_5$.

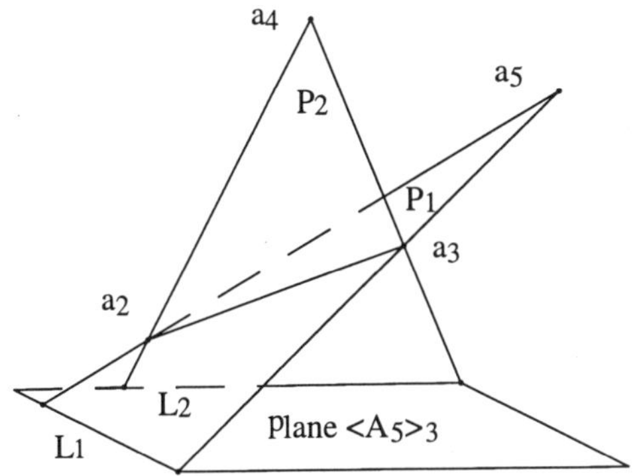

Fig. 6.1   Proof of Proposition 6.54.

If $\mathbf{a}_4\mathbf{a}_5$ is the only line among $\{\mathbf{a}_i\mathbf{a}_j \mid 2 \leq i < j \leq 5\}$ that has non-null intersection with plane $\langle \mathbf{A}_5 \rangle_3$, then for lines $\mathbf{a}_2\mathbf{a}_3, \mathbf{a}_2\mathbf{a}_5, \mathbf{a}_3\mathbf{a}_5$ to intersect either L_1 or L_2, it is necessary that they meet the same line, say L_1. Then $\mathbf{a}_2\mathbf{a}_4$ and $\mathbf{a}_3\mathbf{a}_4$ cannot meet line L_1 any more. They all meet L_2, indicating that $\mathbf{a}_4$ cannot be in plane $\langle \mathbf{A}_5 \rangle_3$. Denote by P_1 the plane determined by $\mathbf{a}_5$ and L_1, and by P_2 the plane determined by $\mathbf{a}_4$ and L_2. Then $\mathbf{a}_2, \mathbf{a}_3$ are both on the line of intersection $P_1 \cap P_2$. By the same argument, $\mathbf{a}_1$ is also on this line, contradicting with the linear independence assumption of vectors $\mathbf{a}_1, \mathbf{a}_2, \mathbf{a}_3$. $\qquad\square$

Proposition 6.55. [5-tuple compression by intersections with lines] There is a partition $(\mathbf{A}_{4(1)}, \mathbf{A}_{4(2)})$ of blade $\mathbf{A}_4 = \mathbf{a}_2 \wedge \mathbf{a}_3 \wedge \mathbf{a}_4 \wedge \mathbf{a}_5$ of shape $(2, 2)$, such that

(1) vector $\mathbf{A}_{4(2)} \vee (\mathbf{a}_1 \wedge \mathbf{A}_{4(1)}) \in \mathcal{W}^4$ is invertible;
(2) if denoting the two vectors in $\mathbf{A}_{4(2)}$ by $\mathbf{a}'_4, \mathbf{a}'_5$, and setting $\mathbf{a}_2\mathbf{a}_3\mathbf{a}_4\mathbf{a}_5 = \mathbf{b}'_1\mathbf{b}'_2\mathbf{a}'_4\mathbf{a}'_5$, there are the following 3-compressed vectors:

$$\mathbf{b}_3^{-1} = \mathbf{a}'_5{}^{-1}\mathbf{a}'_4{}^{-1}\{(\mathbf{a}_1 \wedge \mathbf{b}'_1 \wedge \mathbf{b}'_2) \vee \mathbf{A}_{4(2)}\},$$
$$\mathbf{b}'_3 = \mathbf{a}'_4\mathbf{a}'_5\mathbf{b}_3^{-1};$$

(3) $\mathbf{A}_5 = \mathbf{B}_4\mathbf{b}_3$, where $\mathbf{B}_4 = \mathbf{a}_1\mathbf{b}'_1\mathbf{b}'_2\mathbf{b}'_3$ and $\langle \mathbf{B}_4 \rangle_4 = 0$.

In Proposition 6.55, $\mathbf{A}_5$ may not be compressed in the last step, if versor $\mathbf{B}_4$ is incompressible. We have the following geometric characterization of the result of Proposition 6.55:

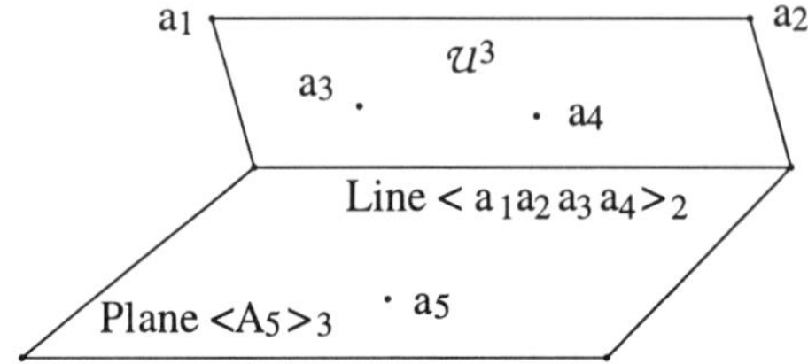

Fig. 6.2   Result of Proposition 6.55.

Proposition 6.56. In $\mathbf{A}_5 = \mathbf{a}_1\mathbf{a}_2\mathbf{a}_3\mathbf{a}_4\mathbf{a}_5$, assume that no three vectors are linearly dependent, that the first four vectors span a 3D subspace $\mathcal{U}^3$, and that the five vectors span a 4D vector space $\mathcal{W}^4$. Then in the 3D projective space $\mathcal{W}^4$, point $\mathbf{a}_5$ is in plane $\langle \mathbf{A}_5 \rangle_3$, and points $\mathbf{a}_1, \mathbf{a}_2, \mathbf{a}_3, \mathbf{a}_4$ are outside the plane. Line $\langle \mathbf{a}_1\mathbf{a}_2\mathbf{a}_3\mathbf{a}_4 \rangle_2$ is the intersection of planes $\mathcal{U}^3$ and $\langle \mathbf{A}_5 \rangle_3$.

Proof.   In $\Lambda(\mathcal{W}^4)$,

$$[\langle \mathbf{A}_5 \rangle_3\mathbf{a}_5] = [\mathbf{A}_5\mathbf{a}_5] = \mathbf{a}_5^2[\mathbf{a}_1\mathbf{a}_2\mathbf{a}_3\mathbf{a}_4] = 0,$$
$$[\langle \mathbf{A}_5 \rangle_3\mathbf{a}_4] = [\mathbf{A}_5\mathbf{a}_4] = [\mathbf{a}_1\mathbf{a}_2\mathbf{a}_3\{2(\mathbf{a}_4 \cdot \mathbf{a}_5)\mathbf{a}_4 - \mathbf{a}_4^2\mathbf{a}_5\}] = -\mathbf{a}_4^2[\mathbf{a}_1\mathbf{a}_2\mathbf{a}_3\mathbf{a}_5] \neq 0.$$

Similarly, $[\langle \mathbf{A}_5 \rangle_3\mathbf{a}_i] \neq 0$ for $i = 1, 2, 3$. $\qquad\square$

6.2.3   *m-tuple compression*

The m-tuple compression by intersections with lines for $m = 3, 4, 5$ can be extended to arbitrary integer m.

Lemma 6.57. For $m \geq 4$ and $\mathbf{A}_m = \mathbf{a}_1\mathbf{a}_2 \cdots \mathbf{a}_m$, assume that $\mathbf{a}_2, \ldots, \mathbf{a}_m$ span an $(m - 1)$D vector space $\mathcal{W}^{m-1}$ whose subspace $\langle \mathbf{A}_m \rangle_{m-2}$ is not null. Assume $\mathbf{a}_1 \in \mathcal{W}^{m-1}$. Then

(i) If $\mathbf{A}_m$ can be compressed to $\mathbf{b}_1\mathbf{b}_2\cdots\mathbf{b}_{m-2}$, and $\mathbf{a}_{m-1}\mathbf{a}_m\mathbf{b}_{m-2}$ is a 3-compressed vector, then $\mathbf{a}_{m-1}\mathbf{a}_m\mathbf{b}_{m-2}$ equals $(\mathbf{a}_1\wedge\cdots\wedge\mathbf{a}_{m-2})\vee(\mathbf{a}_{m-1}\wedge\mathbf{a}_m)$ up to scale.

(ii) For any permutation σ of indices $2,3,\ldots,m$, $(\mathbf{a}_{\sigma(m-1)}\wedge\mathbf{a}_{\sigma(m)})\vee(\mathbf{a}_1\wedge\mathbf{a}_{\sigma(2)}\wedge\cdots\wedge\mathbf{a}_{\sigma(m-2)})$ is a null vector if and only if so is $(\mathbf{a}_{\sigma(m-1)}\wedge\mathbf{a}_{\sigma(m)})\vee\langle\mathbf{A}_m\rangle_{m-2}$. The two points are identical if and only if they are both null.

(iii) Hyperplane $\langle\mathbf{A}_m\rangle_{m-2}$ meets one of the lines $\mathbf{a}_i\mathbf{a}_j$ for $2\le i<j\le m$ at an invertible point, and meets two of the lines $\mathbf{a}_i\mathbf{a}_j$ for $1\le i<j\le m$ at invertible points.

Proof. Much the same as the $m=5$ case. $\qquad\square$

The following proposition is a direct corollary.

Proposition 6.58. [m-tuple compression by intersections with lines] With the same hypothesis as Lemma 6.57, there is a permutation σ of indices $2,3,\ldots,m$, such that

(1) vector $(\mathbf{a}_{\sigma(m-1)}\wedge\mathbf{a}_{\sigma(m)})\vee(\mathbf{a}_1\wedge\mathbf{a}_{\sigma(2)}\wedge\cdots\wedge\mathbf{a}_{\sigma(m-2)})\in\mathcal{W}^{m-1}$ is invertible;

(2) by setting $\mathbf{a}_2\mathbf{a}_3\cdots\mathbf{a}_m=\mathbf{b}_1'\mathbf{b}_2'\cdots\mathbf{b}_{m-3}'\mathbf{a}_{\sigma(m-1)}\mathbf{a}_{\sigma(m)}$, there are the following 3-compressed vectors:

$$\mathbf{b}_{m-2}^{-1}=\mathbf{a}_{\sigma(m)}^{-1}\mathbf{a}_{\sigma(m-1)}^{-1}\{(\mathbf{a}_1\wedge\mathbf{b}_1'\wedge\mathbf{b}_2'\wedge\cdots\wedge\mathbf{b}_{m-3}')\vee(\mathbf{a}_{\sigma(m-1)}\wedge\mathbf{a}_{\sigma(m)})\},$$
$$\mathbf{b}_{m-2}'=\mathbf{a}_{\sigma(m-1)}\mathbf{a}_{\sigma(m)}\mathbf{b}_{m-2}^{-1};$$

(3) $\mathbf{A}_m=\mathbf{B}_{m-1}\mathbf{b}_{m-2}$, where $\mathbf{B}_{m-1}=\mathbf{a}_1\mathbf{b}_1'\mathbf{b}_2'\cdots\mathbf{b}_{m-2}'$ satisfies $\langle\mathbf{B}_{m-1}\rangle_{m-1}=0$, and may be further compressed by i-tuple compressions for i ranging from 3 to $m-1$.

(4) $\mathbf{b}_{m-2}$ is in subspace $\langle\mathbf{A}_m\rangle_{m-2}$, and $\mathbf{a}_1,\mathbf{b}_1',\mathbf{b}_2',\ldots\mathbf{b}_{m-2}'$ are outside the subspace. If $\langle\mathbf{B}_{m-1}\rangle_{m-3}\ne0$, then $\mathbf{a}_1,\mathbf{b}_1',\mathbf{b}_2',\ldots\mathbf{b}_{m-2}'$ span an $(m-2)$D subspace whose intersection with subspace $\langle\mathbf{A}_m\rangle_{m-2}$ is $\langle\mathbf{B}_{m-1}\rangle_{m-3}$.

Algorithm 6.59. Symbolic versor compression by intersections with lines.

Input: Versor $\mathbf{A}$, dimension and signature of the base space $\mathcal{V}^n$.

Output: The versor after compression.

Step 1. Do 3-tuple compressions to $\mathbf{A}$ repeatedly.

Step 2. If the length of $\mathbf{A}$ is less than three, return $\mathbf{A}$ and exit, else denote the first $m-1$ vectors by $\mathbf{A}_{m-1}$.

For m from 4 to the length of $\mathbf{A}$, do the following steps:

Step 3. Find all vectors $\mathbf{x}$ in $\mathbf{A}$ such that $\langle\mathbf{A}_{m-1}\mathbf{x}\rangle_m\ne0$. Denote the set of such vectors by S_m. If there is no such vector then set $m:=m+1$ and go back to the beginning of Step 3.

For any element $\mathbf{x}$ of S_m, do the following steps:

Step 4. Do m-tuple compression to $\mathbf{A}_{m-1}\mathbf{x}$. If it is changed into $\mathbf{B}_{m-1}\mathbf{b}_{m-2}$ by Proposition 6.58, do versor compression to $\mathbf{B}_{m-1}$.

Step 5. If the length of $\mathbf{B}_{m-1}$ is strictly reduced, then

    (1) move $\mathbf{x}$ to the m-th position in $\mathbf{A}$ by Proposition 6.35. Denote the subsequence from the $(m+1)$-th position to the end of $\mathbf{A}$ by $\mathbf{C}$.

    (2) Set $\mathbf{A} := \mathbf{B}_{m-1}\mathbf{b}_{m-2}\mathbf{C}$.

    (3) Go to Step 1.

The above algorithm for m-tuple compression is not always successful when $m > 5$. Proposition 6.37 for $m = 4$ and Proposition 6.47 for $m = 5$ cannot be extended to $m > 5$.

Proposition 6.60. Only when $\mathcal{V}^n$ is (anti-)Euclidean or (anti-)Minkowski can the above algorithm compress any versor to a rigid one.

By Proposition 6.58 and Theorem 6.65 to be introduced in the next section, for an odd non-rigid versor $\mathbf{A}_{2k+1}$ whose vectors span a $(2k)$D vector space, $\mathbf{A}_{2k+1}$ can be changed into another form $\mathbf{b}_1\mathbf{b}_2\cdots\mathbf{b}_{2k+1}$ where the first $2k$ vectors span a $(2k-1)$D subspace. For an even non-rigid versor $\mathbf{A}_{2k}$ whose vectors span a $(2k-1)$D vector space and whose subspace $\langle\mathbf{A}_{2k}\rangle_{2k-2}$ is not null, it can be changed into another form $\mathbf{b}_1\mathbf{b}_2\cdots\mathbf{b}_{2k}$ where the first $2k-2$ vectors span a $(2k-3)$D subspace.

Therefore, the obstruction of the compression of a non-rigid versor comes from the occurrence of a rotor $\mathbf{A}_{2k}$ such that $\langle\mathbf{A}_{2k}\rangle_{2k-2}$ is null. Such a rotor is called a *parabolic* or *hyperbolic* rotor. It is incompressible, and is the topic of the next section.

6.3 Obstructions to versor compression*

Clifford monomial compression is a transformation in Geometric Algebra. During the compression of a Clifford monomial, the underlying Grassmann structure of the monomial is invariant. In particular, the multivector representation of a parabolic or hyperbolic rotor in Grassmann algebra is invariant under the change of representative invertible vectors of the rotor. This invariant structure or representation is the key to solving the versor compression problem.

6.3.1 *Almost null space*

Definition 6.61. An nD inner-product space is said to be *almost null*, if it is not null but has a null $(n-1)$D subspace. It is said to be of *rank r*, if its inner product matrix has rank r.

The following lemma can be easily verified.

Lemma 6.62. An nD almost null space is either one of $\mathbb{R}^{1,0,n-1}, \mathbb{R}^{0,1,n-1}, \mathbb{C}^{1,n-1}$, which are of rank 1, or one of $\mathbb{R}^{1,1,n-2}, \mathbb{C}^{2,n-2}$, which are of rank 2. An nD almost null space has at most two null $(n-1)$D subspaces.

Proposition 6.63. Let $\mathbf{b}_1, \mathbf{b}_2 \ldots, \mathbf{b}_{k+1}$ be nonzero vectors spanning a kD vector space. Let $\langle \mathbf{b}_1 \mathbf{b}_2 \cdots \mathbf{b}_{k+1} \rangle_{k-1} = 0$. Then

(1) at least one of the $k+1$ vectors is null. In particular, if every k of the $k+1$ vectors $\mathbf{b}_i$ are linearly independent, then all the $k+1$ vectors are null.
(2) If $\mathbf{b}_2 \wedge \mathbf{b}_3 \wedge \cdots \wedge \mathbf{b}_{k+1} = 0$, then $\langle \mathbf{b}_2 \mathbf{b}_3 \cdots \mathbf{b}_{k+1} \rangle_{k-2} = 0$.

Proof. (1) Without loss of generality, assume that $\mathbf{b}_1, \mathbf{b}_2, \ldots, \mathbf{b}_k$ are linearly independent. In $\Lambda(\mathbf{b}_1 \wedge \mathbf{b}_2 \wedge \cdots \wedge \mathbf{b}_k)$, denote $[\check{\mathbf{b}}_i] = [\mathbf{b}_1 \mathbf{b}_2 \cdots \check{\mathbf{b}}_i \cdots \mathbf{b}_{k+1}]$, where $1 \le i \le k$.

Using Cramer's rule

$$[\mathbf{b}_1 \mathbf{b}_2 \cdots \mathbf{b}_k] \mathbf{b}_{k+1} = \sum_{i=1}^{k} (-1)^{k+i} [\check{\mathbf{b}}_i] \mathbf{b}_i \tag{6.3.1}$$

to eliminate $\mathbf{b}_{k+1}$ from versor $\mathbf{b}_1 \mathbf{b}_2 \cdots \mathbf{b}_{k+1}$, then using (6.1.38) to expand the $(k-1)$-graded part of the versor, we get

$$\begin{aligned}
&\langle \mathbf{b}_1 \mathbf{b}_2 \cdots \mathbf{b}_{k+1} \rangle_{k-1} \\
&= \sum_{i=1}^{k} (-1)^{k+i} [\check{\mathbf{b}}_i] \langle \mathbf{b}_1 \mathbf{b}_2 \cdots \mathbf{b}_k \mathbf{b}_i \rangle_{k-1} \\
&= \sum_{j=1}^{k} \mathbf{b}_1 \wedge \mathbf{b}_2 \wedge \cdots \wedge \check{\mathbf{b}}_i \wedge \cdots \wedge \mathbf{b}_k \left\{ \mathbf{b}_j^2 [\check{\mathbf{b}}_j] + \sum_{i=1}^{j-1} 2(-1)^{i+j} (\mathbf{b}_i \cdot \mathbf{b}_j)[\check{\mathbf{b}}_i] \right\}.
\end{aligned}$$

By $\langle \mathbf{b}_1 \mathbf{b}_2 \cdots \mathbf{b}_{k+1} \rangle_{k-1} = 0$ and the linear independence among the first k vectors, we get, for any $1 \le j \le k$,

$$\mathbf{b}_j^2 [\check{\mathbf{b}}_j] + \sum_{i=1}^{j-1} 2(-1)^{i+j} (\mathbf{b}_i \cdot \mathbf{b}_j)[\check{\mathbf{b}}_i] = 0. \tag{6.3.2}$$

If none of the $k+1$ vectors $\mathbf{b}_i$ is null, then when $j = 1$, (6.3.2) becomes $[\check{\mathbf{b}}_1] = 0$; when $j = 2$, (6.3.2) becomes $[\check{\mathbf{b}}_2] = 0$. Continuing this way to $j = k$, we get $[\check{\mathbf{b}}_j] = 0$ for all $1 \le j \le k$. The right side of Cramer's rule (6.3.1) becomes zero, however, the left side is nonzero. Contradiction.

Furthermore, if $[\check{\mathbf{b}}_j] \ne 0$ for any $1 \le j \le k$, then from (6.3.1) we get (6.3.2), and by setting $j = 1$, we get $\mathbf{b}_1^2 = 0$. By symmetry, for any $1 \le i \le k+1$, we can start with the Cramer's rule of $\mathbf{b}_i$ with respect to the other k vectors, first obtain the corresponding equation (6.3.2), and then get $\mathbf{b}_i^2 = 0$.

(2) Since $\mathbf{b}_1$ to $\mathbf{b}_{k+1}$ span a kD space, while $\mathbf{b}_2$ to $\mathbf{b}_{k+1}$ span a $(k-1)$D subspace, it must be that $\mathbf{b}_1$ is linearly independent of the other k vectors. By

$$\begin{aligned}
\langle \mathbf{b}_1 \mathbf{b}_2 \cdots \mathbf{b}_{k+1} \rangle_{k-1} &= \mathbf{b}_1 \wedge \langle \mathbf{b}_2 \mathbf{b}_3 \cdots \mathbf{b}_{k+1} \rangle_{k-2} + \mathbf{b}_1 \cdot \langle \mathbf{b}_2 \mathbf{b}_3 \cdots \mathbf{b}_{k+1} \rangle_k \\
&= \mathbf{b}_1 \wedge \langle \mathbf{b}_2 \mathbf{b}_3 \cdots \mathbf{b}_{k+1} \rangle_{k-2},
\end{aligned}$$

we get that $\langle \mathbf{b}_1 \mathbf{b}_2 \cdots \mathbf{b}_{k+1} \rangle_{k-1} = 0$ if and only if $\langle \mathbf{b}_2 \mathbf{b}_3 \cdots \mathbf{b}_{k+1} \rangle_{k-2} = 0$. $\square$

Definition 6.64. Let $\mathbf{A}_{2k} = \mathbf{a}_1 \mathbf{a}_2 \cdots \mathbf{a}_{2k}$, where the $2k$ vectors span a $(2k-1)$D vector space $\mathcal{W}^{2k-1}$, and whose subspace $\langle \mathbf{A}_{2k} \rangle_{2k-2}$ is null. If $\mathcal{W}^{2k-1}$ is of rank 1, then $\mathbf{A}_{2k}$ is called a *parabolic rotor*; if $\mathcal{W}^{2k-1}$ is of rank 2, then $\mathbf{A}_{2k}$ is called a *hyperbolic rotor*.

Theorem 6.65. Let $\mathbf{A}_m = \mathbf{a}_1 \mathbf{a}_2 \cdots \mathbf{a}_m$, where the m vectors span an $(m-1)$D vector space $\mathcal{W}^{m-1}$. If m is odd, then subspace $\langle \mathbf{A}_m \rangle_{m-2}$ is not null.

Proof. Assume that $\langle \mathbf{A}_m \rangle_{m-2}$ is null. Then $\mathcal{W}^{m-1}$ is almost null. There are two cases:

Case 1. $\mathcal{W}^{m-1}$ is of rank 1. We only consider the case $\mathcal{W}^{m-1} = \mathbb{R}^{1,0,m-2}$, as the others are similar.

Let $\mathbf{e}_1, \ldots, \mathbf{e}_{m-1}$ be an orthonormal basis of $\mathbb{R}^{1,0,m-2}$ such that $\mathbf{e}_1^2 = 1$. The coordinate component of $\mathbf{a}_i$ with respect to $\mathbf{e}_1$ is nonzero. Let $\mathbf{a}_i = \mathbf{e}_1 + \lambda_{2i} \mathbf{e}_2 + \lambda_{3i} \mathbf{e}_3 + \cdots + \lambda_{(m-1)i} \mathbf{e}_{m-1}$ for $1 \le i \le m$. Then null blade $\langle \mathbf{A}_m \rangle_{m-2}$ must be identical to $\mathbf{e}_2 \mathbf{e}_3 \cdots \mathbf{e}_{m-1}$, so

$$[(\mathbf{a}_i - \mathbf{a}_j) \langle \mathbf{A}_m \rangle_{m-2}] = 0, \quad \forall 1 \le i < j \le m. \tag{6.3.3}$$

Denote $[\breve{\mathbf{a}}_i] = [\mathbf{a}_1 \cdots \breve{\mathbf{a}}_i \cdots \mathbf{a}_m]$. By (5.4.2) and (5.1.62),

$$
\begin{aligned}
[\mathbf{a}_i \langle \mathbf{A}_m \rangle_{m-2}] &= \sum_{j<i} (-1)^{i+j+1} (\mathbf{a}_i \cdot \mathbf{a}_j) [\mathbf{a}_i (\mathbf{a}_1 \cdots \breve{\mathbf{a}}_j \cdots \breve{\mathbf{a}}_i \cdots \mathbf{a}_m)] \\
&\quad + \sum_{j>i} (-1)^{i+j+1} (\mathbf{a}_i \cdot \mathbf{a}_j) [\mathbf{a}_i (\mathbf{a}_1 \cdots \breve{\mathbf{a}}_i \cdots \breve{\mathbf{a}}_j \cdots \mathbf{a}_m)] \\
&= \left(\sum_{j<i} - \sum_{j>i} \right) (-1)^{j+1} (\mathbf{a}_i \cdot \mathbf{a}_j) [\breve{\mathbf{a}}_j] \\
&= (-1)^i \mathbf{a}_i^2 [\breve{\mathbf{a}}_i] + 2 \sum_{j>i} (-1)^j (\mathbf{a}_i \cdot \mathbf{a}_j) [\breve{\mathbf{a}}_j] \\
&= (-1)^{i+1} \mathbf{a}_i^2 [\breve{\mathbf{a}}_i] + 2 \sum_{j<i} (-1)^{j+1} (\mathbf{a}_i \cdot \mathbf{a}_j) [\breve{\mathbf{a}}_j].
\end{aligned}
\tag{6.3.4}
$$

On one hand, by setting $i = 1, 2$ in the last line of (6.3.4) and using (6.3.3), we get $[\breve{\mathbf{a}}_2] = [\breve{\mathbf{a}}_1]$. Substituting it into the last line of (6.3.4) and increasing the index i, we always get $[\breve{\mathbf{a}}_i] = [\breve{\mathbf{a}}_1]$ for all $2 \le i \le m$. On the other hand, by setting $i = m, 1$ in the last two lines of (6.3.4) respectively, we get $[\breve{\mathbf{a}}_m] = (-1)^m [\breve{\mathbf{a}}_1]$. So m must be even. The Cramer's rule of the m vectors, *i.e.*,

$$\sum_{i=1}^{m} (-1)^{i+1} [\breve{\mathbf{a}}_i] \mathbf{a}_i = 0, \tag{6.3.5}$$

becomes

$$\sum_{i=1}^{m} (-1)^{i+1} \mathbf{a}_i = 0. \tag{6.3.6}$$

Case 2. $\mathcal{W}^{m-1}$ is of rank 2. Again we only consider the real case $\mathbb{R}^{1,1,m-3}$.

Let $\mathbf{e}_1, \ldots, \mathbf{e}_{m-1}$ be a Witt basis of $\mathbb{R}^{1,1,m-3}$ such that $(\mathbf{e}_1, \mathbf{e}_2)$ is a Witt pair. The coordinate components of $\mathbf{a}_i$ with respect to $\mathbf{e}_1, \mathbf{e}_2$ are both nonzero. Let $\mathbf{a}_i = \mathbf{e}_1 + \lambda_i \mathbf{e}_2 + \mu_{3i} \mathbf{e}_3 + \cdots + \mu_{(m-1)i} \mathbf{e}_{m-1}$ for $1 \le i \le m$. Then

$$\mathbf{a}_i^{-1} = \frac{\mathbf{a}_i}{\mathbf{a}_i \cdot \mathbf{a}_i} = -\frac{1}{2}\mathbf{e}_2 - \frac{1}{2\lambda_i}(\mathbf{e}_1 + \mu_{3i}\mathbf{e}_3 + \cdots + \mu_{(m-1)i}\mathbf{e}_{m-1}). \qquad (6.3.7)$$

So for any $1 \le i < j \le m$,

$$\begin{aligned}
(\mathbf{a}_i - \mathbf{a}_j) \wedge (\mathbf{e}_2 \wedge \mathbf{e}_3 \wedge \cdots \wedge \mathbf{e}_{m-1}) &= 0, \\
(\mathbf{a}_i^{-1} - \mathbf{a}_j^{-1}) \wedge (\mathbf{e}_1 \wedge \mathbf{e}_3 \wedge \cdots \wedge \mathbf{e}_{m-1}) &= 0.
\end{aligned} \qquad (6.3.8)$$

Null blade $\langle \mathbf{A}_m \rangle_{m-2}$ must be identical to one of $\mathbf{e}_2 \wedge \mathbf{e}_3 \wedge \cdots \wedge \mathbf{e}_{m-1}$ and $\mathbf{e}_1 \wedge \mathbf{e}_3 \wedge \cdots \wedge \mathbf{e}_{m-1}$ up to scale. There are two cases:

Case 2a. $[(\mathbf{a}_i - \mathbf{a}_j)\langle \mathbf{a}_1 \mathbf{a}_2 \cdots \mathbf{a}_m \rangle_{m-2}] = 0$, for any $1 \le i < j \le m$.

Case 2b. $[(\mathbf{a}_i^{-1} - \mathbf{a}_j^{-1})\langle \mathbf{a}_1^{-1} \mathbf{a}_2^{-1} \cdots \mathbf{a}_m^{-1} \rangle_{m-2}] = 0$, for any $1 \le i < j \le m$.

We only check Case 2a. Substituting the last line of (6.3.4) into (6.3.3), we get, for $1 \le i \le \left[\frac{m+1}{2}\right]$,

$$[\breve{\mathbf{a}}_{2i}] = [\breve{\mathbf{a}}_{2i-1}] = \prod_{j=1}^{i-1} \lambda_{2j} \lambda_{2j+1}^{-1} [\breve{\mathbf{a}}_1] \ne 0. \qquad (6.3.9)$$

Substituting (6.3.9) into (6.3.5), we see that if m is odd, then the $\mathbf{e}_1$-component of the left side of (6.3.5) is $\sum_{i=1}^{m}(-1)^{i+1}[\breve{\mathbf{a}}_i] = [\breve{\mathbf{a}}_m] = 0$. However, by setting $i = m, 1$ in the last two lines of (6.3.4) respectively, we get $(-1)^m \mathbf{a}_m^2 [\breve{\mathbf{a}}_m] = (-1)^2 \mathbf{a}_1^2 [\breve{\mathbf{a}}_1]$, *i.e.*, $[\breve{\mathbf{a}}_m] = -[\breve{\mathbf{a}}_1]\lambda_1/\lambda_m \ne 0$. Contradiction. $\qquad \square$

Corollary 6.66. If $\mathbf{A}_{2k}$ is parabolic or hyperbolic, then any $2k - 1$ vectors in $\mathbf{A}_{2k}$ are linearly independent.

6.3.2 *Parabolic rotors*

Parabolic rotors are much simpler in structure than hyperbolic rotors. The following theorem characterizes the Grassmann structure of a parabolic rotor.

Theorem 6.67. The following statements are equivalent:

(1) $\mathbf{A}_{2k} = \mathbf{a}_1 \mathbf{a}_2 \cdots \mathbf{a}_{2k}$ is parabolic.
(2) For $\mathbf{A}_{2k-1} = \mathbf{a}_1 \mathbf{a}_2 \cdots \mathbf{a}_{2k-1}$, rotor $\mathbf{a}_{2k} \mathbf{A}_{2k-1}$ is parabolic.
(3) $\langle \mathbf{A}_{2k} \rangle \mathbf{a}_{2k}^{-1} = \langle \mathbf{A}_{2k-1} \rangle_1$, *i.e.*, $\langle \mathbf{A}_{2k-1} \rangle_1 \wedge \mathbf{a}_{2k} = 0$.
(4) $\mathbf{A}_{2k-1} \wedge \mathbf{a}_{2k} = 0$.
(5) $\mathbf{A}_{2k} = \mathbf{a}_{2k} \mathbf{A}_{2k-1}$.
(6) $\mathbf{A}_{2k} = \mathbf{A}_{2k-1} \cdot \mathbf{a}_{2k}$.
(7) $\mathbf{A}_{2k-1} = \mathbf{A}_{2k} \wedge \mathbf{a}_{2k}^{-1}$.

Proof. We only prove the theorem under the assumption that the vector space $\mathcal{W}^{2k-1}$ spanned by the $2k$ vectors $\mathbf{a}_i$ has rank 1. After the properties of hyperbolic versors are established in the next subsection, the validity of the theorem in any almost null space $\mathcal{W}^{2k-1}$ will be a direct corollary.

We use the same notation as in the proof of Theorem 6.65, and take $\mathcal{W}^{2k-1}$ as $\mathbb{R}^{1,0,2k-2}$. Subspace $\langle \mathbf{A}_{2k} \rangle_{2k-2}$ is null if and only if (6.3.6) holds. So the equivalence between statements (1) and (2) is obvious. Statements (4) to (7) are obviously equivalent to each other. Below we only prove the necessity of the conditions (3) and (4) for $\mathbf{A}_{2k}$ to be parabolic; the sufficiency of each condition is easy to prove.

In any $\langle \mathbf{a}_{j_1} \mathbf{a}_{j_2} \cdots \mathbf{a}_{j_{2l}} \rangle$, all nonzero contributions come from the $\mathbf{e}_1$-components of the vectors, so

$$\langle \mathbf{a}_{j_1} \mathbf{a}_{j_2} \cdots \mathbf{a}_{j_{2l}} \rangle = \langle \underbrace{\mathbf{e}_1 \mathbf{e}_1 \cdots \mathbf{e}_1}_{2l} \rangle = 1. \tag{6.3.10}$$

Then statement (3) is just (6.3.6).

By (6.1.38) and (6.3.10), for any $1 \le i \le 2k - 1$,

$$\mathbf{a}_i \mathbf{a}_{i+1} \cdots \mathbf{a}_{2k-1} \mathbf{a}_i = (-1)^{i+1} \mathbf{a}_{i+1} \mathbf{a}_{i+2} \cdots \mathbf{a}_{2k-1} + 2 \sum_{j=i+1}^{2k-1} (-1)^{j+1} \mathbf{a}_i \cdots \breve{\mathbf{a}}_j \cdots \mathbf{a}_{2k-1},$$

$$\mathbf{a}_i \mathbf{a}_1 \mathbf{a}_2 \cdots \mathbf{a}_i = (-1)^{i+1} \mathbf{a}_1 \mathbf{a}_2 \cdots \mathbf{a}_{i-1} + 2 \sum_{j=1}^{i-1} (-1)^{j+1} \mathbf{a}_1 \cdots \breve{\mathbf{a}}_j \cdots \mathbf{a}_i.$$

By these formulas and (6.3.6), statement (4) can be derived as follows:

$$\begin{aligned} 2(\mathbf{a}_1 \cdots \mathbf{a}_{2k-1}) \wedge \mathbf{a}_{2k} &= \mathbf{a}_1 \cdots \mathbf{a}_{2k-1} \mathbf{a}_{2k} - \mathbf{a}_{2k} \mathbf{a}_1 \cdots \mathbf{a}_{2k-1} \\ &= \sum_{i=1}^{2k-1} (-1)^{i+1} (\mathbf{a}_1 \mathbf{a}_2 \cdots \mathbf{a}_{2k-1} \mathbf{a}_i - \mathbf{a}_i \mathbf{a}_1 \cdots \mathbf{a}_{2k-1}) \\ &= 2(\sum_{j>i} - \sum_{j<i})(-1)^{i+j} \mathbf{a}_1 \cdots \breve{\mathbf{a}}_j \cdots \mathbf{a}_{2k-1} \\ &= 0. \end{aligned}$$

$\square$

In a general vector space $\mathcal{V}^n$, let $\mathbf{e}_1, \mathbf{e}_2, \ldots, \mathbf{e}_n$ be a fixed basis. For any $1 \le i \le n$, the $\mathbf{e}_i$-*component* of a multivector $\mathbf{B} \in \Lambda(\mathcal{V}^n)$ refers to the component of $\mathbf{B}$ in the subspace $\mathbf{e}_i \wedge \Lambda(\mathcal{V}^n)$. $\mathbf{B}$ has no $\mathbf{e}_i$-component if and only if $\mathbf{e}_i \wedge (\mathbf{B} - \langle \mathbf{B} \rangle) = 0$.

Denote $\mathbf{I}_n = \mathbf{e}_1 \wedge \mathbf{e}_2 \wedge \cdots \wedge \mathbf{e}_n$. Then any multivector $\mathbf{B} \in \Lambda(\mathbf{I}_n)$ is of the form

$$\mathbf{B} = \sum_{\vdash \mathbf{I}_n} \lambda_{\mathbf{I}_{n(1)}} \mathbf{I}_{n(1)}, \tag{6.3.11}$$

where $\lambda_{\mathbf{I}_{n(1)}} \in \mathbb{K}$. The $\mathbf{e}_i$-component of $\mathbf{B}$ is just

$$\sum_{\vdash \mathbf{I}_n, \, \mathbf{e}_i \in \mathbf{I}_{n(1)}} \lambda_{\mathbf{I}_{n(1)}} \mathbf{I}_{n(1)}. \tag{6.3.12}$$

Corollary 6.68. If $\mathbf{A}_{2k}$ is a parabolic rotor, then $\mathbf{A}_{2k} \in \Lambda(\langle \mathbf{A}_{2k} \rangle_{2k-2})$.

Proof. Let $\mathcal{W}^{2k-1}$ be the $(2k-1)$D space spanned by the $2k$ vectors $\mathbf{a}_1$ to $\mathbf{a}_{2k}$ in $\mathbf{A}_{2k}$. A multivector $\mathbf{B} \in \Lambda(\mathcal{W}^{2k-1})$ is in $\Lambda(\langle\mathbf{A}_{2k}\rangle_{2k-2})$ if and only if it does not have any $\mathbf{e}_1$-component, *i.e.*, if and only if $(\mathbf{B} - \langle\mathbf{B}\rangle)\cdot\mathbf{a}_{2k} = 0$. By statement (4) of Theorem 6.67, for any $1 \le i \le k-1$,

$$\langle\mathbf{A}_{2k}\rangle_{2k-2i}\cdot\mathbf{a}_{2k} = (\langle\mathbf{A}_{2k-1}\rangle_{2k-2i+1}\cdot\mathbf{a}_{2k})\cdot\mathbf{a}_{2k} = \langle\mathbf{A}_{2k-1}\rangle_{2k-2i+1}\cdot(\mathbf{a}_{2k}\wedge\mathbf{a}_{2k}) = 0,$$

so $\mathbf{A}_{2k} \in \Lambda(\langle\mathbf{A}_{2k}\rangle_{2k-2})$. $\qquad\square$

Consider an odd non-rigid versor $\mathbf{V}$ whose compression is obstructed. By Proposition 6.58, there exist invertible vectors $\mathbf{a}_1, \mathbf{a}_2, \ldots, \mathbf{a}_{2k+1}$ such that $\mathbf{A}_{2k} = \mathbf{a}_1\mathbf{a}_2\cdots\mathbf{a}_{2k}$ is a parabolic or hyperbolic rotor, and $\mathbf{V} = \mathbf{A}_{2k}\mathbf{a}_{2k+1}$. If $\mathbf{V}$ itself is compressible, say $\mathbf{V} = \mathbf{b}_1\mathbf{b}_2\cdots\mathbf{b}_{2k-1}$, then by denoting $\mathbf{b}_{2k} = \mathbf{a}_{2k+1}^{-1}$, we have

$$\mathbf{A}_{2k} = \mathbf{b}_1\mathbf{b}_2\cdots\mathbf{b}_{2k} = \mathbf{a}_1\mathbf{a}_2\cdots\mathbf{a}_{2k}. \tag{6.3.13}$$

So if $\mathbf{V}$ is compressible, then $\mathbf{A}_{2k}$ has two equal representations, either by vectors $\mathbf{a}_1$ to $\mathbf{a}_{2k}$, or by vectors $\mathbf{b}_1$ to $\mathbf{b}_{2k}$. The $2k$ vectors $\mathbf{a}_i$ span a $(2k-1)$D subspace $\mathcal{W}^{2k-1}$, while the $2k$ vectors $\mathbf{b}_j$ span another $(2k-1)$D subspace $\langle\mathbf{A}_{2k}\mathbf{b}_{2k}^{-1}\rangle_{2k-1}$. The intersection of the two subspaces is the null $(2k-2)$D subspace $\langle\mathbf{A}_{2k}\rangle_{2k-2}$. Figure 6.2 shows the $k = 2$ case.

The compressibility of $\mathbf{A}_{2k}\mathbf{b}_{2k}^{-1}$ is equivalent to the existence and constructibility of another form $\mathbf{b}_1\mathbf{b}_2\cdots\mathbf{b}_{2k}$ of $\mathbf{A}_{2k}$, when given an initial invertible vector $\mathbf{b}_{2k}$ outside the subspace spanned by the $2k$ vectors $\mathbf{a}_i$ in $\mathbf{A}_{2k}$.

Proposition 6.69. In $\mathcal{G}(\mathcal{V}^n)$, if versor $\mathbf{A}_{2k} = \mathbf{a}_1\mathbf{a}_2\cdots\mathbf{a}_{2k}$ is parabolic, vector $\mathbf{b}_{2k}$ has the same signature with vector $\mathbf{a}_{2k}$, and $\mathbf{b}_{2k}\cdot\langle\mathbf{A}_{2k}\rangle_{2k-2} = 0$, then versor $\mathbf{A}_{2k}\mathbf{b}_{2k}$ is compressible.

Proof. Since subspace $\langle\mathbf{A}_{2k}\rangle_{2k-2}$ is null, and $\langle\mathbf{A}_{2k}\mathbf{b}_{2k}\rangle_{2k-1} = \langle\mathbf{A}_{2k}\rangle_{2k-2}\wedge\mathbf{b}_{2k}$, subspace $\langle\mathbf{A}_{2k}\mathbf{b}_{2k}\rangle_{2k-1}$ must have rank 1. Since both $\mathbf{a}_{2k}$ and $\mathbf{b}_{2k}$ are orthogonal to subspace $\langle\mathbf{A}_{2k}\rangle_{2k-2}$, and $\mathbf{a}_{2k}^2 = \mathbf{b}_{2k}^2$ by rescaling, there exists an orthogonal transformation g of $\mathcal{V}^n$ that changes $\mathbf{a}_{2k}$ to $\mathbf{b}_{2k}$ while fixing every vector in subspace $\langle\mathbf{A}_{2k}\rangle_{2k-2}$. Under this transformation, multivector $\mathbf{A}_{2k} \in \Lambda(\langle\mathbf{A}_{2k}\rangle_{2k-2})$ is invariant, so

$$\mathbf{A}_{2k} = g(\mathbf{A}_{2k}) = g(\mathbf{a}_1)g(\mathbf{a}_2)\cdots g(\mathbf{a}_{2k}) = g(\mathbf{a}_1)g(\mathbf{a}_2)\cdots g(\mathbf{a}_{2k-1})\mathbf{b}_{2k}. \tag{6.3.14}$$

$\square$

In fact, the Grassmann structure of a parabolic rotor $\mathbf{A}_{2k}$ can be computed explicitly, so is the equivalent representation (6.3.14) of the rotor.

Lemma 6.70. In an almost null space of rank 1, let vector $\mathbf{a}$ be invertible, and let vectors $\mathbf{d}_1, \ldots, \mathbf{d}_r$ be null and orthogonal to $\mathbf{a}$. Denote $\mathbf{D}_r = \mathbf{d}_1\mathbf{d}_2\cdots\mathbf{d}_r$. Then

$$(\mathbf{a}+\mathbf{d}_1)(\mathbf{a}+\mathbf{d}_2)\cdots(\mathbf{a}+\mathbf{d}_r) = \sum_{i=0}^{r}\mathbf{a}^{r-i}\sum_{(r-i,i)\vdash\mathbf{D}_r}\mathbf{D}_{r(2)}. \tag{6.3.15}$$

Proof. Induction on r. When $r = 1$, the conclusion is trivial. Assume that the conclusion holds for r. Then for $r + 1$,

$$(\mathbf{a} + \mathbf{d}_1)(\mathbf{a} + \mathbf{d}_2) \cdots (\mathbf{a} + \mathbf{d}_r)(\mathbf{a} + \mathbf{d}_{r+1})$$

$$= \sum_{i=0}^{r} \mathbf{a}^{r-i} \sum_{(r-i,i) \vdash \mathbf{D}_r} \mathbf{D}_{r\,(2)}(\mathbf{a} + \mathbf{d}_{r+1})$$

$$= \sum_{i=0}^{r} (-1)^i \mathbf{a}^{r-i+1} \sum_{(r-i,i) \vdash \mathbf{D}_r} \mathbf{D}_{r\,(2)} + \sum_{i=0}^{r} \mathbf{a}^{r-i} \sum_{(r-i,i) \vdash \mathbf{D}_r} \mathbf{D}_{r\,(2)} \mathbf{d}_{r+1}$$

$$= \sum_{i=0}^{r} \mathbf{a}^{r-i+1} \sum_{\substack{(r+1-i,i) \vdash \mathbf{D}_{r+1}, \\ \mathbf{d}_{r+1} \in \mathbf{D}_{r+1\,(1)}}} \mathbf{D}_{r+1\,(2)} + \sum_{i=0}^{r} \mathbf{a}^{r-i} \sum_{\substack{(r-i,i+1) \vdash \mathbf{D}_{r+1}, \\ \mathbf{d}_{r+1} \in \mathbf{D}_{r+1\,(2)}}} \mathbf{D}_{r+1\,(2)}$$

$$= \sum_{i=0}^{r} \mathbf{a}^{r+1-i} \sum_{\substack{(r+1-i,i) \vdash \mathbf{D}_{r+1}, \\ \mathbf{d}_{r+1} \in \mathbf{D}_{r+1\,(1)}}} \mathbf{D}_{r+1\,(2)} + \sum_{i=1}^{r+1} \mathbf{a}^{r+1-i} \sum_{\substack{(r+1-i,i) \vdash \mathbf{D}_{r+1}, \\ \mathbf{d}_{r+1} \in \mathbf{D}_{r+1\,(2)}}} \mathbf{D}_{r+1\,(2)}$$

$$= \sum_{i=0}^{r+1} \mathbf{a}^{r+1-i} \sum_{(r+1-i,i) \vdash \mathbf{D}_{r+1}} \mathbf{D}_{r+1\,(2)}.$$

$\square$

Corollary 6.71. In parabolic rotor $\mathbf{A}_{2k} \in \mathcal{G}(\mathcal{V}^n)$, let $\mathbf{a}_i = \mathbf{a}_{2k} + \mathbf{d}_i$ for $1 \leq i \leq 2k - 1$, where the $\mathbf{d}_i$ are null and orthogonal to $\mathbf{a}_{2k}$. Let $\mathbf{D}_{2k-1} = \mathbf{d}_1 \mathbf{d}_2 \cdots \mathbf{d}_{2k-1}$. Then

$$\mathbf{A}_{2k} = \sum_{i=0}^{k-1} (\mathbf{a}_{2k}^2)^{k-i} \sum_{(2i,2k-2i-1) \vdash \mathbf{D}_{2k-1}} \mathbf{D}_{2k-1\,(1)}. \tag{6.3.16}$$

Furthermore, if vector $\mathbf{b}_{2k} \in \mathcal{V}^n$ satisfies $\mathbf{b}_{2k}^2 = \mathbf{a}_{2k}^2$ and $\mathbf{b}_{2k} \cdot \langle \mathbf{A}_{2k} \rangle_{2k-2} = 0$, then

$$\mathbf{A}_{2k} = (\mathbf{b}_{2k} + \mathbf{d}_1)(\mathbf{b}_{2k} + \mathbf{d}_2) \cdots (\mathbf{b}_{2k} + \mathbf{d}_{2k-1})\mathbf{b}_{2k}. \tag{6.3.17}$$

6.3.3 *Hyperbolic rotors*

Consider an almost null space $\mathcal{W}^{2k-1}$ of rank 2. Fix a Witt basis $\mathbf{e}_1, \mathbf{e}_2, \ldots, \mathbf{e}_{2k-1}$ of $\mathcal{W}^{2k-1}$, where $(\mathbf{e}_1, \mathbf{e}_2)$ is a Witt pair. For invertible vectors $\mathbf{a}_i$, assume that their coordinates are

$$\mathbf{a}_i = \mathbf{e}_1 + \lambda_i \mathbf{e}_2 + \mu_{3i} \mathbf{e}_3 + \cdots + \mu_{(2k-1)i} \mathbf{e}_{2k-1}. \tag{6.3.18}$$

Then the inner product $\mathbf{a}_i \cdot (\mathbf{a}_{j_1} - \mathbf{a}_{j_2}) = \lambda_{j_2} - \lambda_{j_1}$ is independent of the index i.

A hyperbolic rotor $\mathbf{A}_{2k}$ whose $(2k - 2)$-graded part has no $\mathbf{e}_1$-component (or no $\mathbf{e}_2$-component), is said to be $\mathbf{e}_2$-*typed* (or $\mathbf{e}_1$-*typed*). The following lemma can be easily verified.

Lemma 6.72. Let $\mathbf{a}_1, \mathbf{a}_2, \ldots, \mathbf{a}_{2k-1}$ be linearly independent invertible vectors of the form (6.3.18). Then $\{\mathbf{a}_{2i} - \mathbf{a}_{2i-1}, \mathbf{a}_{2i} - \mathbf{a}_{2i+1} \mid 1 \leq i \leq k - 1\}$ is a basis of the null

$(2k-2)$D subspace spanned by $\mathbf{e}_2, \mathbf{e}_3, \ldots, \mathbf{e}_{2k-1}$, and $\{\mathbf{a}_{2i}^{-1} - \mathbf{a}_{2i-1}^{-1}, \mathbf{a}_{2i}^{-1} - \mathbf{a}_{2i+1}^{-1} \mid 1 \le i \le k-1\}$ is a basis of the null $(2k-2)$D subspace spanned by $\mathbf{e}_1, \mathbf{e}_3, \ldots, \mathbf{e}_{2k-1}$.

Lemma 6.73. For any invertible vectors $\mathbf{a}_i \in \mathcal{W}^{2k-1}$,

$$\langle \mathbf{a}_1 \mathbf{a}_2 \cdots \mathbf{a}_{2l} \rangle = -(-2)^{l-1}(\lambda_1 \lambda_3 \cdots \lambda_{2l-1} + \lambda_2 \lambda_4 \cdots \lambda_{2l}), \qquad (6.3.19)$$

$$\langle \mathbf{e}_1 \mathbf{a}_1 \mathbf{a}_2 \cdots \mathbf{a}_{2l-1} \rangle = -(-2)^{l-1} \lambda_1 \lambda_3 \cdots \lambda_{2l-1}, \qquad (6.3.20)$$

$$\langle \mathbf{e}_2 \mathbf{a}_1 \mathbf{a}_2 \cdots \mathbf{a}_{2l-1} \rangle = -(-2)^{l-1} \lambda_2 \lambda_4 \cdots \lambda_{2l-2}. \qquad (6.3.21)$$

Proof. We only prove (6.3.19). The other two equalities can be proved similarly.

Induction on l. When $l = 1$, (6.3.19) is obvious. Assume that the equality holds for $l - 1$. Then for l,

$$\langle \mathbf{a}_1 \mathbf{a}_2 \cdots \mathbf{a}_{2l} \rangle = \sum_{i=2}^{2l} (-1)^i (\mathbf{a}_1 \cdot \mathbf{a}_i) \langle \mathbf{a}_2 \mathbf{a}_3 \cdots \check{\mathbf{a}}_i \cdots \mathbf{a}_{2l} \rangle$$

$$= (-2)^{l-2} \Big\{ \sum_{i=1}^{l} (\lambda_1 + \lambda_{2i})((\lambda_3 \lambda_5 \cdots \lambda_{2i-1})(\lambda_{2i+2} \lambda_{2i+4} \cdots \lambda_{2l})$$

$$+ (\lambda_2 \lambda_4 \cdots \lambda_{2i-2})(\lambda_{2i+1} \lambda_{2i+3} \cdots \lambda_{2l-1}))$$

$$- \sum_{i=1}^{l-1} (\lambda_1 + \lambda_{2i+1})((\lambda_3 \lambda_5 \cdots \lambda_{2i-1})(\lambda_{2i+2} \lambda_{2i+4} \cdots \lambda_{2l})$$

$$+ (\lambda_2 \lambda_4 \cdots \lambda_{2i})(\lambda_{2i+3} \lambda_{2i+5} \cdots \lambda_{2l-1})) \Big\}$$

$$= -(-2)^{l-1}(\lambda_1 \lambda_3 \cdots \lambda_{2l-1} + \lambda_2 \lambda_4 \cdots \lambda_{2l}).$$

$\square$

Definition 6.74. For a hyperbolic rotor $\mathbf{A}_{2k} = \mathbf{a}_1 \mathbf{a}_2 \cdots \mathbf{a}_{2k}$ whose vectors span a $(2k-1)$D vector space $\mathcal{W}^{2k-1}$, its $\mathbf{e}_i$-*typed characteristic vector*, denoted by $\mathbf{c}_i$, where $i = 1, 2$, is defined as a vector in the radical $\mathrm{rad}(\mathcal{W}^{2k-1})$:

$$\begin{aligned}
\mathbf{c}_1 &= \sum_{i=1}^{k} (\mathbf{a}_{2i} - \mathbf{a}_{2i+1})\,(\prod_{j=2}^{i} \lambda_{2j-1})\,(\prod_{j=i+1}^{k} \lambda_{2j}), \\
\mathbf{c}_2 &= \sum_{i=1}^{k} (\mathbf{a}_{2i} - \mathbf{a}_{2i-1})\,(\prod_{j=1}^{i-1} \lambda_{2j})\,(\prod_{j=i}^{k-1} \lambda_{2j+1}).
\end{aligned} \qquad (6.3.22)$$

More explicit expressions of $\mathbf{c}_1, \mathbf{c}_2$ can be obtained. $\mathrm{rad}(\mathcal{W}^{2k-1})$ is just the subspace spanned by basis vectors $\mathbf{e}_3, \ldots, \mathbf{e}_{2k-1}$. Let $\mathbf{a}_i = \mathbf{e}_1 + \lambda_i \mathbf{e}_2 + \mathbf{d}_i$, where $\mathbf{d}_i \in \mathrm{rad}(\mathcal{W}^{2k-1})$. By direct computing, we get

$$\begin{aligned}
\mathbf{c}_1 &= \mathbf{e}_2\,(\prod_{j=1}^{k} \lambda_{2j} - \prod_{j=1}^{k} \lambda_{2j-1}) + \sum_{i=1}^{k} (\mathbf{d}_{2i} - \mathbf{d}_{2i+1})\,(\prod_{j=2}^{i} \lambda_{2j-1})\,(\prod_{j=i+1}^{k} \lambda_{2j}), \\
\mathbf{c}_2 &= \mathbf{e}_2\,(\prod_{j=1}^{k} \lambda_{2j} - \prod_{j=1}^{k} \lambda_{2j-1}) + \sum_{i=1}^{k} (\mathbf{d}_{2i} - \mathbf{d}_{2i-1})\,(\prod_{j=1}^{i-1} \lambda_{2j})\,(\prod_{j=i}^{k-1} \lambda_{2j+1}).
\end{aligned}$$

That their coordinate components with respect to $\mathbf{e}_2$ are equal to zero is the result of the following proposition.

Proposition 6.75. For hyperbolic rotor $\mathbf{A}_{2k}$,

$$\prod_{j=1}^{k} \lambda_{2j} = \prod_{j=1}^{k} \lambda_{2j-1}, \tag{6.3.23}$$

$$|\langle \mathbf{a}_1 \mathbf{a}_2 \cdots \mathbf{a}_{2k}\rangle| = |\mathbf{a}_1|\,|\mathbf{a}_2| \cdots |\mathbf{a}_{2k}|. \tag{6.3.24}$$

Furthermore, the rotor is $\mathbf{e}_2$-typed (or $\mathbf{e}_1$-typed) if and only if it satisfies $\mathbf{c}_2 = 0$ (or $\mathbf{c}_1 = 0$).

Proof. If $\mathbf{A}_{2k}$ is $\mathbf{e}_2$-typed, set $m = 2k$ in (6.3.4), and substitute the last two lines of (6.3.4) for $i = m, 1$ respectively into (6.3.3). The result is $[\breve{\mathbf{a}}_{2k}] = \lambda_1 \lambda_{2k}^{-1}[\breve{\mathbf{a}}_1]$, by which and by (6.3.9) we get (6.3.23).

By (6.3.19),

$$\langle \mathbf{a}_1 \mathbf{a}_2 \cdots \mathbf{a}_{2k}\rangle = \prod_{j=1}^{k} (-2\lambda_{2j-1}) = \mathbf{a}_1^2 \mathbf{a}_3^2 \cdots \mathbf{a}_{2k-1}^2 = \mathbf{a}_2^2 \mathbf{a}_4^2 \cdots \mathbf{a}_{2k}^2, \tag{6.3.25}$$

then (6.3.24) follows.

Substituting (6.3.9) into (6.3.5), we get

$$\sum_{i=1}^{k} (\mathbf{a}_{2i} - \mathbf{a}_{2i-1})\left(\prod_{j=1}^{i-1} \lambda_{2j}\lambda_{2j+1}^{-1}\right) = 0. \tag{6.3.26}$$

Multiplying both sides with $\prod_{j=1}^{k-1} \lambda_{2j+1}$, we get $\mathbf{c}_2 = 0$.

If $\mathbf{A}_{2k}$ is $\mathbf{e}_1$-typed, by (6.3.7), if replacing every $\mathbf{a}_i$ by $-2\mathbf{a}_i^{-1}$, and replacing every λ_i by λ_i^{-1}, we get the corresponding results from those of $\mathbf{e}_2$-typed hyperbolic rotors. Hence (6.3.23), (6.3.19) and (6.3.24) are still the same.

With the understanding that $\mathbf{a}_{2k+1} = \mathbf{a}_1$ and $\lambda_{2k+1} = \lambda_1$, the above replacements change the left side of (6.3.26) into

$$\sum_{i=1}^{k} (-2\mathbf{a}_{2i}^{-1} + 2\mathbf{a}_{2i-1}^{-1})\left(\prod_{j=1}^{i-1} \lambda_{2j}^{-1}\lambda_{2j+1}\right)$$

$$= -\mathbf{a}_1\lambda_1^{-1} + \mathbf{a}_{2k}\lambda_{2k}^{-1}\prod_{j=1}^{k-1} \lambda_{2j}^{-1}\lambda_{2j+1} + \sum_{i=1}^{k-1}(\mathbf{a}_{2i} - \mathbf{a}_{2i+1})\lambda_{2i}^{-1}\prod_{j=1}^{i-1} \lambda_{2j}^{-1}\lambda_{2j+1} \tag{6.3.27}$$

$$= \sum_{i=1}^{k}(\mathbf{a}_{2i} - \mathbf{a}_{2i+1})\lambda_{2i}^{-1}\prod_{j=1}^{i-1} \lambda_{2j}^{-1}\lambda_{2j+1}.$$

Multiplying both sides with $\prod_{j=1}^{k} \lambda_{2j}$, we get $\mathbf{c}_1 = 0$. $\square$

Corollary 6.76. $\mathbf{A}_{2k} = \mathbf{a}_1 \mathbf{a}_2 \cdots \mathbf{a}_{2k}$ is a hyperbolic rotor if and only if so is $\mathbf{a}_{2k}\mathbf{A}_{2k-1}$, where $\mathbf{A}_{2k-1} = \mathbf{a}_1 \mathbf{a}_2 \cdots \mathbf{a}_{2k-1}$. If $\mathbf{A}_{2k}$ is $\mathbf{e}_2$-typed (or $\mathbf{e}_1$-typed), then $\mathbf{a}_{2k}\mathbf{A}_{2k-1}$ is $\mathbf{e}_1$-typed (or $\mathbf{e}_2$-typed).

Proof. If $\mathbf{a}_1\mathbf{a}_2\cdots\mathbf{a}_{2k}$ is shifted to $\mathbf{a}_2\mathbf{a}_3\cdots\mathbf{a}_{2k}\mathbf{a}_1$, then $\mathbf{c}_2, \mathbf{c}_1$ are changed into $-\mathbf{c}_1, -\lambda_1\lambda_2^{-1}\mathbf{c}_2$ respectively. By Proposition 6.75, the shift interchanges the two kinds of hyperbolic rotors. $\square$

Proposition 6.77. For any hyperbolic rotor, $\mathbf{c}_1 + \mathbf{c}_2 \neq 0$.

Proof. By the fact that vectors $\mathbf{a}_1, \mathbf{a}_2, \ldots, \mathbf{a}_{2k-1}$ are linearly independent, it is easy to prove that $\lambda_{2k}\mathbf{c}_2 - \lambda_1\mathbf{c}_1$, being a linear combination of $\mathbf{a}_1, \mathbf{a}_2, \ldots, \mathbf{a}_{2k-1}$, equals zero if and only if $\lambda_i = 1$ for all $1 \leq i \leq 2k$. However, if the latter condition holds, then by (6.3.18), $\mathbf{a}_1, \mathbf{a}_2, \ldots, \mathbf{a}_{2k-1}$ are linearly dependent. This contradiction shows that $\lambda_{2k}\mathbf{c}_2 \neq \lambda_1\mathbf{c}_1$. For a hyperbolic rotor, one of $\mathbf{c}_1, \mathbf{c}_2$ equals zero, so the other must be nonzero. $\square$

Proposition 6.78. For hyperbolic rotor $\mathbf{A}_{2k}$,

$$\mathbf{a}_1^{-1}\langle\mathbf{A}_{2k}\rangle = \langle\mathbf{a}_2\mathbf{a}_3\cdots\mathbf{a}_{2k}\rangle_1 + (-2)^{k-2}(\mathbf{c}_1 + \mathbf{c}_2). \tag{6.3.28}$$

In other words, $\mathbf{a}_1$ differs from $\langle\mathbf{a}_2\mathbf{a}_3\cdots\mathbf{a}_{2k}\rangle_1$ up to scale by a nonzero component in $\mathrm{rad}(\mathcal{W}^{2k-1})$.

Proof. For any $1 \leq l \leq k-1$,

$$\langle\mathbf{a}_1\mathbf{a}_2\cdots\mathbf{a}_{2l+1}\rangle_1 = -\sum_{i=1}^{l}\mathbf{a}_{2i}\langle\mathbf{a}_1\cdots\breve{\mathbf{a}}_{2i}\cdots\mathbf{a}_{2l+1}\rangle + \sum_{i=1}^{l+1}\mathbf{a}_{2i-1}\langle\mathbf{a}_1\cdots\breve{\mathbf{a}}_{2i-1}\cdots\mathbf{a}_{2l+1}\rangle$$

$$= (-2)^{l-1}\{\sum_{i=1}^{l}\mathbf{a}_{2i}(\lambda_1\lambda_3\cdots\lambda_{2i-1}\lambda_{2i+2}\lambda_{2i+4}\cdots\lambda_{2l}$$

$$+\lambda_2\lambda_4\cdots\lambda_{2i-2}\lambda_{2i+1}\lambda_{2i+3}\cdots\lambda_{2l+1})$$

$$-\sum_{i=1}^{l+1}\mathbf{a}_{2i-1}(\lambda_1\lambda_3\cdots\lambda_{2i-3}\lambda_{2i}\lambda_{2i+2}\cdots\lambda_{2l}$$

$$+\lambda_2\lambda_4\cdots\lambda_{2i-2}\lambda_{2i+1}\lambda_{2i+3}\cdots\lambda_{2l+1})\}$$

$$= (-2)^{l-1}\{(\sum_{i=1}^{l}(\mathbf{a}_{2i} - \mathbf{a}_{2i-1})\lambda_2\lambda_4\cdots\lambda_{2i-2}\lambda_{2i+1}\lambda_{2i+3}\cdots\lambda_{2l+1})$$

$$+(\sum_{i=1}^{l}(\mathbf{a}_{2i} - \mathbf{a}_{2i+1})\lambda_1\lambda_3\cdots\lambda_{2i-1}\lambda_{2i+2}\lambda_{2i+4}\cdots\lambda_{2l})$$

$$-(\mathbf{a}_1 + \mathbf{a}_{2l+1})\lambda_2\lambda_4\cdots\lambda_{2l}\}. \tag{6.3.29}$$

On one hand, by setting $l = k$ and $\mathbf{a}_{2l+1} = \mathbf{a}_1$ in (6.3.29), and using (6.3.23), (6.3.25), we get

$$\langle\mathbf{a}_1\mathbf{a}_2\cdots\mathbf{a}_{2k}\mathbf{a}_1\rangle_1 = (-2)^{k-1}\lambda_1(\mathbf{c}_1 + \mathbf{c}_2) + (-2)^k\mathbf{a}_1\lambda_2\lambda_4\cdots\lambda_{2k}$$

$$= (-2)^{k-2}\mathbf{a}_1^2(\mathbf{c}_1 + \mathbf{c}_2) + \mathbf{a}_1\langle\mathbf{a}_1\mathbf{a}_2\cdots\mathbf{a}_{2k}\rangle.$$

On the other hand, by (6.1.39), we get

$$\langle\mathbf{a}_1\mathbf{a}_2\cdots\mathbf{a}_{2k}\mathbf{a}_1\rangle_1 = 2\,\mathbf{a}_1 \wedge \langle\mathbf{a}_1\mathbf{a}_2\cdots\mathbf{a}_{2k}\rangle - \mathbf{a}_1^2\langle\mathbf{a}_2\mathbf{a}_3\cdots\mathbf{a}_{2k}\rangle_1.$$

Combining the two results, we get (6.3.28). $\square$

Corollary 6.79. If hyperbolic rotor $\mathbf{A}_{2k}$ is $\mathbf{e}_2$-typed, then

$$\langle \mathbf{a}_2\mathbf{a}_3\cdots\mathbf{a}_{2k}\rangle_1 - \langle \mathbf{A}_{2k}\rangle\mathbf{a}_1^{-1} = \langle \mathbf{A}_{2k}\rangle\mathbf{a}_2^{-1} - \langle \mathbf{a}_3\mathbf{a}_4\cdots\mathbf{a}_{2k}\mathbf{a}_1\rangle_1. \tag{6.3.30}$$

If $\mathbf{A}_{2k}$ is $\mathbf{e}_1$-typed hyperbolic, then

$$\lambda_1(\langle \mathbf{a}_2\mathbf{a}_3\cdots\mathbf{a}_{2k}\rangle_1 - \langle \mathbf{A}_{2k}\rangle\mathbf{a}_1^{-1}) = \lambda_2(\langle \mathbf{A}_{2k}\rangle\mathbf{a}_2^{-1} - \langle \mathbf{a}_3\mathbf{a}_4\cdots\mathbf{a}_{2k}\mathbf{a}_1\rangle_1). \tag{6.3.31}$$

Proposition 6.80. In hyperbolic rotor $\mathbf{A}_{2k} = \mathbf{a}_1\mathbf{a}_2\cdots\mathbf{a}_{2k}$, denote $\mathbf{A}_{2k-1} = \mathbf{a}_1\mathbf{a}_2\cdots\mathbf{a}_{2k-1}$. Then $\langle \mathbf{A}_{2k-1}\rangle_1 \wedge \mathbf{A}_{2k-1} = 0$.

Proof. First, by (6.1.38),

$$2\langle \mathbf{A}_{2k-1}\rangle_1 \wedge \mathbf{A}_{2k-1}$$
$$= \sum_{i=1}^{2k-1}(-1)^{i+1}\langle \mathbf{a}_1\cdots\breve{\mathbf{a}}_i\cdots\mathbf{a}_{2k-1}\rangle(\mathbf{a}_i\mathbf{a}_1\mathbf{a}_2\cdots\mathbf{a}_{2k-1} - \mathbf{a}_1\mathbf{a}_2\cdots\mathbf{a}_{2k-1}\mathbf{a}_i)$$
$$= 2(\sum_{1\leq j<i\leq 2k-1} - \sum_{1\leq i<j\leq 2k-1})(-1)^{i+j}(\mathbf{a}_i\cdot\mathbf{a}_j)\langle \mathbf{a}_1\cdots\breve{\mathbf{a}}_i\cdots\mathbf{a}_{2k-1}\rangle\mathbf{a}_1\cdots\breve{\mathbf{a}}_j\cdots\mathbf{a}_{2k-1}$$
$$= 2\sum_{j=1}^{2k-1}(-1)^j\mathbf{a}_1\cdots\breve{\mathbf{a}}_j\cdots\mathbf{a}_{2k-1}\{(\sum_{i=j+1}^{2k-1} - \sum_{i=1}^{j-1})(-1)^i(\mathbf{a}_i\cdot\mathbf{a}_j)\langle \mathbf{a}_1\cdots\breve{\mathbf{a}}_i\cdots\mathbf{a}_{2k-1}\rangle\}.$$

Second, we prove that for any $1 \leq j \leq 2k-1$,

$$f(j) := (\sum_{i=j+1}^{2k-1} - \sum_{i=1}^{j-1})(-1)^{i+1}(\mathbf{a}_i\cdot\mathbf{a}_j)\langle \mathbf{a}_1\cdots\breve{\mathbf{a}}_i\cdots\mathbf{a}_{2k-1}\rangle = 0. \tag{6.3.32}$$

When $j = 2p$ is even, then

$$-(-2)^{2-k}f(2p) = \{(\sum_{i=p+1}^{k-1} - \sum_{i=1}^{p-1})(\lambda_{2i} + \lambda_{2p})(\lambda_1\lambda_3\cdots\lambda_{2i-1}\lambda_{2i+2}\lambda_{2i+4}\lambda_{2k-2}$$
$$+\lambda_2\lambda_4\cdots\lambda_{2i-2}\lambda_{2i+1}\lambda_{2i+3}\lambda_{2k-1})\}$$
$$-\{(\sum_{i=p+1}^{k} - \sum_{i=1}^{p})(\lambda_{2i-1} + \lambda_{2p})(\lambda_1\lambda_3\cdots\lambda_{2i-3}\lambda_{2i}\lambda_{2i+2}\lambda_{2k-2}$$
$$+\lambda_2\lambda_4\cdots\lambda_{2i-2}\lambda_{2i+1}\lambda_{2i+3}\lambda_{2k-1})\}$$
$$= 0.$$

When $j = 2p - 1$ is odd, then

$$-(-2)^{2-k}f(2p-1) = \{(\sum_{i=p}^{k-1} - \sum_{i=1}^{p-1})(\lambda_{2i} + \lambda_{2p-1})(\lambda_1\lambda_3\cdots\lambda_{2i-1}\lambda_{2i+2}\lambda_{2i+4}\lambda_{2k-2}$$
$$+\lambda_2\lambda_4\cdots\lambda_{2i-2}\lambda_{2i+1}\lambda_{2i+3}\lambda_{2k-1})\}$$
$$-\{(\sum_{i=p+1}^{k} - \sum_{i=1}^{p-1})(\lambda_{2i-1} + \lambda_{2p-1})(\lambda_1\lambda_3\cdots\lambda_{2i-3}\lambda_{2i}\lambda_{2i+2}\lambda_{2k-2}$$
$$+\lambda_2\lambda_4\cdots\lambda_{2i-2}\lambda_{2i+1}\lambda_{2i+3}\lambda_{2k-1})\}$$
$$= 0.$$

    $\square$

Theorem 6.81. Let $\mathbf{A}_{2k}$ be a hyperbolic rotor. Then

$$\mathbf{A}_{2k} = \begin{cases} 2\lambda_1\mathbf{e}_2 \cdot (\mathbf{a}_2\mathbf{a}_3\cdots\mathbf{a}_{2k}), & \text{if the rotor is } \mathbf{e}_2\text{-typed};\\ 2\mathbf{e}_1 \cdot (\mathbf{a}_2\mathbf{a}_3\cdots\mathbf{a}_{2k}), & \text{if the rotor is } \mathbf{e}_1\text{-typed}. \end{cases} \tag{6.3.33}$$

Proof. We only prove the statement for $\mathbf{e}_2$-typed rotor. For $2 \leq i \leq k$, denote $\tau_i = \prod_{j=1}^{i-1} \lambda_{2j}\lambda_{2j+1}^{-1}$. By (6.3.26),

$$\mathbf{a}_1 = \mathbf{a}_2 + \sum_{i=2}^{k}(\mathbf{a}_{2i} - \mathbf{a}_{2i-1})\tau_i. \tag{6.3.34}$$

Substituting it into $\mathbf{A}_{2k}$, and using the following formulas from (6.1.38):

$$\mathbf{a}_{2i}\mathbf{a}_2\mathbf{a}_3\cdots\mathbf{a}_{2i} = \ \mathbf{a}_{2i}^2\mathbf{a}_2\mathbf{a}_3\cdots\mathbf{a}_{2i-1} + 2\sum_{j=2}^{2i-1}(-1)^j(\mathbf{a}_{2i}\cdot\mathbf{a}_j)\,\mathbf{a}_2\mathbf{a}_3\cdots\breve{\mathbf{a}}_j\cdots\mathbf{a}_{2i},$$

$$\mathbf{a}_{2i-1}\mathbf{a}_2\mathbf{a}_3\cdots\mathbf{a}_{2i-1} = -\mathbf{a}_{2i-1}^2\mathbf{a}_2\mathbf{a}_3\cdots\mathbf{a}_{2i-2}$$
$$+2\sum_{j=2}^{2i-2}(-1)^j(\mathbf{a}_{2i-1}\cdot\mathbf{a}_j)\mathbf{a}_2\mathbf{a}_3\cdots\breve{\mathbf{a}}_j\cdots\mathbf{a}_{2i-1},$$

we get

$$\mathbf{A}_{2k} = \mathbf{a}_2^2\mathbf{a}_3\mathbf{a}_4\cdots\mathbf{a}_{2k} + \sum_{i=2}^{k}\tau_i(\mathbf{a}_{2i}^2\mathbf{a}_2\mathbf{a}_3\cdots\breve{\mathbf{a}}_{2i}\cdots\mathbf{a}_{2k} + \mathbf{a}_{2i-1}^2\mathbf{a}_2\mathbf{a}_3\cdots\breve{\mathbf{a}}_{2i-1}\cdots\mathbf{a}_{2k})$$
$$+2\sum_{j=2}^{2i-2}(-1)^j\sum_{i=2}^{k}\tau_i(\mathbf{a}_j\cdot(\mathbf{a}_{2i} - \mathbf{a}_{2i-1}))\,\mathbf{a}_2\mathbf{a}_3\cdots\breve{\mathbf{a}}_j\cdots\mathbf{a}_{2k}$$
$$-2\sum_{i=2}^{k}\tau_i(\mathbf{a}_{2i-1}\cdot\mathbf{a}_{2i})\,\mathbf{a}_2\mathbf{a}_3\cdots\breve{\mathbf{a}}_{2i-1}\cdots\mathbf{a}_{2k}$$
$$= -2\lambda_2\mathbf{a}_3\mathbf{a}_4\cdots\mathbf{a}_{2k} - 2\sum_{j=2}^{k}\tau_j\lambda_{2j}(\mathbf{a}_2\mathbf{a}_3\cdots\breve{\mathbf{a}}_{2j}\cdots\mathbf{a}_{2k} - \mathbf{a}_2\mathbf{a}_3\cdots\breve{\mathbf{a}}_{2j-1}\cdots\mathbf{a}_{2k})$$
$$+2\sum_{j=1}^{k-1}\mathbf{a}_2\mathbf{a}_3\cdots\breve{\mathbf{a}}_{2j}\cdots\mathbf{a}_{2k}\sum_{i=j+1}^{k}\tau_i(\lambda_{2i-1} - \lambda_{2i})$$
$$-2\sum_{j=1}^{k-2}\mathbf{a}_2\mathbf{a}_3\cdots\breve{\mathbf{a}}_{2j+1}\cdots\mathbf{a}_{2k}\sum_{i=j+2}^{k}\tau_i(\lambda_{2i-1} - \lambda_{2i}). \tag{6.3.35}$$

Multiplying (6.3.35) with $\lambda_3\lambda_5\cdots\lambda_{2k-1}$, and using the identity

$$\lambda_3\lambda_5\cdots\lambda_{2k-1}\sum_{i=j+1}^{k}\tau_i(\lambda_{2i-1} - \lambda_{2i}) = \lambda_2\lambda_4\cdots\lambda_{2j}\lambda_{2j+1}\lambda_{2j+3}\cdots\lambda_{2k-1} - \lambda_2\lambda_4\cdots\lambda_{2k},$$

we get

$$\frac{\lambda_3\lambda_5\cdots\lambda_{2k-1}}{2\,\lambda_2\lambda_4\cdots\lambda_{2k}}\mathbf{A}_{2k} = -\sum_{j=2}^{k-1}\mathbf{a}_2\mathbf{a}_3\cdots\breve{\mathbf{a}}_{2j}\cdots\mathbf{a}_{2k} + \sum_{j=1}^{k-2}\mathbf{a}_2\mathbf{a}_3\cdots\breve{\mathbf{a}}_{2j+1}\cdots\mathbf{a}_{2k}$$
$$-\mathbf{a}_3\mathbf{a}_4\cdots\mathbf{a}_{2k} - \mathbf{a}_2\mathbf{a}_3\cdots\mathbf{a}_{2k-1} + \mathbf{a}_2\mathbf{a}_3\cdots\mathbf{a}_{2k-2}\mathbf{a}_{2k}$$

$$= \sum_{j=2}^{2k}(-1)^{j+1}\mathbf{a}_2\mathbf{a}_3\cdots\breve{\mathbf{a}}_j\cdots\mathbf{a}_{2k}$$
$$= \mathbf{e}_2\cdot(\mathbf{a}_2\mathbf{a}_3\cdots\mathbf{a}_{2k}).$$

$\square$

Corollary 6.82. Let $\mathbf{A}_{2k}$ be a hyperbolic rotor. Then $\mathbf{A}_{2k}\in\Lambda(\langle\mathbf{A}_{2k}\rangle_{2k-2})$.

Proof. By symmetry we only need to consider the $\mathbf{e}_2$-typed case. For a multi-vector $\mathbf{B}\in\Lambda(\mathcal{W}^{2k-1})$ with no scalar part, it has no $\mathbf{e}_1$-component if and only if $\mathbf{e}_2\cdot\mathbf{B}=0$. For $\mathbf{A}_{2k}$, we have

$$\mathbf{e}_2\cdot(\mathbf{A}_{2k}-\langle\mathbf{A}_{2k}\rangle) = -2\lambda_1\mathbf{e}_2\cdot(\mathbf{e}_2\cdot(\mathbf{a}_2\mathbf{a}_3\cdots\mathbf{a}_{2k})-\langle\mathbf{e}_2\mathbf{a}_2\mathbf{a}_3\cdots\mathbf{a}_{2k}\rangle)$$
$$= -2\lambda_1(\mathbf{e}_2\wedge\mathbf{e}_2)\cdot(\mathbf{a}_2\mathbf{a}_3\cdots\mathbf{a}_{2k})$$
$$= 0.$$

$\square$

Proposition 6.83. If $\mathbf{A}_{2k}$ is a parabolic or hyperbolic rotor, then $\langle\mathbf{A}_{2k}\rangle_{2k-2i}\neq 0$ for any $1\leq i\leq k$.

Proof. Denote $\mathbf{B}_{2k-1} = \mathbf{a}_2\mathbf{a}_3\cdots\mathbf{a}_{2k}$. If $\mathbf{A}_{2k}$ is a parabolic rotor, then $\mathbf{A}_{2k} = \mathbf{e}_1\cdot\mathbf{B}_{2k-1}$. By the linear independence among vectors $\mathbf{a}_2,\mathbf{a}_3,\ldots,\mathbf{a}_{2k}$, when $i<k$,

$$\langle\mathbf{A}_{2k}\rangle_{2k-2i} = \mathbf{e}_1\cdot\langle\mathbf{B}_{2k-1}\rangle_{2k-1-2(i-1)}$$
$$= \sum_{(2k-2i,\,2i-1)\vdash\mathbf{B}_{2k-1}}\langle\mathbf{e}_1\mathbf{B}_{2k-1(2)}\rangle\langle\mathbf{B}_{2k-1(1)}\rangle_{2k-2i} \tag{6.3.36}$$
$$= \sum_{(2k-2i,\,2i-1)\vdash\mathbf{B}_{2k-1}}\langle\mathbf{B}_{2k-1(1)}\rangle_{2k-2i}\neq 0.$$

When $i=k$, $\langle\mathbf{A}_{2k}\rangle = 1\neq 0$.

If $\mathbf{A}_{2k}$ is a hyperbolic rotor, we only consider the $\mathbf{e}_2$-typed case. Denote

$$\lambda_+(\mathbf{a}_{j_1}\mathbf{a}_{j_2}\cdots\mathbf{a}_{j_{2i-1}}) := \begin{cases} \lambda_{j_2}\lambda_{j_4}\cdots\lambda_{j_{2i-2}}, & \text{if } 2\leq i\leq k; \\ 1, & \text{if } i=1. \end{cases} \tag{6.3.37}$$

Here $\{j_2,j_4,\ldots,j_{2i-2}\}\subset\{j_1,j_2,\ldots,j_{2i-1}\}$. Then if $i<k$,

$$\langle\mathbf{A}_{2k}\rangle_{2k-2i} = 2\lambda_1\mathbf{e}_2\cdot\langle\mathbf{B}_{2k-1}\rangle_{2k-1-2(i-1)}$$
$$= 2\lambda_1\sum_{(2k-2i,\,2i-1)\vdash\mathbf{B}_{2k-1}}\langle\mathbf{e}_2\mathbf{B}_{2k-1(2)}\rangle\langle\mathbf{B}_{2k-1(1)}\rangle_{2k-2i} \tag{6.3.38}$$
$$= \sum_{(2k-2i,\,2i-1)\vdash\mathbf{B}_{2k-1}}(-2)^i\lambda_1\lambda_+(\mathbf{B}_{2k-1(2)})\langle\mathbf{B}_{2k-1(1)}\rangle_{2k-2i}\neq 0.$$

If $i=k$, $\langle\mathbf{A}_{2k}\rangle = (-2)^k\lambda_1\lambda_+(\mathbf{B}_{2k-1})\neq 0$.

$\square$

Corollary 6.84. Let $\mathbf{A}_{2k}$ be an $\mathbf{e}_2$-typed (or $\mathbf{e}_1$-typed) hyperbolic rotor. Then for any $1\leq i\leq k-1$, the $\mathbf{e}_2$-component (or $\mathbf{e}_1$-component) of $\langle\mathbf{A}_{2k}\rangle_{2k-2i}$ is nonzero.

Proof. We only consider the $\mathbf{e}_2$-typed case. Denote

$$\lambda_-(\mathbf{a}_{j_1}\mathbf{a}_{j_2}\cdots\mathbf{a}_{j_{2i}}) := \lambda_{j_1}\lambda_{j_3}\cdots\lambda_{j_{2i-1}}, \tag{6.3.39}$$

where $1 \le i \le k$, and $\{j_1, j_3, \ldots, j_{2i-1}\} \subset \{j_1, j_2, \ldots, j_{2i}\}$. By (6.3.38), with the understanding that $\lambda(\mathbf{a}_i)$ denotes λ_i,

$$\mathbf{e}_1 \cdot \langle \mathbf{A}_{2k}\rangle_{2k-2i}$$

$$= \sum_{(2i-1,2k-2i)\vdash\mathbf{B}_{2k-1}} (-2)^i \lambda_1 \lambda_+(\mathbf{B}_{2k-1\,(1)})\, \mathbf{e}_1 \cdot \langle\mathbf{B}_{2k-1\,(2)}\rangle_{2k-2i}$$

$$= \sum_{(2i-1,1,2k-2i-1)\vdash\mathbf{B}_{2k-1}} -(-2)^i \lambda_1 \lambda_+(\mathbf{B}_{2k-1\,(1)})\, \lambda(\mathbf{B}_{2k-1\,(2)})\, \langle\mathbf{B}_{2k-1\,(3)}\rangle_{2k-2i-1}$$

$$= \sum_{(2i,2k-2i-1)\vdash\mathbf{B}_{2k-1}} (-2)^i \lambda_1 \{\lambda_-(\mathbf{B}_{2k-1\,(1)}) - \lambda_+(\mathbf{B}_{2k-1\,(1)})\}\langle\mathbf{B}_{2k-1\,(2)}\rangle_{2k-2i-1}.$$

$$\tag{6.3.40}$$

So $\mathbf{e}_1 \cdot \langle\mathbf{A}_{2k}\rangle_{2k-2i} = 0$ if and only if $\lambda_{j_1}\lambda_{j_3}\cdots\lambda_{j_{2i-1}} = \lambda_{j_2}\lambda_{j_4}\cdots\lambda_{j_{2i}}$ for any subsequence $j_1, j_2, \ldots, j_{2i}$ of indices $2, 3, \ldots, 2k$.

However, if the latter condition is satisfied, then it is easy to deduce that $\lambda_2 = \lambda_3 = \ldots = \lambda_{2k}$, which leads to the linear dependence of $\mathbf{a}_2, \mathbf{a}_3, \ldots, \mathbf{a}_{2k}$. Contradiction. $\qquad\square$

So far the Grassmann structures of a parabolic rotor and a hyperbolic rotor are more or less clear. Still it is not clear how the Grassmann structures as invariants of versor compression determine the compressibility of a versor $\mathbf{A}_{2k}\mathbf{b}_{2k}^{-1}$, where $\mathbf{A}_{2k}$ is a parabolic or hyperbolic rotor in $\mathcal{G}(\mathcal{V}^n)$, and $\mathbf{b}_{2k}$ is a general invertible vector in $\mathcal{V}^n$. It remains an open problem for further investigation.

6.3.4 *Maximal grade conjectures*

A parabolic or hyperbolic rotor $\mathbf{A}_{2k}$ discloses a sharp difference between a Grassmann space and the corresponding Clifford space. As a multivector, $\mathbf{A}_{2k}$ belongs to $\mathcal{CL}(\langle\mathbf{A}_{2k}\rangle_{2k-2}) = \Lambda(\langle\mathbf{A}_{2k}\rangle_{2k-2})$, but as a Clifford monomial, it does not belong to $\mathcal{G}(\langle\mathbf{A}_{2k}\rangle_{2k-2})$.

There are several interesting problems inspired by the results of the Grassmann structures of the two rotors, each on a fundamental aspect of the Grassmann structure of Clifford algebra. In this subsection, we probe on two such problems.

All vectors $\mathbf{a}_i$ below are assumed to be nonzero, but can be null.

Definition 6.85. The *maximal grade* (or *minimal grade*) of a multivector $\mathbf{A}$, denoted by $\langle\mathbf{A}\rangle_{\max}$ (or $\langle\mathbf{A}\rangle_{\min}$), is the maximal (or minimal) integer r such that $\langle\mathbf{A}\rangle_r \ne 0$.

By Corollaries 6.68 and 6.82, if $\mathbf{A}_{2k}$ is a parabolic or hyperbolic rotor, then $\mathbf{A}_{2k} \in \Lambda(\langle\mathbf{A}_{2k}\rangle_{\max})$. There are more such instances. If $\mathbf{A}_k$ is a rigid versor, then obviously $\mathbf{A}_k \in \Lambda(\langle\mathbf{A}_k\rangle_k)$.

Proposition 6.86. If $\mathcal{V}^n$ is (anti-)Euclidean or (anti-)Minkowski, then for any Clifford monomial $\mathbf{A}_k \in \mathcal{G}(\mathcal{V}^n)$, $\langle \mathbf{A}_k \rangle_{\max}$ is a blade, and $\mathbf{A}_k \in \Lambda(\langle \mathbf{A}_k \rangle_{\max})$.

Proof. Any versor in $\mathcal{G}(\mathcal{V}^n)$ can be compressed to a rigid one, and almost all vectors in $\mathcal{V}^n$ are invertible. By the multilinearity of the geometric product and the compactness of the projective variety of blades, any Clifford monomial can be approximated by a sequence of versors. By the continuity of the geometric product, we get the conclusion. $\qquad\square$

Corollary 6.87. If $\mathcal{V}^n$ is a 4D or 5D (anti-)Euclidean or (anti-)Minkowski space, then any even Clifford monomial $\mathbf{A}_{2k} \in \mathcal{G}(\mathcal{V}^n)$ has the following Grassmann structure:

$$2 \langle \mathbf{A}_{2k} \rangle \langle \mathbf{A}_{2k} \rangle_4 = \langle \mathbf{A}_{2k} \rangle_2 \wedge \langle \mathbf{A}_{2k} \rangle_2. \tag{6.3.41}$$

In particular, if $\langle \mathbf{A}_{2k} \rangle \langle \mathbf{A}_{2k} \rangle_4 \neq 0$, then $\langle \mathbf{A}_{2k} \rangle_2$ is not a blade.

Proof. As a Clifford monomial, $\mathbf{A}_{2k}$ satisfies

$$\begin{aligned}
\mathbf{A}_{2k} \mathbf{A}_{2k}^\dagger &= ((\langle \mathbf{A}_{2k} \rangle_4 + \langle \mathbf{A}_{2k} \rangle_2 + \langle \mathbf{A}_{2k} \rangle)(\langle \mathbf{A}_{2k} \rangle_4 - \langle \mathbf{A}_{2k} \rangle_2 + \langle \mathbf{A}_{2k} \rangle)) \\
&= \langle \mathbf{A}_{2k} \rangle_4^2 - \langle \mathbf{A}_{2k} \rangle_2^2 + \langle \mathbf{A}_{2k} \rangle^2 + 2 \left[\langle \mathbf{A}_{2k} \rangle_2, \langle \mathbf{A}_{2k} \rangle_4 \right] + 2 \langle \mathbf{A}_{2k} \rangle \langle \mathbf{A}_{2k} \rangle_4 \ \in \mathbb{R}.
\end{aligned}$$

Since $\langle \mathbf{A}_{2k} \rangle_4$ is a blade, $\langle \mathbf{A}_{2k} \rangle_4^2 \in \mathbb{R}$. Condition $\langle \mathbf{A}_{2k} \mathbf{A}_{2k}^\dagger \rangle_2 = 0$ holds trivially. Condition $\langle \mathbf{A}_{2k} \mathbf{A}_{2k}^\dagger \rangle_4 = 0$ is equivalent to

$$\left[\langle \mathbf{A}_{2k} \rangle_2, \langle \mathbf{A}_{2k} \rangle_4 \right] + \langle \mathbf{A}_{2k} \rangle \langle \mathbf{A}_{2k} \rangle_4 = \frac{1}{2} \langle \mathbf{A}_{2k} \rangle_2 \wedge \langle \mathbf{A}_{2k} \rangle_2. \tag{6.3.42}$$

Since $\langle \mathbf{A}_{2k} \rangle_2 \in \Lambda(\langle \mathbf{A}_{2k} \rangle_4)$, we have $\left[\langle \mathbf{A}_{2k} \rangle_2, \langle \mathbf{A}_{2k} \rangle_4 \right] = 0$, and then (6.3.41). $\qquad\square$

Proposition 6.88. Let $\mathbf{A}_k = \mathbf{a}_1 \mathbf{a}_2 \cdots \mathbf{a}_k$, where $\mathbf{a}_i \in \mathcal{V}^n$. Let the maximal grade of $\mathbf{A}_k$ be m. Then $\langle \mathbf{A}_k \rangle_m$ is a blade if $m \leq 2$ or $k \leq 6$, and $\mathbf{A}_k \in \Lambda(\langle \mathbf{A}_k \rangle_m)$ if $m \leq 2$ or $k \leq 5$.

Proof. When $m = 0$ or 1, the conclusion is trivial. When $m = k$ or $k - 2$, $\langle \mathbf{A}_k \rangle_m$ is obviously a blade. When $m = 2$, since

$$\mathbf{A}_{2l} \mathbf{A}_{2l}^\dagger = ((\langle \mathbf{A}_{2l} \rangle_2 + \langle \mathbf{A}_{2l} \rangle)(-\langle \mathbf{A}_{2l} \rangle_2 + \langle \mathbf{A}_{2l} \rangle)) = -\langle \mathbf{A}_{2l} \rangle_2^2 + \langle \mathbf{A}_{2l} \rangle^2 \in \mathbb{R},$$

we have $\langle \mathbf{A}_{2l} \rangle_2 \wedge \langle \mathbf{A}_{2l} \rangle_2 = 0$, so $\langle \mathbf{A}_{2l} \rangle_2$ is a blade. This proves the first part of the proposition.

For the second part, when $m = 0, 1, 2, k$, the conclusion is trivial. When $k = 5$ and $m = 3$, by (6.2.19), $\langle \mathbf{A}_5 \rangle_1 \in \langle \mathbf{A}_5 \rangle_3$. $\qquad\square$

Maximal (or minimal) grade blade conjecture: For any Clifford monomial $\mathbf{A}_k$ in $\mathcal{CL}(\mathcal{V}^n)$, $\langle \mathbf{A}_k \rangle_{\max}$ is a blade; or equivalently, $\langle \mathbf{A}_k \rangle_{\min}$ is a blade.

Maximal grade reduction conjecture: For any Clifford monomial $\mathbf{A}_k$ in $\mathcal{CL}(\mathcal{V}^n)$, $\mathbf{A}_k \in \Lambda(\langle \mathbf{A}_k \rangle_{\max})$.

The two conjectures are closely related to each other. If $\mathcal{V}^n$ is a degenerate inner-product space, then we can extend $\mathcal{V}^n$ to a larger inner-product space that is not degenerate. Without loss of generality, we assume that $\mathcal{V}^n$ is nondegenerate. Let the maximal grade of $\mathbf{A}_k$ be $k - 2m$. Below we always assume that $\langle \mathbf{A}_k \rangle_{k-2m}$ is a blade, and investigate whether or not $\langle \mathbf{A}_k \mathbf{a}_{k+1} \rangle_{\max}$ is a blade for vector $\mathbf{a}_{k+1} \in \mathcal{V}^n$.

Lemma 6.89. If $\mathbf{a}_{k+1}^2 \neq 0$, then $\langle \mathbf{A}_k \mathbf{a}_{k+1} \rangle_{k-2m-1}$ is a blade if and only if $\mathbf{a}_{k+1} \wedge \langle \mathbf{A}_k \rangle_{k-2m-2} \in \Lambda(\langle \mathbf{A}_k \rangle_{k-2m})$.

Proof. If $\mathbf{a}_{k+1} \wedge \langle \mathbf{A}_k \rangle_{k-2m} \neq 0$, then $\langle \mathbf{A}_k \mathbf{a}_{k+1} \rangle_{\max} = \langle \mathbf{A}_k \rangle_{k-2m} \wedge \mathbf{a}_{k+1}$ is obviously a blade. So we assume $\mathbf{a}_{k+1} \in \langle \mathbf{A}_k \rangle_{k-2m}$. Then

$$\langle \mathbf{A}_k \mathbf{a}_{k+1} \rangle_{k-2m-1} = \langle \mathbf{A}_k \rangle_{k-2m} \cdot \mathbf{a}_{k+1} + \langle \mathbf{A}_k \rangle_{k-2m-2} \wedge \mathbf{a}_{k+1}, \tag{6.3.43}$$

from which we get the sufficient statement.

Now we prove the necessity statement. Obviously $\langle \mathbf{A}_k \rangle_{k-2m} \cdot \mathbf{a}_{k+1}$ is a blade. Since $\mathbf{a}_{k+1}^2 \neq 0$, we have $\langle \mathbf{A}_k \rangle_{k-2m} \cdot \mathbf{a}_{k+1} \neq 0$. If $\langle \mathbf{A}_k \rangle_{k-2m-2} \wedge \mathbf{a}_{k+1} = 0$, then the conclusion is trivial. We further assume $\langle \mathbf{A}_k \rangle_{k-2m-2} \wedge \mathbf{a}_{k+1} \neq 0$.

Let $\langle \mathbf{A}_k \mathbf{a}_{k+1} \rangle_{k-2m-1} = \mathbf{b}_1 \wedge \mathbf{b}_2 \wedge \cdots \wedge \mathbf{b}_{k-2m-1}$. If $\mathbf{b}_i \cdot \mathbf{a}_{k+1} \neq 0$ for some i, by rescaling the $\mathbf{b}_i$ appropriately and reordering them in the blade $\mathbf{b}_1 \wedge \mathbf{b}_2 \wedge \cdots \wedge \mathbf{b}_{k-2m-1}$, we can assume that there is an integer $1 \leq l \leq k - 2m - 1$, such that

$$\mathbf{b}_i = \begin{cases} \mathbf{b}_i' + \mathbf{a}_{k+1}, & \text{if } 1 \leq i \leq l, \\ \mathbf{b}_i', & \text{if } l+1 \leq i \leq k - 2m - 1, \end{cases} \tag{6.3.44}$$

where $\mathbf{b}_i' \cdot \mathbf{a}_{k+1} = 0$. Then

$$\begin{aligned}
\langle \mathbf{A}_k \mathbf{a}_{k+1} \rangle_{k-2m-1} &= (\mathbf{a}_{k+1} + \mathbf{b}_1') \wedge (\mathbf{a}_{k+1} + \mathbf{b}_2') \wedge \cdots \wedge (\mathbf{a}_{k+1} + \mathbf{b}_l') \\
&\qquad \wedge \mathbf{b}_{l+1}' \wedge \mathbf{b}_{l+2}' \wedge \cdots \wedge \mathbf{b}_{k-2m-1}' \\
&= \mathbf{b}_1' \wedge \mathbf{b}_2' \wedge \cdots \wedge \mathbf{b}_{k-2m-1}' \\
&\quad + \mathbf{a}_{k+1} \wedge (\mathbf{b}_2' - \mathbf{b}_1') \wedge (\mathbf{b}_3' - \mathbf{b}_1') \wedge \cdots \wedge (\mathbf{b}_l' - \mathbf{b}_1') \\
&\qquad \wedge \mathbf{b}_{l+1}' \wedge \mathbf{b}_{l+2}' \wedge \cdots \wedge \mathbf{b}_{k-2m-1}'.
\end{aligned} \tag{6.3.45}$$

We have

$$\langle \mathbf{A}_k \rangle_{k-2m} \cdot \mathbf{a}_{k+1} = \mathbf{b}_1' \wedge \mathbf{b}_2' \wedge \cdots \wedge \mathbf{b}_{k-2m-1}',$$

$$\begin{aligned}
\langle \mathbf{A}_k \rangle_{k-2m-2} \wedge \mathbf{a}_{k+1} &= \mathbf{a}_{k+1} \wedge (\mathbf{b}_2' - \mathbf{b}_1') \wedge (\mathbf{b}_3' - \mathbf{b}_1') \wedge \cdots \wedge (\mathbf{b}_l' - \mathbf{b}_1') \\
&\qquad \wedge \mathbf{b}_{l+1}' \wedge \mathbf{b}_{l+2}' \wedge \cdots \wedge \mathbf{b}_{k-2m-1}',
\end{aligned}$$

so $\langle \mathbf{A}_k \rangle_{k-2m-2} \wedge \mathbf{a}_{k+1}$ is a subspace of $(\langle \mathbf{A}_k \rangle_{k-2m} \cdot \mathbf{a}_{k+1}) \wedge \mathbf{a}_{k+1} = \mathbf{a}_{k+1}^2 \langle \mathbf{A}_k \rangle_{k-2m}$. $\square$

Corollary 6.90. If subspace $\langle \mathbf{A}_k \rangle_{k-2m}$ is not null, then $\langle \mathbf{A}_k \mathbf{a}_{k+1} \rangle_{k-2m-1}$ is a blade for all vectors $\mathbf{a}_{k+1} \in \langle \mathbf{A}_k \rangle_{k-2m}$ if and only if $\langle \mathbf{A}_k \rangle_{k-2m-2} \in \Lambda(\langle \mathbf{A}_k \rangle_{k-2m})$.

Proof. Since subspace $\langle \mathbf{A}_k \rangle_{k-2m}$ is not null, almost all vectors in it are invertible. If $\langle \mathbf{A}_k \rangle_{k-2m-2}$ has any nonzero component outside $\Lambda(\langle \mathbf{A}_k \rangle_{k-2m})$, then for almost all invertible vectors $\mathbf{a}_{k+1} \in \langle \mathbf{A}_k \rangle_{k-2m}$, the component of $\mathbf{a}_{k+1} \wedge \langle \mathbf{A}_k \rangle_{k-2m-2}$ outside $\Lambda(\langle \mathbf{A}_k \rangle_{k-2m})$ is nonzero. $\square$

6.4 Clifford coalgebra, Clifford summation and factorization*

In Chapter 2, Grassmann-Cayley coalgebra is proposed to make Cayley expansions by changing the meet products and the outer products of meet products into outer coproducts and meet coproducts respectively. In Chapter 5, when expanding graded Clifford monomials, various partitions of the monomials are needed to replace long geometric products by outer products with angular bracket coefficients. Such partitions naturally lead to the concepts of *geometric coproduct* and *Clifford coalgebra*.

Definition 6.91. Let $\mathbf{A}_r = \mathbf{a}_1 \mathbf{a}_2 \cdots \mathbf{a}_r$, where $\mathbf{a}_i \in \mathcal{V}^n$. For any $0 \le t \le r$,

$$\Delta_t(\mathbf{A}_r) := \sum_{(t,r-t)\vdash \mathbf{A}_r} \mathsf{cl}\,(\mathbf{A}_{r\,(1)}) \otimes \mathsf{cl}\,(\mathbf{A}_{r\,(2)}) \tag{6.4.1}$$

is called the *t-th part* of the *Clifford coproduct* of $\mathbf{A}_r$, where "cl" is the geometric product operator defined in (5.2.1). When $t > r$, define $\Delta_t(\mathbf{A}_r) = 0$.

The above definition is correct only when $\Delta_t(\mathbf{A}_r)$ is independent of the representative vectors of Clifford monomial $\mathbf{A}_r$. The following lemma establishes this property.

Lemma 6.92. [Fundamental lemma on Clifford coproduct] Let $r > t > 0$ be integers. Let $\mathbf{A}_r = \mathbf{a}_1 \mathbf{a}_2 \cdots \mathbf{a}_r$. Then

$$\Delta_t(\mathbf{A}_r) = \sum_{l=0}^{[\frac{r-t}{2}]+[\frac{t}{2}]} \sum_{j=\max(0,l-[\frac{r-t}{2}])}^{\min(l,[\frac{t}{2}])} C_l^j\, \Delta_{t-2j}^{\wedge}(\langle \mathbf{A}_r\rangle_{r-2l}). \tag{6.4.2}$$

Proof.   By (5.4.2) and (5.4.14),

$$\sum_{(t,r-t)\vdash \mathbf{A}_r} \mathbf{A}_{r\,(1)} \otimes \mathbf{A}_{r\,(2)}$$

$$= \sum_{(t,r-t)\vdash \mathbf{A}_r} \sum_{k=0}^{[\frac{t}{2}]} \sum_{j=0}^{[\frac{r-t}{2}]} \langle \mathbf{A}_{r\,(1)}\rangle_{t-2k} \otimes \langle \mathbf{A}_{r\,(2)}\rangle_{r-t-2j}$$

$$= \sum_{k=0}^{[\frac{t}{2}]} \sum_{j=0}^{[\frac{r-t}{2}]} \sum_{(2k,t-2k,2j,r-t-2j)\vdash \mathbf{A}_r} \langle \mathbf{A}_{r\,(1)}\rangle \langle \mathbf{A}_{r\,(3)}\rangle \langle \mathbf{A}_{r\,(2)}\rangle_{t-2k} \otimes \langle \mathbf{A}_{r\,(4)}\rangle_{r-t-2j}$$

$$= \sum_{k=0}^{[\frac{t}{2}]} \sum_{j=0}^{[\frac{r-t}{2}]} \sum_{(2k,2j,t-2k,r-t-2j)\vdash \mathbf{A}_r} \langle \mathbf{A}_{r\,(1)}\rangle \langle \mathbf{A}_{r\,(2)}\rangle \langle \mathbf{A}_{r\,(3)}\rangle_{t-2k} \otimes \langle \mathbf{A}_{r\,(4)}\rangle_{r-t-2j}$$

$$= \sum_{k=0}^{[\frac{t}{2}]} \sum_{j=0}^{[\frac{r-t}{2}]} \sum_{(2k+2j,r-2k-2j)\vdash \mathbf{A}_r} C_{k+j}^k \langle \mathbf{A}_{r\,(1)}\rangle \Delta_{t-2k}^{\wedge}(\mathbf{A}_{r\,(2)})$$

$$= \sum_{l=0}^{[\frac{r-t}{2}]+[\frac{t}{2}]} \sum_{k=\max(0,l-[\frac{r-t}{2}])}^{\min(l,[\frac{t}{2}])} C_l^k\, \Delta_{t-2k}^{\wedge}(\langle \mathbf{A}_r\rangle_{r-2l}).$$

$\square$

Definition 6.93. The *Clifford coproduct* defined on $\mathcal{G}(\mathcal{V}^n)$, also called the *geometric coproduct*, is the linear extension of the following *Clifford coproduct* of a Clifford monomial:

$$\Delta = \sum_{t=0}^{\infty} \Delta_t : \quad \mathcal{G}(\mathcal{V}^n) \longrightarrow \mathcal{G}(\mathcal{V}^n) \,\hat{\otimes}\, \mathcal{G}(\mathcal{V}^n). \tag{6.4.3}$$

The *Clifford coalgebra* over $\mathcal{V}^n$ is the graded Clifford space $\mathcal{G}(\mathcal{V}^n)$ together with the Clifford coproduct and the counit map $\langle\ \rangle\colon \mathbf{A} \mapsto \langle\mathbf{A}\rangle$ for $\mathbf{A} \in \mathcal{G}(\mathcal{V}^n)$.

The following proposition can be easily proved.

Proposition 6.94. Δ is a homomorphism of Clifford algebras.

By (5.4.17), the Clifford coproduct has the following geometric interpretation, when composed with two $\mathbb{Z}$-grading operators and one deficit bracket operator: for any Clifford monomial $\mathbf{A}_r$ of length r, any $0 \leq i \leq [\frac{t}{2}]$ and $0 \leq j \leq [\frac{r-t}{2}]$,

$$\sum_{(t,r-t)\vdash\mathbf{A}_r} [\langle\mathbf{A}_{r\,(1)}\rangle_{t-2i}\mathbf{U}_{n-t+2i}] \,\langle\mathbf{A}_{r\,(2)}\rangle_{r-t-2j}$$

$$= \sum_{(t,r-t)\vdash\mathbf{A}_r} (\langle\mathbf{A}_{r\,(1)}\rangle_{t-2i} \vee \mathbf{U}_{n-t+2i})\langle\mathbf{A}_{r\,(2)}\rangle_{r-t-2j} \tag{6.4.4}$$

$$= C_{i+j}^i \langle\mathbf{A}_r\rangle_{r-2(i+j)} \vee \mathbf{U}_{n-t+2i}.$$

Recall that in Chapter 5, when expanding square brackets, we proposed an expansion formula (5.4.24) that can produce much fewer terms than the general formula (5.4.2). That formula can also be written in the form (5.4.26) which is of the following "summation by part" style:

$$\sum_{\vdash\mathbf{A}} \langle\mathbf{A}_{(1)}\rangle\, [\mathbf{A}_{(2)}\mathbf{B}] = \sum_{\vdash\mathbf{B}} \lambda_{\vdash}\, \langle\mathbf{B}_{(1)}\rangle\, [\mathbf{A}\mathbf{B}_{(2)}], \tag{6.4.5}$$

where $\lambda_{\vdash}$ denotes a scalar depending only on the partition. A formula of this form is called a *Clifford summation by part*, or in short, *Clifford summation*.

Clifford coalgebra is an important tool in making Clifford summations. In particular, when only one Clifford monomial is involved in a Clifford summation, *e.g.*, in (6.4.5), $\mathbf{B} = 1$ is a monomial of length zero, then the Clifford summation becomes *Clifford factorization*, the inverse of Clifford expansion.

In the following subsections, we investigate various Clifford summations and the Clifford factorizations induced by them, by categorizing the Clifford summations according to the number of participating Clifford monomials and the types of $\mathbb{Z}$-grading operators. Symbols $\mathbf{a}_i$ denote arbitrary vectors in $\mathcal{V}^n$.

6.4.1   One Clifford monomial

Lemma 6.95. Let $\mathbf{A}_{s+t+2k+2l} = \mathbf{a}_1\mathbf{a}_2\cdots\mathbf{a}_{s+t+2k+2l}$ for any $s,t > 0$ and $k,l \geq 0$. Then

$$\sum_{(s+2k,t+2l)\vdash\mathbf{A}_{s+t+2k+2l}} \langle \mathbf{A}_{s+t+2k+2l\,(1)}\rangle_s \cdot \langle \mathbf{A}_{s+t+2k+2l\,(2)}\rangle_t = 0. \tag{6.4.6}$$

Proof. Direct from (5.4.2) and (5.1.22).   □

Theorem 6.96. For any $s,t,k,l \geq 0$, let $u = s+t+2k+2l$. Let $\mathbf{A}_u = \mathbf{a}_1\mathbf{a}_2\cdots\mathbf{a}_u$. Then

$$\sum_{(s+2k,t+2l)\vdash\mathbf{A}_u} \langle \mathbf{A}_{u(1)}\rangle_s \langle \mathbf{A}_{u(2)}\rangle_t = C_{k+l}^k C_{s+t}^s \langle \mathbf{A}_u\rangle_{s+t}. \tag{6.4.7}$$

Proof. By (6.4.6),

$$\sum_{(s+2k,t+2l)\vdash\mathbf{A}_u} \langle \mathbf{A}_{u(1)}\rangle_s \langle \mathbf{A}_{u(2)}\rangle_t = \sum_{(2k,s,2l,t)\vdash\mathbf{A}_u} \langle \mathbf{A}_{u(1)}\rangle \langle \mathbf{A}_{u(3)}\rangle \langle \mathbf{A}_{u(2)}\rangle_s \langle \mathbf{A}_{u(4)}\rangle_t$$

$$= \sum_{(2k,2l,s,t)\vdash\mathbf{A}_u} \langle \mathbf{A}_{u(1)}\rangle \langle \mathbf{A}_{u(2)}\rangle \langle \mathbf{A}_{u(3)}\rangle_s \wedge \langle \mathbf{A}_{u(4)}\rangle_t$$

$$= C_{k+l}^k C_{s+t}^s \sum_{(2k+2l,s+t)\vdash\mathbf{A}_u} \langle \mathbf{A}_{u(1)}\rangle \langle \mathbf{A}_{u(2)}\rangle_{s+t}$$

$$= C_{k+l}^k C_{s+t}^s \langle \mathbf{A}_u\rangle_{s+t}.$$

     □

Theorem 6.97. For $\mathbf{A}_r = \mathbf{a}_1\mathbf{a}_2\cdots\mathbf{a}_r$,

$$\sum_{(t,r-t)\vdash\mathbf{A}_r} \langle \mathbf{A}_{r(1)}\mathbf{A}_{r(2)}\rangle_{r-2l} = c(r,t,l)\langle \mathbf{A}_r\rangle_{r-2l}, \tag{6.4.8}$$

where

$$c(r,t,l) = \sum_{i=\max(0,l+[\frac{t-r+1}{2}])}^{\min(l,[\frac{t}{2}])} C_l^i C_{r-2l}^{t-2i} \tag{6.4.9}$$

is the coefficient of x^t in the polynomial $(1+x^2)^l(1+x)^{r-2l}$.

Proof. By (6.4.2) and (6.4.7),

$$\mathsf{cl}\circ\Delta_t(\mathbf{A}_r) = \sum_{l=0}^{[\frac{r-t}{2}]+[\frac{t}{2}]} \sum_{j=\max(0,l-[\frac{r-t}{2}])}^{\min(l,[\frac{t}{2}])} C_l^j \; \mathsf{cl}\circ\Delta_{t-2j}^{\wedge}(\langle \mathbf{A}_r\rangle_{r-2l})$$

$$= \sum_{l=0}^{[\frac{r-t}{2}]+[\frac{t}{2}]} \sum_{j=\max(0,l-[\frac{r-t}{2}])}^{\min(l,[\frac{t}{2}])} C_l^j C_{r-2l}^{t-2j} \langle \mathbf{A}_r\rangle_{r-2l}.$$

     □

Corollary 6.98. For $\mathbf{A}_{2l} = \mathbf{a}_1 \mathbf{a}_2 \cdots \mathbf{a}_{2l}$,

$$\sum_{(t,2l-t)\vdash\mathbf{A}_{2l}} \langle \mathbf{A}_{2l(1)}\mathbf{A}_{2l(2)}\rangle = \begin{cases} 0, & \text{if } t \text{ is odd}; \\ C_l^k \langle \mathbf{A}_{2l}\rangle, & \text{if } t = 2k. \end{cases} \tag{6.4.10}$$

Proof. If $t = 2k - 1$, then by (6.4.9), $c(2l, t, l) = \sum_{i=k}^{k-1} C_l^i = 0$. If $t = 2k$, then $c(2l, t, l) = \sum_{i=k}^{k} C_l^i = C_l^k$. $\qquad\square$

6.4.2 *Two Clifford monomials*

Recall the classical definition of combinatorial number C_k^l: for any $0 \le l \le k$,

$$C_k^l = \frac{k!}{l!(k-l)!}. \tag{6.4.11}$$

We extend it to $k, l \in \mathbb{Z}$ so that the following rules are always satisfied:

- For any $l > k$, $C_k^l = 0$.
- For any $k, l \in \mathbb{Z}$,

$$C_k^l = C_{k-1}^l + C_{k-1}^{l-1}. \tag{6.4.12}$$

The C_k^l for $k, l \in \mathbb{Z}$ are called *extended combinatorial numbers*.

Lemma 6.99.

(1) For any $l < 0 \le k$, $C_k^l = 0$.
(2) For any $l \ge k > 0$, $C_{-k}^{-l} = (-1)^{k+l} C_{l-1}^{k-1}$.

Proof. For any $k \ge 0$, from $C_k^{-1} + C_k^0 = C_{k+1}^0$, we get $C_k^{-1} = 0$. Assume that for some $l > 1$, $C_k^{-(l-1)} = 0$ for any $k \ge 0$. From $C_k^{-l} + C_k^{-(l-1)} = C_{k+1}^{-(l-1)}$, we get $C_k^{-l} = 0$. This proves (1).

To prove (2), first we prove $C_{-1}^{-k} = (-1)^{k+1}$ for any $k > 0$. From $C_{-1}^{-1} + C_{-1}^0 = C_0^0$, we get $C_{-1}^{-1} = 1$. Assume that $C_{-1}^{-(k-1)} = (-1)^k$ for some $k > 1$. From $C_{-1}^{-k} + C_{-1}^{-(k-1)} = C_0^{-(k-1)} = 0$, we get $C_{-1}^{-k} = (-1)^{k+1}$.

Now assume that for some $k > 1$, $C_{-(k-1)}^{-l} = (-1)^{k+l-1} C_{l-1}^{k-2}$ for any $l \ge k$.

From $C_{-k}^{-k-1} + C_{-k}^{-k} = C_{-(k-1)}^{-k}$, we get $C_{-k}^{-(k+1)} = -C_{k-1}^{k-2} - 1 = -k$. So (2) is true for $l = k + 1$. Assume that for some $l > k$, $C_{-k}^{-(l-1)} = (-1)^{k+l-1} C_{l-2}^{k-1}$. From $C_{-k}^{-l} + C_{-k}^{-(l-1)} = C_{-(k-1)}^{-(l-1)}$, we get $C_{-k}^{-l} = (-1)^{k+l} C_{l-2}^{k-2} + (-1)^{k+l} C_{l-2}^{k-1} = (-1)^{k+l} C_{l-1}^{k-1}$. $\qquad\square$

The following theorem is dual to Theorem 5.79.

Theorem 6.100. [$(0,0)$-typed summation] Let $\mathbf{A}_m = \mathbf{a}_1\mathbf{a}_2\cdots\mathbf{a}_m$ and $\mathbf{B}_{2k+2l-m} = \mathbf{a}_{m+1}\mathbf{a}_{m+2}\cdots\mathbf{a}_{2k+2l}$. Then for any $k,l > 0$ and any $0 \le m \le 2l$,

$$\sum_{(2l-m,2k)\vdash \mathbf{B}_{2k+2l-m}} \langle \mathbf{A}_m \mathbf{B}_{2k+2l-m\,(1)} \rangle \langle \mathbf{B}_{2k+2l-m\,(2)} \rangle$$
$$= \sum_{i=0}^{[\frac{m}{2}]} \sum_{(2i,m-2i)\vdash \mathbf{A}_m} C^{k-i}_{k+l-m} \langle \mathbf{A}_{m\,(1)} \rangle \langle \mathbf{A}_{m\,(2)} \mathbf{B}_{2k+2l-m} \rangle. \tag{6.4.13}$$

where the C^{k-i}_{k+l-m} are extended combinatorial numbers.

For example, set $m = k = l = 2$, and denote $\mathbf{B}_6 = \mathbf{b}_1\mathbf{b}_2\cdots\mathbf{b}_6$ in (6.4.13):

$$\langle \mathbf{a}_1\mathbf{a}_2\mathbf{b}_1\mathbf{b}_2 \rangle \langle \mathbf{b}_3\mathbf{b}_4\mathbf{b}_5\mathbf{b}_6 \rangle - \langle \mathbf{a}_1\mathbf{a}_2\mathbf{b}_1\mathbf{b}_3 \rangle \langle \mathbf{b}_2\mathbf{b}_4\mathbf{b}_5\mathbf{b}_6 \rangle + \cdots$$
$$+ \langle \mathbf{a}_1\mathbf{a}_2\mathbf{b}_5\mathbf{b}_6 \rangle \langle \mathbf{b}_1\mathbf{b}_2\mathbf{b}_3\mathbf{b}_4 \rangle \text{ (15 terms)}$$
$$= \langle \mathbf{a}_1\mathbf{a}_2\mathbf{b}_1\mathbf{b}_2\mathbf{b}_3\mathbf{b}_4\mathbf{b}_5\mathbf{b}_6 \rangle + 2\,(\mathbf{a}_1 \cdot \mathbf{a}_2) \langle \mathbf{b}_1\mathbf{b}_2\mathbf{b}_3\mathbf{b}_4\mathbf{b}_5\mathbf{b}_6 \rangle.$$

Proof. When $m = 0$, (6.4.13) is just (5.4.10). When $m = 1$, by (5.4.3) and (5.4.10),

$$\sum_{(2l-1,2k)\vdash \mathbf{B}_{2k+2l-1}} \langle \mathbf{a}_1 \mathbf{B}_{2k+2l-1\,(1)} \rangle \langle \mathbf{B}_{2k+2l-1\,(2)} \rangle$$

$$= \sum_{(1,2l-2,2k)\vdash \mathbf{B}_{2k+2l-1}} (\mathbf{a}_1 \cdot \mathbf{B}_{2k+2l-1\,(1)}) \langle \mathbf{B}_{2k+2l-1\,(2)} \rangle \langle \mathbf{B}_{2k+2l-1\,(3)} \rangle$$

$$= C^k_{k+l-1} \sum_{(1,2l+2k-2)\vdash \mathbf{B}_{2k+2l-1}} (\mathbf{a}_1 \cdot \mathbf{B}_{2k+2l-1\,(1)}) \langle \mathbf{B}_{2k+2l-1\,(2)} \rangle$$

$$= C^k_{k+l-1} \langle \mathbf{a}_1 \mathbf{B}_{2k+2l} \rangle.$$

Assume that (6.4.13) holds for $0 \le m \le r-1$. By (6.1.68), (5.4.17) and (6.1.69), for $m = r \ge 2$,

$$\sum_{(2l-r,2k)\vdash \mathbf{B}_{2k+2l-r}} \langle \mathbf{A}_r \mathbf{B}_{2k+2l-r\,(1)} \rangle \langle \mathbf{B}_{2k+2l-r\,(2)} \rangle$$

$$= \sum_{(2l-r,2k)\vdash \mathbf{B}_{2k+2l-r}} \langle \mathbf{A}_r \rangle_r \cdot \langle \mathbf{B}_{2k+2l-r\,(1)} \rangle_r \langle \mathbf{B}_{2k+2l-r\,(2)} \rangle$$

$$+ \sum_{i=1}^{[\frac{r}{2}]} \sum_{\substack{(2i,r-2i)\vdash \mathbf{A}_r, \\ (2l-r,2k)\vdash \mathbf{B}_{2k+2l-r}}} (-1)^{i+1} \langle \mathbf{A}_{r\,(1)} \rangle \langle \mathbf{B}_{2k+2l-r\,(2)} \rangle \langle \mathbf{A}_{r\,(2)} \mathbf{B}_{2k+2l-r\,(1)} \rangle$$

$$= C^k_{k+l-r} \langle \mathbf{A}_r \rangle_r \cdot \langle \mathbf{B}_{2k+2l-r} \rangle_r + \sum_{i=1}^{[\frac{r}{2}]} \sum_{(2i,r-2i)\vdash \mathbf{A}_r} \sum_{j=0}^{[\frac{r}{2}]-i} \sum_{(2j,r-2i-2j)\vdash \mathbf{A}_{r\,(2)}}$$

$$(-1)^{i+1} C^{k-j}_{k+l-r+i} \langle \mathbf{A}_{r\,(1)} \rangle \langle \mathbf{A}_{r\,(21)} \rangle \langle \mathbf{A}_{r\,(22)} \mathbf{B}_{2k+2l-r} \rangle$$

$$= C_{k+l-r}^k \langle \mathbf{A}_r \mathbf{B}_{2k+2l-r} \rangle - C_{k+l-r}^k \sum_{i=1}^{[\frac{r}{2}]} \sum_{(2i,r-2i)\vdash \mathbf{A}_r} (-1)^{i+1} \langle \mathbf{A}_{r(1)} \rangle \langle \mathbf{A}_{r(2)} \mathbf{B}_{2k+2l-r} \rangle$$

$$+ \sum_{i=1}^{[\frac{r}{2}]} \sum_{j=0}^{[\frac{r}{2}]-i} \sum_{(2i,2j,r-2i-2j)\vdash \mathbf{A}_r} (-1)^{i+1} C_{k+l-r+i}^{k-j} \langle \mathbf{A}_{r(1)} \rangle \langle \mathbf{A}_{r(2)} \rangle \langle \mathbf{A}_{r(3)} \mathbf{B}_{2k+2l-r} \rangle$$

$$= C_{k+l-r}^k \langle \mathbf{A}_r \mathbf{B}_{2k+2l-r} \rangle + \sum_{h=1}^{[\frac{r}{2}]} \sum_{(2h,r-2h)\vdash \mathbf{A}_r} \langle \mathbf{A}_{r(1)} \rangle \langle \mathbf{A}_{r(2)} \mathbf{B}_{2k+2l-r} \rangle$$
$$\left\{ (-1)^h C_{k+l-r}^k + \sum_{i=1}^{h} (-1)^{i+1} C_h^i C_{k+l-r+i}^{k-h+i} \right\}.$$

So (6.4.13) is reduced to the following identity: for any $h > 0$,

$$(-1)^h C_{k+l-r}^k + \sum_{i=1}^{h} (-1)^{i+1} C_h^i C_{k+l-r+i}^{k-h+i} = C_{k+l-r}^{k-h}. \tag{6.4.14}$$

We prove (6.4.14) by induction. When $h = 1$, (6.4.14) is just $-C_{k+l-r}^k + C_{k+l-r+1}^k = C_{k+l-r}^{k-1}$, which is true by (6.4.12). Assume that (6.4.14) is true for $h = j - 1$. When $h = j$, by (6.4.12) and induction hypothesis,

$$(-1)^h C_{k+l-r}^k + \sum_{i=1}^{h} (-1)^{i+1} C_h^i C_{k+l-r+i}^{k-h+i}$$
$$= (-1)^h C_{k+l-r}^k + \sum_{i=1}^{h} (-1)^{i+1} \left(C_{h-1}^i C_{k+l-r+i+1}^{k-h+i+1} - C_{h-1}^i C_{k+l-r+i}^{k-h+i+1} \right.$$
$$\left. + C_{h-1}^{i-1} C_{k+l-r+i+1}^{k-h+i+1} - C_{h-1}^{i-1} C_{k+l-r+i}^{k-h+i+1} \right)$$

$$= (-1)^h C_{k+l-r}^k + \sum_{i=1}^{h-1} (-1)^{i+1} C_{h-1}^i C_{k+(l-r+1)+i}^{k-(h-1)+i} - \sum_{i=1}^{h-1} (-1)^{i+1} C_{h-1}^i C_{k+l-r+i}^{k-(h-1)+i}$$
$$- \sum_{i=0}^{h-1} (-1)^{i+1} C_{h-1}^i C_{(k+1)+(l-r+1)+i}^{(k+1)-(h-1)+i} + \sum_{i=0}^{h-1} (-1)^{i+1} C_{h-1}^i C_{(k+1)+(l-r)+i}^{(k+1)-(h-1)+i}$$

$$= C_{k+l-r+1}^{k-(h-1)} + (-1)^h C_{k+l-r+1}^k - C_{k+l-r}^{k-(h-1)} - (-1)^h C_{k+l-r+2}^{k+1} + (-1)^h C_{k+l-r+1}^{k+1}$$

$$= C_{k+l-r}^{k-h}.$$

$\square$

Theorem 6.101. [$(0, n)$-typed summation] Let $k, l \geq 0$, and let $u = n + 2k + 2l$. For any $0 \leq m \leq 2k$, let $\mathbf{A}_m = \mathbf{a}_1 \mathbf{a}_2 \cdots \mathbf{a}_m$ and $\mathbf{B}_{u-m} = \mathbf{a}_{m+1} \mathbf{a}_{m+2} \cdots \mathbf{a}_u$. Define

$$T_{k,l}(\mathbf{A}_m) = \sum_{(2k-m,u-2k)\vdash \mathbf{B}_{u-m}} \langle \mathbf{A}_m \mathbf{B}_{u-m(1)} \rangle \langle \mathbf{B}_{u-m(2)} \rangle_n. \tag{6.4.15}$$

Then

$$T_{k,l}(\mathbf{A}_m) = \begin{cases} 0, & \text{if } m \text{ is odd;} \\ C^l_{k+l-i}\langle\mathbf{A}_m\rangle\,\langle\mathbf{B}_{u-m}\rangle_n, & \text{if } m = 2i. \end{cases} \tag{6.4.16}$$

Proof. By (5.4.17),

$$\begin{aligned}
T_{k,l}(\mathbf{A}_m) &= \sum_{(2k-m,\,u-2k)\vdash\mathbf{B}_{u-m}} \sum_{i=0}^{[\frac{m}{2}]} (\langle\mathbf{A}_m\rangle_{m-2i} \cdot \langle\mathbf{B}_{u-m\,(1)}\rangle_{m-2i})\,\langle\mathbf{B}_{u-m\,(2)}\rangle_n \\
&= \sum_{i=0}^{[\frac{m}{2}]} C^{k-m+i}_{k+l-m+i}\,\langle\langle\mathbf{A}_m\rangle_{m-2i} \cdot \langle\mathbf{B}_{u-m}\rangle_{n+m-2i}\rangle_n \\
&= \begin{cases} 0, & \text{if } m \text{ is odd;} \\ C^l_{k+l-h}\,\langle\mathbf{A}_{2h}\rangle\,\langle\mathbf{B}_{u-2h}\rangle_n, & \text{if } m = 2h. \end{cases}
\end{aligned}$$

$\square$

Theorem 6.102. [(n,n)-typed summation] Let $k, l \geq 0$, and let $u = 2n + 2k + 2l$. For any $0 \leq m \leq n + 2l$, let $\mathbf{A}_m = \mathbf{a}_1\mathbf{a}_2\cdots\mathbf{a}_m$ and $\mathbf{B}_{u-m} = \mathbf{a}_{m+1}\mathbf{a}_{m+2}\cdots\mathbf{a}_u$. Define

$$U_{k,l}(\mathbf{A}_m) = \sum_{(n+2k-m,\,n+2l)\vdash\mathbf{B}_{u-m}} \langle\mathbf{A}_m\mathbf{B}_{u-m\,(1)}\rangle_n\,\langle\mathbf{B}_{u-m\,(2)}\rangle_n. \tag{6.4.17}$$

Then

$$U_{k,l}(\mathbf{A}_m) = \begin{cases} 0, & \text{if either } m < n, \text{ or } m - n \text{ is odd;} \\ C^l_{k+l-i}\langle\mathbf{A}_m\rangle_n\,\langle\mathbf{B}_{u-m}\rangle_n, & \text{if } m - n = 2i. \end{cases}$$

$$\tag{6.4.18}$$

Proof. By (5.4.2) and (6.4.6),

$$\begin{aligned}
U_{k,l}(\mathbf{A}_m) &= \sum_{(n+2k-m,\,n+2l)\vdash\mathbf{B}_{u-m}} \sum_{i=0}^{[\frac{m}{2}]} (\langle\mathbf{A}_m\rangle_{m-2i} \wedge \langle\mathbf{B}_{u-m\,(1)}\rangle_{n-m+2i})\,\langle\mathbf{B}_{u-m\,(2)}\rangle_n \\
&= \sum_{i=\max(0,\,[\frac{m-n+1}{2}])} \sum_{\substack{(n+2k-m,\,n+2l) \\ \vdash\mathbf{B}_{u-m}}}^{[\frac{m}{2}]} \langle\mathbf{A}_m\rangle_{m-2i} \cdot (\langle\mathbf{B}_{u-m\,(1)}\rangle_{n-m+2i} \cdot \langle\mathbf{B}_{u-m\,(2)}\rangle_n) \\
&= \begin{cases} 0, & \text{if } m - n \text{ is either odd or negative;} \\ C^l_{k+l-h}\,\langle\mathbf{A}_m\rangle_n\,\langle\mathbf{B}_{u-m}\rangle_n, & \text{if } m - n = 2h. \end{cases}
\end{aligned}$$

$\square$

6.4.3 *Three Clifford monomials*

The following theorem on the Clifford summation of three Clifford monomials is a corollary of (6.4.16), (6.4.18), and Lemma 6.105 to be introduced in this subsection.

Theorem 6.103. Let $\mathbf{A}_i = \mathbf{a}_1\mathbf{a}_2\cdots\mathbf{a}_i$, $\mathbf{B}_j = \mathbf{b}_1\mathbf{b}_2\cdots\mathbf{b}_j$, and $\mathbf{C}_u = \mathbf{c}_1\mathbf{c}_2\cdots\mathbf{c}_u$, where the $\mathbf{a}_p, \mathbf{b}_q, \mathbf{c}_r$ are vectors in $\mathcal{V}^n$.

(1) If $u = n + 2k + 2l - i - j$, where $i \leq 2k$ and $j \leq 2l$, then

$$\sum_{(2k-i,\,n+2l-j)\vdash\mathbf{C}_u} \langle \mathbf{A}_i\mathbf{C}_{u(1)}\rangle\,[\mathbf{B}_j\mathbf{C}_{u(2)}] = \sum_{m=[\frac{i+1}{2}]}^{\min(k,\,[\frac{i+j}{2}])} \sum_{(2m-i,\,i+j-2m)\vdash\mathbf{B}_j}$$

$$(-1)^{m+ij+\frac{i(i-1)}{2}}\, C_{k+l-m}^{l}\,\langle \mathbf{A}_i\mathbf{B}_{j\,(1)}^{\dagger}\rangle\,[\mathbf{B}_{j\,(2)}\mathbf{C}_u]. \qquad (6.4.19)$$

(2) If $u = 2n + 2k + 2l - i - j$, where $i \leq n + 2k$ and $j \leq n + 2l$, then

$$\sum_{(n+2k-i,\,n+2l-j)\vdash\mathbf{C}_u} [\mathbf{A}_i\mathbf{C}_{u(1)}]\,[\mathbf{B}_j\mathbf{C}_{u(2)}] = \sum_{m=\max(0,\,[\frac{i+1-n}{2}])}^{\min(k,\,[\frac{i+j-n}{2}])} \sum_{\substack{(n+2m-i,\\ i+j-n-2m)\vdash\mathbf{B}_j}}$$

$$(-1)^{m+j(n-i)+\frac{(n-i)(n-i+1)}{2}}\, C_{k+l-m}^{l}\,[\mathbf{A}_i\mathbf{B}_{j\,(1)}^{\dagger}]\,[\mathbf{B}_{j\,(2)}\mathbf{C}_u]. \qquad (6.4.20)$$

In particular, (6.4.20) equals zero when $i + j < n$.

Example 6.104. Set $n = 4$, $i = j = k = l = 2$, $u = 8$ in (6.4.19):

$$\langle\mathbf{a}_1\mathbf{a}_2\mathbf{c}_1\mathbf{c}_2\rangle[\mathbf{b}_1\mathbf{b}_2\mathbf{c}_3\mathbf{c}_4\mathbf{c}_5\mathbf{c}_6\mathbf{c}_7\mathbf{c}_8] - \langle\mathbf{a}_1\mathbf{a}_2\mathbf{c}_1\mathbf{c}_3\rangle[\mathbf{b}_1\mathbf{b}_2\mathbf{c}_2\mathbf{c}_4\mathbf{c}_5\mathbf{c}_6\mathbf{c}_7\mathbf{c}_8] + \cdots$$
$$+ \langle\mathbf{a}_1\mathbf{a}_2\mathbf{c}_7\mathbf{c}_8\rangle[\mathbf{b}_1\mathbf{b}_2\mathbf{c}_1\mathbf{c}_2\mathbf{c}_3\mathbf{c}_4\mathbf{c}_5\mathbf{c}_6]\ (28\ \text{terms})$$
$$= 3\,(\mathbf{a}_1\cdot\mathbf{a}_2)\,[\mathbf{b}_1\mathbf{b}_2\mathbf{c}_1\mathbf{c}_2\mathbf{c}_2\mathbf{c}_4\mathbf{c}_5\mathbf{c}_6\mathbf{c}_7\mathbf{c}_8] - \langle\mathbf{a}_1\mathbf{a}_2\mathbf{b}_2\mathbf{b}_1\rangle\,[\mathbf{c}_1\mathbf{c}_2\mathbf{c}_2\mathbf{c}_4\mathbf{c}_5\mathbf{c}_6\mathbf{c}_7\mathbf{c}_8].$$

Set $n = u = 4$, $i = j = 2$, $k = l = 0$ in (6.4.20):

$$[\mathbf{a}_1\mathbf{a}_2\mathbf{c}_1\mathbf{c}_2]\,[\mathbf{b}_1\mathbf{b}_2\mathbf{c}_3\mathbf{c}_4] - [\mathbf{a}_1\mathbf{a}_2\mathbf{c}_1\mathbf{c}_3]\,[\mathbf{b}_1\mathbf{b}_2\mathbf{c}_2\mathbf{c}_4] + [\mathbf{a}_1\mathbf{a}_2\mathbf{c}_1\mathbf{c}_4]\,[\mathbf{b}_1\mathbf{b}_2\mathbf{c}_2\mathbf{c}_3]$$
$$+ [\mathbf{a}_1\mathbf{a}_2\mathbf{c}_2\mathbf{c}_3]\,[\mathbf{b}_1\mathbf{b}_2\mathbf{c}_1\mathbf{c}_4] - [\mathbf{a}_1\mathbf{a}_2\mathbf{c}_2\mathbf{c}_4]\,[\mathbf{b}_1\mathbf{b}_2\mathbf{c}_1\mathbf{c}_3] + [\mathbf{a}_1\mathbf{a}_2\mathbf{c}_3\mathbf{c}_4]\,[\mathbf{b}_1\mathbf{b}_2\mathbf{c}_1\mathbf{c}_2]$$
$$= [\mathbf{a}_1\mathbf{a}_2\mathbf{b}_1\mathbf{b}_2]\,[\mathbf{c}_1\mathbf{c}_2\mathbf{c}_3\mathbf{c}_4].$$

Set $n = i = 4$, $j = 2$, $k = l = 1$, $u = 6$ in (6.4.20):

$$[\mathbf{a}_1\mathbf{a}_2\mathbf{a}_3\mathbf{a}_4\mathbf{c}_1\mathbf{c}_2]\,[\mathbf{b}_1\mathbf{b}_2\mathbf{c}_3\mathbf{c}_4\mathbf{c}_5\mathbf{c}_6] - [\mathbf{a}_1\mathbf{a}_2\mathbf{a}_3\mathbf{a}_4\mathbf{c}_1\mathbf{c}_3]\,[\mathbf{b}_1\mathbf{b}_2\mathbf{c}_2\mathbf{c}_4\mathbf{c}_5\mathbf{c}_6] + \cdots$$
$$+ [\mathbf{a}_1\mathbf{a}_2\mathbf{a}_3\mathbf{a}_4\mathbf{c}_5\mathbf{c}_6]\,[\mathbf{b}_1\mathbf{b}_2\mathbf{c}_1\mathbf{c}_2\mathbf{c}_3\mathbf{c}_4]\ (15\ \text{terms})$$
$$= [\mathbf{a}_1\mathbf{a}_2\mathbf{a}_3\mathbf{a}_4]\,[\mathbf{b}_1\mathbf{b}_2\mathbf{c}_1\mathbf{c}_2\mathbf{c}_3\mathbf{c}_4\mathbf{c}_5\mathbf{c}_6] - [\mathbf{a}_1\mathbf{a}_2\mathbf{a}_3\mathbf{a}_4\mathbf{b}_2\mathbf{b}_1]\,[\mathbf{c}_1\mathbf{c}_2\mathbf{c}_3\mathbf{c}_4\mathbf{c}_5\mathbf{c}_6].$$

Lemma 6.105. Let k, l, s, t be nonnegative integers, and let $u = s + t + 2k + 2l$. Let $\mathbf{A}_u = \mathbf{a}_1\mathbf{a}_2\cdots\mathbf{a}_u$. For any $i \leq t + 2l$ and $j \leq s + 2k$, let $v = s + 2k - i$, $w = t + 2l - j$. For any partition $(\mathbf{A}_{u(1)}, \mathbf{A}_{u(2)}, \mathbf{A}_{u(3)})$ of $\mathbf{A}_u$ of shape $(i, j, u - i - j)$, define

$$\mathbf{W}(\mathbf{A}_{u(1)}, \mathbf{A}_{u(2)}) = \sum_{(v,w)\vdash\mathbf{A}_{u(3)}} \langle\mathbf{A}_{u(1)}\mathbf{A}_{u(31)}\rangle_s\,\langle\mathbf{A}_{u(2)}\mathbf{A}_{u(32)}\rangle_t. \qquad (6.4.21)$$

For $\mathbf{A}_j = \mathbf{a}_1\mathbf{a}_2\cdots\mathbf{a}_j$, $\mathbf{B}_i = \mathbf{a}_{j+1}\mathbf{a}_{j+2}\cdots\mathbf{a}_{j+i}$, $\mathbf{C}_{u-i-j} = \mathbf{a}_{j+i+1}\mathbf{a}_{j+i+2}\cdots\mathbf{a}_u$:

(1) If $i + j \leq s + 2k$, then

$$\mathbf{W}\left(\mathbf{B}_i, \mathbf{A}_j\right) = \sum_{r=0}^{j} \sum_{(r,j-r)\vdash \mathbf{A}_j} (-1)^{j(s+i)+\frac{r(r-1)}{2}} \mathbf{W}(\mathbf{B}_i \mathbf{A}_{j\,(1)}^{\dagger}, 1). \qquad (6.4.22)$$

(2) If $i + j > s + 2k$, then

$$\begin{aligned}
\mathbf{W}\left(\mathbf{B}_i, \mathbf{A}_j\right) = {} & \sum_{r=0}^{v-1} \sum_{(r,j-r)\vdash \mathbf{A}_j} (-1)^{j(s+i)+\frac{r(r-1)}{2}} \mathbf{W}(\mathbf{B}_i \mathbf{A}_{j\,(1)}^{\dagger}, 1) \\
& + \sum_{(v,j-v)\vdash \mathbf{A}_j} (-1)^{j+k-1+\frac{(s-i)(s-i-1)}{2}} \langle \mathbf{B}_i \mathbf{A}_{j\,(1)}^{\dagger} \rangle_s \, \langle \mathbf{A}_{j\,(2)} \mathbf{C}_{u-i-j} \rangle_t.
\end{aligned}$$
$$(6.4.23)$$

Proof. First, we prove that for any $0 \leq h \leq \min(j, v)$, by denoting $\mathbf{D}_h = \mathbf{a}_{j-h+1}\mathbf{a}_{j-h+2}\cdots\mathbf{a}_j$,

$$\mathbf{W}\left(\mathbf{B}_i, \mathbf{A}_j\right) = \sum_{r=0}^{h} \sum_{(r,h-r)\vdash \mathbf{D}_h} (-1)^{h(s+i)+\frac{r(r-1)}{2}} \mathbf{W}(\mathbf{B}_i \mathbf{D}_{h\,(1)}^{\dagger}, \mathbf{A}_{j-h}). \qquad (6.4.24)$$

When $h = 1$, (6.4.24) can be proved as follows:

$$\begin{aligned}
\mathbf{W}\left(\mathbf{B}_i, \mathbf{A}_{j-1}\right) = {} & \sum_{(v,u-i-j-v+1)\vdash \mathbf{a}_j \mathbf{C}_{u-i-j}} \langle \mathbf{B}_i(\mathbf{a}_j \mathbf{C}_{u-i-j})_{(1)} \rangle_s \, \langle \mathbf{A}_{j-1}(\mathbf{a}_j \mathbf{C}_{u-i-j})_{(2)} \rangle_t \\[2mm]
= {} & (-1)^{s+i} \sum_{(v,u-i-j-v)\vdash \mathbf{C}_{u-i-j}} \langle \mathbf{B}_i \mathbf{C}_{u-i-j\,(1)} \rangle_s \, \langle \mathbf{A}_j \mathbf{C}_{u-i-j\,(2)} \rangle_t \\[2mm]
& + \sum_{(v-1,u-i-j-v+1)\vdash \mathbf{C}_{u-i-j}} \langle \mathbf{B}_i \mathbf{a}_j \mathbf{C}_{u-i-j\,(1)} \rangle_s \langle \mathbf{A}_{j-1} \mathbf{C}_{u-i-j\,(2)} \rangle_t \\[2mm]
= {} & (-1)^{s+i} \mathbf{W}(\mathbf{B}_i, \mathbf{A}_j) + \mathbf{W}(\mathbf{B}_i \mathbf{a}_j, \mathbf{A}_{j-1}).
\end{aligned}$$

Assume that (6.4.24) is true for $h = m - 1$, i.e.,

$$\mathbf{W}\left(\mathbf{B}_i, \mathbf{A}_j\right) = \sum_{r=0}^{m-1} \sum_{(r,m-r-1)\vdash \mathbf{D}_{m-1}} (-1)^{(m-1)(s+i)+\frac{r(r-1)}{2}} \mathbf{W}(\mathbf{B}_i \mathbf{D}_{m-1\,(1)}^{\dagger}, \mathbf{A}_{j-m+1}).$$
$$(6.4.25)$$

Substituting

$$\mathbf{W}(\mathbf{B}_i \mathbf{D}_{m-1\,(1)}^{\dagger}, \mathbf{A}_{j-m+1}) = (-1)^{s+i+r}\{\mathbf{W}(\mathbf{B}_i \mathbf{D}_{m-1\,(1)}^{\dagger}, \mathbf{A}_{j-m}) \\ -\mathbf{W}(\mathbf{B}_i \mathbf{D}_{m-1\,(1)}^{\dagger} \mathbf{a}_{j-m+1}, \mathbf{A}_{j-m})\} \qquad (6.4.26)$$

into (6.4.25), and denoting $\mu = (-1)^{m(s+i)+\frac{r(r-1)}{2}}$, we get

$$\begin{aligned}
\mathbf{W}(\mathbf{B}_i, \mathbf{A}_j) = {} & \Big(\sum_{\substack{r=0 \\ (r,m-r)\vdash \mathbf{D}_m, \\ \mathbf{a}_{j-m+1}\notin \mathbf{D}_{m\,(1)}}} + \sum_{\substack{r=1 \\ (r,m-r)\vdash \mathbf{D}_m, \\ \mathbf{a}_{j-m+1}\in \mathbf{D}_{m\,(1)}}}^{m} \Big) \mu \mathbf{W}(\mathbf{B}_i \mathbf{D}_{m\,(1)}^{\dagger}, \mathbf{A}_{j-m}) \\[2mm]
= {} & \sum_{r=0}^{m} \sum_{(r,m-r)\vdash \mathbf{D}_m} \mu \mathbf{W}(\mathbf{B}_i \mathbf{D}_{m\,(1)}^{\dagger}, \mathbf{A}_{j-m}).
\end{aligned}$$

This proves (6.4.24).

Second, when $j \leq v$, setting $h = j$ in (6.4.24), we get (6.4.22). When $j > v$, setting $h = v$ in (6.4.24), we get

$$\mathbf{W}(\mathbf{B}_i, \mathbf{A}_j) = \sum_{r=0}^{v} \sum_{(r,v-r)\vdash \mathbf{D}_v} (-1)^{s+i+\frac{r(r-1)}{2}} \mathbf{W}(\mathbf{B}_i \mathbf{D}_{v\,(1)}^\dagger, \mathbf{A}_{j-v}). \tag{6.4.27}$$

Third, we prove that for any $0 \leq h \leq j - v$,

$$\mathbf{W}(\mathbf{B}_i, \mathbf{A}_j) = \sum_{r=0}^{v-1} \sum_{(r,v+h-r)\vdash \mathbf{D}_{v+h}} (-1)^{(h-1)(s+i)+\frac{r(r-1)}{2}} \mathbf{W}(\mathbf{B}_i \mathbf{D}_{v+h\,(1)}^\dagger, \mathbf{A}_{j-v-h})$$

$$+ \sum_{(v,h)\vdash \mathbf{D}_{v+h}} (-1)^{h(s+i)+\frac{v(v-1)}{2}} \langle \mathbf{B}_i \mathbf{D}_{v+h\,(1)}^\dagger \rangle_s \langle \mathbf{A}_{j-v-h} \mathbf{D}_{v+h\,(2)} \mathbf{C}_{u-i-j} \rangle_t.$$

$$\tag{6.4.28}$$

The case $h = 0$ is just (6.4.27). Assume that (6.4.28) is true for $h = m$. Substituting (6.4.26) into (6.4.28) for $h = m$, we get that (6.4.28) is true for $h = m + 1$. This proves (6.4.28).

Last, by setting $h = j - v$, we get (6.4.23). $\qquad\square$

6.4.4 *Clifford coproduct of blades*

Let $\mathbf{A}_r$ be an r-blade. Then there exist mutually orthogonal vectors $\mathbf{a}_1, \mathbf{a}_2, \ldots, \mathbf{a}_r$ such that $\mathbf{A}_r = \mathbf{a}_1 \mathbf{a}_2 \cdots \mathbf{a}_r$. The Clifford coproduct of $\mathbf{A}_r$ is identical to the outer coproduct of $\mathbf{A}_r$.

In Clifford coalgebra, the result of the Clifford coproduct of a blade or the geometric product of several blades, can be multiplied using the geometric product after being composed with various $\mathbb{Z}$-grading operators. We first consider the case where only one blade participates in the Clifford coproduct.

Theorem 6.106. Let $\mathbf{A}_r, \mathbf{B}_s$ be blades of grade r, s respectively.

(1) For any $0 \leq t \leq r$ and any $0 \leq l \leq \min(t, r + s - t)$,

$$\sum_{(t,r-t)\vdash \mathbf{A}_r} \langle \mathbf{A}_{r\,(1)}(\mathbf{A}_{r\,(2)} \wedge \mathbf{B}_s) \rangle_{r+s-2l} = C_{r-l}^{t-l} \langle \mathbf{A}_r \mathbf{B}_s \rangle_{r+s-2l}. \tag{6.4.29}$$

(2) For any $0 \leq t \leq s$ and any $0 \leq l \leq \min(t, r + s - t)$,

$$\sum_{(s-t,t)\vdash \mathbf{B}_s} \langle (\mathbf{A}_r \wedge \mathbf{B}_{s\,(1)}) \mathbf{B}_{s\,(2)} \rangle_{r+s-2l} = C_{s-l}^{t-l} \langle \mathbf{A}_r \mathbf{B}_s \rangle_{r+s-2l}. \tag{6.4.30}$$

Proof. By (5.4.33) and (5.4.13),

$$\sum_{(t,r-t)\vdash \mathbf{A}_r} \langle \mathbf{A}_{r\,(1)}(\mathbf{A}_{r\,(2)} \wedge \mathbf{B}_s) \rangle_{r+s-2l}$$

$$= \sum_{i=0}^{l} \sum_{\substack{(t-l,i,l-i,l-i,r-t-l+i)\vdash \mathbf{A}_r, \\ (i,s-i)\vdash \mathbf{B}_s}} (-1)^{i(r-t-l+i)} \langle \mathbf{A}_{r\,(2)} \mathbf{B}_{s\,(1)} \rangle \langle \mathbf{A}_{r\,(3)} \mathbf{A}_{r\,(4)} \rangle \mathbf{A}_{r\,(1)} \wedge \mathbf{A}_{r\,(5)} \wedge \mathbf{B}_{s\,(2)}$$

$$= \sum_{i=0}^{l} \sum_{\substack{(i,r-2l+i,l-i,l-i)\vdash \mathbf{A}_r, \\ (i,s-i)\vdash \mathbf{B}_s}} (-1)^{i(r-i)} C_{r-2l+i}^{t-l} \langle \mathbf{A}_{r(1)} \mathbf{B}_{s(1)} \rangle \langle \mathbf{A}_{r(3)} \mathbf{A}_{r(4)} \rangle \mathbf{A}_{r(2)} \wedge \mathbf{B}_{s(2)}.$$

$$(6.4.31)$$

By (6.4.6), in the result of (6.4.31), only the term corresponding to $i = l$ is nonzero. So

$$\sum_{(t,r-t)\vdash \mathbf{A}_r} \langle \mathbf{A}_{r(1)} (\mathbf{A}_{r(2)} \wedge \mathbf{B}_s) \rangle_{r+s-2l} = (-1)^{l(r-l)} C_{r-l}^{t-l} \sum_{\substack{(l,r-l)\vdash \mathbf{A}_r, \\ (l,s-l)\vdash \mathbf{B}_s}} \langle \mathbf{A}_{r(1)} \mathbf{B}_{s(1)} \rangle \mathbf{A}_{r(2)} \wedge \mathbf{B}_{s(2)}$$

$$= C_{r-l}^{t-l} \langle \mathbf{A}_r \mathbf{B}_s \rangle_{r+s-2l}.$$

(6.4.30) can be derived from (6.4.29) as follows:

$$C_{s-l}^{t-l} \langle \mathbf{A}_r \mathbf{B}_s \rangle_{r+s-2l} = (-1)^{rs-l} C_{s-l}^{t-l} \langle \mathbf{B}_s \mathbf{A}_r \rangle_{r+s-2l}$$

$$= (-1)^{rs-l} \sum_{(t,s-t)\vdash \mathbf{B}_s} \langle \mathbf{B}_{s(1)} (\mathbf{B}_{s(2)} \wedge \mathbf{A}_r) \rangle_{t+(s-t+r)-2l}$$

$$= (-1)^{rs-l+t(s-t+r)-l} \sum_{(t,s-t)\vdash \mathbf{B}_s} \langle (\mathbf{B}_{s(2)} \wedge \mathbf{A}_r) \mathbf{B}_{s(1)} \rangle_{r+s-2l}$$

$$= \sum_{(s-t,t)\vdash \mathbf{B}_s} \langle (\mathbf{A}_r \wedge \mathbf{B}_{s(1)}) \mathbf{B}_{s(2)} \rangle_{r+s-2l}.$$

$$\square$$

Corollary 6.107. For any r-blade $\mathbf{A}_r$ and s-blade $\mathbf{B}_s$, for any $0 < t < r$,

$$\sum_{(t,r-t)\vdash \mathbf{A}_r} \mathbf{A}_{r(1)} \cdot (\mathbf{A}_{r(2)} \wedge \mathbf{B}_s) = \begin{cases} 0, & \text{if } \dfrac{r+s}{2} < t < r; \\ \langle \mathbf{A}_r \mathbf{B}_s \rangle_{r+s-2t}, & \text{if } 1 \leq t \leq \dfrac{r+s}{2}. \end{cases} \quad (6.4.32)$$

As a special case of Corollary 6.107, we have

Proposition 6.108. [70] [Seidel's identity] Let $\mathbf{A}_r = \mathbf{C}_t \wedge \mathbf{A}'_{r-t}$ and $\mathbf{B}_s = \mathbf{C}_t \wedge \mathbf{B}'_{s-t}$, where $\mathbf{A}'_{r-t}, \mathbf{B}'_{s-t}, \mathbf{C}_t$ are blades of grade $r-t, s-t, t$ respectively. Then

$$\langle \mathbf{A}_r \mathbf{B}_s \rangle_{r+s-2t} = \mathbf{C}_t \cdot (\mathbf{A}'_{r-t} \wedge \mathbf{B}_s). \quad (6.4.33)$$

Theorem 6.109. Let $\mathbf{A}_r, \mathbf{B}_s$ be blades of grade r, s respectively. Then $\langle \mathbf{A}_r \mathbf{B}_s \rangle_{\max}$ and $\langle \mathbf{A}_r \mathbf{B}_s \rangle_{\min}$ are both blades. Furthermore, $\mathbf{A}_r \mathbf{B}_s \in \Lambda(\langle \mathbf{A}_r \mathbf{B}_s \rangle_{\max})$. Geometrically, $\langle \mathbf{A}_r \mathbf{B}_s \rangle_{\max}$ represents the orthogonal complement of subspace $\mathbf{A}_r \sqcup \mathbf{B}_s$ in space $\mathbf{A}_r \sqcap \mathbf{B}_s$.

Proof. We assume that $\mathbf{A}_r \mathbf{B}_s \neq 0$. In (6.4.33), let $\mathbf{C}_t = \mathbf{A}_r \sqcup \mathbf{B}_s$. Then $\mathbf{C}_t$ does not contain any vector in either $\mathrm{rad}(\mathbf{A}_r)$ or $\mathrm{rad}(\mathbf{B}_s)$.

Obviously $r + s - 2t$ is the maximal grade of which $\mathbf{A}_r \mathbf{B}_s$ possibly has nonzero component. We only need to prove that $\mathbf{C}_t \cdot (\mathbf{A}'_{r-t} \wedge \mathbf{B}_s) \neq 0$. By Witt Theorem, we only need to consider the following three cases:

Case a. $\mathbf{C}_t = \mathbf{a}$ is an invertible vector. $\mathbf{A}'_{r-1}, \mathbf{B}'_{s-1}$ can be chosen to be orthogonal to $\mathbf{a}$. So $\mathbf{A}_r = \mathbf{a}\mathbf{A}'_{r-1}$, $\mathbf{B}_s = \mathbf{a}\mathbf{B}'_{s-1}$, and

$$\begin{aligned}
\mathbf{C}_t \cdot (\mathbf{A}'_{r-1} \wedge \mathbf{B}_s) &= (-1)^{r-1}\mathbf{a}^2 \mathbf{A}'_{r-1} \wedge \mathbf{B}'_{s-1} \neq 0, \\
\mathbf{A}_r \mathbf{B}_s &= (-1)^{r-1}\mathbf{a}^2 \mathbf{A}'_{r-1}\mathbf{B}'_{s-1}.
\end{aligned} \tag{6.4.34}$$

Case b. $\mathbf{C}_t = \mathbf{a} \wedge \mathbf{b}$, where $(\mathbf{a}, \mathbf{b})$ is a Witt pair. $\mathbf{A}'_{r-2}, \mathbf{B}'_{s-2}$ can be chosen to be orthogonal to both $\mathbf{a}$ and $\mathbf{b}$. So $\mathbf{A}_r = (\mathbf{a} \wedge \mathbf{b})\mathbf{A}'_{r-2}$, $\mathbf{B}_s = (\mathbf{a} \wedge \mathbf{b})\mathbf{B}'_{s-2}$, and

$$\begin{aligned}
\mathbf{C}_t \cdot (\mathbf{A}'_{r-1} \wedge \mathbf{B}_s) &= \mathbf{A}'_{r-2} \wedge \mathbf{B}'_{s-2} \neq 0, \\
\mathbf{A}_r \mathbf{B}_s &= \mathbf{A}'_{r-2}\mathbf{B}'_{s-2}.
\end{aligned} \tag{6.4.35}$$

Case c. $\mathbf{C}_t = \mathbf{a}$ is null, $\mathbf{a} \cdot \mathbf{A}'_{r-1} \neq 0$, and $\mathbf{a} \cdot \mathbf{B}'_{s-1} \neq 0$.

There exists a vector $\mathbf{b}' \in \mathbf{B}'_{s-1}$ such that $\mathbf{a} \cdot \mathbf{b}' \neq 0$. The 2D subspace $\mathbf{a} \wedge \mathbf{b}'$ is nondegenerate, so there is a vector $\mathbf{b} \in \mathbf{a} \wedge \mathbf{b}'$ such that $(\mathbf{a}, \mathbf{b})$ is a Witt pair. Then $\mathbf{b} \in \mathbf{B}'_{s-1}$ but $\mathbf{b} \notin \mathbf{A}'_{r-1}$, otherwise the case is identical to Case b.

Extend $(\mathbf{a}, \mathbf{b})$ to a Witt basis of $\mathbf{B}_s$, and then further extend it to a Witt basis of $\mathcal{V}^n$. With respect to this basis, blades $\mathbf{A}_r$ and $\mathbf{B}_s$ have the following decompositions that are similar to (6.3.45):

$$\begin{aligned}
\mathbf{A}_r &= \mathbf{a} \wedge (\mathbf{C}_{r-1} + \mathbf{b} \wedge \mathbf{D}_{r-2}) = \mathbf{a}\mathbf{C}_{r-1} + (\mathbf{a} \wedge \mathbf{b})\mathbf{D}_{r-2}), \\
\mathbf{B}_s &= \qquad \mathbf{a} \wedge \mathbf{b} \wedge \mathbf{E}_{s-2} \qquad = (\mathbf{a} \wedge \mathbf{b})\mathbf{E}_{s-2}.
\end{aligned} \tag{6.4.36}$$

Here blades $\mathbf{C}_{r-1}, \mathbf{D}_{r-2}, \mathbf{E}_{s-2}$ are orthogonal to both $\mathbf{a}$ and $\mathbf{b}$, moreover, $\mathbf{C}_{r-1} \neq 0$, and $\mathbf{D}_{r-2} \in \Lambda(\mathbf{C}_{r-1})$. Then

$$\begin{aligned}
\mathbf{C}_t \cdot (\mathbf{A}'_{r-1} \wedge \mathbf{B}_s) &= \mathbf{a} \wedge \mathbf{C}_{r-1} \wedge \mathbf{E}_{s-2} \neq 0, \\
\mathbf{A}_r \mathbf{B}_s &= \mathbf{a}\mathbf{C}_{r-1}\mathbf{E}_{s-2} + \mathbf{D}_{r-2}\mathbf{E}_{s-2}.
\end{aligned} \tag{6.4.37}$$

If $\mathcal{V}^n$ is nondegenerate, by

$$\langle \mathbf{A}_r \mathbf{B}_s \rangle_{\min} = (\langle \mathbf{A}_r (\mathbf{B}_s^{\sim}) \rangle_{\max})^{-\sim}, \tag{6.4.38}$$

$\langle \mathbf{A}_r \mathbf{B}_s \rangle_{\min}$ is a blade. If $\mathcal{V}^n$ is degenerate, then $\mathcal{V}^n$ can be extended to a larger inner-product space $\mathcal{W}^m$ that is nondegenerate. With respect to the dual operator in $\mathcal{CL}(\mathcal{W}^m)$, (6.4.38) remains valid. $\qquad\square$

Theorem 6.110. For any r-blade $\mathbf{A}_r$ and s-blade $\mathbf{B}_s$, for any $0 \leq t \leq r$, define

$$\mathbf{P} = \sum_{(t,r-t) \vdash \mathbf{A}_r} \mathbf{A}_{r(1)}(\mathbf{A}_{r(2)} \cdot \mathbf{B}_s).$$

Then if $0 \leq t \leq r - s$,

$$\mathbf{P} = C_{r-s}^t \mathbf{A}_r \cdot \mathbf{B}_s. \tag{6.4.39}$$

If $\max(0, r - s) \leq t < r$, then for any $0 \leq l \leq \min(t, s - r + t)$,

$$\langle \mathbf{P} \rangle_{s-r+2t-2l} = C_{r+l-t}^l \langle \mathbf{A}_r \mathbf{B}_s \rangle_{s-r+2t-2l}. \tag{6.4.40}$$

Proof. By (6.4.6), if $r - t \geq s$, then $\mathbf{P} = \langle \mathbf{P} \rangle_{r-s}$. So

$$
\langle \mathbf{P} \rangle_{r-s} = \sum_{(t,r-t) \vdash \mathbf{A}_r} \mathbf{A}_{r(1)} \wedge (\mathbf{A}_{r(2)} \cdot \mathbf{B}_s)
$$

$$
= \sum_{(t,r-t-s,s) \vdash \mathbf{A}_r} \mathbf{A}_{r(1)} \wedge \mathbf{A}_{r(2)} (\mathbf{A}_{r(3)} \cdot \mathbf{B}_s)
$$

$$
= \sum_{(r-s,s) \vdash \mathbf{A}_r} C^t_{r-s} \, \mathbf{A}_{r(1)} (\mathbf{A}_{r(2)} \cdot \mathbf{B}_s)
$$

$$
= C^t_{r-s} \, \mathbf{A}_r \cdot \mathbf{B}_s.
$$

When $r - t \leq s$, by (5.4.37),

$$
\langle \mathbf{P} \rangle_{s-r+2t-2l} = \sum_{\substack{(t-l,l,r-t) \vdash \mathbf{A}_r, \\ (r-t,l,s-r+t-l) \vdash \mathbf{B}_s}} \langle \mathbf{A}_{r(3)} \mathbf{B}_{s(1)} \rangle \langle \mathbf{A}_{r(2)} \mathbf{B}_{s(2)} \rangle \mathbf{A}_{r(1)} \wedge \mathbf{B}_{s(3)}
$$

$$
= \sum_{\substack{(t-l,r-t+l) \vdash \mathbf{A}_r, \\ (r-t+l,s-r+t-l) \vdash \mathbf{B}_s}} C^l_{r+l-t} \langle \mathbf{A}_{r(2)} \mathbf{B}_{s(1)} \rangle \mathbf{A}_{r(1)} \wedge \mathbf{B}_{s(2)}
$$

$$
= C^l_{r+l-t} \, \langle \mathbf{A}_r \mathbf{B}_s \rangle_{r+s-2(r+l-t)}.
$$

$$\square$$

Theorem 6.111. For any r-blade $\mathbf{A}_r$ and s-blade $\mathbf{B}_s$, for any $0 \leq t \leq r$,

$$
\sum_{(t,r-t) \vdash \mathbf{A}_r} \langle \mathbf{A}_{r(1)} \mathbf{B}_s \mathbf{A}_{r(2)} \rangle_{r+s-2l} = (-1)^{st} b(r,t,l) \langle \mathbf{B}_s \mathbf{A}_r \rangle_{r+s-2l}, \tag{6.4.41}
$$

where

$$
b(r,t,l) = \sum_{i=\max(0,t+l-r)}^{\min(t,l)} (-1)^i C^i_l C^{t-i}_{r-l} \tag{6.4.42}
$$

is the coefficient of x^t in the polynomial $(1-x)^l (1+x)^{r-l}$.

Proof. By (6.4.6), the inner product of any two vectors in $\mathbf{A}_r$ cannot occur in the Clifford expansion of the left side of (6.4.41), so

$$
\sum_{(t,r-t) \vdash \mathbf{A}_r} \langle \mathbf{A}_{r(1)} \mathbf{B}_s \mathbf{A}_{r(2)} \rangle_{r+s-2l}
$$

$$
= \sum_{i=\max(0,l+t-r)}^{\min(t,l)} \sum_{\substack{(t-i,i,l-i,r-t-l+i) \vdash \mathbf{A}_r, \\ (i,s-l,l-i) \vdash \mathbf{B}_s}} \langle \mathbf{A}_{r(2)} \mathbf{B}_{s(1)} \rangle \langle \mathbf{A}_{r(3)} \mathbf{B}_{s(3)} \rangle \mathbf{A}_{r(1)} \wedge \mathbf{B}_{s(2)} \wedge \mathbf{A}_{r(4)}
$$

$$
= \sum_{i=\max(0,l+t-r)}^{\min(t,l)} (-1)^{ts+i} \sum_{\substack{(l-i,i,t-i,r-t-l+i) \vdash \mathbf{A}_r, \\ (s-l,i,l-i) \vdash \mathbf{B}_s}} \langle \mathbf{A}_{r(2)} \mathbf{B}_{s(2)} \rangle \langle \mathbf{A}_{r(1)} \mathbf{B}_{s(3)} \rangle \mathbf{B}_{s(1)} \wedge \mathbf{A}_{r(3)} \wedge \mathbf{A}_{r(4)}
$$

$$= \sum_{i=\max(0,l+t-r)}^{\min(t,l)} (-1)^{ts+i} C_{r-l}^{t-i} \sum_{\substack{(l-i,i,r-l)\vdash \mathbf{A}_r, \\ (s-l,i,l-i)\vdash \mathbf{B}_s}} \langle \mathbf{A}_{r(2)}\mathbf{B}_{s(2)}\rangle \langle \mathbf{A}_{r(1)}\mathbf{B}_{s(3)}\rangle \mathbf{B}_{s(1)} \wedge \mathbf{A}_{r(3)}$$

$$= \sum_{i=\max(0,l+t-r)}^{\min(t,l)} (-1)^{ts+i} C_{r-l}^{t-i} C_l^i \sum_{\substack{(l,r-l)\vdash \mathbf{A}_r, \\ (s-l,l)\vdash \mathbf{B}_s}} \langle \mathbf{A}_{r(1)}\mathbf{B}_{s(2)}\rangle \, \mathbf{B}_{s(1)} \wedge \mathbf{A}_{r(2)}$$

$$= (-1)^{st} b(r,t,l) \langle \mathbf{B}_s \mathbf{A}_r \rangle_{r+s-2l}.$$

$\qquad\qquad\qquad\qquad\qquad\qquad\qquad\qquad\qquad\qquad\qquad\qquad\qquad\qquad$ $\square$

Next we consider the case where two blades participate in the Clifford coproduct. By Proposition 6.94,

$$\Delta(\mathbf{A}_r \mathbf{B}_s) = \Delta(\mathbf{A}_r)\Delta(\mathbf{B}_s), \tag{6.4.43}$$

where the juxtaposition on the right side denotes the geometric product in the twisted tensor product $\mathcal{G}(\mathcal{V}^n) \hat{\otimes} \mathcal{G}(\mathcal{V}^n)$.

As an example, consider the following nested summation:

$$\mathbf{Q} = \sum_{\substack{(r-r',r')\vdash \mathbf{A}_r, \\ (s',s-s')\vdash \mathbf{B}_s}} \langle \mathbf{A}_{r(1)}(\mathbf{A}_{r(2)} \wedge \mathbf{B}_{s(1)})\mathbf{B}_{s(2)}\rangle_{r+s-2l}. \tag{6.4.44}$$

Theorem 6.112. For any r-blade $\mathbf{A}_r$ and s-blade $\mathbf{B}_s$, for any $0 \le r' \le r$, any $0 \le s' \le s$, and any $0 \le l \le \min(r,s)$,

$$\mathbf{Q} = d(r,r',s,s',l) \, \langle \mathbf{A}_r \mathbf{B}_s \rangle_{r+s-2l}, \tag{6.4.45}$$

where

$$d(r,r',s,s',l) = \sum_{\substack{i_1+i_2=i_3+i_4=l,\, i_1 \le r-r',\, i_2 \le r', \\ i_3 \le \min(l-i_2,s'),\, i_4 \le s-s'}} C_{r-l}^{r'-i_2} C_l^{i_3} C_l^{i_1} C_{s-l}^{s'-i_3}$$

$$\tag{6.4.46}$$

$$= \sum_{i_2=\max(0,l-r+r')}^{\min(l,r')} \sum_{i_3=\max(0,l-s+s')}^{\min(l-i_2,s')} C_{r-l}^{r'-i_2} C_l^{i_3} C_l^{i_2} C_{s-l}^{s'-i_3}.$$

Proof. By (5.4.46), (6.4.6) and (5.4.13),

$$\mathbf{Q} = \sum_{\substack{i_1+i_2+i_3+i_4=2l, \\ i_1 \le \min(l,r-r'),\, i_2 \le r', \\ i_3 \le \min(l-i_2,s'),\, i_4 \le \min(l,s-s')}} \sum_{\substack{((i_1,r-r'-i_1,i_2,r'-i_2), \\ (i_3,s'-i_3,i_4,s-s'-i_4)) \\ \vdash \mathbf{A}_r \mathbf{B}_s}} \langle \mathbf{A}_{r(1)}^{(1)} \mathbf{A}_{r(3)}^{(2)} \mathbf{B}_{s(1)}^{(3)} \mathbf{B}_{s(3)}^{(4)} \rangle \, \mathbf{A}_{r(2)}^{(5)} \wedge \mathbf{A}_{r(4)}^{(6)} \wedge \mathbf{B}_{s(2)}^{(7)} \wedge \mathbf{B}_{s(4)}^{(8)}$$

$$= \sum_{\substack{i_1+i_2+i_3+i_4=2l, \\ i_1 \le \min(l,r-r'),\, i_2 \le r', \\ i_3 \le \min(l-i_2,s'),\, i_4 \le \min(l,s-s')}} \sum_{\substack{((r-r'-i_1,r'-i_2,i_1,i_2), \\ (i_3,i_4,s'-i_3,s-s'-i_4)) \\ \vdash \mathbf{A}_r \mathbf{B}_s}} \langle (\mathbf{A}_{r(3)}^{(1)} \wedge \mathbf{A}_{r(4)}^{(2)})(\mathbf{B}_{s(1)}^{(3)} \wedge \mathbf{B}_{s(2)}^{(4)}) \rangle \, \mathbf{A}_{r(1)}^{(5)} \wedge \mathbf{A}_{r(2)}^{(6)} \wedge \mathbf{B}_{s(3)}^{(7)} \wedge \mathbf{B}_{s(4)}^{(8)}$$

$$= \sum_{\substack{i_1+i_2+i_3+i_4=2l, \\ i_1\leq\min(l,r-r'),i_2\leq r', \\ i_3\leq\min(l-i_2,s'),i_4\leq\min(l,s-s')}} \sum_{\substack{((r-r'-i_1,r'-i_2,i_1,i_2), \\ (i_3,i_4,s'-i_3,s-s'-i_4)) \\ \vdash\mathbf{A}_r\mathbf{B}_s}} \langle(\mathbf{A}_{r\,(3)}^{(3)}\wedge\mathbf{A}_{r\,(4)}^{(4)})(\mathbf{B}_{s\,(1)}^{(5)}\wedge\mathbf{B}_{s\,(2)}^{(6)})\rangle\ \mathbf{A}_{r\,(1)}^{(1)}\wedge\mathbf{A}_{r\,(2)}^{(2)}\wedge\mathbf{B}_{s\,(3)}^{(7)}\wedge\mathbf{B}_{s\,(4)}^{(8)}$$

$$= \sum_{\substack{i_1+i_2=i_3+i_4=l, \\ i_1\leq r-r',i_2\leq r', \\ i_3\leq\min(l-i_2,s'),i_4\leq s-s'}} \sum_{\substack{(r-l,l)\vdash\mathbf{A}_r, \\ (l,s-l)\vdash\mathbf{B}_s}} C_{r-l}^{r'-i_2}C_l^{i_3}C_l^{i_1}C_{s-l}^{s'-i_3}\langle\mathbf{A}_{r\,(2)}\mathbf{B}_{s\,(1)}\rangle\,\mathbf{A}_{r\,(1)}\wedge\mathbf{B}_{s\,(2)}.$$

$\square$

6.5   Clifford bracket algebra

Graded inner product algebra is an algebra of advanced invariants for the orthogonal geometry of $\mathcal{V}^n$. While a graded inner-product provides the cosine of an angle formed by two high-dimensional linear subspaces, its grade is restricted to within n. The Pfaffian operator "$\langle\ \rangle$" does not have such a limitation upon the length of its contents, nor does the square bracket operator "$[\ ]$" defined by (5.4.16). They naturally extend the graded inner-product bracket algebra to an algebra of higher-level invariants, called *Clifford bracket algebra*.

Definition 6.113. [Definition of Clifford bracket algebra] Let $\mathbf{a}_1,\ldots,\mathbf{a}_m$ be symbols, called *atomic vectors*, and let $m \geq n$.

- Let the $[\mathbf{a}_{i_1}\cdots\mathbf{a}_{i_{n+2l}}]$ be indeterminates over $\mathbb{K}$ for each ordered sequence $1 \leq i_1,\ldots,i_{n+2l} \leq m$, called *square brackets*, where $l \geq 0$ is arbitrary.
- Let the $\langle\mathbf{a}_{j_1}\cdots\mathbf{a}_{j_{2l+2}}\rangle$ be indeterminates over $\mathbb{K}$ for each ordered sequence $1 \leq i_1,\ldots,i_{2l+2} \leq m$, called *angular brackets*, where $l \geq 0$ is arbitrary.
- For any $1 \leq r \leq n$, let the $(\mathbf{a}_{j_1}\ldots\mathbf{a}_{j_r}\,|\,\mathbf{a}_{k_1}\ldots\mathbf{a}_{k_r})$ be indeterminates over $\mathbb{K}$ for every ordered pair of sequences $1 \leq j_1,\ldots,j_r \leq m$ and $1 \leq k_1,\ldots,k_r \leq m$, called *r-graded inner products*. When $r = 1$, $(\mathbf{a}_j\,|\,\mathbf{a}_k)$ is identified with $\langle\mathbf{a}_j\mathbf{a}_k\rangle$.

The nD *Clifford bracket algebra* generated by the atomic vectors, is the quotient of the polynomial ring generated by the square brackets, angular brackets and graded inner products, modulo the ideal generated by the following syzygies from Definitions 5.35 and 5.37:

B1, B2, IS, IGP, BL, GI1, GI2, GI3, GIL,

together with the following two new syzygies:

AB. (angular bracket expansion)

$$\langle\mathbf{a}_{i_1}\mathbf{a}_{i_2}\cdots\mathbf{a}_{i_{2l}}\rangle - \sum_{j=2}^{2l}(-1)^j\langle\mathbf{a}_{i_1}\mathbf{a}_{i_j}\rangle\langle\mathbf{a}_{i_2}\mathbf{a}_{i_3}\cdots\breve{\mathbf{a}}_{i_j}\cdots\mathbf{a}_{i_{2l}}\rangle; \tag{6.5.1}$$

SB. (square bracket expansion) with the notation $\mathbf{A}_{n+2l} = \mathbf{a}_{i_1}\mathbf{a}_{i_2}\cdots\mathbf{a}_{i_{n+2l}}$,

$$[\mathbf{A}_{n+2l}] - \sum_{(2l,n)\vdash \mathbf{A}_{n+2l}} \langle \mathbf{A}_{n+2l\,(1)}\rangle\,[\mathbf{A}_{n+2l\,(2)}]. \tag{6.5.2}$$

In developing Clifford expansion theory and Geometric Algebra techniques, we have also developed a lot of formulas related to the angular bracket and the square bracket, using the geometric product, the inner product, the outer product, the meet product, the dual operator, and various grading operators. Now that these products and operators do not belong to Clifford bracket algebra, they pose a threat against the validity of the formulas in the setting of Clifford bracket algebra.

In the remaining part of this section, we show that all the identities in Clifford algebra involving only the two kinds of brackets, can be derived from the defining syzygies of Clifford bracket algebra.

In the statements of the following propositions, basic symmetry syzygies B1, B2, IS are not taken into account, because they are always automatically applied. As no graded inner product occurs in the propositions, syzygies GI1, GI2, GI3, GIL do not occur. So only four of the defining syzygies are left: IGP, BL, AB, SB.

Proposition 6.114. The following relations are in the ideal generated by AB:

$$\langle \mathbf{a}_1\mathbf{a}_2\cdots\mathbf{a}_{2l}\rangle = \langle \mathbf{a}_{2l}\mathbf{a}_1\mathbf{a}_2\cdots\mathbf{a}_{2l-1}\rangle = \langle \mathbf{a}_{2l}\mathbf{a}_{2l-1}\cdots\mathbf{a}_1\rangle. \tag{6.5.3}$$

Proof. When $l = 1$, (6.5.3) is just the symmetry relation IS. Assume that (6.5.3) holds for $l = k - 1$. For $l = k$, denote $\mathbf{B}_{2k-2} = \mathbf{a}_2\mathbf{a}_3\cdots\mathbf{a}_{2k-1}$. By syzygy AB, the first equality in (6.5.3) can be generated as follows:

$$\langle \mathbf{a}_1\mathbf{a}_2\cdots\mathbf{a}_{2k}\rangle$$

$$= (\mathbf{a}_1\cdot\mathbf{a}_{2k})\langle\mathbf{B}_{2k-2}\rangle + \sum_{(1,2k-3)\vdash\mathbf{B}_{2k-2}} (\mathbf{a}_1\cdot\mathbf{B}_{2k-2\,(1)})\,\langle\mathbf{B}_{2k-2\,(2)}\mathbf{a}_{2k}\rangle$$

$$= (\mathbf{a}_1\cdot\mathbf{a}_{2k})\langle\mathbf{B}_{2k-2}\rangle + \sum_{(1,2k-4,1)\vdash\mathbf{B}_{2k-2}} (\mathbf{a}_1\cdot\mathbf{B}_{2k-2\,(1)})\,\langle\mathbf{B}_{2k-2\,(2)}\rangle\,(\mathbf{a}_{2k}\cdot\mathbf{B}_{2k-2\,(3)})$$

$$= (\mathbf{a}_1\cdot\mathbf{a}_{2k})\langle\mathbf{B}_{2k-2}\rangle + \sum_{(2k-3,1)\vdash\mathbf{B}_{2k-2}} \langle\mathbf{a}_1\mathbf{B}_{2k-2\,(1)}\rangle\,(\mathbf{a}_{2k}\cdot\mathbf{B}_{2k-2\,(2)})$$

$$= \sum_{(1,2k-2)\vdash\mathbf{a}_1\mathbf{B}_{2k-2}} (\mathbf{a}_{2k}\cdot\mathbf{B}_{2k-2\,(1)})\,\langle\mathbf{B}_{2k-2\,(2)}\rangle$$

$$= \langle\mathbf{a}_{2k}\mathbf{a}_1\mathbf{B}_{2k-2}\rangle.$$

The second equality can be generated as follows:

$$\langle\mathbf{a}_{2k}\mathbf{a}_{2k-1}\cdots\mathbf{a}_1\rangle = \sum_{i=1}^{2k-1} (-1)^{2k-1-i}(\mathbf{a}_{2k}\cdot\mathbf{a}_i)\,\langle\mathbf{a}_{2k-1}\cdots\check{\mathbf{a}}_i\cdots\mathbf{a}_1\rangle$$

$$= \sum_{i=1}^{2k-1} (-1)^{i+1} (\mathbf{a}_{2k} \cdot \mathbf{a}_i) \langle \mathbf{a}_1 \cdots \check{\mathbf{a}}_i \cdots \mathbf{a}_{2k-1} \rangle$$

$$= \langle \mathbf{a}_{2k} \mathbf{a}_1 \cdots \mathbf{a}_{2k-1} \rangle$$

$$= \langle \mathbf{a}_1 \cdots \mathbf{a}_{2k-1} \mathbf{a}_{2k} \rangle.$$

$\square$

Proposition 6.115. The following relations are generated by AB, SB and IGP:

$$[\mathbf{a}_1 \mathbf{a}_2 \cdots \mathbf{a}_{n+2l}] = (-1)^{n-1} [\mathbf{a}_{n+2l} \mathbf{a}_1 \cdots \mathbf{a}_{n+2l-1}]$$
$$= (-1)^{\frac{n(n-1)}{2}} [\mathbf{a}_{n+2l} \mathbf{a}_{n+2l-1} \cdots \mathbf{a}_1]. \tag{6.5.4}$$

Proof. Denote $\mathbf{A} = \mathbf{a}_1 \mathbf{a}_2 \cdots \mathbf{a}_{n+2l-1}$ and $\mathbf{B} = \mathbf{a}_1 \mathbf{a}_2 \cdots \mathbf{a}_{n+2l}$. By syzygy SB, on one hand,

$$[\mathbf{A}\mathbf{a}_{n+2l}] = \sum_{(2l,n-1)\vdash\mathbf{A}} \langle \mathbf{A}_{(1)} \rangle [\mathbf{A}_{(2)}\mathbf{a}_{n+2l}] + \sum_{(2l-1,n)\vdash\mathbf{A}} (-1)^n \langle \mathbf{A}_{(1)}\mathbf{a}_{n+2l} \rangle [\mathbf{A}_{(2)}]$$

$$= \sum_{(2l,n-1)\vdash\mathbf{A}} (-1)^{n-1} \langle \mathbf{A}_{(1)} \rangle [\mathbf{a}_{n+2l}\mathbf{A}_{(2)}] - \sum_{(2l-1,n)\vdash\mathbf{A}} (-1)^{n-1} \langle \mathbf{A}_{(1)}\mathbf{a}_{n+2l} \rangle [\mathbf{A}_{(2)}];$$

on the other hand,

$$[\mathbf{a}_{n+2l}\mathbf{A}] = \sum_{(2l,n-1)\vdash\mathbf{A}} \langle \mathbf{A}_{(1)} \rangle [\mathbf{a}_{n+2l}\mathbf{A}_{(2)}] + \sum_{(2l-1,n)\vdash\mathbf{A}} \langle \mathbf{a}_{n+2l}\mathbf{A}_{(1)} \rangle [\mathbf{A}_{(2)}].$$

By the above two equalities, together with (6.5.3) and syzygies AB, IGP, the first equality in (6.5.4) can be derived as follows:

$$[\mathbf{a}_{n+2l}\mathbf{A}] - (-1)^{n-1}[\mathbf{A}\mathbf{a}_{n+2l}] = \sum_{(2l-1,n)\vdash\mathbf{A}} 2 \langle \mathbf{A}_{(1)}\mathbf{a}_{n+2l} \rangle [\mathbf{A}_{(2)}]$$

$$= \sum_{(1,2l-2,n)\vdash\mathbf{A}} 2 (\mathbf{a}_{n+2l} \cdot \mathbf{A}_{(1)}) \langle \mathbf{A}_{(2)} \rangle [\mathbf{A}_{(3)}]$$

$$= \sum_{(1,n,2l-2)\vdash\mathbf{A}} 2 (\mathbf{a}_{n+2l} \cdot \mathbf{A}_{(1)}) [\mathbf{A}_{(2)}] \langle \mathbf{A}_{(3)} \rangle$$

$$= \sum_{(n+1,2l-2)\vdash\mathbf{A}} 2\langle \mathbf{A}_{(2)} \rangle \sum_{(1,n)\vdash\mathbf{A}_{(1)}} \mathbf{a}_{n+2l} \cdot \mathbf{A}_{(11)} [\mathbf{A}_{(12)}]$$

$$= 0.$$

For the second equality, by syzygy SB and (6.5.3),

$$[\mathbf{a}_{n+2l}\mathbf{a}_{n+2l-1} \cdots \mathbf{a}_1] = \sum_{(2l,n)\vdash\mathbf{B}} \langle \mathbf{B}^{\dagger}_{(1)} \rangle [\mathbf{B}^{\dagger}_{(2)}]$$

$$= \sum_{(2l,n)\vdash\mathbf{B}} (-1)^{\frac{n(n-1)}{2}} \langle \mathbf{B}_{(1)} \rangle [\mathbf{B}_{(2)}]$$

$$= (-1)^{\frac{n(n-1)}{2}} [\mathbf{B}].$$

It is independent of IGP. $\square$

Proposition 6.116.

(1) For any $2 \leq k \leq 2l$, the following relation is generated by AB:

$$\langle \mathbf{a}_1 \mathbf{a}_2 \cdots \mathbf{a}_{2l} \rangle + (-1)^k \langle \mathbf{a}_2 \mathbf{a}_3 \cdots \mathbf{a}_k \mathbf{a}_1 \mathbf{a}_{k+1} \cdots \mathbf{a}_{2l} \rangle \\ = 2 \sum_{i=2}^{k} (-1)^i (\mathbf{a}_1 \cdot \mathbf{a}_i) \langle \mathbf{a}_2 \mathbf{a}_3 \cdots \breve{\mathbf{a}}_i \cdots \mathbf{a}_{2l} \rangle. \tag{6.5.5}$$

(2) For any $2 \leq k \leq n + 2l$, the following relation is generated by AB and SB:

$$[\mathbf{a}_1 \mathbf{a}_2 \cdots \mathbf{a}_{n+2l}] + (-1)^k [\mathbf{a}_2 \mathbf{a}_3 \cdots \mathbf{a}_k \mathbf{a}_1 \mathbf{a}_{k+1} \cdots \mathbf{a}_{n+2l}] \\ = 2 \sum_{i=2}^{k} (-1)^i (\mathbf{a}_1 \cdot \mathbf{a}_i) [\mathbf{a}_2 \mathbf{a}_3 \cdots \breve{\mathbf{a}}_i \cdots \mathbf{a}_{n+2l}]. \tag{6.5.6}$$

Proof. (1) Denote $(\mathbf{a}_1 \mathbf{a}_2 \,|\, \mathbf{a}_3 \mathbf{a}_4) = (\mathbf{a}_1 \cdot \mathbf{a}_3)(\mathbf{a}_2 \cdot \mathbf{a}_4) - (\mathbf{a}_2 \cdot \mathbf{a}_3)(\mathbf{a}_1 \cdot \mathbf{a}_4)$. For $k = 2$, by syzygy AB,

$$\begin{aligned}
& \langle \mathbf{a}_1 \mathbf{a}_2 \mathbf{a}_3 \cdots \mathbf{a}_{2l} \rangle + \langle \mathbf{a}_2 \mathbf{a}_1 \mathbf{a}_3 \cdots \mathbf{a}_{2l} \rangle \\
= & \sum_{3 \leq i < j \leq 2l} (-1)^{i+j} \langle \mathbf{a}_3 \cdots \breve{\mathbf{a}}_i \cdots \breve{\mathbf{a}}_j \cdots \mathbf{a}_{2l} \rangle (\mathbf{a}_1 \wedge \mathbf{a}_2) \cdot (\mathbf{a}_i \wedge \mathbf{a}_j) \\
& + \sum_{3 \leq i < j \leq 2l} (-1)^{i+j} \langle \mathbf{a}_3 \cdots \breve{\mathbf{a}}_i \cdots \breve{\mathbf{a}}_j \cdots \mathbf{a}_{2l} \rangle (\mathbf{a}_2 \wedge \mathbf{a}_1) \cdot (\mathbf{a}_i \wedge \mathbf{a}_j) \\
& + 2 \mathbf{a}_1 \cdot \mathbf{a}_2 \langle \mathbf{a}_3 \cdots \mathbf{a}_{2l} \rangle \\
= & \; 2 \mathbf{a}_1 \cdot \mathbf{a}_2 \langle \mathbf{a}_3 \cdots \mathbf{a}_{2l} \rangle.
\end{aligned}$$

So (6.5.5) is true for $k = 2$. Assume that it is true for $k = r - 1$. For $k = r$, by induction hypothesis and (6.5.3),

$$\begin{aligned}
& \langle \mathbf{a}_1 \mathbf{a}_2 \cdots \mathbf{a}_{2l} \rangle + (-1)^r \langle \mathbf{a}_2 \mathbf{a}_3 \cdots \mathbf{a}_r \mathbf{a}_1 \mathbf{a}_{r+1} \cdots \mathbf{a}_{2l} \rangle \\
= & \langle \mathbf{a}_1 \mathbf{a}_2 \cdots \mathbf{a}_{2l} \rangle + \langle \mathbf{a}_2 \mathbf{a}_1 \cdots \mathbf{a}_{2l} \rangle - \langle \mathbf{a}_2 \mathbf{a}_1 \cdots \mathbf{a}_{2l} \rangle + (-1)^r \langle \mathbf{a}_2 \mathbf{a}_3 \cdots \mathbf{a}_r \mathbf{a}_1 \mathbf{a}_{r+1} \cdots \mathbf{a}_{2l} \rangle \\
= & \; 2 \mathbf{a}_1 \cdot \mathbf{a}_2 \langle \mathbf{a}_3 \cdots \mathbf{a}_{2l} \rangle - \langle \mathbf{a}_1 \mathbf{a}_3 \cdots \mathbf{a}_{2l} \mathbf{a}_2 \rangle + (-1)^r \langle \mathbf{a}_3 \cdots \mathbf{a}_r \mathbf{a}_1 \mathbf{a}_{r+1} \cdots \mathbf{a}_{2l} \mathbf{a}_2 \rangle \\
= & \; 2 \mathbf{a}_1 \cdot \mathbf{a}_2 \langle \mathbf{a}_3 \cdots \mathbf{a}_{2l} \rangle - 2 \sum_{i=3}^{r} (-1)^{i+1} \mathbf{a}_1 \cdot \mathbf{a}_i \langle \mathbf{a}_3 \cdots \breve{\mathbf{a}}_i \cdots \mathbf{a}_{2l} \mathbf{a}_2 \rangle \\
= & \; 2 \sum_{i=2}^{r} (-1)^i \mathbf{a}_1 \cdot \mathbf{a}_i \langle \mathbf{a}_2 \cdots \breve{\mathbf{a}}_i \cdots \mathbf{a}_{2l} \rangle.
\end{aligned}$$

(2) When $k = 2$, let $\mathbf{B} = \mathbf{a}_3 \mathbf{a}_4 \cdots \mathbf{a}_{n+2l}$. By syzygy SB and (6.5.5),

$$[\mathbf{a}_1\mathbf{a}_2\mathbf{a}_3\cdots\mathbf{a}_{n+2l}] + [\mathbf{a}_2\mathbf{a}_1\mathbf{a}_3\cdots\mathbf{a}_{n+2l}]$$

$$= \sum_{(2l,n-2)\vdash\mathbf{B}} \langle\mathbf{B}_{(1)}\rangle([\mathbf{a}_1\mathbf{a}_2\mathbf{B}_{(2)}] + [\mathbf{a}_2\mathbf{a}_1\mathbf{B}_{(2)}]) + \sum_{(2l-1,n-1)\vdash\mathbf{B}} \{-\langle\mathbf{a}_1\mathbf{B}_{(1)}\rangle[\mathbf{a}_2\mathbf{B}_{(2)}]$$

$$+ \langle\mathbf{a}_2\mathbf{B}_{(1)}\rangle[\mathbf{a}_1\mathbf{B}_{(2)}] - \langle\mathbf{a}_2\mathbf{B}_{(1)}\rangle[\mathbf{a}_1\mathbf{B}_{(2)}] + \langle\mathbf{a}_1\mathbf{B}_{(1)}\rangle[\mathbf{a}_2\mathbf{B}_{(2)}]\}$$

$$+ \sum_{(2l-2,n)\vdash\mathbf{B}} (\langle\mathbf{a}_1\mathbf{a}_2\mathbf{B}_{(1)}\rangle + \langle\mathbf{a}_2\mathbf{a}_1\mathbf{B}_{(1)}\rangle)\,[\mathbf{B}_{(2)}]$$

$$= \sum_{(2l-2,n)\vdash\mathbf{B}} 2(\mathbf{a}_1\cdot\mathbf{a}_2)\,\langle\mathbf{B}_{(1)}\rangle\,[\mathbf{B}_{(2)}]$$

$$= 2(\mathbf{a}_1\cdot\mathbf{a}_2)\,[\mathbf{B}].$$

So (6.5.6) is true for $k=2$. Assume that it is true for $k=r-1$. For $k=r$, by induction hypothesis and (6.5.4),

$$[\mathbf{a}_1\mathbf{a}_2\cdots\mathbf{a}_{n+2l}] + (-1)^r[\mathbf{a}_2\mathbf{a}_3\cdots\mathbf{a}_r\mathbf{a}_1\mathbf{a}_{r+1}\cdots\mathbf{a}_{n+2l}]$$

$$= [\mathbf{a}_1\mathbf{a}_2\cdots\mathbf{a}_{n+2l}] + [\mathbf{a}_2\mathbf{a}_1\cdots\mathbf{a}_{n+2l}] - [\mathbf{a}_2\mathbf{a}_1\cdots\mathbf{a}_{n+2l}]$$
$$+ (-1)^{n+r-1}[\mathbf{a}_3\cdots\mathbf{a}_r\mathbf{a}_1\mathbf{a}_{r+1}\cdots\mathbf{a}_{n+2l}\mathbf{a}_2]$$

$$= 2\,\mathbf{a}_1\cdot\mathbf{a}_2[\mathbf{a}_3\cdots\mathbf{a}_{n+2l}] - (-1)^{n-1}[\mathbf{a}_1\mathbf{a}_3\cdots\mathbf{a}_{n+2l}\mathbf{a}_2]$$
$$+ (-1)^{n+r-1}[\mathbf{a}_3\cdots\mathbf{a}_r\mathbf{a}_1\mathbf{a}_{r+1}\cdots\mathbf{a}_{n+2l}\mathbf{a}_2]$$

$$= 2\,\mathbf{a}_1\cdot\mathbf{a}_2[\mathbf{a}_3\cdots\mathbf{a}_{2l}] + 2\sum_{i=3}^{r}(-1)^{n+i+1}\mathbf{a}_1\cdot\mathbf{a}_i[\mathbf{a}_3\cdots\breve{\mathbf{a}}_i\cdots\mathbf{a}_{n+2l}\mathbf{a}_2]$$

$$= 2\sum_{i=2}^{r}(-1)^i\mathbf{a}_1\cdot\mathbf{a}_i[\mathbf{a}_2\cdots\breve{\mathbf{a}}_i\cdots\mathbf{a}_{n+2l}].$$

$\square$

Corollary 6.117. The following relations are generated by AB and SB:

$$\langle\mathbf{a}_1\mathbf{a}_1\mathbf{a}_2\mathbf{a}_3\cdots\mathbf{a}_{2l-1}\rangle = \mathbf{a}_1^2\langle\mathbf{a}_2\mathbf{a}_3\cdots\mathbf{a}_{2l-1}\rangle,$$
$$[\mathbf{a}_1\mathbf{a}_1\mathbf{a}_2\mathbf{a}_3\cdots\mathbf{a}_{2l-1}] = \mathbf{a}_1^2[\mathbf{a}_2\mathbf{a}_3\cdots\mathbf{a}_{n+2l-1}]. \tag{6.5.7}$$

Proposition 6.118. Denote $\mathbf{A}_p = \mathbf{a}_1\mathbf{a}_2\cdots\mathbf{a}_p$ and $\mathbf{B}_q = \mathbf{b}_1\mathbf{b}_2\cdots\mathbf{b}_q$.

(1) [IGP relation for Clifford monomials, see (6.4.16)] For any $1\le i\le k$, and any $l\ge 0$, let

$$\mathbf{A} = \mathbf{A}_{2i-1}, \quad \mathbf{A}' = \mathbf{A}_{2i}, \quad \mathbf{B} = \mathbf{B}_{n+2k+2l-2i+1}, \quad \mathbf{B}' = \mathbf{B}_{n+2k+2l-2i}.$$

Then the following relations are generated by AB, SB and IGP:

$$\sum_{(2k-2i+1,n+2l)\vdash\mathbf{B}} \langle\mathbf{A}\mathbf{B}_{(1)}\rangle[\mathbf{B}_{(2)}] = 0,$$

$$\sum_{(2k-2i,n+2l)\vdash\mathbf{B}'} \langle\mathbf{A}'\mathbf{B}'_{(1)}\rangle[\mathbf{B}'_{(2)}] = C_{k+l-i}^{l}\langle\mathbf{A}'\rangle\,[\mathbf{B}']. \tag{6.5.8}$$

(2) [GP relation for Clifford monomials, see (6.4.18)] For any $k, l \geq 0$, any $0 \leq m < n$, any $1 \leq j \leq l$, and any $0 \leq i \leq l$, let

$$\mathbf{A} = \mathbf{A}_m, \qquad \mathbf{A}' = \mathbf{A}_{n+2j-1}, \qquad \mathbf{A}'' = \mathbf{A}_{n+2i},$$
$$\mathbf{B} = \mathbf{B}_{2n+2k+2l-m}, \quad \mathbf{B}' = \mathbf{B}_{n+2k+2l-2j+1}, \quad \mathbf{B}'' = \mathbf{B}_{n+2k+2l-2i}.$$

Then the following relations are generated by AB, SB, IGP and BL:

$$\sum_{(n+2k-m,\,n+2l)\vdash\mathbf{B}} [\mathbf{A}\mathbf{B}_{(1)}][\mathbf{B}_{(2)}] = 0,$$

$$\sum_{(2k-2j+1,\,n+2l)\vdash\mathbf{B}'} [\mathbf{A}'\mathbf{B}'_{(1)}][\mathbf{B}'_{(2)}] = 0, \tag{6.5.9}$$

$$\sum_{(2k-2i,\,n+2l)\vdash\mathbf{B}''} [\mathbf{A}''\mathbf{B}''_{(1)}][\mathbf{B}''_{(2)}] = C^l_{k+l-i}[\mathbf{A}''][\mathbf{B}''].$$

In fact, all the identities developed in this chapter and the previous chapter on the two kinds of brackets can be derived similarly. Similar to the definitions of the outer product and Grassmann-Cayley algebra by the brackets of atomic vectors and dummy vectors in Chapter 2, any composition of the geometric product and a $\mathbb{Z}$-grading operator, *e.g.*, $\langle \mathbf{a}_1 \mathbf{a}_2 \cdots \mathbf{a}_k \rangle_{k-2l}$ for atomic vectors $\mathbf{a}_i$, can be defined using dummy vectors and deficit brackets. Then the geometric product $\mathbf{a}_1 \mathbf{a}_2 \cdots \mathbf{a}_k$ can be defined by

$$\mathbf{a}_1 \mathbf{a}_2 \cdots \mathbf{a}_k = \sum_{l=0}^{[\frac{k}{2}]} \langle \mathbf{a}_1 \mathbf{a}_2 \cdots \mathbf{a}_k \rangle_{k-2l}. \tag{6.5.10}$$

In this way, the whole Clifford algebra generated by atomic vectors $\mathbf{a}_i$ can be derived from the Clifford bracket algebra, if the latter is supplemented with sufficiently many dummy vectors.

In a general Clifford bracket algebra, the polynomials of syzygies AB and SB are too large, leading to middle expression swell very quickly if they are used directly in algebraic manipulations. To reduce the expression size, we need a special Clifford bracket algebra based on the *homogeneous model* of Euclidean geometry, called *null bracket algebra* [125]. The homogeneous model will be introduced in the next chapter.

Chapter 7

Euclidean Geometry and Conformal Grassmann-Cayley Algebra

Euclidean geometry contains, besides projective structure and inner-product structure, an affine structure, or structure of translation. The Cartesian model of Euclidean geometry is inconvenient in representing the latter structure. The homogeneous coordinates model of Euclidean geometry can represent the affine structure effectively, but cannot make the representation fully compatible with the inner-product structure. It is the conformal model of Euclidean geometry, developed by F.L. Wachter, S. Lie, *et al.* in the 19th century, that provides a representation fully compatible with both structures. Lie further developed a model for the geometry of oriented spheres and hyperplanes based on the conformal model, called the Lie model of contact geometry.

In the conformal model, the primitive geometric objects include not only points, lines and planes, but also circles and spheres of various dimensions. The Grassmann-Cayley algebra on the intersections and extensions of these geometric objects is called *conformal Grassmann-Cayley algebra*. It extends the classical Grassmann-Cayley algebra of linear geometric objects to include nonlinear geometric objects of constant curvature. It can be further developed into an algebra of oriented geometric objects of constant curvature via the Lie model.

This chapter introduces various models of Euclidean geometry and develops their Grassmann-Cayley algebraic representations, with emphasis on the conformal model and the Lie model.

7.1 Homogeneous coordinates and Cartesian coordinates

7.1.1 *Affine space and affine Grassmann-Cayley algebra*

Let $\mathcal{V}^n$ be a vector space, and let $\mathcal{W}^{n-1}$ be a fixed $(n-1)$D subspace of $\mathcal{V}^n$. In the $(n-1)$D projective space $\mathcal{V}^n$, the $(n-1)$D *affine space* $\mathbb{A}^{n-1}$ whose *hyperplane at infinity* is $\mathcal{W}^{n-1}$, is composed of all the projective points not in subspace $\mathcal{W}^{n-1}$:

$$\mathbb{A}^{n-1} := \mathcal{V}^n - \mathcal{W}^{n-1}.$$

A projective point in $\mathbb{A}^{n-1}$ is called an *affine point*, and a projective point in

339

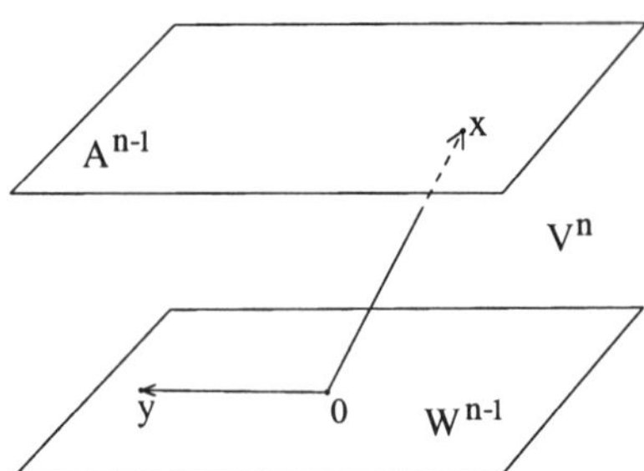

Fig. 7.1   Affine space $\mathbb{A}^{n-1}$. $\mathbf{x}$: affine point; $\mathbf{y}$: point at infinity.

$\mathcal{W}^{n-1}$ is called a *point at infinity*. An rD *affine subspace* refers to the intersection of an rD projective subspace with $\mathbb{A}^{n-1}$. An rD *subspace at infinity* refers to the intersection of an rD projective subspace with $\mathcal{W}^{n-1}$.

The affine structure can be represented in either the Grassmann-Cayley algebra or the inner-product Grassmann algebra over $\mathcal{V}^n$, by specifying either a meet product with a fixed pseudoscalar representing the hyperplane at infinity, or an inner product with a fixed vector representing the *normal direction* of the affine space.

In $\Lambda(\mathcal{V}^n)$, let $\mathbf{I}_{n-1}$ be a fixed $(n-1)$-blade representing the hyperplane at infinity $\mathcal{W}^{n-1}$. For any r-blade $\mathbf{A}_r \in \Lambda(\mathcal{V}^n)$, $\mathbf{A}_r \vee \mathbf{I}_{n-1}$ represents the *space at infinity* of the $(r-1)$D affine subspace which is the intersection of projective subspace $\mathbf{A}_r$ with affine space $\mathbb{A}^{n-1}$. The affine subspace can be compactly represented by the pair $(\mathbf{A}_r, \mathbf{A}_r \vee \mathbf{I}_{n-1})$, and the whole affine space can be represented by $(\mathcal{V}^n, \mathbf{I}_{n-1})$.

Definition 7.1. The *boundary operator* of affine space $(\mathcal{V}^n, \mathbf{I}_{n-1})$, denoted by "$\partial$", is a linear operator in $\Lambda(\mathcal{V}^n)$ defined by

$$\partial(\mathbf{A}) := \mathbf{A} \vee \mathbf{I}_{n-1}, \quad \forall \mathbf{A} \in \Lambda(\mathcal{V}^n). \tag{7.1.1}$$

$\partial(\mathbf{A})$ is called the *boundary* of $\mathbf{A}$.

For example, the boundary of $\mathcal{V}^n$ is $\mathbf{I}_{n-1}$. The boundary of a scalar is zero. The boundary of a vector is zero if and only if it is a point at infinity. In general, the boundary of a multivector $\mathbf{A}$ is zero if and only if it is "at infinity": $\mathbf{A} \in \Lambda(\mathbf{I}_{n-1})$.

Proposition 7.2. $\partial \circ \partial = 0$. Furthermore, for any vectors $\mathbf{a}_i \in \mathcal{V}^n$,

$$\partial(\mathbf{a}_1 \wedge \cdots \wedge \mathbf{a}_r) = \sum_{i=1}^{r}(-1)^{i+1}\partial(\mathbf{a}_i)\,\mathbf{a}_1 \wedge \cdots \breve{\mathbf{a}}_i \wedge \cdots \wedge \mathbf{a}_r. \tag{7.1.2}$$

Proof.   Direct from the cograded anticommutativity of the meet product, and the shuffle formula (2.3.32). $\square$

Corollary 7.3. ∂ is a *graded derivative* with respect to the outer product, *i.e.*, for any r-vector $\mathbf{A}_r$ and s-vector $\mathbf{B}_s$,

$$\partial(\mathbf{A}_r \wedge \mathbf{B}_s) = \partial(\mathbf{A}_r) \wedge \mathbf{B}_s + (-1)^r \mathbf{A}_r \wedge \partial(\mathbf{B}_s). \tag{7.1.3}$$

Furthermore, it has the following property related to the meet product:

$$\partial(\mathbf{A}_r \vee \mathbf{B}_s) = \mathbf{A}_r \vee \partial(\mathbf{B}_s) = (-1)^{n-s}\partial(\mathbf{A}_r) \vee \mathbf{B}_s. \tag{7.1.4}$$

Definition 7.4. Two affine subspaces $(\mathbf{A}_r, \partial(\mathbf{A}_r))$ and $(\mathbf{B}_s, \partial(\mathbf{B}_s))$ are said to be *parallel*, if the intersection of the two linear subspaces $\mathbf{A}_r, \mathbf{B}_s$ is a linear subspace at infinity. Two affine subspaces of the same dimension are said to be *translational*, if they have the same space at infinity. A *parallel projection* in rD direction $\mathbf{A}_r$, is an rD perspective projection whose center $\mathbf{A}_r$ is at infinity.

For example, for lines $\mathbf{12}, \mathbf{34}$ in the plane, they are parallel (or equivalently, translational) if and only if in $\Lambda(\mathcal{V}^3)$, $\partial((\mathbf{1} \wedge \mathbf{2}) \vee (\mathbf{3} \wedge \mathbf{4})) = 0$. For plane $\mathbf{123}$ and line $\mathbf{45}$ in space, they are parallel if and only if in $\Lambda(\mathcal{V}^4)$, $\partial((\mathbf{1} \wedge \mathbf{2} \wedge \mathbf{3}) \vee (\mathbf{4} \wedge \mathbf{5})) = 0$.

A vector in $\mathbf{I}_{n-1}$ is called a *displacement vector*, or *translational vector*. All displacement vectors form a vector subspace of $\mathcal{V}^n$, called the *space of displacements*. The difference between a translational vector and a point at infinity lies in the effect of the scaling: it changes a translational vector while leaving the corresponding point at infinity invariant.

Example 7.5. Any two affine points are translational. For points $\mathbf{x}, \mathbf{y}$, the translational vector from $\mathbf{x}$ to $\mathbf{y}$ is

$$\frac{\partial(\mathbf{x} \wedge \mathbf{y})}{\partial(\mathbf{x})\partial(\mathbf{y})}, \tag{7.1.5}$$

which is independent of the scaling of $\mathbf{x}, \mathbf{y}$. Indeed,

$$\frac{\mathbf{x}}{\partial(\mathbf{x})} + \frac{\partial(\mathbf{x} \wedge \mathbf{y})}{\partial(\mathbf{x})\partial(\mathbf{y})} = \frac{\partial(\mathbf{y})\mathbf{x} + \partial(\mathbf{x})\mathbf{y} - \partial(\mathbf{y})\mathbf{x}}{\partial(\mathbf{x})\partial(\mathbf{y})} = \frac{\mathbf{y}}{\partial(\mathbf{y})}.$$

Definition 7.6. The *affine addition*, or *barycenter*, of two affine points $\mathbf{x}, \mathbf{y}$, is

$$\frac{\mathbf{x}}{\partial(\mathbf{x})} + \frac{\mathbf{y}}{\partial(\mathbf{y})}. \tag{7.1.6}$$

The *translation*, or *displacement*, of an affine point $\mathbf{x}$ by a translational vector $\mathbf{t}$ is

$$\frac{\mathbf{x}}{\partial(\mathbf{x})} + \mathbf{t}. \tag{7.1.7}$$

The two concepts can be extended to more than two affine points and more than one translational vector respectively.

Traditionally, people set $\partial(\mathbf{x}) = 1$ if $\mathbf{x}$ is an affine point, and take $\mathbb{A}^{n-1}$ as the set composed of all such vectors in $\mathcal{V}^n$. This inhomogeneous representation of affine space is traditionally called the *homogeneous coordinates* representation. Indeed, let $\mathbf{e}_0, \mathbf{e}_1, \ldots, \mathbf{e}_{n-1}$ be a basis of $\mathcal{V}^n$, and let $\mathbf{I}_{n-1} = \mathbf{e}_1 \wedge \mathbf{e}_2 \wedge \cdots \wedge \mathbf{e}_{n-1}$, then any affine point $\mathbf{x}$ has homogeneous coordinates $(1, x_1, \ldots, x_{n-1})$.

In the homogeneous coordinates representation, the affine addition of r points $\mathbf{x}_1, \mathbf{x}_2, \ldots, \mathbf{x}_r$ becomes

$$\frac{\mathbf{x}_1 + \mathbf{x}_2 + \cdots + \mathbf{x}_r}{r}. \tag{7.1.8}$$

More generally, an *affine combination* of the r points is a linear combination of the form

$$\lambda_1 \mathbf{x}_1 + \lambda_2 \mathbf{x}_2 + \cdots + \lambda_r \mathbf{x}_r, \quad \text{where} \quad \lambda_1 + \lambda_2 + \cdots + \lambda_r = 1. \tag{7.1.9}$$

An *affine basis* of $\mathcal{V}^n$ is a basis composed of n affine points. A *Cartesian frame*, or *Cartesian basis*, of $\mathcal{V}^n$, is a basis composed of one affine point, called the *origin*, and $n-1$ translational vectors. In the homogeneous coordinates representation, any affine point is either an affine combination of affine basis vectors, or the origin plus a linear combination of translational basis vectors in a Cartesian frame.

Three affine points $\mathbf{1}, \mathbf{2}, \mathbf{3}$ are *collinear* if and only if $\mathbf{1} \wedge \mathbf{2} \wedge \mathbf{3} = 0$ in $\Lambda(\mathcal{V}^n)$, or equivalently, $\partial(\mathbf{1} \wedge \mathbf{2} \wedge \mathbf{3}) = 0$ in $\Lambda(\mathbf{I}_{n-1})$. In general, r affine points $\mathbf{1}, \mathbf{2}, \ldots, \mathbf{r}$ are *affinely independent*, i.e., they are linearly independent as vectors in $\mathcal{V}^n$, if and only if in $\Lambda(\mathbf{I}_{n-1})$,

$$\partial(\mathbf{1} \wedge \mathbf{2} \wedge \cdots \wedge \mathbf{r}) \neq 0. \tag{7.1.10}$$

Lemma 7.7. For any vector $\mathbf{a} \in \mathcal{V}^n$, any r-blade $\mathbf{A}_r$ and s-blade $\mathbf{B}_s$ in $\Lambda(\mathcal{V}^n)$,

$$\mathbf{a} \wedge \partial(\mathbf{a} \wedge \mathbf{A}_r) = \partial(\mathbf{a}) \, \mathbf{a} \wedge \mathbf{A}_r,$$
$$\partial(\mathbf{a})\mathbf{A}_r = \mathbf{a} \wedge \partial(\mathbf{A}_r), \quad \text{if } \mathbf{a} \in \mathbf{A}_r, \tag{7.1.11}$$
$$\partial(\mathbf{a} \wedge \mathbf{A}_r) \wedge \partial(\mathbf{a} \wedge \mathbf{B}_s) = \partial(\mathbf{a}) \, \partial(\mathbf{a} \wedge \mathbf{A}_r \wedge \mathbf{B}_s).$$

Proof. Only the last equality needs proof.

$$\begin{aligned}
& \partial(\mathbf{a} \wedge \mathbf{A}_r) \wedge \partial(\mathbf{a} \wedge \mathbf{B}_s) \\
&= (\partial(\mathbf{a}) \, \mathbf{A}_r - \mathbf{a} \wedge \partial(\mathbf{A}_r)) \wedge (\partial(\mathbf{a}) \, \mathbf{B}_s - \mathbf{a} \wedge \partial(\mathbf{B}_s)) \\
&= \partial(\mathbf{a})^2 \, \mathbf{A}_r \wedge \mathbf{B}_s - \partial(\mathbf{a}) \, \mathbf{a} \wedge (\partial(\mathbf{A}_r) \wedge \mathbf{B}_s + (-1)^r \mathbf{A}_r \wedge \partial(\mathbf{B}_s)) \\
&= \partial(\mathbf{a})^2 \, \mathbf{A}_r \wedge \mathbf{B}_s - \partial(\mathbf{a}) \, \mathbf{a} \wedge \partial(\mathbf{A}_r \wedge \mathbf{B}_s) \\
&= \partial(\mathbf{a}) \, \partial(\mathbf{a} \wedge \mathbf{A}_r \wedge \mathbf{B}_s).
\end{aligned}$$

$\square$

Corollary 7.8. Affine points $\mathbf{1}, \mathbf{2}, \ldots, \mathbf{r}$ are affinely independent if and only if in $\mathbf{I}_{n-1}$, vectors $\partial(\mathbf{1} \wedge \mathbf{i})$ for $\mathbf{i} = \mathbf{2}, \mathbf{3}, \ldots, \mathbf{r}$ are linearly independent.

Proof. Direct from the identity that for any multivectors $\mathbf{A}_1, \ldots, \mathbf{A}_r$,

$$\partial(\mathbf{a} \wedge \mathbf{A}_1) \wedge \cdots \wedge \partial(\mathbf{a} \wedge \mathbf{A}_r) = \partial(\mathbf{a})^{r-1} \partial(\mathbf{a} \wedge \mathbf{A}_1 \wedge \cdots \wedge \mathbf{A}_r). \tag{7.1.12}$$

$\square$

The $(r-1)$D affine subspace generated by affinely independent points $\mathbf{1}, \mathbf{2}, \ldots, \mathbf{r}$ is represented by the pair $(\mathbf{A}_r, \partial(\mathbf{A}_r))$, where $\mathbf{A}_r = \mathbf{1} \wedge \mathbf{2} \wedge \cdots \wedge \mathbf{r}$, in the sense that a point $\mathbf{x} \in \mathbb{A}^{n-1}$ is in the affine subspace if and only if $\mathbf{x} \wedge \mathbf{A}_r = 0$ in $\Lambda(\mathcal{V}^n)$, or equivalently,

$$\frac{\mathbf{x}}{\partial(\mathbf{x})} \wedge \partial(\mathbf{A}_r) = \mathbf{A}_r. \tag{7.1.13}$$

r-blade $\mathbf{A}_r$ is called the *moment* of the affine subspace, and $\partial(\mathbf{A}_r)$ is called the $(r-1)$D *direction*. So $(\mathbf{A}_r, \partial(\mathbf{A}_r))$ is called the *moment-direction* representation of the affine subspace.

An *affine transformation* in $\mathbb{A}^{n-1}$ refers to a linear transformation in $\mathcal{V}^n$ leaving the space at infinity $\mathcal{W}^{n-1}$ invariant, or equivalently, commuting with the boundary operator. *Affine geometry* studies the properties of $\mathbb{A}^{n-1}$ that are invariant under affine transformations, called *affine invariants*.

For example, projective invariants are also affine invariants. In particular, all brackets are affine invariants. The boundary of a vector in $\mathcal{V}^n$ is another affine invariant. Brackets and boundaries of vectors are basic affine invariants, and generate all other affine invariants [205].

There are also rational affine invariants. For example, for collinear points $\mathbf{1}, \mathbf{2}, \mathbf{3}$, the ratio $[\mathbf{12}]/[\mathbf{13}]$ is an affine invariant because it is a projective invariant. It is not an *absolute* affine invariant because it changes with the scaling of $\mathbf{2}$ and $\mathbf{3}$. Instead, the following ratio

$$\frac{[\mathbf{12}]}{[\mathbf{13}]} \frac{\partial(\mathbf{3})}{\partial(\mathbf{2})}, \tag{7.1.14}$$

called the *simple ratio* of point $\mathbf{1}$ with respect to collinear points $\mathbf{2}$ and $\mathbf{3}$, is an absolute affine invariant.

Definition 7.9. The *affine Grassmann-Cayley algebra* over $\mathbb{A}^{n-1} = (\mathcal{V}^n, \mathbf{I}_{n-1})$, is the Grassmann-Cayley algebra $\Lambda(\mathcal{V}^n)$ further equipped with the boundary operator "∂", such that $\partial(\mathbf{x}) \neq 0$ if and only if $\mathbf{x} \in \mathbb{A}^{n-1}$.

Definition 7.10. Let $\mathbf{a}_1, \ldots, \mathbf{a}_m$ be symbols, called *atomic vectors*, and let $m \geq n$.

- Let the $[\mathbf{a}_{i_1} \cdots \mathbf{a}_{i_n}]$ be indeterminates over $\mathbb{K}$ for each sequence of indices $1 \leq i_1, \ldots, i_n \leq m$, called *brackets*.
- Let the $\partial(\mathbf{a}_j)$ be indeterminates over $\mathbb{K}$ for each $1 \leq j \leq m$, called *boundaries of vectors*.

The nD *affine bracket algebra* generated by the $\mathbf{a}$'s, is the quotient of the polynomial ring generated by the brackets and boundaries of vectors, modulo the ideal generated by the syzygies B1, B2, GP in Definition 2.11, together with the following *affine Grassmann-Plücker syzygy* (AGP):

AGP. $\displaystyle\sum_{i=1}^{n+1} (-1)^{i+1} \partial(\mathbf{a}_{j_i}) [\mathbf{a}_{j_1} \ldots \breve{\mathbf{a}}_{j_i} \ldots \mathbf{a}_{j_{n+1}}].$

If $\mathcal{V}^n$ is endowed with a nondegenerate inner product, then by setting $\mathbf{e}_0 = \mathbf{I}^{\sim}_{n-1}$, we have

$$\partial(\mathbf{A}) = \mathbf{e}_0 \cdot (\mathbf{A} - \langle \mathbf{A} \rangle), \quad \forall \mathbf{A} \in \Lambda(\mathcal{V}^n). \tag{7.1.15}$$

Proposition 7.11. In $\mathcal{CL}(\mathcal{V}^n)$, ∂ is a graded derivative with respect to the geometric product: for any r-vector $\mathbf{A}_r$ and s-vector $\mathbf{B}_s$,

$$\partial(\mathbf{A}_r \mathbf{B}_s) = \partial(\mathbf{A}_r)\mathbf{B}_s + (-1)^r \mathbf{A}_r \partial(\mathbf{B}_s). \tag{7.1.16}$$

Furthermore, it has the following property related to the inner product:

$$\partial(\mathbf{A}_r \cdot \mathbf{B}_s) = \begin{cases} \partial(\mathbf{A}_r) \cdot \mathbf{B}_s, & \text{if } r > s, \\ 0, & \text{if } r = s, \\ (-1)^r \mathbf{A}_r \cdot \partial(\mathbf{B}_s), & \text{if } r < s. \end{cases} \tag{7.1.17}$$

7.1.2   *The Cartesian model of Euclidean space*

An nD *Euclidean space*, denoted by $\mathbb{E}^n$, is an affine space in $\mathcal{V}^{n+1}$ whose space of displacements is a Euclidean inner-product space. A *Euclidean transformation* is an affine transformation preserving the inner product in the space of displacements. *Euclidean geometry* studies the properties of $\mathbb{E}^n$ that are invariant under Euclidean transformations, called *Euclidean invariants*.

For example, the *distance* between two points $\mathbf{x}, \mathbf{y}$ is defined as the magnitude of their displacement vector in $\mathcal{V}^{n+1}$:

$$d_{\mathbf{xy}} := \frac{|\partial(\mathbf{x} \wedge \mathbf{y})|}{|\partial(\mathbf{x})|\,|\partial(\mathbf{y})|}. \tag{7.1.18}$$

By setting $\partial(\mathbf{x}) = \partial(\mathbf{y}) = 1$, we get the familiar expression $d_{\mathbf{xy}} = |\mathbf{x} - \mathbf{y}|$.

Traditionally people set the origin of a Cartesian frame to be the zero vector, and identify $\mathbb{E}^n$ with the space of displacements $\mathbb{R}^n$. This is the *Cartesian model* of Euclidean space. It has a special point represented by the zero vector, which is convenient for algebraic computation but not so for geometric explanation.

In the Cartesian model, a vector represents both a point and a direction. A line passing through points $\mathbf{1}, \mathbf{2}$ can be represented by the pair of points. However, this representation is far from being unique: the pair can be replaced by any two other points on the line. The moment-direction representation $(\mathbf{1} \wedge \mathbf{2}, \mathbf{2} - \mathbf{1})$, on the contrary, is unique up to scale: $(\mathbf{1} \wedge \mathbf{2}, \mathbf{2} - \mathbf{1})$ and $(\mathbf{1}' \wedge \mathbf{2}', \mathbf{2}' - \mathbf{1}')$ represent the same line if and only if $(\mathbf{1}' \wedge \mathbf{2}', \mathbf{2}' - \mathbf{1}') = \lambda(\mathbf{1} \wedge \mathbf{2}, \mathbf{2} - \mathbf{1}) = (\lambda \mathbf{1} \wedge \mathbf{2}, \lambda(\mathbf{2} - \mathbf{1}))$ for some scalar $\lambda \neq 0$.

In general, the $(r-1)$D affine subspace generated by points $\mathbf{1}, \mathbf{2}, \ldots, \mathbf{r} \in \mathbb{R}^n$ is represented by the moment-direction $(\mathbf{A}_r, \partial(\mathbf{A}_r))$, where $\mathbf{A}_r = \mathbf{1} \wedge \mathbf{2} \wedge \cdots \wedge \mathbf{r}$, and

$$\partial(\mathbf{A}_r) = (\mathbf{2} - \mathbf{1}) \wedge (\mathbf{3} - \mathbf{1}) \wedge \cdots \wedge (\mathbf{r} - \mathbf{1}) = \sum_{i=1}^{r} (-1)^{i+1} \mathbf{1} \wedge \cdots \wedge \check{\mathbf{i}} \wedge \cdots \wedge \mathbf{r}, \tag{7.1.19}$$

such that any point $\mathbf{a} \in \mathbb{R}^n$ is in the affine subspace if and only if

$$\mathbf{a} \wedge \partial(\mathbf{A}_r) = \mathbf{A}_r. \tag{7.1.20}$$

In particular, if $r = 1$, then point $\mathbf{a} \in \mathbb{R}^n$ has the moment-direction representation $(\mathbf{a}, 1)$. In other words, $\partial(\mathbf{a}) = 1$ for any point $\mathbf{a}$ in the Cartesian model.

It must be pointed out that in the Cartesian model, $\partial(\mathbf{A}_r)$ is just a short-hand notation. It always refers to the blade defined by (7.1.19) when $r > 1$, which is in fact a blade in $\Lambda(\mathbf{A}_r)$ if $\mathbf{A}_r \neq 0$; it refers to r when $r \leq 1$. *In the Cartesian model, $\partial(\mathbf{A}_r)$ does not need to be equal to zero if $\mathbf{A}_r = 0$.* The reason is that the direction is invariant under the change of origin, in particular, it remains the same no matter if the origin is in $\mathbb{R}^n$ or not. The moment, on the contrary, always changes with the origin.

Example 7.12. In the Euclidean plane $\mathbb{R}^2$, the outer product of any three points is always zero. The signed area of triangle **123** can only be represented by

$$\frac{1}{2}[\partial(\mathbf{1} \wedge \mathbf{2} \wedge \mathbf{3})] = \frac{1}{2}([\mathbf{12}] - [\mathbf{13}] + [\mathbf{23}]). \tag{7.1.21}$$

So in the Cartesian model, the constraint that three points are collinear is represented by a 3-termed bracket polynomial equation.

Example 7.13. In $\mathbb{R}^n$, the constraint that two lines **12**, **34** are perpendicular is represented by a 4-termed polynomial equation of inner products of vectors:

$$\partial(\mathbf{1} \wedge \mathbf{2}) \cdot \partial(\mathbf{3} \wedge \mathbf{4}) = (\mathbf{2} - \mathbf{1}) \cdot (\mathbf{4} - \mathbf{3}) = 0. \tag{7.1.22}$$

One may argue that the inner product in (7.1.22) should not be expanded, so that the representation still remains 1-termed. In algebraic computation based on multilinearity properties, such an expansion is unavoidable. Although setting one of the four points to be the origin can reduce the number of terms by two, there is only one origin, so at most one vector can be set to zero. For general algebraic computation, the simplification brought about by the origin is negligible.

Example 7.14. Lines **12**, **34** are parallel if and only if they have the same direction. In $\mathbb{R}^2$, this constraint is represented by

$$(\mathbf{2} - \mathbf{1}) \wedge (\mathbf{4} - \mathbf{3}) = \mathbf{1} \wedge \mathbf{3} - \mathbf{2} \wedge \mathbf{3} - \mathbf{1} \wedge \mathbf{4} + \mathbf{2} \wedge \mathbf{4} = 0. \tag{7.1.23}$$

If the two lines are not parallel, then their intersection can be derived as follows: in the affine GC algebra $\Lambda(\mathcal{V}^3)$, by setting $\partial(\mathbf{x}) = 1$ for any affine point $\mathbf{x}$, the intersection is

$$
\begin{aligned}
\frac{(\mathbf{1} \wedge \mathbf{2}) \vee (\mathbf{3} \wedge \mathbf{4})}{\partial((\mathbf{1} \wedge \mathbf{2}) \vee (\mathbf{3} \wedge \mathbf{4}))} &= \frac{[\mathbf{134}]\mathbf{2} - [\mathbf{234}]\mathbf{1}}{[\mathbf{134}] - [\mathbf{234}]} \\
&= \frac{[\partial(\mathbf{1} \wedge \mathbf{3} \wedge \mathbf{4})]\mathbf{2} - [\partial(\mathbf{2} \wedge \mathbf{3} \wedge \mathbf{4})]\mathbf{1}}{[\partial(\mathbf{1} \wedge \mathbf{3} \wedge \mathbf{4})] - [\partial(\mathbf{2} \wedge \mathbf{3} \wedge \mathbf{4})]} \\
&= \frac{([\mathbf{13}] - [\mathbf{14}] + [\mathbf{34}])\mathbf{2} - ([\mathbf{23}] - [\mathbf{24}] + [\mathbf{34}])\mathbf{1}}{[\mathbf{13}] - [\mathbf{14}] - [\mathbf{23}] + [\mathbf{24}]}.
\end{aligned}
\tag{7.1.24}
$$

In the Cartesian model $\mathbb{R}^2$, the above result is still valid, and remains 10-termed.

We see that in the Cartesian model, the outer product, inner product and meet product in the affine GC algebra $\Lambda(\mathcal{V}^{n+1})$ of $\mathbb{E}^n$ are represented by complicated (rational) polynomials of rather poor geometric meaning. Contrary to the naive idea that the introduction of zero vector to represent a specific point should simplify algebraic representations and manipulations, such an introduction generally makes symbolic computing based on invariants much more complicated.

The homogeneous coordinates model of Euclidean geometry, and its homogeneous version where for any affine point $\mathbf{x}$, it is $\partial(\mathbf{x}) \neq 0$ instead of $\partial(\mathbf{x}) = 1$, both lead to significant simplifications in symbolic computing. However, in this model, the inner product has Euclidean geometric meaning only when restricted to the space of displacements. In computing by multilinearity, if we expand the inner product such as the one in (7.1.22), we get an expression of four terms, none of which is an invariant, no matter how we select the signature for the 1D subspace orthogonal to the space at infinity in $\mathcal{V}^{n+1}$.

Example 7.15. For two affine points $\mathbf{x}, \mathbf{y}$ in $(\mathcal{V}^{n+1}, \mathbf{I}_n)$, where $\partial(\mathbf{x}) = \partial(\mathbf{y}) = 1$, if their inner product $\mathbf{x} \cdot \mathbf{y}$ is a Euclidean invariant, it should be unchanged by the displacement along vector $\mathbf{t} \in \mathbf{I}_n$:

$$(\mathbf{x} + \mathbf{t}) \cdot (\mathbf{y} + \mathbf{t}) = \mathbf{x} \cdot \mathbf{y}, \quad i.e., \quad \mathbf{t} \cdot (\mathbf{x} + \mathbf{y} + \mathbf{t}) = 0. \qquad (7.1.25)$$

Since points $\mathbf{x}, \mathbf{y}$ are arbitrary, it must be that $\mathbf{t} \in \mathrm{rad}(\mathcal{V}^{n+1})$. Since $\mathbf{t} \in \mathbf{I}_n$ is arbitrary, $\mathbf{I}_n$ must be the radical of $\mathcal{V}^{n+1}$, contradicting with the requirement that it has Euclidean signature.

The homogeneous coordinates model disables Euclidean invariant computing, while the Cartesian model is very inconvenient for symbolic computing. More advanced algebraic models are needed.

7.2   The conformal model and the homogeneous model

F.L. Wachter (1792-1817), a student of Gauss, once discovered that a special sphere in non-Euclidean geometry called *horosphere*, whose center is at infinity, is equipped with a Euclidean distance structure. In his Ph.D. dissertation, S. Lie (1872) proposed what is nowadays called *Lie sphere geometry*, where a point in the Euclidean space is taken as a sphere of radius zero, and where an oriented sphere or hyperplane in $\mathbb{E}^n$ is represented by a null vector in space $\mathbb{R}^{n+1,2}$. By discarding the orientations, the dimension of the embedding space can be reduced by one, leading to a representation of spheres and hyperplanes of $\mathbb{E}^n$ by positive vectors in the Minkowski space $\mathbb{R}^{n+1,1}$, while points in $\mathbb{E}^n$ are still represented by null vectors. The model with embedding space $\mathbb{R}^{n+1,2}$ is called the *Lie model*, and will be investigated in Section 7.5. The model with embedding space $\mathbb{R}^{n+1,1}$ is called the *conformal model*.

7.2.1   *The conformal model*

We start from the Euclidean plane. A circle with center (x_0, y_0) and radius ρ has the following equation in the Cartesian coordinates of $\mathbb{R}^2$:

$$(x - x_0)^2 + (y - y_0)^2 = \rho^2. \tag{7.2.1}$$

Expanding the squares, we get

$$xx_0 + yy_0 - \frac{x_0^2 + y_0^2 - \rho^2}{2} - \frac{x^2 + y^2}{2} = 0. \tag{7.2.2}$$

(7.2.2) suggests that we can represent the circle by a 4D vector $(x_0, y_0, (x_0^2 + y_0^2 - \rho^2)/2, 1)$, and by taking point $(x, y) \in \mathbb{R}^2$ as the circle of center (x, y) and radius 0, we can represent the point by another 4D vector $(x, y, (x^2 + y^2)/2, 1)$, such that the left side of (7.2.2) is the inner product of the two 4D vectors. The components of the two 4D vectors are called the *generalized homogeneous coordinates* of the circle and the point respectively.

Denote the basis vectors of the 4D space by $\mathbf{e}_1, \mathbf{e}_2, \mathbf{e}_3, \mathbf{e}_4$. The left side of (7.2.2) becomes

$$\begin{aligned}
&(x_0\mathbf{e}_1 + y_0\mathbf{e}_2 + \frac{x_0^2 + y_0^2 - \rho^2}{2}\mathbf{e}_3 + \mathbf{e}_4) \cdot (x\mathbf{e}_1 + y\mathbf{e}_2 + \frac{x^2 + y^2}{2}\mathbf{e}_3 + \mathbf{e}_4) \\
&= xx_0 + yy_0 - \frac{x_0^2 + y_0^2 - \rho^2}{2} - \frac{x^2 + y^2}{2},
\end{aligned} \tag{7.2.3}$$

for any parameters x, y, x_0, y_0, ρ. A simple computation shows that

$$\begin{aligned}
&\mathbf{e}_1^2 = \mathbf{e}_2^2 = 1, \\
&\mathbf{e}_3^2 = \mathbf{e}_4^2 = 0, \\
&\mathbf{e}_1 \cdot \mathbf{e}_2 = \mathbf{e}_1 \cdot \mathbf{e}_3 = \mathbf{e}_1 \cdot \mathbf{e}_4 = \mathbf{e}_2 \cdot \mathbf{e}_3 = \mathbf{e}_2 \cdot \mathbf{e}_4 = 0, \\
&\mathbf{e}_3 \cdot \mathbf{e}_4 = -1.
\end{aligned} \tag{7.2.4}$$

Therefore, the 4D embedding space is Minkowski with Witt basis $\mathbf{e}_1, \mathbf{e}_2, \mathbf{e}_3, \mathbf{e}_4$, where $(\mathbf{e}_3, \mathbf{e}_4)$ is a Witt pair. Traditionally people write $\mathbf{e} = \mathbf{e}_3$ and $\mathbf{e}_0 = \mathbf{e}_4$. The mapping $\mathbf{f} : \mathbb{R}^2 \mapsto \mathbb{R}^{3,1}$ defined by

$$\mathbf{f}(\mathbf{x}) = \mathbf{e}_0 + \mathbf{x} + \frac{\mathbf{x}^2}{2}\mathbf{e}, \tag{7.2.5}$$

where $\mathbb{R}^2$ is the 2D plane spanned by $\mathbf{e}_1, \mathbf{e}_2$, has the following properties:

(1) For any two points $\mathbf{x}, \mathbf{y} \in \mathbb{R}^2$,

$$\mathbf{f}(\mathbf{x}) \cdot \mathbf{f}(\mathbf{y}) = \mathbf{x} \cdot \mathbf{y} - \frac{x^2 + y^2}{2} = -\frac{|\mathbf{x} - \mathbf{y}|^2}{2} = -\frac{d_{\mathbf{xy}}^2}{2}. \tag{7.2.6}$$

So in the conformal model, the inner product of two points has intrinsic geometric meaning, and is a basic Euclidean invariant.

In particular if $\mathbf{x} = \mathbf{y}$, then $\mathbf{f}(\mathbf{x})^2 = 0$. So $\mathbf{f}(\mathbf{x})$ is a null vector. Furthermore, $\mathbf{f}(\mathbf{x}) \cdot \mathbf{e} = \mathbf{e}_0 \cdot \mathbf{e} = -1$. The set

$$\mathcal{N}_\mathbf{e} = \{\mathbf{a} \in \mathbb{R}^{3,1} \,|\, \mathbf{a}^2 = 0, \, \mathbf{a} \cdot \mathbf{e} = -1\} \tag{7.2.7}$$

is the image space of $\mathbf{f}$. It contains up to scale all null vectors in $\mathbb{R}^{3,1}$ except $\mathbf{e}$.

(2) For any two points $\mathbf{x}, \mathbf{y} \in \mathbb{R}^2$,

$$(\mathbf{f}(\mathbf{x}) - \mathbf{f}(\mathbf{y}))^2 = \mathbf{f}(\mathbf{x})^2 + \mathbf{f}(\mathbf{y})^2 - 2\,\mathbf{f}(\mathbf{x}) \cdot \mathbf{f}(\mathbf{y}) = d_{\mathbf{xy}}^2. \tag{7.2.8}$$

So $\mathbf{f}$ is an isometry from $\mathbb{R}^2$ to $\mathcal{N}_{\mathbf{e}}$, and is invertible. The inverse map is the orthogonal projection onto $\mathbb{R}^2$:

$$\mathbf{f}^{-1}(\mathbf{a}) = P_{\mathbf{e}_1 \wedge \mathbf{e}_2}(\mathbf{a}) = P_{\mathbf{e} \wedge \mathbf{e}_0}^{\perp}(\mathbf{a}), \quad \forall \mathbf{a} \in \mathcal{N}_{\mathbf{e}}. \tag{7.2.9}$$

(3) $\mathbf{f}(0) = \mathbf{e}_0$, so the origin of $\mathbb{R}^2$ corresponds to $\mathbf{e}_0$. In (7.2.5), when $\mathbf{x}^2$ tends to infinity, $2\mathbf{f}(\mathbf{x})/\mathbf{x}^2$ tends to $\mathbf{e}$. So $\mathbf{e}$ is a point at infinity. It is obtained by pinching the line at infinity of the 2D affine plane to a single point, so that the result is a compact topological space homeomorphic to a 2D sphere. This extraneous point outside the plane is called the *conformal point at infinity* of the Euclidean plane.

The set $\mathcal{N}_{\mathbf{e}}$ is a Euclidean distance space but not a Euclidean space, because it is not affine: for two different null vectors $\mathbf{a}, \mathbf{b} \in \mathcal{N}_{\mathbf{e}}$, the 2D plane $\mathbf{a} \wedge \mathbf{b}$ is Minkowski, so $\mathbf{a}, \mathbf{b}$ are the only two null vectors up to scale, and $\lambda \mathbf{a} + \mu \mathbf{b} \notin \mathcal{N}_{\mathbf{e}}$ for all $\lambda \mu \neq 0$.

From this aspect, the conformal model is very unusual: it represents Euclidean geometry by a non-affine model in a non-Euclidean space of two more dimensions through a nonlinear isometry (7.2.5). It is much more complicated than both the homogeneous coordinates model and the Cartesian model of Euclidean geometry. What is amazing is that, this complexity in algebraic structure does not lead to the complication of symbolic manipulations; on the contrary, it brings about highly unusual simplifications for Euclidean geometric computing.

(4) We come back to circles in $\mathbb{R}^2$. We have seen that a circle of center $\mathbf{x}_0 \in \mathbb{R}^2$ and radius ρ can be represented by null vector

$$\mathbf{s} = \mathbf{x}_0 + \frac{\mathbf{x}_0^2 - \rho^2}{2}\mathbf{e} + \mathbf{e}_0 = \mathbf{f}(\mathbf{x}_0) - \frac{\rho^2}{2}\mathbf{e}, \tag{7.2.10}$$

and a point $\mathbf{x} \in \mathbb{R}^2$ is on the circle if and only if $\mathbf{f}(\mathbf{x}) \cdot \mathbf{s} = 0$. In fact, for any point $\mathbf{x} \in \mathbb{R}^2$,

$$\mathbf{f}(\mathbf{x}) \cdot \mathbf{s} = \mathbf{f}(\mathbf{x}) \cdot \mathbf{f}(\mathbf{x}_0) - \frac{\rho^2}{2}\mathbf{f}(\mathbf{x}) \cdot \mathbf{e} = \frac{\rho^2 - d_{\mathbf{xx}_0}^2}{2}. \tag{7.2.11}$$

So $\mathbf{x}$ is inside, on, or outside the circle if and only if $\mathbf{f}(\mathbf{x}) \cdot \mathbf{s} > 0,\ = 0,$ or < 0 respectively.

(5) By (7.2.10), $\mathbf{s}^2 = \rho^2 > 0$, so $\mathbf{s}$ is a positive vector. Furthermore, $\mathbf{s} \cdot \mathbf{e} = -1$.

Conversely, any positive vector $\mathbf{s}$ satisfying $\mathbf{s} \cdot \mathbf{e} = -1$ represents a circle: the radius is $\sqrt{\mathbf{s}^2}$ and the center in $\mathbb{R}^2$ is

$$\mathbf{f}^{-1}\!\left(\mathbf{s} + \frac{\mathbf{s}^2}{2}\mathbf{e}\right) = P_{\mathbf{e} \wedge \mathbf{e}_0}^{\perp}(\mathbf{s}) = ((\mathbf{s} \wedge \mathbf{e} \wedge \mathbf{e}_0) \cdot \mathbf{e}) \cdot \mathbf{e}_0 = \mathbf{e}_0 \cdot (\mathbf{s} \wedge \mathbf{e}) - \mathbf{e}_0. \tag{7.2.12}$$

(6) A positive vector $\mathbf{s}$ such that $\mathbf{s} \cdot \mathbf{e} = 0$ represents a line in the plane. A line can be taken as a circle passing through the conformal point at infinity. We first check the line normal to unit vector $\mathbf{n} \in \mathbb{R}^2$ and to which the signed distance

from the origin along direction $\mathbf{n}$ is δ. A point $\mathbf{x} \in \mathbb{R}^2$ is on the line if and only if $\mathbf{x} \cdot \mathbf{n} = \delta$. Replacing $\mathbf{x}$ by $\mathbf{f}(\mathbf{x}) - \mathbf{e}_0 - \mathbf{x}^2\mathbf{e}/2$, we get

$$\mathbf{f}(\mathbf{x}) \cdot (\mathbf{n} + \delta\mathbf{e}) = 0. \tag{7.2.13}$$

Denote $\mathbf{s} = \mathbf{n} + \delta\mathbf{e}$, then $\mathbf{s}^2 = \mathbf{n}^2 = 1$, and $\mathbf{s} \cdot \mathbf{e} = 0$. Conversely, any positive unit vector $\mathbf{s}$ orthogonal to $\mathbf{e}$ represents a line whose signed distance to the origin is $\mathbf{e}_0 \cdot \mathbf{s}$ along the unit normal

$$\mathbf{s} + (\mathbf{e}_0 \cdot \mathbf{s})\mathbf{e} = \mathbf{e}_0 \cdot (\mathbf{s} \wedge \mathbf{e}). \tag{7.2.14}$$

The conformal model of the Euclidean plane can be directly extended to nD Euclidean space.

Definition 7.16. The *conformal model* of nD Euclidean geometry is the set

$$\mathcal{N}_\mathbf{e} := \{\mathbf{x} \in \mathbb{R}^{n+1,1} \,|\, \mathbf{x} \cdot \mathbf{x} = 0, \ \mathbf{x} \cdot \mathbf{e} = -1\}, \tag{7.2.15}$$

where $\mathbf{e}$ is a fixed null vector in $\mathbb{R}^{n+1,1}$, called the *conformal point at infinity* of the model, together with the following isometry:

Fix a vector $\mathbf{e}_0 \in \mathcal{N}_\mathbf{e}$, and denote the orthogonal complement of the 2D plane $\mathbf{e} \wedge \mathbf{e}_0$ in $\mathbb{R}^{n+1,1}$ by $\mathbb{R}^n$. Then

$$\mathbf{f}(\mathbf{x}) := \mathbf{e}_0 + \mathbf{x} + \frac{\mathbf{x}^2}{2}\mathbf{e}, \quad \forall \mathbf{x} \in \mathbb{R}^n \tag{7.2.16}$$

is an isometry from $\mathbb{R}^n$ onto $\mathcal{N}_\mathbf{e}$. Its inverse is $P^\perp_{\mathbf{e}\wedge\mathbf{e}_0}$. The origin of $\mathbb{R}^n$ corresponds to $\mathbf{e}_0$. $\mathbf{f}$ is called the *formalization map* of the conformal model.

The set of null vectors in $\mathbb{R}^{n+1,1}$, denoted by $\mathcal{N}$, is called the nD *projective null cone* when being considered in the $(n+1)$D projective space $\mathbb{R}^{n+1,1}$. Topologically, the projective null cone is homeomorphic to nD sphere. Since two null vectors in $\mathbb{R}^{n+1,1}$ are orthogonal if and only if they are equal up to scale, when removing a projective point from the projective null cone, the result $\mathcal{N}_\mathbf{e}$ is homeomorphic to $\mathbb{R}^n$. In fact it is isometric to $\mathbb{R}^n$ as a distance space, the distance being defined by

$$|\mathbf{a} - \mathbf{b}| = \sqrt{(\mathbf{a} - \mathbf{b})^2}, \quad \forall \mathbf{a}, \mathbf{b} \in \mathcal{N}_\mathbf{e}. \tag{7.2.17}$$

When the origin moves from $\mathbf{e}_0$ to $\mathbf{f}(\mathbf{o})$, where $\mathbf{o} \in \mathbb{R}^n = (\mathbf{e}\wedge\mathbf{e}_0)^\sim$, for any point $\mathbf{x} \in \mathbb{R}^n$, null vector $\mathbf{f}(\mathbf{x})$ represents the following point $\mathbf{y}$ in the nD Euclidean space $(\mathbf{e} \wedge \mathbf{f}(\mathbf{o}))^\sim$:

$$\mathbf{y} = P^\perp_{\mathbf{e}\wedge\mathbf{f}(\mathbf{o})}(\mathbf{f}(\mathbf{x})) = \mathbf{x} - \mathbf{o} + (\mathbf{o} \cdot (\mathbf{x} - \mathbf{o}))\mathbf{e}. \tag{7.2.18}$$

The orthogonal projection of $\mathbf{y}$ back to $\mathbb{R}^n = (\mathbf{e} \wedge \mathbf{e}_0)^\sim$ is $\mathbf{x} - \mathbf{o}$. As expected, a change of the origin $\mathbf{e}_0$ induces a translation in $\mathbb{R}^n$ along the vector $-\mathbf{o}$ from the new origin to the old one.

When the conformal point at infinity rescales from $\mathbf{e}$ to $\lambda\mathbf{e}$, a point $\mathbf{x} \in \mathbb{R}^n$ is changed into $\lambda^{-1}\mathbf{x} \in \mathbb{R}^n$, because

$$\mathbf{f}(\mathbf{x}) = \mathbf{x} + \mathbf{e}_0 + \frac{\mathbf{x}^2}{2}\mathbf{e} = \lambda(\lambda^{-1}\mathbf{x} + \lambda^{-1}\mathbf{e}_0 + \frac{(\lambda^{-1}\mathbf{x})^2}{2}\lambda\mathbf{e}). \tag{7.2.19}$$

This is a dilation in the Euclidean space centered at the origin. The dilation property can also be seen from

$$\mathbf{a} \cdot \mathbf{b} = -\frac{d_{\mathbf{xy}}^2}{2}(\mathbf{a} \cdot \mathbf{e})(\mathbf{b} \cdot \mathbf{e}) = -\frac{d_{(\lambda^{-1}\mathbf{x})(\lambda^{-1}\mathbf{y})}^2}{2}(\mathbf{a} \cdot (\lambda\mathbf{e}))(\mathbf{b} \cdot (\lambda\mathbf{e})), \qquad (7.2.20)$$

where $\mathbf{a} = \mathbf{f}(\mathbf{x})$ and $\mathbf{b} = \mathbf{f}(\mathbf{y})$.

When $\mathbf{e}, \mathbf{e}_0$ are interchanged but each vector in $(\mathbf{e} \wedge \mathbf{e}_0)^{\sim}$ is invariant, by linear extension we get an orthogonal transformation T in $\mathbb{R}^{n+1,1}$. For any vector $\mathbf{x} \in \mathbb{R}^n$, $\mathbf{f}(\mathbf{x})$ is changed into

$$\mathbf{e} + \mathbf{x} + \frac{\mathbf{x}^2}{2}\mathbf{e}_0 = \frac{\mathbf{x}^2}{2}\left(\mathbf{e}_0 + 2\mathbf{x}^{-1} + \frac{(2\mathbf{x}^{-1})^2}{2}\mathbf{e}\right) = \frac{\mathbf{x}^2}{2}\mathbf{f}(2\mathbf{x}^{-1}), \qquad (7.2.21)$$

i.e., $\mathbf{x}$ is mapped to $2\mathbf{x}^{-1}$. This is the *inversion* with respect to the sphere in $\mathbb{R}^n$ centered at the origin and with radius $\sqrt{2}$.

Definition 7.17. In $\mathbb{E}^n$, the *inversion* with respect to a sphere of center $\mathbf{c}$ and radius ρ, called the *invariant sphere* of the inversion, is the transformation in $\mathbb{E}^n \cup \{\mathbf{e}\}$ ($\mathbf{e}$ being the conformal point at infinity) that interchanges point $\mathbf{c}$ and the conformal point at infinity, and changes any point $\mathbf{x}$ to a point on line $\mathbf{cx}$, such that vectors $\overrightarrow{\mathbf{cx}}, \overrightarrow{\mathbf{cy}}$ have the same direction, and

$$d_{\mathbf{cx}}d_{\mathbf{cy}} = \rho^2. \qquad (7.2.22)$$

An inversion fixes every point on the invariant sphere. It also fixes every ray starting from the center of the invariant sphere. Two points are said to be *in inversion* with respect to a sphere, if they are interchanged by the inversion with respect to the sphere.

When $\mathbf{e} \wedge \mathbf{e}_0$ is replaced by another Minkowski plane $\mathbf{e}' \wedge \mathbf{e}_0'$, a point $\mathbf{x} \in \mathbb{R}^n = (\mathbf{e} \wedge \mathbf{e}_0)^{\sim}$ is changed into another point $\mathbf{y}'$ in $(\mathbf{e}' \wedge \mathbf{e}_0')^{\sim}$. Denote the orthogonal projection of $\mathbf{y}'$ back to $\mathbb{R}^n = (\mathbf{e} \wedge \mathbf{e}_0)^{\sim}$ by $\mathbf{y}$. Then $\mathbf{x} \mapsto \mathbf{y}$ for all vectors $\mathbf{x} \in \mathbb{R}^n$ is a *conformal transformation* in $\mathbb{R}^n \cup \{\mathbf{e}\}$.

A *conformal transformation* in $\mathbb{E}^n$, also called *Möbius transformation*, is an angle-preserving diffeomorphism of $\mathbb{E}^n \cup \{\mathbf{e}\}$. *Conformal geometry*, or *Möbius geometry*, studies the properties of $\mathbb{E}^n \cup \{\mathbf{e}\}$ that are invariant under conformal transformations.

It is a classical result [22] that any orthogonal transformation of $\mathbb{R}^{n+1,1}$ induces a conformal transformation in $\mathbb{E}^n \cup \{\mathbf{e}\}$ via the conformal model, and conversely, any conformal transformation can be generated in this way. This justifies the name "conformal" of the model.

7.2.2   *Vectors of different signatures*

In the conformal model, a null vector represents a point or the conformal point at infinity, while a positive vector represents a sphere or hyperplane in $\mathbb{E}^n$. By the term "sphere" without specifying its dimension from the context, we mean an $(n-1)$D "hyper-sphere" in $\mathbb{E}^n$. A sphere with center $\mathbf{c}$ and radius ρ is usually denoted by the pair $(\mathbf{c}, \rho)$.

Let $\mathbf{s}$ be a positive vector in $\mathbb{R}^{n+1,1}$, then it represents a sphere or hyperplane in the sense that a point represented by a null vector $\mathbf{a}$ is on it if and only if $\mathbf{a} \cdot \mathbf{s} = 0$. It represents a hyperplane if and only if $\mathbf{e} \cdot \mathbf{s} = 0$. In other words, a hyperplane passes through the conformal point at infinity, but not so for a sphere. The representation by positive vectors is unique up to scale.

The following are Euclidean geometric explanations of the inner products of positive vectors and null vectors in the conformal model, where a point is identified with the null vector in $\mathcal{N}_e$ representing it:

- For two points $\mathbf{c}_1$ and $\mathbf{c}_2$,

$$\mathbf{c}_1 \cdot \mathbf{c}_2 = -\frac{d_{\mathbf{c}_1\mathbf{c}_2}^2}{2}. \tag{7.2.23}$$

- For point $\mathbf{c}$ and hyperplane $\mathbf{n} + \delta\mathbf{e}$,

$$\mathbf{c} \cdot (\mathbf{n} + \delta\mathbf{e}) = \mathbf{c} \cdot \mathbf{n} - \delta. \tag{7.2.24}$$

The inner product is positive, zero, or negative, if and only if the vector from the hyperplane to the point is along $\mathbf{n}$, zero, or along $-\mathbf{n}$ respectively. Its absolute value equals the distance between the point and the hyperplane.

- For point $\mathbf{c}_1$ and sphere $\mathbf{c}_2 - \rho^2\mathbf{e}/2$,

$$\mathbf{c}_1 \cdot (\mathbf{c}_2 - \frac{\rho^2}{2}\mathbf{e}) = \frac{\rho^2 - d_{\mathbf{c}_1\mathbf{c}_2}^2}{2}. \tag{7.2.25}$$

The inner product is positive, zero, or negative, if and only if the point is inside, on, or outside the sphere respectively. Let $d_{\max}$ and $d_{\min}$ be respectively the maximal distance and minimal distance between point $\mathbf{c}_1$ and the points on the sphere. Then the absolute value of (7.2.25) equals $d_{\max}d_{\min}/2$.

- For two hyperplanes $\mathbf{n}_1 + \delta_1\mathbf{e}$ and $\mathbf{n}_2 + \delta_2\mathbf{e}$,

$$(\mathbf{n}_1 + \delta_1\mathbf{e}) \cdot (\mathbf{n}_2 + \delta_2\mathbf{e}) = \mathbf{n}_1 \cdot \mathbf{n}_2. \tag{7.2.26}$$

- For hyperplane $\mathbf{n} + \delta\mathbf{e}$ and sphere $\mathbf{c} - \rho^2\mathbf{e}/2$,

$$(\mathbf{n} + \delta\mathbf{e}) \cdot (\mathbf{c} - \frac{\rho^2}{2}\mathbf{e}) = \mathbf{n} \cdot \mathbf{c} - \delta. \tag{7.2.27}$$

It is positive, zero, or negative, if and only if the vector from the hyperplane to the center $\mathbf{c}$ of the sphere is along $\mathbf{n}$, zero, or along $-\mathbf{n}$ respectively. Its absolute value equals the distance between the center and the hyperplane.

- For two spheres $\mathbf{c}_1 - \rho_1^2\mathbf{e}/2$ and $\mathbf{c}_2 - \rho_2^2\mathbf{e}/2$,

$$(\mathbf{c}_1 - \frac{\rho_1^2}{2}\mathbf{e}) \cdot (\mathbf{c}_2 - \frac{\rho_2^2}{2}\mathbf{e}) = \frac{\rho_1^2 + \rho_2^2 - d_{\mathbf{c}_1\mathbf{c}_2}^2}{2}. \tag{7.2.28}$$

It is zero if the two spheres are *perpendicular* to each other, *i.e.*, they intersect at 90°. When the two spheres intersect, (7.2.28) equals the cosine of the angle of intersection multiplied by $\rho_1\rho_2$.

When $\mathbf{s}_1, \mathbf{s}_2$ are positive vectors, the scalar $\mathbf{s}_1 \cdot \mathbf{s}_2 / |\mathbf{s}_1||\mathbf{s}_2|$ is called the *inversive product* of the two spheres or hyperplanes [95].

Proposition 7.18. Let $\mathbf{s}_1, \mathbf{s}_2$ be two positive vectors in $\mathbb{R}^{n+1,1}$ representing two intersecting spheres and planes in $\mathbb{R}^n$. Let $\mathbf{a}$ be a point at their intersection. For $i = 1, 2$, when $\mathbf{s}_i$ represents a sphere, let $\mathbf{s}_i = \mathbf{c}_i - \rho_i^2 \mathbf{e}/2$, and let $\mathbf{m}_i$ be the outward unit normal direction of the sphere at point $\mathbf{a}$; when $\mathbf{s}_i$ represents a hyperplane, let $\mathbf{s}_i = \mathbf{n}_i + \delta_i \mathbf{e}$, and let $\mathbf{m}_i = -\mathbf{n}_i$. Then

$$\frac{\mathbf{s}_1 \cdot \mathbf{s}_2}{|\mathbf{s}_1||\mathbf{s}_2|} = \mathbf{m}_1 \cdot \mathbf{m}_2. \tag{7.2.29}$$

Proof. Direct from (7.2.27), (7.2.28), and the expression $\mathbf{m}_i = \rho_i^{-1}(\mathbf{f}^{-1}(\mathbf{a}) - \mathbf{f}^{-1}(\mathbf{c}_i))$ for $\mathbf{s}_i = \mathbf{c}_i - \rho_i^2 \mathbf{e}/2$. $\qquad \square$

Below we consider the geometric meaning of a negative vector $\mathbf{s} \in \mathbb{R}^{n+1,1}$. Since $\mathbf{s}^2 < 0$, $\mathbf{s} \cdot \mathbf{e} \neq 0$. Let $\mathbf{a} = -\mathbf{s}/(\mathbf{s} \cdot \mathbf{e})$. In the Minkowski plane spanned by vectors $\mathbf{a}$ and $\mathbf{e}$,

$$\mathbf{c} = \mathbf{a} + \frac{\mathbf{a}^2}{2}\mathbf{e} \tag{7.2.30}$$

is the unique null vector in $\mathcal{N}_\mathbf{e}$. Let

$$\begin{aligned} \mathbf{x} &= \mathbf{f}^{-1}(\mathbf{c}) = P^{\perp}_{\mathbf{e} \wedge \mathbf{e}_0}(\mathbf{a}) \in \mathbb{R}^n, \\ \rho &= |\mathbf{a}| = |P_\mathbf{s}(\mathbf{e})|^{-1}. \end{aligned} \tag{7.2.31}$$

Then sphere $(\mathbf{c}, \rho)$ has positive-vector representation

$$\mathbf{s} \cdot (\mathbf{e} \wedge \mathbf{s}). \tag{7.2.32}$$

The space $\mathbf{s}^{\sim}$ is Euclidean and is composed of positive vectors, *i.e.*, spheres and hyperplanes. A sphere $\mathbf{f}(\mathbf{o}') - \rho'^2 \mathbf{e}/2$ is in $\mathbf{s}^{\sim}$ if and only if $\rho'^2 = d_{\mathbf{o}'\mathbf{x}}^2 + |\mathbf{a}|^2$. A hyperplane $\mathbf{n} + \delta \mathbf{e}$ is in $\mathbf{s}^{\sim}$ if and only if $\mathbf{n} \cdot \mathbf{x} = \delta$. So negative vector $\mathbf{s}$ represents all the spheres and hyperplanes each containing an $(n-2)$D great sphere of the sphere $(\mathbf{c}, \rho)$, in the sense that for any positive vector $\mathbf{t}$, $\mathbf{s} \cdot \mathbf{t} = 0$ if and only if the sphere or hyperplane represented by vector $\mathbf{t}$ intersects sphere $(\mathbf{c}, \rho)$ at an $(n-2)$D great sphere on the latter sphere, see Figure 7.2(a).

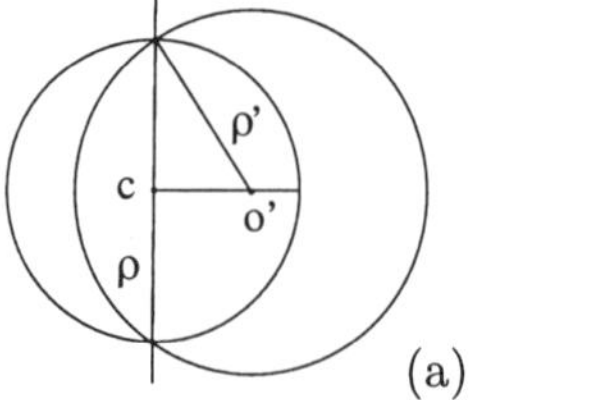

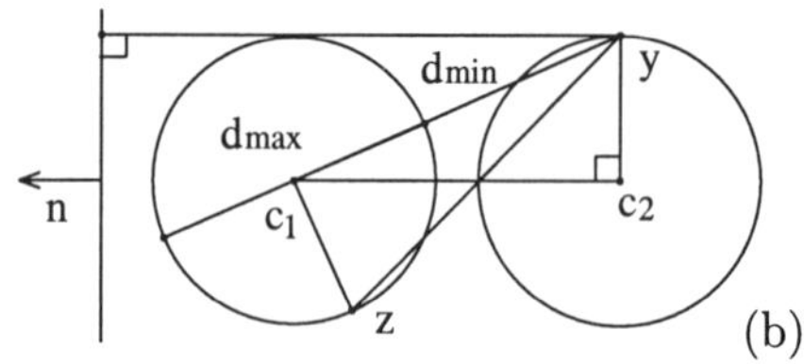

Fig. 7.2 Geometric interpretation of negative vectors in the conformal model.

Negative vector $\mathbf{s}$ after rescaling, can be written in the following *standard form*:

$$\mathbf{s} = \mathbf{c} + \frac{\rho^2}{2}\mathbf{e}. \tag{7.2.33}$$

The Euclidean geometric meaning of the inner product of a negative vector with another vector in $\mathbb{R}^{n+1,1}$ is as follows:

(1) Let $\mathbf{c}_1, \mathbf{c}_2$ be points represented by null vectors in $\mathcal{N}_\mathbf{e}$. Then

$$\mathbf{c}_1 \cdot \left(\mathbf{c}_2 + \frac{\rho^2}{2}\mathbf{e}\right) = -\frac{d^2_{\mathbf{c}_1\mathbf{c}_2} + \rho^2}{2}. \tag{7.2.34}$$

As shown in Figure 7.2(b), let $\mathbf{y}$ be any point on sphere $(\mathbf{c}_2, \rho)$ satisfying $\mathbf{c}_2\mathbf{y} \perp \mathbf{c}_1\mathbf{c}_2$, then (7.2.34) equals $-d^2_{\mathbf{c}_1\mathbf{y}}/2$.

(2) Let $\mathbf{n}$ be a unit vector in $\mathbb{R}^n$, then

$$(\mathbf{n} + \delta\mathbf{e}) \cdot \left(\mathbf{c} + \frac{\rho^2}{2}\mathbf{e}\right) = \mathbf{c} \cdot \mathbf{n} - \delta. \tag{7.2.35}$$

Let $\mathbf{y}$ be any point on sphere $(\mathbf{c}_2, \rho)$ such that $\mathbf{c}_2\mathbf{y} \perp \mathbf{n}$, then (7.2.35) equals the signed distance from hyperplane $\mathbf{n} + \delta\mathbf{e}$ to point $\mathbf{y}$.

(3) Let $\mathbf{c}_1, \mathbf{c}_2 \in \mathcal{N}_\mathbf{e}$, then

$$\left(\mathbf{c}_1 - \frac{\rho_1^2}{2}\mathbf{e}\right) \cdot \left(\mathbf{c}_2 + \frac{\rho_2^2}{2}\mathbf{e}\right) = \frac{\rho_1^2 - \rho_2^2 - d^2_{\mathbf{c}_1\mathbf{c}_2}}{2}. \tag{7.2.36}$$

Let $\mathbf{y}$ be any point on sphere $(\mathbf{c}_2, \rho)$ such that $\mathbf{c}_2\mathbf{y} \perp \mathbf{c}_1\mathbf{c}_2$, and let $d_{\max}, d_{\min}$ be respectively the maximal signed distance and the minimal signed distance from point $\mathbf{y}$ to points on sphere $\mathbf{c}_1 - \rho_1^2\mathbf{e}/2$, then (7.2.36) equals $-d_{\max}d_{\min}/2$.

(4) Let $\mathbf{c}_1, \mathbf{c}_2 \in \mathcal{N}_\mathbf{e}$, then

$$\left(\mathbf{c}_1 + \frac{\rho_1^2}{2}\mathbf{e}\right) \cdot \left(\mathbf{c}_2 + \frac{\rho_2^2}{2}\mathbf{e}\right) = -\frac{\rho_1^2 + \rho_2^2 + d^2_{\mathbf{c}_1\mathbf{c}_2}}{2}. \tag{7.2.37}$$

Let $\mathbf{y}$ be any point on sphere $(\mathbf{c}_2, \rho_2)$ such that $\mathbf{c}_2\mathbf{y} \perp \mathbf{c}_1\mathbf{c}_2$, and let $\mathbf{z}$ be any point on sphere $(\mathbf{c}_1, \rho_1)$ satisfying $\mathbf{c}_1\mathbf{z} \perp \mathbf{c}_1\mathbf{y}$. Then (7.2.37) equals $-d^2_{\mathbf{y}\mathbf{z}}/2$.

7.2.3 *The homogeneous model*

From the definition of the conformal model, it is clear that the model depends on the choice of the origin $\mathbf{e}_0$. For invariant symbolic computing, we need to revise the model by getting rid of the origin. The result is a new model called the *homogeneous model*.

Definition 7.19. [115] The *homogeneous model* of nD Euclidean geometry is a pair $(\mathcal{N}, \mathbf{e})$, where $\mathcal{N}$ is the set of null vectors in $\mathbb{R}^{n+1,1}$, and $\mathbf{e} \in \mathcal{N}$ represents the conformal point at infinity. A vector $\mathbf{a} \in \mathcal{N}$ represents a Euclidean point if and only if $\mathbf{a} \cdot \mathbf{e} \neq 0$. Two vectors in $\mathcal{N}$ represent the same point if and only if they differ by scale.

The homogeneous model has a representation space that is homeomorphic to nD sphere, while in the conformal model, the representation space $\mathcal{N}_\mathbf{e}$ is homeomorphic to $\mathbb{R}^n$. Since it does not require any origin, the homogeneous model does not need a formalization map onto the standard Euclidean space $\mathbb{R}^n$. As a consequence, the representations and computations in the homogeneous model are all homogeneous.

The homogeneous model provides a conformal description of Euclidean geometry, *i.e.*, the model is conformal instead of isometric. By setting $\mathbf{a} \cdot \mathbf{e} = -1$ for any null vector $\mathbf{a}$ not equal to $\mathbf{e}$ up to scale, we obtain an *inhomogeneous representation* that is still origin-free. The homogeneous representation is very convenient for symbolic computation, while the inhomogeneous representation is suitable for geometric interpretation.

Example 7.20. In the homogeneous model, the sphere with center $\mathbf{o}$ (null vector) and radius ρ is represented by

$$\mathbf{o} + \frac{\rho^2 \mathbf{o} \cdot \mathbf{e}}{2} \mathbf{e}; \tag{7.2.38}$$

the sphere with center $\mathbf{o}$ (null vector) and through point $\mathbf{a}$ (null vector) is represented by

$$\mathbf{a} \cdot (\mathbf{e} \wedge \mathbf{o}); \tag{7.2.39}$$

the hyperplane with normal $\mathbf{n} \in \mathbf{e}^\sim$ and through point $\mathbf{a}$ (null vector), where $\mathbf{n}$ is a positive vector but is not necessary to be of unit magnitude, is represented by

$$\mathbf{a} \cdot (\mathbf{e} \wedge \mathbf{n}). \tag{7.2.40}$$

(7.2.38) is quadratic with respect to $\mathbf{e}$, so under the dilation $\mathbf{e} \mapsto \lambda \mathbf{e}$, the radius ρ has to be scaled by λ^{-1} in order to keep (7.2.38) invariant. (7.2.39) is homogeneous with respect to $\mathbf{e}$; it is a dilation-invariant representation while (7.2.38) is not.

7.3 Positive-vector representations of spheres and hyperplanes

Positive vectors in the conformal model represent spheres and hyperplanes in $\mathbb{E}^n$, and the representation is unique up to scale. The outer product of a sequence of positive vectors can represent both the pencil of spheres and hyperplanes spanned by the vectors, and the intersection of the spheres and hyperplanes represented by the vectors.

Definition 7.21. For $r > 1$, let there be a set of spheres and hyperplanes $f_1 = 0, \ldots, f_r = 0$, where $f_i = 0$ is a polynomial equation in Cartesian coordinates representing the i-th sphere or hyperplane. Assume that the polynomials f_i are linearly independent. The rD *pencil of spheres and hyperplanes* spanned by the r spheres and hyperplanes, is the set of spheres and hyperplanes whose equations are of the form $\lambda_1 f_1 + \ldots + \lambda_r f_r = 0$, where the λ_i are parameters. If the dimension of a pencil is not specified, the pencil is assumed to have default dimension two.

In the conformal model, an rD pencil is an rD vector subspace of $\mathbb{R}^{n+1,1}$ spanned by r linearly independent positive vectors $\mathbf{s}_1, \ldots, \mathbf{s}_r$, so it can be represented by blade $\mathbf{s}_1 \wedge \cdots \wedge \mathbf{s}_r$.

7.3.1 *Pencils of spheres and hyperplanes*

Pencils are classified by the signatures of their representing blades. The classification and representations of rD pencils are much the same with those of 2D pencils [110], so we only investigate 2D pencils.

Take the 2D geometry as an example. In $\mathbb{R}^{3,1}$, the 2D subspace $\mathbf{B}_2 = (\mathbf{s}_1 \wedge \mathbf{s}_2)^{\sim}$ has three possible signatures:

(1) Minkowski: $(\mathbf{s}_1 \wedge \mathbf{s}_2)^2 < 0$. $\mathbf{B}_2$ has two different null vectors up to scale.

If $\mathbf{e} \in \mathbf{B}_2$, then both $\mathbf{s}_1, \mathbf{s}_2$ are lines. Besides $\mathbf{e}$, the other null vector $\mathbf{a} \in \mathbf{B}_2$ is the point of intersection of the two lines. Pencil $\mathbf{s}_1 \wedge \mathbf{s}_2$ is called a *concurrent pencil*. Any positive vector in the pencil is a line passing through point $\mathbf{a}$. To express the point of intersection $\mathbf{a}$ explicitly, however, the three vectors $\mathbf{e}, \mathbf{s}_1, \mathbf{s}_2$ are insufficient, and a fourth vector needs to be introduced.

If $\mathbf{e} \notin \mathbf{B}_2$, then at least one of $\mathbf{s}_1, \mathbf{s}_2$ is a circle. The two circles $\mathbf{s}_1, \mathbf{s}_2$, or one circle and one line, intersect at the two points $\mathbf{a}, \mathbf{b}$ corresponding to the two null 1D subspaces of $\mathbf{B}_2$. Pencil $\mathbf{s}_1 \wedge \mathbf{s}_2$ is called a *secant pencil*. Any point or line in the pencil passes through the two points.

The secant pencil contains only one line. It is line $\mathbf{ab}$, whose positive-vector representation is $\mathbf{e} \cdot (\mathbf{s}_1 \wedge \mathbf{s}_2)$. The circle containing points $\mathbf{a}, \mathbf{b}$ as a pair of antipodal points is $P_{\mathbf{s}_1 \wedge \mathbf{s}_2}(\mathbf{e})$. It is also in the pencil.

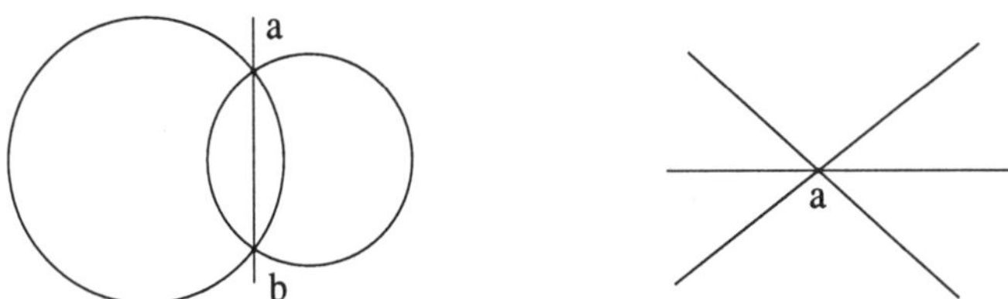

Fig. 7.3　Secant pencil (left) and concurrent pencil (right).

(2) Degenerate: $(\mathbf{s}_1 \wedge \mathbf{s}_2)^2 = 0$. $\mathbf{B}_2$ has only one null vector up to scale, denoted by $\mathbf{a}$.

If $\mathbf{a} = \mathbf{e}$ up to scale, then both $\mathbf{s}_1, \mathbf{s}_2$ are lines, and the two lines are parallel. Pencil $\mathbf{s}_1 \wedge \mathbf{s}_2$ is called a *parallel pencil*. Any positive vector in the pencil is a line with the same normal direction $\mathbf{e}_0 \cdot (\mathbf{s}_1 \wedge \mathbf{s}_2)$.

If $\mathbf{a} \neq \mathbf{e}$ up to scale, then at least one of $\mathbf{s}_1, \mathbf{s}_2$ is a circle. The two circles $\mathbf{s}_1, \mathbf{s}_2$, or one circle and one line, are tangent to each other at point $\mathbf{a} = P_{\mathbf{s}_1}^{\perp}(\mathbf{s}_2)$. Pencil $\mathbf{s}_1 \wedge \mathbf{s}_2$ is called a *tangent pencil*.

The tangent pencil contains only one line. It is line $\mathbf{e} \cdot (\mathbf{s}_1 \wedge \mathbf{s}_2)$. Any circle in the pencil is tangent to the line at the common point of tangency $\mathbf{a}$. All the circles and lines in the pencil are perpendicular to line $(\mathbf{e} \wedge \mathbf{s}_1 \wedge \mathbf{s}_2)^{\sim}$.

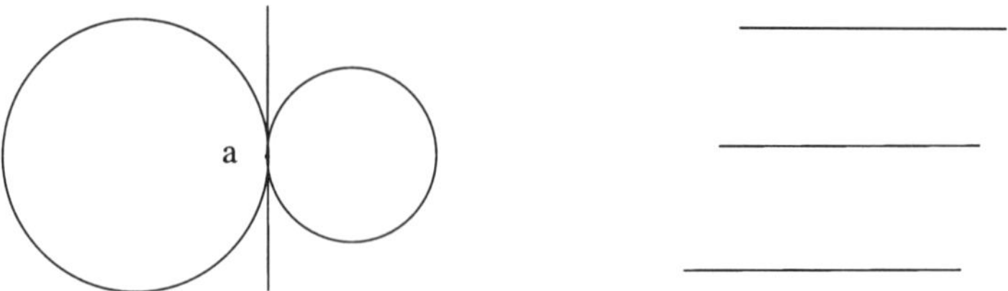

Fig. 7.4   Tangent pencil (left) and parallel pencil (right).

(3) Euclidean: $(\mathbf{s}_1 \wedge \mathbf{s}_2)^2 > 0$. Since $\mathbf{B}_2$ has no null vector, circles $\mathbf{s}_1, \mathbf{s}_2$, or one circle and one line, do not intersect.

As the dual of $\mathbf{B}_2$, blade $\mathbf{s}_1 \wedge \mathbf{s}_2$ is Minkowski, and has two null 1D subspaces. Let $(\mathbf{a}, \mathbf{b})$ be a Witt pair in plane $\mathbf{s}_1 \wedge \mathbf{s}_2$. Without loss of generality, let $\mathbf{a} = \mathbf{e}_0$ be the origin, and let $\mathbf{s}_i = \mathbf{f}(\mathbf{c}_i) - \rho_i^2 \mathbf{e}/2$ for $i = 1, 2$, where $\mathbf{c}_i \in \mathbb{R}^2$.

If $\mathbf{b} = \mathbf{e}$, from $\mathbf{s}_i \wedge \mathbf{e} \wedge \mathbf{e}_0 = 0$ we get $\mathbf{c}_i = 0$. So circles $\mathbf{s}_1, \mathbf{s}_2$ have the same center. Pencil $\mathbf{s}_1 \wedge \mathbf{s}_2$ is called a *concentric pencil*. Any positive vector in the pencil is a circle with the same center $\mathbf{s}_1 \mathbf{e} \mathbf{s}_1$ (null vector).

If $\mathbf{b} = \mathbf{f}(\mathbf{x})$ up to scale, where $\mathbf{x} \in \mathbb{R}^2 - \{0\}$, then from $\mathbf{s}_i \wedge \mathbf{e}_0 \wedge \mathbf{f}(\mathbf{x}) = 0$ we get $\mathbf{x}^2 \mathbf{c}_i = (\mathbf{c}_i^2 - \rho_i^2)\mathbf{x}$. So $\mathbf{x}\mathbf{c}_i = \mathbf{c}_i^2 - \rho_i^2$, and

$$-\mathbf{c}_i(\mathbf{x} - \mathbf{c}_i) = \rho_i^2. \tag{7.3.1}$$

By (7.2.22), the origin and point $\mathbf{x}$ are in inversion with respect to each circle $\mathbf{s}_i$. They are called the *Poncelet points* of the pencil. Pencil $\mathbf{s}_1 \wedge \mathbf{s}_2$ is called a *Poncelet pencil*.

The Poncelet pencil contains only one line. It is the perpendicular bisector of line segment $\mathbf{a}\mathbf{b}$, and has positive-vector representation $\mathbf{e} \cdot (\mathbf{a} \wedge \mathbf{b})$. All the circles in the pencil have the property that the two Poncelet points $\mathbf{a}, \mathbf{b}$ are in inversion with respect to each of them.

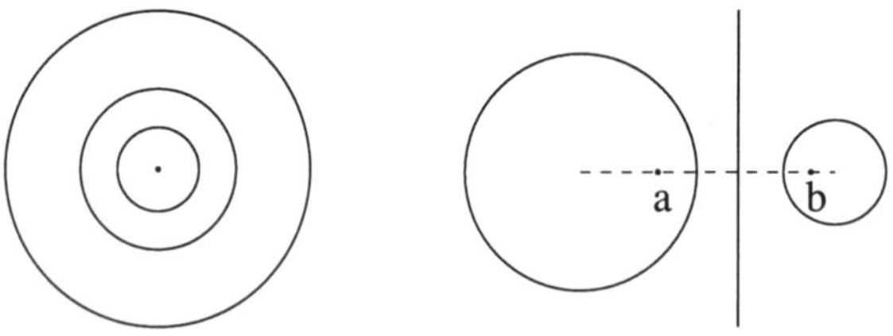

Fig. 7.5   Concentric pencil (left) and Poncelet pencil (right).

Proposition 7.22. In the following statements, a point is identified with a null vector representing it:

(i) The secant pencil passing through points $\mathbf{a}, \mathbf{b}$ is $\mathbf{a} \wedge \mathbf{b}$.

(ii) The concurrent pencil passing through point $\mathbf{a}$ is $\mathbf{e} \wedge \mathbf{a}$.

(iii) The parallel pencil containing line $\mathbf{s}$ is $\mathbf{e} \wedge \mathbf{s}$.

(iv) The parallel pencil with common normal direction $\mathbf{n} \in \mathbb{R}^2$ is $\mathbf{e} \wedge \mathbf{n}$.

(v) The tangent pencil with common point of tangency $\mathbf{a}$, and containing circle (or line) $\mathbf{s}$, is $\mathbf{a} \wedge \mathbf{s}$.

(vi) The tangent pencil with common point of tangency $\mathbf{a}$ and common perpendicular line $\mathbf{s}$, is $(\mathbf{a} \wedge \mathbf{s})^{\sim}$.

(vii) The tangent line of circle $\mathbf{s}$ at the point of tangency $\mathbf{a}$ is $\mathbf{e} \cdot (\mathbf{a} \wedge \mathbf{s})$.

(viii) The concentric pencil centered at point $\mathbf{a}$ is $\mathbf{e} \wedge \mathbf{a}$.

(ix) The Poncelet pencil with Poncelet points $\mathbf{a}, \mathbf{b}$ is $\mathbf{a} \wedge \mathbf{b}$.

(x) For any point (including the conformal point at infinity) $\mathbf{a}$, the circle or line in pencil $\mathbf{s}_1 \wedge \mathbf{s}_2$ and through point $\mathbf{a}$ exists if and only if vector $\mathbf{a} \cdot (\mathbf{s}_1 \wedge \mathbf{s}_2)$ is positive. If it exists, the circle or line is represented by the vector.

Proof. Only (vi) needs proof. The line through point $\mathbf{a}$ and perpendicular to line $\mathbf{s}$ is $(\mathbf{e} \wedge \mathbf{a} \wedge \mathbf{s})^{\sim}$. By (v), the pencil is $\mathbf{a} \wedge (\mathbf{e} \wedge \mathbf{a} \wedge \mathbf{s})^{\sim} = \langle \mathbf{a}(\mathbf{e} \wedge \mathbf{a})\mathbf{s} \rangle_{\widetilde{2}} = \mathbf{a} \cdot \mathbf{e}(\mathbf{a} \wedge \mathbf{s})^{\sim}$. $\qquad\square$

7.3.2 *Positive-vector representation*

Let $\mathbf{s}_1, \mathbf{s}_2$ be two linearly independent positive vectors, and let $\mathbf{a}$ be a null vector. Since $\mathbf{a} \cdot (\mathbf{s}_1 \wedge \mathbf{s}_2) = (\mathbf{a} \cdot \mathbf{s}_1)\mathbf{s}_2 - (\mathbf{a} \cdot \mathbf{s}_2)\mathbf{s}_1 = 0$ if and only if $\mathbf{a} \cdot \mathbf{s}_1 = \mathbf{a} \cdot \mathbf{s}_2 = 0$, blade $\mathbf{s}_1 \wedge \mathbf{s}_2$ can also represent the intersection of spheres or hyperplanes $\mathbf{s}_1, \mathbf{s}_2$, in the sense that point $\mathbf{a}$ is on the intersection if and only if $\mathbf{a} \cdot (\mathbf{s}_1 \wedge \mathbf{s}_2) = 0$. If the two spheres or hyperplanes intersect, their intersection is an $(n-2)$D sphere or plane, and can be represented by blade $\mathbf{s}_1 \wedge \mathbf{s}_2$, which has Euclidean signature.

In the general case, if the intersection of r spheres or hyperplanes $\mathbf{s}_1, \ldots, \mathbf{s}_r$ exists, where $1 \leq r \leq n$, it can be represented by $\mathbf{s}_1 \wedge \cdots \wedge \mathbf{s}_r$ in the sense that a point $\mathbf{a}$ is on the intersection if and only if $\mathbf{a} \cdot (\mathbf{s}_1 \wedge \cdots \wedge \mathbf{s}_r) = 0$. This representation of iD spheres and planes for all $0 \leq i \leq n-1$, is called the *positive-vector representation*. In this representation, blade $\mathbf{s}_1 \wedge \cdots \wedge \mathbf{s}_r$ represents an $(n-r)$D sphere or plane if and only if it is Euclidean.

Definition 7.23. In $\mathbb{E}^n$, a *0D plane*, or *0D line*, is a pair $(\mathbf{a}, \mathbf{e})$ where $\mathbf{a}$ is a point and $\mathbf{e}$ is the conformal point at infinity. A *0D sphere*, or *0D circle*, is a pair of distinct points. The *center* of a 0D sphere is the midpoint of the line segment between the two points of the 0D sphere.

Definition 7.24. For an rD sphere in $\mathbb{E}^n$, where $0 \leq r \leq n-1$, its *supporting plane* is the $(r+1)$D affine plane in $\mathbb{E}^n$ that contains the rD sphere. Its *extensive sphere* is the $(n-1)$D sphere having the same center and radius with the rD sphere. The *supporting plane* of an rD plane is the rD plane itself.

For example, the supporting plane of a 0D sphere is the line passing through the two points of the 0D sphere; the supporting plane of an $(n-1)$D sphere is $\mathbb{E}^n$ itself.

Definition 7.25. In $\mathbb{E}^n$, for any $0 \le r, s \le n-1$,

- when $r, s > 0$, then if an rD sphere and an sD sphere intersect, they are said to be *perpendicular* if at any point of their intersection, their tangent spaces are perpendicular;
- when $r = 0$, a 0D sphere and an sD sphere are said to be *perpendicular*, if the center of the sD sphere is collinear with the two points of the 0D sphere, and the two extensive spheres of the 0D sphere and the sD sphere respectively are perpendicular to each other;
- when $s > 0$, an rD sphere and an sD plane are said to be *perpendicular*, if the center of the rD sphere lies in the sD plane, and the sD plane is either perpendicular to the supporting plane of the rD sphere, or inside the latter plane as a proper affine subspace;
- when $s = 0$, an rD sphere and a 0D plane are said to be *perpendicular*, if the point of the 0D plane is the center of the rD sphere.

Example 7.26. Let the $(\mathbf{f}(\mathbf{a}_i), \mathbf{f}(\mathbf{b}_i))$ for $i = 1, 2$ be two 0D spheres in $\mathbb{R}^n$. The center and radius of sphere $(\mathbf{a}_i, \mathbf{b}_i)$ are denoted by $(\mathbf{o}_i, \rho_i)$ respectively, where $\mathbf{o}_i \in \mathbb{R}^n$. By definition, the two spheres are perpendicular if and only if the four points $\mathbf{a}_1, \mathbf{a}_2, \mathbf{b}_1, \mathbf{b}_2$ are collinear and

$$d_{\mathbf{o}_1\mathbf{o}_2}^2 = \rho_1^2 + \rho_2^2. \tag{7.3.2}$$

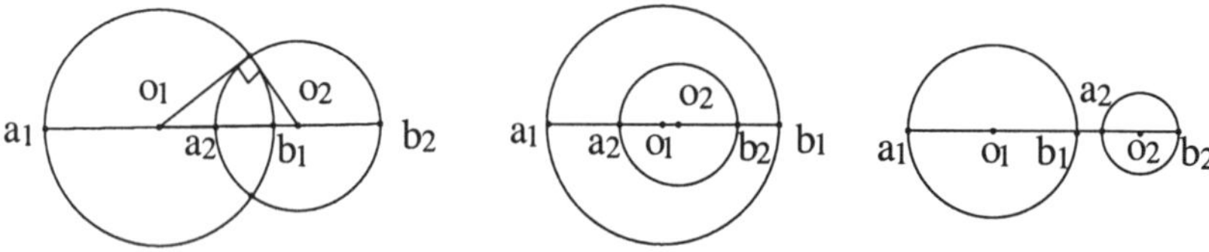

Fig. 7.6   Perpendicularity of two 0D spheres.

(7.3.2) can be expressed as an equality of the distances among the four points.

The extensive spheres of the two 0D spheres have three possible relations: intersecting, inclusive, and exclusive. We consider only the intersecting case, because the formulas to be derived below are valid for the other two cases as well.

Denote $d = d_{\mathbf{o}_1\mathbf{o}_2}$. As shown in the first of Figure 7.6, if $\mathbf{e}_1$ is the unit vector on line $\mathbf{o}_1\mathbf{o}_2$ in the direction from $\mathbf{o}_1$ to $\mathbf{o}_2$, then

$$
\begin{aligned}
\overrightarrow{\mathbf{a}_1\mathbf{b}_1} &= 2\rho_1\mathbf{e}_1, \\
\overrightarrow{\mathbf{a}_2\mathbf{b}_2} &= 2\rho_2\mathbf{e}_1, \\
\overrightarrow{\mathbf{o}_1\mathbf{a}_2} &= (d - \rho_2)\mathbf{e}_1, \\
\overrightarrow{\mathbf{b}_1\mathbf{o}_2} &= (d - \rho_1)\mathbf{e}_1, \\
\overrightarrow{\mathbf{a}_1\mathbf{b}_2} &= (d + \rho_1 + \rho_2)\mathbf{e}_1, \\
\overrightarrow{\mathbf{a}_2\mathbf{b}_1} &= (\rho_1 + \rho_2 - d)\mathbf{e}_1, \\
\overrightarrow{\mathbf{a}_1\mathbf{a}_2} &= \rho_1\mathbf{e}_1 + \overrightarrow{\mathbf{o}_1\mathbf{a}_2} = (d + \rho_1 - \rho_2)\mathbf{e}_1, \\
\overrightarrow{\mathbf{b}_1\mathbf{b}_2} &= \rho_2\mathbf{e}_1 + \overrightarrow{\mathbf{b}_1\mathbf{o}_2} = (d - \rho_1 + \rho_2)\mathbf{e}_1.
\end{aligned}
$$

So (7.3.2) is equivalent to

$$
\overrightarrow{\mathbf{a}_1\mathbf{b}_2}\,\overrightarrow{\mathbf{a}_2\mathbf{b}_1} = \overrightarrow{\mathbf{a}_1\mathbf{a}_2}\,\overrightarrow{\mathbf{b}_1\mathbf{b}_2} = \frac{1}{2}\overrightarrow{\mathbf{a}_1\mathbf{b}_1}\,\overrightarrow{\mathbf{a}_2\mathbf{b}_2}. \tag{7.3.3}
$$

Proposition 7.27. Let $n > 1$. For spheres and hyperplanes $\mathbf{t}, \mathbf{s}_1, \mathbf{s}_2, \ldots, \mathbf{s}_r$, where $1 \le r \le n$, let blade $\mathbf{A}_r = \mathbf{s}_1 \wedge \cdots \wedge \mathbf{s}_r$ be Euclidean.

(1) When $r < n$, $\mathbf{A}_r \cdot \mathbf{t} = 0$ if and only if as spheres and planes geometrically, $\mathbf{A}_r$ and $\mathbf{t}$ intersect and are perpendicular to each other.

(2) When $r = n$, if $\mathbf{e} \cdot \mathbf{A}_n = 0$, then $\mathbf{A}_n \cdot \mathbf{t} = 0$ if and only if $\mathbf{t}$ is a sphere perpendicular to 0D plane $\mathbf{A}_n$, *i.e.*, the $\mathbf{s}_i$ for $1 \le i \le n$ are n hyperplanes meeting at the center of sphere $\mathbf{t}$; if $\mathbf{e} \cdot \mathbf{A}_n \ne 0$, then $\mathbf{A}_n \cdot \mathbf{t} = 0$ if and only if $\mathbf{A}_n$ is a 0D sphere perpendicular to sphere or hyperplane $\mathbf{t}$.

Proof. When $r = 1$, by (7.2.29), all the statements are true, so we assume $r > 1$. We only prove the necessity statements. The sufficiency statements can be proved similarly. By the hypothesis, $\mathbf{A}_r \cdot \mathbf{t} = 0$, *i.e.*, $\mathbf{s}_i \cdot \mathbf{t} = 0$ for every $1 \le i \le r$.

(1) When $r < n$, since $\mathbf{A}_r \cdot \mathbf{t} = 0$, blade $\mathbf{A}_r \wedge \mathbf{t}$ is Euclidean, so it represents the $(n - r - 1)$D intersection of the spheres and hyperplanes $\mathbf{t}, \mathbf{s}_1, \ldots, \mathbf{s}_r$. In particular, $\mathbf{A}_r$ and $\mathbf{t}$ intersect.

Case 1.1. If $\mathbf{e} \cdot (\mathbf{A}_r \wedge \mathbf{t}) = 0$, then $\mathbf{t}$ and all the $\mathbf{s}_i$ are hyperplanes, and obviously plane $\mathbf{A}_r$ and hyperplane $\mathbf{t}$ are perpendicular.

Case 1.2. If $\mathbf{e} \cdot \mathbf{t} \ne 0$ but $\mathbf{e} \cdot \mathbf{A}_r = 0$, then all the $\mathbf{s}_i$ are hyperplanes, and $\mathbf{t}$ is a sphere whose center is on every hyperplane $\mathbf{s}_i$, so the center must be on the intersection $\mathbf{A}_r$ of the hyperplanes.

Case 1.3. If $\mathbf{e} \cdot \mathbf{t} = 0$ but $\mathbf{e} \cdot \mathbf{A}_r \ne 0$, we can choose $\mathbf{s}_i \in \mathbf{A}_r$ such that $\mathbf{e} \cdot \mathbf{s}_i \ne 0$ for all $1 \le i \le r$. The centers of the spheres $\mathbf{s}_i$ are all on hyperplane $\mathbf{t}$. Since the intersection of an sD sphere and an $(n - 1)$D sphere is an $(s - 1)$D sphere whose center is on the line through the centers of the two spheres, the center of $(n - r)$D sphere $\mathbf{A}_r$ must be on hyperplane $\mathbf{t}$. Since the the supporting plane $\mathbf{e} \cdot \mathbf{A}_r$ of the $(n - r)$D sphere is perpendicular to hyperplane $\mathbf{t}$, the $(n - r)$D sphere and the hyperplane are perpendicular.

Case 1.4. If both $\mathbf{e} \cdot \mathbf{t}$ and $\mathbf{e} \cdot \mathbf{A}_r$ are nonzero, by

$$\mathbf{A}_r^{-1} = \frac{(\mathbf{e} \cdot \mathbf{A}_r) \wedge P_{\mathbf{A}_r}(\mathbf{e})}{(\mathbf{e} \cdot \mathbf{A}_r)^2} = \frac{(\mathbf{e} \cdot \mathbf{A}_r) P_{\mathbf{A}_r}(\mathbf{e})}{(\mathbf{e} \cdot \mathbf{A}_r)^2}, \tag{7.3.4}$$

we get that the center $\mathbf{o}'$ of the $(n-1)$D sphere $P_{\mathbf{A}_r}(\mathbf{e})$ is also the center of the $(n-r)$D sphere $\mathbf{A}_r$, and is on the supporting plane $\mathbf{e} \cdot \mathbf{A}_r$ of $(n-r)$D sphere $\mathbf{A}_r$. By (7.3.4),

$$\begin{aligned}
0 &= (\mathbf{e} \cdot \mathbf{A}_r)^2(\mathbf{A}_r^{-1} \cdot \mathbf{t}) \\
&= ((\mathbf{e} \cdot \mathbf{A}_r) \wedge P_{\mathbf{A}_r}(\mathbf{e})) \cdot \mathbf{t} \\
&= \mathbf{e} \cdot \mathbf{A}_r(P_{\mathbf{A}_r}(\mathbf{e}) \cdot \mathbf{t}) + ((\mathbf{e} \cdot \mathbf{A}_r) \cdot \mathbf{t}) \wedge P_{\mathbf{A}_r}(\mathbf{e}),
\end{aligned}$$

so

$$P_{\mathbf{A}_r}(\mathbf{e}) \cdot \mathbf{t} = 0, \qquad (\mathbf{e} \cdot \mathbf{A}_r) \cdot \mathbf{t} = 0. \tag{7.3.5}$$

By Case 1.2, the center $\mathbf{o}$ of sphere $\mathbf{t}$ is on plane $\mathbf{e} \cdot \mathbf{A}_r$. Since both centers $\mathbf{o}, \mathbf{o}'$ are on plane $\mathbf{e} \cdot \mathbf{A}_r$, spheres $\mathbf{t}$ and $\mathbf{A}_r$ are perpendicular if and only if the intersection of sphere $\mathbf{t}$ with plane $\mathbf{e} \cdot \mathbf{A}_r$ is perpendicular to sphere $\mathbf{A}_r$, as the latter is completely situated in the plane. By (7.3.5), spheres $P_{\mathbf{A}_r}(\mathbf{e})$ and $\mathbf{t}$ are perpendicular, so they keep the perpendicularity when restricted to any $(n-r+1)$D plane through their centers, and in particular, when restricted to the plane $\mathbf{e} \cdot \mathbf{A}_r$. The restriction of sphere $P_{\mathbf{A}_r}(\mathbf{e})$ to the plane is exactly $\mathbf{A}_r$.

(2) When $r = n$, $\mathbf{A}_n^{\sim}$ is a Minkowski plane, and has two different null vectors up to scale: $\mathbf{a}, \mathbf{b}$.

Case 2.1. If one of $\mathbf{a}, \mathbf{b}$, say $\mathbf{a}$, equals $\mathbf{e}$ up to scale, then all the $\mathbf{s}_i$ are hyperplanes. Since $\mathbf{t} \cdot \mathbf{A}_n = 0$, by the signature of $\mathbb{R}^{n+1,1}$, $\mathbf{e} \cdot \mathbf{t} \neq 0$. Since the center of sphere $\mathbf{t}$ is on every hyperplane $\mathbf{s}_i$, it must be the point of intersection $\mathbf{a}$ of the n hyperplanes.

Case 2.2. If $\mathbf{e} \cdot \mathbf{A}_n \neq 0$, then $\mathbf{A}_n$ represents a 0D sphere. If furthermore $\mathbf{e} \cdot \mathbf{t} = 0$, then the arguments in Case 1.3 are still valid, and it can be shown that the midpoint of line segment $\mathbf{ab}$ is on hyperplane $\mathbf{t}$, and the line segment is perpendicular to hyperplane $\mathbf{t}$.

Case 2.3. If both $\mathbf{e} \cdot \mathbf{A}_n \neq 0$ and $\mathbf{e} \cdot \mathbf{t} \neq 0$, then similar to Case 1.4, it can be shown that sphere $\mathbf{t}$ passes through the midpoint of line segment $\mathbf{ab}$ and is perpendicular to the line segment. $\qquad\qquad\qquad\qquad\qquad\qquad\qquad\qquad\quad\square$

Definition 7.28. For two spheres with centers $\mathbf{c}_1, \mathbf{c}_2$ and radii ρ_1, ρ_2 respectively, they are said to be *near* if $d_{\mathbf{c}_1\mathbf{c}_2}^2 < \rho_1^2 + \rho_2^2$; they are said to be *far* if $d_{\mathbf{c}_1\mathbf{c}_2}^2 > \rho_1^2 + \rho_2^2$.

Remark: By Pythagorean Theorem, the two spheres are perpendicular if and only if $d_{\mathbf{c}_1\mathbf{c}_2}^2 = \rho_1^2 + \rho_2^2$.

Let $\mathbf{s}_1, \mathbf{s}_2, \mathbf{s}_3$ be either null or positive vectors in $\mathbb{R}^{n+1,1}$. The integer

$$\tau = \text{sign of scalar } (\mathbf{s}_1 \cdot \mathbf{s}_2)(\mathbf{s}_2 \cdot \mathbf{s}_3)(\mathbf{s}_3 \cdot \mathbf{s}_1), \tag{7.3.6}$$

is invariant when the $\mathbf{s}_i$ are rescaled. It is called the *discriminant* of the three vectors. Below we consider the geometric meaning of $\tau = -1$. The geometric meaning of $\tau = 1$ can be discussed similarly.

The following classification is easy to verify.

Proposition 7.29.

(1) When the three s_i are all null vectors, then $\tau = -1$ is always true, as long as no two of them are equal up to scale.

(2) When $s_1 = e$, and s_2 is a point (null vector), and s_3 is positive, then $\tau = -1$ if and only if s_3 is a sphere and point s_2 is outside the sphere.

(3) When $s_1 = e$, and s_2, s_3 are positive, then $\tau = -1$ if and only if s_2, s_3 are far spheres.

(4) When s_1, s_2 are points, and s_3 is positive, then $\tau = -1$ if and only if points s_1, s_2 are on the same side of sphere or hyperplane s_3.

(5) When s_1 is a point, and s_2, s_3 are parallel hyperplanes, then $\tau = -1$ if and only if s_1 is between hyperplanes s_2, s_3. When s_1 is a sphere, then $\tau = -1$ if and only if the center of the sphere is between the two hyperplanes.

(6) When s_1 is a point, and s_2, s_3 are intersecting hyperplanes, then $\tau = -1$ if and only if s_1 is in the wedge domain of acute angle formed by the two hyperplanes. When s_1 is a sphere, the same conclusion holds for the center of the sphere.

(7) When s_1 is a point outside sphere s_2, and s_3 is a hyperplane, then $\tau = -1$ if and only if point s_1 and the center of s_2 are on the same side of hyperplane s_3.

(8) When s_1 is a point inside sphere s_2, and s_3 is a hyperplane, then $\tau = -1$ if and only if point s_1 and the center of s_2 are on different sides of hyperplane s_3.

(9) When s_1 is a point, and s_2, s_3 are far spheres, then $\tau = -1$ if and only if point s_1 is either inside both spheres or outside both spheres.

(10) When s_1 is a point, and s_2, s_3 are near spheres, then $\tau = -1$ if and only if point s_1 is inside one sphere but outside the other.

(11) When s_1, s_2, s_3 are all hyperplanes, then $\tau = -1$ if and only if they intersect pairwise, and in the 3D subspace formed by the normal directions n_1, n_2, n_3 of the three hyperplanes, for any two directions, say n_1 and n_2, the orthogonal projection of the third direction n_3 into the 2D plane formed by n_1, n_2 always lies in the wedge region of obtuse angle formed by 1D lines n_1, n_2.

(12) When s_1 is a hyperplane, and s_2, s_3 are far spheres, then $\tau = -1$ if and only if the two spheres are on the same side of the hyperplane.

(13) When s_1 is a hyperplane, and s_2, s_3 are near spheres, then $\tau = -1$ if and only if the two spheres are on different sides of the hyperplane.

(14) When s_1, s_2, s_3 are all spheres, then $\tau = -1$ if and only if either all three spheres are far from each other, or two spheres are far from each other but both near to the third.

Corollary 7.30. Let s_1, s_2 be two non-intersecting spheres.

- They are *inclusive*, *i.e.*, one is contained in the open ball bordered by the other, if and only if they do not intersect and are near, *i.e.*, $(s_1 \wedge s_2)^2 > 0$ and $(e \cdot s_1)(e \cdot s_2)(s_1 \cdot s_2) > 0$.

- They are *exclusive*, *i.e.*, they are strictly separated from each other by a hyperplane, if and only if they do not intersect and are far, *i.e.*, $(\mathbf{s}_1 \wedge \mathbf{s}_2)^2 > 0$ and $(\mathbf{e} \cdot \mathbf{s}_1)(\mathbf{e} \cdot \mathbf{s}_2)(\mathbf{s}_1 \cdot \mathbf{s}_2) < 0$.

Example 7.31. [38] A *canal surface* in $\mathbb{R}^3$ is an envelope of a 1-parameter family of spheres with varying radii. In the rational parametrization of canal surfaces, the conformal model can provide simple algebraic expressions.

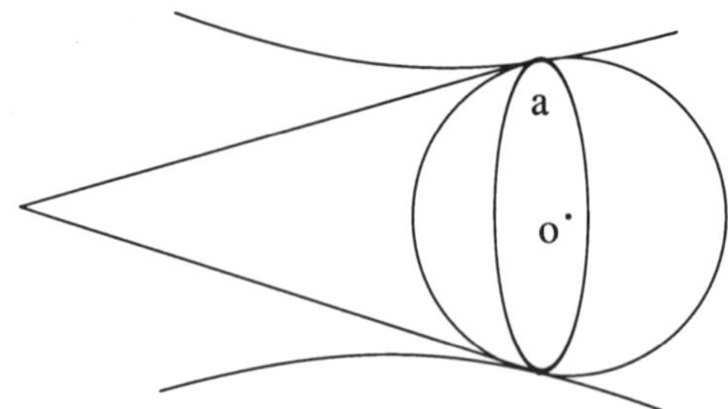

Fig. 7.7   Canal surface in computer-aided geometric design.

Let $\mathbf{o}(t) = (x(t), y(t), z(t))$ be the center of the 1-parameter sphere, and let $\rho(t)$ be the radius, where t is the parameter. The surface has positive-vector representation $\mathbf{c} = \mathbf{c}(t) = \mathbf{f}(\mathbf{o}) - \rho^2 \mathbf{e}/2$. Let $\mathbf{c}' = \mathbf{c}'(t)$ be the derivative with respect to t. Let $\mathbf{a}$ (null vector) be any point on the surface. By the enveloping property of the surface,

$$\begin{cases} \mathbf{a} \cdot \mathbf{c} = 0, \\ \mathbf{a} \cdot \mathbf{c}' = 0, \end{cases} \quad i.e., \quad \mathbf{a} \cdot (\mathbf{c} \wedge \mathbf{c}') = 0.$$

Thus the surface is formed by the pencil of spheres $\mathbf{c} \wedge \mathbf{c}'$. The canal surface is said to be *nondegenerate*, if the pencil does not contain any "bad" spheres, *i.e.*, null vectors and negative vectors. Direct computation gives

$$(\mathbf{c} \wedge \mathbf{c}')^2 = (\mathbf{c} \cdot \mathbf{c}')^2 - \mathbf{c}^2 \mathbf{c}'^2 = \rho^2(\rho'^2 - \mathbf{o}'^2). \tag{7.3.7}$$

So $\mathbf{c} \wedge \mathbf{c}'$ has Euclidean signature if and only if

$$\mathbf{o}'^2 - \rho'^2 > 0. \tag{7.3.8}$$

This is the *nondegeneracy condition* of the canal surface.

When $\mathbf{c} \wedge \mathbf{c}'$ is Minkowski, the surface has a swallowtail; when $\mathbf{c} \wedge \mathbf{c}'$ is degenerate, the surface has a cone vertex.

7.4   Conformal Grassmann-Cayley algebra

In the previous section, a sphere or hyperplane in $\mathbb{E}^n$ is represented by a positive vector. Alternatively, a sphere or hyperplane can be represented by the dual of a positive vector – a Minkowski $(n+1)$-blade, and accordingly, an rD sphere or plane can be represented by a Minkowski $(r+2)$-blade.

Geometrically, since the Minkowski blade representation is based on null vectors instead of positive ones, it describes the geometry of points instead of spheres and hyperplanes. Algebraically, this representation has the property that the grade of a Minkowski blade representing a sphere or hyperplane is independent of the dimension n of Euclidean space $\mathbb{E}^n$, in sharp contrast to the positive-vector representation. The dimension-free property is a big advantage, by which the Grassmann structure of projective incidence geometry can be directly extended to Euclidean incidence geometry.

By *Euclidean incidence relations* we mean the geometric relations including collinearity, cocircularity, parallelism, perpendicularity and tangency among points, lines, circles, planes and spheres of various dimensions in a Euclidean space. *Euclidean incidence geometry* studies the properties of Euclidean incidence relations. The Minkowski blade representation enables the Grassmann-Cayley algebra of projective geometry to be extended to the conformal model of Euclidean geometry, such that the outer product, the meet product and other related products in GC algebra have very nice Euclidean geometric interpretations.

The GC algebra $\Lambda(\mathbb{R}^{n+1,1})$, further equipped with the inner product, the boundary operator, and the Euclidean geometric interpretations of its old and new products and operators derived from the Minkowski blade representation, is called the *conformal Grassmann-Cayley algebra* over $\mathbb{E}^n$. From now on, we always use the Minkowski blade representation in the conformal model and the homogeneous model.

7.4.1 *Geometry of Minkowski blades*

Any nonzero Minkowski $(r+2)$-blade $\mathbf{A}_{r+2}$ in $\Lambda(\mathbb{R}^{n+1,1})$ represents an rD sphere or plane. It represents a plane if and only if it contains the conformal point at infinity, *i.e.*, $\mathbf{e} \wedge \mathbf{A}_{r+2} = 0$. The representation is unique up to scale.

For $0 \le r \le n-1$, the rD sphere passing through $r+2$ affinely independent points $\mathbf{a}_1, \ldots, \mathbf{a}_{r+2}$ represented by null vectors, is represented by $\mathbf{A}_{r+2} = \mathbf{a}_1 \wedge \cdots \wedge \mathbf{a}_{r+2}$, in the sense that a point $\mathbf{b}$ (null vector) is on the sphere if and only if $\mathbf{b} \wedge \mathbf{A}_{r+2} = 0$. The rD plane passing through $r+1$ affinely independent points $\mathbf{a}_1, \ldots, \mathbf{a}_{r+1}$ is represented by $\mathbf{B}_{r+2} = \mathbf{e} \wedge \mathbf{a}_1 \wedge \cdots \wedge \mathbf{a}_{r+1}$ in the same sense. Alternatively, the rD plane passing through $r+2$ points $\mathbf{a}_1, \ldots, \mathbf{a}_{r+2}$ in which no $r+1$ points are on the same $(r-1)$D sphere or plane, can be represented by $\mathbf{A}_{r+2}$.

(1) When $r = n-1$, if $(n+1)$-blade $\mathbf{A}_{n+1}$ represents the sphere of center $\mathbf{c} \in \mathcal{N}_\mathbf{e}$ and radius ρ, then $\mathbf{c} - \rho^2 \mathbf{e}/2$ equals $\mathbf{A}_{n+1}^\sim$ up to scale. Since $\mathbf{e} \cdot (\mathbf{c} - \rho^2 \mathbf{e}/2) = -1$ while $\mathbf{e} \cdot \mathbf{A}_{n+1}^\sim = [\mathbf{e}\mathbf{A}_{n+1}]$, we have

$$\mathbf{c} - \frac{\rho^2}{2}\mathbf{e} = -\frac{\mathbf{A}_{n+1}^\sim}{[\mathbf{e}\mathbf{A}_{n+1}]}. \tag{7.4.1}$$

Squaring both sides, we get

$$\rho^2 = -\frac{\mathbf{A}_{n+1}\rfloor\mathbf{A}_{n+1}}{[\mathbf{e}\mathbf{A}_{n+1}]^2}. \tag{7.4.2}$$

Substituting it into (7.4.1), we get, up to scale,

$$\mathbf{c} = 2\,[\mathbf{e}\mathbf{A}_{n+1}]\mathbf{A}^{\sim}_{n+1} + (\mathbf{A}_{n+1}\rfloor\mathbf{A}_{n+1})\mathbf{e} = \mathbf{A}^{\sim}_{n+1}\mathbf{e}\mathbf{A}^{\sim}_{n+1}, \tag{7.4.3}$$

or in short, $\mathbf{c} = \mathbf{A}^{\sim}_{n+1}$ mod $\mathbf{e}$ up to scale.

Example 7.32. When $n = 2$, circle **123** is represented by $\mathbf{1} \wedge \mathbf{2} \wedge \mathbf{3}$, and the center is represented by

$$[\mathbf{e123}](\mathbf{1} \wedge \mathbf{2} \wedge \mathbf{3})^{\sim} + (\mathbf{1} \cdot \mathbf{2})(\mathbf{1} \cdot \mathbf{3})(\mathbf{2} \cdot \mathbf{3})\mathbf{e}. \tag{7.4.4}$$

The squared radius is

$$\rho^2 = -2\frac{(\mathbf{1} \cdot \mathbf{2})(\mathbf{1} \cdot \mathbf{3})(\mathbf{2} \cdot \mathbf{3})}{[\mathbf{e123}]^2}. \tag{7.4.5}$$

If $\mathbf{A}_{n+1}$ represents hyperplane $\mathbf{n} + \delta\mathbf{e}$, then

$$\mathbf{n} + \delta\mathbf{e} = \frac{\mathbf{A}^{\sim}_{n+1}}{|\mathbf{A}_{n+1}|}, \tag{7.4.6}$$

or in short, $\mathbf{n} = \mathbf{A}^{\sim}_{n+1}$ mod $\mathbf{e}$ up to scale. Making inner product with $\mathbf{e}_0$ at both sides of (7.4.6), we get

$$\delta = -\frac{[\mathbf{e}_0\mathbf{A}_{n+1}]}{|\mathbf{A}_{n+1}|}, \tag{7.4.7}$$

and up to a positive scale,

$$\mathbf{n} = -(\mathbf{e}_0 \cdot \mathbf{e})\mathbf{A}^{\sim}_{n+1} + (\mathbf{e}_0 \cdot \mathbf{A}^{\sim}_{n+1})\mathbf{e} = -(\mathbf{e}_0 \wedge (\mathbf{e} \cdot \mathbf{A}_{n+1}))^{\sim}. \tag{7.4.8}$$

Example 7.33. When $n = 2$, line **12** is represented by $\mathbf{e} \wedge \mathbf{1} \wedge \mathbf{2}$. The direction of the line is $\mathbf{e} \cdot (\mathbf{1} \wedge \mathbf{2})$, and the normal direction is

$$(\mathbf{e} \wedge \mathbf{e}_0 \wedge (\mathbf{e} \cdot (\mathbf{1} \wedge \mathbf{2})))^{\sim} = (\mathbf{e} \wedge \mathbf{1} \wedge \mathbf{2})^{\sim} - [\mathbf{e}\mathbf{e}_0\mathbf{12}]\mathbf{e}. \tag{7.4.9}$$

The signed distance from a point **3** to line **12** along the normal direction is

$$\frac{[\mathbf{e123}]}{|\mathbf{e} \wedge \mathbf{1} \wedge \mathbf{2}|}. \tag{7.4.10}$$

(2) For general r, the rD sphere on $(r + 1)$D affine plane $\mathbf{A}_{r+3}$, with center $\mathbf{o}$ (null vector) and radius ρ, is the intersection of plane $\mathbf{A}_{r+3}$ and sphere $(\mathbf{o}-\rho^2\mathbf{e}/2)^{\sim}$, and has Minkowski blade representation

$$\mathbf{A}_{r+3} \vee (\mathbf{o} - \frac{\rho^2}{2}\mathbf{e})^{-\sim} = (\mathbf{o} - \frac{\rho^2}{2}\mathbf{e}) \cdot \mathbf{A}_{r+3} = (\mathbf{o} - \frac{\rho^2}{2}\mathbf{e})\mathbf{A}_{r+3}. \tag{7.4.11}$$

The rD sphere on $(r + 1)$D affine plane $\mathbf{A}_{r+3}$, with center $\mathbf{o}$ and through point $\mathbf{a}$, according to (7.2.39), has Minkowski blade representation

$$\mathbf{A}_{r+3} \vee (\mathbf{a} \cdot (\mathbf{e} \wedge \mathbf{o}))^{-\sim} = (\mathbf{a} \cdot (\mathbf{e} \wedge \mathbf{o})) \cdot \mathbf{A}_{r+3} = (\mathbf{a} \cdot (\mathbf{e} \wedge \mathbf{o}))\mathbf{A}_{r+3}. \tag{7.4.12}$$

The rD plane on $(r+1)$D plane $\mathbf{A}_{r+3}$, with normal direction $\mathbf{n}$ and passing through point $\mathbf{a}$, is the intersection of plane $\mathbf{A}_{r+3}$ and hyperplane $(\mathbf{a} \cdot (\mathbf{e} \wedge \mathbf{n}))^{\sim}$, and by (7.2.40), has Minkowski blade representation

$$\mathbf{A}_{r+3} \vee (\mathbf{a} \cdot (\mathbf{e} \wedge \mathbf{n}))^{-\sim} = (\mathbf{a} \cdot (\mathbf{e} \wedge \mathbf{n})) \cdot \mathbf{A}_{r+3} = (\mathbf{a} \cdot (\mathbf{e} \wedge \mathbf{n}))\mathbf{A}_{r+3}. \qquad (7.4.13)$$

(3) The rD affine plane passing through affinely independent points $\mathbf{x}_1, \ldots, \mathbf{x}_{r+1}$ in $\mathbb{R}^n$ is

$$\mathbf{A}_{r+2} = \mathbf{e} \wedge \mathbf{f}(\mathbf{x}_1) \wedge \cdots \wedge \mathbf{f}(\mathbf{x}_{r+1}) = \mathbf{e} \wedge \mathbf{B}_{r+1} + \mathbf{e} \wedge \mathbf{e}_0 \wedge \partial(\mathbf{B}_{r+1}), \qquad (7.4.14)$$

where $\mathbf{B}_{r+1} = \mathbf{x}_1 \wedge \cdots \wedge \mathbf{x}_{r+1}$. This is just the moment-direction representation of the rD affine plane. In vector space $\mathbf{B}_{r+1}$, the normal direction of subspace $\partial(\mathbf{B}_{r+1})$ is $\partial(\mathbf{B}_{r+1}) \cdot \mathbf{B}_{r+1}^{-1}$.

The rD sphere determined by $r+2$ points $\mathbf{x}_1, \ldots, \mathbf{x}_{r+2} \in \mathbb{R}^n$ is

$$\begin{aligned}
\mathbf{A}_{r+2} &= \mathbf{f}(\mathbf{x}_1) \wedge \cdots \wedge \mathbf{f}(\mathbf{x}_{r+2}) \\
&= \mathbf{B}_{r+2} + \mathbf{e}_0 \wedge \partial(\mathbf{B}_{r+2}) + \frac{1}{2}\mathbf{e} \wedge \mathbf{C}_{r+1} + \frac{1}{2}\mathbf{e} \wedge \mathbf{e}_0 \wedge \mathbf{D}_r,
\end{aligned} \qquad (7.4.15)$$

where

$$\begin{cases}
\mathbf{B}_{r+2} = \mathbf{x}_1 \wedge \cdots \wedge \mathbf{x}_{r+2}, \\
\mathbf{C}_{r+1} = \displaystyle\sum_{i=1}^{r+2} (-1)^{i+1}\mathbf{x}_i^2\, \mathbf{x}_1 \wedge \cdots \wedge \check{\mathbf{x}}_i \wedge \cdots \wedge \mathbf{x}_{r+2}, \\
\mathbf{D}_r = \displaystyle\sum_{1 \leq i < j \leq r+2} (-1)^{i+j+1}(\mathbf{x}_i^2 - \mathbf{x}_j^2)\, \mathbf{x}_1 \wedge \cdots \wedge \check{\mathbf{x}}_i \wedge \cdots \wedge \check{\mathbf{x}}_j \wedge \cdots \wedge \mathbf{x}_{r+2}.
\end{cases} \qquad (7.4.16)$$

When interchanging $\mathbf{e}, \mathbf{e}_0$ as in (7.2.21), or equivalently, by writing

$$\mathbf{f}(\mathbf{x}) = \frac{\mathbf{x}^2}{2}\left(\mathbf{e} + 2\mathbf{x}^{-1} + \frac{(2\mathbf{x}^{-1})^2}{2}\mathbf{e}_0\right), \qquad (7.4.17)$$

we get another expansion of $\mathbf{f}(\mathbf{x}_1) \wedge \cdots \wedge \mathbf{f}(\mathbf{x}_{r+2})$. Comparing the two results, we obtain

$$\mathbf{C}_{r+1} = \mathbf{x}_1^2\mathbf{x}_2^2\cdots\mathbf{x}_{r+2}^2\, \partial(\mathbf{x}_1^{-1} \wedge \mathbf{x}_2^{-1} \wedge \cdots \wedge \mathbf{x}_{r+2}^{-1}), \qquad (7.4.18)$$

where "∂" is the boundary operator (7.1.19) in the Cartesian model.

By Proposition 2.87, if $\mathbf{B}_{r+2} \neq 0$, then in $\Lambda(\mathbf{B}_{r+2})$,

$$\mathbf{D}_r = -\frac{\partial(\mathbf{B}_{r+2}) \vee \mathbf{C}_{r+1}}{[\mathbf{B}_{r+2}]}. \qquad (7.4.19)$$

By (7.4.15), the supporting plane of rD sphere $\mathbf{A}_{r+2}$ has moment-direction representation $(\mathbf{B}_{r+2}, \partial(\mathbf{B}_{r+2}))$. Comparing (7.4.15) with (7.4.11), we only need to find out how the center $\mathbf{o} \in \mathbb{R}^n$ and radius ρ of the sphere are related to the four components $\mathbf{B}_{r+2}, \partial(\mathbf{B}_{r+2}), \mathbf{C}_{r+1}, \mathbf{D}_r$ in (7.4.15).

By (7.4.11), rD sphere $\mathbf{A}_{r+2}$ has another representation

$$-\left(\mathbf{f}(\mathbf{o}) - \frac{\rho^2}{2}\mathbf{e}\right) \cdot (\mathbf{e} \wedge \mathbf{f}(\mathbf{x}_1) \wedge \cdots \wedge \mathbf{f}(\mathbf{x}_{r+2})) \qquad (7.4.20)$$

that equals (7.4.15). Expanding the inner product in (7.4.20) after substituting (7.4.15) into it, we get

$$-\left(\mathbf{f}(\mathbf{o}) - \frac{\rho^2}{2}\mathbf{e}\right) \cdot (\mathbf{e} \wedge \mathbf{B}_{r+2} + \mathbf{e} \wedge \mathbf{e}_0 \wedge \partial(\mathbf{B}_{r+2}))$$

$$= \mathbf{B}_{r+2} + \mathbf{e}_0 \wedge \partial(\mathbf{B}_{r+2}) + \mathbf{e} \wedge \left(\mathbf{o} \cdot \mathbf{B}_{r+2} + \frac{\rho^2 - \mathbf{o}^2}{2}\partial(\mathbf{B}_{r+2})\right) \tag{7.4.21}$$

$$-\mathbf{e} \wedge \mathbf{e}_0 \wedge (\mathbf{o} \cdot \partial(\mathbf{B}_{r+2})).$$

Since $\mathbf{o}$ is on the $(r+1)$D affine plane $(\mathbf{B}_{r+2}, \partial(\mathbf{B}_{r+2}))$, it is completely determined by $\mathbf{o} \cdot \partial(\mathbf{B}_{r+2})$, or equivalently, by its projection into $\partial(\mathbf{B}_{r+2})$. After $\mathbf{o}$ is determined, ρ can be obtained by comparing (7.4.21) with (7.4.15):

$$\begin{cases} \mathbf{C}_{r+1} = (\rho^2 - \mathbf{o}^2)\partial(\mathbf{B}_{r+2}) + 2\,\mathbf{o} \cdot \mathbf{B}_{r+2} = (\rho^2 + \mathbf{o}^2)\partial(\mathbf{B}_{r+2}) - 2\,\mathbf{o} \wedge (\mathbf{o} \cdot \partial(\mathbf{B}_{r+2})), \\[2mm] \mathbf{D}_r \quad = -2\,\mathbf{o} \cdot \partial(\mathbf{B}_{r+2}) \qquad\qquad\quad = -\dfrac{2}{\rho^2 - \mathbf{o}^2}\mathbf{o} \cdot \mathbf{C}_{r+1}. \end{cases}$$
$$\tag{7.4.22}$$

(4) Consider the geometric meaning of vectors of different signatures in a Minkowski blade.

Let $\mathbf{a}$ be a nonzero vector in $\mathbb{R}^{n+1,1}$, and let $\mathbf{A}_{r+2}$ be a Minkowski blade of grade $r+2$, where $0 \leq r \leq n-1$. If $\mathbf{a}$ is a null vector, then $\mathbf{a} \wedge \mathbf{A}_{r+2} = 0$ if and only if the point or conformal point at infinity represented by $\mathbf{a}$ is on the rD sphere or plane $\mathbf{A}_{r+2}$. If $\mathbf{a}$ is a positive vector, then $\mathbf{a}^\sim$ represents a sphere or hyperplane, and $\mathbf{a} \wedge \mathbf{A}_{r+2} = 0$ if and only if $\mathbf{A}_{r+2} \cdot \mathbf{a}^\sim = 0$, *i.e.*, if and only if the rD sphere or plane $\mathbf{A}_{r+2}$ intersects and is perpendicular to the sphere or hyperplane $\mathbf{a}^\sim$.

If $\mathbf{a}$ is a negative vector, by (7.2.32), when $r > 0$, then $\mathbf{a} \wedge \mathbf{A}_{r+2} = \mathbf{A}_{r+2}^\sim \cdot \mathbf{a} = 0$ if and only if rD sphere or plane $\mathbf{A}_{r+2}$ passes through an $(r-1)$D great sphere of the $(n-1)$D sphere $(\mathbf{a} \cdot (\mathbf{e} \wedge \mathbf{a}))^\sim$. *A negative vector can be taken as the set of great spheres of dimension ranging from 0 to $n-1$ in an $(n-1)D$ sphere.*

When $r = 0$, then Minkowski 2-blade $\mathbf{A}_2$ has two different null vectors up to scale. For negative vector $\mathbf{a}$, $\mathbf{a} \wedge \mathbf{A}_2 = 0$ if and only if (i) when $\mathbf{A}_2$ is a 0D sphere, then the 0D sphere is composed of two antipodal points on sphere $(\mathbf{a} \cdot (\mathbf{e} \wedge \mathbf{a}))^\sim$, (ii) when $\mathbf{A}_2 = \mathbf{e} \wedge \mathbf{x}$, then point $\mathbf{x}$ is the center of sphere $(\mathbf{a} \cdot (\mathbf{e} \wedge \mathbf{a}))^\sim$.

Example 7.34. When $n = 2$, for positive vector $\mathbf{a}$, negative vector $\mathbf{a}'$, and null vector $\mathbf{b}$ in $\mathbb{R}^{3,1}$,

- $\mathbf{e} \wedge \mathbf{a} \wedge \mathbf{b}$ is the line passing through point $\mathbf{b}$ and perpendicular to circle or line $\mathbf{a}^\sim$;
- $\mathbf{e} \wedge \mathbf{a}' \wedge \mathbf{b}$ is the line passing through both point $\mathbf{b}$ and the center of circle $(\mathbf{a}' \wedge (\mathbf{e} \cdot \mathbf{a}'))^\sim$.

(5) Consider the geometric meaning of brackets, or equivalently, Minkowski $(n+2)$-blades in $\Lambda(\mathbb{R}^{n+1,1})$.

Proposition 7.35. Let there be an nD simplex in $\mathbb{E}^n$ whose vertices are $\mathbf{a}_1, \ldots, \mathbf{a}_{n+1}$, and let $S_{\mathbf{a}_1 \ldots \mathbf{a}_{n+1}}$ be its signed volume. Let the sphere circumscribing the simplex be of center $\mathbf{o}$ and radius ρ, and let $\mathbf{a}_{n+2}$ be a point in $\mathbb{E}^n$. Then in the conformal model, with the same symbols denoting the null vectors representing the points,

$$[\mathbf{e}\mathbf{a}_1 \ldots \mathbf{a}_{n+1}] = n! \, S_{\mathbf{a}_1 \ldots \mathbf{a}_{n+1}},$$

$$[\mathbf{a}_1 \ldots \mathbf{a}_{n+2}] = (-1)^n \frac{n!}{2}(\rho^2 - d^2_{\mathbf{o}\mathbf{a}_{n+2}}) S_{\mathbf{a}_1 \ldots \mathbf{a}_{n+1}}. \tag{7.4.23}$$

As a corollary, any $n+2$ points in $\mathbb{E}^n$ are on the same sphere or hyperplane if and only if in $\Lambda(\mathbb{R}^{n+1,1})$, the bracket composed of their representative null vectors equals zero.

Proof. The first equality is direct from (7.4.14). For the second equality, by (7.4.1),

$$\frac{[\mathbf{a}_1 \ldots \mathbf{a}_{n+2}]}{[\mathbf{e}\mathbf{a}_1 \ldots \mathbf{a}_{n+1}]} = (-1)^{n+1} \mathbf{a}_{n+2} \cdot \frac{(\mathbf{a}_1 \wedge \cdots \wedge \mathbf{a}_{n+1})^{\sim}}{[\mathbf{e}\mathbf{a}_1 \ldots \mathbf{a}_{n+1}]} = (-1)^n \mathbf{a}_{n+2} \cdot \left(\mathbf{o} - \frac{\rho^2}{2}\mathbf{e}\right)$$

$$= (-1)^n \frac{\rho^2 - d^2_{\mathbf{o}\mathbf{a}_{n+2}}}{2}.$$

$\square$

For points $\mathbf{x}_1, \cdots, \mathbf{x}_r \in \mathbb{R}^n$, where $2 \le r \le n+1$,

$$(\mathbf{e}\,\mathbf{f}(\mathbf{x}_1) \ldots \mathbf{f}(\mathbf{x}_r) \,|\, \mathbf{e}\,\mathbf{f}(\mathbf{x}_1) \ldots \mathbf{f}(\mathbf{x}_r)) = \begin{vmatrix} 0 & -1 & -1 & \cdots & -1 \\ -1 & 0 & -d^2_{\mathbf{x}_1\mathbf{x}_2}/2 & \cdots & -d^2_{\mathbf{x}_1\mathbf{x}_r}/2 \\ -1 & -d^2_{\mathbf{x}_2\mathbf{x}_1}/2 & 0 & \cdots & -d^2_{\mathbf{x}_2\mathbf{x}_r}/2 \\ \vdots & \vdots & \vdots & \ddots & \vdots \\ -1 & -d^2_{\mathbf{x}_r\mathbf{x}_1}/2 & -d^2_{\mathbf{x}_r\mathbf{x}_2}/2 & \cdots & 0 \end{vmatrix}. \tag{7.4.24}$$

The right side is called the *Cayley-Menger* determinant [162] of the r points. When the conformal point at infinity on the left side of (7.4.24) is replaced by a point, we get

$$(\mathbf{f}(\mathbf{x}_1) \cdots \mathbf{f}(\mathbf{x}_r) \,|\, \mathbf{f}(\mathbf{x}_1) \cdots \mathbf{f}(\mathbf{x}_r)) = \det(\mathbf{f}(\mathbf{x}_i) \cdot \mathbf{f}(\mathbf{x}_j))_{r \times r} = \det(-d^2_{\mathbf{x}_i\mathbf{x}_j}/2)_{r \times r}.$$

Proposition 7.36. [Ptolemy's Theorem] Let $\mathbf{x}_1, \cdots, \mathbf{x}_{n+2}$ be points in $\mathbb{E}^n$, then they are on a sphere or hyperplane if and only if $\det(d^2_{\mathbf{x}_i\mathbf{x}_j})_{(n+2)\times(n+2)} = 0$.

(6) Consider the geometric meaning of a Minkowski 2-blades $\mathbf{A}_2$.

$\mathbf{A}_2$ represents a 0D sphere or plane. When representing a 0D plane, then up to scale,

$$\mathbf{A}_2 = \mathbf{e} \wedge \mathbf{f}(\mathbf{x}) = \mathbf{e} \wedge \mathbf{e}_0 + \mathbf{e} \wedge \mathbf{x}. \tag{7.4.25}$$

Definition 7.37. In the conformal model of $\mathbb{R}^n$, the *affine representation* of a point $\mathbf{x} \in \mathbb{R}^n$ refers to the Minkowski 2-blade representation (7.4.25) of 0D plane $(\mathbf{e}, \mathbf{x})$.

It can be easily verified that for any $\mathbf{x}, \mathbf{y} \in \mathbb{R}^n$ and $\lambda \in \mathbb{R}$,

$$\mathbf{e} \wedge \mathbf{f}(\lambda\mathbf{x} + (1 - \lambda)\mathbf{y}) = \lambda\mathbf{e} \wedge \mathbf{f}(\mathbf{x}) + (1 - \lambda)\mathbf{e} \wedge \mathbf{f}(\mathbf{y}).$$

So the set

$$\mathbf{e} \wedge \mathcal{N}_\mathbf{e} = \{\mathbf{e} \wedge \mathbf{a} \,|\, \mathbf{a} \in \mathcal{N}_\mathbf{e}\} \tag{7.4.26}$$

is an nD affine space whose space at infinity is

$$\mathbf{e} \wedge \mathbf{e}^\perp := \{\mathbf{e} \wedge \mathbf{a} \,|\, \mathbf{a} \in \mathbb{R}^{n+1,1}, \, \mathbf{a} \cdot \mathbf{e} = 0\}. \tag{7.4.27}$$

The homogeneous form of affine space $\mathbf{e} \wedge \mathcal{N}_\mathbf{e}$ is

$$\mathbf{e} \wedge \mathcal{N} := \{\lambda\mathbf{e} \wedge \mathbf{a} \,|\, \mathbf{a} \in \mathcal{N}_\mathbf{e}, \, \lambda \in \mathbb{R} - \{0\}\}. \tag{7.4.28}$$

Definition 7.38. The *affine Grassmann-Cayley algebra upon the conformal model* of $\mathbb{E}^n$, refers to the Grassmann space over the vector space $\mathbf{e} \wedge \mathbb{R}^{n+1,1} = \{\mathbf{e} \wedge \mathbf{a} \,|\, \mathbf{a} \in \mathbb{R}^{n+1,1}\}$, equipped with the outer product (2.3.16), *i.e.*,

$$(\mathbf{e} \wedge \mathbf{A}) \wedge_\mathbf{e} (\mathbf{e} \wedge \mathbf{B}) := \mathbf{e} \wedge \mathbf{A} \wedge \mathbf{B}, \quad \forall \mathbf{A}, \mathbf{B} \in \Lambda(\mathbb{R}^{n+1,1}), \tag{7.4.29}$$

the same meet product "$\vee$" as the GC algebra over $\mathbb{R}^{n+1,1}$, and the boundary operator

$$\partial(\mathbf{A}) := -\mathbf{e} \cdot \mathbf{A}, \quad \forall \mathbf{A} \in \Lambda(\mathbb{R}^{n+1,1}). \tag{7.4.30}$$

Proposition 7.39. Let $\mathbf{A}_r$ be an $(r-2)$D plane, where $3 \leq r \leq n+1$, and let $\mathbf{a}$ (null vector) be a point. The foot drawn from $\mathbf{a}$ to the $(r-2)$D plane has affine representation $\mathbf{e} \wedge P_{\mathbf{A}_r}(\mathbf{a})$.

Proof. First, no null vector is orthogonal to a Minkowski blade, so $P_{\mathbf{A}_r}(\mathbf{a}) \neq 0$. Second, if $\mathbf{e} \cdot P_{\mathbf{A}_r}(\mathbf{a}) = 0$, then $P_{\mathbf{A}_r}(\mathbf{a})$ must be positive. Since $\mathbf{a}$ is null, $P_{\mathbf{A}_r}^\perp(\mathbf{a})$ is negative. However,

$$\mathbf{e} \cdot P_{\mathbf{A}_r}^\perp(\mathbf{a}) = \mathbf{e} \cdot ((\mathbf{a} \wedge \mathbf{A}_r) \cdot \mathbf{A}_r^{-1}) = (-1)^r (\mathbf{e} \wedge \mathbf{A}_r^{-1}) \cdot (\mathbf{a} \wedge \mathbf{A}_r) = 0.$$

It contradicts with the fact that a null vector cannot be orthogonal to a negative vector in a Minkowski space. This proves $\mathbf{e} \cdot P_{\mathbf{A}_r}(\mathbf{a}) \neq 0$, so $\mathbf{e} \wedge P_{\mathbf{A}_r}(\mathbf{a})$ is a Minkowski blade in $\Lambda(\mathbf{A}_r)$.

By

$$(\mathbf{e} \wedge P_{\mathbf{A}_r}(\mathbf{a}) \wedge \mathbf{a}) \cdot \mathbf{A}_r = \mathbf{e} \cdot \{((\mathbf{a} \cdot \mathbf{A}_r)\mathbf{A}_r^{-1}) \cdot (\mathbf{a} \cdot \mathbf{A}_r)\} = 0,$$

line $\mathbf{e} \wedge P_{\mathbf{A}_r}(\mathbf{a}) \wedge \mathbf{a}$ is perpendicular to $(r-2)$D plane $\mathbf{A}_r$. $\square$

Definition 7.40. A 1D *tangent direction* in the conformal model of $\mathbb{E}^n$, refers to a 2-blade in the space at infinity $\mathbf{e} \wedge \mathbf{e}^\perp$ of the nD affine space $\mathbf{e} \wedge \mathcal{N}_\mathbf{e}$. Any 1D tangent direction is of the form $\mathbf{e} \wedge \mathbf{a} = \mathbf{e}\mathbf{a}$, where $\mathbf{a} \in \mathbf{e}^\sim$ is a positive vector. For any point $\mathbf{b}$ in $\mathbb{E}^n$ represented by a null vector in $\mathcal{N}_\mathbf{e}$, the *tangent direction* of $\mathbf{e} \wedge \mathbf{a}$ at *base point* $\mathbf{b}$, refers to the orthogonal rejection of vector $\mathbf{a}$ from $\mathbf{e} \wedge \mathbf{b}$:

$$P_{\mathbf{e}\wedge\mathbf{b}}^\perp(\mathbf{a}) = (\mathbf{a} \wedge \mathbf{e} \wedge \mathbf{b})(\mathbf{e} \wedge \mathbf{b}) = -\mathbf{b} \cdot (\mathbf{e} \wedge \mathbf{a}). \tag{7.4.31}$$

All tangent directions at the base point form the *tangent space* $(\mathbf{e} \wedge \mathbf{b})^\sim$ of the conformal model at base point $\mathbf{b}$, which is an nD Euclidean vector space.

When point $\mathbf{b}$ varies in $\mathcal{N}_{\mathbf{e}}$, so does the tangent direction $\mathbf{b} \cdot (\mathbf{e} \wedge \mathbf{a})$ of $\mathbf{e} \wedge \mathbf{a}$ at base point $\mathbf{b}$, but the length of the tangent direction is invariant:

$$|\mathbf{b} \cdot (\mathbf{e} \wedge \mathbf{a})| = |\mathbf{a} + (\mathbf{a} \cdot \mathbf{b})\mathbf{e}| = |\mathbf{a}|. \tag{7.4.32}$$

In fact, for any two points $\mathbf{b}, \mathbf{c}$, the tangent spaces based on them are isometric canonically: any $\mathbf{x} \in (\mathbf{e} \wedge \mathbf{b})^{\sim}$ corresponds to $P^{\perp}_{\mathbf{e} \wedge \mathbf{c}}(\mathbf{x}) \in (\mathbf{e} \wedge \mathbf{c})^{\sim}$, and *vice versa*.

An rD *tangent direction* of the conformal model is of the form $\mathbf{e} \wedge \mathbf{a}_1 \wedge \cdots \wedge \mathbf{a}_r$, where the $\mathbf{a}_i$ are in $\mathbf{e}^{\sim}$. Its dual is the $(n-r)D$ direction $(\mathbf{e} \wedge \mathbf{a}_1 \wedge \cdots \wedge \mathbf{a}_r)^{\sim}$. The two high-dimensional directions are completely orthogonal to each other, and form an orthogonal decomposition of the tangent space at each base point of $\mathbb{E}^n$.

Example 7.41. When $n = 2$, let $\mathbf{b}$ be a point (null vector), and let $\mathbf{e} \wedge \mathbf{a}$ be a 1D tangent direction. The line passing through point $\mathbf{b}$ and following direction $\mathbf{e} \wedge \mathbf{a}$ is $(\mathbf{e} \wedge \mathbf{a}) \wedge \mathbf{b}$. The line passing through point $\mathbf{b}$ and parallel to line $\mathbf{a}^{\sim}$ is $(\mathbf{e} \wedge \mathbf{a})^{\sim} \wedge \mathbf{b}$.

(7) The outer product of a Minkowski blade with any other blade in $\Lambda(\mathbb{R}^{n+1,1})$, if nonzero, must still be a Minkowski blade. Let $\mathbf{A}_r$ be a Minkowski r-blade, and let $\mathbf{B}_s$ be an arbitrary s-blade, such that $r, s \geq 2$, $r + s < n + 2$, and $\mathbf{A}_r \wedge \mathbf{B}_s \neq 0$. Consider the geometric meaning of $\mathbf{A}_r \wedge \mathbf{B}_s$.

When $r = s = 2$, let $\mathbf{B}_2 = \mathbf{b} \wedge \mathbf{c}$. If $\mathbf{A}_2 = \mathbf{e} \wedge \mathbf{a}$, where $\mathbf{a}$ is null, then

- if $\mathbf{B}_2$ is Minkowski, let $\mathbf{b}, \mathbf{c}$ be null, then $\mathbf{A}_2 \wedge \mathbf{B}_2$ is the 2D affine plane of $\mathbb{E}^n$ determined by three points $\mathbf{a}, \mathbf{b}, \mathbf{c}$;

- if $\mathbf{B}_2$ is degenerate, let $\mathbf{b}$ be null, then $\mathbf{A}_2 \wedge \mathbf{B}_2$ is the 2D affine plane passing through points $\mathbf{a}, \mathbf{b}$ and 1D direction $\mathbf{e} \wedge (\mathbf{e} \cdot (\mathbf{b} \wedge \mathbf{c}))$;

- if $\mathbf{B}_2$ is Euclidean, then $\mathbf{A}_2 \wedge \mathbf{B}_2$ is the 2D plane passing through point $\mathbf{a}$ and perpendicular to spheres or hyperplanes $\mathbf{b}^{\sim}$ and $\mathbf{c}^{\sim}$.

If $\mathbf{A}_2 = \mathbf{a} \wedge \mathbf{a}'$, where $\mathbf{a}, \mathbf{a}'$ are null vectors representing two points, then

- when $\mathbf{B}_2$ is Minkowski, let $\mathbf{B}_2 = \mathbf{b} \wedge \mathbf{b}'$, where $\mathbf{b}, \mathbf{b}'$ are null, then $\mathbf{A}_2 \wedge \mathbf{B}_2$ is the 2D sphere or plane through four points $\mathbf{a}, \mathbf{a}', \mathbf{b}, \mathbf{b}'$ (possibly including the conformal point at infinity).

- When $\mathbf{B}_2$ is degenerate, let $\mathbf{b}$ be the null vector unique up to scale in $\mathbf{B}_2$, then $\mathbf{A}_2 \wedge \mathbf{B}_2$ is the 2D sphere or plane through points $\mathbf{a}, \mathbf{a}', \mathbf{b}$ and the tangent direction $\langle \mathbf{ecb} \rangle_1$ at $\mathbf{b}$.

  The tangent direction is derived as follows. First, $\mathbf{e} \cdot (\mathbf{a} \wedge \mathbf{b})$ is a tangent direction at some base point in $\mathbb{E}^n$, and $\mathbf{e} \wedge (\mathbf{e} \cdot (\mathbf{b} \wedge \mathbf{c}))$ is the bivector representation of the tangent direction. Second, $\mathbf{b} \cdot (\mathbf{e} \wedge (\mathbf{e} \cdot (\mathbf{b} \wedge \mathbf{c})))$ is the tangent direction corresponding to base point $\mathbf{b}$. Third, direct computing gives

$$\mathbf{b} \cdot (\mathbf{e} \wedge (\mathbf{e} \cdot (\mathbf{b} \wedge \mathbf{c}))) = (\mathbf{b} \cdot \mathbf{e})\,\mathbf{e} \cdot (\mathbf{b} \wedge \mathbf{c}) - (\mathbf{b} \cdot \mathbf{e})(\mathbf{b} \cdot \mathbf{c})\mathbf{e} = -(\mathbf{b} \cdot \mathbf{e})\,\langle \mathbf{ecb} \rangle_1.$$

- When $\mathbf{B}_2$ is Euclidean, $\mathbf{A}_2 \wedge \mathbf{B}_2$ is the 2D sphere or plane passing through points $\mathbf{a}, \mathbf{a}'$ and perpendicular to spheres or hyperplanes $\mathbf{b}^{\sim}$ and $\mathbf{c}^{\sim}$.

When $r > 2$ or $s > 2$, the geometric explanation of $\mathbf{A}_r \wedge \mathbf{B}_s$ is similar.

7.4.2   *Inner product of Minkowski blades*

Firstly, consider the geometric meaning of the dual $\mathbf{A}_r^{\sim}$ of a Minkowski blade $\mathbf{A}_r$. It is the inner product of an r-blade and an $(n+2)$-blade in $\Lambda(\mathbb{R}^{n+1,1})$, both of which are Minkowski. Blade $\mathbf{A}_r^{\sim}$ is Euclidean, so it can only be interpreted in terms of tangent directions.

- When $r = 2$ and $\mathbf{A}_2 = \mathbf{e} \wedge \mathbf{a}$, where $\mathbf{a}$ is null, $\mathbf{A}_2^{\sim}$ is the tangent space of the conformal model of $\mathbb{E}^n$ at base point $\mathbf{a}$.

- When $r = 2$ and $\mathbf{e} \wedge \mathbf{A}_2 \neq 0$, blade $\mathbf{e} \wedge \mathbf{A}_2^{\sim}$ is Minkowski. Let $\mathbf{A}_2 = \mathbf{a} \wedge \mathbf{b}$, where $\mathbf{a}, \mathbf{b}$ are null vectors, then $\mathbf{e} \wedge \mathbf{A}_2^{\sim} = (\mathbf{e} \cdot (\mathbf{a} \wedge \mathbf{b}))^{\sim}$ is the perpendicular bisecting hyperplane of line segment $\mathbf{ab}$.

- When $2 < r < n+2$ and $\mathbf{e} \wedge \mathbf{A}_r = 0$, $\mathbf{A}_r^{\sim}$ is the *normal space* of the $(r-2)$D plane $\mathbf{A}_r$, *i.e.*, the space composed of normal directions of the $(r-2)$D plane.

- When $2 < r < n+1$ and $\mathbf{e} \wedge \mathbf{A}_{n+1} \neq 0$, $\mathbf{e} \wedge \mathbf{A}_r^{\sim}$ is the $(n+1-r)$D plane passing through the center of $(r-2)$D sphere $\mathbf{A}_r$ and perpendicular to the supporting plane of the $(r-2)$D sphere.

- When $r = n+1$ and $\mathbf{e} \wedge \mathbf{A}_{n+1} \neq 0$, $\mathbf{e} \wedge \mathbf{A}_{n+1}^{\sim}$ is the affine representation of the center of sphere $\mathbf{A}_{n+1}$.

Secondly, consider the geometric meaning of the inner product of a null vector $\mathbf{a}$ and a Minkowski blade $\mathbf{A}_r$, where $2 \leq r \leq n+1$.

Obviously, $\mathbf{a} \cdot \mathbf{A}_r \neq 0$. Since $\mathbf{a} \cdot (\mathbf{a} \cdot \mathbf{A}_r) = 0$, blade $\mathbf{a} \cdot \mathbf{A}_r$ is either degenerate or Euclidean, and is degenerate if and only if $\mathbf{a}$ is in it, or equivalently, if and only if $\mathbf{a} \cdot (\mathbf{a} \wedge \mathbf{A}_r) = 0$.

When $\mathbf{a} = \mathbf{e}$, if $\mathbf{A}_r = \mathbf{f}(\mathbf{x}_1) \wedge \cdots \wedge \mathbf{f}(\mathbf{x}_r)$, where the $\mathbf{x}_i \in \mathbb{R}^n$, then by (7.4.15),

$$\mathbf{e} \cdot \mathbf{A}_r = -\partial(\mathbf{B}_r) + \frac{1}{2}\mathbf{e} \wedge \mathbf{D}_{r-2}, \qquad (7.4.33)$$

where $\mathbf{B}_r = \mathbf{x}_1 \wedge \cdots \wedge \mathbf{x}_r$. If $\mathbf{A}_r = \mathbf{e} \wedge \mathbf{f}(\mathbf{x}_1) \wedge \cdots \wedge \mathbf{f}(\mathbf{x}_{r-1})$, then by (7.4.14), $\mathbf{e} \cdot \mathbf{A}_r = \mathbf{e} \wedge \partial(\mathbf{B}_{r-1})$, where $\mathbf{B}_{r-1} = \mathbf{x}_1 \wedge \cdots \wedge \mathbf{x}_{r-1}$.

If $\mathbf{A}_r$ is an $(r-2)$D sphere, then $\mathbf{e} \cdot (\mathbf{e} \wedge \mathbf{A}_r)$ is the space at infinity of the supporting affine plane of the $(r-2)$D sphere; if $\mathbf{A}_r$ is an $(r-2)$D plane, then when $r > 2$, $\mathbf{e} \cdot \mathbf{A}_r$ is the space at infinity of the $(r-2)$D plane, when $r = 2$, $\mathbf{e} \cdot \mathbf{A}_r$ is $\mathbf{e}$ up to scale.

When $\mathbf{a} \in \mathcal{N}_{\mathbf{e}}$, without loss of generality, take $\mathbf{a} = \mathbf{e}_0$. When $\mathbf{e}_0 \cdot \mathbf{A}_r$ is degenerate, if $r > 2$, then $\mathbf{e}_0 \cdot \mathbf{A}_r$ is point $\mathbf{e}_0$ together with the $(r-2)$D tangent subspace at $\mathbf{e}_0$ represented by $(\mathbf{e} \wedge \mathbf{e}_0) \cdot \mathbf{A}_r$; if $r = 2$, then $\mathbf{e}_0 \cdot \mathbf{A}_r$ is $\mathbf{e}_0$ up to scale.

If $\mathbf{A}_r = \mathbf{f}(\mathbf{x}_1) \wedge \cdots \wedge \mathbf{f}(\mathbf{x}_r)$, again by (7.4.15),

$$\mathbf{e}_0 \cdot \mathbf{A}_r = -\frac{1}{2}(\mathbf{C}_{r-1} + \mathbf{e}_0 \wedge \mathbf{D}_{r-2}). \qquad (7.4.34)$$

If $\mathbf{D}_{r-2} \neq 0$, denote $\mathbf{y} = \mathbf{C}_{r-1} \cdot \mathbf{D}_{r-2}^{-1}$, then $\mathbf{C}_{r-1} = \mathbf{y} \wedge \mathbf{D}_{r-2}$, and

$$\mathbf{e} \wedge (\mathbf{e}_0 \cdot \mathbf{A}_r) = -\frac{1}{2}\mathbf{e} \wedge \mathbf{f}(\mathbf{y}) \wedge \mathbf{D}_{r-2}. \qquad (7.4.35)$$

In fact, by (7.4.22), if $\mathbf{A}_r$ is an $(r-2)$D sphere with center $\mathbf{o} \in \mathbb{R}^n$ and radius ρ, then

$$\mathbf{y} = (\mathbf{D}_{r-2}\mathbf{C}_{r-1}^{-1})^{-1} = \frac{\mathbf{o}^2 - \rho^2}{2}(P_{\mathbf{C}_{r-1}}(\mathbf{o}))^{-1}. \tag{7.4.36}$$

Similarly, if $\mathbf{A}_r = \mathbf{e} \wedge \mathbf{f}(\mathbf{x}_1) \wedge \cdots \wedge \mathbf{f}(\mathbf{x}_{r-1})$, again by (7.4.14),

$$\mathbf{e}_0 \cdot \mathbf{A}_r = -(\mathbf{B}_{r-1} + \mathbf{e}_0 \wedge \partial(\mathbf{B}_{r-1})). \tag{7.4.37}$$

So $\mathbf{e} \wedge (\mathbf{e}_0 \cdot \mathbf{A}_r) = -\mathbf{A}_r$.

When $\mathbf{e}_0 \cdot \mathbf{A}_r$ is Euclidean, if $(\mathbf{e} \wedge \mathbf{e}_0) \cdot \mathbf{A}_r = 0$, then $\mathbf{e}_0 \cdot \mathbf{A}_r$ is an $(r-1)$D tangent subspace at $\mathbf{e}_0$; if $(\mathbf{e} \wedge \mathbf{e}_0) \cdot \mathbf{A}_r \neq 0$, then if $r = 2$, $\mathbf{e} \wedge (\mathbf{e}_0 \cdot \mathbf{A}_r)$ is a 0D plane, if $r > 2$, $\mathbf{e} \wedge (\mathbf{e}_0 \cdot \mathbf{A}_r)$ is an $(r-2)$D plane with $(r-2)$D direction $(\mathbf{e} \wedge \mathbf{e}_0) \cdot \mathbf{A}_r$, and is just $\mathbf{A}_r$ itself if the latter is an $(r-2)$D plane.

Thirdly, consider the geometric meaning of the inner product of two Minkowski blades of the same grade.

The simplest case is grade two. If $\mathbf{A}_2 = \mathbf{e} \wedge \mathbf{a}$, $\mathbf{B}_2 = \mathbf{e} \wedge \mathbf{b}$, where $\mathbf{a}, \mathbf{b} \in \mathcal{N}_\mathbf{e}$, then $(\mathbf{e} \wedge \mathbf{a}) \cdot (\mathbf{e} \wedge \mathbf{b}) = 1$. If $\mathbf{B}_2 = \mathbf{b} \wedge \mathbf{b}'$, where $\mathbf{b}, \mathbf{b}' \in \mathcal{N}_\mathbf{e}$, then

$$(\mathbf{e} \wedge \mathbf{a}) \cdot (\mathbf{b} \wedge \mathbf{b}') = \frac{d_{\mathbf{ab}}^2 - d_{\mathbf{ab}'}^2}{2}, \tag{7.4.38}$$

which equals zero if and only if $d_{\mathbf{ab}} = d_{\mathbf{ab}'}$. If $\mathbf{A}_2 = \mathbf{a} \wedge \mathbf{a}'$, where $\mathbf{a}, \mathbf{a}' \in \mathcal{N}_\mathbf{e}$, then

$$(\mathbf{a} \wedge \mathbf{a}') \cdot (\mathbf{b} \wedge \mathbf{b}') = \frac{d_{\mathbf{a}'\mathbf{b}}^2 d_{\mathbf{ab}'}^2 - d_{\mathbf{ab}}^2 d_{\mathbf{a}'\mathbf{b}'}^2}{4}, \tag{7.4.39}$$

which equals zero if and only if

$$d_{\mathbf{ab}'}d_{\mathbf{a}'\mathbf{b}} = d_{\mathbf{ab}}d_{\mathbf{a}'\mathbf{b}'}. \tag{7.4.40}$$

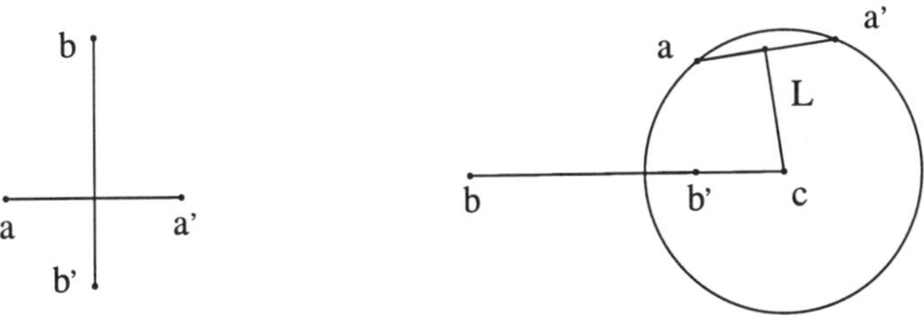

Fig. 7.8　Geometric meaning of $(\mathbf{a} \wedge \mathbf{a}') \cdot (\mathbf{b} \wedge \mathbf{b}') = 0$.

If $d_{\mathbf{ab}} = 0$, then $d_{\mathbf{ab}'} \neq 0$ and $d_{\mathbf{a}'\mathbf{b}} \neq 0$, so (7.4.40) cannot hold. Assume that (7.4.40) holds. Then the distance between any two of the four points is nonzero. Let $d_{\mathbf{ab}} : d_{\mathbf{ab}'} = \lambda > 0$. Let $\mathbf{a}, \mathbf{a}', \mathbf{b}, \mathbf{b}'$ represent points $\mathbf{x}, \mathbf{x}', \mathbf{y}, \mathbf{y}' \in \mathbb{R}^n$ respectively. Set $\mathbf{y}'$ as the origin of $\mathbb{R}^n$. Then (7.4.40) is equivalent to

$$\begin{cases} (\mathbf{x} - \mathbf{y})^2 = \lambda^2 \mathbf{x}^2, \\ (\mathbf{x}' - \mathbf{y})^2 = \lambda^2 \mathbf{x}'^2, \end{cases} \quad i.e., \quad (\mathbf{x} - \frac{\mathbf{y}}{1-\lambda^2})^2 = (\mathbf{x}' - \frac{\mathbf{y}}{1-\lambda^2})^2 = (\frac{\lambda \mathbf{y}}{1-\lambda^2})^2. \tag{7.4.41}$$

If $\lambda = 1$, then $\mathbf{x}, \mathbf{x}'$ are both on the perpendicular bisecting hyperplane of line segment $\mathbf{yy}'$. If $\lambda \neq 1$, then $\mathbf{x}, \mathbf{x}'$ are both on the sphere with center $\mathbf{c} = (\mathbf{y} - \lambda^2 \mathbf{y}')/(1 - \lambda^2)$ and radius $\rho = \lambda d_{\mathbf{cy}'}$. Since $d_{\mathbf{cy}} d_{\mathbf{cy}'} = \rho^2$, and $\mathbf{c}$ is on line $\mathbf{yy}'$ but outside the line segment between $\mathbf{y}, \mathbf{y}'$, points $\mathbf{y}, \mathbf{y}'$ must be in inversion with respect to sphere $(\mathbf{c}, \rho)$.

Proposition 7.42. For any four points $\mathbf{a}, \mathbf{a}', \mathbf{b}, \mathbf{b}'$ in $\mathbb{E}^n$, let L be the perpendicular bisecting hyperplane of line segment $\mathbf{aa}'$. Then in the homogeneous model of $\mathbb{E}^n$, $(\mathbf{a} \wedge \mathbf{a}') \cdot (\mathbf{b} \wedge \mathbf{b}') = 0$ if and only if either (1) points $\mathbf{b}, \mathbf{b}'$ are both on L, or (2) L intersects line $\mathbf{bb}'$ at a point $\mathbf{c}$, such that points $\mathbf{b}, \mathbf{b}'$ are in inversion with respect to the sphere with center $\mathbf{c}$ and through points $\mathbf{a}, \mathbf{a}'$.

Comparing (7.4.40) with (7.3.3), we get

Corollary 7.43. For pairwise different collinear points $\mathbf{a}, \mathbf{a}', \mathbf{b}, \mathbf{b}'$ in $\mathbb{E}^n$, 0D spheres $(\mathbf{a}, \mathbf{a}')$ and $(\mathbf{b}, \mathbf{b}')$ are perpendicular if and only if the two points of 0D sphere $(\mathbf{a}, \mathbf{a}')$ are in inversion with respect to 0D sphere $(\mathbf{b}, \mathbf{b}')$.

The next simplest case is grade $n + 1$. It is already considered in Section 7.3 in the positive-vector representation dual to the Minkowski blade representation.

In the general case where $3 \leq r \leq n$, there are three cases:

(i) If $\mathbf{A}_r, \mathbf{B}_r$ are both planes, by (7.4.14), let

$$\begin{aligned}
\mathbf{A}_r &= \mathbf{e} \wedge \mathbf{A}_{r-1} + \mathbf{e} \wedge \mathbf{e}_0 \wedge \partial(\mathbf{A}_{r-1}), \\
\mathbf{B}_r &= \mathbf{e} \wedge \mathbf{B}_{r-1} + \mathbf{e} \wedge \mathbf{e}_0 \wedge \partial(\mathbf{B}_{r-1}),
\end{aligned} \tag{7.4.42}$$

where $\mathbf{A}_{r-1}, \mathbf{B}_{r-1} \in \Lambda^{r-1}(\mathbb{R}^n)$, then $\mathbf{A}_r \cdot \mathbf{B}_r = \partial(\mathbf{A}_{r-1}) \cdot \partial(\mathbf{B}_{r-1})$ is the inner product of the two Euclidean $(r-2)$-blades representing the tangent spaces of the two planes.

(ii) If $\mathbf{A}_r$ is a plane and $\mathbf{B}_r$ is a sphere, by (7.4.11) and (7.4.14), let

$$\begin{aligned}
\mathbf{A}_r &= \mathbf{e} \wedge \mathbf{A}_{r-1} + \mathbf{e} \wedge \mathbf{e}_0 \wedge \partial(\mathbf{A}_{r-1}), \\
\mathbf{B}_r &= (\mathbf{f}(\mathbf{o}) - \frac{\rho^2}{2}\mathbf{e})(\mathbf{e} \wedge \mathbf{B}'_r + \mathbf{e} \wedge \mathbf{e}_0 \wedge \partial(\mathbf{B}'_r)),
\end{aligned} \tag{7.4.43}$$

where $\mathbf{A}_{r-1} \in \Lambda^{r-1}(\mathbb{R}^n)$, $\mathbf{B}'_r \in \Lambda^r(\mathbb{R}^n)$, and $\mathbf{o} \in \mathbb{R}^n$ is a point in $(r-1)$D affine plane $(\mathbf{B}'_r, \partial(\mathbf{B}'_r))$. Let $\mathbf{x} \in \mathbb{R}^n$ be any point in $(r-2)$D affine plane $(\mathbf{A}_{r-1}, \partial(\mathbf{A}_{r-1}))$. Then the geometric meaning of $\mathbf{A}_r \cdot \mathbf{B}_r$ can be derived as follows:

$$\begin{aligned}
\mathbf{A}_r \cdot \mathbf{B}_r &= (\mathbf{A}_r \wedge \mathbf{f}(\mathbf{o})) \cdot (\mathbf{e} \wedge \mathbf{B}'_r + \mathbf{e} \wedge \mathbf{e}_0 \wedge \partial(\mathbf{B}'_r)) \\
&= (-1)^{r-1}(\mathbf{e} \wedge \mathbf{o} \wedge \mathbf{A}_{r-1} + \mathbf{e} \wedge \mathbf{e}_0 \wedge (\mathbf{A}_{r-1} - \mathbf{o} \wedge \partial(\mathbf{A}_{r-1}))) \\
&\qquad\qquad\qquad\qquad \cdot (\mathbf{e} \wedge \mathbf{B}'_r + \mathbf{e} \wedge \mathbf{e}_0 \wedge \partial(\mathbf{B}'_r)) \\
&= (-1)^{r-1}(\mathbf{A}_{r-1} - \mathbf{o} \wedge \partial(\mathbf{A}_{r-1})) \cdot \partial(\mathbf{B}'_r) \\
&= (-1)^{r-1}((\mathbf{x} - \mathbf{o}) \wedge \partial(\mathbf{A}_{r-1})) \cdot \partial(\mathbf{B}'_r).
\end{aligned} \tag{7.4.44}$$

(iii) If $\mathbf{A}_r, \mathbf{B}_r$ are both spheres, by (7.4.11), let

$$\begin{aligned}
\mathbf{A}_r &= (\mathbf{e} \wedge \mathbf{A}'_r + \mathbf{e} \wedge \mathbf{e}_0 \wedge \partial(\mathbf{A}'_r))(\mathbf{f}(\mathbf{o}_a) - \frac{\rho_a^2}{2}\mathbf{e}), \\
\mathbf{B}_r &= (\mathbf{f}(\mathbf{o}_b) - \frac{\rho_b^2}{2}\mathbf{e})(\mathbf{e} \wedge \mathbf{B}'_r + \mathbf{e} \wedge \mathbf{e}_0 \wedge \partial(\mathbf{B}'_r)),
\end{aligned} \tag{7.4.45}$$

where $\mathbf{A}'_r, \mathbf{B}'_r \in \Lambda^r(\mathbb{R}^n)$, and $\mathbf{o}_a, \mathbf{o}_b \in \mathbb{R}^n$ are points on affine planes $(\mathbf{A}'_r, \partial(\mathbf{A}'_r))$, $(\mathbf{B}'_r, \partial(\mathbf{B}'_r))$ respectively.

The computing of $\mathbf{A}_r \cdot \mathbf{B}_r$ is very complicated in the inner-product Grassmann algebra $\Lambda(\mathbb{R}^{n+1,1})$, but much easier in the Geometric Algebra $\mathcal{G}(\mathbb{R}^{n+1,1})$. In the latter algebra, the computing is as follows:

$$\mathbf{A}_r \cdot \mathbf{B}_r = \langle (\mathbf{e}\mathbf{A}'_r + (\mathbf{e} \wedge \mathbf{e}_0)\partial(\mathbf{A}'_r))(\mathbf{e}_0 + \mathbf{o}_a + \frac{\mathbf{o}_a^2 - \rho_a^2}{2}\mathbf{e})$$
$$(\mathbf{e}_0 + \mathbf{o}_b + \frac{\mathbf{o}_b^2 - \rho_b^2}{2}\mathbf{e})(\mathbf{e}\mathbf{B}'_r + (\mathbf{e} \wedge \mathbf{e}_0)\partial(\mathbf{B}'_r))\rangle$$

$$= \langle \mathbf{e}\mathbf{A}'_r(\mathbf{e}_0 + \mathbf{o}_a)(\mathbf{e}_0 + \mathbf{o}_b)(\mathbf{e} \wedge \mathbf{e}_0)\partial(\mathbf{B}'_r)\rangle$$
$$+ \langle (\mathbf{e} \wedge \mathbf{e}_0)\partial(\mathbf{A}'_r)(\mathbf{e}_0 + \mathbf{o}_a)(\mathbf{e}_0 + \mathbf{o}_b)\mathbf{e}\mathbf{B}'_r\rangle$$
$$+ \langle (\mathbf{e} \wedge \mathbf{e}_0)\partial(\mathbf{A}'_r)(\mathbf{o}_a\mathbf{o}_b + \frac{\mathbf{o}_a^2 - \rho_a^2}{2}\mathbf{e}\mathbf{e}_0 + \frac{\mathbf{o}_b^2 - \rho_b^2}{2}\mathbf{e}_0\mathbf{e})(\mathbf{e} \wedge \mathbf{e}_0)\partial(\mathbf{B}'_r)\rangle$$

$$= -\langle \mathbf{e}\mathbf{A}'_r\mathbf{e}_0(\mathbf{o}_b - \mathbf{o}_a)\partial(\mathbf{B}'_r)\rangle + \langle \partial(\mathbf{A}'_r)\mathbf{e}_0(\mathbf{o}_b - \mathbf{o}_a)\mathbf{e}\mathbf{B}'_r\rangle$$
$$+ \langle \partial(\mathbf{A}'_r)\mathbf{o}_a\mathbf{o}_b\partial(\mathbf{B}'_r)\rangle - \frac{\mathbf{o}_a^2 - \rho_a^2 + \mathbf{o}_b^2 - \rho_b^2}{2}\langle \partial(\mathbf{A}'_r)\partial(\mathbf{B}'_r)\rangle$$

$$= (-1)^r\langle \mathbf{A}'_r(\mathbf{o}_b - \mathbf{o}_a)\partial(\mathbf{B}'_r)\rangle + \langle \partial(\mathbf{A}'_r)(\mathbf{o}_b - \mathbf{o}_a)\mathbf{B}'_r\rangle$$
$$+ \langle \partial(\mathbf{A}'_r)\mathbf{o}_a\mathbf{o}_b\partial(\mathbf{B}'_r)\rangle - \frac{\mathbf{o}_a^2 - \rho_a^2 + \mathbf{o}_b^2 - \rho_b^2}{2}\langle \partial(\mathbf{A}'_r)\partial(\mathbf{B}'_r)\rangle.$$

$$(7.4.46)$$

Direct expansion of the geometric product by multilinearity in the first line of (7.4.46) produces a lot of terms. Using the shift symmetry of the angular bracket, and the (anti-)commutativities of $\mathbf{e}$ and $\mathbf{e}\wedge\mathbf{e}_0$ with elements in $\Lambda(\mathbb{R}^n)$, the expansion can be significantly simplified. For example, the first term $\mathbf{e}\mathbf{A}'_r$ in the first expression $\mathbf{e}\mathbf{A}'_r + (\mathbf{e}\wedge\mathbf{e}_0)\partial(\mathbf{A}'_r)$, and the first term $\mathbf{e}\mathbf{B}'_r$ in the last expression $\mathbf{e}\mathbf{B}'_r + (\mathbf{e}\wedge\mathbf{e}_0)\partial(\mathbf{B}'_r)$, cannot be in the same angular bracket because

$$\langle \mathbf{e}\mathbf{A}'_r \ldots \mathbf{e}\mathbf{B}'_r\rangle = \langle \mathbf{e}\mathbf{B}'_r\mathbf{e}\mathbf{A}'_r \ldots \rangle = (-1)^r\langle \mathbf{B}'_r\mathbf{e}\mathbf{e}\mathbf{A}'_r \ldots \rangle = 0.$$

As a second example, the multilinear expansion generates a term of the form $\langle (\mathbf{e} \wedge \mathbf{e}_0)\partial(\mathbf{A}'_r) \ldots (\mathbf{e} \wedge \mathbf{e}_0)\partial(\mathbf{B}'_r)\rangle$. Since the two $\mathbf{e} \wedge \mathbf{e}_0$'s can be moved together and canceled, the result is

$$\langle \partial(\mathbf{A}'_r)(\mathbf{e}_0 + \mathbf{o}_a + \frac{\mathbf{o}_a^2 - \rho_a^2}{2}\mathbf{e})(\mathbf{e}_0 + \mathbf{o}_b + \frac{\mathbf{o}_b^2 - \rho_b^2}{2}\mathbf{e})\partial(\mathbf{B}'_r)\rangle. \qquad (7.4.47)$$

In (7.4.47), terms containing $\mathbf{e}\mathbf{o}_a, \mathbf{e}_0\mathbf{o}_a$, *etc.*, do not contribute to the angular bracket, so only three terms are left after (7.4.47) is further expanded.

We continue the computing of (7.4.46). By $\mathbf{o}_a \wedge \partial(\mathbf{A}'_r) = \mathbf{A}'_r$ and $\mathbf{o}_b \wedge \partial(\mathbf{B}'_r) = \mathbf{B}'_r$, and using ungrading techniques in Geometric Algebra, we get

$$(-1)^r\langle \mathbf{A}'_r(\mathbf{o}_b - \mathbf{o}_a)\partial(\mathbf{B}'_r)\rangle + \langle \partial(\mathbf{A}'_r)(\mathbf{o}_b - \mathbf{o}_a)\mathbf{B}'_r\rangle$$

$$= -\langle (\partial(\mathbf{A}'_r) \wedge \mathbf{o}_a)(\mathbf{o}_b - \mathbf{o}_a)\partial(\mathbf{B}'_r)\rangle + \langle \partial(\mathbf{A}'_r)(\mathbf{o}_b - \mathbf{o}_a)(\mathbf{o}_b \wedge \partial(\mathbf{B}'_r))\rangle$$

$$= \frac{1}{2}\{-\langle \partial(\mathbf{A}'_r)\mathbf{o}_a(\mathbf{o}_b - \mathbf{o}_a)\partial(\mathbf{B}'_r)\rangle + (-1)^r\langle \mathbf{o}_a\partial(\mathbf{A}'_r)(\mathbf{o}_b - \mathbf{o}_a)\partial(\mathbf{B}'_r)\rangle$$

$$+ \langle \partial(\mathbf{A}'_r)(\mathbf{o}_b - \mathbf{o}_a)\mathbf{o}_b\partial(\mathbf{B}'_r)\rangle + (-1)^{r-1}\langle \partial(\mathbf{A}'_r)(\mathbf{o}_b - \mathbf{o}_a)\partial(\mathbf{B}'_r)\mathbf{o}_b\rangle\}$$

$$= -\langle\partial(\mathbf{A}_r')\mathbf{o}_a\mathbf{o}_b\partial(\mathbf{B}_r')\rangle + \frac{\mathbf{o}_a^2 + \mathbf{o}_b^2}{2}\langle\partial(\mathbf{A}_r')\partial(\mathbf{B}_r')\rangle + (-1)^r\langle\mathbf{o}_a\partial(\mathbf{A}_r')\mathbf{o}_b\partial(\mathbf{B}_r')\rangle$$
$$+(-1)^{r-1}\frac{1}{2}(\langle\mathbf{o}_a\partial(\mathbf{A}_r')\mathbf{o}_a\partial(\mathbf{B}_r')\rangle + \langle\mathbf{o}_b\partial(\mathbf{A}_r')\mathbf{o}_b\partial(\mathbf{B}_r')\rangle).$$

Substituting it into (7.4.46), we get

$$\mathbf{A}_r \cdot \mathbf{B}_r = \frac{\rho_a^2 + \rho_b^2}{2}\langle\partial(\mathbf{A}_r')\partial(\mathbf{B}_r')\rangle + (-1)^r\langle\mathbf{o}_a\partial(\mathbf{A}_r')\mathbf{o}_b\partial(\mathbf{B}_r')\rangle$$
$$+(-1)^{r-1}\frac{1}{2}(\langle\mathbf{o}_a\partial(\mathbf{A}_r')\mathbf{o}_a\partial(\mathbf{B}_r')\rangle + \langle\mathbf{o}_b\partial(\mathbf{A}_r')\mathbf{o}_b\partial(\mathbf{B}_r')\rangle). \tag{7.4.48}$$

Choose $\mathbf{o}_a$ as the origin, *i.e.*, set $\mathbf{o}_a = 0$. By commutation, (7.4.48) becomes

$$2\,\mathbf{A}_r \cdot \mathbf{B}_r = (\rho_a^2 + \rho_b^2)\partial(\mathbf{A}_r') \cdot \partial(\mathbf{B}_r') + (-1)^{r-1}\langle\mathbf{o}_b\partial(\mathbf{A}_r')\mathbf{o}_b\partial(\mathbf{B}_r')\rangle$$
$$= (\rho_a^2 + \rho_b^2)\partial(\mathbf{A}_r') \cdot \partial(\mathbf{B}_r') - \langle\mathbf{o}_b\partial(\mathbf{A}_r')\partial(\mathbf{B}_r')\mathbf{o}_b\rangle$$
$$+ 2\langle\mathbf{o}_b\partial(\mathbf{A}_r')(\partial(\mathbf{B}_r')\wedge\mathbf{o}_b)\rangle \tag{7.4.49}$$
$$= (\rho_a^2 + \rho_b^2 - \mathbf{o}_b^2)\partial(\mathbf{A}_r') \cdot \partial(\mathbf{B}_r') + 2\,(\mathbf{o}_b \wedge \partial(\mathbf{A}_r')) \cdot (\partial(\mathbf{B}_r') \wedge \mathbf{o}_b)$$
$$= \partial(\mathbf{A}_r') \cdot ((\rho_a^2 + \rho_b^2 - \mathbf{o}_b^2)\partial(\mathbf{B}_r') + 2\,\mathbf{o}_b \cdot \mathbf{B}_r').$$

To clarify the geometric meaning of the result of (7.4.49), we need to eliminate one of $\mathbf{B}_r'$ and $\partial(\mathbf{B}_r')$.

If $\mathbf{B}_r' = 0$, *i.e.*, if the center of sphere $\mathbf{A}_r$ is on the supporting affine plane of $(r-2)$D sphere $\mathbf{B}_r$, then

$$\mathbf{A}_r \cdot \mathbf{B}_r = \frac{\rho_a^2 + \rho_b^2 - d_{\mathbf{o}_a\mathbf{o}_b}^2}{2}\,\partial(\mathbf{A}_r') \cdot \partial(\mathbf{B}_r'), \tag{7.4.50}$$

which equals zero if and only if either the supporting planes of the two $(r-2)$D spheres are perpendicular, or the extensive spheres of the two $(r-2)$D spheres are perpendicular.

If $\mathbf{B}_r' \neq 0$, from $P_{\partial(\mathbf{B}_r')}^{\perp}(\mathbf{o}_b)\,\partial(\mathbf{B}_r') = \mathbf{B}_r'$ we get $\partial(\mathbf{B}_r') = (P_{\partial(\mathbf{B}_r')}^{\perp}(\mathbf{o}_b))^{-1} \cdot \mathbf{B}_r'$, so

$$\mathbf{A}_r \cdot \mathbf{B}_r = \{\partial(\mathbf{A}_r') \wedge (\frac{\rho_a^2 + \rho_b^2 - \mathbf{o}_b^2}{2}(P_{\partial(\mathbf{B}_r')}^{\perp}(\mathbf{o}_b))^{-1} + \mathbf{o}_b)\} \cdot \mathbf{B}_r', \tag{7.4.51}$$

which equals zero if and only if either the following point $\mathbf{x}$ is on the supporting plane of $(r-2)$D sphere $\mathbf{A}_r$, or the rD plane containing both $(r-2)$D sphere $\mathbf{B}_r$ and point $\mathbf{o}_a$ is perpendicular to the rD plane containing both $(r-2)$D sphere $\mathbf{A}_r$ and point $\mathbf{x}$:

$$\mathbf{x} = \frac{\rho_a^2 + \rho_b^2 - d_{\mathbf{o}_a\mathbf{o}_b}^2}{2}(P_{\partial(\mathbf{B}_r')}^{\perp}(\mathbf{o}_b - \mathbf{o}_a))^{-1} + \mathbf{o}_b - \mathbf{o}_a. \tag{7.4.52}$$

Finally, assuming that $\mathbf{A}_r, \mathbf{B}_s$ are Minkowski blades of grade r, s respectively, where $2 \leq r < s \leq n+1$, we consider the geometric meaning of their inner product.

Since $\mathbf{A}_r \cdot \mathbf{B}_s = 0$ if and only if $\mathbf{A}_r \cdot \mathbf{D}_r = 0$ for any Minkowski r-blade $\mathbf{D}_r \in \Lambda(\mathbf{B}_s)$, and the geometric meaning of $\mathbf{A}_r \cdot \mathbf{D}_r = 0$ is already obtained, we can assume that $\mathbf{A}_r \cdot \mathbf{B}_s \neq 0$. Since $\mathbf{A}_r \cdot \mathbf{B}_s = \mathbf{B}_s \vee \mathbf{A}_r^{-\sim}$, we only need to consider the meet product of a Minkowski blade $\mathbf{A}_r$ and a Euclidean blade $\mathbf{C}_t$, where $r + t > n + 2$. The result has only one possible signature: Euclidean.

If $\mathbf{e} \cdot (\mathbf{A}_r \vee \mathbf{C}_t) = 0$, then $\mathbf{A}_r \vee \mathbf{C}_t$ represents an $(r + t - n - 2)$D direction in the space at infinity of the supporting plane of $(r - 2)$D sphere or plane $\mathbf{A}_r$. If $\mathbf{e} \cdot (\mathbf{A}_r \vee \mathbf{C}_t) \neq 0$, then $\mathbf{e} \cdot (\mathbf{e} \wedge (\mathbf{A}_r \vee \mathbf{C}_t))$ represents an $(r + t - n - 3)$D direction in the space at infinity of the supporting plane of $\mathbf{A}_r$.

We see that for $r \neq s$, the inner product $\mathbf{A}_r \cdot \mathbf{B}_s$ of two Minkowski blades $\mathbf{A}_r, \mathbf{B}_s$, if nonzero, does not have good interpretation in Euclidean geometry. The reason is that the inner product results in a blade of Euclidean signature, so it can only be interpreted as a high dimensional direction in the space at infinity. On the contrary, the inner product of a Minkowski blade and a Euclidean blade, or equivalently via the dual operator, the meet product of two Minkowski blades, has much better Euclidean geometric interpretation, as to be investigated in the next subsection.

7.4.3 *Meet product of Minkowski blades*

Let $\mathbf{A}_r, \mathbf{B}_s$ be two Minkowski blades in $\Lambda(\mathbb{R}^{n+1,1})$, where $2 \leq r, s \leq n + 1$ and $r + s > n + 2$. Their meet product $\mathbf{A}_r \vee \mathbf{B}_s$, if nonzero, is a blade that has four possible signatures.

Case 1. $\mathbf{A}_r \vee \mathbf{B}_s$ is Minkowski.

Then $r + s > n + 3$, the two spheres or planes intersect, and $\mathbf{A}_r \vee \mathbf{B}_s$ is their intersection. If $r + s = n + 4$, the intersection is either a 0D circle or the affine representation of a point.

Example 7.44. When $n = 2$, for any points $\mathbf{1}, \mathbf{2}, \mathbf{1}', \mathbf{2}'$ in the plane (null vectors in $\mathbb{R}^{3,1}$), let $\mathbf{a}$ (null vector) be the point of intersection of lines $\mathbf{12}$ and $\mathbf{1}'\mathbf{2}'$. Then

$$\mathbf{e} \wedge \mathbf{a} = (\mathbf{e} \wedge \mathbf{1} \wedge \mathbf{2}) \vee (\mathbf{e} \wedge \mathbf{1}' \wedge \mathbf{2}') = \mathbf{e} \wedge ([\mathbf{e11}'\mathbf{2}']\mathbf{2} - [\mathbf{e21}'\mathbf{2}']\mathbf{1}). \tag{7.4.53}$$

Let

$$\mathbf{a} = [\mathbf{e11}'\mathbf{2}']\mathbf{2} - [\mathbf{e21}'\mathbf{2}']\mathbf{1} + \lambda \mathbf{e}, \tag{7.4.54}$$

where λ is an indeterminate. λ can be determined by $\mathbf{a}^2 = 0$, and the result is

$$\lambda = \frac{\mathbf{1} \cdot \mathbf{2}[\mathbf{e11}'\mathbf{2}'][\mathbf{e21}'\mathbf{2}']}{\mathbf{e} \cdot \mathbf{2}[\mathbf{e11}'\mathbf{2}'] - \mathbf{e} \cdot \mathbf{1}[\mathbf{e21}'\mathbf{2}']}. \tag{7.4.55}$$

The denominator of (7.4.55) is a bracket binomial that can be written as a long bracket as follows. By (6.1.38), for null vector $\mathbf{e}$,

$$\frac{1}{2}\mathbf{e12e} = (\mathbf{e} \cdot \mathbf{1})\mathbf{2e} - (\mathbf{e} \cdot \mathbf{2})\mathbf{1e},$$

so

$$\frac{1}{2}[\mathbf{e12e1}'\mathbf{2}'] = \mathbf{e} \cdot \mathbf{2}[\mathbf{e11}'\mathbf{2}'] - \mathbf{e} \cdot \mathbf{1}[\mathbf{e21}'\mathbf{2}']. \tag{7.4.56}$$

Substituting it into (7.4.55), and then into (7.4.54), we get, up to scale,

$$\mathbf{a} = \frac{1}{2}[\mathbf{e12e1}'\mathbf{2}']([\mathbf{e11}'\mathbf{2}']\mathbf{2} - [\mathbf{e21}'\mathbf{2}']\mathbf{1}) + \mathbf{1} \cdot \mathbf{2}[\mathbf{e11}'\mathbf{2}'][\mathbf{e21}'\mathbf{2}']\mathbf{e}. \tag{7.4.57}$$

The geometric meaning of $[\mathbf{e}12\mathbf{e}1'2']$ is obvious from (7.4.56): it is four times the signed area of quadrilateral $1'12'2$. Two lines $12, 1'2'$ are parallel if and only if $[\mathbf{e}12\mathbf{e}1'2'] = 0$. By the shift symmetry of the square bracket, we also have

$$\frac{1}{2}[\mathbf{e}1'2'\mathbf{e}12] = -\frac{1}{2}[\mathbf{e}12\mathbf{e}1'2'] = \mathbf{e} \cdot 2'[\mathbf{e}121'] - \mathbf{e} \cdot 1'[\mathbf{e}122']. \tag{7.4.58}$$

Consider the barycentric coordinates of the intersection (7.4.57) with respect to the affine basis $1, 2$ of line 12. Let $[\mathbf{e}12]$ denote a deficit bracket in $\Lambda(\mathbb{R}^{3,1})$. By the Cramer's rule $[\mathbf{e}12]\mathbf{a} = [12\mathbf{a}]\mathbf{e} - [\mathbf{e}2\mathbf{a}]1 + [\mathbf{e}1\mathbf{a}]2$,

$$\frac{\mathbf{a}}{\partial(\mathbf{a})} = \frac{\mathbf{a}}{-\mathbf{e} \cdot \mathbf{a}} = \frac{[\mathbf{e}2\mathbf{a}]1 - [\mathbf{e}1\mathbf{a}]2 - [12\mathbf{a}]\mathbf{e}}{\mathbf{e} \cdot \mathbf{a}}.$$

By this and (7.4.57), we get the the barycentric coordinates of $\mathbf{a}$ with respect to $1, 2$:

$$\begin{aligned}
\left(\frac{[\mathbf{e}2\mathbf{a}]}{\mathbf{e} \cdot \mathbf{a}[\mathbf{e}12]}, \ \frac{-[\mathbf{e}1\mathbf{a}]}{\mathbf{e} \cdot \mathbf{a}[\mathbf{e}12]}\right) &= \left(2\frac{[\mathbf{e}21'2']}{[\mathbf{e}12\mathbf{e}1'2']}, \ -2\frac{[\mathbf{e}11'2']}{[\mathbf{e}12\mathbf{e}1'2']}\right) \\
&= \left(2\frac{\langle \mathbf{e}21'2'\rangle_4}{\langle \mathbf{e}12\mathbf{e}1'2'\rangle_4}, \ -2\frac{\langle \mathbf{e}11'2'\rangle_4}{\langle \mathbf{e}12\mathbf{e}1'2'\rangle_4}\right).
\end{aligned} \tag{7.4.59}$$

The last form of (7.4.59) is valid not only for the intersection of two coplanar lines, but also for a foot of perpendicular of two noncoplanar lines in space. Let $12, 1'2'$ be noncoplanar lines in $\mathbb{E}^3$. Let their common perpendicular be $\mathbf{aa}'$, where $\mathbf{a}, \mathbf{a}'$ are the feet on lines $12, 1'2'$ respectively. Using the same method of indeterminate coefficients, we get the barycentric coordinates of $\mathbf{a}$ with respect to $1, 2$ as follows:

$$\left(2\langle \mathbf{e}21'2'\rangle_4 \cdot \langle \mathbf{e}12\mathbf{e}1'2'\rangle_4^{-1}, \ -2\langle \mathbf{e}11'2'\rangle_4 \cdot \langle \mathbf{e}12\mathbf{e}1'2'\rangle_4^{-1}\right). \tag{7.4.60}$$

When lines $12, 1'2'$ are coplanar, then $\mathbf{a} = \mathbf{a}'$ is their point of intersection, and (7.4.60) is identical to (7.4.59).

Case 2. $\mathbf{A}_r \vee \mathbf{B}_s$ is degenerate.

If $r + s = n + 3$, then if $\mathbf{A}_r \vee \mathbf{B}_s = \mathbf{e}$ up to scale, planes $\mathbf{A}_r, \mathbf{B}_s$ do not intersect, else they intersect only at the point represented by null vector $\mathbf{A}_r \vee \mathbf{B}_s$. For example when $n = 3$, circles 123 and 145 intersect only at point 1 if the five points $1, 2, 3, 4, 5$ are neither coplanar nor cospherical in space.

If $r + s > n + 3$, then if $\mathbf{e} \in \mathbf{A}_r \vee \mathbf{B}_s$, planes $\mathbf{A}_r$ and $\mathbf{B}_s$ are parallel, and $\mathbf{A}_r \vee \mathbf{B}_s$ is their common directions of dimension $r + s - n - 3$. If $\mathbf{e} \notin \mathbf{A}_r \vee \mathbf{B}_s$, let $\mathbf{a}$ be the null vector in $\mathbf{A}_r \vee \mathbf{B}_s$ which is unique up to scale, then the two spheres or planes $\mathbf{A}_r$ and $\mathbf{B}_s$ are tangent to each other at point $\mathbf{a}$, in the sense that they have a common tangent subspace $\mathbf{e} \cdot (\mathbf{A}_r \vee \mathbf{B}_s)$ of dimension $r + s - n - 3$ at their unique common point $\mathbf{a}$. This is only a *partial tangency* except for the special case where at least one of r, s equals $n + 1$. For any $r, s < n + 1$, since $r + s - n - 3 < \min(r, s) - 2$, $\mathbf{e} \cdot (\mathbf{A}_r \vee \mathbf{B}_s)$ is a proper subspace of both tangent spaces of $\mathbf{A}_r$ and $\mathbf{B}_s$ at $\mathbf{a}$.

For example, when $n = 4$, if two 2D spheres $\mathbf{A}_4, \mathbf{B}_4$ intersect at only one point, their 2D tangent planes at the intersection may be identical, or intersecting at a 1D subspace, or not intersecting. In the first case, $\mathbf{A}_4 \vee \mathbf{B}_4 = 0$, because $\mathbf{A}_4, \mathbf{B}_4$ have a common 3D subspace. In the last case, $\mathbf{A}_4, \mathbf{B}_4$ have a common null 1D subspace but have no common positive vector, contradicting with the fact that $\mathbf{A}_4 \vee \mathbf{B}_4$ is a 2D subspace of a Minkowski space. In the middle case, the two 2D spheres have only one common tangent direction at the tangent point, and are partially tangent to each other.

Case 3. $\mathbf{A}_r \vee \mathbf{B}_s$ is anti-Euclidean.

Then $r + s = n + 3$, spheres or planes $\mathbf{A}_r, \mathbf{B}_s$ do not intersect. Let $\mathbf{a} = \mathbf{A}_r \vee \mathbf{B}_s$. Then $\mathbf{A}_r$ passes through a $(r-3)$D great sphere $\mathbf{A}'_{r-1}$ of sphere $(\mathbf{a} \cdot (\mathbf{e} \wedge \mathbf{a}))^\sim$, and $\mathbf{B}_s$ passes through another $(s-3)$D great sphere $\mathbf{B}'_{s-1}$ of the same sphere.

For example, when $n = 3$, let $\mathbf{x}_1, \mathbf{x}_2$ be linearly independent unit vectors in $\mathbb{R}^3$, and let $\mathbf{y}_1, \mathbf{y}_2$ be non-unit vectors in $\mathbb{R}^3$. For $i = 1, 2$, the two circles each passing through three points $\mathbf{x}_i, -\mathbf{x}_i$ and $\mathbf{y}_i$ are

$$\mathbf{C}_i = \mathbf{f}(\mathbf{x}_i) \wedge \mathbf{f}(-\mathbf{x}_i) \wedge \mathbf{f}(\mathbf{y}_i) = -(2\mathbf{e}_0 + \mathbf{e}) \wedge \mathbf{x}_i \wedge \mathbf{f}(\mathbf{y}_i).$$

So

$$\begin{aligned}
\mathbf{C}_1 \vee \mathbf{C}_2 &= [(2\mathbf{e}_0 + \mathbf{e})\mathbf{x}_1\mathbf{f}(\mathbf{y}_1)\mathbf{x}_2\mathbf{f}(\mathbf{y}_2)](2\mathbf{e}_0 + \mathbf{e}) \\
&= [\mathbf{e}\mathbf{e}_0\mathbf{x}_1\mathbf{x}_2((1 - \mathbf{y}_2^2)\mathbf{y}_1 - (1 - \mathbf{y}_1^2)\mathbf{y}_2)](2\mathbf{e}_0 + \mathbf{e})
\end{aligned}$$

is a negative vector as long as $(1 - \mathbf{y}_2^2)\mathbf{y}_1 - (1 - \mathbf{y}_1^2)\mathbf{y}_2 \notin \mathbf{x}_1 \wedge \mathbf{x}_2$.

Definition 7.45. An rD sphere $\mathbf{A}$ and an sD sphere $\mathbf{B}$ in $\mathbb{E}^3$, where $0 \leq r, s \leq n-1$, are said to be *knotted*, if (1) they do not intersect, (2) sphere $\mathbf{B}$ intersects the $(r+1)$D supporting plane $\mathbf{A}'$ of sphere $\mathbf{A}$ at a point inside sphere $\mathbf{A}$ and at the other point outside sphere $\mathbf{A}$, (3) sphere $\mathbf{A}$ intersects the $(s+1)$D supporting plane $\mathbf{B}'$ of sphere $\mathbf{B}$ at a point inside sphere $\mathbf{B}$ and at the other point outside sphere $\mathbf{B}$.

An rD plane $\mathbf{A}$ and an sD sphere $\mathbf{B}$ are said to be *knotted*, if the intersection of plane $\mathbf{A}$ and the supporting plane $\mathbf{B}'$ of sphere $\mathbf{B}$ is a point inside sphere $\mathbf{B}$.

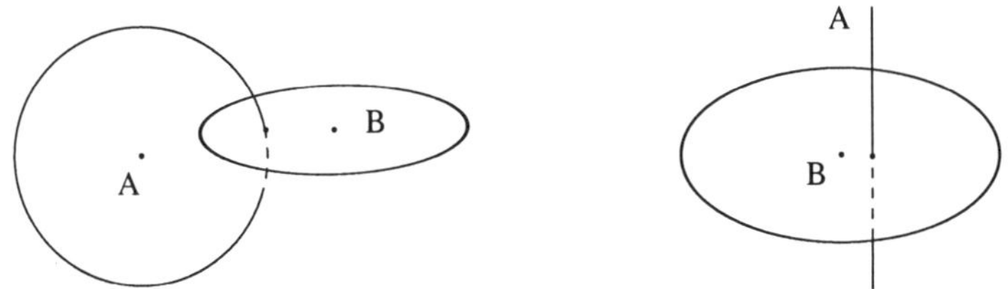

Fig. 7.9   Two knotted spheres (left); a knotted pair of sphere and plane (right).

In particular, a point is knotted with an $(n-1)$D sphere if it is inside the sphere; a 0D sphere is knotted with an $(n-1)$D sphere if one point of the 0D sphere is inside the $(n-1)$D sphere while the other point is outside.

Proposition 7.46. In a $(r+1)$D affine plane $\mathbf{A}_{r+3}$,

(1) two points $\mathbf{a}, \mathbf{a}'$ (null vectors) are on the same side (or on different sides) of an rD sphere $\mathbf{B}_{r+2}$ in the $(r+1)$D plane, if and only if vector $(\mathbf{a} \wedge \mathbf{a}') \cdot (\mathbf{B}_{r+2} \cdot \mathbf{A}_{r+3})$ is positive (or negative);

(2) point $\mathbf{a}$ is outside (or inside) sphere $\mathbf{B}_{r+2}$ if and only if vector $(\mathbf{e} \wedge \mathbf{a}) \cdot (\mathbf{B}_{r+2} \cdot \mathbf{A}_{r+3})$ is positive (or negative).

Proof. We only need to prove the first statement. By Proposition 7.29, two points $\mathbf{a}, \mathbf{a}'$ are on the same side of a sphere $\mathbf{b}^\sim$ (or on different sides), if and only if the discriminant $\tau = (\mathbf{a} \cdot \mathbf{a}')(\mathbf{a} \cdot \mathbf{b})(\mathbf{a}' \cdot \mathbf{b}) < 0$ (or $\tau > 0$). By the identity

$$(\mathbf{a} \wedge \mathbf{a}' \wedge \mathbf{b})^2 = - \begin{vmatrix} 0 & \mathbf{a} \cdot \mathbf{a}' & \mathbf{a} \cdot \mathbf{b} \\ \mathbf{a} \cdot \mathbf{a}' & 0 & \mathbf{a}' \cdot \mathbf{b} \\ \mathbf{a} \cdot \mathbf{b} & \mathbf{a}' \cdot \mathbf{b} & \mathbf{b}^2 \end{vmatrix} = -2(\mathbf{a} \cdot \mathbf{a}')(\mathbf{a} \cdot \mathbf{b})(\mathbf{a}' \cdot \mathbf{b}) + \mathbf{b}^2 (\mathbf{a} \cdot \mathbf{a}')^2,$$

we get the following expression of the discriminant:

$$\tau = \frac{1}{2}(\mathbf{b}^2 (\mathbf{a} \wedge \mathbf{a}')^2 - (\mathbf{a} \wedge \mathbf{a}' \wedge \mathbf{b})^2) = -\frac{1}{2}((\mathbf{a} \wedge \mathbf{a}') \cdot \mathbf{b})^2. \tag{7.4.61}$$

For sphere $\mathbf{B}_{r+2}$ in plane $\mathbf{A}_{r+3}$, its positive-vector representation in $\mathbf{A}_{r+3}$ is $\mathbf{b} = \mathbf{B}_{r+2} \cdot \mathbf{A}_{r+3}$. By (7.4.61), we get the conclusion. $\qquad\square$

Theorem 7.47. For $2 \leq r \leq n+1$, let $\mathbf{A}_r$ be a $(r-2)$D sphere, and $\mathbf{B}_{n-r+3}$ be an $(n-r+1)$D sphere or plane. Then $\mathbf{A}_r, \mathbf{B}_{n-r+3}$ are knotted if and only if vector $\mathbf{A}_r \vee \mathbf{B}_{n-r+3}$ is negative.

Proof. First, consider the case where vector $\mathbf{c} = \mathbf{A}_r \vee \mathbf{B}_{n-r+3}$ is negative. If both $\mathbf{A}_r, \mathbf{B}_{n-r+3}$ are spheres, the intersection of the supporting plane $\mathbf{e} \wedge \mathbf{A}_r$ of sphere $\mathbf{A}_r$ with sphere $\mathbf{B}_{n-r+3}$ is represented by the 2-blade $(\mathbf{e} \wedge \mathbf{A}_r) \vee \mathbf{B}_{n-r+3}$. The 2-blade is Minkowski, because it contains negative vector $\mathbf{c}$. This proves that the intersection of plane $\mathbf{e} \wedge \mathbf{A}_r$ and sphere $\mathbf{B}_{n-r+3}$ is a 0D sphere.

By Proposition 7.46 and by symmetry consideration, we only need to prove that the following vector is negative:

$$\begin{aligned} \mathbf{d} &= ((\mathbf{e} \wedge \mathbf{A}_r) \vee \mathbf{B}_{n-r+3}) \cdot (\mathbf{A}_r \cdot (\mathbf{e} \wedge \mathbf{A}_r)) \\ &= (-1)^r \mathbf{A}_r^2 \, (\mathbf{B}_{n-r+3}^\sim \cdot (\mathbf{e} \wedge \mathbf{A}_r)) \cdot ((\mathbf{e} \wedge \mathbf{A}_r)\mathbf{A}_r^{-1}) \\ &= (-1)^r \mathbf{A}_r^2 \sum_{(r-2,2) \vdash \mathbf{A}_r} (\mathbf{B}_{n-r+3}^\sim \cdot (\mathbf{e} \wedge \mathbf{A}_{r\,(1)})) \, (\mathbf{A}_{r\,(2)} \cdot P_{\mathbf{A}_r}^\perp(\mathbf{e})) \\ &\qquad\qquad\qquad - \mathbf{A}_r^2 \, (\mathbf{e} \wedge (\mathbf{B}_{n-r+3}^\sim \cdot \mathbf{A}_r)) \cdot P_{\mathbf{A}_r}^\perp(\mathbf{e}) \\ &= -\mathbf{A}_r^2 \, (\mathbf{e} \wedge \mathbf{c}) \cdot P_{\mathbf{A}_r}^\perp(\mathbf{e}) \\ &= \mathbf{A}_r^2 \, (\mathbf{e} \cdot P_{\mathbf{A}_r}^\perp(\mathbf{e})) \mathbf{c} \\ &= \mathbf{A}_r^2 \, (P_{\mathbf{A}_r}^\perp(\mathbf{e}))^2 \mathbf{c}. \end{aligned} \tag{7.4.62}$$

Since $\mathbf{A}_r$ is a sphere, $\mathbf{e} \wedge \mathbf{A}_r \neq 0$, so $P_{\mathbf{A}_r}^\perp(\mathbf{e})$ is a nonzero vector in Euclidean vector space $\mathbf{A}_r^\sim$, and $(P_{\mathbf{A}_r}^\perp(\mathbf{e}))^2 \neq 0$. Then $\mathbf{c}, \mathbf{d}$ have the same signature. If $\mathbf{A}_r$ is a sphere but $\mathbf{B}_{n-r+3}$ is a plane, the proof is still valid.

Second, if vector $\mathbf{c}$ is null, then spheres or planes $\mathbf{A}_r, \mathbf{B}_{n-r+3}$ have a common point $\mathbf{c}$, so they cannot be knotted.

Third, if $\mathbf{c}$ is positive, so is $\mathbf{d}$ by (7.4.62). Even if $(\mathbf{e} \wedge \mathbf{A}_r) \vee \mathbf{B}_{n-r+3}$ is a 0D sphere, by Proposition 7.46, the two points of the 0D sphere cannot be separated by sphere $\mathbf{A}_r$ in plane $\mathbf{e} \wedge \mathbf{A}_r$. So $\mathbf{A}_r, \mathbf{B}_{n-r+3}$ cannot be knotted. $\qquad\square$

Case 4. $\mathbf{A}_r \vee \mathbf{B}_s$ is Euclidean.

Then spheres or planes $\mathbf{A}_r, \mathbf{B}_s$ do not intersect. Blade $(\mathbf{A}_r \vee \mathbf{B}_s)^\sim$ is Minkowski, because $r + s - n - 2 \leq n$.

If $\mathbf{e} \in (\mathbf{A}_r \vee \mathbf{B}_s)^\sim$, then $(\mathbf{A}_r \vee \mathbf{B}_s)^\sim$ is a plane. If $\mathbf{A}_r$ is a sphere, its center $\mathbf{c}$ (null vector) must be in plane $(\mathbf{A}_r \vee \mathbf{B}_s)^\sim$, because if ρ is the radius then $\mathbf{c} - \rho^2 \mathbf{e}/2 \in \mathbf{A}_r^\sim \subset (\mathbf{A}_r \vee \mathbf{B}_s)^\sim$. If $\mathbf{A}_r$ is a plane, all its normal directions are in plane $(\mathbf{A}_r \vee \mathbf{B}_s)^\sim$.

If $\mathbf{e} \notin (\mathbf{A}_r \vee \mathbf{B}_s)^\sim$, then $(\mathbf{A}_r \vee \mathbf{B}_s)^\sim$ is a sphere. By Proposition 8.11 to be proved in Chapter 8, sphere $(\mathbf{A}_r \vee \mathbf{B}_s)^\sim$ is invariant under the inversion or mirror reflection with respect to any of $\mathbf{A}_r, \mathbf{B}_s$. It is called the *Poncelet sphere* of $\mathbf{A}_r, \mathbf{B}_s$.

Definition 7.48. Two spheres or planes of dimension between 0 and $n - 1$ in $\mathbb{E}^n$ are said to be *separated*, if they neither intersect, nor are tangent to each other, nor are knotted.

As a corollary of Theorem 7.47, we have

Proposition 7.49. For $2 \leq r \leq n + 1$ and Minkowski blades $\mathbf{A}_r, \mathbf{B}_{n-r+3}$ of grade $r, n - r + 3$ respectively, spheres or planes $\mathbf{A}_r, \mathbf{B}_{n-r+3}$ are separated if and only if $\mathbf{A}_r \vee \mathbf{B}_{n-r+3}$ is Euclidean.

For Minkowski blades $\mathbf{A}_r, \mathbf{B}_s$, their total meet product provides a complete classification of all possible incidence relations between the two spheres or planes.

Definition 7.50. The *extension* of two spheres or planes of dimension r, s respectively, refers to the plane or sphere of lowest dimension that contains both of them.

Definition 7.51. For $0 \leq r, s \leq n - 1$, $0 \leq t \leq \min(r, s)$ and $u = r + s - t$, an rD sphere and an sD sphere in $\mathbb{E}^n$ are said to be

- *tD spherical intersecting* (or *tD planar intersecting*), if their intersection is a tD sphere, and their extension is a uD sphere (or plane).
- *tD spherical tangent* (or *tD planar tangent*), if they have a unique common point, called the *tangent point*, and they have a common tD tangent subspace at the tangent point, and their extension is a $(u + 1)$D sphere (or plane).
- *uD spherical separated* (or *uD planar separated*), if they are separated, and their extension is a $(u + 1)$D sphere (or plane).

For an rD sphere and an sD plane, their extension can only be a plane, and their tD intersecting, tangent and separated relations can be defined similarly. For an rD plane and an sD plane, the definitions are the same, except that if the two planes are tangent to each other, the tangent point can only be the conformal point at infinity, and they are called *parallel* instead.

Below we investigate in details the case of two circles in $\mathbb{E}^3$. The general case of an rD and an sD sphere or plane in $\mathbb{E}^n$ is similar.

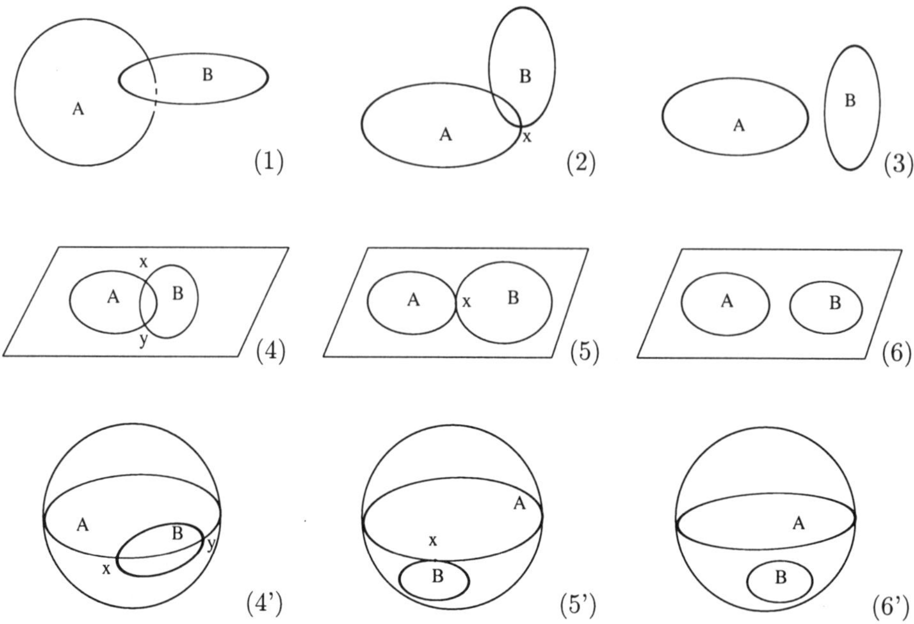

Fig. 7.10 Incidence relations between two circles $\mathbf{A}, \mathbf{B}$ in space: (1) knotted, (2) 0D planar tangent at point $\mathbf{x}$, (3) 2D planar separated, (4) 0D planar intersecting at points $\mathbf{x}, \mathbf{y}$, (5) 1D planar tangent at point $\mathbf{x}$, (6) 1D planar separated, (4') 0D spherical intersecting at points $\mathbf{x}, \mathbf{y}$, (5') 1D spherical tangent at point $\mathbf{x}$, (6') 1D spherical separated.

In $\Lambda(\mathbb{R}^{4,1})$, let $\mathbf{A}_3 = \mathbf{1} \wedge \mathbf{2} \wedge \mathbf{3}$ and $\mathbf{B}_3 = \mathbf{1'} \wedge \mathbf{2'} \wedge \mathbf{3'}$ be two circles each passing through three points. Their total meet product which is defined by (2.4.17), is

$$\begin{aligned}
\mathbf{A}_3 \bar{\vee} \mathbf{B}_3 = \quad & \mathbf{1'} \otimes \mathbf{A}_3 \wedge \mathbf{2'} \wedge \mathbf{3'} - \mathbf{2'} \otimes \mathbf{A}_3 \wedge \mathbf{1'} \wedge \mathbf{3'} + \mathbf{3'} \otimes \mathbf{A}_3 \wedge \mathbf{1'} \wedge \mathbf{2'} \\
& + \mathbf{1'} \wedge \mathbf{2'} \otimes \mathbf{A}_3 \wedge \mathbf{3'} - \mathbf{1'} \wedge \mathbf{3'} \otimes \mathbf{A}_3 \wedge \mathbf{2'} + \mathbf{2'} \wedge \mathbf{3'} \otimes \mathbf{A}_3 \wedge \mathbf{1'} + \mathbf{B}_3 \otimes \mathbf{A}_3.
\end{aligned}$$
$$(7.4.63)$$

From (7.4.63) we get the $(k-2)$D extensions $\mathbf{E}_k$ for $k = 5, 4, 3$, and $(l-2)$D intersections $\mathbf{S}_l$ for $l = 1, 2, 3$, as follows:

$$\mathbf{E}_5 = [\mathbf{1'}]\mathbf{A}_3 \wedge \mathbf{2'} \wedge \mathbf{3'} - [\mathbf{2'}]\mathbf{A}_3 \wedge \mathbf{1'} \wedge \mathbf{3'} + [\mathbf{3'}]\mathbf{A}_3 \wedge \mathbf{1'} \wedge \mathbf{2'},$$

$$\mathbf{S}_1 = [\mathbf{1232'3'}]\mathbf{1'} - [\mathbf{1231'3'}]\mathbf{2'} + [\mathbf{1231'2'}]\mathbf{3'},$$

$$\mathbf{E}_4 = [\mathbf{1'2'}]\mathbf{A}_3 \wedge \mathbf{3'} - [\mathbf{1'3'}]\mathbf{A}_3 \wedge \mathbf{2'} + [\mathbf{2'3'}]\mathbf{A}_3 \wedge \mathbf{1'},$$

$$\mathbf{S}_2 = [\mathbf{1233'}]\mathbf{1'} \wedge \mathbf{2'} - [\mathbf{1232'}]\mathbf{1'} \wedge \mathbf{3'} + [\mathbf{1231'}]\mathbf{2'} \wedge \mathbf{3'},$$

$$\mathbf{E}_3 = [\mathbf{B}_3]\mathbf{A}_3, \quad \mathbf{S}_3 = [\mathbf{A}_3]\mathbf{B}_3.$$

(i) The two circles are either knotted, or 0D planar tangent, or 2D planar separated, if and only if $\mathbf{S}_1$ is either negative, or null, or positive.

(ii) When $\mathbf{S}_1 = 0$, the two circles are coplanar or cospherical. They are coplanar and cospherical simultaneously if and only if they are identical, or equivalently, if and only if $\mathbf{S}_1 = \mathbf{S}_2 = 0$. When they are identical, the two circles are 1D spherical intersecting.

(iii) Assume that $\mathbf{S}_1 = 0$ but $\mathbf{S}_2 \neq 0$. Then Minkowski blade $\mathbf{E}_4$ represents the common supporting plane or sphere of the two circles, depending on whether or not $\mathbf{e} \in \mathbf{E}_4$.

The two circles are either 0D intersecting, or 1D tangent, or 1D separated, if and only if blade $\mathbf{S}_2$ is either Minkowski, or degenerate, or Euclidean.

Example 7.52. In the homogeneous model of 2D geometry, the foot of perpendicular drawn from point $\mathbf{1}$ to line $\mathbf{23}$ is

$$\mathbf{a} = \frac{\mathbf{2}}{\mathbf{e} \cdot \mathbf{2}\langle\mathbf{e123}\rangle} + \frac{\mathbf{3}}{\mathbf{e} \cdot \mathbf{3}\langle\mathbf{e132}\rangle} - \frac{\mathbf{e}}{2\,(\mathbf{e}\cdot\mathbf{1})(\mathbf{e}\cdot\mathbf{2})(\mathbf{e}\cdot\mathbf{3})}. \tag{7.4.64}$$

(7.4.64) can be derived as follows: the foot $\mathbf{a}$ is the intersection of line $\mathbf{23}$ and the line passing through point $\mathbf{1}$ and following the normal direction $(\mathbf{e} \wedge \mathbf{2} \wedge \mathbf{3})^{\sim}$ of line $\mathbf{23}$:

$$\mathbf{e} \wedge \mathbf{a} = (\mathbf{e} \wedge \mathbf{2} \wedge \mathbf{3}) \vee (\mathbf{e} \wedge \mathbf{1} \wedge (\mathbf{e} \wedge \mathbf{2} \wedge \mathbf{3})^{\sim}).$$

Expanding the meet product by separating the first blade, we get

$$\mathbf{e} \wedge \mathbf{a} = \mathbf{e} \wedge ([\mathbf{e13}(\mathbf{e} \wedge \mathbf{2} \wedge \mathbf{3})^{\sim}]\mathbf{2} - [\mathbf{e12}(\mathbf{e} \wedge \mathbf{2} \wedge \mathbf{3})^{\sim}]\mathbf{3}). \tag{7.4.65}$$

In $\mathcal{CL}(\mathbb{R}^{3,1})$,

$$
\begin{aligned}
[\mathbf{e13}(\mathbf{e} \wedge \mathbf{2} \wedge \mathbf{3})^{\sim}] &= -(\mathbf{e} \wedge \mathbf{1} \wedge \mathbf{3}) \cdot (\mathbf{e} \wedge \mathbf{2} \wedge \mathbf{3}) \\
&= \begin{vmatrix} 0 & \mathbf{e}\cdot\mathbf{2} & \mathbf{e}\cdot\mathbf{3} \\ \mathbf{e}\cdot\mathbf{1} & \mathbf{1}\cdot\mathbf{2} & \mathbf{1}\cdot\mathbf{3} \\ \mathbf{e}\cdot\mathbf{3} & \mathbf{2}\cdot\mathbf{3} & 0 \end{vmatrix} \\
&= \mathbf{e}\cdot\mathbf{3}\{(\mathbf{e}\cdot\mathbf{1})(\mathbf{2}\cdot\mathbf{3}) + (\mathbf{e}\cdot\mathbf{2})(\mathbf{1}\cdot\mathbf{3}) - (\mathbf{e}\cdot\mathbf{3})(\mathbf{1}\cdot\mathbf{2})\} \\
&= \mathbf{e}\cdot\mathbf{3}\,\langle\mathbf{e132}\rangle.
\end{aligned}
$$

Similarly, $[\mathbf{e12}(\mathbf{e} \wedge \mathbf{2} \wedge \mathbf{3})^{\sim}] = -\mathbf{e} \cdot \mathbf{2}\,\langle\mathbf{e123}\rangle$. Substituting them into (7.4.65), we get

$$\mathbf{e} \wedge \mathbf{a} = \mathbf{e} \wedge \{(\mathbf{e}\cdot\mathbf{3})\,\langle\mathbf{e132}\rangle\,\mathbf{2} + (\mathbf{e}\cdot\mathbf{2})\,\langle\mathbf{e123}\rangle\,\mathbf{3}\}. \tag{7.4.66}$$

To obtain a null vector representation of $\mathbf{a}$ from its affine representation (7.4.66), we need the method of indeterminate coefficients. Let $\mathbf{a} = \lambda \mathbf{e} + \mathbf{x}$, where λ is an indeterminate. By $\mathbf{a}^2 = 0$, we get $\lambda = -\mathbf{x}^2/(2\,\mathbf{e}\cdot\mathbf{x})$. In (7.4.66),

$$
\begin{aligned}
\mathbf{x} &= (\mathbf{e}\cdot\mathbf{3})\,\langle\mathbf{e}132\rangle\,\mathbf{2} + (\mathbf{e}\cdot\mathbf{2})\,\langle\mathbf{e}123\rangle\,\mathbf{3}, \\
\mathbf{x}^2 &= 2\,(\mathbf{e}\cdot\mathbf{2})(\mathbf{e}\cdot\mathbf{3})(\mathbf{2}\cdot\mathbf{3})\langle\mathbf{e}123\rangle\,\langle\mathbf{e}132\rangle, \\
\mathbf{e}\cdot\mathbf{x} &= (\mathbf{e}\cdot\mathbf{2})(\mathbf{e}\cdot\mathbf{3})(\langle\mathbf{e}123\rangle + \langle\mathbf{e}132\rangle) \\
&= 2\,(\mathbf{e}\cdot\mathbf{1})(\mathbf{e}\cdot\mathbf{2})(\mathbf{e}\cdot\mathbf{3})(\mathbf{2}\cdot\mathbf{3}),
\end{aligned}
\tag{7.4.67}
$$

then (7.4.64) follows.

We need to explain the geometric meaning of the angular brackets in (7.4.64). In the conformal model of $\mathbb{R}^2$, let $\mathbf{e}_0 = \mathbf{2}$, let $\mathbf{1} = \mathbf{f}(\mathbf{x})$ and $\mathbf{3} = \mathbf{f}(\mathbf{y})$, where $\mathbf{x}, \mathbf{y} \in \mathbb{R}^2$. Then

$$\langle\mathbf{e}123\rangle = \langle\mathbf{exe}_0\mathbf{y}\rangle = -\langle\mathbf{ee}_0\mathbf{xy}\rangle = -(\mathbf{e}\cdot\mathbf{e}_0)(\mathbf{x}\cdot\mathbf{y}) = \mathbf{x}\cdot\mathbf{y} = |\mathbf{x}||\mathbf{y}|\cos\angle(\mathbf{x},\mathbf{y}).$$

So

$$\langle\mathbf{e}123\rangle = d_{12}d_{13}\cos\angle(\overrightarrow{\mathbf{21}}, \overrightarrow{\mathbf{23}}), \tag{7.4.68}$$

where $\overrightarrow{\mathbf{23}}$ denotes the vector in $\mathbb{R}^2$ from point $\mathbf{2}$ to point $\mathbf{3}$.

As another example, we use expression (7.4.64) to prove the following theorem, which is an extension of the classical Simson's Theorem [201].

Example 7.53. [Simson's Triangle Theorem] Let $\mathbf{1}, \mathbf{2}, \mathbf{3}, \mathbf{4}$ be four points in the plane. Draw perpendiculars from $\mathbf{4}$ to the three sides $\mathbf{12}$, $\mathbf{23}$, $\mathbf{31}$ of triangle $\mathbf{123}$, and denote the three feet by $\mathbf{3}'$, $\mathbf{1}'$, $\mathbf{2}'$ respectively. Triangle $\mathbf{1}'\mathbf{2}'\mathbf{3}'$ is called a *Simson triangle*. Prove that in the conformal model,

$$[\mathbf{e1}'\mathbf{2}'\mathbf{3}'] = \frac{[\mathbf{1234}]}{2\rho^2}, \tag{7.4.69}$$

where ρ is the radius of the circumcircle of triangle $\mathbf{123}$.

As a special case of the above theorem, if $\mathbf{1}, \mathbf{2}, \mathbf{3}, \mathbf{4}$ are cocircular, then $[\mathbf{1234}] = 0$, and by (7.4.69), feet $\mathbf{1}', \mathbf{2}', \mathbf{3}'$ are collinear, which is just the classical Simson's theorem.

Proof. By (7.4.64), the three feet have the following affine representations in the homogeneous model:

$$
\begin{aligned}
\mathbf{e}\wedge\mathbf{1}' &= \mathbf{e}\wedge((\mathbf{e}\cdot\mathbf{3})\,\langle\mathbf{e}432\rangle\,\mathbf{2} + (\mathbf{e}\cdot\mathbf{2})\,\langle\mathbf{e}423\rangle)\,\mathbf{3}, \\
\mathbf{e}\wedge\mathbf{2}' &= \mathbf{e}\wedge((\mathbf{e}\cdot\mathbf{3})\,\langle\mathbf{e}431\rangle\,\mathbf{1} + (\mathbf{e}\cdot\mathbf{1})\,\langle\mathbf{e}413\rangle)\,\mathbf{3}, \\
\mathbf{e}\wedge\mathbf{3}' &= \mathbf{e}\wedge((\mathbf{e}\cdot\mathbf{1})\,\langle\mathbf{e}412\rangle\,\mathbf{2} + (\mathbf{e}\cdot\mathbf{2})\,\langle\mathbf{e}421\rangle)\,\mathbf{1}.
\end{aligned}
$$

So

$$
\begin{aligned}
&[\mathbf{e1}'\mathbf{2}'\mathbf{3}'] \\
&= [\mathbf{e}\,\{(\mathbf{e}\cdot\mathbf{3})\langle\mathbf{e}432\rangle\,\mathbf{2} + (\mathbf{e}\cdot\mathbf{2})\langle\mathbf{e}423\rangle\,\mathbf{3}\}\,\{(\mathbf{e}\cdot\mathbf{3})\,\langle\mathbf{e}431\rangle\,\mathbf{1} + (\mathbf{e}\cdot\mathbf{1})\,\langle\mathbf{e}413\rangle\,\mathbf{3}\} \\
&\qquad\qquad\qquad\qquad \{(\mathbf{e}\cdot\mathbf{1})\,\langle\mathbf{e}412\rangle\,\mathbf{2} + (\mathbf{e}\cdot\mathbf{2})\,\langle\mathbf{e}421\rangle\,\mathbf{1}\}] \\
&= (\mathbf{e}\cdot\mathbf{1})(\mathbf{e}\cdot\mathbf{2})(\mathbf{e}\cdot\mathbf{3})[\mathbf{e}123]\,\{\langle\mathbf{e}412\rangle\,\langle\mathbf{e}423\rangle\,\langle\mathbf{e}431\rangle + \langle\mathbf{e}421\rangle\,\langle\mathbf{e}432\rangle\,\langle\mathbf{e}413\rangle\}.
\end{aligned}
\tag{7.4.70}
$$

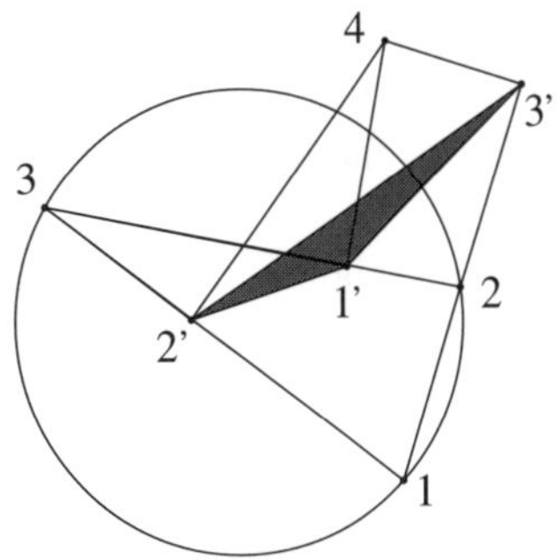
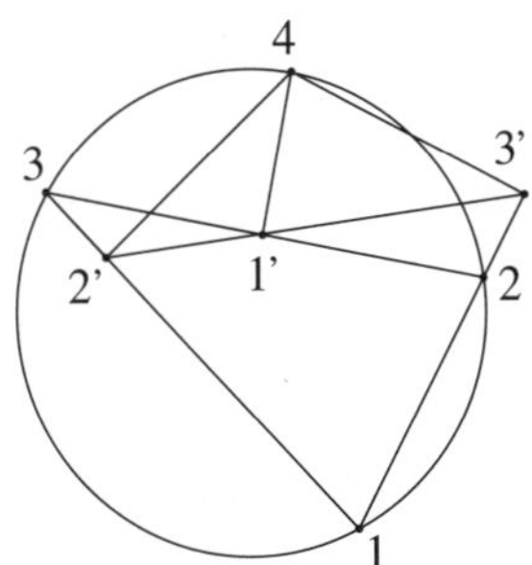

Fig. 7.11　Simson triangle (left); Simson's Theorem (right).

In the conformal model, $[\mathbf{e}\mathbf{1}'\mathbf{2}'\mathbf{3}']$ needs to be divided by $-(\mathbf{e}\cdot\mathbf{1}')(\mathbf{e}\cdot\mathbf{2}')(\mathbf{e}\cdot\mathbf{3}')$. By the last equality in (7.4.67),

$$\mathbf{e}\cdot\mathbf{1}' = 2\,(\mathbf{e}\cdot\mathbf{2})(\mathbf{e}\cdot\mathbf{3})(\mathbf{e}\cdot\mathbf{4})(\mathbf{2}\cdot\mathbf{3}),$$
$$\mathbf{e}\cdot\mathbf{2}' = 2\,(\mathbf{e}\cdot\mathbf{1})(\mathbf{e}\cdot\mathbf{3})(\mathbf{e}\cdot\mathbf{4})(\mathbf{1}\cdot\mathbf{3}),$$
$$\mathbf{e}\cdot\mathbf{3}' = 2\,(\mathbf{e}\cdot\mathbf{1})(\mathbf{e}\cdot\mathbf{2})(\mathbf{e}\cdot\mathbf{4})(\mathbf{1}\cdot\mathbf{2}),$$

so

$$\frac{[\mathbf{e}\mathbf{1}'\mathbf{2}'\mathbf{3}']}{-(\mathbf{e}\cdot\mathbf{1}')(\mathbf{e}\cdot\mathbf{2}')(\mathbf{e}\cdot\mathbf{3}')} = -\frac{[\mathbf{e}\mathbf{123}]\{\langle\mathbf{e412}\rangle\,\langle\mathbf{e423}\rangle\,\langle\mathbf{e431}\rangle + \langle\mathbf{e421}\rangle\,\langle\mathbf{e432}\rangle\,\langle\mathbf{e413}\rangle\}}{8\,(\mathbf{e}\cdot\mathbf{1})(\mathbf{e}\cdot\mathbf{2})(\mathbf{e}\cdot\mathbf{3})(\mathbf{e}\cdot\mathbf{4})^3(\mathbf{1}\cdot\mathbf{2})(\mathbf{1}\cdot\mathbf{3})(\mathbf{2}\cdot\mathbf{3})}.$$

By (7.4.5), when $[\mathbf{1234}]$ is divided by $(\mathbf{e}\cdot\mathbf{1})(\mathbf{e}\cdot\mathbf{2})(\mathbf{e}\cdot\mathbf{3})(\mathbf{e}\cdot\mathbf{4})$, the right side of (7.4.69) becomes

$$-\frac{[\mathbf{1234}][\mathbf{e123}]^2}{4(\mathbf{e}\cdot\mathbf{1})(\mathbf{e}\cdot\mathbf{2})(\mathbf{e}\cdot\mathbf{3})(\mathbf{e}\cdot\mathbf{4})(\mathbf{1}\cdot\mathbf{2})(\mathbf{2}\cdot\mathbf{3})(\mathbf{1}\cdot\mathbf{3})}.$$

So (7.4.69) is changed into the following homogeneous equality:

$$\langle\mathbf{e412}\rangle\,\langle\mathbf{e423}\rangle\,\langle\mathbf{e431}\rangle + \langle\mathbf{e421}\rangle\,\langle\mathbf{e432}\rangle\,\langle\mathbf{e413}\rangle = 2\,(\mathbf{e}\cdot\mathbf{4})^2[\mathbf{1234}][\mathbf{e123}]. \quad (7.4.71)$$

The proof of identity (7.4.71) is very easy. First, expand the left side by angular bracket expansion, and expand the right side by bracket Laplace expansion. Second, now that both sides of (7.4.71) become polynomials of the inner products of five vectors $\mathbf{e},\mathbf{1},\mathbf{2},\mathbf{3},\mathbf{4} \in \mathbb{R}^{3,1}$, the only syzygy relation among the inner products is the inner-product Laplace expansion (5.1.68), *i.e.*,

$$(\mathbf{e}\wedge\mathbf{1}\wedge\mathbf{2}\wedge\mathbf{3}\wedge\mathbf{4})^2 = \begin{vmatrix} 0 & \mathbf{e}\cdot\mathbf{1} & \mathbf{e}\cdot\mathbf{2} & \mathbf{e}\cdot\mathbf{3} & \mathbf{e}\cdot\mathbf{3} \\ \mathbf{e}\cdot\mathbf{1} & 0 & \mathbf{1}\cdot\mathbf{2} & \mathbf{1}\cdot\mathbf{3} & \mathbf{1}\cdot\mathbf{4} \\ \mathbf{e}\cdot\mathbf{2} & \mathbf{1}\cdot\mathbf{2} & 0 & \mathbf{2}\cdot\mathbf{3} & \mathbf{2}\cdot\mathbf{4} \\ \mathbf{e}\cdot\mathbf{3} & \mathbf{1}\cdot\mathbf{3} & \mathbf{2}\cdot\mathbf{3} & 0 & \mathbf{3}\cdot\mathbf{4} \\ \mathbf{e}\cdot\mathbf{4} & \mathbf{1}\cdot\mathbf{4} & \mathbf{2}\cdot\mathbf{4} & \mathbf{3}\cdot\mathbf{4} & 0 \end{vmatrix} = 0. \quad (7.4.72)$$

By polynomial division, (7.4.71) in expanded form is reduced to $0 = 0$ by (7.4.72).

However, for automated geometric reasoning, the task is not to verify an identity, but to deduce the right side of an identity from the left side, *without knowing a priori the right side.* In the current situation, the right side of (7.4.71) is a monomial of basic invariants in Clifford bracket algebra, while the left side is a binomial of advanced invariants. The aim of automated deduction is to factorize the left side in the framework of Clifford bracket algebra.

Below we do the factorization in the algebra of basic invariants: inner-product bracket algebra. Expanding the angular brackets on the left side of (7.4.71) by (5.4.3), we get

$$\langle \mathbf{e412}\rangle\,\langle \mathbf{e423}\rangle\,\langle \mathbf{e431}\rangle + \langle \mathbf{e421}\rangle\,\langle \mathbf{e432}\rangle\,\langle \mathbf{e413}\rangle$$

$$\begin{aligned}
= -2(\mathbf{e}\cdot\mathbf{4})\{&(\mathbf{e}\cdot\mathbf{1})^2(\mathbf{2}\cdot\mathbf{3})(\mathbf{3}\cdot\mathbf{4})(\mathbf{2}\cdot\mathbf{4}) + (\mathbf{e}\cdot\mathbf{2})^2(\mathbf{1}\cdot\mathbf{3})(\mathbf{3}\cdot\mathbf{4})(\mathbf{1}\cdot\mathbf{4})\\
&+(\mathbf{e}\cdot\mathbf{3})^2(\mathbf{1}\cdot\mathbf{2})(\mathbf{2}\cdot\mathbf{4})(\mathbf{1}\cdot\mathbf{4}) - (\mathbf{e}\cdot\mathbf{4})^2(\mathbf{1}\cdot\mathbf{2})(\mathbf{2}\cdot\mathbf{3})(\mathbf{1}\cdot\mathbf{3})\\
&+(\mathbf{e}\cdot\mathbf{1})(\mathbf{e}\cdot\mathbf{2})(\mathbf{3}\cdot\mathbf{4})((\mathbf{1}\cdot\mathbf{2})(\mathbf{3}\cdot\mathbf{4}) - (\mathbf{1}\cdot\mathbf{3})(\mathbf{2}\cdot\mathbf{4}) - (\mathbf{2}\cdot\mathbf{3})(\mathbf{1}\cdot\mathbf{4}))\\
&+(\mathbf{e}\cdot\mathbf{1})(\mathbf{e}\cdot\mathbf{3})(\mathbf{2}\cdot\mathbf{4})((\mathbf{1}\cdot\mathbf{3})(\mathbf{2}\cdot\mathbf{4}) - (\mathbf{1}\cdot\mathbf{2})(\mathbf{3}\cdot\mathbf{4}) - (\mathbf{2}\cdot\mathbf{3})(\mathbf{1}\cdot\mathbf{4}))\\
&+(\mathbf{e}\cdot\mathbf{2})(\mathbf{e}\cdot\mathbf{3})(\mathbf{1}\cdot\mathbf{4})((\mathbf{2}\cdot\mathbf{3})(\mathbf{1}\cdot\mathbf{4}) - (\mathbf{1}\cdot\mathbf{2})(\mathbf{3}\cdot\mathbf{4}) - (\mathbf{1}\cdot\mathbf{3})(\mathbf{2}\cdot\mathbf{4}))\}.
\end{aligned}$$

$$(7.4.73)$$

Denote the polynomial factor in (7.4.73) by λ. Dividing (7.4.72) by polynomial λ, we get that (7.4.72) is just

$$-2\lambda - (\mathbf{e}\cdot\mathbf{4})\,\mu = 0, \qquad (7.4.74)$$

where

$$\begin{aligned}
\mu = \ &2(\mathbf{e}\cdot\mathbf{4})(\mathbf{1}\cdot\mathbf{2})(\mathbf{1}\cdot\mathbf{3})(\mathbf{2}\cdot\mathbf{3})\\
&+(\mathbf{e}\cdot\mathbf{1})(\mathbf{2}\cdot\mathbf{3})((\mathbf{1}\cdot\mathbf{4})(\mathbf{2}\cdot\mathbf{3}) - (\mathbf{2}\cdot\mathbf{4})(\mathbf{1}\cdot\mathbf{3}) - (\mathbf{3}\cdot\mathbf{4})(\mathbf{1}\cdot\mathbf{2}))\\
&+(\mathbf{e}\cdot\mathbf{2})(\mathbf{1}\cdot\mathbf{3})((\mathbf{2}\cdot\mathbf{4})(\mathbf{1}\cdot\mathbf{3}) - (\mathbf{1}\cdot\mathbf{4})(\mathbf{2}\cdot\mathbf{3}) - (\mathbf{3}\cdot\mathbf{4})(\mathbf{1}\cdot\mathbf{2}))\\
&+(\mathbf{e}\cdot\mathbf{3})(\mathbf{1}\cdot\mathbf{2})((\mathbf{3}\cdot\mathbf{4})(\mathbf{1}\cdot\mathbf{2}) - (\mathbf{1}\cdot\mathbf{4})(\mathbf{2}\cdot\mathbf{3}) - (\mathbf{2}\cdot\mathbf{4})(\mathbf{1}\cdot\mathbf{3})).
\end{aligned}$$

Polynomial μ is linear in $\mathbf{e}, \mathbf{4}$ and quadratic in $\mathbf{1}, \mathbf{2}, \mathbf{3}$. The only syzygy relation involving inner products of vectors and having the same degrees as μ in its vector variables is the following bracket Laplace expansion:

$$[\mathbf{e123}][\mathbf{1234}] = -\begin{vmatrix} \mathbf{e}\cdot\mathbf{1} & 0 & \mathbf{1}\cdot\mathbf{2} & \mathbf{1}\cdot\mathbf{3}\\ \mathbf{e}\cdot\mathbf{2} & \mathbf{1}\cdot\mathbf{2} & 0 & \mathbf{2}\cdot\mathbf{3}\\ \mathbf{e}\cdot\mathbf{3} & \mathbf{1}\cdot\mathbf{3} & \mathbf{2}\cdot\mathbf{3} & 0\\ \mathbf{e}\cdot\mathbf{4} & \mathbf{1}\cdot\mathbf{4} & \mathbf{2}\cdot\mathbf{4} & \mathbf{3}\cdot\mathbf{4} \end{vmatrix} = \mu.$$

So $\lambda = (\mathbf{e}\cdot\mathbf{4})[\mathbf{e123}][\mathbf{1234}]$, and the right side of (7.4.71) is deduced. $\qquad\square$

Corollary 7.54. For any null vectors $\mathbf{0}, \mathbf{1}, \mathbf{2}, \mathbf{3}, \mathbf{4} \in \mathbb{R}^{3,1}$,

$$\langle \mathbf{0412}\rangle\,\langle \mathbf{0423}\rangle\,\langle \mathbf{0431}\rangle + \langle \mathbf{0421}\rangle\,\langle \mathbf{0432}\rangle\,\langle \mathbf{0413}\rangle = 2\,(\mathbf{0}\cdot\mathbf{4})^2[\mathbf{0123}][\mathbf{1234}]. \quad (7.4.75)$$

Remark: Example 7.53 shows that algebraic simplifications in inner-product bracket algebra based on basic syzygies of inner products of vectors and brackets are not easy, because a lot of terms are involved. The situation will be changed dramatically when the advanced invariant algebra of Euclidean geometry – *null bracket algebra,* is put to use [125].

7.5   The Lie model of oriented spheres and hyperplanes

A hyperplane has two orientations. Let $\mathbf{x}_1, \ldots, \mathbf{x}_n$ be n affinely independent points in $\mathbb{R}^n$. They generate an affine hyperplane in $\mathbb{R}^n$, and $\mathbf{J}_{n-1} = \partial(\mathbf{x}_1 \wedge \cdots \wedge \mathbf{x}_n)$ defined by (7.1.19) determines an orientation of the hyperplane. Alternatively, the vector $\mathbf{n} = \mathbf{J}_{n-1}^{\sim}$ is normal to the hyperplane, and satisfies $[\mathbf{n}\mathbf{J}_{n-1}] > 0$. It is called the left-handed *normal direction* of the oriented hyperplane, and can be used to indicate the same orientation.

More generally, if $\mathbf{A}_r \neq 0$, then in the moment-direction representation $(\mathbf{A}_r, \partial(\mathbf{A}_r))$ of a $(r-1)$D oriented affine plane, the left-handed *normal direction* of the direction $\partial(\mathbf{A}_r)$ with respect to the moment $\mathbf{A}_r$ is

$$\mathbf{n} = \partial(\mathbf{A}_r) \cdot \mathbf{A}_r^{-1}. \tag{7.5.1}$$

A sphere also has two orientations. Let $\mathbf{x}_1, \ldots, \mathbf{x}_{n+1}$ be $n+1$ affinely independent points in $\mathbb{R}^n$. Then $\mathbf{J}_n = \partial(\mathbf{x}_1 \wedge \cdots \wedge \mathbf{x}_{n+1})$ determines an orientation of the sphere. If $[\mathbf{J}_n] > 0$, the sphere is said to have *positive* orientation; otherwise, it is said to have *negative* orientation.

Let $\mathbf{x}$ be a point on the sphere, and $\mathbf{o}$ be the center of the sphere, then blade $\mathbf{J}_{n-1} = (\mathbf{x} - \mathbf{o}) \cdot \mathbf{J}_n$ determines an orientation of the hyperplane tangent to the sphere at $\mathbf{x}$, called the *induced orientation* of the tangent hyperplane. It satisfies $[(\mathbf{x} - \mathbf{o})\mathbf{J}_{n-1}] = \rho^2[\mathbf{J}_n]$, where ρ is the radius of the sphere.

The normal direction of the tangent hyperplane with the induced orientation is called the *normal direction* of the sphere at $\mathbf{x}$:

$$\mathbf{n} = [\mathbf{J}_n](\mathbf{x} - \mathbf{o}). \tag{7.5.2}$$

$\mathbf{n}$ is in the direction of $\mathbf{x} - \mathbf{o}$ if and only if the sphere has positive orientation. *The two normal directions: inward and outward ones, correspond to the two orientations of the sphere: negative and positive ones, respectively.*

In Section 7.2, we have seen that in the conformal model, a sphere or hyperplane can be represented by a positive vector $\mathbf{s} \in \mathbb{R}^{n+1,1}$, and two such vectors represent the same sphere or hyperplane if and only if they differ by a nonzero scale. We can use $\mathbf{s}$ and $-\mathbf{s}$ to represent the same sphere or hyperplane with opposite orientations, so that two positive vectors represent the same oriented sphere or hyperplane if and only if they differ by a positive scale. In symbolic computation, however, it is difficult to judge whether or not an expression is positive (or negative) definite, and such an inequality representation is awkward.

A better representation is provided by Lie (1872), where an oriented sphere or hyperplane is represented by a null vector, and two null vectors represent the same oriented sphere or hyperplane if and only if they differ by a nonzero scale.

Lie's construction can be described as follows. Minkowski space $\mathbb{R}^{n+1,1}$ is embedded into $\mathbb{R}^{n+1,2}$ as a hyperspace. Let $\mathbf{e}, \mathbf{e}_0, \mathbf{e}_1, \ldots, \mathbf{e}_n$ be a Witt basis of $\mathbb{R}^{n+1,1}$, where $(\mathbf{e}, \mathbf{e}_0)$ is a Witt pair. It extends to a Witt basis of $\mathbb{R}^{n+1,2}$ by attaching a unit negative vector $\mathbf{e}_-$. Null vectors in $\mathbb{R}^{n+1,1}$ are also null vectors in $\mathbb{R}^{n+1,2}$. They are the set

$$\mathcal{N}_0 = \{\mathbf{a} \in \mathbb{R}^{n+1,2} | \mathbf{a} \neq 0,\ \mathbf{a}^2 = 0,\ \mathbf{a} \cdot \mathbf{e}_- = 0\}. \tag{7.5.3}$$

A vector in $\mathcal{N}_0$ represents a point or the conformal point at infinity of $\mathbb{R}^n$.

In $\mathbb{R}^{n+1,1}$, a sphere or hyperplane is represented by a positive vector $\mathbf{s}$. The mapping $\mathbf{g}$: $\mathbb{R}^{n+1,1} \longrightarrow \mathbb{R}^{n+1,2}$ defined by

$$\mathbf{g}(\mathbf{s}) := \mathbf{s} + |\mathbf{s}|\mathbf{e}_-, \tag{7.5.4}$$

maps all such vectors into the following set:

$$\mathcal{N}_* = \{\mathbf{a} \in \mathbb{R}^{n+1,2} | \mathbf{a}^2 = 0,\ \mathbf{a} \cdot \mathbf{e}_- \neq 0\}. \tag{7.5.5}$$

The mapping $\mathbf{g}$ is injective. In particular, if $\mathbf{s} \in \mathbb{R}^{n+1,1}$ is not null, then $\pm\mathbf{s}$ are mapped to two null vectors that are different even modulo scale. The inverse map of $\mathbf{g}$ is $P_{\mathbf{e}_-}^{\perp}$. The preimage of a null vector in $\mathbb{R}^{n+1,2}$ under $\mathbf{g}$ is either a null vector or a positive vector in $\mathbb{R}^{n+1,1}$. $\mathbf{g}$ is called the *formalization map* of the Lie model.

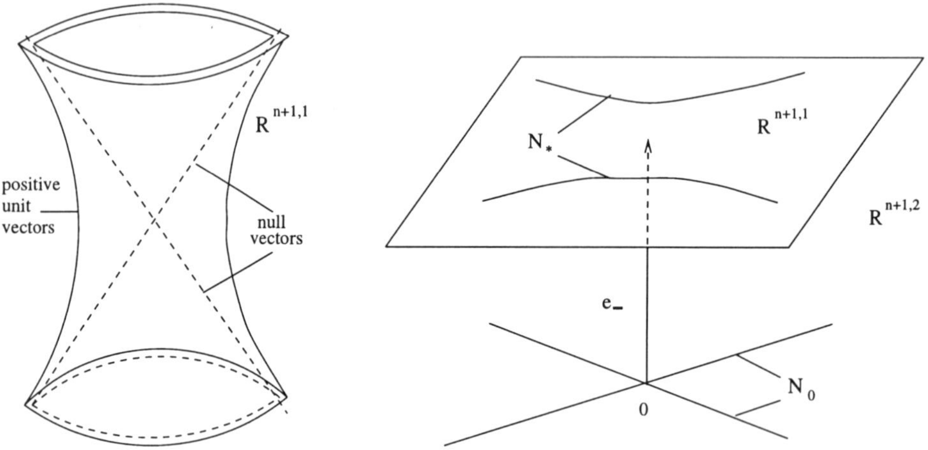

Fig. 7.12   Lie's construction.

Definition 7.55. A *Lie sphere* in $\mathbb{E}^n$ refers to one of the following geometric objects: (1) an oriented sphere, (2) an oriented hyperplane, (3) a point, (4) the conformal point at infinity. A pencil of Lie spheres is called a *Lie pencil*.

Let $\mathbf{E}_3$ be a Minkowski 3-blade in $\mathcal{CL}(\mathbb{R}^{n+1,2})$. Let $\mathbf{e}, \mathbf{e}_0, \mathbf{e}_-$ be a Witt basis of $\mathbf{E}_3$, where $(\mathbf{e}, \mathbf{e}_0)$ is a Witt pair. Then any null vector $\mathbf{s}$ in $\mathbb{R}^{n+1,2}$ represents a Lie sphere in the sense that a point represented by a null vector $\mathbf{a} \in \mathbb{R}^{n+1,2}$ is on the Lie sphere if and only if $\mathbf{a} \cdot \mathbf{s} = 0$. Two null vectors represent the same Lie sphere if and only if they differ by a nonzero scale. Null vector $\mathbf{s} \in \mathbb{R}^{n+1,2}$ represents in $\mathbb{E}^n$:

$$\begin{array}{ll}
\text{the conformal point at infinity,} & \text{if } \mathbf{s} \cdot \mathbf{e} = 0,\ \mathbf{s} \cdot \mathbf{e}_- = 0; \\
\text{a point,} & \text{if } \mathbf{s} \cdot \mathbf{e} \neq 0,\ \mathbf{s} \cdot \mathbf{e}_- = 0; \\
\text{an oriented hyperplane,} & \text{if } \mathbf{s} \cdot \mathbf{e} = 0,\ \mathbf{s} \cdot \mathbf{e}_- \neq 0; \\
\text{an oriented sphere,} & \text{if } \mathbf{s} \cdot \mathbf{e} \neq 0,\ \mathbf{s} \cdot \mathbf{e}_- \neq 0.
\end{array}$$

This algebraic representation of Lie spheres in $\mathbb{E}^n$ is called the *Lie model*.

Definition 7.56. Two oriented hyperplanes or spheres are said to be in *oriented contact*, or simply, in *contact*, if they are tangent to each other, and at the point of tangency they have the same normal direction.

A point or the conformal point at infinity is said to be in *contact* with a Lie sphere, if it is on the Lie sphere.

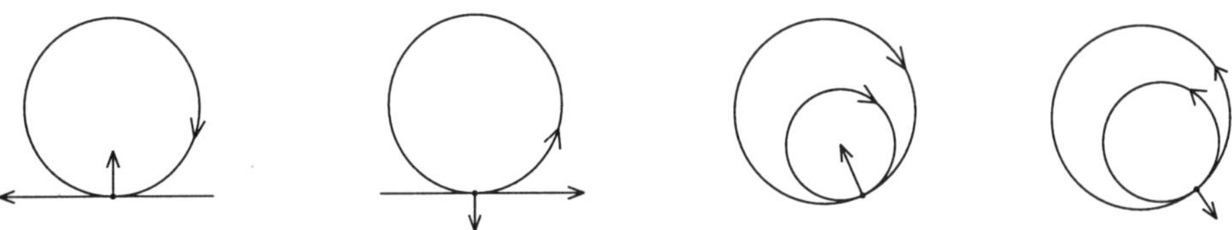

Fig. 7.13　Oriented contact.

7.5.1　*Inner product of Lie spheres*

The following are *standard representations* of Lie spheres in $\mathbb{E}^n$:

- The conformal point at infinity is represented by $\mathbf{e}$.
- Point $\mathbf{c}$ (null vector in $\mathcal{N}_{\mathbf{e}} \subset \mathbb{R}^{n+1,1}$) is still represented by $\mathbf{c}$.
- The hyperplane with unit normal $\mathbf{n} \in \mathbb{R}^n$ and distance δ from the origin along direction $\mathbf{n}$ has two orientations. The oriented hyperplane with normal $\mathbf{n}$ is represented by $\mathbf{n} + \delta\mathbf{e} + \mathbf{e}_-$; the oriented hyperplane with normal $-\mathbf{n}$ is represented by $\mathbf{n} + \delta\mathbf{e} - \mathbf{e}_- = -(-\mathbf{n} - \delta\mathbf{e} + \mathbf{e}_-)$.
- The sphere with center $\mathbf{c}$ (null vector in $\mathcal{N}_{\mathbf{e}}$) and radius $\rho > 0$ has two orientations. The sphere with outward orientation is represented by $\mathbf{c} - \rho^2\mathbf{e}/2 - \rho\mathbf{e}_-$; the sphere with inward orientation is represented by $\mathbf{c} - \rho^2\mathbf{e}/2 + \rho\mathbf{e}_-$.

The last item needs proof. Let $\mathbf{c} = \mathbf{e}_0$ be the origin, let $\rho = 1$, and let $\mathbf{n} \in \mathbb{R}^n$ be a unit vector. The unit sphere with outward orientation is in contact with the oriented hyperplane with unit normal $\mathbf{n}$ and signed distance $\delta = 1$ from the origin. By formula (7.5.5) below, the null vector representations of the sphere and the hyperplane have zero inner product if and only if they have opposite signs in their $\mathbf{e}_-$ components.

Let $\mathbf{s}_1, \mathbf{s}_2$ be null vectors in $\mathbb{R}^{n+1,2}$, and let $\epsilon, \epsilon_1, \epsilon_2 \in \{\pm 1\}$. The following formulas on the inner product $\mathbf{s}_1 \cdot \mathbf{s}_2$ can be easily verified.

(i) If $\mathbf{s}_1 = \mathbf{c}_1$, $\mathbf{s}_2 = \mathbf{c}_2$, where $\mathbf{c}_i \in \mathcal{N}_{\mathbf{e}}$, then

$$\mathbf{s}_1 \cdot \mathbf{s}_2 = -\frac{d^2_{\mathbf{c}_1\mathbf{c}_2}}{2}. \qquad (7.5.1)$$

(ii) If $s_1 = c$, $s_2 = n + \delta e + e_-$, then

$$s_1 \cdot s_2 = c \cdot n - \delta. \tag{7.5.2}$$

(iii) If $s_1 = c_1$, $s_2 = c_2 - \rho^2 e/2 + \epsilon \rho e_-$, then

$$s_1 \cdot s_2 = \frac{\rho^2 - d_{c_1 c_2}^2}{2}. \tag{7.5.3}$$

(iv) If for $i = 1, 2$, $s_i = n_i + \delta_i e + e_-$, then

$$s_1 \cdot s_2 = n_1 \cdot n_2 - 1. \tag{7.5.4}$$

(v) If $s_1 = n + \delta e + e_-$, $s_2 = c - \rho^2 e/2 + \epsilon \rho e_-$, then

$$s_1 \cdot s_2 = c \cdot n - \delta - \epsilon \rho. \tag{7.5.5}$$

(vi) If for $i = 1, 2$, $s_i = c_i - \rho_i^2 e/2 + \epsilon_i \rho_i e_-$, then

$$s_1 \cdot s_2 = \frac{(\rho_1 - \epsilon_1 \epsilon_2 \rho_2)^2 - d_{c_1 c_2}^2}{2}. \tag{7.5.6}$$

Definition 7.57. The *contact distance* between two Lie spheres represented by null vectors s_1, s_2 in the above standard form, is

$$|s_1 - s_2| = \sqrt{2|s_1 \cdot s_2|}. \tag{7.5.7}$$

As a direct corollary of the formulas (7.5.1) to (7.5.6), we have

Proposition 7.58. Two Lie spheres are in contact if and only if their contact distance is zero, or equivalently, if and only if the two null vectors representing them have zero inner product.

Below we compute the explicit form of the contact distance between Lie spheres s_1, s_2 in standard form, using formulas (7.5.1) to (7.5.6).

(1) When s_1, s_2 are both points, $|s_1 - s_2|$ equals the Euclidean distance between them.

(2) When one is a point and the other is a hyperplane, $|s_1 - s_2|$ equals $\sqrt{2}$ times the Euclidean distance between them.

(3) When one is a point and the other is a sphere, let $d_{\max}$ and $d_{\min}$ be respectively the maximal distance and minimal distance between the point and points on the sphere, then $|s_1 - s_2| = \sqrt{d_{\max} d_{\min}}$.

(4) When both are hyperplanes, $|s_1 - s_2|$ equals $2\sin(\theta/2)$, where θ is the angle between their normal directions.

(5) When $s_1 = n + \delta e + e_-$ and $s_2 = c - \rho^2 e/2 + \epsilon \rho e_-$, let $d = c \cdot n - \delta$. Then $|d|$ is the distance between the hyperplane and the center c of the sphere. Let $d_{\max}$ and $d_{\min}$ be respectively the maximal distance and minimal distance between the hyperplane and points on the sphere. There are three cases (Figure 7.14):

- When $d = 0$, then $|s_1 - s_2| = \sqrt{2\rho}$.
- When $\epsilon d < 0$, then $|s_1 - s_2| = \sqrt{2 d_{\max}}$.

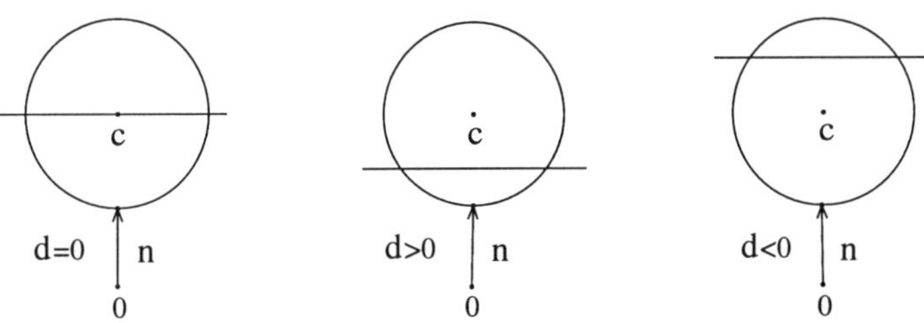

Fig. 7.14   Contact distance between a sphere and a hyperplane.

- When $\epsilon d > 0$, then $|\mathbf{s}_1 - \mathbf{s}_2| = \sqrt{2d_{\min}}$.

Definition 7.59. For two spheres $\mathbf{A}, \mathbf{B}$ in $\mathbb{E}^n$, if there is a hyperplane $\mathbf{C}$ that is tangent to both of them, then when spheres $\mathbf{A}, \mathbf{B}$ are on the same side of $\mathbf{C}$, the distance d^o between the two tangent points is invariant when $\mathbf{C}$ is replaced by any other common tangent hyperplane $\mathbf{C}'$ of spheres $\mathbf{A}, \mathbf{B}$ such that $\mathbf{A}, \mathbf{B}$ are on the same side of $\mathbf{C}'$. d^o is called the *outer tangential distance* between spheres $\mathbf{A}, \mathbf{B}$.

Similarly, if there is a common tangent hyperplane $\mathbf{C}$ of spheres $\mathbf{A}, \mathbf{B}$ such that $\mathbf{A}, \mathbf{B}$ are on different sides of $\mathbf{C}$, the distance d^i between the two tangent points is invariant when $\mathbf{C}$ is replaced by any other tangent hyperplane separating $\mathbf{A}, \mathbf{B}$. d^i is called the *inner tangential distance* between spheres $\mathbf{A}, \mathbf{B}$.

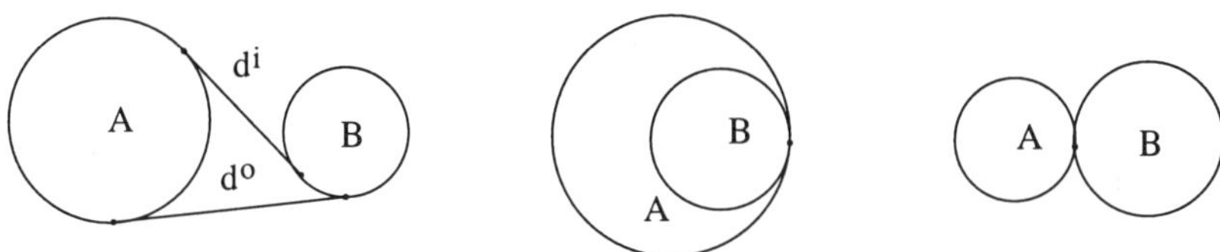

Fig. 7.15   Inner tangential distance d^i and outer tangential distance d^o (left); two inner tangent spheres (middle); two outer tangent spheres (right).

Definition 7.60. Two spheres are said to be *inner tangent* (or *outer tangent*), if their outer tangential distance (or inner tangential distance) exists and equals zero.

Definition 7.61. For two oriented spheres $\mathbf{A}, \mathbf{B}$ in $\mathbb{E}^n$, if they have a common tangent hyperplane $\mathbf{C}$ on which they induce the same orientation, then the *tangential distance* d^t between spheres $\mathbf{A}, \mathbf{B}$ refers to the distance between the two tangent points on hyperplane $\mathbf{C}$.

The following properties are easy to verify:

Proposition 7.62. If spheres $\mathbf{A}, \mathbf{B}$ have the same orientation, then $d^t = d^o$; if they have opposite orientations, then $d^t = d^i$.

(6) Assume that both s_1, s_2 are oriented spheres in standard form.

- If they have the same orientation and are not inclusive, then $|s_1 - s_2|$ equals the outer tangential distance between them. In particular, $|s_1 - s_2| = 0$ if and only if the two spheres are inner tangent to each other.

  The outer tangential distance exists if and only if $(e \cdot s_1)(e \cdot s_2)(s_1 \cdot s_2) \leq 0$.

- If they have different orientations and are exclusive from each other, then $|s_1 - s_2|$ equals the inner tangential distance between them. In particular, the two spheres are outer tangent to each other if and only if $|s_1 - s_2| = 0$.

Proposition 7.63. Let $s_i = c_i - \rho_i^2 e/2 + \epsilon_i \rho_i e_-$ for $i = 1, 2$. Then

$$d_{s_1 s_2}^t = \sqrt{-2 \, s_1 \cdot s_2}. \tag{7.5.8}$$

In particular, $d_{s_1 s_2}^t$ exists if and only if $s_1 \cdot s_2 \leq 0$.

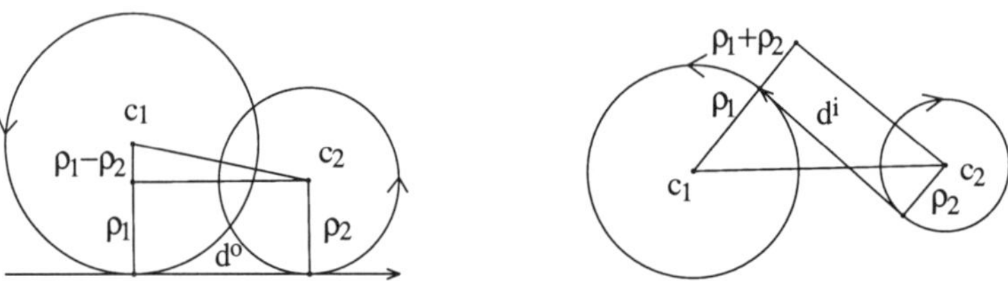

Fig. 7.16   Tangential distance between two oriented spheres.

Proof.   Denote $d = d_{c_1 c_2}$. If $\epsilon_1 \epsilon_2 = 1$, then $d_{s_1 s_2}^t = d_{s_1 s_2}^o$. By (7.5.6) and Figure 7.16 (left), $-2 \, s_1 \cdot s_2 = d^2 - (\rho_1 - \rho_2)^2 = (d_{s_1 s_2}^o)^2$. If $\epsilon_1 \epsilon_2 = -1$, the proof is similar. $\square$

A *contact transformation* is a diffeomorphism in the space of Lie spheres preserving the contact of Lie spheres. *Contact geometry*, also called *Lie sphere geometry*, studies the properties of the space of Lie spheres that are invariant under contact transformations. Any orthogonal transformation in $\mathbb{R}^{n+1,2}$ induces a contact transformation via the Lie model. The converse is also true, and is a classical result.

Historically, since Lie's seminal work on contact geometry (1872), the subject had been actively pursued through the early part of the twentieth century. After this, the subject fell out of favor until in the 1980s, when differential geometers used it to study submanifolds in Euclidean space [35].

Each contact transformation is the composition of a Laguerre transformation and an oriented conformal transformation. *Laguerre transformations* are contact transformations keeping the set of hyperplanes invariant. *Laguerre geometry* is the geometry of Laguerre transformations. An *oriented conformal transformation* is a conformal transformation that either preserves or reverses the orientations of all the oriented spheres and hyperplanes simultaneously. Oriented similarity transformations are the only kind of contact transformations that are both Laguerre and conformal.

In the next subsection, it is to be shown that a general contact transformation maps almost all points to spheres, and a contact transformation mapping almost all points to points is either an orientation-preserving or an orientation-reversing conformal transformation. In classical geometry, it is required that a transformation should map almost all points to points. From this aspect, Lie sphere geometry is not a classical geometry.

7.5.2 *Lie pencils, positive vectors and negative vectors**

Topologically, all Lie spheres in $\mathbb{E}^n$ form an $(n+1)D$ "torus" $\mathbb{RP}^n \times \mathbb{S}^1$, where $\mathbb{RP}^n$ denotes the nD real projective space. The proof is as follows.

When attaching the conformal point at infinity to $\mathbb{E}^n$, the result is an nD sphere $\mathbb{S}^n$. The set of Lie spheres in $\mathbb{E}^n$ is the same as the set of Lie spheres in $\mathbb{S}^n$, so we use the latter as the framework of Lie spheres. For any pair of antipodal points $\mathbf{x}, \mathbf{y}$ on sphere $\mathbb{S}^n$, there is a unique great sphere of $\mathbb{S}^n$ orthogonal to line $\mathbf{xy}$ in $\mathbb{R}^{n+1}$. Fix an orientation of the great sphere, and call it the *initial orientation.*

For any point $\mathbf{z}$ inside line segment $\mathbf{xy}$, when $\mathbf{z}$ moves from $\mathbf{x}$ to $\mathbf{y}$, there is a unique oriented sphere on $\mathbb{S}^n$ centering at $\mathbf{z}$ and whose orientation is the same as the initial orientation when $\mathbf{z}$ passes through the midpoint of $\mathbf{xy}$ (the origin of $\mathbb{R}^{n+1}$). At $\mathbf{z} = \mathbf{x}$ or $\mathbf{y}$, the sphere degenerates to point $\mathbf{x}$ or $\mathbf{y}$. After $\mathbf{z}$ arrives at $\mathbf{y}$, it goes in the reverse direction back to $\mathbf{x}$, and the orientation of the sphere centering at $\mathbf{z}$ is reversed during the return of $\mathbf{z}$ from $\mathbf{y}$ to $\mathbf{x}$. All Lie spheres in $\mathbb{S}^n$ are generated in this way, and for different pairs of antipodal points, the Lie spheres generated are different. So in the bundle of base $\mathbb{RP}^n$ and fibre $\mathbb{S}^1$, the fibres do not intersect. This finishes the proof.

Let $\mathbf{s}$ be a null vector in $\mathbb{R}^{n+1,2}$. Blade $\mathbf{s}^\sim$ has signature $(n, 1, 1)$. It represents the $(n+2)D$ pencil of Lie spheres that are in contact with Lie sphere $\mathbf{s}$, in the sense that a Lie sphere $\mathbf{t}$ is in contact with $\mathbf{s}$ if and only if $\mathbf{t} \wedge \mathbf{s}^\sim = 0$. Such a pencil is called an $(n+2)D$ *contact Lie pencil.* When null vector $\mathbf{s}$ varies in $\mathbb{R}^{n+1,2}$, the topology of $\mathbf{s}^\sim$ is the same.

Topologically, all Lie spheres in $\mathbf{s}^\sim$, where $\mathbf{s}$ is a null vector in $\mathbb{R}^{n+1,2}$, form a closed nD disk.

The proof is as follows. When $\mathbf{s}$ is an oriented sphere, at each point of the sphere there is a 1-parameter family of Lie spheres in contact with $\mathbf{s}$. The space of parameters is $\mathbb{S}^1$, as the contact Lie spheres include the point of contact taken as a sphere of radius zero, and the tangent hyperplane taken as a sphere of infinite radius. The space of all the contact Lie spheres of $\mathbf{s}$ is obtained from the torus $\mathbb{S}^{n-1} \times \mathbb{S}^1$ by first selecting one point from each fibre $\mathbb{S}^1$, and then pinching all such points into a single point. This is because sphere $\mathbf{s}$ is in contact with itself at every point on the sphere, so it is the common point of every fibre.

Cutting the space $\mathbb{S}^{n-1} \times \mathbb{S}^1$ by cutting each fibre at a point on the fibre, and then appending an nD disk to one of the two boundaries of the cutting result, we get a space homeomorphic to a closed nD disk. This finishes the proof.

Below we analyze the geometric meaning of positive vectors and negative vectors in $\mathbb{R}^{n+1,2}$ in terms of pencils of Lie spheres.

Let $\mathbf{s}$ be a positive vector. Then $\mathbf{s} = \mathbf{s}' + \epsilon\lambda\mathbf{e}_-$, where $\mathbf{s}' \in \mathbb{R}^{n+1,1}$ and $\lambda \geq 0$. Since $\mathbf{s}^2 = \mathbf{s}'^2 - \lambda^2 > 0$, $\mathbf{s}'$ is positive. When $\mathbf{s}' = \mathbf{n} + \delta\mathbf{e}$, then $0 \leq \lambda < 1$. When $\mathbf{s}' = \mathbf{c} - \rho^2\mathbf{e}/2$, we can set $\mathbf{s} = \mathbf{s}' + \epsilon\lambda\rho\mathbf{e}_-$, so that $0 \leq \lambda < 1$ is still valid. Set $\lambda = \cos\theta$, where $0 < \theta \leq \pi/2$. When $\theta = \pi/2$, then vector $\mathbf{s}$ loses its $\mathbf{e}_-$ part, *i.e.*, $\mathbf{s} \cdot \mathbf{e}_- = 0$.

The following is a list of inner products between a positive vector $\mathbf{s}$ and a null vector $\mathbf{t}$. In the list, the $\mathbf{n}_i$ are unit vectors in $\mathbb{R}^n$, and the $\mathbf{c}_i \in \mathcal{N}_\mathbf{e}$ are points in the conformal model.

(i) If $\mathbf{s} = \mathbf{n} + \delta\mathbf{e} + \cos\theta\mathbf{e}_-$, $\mathbf{t} = \mathbf{c}$, then

$$\mathbf{s} \cdot \mathbf{t} = \mathbf{c} \cdot \mathbf{n} - \delta. \tag{7.5.9}$$

So $\mathbf{s} \cdot \mathbf{t} = 0$ if and only if point $\mathbf{c}$ is on hyperplane $\mathbf{g}(P^\perp_{\mathbf{e}_-}(\mathbf{s}))$.

(ii) If $\mathbf{s} = \mathbf{n}_1 + \delta_1\mathbf{e} + \cos\theta\mathbf{e}_-$, $\mathbf{t} = \mathbf{n}_2 + \delta_2\mathbf{e} + \mathbf{e}_-$, then

$$\mathbf{s} \cdot \mathbf{t} = \mathbf{n}_1 \cdot \mathbf{n}_2 - \cos\theta. \tag{7.5.10}$$

So $\mathbf{s} \cdot \mathbf{t} = 0$ if and only if the two oriented hyperplanes $\mathbf{g}(P^\perp_{\mathbf{e}_-}(\mathbf{s}))$ and $\mathbf{t}$ intersect at angle θ, or equivalently, their normal directions form an angle θ.

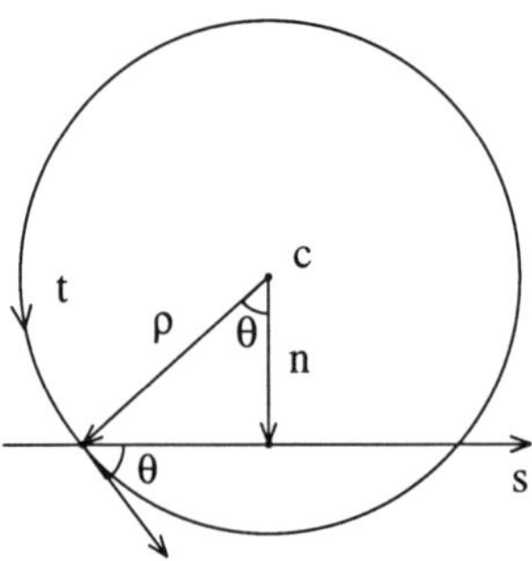

Fig. 7.17   Orthogonality of positive vector $\mathbf{s}$ (sphere) and null vector $\mathbf{t}$ (hyperplane).

(iii) If $\mathbf{s} = \mathbf{n} + \delta\mathbf{e} + \cos\theta\mathbf{e}_-$, $\mathbf{t} = \mathbf{c} - \rho^2\mathbf{e}/2 + \epsilon\rho\mathbf{e}_-$, then

$$\mathbf{s} \cdot \mathbf{t} = \mathbf{c} \cdot \mathbf{n} - \delta - \epsilon\rho\cos\theta. \tag{7.5.11}$$

So $\mathbf{s} \cdot \mathbf{t} = 0$ if and only if oriented sphere $\mathbf{g}(P^\perp_{\mathbf{e}_-}(\mathbf{s}))$ and oriented hyperplane $\mathbf{t}$ intersect at angle θ (Figure 7.17), or more explicitly, the angle formed by the normal directions of $\mathbf{g}(P^\perp_{\mathbf{e}_-}(\mathbf{s}))$ and $\mathbf{t}$ at any point of their intersection is θ.

(iv) If $\mathbf{s} = \mathbf{c}_1 - \rho^2\mathbf{e}/2 + \epsilon\rho\cos\theta\mathbf{e}_-$, $\mathbf{t} = \mathbf{c}_2$, then

$$\mathbf{s}\cdot\mathbf{t} = \frac{\rho^2 - d^2_{\mathbf{c}_1\mathbf{c}_2}}{2}. \tag{7.5.12}$$

So $\mathbf{s}\cdot\mathbf{t} = 0$ if and only if point $\mathbf{c}_2$ is on sphere $\mathbf{g}(P^{\perp}_{\mathbf{e}_-}(\mathbf{s}))$.

(v) If $\mathbf{s} = \mathbf{c} - \rho^2\mathbf{e}/2 + \epsilon\rho\cos\theta\mathbf{e}_-$, $\mathbf{t} = \mathbf{n} + \delta\mathbf{e} + \mathbf{e}_-$, the result of $\mathbf{s}\cdot\mathbf{t}$ is identical to (7.5.11).

(vi) If $\mathbf{s} = \mathbf{c}_1 - \rho_1^2\mathbf{e}/2 + \epsilon_1\rho_1\cos\theta\mathbf{e}_-$, $\mathbf{t} = \mathbf{c}_2 - \rho_2^2\mathbf{e}/2 + \epsilon_2\rho_2\mathbf{e}_-$, then

$$\mathbf{s}\cdot\mathbf{t} = \frac{\rho_1^2 + \rho_2^2 - d^2_{\mathbf{c}_1\mathbf{c}_2}}{2} - \epsilon_1\epsilon_2\rho_1\rho_2\cos\theta = \frac{(\rho_1 - \epsilon_1\epsilon_2\rho_2\cos\theta)^2 + (\rho_2\sin\theta)^2 - d^2_{\mathbf{c}_1\mathbf{c}_2}}{2}.$$
$$\tag{7.5.13}$$

So $\mathbf{s}\cdot\mathbf{t} = 0$ if and only if the two oriented spheres $\mathbf{g}(P^{\perp}_{\mathbf{e}_-}(\mathbf{s}))$ and $\mathbf{g}(P^{\perp}_{\mathbf{e}_-}(\mathbf{t}))$ intersect at angle θ (Figure 7.18): let $\mathbf{x}$ be a point at their intersection, then if $\epsilon_1\epsilon_2 > 0$, $\angle\mathbf{c}_1\mathbf{x}\mathbf{c}_2$ is acute; if $\epsilon_1\epsilon_2 < 0$, $\angle\mathbf{c}_1\mathbf{x}\mathbf{c}_2$ is obtuse.

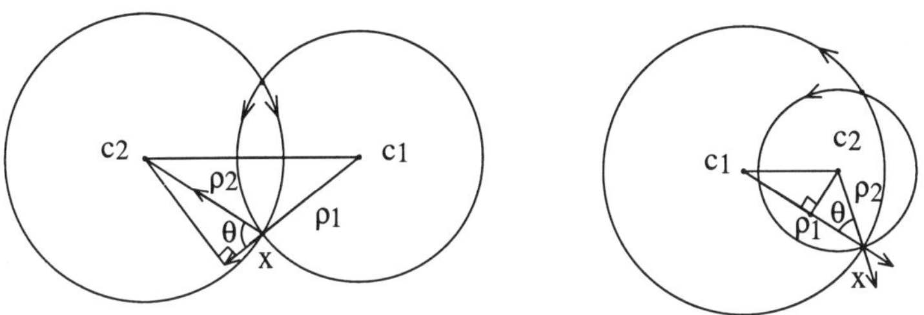

Fig. 7.18   Orthogonality of positive vector $\mathbf{s}$ (sphere) and null vector $\mathbf{t}$ (sphere).

From the above analysis, for a positive vector $\mathbf{s}$ in $\mathbb{R}^{n+1,2}$, its dual $\mathbf{s}^{\sim}$ represents the set of Lie spheres intersecting Lie sphere $\mathbf{g}(P^{\perp}_{\mathbf{e}_-}(\mathbf{s}))$ at a particular angle θ, where $0 < \theta \leq \pi/2$, together with all the points (possibly including the conformal point at infinity) on Lie sphere $\mathbf{g}(P^{\perp}_{\mathbf{e}_-}(\mathbf{s}))$. Such a pencil is called an $(n+2)$D *intersecting Lie pencil* of angle θ. When $\theta = \pi/2$, the pencil is also called an $(n+2)$D *orthogonal Lie pencil*, and is composed of the spheres and hyperplanes perpendicular to $\mathbf{s}$ and with both orientations added.

Topologically, all Lie spheres in $\mathbf{s}^{\sim}$, *where* $\mathbf{s}$ *is a positive vector in* $\mathbb{R}^{n+1,2}$, *form a "cylinder"* $\mathbb{RP}^{n-1} \times [0,1]$.

The proof is as follows. When $\mathbf{s} = \mathbf{c} - \rho^2\mathbf{e}/2 + \epsilon\rho\cos\theta\mathbf{e}_-$, then for every pair of antipodal points $\mathbf{x}, \mathbf{y}$ on sphere $\mathbf{c} - \rho^2\mathbf{e}/2$, for every point $\mathbf{z}$ inside line segment $\mathbf{x}\mathbf{y}$, there is a unique $(n-2)$D sphere $\mathbf{A}$ on sphere $\mathbf{c} - \rho^2\mathbf{e}/2$, centering at $\mathbf{z}$ and perpendicular to line $\mathbf{x}\mathbf{y}$. There is a unique oriented $(n-1)$D sphere $\mathbf{s}'$ intersecting sphere $\mathbf{c} - \rho^2\mathbf{e}/2$ at $\mathbf{A}$ with angle θ. The center of sphere $\mathbf{s}'$ always lies on line $\mathbf{x}\mathbf{y}$.

When $\mathbf{z} = \mathbf{x}$ or $\mathbf{y}$, sphere $\mathbf{s}'$ degenerates to point $\mathbf{x}$ or $\mathbf{y}$. At some point $\mathbf{z} = \mathbf{u}$ inside line segment $\mathbf{x}\mathbf{y}$, the center of sphere $\mathbf{s}'$ is at infinity, and the sphere becomes a hyperplane. Since $\mathbf{s} \wedge \mathbf{s}^{\sim} \neq 0$, sphere $\mathbf{s}'$ cannot be identical to sphere $\mathbf{s}$, so the $[0,1]$ fibres in $\mathbb{RP}^{n-1} \times [0,1]$ do not intersect. This finishes the proof.

Negative vectors are much more complicated in classification. First of all, $\mathbf{e}_-^{\sim}$ represents all points in $\mathbb{E}^n$ together with the conformal point at infinity, which form an nD sphere topologically. $\mathbf{e}_-^{\sim}$ is called the $(n+2)$D *punctual Lie pencil*.

For all other negative vectors $\mathbf{s}$, Lie pencil $\mathbf{s}^{\sim}$ has the same topology as $\mathbf{e}_-^{\sim}$. *Topologically, all Lie spheres in* $\mathbf{s}^{\sim}$, *where* $\mathbf{s}$ *is a negative vector in* $\mathbb{R}^{n+1,2}$, *form an* nD *sphere.*

For any negative vector $\mathbf{s}$ linearly independent of $\mathbf{e}_-$, the signature of vector $P_{\mathbf{e}_-}^{\perp}(\mathbf{s})$ is either (1) positive, or (2) null, or (3) negative, and correspondingly in the conformal model, $P_{\mathbf{e}_-}^{\perp}(\mathbf{s})$ represents either (1) a sphere or hyperplane, or (2) a point or the conformal point at infinity, or (3) all great spheres of a sphere.

The simplest case is $\mathbf{s} = \mathbf{e} + \lambda\mathbf{e}_-$, where $\lambda \neq 0$. For a null vector $\mathbf{t}$, $\mathbf{s} \cdot \mathbf{t} = 0$ if and only if $\mathbf{t}$ is either $\mathbf{e}$ up to scale, or an oriented sphere $\mathbf{c} - \rho^2\mathbf{e}/2 + \epsilon\rho\mathbf{e}_-$ such that $\epsilon\rho = -\lambda^{-1}$. Lie pencil $\mathbf{s}^{\sim}$ contains $\mathbf{e}$ and all the spheres in $\mathbb{E}^n$ having the same radius $|\lambda^{-1}|$ and the same orientation, which is positive if $\lambda < 0$, and negative if $\lambda > 0$. Such a pencil is called an $(n+2)$D *spherical Lie pencil*.

The next simplest case is $\mathbf{s} = \mathbf{c} + \lambda\mathbf{e}_-$, where $\mathbf{c}$ is a point (null vector) and $\lambda \neq 0$. Obviously $\mathbf{c} \in \mathbf{s}^{\sim}$. By

$$(\mathbf{c} + \lambda\mathbf{e}_-) \cdot (\mathbf{n} + \delta\mathbf{e} + \mathbf{e}_-) = \mathbf{c} \cdot \mathbf{n} - \delta - \lambda, \tag{7.5.14}$$

we get that a hyperplane is in $\mathbf{s}^{\sim}$ if and only if the signed distance from the hyperplane to point $\mathbf{c}$ is λ. By

$$(\mathbf{c} + \lambda\mathbf{e}_-) \cdot (\mathbf{c}' - \frac{\rho^2}{2}\mathbf{e} + \epsilon\rho\mathbf{e}_-) = \frac{(\rho - \epsilon\lambda)^2 - d_{\mathbf{cc}'}^2 - \lambda^2}{2}, \tag{7.5.15}$$

which equals zero if and only if $\rho = \epsilon\lambda + \sqrt{d_{\mathbf{cc}'}^2 + \lambda^2}$, we get that when $\mathbf{c} \neq \mathbf{c}'$, for any point $\mathbf{c}'$ there are two spheres centering at $\mathbf{c}'$ and belonging to $\mathbf{s}^{\sim}$, and point $\mathbf{c}$ is in the region sandwiched by the two spheres. Among such spheres, all the spheres surrounding point $\mathbf{c}$ have the same orientation, and all the spheres not surrounding $\mathbf{c}$ have the same opposite orientation. When $\mathbf{c} = \mathbf{c}'$, one sphere degenerates to point $\mathbf{c}$, the other has radius $2|\lambda|$, the smallest among all spheres in $\mathbf{s}^{\sim}$ surrounding $\mathbf{c}$ properly. Such a pencil is called an $(n+2)$D *sandwiching Lie pencil* (Figure 7.19).

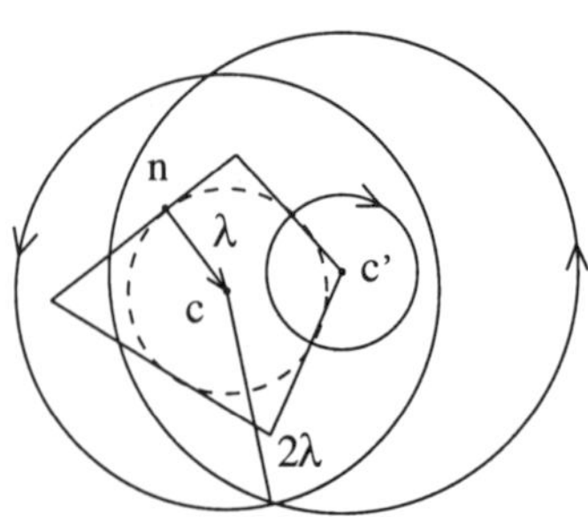

Fig. 7.19 Sandwiching Lie pencil.

The third case is $\mathbf{s} = \mathbf{n} + \delta\mathbf{e} + \lambda\mathbf{e}_-$, where $\mathbf{n}$ is a unit vector in $\mathbb{R}^n$, and $\lambda \neq 0$. Then $\mathbf{s}^\sim$ contains $\mathbf{e}$ and all the points on hyperplane $\mathbf{n} + \delta\mathbf{e}$. Since $\mathbf{s}^2 = 1 - \lambda^2 < 0$, we may assume $\lambda = \cosh t$, where $t > 0$. By

$$(\mathbf{n} + \delta\mathbf{e} + \cosh t\,\mathbf{e}_-) \cdot (\mathbf{n}' + \delta'\mathbf{e} + \mathbf{e}_-) = \mathbf{n} \cdot \mathbf{n}' - \cosh t, \qquad (7.5.16)$$

we get that $\mathbf{s}^\sim$ does not contain any hyperplane. By

$$(\mathbf{n} + \delta\mathbf{e} + \cosh t\,\mathbf{e}_-) \cdot (\mathbf{c} - \frac{\rho^2}{2}\mathbf{e} + \epsilon\rho\mathbf{e}_-) = \mathbf{n} \cdot \mathbf{c} - \delta - \epsilon\rho\cosh t, \qquad (7.5.17)$$

we get that $\mathbf{s}^\sim$ contains all the oriented spheres $\mathbf{c} - \rho^2\mathbf{e}/2 + \epsilon\rho\mathbf{e}_-$ separated from hyperplane $\mathbf{n} + \delta\mathbf{e}$ in such a way that the signed distance $d_\mathbf{c}$ from the hyperplane to the center $\mathbf{c}$ of the sphere satisfies $d_\mathbf{c} = \epsilon\rho\cosh t$.

As shown in Figure 7.20(a), each sphere in $\mathbf{s}^\sim$ has positive orientation if $d_\mathbf{c} < 0$, has negative orientation if $d_\mathbf{c} > 0$, and has distance $d > 0$ from the hyperplane, where

$$\frac{d}{\rho} = 2\sinh^2 \frac{t}{2}. \qquad (7.5.18)$$

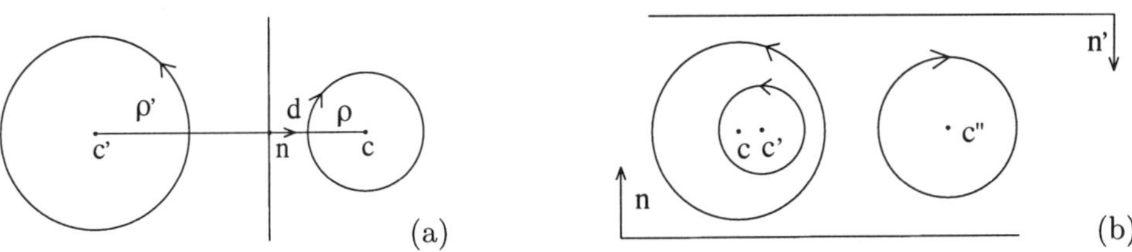

Fig. 7.20   Separating Lie pencil.

The fourth case is $\mathbf{s} = \mathbf{c} - \rho^2\mathbf{e}/2 + \epsilon\rho\cosh t\mathbf{e}_-$. Lie pencil $\mathbf{s}^\sim$ contains all the points on sphere $\mathbf{c} - \rho^2\mathbf{e}/2$. By

$$(\mathbf{c} - \frac{\rho^2}{2}\mathbf{e} + \epsilon\rho\cosh t\mathbf{e}_-) \cdot (\mathbf{n} + \delta\mathbf{e} + \mathbf{e}_-) = \mathbf{c} \cdot \mathbf{n} - \delta - \epsilon\rho\cosh t, \qquad (7.5.19)$$

we get that $\mathbf{s}^\sim$ contains all the oriented hyperplanes separated from sphere $\mathbf{c} - \rho^2\mathbf{e}/2$ in such a way that they satisfy (7.5.18) and $\epsilon d_\mathbf{c} > 0$, where $d_\mathbf{c}$ is the signed distance from the hyperplane to the center $\mathbf{c}$ of the sphere.

Definition 7.64. Let $\mathbf{A}, \mathbf{B}$ be two oriented spheres in $\mathbb{E}^n$. For any point $\mathbf{x}$ (or $\mathbf{y}$) on sphere $\mathbf{A}$ (or $\mathbf{B}$), denote by $\mathbf{n_x}$ (or $\mathbf{n_y}$) the radial direction of the oriented sphere at the point. Then the *maximal* (or *minimal*) *oriented distance* between $\mathbf{A}, \mathbf{B}$, denoted by $d_{\max}$ (or $d_{\min}$), refers to the maximum (or minimum) of the following set of nonnegative values:

$$\{d_{\mathbf{xy}} \mid \mathbf{x} \in \mathbf{A},\ \mathbf{y} \in \mathbf{B},\ \mathbf{n_x} = \mathbf{n_y}\}.$$

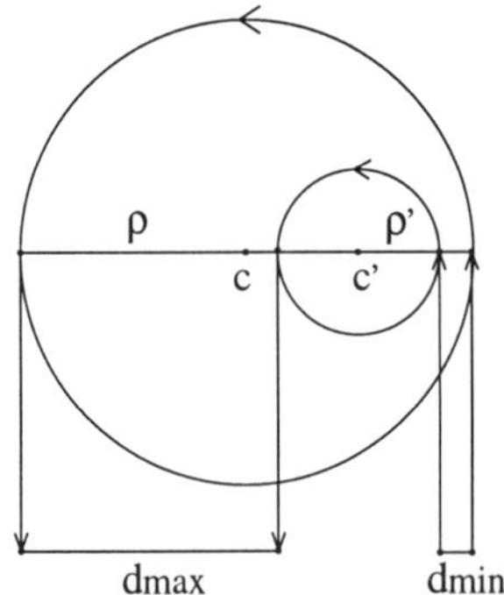 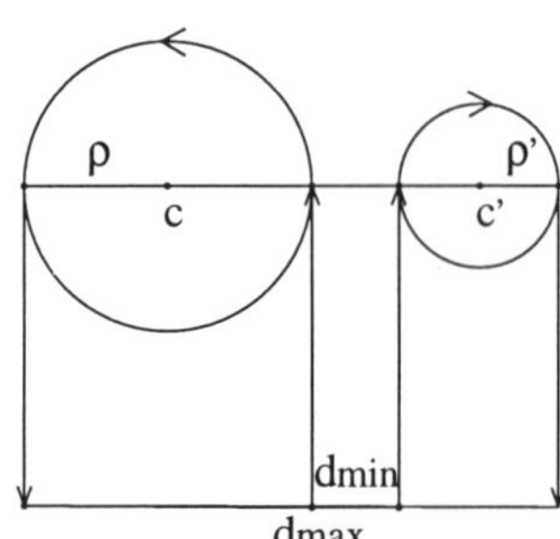

Fig. 7.21   Maximal and minimal oriented distances between two oriented spheres.

Proposition 7.65. If spheres $(\mathbf{c}, \rho)$ and $(\mathbf{c}', \rho')$ are inclusive and have the same orientation, then

$$d_{\max} = |\rho - \rho'| + d_{\mathbf{cc}'}, \quad d_{\min} = |\rho - \rho'| - d_{\mathbf{cc}'}. \tag{7.5.20}$$

If the two spheres are exclusive and have opposite orientations, then

$$d_{\max} = d_{\mathbf{cc}'} + \rho + \rho', \quad d_{\min} = d_{\mathbf{cc}'} - \rho - \rho'. \tag{7.5.21}$$

Proof.   Direct from Figure 7.21. $\square$

For $\mathbf{s} = \mathbf{c} - \rho^2 \mathbf{e}/2 + \epsilon \rho \cosh t \mathbf{e}_-$, by

$$\mathbf{s} \cdot \left(\mathbf{c}' - \frac{\rho'^2}{2}\mathbf{e} + \epsilon'\rho'\mathbf{e}_-\right) = \frac{(\rho - \epsilon\epsilon'\rho'\cosh t)^2 - \rho'^2 \sinh^2 t - d_{\mathbf{cc}'}^2}{2}, \tag{7.5.22}$$

we get that (7.5.22) equals zero if and only if

$$\frac{(\rho - \epsilon\epsilon'\rho')^2 - d_{\mathbf{cc}'}^2}{2\epsilon\epsilon'\rho\rho'} = 2\sinh^2\frac{t}{2}. \tag{7.5.23}$$

As shown in Figure 7.20(b), $\mathbf{s}^{\sim}$ contains all the spheres $(\mathbf{c}', \rho')$ separated from sphere $(\mathbf{c}, \rho)$ in such a way that the following conditions are satisfied:

- if the two spheres $(\mathbf{c}, \rho)$ and $(\mathbf{c}', \rho')$ are inclusive, then they have the same orientation;
- if the two spheres are exclusive, then they have different orientations;
- the maximal and minimal oriented distances between the two spheres satisfy

$$\frac{d_{\max}d_{\min}}{2\rho\rho'} = 2\sinh^2\frac{t}{2}. \tag{7.5.24}$$

The third and fourth cases show that for a positive vector $\mathbf{s}' \in \mathbb{R}^{n+1,1}$, if $\mathbf{s} = \mathbf{s}' + \lambda\mathbf{e}_-$ is negative for some $\lambda \in \mathbb{R}$, then $\mathbf{s}^{\sim}$ is composed of all the points (possibly including the conformal point at infinity) on the sphere or hyperplane $\mathbf{s}'$, together with all the oriented spheres or hyperplanes separated from $\mathbf{s}'$ so as to have a fixed ratio $2\sinh^2(t/2)$ of either distance to ratio or distances to radii, in the sense of either (7.5.18) or (7.5.24). Such a pencil is called an $(n+2)$D *separating Lie pencil* of *separating ratio* $2\sinh^2(t/2)$.

The last case is $\mathbf{s} = \mathbf{c} + \rho^2\mathbf{e}/2 + \lambda\rho\mathbf{e}_-$, where λ is arbitrary. Lie pencil $\mathbf{s}^\sim$ contains no points, nor the conformal point at infinity. Let $\lambda = \epsilon\tan\theta$, where $0 \le \theta < \pi/2$. By

$$\left(\mathbf{c} + \frac{\rho^2}{2}\mathbf{e} + \epsilon\rho\tan\theta\,\mathbf{e}_-\right) \cdot (\mathbf{n} + \delta\mathbf{e} + \mathbf{e}_-) = \mathbf{c}\cdot\mathbf{n} - \delta - \epsilon\rho\tan\theta, \qquad (7.5.25)$$

we get that $\mathbf{s}^\sim$ contains all the oriented hyperplanes whose signed distance d to point $\mathbf{c}$ satisfies $\epsilon d/\rho = \tan\theta$. As shown in Figure 7.22(a), if $\mathbf{x}$ is the foot drawn from point $\mathbf{c}$ to hyperplane $\mathbf{n} + \delta\mathbf{e}$, and $\mathbf{y}$ is any point on sphere $\mathbf{c} - \rho^2\mathbf{e}/2$ satisfying $\mathbf{cy} \perp \mathbf{cx}$, then $\theta = \angle\mathbf{xyc}$, and d has the same sign as ϵ.

By

$$\left(\mathbf{c} + \frac{\rho^2}{2}\mathbf{e} + \epsilon\rho\tan\theta\,\mathbf{e}_-\right) \cdot \left(\mathbf{c}' - \frac{\rho'^2}{2}\mathbf{e} + \epsilon'\rho'\mathbf{e}_-\right) = \frac{(\rho' - \epsilon\epsilon'\rho\tan\theta)^2 - \rho^2\sec^2\theta - d_{\mathbf{cc}'}^2}{2},$$
$$(7.5.26)$$

which equals zero if and only if $\rho' = \sqrt{\rho^2\sec^2\theta + d_{\mathbf{cc}'}^2} + \epsilon\epsilon'\rho\tan\theta$, we get that when $\mathbf{c} \ne \mathbf{c}'$,

- if $\theta = 0$, by setting $\rho' = \sqrt{\rho^2 + d_{\mathbf{cc}'}^2}$, then the two identical spheres $(\mathbf{c}', \rho')$ but with opposite orientations are both in $\mathbf{s}^\sim$;
- if $\theta \ne 0$, then there are two spheres in $\mathbf{s}^\sim$ centering at $\mathbf{c}'$, the bigger one has the same positive or negative orientation as the sign of ϵ, while the smaller one has the opposite orientation, see Figure 7.22(a).

When $\mathbf{c} = \mathbf{c}'$, then $\rho' = \rho(\sec\theta + \epsilon\epsilon'\tan\theta)$,

- if $\theta = 0$, then the two identical spheres $(\mathbf{c}, \rho)$ but with opposite orientations are both in $\mathbf{s}^\sim$;
- if $\theta \ne 0$, then there are two spheres in $\mathbf{s}^\sim$ that are concentric with and sandwich sphere $\mathbf{c} - \rho^2\mathbf{e}/2$, see Figure 7.22(b).

No matter whether $\mathbf{c} = \mathbf{c}'$ or not, when θ approximates $90°$, the radius of the smaller sphere tends to zero, while the radius of the bigger sphere tends to infinity. Pencil $\mathbf{c} + \rho^2\mathbf{e}/2 + \epsilon\rho\tan\theta\,\mathbf{e}_-$ is called an $(n+2)$D *stretching Lie pencil* of *stretching angle* θ.

Summing up, when $\mathbf{s}$ is negative, pencil $\mathbf{s}^\sim$ is $(n+2)$D algebraically, but nD geometrically, and has five different types:

(1) punctual Lie pencil $\mathbf{e}_-^\sim$;
(2) spherical Lie pencil $(\mathbf{e} + \lambda\mathbf{e}_-)^\sim$;
(3) sandwiching Lie pencil $(\mathbf{c} + \lambda\mathbf{e}_-)^\sim$;
(4) separating Lie pencils $(\mathbf{n} + \delta\mathbf{e} + \cosh t\,\mathbf{e}_-)^\sim$ and $(\mathbf{c} - \rho^2\mathbf{e}/2 + \epsilon\rho\cosh t\,\mathbf{e}_-)^\sim$;
(5) stretching Lie pencil $(\mathbf{c} + \rho^2\mathbf{e}/2 + \epsilon\rho\tan\theta\,\mathbf{e}_-)^\sim$.

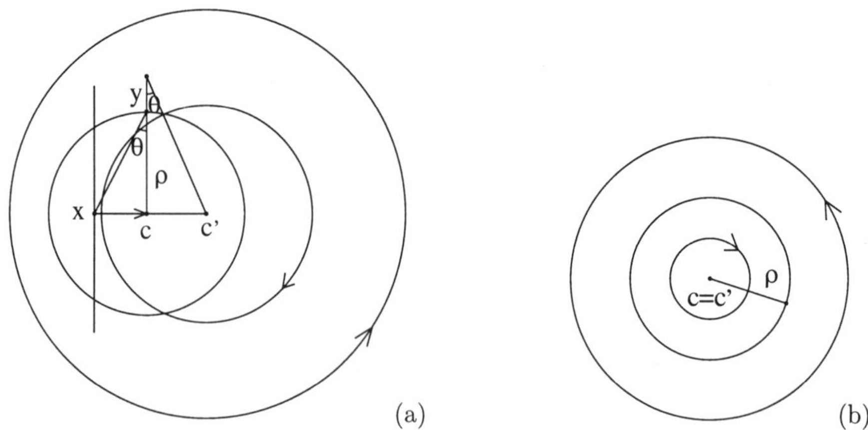

Fig. 7.22   Stretching Lie pencil: (a) $\mathbf{c} \neq \mathbf{c}'$; (b): $\mathbf{c} = \mathbf{c}'$.

Any negative unit vector in $\mathbb{R}^{n+1,2}$ can be obtained from vector $\mathbf{e}_-$ by an orthogonal transformation. Since each contact transformation is induced by an orthogonal transformation of $\mathbb{R}^{n+1,2}$, the images of $\mathbf{e}_-$ under orthogonal transformations can be used to classify contact transformations. Under an orthogonal transformation T of $\mathbb{R}^{n+1,2}$:

(i) If 1D subspace $\mathbf{e}_-$ is invariant, so is the vector space $\mathbb{R}^{n+1,1} = \mathbf{e}_-^{\sim}$, then T induces a conformal transformation in $\mathbb{E}^n$.

(ii) If vector $\mathbf{e}_-$ is changed into $-\mathbf{e}_-$, then T induces the *pure orientation reversing transformation* in $\mathbb{E}^n$: it fixes every point, sphere and hyperplane in $\mathbb{E}^n$; it only reverses the orientation of every sphere and hyperplane.

(iii) If $\mathbf{e}_-$ and $\lambda^{-1}\mathbf{e} + \mathbf{e}_-$ are interchanged, while all the vectors in $(\mathbf{e}_- \wedge \mathbf{e})^{\sim}$ are fixed, then since vector $\mathbf{e}$ is only rescaled, all points in $\mathbb{E}^n$ are changed into spheres having the same radius and orientation with each other.

(iv) If $\mathbf{e}_-$ and $\lambda^{-1}\mathbf{c} + \mathbf{e}_-$ are interchanged, where $\mathbf{c} \in \mathcal{N}_{\mathbf{e}}$, then all points but $\mathbf{c}$ in $\mathbb{E}^n$ are changed into spheres and hyperplanes.

(v) If $\mathbf{e}_-$ and $\mathbf{n} + \delta\mathbf{e} + \cosh t\, \mathbf{e}_-$ are interchanged (up to scale), only points on hyperplane $\mathbf{n} + \delta\mathbf{e}$ remain to be points.

(vi) If $\mathbf{e}_-$ and $\mathbf{c} + \rho^2\mathbf{e}/2 + \epsilon\rho\tan\theta\,\mathbf{e}_-$ are interchanged (up to scale), every point in $\mathbb{E}^n$ is changed into a sphere or hyperplane.

From the above classification, we see that the classical geometric part of contact geometry is oriented conformal geometry: the set of contact transformations changing almost all points in $\mathbb{E}^n$ to points in $\mathbb{E}^n$ is the set of oriented conformal transformations. Contact geometry is not a classical geometry.

7.6 Apollonian contact problem

The Lie model is very efficient in handling contact problems in oriented Euclidean geometry. The classical *Apollonian contact problem* is to construct a circle tangent to three given objects, each of which may be a line, a point, or a circle. In the Lie model, the finding and classification of the solutions to this problem and its generalization to $\mathbb{E}^n$ are very easy.

7.6.1 *1D contact problem*

1D contact geometry is intuitive and almost classical, because its lines and spheres are each composed of one or two points. In $\mathbb{R}^1$:

- A 0D oriented line is a triplet $(\lambda, \mathbf{e}, \epsilon)$, where $\lambda \in \mathbb{R}$ denotes a point on the line, $\mathbf{e}$ is the conformal point at infinity, and $\epsilon \in \{\pm 1\}$ indicates the positive or negative direction (orientation).

  For example, $(\lambda, \mathbf{e}, 1)$ is a 0D line composed of point λ, the conformal point at infinity, and the positive direction.

- A 0D oriented sphere is an ordered pair of distinct points (λ_1, λ_2), whose orientation is the sign of $\lambda_2 - \lambda_1$, where $\lambda_i \in \mathbb{R}$.

  For example, $(\lambda, \lambda+1)$ is a positive sphere, while $(\lambda, \lambda-1)$ is a negative sphere.

In the Lie model $\mathbb{R}^{2,2}$ of $\mathbb{E}^1$, let $\mathbf{e}_1$ be the unit vector of $\mathbb{R}^1 = (\mathbf{e} \wedge \mathbf{e}_0 \wedge \mathbf{e}_-)^\sim$ in the positive direction. Then line $(\lambda, \mathbf{e}, \epsilon)$ is represented by null vector

$$\mathbf{t} = \mathbf{e}_1 + \lambda \mathbf{e} + \epsilon \mathbf{e}_-. \tag{7.6.1}$$

Sphere (λ_1, λ_2) has center $(\lambda_1 + \lambda_2)/2$ and radius $|\lambda_1 - \lambda_2|/2$, and is represented by

$$\mathbf{c} = \mathbf{e}_0 + \frac{\lambda_1 + \lambda_2}{2}\mathbf{e}_1 + \frac{\lambda_1 \lambda_2}{2}\mathbf{e} + \frac{\lambda_2 - \lambda_1}{2}\mathbf{e}_-. \tag{7.6.2}$$

The inner product of line $\mathbf{t}$ and sphere $\mathbf{c}$ is

$$\mathbf{t} \cdot \mathbf{c} = \frac{\lambda_1(1 + \epsilon) + \lambda_2(1 - \epsilon)}{2} - \lambda. \tag{7.6.3}$$

The inner product of two lines is

$$(\mathbf{e}_1 + \lambda_1 \mathbf{e} + \epsilon_1 \mathbf{e}_-) \cdot (\mathbf{e}_1 + \lambda_2 \mathbf{e} + \epsilon_2 \mathbf{e}_-) = 1 - \epsilon_1 \epsilon_2. \tag{7.6.4}$$

The inner product of two null vectors $\mathbf{c}_1, \mathbf{c}_2$ representing two spheres (λ_1, λ_2) and (μ_1, μ_2) is

$$\mathbf{c}_1 \cdot \mathbf{c}_2 = -\frac{(\lambda_1 - \mu_1)(\lambda_2 - \mu_2)}{2}. \tag{7.6.5}$$

For two Lie spheres $\mathbf{s}_1$ and $\mathbf{s}_2$, $\mathbf{s}_1 \cdot \mathbf{s}_2 = 0$ if and only if they are in contact:

- Two lines $(\lambda_1, \mathbf{e}, \epsilon_1)$ and $(\lambda_2, \mathbf{e}, \epsilon_2)$ are in contact if and only if $\epsilon_1 = \epsilon_2$.

- Line $(\lambda, \mathbf{e}, \epsilon)$ and sphere (λ_1, λ_2) are in contact if and only if when $\epsilon = 1$ then $\lambda = \lambda_1$, when $\epsilon = -1$ then $\lambda = \lambda_2$.

- Spheres (λ_1, λ_2) and (μ_1, μ_2) are in contact if and only if either $\lambda_1 = \mu_1$ or $\lambda_2 = \mu_2$.

Based on the above classification, the 1D Apollonian contact problem of constructing a 0D sphere in contact with two points at designated directions is trivial.

7.6.2 *2D contact problem*

In $\mathbb{R}^{3,2}$, let $\mathbf{e}_-, \mathbf{e}, \mathbf{e}_0, \mathbf{e}_1, \mathbf{e}_2$ be a Witt basis, where $(\mathbf{e}, \mathbf{e}_0)$ is a Witt pair, and $\mathbf{e}_1, \mathbf{e}_2$ are positive unit vectors. The Euclidean plane $\mathbf{I}_2 = \mathbf{e}_1 \wedge \mathbf{e}_2$ is denoted by $\mathbb{R}^2$. The orientation of $\mathbb{R}^{3,2}$ is determined by the unit pseudoscalar $\mathbf{I}_5 = \mathbf{e}_- \wedge \mathbf{e} \wedge \mathbf{e}_0 \wedge \mathbf{I}_2$, and $\mathbf{I}_5^{-1} = \mathbf{I}_5^{\dagger} = \mathbf{I}_5$.

Let $\mathbf{s}$ be a null vector in $\mathbb{R}^{3,2}$. Then $(\mathbf{s} \cdot \mathbf{e})(\mathbf{s} \cdot \mathbf{e}_-) < 0$ if and only if $\mathbf{s}$ represents an outward circle, and $(\mathbf{s} \cdot \mathbf{e})(\mathbf{s} \cdot \mathbf{e}_-) > 0$ if and only if $\mathbf{s}$ represents an inward circle.

Proposition 7.66. In the following statements, all points in $\mathbb{E}^2$ are identified with their null vector representations in $\mathcal{N}_\mathbf{e}$:

(1) Let $\mathbf{c}_1, \mathbf{c}_2, \mathbf{c}_3$ be three noncollinear points in $\mathbb{E}^2$. The oriented circle passing through them with the orientation from $\mathbf{c}_1$ to $\mathbf{c}_2$ to $\mathbf{c}_3$, is

$$\mathbf{s} = (\mathbf{c}_1 \wedge \mathbf{c}_2 \wedge \mathbf{c}_3 \wedge \mathbf{e}_-)^{\sim} + |\mathbf{c}_1 \wedge \mathbf{c}_2 \wedge \mathbf{c}_3| \mathbf{e}_-. \tag{7.6.6}$$

(2) The outward circle with center $\mathbf{c}$ and passing through point $\mathbf{a}$ is

$$\mathbf{a} \cdot (\mathbf{e} \wedge \mathbf{c}) + |\mathbf{a} \cdot (\mathbf{e} \wedge \mathbf{c})| \mathbf{e}_-. \tag{7.6.7}$$

(3) Let $\mathbf{c}$ be a point, and let $\mathbf{a}$ be a vector in $\mathbb{R}^2$. The directed line passing through point $\mathbf{c}$ in direction $\mathbf{a}$ is

$$(\mathbf{e} \wedge \mathbf{c} \wedge \mathbf{a} \wedge \mathbf{e}_-)^{\sim} + |\mathbf{e} \wedge \mathbf{c} \wedge \mathbf{a}| \mathbf{e}_-. \tag{7.6.8}$$

(4) Let $\mathbf{c}_1, \mathbf{c}_2$ be two points. The directed line passing through them in the direction from $\mathbf{c}_1$ to $\mathbf{c}_2$, is

$$(\mathbf{e} \wedge \mathbf{c}_1 \wedge \mathbf{c}_2 \wedge \mathbf{e}_-)^{\sim} + |\mathbf{e} \wedge \mathbf{c}_1 \wedge \mathbf{c}_2| \mathbf{e}_-. \tag{7.6.9}$$

Proof.   (1) The positive vector representing the unoriented circle is $(\mathbf{c}_1 \wedge \mathbf{c}_2 \wedge \mathbf{c}_3 \wedge \mathbf{e}_-)^{\sim}$. The magnitude of this vector is $|\mathbf{c}_1 \wedge \mathbf{c}_2 \wedge \mathbf{c}_3|$. Let $\mathbf{c}_i = \mathbf{f}(\mathbf{x}_i)$ for $i = 1, 2, 3$. Vector $\mathbf{s}$ defined by (7.6.6) satisfies

$$\mathbf{e} \cdot \mathbf{s} = [\mathbf{e}_- \mathbf{e} \mathbf{c}_1 \mathbf{c}_2 \mathbf{c}_3] = [\mathbf{e}_- \mathbf{e} \mathbf{e}_0 \partial(\mathbf{x}_1 \wedge \mathbf{x}_2 \wedge \mathbf{x}_3)] = \partial(\mathbf{x}_1 \wedge \mathbf{x}_2 \wedge \mathbf{x}_3) \cdot \mathbf{I}_2^{-1}.$$

Since $(\mathbf{s} \cdot \mathbf{e})(\mathbf{s} \cdot \mathbf{e}_-)$ has the same sign with $-\partial(\mathbf{x}_1 \wedge \mathbf{x}_2 \wedge \mathbf{x}_3) \cdot \mathbf{I}_2^{-1}$, circle $\mathbf{s}$ has the orientation $\partial(\mathbf{x}_1 \wedge \mathbf{x}_2 \wedge \mathbf{x}_3)$ of plane $\mathbb{R}^2$.

(2) The unoriented circle is $\mathbf{a} \cdot (\mathbf{e} \wedge \mathbf{c})$. Since $\mathbf{e} \cdot (\mathbf{a} \cdot (\mathbf{e} \wedge \mathbf{c})) = (\mathbf{e} \cdot \mathbf{a})(\mathbf{e} \cdot \mathbf{c}) > 0$, sphere (7.6.7) is positive.

(3) The left-handed normal direction of the line is $\mathbf{n} = \mathbf{a} \cdot \mathbf{I}_2^{-1} \in \mathbb{R}^2$, because $\mathbf{n} \wedge \mathbf{a}$ has the positive orientation $\mathbf{I}_2$ of the plane. Since

$$\mathbf{n} \cdot (\mathbf{e} \wedge \mathbf{c} \wedge \mathbf{a} \wedge \mathbf{e}_-)^\sim = [\mathbf{ecae}_-\mathbf{n}] = \mathbf{a}^2[\mathbf{e}_-\mathbf{ecI}_2] > 0,$$

vectors $\mathbf{n}$ and $(\mathbf{e} \wedge \mathbf{c} \wedge \mathbf{a} \wedge \mathbf{e}_-)^\sim$ have the same direction.

(4) Direct from (3). □

For two different Lie spheres $\mathbf{s}_1, \mathbf{s}_2$, blade $(\mathbf{s}_1 \wedge \mathbf{s}_2)^\sim$ represents the set of Lie spheres in contact with both $\mathbf{s}_1, \mathbf{s}_2$. It has two possible signatures: $(1, 0, 2)$ and $(2, 1, 0)$.

If $\mathbf{s}_1 \cdot \mathbf{s}_2 = 0$, then $\mathbf{s}_1$ and $\mathbf{s}_2$ are in contact. Blade $(\mathbf{s}_1 \wedge \mathbf{s}_2)^\sim$ has signature $(1, 0, 2)$, and topologically its null vectors form a 1D circle of contact Lie spheres, such that if point $\mathbf{x}$ (possibly being the conformal point at infinity) is where $\mathbf{s}_1, \mathbf{s}_2$ contact each other, then it is also where any two Lie spheres in the 1D circle are in contact. 2-blade $\mathbf{s}_1 \wedge \mathbf{s}_2$ and 3-blade $(\mathbf{s}_1 \wedge \mathbf{s}_2)^\sim$ have the same set of null vectors.

More generally, in $\mathbb{R}^{n+1,2}$, any rD subspace of signature $(r - 2, 0, 2)$ has the property that all its null vectors form a 2D subspace, which is also the set of null vectors in the orthogonal complement of the rD subspace. The set of null vectors in a 2D null subspace of $\mathbb{R}^{n+1,2}$ is called a *parabolic pencil* of Lie spheres. Topologically it is a circle.

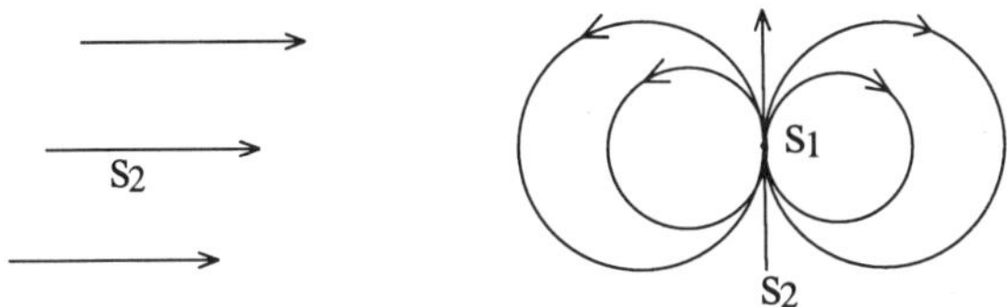

Fig. 7.23   1D parabolic pencils.

We return to the case $n = 2$, which is assumed throughout this subsection.

- If $\mathbf{s}_1 = \mathbf{e}$, and $\mathbf{s}_2$ is a line, then $\mathbf{e} \wedge \mathbf{s}_2$ is all the lines parallel to and in the same direction with $\mathbf{s}_2$.
- If $\mathbf{s}_1$ is a point on line $\mathbf{s}_2$, then $\mathbf{s}_1 \wedge \mathbf{s}_2$ is composed of point $\mathbf{s}_1$, line $\mathbf{s}_2$, and all the oriented circles in contact with line $\mathbf{s}_2$ at point $\mathbf{s}_1$.

In fact, the above are the only two kinds of parabolic pencils.

If $\mathbf{s}_1 \cdot \mathbf{s}_2 \neq 0$, then $\mathbf{s}_1 \wedge \mathbf{s}_2$ is Minkowski, so is $(\mathbf{s}_1 \wedge \mathbf{s}_2)^\sim$. The set of common contact Lie spheres of $\mathbf{s}_1, \mathbf{s}_2$ is topologically a circle. 2-blade $\mathbf{s}_1 \wedge \mathbf{s}_2$ represents a 0D circle of Lie spheres, while 3-blade $(\mathbf{s}_1 \wedge \mathbf{s}_2)^\sim$ represents a 1D circle of Lie spheres. $\mathbf{s}_1 \wedge \mathbf{s}_2$ represents two identical circles or lines but with opposite orientations if and only if $\mathbf{e}_- \wedge \mathbf{s}_1 \wedge \mathbf{s}_2 = 0$.

For example, if $\mathbf{s}$ is an oriented circle, then blade $(\mathbf{e}_- \wedge \mathbf{s})^\sim$ represents all points on the circle. If $\mathbf{s}$ is a line, then $\mathbf{e}_- \wedge \mathbf{s}$ represents the line with its two orientations, and $(\mathbf{e}_- \wedge \mathbf{s})^{\sim}$ represents all points on the line, including the conformal point at infinity.

The following list, together with Figure 7.24, is the classification of all 1D circles of Lie spheres in the plane, in the form of $(\mathbf{s}_1 \wedge \mathbf{s}_2)^{\sim}$ where $\mathbf{s}_1, \mathbf{s}_2$ are two null vectors whose inner product is nonzero.

(1) $\mathbf{s}_1 = \mathbf{e}$, and $\mathbf{s}_2$ is point $\mathbf{a}$;
(2) $\mathbf{s}_1 = \mathbf{e}$, and $\mathbf{s}_2$ is a circle;
(3) $\mathbf{s}_1, \mathbf{s}_2$ are points $\mathbf{a}, \mathbf{b}$ respectively;
(4) $\mathbf{s}_1$ is point $\mathbf{a}$, and $\mathbf{s}_2$ is a line;
(5) $\mathbf{s}_1$ is point $\mathbf{a}$, and $\mathbf{s}_2$ is a circle;
(6) $\mathbf{s}_1, \mathbf{s}_2$ are the same line with opposite directions;
(7) $\mathbf{s}_1, \mathbf{s}_2$ are intersecting lines;
(8) $\mathbf{s}_1, \mathbf{s}_2$ are parallel lines with opposite directions;
(9) $\mathbf{s}_1, \mathbf{s}_2$ are the same circle with opposite orientations;
(10) $\mathbf{s}_1, \mathbf{s}_2$ are inclusive circles;
(11) $\mathbf{s}_1, \mathbf{s}_2$ are intersecting circles;
(12) $\mathbf{s}_1, \mathbf{s}_2$ are exclusive circles.

Let $\mathbf{s}_1, \mathbf{s}_2, \mathbf{s}_3$ be three different Lie spheres in the plane. Then $\mathbf{s}_1 \wedge \mathbf{s}_2 \wedge \mathbf{s}_3 = 0$ if and only if $\mathbf{s}_1, \mathbf{s}_2, \mathbf{s}_3$ belong to a parabolic pencil. Assume that $\mathbf{s}_1 \wedge \mathbf{s}_2 \wedge \mathbf{s}_3 \neq 0$. Then $\mathbf{s}_1 \wedge \mathbf{s}_2 \wedge \mathbf{s}_3$ has three possible signatures: $(2,1,0), (1,1,1)$, or $(1,2,0)$. Correspondingly, $(\mathbf{s}_1 \wedge \mathbf{s}_2 \wedge \mathbf{s}_3)^2 > 0$, $= 0$, or < 0 respectively, and $\mathbf{s}_1 \wedge \mathbf{s}_2 \wedge \mathbf{s}_3$ is called a 3D *Minkowski, degenerate,* or *Euclidean* Lie pencil, respectively. The names come from the fact that blade $(\mathbf{s}_1 \wedge \mathbf{s}_2 \wedge \mathbf{s}_3)^{\sim}$ has three possible signatures: $(1,1,0), (1,0,1), (2,0,0)$.

The solutions of the Apollonian contact problem of Lie spheres $\mathbf{s}_1, \mathbf{s}_2, \mathbf{s}_3$ are just the Lie spheres corresponding to the 1D null subspaces of $(\mathbf{s}_1 \wedge \mathbf{s}_2 \wedge \mathbf{s}_3)^{\sim}$. When $\mathbf{s}_1 \wedge \mathbf{s}_2 \wedge \mathbf{s}_3$ is a Minkowski, degenerate, or Euclidean Lie pencil, the number of 1D null subspaces of $(\mathbf{s}_1 \wedge \mathbf{s}_2 \wedge \mathbf{s}_3)^{\sim}$ is 2, 1, or 0 respectively, which is also the number of solutions of the Apollonian contact problem.

Example 7.67. If $\mathbf{s}_1, \mathbf{s}_2, \mathbf{s}_3$ are three points in the plane, then there is either a circle or a line passing through them, and the circle or line has two possible orientations. Blade $\mathbf{s}_1 \wedge \mathbf{s}_2 \wedge \mathbf{s}_3$ represents a circle or line with two orientations. In comparison, in the conformal model, the blade represents exactly the same circle or line but without any orientation.

Example 7.68. If $\mathbf{s}_1, \mathbf{s}_2, \mathbf{s}_3$ are pairwise intersecting lines, then besides the conformal point at infinity, there exists another common contact Lie sphere $\mathbf{t}$, which is either the inscribed circle or an escribed circle of the triangle formed by the three lines, depending on the orientations of the lines.

In particular, if the three lines concur, then $\mathbf{t}$ is the point where they meet; if $\mathbf{s}_1, \mathbf{s}_2$ are the same line with opposite directions, then $\mathbf{t}$ is the point where lines

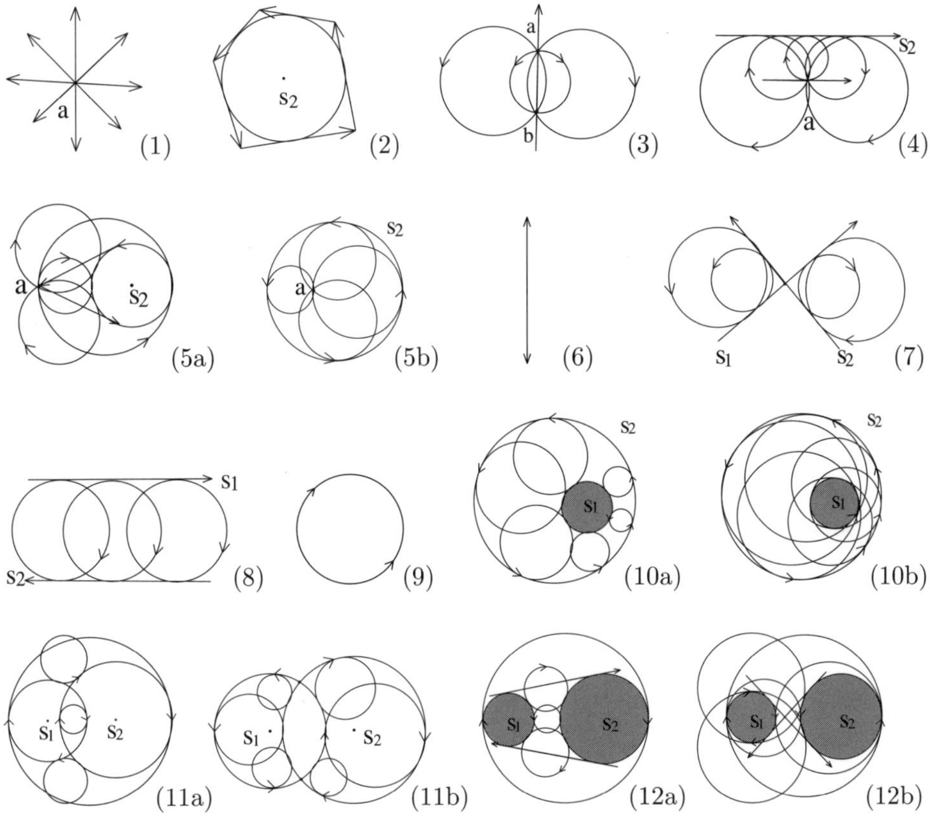

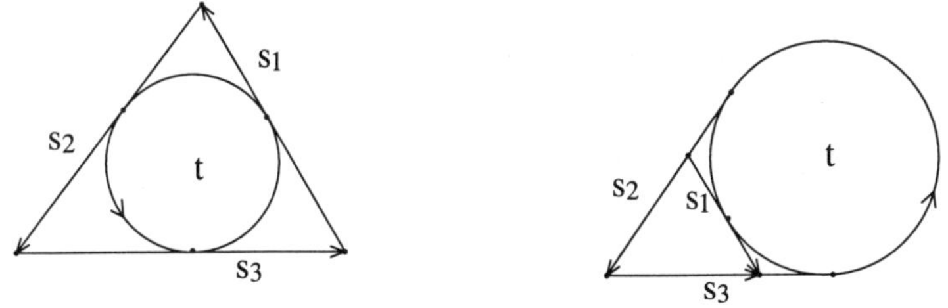

Fig. 7.24   Classification of 3D Minkowski Lie pencils.

s_1, s_3 meet.

Fig. 7.25   Common contact circle of three oriented lines.

Example 7.69. If s_1, s_2 are the same circle or line with opposite orientations, then for any circle s_3 not intersecting s_1, the three Lie spheres have no common contact Lie sphere. If s_3 is in contact with s_1, then the point of contact is the only common contact Lie sphere of s_1, s_2, s_3.

If $s_1 \wedge s_2 \wedge s_3$ has signature $(2,1,0)$ or $(1,2,0)$, it represents a 1D circle of Lie spheres; if it has signature $(1,1,1)$, it represents a pair of parabolic pencils sharing a common Lie sphere, and topologically it is an "8"-shaped curve: the union of two circles by identifying two points, one from each circle.

For example, let $\mathbf{a}$ be a point on line $\mathbf{n} + \delta e$, then $(\mathbf{a} \wedge (\mathbf{n} + \delta e))^{\sim}$ has signature $(1,1,1)$. It represents all the lines and circles perpendicular to line $\mathbf{n} + \delta e$ at point $\mathbf{a}$, each with two orientations, together with point $\mathbf{a}$, which is the Lie sphere shared by the two 1D circles of Lie spheres.

As another example, let $\mathbf{n}_1, \mathbf{n}_2$ be two orthogonal unit vectors in $\mathbb{R}^2$, then $(\mathbf{n}_1 \wedge \mathbf{n}_2)^{\sim}$ is a 3D Euclidean Lie pencil, and topologically it is a circle. The Lie pencil is composed of all the circles centering at the origin, with radii ranging from 0 to infinity, and with both orientations included. When the radius is 0, the circle degenerates to the origin; when the radius is infinite, the circle degenerates to the conformal point at infinity.

Proposition 7.70. In the plane, let s_1, s_2, s_3 be three Lie spheres having exactly two common contact Lie spheres s_4, s_5. Let $\mathbf{T}_3 = s_1 \wedge s_2 \wedge s_3$.

(a) If s_1, s_2, s_3 are all circles, let $s_i = \mathbf{f}(c_i) - \rho_i^2 e/2 + \epsilon_i \rho_i e_-$ for $i = 1, 2, 3$, where $c_i \in \mathbb{R}^2$. Then s_4, s_5 are both lines if and only if

$$\frac{c_1 - c_2}{c_1 - c_3} = \frac{\epsilon_1 \rho_1 - \epsilon_2 \rho_2}{\epsilon_1 \rho_1 - \epsilon_3 \rho_3}. \tag{7.6.10}$$

(b) If at least one of s_1, s_2, s_3 is a circle or point, then s_4, s_5 are circles having opposite orientations if and only if $e \wedge \mathbf{T}_3$, $e_- \wedge \mathbf{T}_3 \neq 0$, but $(e \wedge \mathbf{T}_3) \cdot (e_- \wedge \mathbf{T}_3) = 0$.

(c) If s_1, s_2, s_3 are the three sides $\mathbf{12}, \mathbf{23}, \mathbf{31}$ of triangle $\mathbf{123}$, then the inscribed circle of the triangle is $\mathbf{T}_3 e \mathbf{T}_3$. By changing the directions of the three sides, the three escribed circles can be obtained similarly.

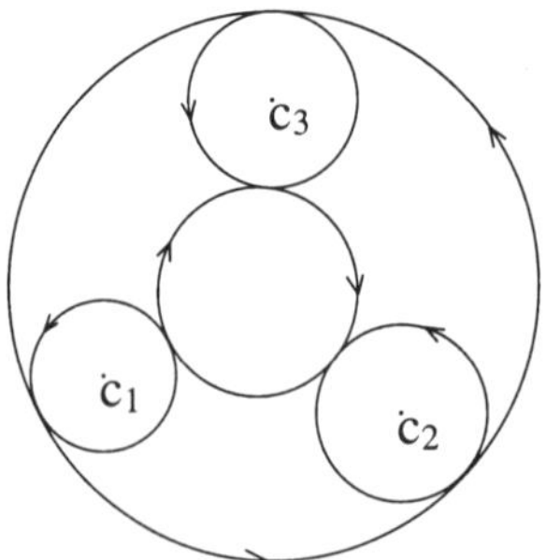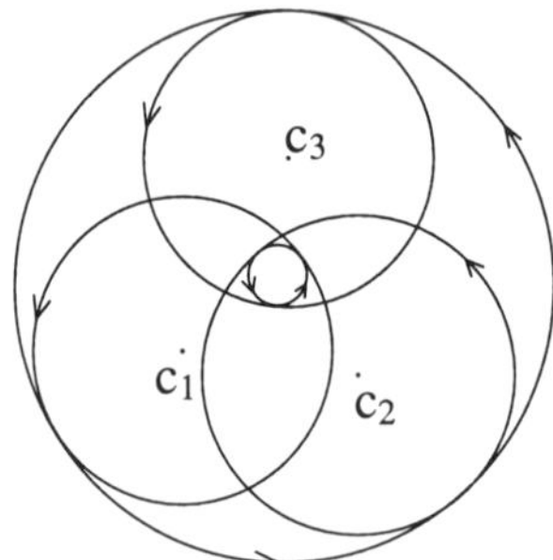

Fig. 7.26   Common contact circles of three oriented circles in the plane.

Proof.   (a) $\mathbf{T}_3^{\sim}$ represents two lines if and only if $e \wedge \mathbf{T}_3 = 0$. Expanding this equality, we get

$$e \wedge e_0 \wedge \partial(c_1 \wedge c_2 \wedge c_3) = 0,$$
$$e \wedge e_- \wedge (\epsilon_1\rho_1 c_2 \wedge c_3 + \epsilon_2\rho_2 c_3 \wedge c_1 + \epsilon_3\rho_3 c_1 \wedge c_2) = 0, \qquad (7.6.11)$$
$$e \wedge e_0 \wedge e_- \wedge \{\epsilon_1\rho_1(c_2 - c_3) + \epsilon_2\rho_2(c_3 - c_1) + \epsilon_3\rho_3(c_1 - c_2)\} = 0.$$

So if s_4, s_5 are both lines, then c_1, c_2, c_3 are collinear, and $\epsilon_1\rho_1(c_2 - c_3) + \epsilon_2\rho_2(c_3 - c_1) + \epsilon_3\rho_3(c_1 - c_2) = 0$. By this and $(c_2 - c_3) + (c_3 - c_1) + (c_1 - c_2) = 0$, we get (7.6.10).

Conversely, if (7.6.10) holds, then c_1, c_2, c_3 are collinear, and the equalities in (7.6.11) can be easily verified.

(b) By the hypotheses, $(s_1 \wedge s_2 \wedge s_3)^\sim$ has signature $(1, 1, 0)$, and $e \notin (s_1 \wedge s_2 \wedge s_3)^\sim$. So $s_4 \cdot s_5 \neq 0$. Let $B_2 = s_4 \wedge s_5 = (s_1 \wedge s_2 \wedge s_3)^\sim$. Then $e \cdot s_4$, $e_- \cdot s_4$ cannot be both zero, and $e \cdot s_5$, $e_- \cdot s_5$ cannot be both zero. Blade B_2 represents two points if and only if $e_- \cdot s_4 = e_- \cdot s_5 = 0$, and represents two lines if and only if $e \cdot s_4 = e \cdot s_5 = 0$.

Assume that B_2 represents neither two points nor two lines. Then $e_- \cdot s_4$, $e_- \cdot s_5$ cannot be both zero, and $e \cdot s_4$, $e \cdot s_5$ cannot be both zero. Furthermore, from

$$(e \wedge T_3) \cdot (e_- \wedge T_3) = \begin{aligned}[t] &(e \cdot B_2) \cdot (e_- \cdot B_2) \\ &= -(s_4 \cdot s_5)\{(e \cdot s_4)(e_- \cdot s_5) + (e \cdot s_5)(e_- \cdot s_4)\}, \end{aligned} \qquad (7.6.12)$$

we get that (7.6.12) equals zero if and only if none of $e \cdot s_4$, $e_- \cdot s_4$, $e \cdot s_5$, $e_- \cdot s_5$ is zero, and $(e \cdot s_4) : (e_- \cdot s_4) = -(e \cdot s_5) : (e_- \cdot s_5)$. The conditions are equivalent to s_4, s_5 being two circles of opposite orientations.

(c) In a Minkowski blade B_2, one null vector x can be determined by the other null vector y up to scale by $x = B_2 y B_2$. $\qquad \square$

When Lie spheres s_1, s_2, s_3 in the plane have a unique common contact Lie sphere, then

$$(s_1 \wedge s_2 \wedge s_3)^2 = -2(s_1 \cdot s_2)(s_1 \cdot s_3)(s_2 \cdot s_3) = 0.$$

So at least one of the $s_i \cdot s_j$ equals zero, but not all of them are zero. Assume that $s_1 \cdot s_2 \neq 0$. Let

$$t = (s_1 \wedge s_2) \cdot (s_1 \wedge s_2 \wedge s_3). \qquad (7.6.13)$$

Then t is a null vector representing the common contact Lie sphere.

Corollary 7.71. If Lie spheres s_1, s_2, s_3 in the plane have a unique common contact Lie sphere t, then t is a point if and only if for any $1 \leq i < j \leq 3$,

$$(e_- \wedge s_i \wedge s_j) \cdot (s_1 \wedge s_2 \wedge s_3) = 0; \qquad (7.6.14)$$

it is a line if and only if for any $1 \leq i < j \leq 3$,

$$(e \wedge s_i \wedge s_j) \cdot (s_1 \wedge s_2 \wedge s_3) = 0. \qquad (7.6.15)$$

If four different Lie spheres s_1, s_2, s_3, s_4 in the plane have a common contact Lie sphere, then vector $\mathbf{x} = (s_1 \wedge s_2 \wedge s_3 \wedge s_4)^{\sim}$ is either zero or null. In both cases $\mathbf{x}^2 = 0$. Conversely, if $\mathbf{x} \neq 0$, it must represent the unique common contact Lie sphere. If $\mathbf{x} = 0$, then if $s_i \wedge s_j \wedge s_k = 0$ for all $1 \leq i < j < k \leq 4$, the four Lie spheres belong to a parabolic pencil, and have infinitely many common contact Lie spheres; if $s_1 \wedge s_2 \wedge s_3 \neq 0$, then any Lie sphere that is in common contact with s_1, s_2, s_3 must also contact s_4. We have proved the following result:

Proposition 7.72. If three Lie spheres s_1, s_2, s_3 in the plane have a common contact Lie sphere, then four Lie spheres s_1, s_2, s_3, s_4 have a common contact Lie sphere if and only if

$$(s_1 \wedge s_2 \wedge s_3 \wedge s_4)^2 = 0. \tag{7.6.16}$$

Corollary 7.73. Four lines in the plane have a common contact Lie sphere other than e if and only if (1) $s_1 \wedge s_2 \wedge s_3 \wedge s_4 = 0$, (2) if three of the four lines are parallel then they have the same direction, (3) if only two of the four lines are parallel then they have opposite directions.

Proof. If the four lines have more than one common contact Lie sphere, then either (a) $s_i \wedge s_j \wedge s_k = 0$ for some $1 \leq i < j < k \leq 4$, or (b) the $s_i \wedge s_j \wedge s_k$ for all $1 \leq i < j < k \leq 4$ are nonzero and equal to each other up to scale.

In both cases, $s_1 \wedge s_2 \wedge s_3 \wedge s_4 = 0$. Furthermore, in (a), lines s_i, s_j, s_k are parallel to each other and are in the same direction; in (b), blade $s_1 \wedge s_2 \wedge s_3$ must be Minkowski, so $s_i \cdot s_j \neq 0$ for any $1 \leq i < j \leq 3$.

This proves the necessity statement. The sufficiency can be proved similarly. $\square$

When the s_i are all points, (7.6.16) can be written as

$$(s_1 \wedge s_2 \wedge s_3 \wedge s_4)^2 = \det(s_i \cdot s_j)_{i,j=1..4} = \frac{1}{16} \det(d^2_{s_i s_j})_{i,j=1..4} = 0.$$

After factorization, we get

Corollary 7.74. [Ptolemy's Theorem] If four points $\mathbf{1}, \mathbf{2}, \mathbf{3}, \mathbf{4}$ are cocircular, then

$$d_{12}d_{34} \pm d_{14}d_{23} \pm d_{13}d_{24} = 0. \tag{7.6.17}$$

When the s_i are all circles, (7.6.16) can be written as $\det(s_i \cdot s_j)_{i,j=1..4} = 0$. After factorization with the aid of (7.5.8), we get

Corollary 7.75. [Casey's Theorem] If four oriented circles $\mathbf{1}, \mathbf{2}, \mathbf{3}, \mathbf{4}$ contact the same oriented circle or oriented line or point, then

$$d^t_{12}d^t_{34} \pm d^t_{14}d^t_{23} \pm d^t_{13}d^t_{24} = 0. \tag{7.6.18}$$

7.6.3 *nD contact problem*

Similar to the case of 2D geometry, in $\Lambda(\mathbb{R}^{n+1,2})$, an r-blade $\mathbf{A}_r$ generated by r null vectors, where $3 \leq r \leq n+1$, has three possible signatures: $(r-1,1,0)$, $(r-2,1,1)$, or $(r-2,2,0)$, and correspondingly the blade is called an rD *Minkowski, degenerate,* or *Euclidean* pencil of Lie spheres in $\mathbb{E}^n$.

The dual space $\mathbf{A}_r^{\sim}$ is the solution space of the Apollonian contact problem of the r Lie spheres represented by the r constitutive null vectors of $\mathbf{A}_r$. Blade $\mathbf{A}_r^{\sim}$ has three possible signatures: $(n-r+2,1,0)$, $(n-r+2,0,1)$, or $(n-r+3,0,0)$, and accordingly the number of solutions of the Apollonian contact problem is the cardinality of $\mathbb{S}^{n-r+1}$, 1, or 0 respectively.

Topologically, for $3 \leq r \leq n+1$, an r-blade $\mathbf{A}_r$ generated by null vectors represents an $(r-2)$D sphere of Lie spheres if it is either Minkowski or Euclidean, and represents a one-point union of a pair of $(r-2)$D spheres of Lie spheres otherwise.

When $r = 2$, $\mathbf{A}_r$ has two possible signatures: $(0,0,2)$ or $(1,1,0)$. In the former case, $\mathbf{A}_r$ is a parabolic pencil representing a 1D circle of Lie spheres; in the latter case, $\mathbf{A}_r$ represents a 0D circle of Lie spheres.

When $r = n + 2$, $\mathbf{A}_{n+2}$ has three possible signatures: $(n+1,1,0)$, $(n,1,1)$, or $(n,2,0)$, and correspondingly, blade $\mathbf{A}_{n+1}^{\sim}$ is a negative, null, or positive vector; topologically, $\mathbf{A}_{n+2}$ represents an nD sphere, disk, or cylinder of Lie spheres.

In particular, if $\mathbf{A}_r$ represents a $(r-2)$D sphere of Lie spheres composed of points only (possibly including the conformal point at infinity), where $2 \leq r \leq n+1$, then $\mathbf{A}_r \in \Lambda(\mathbf{e}_-^{\sim})$. In the conformal model, blade $\mathbf{A}_r$ represents the same $(r-2)$D sphere or plane of points.

The Grassmann-Cayley algebra $\Lambda(\mathbb{R}^{n+1,2})$ can be used to represent and compute the extension and intersection of rD pencils of Lie spheres. The representations and classifications of Lie pencils and the solutions of the corresponding Apollonian contact problems in the previous subsection can be naturally extended to dimension $n > 2$, and the results are similar.

Below we present an example in 3D geometry.

Example 7.76. A *convex polytope* is a compact convex polyhedron enclosed by a finite set of oriented hyperplanes. Let there be a convex polytope of five faces in space. Find the condition for the existence of a sphere inscribed in the polytope.

Let the unit outward normals of the five faces be $\mathbf{n}_1, \mathbf{n}_2, \mathbf{n}_3, \mathbf{n}_4, \mathbf{n}_5$ respectively. Let $\mathbf{a}$ be the point of intersection of the three faces with normals $\mathbf{n}_1$, $\mathbf{n}_2$, $\mathbf{n}_3$, and let δ_4, δ_5 be the signed distances from $\mathbf{a}$ to the faces with normals $\mathbf{n}_4$, $\mathbf{n}_5$, respectively.

Choose $\mathbf{a}$ as the origin of $\mathbb{R}^3$. The five faces can be represented by the following null vectors in $\mathbb{R}^{4,2}$:

$$\mathbf{s}_1 = \mathbf{n}_1 + \mathbf{e}_-,$$
$$\mathbf{s}_2 = \mathbf{n}_2 + \mathbf{e}_-,$$

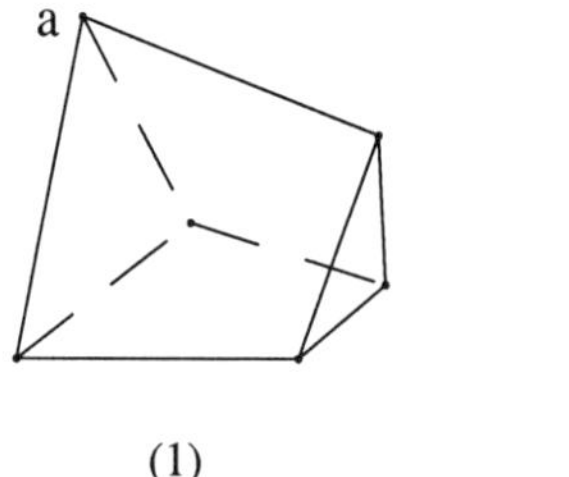
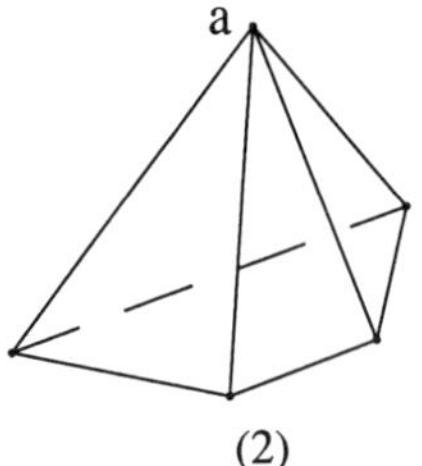

Fig. 7.27   Convex polytope of five faces: (1) 6 vertices; (2) 5 vertices.

$$\begin{aligned}
\mathbf{s}_3 &= \mathbf{n}_3 + \mathbf{e}_-, \\
\mathbf{s}_4 &= \mathbf{n}_4 + \delta_4 \mathbf{e} + \mathbf{e}_-, \\
\mathbf{s}_5 &= \mathbf{n}_5 + \delta_5 \mathbf{e} + \mathbf{e}_-.
\end{aligned} \tag{7.6.19}$$

Corollary 7.73 is the criterion for the existence of a common contact Lie sphere other than $\mathbf{e}$ of $n+2$ hyperplanes in $\mathbb{E}^n$, where $n = 2$. Its extension to the case $n = 3$ is the following:

Proposition 7.77. Five different planes in space represented by null vectors $\mathbf{s}_i \in \mathbb{R}^{4,2}$, where $1 \leq i \leq 5$, have a common contact Lie sphere other than $\mathbf{e}$ if and only if (1) $\mathbf{s}_1 \wedge \mathbf{s}_2 \wedge \mathbf{s}_3 \wedge \mathbf{s}_4 \wedge \mathbf{s}_5 = 0$, (2) if the subspace spanned by the five vectors $\mathbf{s}_i$ is $r\mathrm{D}$ and $r > 2$, then the subspace is not degenerate.

Proof.   Let r-blade $\mathbf{B}_r$ represent the $r\mathrm{D}$ subspace spanned by the five $\mathbf{s}_i$. Then $2 \leq r \leq 5$. We only prove the necessity statement. The sufficiency statement can be proved similarly. Then $r \leq 4$.

Case 1. $r = 2$.

Then $\mathbf{s}_1 \wedge \mathbf{s}_2$ is degenerate, because it has more than two different null vectors up to scale. The five $\mathbf{s}_i$ are in the same parabolic pencil, and all the Lie spheres in the pencil are in contact with the five $\mathbf{s}_i$ simultaneously. That $\mathbf{A}_5 = 0$ is obvious.

Case 2. $r = 3$.

If $\mathbf{s}_1 \wedge \mathbf{s}_2 \wedge \mathbf{s}_3 = 0$, then the three planes are in the same parabolic pencil, and any Lie sphere $\mathbf{t}$ in common contact with them must also be in the pencil. The parabolic pencil can be represented by $\mathbf{e} \wedge \mathbf{t}$. Since $\mathbf{s}_4, \mathbf{s}_5$ are both in contact with $\mathbf{e}$ and $\mathbf{t}$, they must be in the pencil as well. Thus $\mathbf{s}_i \wedge \mathbf{s}_j \wedge \mathbf{s}_k = 0$ for any $1 \leq i < j < k \leq 5$, and $r = 2$ accordingly. Contradiction.

So the $\mathbf{s}_i \wedge \mathbf{s}_j \wedge \mathbf{s}_k$ for all $1 \leq i < j < k \leq 5$ are nonzero and are equal to each other up to scale. Let $\mathbf{B}_3 = \mathbf{s}_1 \wedge \mathbf{s}_2 \wedge \mathbf{s}_3$. Since $\mathbf{B}_3$ does not have any 2D null subspace, neither does $\mathbf{B}_3^{\sim}$. Since each of $\mathbf{B}_3$ and $\mathbf{B}_3^{\sim}$ has at least two different 1D null subspaces, $\mathbf{B}_3$ and $\mathbf{B}_3^{\sim}$ must be both Minkowski. Then 3-blade $\mathbf{B}_3^{\sim}$ has infinitely many different null vectors up to scale, each representing a Lie sphere in contact with the five $\mathbf{s}_i$ simultaneously.

Case 3. $r = 4$.

Since $\mathbf{B}_4$ does not have any 2D null subspace, neither does $\mathbf{B}_4^{\sim}$. Since 2-blade $\mathbf{B}_4^{\sim}$ has at least two different null vectors up to scale, it must be Minkowski. Then $\mathbf{B}_4$ is Minkowski. $\square$

Corollary 7.78. The five faces represented by (7.6.19) surround a convex polytope if and only if $\mathbf{A}_5 = \mathbf{s}_1 \wedge \mathbf{s}_2 \wedge \mathbf{s}_3 \wedge \mathbf{s}_4 \wedge \mathbf{s}_5 = 0$.

Proof. We only prove the necessity statement. By Proposition 7.77, we need only prove the condition (2) there, using the same notation as in the proof of the proposition.

If $r = 3$, then $(\mathbf{s}_1 \wedge \mathbf{s}_2 \wedge \mathbf{s}_3)^2 = 0$ is equivalent to $\mathbf{s}_i \cdot \mathbf{s}_j = 0$ for some $1 \le i \ne j \le 3$. Geometrically, $\mathbf{s}_i \cdot \mathbf{s}_j = 0$ if and only if the two planes are parallel and have the same normal direction. In a convex polyhedron, no two parallel faces have the same outward normal direction. So $\mathbf{s}_i \cdot \mathbf{s}_j \ne 0$ for any $1 \le i < j \le 5$, and $\mathbf{B}_3^2 \ne 0$.

If $r = 4$, by (7.6.19),

$$\begin{aligned}
(\mathbf{s}_1 \wedge \mathbf{s}_2 \wedge \mathbf{s}_3 \wedge \mathbf{s}_4)^2 &= ((\mathbf{n}_1 + \mathbf{e}_-) \wedge (\mathbf{n}_2 + \mathbf{e}_-) \wedge (\mathbf{n}_3 + \mathbf{e}_-) \wedge (\mathbf{n}_4 + \mathbf{e}_-))^2 \\
&= (\mathbf{e}_- \wedge \partial(\mathbf{n}_1 \wedge \mathbf{n}_2 \wedge \mathbf{n}_3 \wedge \mathbf{n}_4))^2 \\
&= (\partial(\mathbf{n}_1 \wedge \mathbf{n}_2 \wedge \mathbf{n}_3 \wedge \mathbf{n}_4))^2.
\end{aligned}$$

It equals zero if and only if $\partial(\mathbf{n}_1 \wedge \mathbf{n}_2 \wedge \mathbf{n}_3 \wedge \mathbf{n}_4) = 0$, *i.e.*, vectors $\mathbf{n}_1, \mathbf{n}_2, \mathbf{n}_3, \mathbf{n}_4$ are on a circle of the unit sphere $\mathbb{S}^2$ of $\mathbb{R}^3$. Since $r = 4$, the five vectors $\mathbf{n}_i$ are on the same circle of $\mathbb{S}^2$. The vector space spanned by the five normals cannot be of dimension two, otherwise the five faces only surround an infinitely long cylinder. So the five normals are not on a great circle of $\mathbb{S}^2$.

Since any non-great circle of $\mathbb{S}^2$ is a section of the sphere by an affine plane not passing through the center of the sphere (the origin of $\mathbb{R}^3$), the endpoints of the five normals must be in the same open half-space bordered by a plane passing through the origin. By the classical theorem in topology [171] that two homotopic mappings from $\mathbb{S}^2$ to $\mathbb{S}^2$ must have the same degree, it is easy to deduce that the five faces cannot surround a closed convex region. This proves $\mathbf{B}_4^2 \ne 0$.

By Proposition 7.77, the five faces must have a common contact Lie sphere $\mathbf{t}$ other than $\mathbf{e}$. It remains to prove that $\mathbf{t}$ can only be a proper sphere. If $\mathbf{t}$ is a point, then the five faces can only surround an infinitely long cone. If $\mathbf{t}$ is a plane, then the five faces must be in the same parabolic pencil $\mathbf{e} \wedge \mathbf{t}$; they are parallel to each other and have the same normal direction, so they can never surround any finite region. $\square$

By Corollary 7.78, to solve the problem in Example 7.77, we only need to compute the explicit form of $\mathbf{A}_5 = 0$. By (7.6.19),

$$\mathbf{A}_5 = \mathbf{e}_- \wedge \mathbf{e} \wedge (\delta_4 \partial(\mathbf{n}_1 \wedge \mathbf{n}_2 \wedge \mathbf{n}_3 \wedge \mathbf{n}_5) - \delta_5 \partial(\mathbf{n}_1 \wedge \mathbf{n}_2 \wedge \mathbf{n}_3 \wedge \mathbf{n}_4)).$$

So $\mathbf{A}_5 = 0$ if and only if

$$\frac{\delta_4}{\delta_5} = \frac{\partial(\mathbf{n}_1 \wedge \mathbf{n}_2 \wedge \mathbf{n}_3 \wedge \mathbf{n}_4)}{\partial(\mathbf{n}_1 \wedge \mathbf{n}_2 \wedge \mathbf{n}_3 \wedge \mathbf{n}_5)}, \tag{7.6.20}$$

whose geometric meaning is as follows: the ratio of the signed distances from the intersection $\mathbf{a}$ of the three faces with normals $\mathbf{n}_1, \mathbf{n}_2, \mathbf{n}_3$, to the two faces with normals $\mathbf{n}_4, \mathbf{n}_5$ respectively, equals the ratio of the signed volumes of the two tetrahedrons whose vertices are the endpoints of vectors $\mathbf{n}_1, \mathbf{n}_2, \mathbf{n}_3, \mathbf{n}_4$, and the endpoints of vectors $\mathbf{n}_1, \mathbf{n}_2, \mathbf{n}_3, \mathbf{n}_5$ respectively, all of which are on the unit sphere of $\mathbb{R}^3$.

Chapter 8

Conformal Clifford Algebra and Classical Geometries

In the previous chapter, we have developed the geometric theory of conformal Grassmann-Cayley algebra. In this chapter, we develop the geometric theory of the geometric products of invertible vectors and Minkowski blades in the conformal model, called *conformal Clifford algebra*.

Conformal Clifford algebra provides surprisingly simple representations for nD conformal transformations. The geometric product of two Minkowski blades representing two geometric objects generates twice the conformal transformation changing one geometric object to the other. The bivector representation of conformal transformations replaces the exponential map from the Lie algebra to the Lie group of conformal transformations by low-degree polynomial maps. The Clifford matrix representation of conformal transformations extends the classical fractional linear form of conformal transformations from 2D to nD.

The Grassmann-Cayley algebra and Clifford algebra established upon the conformal model are called *conformal geometric algebra*. This algebra is not only the covariant algebra for Euclidean incidence geometry, but provides a unified framework for all kinds of classical geometries, including projective, affine, Euclidean, hyperbolic, spherical, elliptic, and conformal geometries. The dual vector algebra of 3D affine objects is also extended to nD by conformal geometric algebra.

8.1 The geometry of positive monomials

In $\mathcal{G}(\mathbb{R}^{n+1,1})$, a *positive monomial* refers to a Clifford monomial whose constituent vectors are positive, an *invertible monomial* refers to a Clifford monomial whose constituent vectors are invertible. A *positive versor* refers to a Clifford monomial that is both positive and invertible.

By Cartan-Dieudonné Theorem, any orthogonal transformation in $\mathbb{R}^{n+1,1}$ is induced by the adjoint action of a versor whose vectors are all positive. Since an orthogonal transformation in $\mathbb{R}^{n+1,1}$ induces a conformal transformation in $\mathbb{E}^n$, positive monomials are generators of conformal transformations, and the geometric theory of positive monomials, or more generally, the geometric theory of invertible

411

monomials in $\mathcal{G}(\mathbb{R}^{n+1,1})$, is then a theory of versor representations of conformal transformations.

8.1.1   *Versors for conformal transformations*

In $\mathbb{R}^{n+1,1}$, any positive vector $\mathbf{s}$ equals either $\mathbf{c} - \rho^2 \mathbf{e}/2$ or $\mathbf{n} + \delta\mathbf{e}$ up to scale. As a versor, vector $\mathbf{s}$ acts on the set $\mathcal{N}$ of all null vectors in $\mathbb{R}^{n+1,1}$ by graded adjoint action (5.3.27).

For any point $\mathbf{a} = \mathbf{f}(\mathbf{x})$, where $\mathbf{x} \in \mathbb{R}^n$, let $\mathbf{b} = Ad_{\mathbf{s}}(\mathbf{a}) = -\mathbf{s}\mathbf{a}\mathbf{s}^{-1}$.

- When $\mathbf{s} = \mathbf{c} - \rho^2 \mathbf{e}/2$, where $\mathbf{c} = \mathbf{f}(\mathbf{o})$ is the center of the sphere, by direct computing,

$$\mathbf{b} = \mathbf{a} - 2(\mathbf{a} \cdot \mathbf{s})\mathbf{s}^{-1} = \frac{d_{\mathbf{ac}}^2}{\rho^2}\mathbf{f}(\mathbf{y}),$$

$$\mathbf{y} = \mathbf{o} + \frac{\rho^2}{d_{\mathbf{ac}}^2}(\mathbf{x} - \mathbf{o}).$$

So $(\mathbf{x} - \mathbf{o})(\mathbf{y} - \mathbf{o}) = \rho^2$, and $\mathbf{x}, \mathbf{y}$ are in inversion with respect to sphere $\mathbf{s}^\sim$.

- When $\mathbf{s} = \mathbf{n} + \delta\mathbf{e}$, by similar computation, we get

$$\mathbf{b} = \mathbf{f}(\mathbf{y}), \quad \mathbf{y} = \mathbf{x} - 2(\mathbf{x} \cdot \mathbf{n} - \delta)\mathbf{n}.$$

So $\mathbf{y}$ is the mirror reflection of $\mathbf{x}$ with respect to hyperplane $\mathbf{s}^\sim$.

Since the group of conformal transformations in $\mathbb{R}^n$ is generated by inversions with respect to spheres and reflections with respect to hyperplanes, it is generated by positive versors in $\mathcal{G}(\mathbb{R}^{n+1,1})$ through the adjoint action on the conformal model. In fact, every versor in $\mathcal{G}(\mathbb{R}^{n+1,1})$ can be changed into a positive versor by dual operations. There are only two kinds of versors fixing every null 1D subspace of $\mathbb{R}^{n+1,1}$: λ and $\lambda \mathbf{I}_{n+2}$, where $\mathbf{I}_{n+2}$ is the unit pseudoscalar in $\Lambda(\mathbb{R}^{n+1,1})$ denoting the orientation of $\mathbb{R}^{n+1,1}$.

Every versor in $\mathcal{G}(\mathbb{R}^{n+1,1})$ generates a conformal transformation in $\mathbb{R}^n$, and conversely, every conformal transformation in $\mathbb{R}^n$ is generated by a versor in $\mathcal{G}(\mathbb{R}^{n+1,1})$. Two versors realize the same conformal transformation if and only if they are equal up to a scalar or pseudoscalar factor.

It is a classical result [150] that any rotor in $\mathcal{G}(\mathbb{R}^{n+1,1})$ can be compressed to a rigid rotor of the form $\mathbf{a}_1 \mathbf{b}_1 \mathbf{a}_2 \mathbf{b}_2 \cdots \mathbf{a}_r \mathbf{b}_r$, where the $\mathbf{a}$'s and $\mathbf{b}$'s are invertible vectors satisfying $\mathbf{a}_i \mathbf{b}_i \mathbf{a}_j \mathbf{b}_j = \mathbf{a}_j \mathbf{b}_j \mathbf{a}_i \mathbf{b}_i$ for any $i \neq j$. Hence it suffices to classifying all rotors generated by two linearly independent unit vectors $\mathbf{a}, \mathbf{b} \in \mathbb{R}^{n+1,1}$.

If blade $\mathbf{a} \wedge \mathbf{b}$ is degenerate, let $\mathbf{I}_2 = \mathbf{a} \wedge \mathbf{b}/(\mathbf{a} \cdot \mathbf{b})$. If $\mathbf{a} \wedge \mathbf{b}$ is nondegenerate, let $\mathbf{I}_2 \in \Lambda^2(\mathbf{a} \wedge \mathbf{b})$ be of unit magnitude. Then there exists $\theta \in \mathbb{R}$ such that

$$\mathbf{ab} = \begin{cases} e^{\mathbf{I}_2\theta} = \cos\theta + \mathbf{I}_2\sin\theta, & \text{if } \mathbf{I}_2^2 = -1; \\ e^{\mathbf{I}_2\theta} = 1 + \mathbf{I}_2\theta, & \text{if } \mathbf{I}_2^2 = 0; \\ e^{\mathbf{I}_2\theta} = \cosh\theta + \mathbf{I}_2\sinh\theta, & \text{if } \mathbf{I}_2^2 = 1 \text{ and } \mathbf{a}^2\mathbf{b}^2 = 1; \\ \mathbf{I}_2 e^{\mathbf{I}_2\theta} = \sinh\theta + \mathbf{I}_2\cosh\theta, & \text{if } \mathbf{I}_2^2 = 1 \text{ and } \mathbf{a}^2\mathbf{b}^2 = -1. \end{cases} \tag{8.1.1}$$

(1) When $\mathbf{I}_2^2 = -1$, then $\mathbf{a}, \mathbf{b}$ are positive, and $\mathbf{I}_2^\sim$ is Minkowski.

If $\mathbf{e} \cdot \mathbf{I}_2 = 0$, then $\mathbf{a}^\sim, \mathbf{b}^\sim$ are two hyperplanes, and $\mathbf{I}_2^\sim$ is their intersection. Vector $\mathbf{e}$ is invariant by the adjoint action of versor $\mathbf{ab}$. Furthermore, versor $\mathbf{ab}$ fixes every point in the $(n-2)$D plane $\mathbf{I}_2^\sim$, and fixes the 2D direction $\mathbf{e} \wedge \mathbf{I}_2$ that is completely orthogonal to the $(n-2)$D plane $\mathbf{I}_2^\sim$. Plane $\mathbf{I}_2^\sim$ is called the *axis* of the rotation.

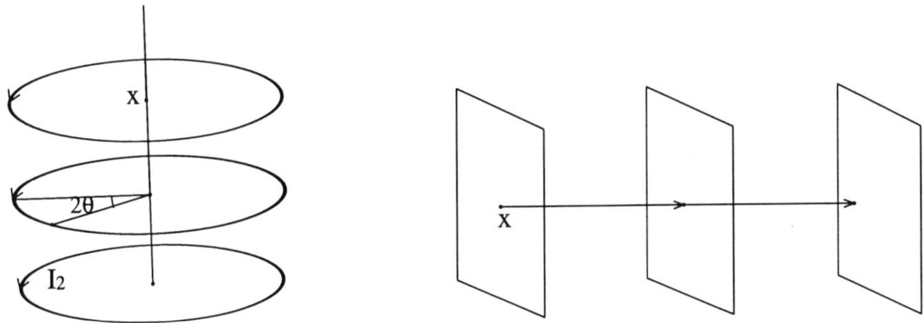

Fig. 8.1   Rotation (left) and translation (right) induced by a rotor of length two.

At every point $\mathbf{x}$ on the axis, there is a unique 2D plane with 2D direction $\mathbf{e} \wedge \mathbf{I}_2$, called a *plane of rotation*. Any point in $\mathbb{E}^n$ is on a unique plane of rotation. On every plane of rotation $\mathbf{B} = \mathbf{e} \wedge \mathbf{x} \wedge \mathbf{I}_2$, where $\mathbf{x}$ is the point on the axis, versor $\mathbf{ab}$ realizes a rotation of angle -2θ in the orientation $\mathbf{I}_2$: by setting $\mathbf{x} = \mathbf{e}_0$, then for any $\mathbf{y} \in \mathbf{I}_2 \subseteq (\mathbf{e} \wedge \mathbf{e}_0)^\sim$,

$$Ad_{\mathbf{ab}}(\mathbf{f}(\mathbf{y})) = e^{\mathbf{I}_2\theta}(\mathbf{e}_0 + \mathbf{y} + \frac{\mathbf{y}^2}{2}\mathbf{e})e^{-\mathbf{I}_2\theta} = \mathbf{e}_0 + e^{\mathbf{I}_2(2\theta)} + \frac{\mathbf{y}^2}{2}\mathbf{e} = \mathbf{f}(\mathbf{y}\cos 2\theta - \mathbf{y}\mathbf{I}_2\sin 2\theta).$$

An equivalent description without resorting to the angle of rotation θ is as follows: let $\mathbf{c}, \mathbf{d}$ be the points of intersection of hyperplanes $\mathbf{a}^\sim, \mathbf{b}^\sim$ respectively with 2D plane $\mathbf{B}$, then versor $\mathbf{ab}$ realizes twice the rotation from point $\mathbf{d}$ to point $\mathbf{c}$ centering at point $\mathbf{x}$ in the direction $\mathbf{I}_2$.

If $\mathbf{e} \cdot \mathbf{I}_2 \neq 0$, then at least one of $\mathbf{a}^\sim, \mathbf{b}^\sim$ is a sphere. Their intersection $\mathbf{I}_2^\sim$ is an $(n-2)$D sphere; every point at the intersection is fixed by versor $\mathbf{ab}$. By an inversion with respect to some sphere $\mathbf{s}^\sim$ whose center is on $(n-2)$D sphere $\mathbf{I}_2^\sim$, the $(n-2)$D sphere is changed into an $(n-2)$D plane. So $\mathbf{ab} = \mathbf{s}(\mathbf{s}^{-1}\mathbf{as})(\mathbf{s}^{-1}\mathbf{bs})\mathbf{s}^{-1}$ is the composition of two inversions and a rotation.

(2) When $\mathbf{I}_2^2 = 0$, then $\mathbf{a}, \mathbf{b}$ are positive.

If $\mathbf{e} \in \mathbf{I}_2$, then it is fixed by versor $\mathbf{ab}$. Furthermore, $\mathbf{e} \cdot \mathbf{a} = \mathbf{e} \cdot \mathbf{b} = 0$, so $\mathbf{a}^\sim, \mathbf{b}^\sim$ are two hyperplanes. Since $\mathbf{e} \wedge \mathbf{a} = \mathbf{e} \wedge \mathbf{b}$ up to scale, the two hyperplanes are parallel, and $\mathbf{e} \wedge \mathbf{a}$ is their common normal direction.

For every point $\mathbf{x}$ on hyperplane $\mathbf{b}^\sim$, line $\mathbf{e} \wedge \mathbf{a} \wedge \mathbf{x}$ is perpendicular to hyperplane $\mathbf{b}^\sim$. The line is fixed by versor $\mathbf{ab}$, while point $\mathbf{x}$ is changed to its mirror reflection with respect to hyperplane $\mathbf{a}^\sim$. The result is a translation that is twice from hyperplane $\mathbf{b}^\sim$ to hyperplane $\mathbf{a}^\sim$.

An equivalent description based on (8.1.1) is as follows. By setting $\mathbf{I}_2 = \mathbf{e} \wedge \mathbf{t}$, where $\mathbf{t} \in (\mathbf{e} \wedge \mathbf{e}_0)^{\sim}$, for any $\mathbf{y} \in (\mathbf{e} \wedge \mathbf{e}_0)^{\sim}$,

$$Ad_{\mathbf{ab}}(\mathbf{f}(\mathbf{y})) = (1 + \theta \mathbf{et})(\mathbf{e}_0 + \mathbf{y} + \frac{\mathbf{y}^2}{2}\mathbf{e})(1 - \theta \mathbf{et}) = \mathbf{f}(\mathbf{y} + 2\theta \mathbf{t}).$$

So $\mathbf{ab}$ realizes the translation by vector $2\theta \mathbf{t}$.

If $\mathbf{e} \notin \mathbf{I}_2$, let $\mathbf{c}$ be the point represented by the null vector in $\mathbf{I}_2$ that is unique up to scale. An inversion with respect to some sphere $\mathbf{s}^{\sim}$ centering at point $\mathbf{c}$ makes $\mathbf{e} \in \mathbf{I}_2$ again. So $\mathbf{ab}$ is the composition of two inversions and a translation.

(3) When $\mathbf{I}_2^2 = 1$, there are three cases:

(3.a) $\mathbf{a}, \mathbf{b}$ are both positive.

If $\mathbf{e} \in \mathbf{I}_2$, then $\mathbf{e} \cdot \mathbf{a}$ and $\mathbf{e} \cdot \mathbf{b}$ are both nonzero, so $\mathbf{a}^{\sim}, \mathbf{b}^{\sim}$ are two spheres. Let $\mathbf{c} \in \mathcal{N}_{\mathbf{e}}$ be the point in $\mathbf{I}_2$ other than $\mathbf{e}$. By rescaling, let $\mathbf{I}_2 = \mathbf{e} \wedge \mathbf{c}$. Then for any $\mathbf{y} \in (\mathbf{e} \wedge \mathbf{c})^{\sim}$,

$$e^{\mathbf{I}_2\theta}\mathbf{e}\,e^{-\mathbf{I}_2\theta} = e^{2\theta(\mathbf{e}\wedge\mathbf{c})}\mathbf{e} = e^{-2\theta}\mathbf{e},$$

$$e^{\mathbf{I}_2\theta}\mathbf{c}\,e^{-\mathbf{I}_2\theta} = e^{2\theta(\mathbf{e}\wedge\mathbf{c})}\mathbf{c} = e^{2\theta}\,\mathbf{c}, \tag{8.1.2}$$

$$e^{\mathbf{I}_2\theta}(\mathbf{c} + \mathbf{y} + \frac{\mathbf{y}^2}{2}\mathbf{e})e^{-\mathbf{I}_2\theta} = e^{2\theta}(\mathbf{c} + \mathbf{y}e^{-2\theta} + \frac{\mathbf{y}^2 e^{-4\theta}}{2}\mathbf{e}).$$

So every vector in $\mathbb{R}^n$ starting from point $\mathbf{c}$ is dilated by scale $e^{-2\theta}$. Point $\mathbf{c}$ is fixed by versor $\mathbf{ab}$. It is called the *center of dilation*. For every point $\mathbf{x} \neq \mathbf{c}$, the line $\mathbf{x} \wedge \mathbf{I}_2$ passing through points $\mathbf{x}, \mathbf{c}$ is fixed by versor $\mathbf{ab}$. Let $\mathbf{y}, \mathbf{z}$ be the points of intersection of spheres $\mathbf{a}^{\sim}, \mathbf{b}^{\sim}$ respectively with line $\mathbf{cx}$. Then versor $\mathbf{ab}$ first dilates vector $\mathbf{cz}$ to vector $\mathbf{cy}$ and then makes the same scaled dilation to vector $\mathbf{cy}$ once again, so it is twice the dilation from point $\mathbf{z}$ to point $\mathbf{y}$ centering at point $\mathbf{c}$.

If $\mathbf{e} \notin \mathbf{I}_2$, by an inversion with respect to a sphere whose center is on 0D sphere $\mathbf{I}_2$, versor $\mathbf{ab}$ can be written as the composition of two inversions and a dilation.

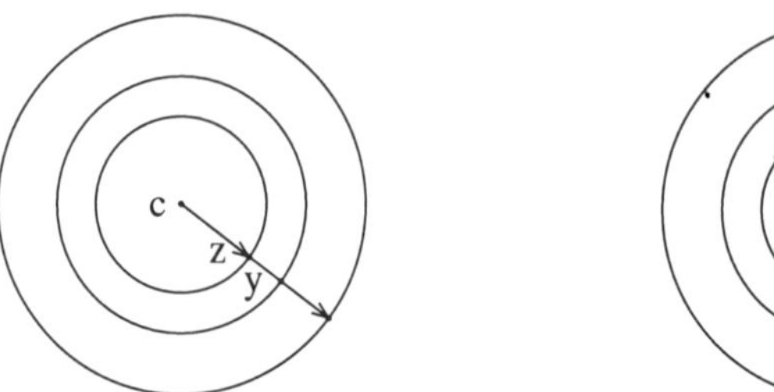

Fig. 8.2   Dilation of positive scale (left), and dilation of negative scale (right).

(3.b) $\mathbf{a}$ is negative, while $\mathbf{b}$ is positive.

If $\mathbf{e} \in \mathbf{I}_2$, let $\mathbf{c} \in \mathcal{N}_{\mathbf{e}}$ be the point in $\mathbf{I}_2$ other than $\mathbf{e}$. Again by rescaling, let $\mathbf{I}_2 = \mathbf{e} \wedge \mathbf{c}$. Since $\mathbf{I}_2\mathbf{e}\mathbf{I}_2^{-1} = -\mathbf{e}$, rotor $\mathbf{I}_2$ induces the *reflection with respect to point* $\mathbf{c}$: any point $\mathbf{x}$ in $\mathbb{E}^n$ is changed to its symmetric point with respect to point $\mathbf{c}$. This transformation is also called the *antipodal transformation* with center $\mathbf{c}$.

By (8.1.1), versor $\mathbf{ab}$ is the composition of an antipodal transformation and a dilation with respect to the same center, so it realizes a dilation of *negative scale* $-e^{-2\theta}$.

If $\mathbf{e} \notin \mathbf{I}_2$, then $\mathbf{ab}$ can be written as the composition of two inversions and a dilation.

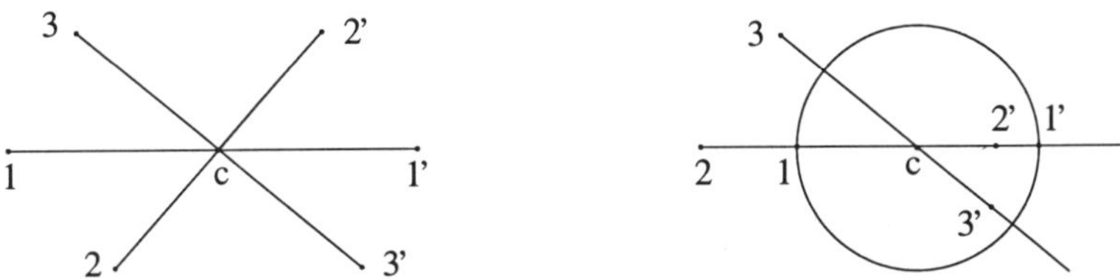

Fig. 8.3   Antipodal transformation (left), and antipodal inversion (right): points $\mathbf{i}, \mathbf{i}'$ are interchanged.

(3.c) $\mathbf{a}, \mathbf{b}$ are both negative.

First consider the versor action of only one negative vector. Let $\mathbf{a} = \mathbf{c}/\rho_1 + \rho_1\mathbf{e}/2$ be a negative unit vector, where $\mathbf{c} \in \mathcal{N}_\mathbf{e}$. Let $\mathbf{s}_1 = \mathbf{c} - \rho_1^2\mathbf{e}/2$. Then

$$\mathbf{a} = \frac{\mathbf{s}_1(\mathbf{e} \wedge \mathbf{c})}{\rho_1\,\mathbf{e} \cdot \mathbf{c}} = -\frac{(\mathbf{e} \wedge \mathbf{c})\mathbf{s}_1}{\rho_1\,\mathbf{e} \cdot \mathbf{c}}. \tag{8.1.3}$$

So $Ad_\mathbf{a}$ is the inversion with respect to sphere $\mathbf{s}_1^\sim$ succeeded (or just the same, preceded) by the reflection with respect to the center of the sphere. This transformation is called the *antipodal inversion* with respect to sphere $\mathbf{s}_1^\sim$.

For two negative unit vectors $\mathbf{a}, \mathbf{b}$, by (8.1.1), $\mathbf{ab}$ is still a dilation of positive scale. If $\mathbf{e} \in \mathbf{I}_2$, then $\mathbf{c}$ is still the center of the dilation. If $\mathbf{e} \notin \mathbf{I}_2$, $\mathbf{ab}$ can be written as the composition of two inversions and a dilation.

Below we seek to represent $\mathbf{ab}$ by a pair of positive vectors. We only consider the case $\mathbf{e} \in \mathbf{I}_2$. Let $\mathbf{I}_2 = \mathbf{e} \wedge \mathbf{c}$, where $\mathbf{c} \in \mathcal{N}_\mathbf{e}$. Since $\mathbf{a}, \mathbf{b}$ are both unit negative vectors in $\mathbf{e} \wedge \mathbf{c}$, they take the form

$$\mathbf{a} = \frac{\mathbf{c}}{\rho_1} + \frac{\rho_1\mathbf{e}}{2}, \qquad \mathbf{b} = \frac{\mathbf{c}}{\rho_2} + \frac{\rho_2\mathbf{e}}{2}. \tag{8.1.4}$$

Let $\mathbf{s}_i = \mathbf{c} - \rho_i^2\mathbf{e}/2$ for $i = 1, 2$. By (8.1.3),

$$\mathbf{ab} = -\frac{\mathbf{s}_1(\mathbf{e} \wedge \mathbf{c})}{\rho_1\,\mathbf{e} \cdot \mathbf{c}}\frac{(\mathbf{e} \wedge \mathbf{c})\mathbf{s}_2}{\rho_2\,\mathbf{e} \cdot \mathbf{c}} = -\frac{\mathbf{s}_1\mathbf{s}_2}{\rho_1\rho_2}. \tag{8.1.5}$$

As an application, given a Minkowski blade $\mathbf{A}_r = \mathbf{a}_1 \wedge \cdots \wedge \mathbf{a}_r \wedge \mathbf{x}$, where $\mathbf{x}$ is a point (null vector) and the $\mathbf{a}_i$ are arbitrary vectors in $\mathbb{R}^{n+1,1}$, *all the non-null vectors among the $\mathbf{a}_i$ can be replaced by either null vectors or tangent directions at point $\mathbf{x}$*, so that $\mathbf{A}_r$ is rewritten as the outer product of several points (null vectors) and a high dimensional tangent directions at point $\mathbf{x}$. This form has much clearer geometric meaning than the original form of $\mathbf{A}_r$.

The rewriting is based on versor action. We use an example to illustrate it.

Example 8.1. Let $\mathbf{A}_3 = \mathbf{a} \wedge \mathbf{x} \wedge \mathbf{y} \in \Lambda(\mathbb{R}^{n+1,1})$, where $\mathbf{x}, \mathbf{y}$ are null vectors representing two points in $\mathbb{E}^n$, and $\mathbf{a}$ is either a positive vector or a negative one. Then $\mathbf{a}$ can be replaced by either a null vector or a tangent vector at point $\mathbf{x}$, under the constraint that the outer product $\mathbf{a} \wedge \mathbf{x} \wedge \mathbf{y}$ is invariant up to scale.

For any null vector $\mathbf{b}$ not orthogonal to $\mathbf{a}$, since blade $\mathbf{a} \wedge \mathbf{b}$ is Minkowski, there is a unique null vector $\mathbf{c} \in \mathbf{a} \wedge \mathbf{b}$ other than $\mathbf{b}$ up to scale. Denote it by $\mathbf{c} = \mathbf{c}(\mathbf{b})$.

First, when $\mathbf{a} \cdot \mathbf{e} \neq 0$, choose $\mathbf{b} = \mathbf{e}$. There are three cases:

(i) If points $\mathbf{c}, \mathbf{x}, \mathbf{y}$ are collinear, then $\mathbf{A}_3 = \mathbf{e} \wedge \mathbf{x} \wedge \mathbf{y}$ up to scale.

(ii) If $\mathbf{c}, \mathbf{x}, \mathbf{y}$ are not collinear, and $\mathbf{a} \cdot \mathbf{x} \neq 0$, then $Ad_{\mathbf{a}}(\mathbf{x}) \in \mathbf{A}_3$ is a null vector that is not equal to $\mathbf{e}$ even up to scale. $\mathbf{A}_3$ represents the circle passing through three different points $\mathbf{x}, \mathbf{y}, Ad_{\mathbf{a}}(\mathbf{x})$.

(iii) If $\mathbf{c}, \mathbf{x}, \mathbf{y}$ are not collinear, and $\mathbf{a} \cdot \mathbf{x} = 0$, then in the GC algebra over the 4D subspace spanned by vectors $\mathbf{e}, \mathbf{c}, \mathbf{x}, \mathbf{y}$, $(\mathbf{e} \wedge \mathbf{c} \wedge \mathbf{x}) \vee (\mathbf{a} \wedge \mathbf{x} \wedge \mathbf{y}) = [\mathbf{ecxy}]\mathbf{a} \wedge \mathbf{x}$. So $\mathbf{x}$ is the point of tangency of line $\mathbf{cx}$ and circle $\mathbf{A}_3$, and $\mathbf{A}_3$ is the circle passing through point $\mathbf{y}$ and tangent to line $\mathbf{cx}$ at point $\mathbf{x}$. Up to scale,

$$\mathbf{A}_3 = ((\mathbf{e} \wedge \mathbf{x})^{-1} \cdot (\mathbf{e} \wedge \mathbf{c} \wedge \mathbf{x})) \wedge \mathbf{x} \wedge \mathbf{y} = -P^{\perp}_{\mathbf{e} \wedge \mathbf{x}}(\mathbf{c}) \wedge \mathbf{x} \wedge \mathbf{y}.$$

Second, when $\mathbf{a} \cdot \mathbf{e} = 0$, then $\mathbf{a}^{\sim}$ represents a hyperplane. Choose point $\mathbf{b}$ arbitrarily. There are also three cases:

(iv) If points $\mathbf{b}, \mathbf{c}, \mathbf{x}, \mathbf{y}$ are either cocircular or collinear, then $\mathbf{A}_3$ represents circle or line $\mathbf{cxy}$.

(v) If the four points are not cocircular, and $\mathbf{a} \cdot \mathbf{x} \neq 0$, then $\mathbf{A}_3$ represents the circle or line passing through three different points $\mathbf{x}, \mathbf{y}, Ad_{\mathbf{a}}(\mathbf{x})$.

(vi) If the four points are not cocircular, and $\mathbf{a} \cdot \mathbf{x} = 0$, then $\mathbf{A}_3$ is the circle or line passing through point $\mathbf{y}$ and tangent to circle $\mathbf{bcx}$ at point $\mathbf{x}$. In the inner-product Grassmann algebra over the 4D subspace spanned by vectors $\mathbf{b}, \mathbf{c}, \mathbf{x}, \mathbf{y}$, the tangent line of circle $\mathbf{bcx}$ at point $\mathbf{x}$ has positive-vector representation

$$\mathbf{d} = \mathbf{e} \cdot (\mathbf{x} \wedge ((\mathbf{b} \wedge \mathbf{c} \wedge \mathbf{x}) \cdot (\mathbf{b} \wedge \mathbf{c} \wedge \mathbf{x} \wedge \mathbf{y}))).$$

So up to scale, $\mathbf{A}_3 = \mathbf{d} \wedge \mathbf{x} \wedge \mathbf{y}$.

Conversely, from the effect of a conformal transformation we can find a versor generating the transformation. The following are some examples.

Example 8.2. The 2D rotation centering at point $\mathbf{c}$, changing point $\mathbf{b}$ to point $\mathbf{a}$ while fixing every point on the $(n-2)$D plane passing through point $\mathbf{c}$ and completely orthogonal to plane $\mathbf{abc}$, is induced by the following rotor in $\mathcal{G}((\mathbf{e} \wedge \mathbf{c})^{\sim})$:

$$((\mathbf{e} \cdot \mathbf{b})P^{\perp}_{\mathbf{e} \wedge \mathbf{c}}(\mathbf{a}) + (\mathbf{e} \cdot \mathbf{a})P^{\perp}_{\mathbf{e} \wedge \mathbf{c}}(\mathbf{b}))P^{\perp}_{\mathbf{e} \wedge \mathbf{c}}(\mathbf{b}) = \mu(\mathbf{e} \wedge \mathbf{c} \wedge \langle \mathbf{aeb} \rangle_1)(\mathbf{e} \wedge \mathbf{c} \wedge \mathbf{b}),$$

where $\mu \in \mathbb{R} - \{0\}$. In particular, the rotation from vector $\mathbf{y} \in \mathbb{R}^n = (\mathbf{e} \wedge \mathbf{c})^{\sim}$ to vector $\mathbf{x} \in \mathbb{R}^n$ is induced by rotor $(\mathbf{x} + \mathbf{y})\mathbf{y}$.

Example 8.3. The translation from point $\mathbf{b}$ to point $\mathbf{a}$ is induced by the following rotor in $\mathcal{G}(\mathbf{e}^\sim)$:

$$(\mathbf{e} \wedge ((\mathbf{e}\cdot\mathbf{a})\mathbf{b} + (\mathbf{e}\cdot\mathbf{b})\mathbf{a}))(\mathbf{e}\wedge\mathbf{b}) = (\mathbf{e}\wedge\langle\mathbf{aeb}\rangle_1)(\mathbf{e}\wedge\mathbf{b}) = -(2 + \mathbf{e}\wedge(\mathbf{a}-\mathbf{b})).$$

In particular, the translation by vector $\mathbf{t} \in \mathbb{R}^n$ is induced by rotor $2 + \mathbf{e}\wedge\mathbf{t} = (2\mathbf{t}^{-1}+\mathbf{e})\mathbf{t}$.

Example 8.4. The dilation centering at point $\mathbf{c}$ and changing point $\mathbf{b}$ to point $\mathbf{a}$, is induced by a rotor in $\mathcal{G}(\mathbf{e}\wedge\mathbf{c})$ of the form $\lambda\mathbf{e}\cdot\mathbf{c} + \mathbf{e}\wedge\mathbf{c}$, where $\lambda \in \mathbb{R}$ is an indeterminate that can be computed as follows.

Geometrically, the dilation changes vector $\overrightarrow{\mathbf{cb}}$ to vector $\overrightarrow{\mathbf{ca}}$ in $\mathbb{R}^n = (\mathbf{e}\wedge\mathbf{c})^\sim$. By (8.1.2), in the inner-product Grassmann algebra over the 3D subspace spanned by vectors $\mathbf{e}, \mathbf{c}, \mathbf{a}$, the dilation changes vector $\mathbf{e}$ to vector

$$\frac{\mathbf{e}\cdot\mathbf{b}[\mathbf{eca}]}{\mathbf{e}\cdot\mathbf{a}[\mathbf{ecb}]}\mathbf{e}.$$

Algebraically, the graded adjoint action of the rotor changes $\mathbf{e}$ to

$$(\lambda\mathbf{e}\cdot\mathbf{c} + \mathbf{e}\wedge\mathbf{c})\mathbf{e}(\lambda\mathbf{e}\cdot\mathbf{c} + \mathbf{e}\wedge\mathbf{c})^{-1} = \frac{\lambda+1}{\lambda-1}\mathbf{e}.$$

So

$$\lambda = -\frac{[\mathbf{ecb}]\mathbf{e}\cdot\mathbf{a} + [\mathbf{eca}]\mathbf{e}\cdot\mathbf{b}}{[\mathbf{ecb}]\mathbf{e}\cdot\mathbf{a} - [\mathbf{eca}]\mathbf{e}\cdot\mathbf{b}} = \frac{[\mathbf{ebc}]\mathbf{e}\cdot\mathbf{a} + [\mathbf{eac}]\mathbf{e}\cdot\mathbf{b}}{[\mathbf{eab}]\mathbf{e}\cdot\mathbf{c}},$$

and the rotor inducing the dilation is

$$[\mathbf{ebc}]\mathbf{e}\cdot\mathbf{a} + [\mathbf{eac}]\mathbf{e}\cdot\mathbf{b} + [\mathbf{eab}]\mathbf{e}\wedge\mathbf{c} = [\mathbf{e}\langle\mathbf{aeb}\rangle_1\mathbf{c}] + [\mathbf{eab}]\mathbf{e}\wedge\mathbf{c}. \qquad (8.1.6)$$

In particular, the reflection with respect to point $\mathbf{c}$ is induced by rotor $\mathbf{e}\wedge\mathbf{c}$.

Example 8.5. A *transversion* in $\mathbb{E}^n$ is defined as the composition of an inversion with respect to a sphere $\mathbf{s}^\sim = (\mathbf{c}-\rho^2\mathbf{e}/2)^\sim$, a translation by vector $\mathbf{t}\in(\mathbf{e}\wedge\mathbf{c})^\sim$, and once again the inversion with respect to $\mathbf{s}^\sim$. A rotor generating the transversion is

$$\mathbf{s}(1+\frac{\mathbf{et}}{2})\mathbf{s} = \lambda(1+\mathbf{c}(\rho^{-2}\mathbf{t})), \qquad (8.1.7)$$

where $\lambda \in \mathbb{R} - \{0\}$. Point $\mathbf{c}$ is invariant under the transversion, and is called the *center of transversion*. $\rho^{-2}\mathbf{t}$ is called the *vector of transversion*.

By a classical theorem of Liouville [130], all orientation-preserving conformal transformations are generated by rotations, translations, dilations of positive scale, and transversions.

Definition 8.6. A *similarity transformation* in $\mathbb{E}^n$ refers to a composition of Euclidean transformations and dilations. A *rigid body motion* refers to an orientation-preserving Euclidean transformation.

For example, all rotations, translations, and dilations of positive scale in $\mathbb{E}^n$, are orientation-preserving similarity transformations. The reflection with respect to a point is orientation-preserving if and only if n is even. The antipodal inversion with respect to a sphere, on the contrary, is orientation-preserving if and only if n is odd.

In $\mathbb{E}^n$, a Euclidean transformation is generated by reflections with respect to hyperplanes, a rigid body motion is generated by rotations and translations. In the conformal model, a conformal transformation is a similarity transformation, or a Euclidean transformation, or a rigid body motion, if and only if it fixes the 1D subspace $\mathbf{e}$, or vector $\mathbf{e}$, or both vector $\mathbf{e}$ and the orientation $\mathbf{I}_{n+2}$ of $\mathbb{R}^{n+1,1}$, respectively.

Having classified all versor generators of length one and two of the conformal group, we are ready to explore the global relationship between the versor group of $\mathcal{G}(\mathbb{R}^{n+1,1})$ and the conformal group of $\mathbb{R}^n$.

The set $\mathcal{N}$ of all null vectors in $\mathbb{R}^{n+1,1}$ has two connected components. In particular, null vectors $\pm\mathbf{a}$ are always in different connected components, as 0 is not a null vector. An orthogonal transformation in $\mathbb{R}^{n+1,1}$ keeping each component of $\mathcal{N}$ invariant is called a *positive orthogonal transformation*. All such transformations form a subgroup $O_+(n+1,1)$ of $O(n+1,1)$, called the *positive orthogonal group*. The orientation-preserving orthogonal transformations of $\mathbb{R}^{n+1,1}$ form another subgroup $SO(n+1,1)$ of $O(n+1,1)$, called the *special orthogonal group*. The intersection of the two subgroups, denoted by $SO_+(n+1,1)$, is called the *Lorentz group*, and its elements are called *Lorentz transformations*. Lorentz transformations are the linear isometries of $\mathbb{R}^{n+1,1}$ connected with the identity transformation.

The negative vectors in $\mathbb{R}^{n+1,1}$ also form two connected components, and each component asymptotes a connected component of $\mathcal{N}$. *Two negative vectors* $\mathbf{a}, \mathbf{b}$ *are in the same connected component if and only if* $\mathbf{a}\cdot\mathbf{b} < 0$. The reason is that because $\mathbf{b}\cdot\mathbf{b} < 0$ and the inner product is continuous, when $\mathbf{a}$ is close to $\mathbf{b}$, $\mathbf{a}\cdot\mathbf{b}$ is always negative.

For each null vector $\mathbf{c}$, the 1D half-space $\{\lambda\mathbf{c}\,|\,\lambda > 0\}$ can be approached by a sequence of negative vectors. For two null vectors of the same component, the two sequences of negative vectors each approaching a 1D half-space generated by one of the two null vectors, belong to the same connected component of negative vectors. By the continuity of the inner product, we have

Proposition 8.7. Any two null vectors $\mathbf{a}, \mathbf{b} \in \mathbb{R}^{n+1,1}$ are in the same connected component if and only if $\mathbf{a}\cdot\mathbf{b} < 0$.

Definition 8.8. A versor in $\mathcal{G}(\mathbb{R}^{n+1,1})$ is said to be *Lorentz* if it is a Clifford monomial where the number of negative vectors is even.

By (8.1.5) and commutations within a versor, any Lorentz versor is equal to a positive versor. They differ only by the representative vectors. Any positive versor induces a positive orthogonal transformation in $\mathbb{R}^{n+1,1}$. To see this we only need to check the graded adjoint action of a positive unit vector $\mathbf{s}$ on a null vector $\mathbf{a}$. By

$$Ad_{\mathbf{s}}(\mathbf{a}) \cdot \mathbf{a} = -\langle \mathbf{sasa} \rangle = -2(\mathbf{a} \cdot \mathbf{s})^2 \leq 0,$$

$Ad_{\mathbf{s}}$ preserves every component of the null vectors.

Proposition 8.9. Any positive rotor induces a unique Lorentz transformation in $\mathbb{R}^{n+1,1}$. Conversely, any Lorentz transformation in $\mathbb{R}^{n+1,1}$ is induced by a positive rotor unique up to scale.

In the conformal model of $\mathbb{E}^n$, any versor $\mathbf{V}$ and its dual $\mathbf{V}^{\sim}$ induce the same conformal transformation. Since $\mathbf{I}_{n+2}$ contains a negative vector, it always interchanges the two connected components of null vectors. So any conformal transformation in $\mathbb{E}^n$ is induced by a positive versor that is unique up to scale.

Any reflection with respect to a hyperplane reverses the orientation of $\mathbb{E}^n$, so does any inversion with respect to a sphere. A conformal transformation is orientation-preserving (or orientation-reversing) if and only if it can be induced by a positive rotor (or positive odd versor).

8.1.2 *Geometric product of Minkowski blades*

In the conformal model, the dual of a Minkowski blade is a Euclidean blade, and the latter is the geometric product of pairwise orthogonal positive vectors. Since a Minkowski blade and its dual induce the same conformal transformation, a geometric product of Minkowski blades has the geometric interpretation of being a versor generator of a conformation transformation.

Below we categorize the geometric products of up to two Minkowski blades in 3D Euclidean geometry. We assume that the generating vectors of any Minkowski blade are null, so that they represent either points or the conformal point at infinity. In this subsection, vectors $\mathbf{a}, \mathbf{b}, \mathbf{c}, \mathbf{a}', \mathbf{b}', \mathbf{c}'$ are always assumed to be null vectors in $\mathcal{N}_{\mathbf{e}}$ representing points in $\mathbb{E}^3$.

First we consider the case of only one Minkowski blade.

Definition 8.10. For $0 \leq r \leq n - 1$, a *reflection* with respect to an rD plane in $\mathbb{E}^n$, refers to the Euclidean transformation induced by the Minkowski blade representation of the rD plane as a versor. An *inversion* with respect to an rD sphere in $\mathbb{E}^n$, refers to the conformal transformation induced by the Minkowski blade representation of the rD sphere as a versor. When $r = n - 1$, they are called *regular reflection* and *regular inversion* respectively, or simply called reflection and inversion by default.

For $s > r$, the reflection with respect to an rD plane fixes every sD plane containing the rD plane. When the reflection is restricted to such an sD plane, it is called a *reflection on* the sD plane. Similarly, the inversion with respect to an rD sphere fixes every sD plane or sphere containing the rD plane. When the inversion is restricted to such an sD plane or sphere, it is called an *inversion on* the sD plane or sphere.

Proposition 8.11. Let $\mathbf{A}_r, \mathbf{B}_s$ be respectively $(r-2)$ and $(s-2)$D spheres or planes in $\mathbb{E}^n$, and let $\mathbf{A}_r \vee \mathbf{B}_s$ be Euclidean. Then sphere or plane $(\mathbf{A}_r \vee \mathbf{B}_s)^\sim$ is invariant under the inversion or mirror reflection with respect to any of $\mathbf{A}_r, \mathbf{B}_s$.

Proof. Direct from $(\mathbf{A}_r \vee \mathbf{B}_s)^\sim = \mathbf{B}_s^\sim \wedge \mathbf{A}_r^\sim$ and the (anti-)commutativity between $\mathbf{B}_s^\sim \wedge \mathbf{A}_r^\sim$ and any of $\mathbf{A}_r^\sim, \mathbf{B}_s^\sim$ in their geometric product. $\qquad\square$

We return to the case of 3D geometry. In $\mathbb{R}^{4,1}$, a Minkowski blade of grade four is either a plane or sphere in space, and induces either an inversion or reflection.

A Minkowski blade of grade two, when taken as a versor, has two possibilities:

(1) $\mathbf{e} \wedge \mathbf{a}$: it generates the reflection with respect to point $\mathbf{a}$. By definition, this is just the reflection with respect to 0D plane $\mathbf{e} \wedge \mathbf{a}$.

(2) $\mathbf{a} \wedge \mathbf{b}$: it generates the inversion with respect to 0D sphere $(\mathbf{a}, \mathbf{b})$, of which the geometric description is as follows (Figure 8.4):

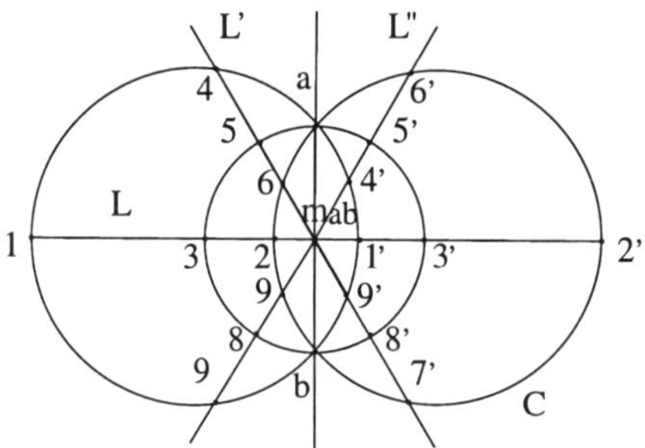

Fig. 8.4   Inversion with respect to 0D sphere $(\mathbf{a}, \mathbf{b})$: points $\mathbf{i}, \mathbf{i}'$ are interchanged.

When restricted to line $\mathbf{ab}$, $Ad_{\mathbf{a} \wedge \mathbf{b}}$ is the regular inversion with respect to circle $(\mathbf{a}, \mathbf{b})$. It interchanges $\mathbf{e}$ and the midpoint $\mathbf{m}_{\mathbf{ab}}$ of line segment $\mathbf{ab}$, but fixes $\mathbf{a}$ and $\mathbf{b}$ respectively.

When restricted to a circle C passing through points $\mathbf{a}, \mathbf{b}$:

- Any line L passing through point $\mathbf{m}_{\mathbf{ab}}$ and perpendicular to line segment $\mathbf{ab}$ is invariant. If C and L are coplanar, then $Ad_{\mathbf{a} \wedge \mathbf{b}}$ interchanges the two points of intersection of C and L.

- Any line L' passing through point $\mathbf{m}_{\mathbf{ab}}$ is changed to its mirror reflection L'' with respect to line $\mathbf{ab}$ in the plane Π supporting lines L' and $\mathbf{ab}$. If circle C lies in plane Π, then the perpendicular bisector L of line segment $\mathbf{ab}$ in plane Π separates the plane into two half-planes, and in each half-plane $Ad_{\mathbf{a} \wedge \mathbf{b}}$ interchanges $L' \cap C$ and $L'' \cap C$.

A Minkowski blade of grade three as a versor also has two possibilities:

(3) $\mathbf{e} \wedge \mathbf{a} \wedge \mathbf{b}$: it generates the *mirror reflection* with respect to line $\mathbf{ab}$, *i.e.*, the $180°$ rotation in space with respect to axis $\mathbf{ab}$. The mirror reflection fixes every plane passing through line $\mathbf{ab}$, and on every such a plane it is a regular reflection.

(4) $\mathbf{a} \wedge \mathbf{b} \wedge \mathbf{c}$: it generates the inversion with respect to circle $\mathbf{abc}$. The geometric description is as follows (Figure 8.5).

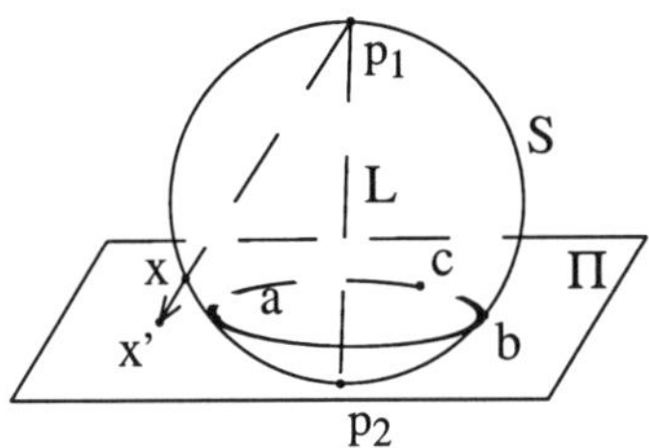

Fig. 8.5   Inversion on a sphere.

On the supporting plane Π of circle $\mathbf{abc}$, $Ad_{\mathbf{a} \wedge \mathbf{b} \wedge \mathbf{c}}$ is a regular inversion, denoted by I. Denote the normal line of plane Π passing through the center of circle $\mathbf{abc}$ by L. Then L intersects every sphere S passing through the circle at two points $\mathbf{p}_1, \mathbf{p}_2$. Do stereographic projection P from point $\mathbf{p}_1$ to plane Π. Then S is projected onto $\Pi \cup \{\mathbf{e}\}$. The inversion on sphere S is just the composition $P^{-1} \circ I \circ P$.

Since any point can be changed into the conformal point at infinity by an inversion, below we consider only the case where any Minkowski blade contains $\mathbf{e}$. The geometric product of such blades always induces a Euclidean transformation, because vector $\mathbf{e}$ is always fixed.

Let $\mathbf{V} = \mathbf{AB}$, where $\mathbf{A}, \mathbf{B}$ are Minkowski blades containing $\mathbf{e}$ and are not equal to each other up to scale. Since the inverse of $\mathbf{AB}$ equals $\mathbf{BA}$ up to scale, without loss of generality, we can assume that the grade of $\mathbf{A}$ is not bigger than that of $\mathbf{B}$. There are all together ten different combinations by grade and by the geometric relationship among the constituent null vectors of the two blades.

(5) Versor

$$\mathbf{V} = (\mathbf{e} \wedge \mathbf{a})(\mathbf{e} \wedge \mathbf{b}) = (\mathbf{e} \cdot \mathbf{a})(\mathbf{e} \cdot \mathbf{b}) + \mathbf{e} \wedge (\mathbf{e} \cdot (\mathbf{a} \wedge \mathbf{b})) \tag{8.1.8}$$

generates twice the translation from point $\mathbf{b}$ to point $\mathbf{a}$.

(6) Versor $\mathbf{V} = (\mathbf{e} \wedge \mathbf{a})(\mathbf{e} \wedge \mathbf{a} \wedge \mathbf{b}) = P^{\perp}_{\mathbf{e} \wedge \mathbf{a}}(\mathbf{b})$ is in fact a positive vector that represents the plane passing through point $\mathbf{a}$ and perpendicular to line $\mathbf{ab}$. $Ad_{\mathbf{V}}$ is the reflection with respect to the plane.

Definition 8.12. (Figure 8.6) In $\mathbb{E}^3$, the *glide mirror reflection* with *gliding vector* $\mathbf{t}$ and mirror plane M, where $\mathbf{t}$ is in the space of displacements of the mirror plane, refers to the transformation that fixes every plane perpendicular to plane M but parallel to vector $\mathbf{t}$, changes every plane parallel to M to its mirror reflection with respect to M, and translates by vector $\mathbf{t}$ every plane perpendicular to both M and $\mathbf{t}$.

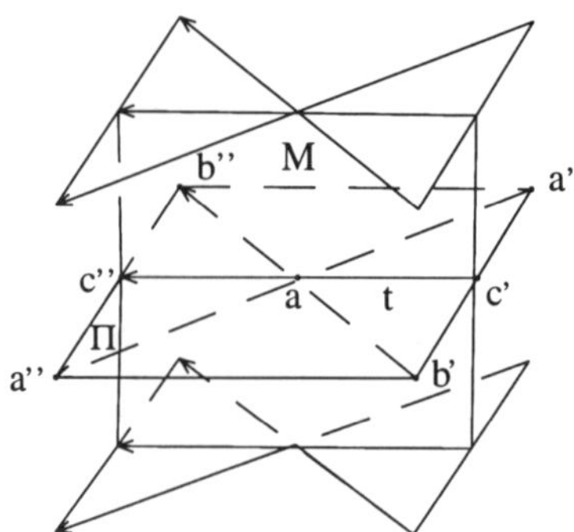

Fig. 8.6   Glide mirror reflection: points $\mathbf{a'}, \mathbf{b'}, \mathbf{c'}$ are changed into points $\mathbf{a''}, \mathbf{b''}, \mathbf{c''}$ respectively.

(7) Let $\mathbf{V} = (\mathbf{e} \wedge \mathbf{a})(\mathbf{e} \wedge \mathbf{a'} \wedge \mathbf{b'})$, where point $\mathbf{a}$ is not on line $\mathbf{a'b'}$.

As shown in Figure 8.6, let $\mathbf{c'}$ be the foot drawn from $\mathbf{a}$ to line $\mathbf{a'b'}$, and let $\mathbf{c''}$ be the reflection of $\mathbf{c'}$ with respect to point $\mathbf{a}$. Let Π be the plane supporting point $\mathbf{a}$ and line $\mathbf{a'b'}$, and let M be the plane passing through line $\mathbf{c'c''}$ and perpendicular to plane Π. Then $Ad_{\mathbf{V}}$ is the glide mirror reflection with gliding vector $\overrightarrow{\mathbf{c'c''}}$ and mirror plane M.

The glide reflection transforms uniformly on every plane parallel to Π. On plane Π, line $\mathbf{c'c''}$ is invariant, and every point on the line undergoes a translation by vector $\overrightarrow{\mathbf{c'c''}}$. Line $\mathbf{a'b'}$ is first translated by vector $\overrightarrow{\mathbf{c'c''}}$, and then flipped over on plane Π, with line $\mathbf{c'c''}$ as the flipping axis.

By setting $\mathbf{a}$ to be the origin of $\mathbb{R}^3$, choosing $\mathbf{a'} = \mathbf{c'}$, setting $\mathbf{t} = \overrightarrow{\mathbf{a'a}}$ and $\mathbf{n} = \overrightarrow{\mathbf{a'b'}}$ in $\mathbb{R}^3$, we can write $\mathbf{e} \wedge \mathbf{a'} \wedge \mathbf{b'}$ as $(\mathbf{e} \wedge \mathbf{a})\mathbf{n} - \mathbf{etn}$. The following is the *normal form* of the versor inducing the glide mirror reflection with gliding vector $2\mathbf{t}$ and mirror plane through the origin and normal to vector $\mathbf{n}$:

$$\mathbf{V} = \mathbf{n}(1 + \mathbf{et}) = (1 + \mathbf{et})\mathbf{n}. \tag{8.1.9}$$

(8) Versor $\mathbf{V} = (\mathbf{e} \wedge \mathbf{a})(\mathbf{e} \wedge \mathbf{a} \wedge \mathbf{b'} \wedge \mathbf{c'})$ is a 2-blade in $\Lambda(\mathbb{R}^{4,1})$ dual to the Minkowski 3-blade representing the line passing through point $\mathbf{a}$ and normal to plane $\mathbf{abc}$. $Ad_{\mathbf{V}}$ is the mirror reflection with respect to the line.

Definition 8.13. The *screw motion*, also called *spiral displacement*, refers to the composition of a rotation and a translation along the axis of the rotation.

When the rotation angle is 180°, the screw motion is called a *glide axial reflection*, where the *gliding vector* is the translational vector, and where the *axis* is the axis of rotation in the screw motion.

(9) Let $\mathbf{V} = (\mathbf{e} \wedge \mathbf{a})(\mathbf{e} \wedge \mathbf{a'} \wedge \mathbf{b'} \wedge \mathbf{c'})$, where point $\mathbf{a}$ is not on plane $\mathbf{a'b'c'}$.

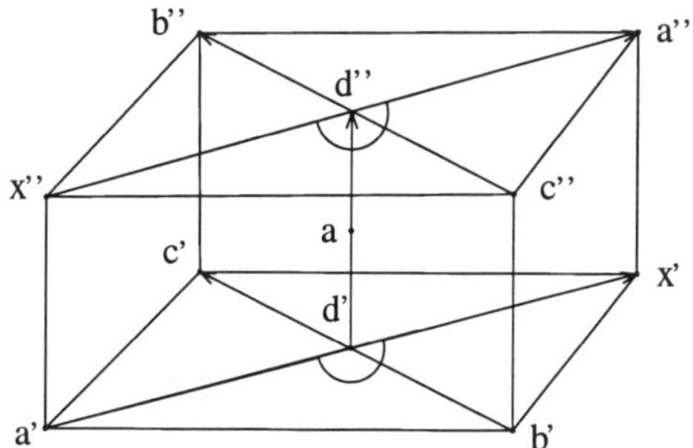

Fig. 8.7   Glide axial reflection: points $\mathbf{a}', \mathbf{b}', \mathbf{c}', \mathbf{d}', \mathbf{x}'$ are changed into points $\mathbf{a}'', \mathbf{b}'', \mathbf{c}'', \mathbf{d}'', \mathbf{x}''$ respectively.

As shown in Figure 8.7, let $\mathbf{d}'$ be the foot drawn from $\mathbf{a}$ to the plane, and let $\mathbf{d}''$ be the reflection of $\mathbf{d}'$ with respect to point $\mathbf{a}$. Then $Ad_{\mathbf{V}}$ is the glide axial reflection with gliding vector $\overrightarrow{\mathbf{d}'\mathbf{d}''}$ and axis $\mathbf{d}'\mathbf{d}''$. It translates by vector $\overrightarrow{\mathbf{d}'\mathbf{d}''}$ every plane that is parallel to plane $\mathbf{a}'\mathbf{b}'\mathbf{c}'$, at the same time makes a 180° rotation on such a plane.

By setting $\mathbf{a}$ to be the origin of $\mathbb{R}^3$, choosing $\mathbf{a}' = \mathbf{d}'$ and $\overrightarrow{\mathbf{a}'\mathbf{b}'} \perp \overrightarrow{\mathbf{a}'\mathbf{c}'}$, and setting $\mathbf{t} = \overrightarrow{\mathbf{a}'\mathbf{a}}$, $l_1 = \overrightarrow{\mathbf{a}'\mathbf{b}'}$, $l_2 = \overrightarrow{\mathbf{a}'\mathbf{c}'}$, $l_1 l_2 = \mathbf{I}_2$ in $\Lambda(\mathbb{R}^3)$, so that $\mathbf{e} \wedge \mathbf{a}' \wedge \mathbf{b}' \wedge \mathbf{c}' = (\mathbf{e} \wedge \mathbf{a})l_1 l_2 - \mathbf{e}\mathbf{t}l_1 l_2$, we get the following *normal form* of the versor inducing the glide axial reflection with gliding vector $2\mathbf{t}$ and axis through the origin:

$$\mathbf{V} = \mathbf{I}_2(1 + \mathbf{e}\mathbf{t}) = (1 + \mathbf{e}\mathbf{t})\mathbf{I}_2. \tag{8.1.10}$$

(10) Let $\mathbf{V} = (\mathbf{e} \wedge \mathbf{a} \wedge \mathbf{b})(\mathbf{e} \wedge \mathbf{a}' \wedge \mathbf{b}')$, where lines $\mathbf{a}\mathbf{b}, \mathbf{a}'\mathbf{b}'$ are coplanar.

If the two lines intersect, let $\mathbf{c}$ be the point of intersection, and let L be the line perpendicular to both lines and passing through point $\mathbf{c}$, then $Ad_{\mathbf{V}}$ is the rotation with axis L and twice the angle of rotation from line $\mathbf{a}'\mathbf{b}'$ to line $\mathbf{a}\mathbf{b}$. If the two lines are parallel, let $\mathbf{t} \in \mathbb{R}^3$ be the translational vector from line $\mathbf{a}'\mathbf{b}'$ to line $\mathbf{a}\mathbf{b}$, then $Ad_{\mathbf{V}}$ is the translation by vector $2\mathbf{t}$.

(11) Let $\mathbf{V} = (\mathbf{e} \wedge \mathbf{a} \wedge \mathbf{b})(\mathbf{e} \wedge \mathbf{a}' \wedge \mathbf{b}')$, where lines $\mathbf{a}\mathbf{b}, \mathbf{a}'\mathbf{b}'$ are noncoplanar.

Let L be the common perpendicular of the two lines, and let $\mathbf{c}, \mathbf{c}'$ be the intersections of L with the two lines respectively. Then $Ad_{\mathbf{V}}$ is the screw motion with translational vector $2\overrightarrow{\mathbf{c}'\mathbf{c}}$ and twice the rotation from vector $\overrightarrow{\mathbf{a}'\mathbf{b}'}$ to vector $\overrightarrow{\mathbf{a}\mathbf{b}}$ having L as the axis.

By setting $\mathbf{c}$ to be the origin of $\mathbb{R}^3$, choosing $\mathbf{a}' = \mathbf{c}'$, and setting $\mathbf{t} = \overrightarrow{\mathbf{a}'\mathbf{a}}$, $l_1 = \overrightarrow{\mathbf{a}\mathbf{b}}$, $l_2 = \overrightarrow{\mathbf{a}'\mathbf{b}'}$, we get the following *normal form* of the versor inducing the screw motion:

$$\mathbf{V} = l_1 l_2(1 + \mathbf{e}\mathbf{t}) = (1 + \mathbf{e}\mathbf{t})l_1 l_2. \tag{8.1.11}$$

The common perpendicular of the two lines is the intersection of two planes $\mathbf{e} \wedge \mathbf{a} \wedge \mathbf{b} \wedge \mathbf{t}$ and $\mathbf{e} \wedge \mathbf{a}' \wedge \mathbf{b}' \wedge \mathbf{t}$. Let $\mathbf{I}_3 = (\mathbf{e} \wedge \mathbf{e}_0)^{\sim}$ in $\Lambda(\mathbb{R}^{4,1})$. Then $\mathbf{t}$ can be expressed up to scale as

$$\mathbf{t} = \{((\mathbf{e} \wedge \mathbf{a} \wedge \mathbf{b}) \vee \mathbf{I}_3) \wedge ((\mathbf{e} \wedge \mathbf{a}' \wedge \mathbf{b}') \vee \mathbf{I}_3)\}\mathbf{I}_3. \tag{8.1.12}$$

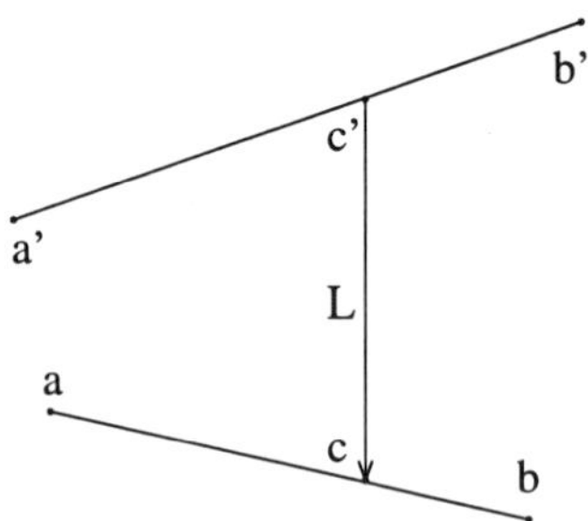

Fig. 8.8   Screw motion.

(8.1.12) requires the use of the origin $\mathbf{e}_0$. We can change it into an origin-free expression as follows. Since $\mathbf{A}_2 = (\mathbf{e} \wedge \mathbf{a} \wedge \mathbf{b})^{\sim}$ is the 2D normal direction of line $\mathbf{ab}$, and $\mathbf{A}'_2 = (\mathbf{e} \wedge \mathbf{a}' \wedge \mathbf{b}')^{\sim}$ is the 2D normal direction of line $\mathbf{a}'\mathbf{b}'$, their 1D intersection should be $\mathbf{t}$ up to scale. In the affine GC algebra $\Lambda(\mathbf{e}^{\sim}) = \Lambda(\mathbf{e} \wedge \mathbb{R}^{4,1})$, up to scale,

$$\mathbf{e} \wedge \mathbf{t} = (\mathbf{e} \wedge \mathbf{A}_2) \vee (\mathbf{e} \wedge \mathbf{A}'_2). \tag{8.1.13}$$

(12) Let line $\mathbf{ab}$ be on plane $\mathbf{a}'\mathbf{b}'\mathbf{c}'$. Let $\mathbf{V} = (\mathbf{e} \wedge \mathbf{a} \wedge \mathbf{b})(\mathbf{e} \wedge \mathbf{a} \wedge \mathbf{b} \wedge \mathbf{c}')$. Then $\mathbf{V}$ is a positive vector representing the plane passing through line $\mathbf{ab}$ and perpendicular to plane $\mathbf{abc}'$. $Ad_{\mathbf{V}}$ is the mirror reflection with respect to the plane.

Definition 8.14. In $\mathbb{E}^3$, an *antipodal rotation*, also called *reflexive rotation*, is the composition of a rotation and a reflection with respect to a point on the axis of the rotation, see Figure 8.9.

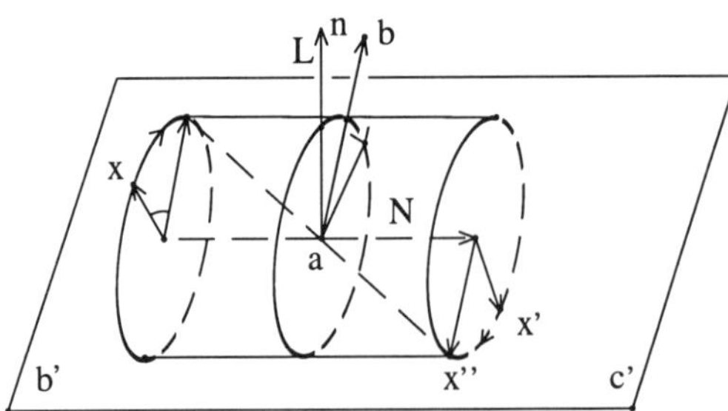

Fig. 8.9   Antipodal rotation: point $\mathbf{x}$ is changed into point $\mathbf{x}'$ by reflection with respect to center $\mathbf{a}$, and then changed into point $\mathbf{x}''$ by rotation with axis N.

(13) Let line $\mathbf{ab}$ intersect plane $\mathbf{a}'\mathbf{b}'\mathbf{c}'$ at point $\mathbf{a}$, and let point $\mathbf{b}$ be outside the plane. Let $\mathbf{V} = (\mathbf{e} \wedge \mathbf{a} \wedge \mathbf{b})(\mathbf{e} \wedge \mathbf{a} \wedge \mathbf{b}' \wedge \mathbf{c}')$.

As shown in Figure 8.9, let L be the line passing through point $\mathbf{a}$ and perpendicular to plane $\mathbf{ab}'\mathbf{c}'$. Let the tangent directions of lines $\mathbf{ab}, \mathbf{ab}', \mathbf{ac}'$ at point $\mathbf{a}$ be $\mathbf{t}, \mathbf{t}_1, \mathbf{t}_2 \in (\mathbf{e} \wedge \mathbf{a})^{\sim}$ respectively. Denote $\mathbf{I}_3 = (\mathbf{e} \wedge \mathbf{a})^{\sim}$. Then $\mathbf{n} = (\mathbf{t}_1 \wedge \mathbf{t}_2)\mathbf{I}_3^{-1}$ is the normal direction of plane $\mathbf{ab}'\mathbf{c}'$, and

$$\mathbf{V} = (\mathbf{e} \wedge \mathbf{a})\mathbf{t}(\mathbf{e} \wedge \mathbf{a})(\mathbf{t}_1 \wedge \mathbf{t}_2) = \mathbf{t}(\mathbf{t}_1 \wedge \mathbf{t}_2) = \mathbf{t}\mathbf{n}\mathbf{I}_3. \tag{8.1.14}$$

Let N be the line in plane $\mathbf{ab'c'}$ passing through point $\mathbf{a}$ and perpendicular to line $\mathbf{ab}$. Then $Ad_{\mathbf{V}}$ is an antipodal rotation: it is the composition of twice the rotation from line L to line $\mathbf{ab}$ with axis N, and the reflection centered at point $\mathbf{a}$.

(14) Let $\mathbf{V} = (\mathbf{e}\wedge\mathbf{a}\wedge\mathbf{b})(\mathbf{e}\wedge\mathbf{a'}\wedge\mathbf{b'}\wedge\mathbf{c'})$, where line $\mathbf{ab}$ is parallel to plane $\mathbf{a'b'c'}$.

As shown in Figure 8.10, let $\mathbf{d'}$ be the foot drawn from $\mathbf{a}$ to plane $\mathbf{a'b'c'}$. Then $Ad_{\mathbf{V}}$ is the glide mirror reflection with gliding vector $2\overrightarrow{\mathbf{d'a}}$ and mirror plane $\mathbf{abd'}$.

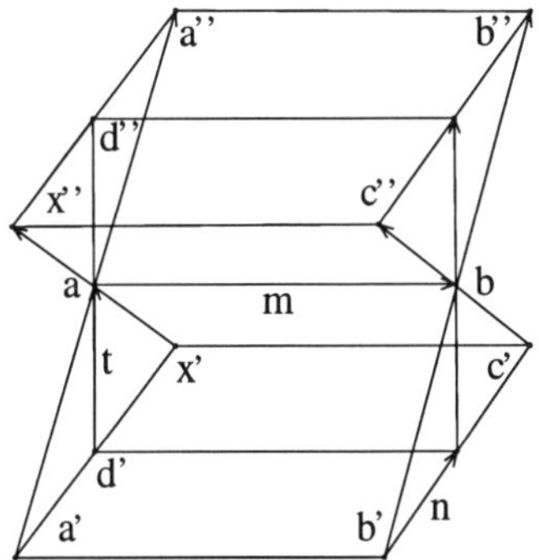

Fig. 8.10 Glide mirror reflection induced by a parallel pair of line and plane: points $\mathbf{a'}, \mathbf{b'}, \mathbf{c'}, \mathbf{d'}, \mathbf{x'}$ are changed into points $\mathbf{a''}, \mathbf{b''}, \mathbf{c''}, \mathbf{d''}, \mathbf{x''}$ respectively.

By setting $\mathbf{a}$ to be the origin of $\mathbb{R}^3$, choosing $\mathbf{a'} = \mathbf{d'}$, and setting $\mathbf{t} = \overrightarrow{\mathbf{a'a}}$, $\mathbf{m} = \overrightarrow{\mathbf{ab}} = \overrightarrow{\mathbf{a'b'}}$, $\mathbf{n} = \overrightarrow{\mathbf{b'c'}}$ in $\mathbb{R}^3$, so that $\mathbf{m}$ is a unit vector, we get the same normal form of the versor inducing the glide mirror reflection as (8.1.9):

$$\mathbf{V} = (\mathbf{e}\wedge\mathbf{a})\mathbf{m}\{(\mathbf{e}\wedge\mathbf{a})\mathbf{mn} - \mathbf{etmn}\} = \mathbf{n}(1+\mathbf{et}) = (1+\mathbf{et})\mathbf{n}. \tag{8.1.15}$$

8.2 Cayley transform and exterior exponential

In this section, we investigate rotors in the conformal model by their Lie algebra generators.

Recall that for a nondegenerate inner-product space $\mathcal{V}^n$, the group of rotors differs from $\mathrm{Spin}(\mathcal{V}^n)$ by a factor $\mathbb{K} - \{0\}$. Since the Lie algebra of the spin group is all bivectors in $\Lambda(\mathcal{V}^n)$, any rotor nearby the identity can be expressed up to scale as the exponential $e^{\mathbf{B}_2}$ of a bivector $\mathbf{B}_2 \in \Lambda^2(\mathcal{V}^n)$. By a classical theorem of Riesz [150], if $\mathcal{V}^n$ is (anti-)Euclidean or (anti-)Minkowski, then any linear isometry of $\mathcal{V}^n$ connected with the identity is induced by a rotor of the exponential form.

As a corollary of Riesz Theorem, any positive rotor in $\mathcal{G}(\mathbb{R}^{n+1,1})$ is in the range of the exponential map, and all orientation-preserving conformal transformations are induced by rotors of the exponential form.

Definition 8.15. In $\Lambda^2(\mathcal{V}^n)$, where $\mathcal{V}^n$ is an inner-product space, if

$$\mathbf{B}_2 = \mathbf{A}_1 + \mathbf{A}_2 + \cdots + \mathbf{A}_r, \tag{8.2.1}$$

where $2r$ is the rank of $\mathbf{B}_2$, the $\mathbf{A}_i$ are 2-blades such that for any $i \neq j$, $\mathbf{A}_i$ and $\mathbf{A}_j$ are completely orthogonal to each other, then (8.2.1) is called a *completely* (or *totally*) *orthogonal decomposition* of bivector $\mathbf{B}_2$.

It is a classical result in the eigenvalue (or spectral) theory of orthogonal transformations [150], that if $\mathcal{V}^n$ is (anti-)Euclidean or (anti-)Minkowski, then any bivector in $\Lambda(\mathcal{V}^n)$ has a completely orthogonal decomposition. The decomposition (8.2.1) is unique if and only if for any $i \neq j$, $\mathbf{A}_i^2 \neq \mathbf{A}_j^2$. For any other nondegenerate real inner-product space, the conclusion is incorrect.

In the current setting of the conformal model of $\mathbb{E}^n$, any bivector $\mathbf{B}_2 \in \Lambda(\mathbb{R}^{n+1,1})$ has a completely orthogonal decomposition as follows:

$$\mathbf{B}_2 = \mathbf{A}_1 + \mathbf{A}_2 + \cdots \mathbf{A}_r = \lambda_1 \mathbf{a}_1 \mathbf{b}_1 + \lambda_2 \mathbf{a}_2 \mathbf{b}_2 + \cdots + \lambda_r \mathbf{a}_r \mathbf{b}_r, \tag{8.2.2}$$

where $\lambda_i \neq 0$, blades $\mathbf{A}_1, \ldots, \mathbf{A}_{r-1}$ are Euclidean, and blade $\mathbf{A}_r$ is either Euclidean, or degenerate, or Minkowski. Correspondingly, for the vectors $\mathbf{a}_i$ and $\mathbf{b}_j$,

(1) $\mathbf{a}_i \cdot \mathbf{b}_j = 0$ for any i, j,
(2) $\mathbf{a}_i \cdot \mathbf{a}_j = 0$ and $\mathbf{b}_i \cdot \mathbf{b}_j = 0$ for any $i \neq j$,
(3) $\mathbf{a}_i^2 = 1$ for any i,
(4) $\mathbf{b}_i^2 = 1$ for any $i \neq r$,
(5) $\mathbf{b}_r^2 = 1$, or 0, or -1.

With respect to the decomposition (8.2.2), the exponential map has the following (hyperbolic) trigonometric function form:

$$\begin{aligned} e^{\mathbf{B}_2} &= e^{\mathbf{A}_1} e^{\mathbf{A}_2} \cdots e^{\mathbf{A}_r} \\ &= (\cos \lambda_1 + \mathbf{a}_1 \mathbf{b}_1 \sin \lambda_1) \cdots (\cos \lambda_{r-1} + \mathbf{a}_{r-1} \mathbf{b}_{r-1} \sin \lambda_{r-1}) \mu, \end{aligned} \tag{8.2.3}$$

where

$$\mu = \begin{cases} \cos \lambda_r + \mathbf{a}_r \mathbf{b}_r \sin \lambda_r, & \text{if } \mathbf{b}_r^2 = 1, \\ 1 + \lambda_r \mathbf{a}_r \mathbf{b}_r, & \text{if } \mathbf{b}_r^2 = 0, \\ \sinh \lambda_r + \mathbf{a}_r \mathbf{b}_r \cosh \lambda_r, & \text{if } \mathbf{b}_r^2 = -1. \end{cases} \tag{8.2.4}$$

Example 8.16. The Lie algebra representation of 3D rigid body motions.

Any rigid body motion in space can be decomposed into a rotation followed by a translation. It can also be decomposed into a translation followed by a rotation. The decomposition is not unique without fixing the axis of rotation. However, there is a unique decomposition, in which the axis of rotation follows exactly the direction of translation. This is the screw motion. So any rigid body motion is a screw motion, and the unique decomposition theorem is known as *Chasles' Theorem* [77].

In the Lie algebra $\Lambda^2(\mathbf{e}^\sim)$ of the spin group of rigid body motions, where the dual operator is in $\Lambda(\mathbb{R}^{4,1})$, the unique decomposition of a rigid body motion corresponds to the unique completely orthogonal decomposition of the bivector $\mathbf{B}_2$ whose exponential map generates the rigid body motion. The uniqueness is because if $\mathbf{B}_2 = \mathbf{C}_1 + \mathbf{C}_2$ is such a decomposition, then $\mathbf{C}_1^2 < 0 = \mathbf{C}_2^2$. Below we compute $\mathbf{C}_1$ and $\mathbf{C}_2$ explicitly.

Let $\mathbf{e}_1, \mathbf{e}_2, \mathbf{e}_3$ be an orthonormal basis of $\mathbb{R}^3$. Then $\Lambda^2(\mathbf{e}^\sim)$ has an orthonormal basis $\mathbf{e}_1\mathbf{e}_2, \mathbf{e}_2\mathbf{e}_3, \mathbf{e}_1\mathbf{e}_3, \mathbf{e}\mathbf{e}_1, \mathbf{e}\mathbf{e}_2, \mathbf{e}\mathbf{e}_3$. Any nonzero bivector in $\Lambda^2(\mathbf{e}^\sim)$ can be written as

$$\mathbf{B}_2 = \frac{\mathbf{I}_2\theta + \mathbf{et}}{2}, \tag{8.2.5}$$

where 2-blade $\mathbf{I}_2 \in \Lambda(\mathbb{R}^3)$ is of unit magnitude, $\theta \in \mathbb{R}$, and $\mathbf{t} \in \mathbb{R}^3$.

If $\theta = 0$, then $e^{\mathbf{B}_2}$ induces the translation by vector $\mathbf{t}$. If $\theta \neq 0$, then

$$\mathbf{C}_1 = \frac{\mathbf{I}_2\theta + \mathbf{e}P_{\mathbf{I}_2}(\mathbf{t})}{2} = \frac{1 - \mathbf{e}(\mathbf{t} \cdot \mathbf{I}_2)\theta^{-1}}{2}\mathbf{I}_2\theta, \qquad \mathbf{C}_2 = \frac{\mathbf{e}P_{\mathbf{I}_2}^\perp(\mathbf{t})}{2}, \tag{8.2.6}$$

and

$$\begin{aligned}
e^{\mathbf{B}_2} &= e^{\mathbf{C}_1}e^{\mathbf{C}_2} \\
&= e^{-\mathbf{e}(\mathbf{t}\cdot\mathbf{I}_2)/(2\theta)}e^{\mathbf{I}_2\theta/2}e^{\mathbf{e}(\mathbf{t}\cdot\mathbf{I}_2)/(2\theta)}e^{\mathbf{e}P_{\mathbf{I}_2}^\perp(\mathbf{t})/2} \\
&= \cos\frac{\theta}{2} + \mathbf{I}_2\sin\frac{\theta}{2} + \frac{1}{\theta}\mathbf{e}P_{\mathbf{I}_2}(\mathbf{t})\sin\frac{\theta}{2} + \frac{1}{2}\mathbf{e}P_{\mathbf{I}_2}^\perp(\mathbf{t})\cos\frac{\theta}{2} + \frac{1}{2}\mathbf{e}P_{\mathbf{I}_2}^\perp(\mathbf{t})\mathbf{I}_2\sin\frac{\theta}{2}.
\end{aligned}$$

$e^{\mathbf{B}_2}$ induces a screw motion with the vector of translation $P_{\mathbf{I}_2}^\perp(\mathbf{t})$, the axis of rotation passing through point $-\mathbf{t} \cdot \mathbf{I}_2/\theta \in \mathbb{R}^3$, and the angle of rotation $-\theta$.

In application, rational polynomial functions are much simpler than exponentials or trigonometric functions. For the special orthogonal group $SO(p,q)$, whose Lie algebra $so(p,q)$ is the set of *antisymmetric linear transformations* g in $\mathbb{R}^{p,q}$, *i.e.*, $g(\mathbf{x}) \cdot \mathbf{y} = \mathbf{x} \cdot g(\mathbf{y})$ for all $\mathbf{x}, \mathbf{y} \in \mathbb{R}^{p,q}$, besides the exponential map, there is also a classical rational polynomial map from the Lie algebra to the Lie group, called *Cayley transform* [130].

Definition 8.17. The mapping

$$\begin{aligned}
so(p,q) &\longrightarrow SO(p,q) \\
g &\longmapsto (I_{\mathbb{R}^{p,q}} + g)(I_{\mathbb{R}^{p,q}} - g)^{-1}, \text{ where } I_{\mathbb{R}^{p,q}} - g \text{ is invertible},
\end{aligned} \tag{8.2.7}$$

is called the *Cayley transform* from $so(p,q)$ to $SO(p,q)$. The mapping is injective but generally not surjective.

In the conformal model of $\mathbb{E}^3$, any orientation-preserving conformal transformation is induced by a Lorentz transformation in $\mathbb{R}^{4,1}$, which in turn is induced by a positive rotor of exponential form in $\mathcal{G}(\mathbb{R}^{4,1})$. A natural idea is to consider simplifying the exponential map by a fractional linear map in the setting of $\mathcal{CL}(\mathbb{R}^{4,1})$, similar to the matrix transformation (8.2.7).

Definition 8.18. [130] The following mapping C:

$$\begin{aligned}
\Lambda^2(\mathbb{R}^{4,1}) &\longrightarrow \mathcal{G}(\mathbb{R}^{4,1}) \\
\mathbf{B}_2 &\longmapsto (1 + \mathbf{B}_2)(1 - \mathbf{B}_2)^{-1}, \text{ where } 1 - \mathbf{B}_2 \text{ is invertible},
\end{aligned} \tag{8.2.8}$$

is called the *Cayley transform* from Lie algebra $\Lambda^2(\mathbb{R}^{4,1})$ to the group of rotors in $\mathcal{G}(\mathbb{R}^{4,1})$.

Proposition 8.19. (1) If $\mathbf{B}_2$ is a blade, then $1 - \mathbf{B}_2$ is not invertible if and only if $\mathbf{B}_2$ is Minkowski and of unit magnitude. $1 - \mathbf{B}_2$ is invertible if and only if $1 + \mathbf{B}_2$ is invertible. When $1 - \mathbf{B}_2$ is invertible,

$$(1 - \mathbf{B}_2)^{-1} = \frac{1 + \mathbf{B}_2}{1 - \mathbf{B}_2^2}, \tag{8.2.9}$$

and in the notation of (8.2.2), $C(\mathbf{B}_2)$ is the Lorentz rotor $(\mathbf{a}_1(\mathbf{a}_1 + \lambda_1 \mathbf{b}_1))^2$.

(2) If $\mathbf{B}_2$ is not a blade, then in the notation of (8.2.2), $1 - \mathbf{B}_2$ is always invertible,

$$(1 - \mathbf{B}_2)^{-1} = \frac{(1 + \mathbf{A}_1 - \mathbf{A}_2)(1 - \mathbf{A}_1 + \mathbf{A}_2)(1 + \mathbf{A}_1 + \mathbf{A}_2)}{1 + \mathbf{A}_1^4 + \mathbf{A}_2^4 - 2\mathbf{A}_1^2 - 2\mathbf{A}_2^2 - 2\mathbf{A}_1^2\mathbf{A}_2^2}, \tag{8.2.10}$$

and $C(\mathbf{B}_2)$ is the positive rotor $\mathbf{a}_1(x\mathbf{a}_1 + \mathbf{b}_1)\mathbf{a}_2(y\mathbf{a}_2 + \mathbf{b}_2)$, where

$$\begin{cases} x = \dfrac{1 - \lambda_1^2 + \lambda_2^2}{2\lambda_1}, & y = \dfrac{1 - \lambda_2^2 + \lambda_1^2}{2\lambda_2}, & \text{if } \mathbf{b}_2^2 = 1; \\[2ex] x = \dfrac{1 - \lambda_1^2}{2\lambda_1}, & y = \dfrac{1 + \lambda_1^2}{2\lambda_2}, & \text{if } \mathbf{b}_2^2 = 0; \\[2ex] x = \dfrac{1 - \lambda_1^2 - \lambda_2^2}{2\lambda_1}, & y = \dfrac{1 + \lambda_1^2 + \lambda_2^2}{2\lambda_2}, & \text{if } \mathbf{b}_2^2 = -1. \end{cases} \tag{8.2.11}$$

Proof.   (1) If $\mathbf{B}_2^2 \neq 1$, then (8.2.9) is obvious. If $\mathbf{B}_2^2 = 1$, then $1 - \mathbf{B}_2 = \mathbf{a}_1(\mathbf{a}_1 - \mathbf{b}_1)$. Since $\mathbf{a}_1 - \mathbf{b}_1$ is null, $1 - \mathbf{B}_2$ is not invertible.

(2) The *discriminant* for the invertibility of $1 - \mathbf{B}_2$ is

$$\Delta_{\mathbf{B}_2} = (1 + \mathbf{A}_1 + \mathbf{A}_2)(1 + \mathbf{A}_1 - \mathbf{A}_2)(1 - \mathbf{A}_1 + \mathbf{A}_2)(1 - \mathbf{A}_1 - \mathbf{A}_2). \tag{8.2.12}$$

$1 - \mathbf{B}_2$ is invertible if and only if $\Delta_{\mathbf{B}_2} \neq 0$. In the notation of (8.2.2),

$$\begin{aligned} \Delta_{\mathbf{B}_2} &= 1 + \mathbf{A}_1^4 + \mathbf{A}_2^4 - 2\mathbf{A}_1^2 - 2\mathbf{A}_2^2 - 2\mathbf{A}_1^2\mathbf{A}_2^2 \\ &= \begin{cases} (1 + (\lambda_1 - \lambda_2)^2)(1 + (\lambda_1 + \lambda_2)^2), & \text{if } \mathbf{b}_2^2 = 1 \\ (1 + \lambda_1^2)^2, & \text{if } \mathbf{b}_2^2 = 0 \\ (\lambda_1^2 + (1 + \lambda_2)^2)(\lambda_1^2 + (1 - \lambda_2)^2), & \text{if } \mathbf{b}_2^2 = -1 \end{cases} \\ &> 0, \end{aligned} \tag{8.2.13}$$

so $1 - \mathbf{B}_2$ is invertible. From

$$\begin{aligned} &(1 + \mathbf{A}_1 + \mathbf{A}_2)(1 + \mathbf{A}_1 - \mathbf{A}_2)(1 - \mathbf{A}_1 + \mathbf{A}_2)(1 + \mathbf{A}_1 + \mathbf{A}_2) \\ &= (1 + \mathbf{A}_1^2 - \mathbf{A}_2^2 + 2\mathbf{A}_1)(1 + \mathbf{A}_2^2 - \mathbf{A}_1^2 + 2\mathbf{A}_2), \end{aligned}$$

we get (8.2.11). That the vector $y\mathbf{a}_2 + \mathbf{b}_2$ is positive is easy to verify. $\quad\square$

Theorem 8.20. The domain of definition of Cayley transform C is all bivectors except the Minkowski blades of unit magnitude, and is a set $\mathbb{R}^{10} - V^5$, where V^5 is a 5D algebraic variety in $\mathbb{R}^{10}$. The image space of C modulo scale is all positive rotors except those of the form $\mathbf{a}_1\mathbf{a}_2\mathbf{a}_3\mathbf{a}_4$, where the $\mathbf{a}_i$ are pairwise orthogonal positive vectors. Geometrically, the image space modulo scale is composed of positive rotors generating all orientation-preserving conformal transformations except the antipodal inversions; topologically, it is the remainder of the Lorentz group of $\mathbb{R}^{4,1}$, which is a 10D connected Lie group, after removal of a 4D open disk.

Proof. All bivectors in $\mathbb{R}^{4,1}$ form a 10D real vector space. The Minkowski 2-blades in $\Lambda(\mathbb{R}^{4,1})$ form a Grassmann variety $O(4,1)/(O(1,1) \times O(3))$, whose dimension is $10 - 1 - 3 = 6$. The unit Minkowski 2-blades form a 5D algebraic variety.

For the image space of C, when $\mathbf{B}_2 \in \Lambda^2(\mathbb{R}^{4,1})$ is a nonzero blade, it is easy to see that the set $\{C(\lambda\mathbf{B}_2) \mid \lambda \in \mathbb{R},\ \lambda^2\mathbf{B}_2^2 \neq 1\}$ modulo scale is composed of all rotors in $\mathcal{G}(\mathbf{B}_2)$ of the form $e^{\mathbf{B}_2\theta}$.

When $\mathbf{B}_2$ is not a blade, by (8.2.11), $\lambda_1 x + \lambda_2 y = 1$. If $\mathbf{b}_2^2 = 1$, by (8.2.11), x can take any value in $\mathbb{R}$ by varying the two λ's. When $x \neq 0$ is fixed, then

$$y = \frac{1 - \lambda_1 x}{\lambda_2} = \frac{1 + x^2 \pm x\sqrt{1 + x^2 + \lambda_2^2}}{\lambda_2},$$

so y can take any value in $\mathbb{R}$ by varying λ_2. When $x = 0$, then y can take any value in $\mathbb{R} - \{0\}$. If $\mathbf{a}_1\mathbf{b}_1 + x$ is rotor $e^{\theta_1\mathbf{a}_1\mathbf{b}_1}$, then $x = \operatorname{ctan}\theta_1$; if $\mathbf{a}_2\mathbf{b}_2 + y$ is rotor $e^{\theta_2\mathbf{a}_2\mathbf{b}_2}$, then $y = \operatorname{ctan}\theta_2$. So all rotors of the form $e^{\theta_1\mathbf{A}_1}e^{\theta_2\mathbf{A}_2}$ up to scale, where $e^{\theta_i\mathbf{A}_i} \neq 1$, are in the image space of Cayley transform, except for $\mathbf{A}_1\mathbf{A}_2$.

If $\mathbf{b}_2^2 = 0$, by similar argument we get that all rotors of the form $e^{\theta_1\mathbf{A}_1}e^{\theta_2\mathbf{A}_2}$ up to scale, where $e^{\theta_i\mathbf{A}_i} \neq 1$, are in the image space of Cayley transform.

If $\mathbf{b}_2^2 = -1$, by (8.2.11), y can take any value in $\mathbb{R} - [-1, 1]$. When $y > 1$ is fixed, then

$$x = \frac{1 - y^2 \pm y\sqrt{y^2 - 1 - \lambda_1^2}}{\lambda_1},$$

so x can take any value in $\mathbb{R}$ by varying $\lambda \in [-\sqrt{y^2 - 1}, \sqrt{y^2 - 1}]$. Similarly, when $y < -1$ is fixed, x can take any value in $\mathbb{R}$. If $\mathbf{a}_1\mathbf{b}_1 + x$ is rotor $e^{\theta_1\mathbf{a}_1\mathbf{b}_1}$, then $x = \operatorname{ctan}\theta_1$; if $\mathbf{a}_2\mathbf{b}_2 + y$ is rotor $e^{\theta_2\mathbf{a}_2\mathbf{b}_2}$, then $y = \operatorname{ctanh}\theta_2 > 1$ or < -1. So all rotors of the form $e^{\theta_1\mathbf{A}_1}e^{\theta_2\mathbf{A}_2}$ up to scale, where $e^{\theta_i\mathbf{A}_i} \neq 1$, are in the image space of Cayley transform. $\qquad\square$

Corollary 8.21. All orientation-preserving similarity transformations in $\mathbb{E}^3$ can be induced by bivectors in $\Lambda(\mathbb{R}^{4,1})$ through Cayley transform and graded adjoint action.

Given a bivector $\mathbf{B}_2$ that is neither a blade nor in the form of a completely orthogonal decomposition, the inverse of $1 - \mathbf{B}_2$ can be determined as follows:

$$(1 + \mathbf{B}_2)(1 - \mathbf{B}_2) = 1 - \langle\mathbf{B}_2^2\rangle - \langle\mathbf{B}_2^2\rangle_4,$$

$$(1 + \mathbf{B}_2)(1 - \mathbf{B}_2)(1 - \langle\mathbf{B}_2^2\rangle + \langle\mathbf{B}_2^2\rangle_4) = 1 - 2\langle\mathbf{B}_2^2\rangle + \langle\mathbf{B}_2^2\rangle^2 - \langle\mathbf{B}_2^2\rangle_4^2 = \Delta_{\mathbf{B}_2}.$$

By (8.2.13), $\Delta_{\mathbf{B}_2} \neq 0$, so $1 - \mathbf{B}_2$ is invertible, with inverse

$$(1 - \mathbf{B}_2)^{-1} = \Delta^{-1}(1 + \mathbf{B}_2)(1 - \langle\mathbf{B}_2^2\rangle + \langle\mathbf{B}_2^2\rangle_4). \tag{8.2.14}$$

In fact, (8.2.14) is valid for all bivectors in the domain of definition of C.

Proposition 8.22. For any $\mathbf{B}_2 \in \Lambda^2(\mathbb{R}^{4,1})$ such that $\mathbf{B}_2^2 \neq 1$, the following equality holds up to scale:

$$C(\mathbf{B}_2) = (1 + \mathbf{B}_2)^2(1 - \mathbf{B}_2 \cdot \mathbf{B}_2 + \mathbf{B}_2 \wedge \mathbf{B}_2). \tag{8.2.15}$$

(8.2.15) can be used as an alternative definition of Cayley transform. From this aspect, Cayley transform is just a polynomial of degree 4 in $\mathbf{B}_2$, with values in the group of positive rotors of $\mathcal{G}(\mathbb{R}^{4,1})$.

In the following, we compute the "inverse" of Cayley transform by finding all the preimages of a rotor in its range. Given a positive rotor $\mathbf{A}$ such that $\mathbf{A} \neq 1$ up to scale, let $\mathbf{B}_2$ be a bivector whose Cayley transform equals $\mathbf{A}$ up to scale. From $(1 + \mathbf{B}_2)(1 - \mathbf{B}_2)^{-1} = \lambda\mathbf{A}$, we get

$$\begin{aligned}
\langle(\lambda\mathbf{A} - 1)(\lambda\mathbf{A} + 1)^{-1}\rangle &= 0, \\
\langle(\lambda\mathbf{A} - 1)(\lambda\mathbf{A} + 1)^{-1}\rangle_4 &= 0, \\
\langle(\lambda\mathbf{A} - 1)(\lambda\mathbf{A} + 1)^{-1}\rangle_2 &= \mathbf{B}_2.
\end{aligned} \tag{8.2.16}$$

If $\langle\mathbf{A}\rangle_4 = 0$, then the *discriminant* for the invertibility of $\lambda\mathbf{A} + 1$, denoted by Δ, is

$$\begin{aligned}
\Delta &= (\lambda\mathbf{A} + 1)(1 + \lambda\langle\mathbf{A}\rangle - \lambda\langle\mathbf{A}\rangle_2) \\
&= 1 + 2\lambda\langle\mathbf{A}\rangle + \lambda^2(\langle\mathbf{A}\rangle^2 - \langle\mathbf{A}\rangle_2^2) \\
&= 1 + \lambda(\mathbf{A} + \mathbf{A}^\dagger) + \lambda^2\mathbf{A}\mathbf{A}^\dagger \\
&= (1 + \lambda\mathbf{A})(1 + \lambda\mathbf{A}^\dagger).
\end{aligned} \tag{8.2.17}$$

It equals zero if and only if $\lambda\mathbf{A} + 1$ is not invertible.

There are at most two values of λ such that $\Delta = 0$. When $\Delta \neq 0$,

$$(\lambda\mathbf{A} + 1)^{-1} = \Delta^{-1}(1 + \lambda\langle\mathbf{A}\rangle - \lambda\langle\mathbf{A}\rangle_2).$$

From the first equation of (8.2.16) we get $\lambda^2 = (\mathbf{A}\mathbf{A}^\dagger)^{-1}$, so $\Delta = 2(1 + \lambda\langle\mathbf{A}\rangle)$.

If $\langle\mathbf{A}\rangle = 0$, then $\Delta = 2$. If $\langle\mathbf{A}\rangle \neq 0$, then $\Delta = 0$ if and only if $\langle\mathbf{A}\rangle_2^2 = 0$ and $\lambda = -\langle\mathbf{A}\rangle^{-1}$. So for $\langle\mathbf{A}\rangle_4 = 0$, if $\langle\mathbf{A}\rangle_2$ is degenerate, then (8.2.16) has a unique solution $\lambda = \langle\mathbf{A}\rangle^{-1}$; else, (8.2.16) has two solutions $\lambda = \pm\sqrt{(\mathbf{A}\mathbf{A}^\dagger)^{-1}}$. From the last equation of (8.2.16), we get

$$\mathbf{B}_2 = \frac{\lambda\langle\mathbf{A}\rangle_2}{1 + \lambda\langle\mathbf{A}\rangle} = \begin{cases} \dfrac{\mathbf{A} - \mathbf{A}^\dagger}{2(\mathbf{A} + \mathbf{A}^\dagger)}, & \text{if } \langle\mathbf{A}\rangle_2^2 = 0, \\[2ex] \dfrac{\mathbf{A} - \mathbf{A}^\dagger}{\mathbf{A} + \mathbf{A}^\dagger \pm 2\sqrt{\mathbf{A}\mathbf{A}^\dagger}}, & \text{otherwise.} \end{cases} \tag{8.2.18}$$

Notice that if $\mathbf{B}_2$ has two solutions $\mathbf{B}', \mathbf{B}''$, then $\mathbf{B}'' = \mathbf{B}'^{-1}$.

If $\langle\mathbf{A}\rangle_4 \neq 0$ but $\langle\mathbf{A}\rangle = 0$, for $\mathbf{A}$ to be within the range of Cayley transform, $\langle\mathbf{A}\rangle_2$ must be a nonzero blade. The discriminant Δ for the invertibility of $\lambda\mathbf{A} + 1$ is

$$\begin{aligned}
\Delta &= (1 + \lambda\mathbf{A})(1 + \lambda\langle\mathbf{A}\rangle_2 - \lambda\langle\mathbf{A}\rangle_4)(1 - \lambda\langle\mathbf{A}\rangle_2 + \lambda\langle\mathbf{A}\rangle_4)(1 - \lambda\langle\mathbf{A}\rangle_2 - \lambda\langle\mathbf{A}\rangle_4) \\
&= 1 + \lambda^4(\langle\mathbf{A}\rangle_2^2 - \langle\mathbf{A}\rangle_4^2)^2 - 2\lambda^2(\langle\mathbf{A}\rangle_2^2 + \langle\mathbf{A}\rangle_4^2) \\
&= 1 + \lambda^4(\mathbf{A}\mathbf{A}^\dagger)^2 - \lambda^2(\mathbf{A}^2 + \mathbf{A}^{\dagger 2}).
\end{aligned}$$

There are at most four values of λ such that $\Delta = 0$. When $\Delta \neq 0$,

$$(1 + \lambda\mathbf{A})^{-1} = \Delta^{-1}(1 + \lambda\langle\mathbf{A}\rangle_2 - \lambda\langle\mathbf{A}\rangle_4)(1 - \lambda\langle\mathbf{A}\rangle_2 + \lambda\langle\mathbf{A}\rangle_4)(1 - \lambda\langle\mathbf{A}\rangle_2 - \lambda\langle\mathbf{A}\rangle_4).$$

Substituting it into (8.2.16), we get $\lambda^2 = (\mathbf{A}\mathbf{A}^\dagger)^{-1}$, so $\Delta = -4\lambda^2 \langle \mathbf{A} \rangle_2^2$.

When 2-blade $\langle \mathbf{A} \rangle_2$ is degenerate, then $\Delta = 0$. Furthermore, $\langle \mathbf{A} \rangle_4^2 > 0$ because rotor $\mathbf{A}$ is positive. So blade $\langle \mathbf{A} \rangle_4$ must be Euclidean. However, no Euclidean space allows a degenerate subspace. This contradiction shows that $\langle \mathbf{A} \rangle_2$ is nondegenerate. So $\Delta \neq 0$, and

$$\begin{aligned}
\mathbf{B}_2 &= -4\Delta^{-1}\lambda\langle \mathbf{A} \rangle_2 (\lambda\langle \mathbf{A} \rangle_4 - 1) \\
&= ((\langle \mathbf{A} \rangle_4 - \lambda^{-1})(\langle \mathbf{A} \rangle_2)^{-1} \\
&= \frac{\mathbf{A} - \mathbf{A}^\dagger}{\mathbf{A} + \mathbf{A}^\dagger \pm 2\sqrt{\mathbf{A}\mathbf{A}^\dagger}}.
\end{aligned} \tag{8.2.19}$$

If both $\langle \mathbf{A} \rangle_4$ and $\langle \mathbf{A} \rangle$ are nonzero, by (6.3.41), $\langle \mathbf{A} \rangle_2$ is not a blade, and

$$2\langle \mathbf{A} \rangle \langle \mathbf{A} \rangle_4 = \langle \mathbf{A} \rangle_2^2 - \langle\langle \mathbf{A} \rangle_2^2 \rangle. \tag{8.2.20}$$

Using $\mathbf{A}\mathbf{A}^\dagger = \langle \mathbf{A} \rangle^2 + \langle \mathbf{A} \rangle_4^2 - \langle\langle \mathbf{A} \rangle_2^2 \rangle$, we get the discriminant Δ for the invertibility of $\lambda \mathbf{A} + 1$ by the following steps:

$$\begin{aligned}
\mathbf{D} :&= (\lambda\mathbf{A} + 1)(1 + \lambda\langle \mathbf{A} \rangle + \lambda\langle \mathbf{A} \rangle_4 - \lambda\langle \mathbf{A} \rangle_2) \\
&= 1 + \lambda^2\langle \mathbf{A} \rangle^2 + \lambda^2\langle \mathbf{A} \rangle_4^2 + 2\lambda\langle \mathbf{A} \rangle - \lambda^2\langle\langle \mathbf{A} \rangle_2^2 \rangle + 2\lambda\langle \mathbf{A} \rangle_4 \\
&= 1 + \lambda^2\mathbf{A}\mathbf{A}^\dagger + 2\lambda\langle \mathbf{A} \rangle + 2\lambda\langle \mathbf{A} \rangle_4,
\end{aligned}$$

$$\begin{aligned}
\Delta &= \mathbf{D}(1 + \lambda^2\mathbf{A}\mathbf{A}^\dagger + 2\lambda\langle \mathbf{A} \rangle - 2\lambda\langle \mathbf{A} \rangle_4) \\
&= (1 + \lambda^2\mathbf{A}\mathbf{A}^\dagger + 2\lambda\langle \mathbf{A} \rangle)^2 - 4\lambda^2\langle \mathbf{A} \rangle_4^2 \\
&= \lambda^4(\mathbf{A}\mathbf{A}^\dagger)^2 + 4\lambda^3\langle \mathbf{A} \rangle\mathbf{A}\mathbf{A}^\dagger + 2\lambda^2(\mathbf{A}\mathbf{A}^\dagger + 2\langle \mathbf{A} \rangle^2 - 2\langle \mathbf{A} \rangle_4^2) + 4\lambda\langle \mathbf{A} \rangle + 1.
\end{aligned}$$

When $\Delta \neq 0$,

$$(\lambda\mathbf{A} + 1)^{-1} = \Delta^{-1}(1 + \lambda(\langle \mathbf{A} \rangle + \langle \mathbf{A} \rangle_4 - \langle \mathbf{A} \rangle_2))(1 + \lambda^2\mathbf{A}\mathbf{A}^\dagger + 2\lambda(\langle \mathbf{A} \rangle - \langle \mathbf{A} \rangle_4)).$$

Substituting it into (8.2.16), we get $\lambda^2 = (\mathbf{A}\mathbf{A}^\dagger)^{-1}$, so $\Delta = 4\{(1 + \lambda\langle \mathbf{A} \rangle)^2 - \lambda^2\langle \mathbf{A} \rangle_4^2\}$.

If $\Delta = 0$, then

$$\langle \mathbf{A} \rangle_4^2 = (\lambda^{-1} + \langle \mathbf{A} \rangle)^2 > 0, \tag{8.2.21}$$

so $\langle \mathbf{A} \rangle_4$ is Euclidean. By $\lambda^{-2} = \mathbf{A}\mathbf{A}^\dagger = \langle \mathbf{A} \rangle^2 + \langle \mathbf{A} \rangle_4^2 - \langle\langle \mathbf{A} \rangle_2^2 \rangle$, (8.2.21) can be written as

$$\lambda^{-1} = \frac{\langle\langle \mathbf{A} \rangle_2^2 \rangle}{2\langle \mathbf{A} \rangle} - \langle \mathbf{A} \rangle. \tag{8.2.22}$$

Substituting it into (8.2.21), we get

$$\langle\langle \mathbf{A} \rangle_2^2 \rangle^2 = 4\langle \mathbf{A} \rangle^2\langle \mathbf{A} \rangle_4^2. \tag{8.2.23}$$

Substituting $\langle \mathbf{A} \rangle_4 = \langle \mathbf{A} \rangle_2 \wedge \langle \mathbf{A} \rangle_2 / (2\langle \mathbf{A} \rangle)$ into (8.2.23), we get

$$(\langle \mathbf{A} \rangle_2 \cdot \langle \mathbf{A} \rangle_2)^2 = (\langle \mathbf{A} \rangle_2 \wedge \langle \mathbf{A} \rangle_2)^2. \tag{8.2.24}$$

The following argument gives an interpretation of condition (8.2.24). By completely orthogonal decomposition, $\langle \mathbf{A} \rangle_2 = \mathbf{C}_1 + \mathbf{C}_2$ where $\mathbf{C}_1, \mathbf{C}_2$ are Euclidean

2-blades in $\Lambda(\langle\mathbf{A}\rangle_4)$. So $\langle\mathbf{A}\rangle_2 \cdot \langle\mathbf{A}\rangle_2 = \mathbf{C}_1^2 + \mathbf{C}_2^2 < 0$, and $\langle\mathbf{A}\rangle_2 \wedge \langle\mathbf{A}\rangle_2 = 2\mathbf{C}_1\mathbf{C}_2$. Substituting them into (8.2.24), we get $(\mathbf{C}_1^2 + \mathbf{C}_2^2)^2 - 4\mathbf{C}_1^2\mathbf{C}_2^2 = (\mathbf{C}_1^2 - \mathbf{C}_2^2)^2 = 0$, *i.e.*, $\mathbf{C}_1^2 = \mathbf{C}_2^2$. When this happens, the completely orthogonal decomposition is not unique.

Definition 8.23. A bivector is said to be *entangled*, or *coherent*, if in its completely orthogonal decomposition there are two components having equal square.

(8.2.24) is the criterion for bivector $\langle\mathbf{A}\rangle_2 \in \Lambda^2(\mathbb{R}^{4,1})$ to be entangled. If $\langle\mathbf{A}\rangle_2$ is not entangled, then $\Delta \neq 0$ for both $\lambda = \pm\sqrt{(\mathbf{A}\mathbf{A}^\dagger)^{-1}}$, and

$$
\begin{aligned}
\mathbf{B}_2 &= 4\Delta^{-1}\lambda\langle\mathbf{A}\rangle_2(1 + \lambda(\langle\mathbf{A}\rangle - \langle\mathbf{A}\rangle_4)) \\
&= \frac{\lambda\langle\mathbf{A}\rangle_2}{1 + \lambda(\langle\mathbf{A}\rangle + \langle\mathbf{A}\rangle_4)} \\
&= \frac{\mathbf{A} - \mathbf{A}^\dagger}{\mathbf{A} + \mathbf{A}^\dagger \pm 2\sqrt{\mathbf{A}\mathbf{A}^\dagger}}.
\end{aligned}
\tag{8.2.25}
$$

If $\langle\mathbf{A}\rangle_2$ is entangled, then (8.2.23) holds. So $\langle\langle\mathbf{A}\rangle_2^2\rangle = -2\,|\langle\mathbf{A}\rangle\,\langle\mathbf{A}\rangle_4|$ because it is negative. Then $|\mathbf{A}| = \sqrt{\mathbf{A}\mathbf{A}^\dagger} = |\langle\mathbf{A}\rangle| + |\langle\mathbf{A}\rangle_4|$. In order to have $\Delta \neq 0$, the only choice of λ, by (8.2.22), is

$$
\lambda^{-1} = \langle\mathbf{A}\rangle - \frac{\langle\langle\mathbf{A}\rangle_2^2\rangle}{2\langle\mathbf{A}\rangle} = \langle\mathbf{A}\rangle + \frac{|\langle\mathbf{A}\rangle\,\langle\mathbf{A}\rangle_4|}{\langle\mathbf{A}\rangle}.
\tag{8.2.26}
$$

So

$$
\mathbf{B}_2 = \frac{\mathbf{A} - \mathbf{A}^\dagger}{\mathbf{A} + \mathbf{A}^\dagger + 2\sqrt{\mathbf{A}\mathbf{A}^\dagger}\,\langle\mathbf{A}\rangle/|\langle\mathbf{A}\rangle|} = \frac{\langle\mathbf{A}\rangle_2}{\langle\mathbf{A}\rangle_4 + 2\langle\mathbf{A}\rangle + |\langle\mathbf{A}\rangle\langle\mathbf{A}\rangle_4|/\langle\mathbf{A}\rangle}.
\tag{8.2.27}
$$

Theorem 8.24. A positive rotor $\mathbf{A}$ in the range of Cayley transform has exactly one bivector preimage if and only if either it is in $\Lambda(\mathbf{C}_2)$ where $\mathbf{C}_2$ is a 2-blade of degenerate signature, or its bivector part is entangled. The unique solution is $\langle\mathbf{A}\rangle_2/(\langle\mathbf{A}\rangle_4 + 2\langle\mathbf{A}\rangle + |\langle\mathbf{A}\rangle\langle\mathbf{A}\rangle_4|/\langle\mathbf{A}\rangle)$. Any other positive rotor $\mathbf{A}$ in the range of Cayley transform has two bivector preimages, and they are inverse to each other: $\langle\mathbf{A}\rangle_2/(\langle\mathbf{A}\rangle_4 + \langle\mathbf{A}\rangle \pm |\mathbf{A}|)$.

In particular, any orientation-preserving similarity transformation which is not a translation is induced by the Cayley transform of exactly two bivectors. A translation is induced by a unique bivector.

Example 8.25. In $\mathcal{CL}(\mathbb{R}^{4,1})$, let $\mathbf{A} = e^{\mathbf{I}_2\frac{\theta}{2}}$, where $\mathbf{I}_2 \in \Lambda(e^\sim)$ is a Euclidean 2-blade of unit magnitude such that $\mathbf{I}_2^\sim$ is the axis of rotation, and $-\theta$ is the angle of rotation. By (8.2.18),

$$
\mathbf{B}_2 = \frac{e^{\mathbf{I}_2\frac{\theta}{2}} - e^{-\mathbf{I}_2\frac{\theta}{2}}}{e^{\mathbf{I}_2\frac{\theta}{2}} + e^{-\mathbf{I}_2\frac{\theta}{2}} + 2} = \mathbf{I}_2\tan\frac{\theta}{4}, \qquad \mathbf{B}_2^{-1} = -\mathbf{I}_2/\tan\frac{\theta}{4},
\tag{8.2.28}
$$

both generate $\mathbf{A}$ by Cayley transform.

While the bivector representation of a rotation via the exponential map is a half-angle representation, the bivector representation via the Cayley transform is a quarter-angle representation. The former is exponential while the latter is both fractional linear and quartic. Furthermore, the exponential map has infinitely many bivector preimages $\mathbf{I}_2(\theta/2 + k\pi)$ for all $k \in \mathbb{Z}$, while the Cayley transform has only two preimages: $\mathbf{I}_2 \tan(\theta/4)$ and its inverse.

Example 8.26. In $\mathcal{CL}(\mathbb{R}^{4,1})$, let $\mathbf{A} = 1 + \mathbf{e}\mathbf{t}/2$ where $\mathbf{t} \in \mathbf{e}^\sim$ is a positive vector. Then $\mathbf{A}$ generates the translation along direction $\mathbf{e} \wedge \mathbf{t}$ with distance $|\mathbf{t}|$. By (8.2.18),

$$\mathbf{B}_2 = \frac{\mathbf{e}\mathbf{t}}{4} \tag{8.2.29}$$

generates $\mathbf{A}$ by Cayley transform.

Cayley transform provides a quarter-distance bivector representation of the translation, while the exponential map provides a half-distance bivector representation.

Example 8.27. Let $\mathbf{A} = e^{\frac{\theta}{2}\mathbf{e}\wedge\mathbf{a}}$, where $\theta \in \mathbb{R}$ and $\mathbf{a} \in \mathcal{N}_\mathbf{e}$ represents a point. Rotor $\mathbf{A}$ generates the dilation centering at $\mathbf{a}$ and with scale $e^{-\theta}$. Denote $\mathbf{I}_2 = \mathbf{e} \wedge \mathbf{a}$. By (8.2.18),

$$\mathbf{B}_2 = \frac{e^{\mathbf{I}_2\frac{\theta}{2}} - e^{-\mathbf{I}_2\frac{\theta}{2}}}{e^{\mathbf{I}_2\frac{\theta}{2}} + e^{-\mathbf{I}_2\frac{\theta}{2}} + 2} = \mathbf{I}_2 \tanh\frac{\theta}{4}, \qquad \mathbf{B}_2^{-1} = \mathbf{I}_2/\tanh\frac{\theta}{4}, \tag{8.2.30}$$

both generate $\mathbf{A}$ by Cayley transform.

Cayley transform provides a quarter-scale bivector representation of the dilation, while the exponential map provides a half-scale bivector representation.

Cayley transform (8.2.15) is quartic with respect to the bivectors in its domain of definition. There is another classical transform in $\mathcal{CL}(\mathbb{R}^{4,1})$ from bivectors to rotors. It is quadratic with respect to the bivectors within its domain, and is equivalent to the Cayley transform (8.2.7) from antisymmetric matrices to special orthogonal matrices, but not equivalent to the Cayley transform (8.2.8) from bivectors to rotors. This is the *exterior exponential* established by Lipschitz in 1880-1886.

Definition 8.28. Let $\mathcal{V}^n$ be a vector space over $\mathbb{K}$. The *exterior exponential*, or *outer exponential*, is the following map from $\Lambda^2(\mathcal{V}^n)$ to $\Lambda(\mathcal{V}^n)$:

$$e^{\wedge\mathbf{B}_2} = 1 + \mathbf{B}_2 + \frac{\mathbf{B}_2 \wedge \mathbf{B}_2}{2!} + \cdots + \frac{\overbrace{\mathbf{B}_2 \wedge \mathbf{B}_2 \wedge \cdots \wedge \mathbf{B}_2}^{r}}{r!}, \tag{8.2.31}$$

where $2r$ is the rank of bivector $\mathbf{B}_2$.

The exterior exponential has two obvious properties: first, the scalar part of $e^{\wedge\mathbf{B}_2}$ is 1; second, the mapping is injective because the bivector part of $e^{\wedge\mathbf{B}_2}$ is $\mathbf{B}_2$.

If $\mathbf{B}_2$ is in the form (8.2.2) of completely orthogonal decomposition, then

$$e^{\wedge\mathbf{B}_2} = (1 + \mathbf{A}_1)(1 + \mathbf{A}_2)\cdots(1 + \mathbf{A}_r).$$

So $e^{\wedge \mathbf{B}_2}$ is invertible if and only if each $\mathbf{A}_i$ is not a Minkowski blade of unit magnitude. If $e^{\wedge \mathbf{B}_2}$ is invertible, then it is a rotor, because each $1 + \mathbf{A}_i$ is a rotor; furthermore,

$$(e^{\wedge \mathbf{B}_2})^{-1} = \frac{e^{\wedge(-\mathbf{B}_2)}}{e^{\wedge \mathbf{B}_2} e^{\wedge(-\mathbf{B}_2)}}.$$

Theorem 8.29. [130] [Lipschitz Theorem] When $\mathcal{V}^n$ is (anti-)Euclidean or (anti-)Minkowski, any antisymmetric linear transformation f of $\mathcal{V}^n$ and its Cayley transform $g = (I_{\mathcal{V}^n} + f)(I_{\mathcal{V}^n} - f)^{-1}$ can be represented respectively by a bivector $\mathbf{B}_2 \in \Lambda^2(\mathcal{V}^n)$ and its exterior exponential $e^{\wedge \mathbf{B}_2}$ as follows:

$$\begin{aligned} f(\mathbf{x}) &= \mathbf{B}_2 \cdot \mathbf{x}, \\ g(\mathbf{x}) &= Ad_{e^{\wedge \mathbf{B}_2}}(\mathbf{x}), \ \ \forall \mathbf{x} \in \mathcal{V}^n. \end{aligned} \tag{8.2.32}$$

Below we assume that $\mathbf{B}_2$ is in the form of (8.2.2) and $e^{\wedge \mathbf{B}_2}$ is invertible, and analyze the range of the exterior exponential by restricting it to the conformal model setting $\mathbb{R}^{4,1}$ of 3D geometry.

If $\mathbf{B}_2$ is a nonzero blade, then $e^{\wedge \mathbf{B}_2} = 1 + \lambda_1 \mathbf{a}_1 \mathbf{b}_1$. When λ_1 varies, the range of $e^{\wedge \mathbf{B}_2}$ modulo scale contains all rotors in $\Lambda(\mathbf{a}_1 \wedge \mathbf{b}_1)$ whose 0-graded part and 2-graded part are both nonzero.

If $\mathbf{B}_2$ is not a blade, then

$$e^{\wedge \mathbf{B}_2} = (1 + \lambda_1 \mathbf{a}_1 \mathbf{b}_1)(1 + \lambda_2 \mathbf{a}_2 \mathbf{b}_2) = 1 + \lambda_1 \mathbf{a}_1 \mathbf{b}_1 + \lambda_2 \mathbf{a}_2 \mathbf{b}_2 + \lambda_1 \lambda_2 \mathbf{a}_1 \mathbf{b}_1 \mathbf{a}_2 \mathbf{b}_2, \tag{8.2.33}$$

whose 0-graded part and 4-graded part are both nonzero. By (6.3.41), that the 2-graded part is not a blade is a direct consequence. The range of $e^{\wedge \mathbf{B}_2}$ modulo scale is all rotors whose 0-graded part and 4-graded part are both nonzero.

Proposition 8.30. When the range of the exterior exponential is restricted to rotors in $\mathcal{G}(\mathbb{R}^{4,1})$, the domain of definition is all bivectors in $\Lambda(\mathbb{R}^{4,1})$ satisfying

$$(\mathbf{B}_2 \wedge \mathbf{B}_2)^2 \neq 4\,(\mathbf{B}_2 \cdot \mathbf{B}_2 - 1), \tag{8.2.34}$$

and is a set $\mathbb{R}^{10} - V^9$, where V^9 is a 9D algebraic variety in $\mathbb{R}^{10}$. The image space modulo scale is all rotors whose scalar part is nonzero; topologically, it is the remainder of the special orthogonal group $SO(4,1)$, which is a 10D Lie group with two connected components, after removal of a 9D closed subset.

Proof. By

$$e^{\wedge \mathbf{B}_2} = 1 + \mathbf{B}_2 + \frac{\mathbf{B}_2 \wedge \mathbf{B}_2}{2}, \tag{8.2.35}$$

we get $e^{\wedge \mathbf{B}_2} e^{\wedge(-\mathbf{B}_2)} = 1 - \mathbf{B}_2 \cdot \mathbf{B}_2 + (\mathbf{B}_2 \wedge \mathbf{B}_2)^2/4$, and (8.2.34) follows. $\qquad \square$

Similar to the exponential map, the exterior exponential provides half-scaled bivector representations for rotations, translations and dilations.

Given a rotor $\mathbf{A}$ in the image space of the exterior exponential, let $\mathbf{B}_2$ be a bivector whose exterior exponential equals $\mathbf{A}$ up to scale. Then

$$1 + \mathbf{B}_2 + \frac{\mathbf{B}_2 \wedge \mathbf{B}_2}{2} = \frac{\mathbf{A}}{\langle \mathbf{A} \rangle}, \tag{8.2.36}$$

so $\mathbf{B}_2 = \langle \mathbf{A} \rangle_2 / \langle \mathbf{A} \rangle$.

Example 8.31. Let $e^{\mathbf{I}_2 \frac{\theta}{2}}$ be a rotation with axis $\mathbf{I}_2^{\sim}$ and angle $-\theta \neq \pi \bmod 2\pi$. Then up to scale,

$$e^{\mathbf{I}_2 \frac{\theta}{2}} = e^{\wedge \mathbf{I}_2 \tan(\frac{\theta}{2})}. \tag{8.2.37}$$

Let $e^{\frac{1}{2}\mathbf{e}\mathbf{t}} = 1 + \mathbf{e}\mathbf{t}/2$ be a translation by vector $\mathbf{t}$. Then

$$e^{\frac{1}{2}\mathbf{e}\mathbf{t}} = e^{\wedge \frac{1}{2}\mathbf{e}\mathbf{t}}. \tag{8.2.38}$$

Let $e^{\frac{\theta}{2}\mathbf{I}_2}$ be a dilation of scale $e^{-\theta}$ centering at point $\mathbf{I}_2$ (affine representation). Then up to scale,

$$e^{\mathbf{I}_2 \frac{\theta}{2}} = e^{\wedge \mathbf{I}_2 \tanh(\frac{\theta}{2})}. \tag{8.2.39}$$

Let $\mathbf{I}_2 e^{\frac{\theta}{2}\mathbf{I}_2}$ be a dilation of scale $-e^{-\theta} \neq -1$. Then up to scale,

$$\mathbf{I}_2 e^{\mathbf{I}_2 \frac{\theta}{2}} = e^{\wedge \mathbf{I}_2 \operatorname{ctanh}(\frac{\theta}{2})}. \tag{8.2.40}$$

The image space of the exterior exponential contains both Lorentz and non-Lorentz rotors, the map is injective and quadratic. Algebraically, exterior exponential is much simpler than Cayley transform.

However, exterior exponential has two severe drawbacks: first, the domain of definition is decomposed into several disconnected regions, which blocks the construction of large-scope bivector parameters in the design of continuous conformal transformations; second, the image space is also decomposed into several disconnected regions, making it impossible to represent rotors of large-scale continuous conformal transformations.

8.3 Twisted Vahlen matrices and Vahlen matrices

In this section, we introduce the classical work of Vahlen (1902) on representing elements of $\mathcal{CL}(\mathbb{R}^{n+1,1})$ by 2×2 matrices whose components are in $\mathcal{CL}(\mathbb{R}^n)$. This work has great impact on the development of Clifford analysis [5], [6], because it leads to the nD extension of fractional linear transformations as representations of Möbius transformations in $\mathbb{E}^n$.

Fix the Witt pair $(\mathbf{e}, \mathbf{e}_0)$ in the conformal model of $\mathbb{R}^n$. By $\mathbf{e}\mathbf{e}_0\mathbf{e} = -2\mathbf{e}$ and $\mathbf{e}_0\mathbf{e}\mathbf{e}_0 = -2\mathbf{e}_0$, any Clifford monomial $\mathbf{M} = (\mathbf{a}_1 + \lambda_1\mathbf{e} + \mu_1\mathbf{e}_0) \cdots (\mathbf{a}_r + \lambda_r\mathbf{e} + \mu_r\mathbf{e}_0)$, where $\mathbf{a}_i \in \mathbb{R}^n$ and $\lambda_i, \mu_i \in \mathbb{R}$, after multilinear expansion, is changed into the following form:

$$\mathbf{M} = -\frac{\mathbf{A}}{2}\mathbf{e}\mathbf{e}_0 - \frac{\mathbf{B}}{2}\mathbf{e} + \mathbf{C}\mathbf{e}_0 - \frac{\mathbf{D}}{2}\mathbf{e}_0\mathbf{e}, \tag{8.3.1}$$

where $\mathbf{A}, \mathbf{B}, \mathbf{C}, \mathbf{D} \in \mathcal{CL}(\mathbb{R}^n)$. By linearity, any element $\mathbf{M} \in \mathcal{CL}(\mathbb{R}^{n+1,1})$ has the unique decomposition (8.3.1).

The orthogonal decomposition of vector space

$$\mathbb{R}^{n+1,1} = \mathbb{R}^n \oplus (\mathbf{e} \wedge \mathbf{e}_0) \tag{8.3.2}$$

induces a natural isomorphism of Clifford algebras: $\mathcal{CL}(\mathbb{R}^{n+1,1}) \simeq \mathcal{CL}(\mathbb{R}^n) \hat{\otimes} \mathcal{CL}(\mathbf{e} \wedge \mathbf{e}_0)$. Since $\mathbf{e} \wedge \mathbf{e}_0$ is Minkowski, by Theorem 5.53, $\mathcal{CL}(\mathbf{e} \wedge \mathbf{e}_0)$ is isomorphic to $M_{2\times 2}(\mathbb{R})$. The following correspondence of bases provides such an isomorphism:

$$
\begin{aligned}
1 &= \begin{pmatrix} 1 & 0 \\ 0 & 1 \end{pmatrix}, &
\mathbf{e} &= \begin{pmatrix} 0 & -2 \\ 0 & 0 \end{pmatrix}, &
\mathbf{e}_0 &= \begin{pmatrix} 0 & 0 \\ 1 & 0 \end{pmatrix}, \\[2mm]
\mathbf{e}\mathbf{e}_0 &= \begin{pmatrix} -2 & 0 \\ 0 & 0 \end{pmatrix}, &
\mathbf{e}_0\mathbf{e} &= \begin{pmatrix} 0 & 0 \\ 0 & -2 \end{pmatrix}, &
\mathbf{e} \wedge \mathbf{e}_0 &= \begin{pmatrix} -1 & 0 \\ 0 & 1 \end{pmatrix}.
\end{aligned}
\tag{8.3.3}
$$

Definition 8.32. The 2×2 *twisted Clifford matrix algebra* over $\mathbb{R}^n$, denoted by $\hat{M}_{2\times 2}(\mathcal{CL}(\mathbb{R}^n))$, is the linear space of 2×2 matrices whose components are in $\mathcal{CL}(\mathbb{R}^n)$, equipped with the *twisted multiplication* defined as follows: for any matrices $\mathbf{M}_1, \mathbf{M}_2 \in \hat{M}_{2\times 2}(\mathcal{CL}(\mathbb{R}^n))$,

$$
\mathbf{M}_1\mathbf{M}_2 = \begin{pmatrix} \mathbf{A} & \mathbf{B} \\ \mathbf{C} & \mathbf{D} \end{pmatrix}\begin{pmatrix} \mathbf{A}' & \mathbf{B}' \\ \mathbf{C}' & \mathbf{D}' \end{pmatrix} := \begin{pmatrix} \mathbf{A}\mathbf{A}' + \mathbf{B}\hat{\mathbf{C}}' & \mathbf{A}\mathbf{B}' + \mathbf{B}\hat{\mathbf{D}}' \\ \mathbf{C}\hat{\mathbf{A}}' + \mathbf{D}\mathbf{C}' & \mathbf{C}\hat{\mathbf{B}}' + \mathbf{D}\mathbf{D}' \end{pmatrix}.
\tag{8.3.4}
$$

In each component on the right side of (8.3.4), the overhat (grade involution) is always added to the element of the second matrix that is not in the same row with the corresponding element of the first matrix multiplied with it. For example, in the first component $\mathbf{A}\mathbf{A}' + \mathbf{B}\hat{\mathbf{C}}'$, $\mathbf{A}, \mathbf{A}'$ are each in the first row of the corresponding matrix, while $\mathbf{B}, \mathbf{C}'$ are in different rows, so the overhat is added to $\mathbf{C}'$.

The isomorphism of Clifford algebras $\mathcal{CL}(\mathbb{R}^{n+1,1}) \simeq \hat{M}_{2\times 2}(\mathcal{CL}(\mathbb{R}^n))$ is a combination of (8.3.1) and (8.3.3):

$$
\mathbf{M} = -\frac{\mathbf{A}}{2}\mathbf{e}\mathbf{e}_0 - \frac{\mathbf{B}}{2}\mathbf{e} + \mathbf{C}\mathbf{e}_0 - \frac{\mathbf{D}}{2}\mathbf{e}_0\mathbf{e} = \begin{pmatrix} \mathbf{A} & \mathbf{B} \\ \mathbf{C} & \mathbf{D} \end{pmatrix}.
\tag{8.3.5}
$$

$\hat{M}_{2\times 2}(\mathcal{CL}(\mathbb{R}^n))$ is a representation of $\mathcal{CL}(\mathbb{R}^{n+1,1})$ induced by the orthogonal decomposition (8.3.2).

Let $\mathbf{I}_n$ be the unit pseudoscalar representing the orientation of $\mathbb{R}^n$. The following formulas can be easily derived from (8.3.5):

$$
\begin{aligned}
\begin{pmatrix} \mathbf{A} & \mathbf{B} \\ \mathbf{C} & \mathbf{D} \end{pmatrix}^{\dagger} &= \begin{pmatrix} \mathbf{D}^{\dagger} & \overline{\mathbf{B}} \\ \overline{\mathbf{C}} & \mathbf{A}^{\dagger} \end{pmatrix}, &
\widehat{\begin{pmatrix} \mathbf{A} & \mathbf{B} \\ \mathbf{C} & \mathbf{D} \end{pmatrix}} &= \begin{pmatrix} \hat{\mathbf{A}} & -\hat{\mathbf{B}} \\ -\hat{\mathbf{C}} & \hat{\mathbf{D}} \end{pmatrix}, \\[3mm]
\overline{\begin{pmatrix} \mathbf{A} & \mathbf{B} \\ \mathbf{C} & \mathbf{D} \end{pmatrix}} &= \begin{pmatrix} \overline{\mathbf{D}} & -\mathbf{B}^{\dagger} \\ -\mathbf{C}^{\dagger} & \overline{\mathbf{A}} \end{pmatrix}, &
\begin{pmatrix} \mathbf{A} & \mathbf{B} \\ \mathbf{C} & \mathbf{D} \end{pmatrix}^{\sim} &= \begin{pmatrix} -\mathbf{A}\mathbf{I}_n^{-1} & \mathbf{B}\mathbf{I}_n^{-1} \\ -\mathbf{C}\mathbf{I}_n^{-1} & \mathbf{D}\mathbf{I}_n^{-1} \end{pmatrix}.
\end{aligned}
\tag{8.3.6}
$$

By Proposition 2.87, any r-blade in $\Lambda(\mathbb{R}^{n+1,1})$ is of the form

$$
\begin{pmatrix} \mathbf{A}_r + \mathbf{D}_{r-2} & \mathbf{B}_{r-1} \\ \mathbf{C}_{r-1} & \mathbf{A}_r - \mathbf{D}_{r-2} \end{pmatrix},
$$

where $\mathbf{A}_r, \mathbf{B}_{r-1}, \mathbf{C}_{r-1}, \mathbf{D}_{r-2} \in \Lambda(\mathbb{R}^n)$ are blades of grade $r, r-1, r-1, r-2$ respectively. In particular, any vector $\mathbf{a} \in \mathbb{R}^{n+1,1}$ corresponds to the following matrix:

$$\begin{pmatrix} \mathbf{x} & \alpha \\ \beta & \mathbf{x} \end{pmatrix}, \tag{8.3.7}$$

where $\mathbf{x} = P^{\perp}_{\mathbf{e} \wedge \mathbf{e}_0}(\mathbf{a})$, $\alpha = 2\mathbf{a} \cdot \mathbf{e}_0$, $\beta = -\mathbf{a} \cdot \mathbf{e}$. Null vector $\mathbf{f}(\mathbf{x})$, positive vectors $\mathbf{n} + \delta \mathbf{e}$ and $\mathbf{f}(\mathbf{x}) - \rho^2 \mathbf{e}/2$, and negative vector $\mathbf{f}(\mathbf{x}) + \rho^2 \mathbf{e}/2$, correspond respectively to the following matrices:

$$\begin{pmatrix} \mathbf{x} & -\mathbf{x}^2 \\ 1 & \mathbf{x} \end{pmatrix}, \quad \begin{pmatrix} \mathbf{n} & -2 \\ 0 & \mathbf{n} \end{pmatrix}, \quad \begin{pmatrix} \mathbf{x} & -\mathbf{x}^2 + \rho^2 \\ 1 & \mathbf{x} \end{pmatrix}, \quad \begin{pmatrix} \mathbf{x} & -\mathbf{x}^2 - \rho^2 \\ 1 & \mathbf{x} \end{pmatrix}. \tag{8.3.8}$$

Of particular interest are the matrices corresponding to versors in $\mathcal{CL}(\mathbb{R}^{n+1,1})$. We first take a look at some examples. Let $\mathbf{I}_2 \in \Lambda^2(\mathbb{R}^n)$ and $\mathbf{t} \in \mathbb{R}^n$.

- The rotor of rotation $e^{\theta \mathbf{I}_2/2}$ corresponds to $\begin{pmatrix} e^{\theta \mathbf{I}_2/2} & 0 \\ 0 & e^{\theta \mathbf{I}_2/2} \end{pmatrix}$.

- The rotor of dilation $e^{\theta \mathbf{e} \wedge \mathbf{e}_0/2}$ corresponds to $\begin{pmatrix} e^{-\theta/2} & 0 \\ 0 & e^{\theta/2} \end{pmatrix}$.

- The rotor of dilation $(\mathbf{e} \wedge \mathbf{e}_0)e^{\theta \mathbf{e} \wedge \mathbf{e}_0/2}$ corresponds to $\begin{pmatrix} -e^{-\theta/2} & 0 \\ 0 & e^{\theta/2} \end{pmatrix}$.

- The rotor of translation $1 + \mathbf{e}\mathbf{t}/2$ corresponds to $\begin{pmatrix} 1 & \mathbf{t} \\ 0 & 1 \end{pmatrix}$.

- The rotor of transversion $1 - \mathbf{e}_0\mathbf{t}$ corresponds to $\begin{pmatrix} 1 & 0 \\ \mathbf{t} & 1 \end{pmatrix}$.

Definition 8.33. A 2×2 matrix $\mathbf{M} = \begin{pmatrix} \mathbf{A} & \mathbf{B} \\ \mathbf{C} & \mathbf{D} \end{pmatrix}$ over $\mathcal{CL}(\mathbb{R}^n)$ is called a *twisted Vahlen matrix*, if

(1) $\mathbf{A}, \mathbf{B}, \mathbf{C}, \mathbf{D}$ are either versors or zero;
(2) $\mathbf{A}\mathbf{B}^\dagger, \mathbf{B}\mathbf{D}^\dagger, \mathbf{D}\mathbf{C}^\dagger, \mathbf{C}\mathbf{A}^\dagger$ are vectors;
(3) $\Delta = \mathbf{A}\mathbf{D}^\dagger + \mathbf{B}\mathbf{C}^\dagger$ is a nonzero scalar.

In the above definition, condition (1) guarantees $\mathbf{B}^\dagger \mathbf{A} = \mathbf{A}^{-1}(\mathbf{A}\mathbf{B}^\dagger)\mathbf{A} \in \mathbb{R}^n$ if $\mathbf{A}\mathbf{B}^\dagger \in \mathbb{R}^n$. So in condition (2), $\mathbf{A}\mathbf{B}^\dagger$ can be replaced by any of $\mathbf{B}\mathbf{A}^\dagger, \mathbf{B}^\dagger\mathbf{A}, \mathbf{A}^\dagger\mathbf{B}$, and so for the other three elements in condition (2). Furthermore, by conditions (1) and (2),

$$\begin{pmatrix} \mathbf{A} & \mathbf{B} \\ \mathbf{C} & \mathbf{D} \end{pmatrix} \begin{pmatrix} \mathbf{A} & \mathbf{B} \\ \mathbf{C} & \mathbf{D} \end{pmatrix}^\dagger = \begin{pmatrix} \mathbf{A}\mathbf{D}^\dagger + \mathbf{B}\mathbf{C}^\dagger & \mathbf{A}\overline{\mathbf{B}} + \mathbf{B}\overline{\mathbf{A}} \\ \mathbf{C}\overline{\mathbf{D}} + \mathbf{D}\overline{\mathbf{C}} & \mathbf{C}\mathbf{B}^\dagger + \mathbf{D}\mathbf{A}^\dagger \end{pmatrix}$$
$$= \begin{pmatrix} \mathbf{A}\mathbf{D}^\dagger + \mathbf{B}\mathbf{C}^\dagger & 0 \\ 0 & (\mathbf{A}\mathbf{D}^\dagger + \mathbf{B}\mathbf{C}^\dagger)^\dagger \end{pmatrix}, \tag{8.3.9}$$

so condition (3) is equivalent to $\mathbf{M}\mathbf{M}^\dagger$ being a nonzero scalar.

Proposition 8.34. In twisted Vahlen matrix $\mathbf{M}$, each of $\mathbf{A}\mathbf{D}^\dagger, \mathbf{B}\mathbf{C}^\dagger, \mathbf{A}^\dagger\mathbf{D}, \mathbf{B}^\dagger\mathbf{C}$ is a rotor that can be compressed to within length two.

Proof.   If $\mathbf{C} \neq 0$, then $\mathbf{AD}^\dagger = (\mathbf{C}^\dagger\mathbf{C})^{-1}(\mathbf{AC}^\dagger)(\mathbf{CD}^\dagger)$ becomes a rotor of length two. If $\mathbf{C} = 0$, then $\mathbf{AD}^\dagger = \Delta$ is a nonzero scalar. $\square$

In twisted Vahlen matrix $\mathbf{M}$, (i) if $\mathbf{A} \neq 0$ and $\mathbf{B} \neq 0$, by denoting

$$\mathbf{A}^\dagger\mathbf{B} = \mathbf{b}, \quad \mathbf{A}^\dagger\mathbf{C} = \mathbf{c}, \quad \mathbf{B}^\dagger\mathbf{D} = \mathbf{d}, \quad \mathbf{A}^\dagger\mathbf{A} = \lambda^{-1}, \quad \mathbf{B}^\dagger\mathbf{B} = \mu^{-1},$$

the matrix can be written as

$$\begin{pmatrix} \mathbf{A} & \mathbf{B} \\ \mathbf{C} & \mathbf{D} \end{pmatrix} = \mathbf{A} \otimes \begin{pmatrix} 1 & \lambda\mathbf{b} \\ \lambda\mathbf{c} & \lambda\mu\mathbf{bd} \end{pmatrix},$$

where $\mathbf{d}$ satisfies

$$\mathbf{d} = \lambda^{-1}\mu^{-1}\Delta\mathbf{b}^{-1} - \lambda\mu^{-1}\mathbf{bcb}^{-1}. \tag{8.3.10}$$

By (8.3.10), $\lambda\mu\mathbf{bd} = \Delta - \lambda^2\mathbf{cb}$, so matrix $\mathbf{M}$ can be written as

$$\begin{pmatrix} \mathbf{A} & \mathbf{B} \\ \mathbf{C} & \mathbf{D} \end{pmatrix} = \mathbf{A} \otimes \begin{pmatrix} 1 & \lambda\mathbf{b} \\ \lambda\mathbf{c} & \Delta - \lambda^2\mathbf{cb} \end{pmatrix}. \tag{8.3.11}$$

(ii) If $\mathbf{A} \neq 0$ but $\mathbf{B} = 0$, matrix $\mathbf{M}$ can be written as

$$\begin{pmatrix} \mathbf{A} & \mathbf{B} \\ \mathbf{C} & \mathbf{D} \end{pmatrix} = \mathbf{A} \otimes \begin{pmatrix} 1 & 0 \\ \lambda\mathbf{c} & \lambda\Delta \end{pmatrix}. \tag{8.3.12}$$

(iii) If $\mathbf{A} = 0$, then $\mathbf{B} \neq 0$, and matrix $\mathbf{M}$ can be written as

$$\begin{pmatrix} \mathbf{A} & \mathbf{B} \\ \mathbf{C} & \mathbf{D} \end{pmatrix} = \mathbf{B} \otimes \begin{pmatrix} 0 & 1 \\ \mu\Delta & \mu\mathbf{d} \end{pmatrix}. \tag{8.3.13}$$

Theorem 8.35. Any twisted Vahlen matrix corresponds via (8.3.5) to the geometric product of a versor in $\mathcal{G}(\mathbb{R}^n)$ and another rotor of length at most two in $\mathcal{G}(\mathbb{R}^{n+1,1})$.

Proof.   The matrix part of the tensor product on the right side of (8.3.11), when multiplied by 2, corresponds to

$$-\mathbf{e}\mathbf{e}_0 - \lambda\mathbf{be} + 2\lambda\mathbf{ce}_0 + (\lambda^2\mathbf{cb} - \Delta)\mathbf{e}_0\mathbf{e}. \tag{8.3.14}$$

When $\mathbf{c} = \nu\mathbf{b}$ for some $\nu \in \mathbb{R}$, then (8.3.14) is in $\Lambda(\mathbf{b} \wedge \mathbf{e} \wedge \mathbf{e}_0)$. Since $\mathbf{MM}^\dagger$ is a nonzero scalar, (8.3.14) is a rotor of length two. When $\mathbf{c}, \mathbf{b}$ are linearly independent, let

$$\mathbf{G} = (\Delta - 1 - \lambda^2\mathbf{bc})(-\mathbf{e}\mathbf{e}_0 - \lambda\mathbf{be} + 2\lambda\mathbf{ce}_0 + (\lambda^2\mathbf{cb} - \Delta)\mathbf{e}_0\mathbf{e}).$$

It is easy to verify that $\langle\mathbf{G}\rangle_4 = 0$, $\langle\mathbf{G}\rangle_2 \neq 0$ and $\langle\mathbf{G}\rangle_2 \wedge \langle\mathbf{G}\rangle_2 = 0$. So the maximal graded part of $\mathbf{G}$ is a nonzero 2-blade. Since $\mathbf{MM}^\dagger$ is a nonzero scalar, so is $\mathbf{GG}^\dagger$. Thus, $\mathbf{G}$ must be a rotor of length two, and (8.3.14) up to scale is the geometric product of rotor $\Delta - 1 - \lambda^2\mathbf{bc} \in \mathcal{CL}(\mathbb{R}^n)$ and rotor $\mathbf{G}$.

In (8.3.12), the matrix part on the right side corresponds to

$$\frac{1 + \lambda\Delta}{2} + ((\frac{\lambda\Delta}{2} - 1)\mathbf{e} + \lambda\mathbf{c}) \wedge \mathbf{e}_0. \tag{8.3.15}$$

Since $\mathbf{MM}^\dagger$ is a nonzero scalar, (8.3.15) is invertible, and is a rotor of length at most two.

In (8.3.13), the matrix part on the right side corresponds to

$$(\mu\Delta\mathbf{e}_0 - \frac{\mathbf{e}}{2})(1 - \frac{\mu\mathbf{d}}{2}(\mu\Delta\mathbf{e}_0 - \frac{\mathbf{e}}{2})^{-1}\mathbf{e}_0\mathbf{e}) = (\mu\Delta\mathbf{e}_0 - \frac{\mathbf{e}}{2})(1 - \frac{\mathbf{de}}{2\Delta})$$
$$= -\mathbf{d}(\mu\Delta\mathbf{e}_0 - \frac{\mathbf{e}}{2})(\mathbf{d}^{-1} - \frac{\mathbf{e}}{2\Delta}), \tag{8.3.16}$$

which is the geometric product of a vector in $\mathbb{R}^n$ and a rotor in $\mathcal{CL}(\mathbb{R}^{n+1,1})$. $\square$

Theorem 8.36. Any versor in $\mathcal{CL}(\mathbb{R}^{n+1,1})$ corresponds via (8.3.5) to a twisted Vahlen matrix.

Proof. Let $\mathbf{M}$ be a versor. When $\mathbf{M}$ is a vector, then it is of the form (8.3.7), and is a twisted Vahlen matrix if and only if it is neither zero nor null. To prove the theorem by induction, we need only prove that for any versor $\mathbf{M}$ and invertible vector $\mathbf{M}'$,

$$\mathbf{MM}' = \begin{pmatrix} \mathbf{A} & \mathbf{B} \\ \mathbf{C} & \mathbf{D} \end{pmatrix} \begin{pmatrix} \mathbf{x} & \alpha \\ \beta & \mathbf{x} \end{pmatrix} = \begin{pmatrix} \mathbf{Ax} + \beta\mathbf{B} & \alpha\mathbf{A} - \mathbf{Bx} \\ -\mathbf{Cx} + \beta\mathbf{D} & \alpha\mathbf{C} + \mathbf{Dx} \end{pmatrix} \tag{8.3.17}$$

is a twisted Vahlen matrix.

By (8.3.11) to (8.3.13), we only need to consider three cases:

$$(1)\ \mathbf{M} = \begin{pmatrix} 1 & \lambda\mathbf{b} \\ \lambda\mathbf{c} & \Delta - \lambda^2\mathbf{cb} \end{pmatrix}, \quad (2)\ \mathbf{M} = \begin{pmatrix} 1 & 0 \\ \lambda\mathbf{c} & \lambda\Delta \end{pmatrix}, \quad (3)\ \mathbf{M} = \begin{pmatrix} 0 & 1 \\ \mu\Delta & \mu\mathbf{d} \end{pmatrix}.$$

The corresponding matrices $\mathbf{MM}'$ are respectively

$$(1)\ \begin{pmatrix} \mathbf{x} + \lambda\beta\mathbf{b} & \alpha - \lambda\mathbf{bx} \\ \Delta\beta - \lambda\mathbf{c}(\mathbf{x} + \lambda\beta\mathbf{b}) & \Delta\mathbf{x} + \lambda\mathbf{c}(\alpha - \lambda\mathbf{bx}) \end{pmatrix},$$

$$(2)\ \begin{pmatrix} \mathbf{x} & \alpha \\ \lambda(\Delta\beta - \mathbf{cx}) & \lambda(\Delta\mathbf{x} + \alpha\mathbf{c}) \end{pmatrix},$$

$$(3)\ \begin{pmatrix} \beta & -\mathbf{x} \\ \mu(\beta\mathbf{d} - \Delta\mathbf{x}) & \mu(\Delta\alpha + \mathbf{dx}) \end{pmatrix},$$

which can be easily verified to be twisted Vahlen matrices. $\square$

Theorem 8.37. [Twisted version of Vahlen's Theorem] Any twisted Vahlen matrix $\mathbf{M}$ generates the following conformal transformation in $\mathbb{R}^n$:

$$\mathbf{x} \longmapsto \mathbf{M}(\mathbf{x}) = (\mathbf{Ax} + \mathbf{B})(\hat{\mathbf{C}}\mathbf{x} + \hat{\mathbf{D}})^{-1}, \quad \forall\,\mathbf{x} \in \mathbb{R}^n. \tag{8.3.18}$$

Conversely, any conformal transformation in $\mathbb{R}^n$ has such a *twisted fractional linear representation*.

Proof.   Vector $\mathbf{f}(\mathbf{x})$ corresponds to the first matrix in (8.3.8). The graded adjoint action of versor $\mathbf{M}$ on $\mathbf{f}(\mathbf{x})$ is

$$
\begin{pmatrix} \mathbf{A} & \mathbf{B} \\ \mathbf{C} & \mathbf{D} \end{pmatrix}
\begin{pmatrix} \mathbf{x} & -\mathbf{x}^2 \\ 1 & \mathbf{x} \end{pmatrix}
\overline{\begin{pmatrix} \mathbf{A} & \mathbf{B} \\ \mathbf{C} & \mathbf{D} \end{pmatrix}}
$$

$$
= \begin{pmatrix}
\mathbf{AxD}^\dagger + \mathbf{BD}^\dagger + \mathbf{x}^2\mathbf{A}\overline{\mathbf{C}} + \mathbf{Bx}\overline{\mathbf{C}} & -\mathbf{AxB}^\dagger - \mathbf{BB}^\dagger - \mathbf{x}^2\mathbf{AA}^\dagger - \mathbf{Bx}\mathbf{A}^\dagger \\
-\mathbf{Cx}\overline{\mathbf{D}} + \mathbf{D}\overline{\mathbf{D}} + \mathbf{x}^2\mathbf{CC}^\dagger - \mathbf{Dx}\mathbf{C}^\dagger & \mathbf{Cx}\overline{\mathbf{B}} - \mathbf{D}\overline{\mathbf{B}} - \mathbf{x}^2\mathbf{C}\overline{\mathbf{A}} + \mathbf{Dx}\overline{\mathbf{A}}
\end{pmatrix}
$$

$$
= \begin{pmatrix}
(\mathbf{Ax}+\mathbf{B})(\mathbf{D}^\dagger + \mathbf{x}\overline{\mathbf{C}}) & -(\mathbf{Ax}+\mathbf{B})(\mathbf{B}^\dagger + \mathbf{x}\mathbf{A}^\dagger) \\
-(\mathbf{Cx}-\mathbf{D})(\overline{\mathbf{D}} - \mathbf{x}\mathbf{C}^\dagger) & (\mathbf{Cx}-\mathbf{D})(\overline{\mathbf{B}} - \mathbf{x}\overline{\mathbf{A}})
\end{pmatrix}.
$$

So $\mathbf{M}$ changes vector $\mathbf{x}$ to vector

$$
-\frac{(\mathbf{Ax}+\mathbf{B})(\mathbf{D}^\dagger + \mathbf{x}\overline{\mathbf{C}})}{(\mathbf{Cx}-\mathbf{D})(\overline{\mathbf{D}} - \mathbf{x}\mathbf{C}^\dagger)} = -(\mathbf{Ax}+\mathbf{B})(\mathbf{D}^\dagger + \mathbf{x}\overline{\mathbf{C}})\widehat{(\mathbf{D}^\dagger + \mathbf{x}\overline{\mathbf{C}})}^{-1}(\mathbf{Cx}-\mathbf{D})^{-1}
$$

$$
= -(\mathbf{Ax}+\mathbf{B})\widehat{(\mathbf{Cx}-\mathbf{D})}^{-1}
$$

$$
= (\mathbf{Ax}+\mathbf{B})(\hat{\mathbf{C}}\mathbf{x}+\hat{\mathbf{D}})^{-1}.
$$

$\square$

Corollary 8.38. Any versor in $\mathcal{G}(\mathbb{R}^{n+1,1})$, when multiplied by a suitable rotor of length within two using the geometric product, can be changed into a versor in $\mathcal{G}(\mathbb{R}^n)$. Equivalently, any conformal transformation in $\mathbb{R}^n$, when composed with a suitable orthogonal transformation in $\mathbb{R}^n$, becomes either a translation, or a dilation, or a transversion.

The twisted multiplication (8.3.4) is just the geometric product in $\mathcal{CL}(\mathbb{R}^{n+1,1})$ under the decomposition (8.3.1) with respect to the partial Witt basis $\mathbf{e}, \mathbf{e}_0$ of $\mathbb{R}^{n+1,1}$. When $\mathcal{CL}(\mathbb{R}^n)$ is represented by a matrix algebra, the twisted matrix multiplication is very inconvenient, and needs to be revised to usual matrix multiplication. The work was done by Vahlen in 1902.

Definition 8.39. The *algebra of 2×2 Clifford matrices* over $\mathcal{CL}(\mathbb{R}^n)$, denoted by $M_{2\times 2}(\mathcal{CL}(\mathbb{R}^n))$, is the linear space of matrices of the form $\mathbf{M} = \begin{pmatrix} \mathbf{A} & \mathbf{B} \\ \mathbf{C} & \mathbf{D} \end{pmatrix}$, where $\mathbf{A}, \mathbf{B}, \mathbf{C}, \mathbf{D} \in \mathcal{CL}(\mathbb{R}^n)$, equipped with the usual matrix multiplication

$$
\begin{pmatrix} \mathbf{A} & \mathbf{B} \\ \mathbf{C} & \mathbf{D} \end{pmatrix}
\begin{pmatrix} \mathbf{A}' & \mathbf{B}' \\ \mathbf{C}' & \mathbf{D}' \end{pmatrix}
= \begin{pmatrix} \mathbf{AA}' + \mathbf{BC}' & \mathbf{AB}' + \mathbf{BD}' \\ \mathbf{CA}' + \mathbf{DC}' & \mathbf{CB}' + \mathbf{DD}' \end{pmatrix}. \tag{8.3.19}
$$

Let $\mathbf{e}_1, \mathbf{e}_2, \ldots, \mathbf{e}_n$ be an orthonormal basis of $\mathbb{R}^n$. By (5.3.25), $\mathcal{CL}(\mathbb{R}^{n+1,1}) \simeq M_{2\times 2}(\mathcal{CL}(\mathbb{R}^n))$ under the correspondence between the Witt basis $\mathbf{e}, \mathbf{e}_0, \mathbf{e}_1, \mathbf{e}_2, \ldots, \mathbf{e}_n$ of $\mathbb{R}^{n+1,1}$ and the following basis of Clifford matrices:

$$
\begin{pmatrix} 0 & -2 \\ 0 & 0 \end{pmatrix}, \;
\begin{pmatrix} 0 & 0 \\ 1 & 0 \end{pmatrix}, \;
\begin{pmatrix} \mathbf{e}_1 & 0 \\ 0 & -\mathbf{e}_1 \end{pmatrix}, \;
\begin{pmatrix} \mathbf{e}_2 & 0 \\ 0 & -\mathbf{e}_2 \end{pmatrix}, \; \ldots, \;
\begin{pmatrix} \mathbf{e}_n & 0 \\ 0 & -\mathbf{e}_n \end{pmatrix}.
$$

Under the above correspondence, $\mathbf{A} \in \mathcal{CL}(\mathbb{R}^n)$ corresponds to $\mathrm{diag}(\mathbf{A}, \hat{\mathbf{A}})$, and any twisted Clifford matrix corresponds to a unique Clifford matrix as follows:

$$\text{twisted Clifford matrix } \begin{pmatrix} \mathbf{A} & \mathbf{B} \\ \mathbf{C} & \mathbf{D} \end{pmatrix} \longleftrightarrow \text{Clifford matrix } \begin{pmatrix} \mathbf{A} & \mathbf{B} \\ \hat{\mathbf{C}} & \hat{\mathbf{D}} \end{pmatrix}. \qquad (8.3.20)$$

Example 8.40. By $\mathcal{CL}(\mathbb{R}^2) \cong M_{2\times 2}(\mathbb{R})$ and $\mathcal{CL}(\mathbb{R}^3) \cong M_{2\times 2}(\mathbb{C})$, $\mathcal{CL}(\mathbb{R}^{3,1})$ and $\mathcal{CL}(\mathbb{R}^{4,1})$ are isomorphic to the algebra of 4×4 real matrices and the real algebra of 4×4 complex matrices respectively.

All the previous results presented in the form of twisted Clifford matrices can be translated easily into Clifford matrices.

Definition 8.41. A 2×2 matrix $\mathbf{M} = \begin{pmatrix} \mathbf{A} & \mathbf{B} \\ \mathbf{C} & \mathbf{D} \end{pmatrix}$ over $\mathcal{CL}(\mathbb{R}^n)$ is called a *Vahlen matrix*, if

(1) $\mathbf{A}, \mathbf{B}, \mathbf{C}, \mathbf{D}$ are either versors or zero;
(2) $\mathbf{AB}^\dagger, \mathbf{BD}^\dagger, \mathbf{DC}^\dagger, \mathbf{CA}^\dagger$ are vectors;
(3) $\Delta = \mathbf{AD}^\dagger - \mathbf{BC}^\dagger$ is a nonzero scalar.

Theorem 8.42. [Vahlen's Theorem] Any Vahlen matrix $\mathbf{M}$ generates the following conformal transformation in $\mathbb{R}^n$:

$$\mathbf{x} \longmapsto \mathbf{M}(\mathbf{x}) = (\mathbf{Ax} + \mathbf{B})(\mathbf{Cx} + \mathbf{D})^{-1}, \quad \forall \mathbf{x} \in \mathbb{R}^n; \qquad (8.3.21)$$

and any conformal transformation has such a *fractional linear representation*.

Vahlen's Theorem has been extended to $\mathbb{R}^{p,q}$, where $p, q \neq 0$ [44], [64], [132].

Definition 8.43. A 2×2 matrix $\mathbf{M} = \begin{pmatrix} \mathbf{A} & \mathbf{B} \\ \mathbf{C} & \mathbf{D} \end{pmatrix}$ over $\mathcal{CL}(\mathbb{R}^{p,q})$, where both p, q are nonzero, is called a *Vahlen matrix* if

(1) $\mathbf{A}, \mathbf{B}, \mathbf{C}, \mathbf{D}$ are either Clifford monomials or zero;
(2) $\mathbf{A}^\dagger\mathbf{B}, \mathbf{BD}^\dagger, \mathbf{D}^\dagger\mathbf{C}, \mathbf{CA}^\dagger$ are vectors;
(3) $\Delta = \mathbf{AD}^\dagger - \mathbf{BC}^\dagger$ is a nonzero scalar.

Theorem 8.44. [145] When both p, q are nonzero, the Vahlen matrices are the group of versors in $\mathcal{CL}(\mathbb{R}^{p+1,q+1})$ under the isomorphism $M_{2\times 2}(\mathcal{CL}(\mathbb{R}^{p,q})) \simeq \mathcal{CL}(\mathbb{R}^{p+1,q+1})$ from (5.3.20). Any Vahlen matrix $\mathbf{M}$ generates the following conformal transformation in $\mathbb{R}^{p,q}$:

$$\mathbf{x} \longmapsto \mathbf{M}(\mathbf{x}) = (\mathbf{Ax} + \mathbf{B})(\mathbf{Cx} + \mathbf{D})^{-1}, \quad \forall \mathbf{x} \in \mathbb{R}^{p,q}; \qquad (8.3.22)$$

and any conformal transformation has such a *fractional linear representation*.

8.4 Affine geometry with dual Clifford algebra

By Definition 7.38, we have obtained from the conformal model of $\mathbb{E}^n$ the homogeneous coordinates representation of its affine structure, by the perspective projection $\mathbf{f}(\mathbf{x}) \mapsto \mathbf{e} \wedge \mathbf{f}(\mathbf{x})$.

The affine space $\mathbf{e} \wedge \mathcal{N}_{\mathbf{e}}$ is composed of affine planes of dimension ranging from 0 to n. A 0D plane is an affine point $\mathbf{x} \in \mathbb{R}^n$ in its affine representation $\mathbf{e} \wedge \mathbf{f}(\mathbf{x}) = \mathbf{e} \wedge \mathbf{e}_0 + \mathbf{e} \wedge \mathbf{x}$. By (7.4.14), an rD plane determined by $r + 1$ affinely independent points $\mathbf{x}_1, \dots, \mathbf{x}_{r+1} \in \mathbb{R}^n$ is represented by the $(r + 2)$-blade

$$\mathbf{e}(\mathbf{x}_1 \wedge \cdots \wedge \mathbf{x}_{r+1}) + (\mathbf{e} \wedge \mathbf{e}_0)\partial(\mathbf{x}_1 \wedge \cdots \wedge \mathbf{x}_{r+1}),$$

which is just the moment-direction representation.

By a second perspective projection with center $\mathbf{e}_0$, we obtain the Cartesian model of $\mathbb{E}^n$:

$$\mathbf{f}(\mathbf{x}) \longmapsto \mathbf{e} \wedge \mathbf{e}_0 \wedge \mathbf{f}(\mathbf{x}) = \mathbf{x}(\mathbf{e} \wedge \mathbf{e}_0) \in \mathbf{e} \wedge \mathbf{e}_0 \wedge \mathbb{R}^n, \quad \forall \mathbf{x} \in \mathbb{R}^n. \qquad (8.4.1)$$

As in Definition 7.38, vector space $\mathbf{e} \wedge \Lambda(\mathbb{R}^{n+1,1}) = (\mathbf{e} \wedge \Lambda(\mathbb{R}^n)) \oplus (\mathbf{e} \wedge \mathbf{e}_0 \wedge \Lambda(\mathbb{R}^n))$ equipped with the outer product "$\wedge_{\mathbf{e}}$" and the meet product in $\Lambda(\mathbb{R}^{n+1,1})$ is a GC algebra. However, this vector space is not closed under the geometric product. The following *dual mapping* can change the vector space into one that is closed under the geometric product, hence closed under all the products induced from the geometric product.

Definition 8.45. Let $(\mathbf{e}, \mathbf{e}_0)$ be a Witt pair in $\mathbb{R}^{n+1,1}$. The *dual mapping* with respect to $\mathbf{e} \wedge \mathbf{e}_0$ in $\mathcal{CL}(\mathbb{R}^{n+1,1})$, refers to the following linear involution in $\mathcal{CL}(\mathbb{R}^{n+1,1})$:

$$\mathbf{E} : \mathbf{A} \longmapsto \mathbf{A}(\mathbf{e} \wedge \mathbf{e}_0), \quad \forall \mathbf{A} \in \mathcal{CL}(\mathbb{R}^{n+1,1}). \qquad (8.4.2)$$

Notice that the dual mapping is neither an isomorphism nor an anti-isomorphism in Clifford algebra $\mathcal{CL}(\mathbb{R}^{n+1,1})$. Under the dual mapping, vector space $\mathbf{e} \wedge \Lambda(\mathbb{R}^{n+1,1})$ is mapped onto the space

$$\Lambda(\mathbf{e} \oplus \mathbb{R}^n) = (\mathbf{e} \wedge \Lambda(\mathbb{R}^n)) \oplus \Lambda(\mathbb{R}^n). \qquad (8.4.3)$$

The latter is obviously closed under the geometric product in $\mathcal{CL}(\mathbb{R}^{n+1,1})$.

Similarly, vector space $\mathbf{e} \wedge \mathbf{e}_0 \wedge \Lambda(\mathbb{R}^n)$ is a Grassmann space, but is not closed under the geometric product in $\Lambda(\mathbb{R}^{n+1,1})$. The dual mapping changes vector space $\mathbf{e} \wedge \mathbf{e}_0 \wedge \Lambda(\mathbb{R}^n)$ to $\Lambda(\mathbb{R}^n)$, and the latter is obviously a Clifford algebra.

Definition 8.46. Vector space $\Lambda(\mathbf{e} \oplus \mathbb{R}^n)$ equipped with the geometric product of $\mathcal{CL}(\mathbb{R}^n)$ which is further extended linearly by

$$\begin{aligned} \mathbf{e}^2 &= 0, \\ \mathbf{e}\mathbf{A} &= \hat{\mathbf{A}}\mathbf{e}, \quad \forall \mathbf{A} \in \mathcal{CL}(\mathbb{R}^n), \end{aligned} \qquad (8.4.4)$$

is called the *dual Clifford algebra* over $\mathbb{R}^n$. Its elements are called *dual multivectors*, or *dual Clifford numbers*.

Definition 8.47. The *dual involution* in $\Lambda(\mathbf{e} \oplus \mathbb{R}^n)$, denoted by the overhead tilde symbol, is defined as the following linear involution:

$$\widetilde{\mathbf{e}} = -\mathbf{e},$$
$$\widetilde{\mathbf{A}} = \mathbf{A}, \ \forall \mathbf{A} \in \Lambda(\mathbb{R}^n), \tag{8.4.5}$$
$$\widetilde{\mathbf{eA}} = -\mathbf{eA}.$$

Proposition 8.48. Let $\mathbf{E}$ be the dual mapping (8.4.2). Then for any $\mathbf{A} \in \Lambda(\mathbf{e} \oplus \mathbb{R}^n)$,
$$\widetilde{\mathbf{A}} = (\mathbf{e} \wedge \mathbf{e}_0)\mathbf{E}(\mathbf{A}) = Ad_{\mathbf{e} \wedge \mathbf{e}_0}(\mathbf{A}). \tag{8.4.6}$$

Proof. Direct from the fact that $\mathbf{e} \wedge \mathbf{e}_0$ commutes with everything in $\Lambda(\mathbb{R}^n)$, but anticommutes with $\mathbf{e}$. $\qquad\square$

The following result discloses the influence of the dual mapping upon the geometric product in $\mathcal{CL}(\mathbb{R}^{n+1,1})$.

Corollary 8.49. Let $\mathbf{E}$ be the dual mapping (8.4.2). For any $\mathbf{A}, \mathbf{B} \in \mathbf{e} \wedge \Lambda(\mathbb{R}^{n+1,1})$,
$$\mathbf{E}(\mathbf{A})\widetilde{\mathbf{E}(\mathbf{B})} = \mathbf{A}\mathbf{B}. \tag{8.4.7}$$

Notations.

- In $\Lambda(\mathbf{e} \oplus \mathbb{R}^n)$, the dual operator with respect to the unit pseudoscalar $\mathbf{I}_n$ of $\mathcal{CL}(\mathbb{R}^n)$ is denoted by "$\perp$": for any dual multivector $\mathbf{A}$,
$$\mathbf{A}^{\perp} := \mathbf{A}\mathbf{I}_n^{-1}. \tag{8.4.8}$$
- The dual operator with respect to an orthonormal basis of $\mathbf{e} \oplus \mathbb{R}^n$ of the form $\mathbf{e}, \mathbf{e}_1, \mathbf{e}_2, \ldots, \mathbf{e}_n$, where the $\mathbf{e}_i$ for all $1 \leq i \leq n$ are in $\mathbb{R}^n$, is independent of the selected orthonormal basis of $\mathbb{R}^n$, and denoted by "$*$" as usual.
- The dual operator in $\Lambda(\mathbb{R}^{n+1,1})$ is still denoted by "$\sim$".

Corollary 8.50. The dual involution commutes with the geometric product, the graded involution, the reversion, the inversion, the conjugate (5.2.8), various grading operators, the dual mapping (8.4.2), and the above three kinds of dual operators.

The advantage of the dual multivector representation of affine planes is that there is only one null vector involved, which reduces the dimension of the base vector space by one. Dual multivectors are generalizations of dual quaternions, and dual Clifford algebra extends dual vector algebra to n dimensions.

We first take a look at the classical dual quaternions. Let $\mathbf{e}_1, \mathbf{e}_2, \mathbf{e}_3$ be an orthonormal basis of $\mathbb{R}^3$. Then $1, \mathbf{e}_1\mathbf{e}_2, \mathbf{e}_2\mathbf{e}_3, \mathbf{e}_1\mathbf{e}_3$ generate the quaternions. Obviously $\mathbf{e}$ commutes with the generators. Denote
$$\epsilon = \mathbf{e} \wedge \mathbf{I}_3. \tag{8.4.9}$$
Then ϵ also commutes with the generators. The algebra of dual quaternions is isomorphic to the even subalgebra of $\mathcal{CL}(\epsilon)$.

A vector $\mathbf{a} = x\mathbf{e}_1 + y\mathbf{e}_2 + z\mathbf{e}_3 \in \mathbb{R}^3$ corresponds to a quaternion $\mathbf{a}^{\perp} = x\mathbf{e}_2\mathbf{e}_3 + y\mathbf{e}_3\mathbf{e}_1 + z\mathbf{e}_1\mathbf{e}_2$. The correspondence between the dual multivector representation and the dual quaternion representation of affine subspaces is as follows:

- point $\mathbf{a} \in \mathbb{R}^3$ is represented by

$$(\mathbf{e} \wedge \mathbf{f}(\mathbf{a}))(\mathbf{e} \wedge \mathbf{e}_0) = 1 + \mathbf{e} \wedge \mathbf{a} = 1 + \epsilon \mathbf{a}^\perp; \qquad (8.4.10)$$

- the line with direction $\mathbf{l} \in \mathbb{R}^3$ and through point $\mathbf{a} \in \mathbb{R}^3$ is represented by

$$(\mathbf{e} \wedge \mathbf{f}(\mathbf{a}) \wedge \mathbf{l})(\mathbf{e} \wedge \mathbf{e}_0)\mathbf{I}_3^{-1} = \mathbf{l}^\perp - \epsilon(\mathbf{a} \wedge \mathbf{l}) = \mathbf{l}^\perp + \epsilon[\mathbf{a}^\perp, \mathbf{l}^\perp]; \qquad (8.4.11)$$

- the plane normal to vector $\mathbf{n} \in \mathbb{R}^3$ and through point $\mathbf{a} \in \mathbb{R}^3$ is represented by

$$-(\mathbf{f}(\mathbf{a}) \cdot (\mathbf{e} \wedge \mathbf{n}))^{\sim}(\mathbf{e} \wedge \mathbf{e}_0) = (\mathbf{n} + (\mathbf{n} \cdot \mathbf{a})\mathbf{e})\mathbf{I}_3^{-1} = \mathbf{n}^\perp + \epsilon(\mathbf{n}^\perp \cdot \mathbf{a}^\perp). \qquad (8.4.12)$$

In the dual vector algebra over $\mathbb{R}^3$, there is a classical result (5.3.13) on the geometric relationship between two spatial lines, one of which passes through point $\mathbf{a}$ in direction $\mathbf{l}_a$, and the other passes through point $\mathbf{b}$ in direction $\mathbf{l}_b$. In dual Clifford algebra, the result becomes, by (8.4.11),

$$\begin{aligned}
&(\mathbf{l}_a^\perp + \epsilon[\mathbf{a}^\perp, \mathbf{l}_a^\perp])(\mathbf{l}_b^\perp + \epsilon[\mathbf{b}^\perp, \mathbf{l}_b^\perp]) \\
&= (\mathbf{e} \wedge \mathbf{f}(\mathbf{a}) \wedge \mathbf{l}_a)(\mathbf{e} \wedge \mathbf{e}_0)\mathbf{I}_3^{-1}(\mathbf{e} \wedge \mathbf{f}(\mathbf{b}) \wedge \mathbf{l}_b)(\mathbf{e} \wedge \mathbf{e}_0)\mathbf{I}_3^{-1} \\
&= (\mathbf{e} \wedge \mathbf{f}(\mathbf{a}) \wedge \mathbf{l}_a)(\mathbf{e} \wedge \mathbf{e}_0)\mathbf{I}_3^{-1}(\mathbf{e} \wedge \mathbf{e}_0)\mathbf{I}_3^{-1}(\mathbf{e} \wedge \mathbf{f}(\mathbf{b}) \wedge \mathbf{l}_b) \\
&= -\{(\mathbf{e} \wedge \mathbf{f}(\mathbf{a}) \wedge \mathbf{l}_a)(\mathbf{e} \wedge \mathbf{e}_0)\}\{(\mathbf{e} \wedge \mathbf{e}_0)(\mathbf{e} \wedge \mathbf{f}(\mathbf{b}) \wedge \mathbf{l}_b)\} \\
&= -\{\mathbf{l}_a + \mathbf{e}(\mathbf{a} \wedge \mathbf{l}_a)\}\{\mathbf{l}_b - \mathbf{e}(\mathbf{b} \wedge \mathbf{l}_b)\} \\
&= -\{\mathbf{l}_a \cdot \mathbf{l}_b + \mathbf{e}((\mathbf{a} - \mathbf{b}) \wedge \mathbf{l}_a \wedge \mathbf{l}_b) + \mathbf{l}_a \wedge \mathbf{l}_b \\
&\qquad + (\mathbf{l}_a \cdot \mathbf{l}_b)\mathbf{e}(\mathbf{a} - \mathbf{b}) - \mathbf{e}((\mathbf{a} \cdot \mathbf{l}_b)\mathbf{l}_a - (\mathbf{b} \cdot \mathbf{l}_a)\mathbf{l}_b)\}.
\end{aligned} \qquad (8.4.13)$$

In the nD extension $\Lambda(\mathbf{e} \oplus \mathbb{R}^n)$ of dual quaternions, the rD plane passing through point $\mathbf{a} \in \mathbb{R}^n$ with rD direction $\mathbf{B}_r \in \Lambda^r(\mathbb{R}^n)$ is represented by $(1 + \mathbf{ea}) \wedge \mathbf{B}_r$. Point $\mathbf{a}$ is represented by $1 + \mathbf{ea}$, and the origin is represented by 1. This representation highly depends on the choice of the origin in $\mathbb{E}^n$.

The extension of (8.4.13) to $\mathbb{R}^n$ is as follows. Let $\mathbf{A}_r, \mathbf{B}_s$ be respectively an r-blade and an s-blade in $\Lambda(\mathbb{R}^n)$. Then in the conformal model,

$$\begin{aligned}
&(\mathbf{e} \wedge \mathbf{f}(\mathbf{a}) \wedge \mathbf{A}_r)(\mathbf{e} \wedge \mathbf{f}(\mathbf{b}) \wedge \mathbf{B}_s) \\
&= \{(\mathbf{e} \wedge \mathbf{f}(\mathbf{a}) \wedge \mathbf{A}_r)(\mathbf{e} \wedge \mathbf{e}_0)\}\{(\mathbf{e} \wedge \mathbf{e}_0)(\mathbf{e} \wedge \mathbf{f}(\mathbf{b}) \wedge \mathbf{B}_s)\} \\
&= \{\mathbf{A}_r + \mathbf{e}(\mathbf{a} \wedge \mathbf{A}_r)\}\{\mathbf{B}_s + \widetilde{\mathbf{e}(\mathbf{b} \wedge \mathbf{B}_s)}\} \\
&= \{\mathbf{A}_r + \mathbf{e}(\mathbf{a} \wedge \mathbf{A}_r)\}\{\mathbf{B}_s - \mathbf{e}(\mathbf{b} \wedge \mathbf{B}_s)\} \\
&= \mathbf{A}_r\mathbf{B}_s - \mathbf{e}\hat{\mathbf{A}}_r(\mathbf{b} \wedge \mathbf{B}_s) + \mathbf{e}(\mathbf{a} \wedge \mathbf{A}_r)\mathbf{B}_s.
\end{aligned} \qquad (8.4.14)$$

By Lemma 8.49, the dual mapping changes the geometric product of $\mathcal{CL}(\mathbb{R}^{n+1,1})$ into the following *nonassociative product* in dual Clifford algebra $\Lambda(\mathbf{e} \oplus \mathbb{R}^n)$, which truly reflects the Euclidean geometric relationship between two affine objects represented by dual multivectors.

Definition 8.51. For any two dual multivectors $\mathbf{A}, \mathbf{B}$, their *nonassociative product* is $\mathbf{A}\widetilde{\mathbf{B}}$.

Definition 8.52. The *dual adjoint action* of a versor $\mathbf{V} \in \Lambda(\mathbf{e} \oplus \mathbb{R}^n)$ on a dual multivector $\mathbf{A}$, denoted by $\widetilde{Ad}_{\mathbf{V}}(\mathbf{A})$, is defined by

$$\widetilde{Ad}_{\mathbf{V}}(\mathbf{A}) := \mathbf{V}\mathbf{A}\widetilde{\mathbf{V}}^{-1}. \tag{8.4.15}$$

Let $\mathbf{V} = \mathbf{A} + \mathbf{e}\mathbf{B}$ be a versor in $\Lambda(\mathbf{e} \oplus \mathbb{R}^n)$, where $\mathbf{A}, \mathbf{B} \in \Lambda(\mathbb{R}^n)$. As in (6.3.44), by versor compression there exist invertible vectors $\mathbf{a}_i \in \mathbb{R}^n$ such that up to scale,

$$\mathbf{V} = (\mathbf{a}_1 + \mathbf{e})(\mathbf{a}_2 + \mathbf{e}) \cdots (\mathbf{a}_r + \mathbf{e})\mathbf{a}_{r+1}\mathbf{a}_{r+2} \cdots \mathbf{a}_{r+s}, \tag{8.4.16}$$

where $r, s \geq 0$, and where the $r + s$ constituent vectors in $\mathbf{e} \oplus \mathbb{R}^n$ are linearly independent. Then

$$\mathbf{A} = (\mathbf{a}_1 \wedge \mathbf{a}_2 \wedge \cdots \wedge \mathbf{a}_r)\mathbf{a}_{r+1}\mathbf{a}_{r+2} \cdots \mathbf{a}_{r+s},$$
$$\mathbf{B} = \partial(\mathbf{a}_1 \wedge \mathbf{a}_2 \wedge \cdots \wedge \mathbf{a}_r)\mathbf{a}_{r+1}\mathbf{a}_{r+2} \cdots \mathbf{a}_{r+s}.$$

When each $\mathbf{e}$ in (8.4.16) is replaced by $-\mathbf{e}$, then $\mathbf{V}$ is changed into $\widetilde{\mathbf{V}}$, which is still a versor:

$$\widetilde{\mathbf{V}} = (\mathbf{a}_1 - \mathbf{e})(\mathbf{a}_2 - \mathbf{e}) \cdots (\mathbf{a}_r - \mathbf{e})\mathbf{a}_{r+1}\mathbf{a}_{r+2} \cdots \mathbf{a}_{r+s} = \mathbf{A} - \mathbf{e}\mathbf{B}.$$

The dual adjoint action is neither a Grassmann algebraic homomorphism nor a Clifford algebraic homomorphism in $\Lambda(\mathbf{e} \oplus \mathbb{R}^n)$. It is related to the adjoint action in $\Lambda(\mathbf{e} \oplus \mathbb{R}^n)$ as follows:

Proposition 8.53. For any versor $\mathbf{V} \in \Lambda(\mathbf{e} \oplus \mathbb{R}^n)$ and any dual multivectors $\mathbf{A}, \mathbf{B}$,

$$\mathbf{V}(\mathbf{A}\widetilde{\mathbf{B}})\mathbf{V}^{-1} = \widetilde{Ad}_{\mathbf{V}}(\mathbf{A})\,\widetilde{\widetilde{Ad}_{\mathbf{V}}(\mathbf{B})}. \tag{8.4.17}$$

Proof. By (8.4.6), the left side of (8.4.17) equals $\mathbf{V}\mathbf{A}(\mathbf{e} \wedge \mathbf{e}_0)\mathbf{B}(\mathbf{e} \wedge \mathbf{e}_0)\mathbf{V}^{-1}$. Similarly, the right side of (8.4.17) equals

$$\mathbf{V}\mathbf{A}\widetilde{\mathbf{V}}^{-1}(\mathbf{e} \wedge \mathbf{e}_0)\mathbf{V}\mathbf{B}\widetilde{\mathbf{V}}^{-1}(\mathbf{e} \wedge \mathbf{e}_0) = \mathbf{V}\mathbf{A}(\mathbf{e} \wedge \mathbf{e}_0)\mathbf{V}^{-1}\mathbf{V}\mathbf{B}(\mathbf{e} \wedge \mathbf{e}_0)\mathbf{V}^{-1}$$
$$= \mathbf{V}\mathbf{A}(\mathbf{e} \wedge \mathbf{e}_0)\mathbf{B}(\mathbf{e} \wedge \mathbf{e}_0)\mathbf{V}^{-1}.$$

$\square$

The orthogonal transformations in $\mathbf{e} \oplus \mathbb{R}^n$ are the similarity transformations in $\mathbb{R}^n$. They keep the unique 1D null subspace $\mathbf{e}$ invariant. However, dual Clifford algebra contains only the versors inducing Euclidean transformations. The versors act upon affine objects represented by dual multivectors by the dual adjoint action.

(1) The versor inducing the reflection with respect to hyperplane $(\mathbf{n} + \delta\mathbf{e})^{\perp} = (1 + \mathbf{e}(\delta\mathbf{n})) \wedge (\mathbf{n}\mathbf{I}_n^{-1})$, where $\mathbf{n}$ is a unit vector in $\mathbb{R}^n$, is $\mathbf{V} = \mathbf{n} + \delta\mathbf{e}$. For example, for any $\mathbf{x} \in \mathbb{R}^n$,

$$\widetilde{Ad}_{\mathbf{V}}(1 + \mathbf{e}\mathbf{x}) = (\mathbf{n} + \delta\mathbf{e})(1 + \mathbf{e}\mathbf{x})(\mathbf{n} - \delta\mathbf{e})$$
$$= 1 + \mathbf{e}(2\delta\mathbf{n} - \mathbf{n}\mathbf{x}\mathbf{n})$$
$$= 1 + \mathbf{e}(\mathbf{x} - 2\mathbf{n}(\mathbf{n} \cdot \mathbf{x} - \delta)).$$

(2) The rotor inducing the rotation from vector $\mathbf{b} \in \mathbb{R}^n$ to vector $\mathbf{a} \in \mathbb{R}^n$ centering at the origin of $\mathbb{R}^n$, is $\mathbf{V} = (\mathbf{a} + \mathbf{b})\mathbf{b}$. It satisfies $\widetilde{Ad}_{\mathbf{V}} = Ad_{\mathbf{V}}$.

(3) The rotor inducing the translation by vector $\mathbf{t}$ is $\mathbf{V} = 1 + \mathbf{et}/2$. For example, for the origin of $\mathbb{R}^n$ represented by scalar 1,

$$\widetilde{Ad}_{\mathbf{V}}(1) = (1 + \frac{\mathbf{et}}{2})\, 1\, (1 + \frac{\mathbf{et}}{2}) = 1 + \mathbf{et}.$$

Contrary to the nonassociative product in Definition 8.51, the geometric product in the dual geometric algebra, although associative, reflects the Euclidean geometric relationship of affine objects very "noisily", because of the interference of the origin.

Consider the geometric product of two planes $(1 + \mathbf{ea}) \wedge \mathbf{A}_r$ and $(1 + \mathbf{eb}) \wedge \mathbf{B}_s$:

$$((1 + \mathbf{ea}) \wedge \mathbf{A}_r)((1 + \mathbf{eb}) \wedge \mathbf{B}_s) = \mathbf{A}_r \mathbf{B}_s + \mathbf{e}\hat{\mathbf{A}}_r(\mathbf{b} \wedge \mathbf{B}_s) + \mathbf{e}(\mathbf{a} \wedge \mathbf{A}_r)\mathbf{B}_s. \quad (8.4.18)$$

It is composed of two parts: the *real part* $\mathbf{A}_r \mathbf{B}_s$ which is the relationship between the directions of the two planes, and the *dual part* which is the sum of two relationships: the relationship between the direction $\hat{\mathbf{A}}_r$ and the moment $\mathbf{b} \wedge \mathbf{B}_s$, and the relationship between the moment $\mathbf{a} \wedge \mathbf{A}_r$ and the direction $\mathbf{B}_s$.

More generally, the geometric product of k planes $(1 + \mathbf{ea}_1) \wedge \mathbf{A}_1, \ldots, (1 + \mathbf{ea}_k) \wedge \mathbf{A}_k$ is

$$((1 + \mathbf{ea}_1) \wedge \mathbf{A}_1)((1 + \mathbf{ea}_2) \wedge \mathbf{A}_2) \cdots ((1 + \mathbf{ea}_k) \wedge \mathbf{A}_k)$$
$$= \mathbf{A}_1 \mathbf{A}_2 \cdots \mathbf{A}_k + \mathbf{e}(\mathbf{a}_1 \wedge \mathbf{A}_1)\mathbf{A}_2 \cdots \mathbf{A}_k + \mathbf{e}\hat{\mathbf{A}}_1(\mathbf{a}_2 \wedge \mathbf{A}_2)\mathbf{A}_3 \cdots \mathbf{A}_k + \cdots \quad (8.4.19)$$
$$+ \mathbf{e}\hat{\mathbf{A}}_1 \cdots \hat{\mathbf{A}}_{k-1}(\mathbf{a}_k \wedge \mathbf{A}_k).$$

Below we investigate the influence of the origin in $\mathbb{E}^n$ upon the geometric interpretations of the outer product, the inner product and the meet product in dual Clifford algebra.

(1) The outer product between rD plane $(1 + \mathbf{ea}) \wedge \mathbf{A}_r$ and sD $(1 + \mathbf{eb}) \wedge \mathbf{B}_s$ is

$$((1 + \mathbf{ea}) \wedge \mathbf{A}_r) \wedge ((1 + \mathbf{eb}) \wedge \mathbf{B}_s) = (1 + \mathbf{e}(\mathbf{a} + \mathbf{b})) \wedge \mathbf{A}_r \wedge \mathbf{B}_s. \quad (8.4.20)$$

It is the $(r + s)$D affine plane passing through point $\mathbf{a} + \mathbf{b}$ with $(r + s)$D direction $\mathbf{A}_r \wedge \mathbf{B}_s$.

(2) The dual of rD plane $(1 + \mathbf{ea}) \wedge \mathbf{A}_r$ is

$$((1 + \mathbf{ea}) \wedge \mathbf{A}_r)^{\perp} = (1 + \mathbf{e}(\mathbf{a} \wedge \mathbf{A}_r)^{\perp}(\mathbf{A}_r^{\perp})^{-1})\mathbf{A}_r^{\perp} = (1 + \mathbf{e}P_{\mathbf{A}_r}^{\perp}(\mathbf{a})) \cdot \mathbf{A}_r^{\perp}. \quad (8.4.21)$$

It no longer represents any affine plane. Instead, it is composed of the foot $P_{\mathbf{A}_r}^{\perp}(\mathbf{a})$ drawn from the origin to the plane, and the direction $\mathbf{A}_r^{\perp}$ which is the orthogonal complement of the rD direction of the plane.

(3) Planes $(1 + \mathbf{ea}) \wedge \mathbf{A}_r$ and sD $(1 + \mathbf{eb}) \wedge \mathbf{B}_s$ are perpendicular if and only if $\mathbf{A}_r \cdot \mathbf{B}_s = 0$. When they are not perpendicular, their inner product is

$$((1 + \mathbf{ea}) \wedge \mathbf{A}_r) \cdot ((1 + \mathbf{eb}) \wedge \mathbf{B}_s)$$
$$= \{1 + \mathbf{e}(\hat{\mathbf{A}}_r \cdot (\mathbf{b} \wedge \mathbf{B}_s) + (\mathbf{a} \wedge \mathbf{A}_r) \cdot \mathbf{B}_s)(\mathbf{A}_r \cdot \mathbf{B}_s)^{-1}\} (\mathbf{A}_r \cdot \mathbf{B}_s). \quad (8.4.22)$$

If $r = s$, the result represents a point. If $r < s$, the result is composed of an $(s-r)$-graded real part, an $(s-r+1)$-graded dual part, and an $(s-r-1)$-graded dual part. By writing (8.4.22) as

$$(1 + \mathbf{e}(\hat{\mathbf{A}}_r \cdot (\mathbf{b} \wedge \mathbf{B}_s)))(\mathbf{A}_r \cdot \mathbf{B}_s)^{-1}) \wedge (\mathbf{A}_r \cdot \mathbf{B}_s) + \mathbf{e}((\mathbf{a} \wedge \mathbf{A}_r) \cdot \mathbf{B}_s), \qquad (8.4.23)$$

we get that the sum of the $(s-r)$-graded part and the $(s-r+1)$-graded part represents an $(s-r)$D plane with direction $\mathbf{A}_r \cdot \mathbf{B}_s$. When the $(s-r-1)$-graded part is nonzero, the dual "$\perp$" of the sum of the $(s-r)$-graded part and the $(s-r-1)$-graded part is

$$(\mathbf{A}_r \cdot \mathbf{B}_s)^{\perp} + \mathbf{e}((\mathbf{a} \wedge \mathbf{A}_r) \cdot \mathbf{B}_s)^{\perp} = (1 + \mathbf{e}P^{\perp}_{P_{\mathbf{B}_s}(\mathbf{A}_r)}(P_{\mathbf{B}_s}(\mathbf{a})))(\mathbf{A}_r \cdot \mathbf{B}_s)^{\perp},$$

which represents an $(n-s+r)$D plane with direction $(\mathbf{A}_r \cdot \mathbf{B}_s)^{\perp}$. If $r > s$, the geometric interpretation is similar.

(4) The meet product of two planes $(1+\mathbf{ea}) \wedge \mathbf{A}_r$ and $(1+\mathbf{eb}) \wedge \mathbf{B}_s$ in $\Lambda(\mathbf{e} \oplus \mathbb{R}^n)$, where $r + s \geq n - 1$, and $0 < r, s < n$, is

$$\begin{aligned}((1 + \mathbf{ea}) \wedge \mathbf{A}_r) \vee ((1 + \mathbf{eb}) \wedge \mathbf{B}_s) &= (\mathbf{e} \wedge \mathbf{a} \wedge \mathbf{A}_r) \vee \mathbf{B}_s + \mathbf{A}_r \vee (\mathbf{e} \wedge \mathbf{b} \wedge \mathbf{B}_s) \\ &\quad + (-1)^{n-s-1}\mathbf{e} \wedge ((\mathbf{a} \wedge \mathbf{A}_r) \vee (\mathbf{e} \wedge \mathbf{b} \wedge \mathbf{B}_s)).\end{aligned}$$
$$(8.4.24)$$

For example, when $n = 3$, $r = s = 1$, then

$$(\mathbf{l}_a + \mathbf{e}(\mathbf{a} \wedge \mathbf{l}_a)) \vee (\mathbf{l}_b + \mathbf{e}(\mathbf{b} \wedge \mathbf{l}_b)) = [\mathbf{e}(\mathbf{b} - \mathbf{a})\mathbf{l}_a \mathbf{l}_b] + \mathbf{e}([\mathbf{eal}_a \mathbf{l}_b]\mathbf{b} - [\mathbf{eal}_a \mathbf{b}]\mathbf{l}_b). \quad (8.4.25)$$

When the two lines are not coplanar, their meet product is a point on the line of intersection of the two planes each spanned by the origin and one of the two lines.

(5) The *dual meet product* of two planes $(1 + \mathbf{ea}) \wedge \mathbf{A}_r$ and $(1 + \mathbf{eb}) \wedge \mathbf{B}_s$, where $r + s \geq n - 2$, and $0 < r, s < n$, is defined to be $((1 + \mathbf{eb}) \wedge \mathbf{B}_s)^{\perp} \cdot ((1 + \mathbf{ea}) \wedge \mathbf{A}_r)$.

By $(\mathbf{e}(\mathbf{b} \wedge \mathbf{B}_s)^{\perp}) \cdot \mathbf{A}_r = 0$ and $\mathbf{B}_s^{\perp} \cdot (\mathbf{e}(\mathbf{a} \wedge \mathbf{A}_r)) = (-1)^{n-s}\mathbf{e}(\mathbf{B}_s^{\perp} \cdot (\mathbf{a} \wedge \mathbf{A}_r))$, we get

$$((1 + \mathbf{eb}) \wedge \mathbf{B}_s)^{\perp} \cdot ((1 + \mathbf{ea}) \wedge \mathbf{A}_r) = \mathbf{B}_s^{\perp} \cdot \mathbf{A}_r + (-1)^{n-s}\mathbf{e}(\mathbf{B}_s^{\perp} \cdot (\mathbf{a} \wedge \mathbf{A}_r)). \quad (8.4.26)$$

For example, when $n = 3$, $r = 1$ and $s = 2$, the dual meet product of a line $\mathbf{t} + \mathbf{e}(\mathbf{a} \wedge \mathbf{t})$ and a plane $(\mathbf{n} + \delta\mathbf{e})^{\perp}$ that are not parallel to each other, is

$$-(\mathbf{n} + \delta\mathbf{e}) \cdot (\mathbf{t} + \mathbf{e}(\mathbf{a} \wedge \mathbf{t})) = -(\mathbf{n} \cdot \mathbf{t})(1 + \mathbf{e}(\mathbf{a} - \frac{\mathbf{n} \cdot \mathbf{a}}{\mathbf{n} \cdot \mathbf{t}}\mathbf{t})). \qquad (8.4.27)$$

It represents a point $\mathbf{x} = \mathbf{a} - \mathbf{t}\,(\mathbf{n} \cdot \mathbf{a})/(\mathbf{n} \cdot \mathbf{t})$ on the line. This point satisfies $\mathbf{x} \cdot \mathbf{n} = 0$, so it is the intersection of line $\mathbf{t} + \mathbf{e}(\mathbf{a} \wedge \mathbf{t})$ and the plane passing through the origin and parallel to plane $(\mathbf{n} + \delta\mathbf{e})^{\perp}$.

The geometric interpretations in dual Clifford algebra are rather poor, because the representations strongly depend on the origin. The homogeneous model, on the contrary, is origin-free, whose algebraic representations are so good that a single algebraic identity can be interpreted in any of Euclidean, spherical and hyperbolic geometries as interrelated geometric theorems. The rest of this chapter is devoted to the unified representation aspect of the homogeneous model.

8.5 Spherical geometry and its conformal model

Spherical geometry is another frequently encountered classical geometry. It can be taken as the geometry of unit directions in a Euclidean space. Trigonometric functions are the most often used algebraic language to describe spherical geometric properties. In this section, we establish the *conformal model* of spherical geometry in a way that is similar to the homogeneous coordinates representation of affine geometry.

8.5.1 *The classical model of spherical geometry*

The symbol $\mathbb{S}^n$ denotes the unit sphere of $\mathbb{R}^{n+1}$. Any point on $\mathbb{S}^n$ is represented by a unique unit vector in $\mathbb{R}^{n+1}$. For two points $\mathbf{a}, \mathbf{b} \in \mathbb{S}^n$, their *spherical distance $d_{\mathbf{ab}}$* is defined as a number in the range $[0, \pi]$ such that

$$\cos d_{\mathbf{ab}} = \mathbf{a} \cdot \mathbf{b}. \tag{8.5.1}$$

This is the *classical model* of nD spherical geometry.

Definition 8.54. Two distance functions d_1, d_2 defined on the same distance space are said to be *conjugate*, if for any two pairs of points $\mathbf{a}_1, \mathbf{b}_1$ and $\mathbf{a}_2, \mathbf{b}_2$, $d_1(\mathbf{a}_1, \mathbf{b}_1) = d_1(\mathbf{a}_2, \mathbf{b}_2)$ if and only if $d_2(\mathbf{a}_1, \mathbf{b}_1) = d_2(\mathbf{a}_2, \mathbf{b}_2)$.

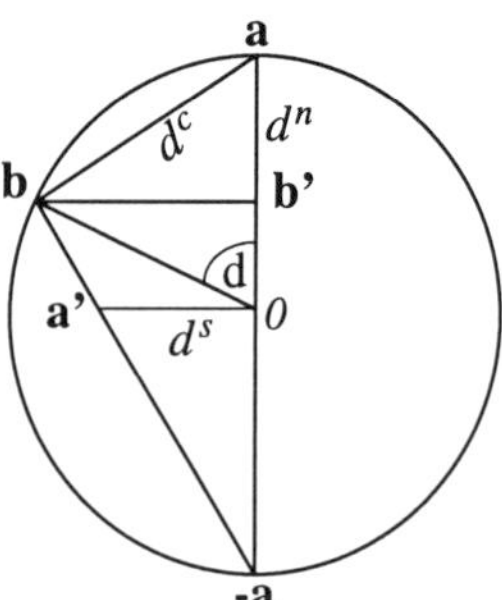

Fig. 8.11   Some conjugate distances in $\mathbb{S}^n$.

The spherical distance has several important conjugate distance functions, as shown in Figure 8.11. The following distance, called *chord distance*, measures the length of the chord between points $\mathbf{a}, \mathbf{b} \in \mathbb{S}^n$:

$$d^c_{\mathbf{ab}} := |\mathbf{a} - \mathbf{b}|. \tag{8.5.2}$$

The following distance, called *normal distance*, measures the distance between point $\mathbf{a}$ and the foot $\mathbf{b}'$ drawn from point $\mathbf{b}$ to the line in $\mathbb{R}^{n+1}$ connecting the origin and point $\mathbf{a}$:

$$d^n_{\mathbf{ab}} := 1 - \mathbf{a} \cdot \mathbf{b}. \tag{8.5.3}$$

The following distance, called *stereographic distance*, measures the distance between the origin and point $\mathbf{a}'$, where $\mathbf{a}'$ is the intersection of line $(-\mathbf{a})\mathbf{b}$ and the hyperplane in $\mathbb{R}^{n+1}$ passing through the origin and perpendicular to line $(-\mathbf{a})\mathbf{a}$:

$$d^s_{\mathbf{ab}} := \frac{|\mathbf{a} \wedge \mathbf{b}|}{1 + \mathbf{a} \cdot \mathbf{b}}. \tag{8.5.4}$$

Some relations among these distances are

$$d^c_{\mathbf{ab}} = 2\sin\frac{d_{\mathbf{ab}}}{2},$$
$$d^n_{\mathbf{ab}} = 1 - \cos d_{\mathbf{ab}}, \tag{8.5.5}$$
$$d^s_{\mathbf{ab}} = \tan\frac{d_{\mathbf{ab}}}{2}.$$

We prefer the normal distance $d^n_{\mathbf{ab}}$ because it is a polynomial function of the Cartesian coordinates of vectors $\mathbf{a}, \mathbf{b}$ in $\mathbb{R}^{n+1}$. The maximal value of the normal distance in $\mathbb{S}^n$ is 2. Two points have normal distance 2 if and only if they are antipodal.

Let $0 < \rho < 2$. A *sphere* of $\mathbb{S}^n$ with center $\mathbf{c} \in \mathbb{S}^n$ and normal radius ρ, is the set of points on $\mathbb{S}^n$ having normal distance ρ with $\mathbf{c}$. Let $\mathbf{x}$ be a point on the sphere. Then $\mathbf{x} \cdot \mathbf{c} = 1 - \rho$. So the sphere is the intersection of $\mathbb{S}^n$ with the affine hyperplane $\{\mathbf{x} \,|\, \mathbf{x} \cdot \mathbf{c} = 1 - \rho\}$ in $\mathbb{R}^{n+1}$. Similarly, for $0 \leq r \leq n - 1$, an rD sphere of $\mathbb{S}^n$ is the intersection of $\mathbb{S}^n$ with an $(r + 1)$D affine plane in $\mathbb{R}^{n+1}$.

An rD *great sphere* is also called an rD *plane* of $\mathbb{S}^n$. It is a sphere with normal radius 1. Any plane is composed of pairs of antipodal points. In Grassmann space $\Lambda(\mathbb{R}^{n+1})$, the rD plane determined by r linearly independent points $\mathbf{x}_1, \ldots, \mathbf{x}_r \in \mathbb{S}^n$ is represented by $\mathbf{x}_1 \wedge \cdots \wedge \mathbf{x}_r$; the rD sphere determined by $r + 1$ linearly independent points $\mathbf{x}_1, \ldots, \mathbf{x}_{r+1} \in \mathbb{S}^n$, has the moment-direction representation $(\mathbf{x}_1 \wedge \cdots \wedge \mathbf{x}_{r+1}, \partial(\mathbf{x}_1 \wedge \cdots \wedge \mathbf{x}_{r+1}))$.

If the normal radius of a sphere is required to be less than 1, then the center of the sphere can be uniquely determined, otherwise it always has two centers. The *interior* of a sphere is the open region of $\mathbb{S}^n$ bordered by the sphere and containing the center. For a sphere on $\mathbb{S}^n$ with center $\mathbf{c}$ and normal radius $\rho < 1$, its interior is $\{\mathbf{x} \,|\, \mathbf{x} \cdot \mathbf{c} > 1 - \rho\}$.

Proposition 8.55. The center of rD sphere $(\mathbf{A}_{r+1}, \partial(\mathbf{A}_{r+1}))$ is $\dfrac{\mathbf{A}_r \cdot (\partial(\mathbf{A}_r))^{-1}}{|\mathbf{A}_r|\,|\partial(\mathbf{A}_r)|^{-1}}$, the normal radius is $1 - |\mathbf{A}_r|/|\partial(\mathbf{A}_r)|$.

Proof. Let $\mathbf{b}$ be the foot drawn from the origin of $\mathbb{R}^{n+1}$ to the rD plane with moment-direction $(\mathbf{A}_{r+1}, \partial(\mathbf{A}_{r+1}))$. Let $\mathbf{a}_1, \ldots, \mathbf{a}_{r+1}$ be $r + 1$ linearly independent points on the given rD sphere. Then $\mathbf{a}_i = \mathbf{b} + \mathbf{c}_i$, where $\mathbf{c}_i \in \partial(\mathbf{A}_{r+1})$. We have

$$\mathbf{A}_{r+1} = (\mathbf{b} + \mathbf{c}_1) \wedge (\mathbf{b} + \mathbf{c}_2) \wedge \cdots \wedge (\mathbf{b} + \mathbf{c}_{r+1}) = \mathbf{b}\,\partial(\mathbf{c}_1 \wedge \mathbf{c}_2 \wedge \cdots \wedge \mathbf{c}_{r+1}).$$

So $\partial(\mathbf{A}_{r+1}) = \partial(\mathbf{c}_1 \wedge \mathbf{c}_2 \wedge \cdots \wedge \mathbf{c}_{r+1})$, and $\mathbf{b} = \mathbf{A}_{r+1}(\partial(\mathbf{A}_{r+1}))^{-1}$. The center of the rD sphere is in the direction of $\mathbf{b}$, and the normal radius is $1 - |\mathbf{b}|$. $\qquad\square$

Definition 8.56. The rD *stereo angle* formed by a sequence of r linearly independent points $\mathbf{a}_1, \mathbf{a}_2, \ldots, \mathbf{a}_r$ on $\mathbb{S}^n$, denoted by $\angle(\mathbf{a}_1, \mathbf{a}_2, \ldots, \mathbf{a}_r)$, is a value in $[-\pi/2, \pi/2]$ such that

$$\sin \angle(\mathbf{a}_1, \mathbf{a}_2, \ldots, \mathbf{a}_r) = [\mathbf{a}_1 \mathbf{a}_2 \cdots \mathbf{a}_r], \tag{8.5.6}$$

where the bracket is in the Grassmann algebra $\Lambda(\mathbf{a}_1 \wedge \mathbf{a}_2 \wedge \cdots \wedge \mathbf{a}_r)$.

Example 8.57. Let $\mathbf{a}, \mathbf{b}, \mathbf{c}$ be three points on $\mathbb{S}^2$. $\angle(\mathbf{a}, \mathbf{b}, \mathbf{c})$ is usually called a *solid angle*.

- $\angle(\mathbf{a}, \mathbf{b}, \mathbf{c}) = 0$ if and only if the three points are collinear.
- $\angle(\mathbf{a}, \mathbf{b}, \mathbf{c}) = \pi/2$ if and only if the three vectors form a positive oriented orthonormal basis of $\mathbb{R}^3$.
- $\angle(\mathbf{a}, \mathbf{b}, \mathbf{c}) = -\pi/2$ if and only if the three vectors form a negative oriented orthonormal basis of $\mathbb{R}^3$.
- $\angle(\mathbf{a}, \mathbf{b}, \mathbf{c})$ is antisymmetric with respect to $\mathbf{a}, \mathbf{b}, \mathbf{c}$.

The nD real projective space $\mathbb{RP}^n$ is $\mathbb{S}^n$ with pairs of antipodal points identified. The distance between two points $\mathbf{a}, \mathbf{b} \in \mathbb{RP}^n$ is $|\mathbf{a} \wedge \mathbf{b}|$. $\mathbb{RP}^n$ equipped with this distance function is called an nD *elliptic space*, and the corresponding geometry is called *elliptic geometry*. This geometry can also be described and studied with Grassmann-Cayley algebra and Clifford algebra.

8.5.2 *The conformal model of spherical geometry*

Since spheres on $\mathbb{S}^n$ have the same representations as affine planes of $\mathbb{R}^{n+1}$ in Grassmann algebra $\Lambda(\mathbb{R}^{n+1})$, it is natural to think of extending the dimension of the surrounding space $\mathbb{R}^{n+1}$ by one, so that any sphere can be represented by a blade. This idea leads to the *conformal model* of spherical geometry.

Definition 8.58. The *conformal model* of nD spherical geometry is the set

$$\mathcal{N}_{\mathbf{p}} := \{\mathbf{x} \in \mathbb{R}^{n+1,1} \,|\, \mathbf{x} \cdot \mathbf{x} = 0, \ \mathbf{x} \cdot \mathbf{p} = -1\}, \tag{8.5.7}$$

where $\mathbf{p}$ is a fixed negative unit vector in $\mathbb{R}^{n+1,1}$, called the *spherical center* of the model, together with the following isometry:

Denote the orthogonal complement of vector $\mathbf{p}$ in $\mathbb{R}^{n+1,1}$ by $\mathbb{R}^{n+1}$, and denote the unit sphere of $\mathbb{R}^{n+1}$ by $\mathbb{S}^n$. Then

$$\mathbf{s}(\mathbf{x}) := \mathbf{p} + \mathbf{x}, \quad \forall \mathbf{x} \in \mathbb{S}^n \tag{8.5.8}$$

is an isometry from $\mathbb{S}^n$ onto $\mathcal{N}_{\mathbf{p}}$, called the *formalization map* of the conformal model of $\mathbb{S}^n$. Its inverse is $P_{\mathbf{p}}^{\perp}$.

In the conformal model, for two points $\mathbf{x}, \mathbf{y} \in \mathbb{S}^n$,

$$\mathbf{s}(\mathbf{x}) \cdot \mathbf{s}(\mathbf{y}) = \mathbf{x} \cdot \mathbf{y} - 1 = -d_{\mathbf{xy}}^n. \tag{8.5.9}$$

A sphere with center $\mathbf{c} \in \mathbb{S}^n$ and normal radius ρ is represented by $(\mathbf{s}(\mathbf{c}) - \rho\mathbf{p})^{\sim}$; a hyperplane with normal direction $\mathbf{n}$ is represented by $\mathbf{n}^{\sim}$. Both blades are Minkowski. Conversely, any Minkowski r-blade $\mathbf{A}_r$ where $1 < r < n+2$, represents an $(r-2)$D sphere of $\mathbb{S}^n$ in the sense that a point $\mathbf{x} \in \mathbb{S}^n$ is on the $(r-2)$D sphere if and only if $\mathbf{s}(\mathbf{x}) \in \mathbf{A}_r$. $\mathbf{A}_r$ represents an $(r-2)$D plane of $\mathbb{S}^n$ if and only if the spherical center $\mathbf{p} \in \mathbf{A}_r$.

The rD plane determined by $r+1$ points $\mathbf{x}_1, \ldots, \mathbf{x}_{r+1}$ of $\mathbb{S}^n$ is represented by $\mathbf{p} \wedge \mathbf{s}(\mathbf{x}_1) \wedge \cdots \wedge \mathbf{s}(\mathbf{x}_{r+1})$. The rD sphere determined by $r+2$ points $\mathbf{x}_1, \ldots, \mathbf{x}_{r+2}$ of $\mathbb{S}^n$ is represented by

$$\mathbf{s}(\mathbf{x}_1) \wedge \cdots \wedge \mathbf{s}(\mathbf{x}_{r+2}) = \mathbf{x}_1 \wedge \cdots \wedge \mathbf{x}_{r+2} + \mathbf{p} \wedge \partial(\mathbf{x}_1 \wedge \cdots \wedge \mathbf{x}_{r+2}). \tag{8.5.10}$$

As expected, (8.5.10) provides for the rD sphere the moment-direction representation of its supporting plane in $\mathbb{R}^{n+1}$.

The *homogeneous model* of nD spherical geometry is a pair $(\mathcal{N}, \mathbf{p})$, where $\mathcal{N}$ is the set of null vectors in $\mathbb{R}^{n+1,1}$, called *projective null cone*, and $\mathbf{p} \in \mathbb{R}^{n+1,1}$ is a negative unit vector representing the spherical center. A vector $\mathbf{a} \in \mathcal{N}$ always satisfies $\mathbf{a} \cdot \mathbf{p} \neq 0$, so it always represents a spherical point. Two vectors in $\mathcal{N}$ represent the same spherical point if and only if they differ by scale.

All the results on points, planes and spheres in the conformal Grassmann-Cayley algebra and conformal Clifford algebra of Euclidean geometry can be transferred to spherical geometry. The unification is based on the identification of the homogeneous model $(\mathcal{N}, \mathbf{p})$ of spherical geometry with the homogeneous model $(\mathcal{N}, \mathbf{e})$ of Euclidean geometry.

In fact, *the stereographic projection from $\mathbb{S}^n$ to $\mathbb{R}^n$ when represented in the homogeneous model, is the identity transformation in the projective null cone $\mathcal{N}$.*

Definition 8.59. Let $\mathbf{a}$ be a fixed point on $\mathbb{S}^n$, called the *north pole*, and let $\mathbb{R}^n$ be the orthogonal complement of vector $\mathbf{a}$ in $\mathbb{R}^{n+1}$, called the *equator plane*. The *stereographic projection* from $\mathbb{S}^n$ to $\mathbb{R}^n$ is the mapping that changes $\mathbf{a}$ to the conformal point at infinity of $\mathbb{R}^n$, changes $-\mathbf{a}$ to the origin of $\mathbb{R}^n$, and changes any other point $\mathbf{x} \in \mathbb{S}^n$ to the intersection $\mathbf{x}'$ of line $\mathbf{a}\mathbf{x}$ with the equator plane.

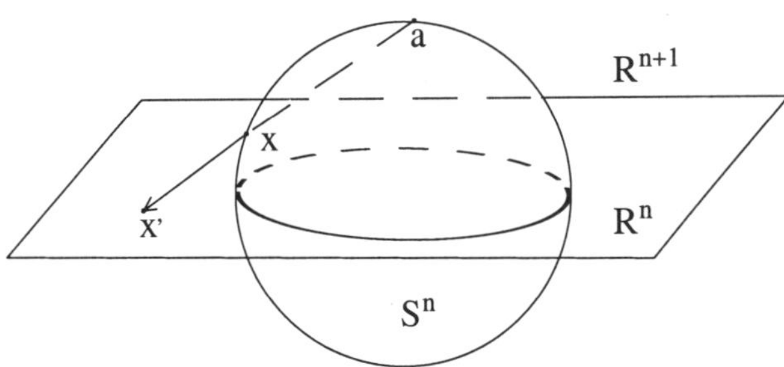

Fig. 8.12   Stereographic projection from $\mathbb{S}^n$ to $\mathbb{R}^n$.

In $\mathbb{R}^{n+1}$, the stereographic projection has a very nice expression:

$$\mathbf{x} \longmapsto \mathbf{x}' = \mathbf{a} + 2(\mathbf{x} - \mathbf{a})^{-1}, \quad \forall \mathbf{x} \in \mathbb{S}^n. \tag{8.5.11}$$

The inverse map has the same expression as (8.5.11).

Since $(\mathbf{x}' - \mathbf{a})(\mathbf{x} - \mathbf{a}) = 2$, the stereographic projection is in fact the inversion with respect to the sphere $(\mathbf{a}, \sqrt{2})$ in $\mathbb{R}^{n+1}$, whose domain of definition is restricted to the unit sphere of $\mathbb{R}^{n+1}$.

In $\mathbb{R}^{n+1,1}$, points in $\mathbb{R}^n$ and $\mathbb{S}^n$ are all represented by null vectors, so they have a natural correspondence. For the spherical center $\mathbf{p}$ and the north pole $\mathbf{a}$, let

$$\mathbf{e} = \mathbf{p} + \mathbf{a}, \quad \mathbf{e}_0 = \frac{\mathbf{p} - \mathbf{a}}{2}. \tag{8.5.12}$$

On one hand, they represent points $\pm\mathbf{a}$ on $\mathbb{S}^n$; on the other hand, they also represent the conformal point at infinity and the origin of $\mathbb{R}^n$. Let $\mathbf{c} \in \mathcal{N}$ but $\mathbf{c} \notin \mathbf{e} \wedge \mathbf{e}_0$. Then $\mathbf{c}$ represents both point $\mathbf{x} = P_{\mathbf{p}}^{\perp}(-\mathbf{c}/\mathbf{c}\cdot\mathbf{p}) \in \mathbb{S}^n$ and point $\mathbf{x}' = P_{\mathbf{e}\wedge\mathbf{e}_0}^{\perp}(-\mathbf{c}/\mathbf{c}\cdot\mathbf{e}) \in \mathbb{R}^n$. The relation (8.5.11) between $\mathbf{x}, \mathbf{x}'$ is easy to verify.

In the conformal model of $\mathbb{R}^n$, blade $\mathbf{p}^{\sim} = (\mathbf{e}_0 + \mathbf{e}/2)^{\sim}$ is Euclidean, whose vectors are positive-vector representations of the set $\mathcal{S}$ of spheres and hyperplanes in $\mathbb{R}^n$ whose intersections with the unit sphere are great spheres of the unit sphere. In the conformal model of $\mathbb{S}^n$, blade $\mathbf{p}^{\sim}$ corresponds to all hyperplanes of $\mathbb{S}^n$ via their positive-vector representations. The stereographic projection (8.5.11) changes the set of hyperplanes of $\mathbb{S}^n$ to the set $\mathcal{S}$; it gives a much clearer (spherical) geometric interpretation of a negative vector in the conformal model setting.

As the stereographic projection is the identity transformation in $\mathcal{N}$, any homogeneous identity on vectors of $\mathcal{N}$ can be explained either as a geometric theorem in spherical geometry, or as a variety of corresponding theorems in Euclidean geometry through different stereographic projections.

8.6 Hyperbolic geometry and its conformal model*

Hyperbolic geometry was discovered by Gauss, Bolyai and Lobachevsky at the beginning of the 19th century. At the end of that century the existence of hyperbolic geometry was proved by Beltrami, Klein and Poincaré by establishing in Euclidean space and Minkowski space various isometric models of this geometry. Poincaré established three models: the hyperboloid model, the disk model and the half-space model. The latter two models are the most often used ones in modern geometry and analysis, because they are established on regions of a Euclidean space.

The hyperboloid model, in fact, has superb algebraic properties because it is *isotropic*: at every point of the model, the metric of the tangent space is the same. As a direct result, a plane in hyperbolic geometry corresponds to a projective plane in a Minkowski space, and a sphere in hyperbolic geometry corresponds to an affine plane in the same Minkowski space.

Since the correspondence of planes and spheres in hyperbolic geometry with planes in Minkowski space is the same as the correspondence of planes and spheres in spherical geometry with planes in Euclidean space, there is a natural similarity between the hyperboloid model of hyperbolic geometry and the classical model of spherical geometry, leading to the conformal model of hyperbolic geometry.

8.6.1 *Poincaré's hyperboloid model of hyperbolic geometry*

Definition 8.60. The set

$$\mathbb{D}^n = \{\mathbf{x} \in \mathbb{R}^{n,1} \,|\, \mathbf{x}^2 = -1\} \tag{8.6.1}$$

has two branches, and they are antipodal to each other, denoted by $\mathbb{H}^n$ and $-\mathbb{H}^n$ respectively. The branch $\mathbb{H}^n$ equipped with the Riemannian metric induced from $\mathbb{R}^{n,1}$, is called the *hyperboloid model* of nD hyperbolic geometry, or simply called the nD *hyperbolic space*. $\mathbb{D}^n$ is called the nD *double-hyperbolic space*.

A hyperbolic *point* refers to a vector in $\mathbb{H}^n$ in the case of hyperbolic geometry, and refers to a vector in $\mathbb{D}^n$ in the case of double-hyperbolic geometry. An rD hyperbolic *plane* is the intersection of a $(r+1)$D vector subspace of $\mathbb{R}^{n,1}$ with $\mathbb{H}^n$ in the case of hyperbolic geometry, and with $\mathbb{D}^n$ in the case of double-hyperbolic geometry. A 1D plane is also called a *line*.

Let $\mathbf{p}, \mathbf{q}$ be two points in $\mathbb{H}^n$ (or $-\mathbb{H}^n$). Since $\mathbf{p}^2 = -1$, by the continuity of the inner product and the fact that $\mathbf{p}, \mathbf{q}$ are on the same branch, we get $\mathbf{p} \cdot \mathbf{q} < 0$. Since $\mathbf{p} \wedge \mathbf{q}$ is Minkowski, $(\mathbf{p} \wedge \mathbf{q})^2 = (\mathbf{p} \cdot \mathbf{q})^2 - 1 > 0$, so $\mathbf{p} \cdot \mathbf{q} < -1$. There exists a unique scalar $d_{\mathbf{pq}} \geq 0$ such that

$$\cosh d_{\mathbf{pq}} = -\mathbf{p} \cdot \mathbf{q}, \tag{8.6.2}$$

and $d_{\mathbf{pq}} = 0$ if and only if $\mathbf{p} = \mathbf{q}$. $d_{\mathbf{pq}}$ is called the *hyperbolic distance* between points $\mathbf{p}, \mathbf{q}$.

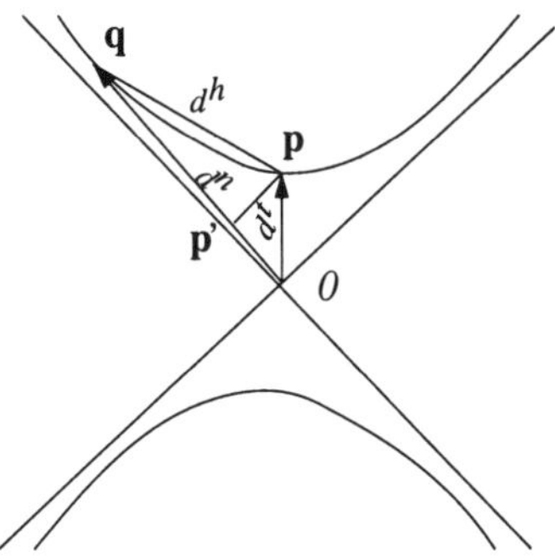

Fig. 8.13   Some conjugate distances in $\mathbb{H}^n$.

Similar to the spherical distance in spherical geometry, the hyperbolic distance also has several important conjugate distance functions.

First, as shown in Figure 8.13, let $\mathbf{p}' = P_{\mathbf{q}}(\mathbf{p})$ be the foot drawn from point $\mathbf{p}$ to the 1D space spanned by vector $\mathbf{q}$. As the 1D space is anti-Euclidean, its distance function is Euclidean. The Euclidean distance between $\mathbf{p}', \mathbf{q}$ is called the *normal distance* between points $\mathbf{p}, \mathbf{q}$:

$$d^n_{\mathbf{pq}} := |\mathbf{q} - P_{\mathbf{q}}(\mathbf{p})| = |\mathbf{q}(1 + \mathbf{p} \cdot \mathbf{q})| = -1 - \mathbf{p} \cdot \mathbf{q}. \tag{8.6.3}$$

Second, vector $\mathbf{p} - \mathbf{p}'$ is orthogonal to vector $\mathbf{q}$ in $\mathbb{R}^{n,1}$, so it is Euclidean. The Euclidean distance between $\mathbf{p}, \mathbf{p}'$ is called the *tangential distance* between points $\mathbf{p}, \mathbf{q}$:

$$d^t_{\mathbf{pq}} := |\mathbf{p} - P_{\mathbf{q}}(\mathbf{p})| = |P_{\mathbf{q}}^{\perp}(\mathbf{p})| = |\mathbf{p} \wedge \mathbf{q}|. \tag{8.6.4}$$

The term "tangential" comes from the fact that, if letting $\mathbf{t}$ be the unit tangent vector of $\mathbb{H}^n$ at point $\mathbf{p}$ such that $\mathbf{t} \cdot \mathbf{p} = 0$ and $\mathbf{q} = \mathbf{p} + \lambda\mathbf{t}$, where $\lambda > 0$, then $\lambda = d^t_{\mathbf{pq}}$. The nD Euclidean space $\mathbf{p}^{\sim}$ is called the *tangent space* of the hyperboloid model at point $\mathbf{p}$, and any vector in it is called a *tangent vector* at base point $\mathbf{p}$.

Third, the following distance is directly induced from $\mathbb{R}^{n,1}$, called the *horo-distance* between points $\mathbf{p}, \mathbf{q}$:

$$d^h_{\mathbf{pq}} := |\mathbf{p} - \mathbf{q}|. \tag{8.6.5}$$

Fourth, the following distance is called the *stereographic distance* between points $\mathbf{p}, \mathbf{q}$:

$$d^s_{\mathbf{pq}} := \tanh \frac{d_{\mathbf{pq}}}{2}. \tag{8.6.6}$$

Some relations among the distances are

$$\begin{aligned}
d^n_{\mathbf{pq}} &= \cosh d_{\mathbf{pq}} - 1, \\
d^t_{\mathbf{pq}} &= \sinh d_{\mathbf{pq}}, \\
d^h_{\mathbf{pq}} &= 2\sinh \frac{d_{\mathbf{pq}}}{2}.
\end{aligned} \tag{8.6.7}$$

In $\mathbb{R}^{n,1}$, negative vectors represent points in $\mathbb{D}^n$. For null vectors, by the argument before Proposition 8.7, there are two branches of null vectors, where each branch can be asymptated by exactly one branch of $\mathbb{D}^n$. Let $\mathcal{N}_{\mathbb{H}^n}$ be the branch asymptated by $\mathbb{H}^n$. The other branch is denoted by $-\mathcal{N}_{\mathbb{H}^n}$. We say vectors in $\mathcal{N}_{\mathbb{H}^n}$ and $\mathbb{H}^n$ are *of the same branch*, and vectors in $-\mathcal{N}_{\mathbb{H}^n}$ and $-\mathbb{H}^n$ are also *of the same branch*.

A 1D null half-space asymptated by $\mathbb{H}^n$ (or $-\mathbb{H}^n$) is called a *point at infinity* of $\mathbb{H}^n$ (or $-\mathbb{H}^n$). The null-vector representation of a point at infinity is unique up to a positive scale. A null vector $\mathbf{u}$ is of the same branch with a point $\mathbf{p} \in \mathbb{D}^n$ if and only if $\mathbf{u} \cdot \mathbf{p} < 0$.

Sometimes there is no need to distinguish between null vectors of different branches, and a 1D null subspace of $\mathbb{R}^{n,1}$ is called an *isotropic point at infinity* of $\mathbb{D}^n$. In classical literature, an isotropic point at infinity is called an *end*. The set of ends is topologically a sphere, called the *isotropic sphere at infinity*. When restricted to $\mathbb{H}^n$, then "end" is a synonym of "point at infinity", and the isotropic sphere at infinity is simply called the *sphere at infinity* of $\mathbb{H}^n$.

An rD hyperbolic *plane* in $\mathbb{D}^n$ determined by $r+1$ points $\mathbf{p}_1,\ldots,\mathbf{p}_{r+1} \in \mathbb{D}^n$, is the intersection of $\mathbb{D}^n$ with the $(r+1)$D vector subspace spanned by the $r+1$ vectors in $\mathbb{R}^{n,1}$. It can be represented by the Minkowski $(r+1)$-blade $\mathbf{A}_{r+1} = \mathbf{p}_1 \wedge \cdots \wedge \mathbf{p}_{r+1}$, in the sense that a point $\mathbf{x} \in \mathbb{D}^n$ is on the plane if and only if $\mathbf{x} \in \mathbf{A}_{r+1}$.

An rD plane in $\mathbb{D}^n$ has two branches. The branch in $\mathbb{H}^n$ (or $-\mathbb{H}^n$) is called an rD plane in $\mathbb{H}^n$ (or $-\mathbb{H}^n$). For rD plane $\mathbf{A}_{r+1}$, the intersection of the $(r+1)$D subspace $\mathbf{A}_{r+1}$ of $\mathbb{R}^{n,1}$ with the sphere at infinity of $\mathbb{D}^n$ (or $\mathbb{H}^n$, or $-\mathbb{H}^n$), is called the sphere at infinity of the rD plane. For example, the sphere at infinity of a line is composed of two ends.

It must be pointed out that a line in hyperbolic geometry is never affine, and any line has two different ends. In contrast, any line in affine geometry has a unique point at infinity, and different lines have different points at infinity. In the conformal model of Euclidean geometry, there is only one point at infinity in the whole model, and all lines share the same point at infinity.

In hyperbolic geometry, the absolute distance between a point and a point at infinity is always infinite. However, the "relative distance" between a point and a point at infinity is meaningful, because the ratio of the two relative distances between a point at infinity and two hyperbolic points collinear with it, is independent of the scaling of the representative null vector of the point at infinity. For a null vector $\mathbf{u}$, the *relative distance* between $\mathbf{u}$ and a point $\mathbf{p} \in \mathbb{D}^n$ is defined by

$$d^r_{\mathbf{up}} := |\mathbf{u} \cdot \mathbf{p}|. \tag{8.6.8}$$

It depends on the scale of $\mathbf{u}$, therefore is said to be relative.

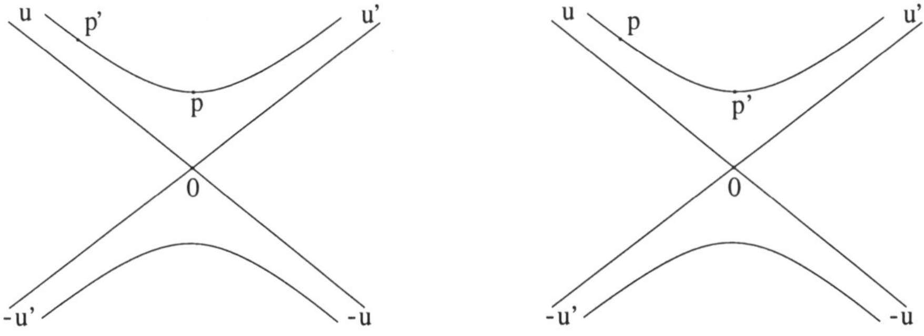

Fig. 8.14   Geometric meaning of $d^r_{\mathbf{up}}$. Left: $\mathbf{p}'$ is between $\mathbf{u}, \mathbf{p}$; right: $\mathbf{p}$ is between $\mathbf{u}, \mathbf{p}'$.

The geometric meaning of $d^r_{\mathbf{up}}$ can be obtained as follows. Let $\mathbf{u}, \mathbf{p}$ be of the same branch of $\mathbb{D}^n$, say $\mathbb{H}^n$. Let $\mathbf{p}' \in \mathbb{H}^n$ be a point on line $\mathbf{u} \wedge \mathbf{p}$ such that $\mathbf{u} \cdot \mathbf{p}' = -1$. Then

$$\mathbf{p}' = \frac{(\mathbf{u} \cdot \mathbf{p})^2 - 1}{2(\mathbf{u} \cdot \mathbf{p})^2}\mathbf{u} - \frac{1}{\mathbf{u} \cdot \mathbf{p}}\mathbf{p}. \tag{8.6.9}$$

From this we get $(\mathbf{u} \cdot \mathbf{p})^2 - 2(\mathbf{p} \cdot \mathbf{p}')(\mathbf{u} \cdot \mathbf{p}) + 1 = 0$, so

$$\mathbf{u} \cdot \mathbf{p} = \mathbf{p} \cdot \mathbf{p}' \pm \sqrt{(\mathbf{p} \cdot \mathbf{p}')^2 - 1} = -\cosh d_{\mathbf{pp}'} \pm \sinh d_{\mathbf{pp}'} = -e^{\pm d_{\mathbf{pp}'}}. \qquad (8.6.10)$$

As shown in Figure 8.14, if $\mathbf{p}'$ is between $\mathbf{u}$ and $\mathbf{p}$, then $1 = d^r_{\mathbf{up}'} < d^r_{\mathbf{up}}$, so $d^r_{\mathbf{up}} = e^{d_{\mathbf{pp}'}}$; if $\mathbf{p}$ is between $\mathbf{u}$ and $\mathbf{p}'$, then $d^r_{\mathbf{up}} = e^{-d_{\mathbf{pp}'}}$.

The *distance between a point* $\mathbf{p}$ *and an* rD *plane* $\mathbf{A}_{r+1}$ in $\mathbb{D}^n$ is defined as the distance between $\mathbf{p}$ and the foot drawn from $\mathbf{p}$ to the rD plane that is on the same branch of $\mathbb{D}^n$ with $\mathbf{p}$. The foot is $P_{\mathbf{A}_{r+1}}(\mathbf{p})/|P_{\mathbf{A}_{r+1}}(\mathbf{p})|$, because $\mathbf{p} = P_{\mathbf{A}_{r+1}}(\mathbf{p}) + P^{\perp}_{\mathbf{A}_{r+1}}(\mathbf{p})$ is negative, $P^{\perp}_{\mathbf{A}_{r+1}}(\mathbf{p}) = (\mathbf{p} \wedge \mathbf{A}_{r+1})\mathbf{A}^{-1}_{r+1}$ is positive, and $\mathbf{p} \cdot P_{\mathbf{A}_{r+1}}(\mathbf{p}) = (P_{\mathbf{A}_{r+1}}(\mathbf{p}))^2 < 0$.

A positive vector $\mathbf{a} \in \mathbb{R}^{n,1}$ has the property that $\mathbf{a}^{\sim}$ corresponds to a Minkowski n-blade, whose intersection as an nD vector subspace of $\mathbb{R}^{n,1}$ with $\mathbb{D}^n$ is a hyperbolic hyperplane. $\mathbf{a}$ is called a *normal direction* of the hyperplane. $\pm \mathbf{a}^{\sim}$ represent the same hyperplane with opposite orientations. The 1D half-space $\{\lambda \mathbf{a} \,|\, \lambda > 0\}$ is called an *imaginary point* of $\mathbb{D}^n$. It can be represented by vector $\mathbf{a}$, and the representation is unique up to a positive scale. The 1D space spanned by vector $\mathbf{a}$ is called an *isotropic imaginary point*, or *ideal point* in classical literature. An ideal point is said to be *on* an rD plane, if as a 1D linear space it is in the $(r+1)D$ linear subspace of $\mathbb{R}^{n,1}$ supporting the rD plane.

Let $\mathbf{a}$ be an ideal point and $\mathbf{p}$ be a point. The distance between $\mathbf{a}$ and $\mathbf{p}$ is defined as the distance between point $\mathbf{p}$ and hyperplane $\mathbf{a}^{\sim}$. Blade $\mathbf{a} \wedge \mathbf{p}$ is always Minkowski, so $(\mathbf{a} \wedge \mathbf{p})^2 = (\mathbf{a} \cdot \mathbf{p})^2 + 1 \geq 1$, and $|\mathbf{a} \wedge \mathbf{p}| - 1 \geq 0$. The equality holds if and only if $\mathbf{a} \cdot \mathbf{p} = 0$, *i.e.*, point $\mathbf{p}$ is on hyperplane $\mathbf{a}^{\sim}$.

The following identities are easy to verify:

$$\begin{aligned}
\cosh d_{\mathbf{ap}} &= |\mathbf{a} \wedge \mathbf{p}|, \\
d^n_{\mathbf{ap}} &= |\mathbf{a} \wedge \mathbf{p}| - 1, \\
d^t_{\mathbf{ap}} &= |\mathbf{a} \cdot \mathbf{p}|, \\
d^h_{\mathbf{ap}} &= 2 \sinh \frac{d_{\mathbf{ap}}}{2}.
\end{aligned} \qquad (8.6.11)$$

For positive unit vector $\mathbf{a} \in \mathbb{R}^{n,1}$, the half space $\{\mathbf{p} \in \mathbb{D}^n | \mathbf{a} \cdot \mathbf{p} > 0\}$ is called the *positive side* of oriented hyperplane $\mathbf{a}^{\sim}$. It is composed of two connected components, one on each branch of $\mathbb{D}^n$, as shown in Figure 8.15. The geometric meaning of $\mathbf{a} \cdot \mathbf{p} > 0$ can be derived as follows. Let $\mathbf{q}$ be the foot drawn from $\mathbf{p}$ to hyperplane $\mathbf{a}^{\sim}$, which is on the same branch of $\mathbb{D}^n$ with $\mathbf{p}$. Let $\mathbf{t}$ be the unit tangent vector of $\mathbb{D}^n$ at point $\mathbf{q}$ such that $\mathbf{p} = \mathbf{q} + \lambda \mathbf{t}$, where $\lambda > 0$. Then $\mathbf{t} = \mathbf{a}$ if and only $\mathbf{a} \cdot \mathbf{p} > 0$.

In hyperbolic geometry, a sphere is a set of points having the same distance with the center of the sphere, and the center can be either a hyperbolic point, or a point at infinity, or an imaginary point, so there are three kinds of spheres. A sphere can be denoted by a pair $(\mathbf{c}, \rho)$, where $\mathbf{c}$ is the center and $\rho > 0$ is the radius.

Case 1. When $\mathbf{c}^2 = -1$, *i.e.*, $\mathbf{c}$ is a point, then if $\mathbf{c} \in \mathbb{H}^n$, the set $\{\mathbf{p} \in \mathbb{H}^n | d^n_{\mathbf{pc}} = \rho\}$ is the *hyperbolic sphere* in $\mathbb{H}^n$ with center $\mathbf{c}$ and normal radius ρ; if $\mathbf{c}$ is in $-\mathbb{H}^n$,

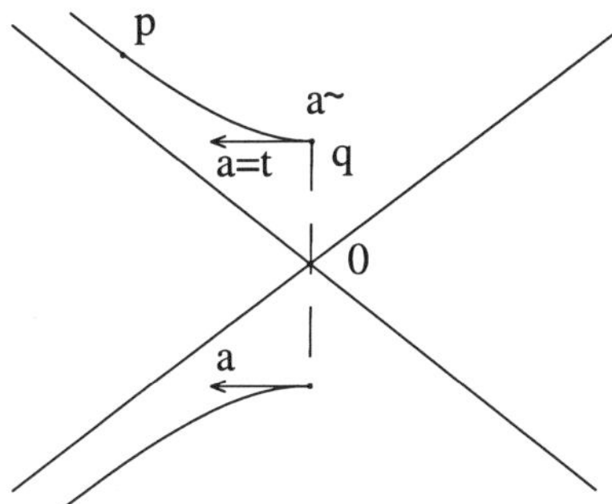

Fig. 8.15   Positive side of oriented hyperplane $\mathbf{a}^{\sim}$.

the set $\{\mathbf{p} \in -\mathbb{H}^n | d^n_{\mathbf{pc}} = \rho\}$ is a hyperbolic sphere in $-\mathbb{H}^n$. A *hyperbolic sphere* in $\mathbb{D}^n$ refers to a sphere in either $\mathbb{H}^n$ or $-\mathbb{H}^n$. Thus, a sphere in $\mathbb{D}^n$ with center $\mathbf{c}$ and normal radius ρ is the set

$$\{\mathbf{p} \in \mathbb{D}^n | \mathbf{p} \cdot \mathbf{c} = -(1 + \rho)\}. \tag{8.6.12}$$

Case 2. When $\mathbf{c}^2 = 0$, *i.e.*, $\mathbf{c}$ is a point at infinity, then if $\mathbf{c}$ is of the same branch with $\mathbb{H}^n$, the set $\{\mathbf{p} \in \mathbb{H}^n | d^r_{\mathbf{pc}} = \rho\}$ is the *horosphere* in $\mathbb{H}^n$ with center $\mathbf{c}$ and relative radius ρ; if $\mathbf{c}$ is of the same branch with $-\mathbb{H}^n$, the set $\{\mathbf{p} \in -\mathbb{H}^n | d^r_{\mathbf{pc}} = \rho\}$ is a horosphere in $-\mathbb{H}^n$. A *horosphere* in $\mathbb{D}^n$ refers to a horosphere in either $\mathbb{H}^n$ or $-\mathbb{H}^n$. A horosphere in $\mathbb{D}^n$ with center $\mathbf{c}$ and relative radius ρ is the set

$$\{\mathbf{p} \in \mathbb{D}^n | \mathbf{p} \cdot \mathbf{c} = -\rho\}. \tag{8.6.13}$$

Case 3. When $\mathbf{c}^2 = 1$, *i.e.*, $\mathbf{c}$ is an imaginary point, then

$$\{\mathbf{p} \in \mathbb{D}^n | \mathbf{p} \cdot \mathbf{c} = -\rho\} \tag{8.6.14}$$

is the *hypersphere* in $\mathbb{D}^n$ with center $\mathbf{c}$ and tangential radius ρ; its intersection with $\mathbb{H}^n$ (or $-\mathbb{H}^n$) is called a *hypersphere* in $\mathbb{H}^n$ (or $-\mathbb{H}^n$). The hyperplane $\mathbf{c}^{\sim}$ is called the *axis* of the hypersphere.

A hypersphere in $\mathbb{D}^n$ has two branches, each in a branch of $\mathbb{D}^n$. They are both on the *negative side* of the oriented axis. A hyperplane can be taken as a hypersphere with zero radius.

The set $\{\mathbf{p} \in \mathbb{D}^n | d^t_{\mathbf{pc}} = \rho\}$ is called the *double-sphere* in $\mathbb{D}^n$ with center $\mathbf{c}$ and tangential radius ρ; its intersection with $\mathbb{H}^n$ (or $-\mathbb{H}^n$) is called a *double-sphere* in $\mathbb{H}^n$ (or $-\mathbb{H}^n$). The hyperplane $\mathbf{c}^{\sim}$ is called the *axis* of the double-sphere. A double-sphere is composed of two hyperspheres symmetric with respect to the axis. A double-sphere in $\mathbb{D}^n$ has four connected components, while a double-sphere in $\mathbb{H}^n$ or $-\mathbb{H}^n$ has two.

A sphere in $\mathbb{D}^n$ has default dimension $n - 1$. By (8.6.12) to (8.6.14), a sphere is the intersection of $\mathbb{D}^n$ with an affine hyperplane in $\mathbb{R}^{n,1}$. In $\Lambda(\mathbb{R}^{n,1})$, a sphere can be represented by the moment-direction representation $(\mathbf{A}_{n+1}, \partial(\mathbf{A}_{n+1}))$ of the

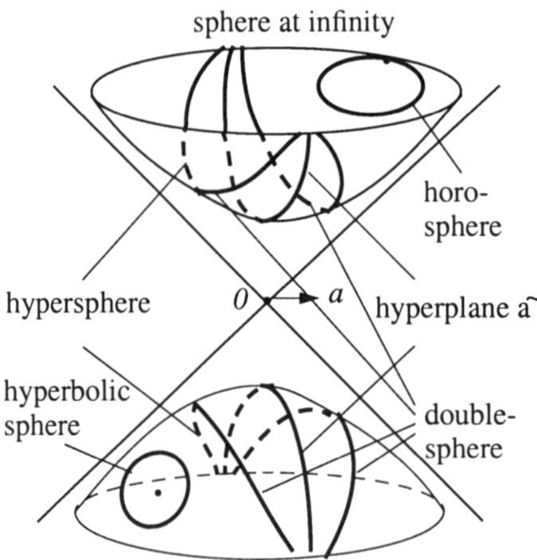

Fig. 8.16   Spheres in hyperbolic geometry.

supporting affine hyperplane, in the sense that a point $\mathbf{p} \in \mathbb{D}^n$ is on the sphere if and only if $\mathbf{p} \wedge \partial(\mathbf{A}_{n+1}) = \mathbf{A}_{n+1}$.

In the general case, an rD sphere is the intersection of $\mathbb{D}^n$ with an $(r + 1)$D affine plane in $\mathbb{R}^{n,1}$. Let $\mathbf{x}_1, \ldots, \mathbf{x}_{r+2} \in \mathbb{H}^n$ be linearly independent in $\mathbb{R}^{n,1}$, then they determine a unique rD sphere in $\mathbb{H}^n$. The rD sphere has the moment-direction representation $(\mathbf{A}_{r+2}, \partial(\mathbf{A}_{r+2}))$, where $\mathbf{A}_{r+2} = \mathbf{x}_1 \wedge \cdots \wedge \mathbf{x}_{r+2}$. While blade $\mathbf{A}_{r+2}$ is always Minkowski, blade $\partial(\mathbf{A}_{r+2})$ has three possible signatures:

(1) If $\partial(\mathbf{A}_{r+2})$ is Euclidean, then vector $\mathbf{b} = \mathbf{A}_{r+2}(\partial(\mathbf{A}_{r+2}))^{-1}$ is negative. $(\mathbf{A}_{r+2}, \partial(\mathbf{A}_{r+2}))$ represents an rD hyperbolic sphere, with center $\mathbf{b}/|\mathbf{b}|$ and normal radius

$$-1 - \frac{\mathbf{A}_{r+2}(\partial(\mathbf{A}_{r+2}))^{-1}}{|\mathbf{A}_{r+2}|\,|\partial(\mathbf{A}_{r+2})|^{-1}} \cdot \mathbf{x}_1 = -1 - \frac{\mathbf{A}_{r+2}(\partial(\mathbf{A}_{r+2})^{\dagger} \wedge \mathbf{x}_1)}{|\mathbf{A}_{r+2}|\,|\partial(\mathbf{A}_{r+2})|} = \frac{|\mathbf{A}_{r+2}|}{|\partial(\mathbf{A}_{r+2})|} - 1,$$

where $\mathbf{x}_1$ is any point on the rD sphere.

(2) If $\partial(\mathbf{A}_{r+2})$ is degenerate, then $\mathbf{b} = \partial(\mathbf{A}_{r+2})\mathbf{A}_{r+2}^{\dagger}$ is the unique null vector in $\partial(\mathbf{A}_{r+2})$ up to scale. For any point $\mathbf{x}_1$ on the sphere,

$$\mathbf{x}_1 \cdot \mathbf{b} = (\mathbf{x}_1 \wedge \partial(\mathbf{A}_{r+2}))\mathbf{A}_{r+2}^{\dagger} = \mathbf{A}_{r+2}\mathbf{A}_{r+2}^{\dagger} = -|\mathbf{A}_{r+2}|^2 < 0,$$

so $(\mathbf{A}_{r+2}, \partial(\mathbf{A}_{r+2}))$ represents an rD horosphere with center $\mathbf{b}$ and relative radius $|\mathbf{A}_{r+2}|^2$.

(3) If $\partial(\mathbf{A}_{r+2})$ is Minkowski, then $\mathbf{b} = \mathbf{A}_{r+2}(\partial(\mathbf{A}_{r+2}))^{-1}$ is a positive vector. Let $\mathbf{x}_1$ be any point on the rD sphere. $(\mathbf{A}_{r+2}, \partial(\mathbf{A}_{r+2}))$ represents an rD hypersphere, with center $-\mathbf{b}/|\mathbf{b}|$ and tangential radius

$$\frac{\mathbf{A}_{r+2}(\partial(\mathbf{A}_{r+2}))^{-1}}{|\mathbf{A}_{r+2}|\,|\partial(\mathbf{A}_{r+2})|^{-1}} \cdot \mathbf{x}_1 = -\frac{\mathbf{A}_{r+2}(\partial(\mathbf{A}_{r+2})^{\dagger} \wedge \mathbf{x}_1)}{|\mathbf{A}_{r+2}|\,|\partial(\mathbf{A}_{r+2})|} = \frac{|\mathbf{A}_{r+2}|}{|\partial(\mathbf{A}_{r+2})|}.$$

8.6.2 *The conformal model of double-hyperbolic geometry*

Since rD spheres in $\mathbb{D}^n$ have the same representations as $(r+1)$D affine planes in $\mathbb{R}^{n,1}$, we can extend the dimension of the embedding space $\mathbb{R}^{n,1}$ of $\mathbb{D}^n$ by one, so that any rD sphere can be represented by a $(r+2)$-blade, just like the homogeneous coordinates representation of affine geometry and the conformal model of spherical geometry. This idea leads to the *conformal model* of double-hyperbolic geometry.

Definition 8.61. The *conformal model* of nD double-hyperbolic geometry is the set

$$\mathcal{N}_{\mathbf{a}} := \{\mathbf{x} \in \mathbb{R}^{n+1,1} \mid \mathbf{x} \cdot \mathbf{x} = 0, \ \mathbf{x} \cdot \mathbf{a} = -1\}, \tag{8.6.15}$$

where $\mathbf{a}$ is a fixed positive unit vector in $\mathbb{R}^{n+1,1}$, called the *hyperbolic center* of the model, together with the following isometry:

Denote the orthogonal complement of vector $\mathbf{a}$ in $\mathbb{R}^{n+1,1}$ by $\mathbb{R}^{n,1}$, and denote the set of unit negative vectors in $\mathbb{R}^{n,1}$ by $\mathbb{D}^n$. Then

$$\mathbf{h}(\mathbf{x}) := -\mathbf{a} + \mathbf{x}, \quad \forall \mathbf{x} \in \mathbb{D}^n \tag{8.6.16}$$

is an isometry from $\mathbb{D}^n$ onto $\mathcal{N}_{\mathbf{a}}$, called the *formalization map* of the conformal model. Its inverse is $P_{\mathbf{a}}^{\perp}$.

In the conformal model, an end or point at infinity is represented by a null vector orthogonal to $\mathbf{a}$; an imaginary point or ideal point is represented by a positive vector orthogonal to $\mathbf{a}$. The sphere at infinity is represented by $\mathbf{a}^{\sim}$, the hyperplane normal to positive vector $\mathbf{n} \in \mathbf{a}^{\sim}$ is represented by $\mathbf{n}^{\sim}$.

For two points $\mathbf{x}, \mathbf{y}$ in $\mathbb{H}^n$,

$$\mathbf{h}(\mathbf{x}) \cdot \mathbf{h}(\mathbf{y}) = 1 + \mathbf{x} \cdot \mathbf{y} = -d_{\mathbf{x}\mathbf{y}}^n. \tag{8.6.17}$$

A direct corollary is the following unified representation of a sphere with center $\mathbf{c}$ and radius ρ, which may be any of the three kinds of spheres from (8.6.12) to (8.6.14), and where $\mathbf{c} \in \mathcal{N}_{\mathbf{a}}$ when $\mathbf{c}$ represents a point. A point $\mathbf{p}$ is on the sphere if and only if $\mathbf{h}(\mathbf{p}) \cdot \mathbf{c} = -\rho$. So *sphere* $(\mathbf{c}, \rho)$ *can be represented by blade* $(\mathbf{c} - \rho\mathbf{a})^{\sim}$, so that a point $\mathbf{p} \in \mathbb{D}^n$ is on the sphere if and only if $\mathbf{h}(\mathbf{p}) \in (\mathbf{c} - \rho\mathbf{a})^{\sim}$. Since $(\mathbf{c} - \rho\mathbf{a})^2 > 0$ for all three kinds of spheres, blade $(\mathbf{c} - \rho\mathbf{a})^{\sim}$ is Minkowski.

In the general case, the rD plane determined by $r+1$ points $\mathbf{x}_1, \ldots, \mathbf{x}_{r+1}$ in $\mathbb{D}^n$ is represented by Minkowski blade $\mathbf{a} \wedge \mathbf{h}(\mathbf{x}_1) \wedge \cdots \wedge \mathbf{h}(\mathbf{x}_{r+1})$. The rD sphere determined by $r+2$ points $\mathbf{x}_1, \ldots, \mathbf{x}_{r+2} \in \mathbb{D}^n$ is represented by Minkowski blade

$$\mathbf{h}(\mathbf{x}_1) \wedge \cdots \wedge \mathbf{h}(\mathbf{x}_{r+2}) = \mathbf{x}_1 \wedge \cdots \wedge \mathbf{x}_{r+2} - \mathbf{a} \wedge \partial(\mathbf{x}_1 \wedge \cdots \wedge \mathbf{x}_{r+2}). \tag{8.6.18}$$

As expected, (8.6.18) provides for the rD sphere the moment-direction representation of its supporting plane in $\mathbb{R}^{n,1}$.

Conversely, let $\mathbf{A}_r$ be a Minkowski r-blade in $\Lambda(\mathbb{R}^{n+1,1})$, where $2 \le r \le n+1$. Then $\mathbf{A}_r$ represents a $(r-2)$D sphere at infinity, or plane, or sphere in $\mathbb{D}^n$. From (8.6.18), we can easily derive the following classification:

- If $\mathbf{a} \cdot \mathbf{A}_r = 0$, then $\mathbf{A}_r$ represents a sphere at infinity.
- If $\mathbf{a} \cdot \mathbf{A}_r$ is Euclidean, then $\mathbf{A}_r$ represents a hyperbolic sphere.
- If $\mathbf{a} \cdot \mathbf{A}_r$ is degenerate, then $\mathbf{A}_r$ represents a horosphere.
- If $\mathbf{a} \cdot \mathbf{A}_r$ is Minkowski, but $\mathbf{a} \wedge \mathbf{A}_r \neq 0$, then $\mathbf{A}_r$ represents a hypersphere.
- If $\mathbf{a} \wedge \mathbf{A}_r = 0$, then $\mathbf{a} \cdot \mathbf{A}_r$ is Minkowski, and $\mathbf{A}_r$ represents a plane.

The *homogeneous model* of nD double-hyperbolic geometry is a pair $(\mathcal{N}, \mathbf{a})$, where $\mathcal{N}$ is the projective null cone in $\mathbb{R}^{n+1,1}$, and $\mathbf{a} \in \mathbb{R}^{n+1,1}$ is a positive unit vector representing the hyperbolic center. A vector $\mathbf{c} \in \mathcal{N}$ represents a hyperbolic point if and only if $\mathbf{c} \cdot \mathbf{a} \neq 0$. Two vectors in $\mathcal{N}$ represent the same hyperbolic point if and only if they differ by scale.

All the results on points, planes and spheres in the conformal Grassmann-Cayley algebras and conformal Clifford algebras of Euclidean geometry and spherical geometry respectively, can be transferred to double-hyperbolic geometry via the identification of the three homogeneous models.

8.6.3 *Poincaré's disk model and half-space model*

Definition 8.62. In the hyperboloid model, let $-\mathbf{p}$ be a fixed point in $-\mathbb{H}^n$, called the *north pole*, and let $\mathbb{R}^n$ be the orthogonal complement of vector $\mathbf{p}$ in $\mathbb{R}^{n,1}$, called the *equator plane*. The *stereographic projection* from $\mathbb{D}^n$ to $\mathbb{R}^n$ is the mapping that changes $-\mathbf{p}$ to the conformal point at infinity of $\mathbb{R}^n$, changes $\mathbf{p}$ to the origin of $\mathbb{R}^n$, and changes any other point $\mathbf{x} \in \mathbb{D}^n$ to the intersection $\mathbf{x}'$ of line $(-\mathbf{p})\mathbf{x}$ with the equator plane.

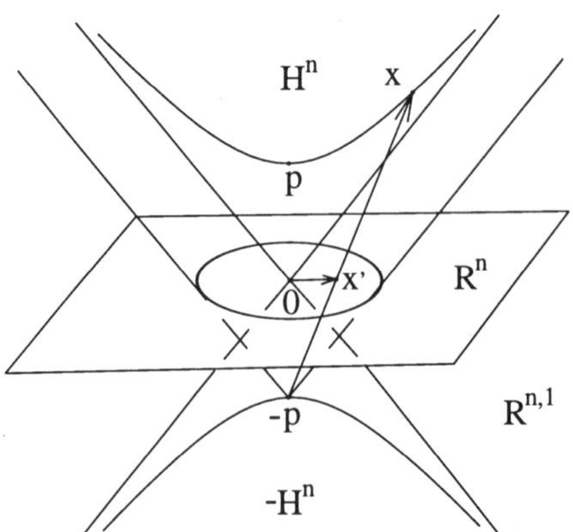

Fig. 8.17   Stereographic projection from $\mathbb{D}^n$ to $\mathbb{R}^n$.

Similar to (8.5.11), in $\mathbb{R}^{n,1}$, the stereographic projection has a very nice inversion-like expression:

$$\mathbf{x} \longmapsto \mathbf{x}' = -\mathbf{p} - 2(\mathbf{x} + \mathbf{p})^{-1}, \quad \forall \mathbf{x} \in \mathbb{D}^n. \tag{8.6.19}$$

Its inverse has the same expression:

$$\mathbf{x}' \longmapsto \mathbf{x} = -\mathbf{p} - 2(\mathbf{x}' + \mathbf{p})^{-1}, \quad \forall \mathbf{x}' \in \mathbb{R}^n - \mathbb{S}^{n-1}, \tag{8.6.20}$$

where $\mathbb{S}^{n-1}$ is the unit sphere of $\mathbb{R}^n$.

The stereographic projection maps $\mathbb{H}^n$ onto the open unit disk of $\mathbb{R}^n$, and maps $-\mathbb{H}^n - \{-\mathbf{p}\}$ onto the complement of the closed unit disk in $\mathbb{R}^n$. It can be extended to the sphere at infinity S_∞ of $\mathbb{D}^n$ naturally: for any end $\mathbf{x} \in S_\infty$, its image under the stereographic projection is the intersection of line $(-\mathbf{p})\mathbf{x}$ with the equator plane:

$$\mathbf{x} \longmapsto \mathbf{x}' = -\mathbf{p} - \frac{\mathbf{x}}{\mathbf{x} \cdot \mathbf{p}}, \quad \forall \mathbf{x} \in S_\infty. \tag{8.6.21}$$

The stereographic projection maps $\mathbb{D}^n \cup S_\infty$ onto $\mathbb{R}^n \cup \{\mathbf{e}\}$.

The image of $\mathbb{H}^n$ under the stereographic projection, when equipped with the Riemannian metric induced by the stereographic projection from $\mathbb{R}^{n,1}$, is called *Poincaré's disk model*.

Just as in spherical geometry, *the stereographic projection from $\mathbb{D}^n$ to $\mathbb{R}^n$ when represented in the homogeneous model, is the identity transformation in the projective null cone $\mathcal{N}$.*

In $\mathcal{N}$, again let $\mathbf{e} = \mathbf{p} + \mathbf{a}$, $\mathbf{e}_0 = (\mathbf{p} - \mathbf{a})/2$ as in (8.5.12). They represent respectively points $\mp\mathbf{p}$ in $\mathbb{D}^n$; they also represent the conformal point at infinity and the origin of $\mathbb{R}^n$. Let $\mathbf{c} \in \mathcal{N}$ but $\mathbf{c} \notin \mathbf{e} \wedge \mathbf{e}_0$. Then $\mathbf{c}$ represents point $\mathbf{x}' = P_{\mathbf{e}\wedge\mathbf{e}_0}^{\perp}(-\mathbf{c}/\mathbf{c} \cdot \mathbf{e}) \in \mathbb{R}^n$; it also represents point $\mathbf{x} = P_{\mathbf{a}}^{\perp}(-\mathbf{c}/\mathbf{c} \cdot \mathbf{a}) \in \mathbb{D}^n$ if $\mathbf{c} \cdot \mathbf{a} \neq 0$, or end $\mathbf{c}$ of $\mathbb{D}^n$ if $\mathbf{c} \cdot \mathbf{a} = 0$. The relations (8.6.19) and (8.6.20) are easy to verify.

In the conformal model of $\mathbb{D}^n$, for a positive vector $\mathbf{s}$, blade $\mathbf{s}^{\sim}$ represents a hyperplane if and only if $\mathbf{s} \cdot \mathbf{a} = 0$. In the conformal model of $\mathbb{R}^n$, $\mathbf{a}^{\sim}$ represents the unit sphere centering at the origin of $\mathbb{R}^n$, because $-\mathbf{a} = \mathbf{e}_0 - \mathbf{e}/2$. So in Poincaré's disk model, the hyperplanes in $\mathbb{D}^n$ are the spheres and hyperplanes in $\mathbb{R}^n$ perpendicular to the unit sphere of $\mathbb{R}^n$.

The metric of Poincaré's disk model can be easily derived from the rescaling of null vectors in $\mathcal{N}$: let $\mathbf{x}, \mathbf{y} \in \mathbb{R}^n$ be two points in the open unit disk of $\mathbb{R}^n$, then

$$\mathbf{x}' = P_{\mathbf{a}}^{\perp}\left(-\frac{f(\mathbf{x})}{f(\mathbf{x}) \cdot \mathbf{a}}\right), \qquad \mathbf{y}' = P_{\mathbf{a}}^{\perp}\left(-\frac{f(\mathbf{y})}{f(\mathbf{y}) \cdot \mathbf{a}}\right) \tag{8.6.22}$$

are the two points in $\mathbb{H}^n$ they represent, whose squared horo-distance is

$$
\begin{aligned}
|\mathbf{x}' - \mathbf{y}'|^2 &= |\left(-\frac{f(\mathbf{x})}{f(\mathbf{x}) \cdot \mathbf{a}} + \frac{f(\mathbf{y})}{f(\mathbf{y}) \cdot \mathbf{a}}\right) \wedge \mathbf{a}|^2 \\
&= \left(\frac{f(\mathbf{x})}{f(\mathbf{x}) \cdot \mathbf{a}} - \frac{f(\mathbf{y})}{f(\mathbf{y}) \cdot \mathbf{a}}\right)^2 \\
&= \frac{-2\,f(\mathbf{x}) \cdot f(\mathbf{y})}{(f(\mathbf{x}) \cdot \mathbf{a})(f(\mathbf{y}) \cdot \mathbf{a})} \\
&= \frac{4\,d_{\mathbf{xy}}^2}{(1 - \mathbf{x}^2)(1 - \mathbf{y}^2)}.
\end{aligned}
\tag{8.6.23}
$$

Now let $\mathbf{y} = \mathbf{x} + \mathbf{e}_1 dx_1 + \mathbf{e}_2 dx_2 + \cdots + \mathbf{e}_n dx_n$, where $\{\mathbf{e}_1, \mathbf{e}_2, \ldots, \mathbf{e}_n\}$ is an orthonormal basis of $\mathbb{R}^n$. By letting dx_i tend to zero for all $1 \le i \le n$, we get from (8.6.23) the metric of Poincaré's disk model at point $\mathbf{x} = (x_1, x_2, \ldots, x_n)^T$:

$$ds^2 = \frac{4(dx_1^2 + dx_2^2 + \cdots + dx_n^2)}{(1 - x_1^2 - x_2^2 - \cdots - x_n^2)^2}. \tag{8.6.24}$$

In establishing Poincaré's disk model, we have chosen the hyperbolic center $\mathbf{a}$ to be a positive vector $-\mathbf{e}_0 + \mathbf{e}/2$ representing a sphere in $\mathbb{R}^n$. We can certainly choose $\mathbf{a}$ to be a positive vector in $\mathbf{e}^\sim$ representing a hyperplane in $\mathbb{R}^n$ instead. This alternative leads to *Poincaré's half-space model.*

Choose two ends $\mathbf{e}, \mathbf{e}_0$ of $\mathbb{D}^n$ such that the two vectors form a Witt pair. Then $\mathbf{a} \in (\mathbf{e} \wedge \mathbf{e}_0)^\sim = \mathbb{R}^n$. The two scalings of null vectors $\mathbf{x} \mapsto -\mathbf{x}/(\mathbf{x} \cdot \mathbf{a})$ and $\mathbf{x} \mapsto -\mathbf{x}/(\mathbf{x} \cdot \mathbf{e})$ for all $\mathbf{x} \in \mathcal{N}$, induce a correspondence between points in $\mathbb{H}^n$ and points in a half space of $\mathbb{R}^n$ bordered by hyperplane $\mathbf{a}^\sim$. The sphere at infinity of $\mathbb{H}^n$ corresponds to the boundary of the half space in $\mathbb{R}^n$. So hyperplanes in $\mathbb{D}^n$ correspond to spheres and hyperplanes in $\mathbb{R}^n$ perpendicular to the bordering hyperplane.

Poincaré's half-space model refers to the image space of the mapping from $\mathbb{H}^n$ to $\mathbb{R}^n$ induced by the above two scalings of null vectors in $\mathcal{N}$, equipped with the metric induced by the mapping from $\mathbb{H}^n$.

The derivation of the metric is also easy. Let $\mathbf{x}, \mathbf{y}$ be two points in the open half space of $\mathbb{R}^n$ bordered by hyperspace $\mathbf{a}^\sim$, where vectors $\mathbf{x}, \mathbf{y}, \mathbf{a} \in \mathbb{R}^n$, and where $\mathbf{x} \cdot \mathbf{a}$ and $\mathbf{y} \cdot \mathbf{a}$ are of the same sign. Let $\mathbf{x}', \mathbf{y}'$ be the two points in $\mathbb{H}^n$ corresponding to the two points $\mathbf{x}, \mathbf{y}$ in $\mathbb{R}^n$ via the identification of their null-vector representations in the homogeneous models. The computing in (8.6.23) is valid till the next to the last step, only the final step becomes

$$|\mathbf{x}' - \mathbf{y}'|^2 = -2 \frac{\mathbf{f}(\mathbf{x}) \cdot \mathbf{f}(\mathbf{y})}{(\mathbf{f}(\mathbf{x}) \cdot \mathbf{a})(\mathbf{f}(\mathbf{y}) \cdot \mathbf{a})} = \frac{d_{\mathbf{xy}}^2}{(\mathbf{a} \cdot \mathbf{x})(\mathbf{a} \cdot \mathbf{y})}. \tag{8.6.25}$$

Let $\mathbf{y} = \mathbf{x} + \mathbf{e}_1 dx_1 + \cdots + \mathbf{e}_{n-1} dx_{n-1} + \mathbf{a}\, dx_n$, where $\{\mathbf{e}_1, \ldots, \mathbf{e}_{n-1}, \mathbf{a}\}$ is an orthonormal basis of $\mathbb{R}^n$. By letting dx_i tend to zero for all $1 \le i \le n$, we get from (8.6.25) the metric of Poincaré's half-space model at point $\mathbf{x} = (x_1, x_2, \ldots, x_n)^T$:

$$ds^2 = \frac{dx_1^2 + dx_2^2 + \cdots + dx_n^2}{x_n^2}. \tag{8.6.26}$$

8.7 Unified algebraic framework for classical geometries

The mapping $\mathbf{x} \mapsto -\mathbf{x}/(\mathbf{x} \cdot \mathbf{c})$ for a fixed nonzero vector $\mathbf{c} \in \mathbb{R}^{n+1,1}$ and for any vector $\mathbf{x} \in \mathcal{N}$, is called the *inhomogenization* of the projective null cone $\mathcal{N}$ with respect to vector $\mathbf{c}$. Its image space is the set

$$\mathcal{N}_{\mathbf{c}} := \{\mathbf{x} \in \mathcal{N} \,|\, \mathbf{x} \cdot \mathbf{c} = -1\}. \tag{8.7.1}$$

Vector $\mathbf{c}$ has three possible signatures: positive, null, and negative, and correspondingly the inhomogenization generates three conformal models for double-hyperbolic, Euclidean, and spherical geometries respectively.

Prior to the inhomogenization, different geometries have the same algebraic representation: a projective point in the homogeneous model represents a point or point at infinity, a Minkowski blade represents a sphere or plane, and an orthogonal transformation of vectors in $\mathcal{N}$ induces a conformal transformation of the points and points at infinity.

So there is a unified algebraic framework that is independent of the inhomogenization. We call this framework the *homogeneous model* of classical geometries. In the homogeneous model, a single algebraic identity can be translated into different geometric theorems in different geometries. Even in the same geometry, a single algebraic identity can have various geometric explanations. It often occurs that one explanation is much simpler than others.

Consequently, proving one geometric theorem by verifying the representational algebraic identity in the homogeneous model, is equivalent to proving all the theorems in classical geometries having the same algebraic representation. Replacing a geometric theorem by an algebraically equivalent but geometrically simpler theorem is a novel way of simplifying geometric theorem proving.

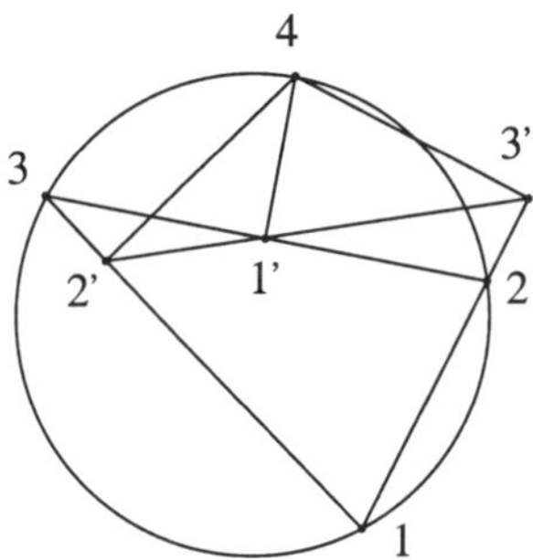

Fig. 8.18   Simson's Theorem.

Example 8.63. In the homogeneous model of 2D Euclidean geometry, the classical Simson's Theorem in Example 7.53 has its hypotheses represented by the following homogeneous bracket equalities and inequalities:

$$
\begin{aligned}
[\mathbf{e123}] &\neq 0 & &\mathbf{1,2,3} \text{ are not collinear,} \\
[\mathbf{4123}] &= 0 & &\mathbf{4,1,2,3} \text{ are cocircular,} \\
\left. \begin{array}{l} (\mathbf{e41'}|\mathbf{e23}) = 0 \\ [\mathbf{e1'23}] = 0 \end{array} \right\} & & &\mathbf{1'} \text{ is the foot drawn from } \mathbf{4} \text{ to line } \mathbf{23,} \\
\left. \begin{array}{l} (\mathbf{e42'}|\mathbf{e13}) = 0 \\ [\mathbf{e12'3}] = 0 \end{array} \right\} & & &\mathbf{2'} \text{ is the foot drawn from } \mathbf{4} \text{ to line } \mathbf{13,} \\
\left. \begin{array}{l} (\mathbf{e43'}|\mathbf{e12}) = 0 \\ [\mathbf{e123'}] = 0 \end{array} \right\} & & &\mathbf{3'} \text{ is the foot drawn from } \mathbf{4} \text{ to line } \mathbf{12.}
\end{aligned}
\tag{8.7.2}
$$

The conclusion is $[\mathbf{e1'2'3'}] = 0$, *i.e.*, $\mathbf{1', 2', 3'}$ are collinear.

In the algebraic representation of the hypotheses, $\mathbf{e}$ is just a null vector picked out from the initial five vectors $\mathbf{e}, \mathbf{1}, \mathbf{2}, \mathbf{3}, \mathbf{4}$. We can certainly explain $\mathbf{e}$ as a point while explain $\mathbf{4}$ as the conformal point at infinity. This is equivalent to interchanging $\mathbf{e}, \mathbf{4}$ in (8.7.2), and the resulting hypotheses are

$$
\begin{array}{ll}
[\mathbf{4123}] \neq 0 & \mathbf{4,1,2,3} \text{ are not collinear,} \\[4pt]
[\mathbf{e123}] = 0 & \mathbf{1,2,3} \text{ are collinear,} \\[4pt]
\left.\begin{array}{l} (\mathbf{e41'}|\mathbf{423}) = 0 \\[4pt] [\mathbf{41'23}] = 0 \end{array}\right\} & \mathbf{41'} \text{ is the diameter of circle } \mathbf{423}, \\[8pt]
\left.\begin{array}{l} (\mathbf{e42'}|\mathbf{413}) = 0 \\[4pt] [\mathbf{412'3}] = 0 \end{array}\right\} & \mathbf{42'} \text{ is the diameter of circle } \mathbf{413}, \\[8pt]
\left.\begin{array}{l} (\mathbf{e43'}|\mathbf{412}) = 0 \\[4pt] [\mathbf{4123'}] = 0 \end{array}\right\} & \mathbf{43'} \text{ is the diameter of circle } \mathbf{412}.
\end{array}
\tag{8.7.3}
$$

The conclusion becomes $[\mathbf{41'2'3'}] = 0$, *i.e.*, $\mathbf{4}, \mathbf{1'}, \mathbf{2'}, \mathbf{3'}$ are cocircular.

In the geometric configuration described by (8.7.3), point $\mathbf{1'}$ has the following obvious *linear construction*, *i.e.*, construction by linear equations: it is the intersection of the line passing through point $\mathbf{2}$ and perpendicular to line $\mathbf{42}$, with the line passing through point $\mathbf{3}$ and perpendicular to line $\mathbf{43}$. Points $\mathbf{2'}, \mathbf{3'}$ have similar linear constructions. In this way, (8.7.3) is equivalent to the following sequence of linear constructions:

$$
\begin{array}{l}
\mathbf{1'}, \mathbf{2'}, \mathbf{3'} \text{ are free points in the plane,} \\[4pt]
\mathbf{4} \text{ is a semifree point in the plane such that if} \\[4pt]
\quad \mathbf{1} \text{ is the foot drawn from } \mathbf{4} \text{ to line } \mathbf{2'3'}, \\[4pt]
\quad \mathbf{2} \text{ is the foot drawn from } \mathbf{4} \text{ to line } \mathbf{1'3'}, \\[4pt]
\quad \mathbf{3} \text{ is the foot drawn from } \mathbf{4} \text{ to line } \mathbf{1'2'}, \\[4pt]
\text{then } \mathbf{4,1,2,3} \text{ are not collinear,} \\[4pt]
\text{and } \mathbf{1,2,3} \text{ are collinear.}
\end{array}
\tag{8.7.4}
$$

What is amazing is that when we interchange the triplets $\mathbf{1}, \mathbf{2}, \mathbf{3}$ and $\mathbf{1'}, \mathbf{2'}, \mathbf{3'}$ in (8.7.4), we get exactly the geometric configuration of the converse theorem of Simson's Theorem. The converse theorem says that for triangle $\mathbf{123}$, if the three feet $\mathbf{1'}, \mathbf{2'}, \mathbf{3'}$ drawn from a point $\mathbf{4}$ to the three sides of the triangle respectively are collinear, then $\mathbf{4}$ is on the circumcircle of the triangle. So Simson's theorem and its converse are algebraically equivalent.

The benefit of this equivalence is that the geometric configuration of the converse theorem is much simpler because no nonlinear object is involved. For automated proving, the converse theorem is much easier to prove, because the nonlinearity of the geometric problem is moved from the hypotheses to the conclusion.

We come back to the theme of theoretical exploration of the homogeneous model of classical geometries. This model needs to be equipped with a distance function to become a distance space. We define the following distance function, called *conformal distance*, in the projective null cone $\mathcal{N}$ of $\mathbb{R}^{n+1,1}$: for fixed nonzero vector $\mathbf{c} \in$

$\mathbb{R}^{n+1,1}$, for any $\mathbf{x}, \mathbf{y} \in \mathcal{N}$,

$$d_{\mathbf{xy}}^{\mathbf{c}} := \left| \frac{\mathbf{x}}{\mathbf{x} \cdot \mathbf{c}} - \frac{\mathbf{y}}{\mathbf{y} \cdot \mathbf{c}} \right| \in \mathbb{R} \cup \{+\infty\}. \tag{8.7.5}$$

(1) When $\mathbf{c}^2 = 0$, then $d^{\mathbf{c}}$ is a Euclidean geometric distance. Only the distance between a point and the conformal point at infinity is infinite.

(2) When $\mathbf{c}^2 = -1$, then $d^{\mathbf{c}}$ is a spherical geometric distance, or more accurately, the chord distance.

(3) When $\mathbf{c}^2 = 1$, then $d^{\mathbf{c}}$ is a hyperbolic geometric distance, or more accurately, the horo-distance when restricted to $\mathbb{H}^n$ or $-\mathbb{H}^n$. The distance between a point and an end is infinite, so is the distance between any two different ends.

For any two nonzero vectors $\mathbf{c}, \mathbf{c}' \in \mathbb{R}^{n+1,1}$, and any two null vectors $\mathbf{x}, \mathbf{y}$,

$$d_{\mathbf{xy}}^{\mathbf{c}'} = \sqrt{\frac{2|\mathbf{x} \cdot \mathbf{y}|}{|\mathbf{x} \cdot \mathbf{c}'||\mathbf{y} \cdot \mathbf{c}'|}} = \sqrt{\frac{2|\mathbf{x} \cdot \mathbf{y}|}{|\mathbf{x} \cdot \mathbf{c}||\mathbf{y} \cdot \mathbf{c}|}} \sqrt{\frac{|\mathbf{x} \cdot \mathbf{c}||\mathbf{y} \cdot \mathbf{c}|}{|\mathbf{x} \cdot \mathbf{c}'||\mathbf{y} \cdot \mathbf{c}'|}} = d_{\mathbf{xy}}^{\mathbf{c}} \sqrt{\frac{|\mathbf{x} \cdot \mathbf{c}||\mathbf{y} \cdot \mathbf{c}|}{|\mathbf{x} \cdot \mathbf{c}'||\mathbf{y} \cdot \mathbf{c}'|}}.$$

So a rescaling in $\mathcal{N}$ is always a conformal map: if ds^2, ds'^2 are the metrics of $\mathcal{N}_{\mathbf{c}}$ and $\mathcal{N}_{\mathbf{c}'}$ respectively, then

$$ds'^2 = \frac{ds^2}{(\mathbf{x} \cdot \mathbf{c}')^2}, \quad \forall \mathbf{x} \in \mathcal{N}_{\mathbf{c}}; \tag{8.7.6}$$

or in symmetric form, $(\mathbf{x} \cdot \mathbf{c}')^2 ds'^2 = (\mathbf{x} \cdot \mathbf{c})^2 ds^2$.

From the coordinate point of view, the upgrade from the homogeneous coordinates model $\mathbb{R}^{n,0,1}$ of nD Euclidean geometry to the conformal model $\mathbb{R}^{n+1,1}$ is the upgrade from homogeneous coordinates to *Euclidean conformal coordinates*, or *generalized homogeneous coordinates*.

A Cartesian frame of affine space $\mathbb{E}^n$ in $\mathbb{R}^{n,0,1}$ is an orthonormal basis $\{\mathbf{e}_0, \mathbf{e}_1, \mathbf{e}_2, \ldots, \mathbf{e}_n\}$, where $\mathbf{e}_0$ is a null vector representing the origin of the frame, and $\mathbf{e}_1, \mathbf{e}_2, \ldots, \mathbf{e}_n$ are an orthonormal basis of $\mathbb{R}^n$. In the conformal model of $\mathbb{R}^n$, a *Euclidean conformal frame* is of the form $\{\mathbf{e}, \mathbf{e}_0, \mathbf{e}_1, \mathbf{e}_2, \ldots, \mathbf{e}_n\}$, where $(\mathbf{e}, \mathbf{e}_0)$ is a Witt pair representing the conformal point at infinity and the origin respectively.

A general *Euclidean conformal frame* in $\mathbb{R}^{n+1,1}$ is independent of the choice of the conformal point at infinity $\mathbf{e}$. It refers to a Witt basis of $\mathbb{R}^{n+1,1}$ of the form

$$\{\mathbf{a}, \mathbf{b}, \mathbf{c}_1, \mathbf{c}_2, \ldots, \mathbf{c}_n\}, \tag{8.7.7}$$

where $(\mathbf{a}, \mathbf{b})$ is a Witt pair. The *Euclidean conformal coordinates* of a point, or sphere, or hyperplane in $\mathbb{E}^n$, are the coordinates of their representational vectors in $\mathbb{R}^{n+1,1}$ with respect to the basis.

In the special case where both $\mathbf{a}, \mathbf{b}$ stand for two points in $\mathbb{E}^n$, then the $\mathbf{c}_i$ stand for pairwise perpendicular spheres or hyperplanes passing through the two points. In some applications, Euclidean conformal coordinates in such a coordinate frame can lead to substantial simplifications.

Similarly, in the conformal model of $\mathbb{S}^n$, a *spherical conformal frame* is of the form $\{\mathbf{p}, \mathbf{e}_0, \mathbf{e}_1, \mathbf{e}_2, \ldots, \mathbf{e}_n\}$, where $\mathbf{p}$ is the spherical center, and $\mathbf{e}_0$ is the north pole of $\mathbb{S}^n$. In the conformal model of $\mathbb{D}^n$, a *hyperbolic conformal frame* is of the form $\{\mathbf{a}, \mathbf{e}_0, \mathbf{e}_1, \mathbf{e}_2, \ldots, \mathbf{e}_n\}$, where $\mathbf{a}$ is the hyperbolic center, and $\mathbf{e}_0$ is the *south pole* of $\mathbb{D}^n$, *i.e.*, the antipodal point of the north pole in $\mathbb{D}^n$, as shown in Figure 8.17.

Definition 8.64. A general *conformal frame* in $\mathbb{R}^{n+1,1}$, also called *conformal basis*, is a basis of the form $\{\mathbf{b}, \mathbf{e}_0, \mathbf{e}_1, \ldots, \mathbf{e}_n\}$, where $\mathbf{b}$ is either a null vector or a unit vector, where $\mathbf{e}_0$ is a null vector satisfying $\mathbf{e}_0 \cdot \mathbf{b} = -1$, and where $\mathbf{e}_1, \mathbf{e}_2, \ldots, \mathbf{e}_n$ are pairwise orthogonal positive unit vectors in $\mathbb{R}^{n+1,1}$ and are orthogonal to both $\mathbf{b}$ and $\mathbf{e}_0$.

For example, a general *spherical conformal frame* in $\mathbb{R}^{n+1,1}$ is independent of the choice of the spherical center $\mathbf{p}$. It refers to a conformal basis of $\mathbb{R}^{n+1,1}$ where the first basis vector is a negative unit vector. A general *hyperbolic conformal frame* in $\mathbb{R}^{n+1,1}$ is also independent of the choice of the hyperbolic center $\mathbf{a}$. It refers to a conformal basis of $\mathbb{R}^{n+1,1}$ where the first basis vector is a positive unit vector.

The *spherical conformal coordinates* or *hyperbolic conformal coordinates* of points, spheres and hyperplanes in $\mathbb{E}^n$, are the coordinates of their representational vectors in $\mathbb{R}^{n+1,1}$ with respect to the given spherical conformal frame or hyperbolic conformal frame. When the geometric space is changed to $\mathbb{S}^n$ or $\mathbb{D}^n$, the Euclidean, spherical, and hyperbolic conformal coordinates can be defined similarly.

So there are two systems in describing a classical geometry. The first system is a conformal distance function $d^{\mathbf{c}}$ given by (8.7.5), together with the distance space defined by $d^{\mathbf{c}}$ on $\mathcal{N}$. This is the defining algebraic system of the geometry by measurement, where the tool of measurement can be double-hyperbolic, or Euclidean, or spherical, depending on the signature of vector $\mathbf{c}$. The second system is a conformal frame of $\mathbb{R}^{n+1,1}$, together with the conformal coordinates of the vectors in $\mathbb{R}^{n+1,1}$ representing points, spheres and hyperplanes in the geometry defined by $d^{\mathbf{c}}$. This is the coordinate descriptive system of the geometry.

The *universal conformal model* of classical metric geometries in coordinate form, is the triplet

$$\mathcal{N}, \quad \{\, d^{\mathbf{c}} \,|\, \mathbf{c} \in \mathbb{R}^{n+1,1} - \{0\} \,\}, \quad \{\, \mathcal{F}_{\mathbf{b}} \,|\, \mathbf{b} \in \mathbb{R}^{n+1,1} - \{0\} \,\}, \tag{8.7.8}$$

where $d^{\mathbf{c}}$ is a conformal distance, and $\mathcal{F}_{\mathbf{b}}$ is the set of conformal frames where the leading basis vector is $\mathbf{b}$.

Any conformal frame $F_{\mathbf{b}} \in \mathcal{F}_{\mathbf{b}}$ determines a *coordinate space* $\mathcal{X}_{\mathbf{b}}$, the latter being the space of coordinates of all vectors in $\mathcal{N}_{\mathbf{b}} = \{\mathbf{x} \in \mathcal{N} \,|\, \mathbf{x} \cdot \mathbf{b} = -1\}$ with respect to the conformal frame. Given a coordinate space $\mathcal{X}_{\mathbf{b}}$ and a conformal distance $d^{\mathbf{c}}$,

- if $\mathbf{b} = \pm\mathbf{c}$, the geometric space $(\mathcal{N}, d^{\mathbf{c}})$ is *canonical* in that the distance function agrees with the metric of the coordinate space induced from $\mathbb{R}^{n+1,1}$;
- else if $\mathbf{b}$ and $\mathbf{c}$ have the same signature, the geometric space is *conformal* in that the distance function differs from the metric of the coordinate space by a scale that is irrelevant to tangent directions;

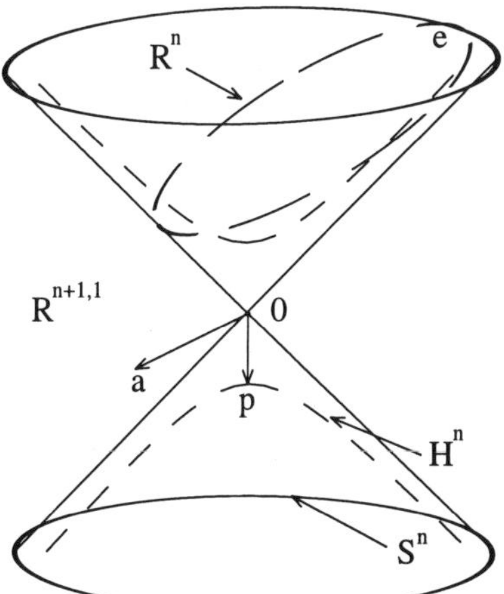

Fig. 8.19　Three kinds of conformal frames of $\mathbb{R}^{n+1,1}$ determine three kinds of coordinate spaces.

- else, the coordinate space is a *conformal representation* of the geometric space. Although the distance and the metric of the coordinate space still have a conformal relation, they define topologically different spaces of points at infinity.

Appendix A

Cayley Expansion Theory for 2D and 3D Projective Geometries

A.1 Cayley expansions of p_{II}

Proof of Proposition 2.59. For $p_{II} = (12 \vee 34 \vee 56)(1'2' \vee 3'4' \vee 5'6')$, item 1 is trivial. First we prove item 3.

3. By symmetry, we only need to consider the expansion (2.5.17), *i.e.*,

$$p_{II} = [134][256]1'2' \vee 3'4' \vee 5'6' - [234][156]1'2' \vee 3'4' \vee 5'6'.$$

Since $p_{II} \neq 0$, none of $[134], [256]$ can be equal to $[234]$ or $[156]$. Assume that there are intergroup like terms. Then $[134][256]$ must be in an expansion result of $1'2' \vee 3'4' \vee 5'6'$, and $1', \ldots, 6'$ must be a permutation of $1, \ldots, 6$. Below we assume $[134][256][234][156] \neq 0$, otherwise there are no intergroup like terms.

If $\{12, 34, 56\} = \{1'2', 3'4', 5'6'\}$, we can assume $\mathbf{i}' = \mathbf{i}$ for $1 \leq i \leq 6$. After combining intergroup like terms, we get three 3-termed expansions of $(12 \vee 34 \vee 56)^2$, which correspond to $([124][356] - [123][456])^2$ and the squares of two other expansions of $12 \vee 34 \vee 56$ respectively.

Assume that $\{12, 34, 56\} \neq \{1'2', 3'4', 5'6'\}$. Without loss of generality, assume that $134 = 1'3'4'$, $256 = 2'5'6'$ up to the order of points. By symmetry, there are only two possibilities:

$$
\begin{array}{ll}
\text{either} & 1 = 1', 3 = 3', 4 = 4' \text{ and } 2 = 5', 6 = 6', 5 = 2', \\
\text{or} & 1 = 3', 4 = 4', 3 = 1' \text{ and } 2 = 5', 6 = 6', 5 = 2'.
\end{array}
\tag{A.1.1}
$$

For the like terms to occur, there are four cases, in which $[234][156]$ is identical respectively to

(1) $[2'3'4'][1'5'6']$,
(2) $[1'2'3'][4'5'6']$,
(3) $[1'2'5'][3'4'6']$,
(4) $[1'2'6'][3'4'5']$.

Case (1): By (A.1.1), $[2'3'4'][1'5'6']$ is identical to either $[534][126]$ or $[514][326]$. Only the latter one can be identical to $[234][156]$, and the additional condition $4 = 6$ is required. The pattern is $(12 \vee 34 \vee 54)(35 \vee 14 \vee 24)$. The

469

result after combining like terms is the same as the result by expanding both meet products into $[\mathbf{234}][\mathbf{145}] - [\mathbf{134}][\mathbf{245}]$.

Case (2): By (A.1.1), $[\mathbf{1'2'3'}][\mathbf{4'5'6'}]$ is identical to $[\mathbf{153}][\mathbf{426}]$, which is also identical to $[\mathbf{234}][\mathbf{156}]$ if $\mathbf{3} = \mathbf{6}$. There are two patterns: $(\mathbf{12} \vee \mathbf{34} \vee \mathbf{53})(\mathbf{15} \vee \mathbf{34} \vee \mathbf{23})$ and $(\mathbf{12} \vee \mathbf{34} \vee \mathbf{53})(\mathbf{35} \vee \mathbf{14} \vee \mathbf{23})$. The results after canceling like terms are identical to those by expanding the two meet products in p_{II} into bracket monomials.

Case (3): By (A.1.1), $[\mathbf{1'2'5'}][\mathbf{3'4'6'}]$ is identical to $[\mathbf{152}][\mathbf{346}]$ or $[\mathbf{352}][\mathbf{146}]$. The former cannot be identical to $[\mathbf{234}][\mathbf{156}]$, while the latter can, if $\mathbf{4} = \mathbf{5}$. The result after canceling like terms is identical to that by expanding the meet products into bracket monomials.

Case (4): By (A.1.1), $[\mathbf{1'2'6'}][\mathbf{3'4'5'}]$ is identical to $[\mathbf{156}][\mathbf{342}]$ or $[\mathbf{356}][\mathbf{142}]$. The latter cannot be identical to $[\mathbf{234}][\mathbf{156}]$, while the former has a pattern $(\mathbf{12} \vee \mathbf{34} \vee \mathbf{56})(\mathbf{15} \vee \mathbf{34} \vee \mathbf{26})$, which corresponds to (2.5.19). If (2.5.19) is a monomial expansion, then the two meet products of p_{II} both have monomial expansions.

2. First, assume that $\{\mathbf{1}, \ldots, \mathbf{6}, \mathbf{1'}, \ldots, \mathbf{6'}\}$ is a set of generic points, and $\mathbf{3} = \mathbf{1}$. If p_{II} has a monomial expansion, since the bracket ring is a UFD, the meet product in (2.5.18) must have a monomial expansion. Thus, expansion (2.5.18) is the shortest.

Among the expansions of p_{II} having no like terms, (2.5.18) is obviously the unique shortest expansion. By the proof of item 3, (2.5.18) is the unique shortest expansion in Cases (1), (2), (3), together with the case $\{\mathbf{12}, \mathbf{34}, \mathbf{56}\} = \{\mathbf{1'2'}, \mathbf{3'4'}, \mathbf{5'6'}\}$. In Case (4), the expansion of $(\mathbf{12} \vee \mathbf{14} \vee \mathbf{56})(\mathbf{15} \vee \mathbf{14} \vee \mathbf{26})$ by (2.5.19) gives the same result as (2.5.18).

Second, let $\mathbf{1}, \ldots, \mathbf{6}, \mathbf{1'}, \ldots, \mathbf{6'}$ be points in the plane. We prove that if the meet product in (2.5.18) has no monomial expansion, then any other expansion of p_{II} has at least two terms. The assumption indicates (a) $\mathbf{1}, \mathbf{2}, \mathbf{4}$ are not collinear, (b) $\mathbf{3}, \mathbf{5}, \mathbf{6}$ are not collinear, (c) $\mathbf{1'}, \mathbf{2'}$ are not on lines $\mathbf{3'4'}, \mathbf{5'6'}$, (d) $\mathbf{3'}, \mathbf{4'}$ are not on lines $\mathbf{1'2'}, \mathbf{5'6'}$, (e) $\mathbf{5'}, \mathbf{6'}$ are not on lines $\mathbf{1'2'}, \mathbf{3'4'}$.

When an expansion does not generate intergroup like terms, the expansion result is obviously longer than the shortest result from (2.5.18). By the proof of item 3, in Cases (1), (2), (3), together with the case $\{\mathbf{12}, \mathbf{34}, \mathbf{56}\} = \{\mathbf{1'2'}, \mathbf{3'4'}, \mathbf{5'6'}\}$, the expansion (2.5.18) is the shortest. In case (4), since the second meet product in p_{II} has no monomial expansion, neither does (2.5.19).

4. The proof is already included in those of items 2 and 3. $\square$

A.2   Cayley expansions of p_{III}

For the expression $p_{III} = [\mathbf{1}(\mathbf{1'2'} \vee \mathbf{3'4'})(\mathbf{1''2''} \vee \mathbf{3''4''})]$, there are two initial outer product expansions. For example, the outer product expansion by distributing $\mathbf{1'2'}$

and $\mathbf{3'4'}$ is

$$p_{III} = [\mathbf{13'4'}]\mathbf{1'2'} \vee \mathbf{1''2''} \vee \mathbf{3''4''} - [\mathbf{11'2'}]\mathbf{3'4'} \vee \mathbf{1''2''} \vee \mathbf{3''4''}. \qquad (A.2.1)$$

There are also four initial meet product expansions. For example, the meet product expansion by separating $\mathbf{1'}$ and $\mathbf{2'}$ is

$$p_{III} = [\mathbf{1'3'4'}]\mathbf{12'} \vee \mathbf{1''2''} \vee \mathbf{3''4''} - [\mathbf{2'3'4'}]\mathbf{11'} \vee \mathbf{1''2''} \vee \mathbf{3''4''}. \qquad (A.2.2)$$

Proposition A.2.1. p_{III} has 46 generic expansions into bracket polynomials.

Proof. Denote initial the meet product expansion result from separating $\mathbf{i,j}$ by $h_{\mathbf{ij}}$. By (A.2.2), $h_{\mathbf{1'2'}}$ has $3 \times 3 = 9$ expansions into bracket polynomials, one of which has brackets $[\mathbf{1''2''3''}]$ and $[\mathbf{1''2''4''}]$, one of which has brackets $[\mathbf{1''3''4''}]$ and $[\mathbf{2''3''4''}]$, and one of which has brackets $[\mathbf{11''2''}]$ and $[\mathbf{13''4''}]$. Similarly, $h_{\mathbf{3'4'}}$ has 9 expansions into bracket polynomials, none of which can be obtained from expanding $h_{\mathbf{1'2'}}$.

$h_{\mathbf{1''2''}}$ has 9 expansions into bracket polynomials, two of which have been obtained from expanding $h_{\mathbf{1'2'}}$ and $h_{\mathbf{3'4'}}$ respectively, one of which has brackets $[\mathbf{11'2'}]$ and $[\mathbf{13'4'}]$. So there are 7 new expansions. Similarly, $h_{\mathbf{3''4''}}$ brings 7 new results.

The outer product expansion separating $\mathbf{1'2'}, \mathbf{3'4'}$ has 9 expansions into bracket polynomials, two of which are old ones from $h_{\mathbf{1''2''}}$ and $h_{\mathbf{3''4''}}$ respectively. So there are 7 new expansions. Similarly, the other outer product expansion brings 7 new results. All together there are $2 \times (9 + 7 + 7) = 46$ expansions into bracket polynomials. $\qquad\square$

Proof of Theorem 2.60. Item 1 is trivial. We start with item 2.

2. We only need to prove the unique shortest statement in item 2. Assume that the meet product on the right of (2.5.21) has only 2-term expansions. Then $\mathbf{1, 1', 1'', 2'', 3'', 4''}$ are generic points, so are $\mathbf{1', 2', 4'}$. p_{III} has two outer product expansions:

$$\begin{aligned}
p_{III} &= [\mathbf{11'4'}]\mathbf{1'2'} \vee \mathbf{1''2''} \vee \mathbf{3''4''} - [\mathbf{11'2'}]\mathbf{1'4'} \vee \mathbf{1''2''} \vee \mathbf{3''4''} & (A.2.3)\\
&= [\mathbf{11''2''}]\mathbf{1'2'} \vee \mathbf{1'4'} \vee \mathbf{3''4''} - [\mathbf{13''4''}]\mathbf{1'2'} \vee \mathbf{1'4'} \vee \mathbf{1''2''}. & (A.2.4)
\end{aligned}$$

By symmetry, p_{III} has one kind of meet product expansion different from (2.5.21):

$$p_{III} = [\mathbf{1''2''3''}]\mathbf{1'2'} \vee \mathbf{1'4'} \vee \mathbf{14''} - [\mathbf{1''2''4''}]\mathbf{1'2'} \vee \mathbf{1'4'} \vee \mathbf{13''}. \qquad (A.2.5)$$

(A.2.3): If one of the brackets $[\mathbf{11'4'}], [\mathbf{11'2'}]$ is zero, then $\mathbf{1} = \mathbf{4'}$ or $\mathbf{2'}$, the expansion is the same with (2.5.21). When both brackets are nonzero, then $[\mathbf{11'4'}]$ cannot occur in any expansion result of $\mathbf{1'4'} \vee \mathbf{1''2''} \vee \mathbf{3''4''}$, so (A.2.3) has no intergroup like terms. It has a 2-termed expansion if and only if $\{\mathbf{2', 4'}\} \subset \{\mathbf{1'', 2'', 3'', 4''}\}$. If $\mathbf{2'4'} = \mathbf{1''2''}$ (or $\mathbf{3''4''}$), then (2.5.21) is a shortest expansion among the expansions of (A.2.3). The only case left is $\mathbf{2'} = \mathbf{1''}$ and $\mathbf{4'} = \mathbf{3''}$, and the corresponding 2-termed expansion is (2.5.22).

(A.2.4): There are no intergroup like terms, because $[\mathbf{13''4''}]$ does no occur in an expansion of $\mathbf{1'2'} \vee \mathbf{1'4'} \vee \mathbf{3''4''}$. The shortest expansions of (A.2.4) have two terms if and only if $\{\mathbf{2'}, \mathbf{4'}\} \subset \{\mathbf{1''}, \mathbf{2''}, \mathbf{3''}, \mathbf{4''}\}$, in which case the expansions of both meet products in (A.2.4) are unique, and the result is identical to an expansion of (2.5.21).

(A.2.5): There are no intergroup like terms, because $[\mathbf{1''2''3''}]$ does not occur in any expansion of $\mathbf{1'2'} \vee \mathbf{1'4'} \vee \mathbf{13''}$. The shortest expansions of (A.2.4) have two terms if and only if both meet products of it can be expanded into bracket monomials, in which case bracket $[\mathbf{1'2'4'}]$ occurs as the common factor, and the expansion result can also be obtained from (2.5.21).

3 – 4. The proofs are similar. (2.5.23) comes from the following identity:

$$(\mathbf{12} \vee \mathbf{34})(\mathbf{12} \vee \mathbf{3'4'}) = (\mathbf{12} \vee \mathbf{34} \vee \mathbf{3'4'})\mathbf{12}. \qquad \text{(A.2.6)}$$

5. If $\mathcal{M}$ has two elements, then there are only two cases for p_{III}: (a) $[\mathbf{1(1'2'} \vee \mathbf{3'4')(1'2''} \vee \mathbf{2'4'')}]$, (b) $[\mathbf{1(1'2'} \vee \mathbf{3'4')(1'2''} \vee \mathbf{3'4'')}]$. If $\mathcal{M}$ has three elements, then there is only one case for p_{III}: (c) $[\mathbf{1(1'2'} \vee \mathbf{3'4')(1'3'} \vee \mathbf{2'4'')}]$. If $\mathcal{M}$ has four elements, the only case for p_{III} is (d) $[\mathbf{1(1'2'} \vee \mathbf{3'4')(1'3'} \vee \mathbf{2'4')}]$.

For the expression in case (d), by symmetry, the outer product expansion gives

$$[\mathbf{1(1'2'}\vee\mathbf{3'4')(1'3'}\vee\mathbf{2'4')}] = [\mathbf{13'4'}]\mathbf{1'2'}\vee\mathbf{1'3'}\vee\mathbf{2'4'} - [\mathbf{11'2'}]\mathbf{3'4'}\vee\mathbf{1'3'}\vee\mathbf{2'4'}, \qquad \text{(A.2.7)}$$

and the meet product expansion gives

$$[\mathbf{1'3'4'}]\mathbf{12'} \vee \mathbf{1'3'} \vee \mathbf{2'4'} - [\mathbf{2'3'4'}]\mathbf{11'} \vee \mathbf{1'3'} \vee \mathbf{2'4'}. \qquad \text{(A.2.8)}$$

The expansions of the two meet products in (A.2.7) are both unique, so the initial outer product expansion always produces

$$[\mathbf{13'4'}][\mathbf{1'2'3'}][\mathbf{1'2'4'}] - [\mathbf{11'2'}][\mathbf{1'3'4'}][\mathbf{2'3'4'}]. \qquad \text{(A.2.9)}$$

The expansions of the two meet products in (A.2.8) are all listed below:

$$\mathbf{11'} \vee \mathbf{1'3'} \vee \mathbf{2'4'} = [\mathbf{11'3'}][\mathbf{1'2'4'}] = [\mathbf{11'2'}][\mathbf{1'3'4'}] + [\mathbf{11'4'}][\mathbf{1'2'3'}],$$
$$\mathbf{12'} \vee \mathbf{1'3'} \vee \mathbf{2'4'} = [\mathbf{12'4'}][\mathbf{1'2'3'}] = [\mathbf{12'3'}][\mathbf{1'2'4'}] - [\mathbf{11'2'}][\mathbf{2'3'4'}].$$

So the results from the initial meet product expansion are either (A.2.9) or 3-termed ones. None of the expansions of expression (d) is a factored one.

Expressions (a) and (b) have the following binomial expansions:

$$[\mathbf{1(1'2'} \vee \mathbf{3'4')(1'2''} \vee \mathbf{2'4'')}] = [\mathbf{12'4''}][\mathbf{1'2'2''}][\mathbf{1'3'4'}] - [\mathbf{11'2''}][\mathbf{1'2'4''}][\mathbf{2'3'4'}],$$
$$[\mathbf{1(1'2'} \vee \mathbf{3'4')(1'2''} \vee \mathbf{3'4'')}] = [\mathbf{13'4'}][\mathbf{1'2'2''}][\mathbf{1'3'4''}] - [\mathbf{11'2'}][\mathbf{1'3'2''}][\mathbf{3'4'4''}].$$

Setting $\mathbf{2''} = \mathbf{3'}$ in (a), we get (c). Setting $\mathbf{4''} = \mathbf{4}$ in (c), we get (d). Setting $\mathbf{2''} = \mathbf{4'}$ and $\mathbf{4''} = \mathbf{2'}$ in (b), we get (d). Since (d) has no factored expansion, neither do (a), (b) and (c).

In the end, if $\mathcal{M}$ has only one element, then the expansions of p_{III} have either three or four terms. If $\mathcal{M}$ is empty, the expansions of p_{III} are all 4-termed. In both cases, p_{III} has no factored expansion, because (d) does not have any. $\square$

Theorem A.2.2. Semifree expansions of $p_{III} = [\mathbf{1(1'2'} \vee \mathbf{3'4')(1''2''} \vee \mathbf{3''4'')}]$:

(1) (Trivially zero) If one of the following conditions is satisfied, p_{III} is trivially zero: (a) one of the pairs, $\mathbf{1'2'}, \mathbf{3'4'}, \mathbf{1''2''}, \mathbf{3''4''}$, is two identical points; (b) one of the 4-tuples, $\{\mathbf{1'}, \mathbf{2'}, \mathbf{3'}, \mathbf{4'}\}$ and $\{\mathbf{1''}, \mathbf{2''}, \mathbf{3''}, \mathbf{4''}\}$, is four collinear points; (c) the four lines, $\mathbf{1'2'}, \mathbf{3'4'}, \mathbf{1''2''}, \mathbf{3''4''}$, are concurrent; (d) point $\mathbf{1}$ is identical to one of $\mathbf{1'2'} \cap \mathbf{3'4'}$ and $\mathbf{1''2''} \cap \mathbf{3''4''}$.

In the following items, assume that p_{III} is not trivially zero.

(2) (Inner intersection) If $\mathbf{1'}, \mathbf{2'}, \mathbf{3'}$ are collinear, point $\mathbf{3'}$ is called an *inner intersection* in p_{III}. The following is a factored expansion:
$$[\mathbf{1}(\mathbf{1'2'} \vee \mathbf{3'4'})(\mathbf{1''2''} \vee \mathbf{3''4''})] = [\mathbf{1'2'4'}]\mathbf{13'} \vee \mathbf{1''2''} \vee \mathbf{3''4''}. \tag{A.2.10}$$
If the meet product on the right side of (A.2.10) has a monomial expansion, then (A.2.10) produces a shortest expansion of p_{III}; else, (A.2.10) leads to a shortest expansion of p_{III} under the additional condition that none of the points $\mathbf{1'}, \mathbf{2'}, \mathbf{4'}$ is on two of the three lines $\mathbf{13'}, \mathbf{1''2''}, \mathbf{3''4''}$.

(3) (Outer intersection) If $\mathbf{1''} = \mathbf{1'2'} \cap \mathbf{3'4'}$, it is called an *outer intersection* in p_{III}. Then
$$\begin{aligned}[\mathbf{1}(\mathbf{1'2'} \vee \mathbf{3'4'})(\mathbf{1''2''} \vee \mathbf{3''4''})] = {} & [\mathbf{11''2''}]\mathbf{1'2'} \vee \mathbf{3'4'} \vee \mathbf{3''4''} \\ = {} & -[\mathbf{1''3''4''}]\mathbf{12''} \vee \mathbf{1'2'} \vee \mathbf{3'4'}. \end{aligned} \tag{A.2.11}$$
If one of the expressions on the right side of (A.2.11) can be expanded into a bracket monomial, then it leads to a shortest expansion of p_{III}; else, the shortest expansions of p_{III} are 2-termed if p_{III} has no inner intersection.

(4) (Double line) If $\mathbf{1'}, \mathbf{2'}, \mathbf{1''}, \mathbf{2''}$ are collinear, line $\mathbf{1'2'}$ is called a *double line* of p_{III}. The following are two factored expansions, and lead to at least one shortest expansion of p_{III} into a bracket polynomial:
$$\begin{aligned}[\mathbf{1}(\mathbf{1'2'} \vee \mathbf{3'4'})(\mathbf{1''2''} \vee \mathbf{3''4''})] = {} & [\mathbf{11'2'}]\mathbf{1''2''} \vee \mathbf{3'4'} \vee \mathbf{3''4''} \\ = {} & [\mathbf{11''2''}]\mathbf{1'2'} \vee \mathbf{3'4'} \vee \mathbf{3''4''}. \end{aligned} \tag{A.2.12}$$

(5) (Recursion of 1) If $\mathbf{1}, \mathbf{1'}, \mathbf{2'}$ are collinear, we say point $\mathbf{1}$ *recurs*. Then
$$[\mathbf{1}(\mathbf{1'2'} \vee \mathbf{3'4'})(\mathbf{1''2''} \vee \mathbf{3''4''})] = [\mathbf{13'4'}]\mathbf{1'2'} \vee \mathbf{1''2''} \vee \mathbf{3''4''}. \tag{A.2.13}$$
If the meet product on the right side of (A.2.13) has a monomial expansion, then (A.2.13) leads to a shortest expansion of p_{III}; else, it leads to a shortest expansion of p_{III} under the additional condition that p_{III} has no double line, and none of the points $\mathbf{1}, \mathbf{3'}, \mathbf{4'}$ is on two of the three lines $\mathbf{1'2'}, \mathbf{1''2''}, \mathbf{3''4''}$.

(6) If p_{III} has neither inner intersection, nor outer intersection, nor double line, and point $\mathbf{1}$ does not recur, then any expansion of p_{III} has at least two terms.

(7) Under the hypotheses in the previous item, there are only three cases in which p_{III} has factored expansions:

(a) (Diagonal) If $\mathbf{1}, \mathbf{1''}, \mathbf{3''}$ are collinear, we say $\mathbf{1}$ is on *diagonal* $\mathbf{1''3''}$. Then
$$\begin{aligned}& [\mathbf{1}(\mathbf{1''4''} \vee \mathbf{3''4'})(\mathbf{1''2''} \vee \mathbf{3''4''})] \\ = {} & [\mathbf{1''3''4''}](\mathbf{14'} \vee \mathbf{1''2''} \vee \mathbf{3''4''} - [\mathbf{12''3''}][\mathbf{1''4''4'}]) \\ = {} & [\mathbf{1''3''4''}]([\mathbf{11''4'}][\mathbf{2''3''4''}] + \mathbf{12''} \vee \mathbf{1''4''} \vee \mathbf{3''4'}). \end{aligned} \tag{A.2.14}$$

(b) (Quadrilateral) If $1 = 1''3'' \cap 2''4''$, then

$$[1(1''4'' \vee 3'4')(1''2'' \vee 3''4'')] = [11''4'']([1''3'4'][2''3''4''] + [1''2''3''][4''3'4']). \quad (A.2.15)$$

(c) (Complete quadrilateral) If $1 = 1''3'' \cap 2''4''$, then

$$[1(1'4'' \vee 2''3'')(1''2'' \vee 3''4'')] = 2[11''4''][1''2''3''][2''3''4'']. \quad (A.2.16)$$

Proof. Item 1 is trivial.

2. If $p_{III} = 0$, one of the two factors in (A.2.10) must be zero. Since (A.2.10) has at most two terms, we only need to prove that if (A.2.10) has only 2-termed expansions, then any other expansion of p_{III} has at least two terms. This hypothesis indicates that (i) $1', 2', 4'$ are not collinear, (ii) $1 \neq 3'$, (iii) $1, 3'$ are not on lines $1''2'', 3''4''$, (iv) $1'', 2'', 3'', 4''$ are not on line $13'$, (v) $1'', 2''$ are not on line $3''4''$, (vi) $3'', 4''$ are not on line $1''2''$.

By symmetry, the meet product expansion of p_{III} also gives two other results:

$$p_{III} = \ [1'3'4']12' \vee 1''2'' \vee 3''4'' - [2'3'4']11' \vee 1''2'' \vee 3''4'' \quad (A.2.17)$$

$$= -[1''2''4'']13'' \vee 1'2' \vee 3'4' + [1''2''3'']14'' \vee 1'2' \vee 3'4'. \quad (A.2.18)$$

The outer product expansion gives two more results:

$$p_{III} = \ [13'4']1'2' \vee 1''2'' \vee 3''4'' - [11'2']3'4' \vee 1''2'' \vee 3''4'' \quad (A.2.19)$$

$$= -[13''4'']1'2' \vee 3'4' \vee 1''2'' + [11''2'']1'2' \vee 3'4' \vee 3''4''. \quad (A.2.20)$$

(A.2.17): (a) If $3' = 1'$ or $2'$, the expansion gives the same result as (A.2.10), so we can assume that $3' \neq 1', 2'$. Then $[1'3'4'], [2'3'4'] \neq 0$.

(b) If $1 = 2'$, the expansion gives $[2'3'4']1'2' \vee 1''2'' \vee 3''4''$, which has a monomial expansion if and only if $1'$ is on either $1''2''$ or $3''4''$. This is forbidden by the assumption in item 1, because at the same time $1'$ is on line $13'$. So we can assume that $1 \neq 2'$.

(c) In $12' \vee 1''2'' \vee 3''4''$, the three lines are pairwise distinct. By the assumption in item 1, $2' \neq 1''2'' \cap 3''4''$, so $12' \vee 1''2'' \vee 3''4'' \neq 0$. Similarly, in (A.2.17), the other meet product $11' \vee 1''2'' \vee 3''4'' \neq 0$.

Based on the above arguments, any expansion of (A.2.17) into a bracket polynomial has at least two terms. Since $3' \neq 1, 1'', 2'', 3'', 4''$, bracket $[2'3'4']$ cannot occur in any expansion of $12' \vee 1''2'' \vee 3''4''$, so there are no intergroup like terms.

(A.2.18): Since $13'' \vee 1'2' \vee 3'4' = [13''3'][1'2'4'] \neq 0$, and $14'' \vee 1'2' \vee 3'4' = [14''3'][1'2'4'] \neq 0$, (A.2.18) has at least two nonzero terms. We only need to prove that there are no intergroup like terms in the expansions.

For $[1''2''3'']$ to occur in an expansion of $13'' \vee 1'2' \vee 3'4'$, there are only two possibilities: it is identical to either $[1'2'4']$ or $[1'2'3'']$ up to sign. Since $3'$ is not on line $1''2''$, we have $1'2' \neq 1''2''$. Thus both are impossible.

(A.2.19): (a) If $[11'2'] = 0$, then $[13'4'] \neq 0$. Since $1', 2'$ cannot be on any of the lines $1''2'', 3''4''$, the meet product $1'2' \vee 1''2'' \vee 3''4''$ neither is zero nor has any monomial expansion. So we can assume that $[11'2'], [13'4'] \neq 0$.

(b) Since $1', 2', 4' \neq 1''2'' \cap 3''4''$, the two meet products in (A.2.19) are both nonzero.

(c) Since 1 is not on any of the lines $1'2', 3'4', 1''2'', 3''4''$, (A.2.19) does not have any intergroup like terms when expanded into a bracket polynomial.

Based on the above arguments, we conclude that any expansion of (A.2.19) into a bracket polynomial has at least two terms.

(A.2.20): Since $1'2' \vee 3'4' \vee 1''2'' = [1'2'4'][3'1''2''] \neq 0$, and $1'2' \vee 3'4' \vee 3''4'' = [1'2'4'][3'3''4''] \neq 0$, (A.2.17) has at least two nonzero terms. We only need to consider those expansions with intergroup like terms.

For $[11''2'']$ to occur in an expansion of $1'2' \vee 3'4' \vee 1''2''$, there are three possibilities:

(a) one of $1', 2', 4'$ equals 1;
(b) $1'2'1''$ or $1'2'2''$ equals $11''2''$;
(c) $1'2'4' = 11''2''$.

In (b), $2''$ or $1''$ equals one of $1', 2'$, and $13'$ is just line $1'2'$, so $13' \cap 1''2''$ equals one of $1', 2'$, contradicting with the assumption of item 2. In (c), by symmetry we only need to consider the case where $1' = 1$, $2' = 1''$ and $4' = 2''$. Then $13'$ is just line $1'2'$, whose intersection with $1''2''$ is $2'$, again contradicting with the assumption of item 2.

So there are only two cases to consider for (A.2.20):

(a1) $1 = 1'$,
(a2) $1 = 4'$.

In (a1), after canceling a pair of like terms in the expansion result, we get $[13'4']([11''2''][2'3''4''] - [2'1''2''][13''4''])$, which is a monomial only when $2'$ is on lines $1''2''$ or $3''4''$. This is forbidden by the assumption of item 2, because $2'$ is at the same time on line $13'$. In (a2), after canceling a pair of like terms in the expansion result, we get $[11'2']([11''2''][3'3''4''] - [13''4''][3'1''2''])$, which has no more like terms and all brackets of which are nonzero. The conclusion is that all expansion results from (A.2.20) with intergroup like terms have at least two terms left after the like terms are combined.

$3 - 5$. The proofs are similar.

6. Let $\mathcal{M}' = \{1'', 2'', 3'', 4''\}$. By symmetry we only need to consider the two expansions (A.2.1) and (A.2.2) of p_{III}. If an expansion starting from (A.2.1) or (A.2.2) does not have intergroup like terms, then since the meet products in (A.2.1) and (A.2.2) are all nonzero, the expansion produces at least two terms. So we only need to consider those expansions having like terms.

(A.2.1) does not have intergroup like terms, because 1 does not recur. In (A.2.2), if an expansion of $12' \vee 1''2'' \vee 3''4''$ produces $[2'3'4']$, the bracket must be formed by three points in $\mathcal{M}'$. Similarly, if an expansion of $11' \vee 1''2'' \vee 3''4''$ produces $[1'3'4']$, the bracket must also be formed by three points in $\mathcal{M}'$. Thus $1', 2', 3', 4'$ must be a permutation of $1'', 2'', 3'', 4''$. There is only one pattern allowed: $1'', 2'', 3'', 4'' =$

$1', 3', 2', 4'$ respectively. Expanding p_{III} by an initial meet product expansion and combining like terms, we get a bracket polynomial of three terms, which does not have any bracket factor.

7. The only case in which an expansion of p_{III} has like terms is already considered in the proof of the previous item. Below we simply assume that the expansions under consideration have no like terms. We establish the following conclusions:

(a). Any two expansions of $\mathbf{1'2'} \vee \mathbf{1''2''} \vee \mathbf{3''4''}$ and $\mathbf{3'4'} \vee \mathbf{1''2''} \vee \mathbf{3''4''}$ respectively do not have common bracket factors.

(b). If two expansions of $\mathbf{13'} \vee \mathbf{1''2''} \vee \mathbf{3''4''}$ and $\mathbf{14'} \vee \mathbf{1''2''} \vee \mathbf{3''4''}$ respectively, have a common bracket factor, then $\mathbf{1} = \mathbf{1''3''} \cap \mathbf{2''4''}$ and $\{\mathbf{3'}, \mathbf{4'}\} \subset \mathcal{M}'$.

(c). If an expansion of $\mathbf{12'} \vee \mathbf{1''2''} \vee \mathbf{3''4''}$ produces a bracket factor $[\mathbf{2'3'4'}]$, then the bracket must be formed by three points in $\mathcal{M}'$, and the fourth point in $\mathcal{M}'$ must be collinear with $\mathbf{1}, \mathbf{2'}$.

These conclusions can be proved by comparing all the expansion results of the meet product expressions. The proofs are omitted. By conclusion (a), (A.2.1) has no factored expansion. By conclusion (b), by symmetry we can assume that $\mathbf{1'} = \mathbf{1''}$. Then $\mathbf{2'} = \mathbf{4''}$. The factored expansion is given by (A.2.15). The polynomial factor in (A.2.15) has like terms if and only if $\mathbf{3'4'} = \mathbf{2''3''}$, in which case the expansion is (A.2.16). By conclusion (c), without loss of generality, we assume that $\mathbf{2'3'4'} = \mathbf{1''2''3''}$, and more specifically, assume $\mathbf{2'} = \mathbf{1''}$, $\mathbf{3'} = \mathbf{2''}$, and $\mathbf{4'} = \mathbf{3''}$. Conclusion (c) requires that $\mathbf{4''}, \mathbf{1}, \mathbf{1''}$ be collinear, *i.e.*, $\mathbf{1}$ be on diagonal $\mathbf{1''4''}$. Then p_{III} has only two factored expansions. They can be obtained from the two meet product expansions which separate $\mathbf{3''}, \mathbf{4'}$ and $\mathbf{1''}, \mathbf{2''}$ respectively. $\qquad\square$

A.3 Cayley expansions of p_{IV}

For expression $p_{IV} = [(\mathbf{12} \vee \mathbf{34})(\mathbf{1'2'} \vee \mathbf{3'4'})(\mathbf{1''2''} \vee \mathbf{3''4''})]$, there are three initial outer product expansions. For example, the outer product expansion by distributing $\mathbf{12}, \mathbf{34}$ is

$$p_{IV} = (\mathbf{12} \vee \mathbf{1'2'} \vee \mathbf{3'4'})(\mathbf{34} \vee \mathbf{1''2''} \vee \mathbf{3''4''}) - (\mathbf{34} \vee \mathbf{1'2'} \vee \mathbf{3'4'})(\mathbf{12} \vee \mathbf{1''2''} \vee \mathbf{3''4''}).$$
$$(\text{A.3.1})$$

There are also six initial meet product expansions. For example, the meet product expansion by separating $\mathbf{1}, \mathbf{2}$ is

$$p_{IV} = [\mathbf{134}][\mathbf{2}(\mathbf{1'2'} \vee \mathbf{3'4'})(\mathbf{1''2''} \vee \mathbf{3''4''})] - [\mathbf{234}][\mathbf{1}(\mathbf{1'2'} \vee \mathbf{3'4'})(\mathbf{1''2''} \vee \mathbf{3''4''})].$$
$$(\text{A.3.2})$$

Proposition A.3.3. If all points in p_{IV} are generic, then p_{IV} has 16847 different expansion results into bracket polynomials.

Proof. Denote the initial meet product expansion separating $\mathbf{i}, \mathbf{j}$ by h_{ij}. By (A.3.2), $h_{\mathbf{12}}$ has $46^2 = 2116$ different expansions into bracket polynomials. Similarly, h_{34} has 2116 expansion results, none of which occurs in h_{12}. Starting from h_{12}, there are the following expansion results which are also the results of $h_{1'2'}$:

$$[\mathbf{134}][\mathbf{1'3'4'}]\mathbf{22'} \vee \mathbf{1''2''} \vee \mathbf{3''4''} - [\mathbf{134}][\mathbf{2'3'4'}]\mathbf{21'} \vee \mathbf{1''2''} \vee \mathbf{3''4''}$$
$$-[\mathbf{234}][\mathbf{1'3'4'}]\mathbf{12'} \vee \mathbf{1''2''} \vee \mathbf{3''4''} + [\mathbf{234}][\mathbf{2'3'4'}]\mathbf{11'} \vee \mathbf{1''2''} \vee \mathbf{3''4''}. \tag{A.3.3}$$

(A.3.3) has $3^4 = 81$ different expansions into bracket polynomials. It has only one expansion producing both $[\mathbf{1''3''4''}]$ and $[\mathbf{2''3''4''}]$.

According to the above account, the initial meet product expansions of p_{IV} lead to all together $6 \times 2116 - 3 \times 4 \times 81 + 2 \times 4 \times 1 = 11732$ different expansions into bracket polynomials.

Next consider the initial outer product expansions of p_{IV}. As a corollary of Proposition 2.58, the number of different expansions of (A.3.1) into bracket polynomials is $45^2 = 2025$. The following expansion of (A.3.1) also belongs to $h_{3'4'}$:

$$[\mathbf{123'}][\mathbf{1'2'4'}]\mathbf{34} \vee \mathbf{1''2''} \vee \mathbf{3''4''} - [\mathbf{124'}][\mathbf{1'2'3'}]\mathbf{34} \vee \mathbf{1''2''} \vee \mathbf{3''4''}$$
$$-[\mathbf{343'}][\mathbf{1'2'4'}]\mathbf{12} \vee \mathbf{1''2''} \vee \mathbf{3''4''} + [\mathbf{344'}][\mathbf{1'2'3'}]\mathbf{12} \vee \mathbf{1''2''} \vee \mathbf{3''4''}. \tag{A.3.4}$$

(A.3.4) contains $3^4 = 81$ different expansions into bracket polynomials. It has only one expansion producing both $[\mathbf{1''3''4''}]$ and $[\mathbf{2''3''4''}]$.

There are three initial outer product expansions of p_{IV}. No expansion of p_{IV} into a bracket polynomial can start from two of the three initial outer product expansions.

According to the above arguments, there are $3 \times 4 \times 81 = 972$ expansions into bracket polynomials which belong to an outer product expansion and a meet product expansion of p_{IV}, and there are $3 \times 4 = 12$ different expansions which belong to an outer product expansion and two meet product expansions of p_{IV}. The three outer product expansions contribute $3 \times 2025 - 972 + 12 = 5115$ new expansions into bracket polynomials. The total sum of expansions is $11732 + 5115 = 16847$.

$\square$

Proof of Theorem 2.62. For $p_{IV} = [(\mathbf{12} \vee \mathbf{34})(\mathbf{1'2'} \vee \mathbf{3'4'})(\mathbf{1''2''} \vee \mathbf{3''4''})]$, item 1 is trivial. Item 5 will be proved at the end of the proof.

2.

We only need to prove that (2.5.26) leads to a shortest expansion of p_{IV} into a bracket polynomial if any expansion of factor $p_{III} = [\mathbf{1}(\mathbf{1'2'} \vee \mathbf{3'4'})(\mathbf{1''2''} \vee \mathbf{3''4''})]$ in (2.5.26) into a bracket polynomial produces at least three terms.

- If p_{IV} has any double line, there are two possibilities: either $\mathbf{1'2'} = \mathbf{1''2''}$ or $\mathbf{12} = \mathbf{1'2'}$. In the former case, p_{III} has a double line; in the latter case, $\mathbf{1}$ recurs. So p_{III} has an expansion into a bracket polynomial of at most two terms.
- If p_{IV} has a second inner intersection, say $\mathbf{1'} = \mathbf{3'}$, then p_{III} has an inner intersection, and has an expansion into a bracket polynomial of at most two terms.
- If p_{IV} has a triangle, then $\mathbf{1}$ must be a vertex of the triangle. So $\mathbf{1}$ has to recur in p_{III}.

Based on the above observations, we can assume that p_{IV} has only one inner intersection, has neither double line nor triangle. The shortest expansions of p_{III} into bracket polynomials have three or four terms if and only if $\mathcal{M} = \{\mathbf{1}', \ldots, \mathbf{4}'\} \cap \{\mathbf{1}'', \ldots, \mathbf{4}''\}$ has one element or is empty, and $\mathbf{1}$ is not in $\{\mathbf{1}', \ldots, \mathbf{4}', \mathbf{1}'', \ldots, \mathbf{4}''\}$. By symmetry, there is only one kind of initial meet product expansion of p_{IV} producing a different result from (2.5.26):

$$p_{IV} = [\mathbf{1}'\mathbf{2}'\mathbf{3}'][\mathbf{4}'(\mathbf{12} \vee \mathbf{14})(\mathbf{1}''\mathbf{2}'' \vee \mathbf{3}''\mathbf{4}'')] - [\mathbf{1}'\mathbf{2}'\mathbf{4}'][\mathbf{3}'(\mathbf{12} \vee \mathbf{14})(\mathbf{1}''\mathbf{2}'' \vee \mathbf{3}''\mathbf{4}'')].$$
$$(A.3.5)$$

By symmetry, there are two kinds of initial outer product expansions producing different results from (2.5.26):

$$\begin{aligned} p_{IV} = \quad & (\mathbf{12} \vee \mathbf{14} \vee \mathbf{3}'\mathbf{4}')(\mathbf{1}'\mathbf{2}' \vee \mathbf{1}''\mathbf{2}'' \vee \mathbf{3}''\mathbf{4}'') & (A.3.6) \\ & -(\mathbf{12} \vee \mathbf{14} \vee \mathbf{1}'\mathbf{2}')(\mathbf{3}'\mathbf{4}' \vee \mathbf{1}''\mathbf{2}'' \vee \mathbf{3}''\mathbf{4}'') \\ = \quad & (\mathbf{12} \vee \mathbf{1}'\mathbf{2}' \vee \mathbf{3}'\mathbf{4}')(\mathbf{14} \vee \mathbf{1}''\mathbf{2}'' \vee \mathbf{3}''\mathbf{4}'') & (A.3.7) \\ & -(\mathbf{14} \vee \mathbf{1}'\mathbf{2}' \vee \mathbf{3}'\mathbf{4}')(\mathbf{12} \vee \mathbf{1}''\mathbf{2}'' \vee \mathbf{3}''\mathbf{4}''). \end{aligned}$$

First assume that none of the above expansions has intergroup like terms. In (A.3.5), $[\mathbf{4}'(\mathbf{12}\vee\mathbf{14})(\mathbf{1}''\mathbf{2}''\vee\mathbf{3}''\mathbf{4}'')] = [\mathbf{124}]\mathbf{4}'\mathbf{1}\vee\mathbf{1}''\mathbf{2}''\vee\mathbf{3}''\mathbf{4}''$ is a shortest expansion, so is $[\mathbf{3}'(\mathbf{12} \vee \mathbf{14})(\mathbf{1}''\mathbf{2}'' \vee \mathbf{3}''\mathbf{4}'')] = [\mathbf{124}]\mathbf{3}'\mathbf{1} \vee \mathbf{1}''\mathbf{2}'' \vee \mathbf{3}''\mathbf{4}''$. They lead (A.3.5) to a shortest expansion of p_{IV} into a bracket polynomial. For the same reason, (A.3.6) is led to a shortest expansion of p_{IV} by $\mathbf{12} \vee \mathbf{14} \vee \mathbf{3}'\mathbf{4}' = [\mathbf{124}][\mathbf{13}'\mathbf{4}']$ and $\mathbf{12} \vee \mathbf{14} \vee \mathbf{1}'\mathbf{2}' = [\mathbf{124}][\mathbf{11}'\mathbf{2}']$. In (A.3.7), If $\mathcal{M}$ has one element, the expansion can only lead to bracket polynomials of at least three terms, as the worst case is to let $\mathbf{2}$ be the element in $\mathcal{M}$ and let $\mathbf{4}$ be in $\{\mathbf{1}', \ldots, \mathbf{4}', \mathbf{1}'', \ldots, \mathbf{4}''\}$. Likewise, if $\mathcal{M}$ is empty, then the expansion can only lead to bracket polynomials of at least four terms.

Now we consider the expansions in (A.3.5)–(A.3.7) having intergroup like terms.

(A.3.5): For $[\mathbf{3}'(\mathbf{12} \vee \mathbf{14})(\mathbf{1}''\mathbf{2}'' \vee \mathbf{3}''\mathbf{4}'')]$ to produce $[\mathbf{1}'\mathbf{2}'\mathbf{3}']$ by expansion, one of $\mathbf{2}, \mathbf{4}$ must be $\mathbf{1}'$ or $\mathbf{2}'$, and the two meet products $\mathbf{12} \vee \mathbf{14}$ and $\mathbf{1}''\mathbf{2}'' \vee \mathbf{3}''\mathbf{4}''$ must be expanded by meet product expansions. Since $\mathbf{12} \vee \mathbf{14} = [\mathbf{124}]\mathbf{1}$, bracket $[\mathbf{1}'\mathbf{2}'\mathbf{3}']$ cannot occur in the expansion result. So (A.3.5) has no like terms.

(A.3.6): If there are any like terms, the two meet products $\mathbf{12} \vee \mathbf{14} \vee \mathbf{1}'\mathbf{2}'$ and $\mathbf{12}\vee\mathbf{14}\vee\mathbf{3}'\mathbf{4}'$ must have like terms when expanded. Comparing all their expansions,

$$\mathbf{12} \vee \mathbf{14} \vee \mathbf{1}'\mathbf{2}' = [\mathbf{124}][\mathbf{11}'\mathbf{2}'] = [\mathbf{121}'][\mathbf{142}'] - [\mathbf{122}'][\mathbf{141}'],$$
$$\mathbf{12} \vee \mathbf{14} \vee \mathbf{3}'\mathbf{4}' = [\mathbf{124}][\mathbf{13}'\mathbf{4}'] = [\mathbf{123}'][\mathbf{144}'] - [\mathbf{123}'][\mathbf{144}'],$$

we conclude that there cannot be any like terms between their expansion results.

(A.3.7): By symmetry, we only need to consider the following outer product expansions which produce different results from the meet product expansions, where the four terms of each result are divided into two braced groups according to the two terms of (A.3.7) producing them:

Case 2.1. $\{[\mathbf{11}'\mathbf{2}'][\mathbf{23}'\mathbf{4}']\mathbf{14} \vee \mathbf{1}''\mathbf{2}'' \vee \mathbf{3}''\mathbf{4}'' - [\mathbf{13}'\mathbf{4}'][\mathbf{21}'\mathbf{2}']\mathbf{14} \vee \mathbf{1}''\mathbf{2}'' \vee \mathbf{3}''\mathbf{4}''\}$
$-\{[\mathbf{11}''\mathbf{2}''][\mathbf{23}''\mathbf{4}'']\mathbf{14} \vee \mathbf{1}'\mathbf{2}' \vee \mathbf{3}'\mathbf{4}' - [\mathbf{13}''\mathbf{4}''][\mathbf{21}''\mathbf{2}'']\mathbf{14} \vee \mathbf{1}'\mathbf{2}' \vee \mathbf{3}'\mathbf{4}'\}$.

Case 2.2. $\{[11'2'][23'4']14 \vee 1''2'' \vee 3''4'' - [13'4'][21'2']14 \vee 1''2'' \vee 3''4''\}$
$-\{[122''][1''3''4'']14 \vee 1'2' \vee 3'4' - [121''][2''3''4'']14 \vee 1'2' \vee 3'4'\}.$

Case 2.3. $\{[122'][1'3'4']14 \vee 1''2'' \vee 3''4'' - [121'][2'3'4']14 \vee 1''2'' \vee 3''4''\}$
$-\{[122''][1''3''4'']14 \vee 1'2' \vee 3'4' - [121''][2''3''4'']14 \vee 1'2' \vee 3'4'\}.$

Case 2.4. $\{[11'2'][23'4']14 \vee 1''2'' \vee 3''4'' - [21'2'][13'4']14 \vee 1''2'' \vee 3''4''\}$
$-\{[11'2'][43'4']12 \vee 1''2'' \vee 3''4'' - [41'2'][13'4']12 \vee 1''2'' \vee 3''4''\}.$

Case 2.5. $\{[11'2'][23'4']14 \vee 1''2'' \vee 3''4'' - [21'2'][13'4']14 \vee 1''2'' \vee 3''4''\}$
$-\{[1'2'4'][143']12 \vee 1''2'' \vee 3''4'' - [1'2'3'][144']12 \vee 1''2'' \vee 3''4''\}.$

Case 2.6. $\{[1'2'4'][123']14 \vee 1''2'' \vee 3''4'' - [1'2'3'][124']14 \vee 1''2'' \vee 3''4''\}$
$-\{[1'3'4'][142']12 \vee 1''2'' \vee 3''4'' - [2'3'4'][141']12 \vee 1''2'' \vee 3''4''\}.$

Case 2.1: For $[11'2']$ to be produced from the second group, there are two cases: either $4 = 1'$ or $4 = 2'$. Similarly, for $[23'4']$ to occur in an expansion of the second group, there are two cases: either $2 = 1'$ or $2 = 2'$. By symmetry we consider only $4 = 2'$ and $2 = 1'$. Then the second group cannot produce the product $[11'2'][23'4'] = [124][23'4']$. So there are no like terms.

Case 2.2: For $[23'4']$ to occur in an expansion of the second group, there are two cases: either $2 = 1'$ or $2 = 2'$. By symmetry, assume $2 = 1'$. For $[11'2'] = [122']$ to occur in an expansion of the second group, $2'$ must equal $2''$, which is the only element in $\mathcal{M}$. The result of the expansion after canceling like terms is

$$-[122'][141''][23'4'][2'3''4''] - [124][122'[2'3'4'][1''3''4''] \\ + [121''][2'3''4'']14 \vee 22' \vee 3'4'. \tag{A.3.8}$$

Expanding (A.3.8) produces no like terms. Since 4 can be identified with an element of $\{1'', 2', 3', 4'\}$, any expansion of (A.3.8) has at least three terms.

Case 2.3: By the symmetry of the expansion, we can assume that the first terms of the two groups have like terms when expanded. Then $[1'3'4']$ can only occur in an expansion result of $14 \vee 1'2' \vee 3'4'$, and $[1''3''4'']$ can only occur in an expansion result of $14 \vee 1''2'' \vee 3''4''$. As a consequence, $[122']$ must be identical to $[122'']$. So $2' = 2''$, which is the only element in $\mathcal{M}$. The result of the expansion after canceling like terms is

$$-[122'][141''][1'3'4'][2'3''4''] - [121'][2'3'4']14 \vee 1''2' \vee 3''4'' \\ +[122'][141'][2'3'4'][1''3''4''] + [121''][2'3''4'']14 \vee 1'2' \vee 3'4'. \tag{A.3.9}$$

Expanding (A.3.9) produces no like terms. By analyzing all possibilities of $\{2, 4\} \subset \{1', 2', 3', 4', 1'', 3'', 4''\}$, we conclude that the shortest expansions of (A.3.9) all have three terms.

Case 2.4: The second group produces neither $[21'2']$ nor $[23'4']$ by expansions, so there are no like terms.

Case 2.5: The second group cannot produce bracket $[\mathbf{23'4'}]$ by expansions. There are two cases for $[\mathbf{21'2'}]$ to occur in an expansion result of the second group: either $\mathbf{2} = \mathbf{3'}$ or $\mathbf{2} = \mathbf{4'}$. By symmetry we consider only $\mathbf{2} = \mathbf{4'}$. For like terms to occur between the second term in the first group and the first term in the second group when both are expanded, $[\mathbf{13'4'}] = -[\mathbf{123'}]$ must be in an expansion result of $\mathbf{12} \vee \mathbf{1''2''} \vee \mathbf{3''4''}$. Assume that $\mathbf{1''} = \mathbf{3'}$. This is the only element in $\mathcal{M}$. The result of the expansion after canceling like terms is

$$[\mathbf{21'2'}][\mathbf{3'3''4''}]([\mathbf{123'}][\mathbf{142''}] - [\mathbf{143'}][\mathbf{122''}]) - [\mathbf{124}][\mathbf{1'2'3'}]\mathbf{12} \vee \mathbf{3'2''} \vee \mathbf{3''4''}.$$
$$(\text{A.3.10})$$

(A.3.10) does not produce any like terms when expanded. The minimal number of terms in any expansion result of (A.3.10) is three, which can be reached by setting $\mathbf{4} = \mathbf{2''}$ or $\mathbf{3'}$.

Case 2.6: The second group cannot produce any of the brackets $[\mathbf{1'2'3'}], [\mathbf{1'2'4'}]$ by expansions, so there are no like terms.

This finishes the proof of item 2.

3.

(i) When the shortest expansions of the right side of (2.5.27) are 2-termed, they are obviously the shortest.

(ii) When the shortest expansions of the right side of (2.5.27) are 3-termed, then the two meet products must be identical. By symmetry we can assume that $\mathbf{34} = \mathbf{1''2''}$ and $\mathbf{3'4'} = \mathbf{3''4''}$. Then p_{IV} has three double lines. By symmetry there is only one kind of meet product expansion:

$$[(\mathbf{12} \vee \mathbf{34})(\mathbf{12} \vee \mathbf{3'4'})(\mathbf{34} \vee \mathbf{3'4'})] = \begin{array}{l} [\mathbf{134}][2(\mathbf{12} \vee \mathbf{3'4'})(\mathbf{34} \vee \mathbf{3'4'})] \\ -[\mathbf{234}][1(\mathbf{12} \vee \mathbf{3'4'})(\mathbf{34} \vee \mathbf{3'4'})], \end{array} \quad (\text{A.3.11})$$

and there is only one kind of outer product expansion, which is just (2.5.27).

(A.3.11) has no expansion producing intergroup like terms. By $[2(\mathbf{12}\vee\mathbf{3'4'})(\mathbf{34}\vee \mathbf{3'4'})] = [\mathbf{23'4'}]\mathbf{12} \vee \mathbf{34} \vee \mathbf{3'4'}$ and $[1(\mathbf{12} \vee \mathbf{3'4'})(\mathbf{34} \vee \mathbf{3'4'})] = [\mathbf{13'4'}]\mathbf{12} \vee \mathbf{34} \vee \mathbf{3'4'}$, (2.5.27) leads to a shortest expansion of the right side of (A.3.11).

(iii) When there is another double line, say $\mathbf{34} = \mathbf{1''2''}$, by symmetry we only need to consider the following meet product expansions:

$$p_{IV} = [\mathbf{134}][2(\mathbf{12} \vee \mathbf{3'4'})(\mathbf{34} \vee \mathbf{3''4''})] - [\mathbf{234}][1(\mathbf{12} \vee \mathbf{3'4'})(\mathbf{34} \vee \mathbf{3''4''})] \quad (\text{A.3.12})$$

$$= [\mathbf{13'4'}][2(\mathbf{34} \vee \mathbf{3''4''})(\mathbf{12} \vee \mathbf{34})] - [\mathbf{23'4'}][1(\mathbf{34} \vee \mathbf{3''4''})(\mathbf{12} \vee \mathbf{34})] \quad (\text{A.3.13})$$

$$= [\mathbf{124'}][3'(\mathbf{34} \vee \mathbf{3''4''})(\mathbf{12} \vee \mathbf{34})] - [\mathbf{123'}][4'(\mathbf{34} \vee \mathbf{3''4''})(\mathbf{12} \vee \mathbf{34})]. \quad (\text{A.3.14})$$

Assume that $p_{III} = (\mathbf{12} \vee \mathbf{34} \vee \mathbf{3'4'})(\mathbf{12} \vee \mathbf{34} \vee \mathbf{3''4''})$ has only 4-termed expansions. Then $\{\mathbf{3'}, \mathbf{4'}\} \cap \{\mathbf{3''}, \mathbf{4''}\}$ has at most one element.

When none of the expansions (A.3.12)–(A.3.14) produces intergroup like terms when further expanded into bracket polynomials, then (2.5.27) leads to the shortest bracket polynomial among all these expansions. So we only need to consider those expansions with like terms. For (A.3.12), the result from the expansion with like terms is $([\mathbf{134}][\mathbf{23''4''}] - [\mathbf{234}][\mathbf{13''4''}])\mathbf{12} \vee \mathbf{34} \vee \mathbf{3'4'}$, which is a special expansion of the right side of (2.5.27). For (A.3.13) and (A.3.14), there are no intergroup like terms.

(iv) Below we assume that p_{IV} has only one double line, and the right side of (2.5.27) has only 4-termed expansions. Then $\mathcal{N} = \{\mathbf{3}, \mathbf{4}, \mathbf{3'}, \mathbf{4'}\} \cap \{\mathbf{1''}, \mathbf{2''}, \mathbf{3''}, \mathbf{4''}\}$ has at most three elements. By symmetry, there are three kinds of meet product expansions:

$$\begin{aligned}
p_{IV} = \; & [\mathbf{134}][\mathbf{2}(\mathbf{12} \vee \mathbf{3'4'})(\mathbf{1''2''} \vee \mathbf{3''4''})] & (A.3.15)\\
& -[\mathbf{234}][\mathbf{1}(\mathbf{12} \vee \mathbf{3'4'})(\mathbf{1''2''} \vee \mathbf{3''4''})] \\
= \; & [\mathbf{124}][\mathbf{3}(\mathbf{12} \vee \mathbf{3'4'})(\mathbf{1''2''} \vee \mathbf{3''4''})] & (A.3.16)\\
& -[\mathbf{123}][\mathbf{4}(\mathbf{12} \vee \mathbf{3'4'})(\mathbf{1''2''} \vee \mathbf{3''4''})] \\
= \; & [\mathbf{1''2''4''}][\mathbf{3''}(\mathbf{12} \vee \mathbf{34})(\mathbf{12} \vee \mathbf{3'4'})] & (A.3.17)\\
& -[\mathbf{1''2''3''}][\mathbf{4''}(\mathbf{12} \vee \mathbf{34})(\mathbf{12} \vee \mathbf{3'4'})],
\end{aligned}$$

and there is one outer product expansion leading to different results from (2.5.27):

$$p_{IV} = (\mathbf{12} \vee \mathbf{34} \vee \mathbf{1''2''})(\mathbf{12} \vee \mathbf{3'4'} \vee \mathbf{3''4''}) - (\mathbf{12} \vee \mathbf{34} \vee \mathbf{3''4''})(\mathbf{12} \vee \mathbf{3'4'} \vee \mathbf{1''2''}). \quad (A.3.18)$$

First we assume that none of the above expansions has intergroup like terms.

(A.3.15) and (A.3.17): (2.5.27) leads to a shortest bracket polynomial among the expansions of (A.3.15) and (A.3.17).

(A.3.16): $[\mathbf{3}(\mathbf{12} \vee \mathbf{3'4'})(\mathbf{1''2''} \vee \mathbf{3''4''})]$ and $[\mathbf{4}(\mathbf{12} \vee \mathbf{3'4'})(\mathbf{1''2''} \vee \mathbf{3''4''})]$ are not trivially zero, have neither inner intersection nor double line.

Case 3.4.1. If none of $\mathbf{3}, \mathbf{4}$ occurs in $\{\mathbf{1''}, \mathbf{2''}, \mathbf{3''}, \mathbf{4''}\}$, then expanding (A.3.16) leads to bracket polynomials of at least four terms.

Case 3.4.2. If both $\mathbf{3}$ and $\mathbf{4}$ occur in $\{\mathbf{1''}, \mathbf{2''}, \mathbf{3''}, \mathbf{4''}\}$, by symmetry we can assume that $\mathbf{3} = \mathbf{1''}$ and $\mathbf{4} = \mathbf{3''}$. Then (A.3.16) has the following expansion whose further expansion into a bracket polynomial is the shortest among all the expansions of (A.3.16):

$$-[\mathbf{124}][\mathbf{344''}]\mathbf{12} \vee \mathbf{3'4'} \vee \mathbf{32''} + [\mathbf{123}][\mathbf{342''}]\mathbf{12} \vee \mathbf{3'4'} \vee \mathbf{44''}. \quad (A.3.19)$$

If either $\mathbf{4'} = \mathbf{4''}$ or $\mathbf{3'} = \mathbf{2''}$, then expanding (A.3.19) leads to at least three terms; if none is satisfied, then expanding (A.3.19) leads to four terms.

Case 3.4.3. If $\mathbf{3} = \mathbf{1''}$ but $\mathbf{4} \neq \mathbf{2''}, \mathbf{3''}, \mathbf{4''}$, then (A.3.16) has the following expansion whose further expansion into a bracket polynomial is the shortest among all the expansions of (A.3.16):

$$-[\mathbf{124}][\mathbf{33''4''}]\mathbf{12} \vee \mathbf{3'4'} \vee \mathbf{32''} - [\mathbf{123}][\mathbf{4}(\mathbf{12} \vee \mathbf{3'4'})(\mathbf{32''} \vee \mathbf{3''4''})]. \quad (A.3.20)$$

If $\mathcal{M} = \{\mathbf{3'}, \mathbf{4'}\} \cap \{\mathbf{2''}, \mathbf{3''}, \mathbf{4''}\}$ is empty, then (A.3.20) has at least six terms when expanded into a bracket polynomial. If $\mathcal{M}$ has two elements, by symmetry there is only one case: $\mathbf{3'} = \mathbf{2''}$ and $\mathbf{4'} = \mathbf{3''}$. Then (A.3.20) has a 3-termed expansion. If

$\mathcal{M}$ has one element, any expansion of (A.3.20) leads to a bracket polynomial of at least four terms.

(A.3.18): $\mathbf{12 \vee 34 \vee 1''2''}$ has a monomial expansion if and only if $\{\mathbf{3, 4}\} \cap \{\mathbf{1'', 2''}\}$ has one element. It is obvious that the shortest expansions of (A.3.18) have three or at least four terms if and only if $\{\mathbf{3, 4, 3', 4'}\} \cap \{\mathbf{1'', 2'', 3'', 4''}\}$ has three or at most two elements.

Now we consider expansions producing intergroup like terms.

(A.3.15): There are no like terms, because $[\mathbf{234}]$ cannot occur in any expansion result of $[\mathbf{2(12 \vee 3'4')(1''2'' \vee 3''4'')}]$.

(A.3.16): The only expansion with like terms is by means of

$$[\mathbf{3(12 \vee 3'4')(1''2'' \vee 3''4'')}] = [\mathbf{33'4'}]\mathbf{12 \vee 1''2'' \vee 3''4''} - [\mathbf{123}]\mathbf{3'4' \vee 1''2'' \vee 3''4''},$$
$$[\mathbf{4(12 \vee 3'4')(1''2'' \vee 3''4'')}] = [\mathbf{43'4'}]\mathbf{12 \vee 1''2'' \vee 3''4''} - [\mathbf{124}]\mathbf{3'4' \vee 1''2'' \vee 3''4''},$$

and the result is $([\mathbf{124}][\mathbf{33'4'}] - [\mathbf{123}][\mathbf{43'4'}])\mathbf{12 \vee 1''2'' \vee 3''4''}$, a special expansion of the right side of (2.5.27).

(A.3.17): For $[\mathbf{1''2''3''}]$ to occur in an expansion result of $[\mathbf{3''(12 \vee 34)(12 \vee 3'4')}]$, points $\mathbf{1''}$ and $\mathbf{2''}$ must belong to $\{\mathbf{3, 4}\}$ and $\{\mathbf{3', 4'}\}$ separately. Without loss of generality, assume that $\mathbf{1''} = \mathbf{3}$ and $\mathbf{2''} = \mathbf{3'}$. The result of the expansion after canceling like terms is

$$\begin{aligned}
&-[\mathbf{124}][\mathbf{123'}][\mathbf{33'4''}][\mathbf{34'3''}] - [\mathbf{123}][\mathbf{33'4''}]\mathbf{12 \vee 3'4' \vee 3''4} \\
&+[\mathbf{124}][\mathbf{123'}][\mathbf{33'3''}][\mathbf{34'4''}] + [\mathbf{123}][\mathbf{33'3''}]\mathbf{12 \vee 3'4' \vee 4''4}.
\end{aligned} \tag{A.3.21}$$

Expanding (A.3.21) produces no like terms. If $\mathcal{L} = \{\mathbf{3'', 4''}\} \cap \{\mathbf{4, 4'}\}$ is empty, then expanding (A.3.21) produces at least six terms; else, $\mathcal{L}$ has one element, and expanding (A.3.21) produces at least four terms.

(A.3.18): By symmetry, there are the following expansions which are not covered by the meet product expansions:

Case 3.4.4. $\{[\mathbf{134}][\mathbf{21''2''}]\mathbf{12 \vee 3'4' \vee 3''4''} - [\mathbf{234}][\mathbf{11''2''}]\mathbf{12 \vee 3'4' \vee 3''4''}\}$
     $-\{[\mathbf{124}][\mathbf{33''4''}]\mathbf{12 \vee 3'4' \vee 1''2''} - [\mathbf{123}][\mathbf{43''4''}]\mathbf{12 \vee 3'4' \vee 1''2''}\}.$

Case 3.4.5. $\{[\mathbf{134}][\mathbf{21''2''}]\mathbf{12 \vee 3'4' \vee 3''4''} - [\mathbf{234}][\mathbf{11''2''}]\mathbf{12 \vee 3'4' \vee 3''4''}\}$
     $-\{[\mathbf{123''}][\mathbf{344''}]\mathbf{12 \vee 3'4' \vee 1''2''} - [\mathbf{124''}][\mathbf{343''}]\mathbf{12 \vee 3'4' \vee 1''2''}\}.$

Case 3.4.6. $\{[\mathbf{124}][\mathbf{31''2''}]\mathbf{12 \vee 3'4' \vee 3''4''} - [\mathbf{123}][\mathbf{41''2''}]\mathbf{12 \vee 3'4' \vee 3''4''}\}$
     $-\{[\mathbf{123''}][\mathbf{344''}]\mathbf{12 \vee 3'4' \vee 1''2''} - [\mathbf{124''}][\mathbf{343''}]\mathbf{12 \vee 3'4' \vee 1''2''}\}.$

Case 3.4.7. $\{[\mathbf{121''}][\mathbf{342''}]\mathbf{12 \vee 3'4' \vee 3''4''} - [\mathbf{122''}][\mathbf{341''}]\mathbf{12 \vee 3'4' \vee 3''4''}\}$
     $-\{[\mathbf{123''}][\mathbf{344''}]\mathbf{12 \vee 3'4' \vee 1''2''} - [\mathbf{124''}][\mathbf{343''}]\mathbf{12 \vee 3'4' \vee 1''2''}\}.$

In cases 3.4.4 and 3.4.5, $[\mathbf{134}], [\mathbf{234}]$ cannot occur in the second group. In cases 3.4.6 and 3.4.7, $[\mathbf{343''}], [\mathbf{344''}]$ cannot occur in the first group. So there are no like terms. This finishes the proof of item 3.

4.

(2.5.28) can be obtained by any outer product expansion. Now assume that p_{IV} has neither inner intersection nor double line. The two terms of the bracket binomial factor in the result of (2.5.28) have a bracket in common if and only if the pattern is a quadrilateral one. The two terms of the bracket binomial factor in (2.5.30) are like terms if and only if the pattern is a complete quadrilateral one. (2.5.31) can be obtained by any outer product expansion. The meet product in (2.5.31) has no monomial expansion.

That there are only three further factorizable subpatterns is based on item 5. Since there is no more factored pattern except for the triangle one, if p_{IV} has two triangles, then if the two triangles share a common side, the subpattern is the quadrilateral; if there is no common side, the subpattern is the triangle pair. If p_{IV} has three triangles, the pattern has to be the complete quadrilateral, in which case there are four triangles. There cannot be more than four triangles in p_{IV}.

In the following, assume that p_{IV} is not trivially zero, and has neither inner intersection, nor double line, nor triangle. Then p_{IV} cannot have more than two triple points.

6.

There are only two patterns of p_{IV}:

(1) $[(\mathbf{12} \vee \mathbf{34})(\mathbf{12'} \vee \mathbf{34'})(\mathbf{12''} \vee \mathbf{34''})]$, where $\mathbf{1}, \mathbf{3}$ are triple points and the others are either double or single points;

(2) $[(\mathbf{12} \vee \mathbf{34})(\mathbf{12'} \vee \mathbf{24'})(\mathbf{12''} \vee \mathbf{24''})]$, where $\mathbf{1}, \mathbf{2}$ are triple points and the others are either double or single points.

They have the following binomial outer product expansions:

$$[(\mathbf{12} \vee \mathbf{34})(\mathbf{12'} \vee \mathbf{34'})(\mathbf{12''} \vee \mathbf{34''})] = \begin{aligned}&[\mathbf{123}][\mathbf{134''}][\mathbf{12'2''}][\mathbf{344'}] \\ &+[\mathbf{122'}][\mathbf{134}][\mathbf{132''}][\mathbf{34'4''}],\end{aligned}$$

$$[(\mathbf{12} \vee \mathbf{34})(\mathbf{12'} \vee \mathbf{24'})(\mathbf{12''} \vee \mathbf{24''})] = \begin{aligned}&[\mathbf{122'}][\mathbf{122''}][\mathbf{134}][\mathbf{24'4''}] \\ &-[\mathbf{124'}][\mathbf{124''}][\mathbf{12'2''}][\mathbf{234}].\end{aligned}$$

This finishes the proof of item 6. The two patterns of p_{IV} in this item can be further classified as follows, which will be used in the proof of item 5:

Case 6.1. If there is no double point, then p_{IV} is of the following patterns, where the asterisks represent independent generic points in the plane:

$$[(\mathbf{12} \vee **)(\mathbf{1}*\vee \mathbf{2}*)(\mathbf{1}*\vee \mathbf{2}*)], \quad [(\mathbf{1}*\vee \mathbf{3}*)(\mathbf{1}*\vee \mathbf{3}*)(\mathbf{1}*\vee \mathbf{3}*)]. \tag{A.3.22}$$

Case 6.2. If there is one double point, then p_{IV} is of the following patterns:

$$[(\mathbf{12} \vee \mathbf{3}*)(\mathbf{13} \vee \mathbf{2}*)(\mathbf{1}*\vee \mathbf{2}*)], \quad [(\mathbf{12} \vee \mathbf{3}*)(\mathbf{1}*\vee \mathbf{32})(\mathbf{1}*\vee \mathbf{3}*)]. \tag{A.3.23}$$

Case 6.3. If there are two double points, then p_{IV} is of the following patterns:

$$\begin{aligned}&[(\mathbf{12} \vee \mathbf{34})(\mathbf{13} \vee \mathbf{24})(\mathbf{1}*\vee \mathbf{2}*)], \quad [(\mathbf{12} \vee \mathbf{34})(\mathbf{13} \vee \mathbf{2}*)(\mathbf{1}*\vee \mathbf{24})], \\ &[(\mathbf{12} \vee \mathbf{34})(\mathbf{14} \vee \mathbf{32})(\mathbf{1}*\vee \mathbf{3}*)], \quad [(\mathbf{12} \vee \mathbf{34})(\mathbf{1}*\vee \mathbf{32})(\mathbf{14} \vee \mathbf{3}*)].\end{aligned} \tag{A.3.24}$$

Case 6.4. If there are three double points, then p_{IV} is of the following pattern:

$$[(\mathbf{12} \vee \mathbf{34})(\mathbf{14} \vee \mathbf{35})(\mathbf{15} \vee \mathbf{32})]. \tag{A.3.25}$$

7.

Assume that p_{IV} has only one triple point $\mathbf{1}$.

(i) If p_{IV} has no double point, then it has only one pattern:

$$[(\mathbf{1} * \vee **)(\mathbf{1} * \vee **)(\mathbf{1} * \vee **)], \tag{A.3.26}$$

whose shortest expansions have four terms. If p_{IV} has one double point, there are two patterns:

$$[(\mathbf{12} \vee **)(\mathbf{1} * \vee \mathbf{2}*)(\mathbf{1} * \vee **)], \quad [(\mathbf{1} * \vee \mathbf{2}*)(\mathbf{1} * \vee \mathbf{2}*)(\mathbf{1} * \vee **)]. \tag{A.3.27}$$

Their shortest expansions have three terms.

(ii) If p_{IV} has two double points $\mathbf{2}, \mathbf{3}$, there are the following six patterns:

$$\begin{aligned}
&[(\mathbf{1} * \vee \mathbf{23})(\mathbf{1} * \vee \mathbf{2}*)(\mathbf{1} * \vee \mathbf{3}*)], \quad [(\mathbf{1} * \vee \mathbf{23})(\mathbf{12} \vee **)(\mathbf{1} * \vee \mathbf{3}*)], \\
&[(\mathbf{12} \vee \mathbf{3}*)(\mathbf{1} * \vee \mathbf{2}*)(\mathbf{13} \vee **)], \quad [(\mathbf{12} \vee \mathbf{3}*)(\mathbf{1} * \vee \mathbf{2}*)(\mathbf{1} * \vee \mathbf{3}*)], \\
&[(\mathbf{12} \vee \mathbf{3}*)(\mathbf{13} \vee \mathbf{2}*)(\mathbf{1} * \vee **)], \quad [(\mathbf{12} \vee \mathbf{3}*)(\mathbf{1} * \vee \mathbf{23})(\mathbf{1} * \vee **)].
\end{aligned} \tag{A.3.28}$$

It can be proved that for the last four patterns, the shortest expansions have three terms. For the first two patterns, there are the following binomial expansions:

$$\begin{aligned}
[(\mathbf{12} \vee \mathbf{34})(\mathbf{12}' \vee \mathbf{34}')(\mathbf{12}'' \vee \mathbf{44}'')] =\ &[\mathbf{122}'][\mathbf{134}'][\mathbf{142}''][\mathbf{344}''] \\
&-[\mathbf{122}''][\mathbf{132}'][\mathbf{144}''][\mathbf{344}'], \\
[(\mathbf{12} \vee \mathbf{34})(\mathbf{12}' \vee \mathbf{34}')(\mathbf{14} \vee \mathbf{3}''\mathbf{4}'')] =\ &-[\mathbf{122}'][\mathbf{134}][\mathbf{134}'][\mathbf{43}''\mathbf{4}''] \\
&-[\mathbf{124}][\mathbf{132}'][\mathbf{13}''\mathbf{4}''][\mathbf{344}'].
\end{aligned} \tag{A.3.29}$$

(iii) If p_{IV} has three double points $\mathbf{2}, \mathbf{3}, \mathbf{4}$, there are the following eight patterns:

$$\begin{aligned}
&[(\mathbf{12} \vee \mathbf{34})(\mathbf{13} \vee \mathbf{24})(\mathbf{1} * \vee **)], \quad [(\mathbf{12} \vee \mathbf{34})(\mathbf{13} \vee \mathbf{4}*)(\mathbf{1} * \vee \mathbf{2}*)], \\
&[(\mathbf{12} \vee \mathbf{3}*)(\mathbf{13} \vee \mathbf{4}*)(\mathbf{14} \vee \mathbf{2}*)], \quad [(\mathbf{12} \vee \mathbf{34})(\mathbf{13} \vee \mathbf{2}*)(\mathbf{1} * \vee \mathbf{4}*)], \\
&[(\mathbf{12} \vee \mathbf{34})(\mathbf{1} * \vee \mathbf{23})(\mathbf{14} \vee **)], \quad [(\mathbf{12} \vee \mathbf{34})(\mathbf{1} * \vee \mathbf{23})(\mathbf{1} * \vee \mathbf{4}*)], \\
&[(\mathbf{12} \vee \mathbf{3}*)(\mathbf{1} * \vee \mathbf{24})(\mathbf{1} * \vee \mathbf{34})], \quad [(\mathbf{12} \vee \mathbf{3}*)(\mathbf{14} \vee \mathbf{2}*)(\mathbf{1} * \vee \mathbf{34})].
\end{aligned} \tag{A.3.30}$$

The shortest expansions of the first three patterns each have three terms. The other five patterns belong to the patterns on the left side of (A.3.29).

(iv) If p_{IV} has four double points $\mathbf{2}, \mathbf{3}, \mathbf{4}, \mathbf{5}$, there are the following three patterns:

$$\begin{aligned}
&[(\mathbf{12} \vee \mathbf{3}*)(\mathbf{14} \vee \mathbf{25})(\mathbf{13} \vee \mathbf{45})], \quad [(\mathbf{1} * \vee \mathbf{23})(\mathbf{14} \vee \mathbf{25})(\mathbf{13} \vee \mathbf{45})], \\
&[(\mathbf{1} * \vee \mathbf{23})(\mathbf{14} \vee \mathbf{25})(\mathbf{15} \vee \mathbf{34})].
\end{aligned} \tag{A.3.31}$$

They all belong to the patterns on the left side of (A.3.29).

8.

Let $\mathbf{1}, \mathbf{2}, \mathbf{3}, \mathbf{4}$ be four double points. The 4-2-2 pattern has the following binomial expansions by distributing $\mathbf{12}$ and $\mathbf{34}$:

$$[(12 \vee 34)(13 \vee 3'4')(24 \vee 3''4'')] = \begin{aligned}[t] &[124][134][23''4''][33'4'] \\ &-[123][13'4'][234][43''4''], \end{aligned}$$

$$[(12 \vee 34)(13 \vee 3'4')(22'' \vee 44'')] = \begin{aligned}[t] &[123][13'4'][242''][344''] \\ &+[122''][134][244''][33'4'], \end{aligned}$$

$$[(12 \vee 34)(12' \vee 34')(22'' \vee 44'')] = \begin{aligned}[t] &[122'][134'][234][42''4''] \\ &-[122''][132'][244''][344']. \end{aligned} \tag{A.3.32}$$

The 3-3-2 pattern has the following binomial expansion by separating $\mathbf{1}, \mathbf{2}$:

$$[(12 \vee 56)(13 \vee 46')(24 \vee 36'')] = [134][156][236''][246'] - [136''][146'][234][256]. \tag{A.3.33}$$

If $\mathbf{1}, \mathbf{2}, \mathbf{3}, \mathbf{4}$ are the only double points in p_{IV}, then there are all together thirteen patterns for p_{IV}:

$$\begin{aligned}
&[(12 \vee 34)(13 \vee **)(24 \vee **)], && [(12 \vee 34)(13 \vee **)(2* \vee 4*)], \\
&[(12 \vee 34)(1* \vee 3*)(2* \vee 4*)], && [(12 \vee **)(13 \vee 4*)(24 \vee 3*)], \\
&[(12 \vee 34)(1* \vee 2*)(3* \vee 4*)], && [(12 \vee 34)(13 \vee 24)(** \vee **)], \\
&[(12 \vee 34)(13 \vee 2*)(4* \vee **)], && [(12 \vee 3*)(24 \vee 3*)(1* \vee 4*)], \\
&[(12 \vee 3*)(23 \vee 4*)(14 \vee **)], && [(12 \vee 3*)(23 \vee 4*)(1* \vee 4*)], \\
&[(12 \vee 3*)(34 \vee 2*)(1* \vee 4*)], && [(12 \vee 3*)(14 \vee 2*)(34 \vee **)], \\
&[(12 \vee 3*)(14 \vee 2*)(3* \vee 4*)].
\end{aligned} \tag{A.3.34}$$

It can be proved that the shortest expansions of the above patterns have at least three terms.

If there are five double points $\mathbf{1}, \dots, \mathbf{5}$, they form six patterns of p_{IV}:

$$\begin{aligned}
&[(12 \vee 34)(13 \vee 25)(4* \vee 5*)], && [(12 \vee 34)(1* \vee 25)(3* \vee 45)], \\
&[(12 \vee 34)(13 \vee 25)(45 \vee **)], && [(12 \vee 34)(13 \vee 5*)(24 \vee 5*)], \\
&[(12 \vee 34)(1* \vee 35)(25 \vee 4*)], && [(12 \vee 34)(1* \vee 35)(24 \vee 5*)].
\end{aligned} \tag{A.3.35}$$

The last three patterns belong to the 4-2-2 pattern, and the third pattern belongs to the 3-3-2 pattern. By analyzing all the meet product and outer product expansions, we find that the shortest expansions of the first two patterns have three terms.

If there are six double points $\mathbf{1}, \dots, \mathbf{6}$, they form three patterns of p_{IV}:

$$\begin{aligned}
&[(12 \vee 34)(15 \vee 36)(26 \vee 45)], && [(12 \vee 34)(15 \vee 26)(35 \vee 46)], \\
&[(12 \vee 34)(13 \vee 56)(25 \vee 46)].
\end{aligned} \tag{A.3.36}$$

The first can be called the *Pascal pattern*, as it corresponds to Pascal's Conic Theorem. The three patterns all belong to the 4-2-2 pattern.

If there are three double points $\mathbf{1}, \mathbf{2}, \mathbf{3}$, they form seven patterns of p_{IV}:

$$\begin{aligned}
&[(12 \vee 3*)(13 \vee 2*)(** \vee **)], && [(12 \vee 3*)(13 \vee **)(2* \vee **)], \\
&[(12 \vee 3*)(1* \vee 2*)(3* \vee **)], && [(12 \vee 3*)(1* \vee 3*)(2* \vee **)], \\
&[(1* \vee 2*)(1* \vee 3*)(2* \vee 3*)], && [(12 \vee **)(1* \vee 3*)(2* \vee 3*)], \\
&[(12 \vee **)(13 \vee **)(2* \vee 3*)].
\end{aligned} \tag{A.3.37}$$

These patterns include respectively the following patterns formed by four double points $\mathbf{1, 2, 3, 4}$, by specifying two of the six generic points (asterisks) in each pattern as point $\mathbf{4}$:

$$\begin{aligned}
&[(\mathbf{12} \vee \mathbf{34})(\mathbf{13} \vee \mathbf{24})(** \vee **)], \quad [(\mathbf{12} \vee \mathbf{34})(\mathbf{13} \vee \mathbf{4}*)(\mathbf{2} * \vee **)], \\
&[(\mathbf{12} \vee \mathbf{34})(\mathbf{14} \vee \mathbf{2}*)(\mathbf{3} * \vee **)], \quad [(\mathbf{12} \vee \mathbf{34})(\mathbf{14} \vee \mathbf{3}*)(\mathbf{2} * \vee **)], \\
&[(\mathbf{1} * \vee \mathbf{24})(\mathbf{14} \vee \mathbf{3}*)(\mathbf{2} * \vee \mathbf{3}*)], \quad [(\mathbf{12} \vee \mathbf{4}*)(\mathbf{14} \vee \mathbf{3}*)(\mathbf{2} * \vee \mathbf{3}*)], \\
&[(\mathbf{12} \vee \mathbf{4}*)(\mathbf{13} \vee \mathbf{4}*)(\mathbf{2} * \vee \mathbf{3}*)].
\end{aligned} \tag{A.3.38}$$

Since none of the patterns in (A.3.38) has binomial expansions, nor does any of the patterns in (A.3.37).

If there are two double points $\mathbf{1, 2}$, they form four patterns of p_{IV}:

$$\begin{aligned}
&[(\mathbf{12} \vee **)(\mathbf{1} * \vee \mathbf{2}*)(** \vee **)], \quad [(\mathbf{12} \vee **)(\mathbf{1} * \vee **)(\mathbf{2} * \vee **)], \\
&[(\mathbf{1} * \vee \mathbf{2}*)(\mathbf{1} * \vee \mathbf{2}*)(** \vee **)], \quad [(\mathbf{1} * \vee \mathbf{2}*)(\mathbf{1} * \vee **)(\mathbf{2} * \vee **)].
\end{aligned} \tag{A.3.39}$$

For the first three patterns, the shortest expansions have four terms; for the fourth pattern, the shortest expansions have five terms.

If there is one double point $\mathbf{1}$, there is only one pattern of p_{IV}:

$$[(\mathbf{1} * \vee **)(\mathbf{1} * \vee **)(** \vee **)]. \tag{A.3.40}$$

The shortest expansions of (A.3.40) have six terms.

If there is no double point, any expansion of p_{IV} has eight terms.

5.

If all points in p_{IV} are single, then there is no factored expansion. Below we pick out all the patterns in the proofs of items 6, 7, 8, and verify that none of them has any factored expansion.

(1) Both patterns in (A.3.22) can be *specified* as patterns in (A.3.23), *i.e.*, they become patterns in (A.3.23) when some of their generic points (denoted by asterisks) are specified. So if the patterns in (A.3.23) have no factored expansion, neither do the patterns in (A.3.22).

(2) The two patterns in (A.3.23) can be specified as the following patterns in (A.3.24):

$$[(\mathbf{12} \vee \mathbf{34})(\mathbf{13} \vee \mathbf{24})(\mathbf{1} * \vee \mathbf{2}*)], \quad [(\mathbf{12} \vee \mathbf{34})(\mathbf{1} * \vee \mathbf{32})(\mathbf{14} \vee \mathbf{3}*)].$$

(3) The last pattern in (A.3.24) can be specified as the pattern (A.3.25).

(4) The six patterns in (A.3.28) can be specified as the following patterns in (A.3.23):

$$\begin{aligned}
&[(\mathbf{1} * \vee \mathbf{23})(\mathbf{1} * \vee \mathbf{2}*)(\mathbf{12} \vee \mathbf{3}*)], \quad [(\mathbf{1} * \vee \mathbf{23})(\mathbf{12} \vee \mathbf{3}*)(\mathbf{1} * \vee \mathbf{3}*)], \\
&[(\mathbf{12} \vee \mathbf{3}*)(\mathbf{1} * \vee \mathbf{2}*)(\mathbf{13} \vee \mathbf{2}*)], \quad [(\mathbf{12} \vee \mathbf{3}*)(\mathbf{1} * \vee \mathbf{2}*)(\mathbf{1} * \vee \mathbf{32})], \\
&[(\mathbf{12} \vee \mathbf{3}*)(\mathbf{13} \vee \mathbf{2}*)(\mathbf{1} * \vee \mathbf{2}*)], \quad [(\mathbf{1} * \vee \mathbf{23})(\mathbf{12} \vee \mathbf{3}*)(\mathbf{1} * \vee \mathbf{3}*)].
\end{aligned}$$

(5) The third patterns in (A.3.30) can be specified as the first pattern in (A.3.31). The other seven patterns in (A.3.30) can be specified as the following patterns in

(A.3.24):

$$[(12 \vee 34)(13 \vee 24)(1 * \vee 2*)], \quad [(12 \vee 34)(13 \vee 42)(1 * \vee 2*)],$$
$$[(12 \vee 34)(13 \vee 2*)(1 * \vee 42)],$$
$$[(12 \vee 34)(1 * \vee 23)(14 \vee 3*)], \quad [(12 \vee 34)(14 \vee 23)(1 * \vee 4*)],$$
$$[(12 \vee 3*)(13 \vee 24)(1 * \vee 34)], \quad [(12 \vee 3*)(14 \vee 23)(1 * \vee 34)].$$

(6) The first pattern in (A.3.31) can be specified as the pattern (A.3.25).

(7) The first four patterns in (A.3.34) can be specified as the following patterns in (A.3.35):

$$[(12 \vee 34)(13 \vee 5*)(24 \vee 5*)], \quad [(12 \vee 34)(13 \vee 5*)(25 \vee 4*)],$$
$$[(12 \vee 34)(1 * \vee 35)(25 \vee 4*)], \quad [(12 \vee **)(13 \vee 45)(24 \vee 35)].$$

The last nine patterns in (A.3.34) can be specified as the following patterns in (A.3.30):

$$[(12 \vee 34)(1 * \vee 2*)(31 \vee 4*)], \quad [(12 \vee 34)(13 \vee 24)(1 * \vee **)],$$
$$[(12 \vee 34)(13 \vee 2*)(4 * \vee 1*)], \quad [(12 \vee 3*)(24 \vee 31)(1 * \vee 4*)],$$
$$[(12 \vee 3*)(23 \vee 4*)(14 \vee 3*)], \quad [(12 \vee 34)(23 \vee 4*)(1 * \vee 42)],$$
$$[(12 \vee 3*)(34 \vee 2*)(1 * \vee 4*)], \quad [(12 \vee 3*)(14 \vee 2*)(34 \vee 1*)],$$
$$[(12 \vee 3*)(14 \vee 2*)(31 \vee 4*)].$$

(8) The first five patterns in (A.3.35) can be specified as the following patterns in (A.3.31):

$$[(12 \vee 34)(13 \vee 25)(42 \vee 5*)], \quad [(12 \vee 34)(13 \vee 25)(3 * \vee 45)],$$
$$[(12 \vee 34)(13 \vee 25)(45 \vee 3*)], \quad [(12 \vee 34)(13 \vee 52)(24 \vee 5*)],$$
$$[(12 \vee 34)(14 \vee 35)(25 \vee 4*)].$$

The last pattern in (A.3.35) can be specified as $[(12 \vee 34)(16 \vee 35)(24 \vee 56)]$ in (A.3.36).

(9) The seven patterns in (A.3.37) can be specified as patterns in (A.3.34), which is already used in the proof of item 8.

(10) The four patterns in (A.3.39) can be specified as the following patterns in (A.3.27):

$$[(12 \vee **)(1 * \vee 2*)(1 * \vee **)], \quad [(12 \vee **)(1 * \vee **)(2 * \vee 1*)],$$
$$[(1 * \vee 2*)(1 * \vee 2*)(1 * \vee **)], \quad [(1 * \vee 2*)(1 * \vee **)(2 * \vee 1*)].$$

(11) The pattern in (A.3.40) can be specified as the pattern (A.3.26).

Thus, we only need to consider the pattern (A.3.25), the first three patterns in (A.3.24), the three patterns in (A.3.36), and the last two patterns in (A.3.31). By analyzing all their expansions up to symmetry, we conclude that none of the nine patterns has any factored expansion. $\square$

When there are collinear constraints among the points in p_{IV}, the complete classification of factored expansions and binomial expansions of p_{IV} is extremely difficult. Below are some typical but not exhaustive patterns having factored expansions.

Proposition A.3.4. Some factored semifree expansions of $p_{IV} = [(12 \vee 34)(1'2' \vee 3'4')(1''2'' \vee 3''4'')]$:

(1) (Inner intersection) If $\mathbf{1}, \mathbf{2}, \mathbf{3}$ are collinear, then

$$p_{IV} = [\mathbf{124}][\mathbf{3}(\mathbf{1'2'} \vee \mathbf{3'4'})(\mathbf{1''2''} \vee \mathbf{3''4''})]. \tag{A.3.41}$$

(2) (Outer intersection) If $\mathbf{1'} = \mathbf{12} \cap \mathbf{34}$, then

$$\begin{aligned} p_{IV} &= (\mathbf{12} \vee \mathbf{34} \vee \mathbf{3'4'})(\mathbf{1'2'} \vee \mathbf{1''2''} \vee \mathbf{3''4''}) \\ &= -[\mathbf{1'3'4'}][\mathbf{2'}(\mathbf{12} \vee \mathbf{34})(\mathbf{1''2''} \vee \mathbf{3''4''})]. \end{aligned} \tag{A.3.42}$$

(3) (Double line) If $\mathbf{1}, \mathbf{2}, \mathbf{1'}, \mathbf{2'}$ are collinear, then

$$\begin{aligned} p_{IV} &= (\mathbf{12} \vee \mathbf{34} \vee \mathbf{3'4'})(\mathbf{1'2'} \vee \mathbf{1''2''} \vee \mathbf{3''4''}) \\ &= (\mathbf{1'2'} \vee \mathbf{34} \vee \mathbf{3'4'})(\mathbf{12} \vee \mathbf{1''2''} \vee \mathbf{3''4''}). \end{aligned} \tag{A.3.43}$$

(4) (Generalized triangle) If $\mathbf{1}, \mathbf{1'}, \mathbf{2'}, \mathbf{1''}$ are collinear, and $\mathbf{2}, \mathbf{1''}, \mathbf{2''}$ are collinear, then

$$p_{IV} = [\mathbf{1'2'2''}]([\mathbf{134}][\mathbf{23''4''}][\mathbf{3'4'1''}] - [\mathbf{13'4'}][\mathbf{234}][\mathbf{1''3''4''}]). \tag{A.3.44}$$

The following are some further factorizable generalized triangle patterns:

(a) (Generalized complete quadrilateral) If $\mathbf{1}, \mathbf{3}, \mathbf{1'}, \mathbf{2'}$ are collinear, and $\mathbf{2}, \mathbf{3}, \mathbf{2''}$ are collinear, then

$$[(\mathbf{12} \vee \mathbf{34})(\mathbf{1'2'} \vee \mathbf{24})(\mathbf{32''} \vee \mathbf{14})] = -2[\mathbf{1'2'2''}][\mathbf{124}][\mathbf{134}][\mathbf{234}]. \tag{A.3.45}$$

(b) (Generalized quadrilateral) If $\mathbf{1}, \mathbf{1'}, \mathbf{2'}, \mathbf{1''}$ are collinear, and $\mathbf{2}, \mathbf{1''}, \mathbf{2''}$ are collinear, then

$$\begin{aligned} &[(\mathbf{12} \vee \mathbf{34})(\mathbf{1'2'} \vee \mathbf{24'})(\mathbf{1''2''} \vee \mathbf{14'})] \\ &= [\mathbf{1'2'2''}][\mathbf{124'}]([\mathbf{134}][\mathbf{24'1''}] + [\mathbf{14'1''}][\mathbf{234}]). \end{aligned} \tag{A.3.46}$$

(c) (Generalized triangle pair) If $\mathbf{1}, \mathbf{1'}, \mathbf{2'}, \mathbf{1''}$ are collinear, and $\mathbf{2}, \mathbf{1''}, \mathbf{2''}$ are collinear, then

$$[(\mathbf{12} \vee \mathbf{34})(\mathbf{1'2'} \vee \mathbf{34'})(\mathbf{1''2''} \vee \mathbf{44'})] = [\mathbf{1'2'2''}][\mathbf{344'}]\mathbf{13} \vee \mathbf{24} \vee \mathbf{4'1''}. \tag{A.3.47}$$

(d) (Perspective triangle) If $\mathbf{1}, \mathbf{3}, \mathbf{2''}$ are collinear, then

$$\begin{aligned} &[(\mathbf{12} \vee \mathbf{34})(\mathbf{13} \vee \mathbf{24'})(\mathbf{22''} \vee \mathbf{14'})] \\ &= [\mathbf{123}][\mathbf{124'}]([\mathbf{12''4'}][\mathbf{234}] + [\mathbf{134}][\mathbf{22''4'}]). \end{aligned} \tag{A.3.48}$$

(5) (Perspective pattern) If $\mathbf{1}, \mathbf{3}, \mathbf{2''}$ are collinear, then

$$\begin{aligned} &[(\mathbf{12} \vee \mathbf{34})(\mathbf{13} \vee \mathbf{3'4'})(\mathbf{22''} \vee \mathbf{3''4''})] \\ &= [\mathbf{123}]([\mathbf{134}][\mathbf{23''4''}][\mathbf{2'3'4'}] - [\mathbf{13'4'}][\mathbf{234}][\mathbf{2''3''4''}]). \end{aligned} \tag{A.3.49}$$

(6) (Double perspective pattern) If $\mathbf{2}, \mathbf{3}, \mathbf{1''}$ are collinear, and $\mathbf{1}, \mathbf{2}, \mathbf{2''}$ are collinear, then

$$\begin{aligned} [(\mathbf{12} \vee \mathbf{34})(\mathbf{13} \vee \mathbf{24'})(\mathbf{1''2''} \vee \mathbf{3''4''})] = [\mathbf{123}]&([\mathbf{124'}][\mathbf{1''3''4''}][\mathbf{2'34}] \\ &-[\mathbf{124}][\mathbf{1''34'}][\mathbf{2''3''4''}] \\ &+[\mathbf{2''3''4''}]\mathbf{13} \vee \mathbf{24'} \vee \mathbf{1''4}). \end{aligned} \tag{A.3.50}$$

Remark: The equality (A.3.47) can only be realized by a binomial expansion followed by a degree-3 Cayley factorization. It indicates that if $13 \vee 24 \vee 4'1''$ has a monomial expansion, then this result cannot be obtained from any Cayley expansion of the left side of (A.3.47). However, the result can always be obtained with the aid of collinearity transformations and Cayley factorizations.

Proof of Proposition A.3.4. We only prove items 4–6.

4.

$$
\begin{aligned}
p_{IV} &= [\mathbf{134}][2(1'2' \vee 3'4')(1''2'' \vee 3''4'')] - [\mathbf{234}][1(1'2' \vee 3'4')(1''2'' \vee 3''4'')] \\
&= -[\mathbf{134}][23''4'']1'2' \vee 3'4' \vee 1''2'' - [\mathbf{234}][13'4']1'2' \vee 1''2'' \vee 3''4'' \\
&= [1'2'2'']([\mathbf{134}][23''4''][3'4'1''] - [13'4'][\mathbf{234}][1''3''4'']).
\end{aligned}
$$

(A.3.45) and (A.3.46) are special cases of (A.3.44). (A.3.47) can be proved as follows: by (A.3.44),

$$
[(12\vee34)(1'2'\vee34')(1''2''\vee44')] = [1'2'2'']([\mathbf{134}][244'][34'1''] - [134'][\mathbf{234}][44'1'']).
$$

So

$$
\begin{aligned}
&\quad\; [\mathbf{134}][244'][34'1''] - [134'][\mathbf{234}][44'1''] - [344']13 \vee 24 \vee 4'1'' \\
&= [\mathbf{134}][244'][34'1''] - [134'][\mathbf{234}][44'1''] - [344']([\mathbf{134}][24'1''] + [123][44'1'']) \\
&= [\mathbf{134}]([244'][34'1''] - [344'][24'1'']) - [44'1'']([134'][\mathbf{234}] + [344'][123]) \\
&= [\mathbf{134}][234'][44'1''] - [44'1''][\mathbf{134}][234'] \\
&= 0.
\end{aligned}
$$

(A.3.48) is obtained as follows:

$$
\begin{aligned}
&\quad\; [(12 \vee 34)(13 \vee 24')(22'' \vee 14')] \\
&= (12 \vee 34 \vee 24')(13 \vee 22'' \vee 14') - (12 \vee 34 \vee 13)(24' \vee 22'' \vee 14') \\
&= [123][124']([12''4'][\mathbf{234}] + [\mathbf{134}][22''4']).
\end{aligned}
$$

5.

$$
\begin{aligned}
&\quad\;\; [(12 \vee 34)(13 \vee 3'4')(22'' \vee 3''4'')] \\
&= [23''4''][2''(12 \vee 34)(13 \vee 3'4')] - [2''3''4''][2(12 \vee 34)(13 \vee 3'4')] \\
&= -[23''4''][2''3'4']13 \vee 12 \vee 34 - [\mathbf{234}][2''3''4'']12 \vee 13 \vee 3'4' \\
&= [123]([\mathbf{134}][23''4''][2''3'4'] - [13'4'][\mathbf{234}][2''3''4'']).
\end{aligned}
$$

6.

$$
\begin{aligned}
&\quad\; [(12 \vee 34)(13 \vee 24')(1''2'' \vee 3''4'')] \\
&= [1''3''4''][2''(12 \vee 34)(13 \vee 24')] - [2''3''4''][1''(12 \vee 34)(13 \vee 24')] \\
&= [1''3''4''][2''34][123][124'] - [2''3''4'']([124]1''3 \vee 13 \vee 24' \\
&\qquad\qquad\qquad\qquad\qquad\qquad\quad -[123]1''4 \vee 13 \vee 24') \\
&= [123]([124'][1''3''4''][2''34] - [124][1''34'][2''3''4''] + [2''3''4'']13 \vee 24' \vee 1''4).
\end{aligned}
$$

$\square$

A.4   Cayley expansions of q_I, q_{II} and q_{III}

The Cayley expansions of

$$q_I = [1(1'2' \vee 3'4')\mathbf{A}_{5''6''}]$$

are those of

$$-\lambda_{6''}15'' \vee 1'2' \vee 3'4' - \lambda_{5''}16'' \vee 1'2' \vee 3'4'. \tag{A.4.1}$$

Their classification is similar to that of p_{III}, so the proof of the following proposition is omitted.

Proposition A.4.5. Generic expansions of q_I:

(1) (Trivially zero) q_I is trivially zero if either (a) one of the pairs, $1'2', 3'4'$, is two identical points, or (b) $\{1', 2'\} = \{3', 4'\}$.

Below we assume that q_I is not trivially zero.

(2) (Inner intersection) If $1' = 3'$, the following is the unique shortest expansion:

$$[1(1'2' \vee 1'4')\mathbf{A}_{5''6''}] = [1'2'4'][11'\mathbf{A}_{5''6''}]. \tag{A.4.2}$$

(3) (Double line) If $\{1', 2'\} = \{5'', 6''\}$, the following is the unique shortest expansion:

$$[1(1'2' \vee 3'4')\mathbf{A}_{1'2'}] = [11'2'][3'4'\mathbf{A}_{1'2'}]. \tag{A.4.3}$$

(4) (Recursion of 1) If $1 = 1'$, the following is the unique shortest expansion:

$$[1(12' \vee 3'4')\mathbf{A}_{5''6''}] = [13'4'][12'\mathbf{A}_{5''6''}]. \tag{A.4.4}$$

If $1 = 5''$, the following is the unique expansion leading to the shortest expansion of q_I into a bracket polynomial:

$$[1(1'2' \vee 3'4')\mathbf{A}_{16''}] = -\lambda_1 16'' \vee 1'2' \vee 3'4'. \tag{A.4.5}$$

(5) (Other cases) If q_I is not trivially zero, and has neither inner intersection nor double line, and 1 does not recur, then it has no factored expansion. Its shortest expansions have two, three, or four terms if and only if $\{5'', 6''\} \cap \{1', 2', 3', 4'\}$ has two, one, or no element.

The Cayley expansions of

$$q_{II} = [(12 \vee 34)\mathbf{A}_{5'6'}\mathbf{B}_{5''6''}]$$

are those of

$$\lambda_{6'}\mu_{6''}12 \vee 34 \vee 5'5'' + \lambda_{6'}\mu_{5''}12 \vee 34 \vee 5'6''$$
$$+\lambda_{5'}\mu_{6''}12 \vee 34 \vee 6'5'' + \lambda_{5'}\mu_{5''}12 \vee 34 \vee 6'6''. \tag{A.4.6}$$

The classification is similar to that of p_{IV}.

Proposition A.4.6. Generic expansions of q_{II}:

(1) (Trivially zero) q_{II} is trivially zero if one of the following conditions is satisfied: (a) one of the pairs, $\mathbf{12}, \mathbf{34}$, is two identical points; (b) $\{\mathbf{1}, \mathbf{2}\} = \{\mathbf{3}, \mathbf{4}\}$.

Below we assume that q_{II} is not trivially zero.

(2) (Inner intersection) If $\mathbf{1} = \mathbf{3}$, the following is a shortest expansion:

$$[(\mathbf{12} \vee \mathbf{14})\mathbf{A_{5'6'}B_{5''6''}}] = [\mathbf{124}][\mathbf{1A_{5'6'}B_{5''6''}}]. \tag{A.4.7}$$

(3) (Double line) If $\{\mathbf{5'}, \mathbf{6'}\} = \{\mathbf{5''}, \mathbf{6''}\}$, the following is the unique expansion:

$$[(\mathbf{12} \vee \mathbf{34})\mathbf{A_{5'6'}B_{5'6'}}] = (\lambda_{6'}\mu_{5'} - \lambda_{5'}\mu_{6'})\mathbf{12} \vee \mathbf{34} \vee \mathbf{5'6'}. \tag{A.4.8}$$

If $\{\mathbf{1}, \mathbf{2}\} = \{\mathbf{5'}, \mathbf{6'}\}$, the following is a shortest expansion:

$$[(\mathbf{12} \vee \mathbf{34})\mathbf{A_{12}B_{5''6''}}] = [\mathbf{12B_{5''6''}}][\mathbf{34A_{12}}]. \tag{A.4.9}$$

(4) (Triangle) If $\mathbf{5'} = \mathbf{1}$ and $\{\mathbf{5''}, \mathbf{6''}\} = \{\mathbf{2}, \mathbf{6'}\}$, then $\mathbf{126'}$ is called a *triangle* in q_{II}. The following is a shortest expansion:

$$[(\mathbf{12} \vee \mathbf{34})\mathbf{A_{16'}B_{26'}}] = [\mathbf{126'}](\lambda_1\mu_{6'}[\mathbf{234}] - \lambda_{6'}\mu_2[\mathbf{134}]). \tag{A.4.10}$$

(5) (Other cases) If q_{II} is not trivially zero, and has neither inner intersection, nor double line, nor triangle, then it has no factored expansion. q_{II} has a binomial expansion if and only if it is of the form $[(\mathbf{12}\vee\mathbf{34})\mathbf{A_{13}B_{24}}]$. The unique binomial expansion is

$$[(\mathbf{12} \vee \mathbf{34})\mathbf{A_{13}B_{24}}] = [\mathbf{12A_{13}}][\mathbf{34B_{24}}] - [\mathbf{12B_{24}}][\mathbf{34A_{13}}]. \tag{A.4.11}$$

The Cayley expansions of

$$q_{III} = [(\mathbf{12} \vee \mathbf{34})(\mathbf{1'2'} \vee \mathbf{3'4'})\mathbf{A_{5''6''}}]$$

are those of

$$\lambda_{6''}[\mathbf{5''}(\mathbf{12} \vee \mathbf{34})(\mathbf{1'2'} \vee \mathbf{3'4'})] + \lambda_{5''}[\mathbf{6''}(\mathbf{12} \vee \mathbf{34})(\mathbf{1'2'} \vee \mathbf{3'4'})]. \tag{A.4.12}$$

Proposition A.4.7. Generic expansions of q_{III}:

(1) (Trivially zero) q_{III} is trivially zero if one of the following conditions is satisfied: (a) one of the pairs, $\mathbf{12}, \mathbf{34}$, is a pair of identical points; (b) $\{\mathbf{1}, \mathbf{2}\} = \{\mathbf{3}, \mathbf{4}\}$ or $\{\mathbf{1'}, \mathbf{2'}\} = \{\mathbf{3'}, \mathbf{4'}\}$; (c) $\{\mathbf{12}, \mathbf{34}\} = \{\mathbf{1'2'}, \mathbf{3'4'}\}$.

Below we assume that q_{III} is not trivially zero.

(2) (Inner intersection) If $\mathbf{1} = \mathbf{3}$, the following leads to a shortest expansion of q_{III}:

$$[(\mathbf{12} \vee \mathbf{14})(\mathbf{1'2'} \vee \mathbf{3'4'})\mathbf{A_{5''6''}}] = -[\mathbf{124}]\mathbf{1A_{5''6''}} \vee \mathbf{1'2'} \vee \mathbf{3'4'}. \tag{A.4.13}$$

(3) (Double line) If $\{\mathbf{1}, \mathbf{2}\} = \{\mathbf{1'}, \mathbf{2'}\}$, the following leads to a shortest expansion of q_{III}:

$$[(\mathbf{12} \vee \mathbf{34})(\mathbf{12} \vee \mathbf{3'4'})\mathbf{A_{5''6''}}] = [\mathbf{12A_{5''6''}}]\mathbf{12} \vee \mathbf{34} \vee \mathbf{3'4'}. \tag{A.4.14}$$

If $\{\mathbf{1}, \mathbf{2}\} = \{\mathbf{5''}, \mathbf{6''}\}$, the following leads to a shortest expansion of q_{III}:

$$[(\mathbf{12} \vee \mathbf{34})(\mathbf{1'2'} \vee \mathbf{3'4'})\mathbf{A_{12}}] = [\mathbf{34A_{12}}]\mathbf{12} \vee \mathbf{1'2'} \vee \mathbf{3'4'}. \tag{A.4.15}$$

(4) (Triangle) If $\mathbf{1'} = \mathbf{1}$ and $\{\mathbf{5''}, \mathbf{6''}\} = \{\mathbf{2}, \mathbf{2'}\}$, then $\mathbf{122'}$ is called a *triangle* in q_{III}. The following is a shortest expansion:

$$[(\mathbf{12} \vee \mathbf{34})(\mathbf{12'} \vee \mathbf{3'4'})\mathbf{A_{22'}}] = [\mathbf{122'}](\lambda_{\mathbf{2'}}[\mathbf{13'4'}][\mathbf{234}] + \lambda_{\mathbf{2}}[\mathbf{134}][\mathbf{2'3'4'}]). \quad \text{(A.4.16)}$$

The triangle pattern has one further factorizable subpattern:

(Quadrilateral) $(\mathbf{1234}, \mathbf{14})$ is called a *quadrilateral* in $[(\mathbf{12} \vee \mathbf{34})(\mathbf{13} \vee \mathbf{24})\mathbf{A_{14}}]$. The following is a shortest expansion:

$$[(\mathbf{12} \vee \mathbf{34})(\mathbf{13} \vee \mathbf{24})\mathbf{A_{14}}] = [\mathbf{124}][\mathbf{134}](\lambda_{\mathbf{4}}[\mathbf{123}] - \lambda_{\mathbf{1}}[\mathbf{234}]). \quad \text{(A.4.17)}$$

(5) If q_{III} is not trivially zero, and has neither inner intersection, nor double line, nor triangle, then it has no factored expansion.

In the following, the above hypothesis is always assumed.

(6) (Two triple points) If q_{III} has two triple points, then it has a binomial expansion.

(7) (One triple point and two double points) If q_{III} has only one triple point $\mathbf{1}$, then it has a binomial expansion if and only if there are two double points $\mathbf{2}, \mathbf{3}$ such that q_{III} is of the form $[(\mathbf{14} \vee \mathbf{23})(\mathbf{15} \vee \mathbf{36})\mathbf{A_{12}}]$, where $\mathbf{4}, \mathbf{5}, \mathbf{6}$ are either double or single points.

(8) (Four double points) If q_{III} has no triple point, then it has a binomial expansion if and only if there are four double points $\mathbf{1}, \mathbf{2}, \mathbf{3}, \mathbf{4}$ such that q_{III} is in one of the following forms, where $\mathbf{5}, \mathbf{6}$ are either single or double points:

    (a) (4-2-2 pattern) $[(\mathbf{12} \vee \mathbf{34})(\mathbf{24} \vee \mathbf{56})\mathbf{A_{13}}]$.

    (b) (3-3-2 pattern) $[(\mathbf{12} \vee \mathbf{45})(\mathbf{24} \vee \mathbf{36})\mathbf{A_{13}}]$.

A.5   Cayley expansions of r_I and r_{II}

The expression

$$r_I = \mathbf{12} \vee \mathbf{1'2'3'} \vee \mathbf{1''2''3''}$$

has three different Cayley expansions.

Proposition A.5.8. Generic expansions of r_I:

(1) (Trivially zero) r_I is trivially zero if one of the following conditions is satisfied: (a) one of the tuples, $\mathbf{12}, \mathbf{1'2'3'}, \mathbf{1''2''3''}$, contains two identical points; (b) the two 3-tuples are identical; (c) the 2-tuple is in one of the 3-tuples; (d) the three tuples have a point in common.

In the following, assume that r_I is not trivially zero.

(2) (Recursion of 1) If $\mathbf{1} = \mathbf{1'}$, then

$$\mathbf{12} \vee \mathbf{12'3'} \vee \mathbf{1''2''3''} = [\mathbf{122'3'}][\mathbf{11''2''3''}]. \quad \text{(A.5.1)}$$

(3) (Double line) If $\{\mathbf{1'}, \mathbf{2'}\} = \{\mathbf{1''}, \mathbf{2''}\}$, then

$$\mathbf{12} \vee \mathbf{1'2'3'} \vee \mathbf{1'2'3''} = [\mathbf{1'2'3'3''}][\mathbf{121'2'}]. \qquad (A.5.2)$$

(4) If r_I has neither a point that recurs, nor a double line, then it has no factored expansion.

The expression

$$r_{II} = \mathbf{123} \vee \mathbf{1'2'3'} \vee \mathbf{1''2''3''} \vee \mathbf{1'''2'''3'''} \qquad (A.5.3)$$

has twelve different initial Cayley expansions. The initial meet product expansion which distributes $\mathbf{1}, \mathbf{2}, \mathbf{3}$ towards $\mathbf{1'2'3'}$ is

$$r_{II} = [\mathbf{11'2'3'}]\mathbf{23} \vee \mathbf{1''2''3''} \vee \mathbf{1'''2'''3'''} - [\mathbf{21'2'3'}]\mathbf{13} \vee \mathbf{1''2''3''} \vee \mathbf{1'''2'''3'''}$$
$$+ [\mathbf{31'2'3'}]\mathbf{12} \vee \mathbf{1''2''3''} \vee \mathbf{1'''2'''3'''}. $$
$$(A.5.4)$$

Proposition A.5.9. Generic expansions of r_{II}:

(1) (Trivially zero) If one of the following conditions is satisfied, r_{II} is trivially zero: (a) one of the 3-tuples, $\mathbf{123}, \mathbf{1'2'3'}, \mathbf{1''2''3''}, \mathbf{1'''2'''3'''}$, contains two identical points; (b) two of the four 3-tuples are identical; (c) three of the four 3-tuples have two points in common; (d) the four 3-tuples have a point in common.

In the following, assume that r_{II} is not trivially zero.

(2) (Triple point) If $\mathbf{1} = \mathbf{1'} = \mathbf{1''}$, then $\mathbf{1}$ is called a *triple point* in r_{II}. Then

$$\mathbf{123} \vee \mathbf{12'3'} \vee \mathbf{12''3''} \vee \mathbf{1'''2'''3'''} = [\mathbf{11'''2'''3'''}]\,\mathbf{23} \vee_1 \mathbf{2'3'} \vee_1 \mathbf{2''3''} \qquad (A.5.5)$$

is the unique factored expansion of r_{II}. It also leads to the unique shortest expansion of r_{II} into a bracket polynomial.

(3) (Double line) If $\{\mathbf{1}, \mathbf{2}\} = \{\mathbf{1'}, \mathbf{2'}\}$, then $\mathbf{12}$ is called a *double line* in r_{II}. The following is the unique factored expansion, and also leads to the unique shortest expansion of r_{II}:

$$\mathbf{123} \vee \mathbf{123'} \vee \mathbf{1''2''3''} \vee \mathbf{1'''2'''3'''} = [\mathbf{1233'}]\,\mathbf{12} \vee \mathbf{1''2''3''} \vee \mathbf{1'''2'''3'''}. \qquad (A.5.6)$$

(4) (Other cases) If r_{II} is not trivially zero, and has neither triple point nor double line, then it has no factored expansion. The shortest expansions of r_{II} are 2-termed if and only if among the four 3-tuples of points $\{\mathbf{123}, \mathbf{1'2'3'}, \mathbf{1''2''3''}, \mathbf{1'''2'''3'''}\}$ in r_{II}, there is a 3-tuple whose intersections with the other three 3-tuples are respectively the three elements of the 3-tuple. A corresponding binomial expansion is

$$\mathbf{123} \vee \mathbf{12'3'} \vee \mathbf{22''3''} \vee \mathbf{32'''3'''}$$
$$= [\mathbf{122'3'}][\mathbf{132'''3'''}][\mathbf{232''3''}] - [\mathbf{122''3''}][\mathbf{132'3'}][\mathbf{232'''3'''}]. \qquad (A.5.7)$$

Bibliography

[1] Abhyankar, S.S. *Enumerative Combinatorics of Young Tableaux.* Marcel Dekker, New York, 1988.

[2] Abhyankar, S.S. *Algebraic Geometry for Scientists and Engineers.* American Mathematical Society, Providence, 1990.

[3] Abhyankar, S.S. Invariant Theory and Enumerative Combinatorics of Young Tableaux. In: Mundy, J.L. and Zisserman, A. (*eds.*) *Geometric Invariance in Computer Vision.* MIT Press, Cambridge, Massachusetts, pp. 45-76, 1992.

[4] Ablamowicz, R. and Sobczyk, G. (*eds.*) *Lectures on Clifford Geometric Algebras and Applications.* Birkhäuser, Boston, 2004.

[5] Ahlfors, L.V. Möbius Transformations and Clifford Numbers. In: Chavel, I. and Farkas, H.M. (*eds.*), *Differential Geometry and Complex Analysis,* Springer, Berlin, Heidelberg, 1985.

[6] Ahlfors, L.V. Möbius Transformations in $\mathbb{R}^n$ Expressed through 2×2 Matrices of Clifford Numbers. *Complex Variables* **5**: 215-224, 1986.

[7] Altmann, S.L. *Rotations, Quaternions, and Double Groups.* Oxford University Press, Oxford, 1986.

[8] Angles, P. *Conformal Groups in Geometry and Spin Structures,* Birkhäuser, Boston, 2007.

[9] Anick, D. and Rota, G.-C. Higher-Order Syzygies for the Bracket Algebra and for the Ring of Coordinates of the Grassmannian. *Proc. Nat. Acad. Sci.* **88**(18): 8087-8090, 1991.

[10] Arbarello, E., Cornalba, M., Griffiths, P.A., and Harris, J. *Geometry of Algebraic Curves,* Volume I. Springer, New York, 1985.

[11] Artin, E. *Geometric Algebra.* Interscience, New York, 1957, 1988.

[12] Atiyah, M.F., Bott, R. and Shapiro, A. Clifford Modules. *Topology* **3**: 3-38, suppl. 1, 1964.

[13] Balls, R.S. *A Treatise on the Theory of Screws.* Cambridge University Press, Cambridge, 1900.

[14] Barnabei, M., Brini, A. and Rota, G.-C. On the Exterior Calculus of Invariant Theory. *J. of Algebra* **96**: 120-160, 1985.

[15] Baylis, W.E. (*ed.*) *Clifford (Geometric) Algebra with Applications to Physics, Mathematics, and Engineering,* Birkhäuser Boston, 1996.

[16] Bayro-Corrochano, E. and Lasenby, J. Object Modeling and Motion Analysis Using Clifford Algebra. In: Mohr, R. and Wu, C. (*eds.*), *Proc. Europe-China Workshop on Geometric Modeling and Invariants for Computer Visions,* Xi'an, China, pp. 143-149, 1995.

[17] Bayro-Corrochano, E. and Sobczyk, G. (*eds.*) *Geometric Algebra with Applications in Science and Engineering*, Birkhäuser, Boston, 2001.

[18] Bayro-Corrochano, E. and López-Franco, C. Omnidirectional Vision: Unified Model Using Conformal Geometry. *ECCV'04* (1): 536-548, 2004.

[19] Bayro-Corrochano, E. Robot Perception and Action Using Conformal Geometric Algebra. In: Bayro-Corrochano, E. (*ed.*), *Handbook of Geometric Computing*, Springer, Heidelberg, pp. 405-458, 2005.

[20] Bayro-Corrochano E. and Falcon, L. E. Geometric Algebra of Points, Lines, Planes and Spheres for Computer Vision and Robotics. *Robotica* **23**: 755-770, 2005.

[21] Bayro-Corrochano, E. , Reyes-Lozano, L. and Zamora-Esquivel, J. Conformal Geometric Algebra for Robotic Vision. *Journal of Mathematical Imaging and Vision* **24**(1): 55-81, 2006.

[22] Berger, M. *Geometry I, II*. Springer, 1987.

[23] Bix, R. *Conics and Cubics*. Springer, 1998.

[24] Blaschke, W. *Vorlesungen über Differentialgeometrie und geometrische Grundlagen von Einsteins Relativitätstheorie*. Vol. 3, Springer, Berlin, 1929.

[25] Blaschke, W. *Kinematik und Quaternionen*. VEB Deutscher Verlag der Wissenschaften, Berlin, 1960.

[26] Blumenthal, L.M. *Theory and Applications of Distance Geometry*. Cambridge University Press, Cambridge, 1953.

[27] Bokowski, J. and Sturmfels, B. *Computational Synthetic Geometry*. LNM **1355**, Springer, Berlin, Heidelberg. 1989.

[28] Bravi, P. and Brini, A. Remarks on Invariant Geometric Calculus, Cayley-Grassmann Algebras and Geometric Clifford Algebras. In: Crapo, H. and Senato, D. (*eds.*), *Algebraic Combinatorics and Computer Science*, Springer, Milano, pp. 129-150, 2001.

[29] Brini, A., Regonati, F. and Teolis, A.G.B. Grassmann Geometric Calculus, Invariant Theory and Superalgebras. In: Crapo, H. and Senato D. (*eds.*), *Algebraic Combinatorics and Computer Science*, Springer, Milano, pp. 151-196, 2001.

[30] Brackx, F., Delanghe, D. and Sommen, F. *Clifford Analysis*. Pitman, Boston, 1982.

[31] Buchberger, B. Application of Gröbner Basis in Non-linear Computational Geometry. In: Rice, J. (*ed.*), *Scientific Software*. Springer, New York, 1988.

[32] Caianiello, E. *Combinatorics and Renormalization in Quantum Field Theory*. Benjamin, Reading, 1973.

[33] Cartan, É. Nombres Complexes. In: Molk, J. (red.), *Encyclopédie des Sciences Mathématiques*, Tome **I**, vol. 1, Fasc. 4, art. I5, pp. 329-468, 1908.

[34] Cartan, É. *The Theory of Spinors*, 2nd edition, Hermann, Paris, 1966.

[35] Cecil, T.E. *Lie Sphere Geometry*. Springer, New York, 1992.

[36] Chevalley, C. *The Algebraic Theory of Spinors and Clifford Algebras*, Collected Works, Volume 2. Editors: Cartier, P. and Chevalley, C. Springer, Berlin, Heidelberg, 1997. Originally published by Columbia University Press, New York, 1954.

[37] Chevallier, D.P. Lie Algebras, Modules, Dual Quaternions, and Algebraic Methods in Kinematics. *Machine and Mechanical Theory* **26**: 613-627, 1991.

[38] Cho, H.C., Choi, H.I., Kwon, S.-H., Lee, D.-S., and Wee, N.-S. Clifford Algebra, Lorentzian Geometry, and Rational Parametrization of Canal Surfaces. *Computer Aided Geometric Design* **21**: 327-339, 2004.

[39] Chou, S.C. *Mechanical Geometry Theorem Proving*. D. Reidel, Dordrecht, 1988.

[40] Chou, S.C., Gao, X.S. and Zhang, J.Z. *Machine Proofs in Geometry - Automated Production of Readable Proofs for Geometric Theorems*. World Scientific, Singapore, 1994.

[41] Chou, S.C., Gao, X.S. and Zhang, J.Z. A Deductive Database Approach to Auto-

mated Geometry Theorem Proving and Discovering. *J. Automated Reasoning* **25**(10): 219-246, 2000.

[42] Clifford, W.K. Preliminary Sketch of Bi-quaternions. *Proc. London Math. Society* **4**: 381-395, 1873.

[43] Clifford, W.K. Application of Grassmann's Extensive Algebra. *American Journal of Mathematics* I: 350-358, 1878.

[44] Cnops, J. Vahlen Matrices for Non-definite Metrics. In: Ablamowicz, R. *et al.* (*eds.*), *Clifford Algebras with Numeric and Symbolic Computations*, Birkhäuser, Boston, 1996.

[45] Crapo, H. and Richter-Gebert, J. Automatic Proving of Geometric Theorems. In: White, N. (*ed.*), *Invariant Methods in Discrete and Computational Geometry*, Kluwer Academic Publishers, pp. 107-139, 1994.

[46] Crapo, H. and Senato D. (*eds.*) *Algebraic Combinatorics and Computer Science.* Springer, Milano, 2001.

[47] Crippen, G.M. and Havel, T.F. *Distance Geometry and Molecular Conformation.* Research Studies Press and John Wiley & Sons Inc., Taunton, England, 1988.

[48] Crumeyrolle, A. *Orthogonal and Symplectic Clifford Algebras.* D. Reidel, Dordrecht, Boston, 1990.

[49] Daniilidis, K. Hand-Eye calibration Using Dual Quaternions. *International J. Robotics Research* **18**: 286-298, 1999.

[50] DeConcini, C. and Procesi C. A Characteristic Free Approach to Invariant Theory. *Adv. in Math.* **21**: 330-354, 1976.

[51] DeConcini, C., Eisenbud E. and Procesi C. *Hodge Algebras. Astérisque* **91**, 1982.

[52] Delanghe, R., Sommen, F. and Soucek, V. *Clifford Algebra and Spinor-Valued Functions*, D. Reidel, Dordrecht, 1992.

[53] Dixon, A.C. On the Newtonian Potential. *Quarterly J. of Mathematics* **35**: 283-296, 1904.

[54] Dorst, L. Objects in Contact: Boundary Collisions as Geometric Wave Propagation. In: Bayro-Corrochano, E. and Sobczyk, G. (*eds.*) *Geometric Algebra with Applications in Science and Engineering*, Birkhäuser, Boston, pp. 349-390, 2001.

[55] Dorst, L., Doran, C. and Lasenby, J. (*eds.*) *Applications of Geometric Algebra in Computer Science and Engineering*, Birkhäuser, Boston, 2002.

[56] Dorst, L., Fontijne, D. and Mann, S. *Geometric Algebra for Computer Science*, Morgan Kaufmann Publishers, Elsevier Inc., 2007.

[57] Doubilet, P., Rota, G.-C. and Stein, J. On the Foundations of Combinatorial Theory IX: Combinatorial Methods in Invariant Theory. *Stud. Appl. Math.* **57**: 185-216, 1974.

[58] Dress, A. and Wenzel, D. Grassmann-Plücker Relations and Matroids with Coefficients. *Advances in Mathematics* **86**: 68-110, 1991.

[59] Faugeras, O. and Mourrain, B. On the Geometry and Algebra of the Point and Line Correspondences between n Images. In: *Proc. ICCV'95*, Boston, IEEE Computer Society Press, pp. 951-956, 1995.

[60] Faugeras, O. and Papadopoulo, T. Grassmann-Cayley Algebra for Modeling Systems of Cameras and the Algebraic Equations of the Manifold of Trifocal Tensors. *Phil. Trans. R. Soc. Lond. A* **356**(1740): 1123-1152, 1998.

[61] Faugeras, O. and Luong, Q.-T. *The Geometry of Multiple Images*, MIT Press, Cambridge, Massachusetts, 2001.

[62] Fauser, B. A Treatise on Quantum Clifford Algebra, Univ. of Konstanz, Germany, 2002.

[63] Fevre, S. and Wang, D. Combining Clifford Algebraic Computing and Term-Rewriting for Geometric Theorem Proving. *Fundamenta Informaticae* **39**(1-2): 85-104, 1999.

[64] Fillmore, J.P. and Springer, A. Möbius Groups over General Fields Using Clifford

Algebras Associated with Spheres. *International J. Theoretical Phys.* **29**: 225-246, 1990.

[65] Fontijne, D. and Dorst, L. Modeling 3D Euclidean Geometry - Performance and Elegance of Five Models of 3D Euclidean Geometry in a Ray Tracing Application. *Computer Graphics and Applications* **23**(2): 68-78, 2003.

[66] Gao, X.S. and Wang, D. *Mathematics Mechanization and Applications.* Academic Press, London, 2000.

[67] Gibbs, J.M. Quaternions and the Algebra of Vectors. *Nature* **47**: 463-464, 1893.

[68] Grassmann, H. *Linear Extension Theory (Die Lineale Ausdehnungslehre)*, 1844. Translated by Kannenberg, L.C. in *The Ausdehnungslehre of 1844 and Other Works*, Chicago, La Salle: Open Court Publ., 1995.

[69] Grosshans, F.D., Rota, G.-C. and Stein, J. *Invariant Theory and Superalgebras*, AMS, 1987.

[70] Havel, T.F. Some Examples of the Use of Distances as Coordinates for Euclidean Geometry. *J. Symb. Comp.* **11**: 579-594, 1991.

[71] Havel, T.F. and Dress, A. Distance Geometry and Geometric Algebra, *Found. Phys.* **23**: 1357-1374, 1993.

[72] Havel, T.F. Geometric Algebra and Möbius Sphere Geometry as a Basis for Euclidean Invariant Theory. In: White, N. (*ed.*), *Invariant Methods in Discrete and Computational Geometry*, D. Reidel, Dordrecht, pp. 245-256, 1995.

[73] Havel, T.F. Qubit Logic, Algebra and Geometry. In: Richter-Gebert, J. and Wang, D. (*eds.*), *Automated Deduction in Geometry*, LNAI 2061, Springer, Berlin, Heidelberg, pp. 228-245, 2001.

[74] Havel, T.F., Cory, D.G., Somaroo, S.S. and Tseng, C.H. Geometric Algebra in Quantum Information Processing by Nuclear Magnetic Resonance. In: Bayro-Corrochano, E. and Sobczyk, G. (*eds.*) *Geometric Algebra with Applications in Science and Engineering*, Birkhäuser, Boston, pp. 281-308, 2001.

[75] Helmstetter, J. Algebres de Clifford et algebres de Weyl. *Cahiers Math.* **25**, Montpellier, 1982.

[76] Hestenes, D. *Space-Time Algebra.* Gordon and Breach, New York, 1966.

[77] Hestenes, D. and Sobczyk, G. *Clifford Algebra to Geometric Calculus.* D. Reidel, Dordrecht, Boston, 1984.

[78] Hestenes, D. A Unified Language for Mathematics and Physics. In: Chisholm, J.S.R. and Common A.K. (*eds.*), *Clifford Algebras and their Applications in Mathematical Physics*, Kluwer Academic Publishers, Dordrecht, pp. 1-23, 1986.

[79] Hestenes, D. and Ziegler, R. Projective Geometry with Clifford Algebra, *Acta Appl. Math.* **23**: 25-63, 1991.

[80] Hestenes, D. The Design of Linear Algebra and Geometry, *Acta Appl. Math.* **23**: 65-93, 1991.

[81] Hestenes, D. Invariant Body Kinematics I, II. *Neural Networks* **7**: 65-77, 79-88, 1994.

[82] Hestenes, D. Grassmann's Vision. In: Schubring, G. (*ed.*), *Hermann Gunther Grassmann (1809-1877): Visionary Mathematician, Scientist and Neohumanist Scholar*, Kluwer Academic Publishers, Dordrecht, 1996.

[83] Hestenes, D. *New Foundations for Classical Mechanics.* D. Reidel, Dordrecht, Boston, 2nd edition, 1998.

[84] Hestenes, D., Li, H. and Rockwood, A. New algebraic tools for classical geometry. In: Sommer, G. (*ed.*), *Geometric Computing with Clifford Algebra*, Springer, Heidelberg, pp. 3-26, 2001.

[85] Hestenes, D. Old Wine in New Bottles: A New Algebraic Framework for Computational Geometry. In: Bayro-Corrochano, E. and Sobczyk, G. (*eds.*), *Geometric Algebra*

with Applications in Science and Engineering, Birkhäuser, Boston, pp. 3-17, 2001.

[86] Hestenes, D. and Holt, J. The Crystallographic Space Groups in Geometric Algebra. *Journal of Mathematical Physics* **48**, 023514, 2007.

[87] Hildenbrand, D., Fontijne, D., Perwass, C. and Dorst, L. Geometric Algebra and Its Application to Computer Graphics. Tutorial of *Eurographics 2004*, published by INRIA and Eurographics Assoc., Grenoble, 2004.

[88] Hildenbrand, D. Geometric Computing in Computer Graphics Using Conformal Geometric Algebra. *Computers and Graphics* **29**(5): 795-803, 2005.

[89] Hilbert, D. Über die Theorie der Algebraischen Formen. *Mathematische Ann.* **36**: 473-534, 1890.

[90] Hilbert, D. Über die vollen Invariantensysteme. *Mathematische Ann.* **42**: 313-373, 1893.

[91] Hitzer, E.M.S. Euclidean Geometric Objects in the Clifford Geometric Algebra of {Origin, 3-Space, Infinity}. *Simon Stevin* **11**(5): 653-662, 2004.

[92] Hou, X., Li, H., Wang, D. and Yang, L. "Russian Killer" No. 2: a Challenging Geometric Theorem with Machine vs. Human Proofs. *Math. Intelligencer* **23**(1): 9-15, 2001.

[93] Hodge, W.V.D. and Pedoe, D. *Methods of Algebraic Geometry*. Cambridge University Press, Cambridge, 1953.

[94] Huang, R.Q., Rota, G.-C. and Stein, J. Supersymmetric Bracket Algebra and Invariant Theory, *Acta Appl. Math.* **21**: 193-246, 1990.

[95] Iversen, B. *Hyperbolic Geometry*. Cambridge University Press, Cambridge, 1992.

[96] Kapur, D. Automated Geometric Reasoning: Dixon Resultants, Gröbner Bases and Characteristic Sets. In: Wang, D. (*ed.*), *Automated Deduction in Geometry*, LNAI 1360, Springer, Berlin, Heidelberg, pp. 1-36, 1998.

[97] Klein, F. *Vorlesungen über nichteuklidische Geometrie*. Springer, Berlin, 1928.

[98] Lasenby, A. and Lasenby, J. Surface Evolution and Representation Using Geometric Algebra. In: Cipolla, R. and Martin, R. (*eds.*), *The Mathematics of Surfaces IX*, Springer, London, pp. 144-168, 2000.

[99] Lasenby, A. and Doran, C. *Geometric Algebra for Physicists*. Cambridge University Press, Cambridge, 2003.

[100] Lasenby, A. Conformal Geometry and the Universe, submitted to *Phil. Trans. R. Soc. Lond. A*, 2004.

[101] Lasenby, A. Recent Applications of Conformal Geometric Algebra. In: Li, H. *et al.* (*eds.*), *Computer Algebra and Geometric Algebra with Applications*, LNCS 3519, Springer, Berlin, Heidelberg, pp. 298-328, 2005.

[102] Lawson, H.B. and Michelsohn, M.L. *Spin Geometry*. Princeton University Press, Princeton, 1989.

[103] Li, H. New Explorations in Automated Theorem Proving in Geometries. Ph.D. Thesis, Peking University, Beijing, 1994.

[104] Li, H. Hyperbolic Geometry with Clifford Algebra, *Acta Appl. Math.* **48**(3): 317-358, 1997.

[105] Li, H. and Shi, H. On Erdös' Ten-Point Problem, *Acta Mathematica Sinica New Series*, **13**(2): 221-230, 1997.

[106] Li, H. Some Applications of Clifford Algebra to Geometries. In: Gao, X.S. *et al.* (*eds.*), *Automated Deduction in Geometry*, LNAI 1669, Springer, Berlin, Heidelberg, pp. 156-179, 1998.

[107] Li, H. Vectorial Equation-Solving for Mechanical Geometry Theorem Proving. *J. Automated Reasoning* **25**: 83-121, 2000.

[108] Li, H. The Lie Model for Euclidean Geometry. In: Sommer, G. and Zeevi, Y. (*eds.*),

Algebraic Frames for the Perception-Action Cycle, LNCS 1888, Springer, Berlin, pp. 115-133, 2000.

[109] Li, H. Clifford Algebra Approaches to Automated Geometry Theorem Proving. In: Gao, X.S. and Wang, D. (*eds.*), *Mathematics Mechanization and Applications*, Academic Press, London, pp. 205-230, 2000.

[110] Li, H., Hestenes, D. and Rockwood, A. Generalized Homogeneous Coordinates for Computational Geometry. In: Sommer, G. (*ed.*), *Geometric Computing with Clifford Algebras*, Springer, Heidelberg, pp. 27-60, 2001.

[111] Li, H., Hestenes, D. and Rockwood, A. Spherical Conformal Geometry with Geometric Algebra. In: Sommer, G. (*ed.*), *Geometric Computing with Clifford Algebras*, Springer, Heidelberg, pp. 61-76, 2001.

[112] Li, H., Hestenes, D. and Rockwood, A. A Universal Model for Conformal Geometries of Euclidean, Spherical and Double-Hyperbolic Spaces. In: Sommer, G. (*ed.*), *Geometric Computing with Clifford Algebras*, Springer, Heidelberg, pp. 77-104, 2001.

[113] Li, H. Hyperbolic Conformal Geometry with Clifford Algebra. *International Journal of Theoretical Physics* **40**(1): 79-91, 2001.

[114] Li, H. and Sommer, G. Coordinate-free Projective Geometry for Computer Vision. In: Sommer, G. (*ed.*), *Geometric Computing with Clifford Algebras*, Springer, Heidelberg, pp. 415-454, 2001.

[115] Li, H. Automated Theorem Proving in the Homogeneous Model with Clifford Bracket Algebra. In: Dorst, L. *et al.* (*eds.*), *Applications of Geometric Algebra in Computer Science and Engineering*, Birkhäuser, Boston, pp. 69-78, 2002.

[116] Li, H. Clifford Expansions and Summations. *MM Research Preprints* **21**: 112-154, 2002.

[117] Li, H. and Wu, Y. Automated Short Proof Generation in Projective Geometry with Cayley and Bracket Algebras I, II. *J. Symb. Comput.* **36**(5): 717-762, 763-809, 2003.

[118] Li, H. Clifford Algebras and Geometric Computation. In: Chen, F. and Wang, D. (*eds.*), *Geometric Computation*, World Scientific, Singapore, pp. 221-247, 2004.

[119] Li, H. Automated Geometric Theorem Proving, Clifford Bracket Algebra and Clifford Expansions. In: Qian, T. *et al.* (*eds.*), *Trends in Mathematics: Advances in Analysis and Geometry*, Birkhäuser, Basel, pp. 345-363, 2004.

[120] Li, H. Algebraic Representation, Elimination and Expansion in Automated Geometric Theorem Proving. In: Winkler, F. (*ed.*), *Automated Deduction in Geometry*, LNAI 2930, Springer, Berlin, Heidelberg, pp. 106-123, 2004.

[121] Li, H. Symbolic Computation in the Homogeneous Geometric Model with Clifford Algebra. In: Gutierrez, J. (*ed.*), *Proc. ISSAC 2004*, ACM Press, New York, pp. 221-228, 2004.

[122] Li, H., Zhao, L., Chen, Y. Polyhedral Scene Analysis Combining Parametric Propagation with Calotte Analysis. In: Li, H. *et al.* (*eds.*), *Computer Algebra and Geometric Algebra with Applications*, LNCS 3519, Springer, Berlin, Heidelberg, pp. 383-402, 2005.

[123] Li, H. Geometric Reasoning with Invariant Algebras. In: *Proc. ATCM 2006*, ATCM Inc., Blacksburg, USA, pp. 6-19, 2006.

[124] Li, H. A Recipe for Symbolic Geometric Computing: Long Geometric Product, BREEFS and Clifford Factorization. In: Brown, C.W. (*ed.*), *Proc. ISSAC 2007*, ACM Press, New York, pp. 261-268, 2007.

[125] Li, H. *Symbolic Computational Geometry with Advanced Invariant Algebras.* Manuscript, sequel of *Invariant Algebras and Geometric Reasoning*, finished in September 2007.

[126] Lie, S. Über Komplexe, inbesondere Linien- und Kugelkomplexe, mit Anwendung auf der Theorie der partieller Differentialgleichungen. *Math. Ann.* **5**: 145-208, 209-256,

1872. (Ges. Abh. **2**: 1-121)

[127] Lipschitz, R. *Untersuchungen über die Summen von Quadraten.* Max Cohen und Sohn, Bonn, 1886.

[128] Lopez-Franco, C. and Bayro-Corrochano, E. Omnidirectional Robot Vision Using Conformal Geometric Computing. *J. Mathematical Imaging and Vision* **26**(3): 243-260, 2006.

[129] Lotze, A. Über eine neue Begründung der regressiven Multiplikation extensiver Grössen in einem Hauptgebeit n-ter Stufe, *Jahresbericht d. Deutschen Mathem. Vereinigung* **57**: 102-110, 1955.

[130] Lounesto, P. *Clifford Algebras and Spinors.* Cambridge University Press, Cambridge, 1997.

[131] Mainetti, M. and Yan, C.H. Arguesian Identities in Linear Lattices. *Advances in Mathematics* **144**: 50-93, 1999.

[132] Maks, J. Clifford Algebras and Möbius Transformations. In: Micali, A. *et al.* (*eds.*), *Clifford Algebras and Their Applications in Mathematical Physics*, Kluwer, Dordrecht, pp. 57-63, 1992.

[133] McMillan, T. and White, N. The Dotted Straightening Algorithm. *J. Symb. Comput.* **11**: 471-482, 1991.

[134] Merlet, J.-P. *Parallel Robot.* Kluwer Academic Publishers, Dordrecht, 2000.

[135] Mourrain, B. New Aspects of Geometrical Calculus with Invariants. *MEGA'91*, 1991.

[136] Mourrain, B. and Stolfi, N. Computational Symbolic Geometry. In: White, N. (*ed.*), *Invariant Methods in Discrete and Computational Geometry*, D. Reidel, Dordrecht, Boston, pp. 107-139, 1995.

[137] Mourrain, B. and Stolfi, N. Applications of Clifford Algebras in Robotics. In: Merlet, J.-P. and Ravani, B. (*eds.*), *Computational Kinematics '95*, D. Reidel, Dordrecht, Boston, pp. 141-150, 1995.

[138] Mundy, J.L. and Zisserman, A. (*eds.*) *Geometric Invariance in Computer Vision.* MIT Press, Cambridge, Massachusetts, 1992.

[139] Olver, P.J. *Classical Invariant Theory.* Cambridge University Press, Cambridge, 1999.

[140] Oziewicz, Z. Clifford Coalgebra, and Linear Energy-Momenta Addition. Draft, Univ. Nacional Autonoma de Mexico, October 10, 2001.

[141] Pauli, W. Zur Quantenmechanik des magnetischen Elektrons. *Z. Phys.* **42**: 601-623, 1927.

[142] Peano, G. *Geometric Calculus according to the Ausdehnungslehre of H. Grassmann,* 1888. Translation copyright 1997 by L. Kannenberg.

[143] Penrose, R. and Rindler, R. *Spinors and Space-Time.* Cambridge University Press, Cambridge, 1984.

[144] Perwass, C., Banarer, V. and Sommer, G. Spherical Decision Surfaces Using Conformal Modeling. In: *DAGM 2003*, Magdeburg, LNCS 2781, Springer, Berlin, pp. 9-16, 2003.

[145] Porteous, I.R. *Clifford Algebras and the Classical Groups.* Cambridge University Press, Cambridge, 1995.

[146] Pozo, J.M. and Sobczyk, G. Realizations of the Conformal Group. In: Bayro-Corrochano, E. and Sobczyk, G. (*eds.*), *Geometric Algebra with Applications in Science and Engineering*, Birkhäuser, Boston, pp. 42-60, 2001.

[147] Ratcliffe, J. *Foundations of Hyperbolic Manifolds.* GTM **149**, 2nd edition, Sprinber, Berlin, Heidelberg, 2006.

[148] Richter-Gebert, J. Mechanical Theorem Proving in Projective Geometry. *Annals of Math. and Artificial Intelligence* **13**: 159-171, 1995.

[149] Richter-Gebert, J. *Realization Spaces of Polytopes.* LNM 1643, Springer, Berlin, Heidelberg, 1996.

[150] Riesz, M. *Clifford Numbers and Spinors*, 1958. From lecture notes made in 1957-8, edited by Bolinder, E. and Lounesto, P. Kluwer Academic Publisher, Dordrecht, 1993.

[151] Rosenhahn, B. and Sommer, G. Adaptive Pose Estimation for Different Corresponding Entities, In: van Gool, L. (*ed.*), *Pattern Recognition, 24 Symposium für Mustererkennung*, LNCS 2449, Springer, Berlin, Heidelberg, pp. 265-273, 2002.

[152] Rosenhahn, B. Pose Estimation Revisited. Ph.D. Dissertation, also Technical Report Number 0308, CAU zu Kiel, 2003.

[153] Rosenhahn, B., Perwass, C. and Sommer, G. Free-form Pose Estimation by Using Twist Representations. *Algorithmica* **38**: 91-113, 2004.

[154] Rosenhahn, B. and Sommer, G. Pose Estimation of Free-form Objects. In: *ECCV 2004*, Prag, LNCS 3021, Springer, Berlin, Heidelberg, pp. 414-427, 2004.

[155] Rosenhahn, B., Perwass, Ch. and Sommer, G. Pose Estimation of Free-form Contours. *International Journal of Computer Vision* **62**(3): 267-289, 2005.

[156] Rosenhahn, B. and Sommer, G. Pose Estimation in Conformal Geometric Algebra I, II. *J. Mathematical Imaging and Vision* **22**: 27-48, 49-70, 2005.

[157] Rota, G.-C. and Stein, J. Applications of Cayley Algebras. Academia Nazionale dei Lincei atti dei Convegni Lincei **17**, Colloquio Internazionale sulle Teorie Combinatoire, Tomo **2**, Roma.

[158] Rota, G.-C. and Stein, J. Symbolic Method in Invariant Theory, *Proc. Nat. Acad. Sci.* **83**: 844-847, 1986.

[159] Rota, G.-C. and Sturmfels, B. Introduction to Invariant Theory in Superalgebras. In: Staton, D. (*ed.*), *Invariant Theory and Tableaux*, Springer, New York, pp. 1-35, 1990.

[160] Ryan, J. Conformal Clifford Manifolds Arising in Clifford Analysis. *Proc. Roy. Irish Acad. Sect. A* **85**: 1-23, 1985.

[161] Samuel, P. *Projective Geometry.* Springer, New York, 1988.

[162] Seidel, J.J. Distance-Geometric Development of Two-dimensional Euclidean, Hyperbolic and Spherical Geometry I, II. *Simon Stevin* **29**: 32-50, 65-76, 1952.

[163] Seidel, J.J. Angles and Distances in n-Dimensional Euclidean and Non-Euclidean geometry I-III. *Indag. Math.* **17**, No. 3 & 4, 1952.

[164] Selig, J.M. *Geometrical Methods in Robotics.* Springer, New York, 1996.

[165] Sobczyk, G. and Pozo, J.M. Geometric Algebra in Linear Algebra and Geometry. *Acta Appl. Math.* **71**: 207-244, 2002.

[166] Sommer, G. (*ed.*) *Geometric Computing with Clifford Algebras.* Springer, Berlin, Heidelberg, 2001.

[167] Sommer, G. A Geometric Algebra Approach to Some Problems of Robot Vision. In: Byrnes, J. (*ed.*), *Computational Noncommunicative Algebra and Applications*, Kluwer Acad., NATO Science Series 136, pp. 309-338, 2004.

[168] Sommer, G. Application of Geometric Algebra in Robot Vision. In: Li, H. *et al.* (*eds.*), *Computer Algebra and Geometric Algebra with Applications*, LNCS 3519, Springer, Berlin, Heidelberg, pp. 258-277, 2005.

[169] Sommer, G., Rosenhahn, B. and Perwass, C. Twists - An Operational Representation of Shape. In: Li, H. *et al.* (*eds.*), *Computer Algebra and Geometric Algebra with Applications*, LNCS 3519, Springer, Berlin, Heidelberg, pp. 278-297, 2005.

[170] Staton, D. (*ed.*) *Invariant Theory and Tableaux.* Springer, New York, 1990.

[171] Steenrod, N. *The Topology of Fibre Bundles.* Princeton University Press, Princeton, New Jersey, 1974.

[172] Study, E. *Geometrie der Dynamen.* Leipzig, 1903.

[173] Sturmfels, B. Computing Final Polynomials and Final Syzygies Using Buchberger's Gröbner Basis Method. *Result in Math.* **15**: 551-360, 1989.

[174] Sturmfels, B. and White, N. Gröbner Bases and Invariant Theory. *Advances in Mathematics* **76**: 245-259, 1989.

[175] Sturmfels, B. Computational Algebraic Geometry of Projective Configurations. *J. Symbolic Computation* **11**: 595-618, 1991.

[176] Sturmfels, B. and Whiteley, W. On the Synthetic Factorization of Homogeneous Invariants. *J. Symbolic Computation* **11**: 439-454, 1991.

[177] Sturmfels, B. *Algorithms in Invariant Theory.* Springer, New York, 1993.

[178] Sugihara, K. *Machine Interpretation of Line Drawings.* MIT Press, 1986.

[179] Sweedler, M.E. *Hopf Algebra.* Benjamin, New York, 1969.

[180] Thurston, W. *Three-dimensional Geometry and Topology*, Volume 1. Princeton University Press, Princeton, 1997.

[181] Tolvanen, A., Perwass, C. and Sommer, G. Projective Model for Central Catadioptric Cameras Using Clifford Algebra. In: *DAGM 2005*, LNCS 3663, Springer, Berlin, Heidelberg, pp. 192-199, 2005.

[182] Turnbull, H.W. *Determinants, Matrices and Invariants.* Blackie and Son, Glasgow, 1928.

[183] Turnbull, H.W. and Young, A. Linear Invariants of Ten Quaternary Quadrics. *Trans. Camb. Phil. Soc.* **23**: 265-301, 1926.

[184] Vahlen, K. Über Bewegungen und komplexe Zahlen. *Math. Ann.* **55**: 585-593, 1902.

[185] van der Waerden, B.L. On Clifford Algebras. *Nederl. Akad. Wetensch. Proc. Ser. A* **69**: 78-83, 1966.

[186] Wang, D. Clifford Algebraic Calculus for Geometric Reasoning, with Application to Computer Vision. In: Wang, D. (*ed.*), *Automated Deduction in Geometry*, LNAI 1360, Springer, Berlin, Heidelberg, pp. 115-140, 1996.

[187] Wang, D. Geometric Reasoning with Geometric Algebra. In: Bayro-Corrochano, E. and Sobczyk, G. (*eds.*) *Geometric Algebra with Applications in Science and Engineering*, Birkhäuser, Boston, pp. 89-109, 2001.

[188] Weitzenbök, R. *Invariantentheorie.* Noordhoff, Groningen, 1923.

[189] Weyl, H. *The Classical Groups.* Princeton Mathematics Series, Princeton University Press, Princeton, New Jersey, 1939.

[190] White, N. The Bracket Ring of Combinatorial Geometry I. *Trans. Amer. Math. Soc.* **202**: 79-103, 1975.

[191] White, N. Implementation of the Straightening Algorithm of Classical Invariant Theory. In: Staton, D. (*ed.*), *Invariant Theory and Tableaux*, Springer, New York, pp. 36-45, 1990.

[192] White, N. Multilinear Cayley Factorization. *J. Symb. Comput.* **11**: 421-438, 1991.

[193] White, N. (*ed.*) *Invariant Methods in Discrete and Computational Geometry.* Kluwer, Dordrecht, 1994.

[194] White, N. Geometric Applications of the Grassmann-Cayley Algebra. In: Goodman, J. E. and O'Rourke, J. (*eds.*), *Handbook of Discrete and Computational Geometry*, CRC Press, Florida, 1997.

[195] White, N. Grassmann-Cayley Algebra and Robotics Applications. In: Bayro-Corrochano, E. (*ed.*), *Handbook of Geometric Computing*, Springer, Heidelberg, pp. 629-656, 2005.

[196] Whiteley, W. Invariant Computations for Analytic Projective Geometry. *J. Symb. Comput.* **11**: 549-578, 1991.

[197] Willmore, T.J. *Riemannian Geometry*, Clarendon Press, Oxford, New York, 1993.

[198] Winter, D. *The Structure of Fields.* GTM 16, Springer, New York, 1974.

[199] Witt, E. Theorie der quadratischen Formen in beliebigen Körpern. *J. Reine Angew. Math.* **176**: 31-44, 1937.

[200] Witte, F.M.C. Lightlike Infinity in CGA Models of Spacetime. *J. of Physics A - Mathematics and General* **37**(42): 9965-9973, 2004.

[201] Wu, W.-T. On the Mechanization of Theorem-Proving in Elementary Differential Geometry. *Scientia Sinica* (Math Suppl. I): 94-102, 1979.

[202] Wu, W.-T. *Basic Principles of Mechanical Theorem Proving in Geometries, Volume I: Part of Elementary Geometries.* Science Press, Beijing, 1984; Springer, 1994.

[203] Wu, W.-T. *Mathematics Mechanization.* Science Press and Kluwer Academic, Beijing, 2000.

[204] Wu, W.-T. and Gao, X.S. Automated Reasoning and Equation Solving with the Characteristic Set Method. *J. Computer Science and Technology* **22**(2): 756-764, 2007.

[205] Wu, Y. Bracket Algebra, Affine Bracket Algebra and Automated Geometric Theorem Proving. Ph.D. Dissertation, Acad. Math. and Sys. Sci., Chinese Academy of Sciences, Beijing, 2001.

[206] Xu, R. Clifford Coalgebra and Geometric Theorem Completion. Ph.D. Dissertation, Acad. Math. and Sys. Sci., Chinese Academy of Sciences, Beijing, 2006.

[207] Yaglom, I.M. *Felix Klein and Sophus Lie.* Birkhäuser, Boston, Basel, 1988. Translated by Sosdsinsky, S.

[208] Yang, A.T. and Freudenstein, F. Application of Dual Number Quaternion Algebra to the Analysis of Spatial Mechanisms. *J. Appl. Mech.* **31**: 300-308, 1964.

[209] Young, A. *The Collected Papers of Alfred Young, 1873-1940.* University of Toronto Press, 1977.

[210] Zaharia, M.D. and Dorst, L. Modeling and Visualization of 3D Polygonal Mesh Surfaces using Geometric Algebra. *Computers and Graphics* **28**(4): 519-526, 2004.

[211] Zang, D. and Sommer, G. Signal Modeling for Two-dimensional Image Structures. *J. Visual Communication and Image Representation* **18**(1): 81-99, 2007.

Index